PRACTICE BOOK

ON

QUICKER MATHS

- Miraculous for Banks, LIC, GIC, UTI, SSC, CPO, Management, Railways and other competitive exams
- Stimulating for general use

M. TYRA & K. KUNDAN

BSC PUBLISHING CO. PVT. LTD.

SEEKOSITY BOOKS

LDS Marketing
Daryaganj, N.Delhi-2

BSC PUBLISHING CO. PVT. LTD.

First Edition: 2002
Second Edition: 2013, 2014, 2015
Third Edition: 2021
Reprint Edition- 2026

ISBN: 978-81-904589-5-5

DISCLAIMER

Every effort has been made to avoid errors or omissions in the publication. In spite of this, some errors might have crept in. The publisher shall not be responsible for the same. Any mistake, error or discrepancy noted may be brought to our notice which shall be taken care of in the next edition. It is notified that neither the publisher nor the author or sellers will be responsible for any damage or loss of action of any one, of any kind, in any manner, therefrom.

PRICE: Rs. 570

Distributed by

Lov Dev & Sons

96, Chapter House, Daryaganj, N.Delhi-2
Mob.: 9355100299

Printed at

Rama Printers
F-7/B, Okhla Phase-1, New Delhi-110020

Contents

PREFACE

Magical Book on Quicker Maths has been greatly appreciated by the competitive exam aspirants. But they have always been complaining that this book does not provide sufficient questions for practice. Keeping the problems of the aspirants in mind, we have attempted to prepare this book under the name *Practice Book on Quicker Maths*. We hope this book will also be as helpful as *Magical Book on Quicker Maths*. Remember this book is not a substitute for *Magical Book on Quicker Maths* but a complement to it. In other words, it is a companion volume of *Magical Book on Quicker Maths*.

We advise the aspirants to first go through *Magical Book on Quicker Maths*. Try to understand each and every method, its derivation, application and the concepts involved in it. Only then you can afford to take up this book. As every chapter starts with rules in this book, at first instance you may get the feeling that the same rules as discussed in the previous book have been given. But this has been done deliberately for the convenience of the aspirants. It serves the purpose of revision. You don't need to revise the rules, since you will learn rules here itself by doing practice. We hope that the aspirants will appreciate our unique attempt.

We would like to thank Mr SK Raju, a member of the BSC research team and faculty of BSC Academy (Head Office), who came up with the idea of having such a book among our titles. We would also extend our thanks to all those who have played a role in making the book available to the reader. We are grateful to those in the computer section — Mr Ashok Kumar, Mr Binod Kumar, Mr Omprakash and Mr Pramod — without whom this book would not have been possible. Finally, We would like to thank Mr Pradeep Gupta for printing.

M Tyra
K Kundan

1 Number System

Rule 1

Dividend = (Divisor × Quotient) + Remainder

Illustrative Example

Ex.: A number when divided by 602 leaves remainder 36 and the value of quotient is 5. Find the number.

Soln: By the above formula, we get

Number = $(602 \times 5) + 36 = 3046$

Exercise

1. In a divison sum the quotient is 120, the divisor 456, and the remainder 333, find the dividend.
   a) 55035 b) 55053 c) 50553 d) 55503
2. In a division the quotient is 105, the remainder is 195, the divisor is equal to the sum of the quotient and remainder, what is the dividend?
   a) 31695 b) 36195 c) 31659 d) 31965
3. In a division sum the divisor is 10 times the quotient and 5 times the remainder. What is the dividend, if the remainder be 469?
   a) 5566 b) 5336 c) 5363 d) 3556
4. The quotient arising from the division of a number by 62 is 463 and the remainder is 60, what is the number?
   a) 28766 b) 28566 c) 27866 d) 28676
5. The divisor is 321, the quotient 11 and the remainder 260. Find the dividend.
   a) 3719 b) 3971 c) 3791 d) 3179
6. In a division sum the divisor is 5 times and the quotient is 6 times the remainder which is 73. What is the dividend?
   a) 169943 b) 159963 c) 159943 d) 159953
7. The quotient is 702, the remainder is 24, and the divisor 7 more than the sum of both. What is the dividend?
   a) 514590 b) 541590 c) 514950 d) 514509
8. In a division sum the divisor is 7239, quotient 1308 and remainder 209. By how much should the dividend be increased so that when it is divided by the same divisor a quotient 1311 and a remainder 730 is obtained?
   a) 22238 b) 22283 c) 22338 d) 22233
9. In a division sum, the divisor is 10 times the quotient and 5 times the remainder. If the remainder is 48, the dividend is
   a) 808 b) 5008 c) 5808 d) 8508
10. In a division sum, the divisor is 10 times the quotient and 5 times the remainder. If the remainder is 46, the dividend is
    a) 4236 b) 4306 c) 4336 d) 5336

Answers

1\. b 2. a 3. b 4. a 5. c 6. c 7. a
8\. a 9. c 10. d

Rule 2

$$\textit{Divisor} = \frac{\textit{Dividend} - \textit{Remainder}}{\textit{Quotient}}$$

Illustrative Example

Ex.: On dividing 397246 by a certain number, the quotient is 865 and the remainder is 211. Find the divisor.

Soln: Applying the above formula, we get

$$\text{Divisor} = \frac{397246 - 211}{865} = 459$$

Exercise

1. On dividing 7865321 by a certain number, the quotient is 33612 and the remainder is 113. Find the divisor.
   a) 254 b) 234 c) 284 d) 264
2. The dividend is 3792, the quotient 12 and the remainder 0. Find the divisor.
   a) 316 b) 261 c) 361 d) 136
3. What is the divisor when the dividend is 345, the remainder 5 and the quotient 20?
   a) 27 b) 17 c) 7 d) 37
4. A boy had to divide 76428 by 123. He copied a figure wrong in the divisor and obtained as his quotient 611 with remainder 53. What mistake did he make?
   a) He made no mistake
   b) He copied 133 instead of 123
   c) He copied 125 instead of 123
   d) He copied 213 instead of 123.

5. The quotient arising from the division of 24446 by a certain number is 79 and the remainder is 35, what is the divisor?
a) 309 b) 319 c) 310 d) 379
6. A boy had to divide 49471 by 210. He made some mistake in copying the divisor and obtained as his quotient 246 with a remainder 25. What mistake did he make?
a) He made no mistake b) He put down 120 for 210
c) He put down 102 for 210 d) He put dwn 201 for 210
7. In a division sum the dividend is 57324 and quotient 123. If the remainder is greater than the quotient but less han twice the quotient. Find the divisor.
a) 465 b) 475 c) 645 d) 565

Answers

1. b 2. a 3. b 4. c 5. a 6. d 7. a

Rule 3

A number (Dividend) can be made completely divisible with the help of either of the following methods:

Divisor) Dividend (Quotient

Remainder

Method I: ***By subtracting remainder from dividend. For finding the greatest n-digit number completely divisible by a divisor, this rule is applicable.***

Illustrative Examples

Ex. 1: Find the greatest number of 3 digits, which is exactly divisible by 35.

Soln: The greatest number of 3 digit = 999
On dividing 999 by 35, remainder = 19.
Now, applying the above method,
the required number = dividend – remainder = 999 – 19 = 980

Ex. 2: Find the least number that must be subtracted from 87375, to get a number exactly divisible by 698.

Soln: On dividing 67375 by 698, the remainder is 125. By the above method,
The least number to be subtracted is the remainder from dividend.
∴ the least number to be subtracted = 125.

Method II: ***By adding (divisor – remainder) to dividend. For finding the least n-digit number completely divisible by a divisor, this rule is applicable.***

Illustrative Examples

Ex. 1: What least number must be added to 49123 to get a number exactly divisible by 263.

Soln: On dividing 49123 by 263, the remainder is 205.
By the above method,
The least number to be added to the dividend
= divisor – remainder
= 263 – 205 = 58.
∴ the least number to be added = 58.

Ex. 2: Find the least number of 3 digits, which is exactly divisible by 14.

Soln: The least number of 3 digits = 100
On dividing 100 by 14, remainder = 2
To determine exactly divisible least number, the above method will be applied.
∴ The required number
= Dividend + (Divisor – Remainder)
= 100 + (14 – 2) = 112.

Exercise

1. What least number must be subtracted from 5731625, to get a number exactly divisible by 3546?
a) 1189 b) 1829 c) 1289 d) 1982
2. Find the least number of 5 digits which is exactly divisible by 456.
a) 10456 b) 10424 c) 10032 d) 10023
3. Find the number which is nearest to 68624 and exactly divisible by 587.
a) 68679 b) 69156 c) 68569 d) 68689
4. Find the number nearest to 144759 and exactly divisible by 927.
a) 144906 b) 144612 c) 144169 d) 144621
5. Find the greatest number of 5-digits, which is exactly divisible by 547.
a) 99456 b) 99554 c) 10545 d) 99545
6. What least number must be added to 954131, to get a number exactly divisible by 548?
a) 63 b) 563 c) 485 d) 611
7. What least number be subtracted from 6501 to get a number exactly divisible by 135?
a) 21 b) 12 c) 35 d) 53
8. What least number be added to 5200 to get a number exactly divisible by 180.
a) 160 b) 60 c) 20 d) 180
9. Find the number which is nearest to 6555 and exactly divisible by 21.
a) 6558 b) 6576 c) 6552 d) 6534
10. Find the number which is nearest to 8845 and exactly divisible by 80.
a) 8890 b) 8810 c) 8800 d) 8880
11. What least number must be subtracted from 13601 to get a number exactly divisible by 87.
a) 39 b) 29 c) 27 d) 33
12. What least number must be added to 1056 to get a number exactly divisible by 23.
a) 21 b) 23 c) 2 d) 4
13. The largest number of four digits exactly divisible by 88 is
a) 9856 b) 9944 c) 9988 d) 9994
14. Find the greatest number of five digits exactly divisible by 279.

a) 99882 b) 99720 c) 99782 d) 99982

15. Find the nearest integer to 56100 which is exactly divisible by 456.
a) 56556 b) 56088 c) 56112 d) 56188
16. What is the nearest whole number to one million which is divisible by 537 without remainder?
a) 999894 b) 999994 c) 999984 d) 999948
17. What least number must be added to 2716321 to make it exactly divisible by 3456?
a) 3361 b) 95 c) 105 d) 3316
18. What least number must be subtracted from 2716321 to make it exactly divisible by 3456?
a) 3361 b) 95 c) 85 d) 3613
19. Find the least number of five digits which is exactly divisible by 654.
a) 10190 b) 10654 c) 10464 d) 10644
20. Which least number should be subtracted from 427396 so that the remainder would be divisible by 15?
[BSRB Delhi PO, 2000]
a) 6 b) 1 c) 16 d) 4

Answers

1. c	2. c	3. a	4. b	5. b	6. c	7. a
8. c	9. c	10. d	11. b	12. c	13. b	14. a
15. b	16. a	17. b	18. a	19. c	20. b	

Rule 4

Theorem: ***When two numbers, after being divided by a third number, leave the same remainder, the difference of those two numbers must be perfectly divisible by the third number.***

Illustrative Examples

Ex. 1: 24345 and 33334 are divided by a certain number of three digits and the remainder is the same in both the cases. Find the divisor and the remainder.

Soln: By the above theorem, the difference of 24345 and 33334 must be perfectly divisible by the divisor. We have the difference = 33334 - 24345 = 8989 = 101 × 89
Thus, the three-digit number is 101.
The remainder can be obtained by dividing one of the numbers by 101. If we divide 24345 by 101, the remainder is 4.

Ex. 2: 451 and 607 are divided by a number and we get the same remainder in both the cases. Find all the possible divisors (other than 1).

Soln: By the above theorem:
607 – 451 = 156 is perfectly divisible by those numbers (divisors).
Now, $156 = 2 \times 2 \times 3 \times 13$
Thus, 1-digit numbers = $2, 3, 2 \times 2, 2 \times 3 = 2, 3, 4, 6$
2-digit numbers = 12, 13, 26, 39, 52, 78
3-digit number = 156

Exercise

1. 457213 and 343373 are divided by a certain number of four digits and the remainder is the same in both the cases. Find the divisor.
a) 1423 b) 1432 c) 1422 d) 1433
2. 31593and 23456 are divided by a certain number of three digits and the remainder is the same in both the cases. Find the remainder.
a) 75 b) 66 c) 68 d) 88

Answers

1. a 2. a

Rule 5

To find the product of the two numbers when the sum and the difference of the two numbers are given.
Product of the numbers

$$= \frac{(Sum + Difference)(Sum - Difference)}{4}$$

Illustrative Example

Ex. The sum of two numbers is 14 and their difference is 10. Find the product of the two numbers.

Soln: **Detail Method:** Let the two numbers be x and y, then $x + y = 14$ and $x - y = 10$

Now, we have, $(x+y)^2 = (x-y)^2 + 4xy$

or, $(14)^2 = (10)^2 + 4xy$

$$\therefore xy = \frac{(14)^2 - (10)^2}{4} = \frac{96}{4} = 24$$

Quicker Method: Applying the above formula, we have

$$\text{Product} = \frac{(14+10)(14-10)}{4} = 24$$

Note: The numbers can also be found by the direct formula

$$x = \frac{Sum + Difference}{2} = \frac{14+10}{2} = 12$$

$$y = \frac{Sum - Difference}{2} = \frac{14-10}{2} = 2$$

Exercise

1. The sum of two numbers is 20 and their difference is 10. Find the product of the two numbers.
a) 60 b) 100 c) 80 d) 75
2. The sum of two numbers is 49 and their difference is 3. Find the product of the two numbers.
a) 598 b) 958 c) 589 d) 859
3. The sum of two numbers is 38 and their difference is 4. Find the product of the two numbers.
a) 537 b) 375 c) 357 d) 753
4. The sum of two numbers is 24 and their difference is 18.

Find the product of the two numbers.
a) 54 b) 63 c) 36 d) 64

5. The sum of two numbers is 33 and their difference is 21. Find the product of the two numbers.
a) 162 b) 126 c) 102 d) 216

6. The difference of two numbers is 11 and $\frac{1}{5}$th of their sum is 9. The numbers are: **[RRB Exam 1991]**
a) 31, 20 b) 30, 19 c) 29, 18 d) 28, 17

Answers

1. d 2. a 3. c 4. b 5. a
6. d; **Hint:** See Note.

Rule 6

Ex. If one-fifth of one-third of one-half of number is 15, find the number.

Soln: **Detail Method:** Let the number be x. Then we have,

$$x\left(\frac{1}{5}\right)\left(\frac{1}{3}\right)\left(\frac{1}{2}\right)=15$$

$$\therefore x=15\times5\times3\times2=450$$

Direct Formula:

(*) The required number $=15\left(\frac{5}{1}\right)\left(\frac{3}{1}\right)\left(\frac{2}{1}\right)=450$

Note:(*) The resultant should be multiplied by the reverse of each fraction.

Exercise

1. If one-third of one-sixth of two-third of number is 64, find the number.
a) 1278 b) 1782 c) 1728 d) 3456
2. If one-tenth of one-fourth of one-fifth of number is 10, find the number.
a) 200 b) 2000 c) 500 d) 1000
3. If three-fourth of two-third of two-fifth of one-half of number is 60, find the number.
a) 600 b) 400 c) 650 d) 575
4. If two-fifth of one-third of two-third of number is 16, find the nmber.
a) 160 b) 280 c) 180 d) 190
5. If one-fifth of two-third of one-half of number is 30, find the number.
a) 450 b) 900 c) 950 d) 400
6. Three-fourth of one-fifth of a number is 60. The number is: **[Bank PO Exam, 1990]**
a) 300 b) 400 c) 450 d) 1200
7. Four-fifths of three-eighths of a number is 24. What is 250 per cent of that number? **[BSRB Mumbai, 1998]**
a) 100 b) 160 c) 120 d) 200
8. Two-fifths of thirty per cent of one-fourth of a number is 15. What is 20 per cent of that number? **[BSRB Mumbai 1998]**
a) 90 b) 150 c) 100 d) 120
9. Two-fifths of one-fourth of five eighths of a number is 6. What is 50 per cent of that number? **[BSRB Calcutta PO 1999]**
a) 96 b) 32 c) 24 d) 48
10. If $\frac{4}{9}$ of $\frac{3}{10}$ of $\frac{5}{8}$ of a number is 45, what is the number? **[BSRB Hyderabad PO 1999]**
a) 450 b) 540 c) 560 d) 650
11. Two-thirds of three-fifths of one-eighth of a certain number is 268.50. What is 30 per cent of that number? **[NABARD 1999]**
a) 1611.0 b) 716.0 c) 1342.5 d) 596.60
12. If $\frac{1}{8}$ of $\frac{2}{3}$ of $\frac{4}{5}$ of a number is 12 then 30 per cent of the number will be **[SBI Bank PO 2001]**
a) 48 b) 64 c) 54 d) 42

Answers

1. c 2. b 3. a 4. c 5. a 6. b 7. d
8. c 9. d 10. b 11. a 12. c

Rule 7

The sum of the digits of a two-digit number is S. If the digits are reversed, the number is decreased by N, then the number is given by $5\left[S+\frac{N}{9}\right]+\frac{1}{2}\left[S-\frac{N}{9}\right]$

or

$$5\left[Sum\ of\ digits+\frac{Decrease}{9}\right]+\frac{1}{2}\left[Sum\ of\ digits-\frac{Decrease}{9}\right]$$

Illustrative Example

Ex. The sum of the digits of a two-digit number is 8. If the digits are reversed, the number is decreased by 54. Find the number.

Soln: **Detail Method:** Let the two-digit number be $10x+y$. Then, we have; $x+y=8$ ... (1) and
$10y+x=10x+y-54$

or, $x-y=\frac{54}{9}=6$ .... (2)

From equations (1) and (2)

$x=\frac{8+6}{2}=7$ and $y=1$

$\therefore$ The required number $=7\times10+1=71$

Quicker Method:
The required number =

$$5\left[\text{Sum of digits}+\frac{\text{Decrease}}{9}\right]+\frac{1}{2}\left[\text{Sum of digits}-\frac{\text{Decrease}}{9}\right]$$

$= 5(8+6)+\frac{1}{2}(8-6)=70+1=71$

Exercise

1. The sum of the digits of a two-digit number is 12. If the digits are reversed, the number is decreased by 18. Find the number.
   a) 75 b) 93 c) 84 d) 57
2. The sum of the digits of a two-digit number is 9. If the digits are reversed, the number is decreased by 63. Find the number.
   a) 72 b) 63 c) 54 d) 81
3. The sum of the digits of a two-digit number is 10. If the digits are reversed, the number is decreased by 72. Find the number.
   a) 91 b) 82 c) 73 d) 64
4. The sum of the digits of a two-digit number is 13. If the digits are reversed, the number is decreased by 45. Find the number.
   a) 85 b) 76 c) 49 d) 94
5. The sum of the digits of a two-digit number is 7. If the digits are reversed, the number is decreased by 45. Find the number.
   a) 52 b) 43 c) 61 d) 25
6. A certain number consists of two digits whose sum is 9. If the order of digits is reversed, the new number is 9 less than the original number. The original number is
   a) 45 b) 36 c) 54 d) 63
7. In a two-digit number the digit in the unit's place is more than the digit in the ten's place by 2. If the difference between the number and the number obtained by interchanging the digits is 18. What is the original number. **[SBI Associates PO 1999]**
   a) 46 b) 68 c) 24 d) Data inadequate

Answers

1. a 2. d 3. a 4. d 5. c 6. c

7. d; **Hint:** Let the no. be 10x + y
then y = x + 2 or, y − x = 2 .... (i)
(10y + x) − (10x + y) = 18
or, 9y − 9x = 18
or, y − x = 2 ..... (ii)
From eqn (i) and (ii) we can't get any conclusion.

Rule 8

If the sum of a number and its square is x, then the number is given by $\left[\frac{\sqrt{1+4x}-1}{2}\right]$.

Illustrative Example

Ex.: If the sum of a number and its square is 182, what is the number?

Soln: **Detail Method:** Let the number = x.

Then, $x^2+x=182$

or, $x^2+x-182=0$

or, $x^2+14x-13x-182=0$

or, $x(x+14)-13(x+14)=0$

or, $(x-13)(x+14)=0$

or, $x = 13$ (negative value is neglected).

Quicker Method: Applying the above rule, we have the required answer

$$=\frac{\sqrt{1+182\times4}-1}{2}=\frac{\sqrt{729}-1}{2}=\frac{27-1}{2}=13$$

Exercise

1. If the sum of a number and its square is 240, what is the number?
   a) 15 b) 18 c) 25 d) 22
2. If the sum of a number and its square is 306, what is the number?
   a) 16 b) 18 c) 17 d) 19
3. If the sum of a number and its square is 702, what is the number?
   a) 26 b) 27 c) 28 d) 29
4. If the sum of a number and its square is 1560, what is the number?
   a) 38 b) 37 c) 36 d) 39
5. If the sum of a number and its square is 156, what is the number?
   a) 16 b) 14 c) 12 d) 13
6. If the sum of a number and its square is 210, what is the number?
   a) 12 b) 13 c) 14 d) 16
7. If the sum of a number and its square is 90, what is the number?
   a) 7 b) 8 c) 9 d) 8
8. If the sum of a number and its square is 380, what is the number?
   a) 17 b) 18 c) 19 d) 21
9. If the sum of a number and its square is 342, what is the number?
   a) 14 b) 28 c) 18 d) 23
10. If the sum of a number and its square is 552, what is the number?
    a) 21 b) 22 c) 23 d) 24

Answers

1. a 2. c 3. a 4. d 5. c 6. c 7. c
8. c 9. c 10. c

Rule 9

The sum of the digits of a two-digit number is S. If the digits are reversed, the number is increased by N, then the num-

ber is given by $5\left[S-\frac{N}{9}\right]+\frac{1}{2}\left[S+\frac{N}{9}\right]$ ***or***

$$5\left[\text{Sum of digits}-\frac{\text{Increase}}{9}\right]+\frac{1}{2}\left[\text{Sum of digits}+\frac{\text{Increase}}{9}\right].$$

Illustrative Example

Ex.: The sum of the digits of a two-digit number is 8. If the digits are reversed, the number is increased by 54. Find the number.

Soln: Detail Method:

Let the two digit number be $10x + y$

Then, we have, $x+y=8$ ... (i) and

$10y+x=10x+y+54$

or, $y-x=6$ .... (ii)

From eqn (i) and (ii)

$x=1$ and $y=7$.

$\therefore$ the required number $=1\times 10+7=17$

Quicker Method: Applying the above formula, we have

$$\text{Required number}=5\left[8-\frac{54}{9}\right]+\frac{1}{2}\left[8+\frac{54}{9}\right]$$
$$=10+7=17$$

Exercise

1. The sum of the digits of a two-digit number is 7. If the digits are reversed, the number is increased by 27. Find the number.
   a) 25 b) 34 c) 16 d) None of these
2. The sum of the digits of a two-digit number is 6. If the digits are reversed, the number is increased by 36. Find the number.
   a) 24 b) 15 c) 51 d) 42
3. The sum of the digits of a two-digit number is 9. If the digits are reversed, the number is increased by 63. Find the number.
   a) 27 b) 36 c) 45 d) 18
4. The sum of the digits of a two-digit number is 5. If the digits are reversed, the number is increased by 27. Find the number.
   a) 23 b) 32 c) 14 d) 41
5. A number consists of two digits whose sum is 15. If 9 is added to the number, then the digits change their places. The number is ____.
   a) 69 b) 78 c) 87 d) 96

Answers

1. a 2. b 3. d 4. c 5. b

Rule 10

If x% of a number is n, then y% of z% of that number is given by $\left[\frac{yzn}{x\times 100}\right]$.

Illustrative Example

Ex. If 40% of a number is 360, what will be 15% of 15% of that number?

Soln: Detail Method: Let the number be x. Then we have 40% of $x=360$

$$\therefore x=\frac{360\times 100}{40}=900$$

$$\text{Now, } 15\% \text{ of } x=\frac{15}{100}\times 900=135$$

$$\text{Again, } 15\% \text{ of } 135=\frac{15}{100}\times 135=20.25$$

Quicker Method: Applying the above rule, we have

$$\text{the required answer}=\frac{15\times 15\times 360}{40\times 100}=20.25.$$

Exercise

1. If 90% of a number is 540, what will be 10% of 5% of that number.
   a) 30 b) 3.5 c) 3 d) 35
2. If 35% of a number is 385, what will be 5% of 5% of that number.
   a) 11 b) 5.5 c) 2.5 d) 2.75
3. If 17% of a number is 68, what will be 15% of 25% of that number.
   a) 20 b) 15 c) 35 d) 25
4. If 18% of a number is 144, what will be 12% of 25% of that number.
   a) 8 b) 12 c) 16 d) 24
5. If 39% of a number is 780, what will be 35% of 13% of that number.
   a) 91 b) 52 c) 65 d) 78

Answers

1. c 2. d 3. b 4. d 5. a

Rule 11

If the ratio of the sum and the difference of two numbers is a : b, then the ratio of these two numbers is given by $\left(\frac{a+b}{a-b}\right)$.

Illustrative Example

Ex. The ratio of the sum and the difference of two numbers is 7 : 1. Find the ratio of those two numbers.

Soln: Detail Method: Let the two numbers be x and y. Then we have

$$\frac{x+y}{x-y}=\frac{7}{1}$$

$$\Rightarrow x+y=7x-7y$$

$$\text{or, } 6x=8y \quad \therefore \frac{x}{y}=\frac{8}{6}=\frac{4}{3}=4:3$$

Quicker Method: Applying the above rule, we have

the required ratio $= \frac{7+1}{7-1} = \frac{8}{6} = \frac{4}{3} = 4:3$

Exercise

1. Ratio of the sum and the difference of the two numbers is 5 : 3. Find the ratio of those two numbers.
a) 4 : 1 b) 3 : 2 c) 3 : 1 d) 2 : 1
2. Ratio of the sum and the difference of the two numbers is 9 : 1. Find the ratio of those two numbers.
a) 5 : 3 b) 5 : 4 c) 4 : 1 d) 5 : 2
3. Ratio of the sum and the difference of the two numbers is 7 : 3. Find the ratio of those two numbers.
a) 5 : 2 b) 5 : 3 c) 3 : 2 d) 7 : 4
4. Ratio of the sum and the difference of the two numbers is 2 : 1. Find the ratio of those two numbers.
a) 1 : 2 b) 3 : 2 c) 4 : 3 d) 3 : 1
5. Ratio of the sum and the difference of the two numbers is 13 : 3. Find the ratio of those two numbers.
a) 5 : 8 b) 8 : 3 c) 8 : 5 d) 8 : 7

Answers

1. a 2. b 3. a 4. d 5. c

Rule 12

To find the difference of the two digits of a two-digit number, when the difference between two-digit number and the number obtained by interchanging the digits is given.

Difference of two digits

$$= \frac{\textit{Diff. in original and interchanged number}}{9}$$

Note: We cannot get the sum of two digits.

Illustrative Example

Ex. The difference between a two-digit number and the number obtained by interchanging the digits is 27. What are the sum and the difference of the two digits of the number?

Soln: **Detail Method:** Let the number be $10x + y$. Then we have

$$(10x + y) - (10y + x) = 27$$

or, $9(x - y) = 27 \quad \therefore x - y = \frac{27}{9} = 3$

Thus, the difference is 3, but we cannot get the sum of two digits.

Quicker Method: Applying the above rule, we have

Required answer $= \frac{27}{9} = 3$

Exercise

1. The difference between a two-digit number and the number obtained by interchanging the digits is 18. What is the sum of the two digits of the number?
a) 2 b) 1 c) 9 d) Can't be determined
2. The difference between two-digit number and the number obtained by interchanging the digits is 36. What is the difference of the two digits of the number?
a) 4 b) 3 c) 2 d) 8
3. The difference between two-digit number and the number obtained by interchanging the digits is 63. What is the difference of the two digits of the number?
a) 7 b) 9 c) 8 d) 6
4. The difference between two-digit number and the number obtained by interchanging the digits is 9. What is the difference of the two digits of the number?
a) 2 b) 5 c) 3 d) 1
5. The difference between two-digit number and the number obtained by interchanging the digits is 72. What is the difference of the two digits of the number?
a) 7 b) 9 c) 8 d) Can't be determined
6. The difference between two-digit number and the number obtained by interchanging the digits is 45. What is the difference of the two digits of the number?
a) 6 b) 5 c) 8 d) Can't be determined
7. The difference between the digits of a two-digit number is one-ninth of the difference between the original number and the number obtained by interchanging the positions of the digits. What definitely is the sum of the digits of that number? **[BSRB Mumbai PO, 1998]**
a) 5 b) 14 c) 12 d) Data inadequate
8. The sum of the digits of a two-digit number is $\frac{1}{11}$ of the sum of the number and the number obtained by interchanging the position of the digits. What is the difference between the digits of that number? **[Bank of Baroda PO, 1999]**
a) 3 b) 2 c) 6 d) Data inadequate
9. The difference between a two-digit number and the number obtained by interchanging the position of the digits of that number is 54. What is the sum of the digits of that number? **[BSRB Calcutta PO, 1999]**
a) 6 b) 9 c) 15 d) Data inadequate
10. The sum of the digits of a two-digit number is $\frac{1}{5}$ of the difference between the number and the number obtained by interchanging the positions of the digits. What definitely is the difference between the digits of that number? **[BSRB Chennai PO, 2000]**
a) 5 b) 9 c) 7 d) Data inadequate

Answers

1. d 2. a 3. a 4. d 5. c 6. b

7. d; **Hint:**

$$x - y = \frac{1}{9}\{(10x + y) - (10y + x)\} = \frac{1}{9}(9x - 9y) = x - y$$

8. d; **Hint:** Let, the two no. be xy, ie 10x + y then,

$x+y=\frac{1}{11}[(10x+y)+(10y+x)]=x+y$

Thus we see that the difference of x and y can't be determined.

Hence, the answer is data inadequate.

9. d; **Hint:** See note.

Let the two-digit no. be 10x + y

According to question,

$(10x+y)-(10y+x)=54$

$9x-9y=54 \therefore x-y=6$

10. a; **Hint:** Let the two-digit number be 10x + y

Then, $x+y=\frac{1}{5}(10x+y-10y-x)$

or, $x+y=\frac{9}{5}(x-y)$

or, $4x-14y=0 \Rightarrow \frac{x}{y}=\frac{7}{2}$

Using componendo & dividendo, we have,

$\frac{x+y}{x-y}=\frac{7+2}{7-2}=\frac{9}{5}$ ie $x-y=5K$

Here, K has the only possible value, K = 1.

Because the difference of two single-digit numbers will always be of a single digit.

Rule 13

Ex. The average of 7 consecutive integers is 7. Find the average of the squares of these integers.

Soln: Use the formula: [for odd number of consecutive integers)

Average of squares

$$=\frac{1}{\text{No. of integers}}\times\left[\frac{n_1(n_1+1)(2n_1+1)}{6}-\frac{n_2(n_2+1)(2n_2+1)}{6}\right]$$

Where, $n_1=\text{Average}+\frac{\text{No. of integers}-1}{2}$

and $n_2=\text{Average}-\frac{\text{No. of integers}+1}{2}$

In the above case

$n_1=7+\frac{7-1}{2}=10$

$n_2=7-\frac{7+1}{2}=3$

$$\therefore \text{Average of squares}=\frac{1}{7}\left[\frac{10\times11\times21}{6}-\frac{3(4)(7)}{6}\right]$$

$$=\frac{1}{7}[385-14]=\frac{371}{7}=53$$

Exercise

1. The average of 5 consecutive integers is 4. Find the average of the squares of these integers.
   a) 22.5 b) 45 c) 18 d) Can't be determined
2. The average of 15 consecutive integers is 15. Find the average of the squares of these integers.
   a) 243.66 approx b) 300
   c) 225.4 approx d) 394.26 approx
3. The average of 9 consecutive integers is 9. Find the average of the squares of these integers.
   a) 87 b) $87\frac{2}{3}$ c) 88 d) $85\frac{1}{3}$
4. The average of 7 consecutive integers is 6. Find the average of the squares of these integers.
   a) $46\frac{2}{3}$ b) $46\frac{1}{3}$ c) 40 d) $47\frac{1}{3}$
5. The average of 3 consecutive integers is 3. Find the average of the squares of these integers.
   a) 5 b) $4\frac{1}{3}$ c) $9\frac{2}{3}$ d) None of these

Answers

1. c 2. a 3. b 4. c 5. c

Rule 14

To find the least number which when divided by x_1, x_2 and x_3 leaves the remainders a_1, a_2, and a_3 respectively. $(x_1-a_1)=(x_2-a_2)=(x_3-a_3)$. We have an established method that is given below.

Required least number = (LCM of x_1, x_2 and x_3) − (x_1-a_1) or (x_2-a_2) or (x_3-a_3)

Illustrative Example

Ex.: Find the least number which, when divided by 13, 15 and 19, leaves the remainder 2, 4 and 8 respectively.

Soln: Applying the above rule,

$13-2=15-4=19-18=11$

Now, LCM of 13, 15, 19 = 3705

$\therefore$ the required least number = 3705 − 11 = 3694

Note: Find the least number which, when divided by 13, 15 and 19, leaves the remainders 1, 2, 3 respectively. Can we find the specific solution. No, because $13-1 \neq 15-2 \neq 19-3$

Exercise

1. Find the least number which when divided by 24, 32 and 36 leaves the remainders 19, 27, and 31 respectively.
   a) 288 b) 283 c) 287 d) 285
2. Find the least number which when divided by 12, 21 and 35 leaves the remainders 6, 15, and 29 respectively.

a) 414 b) 418 c) 420 d) 410

3. Find the least number which when divided by 48, 60 and 64 leaves the remainders 38, 50, and 54 respectively.
a) 860 b) 960 c) 950 d) 850

4. Find the least number which when divided by 5, 6, 8, 9 and 12 leaves the remainders 3, 4, 6, 7 and 10 respectively.
a) 360 b) 358 c) 362 d) 258

5. Find the least number which when divided by 9, 10 and 15 leaves the remainders 4, 5, and 10 respectively.
a) 90 b) 95 c) 85 d) 80

Answers

1. b 2. a 3. c 4. b 5. c

Rule 15

To find the greatest number that will divide given numbers say $x_1, x_2, ... x_n$ so as to leave the same remainder in each case, we find the HCF of the positive difference of numbers ie $|x_1 - x_2|, |x_2 - x_3|, ...$ and so on.

Illustrative Example

Ex. Find the greatest number that will divide 55, 127 and 175 so as to leave the same remainder in each case.

Soln: **Detail Method:** Let x be the remainder, then the numbers $(55 - x)$, $(127 - x)$ and $(175 - x)$ must be exactly divisible by the required number.

Now, we know that if two numbers are divisible by a certain number, then their difference is also divisible by that number. Hence, the numbers

$(27-x)-(55-x)$, $(175-x)-(127-x)$ and

$(175-x)-(55-x)$

or, 72, 48 and 120 are also divisible by the required number. HCF of 72, 48 and 120 is 24.

Therefore, the required number is 24.

Quicker Method: If you don't want to go into the details of the method, find the HCF of the positive differences of numbers. It will serve your purpose quickly. For example, in the above case, positive difference of numbers are $(127 - 55 = 72)$, $(175 - 127 = 48)$ and $(175 - 55 = 120)$.

HCF of 72, 48 and 120 is 24

$\therefore$ required number = 24.

Exercise

1. Find the greatest number which is such that when, 12288, 19139 and 28200 are divided by it, the remainders are all the same.
a) 221 b) 212
c) 122 d) 321

2. Find the greatest number which is such that when 76, 151 and 226 are divided by it, the remainders are all alike. Find also the common remainder.
a) 70, 6 b) 71, 5 c) 75, 1 d) 73, 3

3. The greatest number which when divides 99, 123 and 183 leaves the same remainder is
a) 12 b) 24 c) 18 d) 26

4. Find the greatest number which divides 77, 112 and 287 and leaves the same remainder in each case.
a) 35 b) 25 c) 45 d) 15

5. Find the greatest number which divides 95, 195 and 175 and leaves the same remainder in each case.
a) 5 b) 10 c) 20 d) 25

Answers

1. a 2. c 3. a 4. a 5. c

Rule 16

The ratio between a two-digit number and the sum of the digits of that number is a : b. If the digit in the unit's place is n more than the digit in the ten's place, then the number is given by $\left(\frac{9a}{11b-2a}\right)$ n and the digits in unit's place and ten's place are $n\left(\frac{10b-a}{11b-2a}\right)$ and $n\left(\frac{a-b}{11b-2a}\right)$ respectively.

Illustrative Example

Ex.: The ratio between a two-digit number and the sum of the digits of that number is 4 : 1. If the digit in the unit's place is 3 more than the digit in the ten's place, what is the number?

Soln: **Detail Method:** Suppose the two-digit number = $10x + y$

Then we have $\frac{10x+y}{x+y} = \frac{4}{1}$

or, $10x + y = 4x + 4y$

or, $6x = 3y$ or, $2x = y$

or, $x = y - x = 3$ (given) and $y = 6$

$\therefore$ the number is 36.

Quicker Method: Applying the above rule, we have

Required number

$$= \left(\frac{9\times4}{11\times1-2\times4}\right)\times3 = \frac{9\times4}{3}\times3 = 36$$

Exercise

1. The ratio between a two-digit number and the sum of the digits of that number is 5 : 1. If the digit in the unit's place is 1 more than the digit in the ten's place, what is the value of unit's place digit of that number?
a) 4 b) 5 c) 3 d) 7

2. The ratio between a two-digit number and the sum of the digits of that number is 2 : 1. If the digit in the unit's place

is 7 more than the digit in the ten's place. What is the value of ten's place digit of that number?
a) 1 b) 2 c) 3 d) 64

3. The ratio between a two-digit number and the sum of the digits of that number is 3 : 1. If the digit in the unit's place is 5 more than the digit in the ten's place. What is the value of ten's place digit of that number?
a) 4 b) 3 c) 2 d) 1

4. The ratio between a two-digit number and the sum of the digits of that number is 14 : 5. If the digit in the unit's place is 6 more than the digit in the ten's place. What is the sum of the digits of that number?
a) 10 b) 12 c) 13 d) 9

5. The ratio between a two-digit number and the sum of the digits of that number is 4 : 1. If the digit in the unit's place is 4 more than the digit in the ten's place. What is the sum of the digits of that number?
a) 9 b) 10 c) 15 d) 12

Answers

1. b 2. a 3. c 4. a 5. d

Rule 17

To find the remainder when $\left(x^n + K\right)$ is divided by x – 1.

(i) Remainder = 1 + K; when K < x – 1

(ii) Remainder = (1 + Remainder obtained when K is divided by x – 1);
when K > x – 1.

Illustrative Example

Ex.: Find the remainder when $7^{13}+1$ is divided by 6.

Soln: **Detail Method:** See the following binomial expansion

$(x+y)^n =$

$x^n + {}^nC_1x^{n-1}y + {}^nC_2x^{n-2}y^2 + {}^nC_3x^{n-3}y^3 + ... + {}^nC_{n-1}xy^{n-1} + y^n$

We find that each of the terms except the last term $\left(y^n\right)$ contains x. It means each term except y^n is perfectly divisible by x.

Note: y^n may be perfectly divisible by x but we cannot say without knowing the values of x and y.

Following the same logic,

$7^{13} = (6+1)^{13}$ has each term except 1^{13} exactly divisible by 6. Thus, when 7^{13} is divided by 6 we have the remainder $1^{13} = 1$ and hence, when $\left(7^{13}+1\right)$ is divided by 6 the remainder is 1 + 1 = 2.

Quicker Method: Applying the above rule, we have K = 1 and $x - 1 = 6$

ie K < x – 1.

Therefore, we apply rule (i)

∴ required answer = 1 + 1 = 2.

Exercise

1. Find the remainder when $\left(9^{19}+6\right)$ is divided by 8.
a) 2 b) 3 c) 5 d) 7

2. Find the remainder when $\left(7^{13}+8\right)$ is divided by 6.
a) 2 b) 3 c) 9 d) 5

3. Find the remainder when $\left(5^{23}+3\right)$ is divided by 4.
a) 7 b) 4 c) 3 d) 2

4. Find the remainder when $\left(12^{150}+8\right)$ is divided by 11.
a) 19 b) 7 c) 9 d) 8

5. Find the remainder when $\left(25^{625}+241\right)$ is divided by 24.
a) 23 b) 2 c) 1 d) Can't be determined

Answers

1. d 2. b 3. b 4. c 5. b

Rule 18

To find the all possible numbers, when the product of two numbers and their HCF are given, we follow the following method.

Step I: Find the value of $\frac{Product}{(HCF)^2}$.

Step II: Find the possible pairs of value got in step I.

Step III: Multiply the HCF with the pair of prime factors obtained in step II.

Illustrative Example

Ex.: The product of two numbers is 7168 and their HCF is 16. Find the numbers.

Soln: **Step I:** $\frac{7168}{(16)^2} = 28$

Step II: (1, 28), (2, 14), (4, 7)

Step III: (1 × 16, 28 × 16) and (4 × 16, 7 × 16) or (16, 448) and (64, 112)

Note: (2, 14), which are not prime to each other should be rejected.

Exercise

1. The product of two numbers is 286 and their HCF is 12. Find the sum of the numbers.
a) 12 b) 24 c) 36 d) 48

2. The product of two numbers is 3125 and their HCF is 25. Find the sum of the numbers.
a) 75 b) 100 c) 125 d) 50

3. The product of two numbers is 2016 and their HCF is 12. Find the number of all possible pairs of numbers.
a) 1 b) 2 c) 3 d) Can't be determined

4. The product of two numbers is 338 and their HCF is 13. Find the difference of the numbers.
a) 13 b) 26 c) 39 d) 52

Answers

1. c 2. b 3. b 4. a

Rule 19

A number on being divided by d_1 and d_2 successively leaves the remainders r_1 and r_2 respectively. If the number is divided by $d_1 \times d_2$, then the remainder is given by $(d_1 \times r_2 + r_1)$.

Illustrative Example

Ex. A number on being divided by 5 and 7 successively leaves the remainders 2 and 4 respectively. Find the remainder when the same number is divided by $5 \times 7 = 35$.

Soln: Detail Method:

5	A	
7	B	2
	C	4

In the above arrangement, A is the number which, when divided by 5, gives B as a quotient and leaves 2 as a remainder. Again, when B is divided by 7, it gives C as a quotient and 4 as a remainder. For simplicity, we may take $C = 1$.

$\therefore B = 7 \times 1 + 4 = 11$ and $A = 5 \times 11 + 2 = 57$

Now, when 57 is divided by 35, we get 22 as the remainder.

Quicker Method: The required remainder $= d_1 \times r_2 + r_1$

Where, d_1 = the first divisor = 5

r_1 = the first remainder = 2

r_2 = the second remainder = 4

$\therefore$ the required remainder $= 5 \times 4 + 2 = 22$.

Exercise

1. A number on being divided by 12 and 15 successively leaves the remainders 4 and 6 respectively. Find the remainder when the same number is divided by 180.
   a) 46 b) 76 c) 84 d) 18
2. A number on being divided by 5 and 7 successively leaves the remainders 3 and 6 respectively. Find the remainder when the same number is divided by 35.
   a) 33 b) 23 c) 32 d) Can't be determined
3. A number on being divided by 8 and 9 successively leaves the remainders 5 and 7 respectively. Find the remainder when the same number is divided by 72.
   a) 61 b) 8 c) 71 d) 9
4. A number on being divided by 4 and 6 successively leaves the remainders 2 and 3 respectively. Find the remainder when the same number is divided by 24.
   a) 12 b) 10 c) 14 d) 16
5. A number on being divided by 10 and 11 successively leaves the remainders 5 and 7 respectively. Find the remainder when the same number is divided by 110.
   a) 70 b) 98 c) 74 d) 75
6. A number on being divided by 3 and 7 successively leaves the remainders 2 and 5 respectively. Find the sum of digits of the remainder when the same number is divided by 21.
   a) 7 b) 17 c) 8 d) 6

Answers

1. b 2. a 3. a 4. c 5. d 6. c

Rule 20

To find the number of zeros at the end of the product.

We know that zeros are produced only due to the following reasons.

(i) If there is any zero at the end of any multiplicand.

(ii) If 5 or multiple of 5 are multiplied by any even number.

To generalise the above two statements, we may say that

$(2)^m (5)^n$ ***has n zeros if m > n or m zeros if m < n.***

Note: Always lesser value of the exponents of 5 and 2 will be the required answer. Thus, write the product in the form $(2^m \times 5^n \times ...)$

Illustrative Example

Ex.: Find the number of zeros at the end of the products.
$12 \times 18 \times 15 \times 40 \times 25 \times 16 \times 55 \times 105$

Soln: $12 \times 18 \times 15 \times 40 \times 25 \times 16 \times 55 \times 105$

$= 12 \times 18 \times 16 \times 40 \times 15 \times 25 \times 55 \times 105$

$= (2^2 \times 3) \times (2 \times 9) \times (2)^4 \times (2^3 \times 5) \times (5 \times 3) \times (5)^2 \times (5 \times 11) \times (5 \times 21)$

$= 2^{10} \times 5^6 \times ....$

[Since numbers other than 2 and 5 are useless]

Since $10 > 6$, there are 6 zeros at the end of the product.

Note: This is the easiest way to count the number of zeros in the chain of products. By this method, we can easily find that the product of $1 \times 2 \times 3 \times ... \times 100$ contains 24 zeros.

Exercise

1. Find the number of zeros at the end of the product
   $15 \times 16 \times 18 \times 25 \times 35 \times 24 \times 20$
   a) 10 b) 8 c) 5 d) Can't be determined
2. Find the number of zeros at the end of the product
   $5^2 \times 20 \times 2^8 \times 10 \times 16 \times 125$
   a) 15 b) 22 c) 7 d) 8
3. Find the number of zeros at the end of the product
   $50 \times 625 \times 15 \times 10 \times 30$
   a) 10 b) 9 c) 12 d) 3
4. Find the number of zeros at the end of the product

$150 \times 250 \times 625 \times 125 \times 75 \times 20 \times 16$

a) 9 b) 14 c) 23 d) 5

5. Find the number of zeros at the end of the product $70 \times 80 \times 16 \times 64 \times 5^6 \times 13 \times 18 \times 3125$

a) 16 b) 12 c) 10 d) 25

Answers

1. c 2. c 3. d 4. a 5. b

Rule 21

To find the number of different divisors.

Find the prime factors of the number and increase the index of each factor by 1. The continued product of increased indices will give the result including unity and the number itself.

Note: Also see **Rule - 36.**

Illustrative Examples

Ex. 1: Find the number of different divisors of 50, besides unity and the number itself.

Soln: If you solve this problem without knowing the rule, you will take the numbers in succession and check the divisibility. In doing so, you may miss some numbers. It will also take more time.

Different divisors of 50 are: 1, 2, 5, 10, 25, 50

If we exclude 1 and 50, the number of divisors will be 4.

By rule: $50 = 2 \times 5 \times 5 = 2^1 \times 5^2$

$\therefore$ the number of total divisors $= (1+1) \times (2+1)$

$= 2 \times 3 = 6$

or, the number of divisors excluding 1 and 50 $= 6 - 2$

$= 4$

Ex. 2: Find the different divisors of 37800, excluding unity.

Soln: $37800 = 2 \times 2 \times 2 \times 3 \times 3 \times 3 \times 5 \times 5 \times 7$

$= 2^3 \times 3^3 \times 5^2 \times 7^1$

Total no. of divisors $= (3+1)(3+1)(2+1)(1+1) = 96$

$\therefore$ the number of divisors excluding unity $= 96 - 1 = 95$.

Exercise

1. Find the number of different divisors of 307692.

a) 96 b) 12 c) 6 d) 48

2. Find the number of different divisors of 400, besides unity and the number itself.

a) 15 b) 14 c) 13 d) 12

3. Find the number of divisors of 999999, excluding unity.

a) 64 b) 62 c) 63 d) 79

4. Find the number of different divisors of 13231.

a) 64 b) 4 c) 25 d) 5

5. Find the no. of different divisors of 30030, besides unity and the number itself.

a) 64 b) 63 c) 62 d) 60

6. Find the no. of different divisors of 4452.

a) 24 b) 32 c) 16 d) 22

Answers

1. a 2. c 3. c

4. b; **Hint:** $13231 = 131 \times 101$, 131 and 101 are primes

5. c 6. a

Rule 22

To find the number of numbers divisible by a certain integer.

Illustrative Examples

Ex. 1: How many numbers up to 100 are divisible by 6?

Soln: Divide 100 by 6. The quotient obtained is the required number of numbers.

$100 = \underline{16} \times 6 + 4$

Thus, there are 16 numbers.

Ex. 2: How many numbers up to 200 are divisible by 4 and 3 together?

Soln: LCM of 4 and 3 = 12

Now, divide 200 by 12 and the quotient obtained is the required number of numbers.

$200 = \underline{16} \times 12 + 8$

Thus, there are 16 numbers.

Ex. 3: How many numbers between 100 and 300 are divisible by 7?

Soln: Up to 100, there are 14 numbers which are divisible by 7 (since $100 = \underline{14} \times 7 + 2$). Up to 300, there are 42 numbers which are divisible by 7 (since $300 = \underline{42} \times 7 + 6$)

Hence, there are $42 - 14 = 28$ numbers.

Exercise

1. How many numbers up to 150 are divisible by 9?

a) 16 b) 15 c) 10 d) 6

2. How many numbers up to 200 are divisible by 7?

a) 26 b) 22 c) 18 d) 28

3. How many numbers up to 532 are divisible by 15?

a) 25 b) 26 c) 36 d) 35

4. How many numbers up to 300 are divisible by 5 and 7 together?

a) 9 b) 8 c) 10 d) 7

5. How many numbers up to 450 are divisible by 4, 6 and 8 together?

a) 19 b) 18 c) 17 d) 16

6. How many numbers between 50 and 150 are divisible by 8?

a) 24 b) 12 c) 18 d) 8

7. How many numbers between 100 and 200 are divisible by 2 and 8 together?

a) 12 b) 13 c) 9 d) 16

8. How many numbers between 100 and 300 are divisible by 9?

a) 11 b) 13 c) 19 d) 22

Answers

1. a 2. d 3. d 4. b 5. b 6. b 7. b 8. d

Rule 23

The number which when multiplied by x is increased by y is given by $\left(\frac{y}{x-1}\right)$ ***or*** $\left(\frac{\textit{Increased Value}}{\textit{Multiplier}-1}\right)$.

Illustrative Example

Ex. Find the number which when multiplied by 16 is increased by 225.

Soln: **Detail Method:** Let that number be x. Then

$16x - x = 225$

$\therefore x = \frac{225}{15} = 15$

Quicker Method: Applying the above rule, we have

the required number $= \frac{225}{16-1} = \frac{225}{15} = 15$

Exercise

1. Find the number which when multiplied by 36 is increased by 1050.
   a) 30 b) 28 c) 32 d) 35
2. Find the number which when multiplied by 9 is increased by 128.
   a) 12 b) 15 c) 16 d) 18
3. Find the number which when multiplied by 17 is increased by 256.
   a) 12 b) 14 c) 18 d) 16
4. Find the number which when multiplied by 15 is increased by 378.
   a) 26 b) 16 c) 27 d) 28
5. Find the number which when multiplied by 26 is increased by 625.
   a) 26 b) 25 c) 24 d) 27

Answers

1. a 2. c 3. d 4. c 5. b

Rule 24

Theorem: Sum of all the first n natural numbers $= \frac{n(n+1)}{2}$

Illustrative Example

Ex.: Find the value of $1 + 2 + 3 + ... + 105$.

Soln: Reuired sum $= \frac{105(105+1)}{2} = 5565$

Exercise

1. Find the sum of first 45 natural numbers.
   a) 1035 b) 1235 c) 1135 d) 1305
2. Find the sum of natural numbers between 20 and 100.
   a) 4480 b) 4840 c) 4800 d) 4850
3. Find the value of $1 + 2 + 3 + ... + 210$.
   a) 22155 b) 21255 c) 22515 d) 22255
4. Find the value of $1 + 2 + 3 + ... + 62$.
   a) 1953 b) 1395 c) 1593 d) 1359
5. Find the value of
   $(1+2+3+4+...+80)-(1+2+3+...+60)$
   a) 1830 b) 1410 c) 1140 d) 1380

Answers

1. a 2. b 3. a 4. a 5. b

Rule 25

Theorem: Sum of first n odd numbers $= n^2$.

Illustrative Example

Ex.: Find the value of $1 + 3 + 5 + ... + 20$th odd number.

Soln: $20^2 = 400$.

Exercise

1. Find the sum of first 50 odd numbers.
   a) 6250 b) 2500 c) 2520 d) 2450
2. Find the value of
   $(1+3+5+...+80\text{th odd number})-(1+3+5+7+...+30\text{th odd number})$
   a) 5500 b) 6100 c) 5400 d) 7300
3. Find the value of
   $35+37+...+25$th odd number.
   a) 356 b) 336 c) 363 d) 365
4. Find the value of $1+3+5+...+199$
   a) 40000 b) 10000 c) 39601 d) Can't be determined
5. Find the value of $15 + 17 + ... + 51$
   a) 627 b) 676 c) 725 d) None of these
6. $\frac{1+3+5+...+3983}{1992}$ is equal to
   a) 1988 b) 1989 c) 1990 d) 1992

Answers

1. b 2. a

3. b; **Hint:** We have the following formula,

$t_n = a + (n-1)d$

t_n = nth term of the series

a = first term of the series

n = number of numbers

d = common difference

For the case of odd number

$a = 1, d = 2$

$\therefore t_n = 1 + (n-1)2 = 2n - 1$

We apply this formula for solving this question. First we calculate $1 + 3 + 5 + ... + 33$ and then $1 + 2 + 3 + ... + 25$th odd number. For getting required answer, we subtract first from second.

How do we calculate first ie $(1 + 3 + 5 + ... + 33)$?

We have,

$33 = 2n - 1$ [see formula) $\therefore n = 17$

$\therefore 1 + 3 + 5 + ... + 33 = 1 + 3 + 5 + ... +$ 17th odd number.

$= (17)^2 = 289$

4. b 5. a 6. d

Rule 26

Theorem: Sum of first n even numbers = *n* (*n* + 1)

Illustrative Example

Ex.: Find the value of $2 + 4 + 6 + 8 + ... + 100$ (or 50th even number)

Soln: $50 \times (50 + 1) = 2550$

Note: We have the following formula,

$t_n = a + (n-1)d$

where, t_n = nth term

a = first term

n = no. of numbers

d = common difference.

For the case of even numbers

$t_n = 2 + (n-1)2$

$= 2 + 2n - 2 = 2n$

ro, $n = \frac{t_n}{2}$

Exercise

1. Find the value of $2 + 4 + 6 + ... + 100$th even number.
   a) 10000 b) 10100 c) 11000 d) 10101
2. Find the value of $26 + 28 + ... + 28$th even number.
   a) 656 b) 665 c) 566 d) 565
3. Find the value of $2 + 4 + 6 + ... + 1002$.
   a) 251502 b) 250512 c) 215502 d) 255102
4. Find the value of $68 + 70 + ... + 180$
   a) 7608 b) 7680 c) 6078 d) 7068
5. Find the value of $2 + 4 + 6 ... + 56$th even number.
   a) 3912 b) 3192 c) 3219 d) 3129

Answers

1. b 2. a 3. a 4. d 5. b

Rule 27

Theorem: Sum of squares of first n natural numbers

$$= \frac{n(n+1)(2n+1)}{6}.$$

Illustrative Example

Ex.: Find the value of $1^2 + 2^2 + 3^2 + ... + 10^2$

Soln: $1^2 + 2^2 + 3^2 + ... + 10^2 = \frac{10(10+1)(2 \times 10 + 1)}{6}$

$= \frac{10 \times 11 \times 21}{6} = 385$

Exercise

1. Find the value of $1^2 + 2^2 + ... + 25^2$.
   a) 5255 b) 5525 c) 5552 d) 5252
2. Find the value of $25^2 + 26^2 + ... + 50^2$.
   a) 38025 b) 30825 c) 38250 d) 38205
3. Find the value of $1^2 + 2^2 + 3^2 + ... + 16^2$.
   a) 1946 b) 1649 c) 1469 d) 1496
4. Find the value of $2^2 + 3^2 + ... + 24^2$.
   a) 4899 b) 4900 c) 4901 d) 4898
5. Find the value of $1^2 + 2^2 + ... + (30th\ natural\ number)^2$
   a) 9454 b) 9544 c) 9455 d) 9555
6. $(1^2 + 2^2 + 3^2 + ... + 10^2) - (1 + 2 + 3 + ... + 10)$ is equal to
   a) 330 b) 440 c) 550 d) 660
7. If $(1^2 + 2^2 + 3^2 + ... + 10^2) = 385$, then the value of $(2^2 + 4^2 + 6^2 + ... + 20^2)$ is
   a) 770 b) 1540 c) 1155 d) (385×385)

Answers

1. b 2. a 3. d 4. a 5. c 6. a

7. b; **Hint:** $2^2 + 4^2 + ... + 20^2$

$= (1 \times 2)^2 + (2 \times 2)^2 + (2 \times 3)^2 + ... + (2 \times 10)^2$

$= 2^2[1^2 + 2^2 + 3^2 + ... + 10^2]$

$= 4 \times 385 = 1540$

Rule 28

Theorem: Sum of cubes of first n natural numbers

$$= \left[\frac{n(n+1)}{2}\right]^2.$$

Illustrative Example

Ex.: Find the value of $1^3 + 2^3 + ... + 6^3$.

Soln: $\left[\frac{6 \times (6+1)}{2}\right]^2 = (21)^2 = 441$

Exercise

1. Find the value of $1^3 + 2^3 + ... + 12^3$.
   a) 6804 b) 6084 c) 6048 d) 6408
2. Find the value of $2^3 + 3^3 + ... + 16^3$.
   a) 18496 b) 18495 c) 18497 d) 14895
3. Find the value of $8^3 + 9^3 + ... + 15^3$.
   a) 16316 b) 13661 c) 16361 d) 13616
4. Find the value of $1^3 + 2^3 + ... + (10th\ natural\ number)^3$
   a) 3025 b) 3205 c) 3052 d) 3250

5. Find the value of $2^3+3^3+4^3+...+9^3$.
a) 2024 b) 2025 c) 2225 d) 2205
6. Find the value of $3^3+4^3+...+11^3$.
a) 4356 b) 4348 c) 4347 d) 4374

Answers
1. b . 2. b 3. d 4. a 5. a 6. c

Rule 29

In the first n counting numbers, there are $\frac{n}{2}$ odd and $\frac{n}{2}$ even numbers provided n, the number of numbers, is even.

For example, from 1 to 50, there are $\frac{50}{2}=25$ odd numbers and $\frac{50}{2}=25$ even numbers.

Exercise
1. In the first 62 counting numbers, find the number of even numbers.
a) 30 b) 31 c) 32 d) 34
2. From 1 to 78, how many are the odd numbers?
a) 20 b) 38 c) 39 d) 40
3. From 1 to 28, find the number of even numbers.
a) 14 b) 13 c) 12 d) 15
4. From 1 to 100 find the number of even and the number of odd numbers.
a) 50, 50 b) 51, 50 c) 50, 51 d) 49, 50
5. From 1 to 80 how many are the even numbers?
a) 41 b) 42 c) 39 d) 40
6. From 50 to 90, find the number of odd and even numbers.
a) 20, 21 b) 20, 20 c) 21, 22 d) 19, 20

Answers
1. b 2. c 3. a 4. a 5. d 6. a

Rule 30

In the first n counting numbers, if n, the number of numbers, is odd, then there are $\frac{1}{2}(n+1)$ odd numbers and $\frac{1}{2}(n-1)$ even numbers.

For example, from 1 to 51 there are $\frac{51+1}{2}=26$ odd numbers and $\frac{51-1}{2}=25$ even numbers.

Exercise
1. In the first 61 counting numbers, find the number of even numbers.
a) 30 b) 31 c) 32 d) 29
2. From 1 to 31, how many are the odd numbers?
a) 15 b) 16 c) 14 d) 17
3. From 1 to 51, find the number of even and odd numbers.
a) 26, 25 b) 25, 26 c) 24, 25 d) 25, 24
4. From 51 to 91, find the number of even and odd numbers.
a) 20, 21 b) 21, 20 c) 21, 22 d) 19, 20
5. From 51 to 90, find the number of even and odd numbers.
a) 20, 20 b) 21, 20 c) 20, 21 d) 19, 20

Answers
1. a 2. b 3. b 4. a 5. a

Rule 31

The difference between the squares of two consecutive numbers is always an odd number and the difference between the squares of two consecutive numbers is the sum of the two consecutive numbers.
For example, 16 and 25 are squares of 4 and 5 respectively (two consecutive numbers).
∴ Difference = 25 – 16 = 9 (an odd number)
and 5^2-4^2 (Difference) = 5 + 4 = 9

Reasoning: $a^2-b^2=(a-b)(a+b)=a+b\ [\because a-b=1]$

Exercise
a) 24 b) 12 c) 18 d) 8
1. Find the value of 6^2-5^2.
a) 11 b) 9 c) 8 d) 10
2. Find the value of 35^2-34^2.
a) 59 b) 69 c) 70 d) 71
3. Find the value of
$10^2-9^2+8^2-7^2+6^2-5^2+4^2-3^2+2^2-1^2$
a) 50 b) 65 c) 45 d) 55
4. Find the value of
$29^2+35^2+33^2+31^2-34^2-32^2-30^2-28^2$.
a) 250 b) 252 c) 352 d) 342
5. Find the value of 65^2-64^2
a) 129 b) 128 c) 120 d) 125

Answers
1. a 2. b 3. d 4. b 5. a

Rule 32

To find the number in the unit place for odd numbers. When there is an odd digit in the unit place (except 5), multiply the number by itself until you get 1 in the unit place.

$(...1)^n=(...1)$
$(...3)^{4n}=(...1)$
$(...7)^{4n}=(...1)$

$(...9)^{2n} = (...1)$

where n = 1, 2, 3, ...

Illustrative Examples

Ex. 1: What is the number in the unit place in $(729)^{59}$?

Soln: When 729 is multiplied twice, the number in the unit place is 1. In other words, if 729 is multiplied an even number of times, the number in the unit place will be 1. Thus, the number in the unit place in $(729)^{58}$ is 1.

$\therefore (729)^{59} = (729)^{58} \times (729) = (...1) \times (729) = 9$ in the unit place

Ex. 2: Find the number in the unit place in $(623)^{36}$, $(623)^{38}$ and $(623)^{39}$

Soln: When 623 is multiplied twice, the number in the unit place is 9. When it is multiplied 4 times, the number in the unit place is 1. Thus we say that if 623 is multiplied 4n number of times, the number in the unit place will be 1. So,

$(623)^{36} = (623)^{4\times9} = 1$ in the unit place

$(623)^{38} = (623)^{4\times9} \times (623)^2 = (...1) \times (...9) = 9$ in the unit place.

$(623)^{39} = (623)^{4\times9} \times (623)^3 = (...1) \times (...7) = 7$ in the unit place.

Exercise

1. What is the number in the unit place in $(659)^{56}$?
   a) 1 b) 9 c) 6 d) None of these
2. What is the number in the unit place in $(329)^{79}$?
   a) 1 b) 9 c) 7 d) 4
3. What is the number in the unit place in $(147)^{48}$?
   a) 7 b) 6 c) 9 d) 1
4. What is the number in the unit place in $(87)^{90}$?
   a) 1 b) 7 c) 9 d) 3
5. What is the number in the unit place in $(127)^{127}$?
   a) 1 b) 7 c) 3 d) 9
6. What is the number in the unit place in $(5427)^{641}$?
   a) 1 b) 7 c) 9 d) 3
7. What is the number in the unit place in $(6231)^{928}$?
   a) 1 b) 8 c) 3 d) 4
8. What is the number in the unit place in $(543)^{12}$?
   a) 1 b) 3 c) 6 d) 9
9. What is the number in the unit place in $(333)^{74}$?
   a) 1 b) 6 c) 2 d) 9
10. What is the number in the unit place in $(4673)^{721}$?
    a) 1 b) 6 c) 3 d) 9
11. What is the number in the unit place in $(5483)^{843}$?
    a) 1 b) 7 c) 9 d) 3
12. What is the number in the unit place in $(1243)^{76} \times (1547)^{100}$?
    a) 1 b) 2 c) 3 d) 9
13. What is the number in the unit place $(24533)^{761} \times (12349)^{839}$?
    a) 7 b) 1 c) 9 d) 3
14. What is the number in the unit place in $(157)^{157} \times (159)^{159}$?
    a) 3 b) 9 c) 6 d) 1
15. What is the number in the unit place in $(751)^{751} \times (263)^{271} \times (137)^{138} \times (339)^{339}$?
    a) 7 b) 9 c) 1 d) 6

Answers

1. a 2. b

3. d; **Hint:** When 7 is multiplied 4 times, the number in the unit place is 1. ie if 7 is multiplied 4n number of times, the number in the unit place will be 1.

$\therefore (147)^{48} = (147)^{4\times12} = 1$ in the unit place.

4. c; **Hint:** $(87)^{90} = (87)^{4\times22} \times 87 \times 87$

$= (...1) \times (...9) = 9$

5. c 6. b 7. a 8. b 9. d 10. a 11. b

12. a; **Hint:** $(1243)^{76} = (1243)^{4\times19} = (...1)$ in the unit place.

$(1547)^{100} = (1547)^{4\times25} = (...1)$ in the unit place.

13. a; **Hint:** $(24533)^{761} = (24533)^{4\times190} \times (24533)$

$= (...1) \times (...3) = (...3)$ in the unit place

$(12349)^{839} = (12349)^{2\times419} \times (12349) = (...1)(...9) = (...9)$ in the unit place.

14. a

15. a; **Hint:** $(751)^{751} = (...1)$ in the unit place

$(263)^{271} = (263)^{4\times67} \times (263)^3$

$= (...1) \times (...7) = (...7)$ in the unit place

$(137)^{138} = (137)^{4\times34} \times (137)^2 = (...1) \times (...9) = (...9)$ in the unit place

$(339)^{339} = (339)^{2\times169} \times (339) = (...1) \times (...9) = (...9)$ in the unit place.

$\therefore$ required answer $= (...1)(...7)(...9)(...9) = (...7)$ in the unit place.

Rule 33

To find the number in the unit place for even numbers. When there is an even digit in the unit place, multiply the number by itself until you get 6 in the unit place.

$(...2)^{4n} = (...6)$

$(...4)^{2n} = (...6)$

$(...6)^{n} = (...6)$

$(...8)^{4n} = (...6)$; where n = 1, 2, 3, ...

Illustrative Examples

Ex. 1: Find the number in the unit place in $(122)^{20}$, $(122)^{22}$ and $(122)^{23}$.

Soln: $(...2) \times (...2) = ...4$
$(...2) \times (...2) \times (...2) = 8$
$(...2) \times (...2) \times (...2) \times (...2) = ...6$
We know that $(...6) \times (...6) = ...6$
Thus, when (122) is multiplied 4n times, the last digit is 6. Therefore,

$(122)^{20} = (122)^{4\times5} = (...6) = 6$ in the unit place

$(122)^{22} = (122)^{4\times5} \times (122)^2 = (...6) \times (...4) = 4$ in the unit place

$(122)^{23} = (122)^{4\times5} \times (122)^3 = (...6) \times (...8) = 8$ in the unit place.

Ex. 2: Find the number in the unit place in $(98)^{40}$, $(98)^{42}$ and $(98)^{43}$.

Soln: $(98)^4 = (...6)$

$\therefore (98)^{4n} = (...6)$

Thus, $(98)^{40} = (98)^{4\times10} = (...6) = 6$ in the unit place

$(98)^{42} = (98)^{4\times10} \times (98)^2 = (...6) \times (...4) = 4$ in the unit place

$(98)^{43} = (98)^{4\times10} \times (98)^3 = (...6) \times (...2) = 2$ in the unit place

Exercise

1. Find the number in the unit place in $(542)^{540}$.
   a) 6 b) 2 c) 3 d) 9
2. Find the number in the unit place in $(1542)^{541}$.
   a) 2 b) 4 c) 6 d) 8
3. Find the number in the unit place in $(1602)^{602}$.
   a) 2 b) 4 c) 8 d) 6
4. Find the number in the unit place in $(1392)^{91}$.
   a) 2 b) 4 c) 6 d) 8
5. Find the number in the unit place in $(194)^{64}$.
   a) 6 b) 8 c) 2 d) 4
6. Find the number in the unit place in $(5924)^{429}$.
   a) 4 b) 6 c) 8 d) 2
7. Find the number in the unit place in $(216)^{216}$.
   a) 6 b) 4 c) 8 d) 2
8. Find the number in the unit place in $(958)^{116}$.
   a) 4 b) 2 c) 8 d) 6
9. Find the number in the unit place in $(958)^{117}$
   a) 2 b) 4 c) 6 d) 8
10. Find the number in the unit place in $(958)^{118}$.
    a) 4 b) 2 c) 6 d) 8
11. Find the number in the unit place in $(958)^{119}$.
    a) 2 b) 4 c) 6 d) 8
12. Find the number in the unit place in $(1532)^{162} \times (3454)^{161} \times (1236)^{162} \times (5318)^{243}$.
    a) 2 b) 4 c) 6 d) 8
13. Find the number in the unit place in $(4152)^{51} \times (3268)^{67} \times (5913)^{83} \times (6217)^{103}$.
    a) 4 b) 2 c) 6 d) 8

Answers

1. a	2. a	3. b	4. d	5. a	6. a	7. a
8. d	9. d	10. a	11. a	12. a	13. c	

Rule 34

If there is 1, 5 or 6 in the unit place of the given number, then after any times of its multiplication, it will have the same digit in the unit place ie

$(...1)^n = (...1)$

$(...5)^n = (...5)$

$(...6)^n = (...6)$

Illustrative Example

Ex.: Find the number in the unit place in $(621)^{240}$, $(625)^{125}$, $(636)^{36}$

Soln: From the above rule,
$(621)^{240} = (...1)^{240} = 1$ in the unit place

$(625)^{125} = (...5)^{125} = 5$ in the unit place

$(636)^{36} = (...6)^{36} = 6$ in the unit place

Exercise

1. Find the number in the unit place in $(1845)^{145}$.
   a) 5 b) 3 c) 9 d) 1
2. Find the number in the unit place in $(99026)^{1456}$.
   a) 3 b) 9 c) 6 d) 1
3. Find the number in the unit place in $(441)^{441} \times (495)^{126} \times (1536)^{236}$.
   a) 1 b) 5 c) 6 d) 0
4. Find the number in the unit place in $(321)^{321} \times (325)^{326}$.
   a) 1 b) 5 c) 6 d) 8

Answers

1. a 2. c 3. d 4. b

Rule 35

Ex.: What is the number in the unit place when 781, 325, 497 and 243 are multiplied together?

Soln: Multiply all the numbers in the unit place, ie $1 \times 5 \times 7 \times 3$, the result is a number in which 5 is in the unit place.

Exercise

1. Find the number in the unit place in $962 \times 966 \times 454 \times 959$.
   a) 2 b) 4 c) 6 d) 8
2. Find the number in the unit place in $954 \times 9625 \times 43216 \times 15437 \times 12343$.
   a) 0 b) 1 c) 5 d) 6
3. Find the number in the unit place in $14532 \times 14531 \times 243 \times 245 \times 247 \times 249$.
   a) 3 b) 6 c) 4 d) 0
4. Find the number in the unit place in $1431 \times 5343 \times 9645 \times 1489$.
   a) 3 b) 6 c) 0 d) 5

Answers

1. a 2. a 3. d 4. d

Rule 36

If N is a composite number and $N = a^p b^q c^r$ ...
Where a, b, c, ... are different prime numbers and p, q, r are positive integers. Then the number of divisors are (p + 1) (q + 1) (r + 1) ...

Note: This includes unity and the number itself as divisors.

Illustrative Example

Ex.: Find the no. of divisors of 8064.

Soln: $8064 = 2^7 \times 3^1 \times 7^2$

Now, apply the above rule,

Number of divisors $= (7 + 1)(1 + 1)(2 + 1) = 48$

Exercise

1. Find the no. of divisors of 225.
   a) 4 b) 9 c) 8 d) 6
2. Find the no. of divisors of 63504.
   a) 25 b) 32 c) 75 d) 56
3. Find the no. of divisors of 17640, besides unity and itself.
   a) 12 b) 60 c) 72 d) 70
4. Find the no. of divisors of 25200, excluding unity.
   a) 90 b) 89 c) 88 d) 86
5. Find the no. of divisors of 234.
   a) 12 b) 6 c) 2 d) 8
6. Find the no. of divisors of 9000.
   a) 36 b) 48 c) 54 d) 18
7. Find the no. of divisors of 20570, besides unity and itself.
   a) 24 b) 22 c) 21 d) 18
8. Find the no. of divisors of 10000, excluding itself.
   a) 24 b) 25 c) 16 d) 32

Answers

1. b 2. c 3. d 4. b 5. a 6. b
7. b 8. a

Rule 37

Let $N = a^p b^q c^r$ ...., then the sum of the divisors of a number

$$= \frac{a^{p+1}-1}{a-1} \times \frac{b^{q+1}-1}{b-1} \times \frac{c^{r+1}-1}{c-1} \times ...$$

Note: This includes unity and the number itself as divisors.

Illustrative Example

Ex.: Find the sum of the divisors of a number 8064.

Soln: Factorize 8064 into its prime factors.

$8064 = 2^7 \times 3^1 \times 7^2$

Now, apply the above rule

$$\frac{2^{7+1}-1}{2-1} \times \frac{3^{1+1}-1}{3-1} \times \frac{7^{2+1}-1}{7-1}$$

$$= \frac{256-1}{1} \times \frac{9-1}{2} \times \frac{343-1}{6}$$

$= 255 \times 4 \times 57 = 58140.$

Exercise

1. Find the sum of the divisors of a number 225.
   a) 430 b) 403 c) 503 d) 303
2. Find the sum of the divisors of a number 63504.
   a) 213870 b) 231807 c) 213807 d) 213708
3. Find the sum of the divisors of a number 17640.
   a) 66960 b) 66690 c) 96660 d) 69660
4. Find the sum of the divisors of a number 180.
   a) 465 b) 546 c) 564 d) 654

5. Find the sum of the divisors of a number 120.
   a) 360 b) 420 c) 480 d) 630
6. Find the sum of the divisors of a number 64.
   a) 128 b) 127 c) 63 d) 130
7. Find the sum of the divisors of a number 3125.
   a) 3906 b) 3609 c) 3096 d) 3069
8. Find the number and the sum of the divisors of the number 2460 excluding one and itself.
   a) 24, 7056 b) 42, 7056 c) 24, 4594 d) 24, 4595

Answers

1. b 2. c 3. b 4. b 5. a 6. b 7. a

8. d; **Hint:** Sum of the divisors excluding 1 and itself = 7056.

$\therefore$ sum of the divisors including 1 and itself

$= 7056 - (2460 + 1) = 4595.$

Rule 38

If the places of last two digits of a three-digit number are interchanged, a new number greater than the original number by N is obtained, then the difference between the last two digits of that number is given by $\left(\frac{N}{9}\right)$ or $\left(\frac{\text{Difference in two values}}{9}\right)$.

Illustrative Example

Ex.: If the places of last two digits of a three digit number are interchanged, a new number greater than the original number by 54 is obtained. What is the difference between the last two digits of that number? **[NABARD 1999]**

Soln: Detail Method:

Let the three-digit number be $100x + 10y + z$.

According to the question,

$(100x + 10z + y) - (100x + 10y + z) = 54$

or, $9z - 9y = 54$ or $z - y = 6$

Quicker Method: Applying the above rule, we have

the required answer $= \frac{54}{9} = 6$.

Exercise

1. If the places of last two-digits of a three-digit number are interchanged, a new number greater than the original number by 18 is obtained. What is the difference between the last two digits of that number?
   a) 1 b) 2 c) 3 d) 4
2. If the places of last two-digits of a three-digit number are interchanged, a new number greater than the original number by 9 is obtained. What is the difference between the last two digits of that number?
   a) 1 b) 3 c) 4 d) 6
3. If the places of last two-digits of a three-digit number are interchanged, a new number greater than the original number by 27 is obtained. What is the difference between the last two digits of that number?
   a) 1 b) 2 c) 3 d) 4
4. If the places of last two-digits of a three-digit number are interchanged, a new number greater than the original number by 36 is obtained. What is the difference between the last two digits of that number?
   a) 1 b) 2 c) 3 d) 4
5. If the places of last two-digits of a three-digit number are interchanged, a new number greater than the original number by 45 is obtained. What is the difference between the last two digits of that number?
   a) 3 b) 4 c) 5 d) 6
6. If the places of last two-digits of a three-digit number are interchanged, a new number greater than the original number by 63 is obtained. What is the difference between the last two digits of that number?
   a) 7 b) 5 c) 6 d) 8
7. If the places of last two-digits of a three-digit number are interchanged, a new number greater than the original number by 72 is obtained. What is the difference between the last two digits of that number?
   a) 7 b) 5 c) 4 d) 8
8. If the places of last two-digits of a three-digit number are interchanged, a new number greater than the original number by 81 is obtained. What is the difference between the last two digits of that number?
   a) 7 b) 8 c) 9 d) 1

Answers

1. b 2. a 3. c 4. d 5. c 6. a
7. d 8. c

Rule 39

A number is divided by a certain number N_1 and gives a remainder 'R'. If the same number is divided by another number N_2, then the new remainder is obtained by the following method.

"Divide R by N_2 and the remainder obtained in this division will be the new remainder". (Note: Here $N_1 > N_2$ and N_1 is divisible N_2.)

Illustrative Example

Ex.: A number when divided by 899 gives a remainder 63. What remainder will be obtained by dividing the same number by 29.

Soln: Detail Method:

Number = Divisor × Quotient + Remainder = 899 × Quotient + 63

$= 29 \times 31 \times \text{Quotient} + 2 \times 29 + 5$

Therefore, the remainder obtained by dividing the number by 29 is clearly 5.

Quicker Method: Applying the above rule, we have,

$63 \div 29$ i.e. 29) 63 (2
58

5

$\therefore$ required answer = 5

Exercise

1. A number when divided by 221 gives a remainder 43, what remainder will be obtained by dividing the same number by 17?
a) 7 b) 6 c) 8 d) 9
2. A number when divided by 609 gives a remainder 65. What remainder would be obtained by dividing the same number by 29?
a) 6 b) 5 c) 6 d) 7
3. A number when divided by 738 gives a remainder 92. What remainder would be obtained by dividing the same number by 18?
a) 2 b) 1 c) 9 d) 8
4. A number when divided by 1491 gives a remainder 83. What remainder would be obtained by dividing the same number by 21?
a) 21 b) 2 c) 20 d) 18
5. A number when divided by 1092 gives a remainder 60. What remainder would be obtained by dividing the same number by 28?
a) 6 b) 2 c) 5 d) 4
6. A number when divided by 1156 gives a remainder 73. What remainder would be obtained by dividing the same number by 34?
a) 5 b) 17 c) 13 d) 4
7. A number when divided by 1836 gives a remainder 79. What remainder would be obtained by dividing the same number by 36?
a) 7 b) 9 c) 19 d) 14
8. A number when divided by 1207 gives a remainder 85. What remainder would be obtained by dividing the same number by 17?
a) 7 b) 2 c) 0 d) 6
9. A number when divided by 2470 gives a remainder 80. What remainder would be obtained by dividing the same number by 38?
a) 4 b) 18 c) 9 d) 6
10. A number when divided by 1404 gives a remainder 93. What remainder would be obtained by dividing the same number by 39?
a) 4 b) 13 c) 19 d) 15
11. A number when divided by 17, leaves a remainder 5. What remainder would be obtained by dividing the same number by 357?
a) 39 b) 29
c) 21 d) 38

Answers

1. d 2. d 3. a 4. c 5. d 6. a 7. a
8. c 9. a 10. d

11. a; **Hint:** Here we apply "Remainder Rule".

This rule is applicable when the same number (dividend) is divided by two different divisors which are multiples of each other.

Suppose, the larger divisor is N_1, and the smaller divisor is N_2.

Where, $N_1 = KN_2$ and K = any integer > 1.

Now, when the number is divided by N_2, then remainder is R_2 (say) and when the same number is divided by $N_1(=KN_2)$, remainder is R_1 (say).

Then, by the remainder rule, we have the following formula,

$\mathbf{2N_2 + R_2 = R_1}$

In the given question,

$N_2 = 17$ and $KN_2 = 357$ $\therefore K = \frac{357}{17} = 21$

Here, K > 1 an integer. Now, we can apply the remainder rule.

$2N_2 + R_2 = R_1$

or, $2 \times 17 + 5 = R_1$

$\therefore R_1 = 39$

Hence, the required remainder = 39.

Note: All the other questions can also be solved by this rule.

Rule 40

If the sum of two numbers is x and their difference is y, then the difference of their squares is xy.

Illustrative Example

Ex.: The sum of two numbers is 75 and their difference is 20. Find the difference of their squares.

Soln: **Detail Method:** Let the numbers be x and y.

According to the question,

$x + y = 75$ .... (i) and

$x - y = 20$ .... (ii)

Now, multiplying eqn (i) and (ii), we get

$x^2 - y^2$ = Difference of the squares of numbers

$= 75 \times 20 = 1500$

Quicker Method: Applying the above rule, we have, required answer = $75 \times 20 = 1500$

Exercise

1. The sum of two numbers is 100 and their difference is 37. The difference of their squares is

[Clerk's Grade Exam, 1991]

a) 37 b) 100 c) 63 d) 3700

2. The sum of two numbers is 50 and their difference is 6. The difference of their squares is
a) 400 b) 500 c) 350 d) 300
3. The sum of two numbers is 75 and their difference is 9. The difference of their squares is
a) 685 b) 625 c) 675 d) 775
4. The sum of two numbers is 160 and their difference is 39. The difference of their squares is
a) 6420 b) 4620 c) 8420 d) 6240
5. The sum of two numbers is 175 and their difference is 75. The difference of their squares is
a) 13025 b) 13125 c) 13215 d) 13152

Answers

1. d 2. d 3. c 4. d 5. b

Rule 41

If the difference between the squares of two consecutive numbers is x, then the numbers are $\left(\frac{x-1}{2}\right)$ and $\left(\frac{x+1}{2}\right)$.

Illustrative Example

Ex.: The difference between the squares of two consecutive numbers is 37. Find the numbers.

Soln: **Detail Method:** Let the numbers are x and $x+1$
According to the question,

$(x+1)^2 - x^2 = 37$

or, $x^2 + 1 + 2x - x^2 = 37$

or, $2x = 37 - 1 = 36$

$\therefore x = 18$ and $x + 1 = 19$

$\therefore$ numbers are 18, and 19

Quicker Method: Applying the above rule, we have

the required answer $= \frac{37-1}{2}$ and $\frac{37+1}{2} = 18$ and 19

Exercise

1. The difference between the squares of two consecutive numbers is 39. Find the numbers.
a) 19, 20 b) 20, 21 c) 18, 19 d) 17, 18
2. The difference between the squares of two consecutive numbers is 27. Find the numbers.
a) 14, 15 b) 13, 14 c) 15, 16 d) 16, 7
3. The difference between the squares of two consecutive numbers is 35. Find the numbers.
a) 14, 15 b) 15, 16 c) 17, 18 d) 18, 19
4. The difference between the squares of two consecutive numbers is 59. Find the numbers.
a) 29, 30 b) 30, 31 c) 28, 29 d) 27, 28
5. The difference between the squares of two consecutive numbers is 77. Find the numbers.
a) 38, 39 b) 39, 40 c) 40, 41 d) 37, 38

Answers

1. a 2. b 3. c 4. a 5. a

Rule 42

If the two consecutive numbers are x and y, then the difference of their squares is given by x + y.

Illustrative Example

Ex.: Two consecutive numbers are 8 and 9. Find the difference of their squares.

Soln: **Detail Method:**

Required answer $= 9^2 - 8^2 = 81 - 64 = 17$

Quicker Method: Applying the above rule, we have the required answer $= 8 + 9 = 17$

Exercise

1. Two consecutive numbers are 17 and 18. Find the difference of their squares.
a) 36 b) 25 c) 35 d) 34
2. Two consecutive numbers are 75 and 76. Find the difference of their squares.
a) 141 b) 151 c) 131 d) 115
3. Two consecutive numbers are 79 and 80. Find the difference of their squares.
a) 159 b) 169 c) 149 d) 158
4. Two consecutive numbers are 15 and 16. Find the difference of their squares.
a) 31 b) 32 c) 30 d) 21
5. Two consecutive numbers are 26 and 27. Find the difference of their squares.
a) 53 b) 52 c) 43 d) 63

Answers

1. c 2. b 3. a 4. a 5. a

Rule 43

If the sum of two numbers is x and sum of their squres is y, then the

(i) product of numbers is given by $\left(\frac{x^2-y}{2}\right)$ and

(ii) the numbers are $\left(\frac{x-\sqrt{2y-x^2}}{2}\right)$ and $\left(\frac{x+\sqrt{2y-x^2}}{2}\right)$.

Illustrative Example

Ex.: The sum of two numbers is 13 and the sum of their squares is 85. Find the numbers.

Soln: **Detail Method:** Let the numbers be x and y.
According to the question,

$x + y = 13$ .... (i) and $x^2 + y^2 = 85$ .... (ii)

Now, from eqn (i) and eqn (ii), we have

$(x+y)^2 = 169$

or, $x^2+y^2+2xy=169$

or, $2xy=169-85=84$

$\therefore$ $xy=42$ [xy = product of two numbers]

Again,

$(x-y)^2=(x+y)^2-4xy$

$=169-4\times 42=1$

$\therefore$ x − y = 1 .... (iii)

From eqn (i) and eqn (iii) we have,

x = 7 and y = 6

$\therefore$ Numbers are 7 and 6

Quicker Method: Applying the above rule, we have,

required answers $=\frac{13-\sqrt{170-169}}{2}$ and $\frac{13+\sqrt{170-169}}{2}=6$ and 7

Exercise

1. The sum of two numbers is 15 and sum of their squares is 113. The numbers are: **[CDS Exam, 1991]**
   a) 4, 11 b) 5, 10 c) 6, 9 d) 7, 8
2. The sum of two numbers is 25 and sum of their squares is 313. The numbers are:
   a) 12, 13 b) 20, 25 c) 9, 16 d) 21, 4
3. The sum of two numbers is 26 and sum of their squares is 340. The numbers are:
   a) 12, 14 b) 11, 15 c) 9, 17 d) 8, 18
4. The sum of two numbers is 30 and sum of their squares is 458. The numbers are:
   a) 14, 16 b) 12, 18 c) 13, 17 d) 11, 15
5. The sum of two numbers is 14 and sum of their squares is 100. The numbers are:
   a) 6, 8 b) 5, 9 c) 4, 10 d) 3, 11
6. The sum of two numbers is 13 and sum of their squares 89. Find the product of the two numbers.
   a) 40 b) 36 c) 22 d) 30
7. The sum of two numbers is 32 and sum of their squares 514. Find the product of the two numbers.
   a) 510 b) 225 c) 255 d) 355

Answers

1. d 2. a 3. a 4. c 5. a 6. a 7. c

Rule 44

If the sum of squares of two numbers is x and the square of their difference is y, then the product of the two numbers is $\left(\frac{x-y}{2}\right)$.

Illustrative Example

Ex.: The sum of squares of two numbers is 90 and the square of their difference is 46. The product of the two numbers is ____.

Soln: **Detail Method:** Let the numbers be x and y.

According to the question,

$x^2+y^2=90$ ..... (i) and

$(x-y)^2=46$ .... (ii)

From eqn (ii)

$(x-y)^2=46$

or, $x^2+y^2-2xy=46$

or, $90-2xy=46$ [Putting the value of eqn (i)]

or, $xy=\frac{90-46}{2}=22$

$\therefore$ product of two numbers = 22

Quicker Method: Applying the above rule, we have

the required answer $=\frac{90-46}{2}=22$

Exercise

1. The sum of squares of two numbers is 80 and the square of their difference is 36. The product of the two numbers is **[Clerks' Grade Exam, 1991]**
   a) 22 b) 44 c) 58 d) 116
2. The sum of squares of two numbers is 40 and the square of their difference is 20. The product of the two numbers is
   a) 10 b) 20 c) 15 d) 16
3. The sum of squares of two numbers is 95 and the square of their difference is 37. The product of the two numbers is
   a) 18 b) 19 c) 29 d) 27
4. The sum of squares of two numbers is 94 and the square of their difference is 24. The product of the two numbers is
   a) 36 b) 40 c) 30 d) 35
5. The sum of squares of two numbers is 87 and the square of their difference is 25. The product of the two numbers is
   a) 31 b) 35 c) 32 d) 30

Answers

1. a 2. a 3. c 4. c 5. a

Rule 45

If the product of two numbers is x and the sum of their squares is y, then (i) the sum of the two numbers is given by $\sqrt{y+2x}$ ***and (ii) the difference is*** $\sqrt{y-2x}$.

Illustrative Example

Ex.: The product of two numbers is 143. The sum of their squares is 290. Find the sum of the two numbers and also find the difference of the two numbers.

Soln: **Detail Method:** Let the numbers be x and y.

According to the question,

$xy = 143$ and $x^2 + y^2 = 290$

Now,

$(x+y)^2 = x^2 + y^2 + 2xy$

$= 290 + 2 \times 143 = 576$

or, $x + y = \sqrt{576} = 24$

$\therefore$ Sum of the numbers = 24

Again,

$(x-y)^2 = x^2 + y^2 - 2xy$

$= 290 - 286 = 4$

or, $x - y = 2$

$\therefore$ difference of the numbers = 2

Quicker Method: Applying the above rule, we have the sum of the numbers

$= \sqrt{290 + 2 \times 143} = \sqrt{576} = 24$ and

the difference of the numbers

$= \sqrt{290 - 2 \times 143} = \sqrt{4} = 2$

Exercise

1. The product of two numbers is 120. The sum of their squares is 289. The sum of the two numbers is _____.
[Clerks' Grade Exam, 1991]
a) 20 b) 23 c) 169 d) 33
2. The product of two numbers is 48. The sum of their squares is 100. The sum of the two numbers is _____.
a) 14 b) 12 c) 18 d) 24
3. The product of two numbers is 168. The sum of their squares is 340. The sum of the two numbers is _____.
a) 36 b) 24 c) 26 d) 28
4. The product of two numbers is 36. The sum of their squares is 97. The sum of the two numbers is _____.
a) 13 b) 12 c) 15 d) 11
5. The product of two numbers is 35. The sum of their squares is 74. The sum of the two numbers is _____.
a) 13 b) 12 c) 14 d) 17
6. The product of two numbers is 120. The sum of their squares is 289. The difference of the two numbers is _____.
a) 7 b) 9 c) 8 d) 23
7. The product of two numbers is 180. The sum of their squares is 369. The difference of the two numbers is _____.
a) 3 b) 27 c) 5 d) 17
8. The product of two numbers is 224. The sum of their squares is 452. The difference of the two numbers is _____.
a) 30 b) 2 c) 4 d) 15

Answers

1. b 2. a 3. c 4. a 5. b 6. a
7. a 8. b

Rule 46

The denominator of a rational number is 'D' more than its numerator. If the numerator is increased by x and the denominator is decreased by y, we obtain P, then the rational number is given by $\left[\frac{x - P(D-y)}{x + (yP - D)}\right]$.

Illustrative Example

Ex.: The denominator of a rational number is 3 more than its numerator. If the numerator is increased by 7 and the denominator is decreased by 2, we obtain 2. The rational number is _____.

Soln: **Detail Method:** Let the numerator be x and the denominator $= x + 3$.

According to the question,

$\frac{x+7}{x+3-2} = 2$

or, $x + 7 = 2x + 2 \quad \therefore x = 5$

$\therefore$ Numerator = 5 and the denominator = 5 + 3 = 8

$\therefore$ rational number $= \frac{5}{8}$

Quicker Method: Applying the above rule, we have

Required answer $= \frac{7 - 2(3-2)}{7 + (2 \times 2 - 3)} = \frac{5}{8}$

Exercise

1. The numerator of a rational number is 4 less than its denominator. If the numerator is increased by 8 and the denominator is decreased by 2, we obtain 3. Find the rational number.
a) $\frac{7}{11}$ b) $\frac{3}{7}$ c) $\frac{1}{5}$ d) $\frac{5}{9}$
2. The denominator of a rational number is 6 more than its numerator. If the numerator is increased by 9 and the denominator is decreased by 5, we obtain 5. Find the rational number.
a) $\frac{1}{7}$ b) $\frac{2}{8}$ c) $\frac{7}{13}$ d) $\frac{5}{11}$
3. The denominator of a rational number is 3 more than its numerator. If the numerator is increased by 6 and the denominator is decreased by 2, we obtain 2. Find the rational number.
a) $\frac{1}{3}$ b) $\frac{5}{8}$ c) $\frac{7}{10}$ d) $\frac{4}{7}$
4. The denominator of a rational number is 8 more than its numerator. If the numerator is increased by 7 and the denominator is decreased by 8, we obtain 8. Find the

rational number.

a) $\frac{1}{9}$ b) $\frac{2}{10}$ c) $\frac{3}{11}$ d) $\frac{5}{13}$

5. The denominator of a rational number is 2 more than its numerator. If the numerator is increased by 9 and the denominator is decreased by 5, we obtain 7. Find the rational number.

a) $\frac{5}{7}$ b) $\frac{7}{9}$ c) $\frac{9}{11}$ d) $\frac{3}{5}$

6. The denominator of a fraction is 2 more than thrice its numerator. If the numerator as well as denominator is increased by one, the fraction becomes $\frac{1}{3}$. What was the original fraction. **[SBI PO, 1999]**

a) $\frac{4}{13}$ b) $\frac{3}{11}$ c) $\frac{5}{13}$ d) $\frac{5}{11}$

Answers

1. c 2. a 3. d 4. a 5. a

6. b; **Hint:** This type of question may be solved by **hit and trial method.**

First divide the question in different parts. Then start from the answer-choices one-by-one. The choice, which satisfies all the parts of the given question, will be required answer. For example, in the above question we have two parts.

(I) The denominator of a fraction is 2 more than thrice its numerator.

(II) If the numerator as well as denominator is increased by 1, the fraction becomes 1/3.

Both parts will be satisfied by the answer choice (b), hence (b) is the required answer.

Rule 47

When a number 'A' is added to another number 'B' and the total ie (A + B) becomes P% of the number B, then the ratio between A and B is given by $\left(\frac{P-100}{100}\right)$.

Illustrative Example

Ex.: When a number is added to another number the total becomes 150 per cent of the second number. What is the ratio between the first and the second number?

Soln: **Detail Method:**

Let the numbers be x and y.

According to the question,

$x+y=150\%\text{ of }y=\frac{150}{100}y$

or, $x+y=\frac{3}{2}y$ or, $x=\frac{1}{2}y$

$\therefore \frac{x}{y}=\frac{1}{2}$ $\therefore x:y=1:2$

Quicker Method: Applying the above rule, we have

the required ratio $=\frac{150-100}{100}=\frac{1}{2}=1:2$

Note: In case the total ie (A + B) becomes P% of the number A, the ratio between A and B is given by $\left(\frac{100}{P-100}\right)$.

Exercise

1. When a number is added to another number the total becomes $333\frac{1}{3}$ per cent of the first number. What is the ratio between the first and the second number?

a) 3 : 7 b) 7 : 4 c) 7 : 3 d) Data inadequate

2. When a number is added to another number the total becomes $333\frac{1}{3}$ per cent of the second number. What is the ratio between the first and the second number? **[SBI PO 2000]**

a) 3 : 7 b) 7 : 4 c) 7 : 3 d) 4 : 7

3. When a number is added to another number the total becomes 250 per cent of the second number. What is the ratio between the first and the second number?

a) 3 : 2 b) 2 : 3 c) 4 : 3 d) 3 : 4

4. When a number is added to another number the total becomes 175 per cent of the first number. What is the ratio between the first and the second number?

a) 4 : 3 b) 3 : 4 c) 5 : 3 d) 3 : 5

5. When a number is added to another number the total becomes 275 per cent of the first number. What is the ratio between the first and the second number?

a) 4 : 7 b) 7 : 4 c) 3 : 8 d) 8 : 3

6. When a number is added to another number the total becomes 125 per cent of the second number. What is the ratio between the first and the second number?

a) 1 : 4 b) 4 : 1 c) 1 : 2 d) 2 : 1

7. When a number is added to another number the total becomes 375 per cent of the second number. What is the ratio between the first and the second number?

a) 4 : 11 b) 11 : 4 c) 4 : 7 d) 7 : 4

8. When a number is added to another number the total becomes 375 per cent of the first number. What is the ratio between the first and the second number?

a) 4 : 11 b) 11 : 4 c) 4 : 7 d) 7 : 4

9. When a number is added to another number the total becomes 225 per cent of the first number. What is the ratio between the first and the second number?

a) 5 : 4 b) 4 : 5 c) 3 : 4 d) 4 : 3

10. When a number is added to another number the total becomes 225 per cent of the second number. What is the

ratio between the first and the second number?

a) 3 : 4 b) 4 : 3 c) 5 : 4 d) 4 : 5

Answers

1. a 2. c 3. a 4. a 5. a 6. a 7. b
8. a 9. b 10. c

Rule 48

The sum of three consecutive even or odd numbers is P less or more than $\frac{a}{b}$ of Q. Then the middle number is given by

$$\left[\frac{Q\left(\frac{a}{b}\right) \pm P}{3}\right].$$

Note: +ve and –ve sign indicate more and less respectively.

Illustrative Example

Ex: The sum of three consecutive even numbers is 15 less than three-fourth of 60. What is the middle number?

Soln: **Detail Method:**

Let the middile number be x

According to the question,

$$x - 2 + x + x + 2 = \frac{60 \times 3}{4} - 15$$

or, $3x = 30 \therefore x = 10$

$\therefore$ required answer = 10

Quicker Method: Since we have **less** type of question, the above formula will be like

$$\text{Middle number} = \frac{Q\left(\frac{a}{b}\right) - P}{3} = \frac{60 \times \frac{3}{4} - 15}{3} = 10.$$

Exercise

1. The sum of three consecutive even numbers is 14 less than one-fourth of 176. What is the middle number. **[BSRB Mumbai PO, 1998]**
   a) 10 b) 8 c) 6 d) 4
2. The sum of three consecutive odd numbers is 15 more than one fourth of 120. What is the middle number?
   a) 15 b) 13 c) 17 d) 21
3. The sum of three consecutive even numbers is 24 less than one-sixth of 324. What is the middle number?
   a) 12 b) 10 c) 14 d) 20
4. The sum of three consecutive even numbers is 8 less than two-third of 66. What is the middle number?
   a) 10 b) 18 c) 16 d) 12
5. The sum of three consecutive odd numbers is 25 more than two-fifth of 65. What is the middle number?
   a) 15 b) 19 c) 17 d) 21

Answers

1. a 2. a 3. b 4. d 5. c

Rule 49

Two different numbers when divided by the same divisor, leaves remainders x and y respectively, and when their sum is divided by the same divisor, remainder is z, then the divisor is given by (x + y - z). Or,

Divisor = (sum of remainders) - (Remainder when sum is divided)

Illustrative Example

Ex: Two different numbers when divided by the same divisor, left remainders 11 and 21 respectively, and when their sum was divided by the same divisor, remainder was 4. What is the divisor?

Soln: Applying the above rule, we have the required answer $= 11 + 21 - 4 = 28$

Exercise

1. Two different numbers when divided by the same divisor, left remainders 10 and 15 respectively, and when their sum was divided by the same divisor, remainder was 3. What is the divisor?
   a) 22 b) 25 c) 23 d) 21
2. Two different numbers when divided by the same divisor, left remainders 5 and 7 respectively, and when their sum was divided by the same divisor, remainder was 2. What is the divisor?
   a) 11 b) 12 c) 10 d) 9
3. Two different numbers when divided by the same divisor, left remainders 13 and 23 respectively, and when their sum was divided by the same divisor, remainder was 5. What is the divisor?
   a) 32 b) 36 c) 30 d) 31
4. Two different numbers when divided by the same divisor, left remainders 12 and 21 respectively, and when their sum was divided by the same divisor, remainder was 4. What is the divisor?
   a) 28 b) 27 c) 31 d) 29
5. Two different numbers when divided by the same divisor, left remainders 15 and 17 respectively, and when their sum was divided by the same divisor, remainder was 8. What is the divisor?
   a) 24 b) 25 c) 32 d) 42

Answers

1. a 2. c 3. d 4. d 5. a

Rule 50

If the product of two numbers is x and the sum of these two numbers is y, then the numbers are given by $\left(\frac{y + \sqrt{y^2 - 4x}}{2}\right)$

and $\left(\frac{y-\sqrt{y^2-4x}}{2}\right)$.

Illustrative Example

Ex: The product of two numbers is 192 and the sum of these two numbers is 28. What is the smaller of these two numbers?

[BSRB Calcutta PO 1999]

Soln: Detail Method:

Let the numbers be x and y.

$\therefore xy = 192, x + y = 28$ .....(i)

$\therefore (x-y)^2 = (x+y)^2 - 4xy$

$= 784 - 768 = 16$

$\therefore x - y = 4$ ....(ii)

Combining eqn (i) and eqn (ii)

$x = 16$, and $y = 12$

$\therefore$ smaller number = 12.

Quicker Method: Applying the above rule, we have

the required numbers = $\frac{28+\sqrt{28^2-4\times192}}{2}$

$= \frac{28+\sqrt{784-768}}{2} = \frac{28+4}{2}$

= 16 and

$\frac{28-\sqrt{28^2-4\times192}}{2} = \frac{28-4}{2} = \frac{24}{2} = 12$.

$\therefore$ smaller number = 12.

Exercise

1. The product of two numbers is 154 and the sum of these two numbers is 25. Find the difference between the numbers.
   a) 3 b) 4 c) 5 d) 8
2. The product of two numbers is 252 and the sum of these two numbers is 33. Find the greater number.
   a) 21 b) 12 c) 13 d) 23
3. The product of two numbers is 255 and the sum of these two numbers is 32. Find the smaller number.
   a) 17 b) 16 c) 15 d) 13
4. The product of two numbers is 168 and the sum of these two numbers is 26. Find the smaller number.
   a) 12 b) 14 c) 16 d) 18
5. The product of two numbers is 486 and the sum of these two numbers is 45. Find the smaller number.
   a) 12 b) 18 c) 26 d) 34

Answers

1. a 2. a 3. c 4. a 5. b

Rule 51

If the product of two numbes is x and the difference between these two numbers is y, then the numbers are $\left(\frac{\sqrt{y^2+4x}+y}{2}\right)$ ***and*** $\left(\frac{\sqrt{y^2+4x}-y}{2}\right)$.

Illustrative Example

Ex: The product of two numbers is 192 and the difference of these two numbers is 4. What is the greater of these two numbers?

Soln: Detail Method:

Let the numbers is x and y.

$xy = 192$ and $x - y = 4$ ....(i)

$(x+y)^2 = (x-y)^2 + 4xy$

$= (4)^2 + 4\times192 = 784$

$x + y = 28$ ....(ii)

Solving eqn (i) and eqn (ii) we have

$x = 16$ and $y = 12$

$\therefore$ Greater number = 16

Quicker Method: Applying the above rule, we have required answer =

$\frac{\sqrt{y^2+4x}+y}{2} = \frac{\sqrt{784}+4}{2} = \frac{28+4}{2}$

$= \frac{32}{2} = 16$.

Note: $\frac{\sqrt{y^2+4x}+y}{2} > \frac{\sqrt{y^2+4x}-y}{2}$

Exericse

1. The product of two numbers is 221 and the difference of these two numbers is 4. Find the smaller number.
   a) 13 b) 14 c) 16 d) 17
2. The product of two numbers is 198 and the difference of these two numbers is 7. Find the greater number.
   a) 18 b) 15 c) 13 d) 11
3. The product of two numbers is 180 and the difference of these two numbers is 3. Find the sum of the numbers.
   a) 26 b) 25 c) 28 d) 27
4. The product of two numbers is 594 and the difference of these two numbers is 5. Find the sum of the numbers.
   a) 46 b) 39 c) 40 d) 49
5. The product of two numbers is 468 and the difference of these two numbers is 8. Find the sum of the numbers.
   a) 42 b) 44 c) 48 d) 34

Answers

1. a 2. a 3. d 4. d 5. b

Miscellaneous

1. If a fraction's numerator is increased by 1 and the denominator is increased by 2 then the fraction becomes $\frac{2}{3}$. But when the numerator is increased by 5 and the denominator is increased by 1 then the fraction becomes $\frac{5}{4}$. What is the value of the original fraction?

[Bank of Baroda PO, 1999]

a) $\frac{3}{7}$ b) $\frac{5}{8}$ c) $\frac{5}{7}$ d) $\frac{6}{7}$

2. If the numerator of a fraction is increased by 2 and denominator is increased by 3, the fraction becomes 7/9; and if numerator as well as denominator are decreased by 1 the fraction becomes 4/5. What is the original fraction? **[SBI Associates PO, 1999]**

a) $\frac{13}{16}$ b) $\frac{9}{11}$ c) $\frac{5}{6}$

d) $\frac{17}{21}$ e) None of these

3. If the numerator of a fraction is increased by 2 and the denominator is increased by 1, the fraction becomes $\frac{5}{8}$ and if the numerator of the same fraction is increased by 3 and the denominator is increased by 1 the fraction becomes $\frac{3}{4}$. What is the original fraction?

[Guwahati PO Exam, 1999]

a) Data inadequate b) $\frac{2}{7}$ c) $\frac{4}{7}$

d) $\frac{3}{7}$ e) None of these

4. In a two-digit number, the digit at unit place is 1 more than twice of the digit at tens place. If the digit at unit and tens place be interchanged, then the difference between the new number and original number is less than 1 to that of original number. What is the original number? **[BSRB Hyderabad PO, 1999]**

a) 52 b) 73 c) 25
d) 49 e) 37

5. $\frac{1}{5}$ of a number is equal to $\frac{5}{8}$ of the second number. If 35 is added to the first number then it becomes 4 times of second number. What is the value of the second number? **[BSRB Hyderabad PO, 1999]**

a) 125 b) 70 c) 40
d) 25 e) None of these

6. The ratio of two numbers is 3 : 2. If 10 and the sum of the two numbers are added to their product, square of sixteen is obtained. What could be the smaller number?

[NABARD, 1999]

a) 14 b) 12 c) 16
d) 18 e) None of these

7. The numbers x, y, z are such that xy = 96050 and xz = 95625 and y is greater than z by one. Find out the number z. **[NABARD, 1999]**

a) 425 b) 220 c) 525
d) 226 e) 225

8. If the sum of one-half, one-third and one-fourth of a number exceeds the number itself by 4, what could be the number? **[NABARD, 1999]**

a) 24 b) 36 c) 72
d) 84 e) None of these

9. When any number is divided by 12 then dividend becomes $\frac{1}{4}$ of the other number. By how much per cent is first number greater than the second number?

[BSRB Chennai PO, 2000]

a) 200 b) 150 c) 300
d) Data inadequate e) None of these

10. A number gets reduced to its one-third when 48 is substracted from it. What is two-third of that number?

[BSRB Bhopal PO, 2000]

a) 24 b) 72 c) 36
d) 46 e) None of these

11. The sum of three consecutive numbers is given. What is the difference between first and third number?

[BSRB Bhopal PO, 2000]

a) One 2) Three
c) Either one or three d) Two
e) None of these

12. If the two digits of the age of Mr Manoj are reversed then the new age so obtained is the age of his wife. $\frac{1}{11}$ of the sum of their ages is equal to the difference between their ages. If Mr Manoj is elder than his wife then find the difference between their ages.

[BSRB Bangalore PO, 2000]

a) Cannot be determined b) 10 years
c) 8 years d) 7 years
e) 9 years

13. A number is greater than the square of 44 but smaller than the square of 45. If one part of the number is the square of 6 and the number is a multiple of 5, then find the number. **[BSRB Bangalore PO, 2000]**

a) 1940 b) 2080 c) 1980
d) Cannot be determined e) None of these

14. If a number is decreased by 4 and divided by 6 the result

is 9. What would be the result if 3 is subtracted from the number and then it is divided by 5?

[BSRB Delhi PO, 2000]

a) $9\frac{2}{5}$ b) $10\frac{1}{5}$ c) $11\frac{2}{5}$

d) 11 e) None of these

15. A two-digit number is seven times the sum of its digits. If each digit is increased by 2, the number thus obtained is 4 more than six times the sum of its digits. Find the number. **[BSRB Patna PO, 2001]**

a) 42 b) 24 c) 48

d) Data inadequate e) None of these

16. The digit in the units place of a number is equal to the digit in the tens place of half of that number and the digit in the tens place of that number is less than the digit in units place of half of the number by 1. If the sum of the digits of the number is seven, then what is the number? **[SBI Bank PO, 2001]**

a) 52 b) 16 c) 34

d) Data inadequate e) None of these

17. A fraction becomes 4 when 1 is added to both the numerator and denominator, and it becomes 7 when 1 is subtraced from both the numerator and denominator. The numerator of the given fraction is:

a) 2 b) 3 c) 7 d) 15

(NDA Exam 1990)

18. If 1 is added to the denominator of a fraction, the fraction becomes (1/2). If 1 is added to the numerator, the fraction becomes 1. The fraction is:

a) $\frac{4}{7}$ b) $\frac{5}{9}$ c) $\frac{2}{3}$ d) $\frac{10}{11}$

[CDS Exam 1991]

19. The sum of two numbers is twice their difference. If one of the numbers is 10, the other number is:

a) $3\frac{1}{3}$ b) 30 c) 30 or $-3\frac{1}{3}$ d) 30 or $3\frac{1}{3}$

[RRB Exam 1991]

20. $\frac{4}{5}$ of a certain number is 64. Half of that number is:

a) 32 b) 40 c) 80 d) 16

[BSRB Exam 1991]

21. $\frac{1}{4}$ of a number subtracted from $\frac{1}{3}$ of the number gives 12. The number is:

a) 144 b) 120 c) 72 d) 63

[Hotel Management, 1991]

22. If one fifth of a number decreased by 5 is 5, then the number is:

a) 25 b) 50 c) 60 d) 75

[Clerks' Grade Exam 1991]

23. 11 times a number gives 132. The number is

a) 11 b) 12 c) 13.2 d) None of these

[Clerks' Grade Exam 1991]

24. A number is 25 more than its $\frac{2}{5}th$. The number is:

[Clerks' Grade Exam, 1991]

a) $\frac{125}{3}$ b) $\frac{125}{7}$ c) 60 d) 80

25. 24 is divided into two parts such that 7 times the first part added to 5 times the second part makes 146. The first part is:

a) 11 b) 13 c) 16 d) 17

[RRB Exam 1991]

26. $\frac{4}{5}$ of a number exceeds its $\frac{2}{3}$ by 8. The number is:

a) 30 b) 60 c) 90 d) None of these

[RRB Exam 1989]

27. The difference between squares of two numbers is 256000 and sum of the numbers is 1000. The numbers are:

a) 628, 372 b) 600, 400

c) 640, 630 d) None of these

[GICAAO Exam, 1988]

28. Three numbers are in the ratio 3 : 4 : 5. The sum of the largest and the smallest equals the sum of the third and 52. The smallest number is:

a) 20 b) 27 c) 39 d) 52

[Accountants' Exam 1986]

29. A positive number when decreased by 4, is equal to 21 times the reciprocal of the number. The number is:

a) 3 b) 5 c) 7 d) 9

[NDA Exam 1987]

30. The sum of 3 numbers is 68. If the ratio between first and second be 2 : 3 and that between second and third be 5 : 3, then the second number is:

a) 30 b) 20 c) 58 d) 48

[SSC Exam 1986]

31. Two numbers are such that the ratio between them is 3 : 5; but if each is increased by 10, the ratio between them becomes 5 : 7. The numbers are:

a) 3, 5 b) 7, 9 c) 13, 22 d) 15, 25

[RRB Exam. 1989]

32. Divide 50 into two parts so that the sum of their reciprocals is (1/12):

a) 20, 30 b) 24, 26 c) 28, 22 d) 36, 14

[RRB Exam 1988]

33. The sum of seven numbers is 235. The average of the first three is 23 and that of the last three is 42. The fourth number is:

a) 40 b) 126 c) 69 d) 195

[Clerks' Grade Exam. 1991]

34. How many figures (digits) are required to number a book

containing 200 pages?

a) 200 b) 600 c) 492 d) 372

[MBA 1980]

35. In a question, divisor is $\frac{2}{3}$ of the dividend and 2 times the remainder. If the remainder is 5, find the dividend.

a) 15 b) 25 c) 18 d) 24

[SSC 94]

36. A number when divided by 5 leaves a remainder 3. What is the remainder when the square of the same number is divided by 5?

a) 9 b) 3 c) 1 d) 4

[MBA 1990]

37. Assuming that A, B and C are different single-digit numerical values other than what is already used in the following equation, what number C definitely cannot be?

8A2 + 3B5 + C4 - 1271

a) 7 b) 9 c) Either 7 or 9
d) 6 e) None of these

Answers

1. c; Let the fraction be $\frac{x}{y}$ then,

$\frac{x+1}{y+2} = \frac{2}{3}$ or, 3x + 3 = 2y + 4

or, 3x = 2y + 1 ....(i)

Also, we have $\frac{x+5}{y+1} = \frac{5}{4}$

or, 4x + 20 = 5y + 5 or, 4x = 5y - 15 ....(ii)

From (i) and (ii), we get

$\frac{2y+1}{3} = \frac{5y-15}{4}$ or, 8y + 4 = 15y - 45

or, 7y = 49 ∴ y = 7 and $x = \frac{2y+1}{3} = \frac{2 \times 7+1}{3} = 5$

∴ required original fraction = $\frac{x}{y} = \frac{5}{7}$

2. c; Let the numerator and denominator be x and y respectively. Then $\frac{x+2}{y+3} = \frac{7}{9}$

or, $9(x+2) = 7(y+3)$ or, $9x - 7y = 3$ ....(i)

$\frac{x-1}{y-1} = \frac{4}{5}$ or, 5x - 4y = 1 ....(ii)

Solving (i) and (ii), we get x = 5, y = 6

Reqd fraction = 5/6

3. d; Let the original fraction be $\frac{x}{y}$.

Then $\frac{x+2}{y+1} = \frac{5}{8}$ or, 8x - 5y = -11 ....(i)

Again, $\frac{x+3}{y+1} = \frac{3}{4}$ or, 4x - 3y = -9 ....(ii)

solving eqn (i) and (ii), we get

$x = 3$ and y = 7 ∴ fraction = $\frac{3}{7}$

4. e; Let the original number be 10x + y

y = 2x + 1 ....(i)

and (10y + x) - (10x + y) = 10x + y - 1

or, 9y - 9x = 10x + y - 1 or, 19x - 8y = 1 ....(ii)

Putting the value of (i) in equation (ii) we get,

19x - 8(2x + 1) = 1 or, 19x - 16x - 8 = 1

or, 3x = 9 or, x = 3 So, y = 2 × 3 + 1 = 7

∴ original number = 10 × 3 + 7 = 37

5. c; $\frac{1}{5}I = \frac{5}{8}II$ ∴ $\frac{I}{II} = \frac{25}{8}$ ....(i)

$I + 35 = 4II$ or, $\frac{25}{8}II + 35 = 4II$

∴ $II = 40$

6. b; Let the two numbers be 3x and 2x.

According to the question,

$10 + (3x + 2x) + (3x \times 2x) = (16)^2$

or, $6x^2 + 5x - 246 = 0$ or, $6x^2 + 41x - 36x - 246 = 0$

or, 6x(6x + 41) - 6(6x + 41) = 0

or, (6x + 41) (x - 6) = 0

∴ $x = 6$

or $\frac{-41}{6}$ (But -ve value cannot be accepted)

So, x = 6. Hence, smaller number = 2x = 12

7. e; xy = 96050 ...(i) and xz = 95625 ....(ii)

and y - z = 1 ...(iii)

Dividing (i) by (ii), we get $\frac{y}{z} = \frac{96050}{95625} = \frac{3842}{3825}$ ....(iv)

Combining (iii) and (iv), we get z = 225

8. e; Let the number be x.

$\therefore \left(\frac{1}{2} + \frac{1}{3} + \frac{1}{4}\right)x = \left(\frac{6+4+3}{12}\right)x = \frac{13}{12}x$

According to the question,

$\frac{13}{12}x - x = 4$ ∴ x = 48

9. d; Here neither the remainder nor the dividend nor the second number is given, so can't be determined.

10. e; Let the number be x.

then, $x - \frac{x}{3} = 48$ ∴ $\frac{2}{3}x = 48$

11. d; Let the three consecutive numbers be x, x + 1 and x + 2 respectively.
$\therefore$ Diff. between first and third numbers
$= x + 2 - x = 2$

12. e; Let the age of Mr Manoj be (10x + y) yrs.
$\therefore$ His wife's age = (10y + x) years

Then, $(10x + y + 10y + x)\frac{1}{11} = 10x + y - 10y - x$

or, $x + y = 9x - 9y$ or, $8x = 10y$ or, $\frac{x}{y} = \frac{5}{4}$

$\therefore$ x = 5 and y = 4 (because any other multiple of 5 will make x of two digits)
$\therefore$ Diff $= 10x + y - 10y - x = 9x - 9y = 9(x - y)$
$= 9(5 - 4) = 9$ yrs

13. c; Let the number be x.

$44^2 < x < 45^2 \Rightarrow 1936 < x < 2025$ ....(i)

From equation (i), the required number will be any number between 1936 and 2025.
Since one part of the number is the square of 6 means one factor is 36.
$\therefore$ LCM of 36 and 5 = 180
$\therefore$ Number will be multiple of 180 ie $180 \times 11 = 1980$ the only value which satisfies the equation (i)

14. d; Let the number be x

$\therefore \frac{x-4}{6} = 9 \Rightarrow x = 58$ Again, $\frac{x-3}{5} = \frac{58-3}{5} = 11$

15. a; Let the two-digit number be 10x + y.
$10x + y = 7(x + y) \Rightarrow x = 2y$ ....(i)
$10(x + 2) + y + 2 = 6(x + y + 4) + 4$
or, $10x + y + 22 = 6x + 6y + 28 \Rightarrow 4x - 5y = 6$ .....(ii)
Solving equations (i) and (ii), we get x = 4 and y = 2

16. a; Let 1/2 of the no. = 10x + y
and the no. = 10V + W
From the given conditions, W = x and V = y - 1
Thus the no. = 10(y - 1) + x ....(A)
$\therefore 2(10x + y) = 10(y - 1) + x \Rightarrow 8y - 19x = 10$ ....(i)
$V + W = 7 \Rightarrow y - 1 + x = 7$ $\therefore x + y = 8$ ....(ii)
Solving equations (i) and (ii), we get
x = 2 and y = 6 $\therefore$ From equation (A),
Number = 10(y - 1) + x = 52

17. d; Let the required fraction be $\frac{x}{y}$.

Then, $\frac{x+1}{y+1} = 4 \Rightarrow x - 4y = 3$

And, $\frac{x-1}{y-1} = 7 \Rightarrow x - 7y = -6$

Solving (i) and (ii), we get x = 15, y = 3.

18. c; Let the required fraction = $\frac{x}{y}$.

$\therefore \frac{x}{y+1} = \frac{1}{2} \Rightarrow 2x - y = 1$

And, $\frac{x+1}{y} = 1 \Rightarrow x - y = -1$

Solving (i) and (ii), we get x = 2, y = 3.

$\therefore$ The fraction is $\frac{2}{3}$.

19. b; Let the other number = x
$10 + x = 2(x - 10) \Rightarrow x = 30$.

20. b; $\frac{4}{5} \times x = 64 \Rightarrow x = \frac{64 \times 5}{4} = 80$

$\therefore \frac{1}{2} \times x = \frac{1}{2} \times 80 = 40$

21. a; $\left(\frac{1}{3} \times x - \frac{1}{4} \times x\right) = 12 \Rightarrow \frac{1}{12}x = 12 \Rightarrow x = 144$.

22. b; $\left(\frac{1}{5} \text{ of } x\right) - 5 = 5 \Rightarrow \frac{x}{5} = 10 \Rightarrow x = 50$.

23. b; $11x = 132 \Rightarrow x = 12$

24. a; $x - \frac{2}{5}x = 25 \Rightarrow \frac{3x}{5} = 25 \Rightarrow x = \frac{125}{3}$.

25. b; Let these parts be x and (24 - x). Then,
$7x + 5(24 - x) = 146 \Rightarrow x = 13$.
So the first part is 13.

26. b; Let the number be *x*. Then,

$\frac{4}{5}x - \frac{2}{3}x = 8 \Rightarrow \frac{2}{15}x = 8 \Rightarrow x = 60$.

27. a; Let the numbers be x and y. Then,

$x^2 - y^2 = 256000$ and $x + y = 1000$.

$\therefore x - y = \frac{x^2 - y^2}{x + y} = \frac{256000}{1000} = 256$

Solving x + y = 1000, x - y = 256, we get x = 628, y = 372.

28. c; Let the numbers be 3x, 4x and 5x.
$5x + 3x = 4x + 52 \Rightarrow x = 13$.
$\therefore$ smallest number = 3x = 39.

29. c; Let the number be x. Then,

$x - 4 = \frac{21}{x} \Rightarrow x^2 - 4x - 21 = 0 \Rightarrow x = 7$.

$\therefore$ required number = 7

30. a; Let the numbers be x, y, z. Then,

$$\frac{x}{y}=\frac{2}{3},\frac{y}{z}=\frac{5}{3} \Rightarrow 3x=2y \text{ and } 5z=3y.$$

$$\therefore\ y=\frac{3}{2}x,\ z=\frac{3}{5}y=\frac{3}{5}\times\frac{3}{2}x=\frac{9}{10}x$$

$$\therefore\ x+\frac{3}{2}x+\frac{9}{10}x=98 \Rightarrow 34x=680 \Rightarrow x=20.$$

So, second number $=\frac{3}{2}x=\left(\frac{3}{2}\times 20\right)=30$

31. d; Let the numbers be 3x and 5x.

$$\therefore\ \frac{3x+10}{5x+10}=\frac{5}{7} \Rightarrow x=5$$

Hence, the numbers are 15, 25.

32. a; Let the numbers be x and (50 - x). Then,

$$\frac{1}{x}+\frac{1}{50-x}=\frac{1}{12} \Rightarrow \frac{50-x+x}{x(50-x)}=\frac{1}{12}$$

$$\Rightarrow x^2-50x+600=0 \Rightarrow x=30 \text{ or } 20.$$

33. a; $(23\times 3+x+42\times 3)=235 \Rightarrow x=40.$

$\therefore$ fourth number = 40.

34. c; Number of one digit pages from 1 to 9 = 9

Number of two digit pages from 10 to 99 = 90

Number of three digit pages from 100 to 200 = 101

$\therefore$ total number of required figures =

$9\times 1+90\times 2+101\times 3=492$

35. a; According to the question

Divisor $=\frac{2}{3}\times$ dividend and $2\times$ remainder

or, $\frac{2}{3}\times \text{dividend}=2\times 5$

$\therefore$ Dividend $=\frac{2\times 5\times 3}{2}=15$.

36. d; The number is of the form (5x + 3), where x is an integer

$$\therefore\ \frac{(5x+3)^2}{5}=\frac{25x^2+30x+9}{5}=\frac{25x^2}{5}+\frac{30x}{5}+\frac{5+4}{5}$$

$\therefore$ the remainder is 4.

37. e; Since A + B + C = 16

(Possible values of A, B and C are 0, 6, 7 & 9).

Also $A\neq B, B\neq C, A\neq C$.

If C = 6, A + B should be 10, which is not possible.

If C = 9, A + B should be 7, which is also not possible.

If C = 0, A + B should be 16 which is also not possible.

2 HCF and LCM

Rule 1

To find the HCF of two or more numbers

Method-I: Method of Prime Factors

Break the given numbers into prime factors and then find the product of all the prime factors common to all the numbers. The product will be the required HCF.

Illustrative Examples

Ex. 1: Find the HCF of 42 and 70.

Soln: $42 = \underline{2} \times 3 \times \underline{7}$

$70 = \underline{2} \times 5 \times \underline{7}$

$HCF = 2 \times 7 = 14$

Ex. 2: Find the HCF of 1365, 1560 and 1755.

Soln: $1365 = \underline{3} \times \underline{5} \times 7 \times \underline{13}$

$1560 = 2 \times 2 \times 2 \times \underline{3} \times \underline{5} \times \underline{13}$

$1755 = 3 \times 3 \times \underline{3} \times \underline{5} \times \underline{13}$

$HCF = 3 \times 5 \times 13 = 195$

Method II: Method of Division

(a) HCF of two numbers:

Divide the greater number by the smaller number, divide the divisor by the remainder, divide the remainder by the next remainder, and so on until no remainder is left. The last divisor is the required HCF.

Illustrative Example

Ex.: Find the HCF of 42 and 70 by the method of division.

Soln:
```
42) 70 (1
    42
    28) 42 (1
        28
        14) 28 (2
            28
            00
```

∴ HCF = 14

(b) HCF of more than two numbers

Find the HCF of any two of the numbers and then find the HCF of this HCF and the third number and so on. The last HCF will be the required HCF.

Illustrative Example

Ex.: Find the HCF of 1365, 1560 and 1755.

Soln:
```
1365) 1560 (1
      1365
      195) 1365 (7
           1365
           0000
```

Therefore, 195 is the HCF of 1365 and 1560

Again,
```
195) 1755 (9
     1755
     0000
```

∴ the required HCF = 195

Method III: The work of finding the HCF may sometimes be simplified by the following devices:

(i) Any obvious factor which is common to both numbers may be removed before the rule is applied. Care should however be taken to multiply this factor into the HCF of the quotients.

(ii) If one of the numbers has a prime factor not contained in the other, it may be rejected.

(iii) At any stage of the work, any factor of the divisor not contained in the dividend may be rejected. This is because any factor which divides only one of the two cannot be a portion of the required HCF.

Ex.: Find the HCF of 42237 and 75582.

Soln: $42237 = 9 \times 4693$

$75582 = 2 \times 9 \times 4199$

We may reject 2 which is not a common factor (by rule i). But 9 is a common factor. We, therefore, set it aside (by rule ii) and find the HCF of 4199 and 4693.

```
4199) 4693 (1
      4199
       494
```

494 is divisible by 2 but 4199 is not. We, therefore, divide 494 by 2 and proceed with 247 and 4199 (by rule iii).

```
247) 4199 (17
     247
     1729
     1729
        0
```

The HCF of 4199 and 4693 is 247. Hence, the HCF of the original numbers is $247 \times 9 = 2223$.

Note: If the HCF of two numbers be unity, the numbers must be prime to each other.

Exercise

1. Find the HCF of 144 and 192.
2. Find the HCF of 1260 and 2376.
3. Find the HCF of 144, 180 and 192.
4. Find the HCF of 624, 936 and 264.
5. Find the HCF of 3192 and 14280.
6. Find the HCF of 234, 519, 786.
7. Find the HCF of 876, 492, 1824, 1960.
8. Find the HCF of 1794, 2346, 4761.
9. Find the HCF of 2103, 9945, 9216.
10. Find the HCF of 1492, 1942, 2592.
11. What is the HCF of two consecutive numbers?
12. Two vats contain respectively 540 and 720 litres, find the vessel of greatest capacity that will empty off both vats.
    a) 90 litres b) 160 litres c) 180 litres d) 80 litres
13. Two masses of gold weighing 44270 and 72190 grams respectively are each to be made into coins of the same size, what is the weight of the largest possible coin?
    a) 90 gm b) 110 gm c) 10 gm d) 210 gm

Answers

1. 48 2. 36 3. 12 4. 24 5. 168 6. 3 7. 4
8. 69 9. 3 10. 2 11. 1 12. c 13. c

Rule 2

To find the HCF of two or more concrete quantities

First, the quantities should be reduced to the same unit.

Illustrative Example

Ex.: Find the greatest weight which can be contained exactly in 1 kg 235 gm and 3 kg 430 gm

Soln: 1 kg 235 gm = 1235 gm

3 kg 430 gm = 3430 gm

The greatest weight required is the HCF of 1235 and 3430, which will be found to be 5 gm.

Exercise

1. Find the greatest weight which can be contained exactly in 6 kg 7 hg 4 dag 3g and 9 kg 9 dag 7 g.
   a) 11 g b) 27 g c) 12 g d) 17 g
2. Find the greatest weight which can be contained exactly in 3 kg 7 hg 8 dag 1 g and 9 kg 1 hg 5 dag 4 g.
   a) 199 g b) 299 g c) 189 g d) None of these
3. Find the greatest measure which is exactly contained in 10 litres 857 millilitres and 15 litres 87 millilitres.
   a) 140 ml b) 138 ml c) 141 ml d) 142 ml
4. Find the greatest length which can be contained exactly in 10 m 5 dm 2 cm 4 mm and 12 m 7 dm 5 cm 2 mm.
   a) 5 mm b) 7 mm c) 4 mm d) 6 mm
5. What is the greatest length which can be used to measure exactly 3 m 60 cm, 6 m, 8 m 40 cm and 18 m?
   a) 1 m 20 cm b) 1 m 10 cm
   c) 105 cm d) 125 cm
6. A man bought a certain number of mangoes Rs 14.40 P, he gained 44P by selling some of them for Rs 8. Find at least how much mangoes he had left with.
   a) 19 b) 36 c) 38 d) 21
7. What is the largest sum of money which is contained in Rs 6.25 and Rs 7.50 exactly?
   a) 125 P b) 120 P c) 175 P d) 75 P
8. What is the largest sum of money which will divide Rs 12.90 and Rs 9.30 exactly?
   a) Rs 1.30 b) 30 c) Rs 3 d) 130
9. What is the greatest length which can be used to measure exactly 1 m 8 cm, 2m 43 cm, 1 m 35 cm, and 1 m 89 cm?
   a) 26 cm b) 37 cm c) 28 cm d) 27 cm
10. Two bills, one of Rs 27.50 and the other of Rs 13 are to be paid in coins of the same kind. Find the largest coin that can be used.
    a) 50 paise b) 25 paise
    c) Rs 1 d) None of these

Answers

1. a; **Hint:** 6 kg 7 hg 4 dag 3 g = 6743 g
   9 kg 9 dag 7 g = 9097 g.

2. a 3. c 4. c 5. a

6. a; **Hint:** Cost price of all the mangoes = 1440 P
   Cost price of the mangoes sold = Rs 8 – 44P = 756 P.
   Now, the HCF of 1440 P and 756 P = 36 P
   ∴ Highest possible cost price of each mango = 36 P
   Again the cost price of the mangoes left = 1440 P – 756 P = 684 P
   ∴ The minimum number of mangoes left = $684 \div 36 = 19$.

7. a 8. c 9. d 10. a

Rule 3

To find the HCF of decimals

First make (if necessary) the same number of decimal places in all the given numbers, then find their HCF as if they are integers and marks off in the result as many decimal places as there are in each of the numbers.

Illustrative Examples

Ex. 1: Find the HCF of 16.5, 0.45 and 15.

Soln: The given numbers are equivalent to 16.50, 0.45 and 15.00

Step I: First we find the HCF of 1650, 45 and 1500. Which comes to 15.

Step II: The required HCF = 0.15.

Ex. 2: Find the HCF of 1.7, 0.51, and 0.153.

Soln: **Step I:** First we find the HCF of 1700, 510 and 153. Which comes to 17.

Step II: The required HCF = 0.017.

Exercise

1. Find the HCF of 405.9 and 219.
2. Find the HCF of 1.84, 2.3 and 2.76.
3. Find the HCF of 18.4, 23 and 27.6.
4. Find the HCF of 2.8, 0.98, 42, 0.161, 0.0189.
5. Find the HCF of 4.8, 5.4 and 0.06.
6. Find the HCF of 6.16 and 13.
7. Find the HCF of 11.52, 12.96, 14.4, 15.84 and 17.28.

Answers

1. 0.3 2. 0.46 3. 4.6 4. 0.0007 5. 0.6 6. 0.04 7. 14.4

Rule 4

To find the LCM of two or more given numbers.

Method I: Method of Prime Factors

Resolve the given numbers into their prime factors and then find the product of the highest power of all the factors that occur in the given numbers. This product will be the LCM.

Illustrative Example

Ex.: Find the LCM of 18, 24, 60 and 150.

Soln: $18 = 2 \times 3 \times 2 = 2 \times 3^2$ $24 = 2 \times 2 \times 2 \times 3 = 2^3 \times 3$
$60 = 2 \times 2 \times 3 \times 5 = 2^2 \times 3 \times 5$ $150 = 2 \times 3 \times 5 \times 5 = 2 \times 3 \times 5^2$

Here, the prime factors that occur in the given numbers are 2, 3 and 5, and their highest powers are 2^3, 3^2 and 5^2 respectively.

Hence, the required LCM $= 2^3 \times 3^2 \times 5^2 = 1800$

Note: The LCM of two numbers which are prime to each other is their product.

Thus, the LCM of 15 and 17 is $15 \times 17 = 255$

Method II: The LCM of several small numbers can be easily found by the following method:

Write down the given numbers in a line separating them by commas. Divide by any one of the prime numbers 2, 3, 5, 7, etc., which will exactly divide at least any two of the given numbers. Set down the quotients and the undivided numbers in a line below the first. Repeat the process until you get a line of numbers which are prime to one another. The product of all divisors and the numbers in the last line will be the required LCM.

Note: To simplify the work, we may cancel, at any stage of the process, any one of the numbers which is a factor of any other number in the same line.

Illustrative Example

Ex.: Find the LCM of 12, 15, 90, 108, 135, 150.

Soln:

2	12,	15,	90,	108,	135,	150	.... (1)
3			45,	54,	135,	75	.... (2)
3				18,	45,	25	.... (3)
5				6,	15,	25	.... (4)
				6,	3,	5	.... (5)

∴ the required LCM $= 2 \times 3 \times 3 \times 5 \times 6 \times 5 = 2700$

In line (1), 12 and 15 are the factors of 108 and 90 respectively, therefore, 12 and 15 are struck off.

In line (2), 45 is a factor of 135, therefore 45 is struck off.

In line (5), 3 is a factor of 6, therefore 3 is struck off.

Exercise

1. Find the LCM of 40, 36 and 126.
2. Find the LCM of 84, 90 and 120.
3. Determine the LCM of 624 and 936.
4. Find the LCM of 112, 140 and 168.
5. Find the LCM of 75, 250, 225 and 525.
6. Find the LCM of 48, 64, 72, 96 and 108.
7. Find the LCM of 240, 420 and 660.
8. Find the LCM of 12, 15, 24, 52, 55, 60, 77 and 210.
9. Find the LCM of 2184, 2730 and 3360.
10. Find the LCM of 60, 32, 45, 80, 36 and 120.
11. Find the LCM of 91, 65, 75, 39, 77 and 130.
12. Find the LCM of 364, 2520 and 5265.

Answers

1. 2520	2. 2520	3. 1872
4. 1680	5. 15750	6. 1728
7. 18480	8. 120120	9. 43680
10. 1440	11. 150150	12. 294840

Rule 5

To find the LCM of Decimals

First make (if necessary) the same number of decimal places in all the given numbers; then find their LCM as if they were integers, and mark in the result as many decimal places as there are in each of the numbers.

Illustrative Example

Ex.: Find the LCM of 0.6, 9.6 and 0.36.

Soln: The given numbers are equivalent to 0.60, 9.60 and 0.36.

Now, find the LCM of 60, 960 and 36. Which is equal to 2880.

∴ the required LCM = 28.80.

Exercise

1. Find the LCM of 3, 1.2 and 0.06.
2. Find the LCM of 3.75 and 7.25.
3. Find the LCM of 72.12 and 0.03.
4. Find the LCM of 0.02, 0.4, and 0.008.
5. Find the LCM of 1.2, 0.24 and 6.
6. Find the LCM of 1.6, 0.04 and 0.005.
7. Find the LCM of 2.4, 0.36 and 7.2.
8. Find the LCM of 0.08, 0.002 and 0.0001.
9. Find the LCM of 3.9, 6.6 and 8.22.
10. Find the LCM of 0.6, 0.09 and 1.8.
11. Find the LCM of 0.18, 2.4 and 60.
12. Find the LCM of 20, 2.8 and 0.25.

13. Find the LCM of 1.5, 0.25 and 0.075.

Answers

1. 600 2. 108.75 3. 72.12 4. 0.4
5. 6 6. 1.6 7. 7.2 8. 0.08
9. 11754.6 10. 1.8 11. 180
12. 140 13. 1.5

Rule 6

$$\textit{HCF of fractions} = \frac{\textit{HCF of Numerators}}{\textit{LCM of Denominators}}$$

Illustrative Example

Ex.: Find the HCF of $\frac{54}{9}$, $3\frac{9}{17}$ and $\frac{36}{51}$

Soln: Here, $\frac{54}{9} = \frac{6}{1}$, $3\frac{9}{17} = \frac{60}{17}$ and $\frac{36}{51} = \frac{12}{17}$

Thus the fractions are $\frac{6}{1}$, $\frac{60}{17}$ and $\frac{12}{17}$

$\therefore$ HCF $= \frac{\text{HCF of } 6, 60, 12}{\text{LCM of } 1, 17, 17} = \frac{6}{17}$

Note: (i) First express the given fractions in their lowest terms.

(ii) We see that each of the numbers is perfectly divisible by $\frac{6}{17}$.

Exercise

1. Find the HCF of $\frac{3}{4}$, $\frac{5}{6}$ and $\frac{6}{7}$.
2. Find the HCF of 6, $3\frac{3}{4}$ and $\frac{18}{20}$.
3. Find the HCF of $\frac{6}{8}$, $2\frac{1}{2}$ and $\frac{15}{16}$.
4. Find the HCF of $\frac{100}{25}$, $\frac{88}{33}$ and 4.
5. Find the HCF of $1\frac{37}{78}$, $2\frac{54}{65}$ and $5\frac{35}{39}$.
6. What is the greatest length which is contained a whole number of times exactly in both $7\frac{1}{2}$ metres and $4\frac{1}{4}$ metres?

   a) 25 cm b) 26 cm c) 80 cm d) 30 cm

Answers

1. $\frac{1}{84}$ 2. $\frac{3}{20}$ 3. $\frac{1}{16}$ 4. $1\frac{1}{3}$ 5. $\frac{23}{390}$ 6. a

Rule 7

$$\textit{LCM of Fractions} = \frac{\textit{LCM of Numerators}}{\textit{HCF of Denominators}}$$

Illustrative Example

Ex.: Find the LCM of $4\frac{1}{2}$, 3, $10\frac{1}{2}$

Soln: $4\frac{1}{2} = \frac{9}{2}$, $10\frac{1}{2} = \frac{21}{2}$

Thus, the fractions are $\frac{9}{2}$, $\frac{3}{1}$ and $\frac{21}{2}$

$\therefore$ the required LCM

$= \frac{\text{LCM of } 9, 3 \text{ and } 21}{\text{HCF of } 2, 1 \text{ and } 2} = \frac{63}{1} = 63$

Note: (i) First express the fractions in their lowest terms.

(ii) LCM of fractions may be a fraction or an integer.

Exercise

1. Find the LCM of 8, $\frac{16}{20}$ and $1\frac{3}{7}$.
2. Find the LCM of $\frac{6}{8}$, $2\frac{1}{2}$ and $\frac{15}{16}$.
3. Find the LCM of $1\frac{37}{78}$, $2\frac{54}{65}$ and $5\frac{35}{39}$.
4. Find the least number which, when divided by each of the fractions $\frac{4}{5}$, $\frac{3}{10}$, and $\frac{12}{35}$ gives a whole number as quotient in each case.

   a) $2\frac{2}{5}$ b) $3\frac{3}{5}$ c) $2\frac{3}{5}$ d) $3\frac{2}{5}$
5. Four bells commence tolling together, they tall at intervals of 1, $1\frac{1}{4}$, $1\frac{1}{2}$ and $1\frac{3}{4}$ seconds respectively, after what interval will they tall together again?

   a) 2 min 40 seconds b) 1 min 40 seconds
   c) 2 min 45 seconds d) 1 min 45 seconds
6. The circumferences of the fore and hind-wheels of a carriage are $2\frac{2}{5}$ and $3\frac{3}{7}$ metres respectively. A chalk mark is put on the point of contact of each wheel with the ground at any given moment. How far will the carriage have travelled so that their chalk marks may be again on the ground at the same time?

   a) 26 metres b) 24 metres
   c) 42 metres d) 16 metres

7. The circumferences of the wheels of a carriage are $6\frac{3}{14}$ dm and $8\frac{1}{18}$ dm. What is the least distance in which both wheels simultaneously complete an integral number of revolutions? How often will the points of the two wheels which were lowest at the time of starting touch the ground together in 1 kilometre?

Answers

1. 40 2. $7\frac{1}{2}$ 3. $70\frac{10}{13}$ 4. a

5. d; **Hint:** Find LCM of 1, $1\frac{1}{4}$, $1\frac{1}{2}$ and $1\frac{3}{4}$.

6. b; **Hint:** A little reflection will show that chalk marks will touch the ground together for the first time after the wheels have passed over a distance which is the LCM of $2\frac{2}{5}$ metres and $3\frac{3}{7}$ metres. LCM of $\frac{12}{5}$ metres and $\frac{24}{7}$ metres = 24 metres.

7. **Hint:** LCM of $6\frac{3}{14}$ dm and $8\frac{1}{18}$ dm

$$= \frac{435}{2} = 217\frac{1}{2} \text{ dm}$$

and the required no. of revolutions $= \frac{2 \times 10000}{435} = 45.9$

∴ required answer = 45 times (without including the touch at the start.)

Note: See Q. no. 18 of Miscellaneous.

Rule 8

HCF of Numbers × LCM of Numbers = Product of Numbers

Illustrative Example

Ex.: The LCM of two numbers is 2079 and their HCF is 27. If one of the numbers is 189, find the other.

Soln: The required number $= \frac{\text{LCM} \times \text{HCF}}{\text{First Number}}$

$$= \frac{2079 \times 27}{189} = 297.$$

Exercise

1. The LCM of two numbers is 64699, their GCM (or HCF) is 97 and one of the numbers is 2231. Find the other.
   a) 2183 b) 2813 c) 2831 d) 2381
2. The LCM of two numbers is 11781 and their HCF is 119. If one of the numbers is 1071, find the others.
   a) 1309 b) 1903 c) 1039 d) 1390
3. The product of two numbers is 20736 and their HCF is 54. Find their LCM.
   a) 684 b) 468 c) 648 d) 864
4. The product of two numbers is 396 × 576 and their LCM is 6336. Find their HCF.
   a) 36 b) 34 c) 63 d) 43
5. The HCF of two numbers, each consisting of four digits is 103, and their LCM is 19261, find the numbers.
   a) 1133, 1751 b) 1313, 1571 c) 1331, 1751 d) 1133, 1715
6. The HCF and LCM of two numbers are 16 and 192 respectively, one of the numbers is 48, find the other.
   a) 64 b) 46 c) 63 d) 72
7. The HCF and LCM of two numbers are 10 and 30030 respectively, one of the numbers is 770, what is the other?
   a) 380 b) 370 c) 385 d) 390
8. The HCF and LCM of the two numbers are 14 and 3528 respectively. If one number is 504, find the other.
   a) 88 b) 98 c) 84 d) 112
9. The HCF of two numbers is 99 and their LCM is 2772. The numbers are
   a) 198, 1386 b) 198, 297
   c) 297, 495 d) None of these
10. The LCM of two numbers is 14 times their HCF. The sum of LCM and HCF is 600. If one number is 80, then the other is ______.
    a) 160 b) 60 c) 40 d) 280
11. The HCF of two numbrs is 1/5 th of their LCM. If the product of the two numbers is 720, the HCF is ______.
    a) 20 b) 12 c) 15 d) 18
12. Two numbers have 16 as their HCF and 146 as their LCM. Then, one can say that;
    a) Many such pairs of numbers exist.
    b) Only on such pair of numbers exists.
    c) No such pair of numbers exists.
    d) Only two such pairs of numbers exist.
13. The LCM of two numbers is 39780 and their ratio is 13 : 15. Then, the numbers are ______.
    a) 2652, 3060 b) 273, 315 c) 585, 675 d) 2562, 6030

Answers

1. b 2. a 3. d 4. a 5. a 6. a 7. d
8. b

9. a; **Hint:** For this kind of question you have to start from the answers choice.
Try the pair of numbers 198, 1386
The HCF of these numbers is 99

$$\therefore \text{LCM} = \frac{198 \times 1386}{99} = 2772$$

Hence, (a) is the required answer.

10. d; [**Hint:** LCM = 14 HCF

Since LCM + HCF = 600
or, 14 HCF + HCF = 600

or, $HCF = \frac{600}{15} = 40$

or, $LCM = 14 \times 40 = 560$

∴ The other number

$$= \frac{LCM \times HCF}{\text{Given number}} = \frac{560 \times 40}{80} = 280$$

11. b; **Hint:** LCM = 5 HCF
The product of two numbers = LCM × HCF = 720
or, 5HCF × HCF = 720

or, $(HCF)^2 = \frac{720}{5} = 144$ ∴ HCF = 12

12. c; **Hint:** As a rule HCF of the numbers completely divides their LCM. However, 146 is not exactly divisible by 16, so no such pair of numbers will exist.

13. a; **Hint:** Let the numbers be $13x$ and $15x$. Clearly x is their HCF.
Now, as a rule, the product of two numbers = HCF × LCM
or, $13x \times 15x = x \times 39780$

$\therefore x = \frac{39780}{13 \times 15} = 204$

Therefore, numbers are 13 × 204 = 2652 and 15 × 204 = 3060.

Rule 9

To find the greatest number that will exactly divide x, y and z.
Required number = HCF of x, y and z

Illustrative Example

Ex.: What is the greatest number that will exactly divide 1365, 1560 and 1755?

Soln: Applying the above rule, the required greatest number = HCF of 1365, 1560 and 1755 = 195

Exercise

1. What is the greatest number that will exactly divide 96, 528 and 792?
a) 12 b) 48 c) 36 d) 24
2. What is the greatest number that will exactly divide 370 and 592?
a) 37 b) 74 c) 47 d) 73
3. What is the greatest number that will exactly divide 312, 351 and 650?
a) 39 b) 13 c) 26 d) 52
4. What is the greatest number that will exactly divide 48, 168, 324 and 1400?
a) 14 b) 4 c) 16 d) 8
5. What is the greatest number that will exactly divide 1600 and 1420?
a) 40 b) 20 c) 10 d) 30

Answers

1. d 2. b 3. b 4. b 5. b

Rule 10

To find the greatest number that will divide x, y and z leaving remainders a, b and c respectively.
Required number = HCF of (x – a), (y – b) and (z – c)

Illustrative Example

Ex.: What is the greatest number that will divide 38, 45, and 52 and leave as remainders 2, 3 and 4 respectively?

Soln: Applying the above rule, we have,
the required greatest number = HCF of (38 – 2), (45 – 3) and (52 – 4) or 36, 42 and 48 = 6
∴ Ans = 6

Exercise

1. Find the greatest number that will divide 728 and 900, leaving the remainders 8 and 4 respectively.
a) 16 b) 15 c) 14 d) 24
2. What is the greatest number that will divide 2930 and 3250 and will leave as remainders 7 and 11 respectively?
a) 69 b) 59 c) 97 d) 79
3. What is the greatest number that will divide 3460 and 9380 and will leave as remainders 9 and 13 respectively?
a) 943 b) 439 c) 493 d) 349
4. What is the greatest number that will divide 29, 60 and 103 and will leave as remainders 5, 12 and 7 respectively?
a) 24 b) 16 c) 12 d) 14
5. What is the greatest number that will divide 191, 216 and 266 and will leave as remainders 4, 7 and 13 respectively?
a) 22 b) 39 c) 33 d) 11
6. What is the greatest number that will divide 130, 305 and 245 and will leave as remainders 6, 9 and 17 respectively?
a) 4 b) 5 c) 14 d) 24

Answers

1. a 2. d 3. c 4. a 5. d 6. a

Rule 11

To find the least number which is exactly divisible by x, y and z.
Required number = LCM of x, y and z

Illustrative Example

Ex.: Find the least number which is exactly divisible by 8, 12, 15 and 21.

Soln: By the above rule, we have,
The required least number = LCM of 8, 12, 15 and 21 = 840
∴ Ans = 840

Exercise

1. Find the least number which is exactly divisible by 72, 90 and 120.
   a) 260 b) 630 c) 360 d) 620
2. Find the least number which is exactly divisible by 24, 63 and 70.
   a) 5220 b) 2550 c) 5252 d) 2520
3. Find the least number which is exactly divisible by 35, 48 and 56.
   a) 1680 b) 1860 c) 1380 d) 1830
4. Find the least number which is exactly divisible by 15, 55 and 99.
   a) 485 b) 435 c) 495 d) 395
5. Find the least number which is exactly divisible by 52, 63 and 162.
   a) 29484 b) 24984 c) 29488 d) 29448
6. Find the greatest number of 4 digits which is divisible by 48, 60 and 64.
   a) 9600 b) 1960 c) 9620 d) 9610
7. Find the smallest number which is exactly divisible by 999 and 9999.
   a) 1199889 b) 1109989 c) 1109999 d) 1109889
8. What is the smallest number which is exactly divisible by 36, 45, 63 and 80?
   a) 5040 b) 4050 c) 5400 d) 4500
9. Find the least number into which 47601 and 37668 will each divide without remainder.
   a) 13899492 b) 12899492 c) 13899493 d) 13894992
10. Find the least number that can be divided exactly by all numbers upto 13 inclusive.
    a) 360360 b) 306360 c) 360306 d) 363060
11. Find the least number that can be divided exactly by all the odd numbers upto 15 inclusive.
    a) 46046 b) 45450 c) 45045 d) 40545
12. Find the greatest number less than 900, which is divisible by 8, 12 and 28.
    a) 640 b) 480 c) 840 d) 940
13. What is the smallest number which when increased by 3 is divisible by 27, 35, 25 and 21?
    a) 4725 b) 4722 c) 4723 d) 4728
14. What is the least number which when lessened by 5 is divisible by 36, 48, 21 and 28?
    a) 1008 b) 1003 c) 1013 d) 1023
15. What greatest number can be subtracted from 10000, so that the remainder may be divisible by 32, 36, 48 and 54?
    a) 9136 b) 9316 c) 1360 d) 8640
16. What greatest number can be subtracted from 2470 so that the remainder may be divisible by 42, 98 and 105?
    a) 1000 b) 1470 c) 1400 d) 1407
17. Find the greatest number of five digits which is divisible by 32, 36, 40, 42 and 48.
    a) 99720 b) 90702 c) 90720 d) 90730
18. Find the least number of four digits which is divisible by 4, 6, 8 and 10.
    a) 1050 b) 1070 c) 1080 d) 1008

Answers

1. c 2. d 3. a 4. c 5. a

6. a; **Hint:** The least number divisible by 48, 60 and 64 is their LCM, which is 960.
Clearly, any multiple of 960 will be exactly divisible by each of the numbers 48, 60 and 64. But since the required number is not to exceed 10,000, it is $960 \times 10 = 9600$.
The above question could also be worded thus — **"Find the greatest number less than 10000 which is divisible by 48, 60 and 64."**

7. d 8. a 9. a 10. a 11. c 12. c

13. b; **Hint:** The LCM of 27, 35, 25 and 21 = 4725
$\therefore$ the required no. = 4725 − 3 = 4722

14. c; **Hint:** The LCM of 36, 48, 21 and 28 = 1008.
$\therefore$ the required no. = 1008 + 5 = 1013.

15. a; **Hint:** The least number divisible by 32, 36, 48 and 54 is their LCM which is 864.
$\therefore$ the greatest number that should be subtracted from 10000 is 10000 − 864 = 9136.

16. a 17. c 18. c

Rule 12

To find the *least number which when divided by x, y and z leaves the remainders a, b and c respectively. It is always observed that, (x − a) = (y − b) = (z − c) = K (say)*

$\therefore$ Required number = (LCM of x, y and z) − K

Illustrative Example

Ex.: What is the least number which, when divided by 52, leaves 33 as the remainder, and when divided by 78 leaves 59, and when divided by 117 leaves 98 as the respective remainders.

Soln: Since (52 − 33) = 19, (78 − 59) = 19, (117 − 98) = 19
We see that the remainder in each case is less than the divisor by 19. Hence, if 19 is added to the required number, it becomes exactly divisible by 52, 78 and 117. Therefore, the required number is 19 less than the LCM of 52, 78 and 117.
The LCM of 52, 78 and 117 = 468
$\therefore$ The required number = 468 − 19 = 449

Exercise

1. Find the least number which when divided by 24, 32 and 36 leaves the remainders 19, 27 and 31 respectively.
   a) 283 b) 823 c) 382 d) 238
2. Find the greatest number of six digits which on being divided by 6, 7, 8, 9 and 10 leaves 4, 5, 6, 7 and 8 as remainder respectively.
   a) 997920 b) 997918 c) 998918 d) 999918
3. What is the least multiple of 7, which when divided by 2, 3, 4, 5, and 6 leaves the remainders 1, 2, 3, 4 and 5 respec-

tively?

a) 119 b) 126 c) 112 d) Can't be determined

4. Find the least multiple of 23, which when divided by 18, 21 and 24 leaves the remainder 7, 10 and 13 respectively.

a) 3013 b) 3103 c) 3130 d) 3301

5. Find the greatest number of five digits which being divided by 56, 72, 84 and 96 leaves 50, 66, 78 and 90 as remainders respectively.

a) 97887 b) 97878 c) 98778 d) 97788

6. Find the least number which when divided by 12 and 16 will leave the remainders 5 and 9 respectively.

a) 4 b) 41 c) 43 d) 39

7. Find the least number which when divided by 24 and 36 will leave the remainders 14 and 26 respectively.

a) 64 b) 62 c) 59 d) 63

8. Find the least number which when divided by 48, 64, 72, 80, 120 and 140 will leave the remainders 38, 54, 62, 70, 110 and 130 respectively.

a) 21050 b) 20250 c) 21005 d) 20150

9. Find the greatest number of six digits which when divided by 5, 7, 12 and 15 leaves respectively remainders 3, 5, 10 and 13.

a) 999600 b) 999596 c) 999598 d) 999602

Answers

1. a

2. b; **Hint:** The LCM of 6, 7, 8, 9 and 10 = 2520

The greatest number of six digits is 999999.

Dividing 999999 by 2520 we get 2079 as remainder.

Hence the number divisible by 2520 is 999999 – 2079 or 997920.

Since 6 – 4 = 2, 7 – 5 = 2, 8 – 6 = 2, 9 – 7 = 2, 10 – 8 = 2, the remainder in each case is less than the divisor by 2.

∴ the required number = 997920 – 2 = 997918.

3. a; **Hint:** LCM of 2, 3, 4, 5 and 6 = 60

Moreover, the difference between each divisor and the corresponding remainder is the same, which is 1.

∴ required number is of the form (60 K – 1), which is divisible by 7 for the least value of K.

Now, on dividing 60K – 1 by 7,

$$7)\,60K-1\,(8K$$
$$\underline{56K}$$
$$(4K-1)$$

We get (4K – 1) as the remainder.

We find the least positive number K for which (4K – 1) is divisible by 7. By inspection K = 30.

Hence the required number = 4 × 30 – 1 = 119.

4. a 5. c 6. b 7. b 8. d

9. c; **Hint:** LCM of 5, 7, 12 and 15 = 420

The greatest number of 6 digits = 999999

We can break this number into multiple of 420 as 420 × 2380 + 399

Hence, the greatest number of six digits that is exactly divisible by the above number is 420 × 2380 = 999600.

Now, 5 – 3 = 2, 7 – 5 = 2, 12 – 10 = 2, 15 – 13 = 2

Hence, subtracting 2 from this greatest number we shall get the required number which is therefore equal to 999598.

Rule 13

To find the least number which, when divided by x, y and z leaves the same remainder r in each case.

Required number = (LCM of x, y and z) + r

Illustrative Example

Ex.: Find the least number which, upon being divided by 2, 3, 4, 5 and 6 leaves in each case a remainder of 1.

Soln: By the above rule, we have,

Required least number = (LCM of 2, 3, 4, 5 and 6) + 1
= 60 + 1 = 61

Exercise

1. Find the least number which when divided by 12, 21 and 35 will leave in each case the same remainder 6.

a) 426 b) 326 c) 536 d) 436

2. Find the least number which when divided by 18, 24, 30 and 42, will leave in each case the same remainder 1.

a) 2523 b) 2521 c) 2520 d) 2519

3. What is the least number, which when divided by 98 and 105 has in each case 10 as remainder?

a) 1840 b) 1400 c) 1460 d) 1480

4. What is the lowest number which when divided separately by 27, 42, 63 and 84 will in each case leave 21 as remainder?

a) 777 b) 767 c) 707 d) 787

5. What smallest number must be subtracted from 7894135 so that the remainder when divided by 34, 38, 85 and 95 leaves the same remainder 11 in ech case.

a) 6 b) 8 c) 4 d) 3241

6. What is the least multiple of 17, which leaves a remainder of 1, when divided by each of the first twelve integers excepting unity?

a) 27720 b) 138601 c) 138599 d) 27719

7. What is the least multiple of 19, which leaves a remainder of 2, when divided by 8, 12 or 15?

a) 718 b) 724 c) 722 d) 716

8. Find the least number which when divided by 12, 16 and 18, will leave in each case a remainder 5.

a) 139 b) 144 c) 149 d) 154

9. Find the least number which when divided by 12, 18 and 30 gives the same remainder 9 in each case.

a) 189 b) 187 c) 179 d) 198

10. Find the least number which when divided by 128 and 96 will leave in each case the same remainder 5.

a) 289 b) 389 c) 489 d) 398

11. Find the least number of six digits which when divided by 4, 6, 10 and 15, leaves in each case the same remainder 2.

a) 10020 b) 10018
c) 10022 d) Can't be determined

12. Find the least number of six digits which when divided by 5, 8, 12, 16 and 20 leaves a remainder 3 in each case.
a) 100883 b) 100886
c) 100083 d) 190083

13. Find the least multiple of 13 which when divided by 4, 6, 7 and 10 leaves the remainder 2 in each case.
a) 2522 b) 2252 c) 2225 d) 2552

Answers

1. a 2. b 3. d 4. a

5. c; **Hint:** LCM of 34, 38, 85 and 95 is 3230.
Now, divide 7894135 by 3230, we obtain 15 as remainder and 2444 as the quotient.
But, according to the question, the remainder should be 11. Hence, the required smallest number that must be subtracted is $15 - 11 = 4$.

6. b; **Hint:** LCM of first twelve integers excepting unity is 27720.
The required number is of the form (27720K + 1) which leaves remainder 1 in each case.
17) 27720K + 1 (1630K
27710K
10K + 1
Now, on dividing (27720K + 1) by 17,
we get (10 K + 1) as the remainder.
We find the least positive number K for which (10K + 1) is divisible by 17.
By inspection K = 5.
Hence, the required number $= 27720 \times 5 + 1 = 138601$

7. c 8. c 9. a 10. b

11. c; **Hint:** The LCM of 4, 6, 10 and 15 is 60. Now the least number of six digits = 100000. When this is divided by 60, 40 is left as remainder. Also $60 - 40 = 20$, the least number of six digits exactly divisible by each of the above numbers $= 100000 + 20 = 100020$.
∴ the least number of six digits which will leave a remainder 2 when divided by each of the given numbers $= 100020 + 2 = 100022$.

12. c 13. a

Rule 14

To find the greatest number that will divide x, y and z leaving the same remainder 'r' in each case.

Required number = HCF of (x – r), (y – r) and (z – r)

Illustrative Example

Ex.: Find the greatest number which will divide 410, 751 and 1030 so as to leave remainder 7 in each case.

Soln: By the above rule, the required greatest number
= HCF of (410 – 7), (751 – 7) and (1030 – 7) = 31
∴ Ans = 31

Exercise

1. Find the greatest number which will divided 16997 and 64892 so as to leave the remainder 2 in each case.
a) 1455 b) 1544 c) 1545 d) 1554

2. Find the greatest number which will divide 410, 751 and 1030 so as to leave the remainder 7 in each case.
a) 63 b) 31 c) 13 d) 36

3. Find the greatest number which will divide 260, 720 and 1410 so as to leave the remainder 7 in each case.
a) 33 b) 43 c) 32 d) 23

4. Find the greatest number which will divide 369, 449, 689, 5009 and 729 so as to leave the remainder 9 in each case.
a) 42 b) 49 c) 35 d) 40

5. Find the greatest number which will divide 772 and 2778 so as to leave the remainder 5 in each case.
a) 59 b) 69 c) 49 d) 95

6. Find the greatest number that will divide 261, 933 and 1381, leaving the remainder 5 in each case.
a) 31 b) 52 c) 32 d) 42

Answers

1. c 2. b 3. d 4. d 5. a 6. c

Rule 15

To find the greatest number that will divide x, y and z living the same remainder in each case.

Required number = HCF of |(x – y)|, |(y – z)| and |(z – x)|

Note: Here value of remainder will not be given in the question.

Illustrative Example

Ex.: Find the greatest number which is such that when 76, 151 and 226 are divided by it, the remainders are all alike.

Soln: By the above rule, we get
$|(x - y)| = |(76 - 151)| = 75$
$|(y - z)| = |(151 - 226)| = 75$
$|(z - x)| = |(226 - 76)| = 150$
∴ The required greatest number = HCF of 75, 75 and 150 = 75.

Exercise

1. Find the greatest number which is such that when 12288, 19139 and 28200 are divided by it, the remainders are all the same.
a) 222 b) 221 c) 121 d) 122

2. Find the greatest number which is such that when 76, 151 and 226 are divided by it, the remainders are all alike. Find also the common remainder.
a) 57, 2 b) 75, 2 c) 75, 1 d) 57, 1

3. Find a number of three digits which gives the same remainder when it divides 2272 and 875.
a) 172 b) 127 c) 125 d) 137

4. Find the greatest number that will divide 1305, 4665 and

6905 leaving in each case the same remainder. Find also the common remainder.
a) 1210, 158 b) 1120, 158 c) 1120, 185 d) 1210, 185

5. Find the greatest number that will divide 705, 1805 and 1475 leaving in each case the same remainder.
a) 110 b) 120 c) 114 d) 115

6. Which of the following numbers gives the same remainder when it divides 1110 and 864.
a) 123 b) 213 c) 245 d) 132

Answers

1. b 2. c 3. b 4. c 5. a
6. a; **Hint:** Another number is 246, which gives the same remainder when it divides 1110 and 864.

Rule 16

To find the n-digit greatest number which, when divided by x, y and z,
(i) leaves no remainder (ie exactly divisible)
Following stepwise methods are adopted.

Step I: LCM of x, y and z = L
Step II: L) n-digit greatest number (

Remainder (R)
Step III: Required number = n-digit greatest number – R

Illustrative Example

Ex.: Find the greatest number of four digits which, when divided by 12, 15, 20 and 35 leaves no remainder.

Soln: Using the above method, we get,
Step I: LCM of 12, 15, 20 and 35 = 420
Step II: 420) 9999 (23 [∵ 4-digit greatest no. = 9999]
9660
339
Step III: Required number = 9999 – 339 = 9660

(ii) leaves remainder K in each case
Following stepwise method is adopted.
Step I: LCM of x, y and z = L
Step II: L) n-digit greatest number (

Remainder (R)
Step III: Required number = (n-digit greatest number – R) + K

Illustrative Example

Ex.: Find the greatest 4-digit number which, when divided by 12, 18, 21 and 28 leaves a remainder 3 in each case.

Soln: **Step I:** LCM of 12, 18, 21 and 28 = 252
Step II: 252) 9999 (39
9828
171
Step III: The required number = (9999 – 171) + 3 = 9931

Exercise

1. Find the greatest number of three digits which, when divided by 3, 4 and 5 leaves no remainder.
a) 960 b) 860 c) 690 d) 680

2. Find the greatest 3-digit number such that when divided by 3, 4 and 5, it leaves remainder 2 in each case.
a) 122 b) 962 c) 958 d) 118

3. The greatest number of four digits which is divisible by each one of the numbers 12, 18, 21 and 28 is ____.
a) 9848 b) 9864 c) 9828 d) 9636

4. Find the greatest number of five digits which is divisible by 48, 60 and 64.
a) 96000 b) 99940 c) 99840 d) 98940

5. Find the greatest number of 4 digits which, when divided by 16, 24 and 36 leaves 4 as a remainder in each case.
a) 9936 b) 9932 c) 9940 d) 9904

6. Find the greatest number of five digits which when divided by 52, 56, 78, and 91 leaves no remainder.
a) 12264 b) 98280 c) 97280 d) 13264

Answers

1. a 2. b 3. c 4. c 5. c 6. b

Rule 17

To find the n-digit smallest number which, when divided by x, y and z.
(i) leaves no remainder (ie exactly divisible)
Following steps are followed.

***Step I:* LCM of x, y and z = L**
Step II: L) n-digit smallest number (

Remainder (R)
Step III: The required number = n-digit smallest number + (L – R)

Illustrative Example

Ex.: Find the 4-digit smallest number which when divided by 12, 15 20 and 35 leaves no remainder.

Soln: Using the above method
Step I: LCM of 12, 15, 20 and 35 = 420
Step II: 420) 1000 (2
840
160
Step III: The required number = 1000 + (420 – 160) = 1260

(ii) leaves remainder K in each case.
First two steps are the same as in the case of (i)
Step III: Required number = n-digit smallest number + (L – R) + K

Illustrative Example

Ex.: Find the 4-digit smallest number which, when divided by 12, 18, 21 and 28, leaves a remainder 3 in each case.

Soln: By using the above method, we have;

Step I: LCM of 12, 18, 21 and 28 = 252

Step II: 252) 1000 (3
756
———
244

Step III: The required number = 1000 + (252 – 244) + 3
= 1011

Exercise

1. Find the smallest 3-digit number, such that they are exactly divisible by 3, 4 and 5.
   a) 105 b) 120 c) 115 d) 130
2. Find the smallest 3-digit number, such that when divided by 3, 4 and 5, it leaves remainder 2 in each case.
   a) 118 b) 120 c) 122 d) 132
3. The least number of four digits which is divisible by each one of the numbers 12, 18, 21 and 28 is ______.
   a) 1008 b) 1006 c) 1090 d) 1080
4. Find the smallest number of 6 digits, such that when divided by 15, 18 and 27 it leaves 5 as a remainder in each case.
   a) 100270 b) 100275 c) 100005 d) 100095
5. Find the smallest number of 4 digits which, when divided by 4, 8 and 10, leaves 3 as a remainder in each case.
   a) 1040 b) 1008 c) 1043 d) 1084
6. Find the least number of five digits which when divided by 52, 56, 78 and 91 leaves no remainder.
   a) 10920 b) 19020 c) 10290 d) 10820

Answers

1. b 2. c 3. a 4. d 5. c 6. a

Rule 18

To find the least number which on being divided by x, y and z leaves in each case a remiander R, but when divided by N leaves no remainder, following stepwise methods are adopted.

Step I: Find the LCM of x, y and z say (L).

Step II: Required number will be in the form of (LK + R); where K is a positive integer.

Step III: N) L (Quotient (Q)

Remainder (R_0)

$\therefore L = N \times Q + R_0$

Now put the vaue of L into the expression obtained in **step II**.

$\therefore$ required number will be in the form of $(N \times Q + R_0)$ $K + R$

or, $(N \times Q \times K) + (R_0K + R)$

Clearly, $N \times Q \times K$ is always divisible by N.

Step IV: Now make $(R_0 K + R)$ divisible by N by putting the least value of K. Say, 1, 2, 3, 4 ......

Now put the value of K into the expression (LK + R) which will be the required number.

Illustrative Example

Ex.: Find the least number which on being divided by 5, 6, 8, 9, 12 leaves in each case a remainder 1, but when divided by 13 leaves no remainder.

Soln: **Step I:** The LCM of 5, 6, 8, 9, 12 = 360

Step II: The required number = 360K + 1; where K is a positive integer.

Step III: 13) 360 (27
26
———
100
91
———
9

$\therefore 360K + 1 = (13 \times 27 + 9)K + 1$
$= (13 \times 27 \times K) + (9K + 1)$

Step IV: Now this number has to be divisible by 13. Whatever may be the value of K the portion (13 × 27K) is always divisible by 13. Hence we must choose that least value of K which will make (9K + 1) divisible by 13. Putting K equal to 1, 2, 3, 4, 5 etc in succession, we find that K must be 10.

$\therefore$ the required number $= 360 \times K + 1 = 360 \times 10 + 1$
$= 3601.$

Note: The above example could also be worded thus — **"A gardener had a number of shrubs to plant in rows. At first he tried to plant 5 in each row, then 6, then 8, then 9 and then 12 but had always 1 left. On trying 13 he had none left. What is the smallest number of shrubs that he could have had".**

Exercise

1. Find the least number which being divided by 2, 3, 4, 5, 6, leaves in each case a remainder 1, but when divided by 7 leaves no remainder.
   a) 301 b) 201
   c) 302 d) 310
2. Find the least number which when divided by each of the numbers 3, 5, 6, 8, 10 and 12 leaves in each case a remainder 2 but which when divided by 13 leaves no remainder.
   a) 963 b) 692
   c) 269 d) 962
3. A heap of pebbles can be made up exactly into groups of 25, but when made up into groups of 18, 27 and 32, there is in each case a remainder of 11, find the least number of pebbles such a heap can contain.
   a) 775 b) 975 c) 785 d) 875

Answers

1. a 2. d 3. d

Rule 19

There are n numbers. If the HCF of each pair is x and the LCM of all the n numbers is y, then the product of n numbers is given by $\left[(x)^{n-1} \times y\right]$ or

Product of 'n' numbers = (HCF of each pair)$^{n-1}$ × (LCM of n numbers).

Illustrative Example

Ex.: There are 4 numbers. The HCF of each pair is 3 and the LCM of all the 4 numbers is 116. What is the product of 4 numbers?

Soln: Applying the above rule, we have,

the required answer = $(3)^{4-1} \times 116 = 3132$

Exercise

1. There are 4 numbers. The HCF of each pair is 7 and the LCM of all the 4 numbers is 1470. What is the product of 4 numbers?
   a) 504210 b) 502410
   c) 504120 d) Can't be determined
2. There are 3 numbers. The HCF of each pair is 3 and the LCM of all the 3 numbers is 858. What is the product of 3 numbers?
   a) 7722 b) 7272 c) 6622 d) 7822
3. There are 4 numbers. The HCF of each pair is 4 and the LCM of all the 4 numbers is 840. What is the product of 4 numbers?
   a) 35760 b) 53670 c) 35670 d) 53760
4. There are 4 numbers. The HCF of each pair is 5 and the LCM of all the 4 numbers is 2310. What is the product of 4 numbers?
   a) 288750 b) 288570 c) 828570 d) 288650
5. There are 3 numbers. The HCF of each pair is 6 and the LCM of all the 3 numbers is 420. What is the product of 3 numbers?
   a) 15110 b) 15120 c) 15210 d) 25120
6. There are 3 numbers. The HCF of each pair is 2 and the LCM of all the 3 numbers is 210. What is the product of 3 numbers?
   a) 840 b) 480 c) 740 d) 850

Answers

1. a 2. a 3. d 4. a 5. b 6. a

Miscellaneous

1. The numbers 11284 and 7655, when divided by a certain number of three digits, leave the same remainder. Find the number and the remainder.
   a) 119, 15 b) 191, 15 c) 192, 52 d) 191, 51
2. The sum of two numbers is 1215 and their HCF is 81. How many pairs of such numbers can be formed? Find them.
   a) 1 b) 2 c) 3 d) 4
3. The product of two numbers is 7168 and their HCF is 16, find the sum of all possible numbers.
   a) 640 b) 860 c) 460 d) Data inadequate
4. In a long division sum the dividend is 529565 and the successive remainders from the first to the last are 246, 222, 542. Find the divisor and the quotient.
   a) 561, 943 b) 669, 493 c) 516, 943 d) 561, 493
5. In finding HCF of two numbers, the last divisor is 49 and the quotients 17, 3, 2. Find the numbers.
   a) 343, 5929 b) 434, 2959
   c) 433, 5299 d) Can't be determined
6. An inspector of schools wishes to distribute 84 balls and 180 bats equally among a number of boys. Find the greatest number receiving the gift in this way.
   a) 14 b) 15 c) 16 d) 12
7. In a school 391 boys and 323 girls have been divided into the largest possible equal classes, so that there are equal number of boys and girls in each class. What is the number of classes?
   a) 23 girl's classes, 19 boy's classes
   b) 23 boy's classes, 19 girl's classes
   c) 17 boy's classes, 23 girl's classes
   d) 23 boy's classes, 17 girls' classes
8. What least number must be subtracted from 1936 so that the remainder when divided by 9, 10, 15 will leave in each case the same remainder 7.
   a) 46 b) 53 c) 39 d) 44
9. Find the two numbers whose LCM is 1188 and HCF is 9.
   a) 27, 396 b) 9, 27 c) 36, 99 d) Data inadequate
10. Find the sum of three numbers which are prime to one another such that the product of the first two is 437 and that of the last two is 551.
    a) 91 b) 81 c) 71 d) 70
11. Find the number lying between 900 and 1000 which when divided by 38 and 57, leaves in each case a remainder 23.
    a) 935 b) 945 c) 925 d) 955
12. In a long division sum the successive remainders from the first to the last were 312, 383 and 1. If the dividend be 86037, find the divisor and the quotient.
    a) 548, 157 b) 274, 1 c) 1096, 158 d) None of these
13. Among how many children may 429 mangoes and also 715 oranges be equally divided?
    a) 143 b) 15 c) 18 d) 153
14. The product of two numbers is 4928. If 8 be their HCF find how many pairs of such numbers.
    a) 3 b) 4 c) 2 d) 1
15. Five bells begin to toll together and toll respectively at intervals of 6, 7, 8, 9 and 12 seconds. How many times they will toll together in one hour, excluding the one at the start?
    a) 3 b) 5 c) 7 d) 9
16. 21 mango trees, 42 apple trees and 56 orange trees have to be planted in rows such that each row contains the

same number of trees of one variety only. Minimum number of rows in which the above trees may be planted is
a) 15 b) 17 c) 3 d) 20

17. The sum and difference of the LCM and the HCF of two numbers are 592 and 518 respectively. If the sum of two numbers be 296, find the numbers.
a) 111, 185 b) 37, 259
c) Data inadequate d) None of these

18. The circumferences of the fore and hind wheels of a carriage are $6\frac{3}{14}$ metres and $8\frac{1}{18}$ metres respectively. At any given moment, a chalk mark is put on the point of contact of each wheel with the ground. Find the distance travelled by the carriage so that both the chalk marks are again on the ground at the same time.
a) 218 m b) 217.5 m c) 218.25 m d) 217 m

19. A merchant has three kinds of wine; of the first kind 403 gallons, of the second 527 gallons and of the third 589 gallons. What is the least number of full casks of equal size in which this can be stored without mixing?
a) 21 b) 29 c) 33 d) 31

20. Find the least number of square tiles required for a terrace 15.17 m long and 9.02 m broad.
a) 841 b) 714 c) 814 d) None of these

21. Three pieces of timber 24 metres, 28.8 metres and 33.6 metres long have to be divided into planks of the same length. What is the greatest possible length of each plank?
a) 8.4 m b) 4.8 m c) 4.5 m d) 5.4 m

22. Four bells toll at intervals of 6, 8, 12 and 18 minutes respectively. If they start tolling together at 12 a.m.; find after what interval will they toll together and how many times will they toll together in 6 hours.
a) 6 times b) 5 times c) 4 times d) Data inadequate

23. Three persons A, B, C run along a circular path 12 km long. They start their race from the same point and at the same time with a speed of 3 km/hr, 7 km/hr and 13 km/hr respectively. After what time will they meet again?
a) 12 hrs b) 9 hrs c) 24 hrs d) 16 hrs

24. When in each box 5 or 6 dozens of oranges were packed, three dozens were remaining. Therefore, bigger boxes were taken to pack 8 or 9 dozens of oranges. However, still three dozens of oranges remained. What was the least number of dozens of oranges to be packed?
[NABARD, 1999]
a) 216 b) 243 c) 363 d) 435

Answers

1. b; The required number must be a factor of (11284 – 7655) or 3692.
Now $3692 = 19 \times 191$

```
191) 7655 (40
     764
     ---
      15
```

$\therefore$ 191 is the required number, and 15 is the remainder.

2. d; Let the two numbers be 81a and 81b where a and b are two numbers prime to each other.
$\therefore 81a + 81b = 1215$
$\therefore a + b = \frac{1215}{81} = 15$
Now find two numbers, whose sum is 15. The possible pairs are (14, 1); (13, 2); (12, 3); (11, 4); (10, 5); (9, 6); (8, 7). Of these the only pairs of numbers that are prime to each other are (14, 1), (13, 2), (11, 4) and (8, 7).
Hence the required numbers are
$(14 \times 81, 1 \times 81); (13 \times 81, 2 \times 81); (11 \times 81, 4 \times 81); (8 \times 81, 7 \times 81)$
or, (1134, 81); (1053, 162); (891, 324); (648, 567).
So, there are four such pairs.

3. a; Let the numbers be 16a and 16b where a and b are two numbers prime to each other.
$\therefore 16a \times 16b = 7168 \quad \therefore ab = 28$
Now the pairs of numbers whose product is 28, are (28, 1); (14, 2); (7, 4)
14 and 2 which are not prime to each other should be rejected.
Hence the required numbers are
$28 \times 16; 1 \times 16; 7 \times 16; 4 \times 16$
or, 448, 16, 112, 64
Hence the required answer $= 448 + 16 + 112 + 64 = 640$

4. a; On subtracting the remainders 246, 222, 542 from the numbers giving rise to them, the successive partial products will be found to be 5049, 2244, 1683.

```
 ...
   )529565(
    2466
    ----
    2225
    ----
     542
```

Hence the divisor must be a common factor of these three partial product.
Now 561 is their HCF and no smaller factor (for example 51) will serve the purpose, since $5049 \div 51 = 99$ a two-digit number which is absurd.
$\therefore$ the divisor = 561 and the quotient = 943.

5. a; $\because$ The last divisor = 49 and quotient = 2
$\therefore$ dividend $= 49 \times 2 = 98$

```
343) 5929 (17
   98) 343 (3
      49) 98 (2
           ×
```

Now, divisor = 98, quotient $= 98 \times 3 + 49 = 343$
Again divisor = 343, quotient = 17, and

remainder = 98.

$\therefore$ dividend = 343 × 17 + 98 = 5929

Hence the required numbers are 343, 5929.

6. d; Find the HCF of 84 and 180, which is 12 and this is the required answer.

7. b; The largest possible number of persons in a class is given by the HCF of 391 and 323 ie 17.

$\therefore$ No. of classes of boys = $\frac{391}{17} = 23$ and

No. of classes of girls = $\frac{323}{17} = 19$.

8. c; The LCM of 9, 10, 15 = 90

On dividing 1936 by 90, the remainder = 46

But a part of this remainder = 7.

Hence the required number = 46 – 7 = 39.

9. a; Let the two numbers be 9a and 9b where a and b are two numbers prime to each other. The LCM of 9a and 9b is 9ab.

$\therefore$ 9ab = 1188 $\therefore$ ab = 132

Now the possible pairs of factors of 132 are 1 × 132, 2 × 66,. 3 × 44, 6 × 22, 11 × 12. Of these pairs (2, 66) and (6, 22) are not prime to each other, and therefore, not admissible.

Hence the admissible pairs are

1, 132; 3, 44; 4, 33; 11, 12

$\therefore$ a = 1, b = 132; a = 3, b = 44, a = 4, b = 33; a = 11, b = 12.

Hence the required numbers are 9, 9 × 132; 9 × 3, 9 × 44; 9 × 4, 9 × 33; 9 × 11, 9 × 12

or, 9, 1188; 27, 396; 36, 297; 99, 108.

10. c; From the question we see that the second number is a common factor of the two products, and since the numbers are prime to one another, it is their HCF and is, therefore, 19.

$\therefore$ the first number = 437 ÷ 19 = 23 and

the third number = 551 ÷ 19 = 29.

Hence the numbers are 23, 19 and 29.

11. a; The least common multiple of 38 and 57 is 114 and the multiple which is between 900 and 1000 is 912.

Now, 912 + 23 ie, 935 lies between 900 and 1000 and when divided by 38 and 57 leaves in each case 23 as the remainder. Therefore 935 is the number required.

12. b; Since the last but one remainder is 383 and the last figure to be affixed to it is 7, $\because$ the last partial product is 3837 – 1 = 3836. Similarly, the other partial products will be 2740 and 548.

```
548) 86037 (157
     548
     ----
     3123
     2740
     ----
      3837
      3836
      ----
         1
```

The HCF of these three partial produts = 548

$\therefore$ the divisor = 548 or a factor of 548. But the divisor must be greater than each of the partial remainders 312, 383 and 1.

$\therefore$ The divisor is 548; hence the quotient is 157.

13. a; The number of children required must be a common factor of 429 and 715. Now the HCF of 429 and 715 is 143.

$\therefore$ the number of children required must be 143 or a factor of 143. But 143 = 13 × 11.

$\therefore$ the number of children required is 143, 13 or 11.

14. c; Let the numbers be $8x$ and 8y, where x and y are prime to each other. Then, $8x \times 8y = 4928$ or $64xy = 4928$

$\therefore xy = 77 \quad \therefore x = 1$ or 7 and $y = 77$ or 11

$\therefore$ these pairs of required numbers will be (8, 77 × 8) or, (8 × 7, 8 × 11) that is (8, 616) or (56, 88).

15. c; LCM of 6, 7, 8, 9, 12 is 504.

So, the bells will toll together after 504 sec.

In 1 hour, they will toll together = $\left(\frac{60 \times 60}{504}\right)$ times

= 7 times.

16. b; HCF of 21, 42, 56 = 7

Number of rows of mango trees, apple trees and orange trees are $\frac{21}{7} = 3$, $\frac{42}{7} = 6$ and $\frac{56}{7} = 8$

$\therefore$ required number of rows = (3 + 6 + 8) = 17.

17. a; Let the LCM and HCF be h and k respectively.

$\therefore$ h + k = 592 and h – k = 518

Consequently, $h = \frac{592 + 518}{2} = 555$ &

$k = \frac{592 - 518}{2} = 37$.

i.e., LCM = 555 and HCF = 37

Now, let the numbers be 37a and 37b, where a and b are co-primes.

$\therefore$ 37a + 37b = 296 or a + b = 8.

Possible pairs of co-primes, whose sum is 8 are (1, 7) & (3, 5).

$\therefore$ possible pairs of numbers are:

$$\left.\begin{array}{l}(37 \times 1, 37 \times 7) \text{ or } (37, 259) \\ \text{and } (37 \times 3, 37 \times 5) \text{ or } (111, 185)\end{array}\right\}$$

Now, HCF × LCM = 555 × 37 = 20535.

Also, 111 × 185 = 20535, while 37 × 259 ≠ 20535

Hence, the required numbers are 111 and 185.

18. b; The required distance in metres = LCM of $\frac{84}{14}$ and $\frac{145}{18}$

$$= \frac{\text{LCM of } 87 \,\&\, 145}{\text{HCF of } 14 \,\&\, 18} = \left(\frac{435}{2}\right) \text{m} = 217.5 \text{ m}$$

19. d; HCF of 403, 527 and 589 is 31.
$\therefore$ required answer = 31

20. c; Tiles are least, when size of each is largest. So, HCF of 1517 cm and 902 cm gives each side of a tile, which is 41 cm.

$$\therefore \text{ number of tiles} = \left(\frac{1517 \times 902}{41 \times 41}\right) = 814$$

21. b; Find the HCF of 2400 cm, 2880 cm and 3360 cm, which is 480 cm. Hence required answer is 4.8 metres.

22. a; LCM of 6, 8, 12, 18 min = 72 min = 1 hr 12 min.
So, they will toll together after 1 hr 12 min.

In 6 hours, they will toll together

$$= \left(\frac{360}{72} \textit{time} + 1 \textit{ time at the start}\right) = 6 \text{ times}$$

23. a; Time taken by A, B, C to cover 12 km is 4 hours, $\frac{12}{7}$ hours and $\frac{12}{13}$ hours respectively.

LCM of 4, $\frac{12}{7}$ and $\frac{12}{13}$ = 12.

So, they will meet again after 12 hours.

24. c; **Hint:** Required number = (LCM of 5, 6, 8, 9) + 3
= 360 + 3 = 363.

3

Ratio and Proportion

Rule 1

To Find Compound Ratio

Ratios are compounded by multiplying together the antecedents for a new antecedent, and the consequents for a new consequent.

Note:

1. When the ratio 4 : 3 is compounded with itself the resulting ratio is $4^2 : 3^2$. It is called the **duplicate ratio** of 4 : 3. Similarly, $4^3 : 3^3$ is the **triplicate ratio** of 4 : 3.

   $\sqrt{4} : \sqrt{3}$ is called the **subduplicate ratio** of 4 : 3.

   $a^{1/3} : b^{1/3}$ is **subtriplicate ratio** of a and b.
2. The number of times one quantity contains another quantity of the same kind is called the **ratio** of the two quantities. The ratio 2 to 3 is written as 2:3 or $\frac{2}{3}$. 2 and 3 are called **terms of the ratio**. 2 is the **first term** and 3 is the **second term**. The first term of the ratio is called the **antecedent** and the second the **consequent**. In the ratio 2:3, 2 is the **antecedent** and 3 is the **consequent.**
3. If 2:3 be the given ratio, then $\frac{1}{2}:\frac{1}{3}$ or 3:2 is called its **inverse** or **reciproral** ratio.
4. If the antecedent = the consequent, the ratio is called the **ratio of equality**; as 3:3.
5. If the antecedent > the consequent the ratio is called the ratio of **greater inequality**, as 4:3
6. If the antecedent < the consequent, the ratio is called the **ratio of less** inequality, as 3:4

Illustrative Example

Ex.: Find the ratio compounded of the four ratios: 4 : 3, 9 : 13, 26 : 5 and 2 : 15

Soln: The required ratio = $\frac{4\times9\times26\times2}{3\times13\times5\times15}=\frac{16}{25}$

Exercise

1. Find the compound ratio of the following: 1:2, 3:5 and 5:9

   a) $\frac{1}{6}$ b) $\frac{1}{5}$ c) $\frac{1}{3}$ d) $\frac{1}{18}$
2. Find the compound ratio of the following: 5 : 7, 15 : 14 and 98 : 75.

   a) 1:5 b) 1:1 c) 2:1 d) 5:1
3. Find the compound ratio of the following: 5:6, 12:19, 57:60 and 50:31

   a) 13:25 b) 31:25 c) 25:31 d) 25:13
4. Find the subduplicate ratio of 16:25.

   a) 4:5 b) 5:4 c) 256:625 d) 625:256
5. Find the subtriplicate ratio of 343:729.

   a) 5:7 b) 9:7 c) 7:9 d) 7:5
6. Find the duplicate ratio of 14:17.

   a) 196:289 b) 169:256 c) 196:729 d) 576:729
7. Find the triplicate ratio of 3:5.

   a) 27:125 b) 9:25 c) $3^{\frac{1}{3}} : 5^{\frac{1}{3}}$ d) 125:27

Answers

1. a 2. b 3. c 4. a 5. c 6. a 7. a

Rule 2

Theorem: If four quantities be in proportion the product of the extremes is equal to the product of the means.

Let the four quantities 3, 4, 9 and 12 be in proportion.

We have $\frac{3}{4}=\frac{9}{12}$

Multiply each ratio by 4 × 12

$\therefore \frac{3}{4}\times4\times12=\frac{9}{12}\times4\times12$

$\therefore 3\times12=4\times9$

Note:

1. The 4th term can be found by multiplying the 2nd and 3rd terms together and divide the product by the first

term.

2. Consider the two ratios:

1st ratio	2nd ratio
6:18	8:24

Since 6 is one-third of 18, and 8 is one-third of 24, the two ratios are equal. The equatity of ratios is called **proportion**.

The numbers 6, 18, 8 and 24 are said to be in proportion. The proportion may be written as 6:18::8:24 (6 is to 18 as 8 is to 24)

or, 6:18 = 8:24 or $\frac{6}{18}:\frac{8}{24}$

The numbers 6, 18, 8 and 24 are called **trms**. 6 is the **first term**, 18 the **second**, 8 the **third** and 24 the **fourth**. The first and fourth terms ie 6 and 24 are called the **extremes** (end terms), and the second and the third terms, ie, 18 and 8 are called the **means** (middle terms). 24 is called the **fourth** proportional.

Illustrative Example

Ex.: Find the fourth proportional to the numbers 6, 8 and 15.

Soln: If x be the fourth proportional, then $6 : 8 = 15 : x$

$\therefore x = \frac{8 \times 15}{6} = 20$

Exercise

1. Find a fourth proportional to the numbers 6, 8, 9.
   a) 12 b) 7 c) 5 d) 14
2. Find the value of the missing figure in the question given below.
   6 : ? :: 5 : 35
   a) 30 b) 36 c) 42 d) 48
3. Find a fourth proportional to the numbers 12, 14, 24.
   a) 38 b) 36 c) 28 d) 30
4. Find a fourth proportional to the numbers 5, 7, 15.
   a) 21 b) 35 c) 20 d) 30
5. Find a fourth proportional to the numbers 21, 33, 56.
   a) 77 b) 78 c) 88 d) 87
6. Find a fourth proportional to the numbers 45, 60, 72.
   a) 120 b) 96 c) 72 d) 84
7. Find the value of the missing figure in the following.
   ? : 13 :: 35 : 65
   a) 7 b) 9 c) 6 d) 5
8. Find the value of x in the following proportion.
   5 : 15 = 2 : x
   a) 6 b) 3 c) 12 d) 9
9. Find the value of x in the following proportion.
   75 : 3 = x : 9
   a) 125 b) 120 c) 225 d) 220
10. Calculate a fourth proportional to the numbers.
    1, 2 and 3.
    a) 6 b) 1.5 c) 0.6 d) 5
11. Calculate a fourth proportional to the numbers.
    500, 70, and 69
    a) $9\frac{43}{50}$ b) $8\frac{33}{50}$ c) $9\frac{33}{50}$ d) $19\frac{33}{50}$
12. Calculate a fourth proportional to the numbers.
    2.5, 1.5, and 1.5
    a) 0.9 b) 0.89 c) 0.91 d) 0.09
13. What should come in place of the question mark (?) in the following equation?
    $\frac{28}{?} = \frac{?}{112}$
    a) 70 b) 56 c) 48 d) 64

Answers

1. a 2. c 3. c 4. a 5. c 6. b 7. a 8. a 9. c 10. a 11. c 12. a

13. b; **Hint:** Answer = $\sqrt{28 \times 112} = 56$

Rule 3

Theorem: ***Three quantities of the same kind are said to be in continued proportion when the ratio of the first to the second is equal to the ratio of the second to the third. The second quantity is called the mean proportional between the first and the third; and the third quantity is called the third proportional to the first and second.***

Thus, 9, 6 and 4 are in continued proportion for 9 : 6 : : 6 : 4. Hence, 6 is the mean proportional between 9 and 4, and 4 is the third proportional to 9 and 6.

Illustrative Examples

Ex. 1: Find the third proportional to 15 and 20.

Soln: Here, we have to find a fourth proportional to 15, 20 and 20.

If x be the fourth proportional, we have $15 : 20 = 20 : x$

$\therefore x = \frac{20 \times 20}{15} = \frac{80}{3} = 26\frac{2}{3}$

Ex. 2: Find the mean proportional between 3 and 75.

Soln: If x be the required mean proportional, we have $3 : x : : x : 75$

$\therefore x = \sqrt{3 \times 75} = 15$

Note: It is evident that the mean proportional between two numbers is equal to the square root of their product.

Exercise

1. Find a third proportional to the numbers 3 and 6.
   a) 21 b) 1.5 c) 18 d) 12
2. Find a third proportional to the numbers 1.2 and 1.8.
   a) 2.8 b) 2.7 c) 3.2 d) 3.7
3. Find a third proportional to the numbers 225 and 75.
   a) 25 b) 15 c) 35 d) 30

4. Calculate the mean proportional between 3 and 192
a) 24 b) 26 c) 22 d) 28

5. Calculate the mean proportional between $\frac{14}{363}$ and $\frac{21}{128}$

a) $\frac{7}{88}$ b) $\frac{7}{86}$ c) $\frac{9}{88}$ d) $\frac{14}{87}$

6. Calculate the mean proportional between 0.5 and 1922
a) 39 b) 29 c) 31 d) 41

7. Find the third proportional to the numbers 5 and 10
a) 20 b) 2.5 c) 25 d) 30

Answers

1. d 2. b 3. a 4. a 5. a 6. c 7. a

Rule 4

The Rule of Three: ***The method of finding the 4th term of a proportion when the other three are given is called Simple Proportion or the Rule of Three.***

In every question of simple proportion, two of the given terms are of the same kind, and the third term is of the same kind as the required fourth term.

Now we give the rule of arranging the terms in a question of simple proportion.

Rule: I. Denote the quantity to be found by the letter 'x', and set it down as the 4th term.

II. Of the three given quantities, set down that for the third term which is of the same kind as the quantity to be found.

III. Now, consider carefully whether the quantity to be found will be greater or less than the third term; if greater, make the greater of the two remaining quantities the 2nd term, and the other 1st term, but if less, make the less quantity the second term, and the greater the 1st term.

IV. Now, the required value

$$= \frac{\text{Multiplication of means}}{\text{1st term}}$$

Illustrative Examples

Ex. 1: If 15 books cost Rs 35, what do 21 books cost?

Soln: This is an example of direct proportion. Because if the number of books is increased, their cost also increases.

By the Rule of Three:

Step I: ... : ... = ... : Required cost.

Step II: ... : ... = Rs 35 : Required cost

Step III: The required cost will be greater than the given cost; so the greater quanity will come as the 2nd term. Therefore,

15 books : 21 books = Rs 35 : Required cost.

Step IV: ∴ the required cost $= \frac{21 \times 35}{15} =$ Rs 49

Ex. 2: If 15 men can reap a field in 28 days, in how many days will 5 men reap it?

Soln: **Step I:** ... : ... = ... : Required number of days.

Step II: ... : ... = 28 : x

Step III: The required number of days will be more, since 5 men will take more time than 15 men. Therefore, 5 : 15 = 28 : x

Step IV: $x = \frac{15 \times 28}{5} = 84$ days.

Ex. 3: A fort had provisions for 150 men for 45 days. After 10 days, 25 men left the fort. How long will the food last at the same rate for the remaining men?

Soln: The remaining food would last for 150 men for (45 – 10 =) 35 days. But as 25 men have gone out, the remaining food would last for a longer period. Hence, by the **Rule of Three**, we have the following relationhip.

125 men : 150 men = 35 days : the required no. of days.

∴ the required no. of days $= \frac{150 \times 35}{125} = 42$ days.

Compound Proportion or Double Rule of Three

Ex.: If 8 men can reap 80 hectares in 24 days, how many hectares can 36 men reap in 30 days.

Soln: We can resolve this problem into two questions.

1st: If 8 men can reap 80 hectares, how many hectares can 36 men reap?

8 men : 36 men = 80 hectares : the required no. of hectares

∴ the required no. of hectares $= \frac{36 \times 80}{8} = 360$ hectares.

2nd: If 360 hectares can be reaped in 24 days, how many hectares can be reaped in 30 days?

By the Rule of Three

24 days : 30 days = 360 hectares : the required no. of hectares.

∴ the required no. of hectares $= \frac{30 \times 360}{24} = 450$

We observe that the original number of hectares, namely 80, has been changed in the ratio formed by compounding the ratio $\frac{36}{8}$ and $\frac{30}{24}$.

The above question can be solved in a single step. We arrange the figures in the following form:

8 men : 36 men
24 days : 30 days] : 80 hect : the reqd no. of hectares

The reqd no. of hectares

$$= \frac{\text{Multiplication of means}}{\text{Multiplication of 1st terms}} = \frac{80 \times 36 \times 30}{8 \times 24} = 450$$

Exercise

1. If 30 men do a piece of work in 27 days, in what time can 18 men do another piece of work 3 times as great?
   a) 145 days b) 135 days c) 130 days d) 134 days
2. When wheat is Rs 3.90 per kg. 60 men can be fed for 15 days at a certain cost. how many men can be fed for 45 days at the same cost, when wheat is Rs 3 per kg?
   a) 25 men b) 26 men c) 28 men d) 27 men
3. If a family of 7 persons can live on Rs 8400 for 36 days, how long can a family of 9 persons live on Rs 8100?
   a) 27 days b) 37 days c) 36 days d) 24 days
4. If 1000 copies of a book of 13 sheets require 26 reams of paper, how much paper is required for 5000 copies of a book of 17 sheets?
   a) 180 reams b) 170 reams c) 140 reams d) 270 reams
5. 5 horses eat 18 quintals of oats in 9 days, how long at the same rate will 66 quintals last for 15 horses?
   a) 12 days b) 9 days c) 13 days d) 11 days
6. If the carriage of 810 kg for 70 km cost Rs 112.50, what will be the cost of the carriage of 840 kg for a distance of 63 km at half the former rate?
   a) Rs 52 b) Rs 53 c) Rs $52\frac{1}{2}$ d) Rs $50\frac{1}{4}$
7. If 300 men could do a piece of work in 16 days, how many men would do $\frac{1}{4}$ of the same work in 15 days?
   a) 80 men b) 85 men c) 90 men d) 75 men
8. If 27 men take 15 days to mow 225 hectares of grass, how long will 33 men take to mow 165 hectares?
   a) 9 days b) 12 days c) 15 days d) 6 days
9. How many horses would be required to plough 117 hectares of land in 35 days, if 10 horses can plough 13 hectares in 7 days?
   a) 28 horses b) 18 horses c) 24 horses d) 16 horses
10. If 6 men can do a piece of work in 30 days of 9 hours each, how many men will it take to do 10 times the amount of work if they work 25 days of 8 hours each?
    a) 81 men b) 80 men c) 79 men d) 82 men
11. If I can walk a certain distance in 50 days when I rest 9 hours each day, how long will it take me to walk twice as far if I walk twice as fast and rest twice as long each day?
    a) 125 b) 120 c) 124 d) 130
12. A garrison of 2200 men is provisioned for 16 weeks at the rate of 45 dag per diem per man. How many men must leave so that the same provisions may last 24 weeks at 33 dag per diem per man?
    a) 200 b) 2000 c) 120 d) 220
13. A gang of labourers promise to do a piece of work in 10 days, but five of them become absent. If the rest of the gang do the work in 12 days, find the original number of men.
    a) 30 b) 35 c) 40 d) 25
14. A man can walk 600 kilometres in 35 days, resting 9 hours each day. How long will he take to walk 375 kilometres if he rests 10 hours each day and walks $1\frac{1}{2}$ times as fast as before?
    a) $15\frac{5}{8}$ days b) 15 days c) $15\frac{3}{8}$ days d) $15\frac{4}{5}$ days
15. Two gangs of 6 men and 9 men are set to reap two fields of 35 and 45 hectares respectively. The first gang complete their work in 12 days; in how many days will the second gang complete theirs?
    a) $\frac{70}{9}$ days b) $\frac{79}{7}$ days c) $\frac{72}{7}$ days d) $\frac{82}{7}$ days
16. If 10 masons can build a wall 50 metres long in 25 days of 8 hours each, in how many days of 6 hours each will 15 masons build a wall 36 metres long?
    a) 18 days b) 15 days c) 24 days d) 16 days

Answers

1. b 2. b 3. a 4. b 5. d 6. c 7. a
8. a 9. b 10. a

11. a; **Hint:** In the first case I walk (24 – 9) or 15 hours each day.
In the 2nd I walk (24 – 18) or 6 hours each day.
Now, we have the following **proportion**.

$$\left.\begin{matrix}\text{distance} & 1:2\\ \text{rate} & 2:1\\ \text{hours} & 6:15\end{matrix}\right\} :: 50\text{ days} : \text{reqd no. of days}$$

$$\therefore \text{ reqd. no. of days} = \frac{2\times1\times15\times50}{1\times2\times6} = 125$$

12. a; **Hint:** 'Per diem' means per day.
2200 men provisioned for 16 weeks at 45 dag per day per man
? men provisioned for 24 weeks at 33 dag per day per man
We have the following proportion

$$\left.\begin{matrix}\text{weeks} & 24:16\\ \text{dag} & 33:45\end{matrix}\right\} :: 2200\text{ men} : \text{men reqd.}\left\{\begin{matrix}\text{more weeks,}\\ \text{less men}\\ \text{less dag,}\\ \text{more men}\end{matrix}\right.$$

$$\therefore \text{ men required} = \frac{16\times45\times2200}{24\times33} = 2000$$

Hence (2200 - 2000) or 200 men must leave.

13. a; **Hint:** We have at once, $\frac{\text{Original number}}{\text{Original number - 5}} = \frac{12}{10}$
Here the difference of the last two terms 12 and 10 is 2, but the difference of the first two terms is 5.

Now we have the proportion:

diff		diff		men		men
2	:	5	::	12	:	30

∴ original number of men = 30.

14. a 15. c 16. d

Rule 5

Theorem: ***If two numbers are in the ratio of a : b and the sum of these numbers is x, then these numbers will be*** $\frac{ax}{a+b}$ ***and*** $\frac{bx}{a+b}$ ***respectively.***

Illustrative Examples

Ex. 1: Two numbers are in the ratio of 3 : 1. If sum of these two numbers is 440, find the numbers.

Soln: By using the above rule, we have,

a = 3, b = 1, x = 440

$$\therefore \text{ First number} = \frac{ax}{a+b} = \frac{3\times 440}{3+1} = 330$$

$$\text{Second number} = \frac{bx}{a+b} = \frac{1\times 440}{3+1} = 110$$

Ex. 2: The ratio of the number of boys and girls in a school is 2 : 5. If there are 350 students in the school, find the number of girls in the school.

Soln: Here we can consider

a = number of boys = 2

b = Number of girls = 5

x = Total number of students = 350

We have to find number of girls ie b

$$\Rightarrow \frac{bx}{a+b} \Rightarrow \frac{5\times 350}{2+5} = 250 \text{ girls.}$$

Exercise

1. Two numbers are in the ratio of 8:7. If sum of these two numbers is 450, find the numbers.
   a) 210, 240 b) 240, 210 c) 235, 215 d) 215, 235
2. Two numbers are in the ratio of 9:11. If sum of these two numbers is 660, find the difference between the numbers.
   a) 66 b) 56 c) 46 d) 76
3. Two numbers are in the ratio of 4:5. If sum of these two numbers is 27, find the product of the numbers.
   a) 190 b) 180 c) 225 d) 240
4. The ratio of the number of boys and girls in a school is 7:13. If there are 400 students in the school, find the difference of the numbers of girls and boys.
   a) 160 b) 140 c) 260 d) 120
5. The ratio of the rural and urban population of a country is 9:5. If the total population of the country is 560 millions, find the urban population of that country.
   a) 360 millions b) 260 millions
   c) 240 millions d) 200 millions

Answers

1. b 2. a 3. b 4. d 5. d

Rule 6

To find the number of coins.

$$\textit{Number of each type of coins} = \frac{\textit{Amount in rupees}}{\textit{Value of coins in rupees}}$$

Illustrative Examples

Ex. 1: A bag contains an equal number of one rupee, 50 paise and 25 paise coins respectively. If the total value is Rs 35, how many coins of each type are there?

Soln: Here number of each type of coins is same, hence we may write,

Number of each type of coins

$$= \frac{\text{Total amount in Rs}}{\text{Sum of value of each coin}}$$

∴ Number of each type of coins

$$= \frac{35}{1+0.5+0.25} = 20 \text{ coins of each type.}$$

Ex. 2: A bag contains rupee, fifty paise and twenty five paise coins whose values are in the proportion of 2 : 3 : 4. If the total number of coins are 480 find the value of each coin and the total amount in rupees.

Soln: $\text{Number of coins} = \frac{\text{Amount in rupees}}{\text{Value of coins in rupees}}$

$$\therefore \text{ Number of 1 rupee coin} = \frac{2x}{1}$$

$$\text{Number of 50 paise coin} = \frac{3x}{1/2}$$

$$\text{Number of 25 paise coin} = \frac{4x}{1/4}$$

∴ $2x + 6x + 16x = 480$ (given)

∴ $x = 20$

∴ Value of 1 rupee coin = $2x$ = Rs 40

Value of 50 paise coin = $3x$ = Rs 60

Value of 25 paise coin = $4x$ = Rs 80

Value of total amount = Rs 180 in the bag.

Exercise

1. A bag contains rupee, 50-paise and 25-paise coins in the ratio 5:7:9. If the total amount in the bag is Rs 430, find the number of coins of each kind.
   a) 200, 280, 360 b) 280, 200, 360
   c) 360, 280, 200 d) 360, 200, 280

2. In a bag there are coins of 25-paise, 10-paise and 5-paise in the ratio 1:2:3. The value of the money in the bag is Rs 60. Then, the number of 25-paise coin is
a) 10 b) 120 c) 100 d) 20
3. A bag contains rupee, fifty paise and twenty five paise coins whose values are in the proportion of 4:5:6. If the total number of coins are 760, find the number of 50-paise coins.
a) 80 b) 200 c) 480 d) 280
4. A bag contains rupee, 50-paise and 25-paise coins in the ratio 3:4:5. If the total amount in the bag is Rs 625, find the no. of coins of 25-paise.
a) 125 b) 1250 c) 500 d) 1000
5. A bag contains an equal number of one-rupee, 50-paise and 25-paise coins respectively. If the total value is Rs 43.75, how many coins of each type are there?
a) 40 b) 25 c) 35 d) 30
6. A bag contains an equal number of 50-paise, 25-paise, 20 paise and 5-paise coins respectively. If the total value is Rs 40, how many coins of each type are there?
a) 40 b) 25 c) 30 d) 20

Answers

1. a; **Hint:** Ratio among the values of the coins

$$=\frac{5}{1}:\frac{7}{2}:\frac{9}{4}=20:14:9$$

Thus, the value of one-rupee coins $= 430\times\frac{20}{43} =$ Rs 200

the value of 50-paise coins $= 430\times\frac{14}{43} =$ Rs 140

the value of 25-paise coins $= 430\times\frac{9}{43} =$ Rs 90

$\therefore$ the number of one-rupee coins $= 200 \times 1 = 200$
the number of 50-paise coins $= 140 \times 2 = 280$
the number of 25-paise coins $= 90 \times 4 = 360$

2. c 3. b 4. c 5. b 6. a

Rule 7

To find the strength to milk

Strength of milk in the mixture

$$= \frac{\textit{Quantity of Milk}}{\textit{Total Quantity of Mixture}}$$

Illustrative Example

Ex.: One man adds 3 litres of water to 12 litres of milk and another 4 litres of water to 10 litres of milk. What is the ratio of the strengths of milk in the two mixtures?

Soln: Strength of milk in the mixture

$$= \frac{\text{Quantity of Milk}}{\text{Total Quanity of Mixture}}$$

$\therefore$ Strength of milk in the first mixture $= \frac{12}{12+3} = \frac{12}{15}$

Strength of milk in the second mixture $= \frac{10}{10+4} = \frac{10}{14}$

$\therefore$ The ratio of their strengths $= \frac{12}{15}:\frac{10}{14}$

$= 12 \times 14 : 15 \times 10$

$= 28 : 25$

Exercise

1. One man adds 5 litres of water to 15 litres of milk and another 6 litres of water to 12 litres of milk. What is the ratio of the strengths of milk in the two mixtures?
a) 9:8 b) 8:9 c) 7:6 d) 6:7
2. One man adds 2 litres of water to 10 litres of milk and another 3 litres of water to 12 litres of milk. What is the ratio of the strengths of milk in the two mixtures?
a) 10:9 b) 12:11 c) 25:24 d) 26:25
3. One man adds 11 litres of water to 14 litres of milk and another 12 litres of water to 13 litres of milk. What is the ratio of the strengths of milk in the two mixtures?
a) 13:14 b) 14:13 c) 11:13 d) 12:11
4. One man adds 5 litres of water to 8 litres of milk and another 3 litres of water to 10 litres of milk. What is the ratio of the strengths of milk in the two mixtures?
a) 5:4 b) 7:5 c) 4:5 d) 9:10
5. One man adds 6 litres of water to 11 litres of milk and another 9 litres of water to 8 litres of milk. What is the ratio of the strengths of milk in the two mixtures?
a) 2:3 b) 3:2 c) 11:8 d) 8:11

Answers

1. a 2. c 3. b 4. c 5. c

Rule 8

Ex.: Two vessels contain equal quantity of mixtures of milk and water in the ratio 5 : 2 and 6 : 1 respectively. Both the mixtures are now mixed thoroughly. Find the ratio of milk to water in the new mixture so obtained.

Soln: Detail Method: Change the ratios into fractions

	Milk	:	Water
Vessel I	$\frac{5}{7}$		$\frac{2}{7}$
Vessel II	$\frac{6}{7}$		$\frac{1}{7}$

Now, both the mixtures are mixed thoroughly.
Therefore, the ratio of water to milk in the new vessel

$$=\left(\frac{5}{7}+\frac{6}{7}\right):\left(\frac{2}{7}+\frac{1}{7}\right)=\frac{11}{7}:\frac{3}{7}=11:3$$

Quicker Method: See carefully, if the sum of the ratios of milk to water in two vessels is equal [as in the above example $(5+2=6+1)$] then new ratio of milk to water will be the sum of respective ratios of milk to water in two vessels ie $5+6=11$ and $2+1=3$ respectively.

Note: This will not apply if the sum of the ratios are not same in two vessels.

Exercise

1. Two vessels contain equal quantity of mixtures of milk and water in the ratio 3:2 and 4:1 respectively. Both the mixtures are now mixed thoroughly. Find the ratio of milk to water in the new mixture so obtained.
a) 3:7 b) 7:3 c) 1:1 d) 4:3
2. Two vessels contain equal quantity of mixtures of milk and water in the ratio 2:7 and 5:4 respectively. Both the mixtures are now mixed thoroughly. Find the ratio of milk to water in the new mixture so obtained.
a) 1:2 b) 1:1 c) 11:7 d) 7:11
3. Two vessels contain equal quantity of mixtures of milk and water in the ratio 8:5 and 3:10 respectively. Both the mixtures are now mixed thoroughly. Find the ratio of milk to water in the new mixture so obtained.
a) 11:15 b) 15:11 c) 1:1 d) 4:9
4. Two vessels contain equal quantity of mixtures of milk and water in the ratio 8:9 and 12:5 respectively. Both the mixtures are now mixed thoroughly. Find the ratio of milk to water in the new mixture so obtained.
a) 7:10 b) 13:21 c) 21:13 d) 10:7
5. Two vessels contain equal quantity of mixtures of milk and water in the ratio 9:5 and 4:3 respectively. Both the mixtures are now mixed thoroughly. Find the ratio of milk to water in the new mixture so obtained.
a) 17:11 b) 11:17 c) 8:13 d) 13:8

Answers

1. b 2. d 3. a 4. d
5. a; **Hint:** $4:3=8:6$ and $9+5=14$, $8+6=14$.

Rule 9

Theorem: The contents of two vessels containing water and milk are in the ratio $x_1:y_1$ and $x_2:y_2$ are mixed in the ratio x:y. The resulting mixture will have water and milk in the ratio of

$$xx_1(x_2+y_2)+yx_2(x_1+y_1):xy_1(x_2+y_2)+yy_2(x_1+y_1).$$

Illustrative Example

Ex.: The contents of two vessels containing water and milk are in the ratio 1 : 2 and 2 : 5 are mixed in the ratio 1 : 4. The resulting mixture will have water and milk in the ratio ______.

Soln: **Detail Method:** Change the ratios into fractions.

	Water	:	Milk
Vessel I	$\frac{1}{3}$		$\frac{2}{3}$
Vessel II	$\frac{2}{7}$		$\frac{5}{7}$

From Vessel I, $\frac{1}{5}$ is taken and from Vessel II, $\frac{4}{5}$ is taken.

Therefore, the ratio of water to milk in the new vessel

$$=\left(\frac{1}{3}\times\frac{1}{5}+\frac{2}{7}\times\frac{4}{5}\right):\left(\frac{2}{3}\times\frac{1}{5}+\frac{5}{7}\times\frac{4}{5}\right)$$

$$=\left(\frac{1}{15}+\frac{8}{35}\right):\left(\frac{2}{15}+\frac{20}{35}\right)=\frac{31}{105}:\frac{74}{105}=31:74$$

Quicker Method: Applying the above theorem, we have,

the required answer $=1\times1\times(2+5)+4\times2\,(1+2):1\times2(2+5)+4\times5(1+2)$

$=1\times7+8\times3:2\times7+20\times3=31:74$

Exercise

1. The contents of two vessels containing water and milk are in the ratio 2:3 and 4:5 are mixed in the ratio 1:2. The resulting mixture will have water and milk in the ratio ______.
a) 77:58 b) 58:77 c) 68:77 d) 77:68
2. The contents of two vessels containing water and milk are in the ratio 3:4 and 5:4 are mixed in the ratio 1:4. The resulting mixture will have water and milk in the ratio ______.
a) 184:176 b) 167:184 c) 167:148 d) 148:167
3. The contents of two vessels containing water and milk are in the ratio 1:2 and 2:3 are mixed in the ratio 3:4. The resulting mixture will have water and milk in the ratio ______.
a) 13:22 b) 31:22 c) 22:31 d) 22:13
4. The contents of two vessels containing water and milk are in the ratio 2:5 and 4:7 are mixed in the ratio 1:3. The resulting mixture will have water and milk in the ratio ______.
a) 53:101 b) 101:53 c) 35:101 d) 101:35
5. The contents of two vessels containing water and milk are in the ratio 1:3 and 1:6 are mixed in the ratio 1:5. The resulting mixture will have water and milk in the ratio ______.
a) 47:9 b) 74:9 c) 9:141 d) 9:47

Answers

1. b 2. c 3. a 4. a 5. d

Rule 10

Theorem: ***If two numbers are in the ratio of a : b and the difference between these numbers is x, then these numbers will be***

(i) $\frac{ax}{a-b}$ *and* $\frac{bx}{a-b}$ *respectively. (where, a > b)*

(ii) $\frac{ax}{b-a}$ *and* $\frac{bx}{b-a}$ *respectively (where a < b)*

Illustrative Examples

Ex. 1: Two numbers are in the ratio of 9 : 14. If the larger number is 55 more than the smaller number, find the numbers.

Soln: Here rule (ii) will apply because (a < b).

$$\text{Smaller number} = \frac{ax}{b-a} = \frac{9\times 55}{14-9} = 99$$

$$\text{Larger number} = \frac{bx}{b-a} = \frac{14\times 55}{5} = 154$$

Hence, numbers are 99 and 154.

Ex. 2: If income of A, B and C is in the ratio of 3 : 7 : 9 and income of B is Rs 240 more than that of A, find the income of C.

Soln: Here, this rule will also apply.
Ratio of the income of A, B and C = 3 : 7 : 9
∴ Ratio of the income of A and B = 3 : 7
And difference between income of A and B = Rs 240

$$\therefore \text{Income of C} = \frac{cx}{b-a} = \frac{9\times 240}{7-3} = \text{Rs } 540$$

[∵ a = 3, b = 7, and c = 9; and x = Rs 240]

∴ Income of C = Rs 540

Exercise

1. Two numbers are in the ratio of 8 : 5. If the larger number is 27 more than the smaller number, find the sum of the numbers.
   a) 117 b) 118 c) 115 d) 116
2. Two numbers are in the ratio of 4 : 7. If the larger number is 9 more than the smaller number, find the remainder when larger number is divided by the smaller number.
   a) 9 b) 8 c) 6 d) 7
3. Two numbers are in the ratio of 4 : 5. If the larger number is 15 more than the smaller number, find the product of the numbers.
   a) 3500 b) 3000 c) 4500 d) 4550
4. Two numbers are in the ratio of 3 : 1. If the larger number is 12 more than the smaller number. Find the smallest number that should be subtracted from the product of the numbers so that remainder is divisible by sum of the numbers.
   a) 12 b) 18 c) 8 d) 15
5. If income of A, B and C is in the ratio of 2:9:11 and income of B is Rs 280 more than that of A, find the income of C.
   a) Rs 480 b) Rs 440 c) Rs 540 d) Rs 450
6. The prices of a scooter and a television set are in the ratio 3:2. If a scooter costs Rs 6000 more than the television set, the price of the television set is:
   a) Rs 18000 b) Rs 12000 c) Rs 10000 d) Rs 6000
   (Bank PO Exam 1989)
7. The monthly salary of A, B and C is in the proportion 2 : 3 : 5. If C's monthly salary is Rs 1200 more than A's monthly salary, B's annual salary is:
   a) Rs 14400 b) Rs 24000 c) Rs 1200 d) Rs 2000
   (Bank PO Exam 1990)

Answers

1. a 2. a 3. c 4. a 5. b 6. b

7. a; **Hint:** Let the salaries of A, B, C be 2x, 3x and 5x respectively.
Now, 5x − 2x = 1200 ⇒ x = 400.
∴ B's monthly salary = 3x = Rs 1200.
∴ B's annual salary = 120 × 12 = Rs 14400

Rule 11

Theorem: ***If three numbers are in the ratio of a : b : c and the sum of these numbers is x, then these numbers will be***

$\frac{ax}{a+b+c}, \frac{bx}{a+b+c}$ *and* $\frac{cx}{a+b+c}$ *respectively.*

Illustrative Example

Ex.: An amount of Rs 750 is distributed among A, B and C in the ratio of 4 : 5 : 6. What is the share of B.

Soln: $\text{Share of B} = \frac{bx}{a+b+c} = \frac{5\times 750}{4+5+6} = \text{Rs } 250$

Exercise

1. An amount of Rs 950 is distributed among A, B and C in the ratio of 5 : 11 : 3, what is the difference between the share of B and A?
   a) 550 b) 250 c) 200 d) 300
2. An amount of Rs 1250 is distributed among A, B and C in the ratio of 4 : 7 : 14. What is the ratio between the difference of shares of B and A and the difference of shares of C and B?
   a) 7 : 3 b) 2 : 7 c) 3 : 7 d) 7 : 2
3. An amount of Rs 360 is distributed among A, B and C in the ratio of 1 : 2 : 3. Find the HCF of the shares obtained by A, B and C.
   a) 120 b) 60 c) 15 d) 20
4. An amount of Rs 975 is distributed among A, B and C in the ratio of 5 : 7 : 13. What is the share of C?
   a) Rs 509 b) Rs 507 c) Rs 273 d) Rs 237
5. An amount of Rs 1170 is distributed among A, B and C in the ratio of 3 : 5 : 7. What is the share of C?

a) Rs 546 b) Rs 456 c) Rs 576 d) Rs 586

6. An amount of Rs 975 is distributed among A, B and C in the ratio of 3 : 4 : 8. What is the share of C?
a) Rs 520 b) Rs 500 c) Rs 575 d) Rs 530

Answers

1. d 2. c 3. b 4. b 5. a 6. a

Rule 12

Theorem: ***If two alloys A and B contain gold and silver in the ratio of a : b and c : d respectively then a third alloy C formed by mixing A and B in the ratio of x : y will contain***

$$\left[\frac{\frac{ax}{a+b}+\frac{cy}{c+d}}{x+y}\right]\times 100\% \text{ gold and}$$

$$\left[100\%-\left\{\frac{\frac{ax}{a+b}+\frac{cy}{c+d}}{x+y}\right\}\times 100\%\right]$$

$$\textit{or}\ \left[\frac{\frac{bx}{a+b}+\frac{dy}{c+d}}{x+y}\right]\times 100\% \text{ silver.}$$

Illustrative Example

Ex.: Two alloys contain silver and copper in the ratio of 3 : 1 and 5 : 3. In what ratio the two alloys should be added together to get a new alloy having silver and copper in the ratio of 2 : 1?

Soln: Suppose that the two alloys are mixed in the ratio of x : y
Now, applying the above rule,
Percentage quantity of silver in the mixed alloy

$$=\left[\frac{\frac{3x}{4}+\frac{5y}{8}}{x+y}\right]\times 100\%$$

$$=\frac{6x+5y}{8(x+y)}\times 100\% \quad \text{.... (i)}$$

∵ Ratio of silver to copper in the mixd alloy is 2 : 1.
∴ Percentage quantity of silver in the mixed alloy

$$=\frac{2}{3}\times 100\%=\frac{200}{3}\% \quad \text{... (ii)}$$

From the equation (i) and (ii),

$$\frac{6x+5y}{8(x+y)}\times 100\%=\frac{200}{3}\%$$

or, $\frac{6x+5y}{8(x+y)}=\frac{2}{3}$

or, $18x+15y=16x+16y$

or, $2x=y$ or, $\frac{x}{y}=\frac{1}{2}$ or, x : y = 1 : 2

Hence the two alloys should be mixed in the ratio of 1 : 2.

Exercise

1. Two alloys A and B contain gold and silver in the ratio of 1 : 2 and 1 : 3 respectively. A third alloy C is formed by mixing A and B in the ratio of 2 : 3. Find the percentage of silver in the alloy C.
a) $71\frac{2}{3}\%$ b) $28\frac{1}{3}\%$ c) $70\frac{2}{3}\%$ d) $29\frac{1}{3}\%$

2. Two alloys contain silver and copper in the ratio of 2 : 3 and 3 : 4. In what ratio the two alloys should be added together to get a new alloy having silver and copper in the ratio of 1 : 2? Find the percentage of gold in the alloy C.
a) $41\frac{19}{21}\%$ b) $67\frac{13}{21}\%$ c) $32\frac{5}{21}\%$ d) $67\frac{16}{21}\%$

3. Two alloys contain silver and copper in the ratio of 4 : 5 and 3 : 4. In what ratio the two alloys should be added together to get a new alloy having silver and copper in the ratio of 2 : 3? Find the percentage of gold in the alloy C.
a) $\frac{3560}{63}\%$ b) $\frac{3650}{63}\%$ c) $\frac{2470}{63}\%$ d) $\frac{2740}{63}\%$

4. Two alloys contain silver and copper in the ratio of 2 : 1 and 4 : 1. In what ratio the two alloys should be added together to get a new alloy having silver and copper in the ratio of 3 : 1?
a) 5 : 3 b) 3 : 5 c) 2 : 5 d) 5 : 2

5. Two alloys contain silver and copper in the ratio of 2 : 1 and 4 : 1. In what ratio the two alloys should be added together to get a new alloy having silver and copper in the ratio of 3 : 2?
a) 3 : 5 b) 5 : 2
c) 1 : 2 d) Can't be determined

6. A and B are two alloys of gold and copper prepared by mixing metals in proportions 7 : 2 and 7 : 11 respectively. If equal quantities of alloys are melted to form a third alloy C, the proportion of gold and copper in C will be:
a) 5 : 9 b) 5 : 7 c) 7 : 5 d) 9 : 5
(CDS Exam, 1989)

Answers

1. a 2. a 3. d 4. b

5. d; **Hint:** $\left(\dfrac{\dfrac{2x}{3}+\dfrac{4y}{5}}{x+y}\right)\times 100=\dfrac{3}{5}\times 100$ or, $\dfrac{10x+12y}{15(x+y)}=\dfrac{3}{5}$

or, $x+3y=0$

For any positive value of x and y, this equation will not be satisfied.

6. c; **Hint:** In alloy C (when one unit each of A and B is mixed),

$\text{gold}=\left(\dfrac{7}{9}+\dfrac{7}{18}\right)=\dfrac{21}{18}$ and $\text{copper}=\left(\dfrac{2}{9}+\dfrac{11}{18}\right)=\dfrac{15}{18}$.

$\therefore$ Ratio of gold and copper $=\dfrac{21}{18}:\dfrac{15}{18}=7:5$.

Rule 13

Theorem: ***If in a mixture of x litres, milk and water are in the ratio of a : b then the quantities of milk and water in the mixture will be*** $\dfrac{ax}{a+b}$ ***litres and*** $\dfrac{bx}{a+b}$ ***litres respectively.***

Illustrative Example

Ex.: In a mixture of 65 litres milk and water are in the ratio of 3 : 2. What are the quantities of milk and water in the mixture?

Soln: Applying the above rule, we have,

quantity of milk in the mixture $=\dfrac{ax}{a+b}=\dfrac{3\times 65}{5}=39$

quantity of water in the mixture $=\dfrac{bx}{a+b}=\dfrac{2\times 65}{5}=26$

Exercise

1. In a mixture of 64 litres, milk and water are in the ratio of 1 : 3. What is the quantity of milk in the mixture?
   a) 16 b) 48 c) 15 d) 21
2. In a mixture of 99 litres, milk and water are in the ratio of 5 : 6. What is the quantity of water in the mixture?
   a) 54 b) 45 c) 48 d) 44
3. In a mixture of 80 litres, milk and water are in the ratio of 11 : 9. What is the quantity of milk in the mixture?
   a) 54 b) 45 c) 44 d) 36
4. In a mixture of 625 litres, milk and water are in the ratio of 13 : 12. What is the quantity of water in the mixture?
   a) 300 b) 180 c) 350 d) 325
5. In a mixture of 85 litres, milk and water are in the ratio of 14 : 3. What is the quantity of milk in the mixture?
   a) 80 b) 15 c) 70 d) 65

Answers

1. a 2. a 3. c 4. a 5. c

Rule 14

Theorem: ***If an alloy contains A and B metals in the ratio a : b then percentage of A and B metals in the alloy will be***

$\left(\dfrac{a}{a+b}\right)\times 100\%$ ***and*** $\left(100\%-\dfrac{a\times 100\%}{a+b}\right)$ ***or*** $\dfrac{b}{a+b}\times 100\%$

respectively.

Illustrative Example

Ex.: If an alloy contains copper and zinc in the ratio of 7 : 13, what is the percentage quantity of copper in the alloy?

Soln: Applying the above rule, we have,
the percentage quantity of copper in the alloy

$=\left(\dfrac{a}{a+b}\right)\times 100\%=\left(\dfrac{7}{7+13}\right)\times 100=35\%$

Exercise

1. If an alloy contains zinc and silver in the ratio 4 : 1, what is the percentage quantity of silver in the alloy?
   a) 80% b) 30% c) 70% d) 20%
2. If an alloy contains copper and silver in the ratio 3:5, what is the percentage quantity of copper in the alloy?
   a) $62\frac{1}{2}\%$ b) $37\frac{1}{2}\%$ c) $47\frac{1}{2}\%$ d) $52\frac{1}{2}\%$
3. If an alloy contains zinc and silver in the ratio 21 : 4, what is the percentage quantity of silver in the alloy?
   a) 84% b) 16% c) 26% d) 64%
4. If an alloy contains zinc and silver in the ratio 7 : 9, what is the percentage quantity of zinc?
   a) $43\frac{3}{4}\%$ b) $44\frac{3}{4}\%$ c) $45\frac{3}{4}\%$ d) $48\frac{1}{4}\%$

Answers

1. d 2. b 3. b 4. a

Rule 15

Theorem: ***If the ratio between the first and the second quantities is a : b and the ratio between the second and the third quantities is c : d, then the ratio among first, second and third quantities is given by ac : bc : bd.***
The above ratio can be represented diagrammatically as

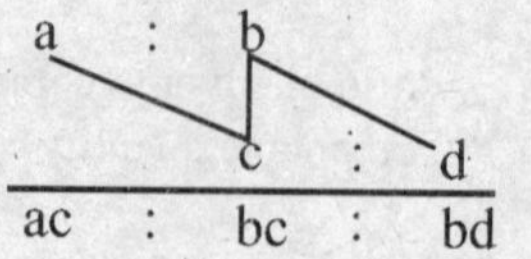

Illustrative Examples

Ex. 1: The sum of three numbers is 98. If the ratio between the first and second be 2 : 3 and that between the second and third be 5 : 8, then find the second number.

Soln: The theorem does not give the direct value of the second number, but we can find the combined ratio of all the three numbers by using the above theorem.
The ratio among the three numbers is

2 : 3
 5 : 8
10 : 15 : 24

$\therefore$ The second number $= \dfrac{98}{10+15+24} \times 15 = 30$

Ex. 2: The ratio of the money with Rita and Sita is 7 : 15 and that with Sita and Kavita is 7 : 16. If Rita has Rs 490, how much money does Kavita have?

Soln: Rita : Sita : Kavita
7 : 15
 7 : 16
49 : 105 : 240
The ratio of money with Rita, Sita and Kavita is 49 : 105 : 240
We see that 49 ≡ Rs 490 $\therefore$ 240 ≡ Rs 2400

Exercise

1. The sum of three numbers is 105. If the ratio between the first and second be 2 : 3 and that between the second and third be 4 : 5, then find the second number.
   a) 35 b) 24 c) 36 d) 45
2. The sum of three numbers is 275. If the ratio between the first and second be 3 : 7 and that between the second and third be 2 : 5, then find the second number.
   a) 30 b) 175 c) 70 d) 80
3. The sum of three numbers is 124. If the ratio between the first and second be 4 : 9 and that between the second and third be 1 : 2. Find the difference between the third number and the sum of first and second number.
   a) 20 b) 72 c) 52 d) 36
4. The sum of three numbers is 230. If the ratio between the first and second be 1 : 3 and that between the second and third be 2 : 5, find the sum of the first and second numbers.
   a) 60 b) 210 c) 170 d) 80
5. The ratio of the money with Sita and Geeta is 3 : 4 and that with Geeta and Guddu is 4 : 5. If Sita has Rs 300, how much money does Guddu have?
   a) 300 b) 400
   c) 500 d) Can't be determined

Answers

1. c 2. c 3. a 4. d 5. c

Rule 16

Theorem: ***If the ratio between the first and the second quantities is a : b; the ratio between the second and the third quantities is c : d and the ratio between the third and the fourth quantities is e : f then the ratio among the first, second, third and fourth quantities is given by***

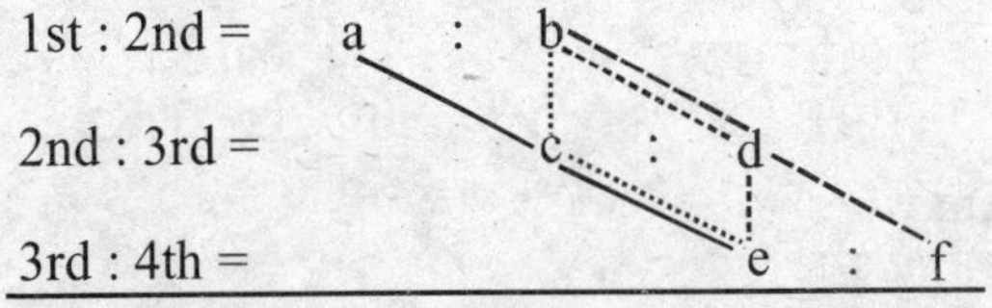

1st : 2nd : 3rd : 4th = ace : bce : bde : bdf

Illustrative Examples

Ex. 1: If A : B = 3 : 4, B : C = 8 : 10 and C : D = 15 : 17
Then find A : B : C : D.

Soln: A : B = 3 : 4
B : C = 8 : 10
C : D = 15 : 17
A : B : C : D = 3 × 8 × 15 : 4 × 8 × 15 : 4 × 10 × 15 : 4 × 10 × 17
= 9 : 12 : 15 : 17

Ex. 2: If A : B = 1 : 2, B : C = 3 : 4, C : D = 2 : 3 and D : E = 3 : 4
Then find A : B : C : D : E.

Soln:

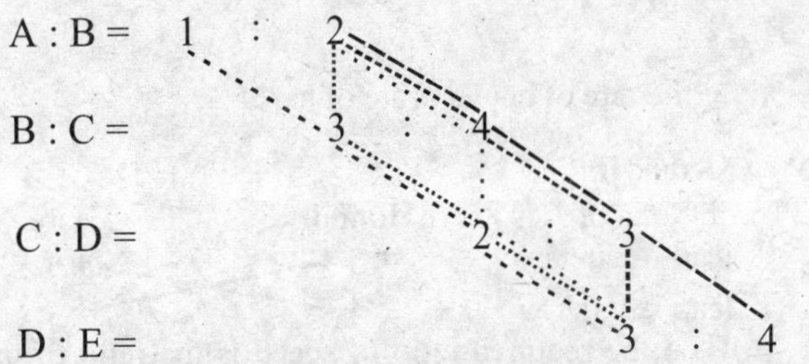

A : B : C : D : E = 1 × 3 × 2 × 3 : 2 × 3 × 2 × 3 : 2 × 4 × 2 × 3 : 2 × 4 × 3 × 3 : 2 × 4 × 3 × 4

= 3 : 6 : 8 : 12 : 16

Exercise

1. If A : B = 1 : 2, B : C = 2 : 3 and C : D = 3 : 4. Find A : B : C : D.
   a) 1 : 2 : 3 : 4 b) 1 : 3 : 4 : 5
   c) 1 : 3 : 5 : 6 d) 1 : 2 : 3 : 5
2. If A : B = 2 : 3, B : C = 3 : 5 and C : D = 9 : 11 Find A : B : C : D.
   a) 54 : 81 : 135 : 156 b) 54 : 81 : 135 : 165
   c) 54 : 18 : 135 : 165 d) Data inadequate
3. If A : B = 2 : 3, B : C = 4 : 5 and C : D = 3 : 7 Find A's share of property of Rs 2100.
   a) Rs 240 b) Rs 340 c) Rs 260 d) Rs 420
4. If A : B = 2 : 3, B : C = 4 : 5, C : D = 6 : 7 and D : E = 8 : 9, then find E's share of property of Rs 34650.
   a) Rs 9450 b) Rs 8400
   c) Rs 7200 d) Data inadequate

5. If A : B = 3 : 4, B : C = 5 : 7 and C : D = 3 : 5, then find A : B : C : D.
   a) 9 : 21 : 12 : 28 b) 45 : 60 : 84 : 140
   c) 9 : 12 : 28 : 21 d) 9 : 12 : 21 : 82

Answers

1. a 2. b 3. a

4. a; **Hint:** A : B : C : D : E = 384 : 576 : 720 : 840 : 945

$$\therefore \text{ share of E} = \frac{945 \times 34650}{384+576+720+840+945}$$
$$= \text{Rs } 9450.$$

5. b

Rule 17

Ex. A hound pursues a hare and takes 5 leaps for every 6 leaps of the hare, but 4 leaps of the hound are equal to 5 leaps of the hare. Compare the rates of the hound and the hare.

Soln: Method I:

4 leaps of hound = 5 leaps of hare

$\therefore$ 5 leaps of hound $= \frac{25}{4}$ leaps of hare

$\therefore$ the rate of hound : rate of hare $= \frac{25}{4} : 6 = 25 : 24$

Method II:

or, Ratio of	Hound	Hare
leap frequency	5	6
leap length	4	5

Then the required ratio of speed is the ratio of the cross-product.

That is, speed of hound : speed of hare
$= 5 \times 5 : 6 \times 4 = 25 : 24.$

Exercise

1. A hound pursues a hare and takes 6 leaps for every 7 leaps of the hare, but 5 leaps of the hound are equal to 6 leaps of the hare. Compare the rates of the hound and the hare.
   a) 36 : 35 b) 35 : 34 c) 34 : 33 d) 33 : 32
2. A hound pursues a hare and takes 3 leaps for every 4 leaps of the hare, but 2 leaps of the hound are equal to 3 leaps of the hare. Compare the rates of the hound and the hare.
   a) 9 : 8 b) 7 : 6 c) 5 : 6 d) 8 : 9
3. A hound pursues a hare and takes 7 leaps for every 8 leaps of the hare, but 5 leaps of the hound are equal to 6 leaps of the hare. Compare the rates of the hound and the hare.
   a) 20 : 19 b) 18 : 17 c) 21 : 20 d) 20 : 21
4. A hound pursues a hare and takes 6 leaps for every 9 leaps of the hare, but 3 leaps of the hound are equal to 5 leaps of the hare. Compare the rates of the hound and the hare.
   a) 30 : 29 b) 6 : 5 c) 10 : 9 d) 11 : 10
5. A hound pursues a hare and takes 5 leaps for every 12 leaps of the hare, but 1 leaps of the hound are equal to 3 leaps of the hare. Compare the rates of the hound and the hare.
   a) 15 : 14 b) 15 : 13 c) 13 : 12 d) 5 : 4

Answers

1. a 2. a 3. c 4. c 5. d

Rule 18

Ex.: A can do a piece of work in 12 days. B is 60% more efficient than A. Find the number of days it takes B to do the same piece of work.

Soln: Method I:

	A	:	B
Efficiency	100	:	160
Days	160	:	100
or	8	:	5

$\therefore$ the number of days taken by B $= \frac{12}{8} \times 5 = \frac{15}{2} = 7\frac{1}{2}$ days.

Method II:

By the rule of fraction: As B is more efficient, it is clear that B will complete the work in less days. So, the number of days (12) should be multiplied by a less-than-one fraction and that fraction is $\frac{100}{100+60}$, i.e., $\frac{100}{160}$. Therefore, our required answer is

$12 \times \frac{100}{160} = \frac{12 \times 5}{8} = \frac{15}{2} = 7\frac{1}{2}$ days.

Note: You are advised to solve such kind of questions by the **"rule of fraction"**.

Exercise

1. A can do a piece of work in 25 days. B is 25% more efficient than A. Find the number of days it takes B to do the same piece of work.
   a) 20 days b) $31\frac{1}{4}$ days c) $33\frac{1}{3}$ days d) 24 days
2. A can do a piece of work in 18 days. B is 20% more efficient than A. Find the number of days it takes B to do the same piece of work.
   a) 16 days b) $22\frac{1}{2}$ days c) 15 days d) 24 days
3. A can do a piece of work in 16 days. B is 20% less efficient than A. Find the number of days it takes B to do

the same piece of work.

a) 20 days b) $12\frac{4}{5}$ days c) 15 days d) 24 days

4. A can do a piece of work in 26 days. B is 30% more efficient than A. Find the number of days it takes B to do the same piece of work.
a) 25 days b) 15 days c) 16 days d) 20 days

5. A can do a piece of work in 15 days. B is 25% more efficient than A. Find the number of days it takes B to do the same piece of work.
a) 25 days b) 20 days c) 18 days d) 24 days

Answers

1. a 2. c

3. a; **Hint:** Required answer $= \frac{100}{100-20} \times 16 = 20$ days

(See **rule of fraction**).

4. d 5. b

Rule 19

Theorem: ***If in x litres mixture of milk and water, the ratio of milk and water is a : b, the quantity of water to be added in order to make this ratio c : d is*** $\frac{x(ad-bc)}{c(a+b)}$.

Illustrative Examples

Ex. 1: In 40 litres mixture of milk and water the ratio of milk and water is 3 : 1. How much water should be added in the mixture so that the ratio of milk to water becomes 2 : 1?

Soln: Solving the above question by the direct formula given in the above theorem:
The quantity of water to be added to get
the required ratio $= \frac{40(3\times1-1\times2)}{(3+1)2} = \frac{40}{8} = 5$ litres.

Note: The above solution can be verified as follows:
In 40 litres of mixture, milk $= \frac{40}{3+1} \times 3 = 30$ litres
and water $= 40 - 30 = 10$ litres.
5 litres water is added; so in the new mixture, milk is 30 litres and water is $10 + 5 = 15$ litres.
Thus, the new ratio is 30 : 15 = 2 : 1. This ratio is the same as given in the question.

Ex. 2: In 30 litres mixture of milk and water, the ratio of milk and water is 7 : 3. Find the quantity of water to be added in the mixture in order to make this ratio 3 : 7.

Soln: Following the same theorem, we have,
the required answer $= \frac{30(7\times7-3\times3)}{3(7+3)} = 40$ litres

Note: The above question is the special case of the above mentioned theorem. Here, we see that the first ratio is reversed in the second case. That is, a : b becomes b : a in the new mixture. Moreover the total quantity of initial mixture equals the denominator $[c(a+b)]$. In this case, the water to be added $= a^2 - b^2$.

Exercise

1. In 44 litres mixture of milk and water the ratio of milk and water is 6 : 5. How much water should be added in the mixture so that the ratio of milk to water becomes 2 : 3?
a) 61 litres b) 16 litres c) 8 litres d) 18 litres

2. In 60 litres mixture of milk and water the ratio of milk and water is 4 : 1. How much water should be added in the mixture so that the ratio of milk to water becomes 3 : 1?
a) 4 litres b) 15 litres c) 16 litres d) 20 litres

3. In 24 litres mixture of milk and water the ratio of milk and water is 9 : 4. How much water should be added in the mixture so that the ratio of milk to water becomes 4 : 9?
a) 25 litres b) 20 litres
c) 30 litres d) Cann't be determined

4. In 153 litres mixture of milk and water the ratio of milk and water is 11 : 6. How much water should be added in the mixture so that the ratio of milk to water becomes 3 : 2?
a) 26 litres b) 13 litres c) 24 litres d) 12 litres

5. In 80 litres mixture of milk and water the ratio of milk and water is 13 : 3. How much water should be added in the mixture so that the ratio of milk to water becomes 5 : 4?
a) 37 litres b) 40 litres c) 36 litres d) 24 litres

6. In 60 litres mixture of milk and water the ratio of milk and water is 7 : 5. How much water should be added in the mixture so that the ratio of milk to water becomes 5 : 7?
a) 42 litres b) 30 litres c) 24 litres d) 36 litres

7. In 28 litres mixture of milk and water the ratio of milk and water is 5 : 2. How much water should be added in the mixture so that the ratio of milk to water becomes 2 : 5?
a) 42 litres b) 32 litres c) 24 litres d) 39 litres

8. In a mixture of 60 litres, the ratio of milk and water is 2 : 1. If the ratio of milk and water is to be 1 : 2, then the amount of water to be further added is:
(NDA Exam 1990)
a) 42 litres b) 56 litres c) 60 litres d) 77 litres

Answers

1. b 2. a 3. c 4. d 5. a 6. c 7. a 8. c

Rule 20

Theorem: ***A mixture contains milk and water in the ratio a : b. If x litres of water is added to the mixture, milk and water become in the ratio a : c. Then the quantity of milk in the mixture is given by*** $\frac{ax}{c-b}$ ***and that of water is given by*** $\frac{bx}{c-b}$.

Illustrative Examples

Ex. 1: A mixture contains milk and water in the ratio of 3 : 2. If 4 litres of water is added to the mixture, milk and water in the mixture become equal. Find the quantities of milk and water in the mixture.

Soln: If we want to solve the above question by the theorem stated above, we will have to change the form of ratios to a : b and a : c. In the above question, the initial ratio is 3 : 2. Thus, to equate the antecedents of the ratio, we write the second ratio as 3 : 3. Then by the above direct formula :

The quantity of milk $= \frac{3\times4}{3-2} = 12$ litres.

and the quantity of water $= \frac{2\times4}{3-2} = 8$ litres.

Ex. 2: A mixture contains milk and water in the ratio of 8 : 3. On adding 3 litres of water, the ratio of milk to water becomes 2 : 1. Find the quantities of milk and water in the mixture.

Soln: To follow the above theorem, we change the ratios in the form a : b and a : c. Then the ratios can be written as 8 : 3 and 8 : 4. Thus, the quantity of milk in the mixture $= \frac{8\times3}{4-3} = 24$ litres

and the quantity of water in the mixture $= \frac{3\times3}{4-3} = 9$ litres.

Exercise

1. A mixture contains milk and water in the ratio of 9 : 4. On adding 4 litres of water, the ratio of milk to water becomes 3 : 2. Find the total quantity of the original mixture.
   a) 26 litres b) 18 litres c) 10 litres d) 30 litres
2. A mixture contains milk and water in the ratio of 4 : 3. On adding 2 litres of water, the ratio of milk to water becomes 8 : 7. Find the total quantity of the final mixture.
   a) 16 litres b) 12 litres c) 28 litres d) 30 litres
3. A mixture contains milk and water in the ratio of 12 : 5. On adding 8 litres of water, the ratio of milk to water becomes 4 : 3. Find the quantity of milk in the mixture.
   a) 24 litres b) 10 litres c) 14 litres d) 16 litres
4. A mixture contains milk and water in the ratio of 6 : 1. On adding 4 litres of water, the ratio of milk to water becomes 6 : 5. Find the quantity of water in the mixture.
   a) 6 litres b) 4 litres c) 1 litre d) 2 litres
5. A mixture contains milk and water in the ratio of 2 : 1. On adding 3 litres of water, the ratio of milk to water becomes 4 : 3. Find the quantity of water in the mixture.
   a) 6 litres b) 12 litres c) 8 litres d) 10 litres

Answers

1. a 2. d 3. a 4. c 5. a

Rule 21

Theorem : ***If two quantities X and Y are in the ratio x : y. Then*** $X+Y : X-Y :: x+y : x-y$

Illustrative Examples

Ex. 1: A sum of money is divided between two persons in the ratio of 3 : 5. If the share of one person is Rs 20 less than that of the other, find the sum.

Soln : By the above theorem : $\frac{Sum}{20} = \frac{3+5}{5-3}$

$\therefore$ Sum $= \frac{8}{2}\times 20 =$ Rs 80

Note: The above question can also be solved as follows (this method is similar to the above theorem):

$5-3 \equiv$ Rs 20 $\quad \therefore 5+3 \equiv \frac{20}{5-3}\times(5+3) =$ Rs 80

Ex. 2: The prices of a scooter and a moped are in the ratio of 9 : 5. If a scooter costs Rs 4200 more than a moped, find the price of the moped.

Soln: Following the method mentioned in the above **note**, we have,

9 - 5 $\equiv$ Rs 4200 $\quad \therefore 5 \equiv \frac{4200}{9-5}\times 5 =$ Rs 5250

Exercise

1. A sum of money is divided between two persons in the ratio of 2 : 9. If the share of one person is Rs 21 less than that of the other, find the sum.
   a) Rs 32 b) Rs 44 c) Rs 33 d) Rs 36
2. A sum of money is divided between two persons in the ratio of 4 : 7. If the share of one person is Rs 12 less than that of the other, find the sum.
   a) Rs 44 b) Rs 36 c) Rs 40 d) Rs 42
3. A sum of money is divided between two persons in the ratio of 5 : 2. If the share of one person is Rs 15 less than that of the other, find the sum.
   a) Rs 25 b) Rs 30 c) Rs 45 d) Rs 35
4. The prices of a pen and a pencil are in the ratio of 5 : 2. If a pen costs Rs 3 more than a pencil, find the price of a pencil.
   a) Rs 5 b) Rs 2 c) Rs 3 d) Rs 4
5. The prices of a scooter and a moped are in the ratio of 11 : 7. If a scooter costs Rs 4400 more than a moped, find the price of the moped.
   a) Rs 7700 b) Rs 6700 c) Rs 7600 d) Rs 5500

Answers

1. c 2. a 3. d 4. b 5. a

Rule 22

Theorem: ***In any two two-dimensional figure, if the corresponding sides are in the ratio a : b, then their areas are in the ratio*** $a^2 : b^2$.

Illustrative Examples

Ex. 1: The sides of a hexagon are enlarged by three times. Find the ratio of the areas of the new and old hexagons.

Soln: Following the above theorem, we see that the ratio of the corresponding sides of the two hexagons is a : b = 1 : 3.
Therefore, the ratio of their areas is given by
$a^2 : b^2 = 1^2 : 3^2 = 1 : 9$

Ex. 2: The ratio of the diagonals of two squares is 2 : 1. Find the ratio of their areas.

Soln: We should follow the same rule when the ratio of diagonals is given instead of the ratio of sides. Thus, the ratio of their areas = $2^2 : 1^2 = 4 : 1$.

Ex. 3: The ratio of the radius (or diameter or circumference) of two circles is 3 : 4. Find the ratio of their areas.

Soln: Following the rule, we have,
ratio of their areas = $3^2 : 4^2 = 9 : 16$.

Note: The above mentioned theorem is true for any two-dimensional figure and for any measuring length related to that figure.

Exercise

1. The ratio of the sides of two squares is 3 : 4. Find the ratio of their areas.
   a) 16 : 9 b) 9 : 16 c) 4 : 3 d) 12 : 16
2. The areas of the two hexagons are in the ratio of 25 : 16. Find the ratio of their sides.
   a) 625 : 256 b) 5 : 3 c) 5 : 6 d) 5 : 4
3. The ratio of circumference of the two circles is 2 : 9. Find the ratio of their areas.
   a) 4 : 9 b) 9 : 4 c) 4 : 81 d) 81 : 4
4. The ratio of diagonals of two squares is 13 : 11. Find the ratio of their areas.
   a) 169 : 121 b) 121 : 169 c) 112 : 196 d) 169 : 112
5. The sides of a hexagon are enlarged by 5 times. Find the ratio of the areas of the old and new hexagons.
   a) 1 : 25 b) 25 : 1
   c) 2 : 25 d) Can't be determined
6. The ratio of the radius of two circles is 5 : 4. Find the ratio of their areas.
   a) 16 : 25 b) 9 : 25 c) 25 : 9 d) 25 : 16
7. The ratio of the perimeters of two squares is 3 : 1. Find the ratio of their areas.
   a) 1 : 9 b) 9 : 1 c) 16 : 1 d) 1 : 16
8. The ratio of the diameters of two circles is 5 : 8. Find the ratio of their areas.
   a) 25 : 64 b) 25 : 46 c) 25 : 81 d) 36 : 81
9. The ratio of the perimeters of two hexagons is 2 : 1. Find the ratio of their areas.
   a) 4 : 1 b) 1 : 4 c) 9 : 4 d) 4 : 9
10. The ratio of diagonals of two squares is 4 : 3. Find the ratio of their areas.
    a) 9 : 16 b) 16 : 25 c) 16 : 9 d) 25 : 16

Answers

1. b 2. d 3. c 4. a 5. b 6. d 7. b
8. a
9. a; **Hint:** See the 'note' in the above mentioned rule.
10. c

Rule 23

Theorem: ***In any two 3-dimensional figure, if the corresponding sides or other measuring lengths are in the ratio a : b, then their volumes are in the ratio*** $a^3 : b^3$.

Illustrative Example

Ex. (a) The sides of two cubes are in the ratio 2 : 1. Find the ratio of their volumes.
(b) Each side of a parallelopipe is doubled find the ratio of volume of old to new parallelopipe.

Soln : (a) The required ratio = $(2)^3 : (1)^3 = 8 : 1$
(b) The required ratio = $(1)^3 : (2)^3 = 1 : 8$

Exercise

1. The diagonals of two cubes are in the ratio of 3 : 4. Find the ratio of their volumes.
   a) 64 : 27 b) 27 : 64 c) 81 : 216 d) 216 : 81
2. The radii of two spheres are in the ratio of 1 : 7. Find the ratio of their volumes.
   a) 1 : 343 b) 343 : 1
   c) 1 : 243 d) Can't be determined
3. The diameters of two sphere are in the ratio of 3 : 8. Find the ratio of their volumes.
   a) 27 : 216 b) 27 : 512 c) 512 : 27 d) 216 : 27
4. Each side of a parallelopipe is increased 4 times. Find the ratio of volume of new to old parallelopipe.
   a) 16 : 1 b) 64 : 1 c) 1 : 64 d) 1 : 16
5. The sides of two cubes are in the ratio 5 : 2. Find the ratio of their volumes.
   a) 4 : 25 b) 8 :.125 c) 125 ; 8 d) 25 : 4

Answers

1. b 2. a 3. b 4. b 5. c

Rule 24

Theorem: The ratio between two numbers is a : b. If each number be increased by x, the ratio becomes c : d. Then, the two numbers are given as $\frac{xa(c-d)}{ad-bc}$ ***and*** $\frac{xb(c-d)}{ad-bc}$***, where*** $c - a \neq d - b$

Illustrative Example

Ex.: The ratio between two numbers is 3 : 4. If each number be increased by 2, the ratio becomes 7 : 9. Find the numbers.

Soln: Following the above theorem, the numbers are

$$\frac{2\times3(7-9)}{3\times9-4\times7} \text{ and } \frac{2\times4(7-9)}{3\times9-4\times7}$$

or, 12 and 16.

Note: The above question may be rewritten as "the ratio between two numbers is 7 : 9. If each number be decreased by 2 the ratio becomes 3 : 4. Find the numbers."

Exercise

1. The ratio between two numbers is 4 : 5. If each number be increased by 3, the ratio becomes 7 : 8. Find the numbers.
   a) 4, 5 b) 8, 10 c) 12, 15 d) 16, 20
2. The ratio between two numbers is 5 : 7. If each number be increased by 11, the ratio becomes 13 : 16. Find the numbers.
   a) 5, 7 b) 15, 21 c) 10, 14 d) 75, 105
3. The ratio between two numbers is 15 : 7. If each number be decreased by 2, the ratio becomes 7 : 3. Find the numbers.
   a) 15, 7 b) 30, 14 c) 45, 21 d) 60, 28
4. The ratio between two numbers is 15 : 11. If each number be decreased by 6, the ratio becomes 3 : 2. Find the numbers.
   a) 30, 22 b) 15, 11 c) 45, 33 d) 60, 44
5. The ratio between two numbers is 2 : 1. If each number be decreased by 7, the ratio becomes 5 : 2. Find the numbers.
   a) 30, 15 b) 24, 12 c) 42, 21 d) 62, 31
6. Two numbers are such that the ratio between them is 3 : 5 but if each is increased by 10, the ratio between them becomes 5 : 7. The numbers are: **[RRB Exam, 1989]**
   a) 3, 5 b) 7, 9 c) 13, 22 d) 15, 25

Answers

1. a 2. b

3. b; **Hint:** The given question may be rewritten as "the ratio between two numbers is 7:3. If each number be **increased** (because we have direct formula for 'increase') by 2 the ratio becomes 15 : 7. Find the numbers."
Here x = 2, a : b = 7 : 3 and c : d = 15 : 7.

4. a 5. c 6. d

Rule 25

Theorem: *The ratio between two numbers is a : b. If each number be increased by x, the ratio becomes c : d and c - a = d - b, then the numbers are given as $\left(\frac{ax}{c-a}\right)$ and $\left(\frac{bx}{c-a}\right)$.*

Illustrative Examples

Ex. 1: The ratio between two numbers is 3 : 4. If each number be increased by 6, the ratio becomes 4 : 5. Find the two numbers.

Soln: The above question may be considered as a special case of the above theorem where c – a = d – b

It is easy to distinguish this type of question. In such a question, there should be a uniform increase in ratio, i.e., the antecedent and consequent is increased by the same value.

In the above question, we see that both the antecedent and the consequent are increased by 1 each, and the numbers are increased by 6. Therefore, we may say that

$1 \equiv 6$

or, $3 \equiv 3\times6 = 18$ and $4 \equiv 4\times6 = 24$

Thus the numbers are 18 and 24.

Note: The above question may be rewritten as :
The ratio of two numbers is 4 : 5. If each of them is decreased by 6, the ratio becomes 3 : 4. Find the two numbers.
Apply the same rule in this case also.

Ex. 2: The students in three classes are in the ratio 2 : 3 : 5. If 20 students are increased in each class, the ratio changes to 4 : 5 : 7. What was the total number of students in the three classes before the increase?

Soln : In the above question also, we see that each term increases by the same value. That is,
$4 - 2 = 5 - 3 = 7 - 5 = 2$. Thus, we have
$2 \equiv 20$

$$\therefore (2+3+5) \equiv \frac{20}{2}\times10 = 100 \text{ students.}$$

Exercise

1. The ratio between two numbers is 3 : 4. If each number be increased by 4, the ratio becomes 5 : 6. Find the two numbers.
   a) 6, 8 b) 12, 16 c) 18, 24 d) 24, 32
2. The ratio between two numbers is 7 : 11. If each number be increased by 6, the ratio becomes 5 : 7. Find the two numbers.
   a) 14, 22 b) 7, 11 c) 21, 33 d) 28, 44
3. The ratio of two numbers is 8 : 13. If each of them is decreased by 15, the ratio becomes 3 : 8. Find the numbers.
   a) 24, 39 b) 16, 26 c) 32, 52 d) 40, 65
4. The ratio between two numbers is 4 : 3. If each number be decreased by 9, the ratio becomes 3 : 2. Find the two numbers.

a) 27, 18 b) 36 : 27 c) 44 : 33 d) 48 : 36

5. The students in three classes are in the ratio 4 : 6 : 9. If 12 students are increased in each class, the ratio changes to 7 : 9 : 12. What was the total number of students in the three classes before the increase?
a) 76 b) 95 c) 100 d) 114

6. The students in three classes are in the ratio 5 : 9 : 13. If 100 students are increased in each class, the ratio changes to 9 : 13 : 17. What was the total number of students in the three classes before the increase?
a) 765 b) 576 c) 675 d) 657

7. The students in three classes are in the ratio of 2 : 4 : 5. If 15 students are taken out from each class, the ratio changes to 1 : 3 : 4. What is the total number of students in the three classes, now?
a) 165 b) 105 c) 120 d) 115

Answers

1. a

2. a; **Hint:** a : b = 7 : 11, x = 6, c : d = 5 : 7 = 5 × 2 : 7 × 2 = 10 : 14
∴ c - a = 10 - 7 = d - b = 14 - 11 = 3

3. a; **Hint:** The given question may be rewritten as "the ratio between two numbers is 3 : 8. If each number be increased by 15, the ratio becomes 8 : 13. Find the numbers."
Here, a : b = 3 : 8, x = 15 and c : d = 8 : b

4. b 5. a 6. c

7. c; **Hint:** The given question may be rewritten as "the students in three classes are in the ratio 1 : 2 : 3. If 15 students are **increased** in each class, the ratio changes to 2 : 4 : 5. What is the total number of students in the three classes, now?"
Here, total no. of students in the three classes before the 15 students from each class are taken out = $\frac{(2+4+5)\times 15}{1} = 165$ students.
According to the question, total no. of students now = 165 - 3 × 15 = 120 students.

Rule 26

Theorem: ***The incomes of two persons are in the ratio a : b and their expenditures are in the ratio c : d. If each of them saves Rs X, then their incomes are given by*** $\frac{Xa(d-c)}{ad-bc}$ ***and*** $\frac{Xb(d-c)}{ad-bc}$.

Illustrative Example

Ex.: The incomes of A and B are in the ratio 3 : 2 and their expenditures are in the ratio 5 : 3. If each saves Rs 2000, what is their income?

Soln: According to the above theorem,

a : b = 3 : 2 (Income)
c : d = 5 : 3 (Expenditure)
X = 2000 (Savings)

Therefore, A's income $= \frac{Xa(d-c)}{ad-bc}$

$= \frac{2000\times 3\times(3-5)}{3\times 3-2\times 5} = \text{Rs } 12{,}000$

and B's income $= \frac{Xb(d-c)}{ad-bc} = \frac{2000\times 2\times(3-5)}{3\times 3-2\times 5}$

= Rs 8,000

Exercise

1. The incomes of Ram and Shyam are in the ratio 4 : 3 and their expenditures are in the ratio 3 : 2. If each saves Rs 2500, what are their incomes?
a) Rs 10000, Rs 7500 b) Rs 12000, Rs 9000
c) Rs 8000, Rs 6000 d) None of these

2. The incomes of A and B are in the ratio 7 : 2 and their expenditures are in the ratio 4 : 1. If each saves Rs 1000, what are their incomes?
a) Rs 6000, Rs 21000 b) Rs 21000, Rs 6000
c) Rs 42000, Rs 12000 d) Rs 12000, Rs 42000

3. The incomes of A and B are in the ratio 4 : 3 and their expenditures are in the ratio 5 : 2. If each saves Rs 4900, what are their incomes?
a) Rs 8400, Rs 6300 b) Rs 10000 Rs 7500
c) Rs 9200, Rs 6900 d) Rs 9600, Rs 7200

4. The incomes of A and B are in the ratio 7 : 5 and their expenditures are in the ratio 5 : 3. If each saves Rs 1600, what are their incomes?
a) Rs 5600, Rs 4000 b) Rs 7000, Rs 5000
c) Rs 14000, Rs 10000 d) Rs 21000, Rs 15000

5. The incomes of A and B are in the ratio 9 : 4 and their expenditures are in the ratio 7 : 3. If each saves Rs 2000, what are their incomes?
a) Rs 90000, Rs 4000 b) Rs 27000, Rs 12000
c) Rs 72000, Rs 16000 d) Rs 72000, Rs 32000

Answers

1. a 2. b 3. a 4. a 5. d

Rule 27

Theorem: ***The incomes of two persons are in the ratio a : b and their expenditures are in the ratio c : d. If each of them saves Rs X, then their expenditures are given by*** $\frac{Xc(b-a)}{ad-bc}$ ***and*** $\frac{Xd(b-a)}{ad-bc}$.

Illustrative Example

Ex.: The incomes of A and B are in the ratio 3 : 2 and their expenditures are in the ratio 5 : 3. If each saves Rs

2000, what are their expenditures?

Soln: According to the above theorem,
a : b = 3 : 2 (income)
c : d = 5 : 3 (expenditure)
X = 2000 (savings)
Therefore,

$$\text{A's expenditure} = \frac{Xc(b-a)}{(ad-bc)} = \frac{2000 \times 5(3-2)}{3 \times 3 - 2 \times 5}$$

$$= \text{Rs } 10000$$

$$\text{B's expenditure} = \frac{Xd(b-a)}{ad-bc} = \frac{2000 \times 3(3-2)}{3 \times 3 - 2 \times 5}$$

$$= \text{Rs } 6000$$

Exercise

1. The incomes of Ram and Shyam are in the ratio 4 : 3 and their expenditures are in the ratio 3 : 2. If each saves Rs 2500, what are their expenditures?
a) Rs 7500, Rs 5000 b) Rs 6300, Rs 4200
c) Rs 6000, Rs 2000 d) Rs 7200 Rs 4800
2. The incomes of A and B are in the ratio 7 : 2 and their expenditures are in the ratio 4 : 1. If each saves Rs 1000, what are their expenditures?
a) Rs 20000, Rs 5000 b) Rs 16000, Rs 4000
c) Rs 12000, Rs 3000 d) Rs 24000, Rs 6000
3. The incomes of A and B are in the ratio 4 : 3 and their expenditures are in the ratio 5 : 2. If each saves Rs 4900, what are their expenditures?
a) Rs 3500, Rs 1400 b) Rs 4000 Rs 1600
c) Rs 4500, Rs 1800 d) Rs 5000, Rs 2000
4. The incomes of A and B are in the ratio 7 : 5 and their expenditures are in the ratio 5 : 3. If each saves Rs 1600, what are their expenditures?
a) Rs 3500, Rs 2100 b) Rs 4000, Rs 2400
c) Rs 4500, Rs 2700 d) Rs 5000, Rs 3000
5. The incomes of A and B are in the ratio 9 : 4 and their expenditures are in the ratio 7 : 3. If each saves Rs 2000, what are their expenditures?
a) Rs 56000, Rs 24000 b) Rs 63000, Rs 27000
c) Rs 49000, Rs 21000 d) Rs 70000, Rs 30000

Answers

1. a 2. a 3. a 4. b 5. d

Rule 28

Theorem: ***The ratio between two numbers is a : b. if each number be increased by x, the ratio becomes c : d. Then,***

$$\textit{sum of two numbers} = \frac{x(a+b)(c-d)}{ad-bc}.$$

Illustrative Example

Ex.: The ratio between two numbers is 3 : 4. If each number be increased by 9, the ratio becomes 18 : 23. Find the sum of the numbers.

Soln: Applying the above rule, we have,

$$\text{sum of two numbers} = \frac{x(a+b)(c-d)}{ad-bc}$$

$$= \frac{9(3+4)(18-23)}{3 \times 23 - 4 \times 18} = \frac{9 \times 7(-5)}{-3} = 105.$$

Exercise

1. The ratio between two numbers is 5 : 4. If each number be increased by 7, the ratio becomes 22 : 19. Find the sum of the numbers.
a) 27 b) 31 c) 41 d) 72
2. The ratio between two numbers is 7 : 15. If each number be increased by 5, the ratio becomes 19 : 35. Find the sum of the numbers.
a) 64 b) 54 c) 44 d) 34
3. The ratio between two numbers is 4 : 9. If each number be increased by 8, the ratio becomes 6 : 11. Find the sum of the numbers.
a) 42 b) 52 c) 62 d) 72
4. The ratio between two numbers is 11 : 15. If each number be increased by 22, the ratio becomes 11 : 13. Find the sum of the numbers.
a) 44 b) 77 c) 65 d) 52
5. The ratio between two numbers is 9 : 2. If each number be increased by 9, the ratio becomes 12 : 5. Find the sum of the numbers.
a) 32 b) 33 c) 31 d) 34

Answers

1. a 2. c 3. b 4. d 5. b

Rule 29

Theorem: ***The ratio between two numbers is a : b. If each number be increased by x, the ratio becomes c : d. Then,***

$$\textit{difference of the two numbers} = \frac{x(a-b)(c-d)}{ad-bc}.$$

Illustrative Example

Ex.: The ratio between two numbers is 3 : 4. If each number be increased by 2, the ratio becomes 7 : 9. Find the difference of the two numbers.

Soln: Applying the above rule, we have

$$\text{difference of the two numbers} = \frac{x(a-b)(c-d)}{ad-bc}$$

$$= \frac{2(3-4)(7-9)}{9 \times 3 - 4 \times 7} = -4$$

ie difference of the two numbers = 4.

Exercise

1. The ratio between two numbers is 9 : 4. If each number

be increased by 3, the ratio becomes 2 : 1. Find the difference of the two numbers.
a) 4 b) 2 c) 15 d) 3

2. The ratio between two numbers is 9 : 5. If each number be increased by 7, the ratio becomes 17 : 11. Find the difference of the two numbers.
a) 10 b) 11 c) 12 d) 14

3. The ratio between two numbers is 4 : 3. If each number be increased by 5, the ratio becomes 5 : 4. Find the difference of the two numbers.
a) 5 b) 13 c) 12 d) 11

4. The ratio between two numbers is 3 : 5. If each number be increased by 10, the ratio becomes 5 : 7. Find the difference of the two numbers.
a) 9 b) 10 c) 12 d) 15

5. The ratio between two numbers is 9 : 4. If each number be increased by 16, the ratio becomes 13 : 8. Find the difference of the two numbers.
a) 9 b) 10 c) 18 d) 20

6. The ratio between two numbers is 7 : 4. If each number be increased by 2, the ratio becomes 5 : 3. Find the difference of the two numbers.
a) 10 b) 12 c) 14 d) 16

7. The ratio between two numbers is 9 : 7. If each number be increased by 6, the ratio becomes 21 : 17. Find the difference of the two numbers.
a) 8 b) 14 c) 16 d) 18

Answers

1. c 2. c 3. a 4. b 5. d 6. b 7. a

Rule 30

Theorem: ***The ratio between two numbers is a : b. If each number be decreased by x, the ratio becomes c : d. Then,***

sum of the two numbers $= \frac{x(a+b)(d-c)}{ad-bc}$.

Illustrative Example

Ex.: The ratio of two numbers is 7 : 9. If each number is decreased by 2, the ratio becomes 3 : 4. Find the sum of the two numbers.

Soln: From the above rule, we have,
sum of the two numbers

$$= \frac{x(a+b)(d-c)}{ad-bc} = \frac{2(7+9)(4-3)}{28-27} = 32$$

Exercise

1. The ratio of two numbers is 7 : 8. If each number is decreased by 2, the ratio becomes 6 : 7. Find the sum of the two numbers.
a) 30 b) 32 c) 28 d) 36

2. The ratio of two numbers is 5 : 9. If each number is decreased by 5, the ratio becomes 5 : 11. Find the sum of the two numbers.
a) 32 b) 36 c) 42 d) 46

3. The ratio of two numbers is 13 : 7. If each number is decreased by 6, the ratio becomes 5 : 2. Find the sum of the two numbers.
a) 38 b) 35 c) 45 d) 40

4. The ratio of two numbers is 9 : 19. If each number is decreased by 2, the ratio becomes 4 : 9. Find the sum of the two numbers.
a) 52 b) 56 c) 62 d) 65

5. The ratio of two numbers is 3 : 2. If each number is decreased by 12, the ratio becomes 2 : 1. Find the sum of the two numbers.
a) 50 b) 60 c) 48 d) 56

6. The ratio of two numbers is 11 : 12. If each number is decreased by 3, the ratio becomes 19 : 21. Find the sum of the two numbers.
a) 46 b) 40 c) 44 d) 64

Answers

1. a 2. c 3. d 4. b 5. b 6. a

Rule 31

Theorem: ***The ratio between two numbers is a : b. If each number be decreased by x, the ratio becomes c : d. Then, the difference of the two numbers*** $= \frac{x(a-b)(d-c)}{ad-bc}$.

Illustrative Example

Ex.: The ratio of two numbers is 7 : 9. If each number is decreased by 2 the ratio becomes 3 : 4. Find the difference of the numbers.

Soln: From the above rule, we have, the difference of the numbers

$$= \frac{x(a-b)(d-c)}{ad-bc} = \frac{2(7-9)(4-3)}{7\times4-9\times3} = -4$$

$\therefore$ difference of the numbers = 4

Exercise

1. The ratio of two numbers is 2 : 3. If each number is decreased by 3 the ratio becomes 3 : 5. Find the difference of the numbers.
a) 5 b) 7 c) 6 d) 4

2. The ratio of two numbers is 14 : 9. If each number is decreased by 8 the ratio becomes 2 : 1. Find the difference of the numbers.
a) 9 b) 10 c) 8 d) 12

3. The ratio of two numbers is 17 : 14. If each number is decreased by 4 the ratio becomes 5 : 4. Find the difference of the numbers.
a) 6 b) 7 c) 8 d) 9

4. The ratio of two numbers is 5 : 9. If each number is decreased by 5 the ratio becomes 1 : 2. Find the difference

of the numbers.
a) 12 b) 18 c) 10 d) 20

5. The ratio of two numbers is 5 : 8. If each number is decreased by 6 the ratio becomes 1 : 2. Find the difference of the numbers.
a) 9 b) 10 c) 12 d) 8

6. The ratio of two numbers is 16 : 5. If each number is decreased by 2 the ratio becomes 15 : 4. Find the difference of the numbers.
a) 15 b) 17 c) 19 d) 22

Answers

1. c 2. b 3. a 4. d 5. a 6. d

Rule 32

Theorem : ***If the ratio of any quantities be a : b : c : d, then the ratio of other quantities which are inversely proportional to that is given by*** $\frac{1}{a}:\frac{1}{b}:\frac{1}{c}:\frac{1}{d}$.

Illustrative Examples

Ex. 1: The speed of three cars are in the ratio 2 : 3 : 4. What is the ratio among the times taken by these cars to travel the same distance?

Soln: We know that speed and time taken are inversely proportional to each other. That is, if speed is more the time taken is less and vice versa. So, we can apply the above theorem in this case.

Hence, ratio of time taken by the three cars = $\frac{1}{2}:\frac{1}{3}:\frac{1}{4}$

Now, multiply each fraction by the LCM of denominators i.e., the LCM of 2, 3, 4, i.e., 12. So, the required ratio is given by $\frac{12}{2}:\frac{12}{3}:\frac{12}{4}=6:4:3$

Ex. 2: The same type of work is assigned to three groups of men. The ratio of persons in the groups is 3 : 4 : 5. Find the ratio of days in which they will complete the work.

Soln: We see that in this case also, man and days are inversely proportional to each other. So, the above rule can be applied in this case also. Therefore, the required ratio is $\frac{1}{3}:\frac{1}{4}:\frac{1}{5}$

Multiplying the above fractions by the LCM of 3, 4 and 5, i.e. 60, we have, $\frac{60}{3}:\frac{60}{4}:\frac{60}{5}=20:15:12$

Exercise

1. The speed of three cars are in the ratio 1 : 3 : 5. What is the ratio among the times taken by these cars to travel the same distance?
a) 15 : 5 : 3 b) 15 : 3 : 5 c) 3 : 5 : 15 d) 5 : 3 : 1

2. The speed of three cars are in the ratio 4 : 6 : 9. What is the ratio among the times taken by these cars to travel the same distance?
a) 16 : 36 : 81 b) 6 : 4 : 9 c) 9 : 6 : 4 d) 81 : 36 : 16

3. The speed of three cars are in the ratio 1 : 2 : 3. What is the ratio among the times taken by these cars to travel the same distance?
a) 3 : 2 : 1 b) 6 : 3 : 2 c) 6 : 5 : 2 d) 6 : 3 : 1

4. A person distributes his pens among four friends A, B, C and D in the ratio $\frac{1}{3}:\frac{1}{4}:\frac{1}{5}:\frac{1}{6}$. What is the minimum number of pens that the person should have?
a) 57 b) 75 c) 67 d) 76

5. The same type of work is assigned to three groups of men. The ratio of persons in the groups is 5 : 4 : 6. Find the ratio of days in which they will complete the work.
a) 15 : 18 : 10 b) 12 : 15 : 10 c) 12 : 18 : 21 d) 6 : 4 : 5

6. The same type of work is assigned to three groups of men. The ratio of persons in the groups is 6 : 7 : 8. Find the ratio of days in which they will complete the work.
a) 28 : 24 : 21 b) 28 : 26 : 21 c) 21 : 24 : 28 d) 8 : 7 : 6

7. The same type of work is assigned to three groups of men. The ratio of persons in the groups is 2 : 4 : 9. Find the ratio of days in which they will complete the work.
a) 18 : 9 : 4 b) 18 : 8 : 5 c) 4 : 9 : 18 d) 9 : 4 : 2

Answers

1. a 2. c 3. b

4. a; **Hint:** LCM of 3, 4, 5 and 6 = 60
So, the pens are distributed among A, B, C and D in the ratio $\frac{1}{3}\times 60:\frac{1}{4}\times 60:\frac{1}{5}\times 60:\frac{1}{6}\times 60$ ie 20 : 15 : 12 : 10.
$\therefore$ total number of pens = $20x+15x+12x+10x=57x$
For minimum number of pens, $x=1$
$\therefore$ the person should have at least 57 pens.

5. b 6. a 7. a

Rule 33

Theorem: ***If the sum of two numbers is A and their difference is a, then the ratio of numbers is given by A + a : A − a.***

Illustrative Example

Ex.: The sum of two numbers is 40 and their difference is 4. What is the ratio of the two numbers?

Soln: Following the above theorem,
the required ratio of numbers = 40 + 4 : 40 − 4
= 44 : 36 = 11 : 9

Exercise

1. The sum of two numbers is 36 and their difference is 6. What is the ratio of the two numbers?
a) 5 : 7 b) 7 : 5 c) 6 : 5 d) 5 : 6

2. The sum of two numbers is 45 and their difference is 5. What is the ratio of the two numbers?
a) 4 : 5 b) 5 : 4 c) 5 : 3 d) 5 : 2
3. The sum of two numbers is 55 and their difference is 9. What is the ratio of the two numbers?
a) 32 : 23 b) 23 : 32
c) 31 : 13 d) 32 : 13
4. The sum of two numbers is 20 and their difference is 2. What is the ratio of the two numbers?
a) 11 : 8 b) 11 : 7 c) 11 : 6 d) 11 : 9
5. The sum of two numbers is 44 and their difference is 4. What is the ratio of the two numbers?
a) 6 : 7 b) 5 : 6 c) 6 : 5 d) 7 : 9
6. The sum of two numbers is 64 and their difference is 8. What is the ratio of the two numbers?
a) 9 : 5 b) 9 : 8 c) 9 : 7 d) 9 : 11
7. The sum of two numbers is 77 and their difference is 7. What is the ratio of the two numbers?
a) 6 : 5 b) 6 : 7 c) 6 : 1 d) 5 : 6

Answers

1. b 2. b 3. a 4. d 5. c 6. c 7. a

Rule 34

Theorem: ***A number which, when added to the terms of the ratio a: b makes it equal to the ratio c : d is*** $\frac{ad-bc}{c-d}$.

Illustrative Example

Ex.: Find the number which, when added to the terms of the ratio 11 : 23 makes it equal to the ratio 4 : 7.

Soln: Following the above rule, we have,
a : b = 11 : 23
c : d = 4 : 7
∴ the required number

$$=\frac{ad-bc}{c-d}=\frac{11\times 7-23\times 4}{4-7}=\frac{(-)15}{(-)3}=5$$

Exercise

1. Find the number which, when added to the terms of the ratio 13 : 28 makes it equal to the ratio 1 : 2.
2. Find the number which, when added to the terms of the ratio 9 : 17 makes it equal to the ratio 3 : 5.
3. Find the number which, when added to the terms of the ratio 7 : 15 makes it equal to the ratio 3 : 5.
4. Find the number which, when added to the terms of the ratio 11 : 29 makes it equal to the ratio 11 : 20.
5. Find the number which, when added to the terms of the ratio 13 : 25 makes it equal to the ratio 3 : 5.
6. Find the number which, when added to the terms of the ratio 17 : 31 makes it equal to the ratio 13 : 20.

Answers

1. 2 2. 3 3. 5 4. 11 5. 5 6. 9

Rule 35

Theorem : ***A number which, when subtracted from the terms of the ratio a: b makes it equal to the ratio c : d is*** $\frac{bc-ad}{c-d}$.

Illustrative Example

Ex.: Find the number which, when subtracted from the terms of the ratio 11 : 23 makes it equal to the ratio 3 : 7.

Soln: Here, a : b = 11 : 23
c : d = 3 : 7
∴ the required number

$$=\frac{bc-ad}{c-d}=\frac{23\times 3-11\times 7}{3-7}=\frac{8}{4}=2.$$

Exercise

1. Find the number which, when subtracted from the terms of the ratio 15 : 17 makes it equal to the ratio 6 : 7.
2. Find the number which, when subtracted from the terms of the ratio 11 : 25 makes it equal to the ratio 4 : 11.
3. Find the number which, when subtracted from the terms of the ratio 13 : 27 makes it equal to the ratio 4 : 11.
4. Find the number which, when subtracted from the terms of the ratio 23 : 33 makes it equal to the ratio 3 : 5.
5. Find the number which, when subtracted from the terms of the ratio 12 : 17 makes it equal to the ratio 2 : 3.
6. Find the number which, when subtracted from the terms of the ratio 5 : 13 makes it equal to the ratio 1 : 5.
7. Find the number which, when subtracted from the terms of the ratio 13 : 37 makes it equal to the ratio 1 : 13.

Answers

1. 3 2. 3 3. 5 4. 8 5. 2 6. 3 7. 11

Rule 36

Ex.: A bucket contains a mixture of two liquids A and B in the proportion 7 : 5. If 9 litres of the mixture is replaced by 9 litres of liquid B, then the ratio of the two liquids becomes 7 : 9. How much of the liquid A was there in the bucket?

Soln: **Detailed Method:**
Suppose the two liquids A and B are 7x litres and 5x litres respectively.
Now, when 9 litres of mixture are taken out,

A remains $7x-9\left(\frac{7}{7+5}\right)=7x-\frac{9\times 7}{12}=\left(7x-\frac{21}{4}\right)$

litres and

B remains $5x-9\left(\frac{5}{7+5}\right)=5x-\frac{9\times 5}{12}=\left(5x-\frac{15}{4}\right)$

litres. Now, when 9 litres of liquid B are added,

$$\left(7x-\frac{21}{4}\right):\left(5x-\frac{15}{4}+9\right)=7:9$$

or, $\dfrac{7x-\frac{21}{4}}{5x-\frac{15}{4}+9}=\dfrac{7}{9}$

or, $63x-\dfrac{189}{4}=35x-\dfrac{105}{4}+63$

$28x=\dfrac{189}{4}-\dfrac{105}{4}+63=21+63=84$

or, $x=\dfrac{84}{28}=3$ $\therefore 7x=7\times3=21$ litres

Quicker Method: If we ignore the intermediate steps, we find a formula which is fast-working as well as easier to remember.

1st ratio = 7 : 5, 2nd ratio = 7 : 9

D = Difference of cross-products of ratios

$=7\times9-7\times5$

$=63-35=28$

Now, the formula is:

Common factor of first ratio

$=\left[\dfrac{\text{Quantity Replaced}}{\text{Sum of terms in 1st ratio}}\right]$

$+\left[\dfrac{\text{Quantity replaced}\times\text{term A in 2nd ratio}}{D}\right]$

$=\left[\dfrac{9}{7+5}\right]+\left[\dfrac{9\times7}{28}\right]=\dfrac{9}{12}+\dfrac{9}{4}=\dfrac{36}{12}=3$

$\therefore$ Quantity of A = 7 × 3 = 21 litres.

Similarly, quantity of B = 5 × 3 = 15 litres.

Exercise

1. A bucket contains a mixture of two liquids A and B in the proportion 5 : 3. If 16 litres of the mixture is replaced by 16 litres of liquid B, then the ratio of the two liquids becomes 3 : 5. How much of the liquid B was there in the bucket?
   a) 25 litres b) 15 litres c) 18 litres d) 24 litres
2. A bucket contains a mixture of two liquids A and B in the proportion 6 : 5. If 33 litres of the mixture is replaced by 33 litres of liquid B, then the ratio of the two liquids becomes 3 : 4. How much of the liquid A was there in the bucket?
   a) 84 litres b) 48 litres c) 70 litres d) 64 litres
3. A bucket contains a mixture of two liquids A and B in the proportion 8 : 3. If 11 litres of the mixture is replaced by 11 litres of liquid B, then the ratio of the two liquids becomes 5 : 2. How much of the liquid B was there in the bucket?
   a) 448 litres b) 484 litres c) 168 litres d) 178 litres
4. A bucket contains a mixture of two liquids A and B in the proportion 4 : 1. If 45 litres of the mixture is replaced by 45 litres of liquid B, then the ratio of the two liquids becomes 2 : 5. How much of the liquid B was there in the bucket? What quantity does the bucket hold?
   a) 56 litres, 70 litres b) 14 litres, 70 litres
   c) 65 litres, 72 litres d) 18 litres, 90 litres
5. A bucket contains a mixture of two liquids A and B in the proportion 5 : 2. If 28 litres of the mixture is replaced by 28 litres of liquid B, then the ratio of the two liquids becomes 3 : 2. How much of the liquid A was there in the bucket?
   a) 50 litres b) 125 litres c) 75 litres d) 100 litres

Answers

1. b 2. a 3. c 4. b 5. b

Rule 37

Ex.: A vessel contains liquids A and B in ratio 5 : 3. If 16 litres of the mixture are removed and the same quantity of liquid B is added, the ratio becomes 3 : 5. What quantity does the vessel hold?

Soln: Detailed Method:

Suppose the vessel contains 5x litres and 3x litres of liquids A and B respectively.

The removed quantity contains $\dfrac{16}{5+3}\times5=10$ litres of A and 16 - 10 = 6 litres of B. Now,

$(5x-10):(3x-6+16)=3:5$

or, $\dfrac{5x-10}{3x+10}=\dfrac{3}{5}$ or, $25x-50=9x+30$

or, $16x=80$ $\therefore x=5$

$\therefore$ The vessel contains $8x=8\times5=40$ litres.

Quicker Method: When the ratio is reversed (i.e. 5 : 3 becomes 3 : 5), we can use the formula:

Total quantity $=\dfrac{(5+3)^2}{5^2-3^2}\times$ Quantity of A in the removed mixture

$=\dfrac{64}{16}\times10=40$ litres.

Note: Since, the liquid B is used as a filler, the quantity of A is used in the formula. If liquid A be the filler the quantity of B is used in the formula.

Exercise

1. A vessel contains liquids A and B in ratio 9 : 5. If 14 litres of the mixture are removed and the same quantity of liquid B is added, the ratio becomes 5 : 9. What quantity does the vessel hold?
   a) 32 litres b) 31 litres c) 31.5 litres d) 33 litres
2. A vessel contains liquids A and B in ratio 5 : 4. If 18 litres

of the mixture are removed and the same quantity of liquid A is added, the ratio becomes 4 : 5. What quantity does the vessel hold?

a) 72 litres b) 90 litres c) 64 litres d) 80 litres

3. A vessel contains liquids A and B in ratio 3 : 2. If 15 litres of the mixture are removed and the same quantity of liquid B is added, the ratio becomes 2 : 3. What quantity does the vessel hold?

a) 30 litres b) 35 litres c) 40 litres d) 45 litres

4. A vessel contains liquids A and B in ratio 3 : 1. If 8 litres of the mixture are removed and the same quantity of liquid B is added, the ratio becomes 1 : 3. What quantity does the vessel hold?

a) 12 litres b) 14 litres c) 16 litres d) 10 litres

5. A vessel contains liquids A and B in ratio 7 : 6. If 26 litres of the mixture are removed and the same quantity of liquid B is added, the ratio becomes 6 : 7. What quantity does the vessel hold?

a) 142 litres b) 172 litres c) 156 litres d) 182 litres

Answers

1. c
2. a; **Hint:** See the 'note'.

The removed quantity contains $\frac{18}{5+4}\times 5=10$ litres of A and 18 - 10 = 8 litres of B. Now,

Total quantity $=\frac{(5+4)^2}{5^2-4^2}\times$ quantity of B in the removed mixture = 72 litres.

3. d 4. a 5. d

Rule 38

Ex.: An employer reduces the number of his employees in the ratio 9 : 8 and increases their wages in the ratio 14 : 15. State whether his bill of total wages increases or decreases, and in what ratio.

Soln: 9 : 8

14 : 15

We know that the total bill = wage per person × no. of total employees.

Therefore, the ratio of change in bill

$= 9\times 14 : 8\times 15 = 126 : 120 = 21 : 20$

The ratio shows that there is a decrease in the bill.

Note: For a detailed method let the no. of employees in two cases = $9x$ & $8x$. Wages in two cases be 14y & 15 y

Initial wage = $9x \times 14y = 126xy$

Changed wage = $8x \times 15y = 120xy$

This shows the decrease in bill and ratio is $126xy : 120xy = 21 : 20$.

Exercise

1. An employer reduces the number of his employees in the ratio 7 : 6 and increases their wages in the ratio 13 : 14. State whether his bill of total wages increases or decreases, and in what ratio?

a) Increase, 13 : 12 b) Decrease, 13 : 12
c) Increase, 14 : 13 d) Decrease, 13 : 14

2. An employer reduces the number of his employees in the ratio 8 : 7 and increases their wages in the ratio 7 : 2. State whether his bill of total wages increases or decreases, and in what ratio?

a) Decrease, 4 : 1 b) Increase, 1 : 4
c) Decrease, 5 : 2 d) Increase, 2 : 5

3. An employer reduces the number of his employees in the ratio 7 : 5 and increases their wages in the ratio 10 : 9. State whether his bill of total wages increases or decreases, and in what ratio?

a) Decrease, 13 : 9 b) Decrease 14 : 9
c) Increase 9 : 14 d) Increase 9 : 13

4. An employer reduces the number of his employees in the ratio 8 : 3 and increases their wages in the ratio 3 : 8. State whether his bill of total wages increases or decreases, and in what ratio?

a) Remains unchanged, 1 : 1 b) Decrease, 3 : 1
c) Decrease, 2 : 1 d) Can't be determined

5. An employer reduces the number of his employees in the ratio 9 : 4 and increases their wages in the ratio 2 : 5. State whether his bill of total wages increases or decreases, and in what ratio?

a) Decrease, 10 : 9 b) Increase, 10 : 9
c) Decrease 9 : 11 d) Increase, 9 : 10

Answers

1. b 2. a 3. b 4. a 5. d

Rule 39

Theorem: ***Two candles of the same height are lighted at the same time. The first is consumed in T_1 hours and the second in T_2 hours. Assuming that each candle burns at a constant rate, the time after which the ratio of first candle to second candle becomes x : y is given by*** $\left[\frac{T_1T_2\left(\frac{x}{y}-1\right)}{\left(\frac{x}{y}\right)T_1-T_2}\right]$ ***hours.***

Illustrative Example

Ex.: Two candles of the same height are lighted at the same time. The first is consumed in 4 hours and the second in 3 hours. Assuming that each candle burns at a constant rate, in how many hours after being lighted, was the first candle twice the height of the second?

Soln: Detail Method: Let the height of candles be h and candle be A and B.

The candle A burns $\frac{1}{4}$ of its height in 1 hour while candle B burns $\frac{1}{3}$ of its height in one hour.

As per question, the height of candle A after x hours be double of height of candle B.

Height of candle A after x hours = $h-\frac{xh}{4}=\frac{4h-xh}{4}$

Height of candle B after x hours = $h-\frac{xh}{3}=\frac{3h-xh}{3}$

As per given question,

$\frac{4h-xh}{4}=\frac{(3h-xh)}{3}\times 2$

or, $12-3x=24-8x$

or, $5x=12$

or, $x=\frac{12}{5}$ hours or 2 hours 24 minutes.

$\therefore$ The required answer is 2 hours 24 minutes.

Quicker Method: Applying the above theorem, we have,

the required answer = $\frac{4\times 3\left(\frac{2}{1}-1\right)}{\frac{2}{1}\times 4-3}=\frac{12}{5}$ hours

= 2 hours 24 minutes.

Note: Here, x : y = 2 : 1

Exercise

1. Two candles of the same height are lighted at the same time. The first is consumed in 8 hours and the second in 6 hours. Assuming that each candle burns at a constant rate, in how many hours after being lighted, the ratio between the first and second candles becomes 2 : 1.
   a) 2 hours 24 minutes b) 4 hours
   c) 1 hour 12 minutes d) 4 hours 48 minutes
2. Two candles of the same height are lighted at the same time. The first is consumed in 5 hours and the second in 4 hours. Assuming that each candle burns at a constant rate, in how many hours after being lighted, the ratio between the first and second candles becomes 3 : 2.
   a) 3 hours b) 3 hours 45 minutes
   c) $\frac{20}{7}$ hours d) Can't be determined
3. Two candles of the same height are lighted at the same time. The first is consumed in 7 hours and the second in 6 hours. Assuming that each candle burns at a constant rate, in how many hours after being lighted, the ratio between the first and second candles becomes 3 : 1.
   a) 5 hours 36 minutes b) 5 hours
   c) 5 hours 60 minutes d) 6 hours
4. Two candles of the same height are lighted at the same time. The first is consumed in 3 hours and the second in 1 hour. Assuming that each candle burns at a constant rate, in how many hours after being lighted, the ratio between the first and second candles become 2 : 1.
   a) 48 minutes b) 1 hour 36 min
   c) 36 minutes d) 60 minutes
5. Two candles of the same height are lighted at the same time. The first is consumed in 7 hours and the second in 3 hours. Assuming that each candle burns at a constant rate, in how many hours after being lighted, the ratio between the first and second candles becomes 3 : 1.
   a) 2 hours b) 2 hours 20 minutes
   c) 3 hours 20 minutes d) 3 hours

Answers

1. d 2. c 3. a 4. c 5. b

Miscellaneous

1. Find the sum of three numbers in the ratio of 3 : 2 : 5, such that the sum of their squares is equal to 1862.
   a) 70 b) 75 c) 69 d) 60
2. A, B, C, D are four quantities of the same kind such that $A:B=3:4, B:C=8:9, C:D=15:16$
   (i) Find the ratio A : D
   a) 5 : 8 b) 8 : 5 c) 4 : 5 d) 5 : 4
   (ii) Compare A, B, C, D.
   a) $4:3:\frac{9}{2}:\frac{24}{5}$ b) $3:4:\frac{9}{2}:\frac{24}{5}$
   c) $4:3:\frac{24}{5}:\frac{9}{2}$ d) $3:4:\frac{24}{5}:\frac{9}{2}$
3. Divide 94 into two parts in such a way that fifth part of the first and eighth part of the second are in the ratio 3 : 4.
   a) 30, 64 b) 20, 44 c) 35, 69 d) 30, 65
4. Divide 1162 into three parts such that 4 times the first may be equal to 5 times the second and 7 times the third. Find the value of smallest part.
   a) 490 b) 492 c) 390 d) 280
5. Divide Rs. 680 among A, B and C such that A gets $\frac{2}{3}$ of what B gets and B gets $\frac{1}{4}$ th of what C gets. What is C's share?
   a) Rs 280 b) Rs 380 c) Rs 480 d) Rs 120

6. If 10% of m is the same as 20% of n, then m : n is equal to:
a) 1 : 2 b) 2 : 1 c) 5 : 1 d) 10 : 1
(C.B.I. Exam. 1990)

7. If A : B = 2 : 3 and B : C = 4 : 5, then C : A is equal to:
a) 15 : 8 b) 12 : 10 c) 8 : 5 d) 8 : 15
(Railway Recruitment Board, 1991)

8. Rs 600 has been divided among A, B and C in such a way that Rs. 40 more than (2/5) of A's share, Rs. 20 more than (2/7) of B's share, Rs. 10 more than (9/17) of C's share, are all equal. A's share is:
a) Rs 280 b) Rs 170 c) Rs 150 d) Rs 200
(Railway Recruitment Board, 1991)

9. 729 ml of a mixture contains milk and water in the ratio 7 : 2. How much more water is to be added to get a new mixture containing milk and water in the ratio 7 : 3?
a) 600 ml b) 710 ml c) 520 ml d) None of these
(Railway Recruitment Board, 1991)

10. Gold is 19 times as heavy as water and copper 9 times as heavy as water. The ratio in which these two metals be mixed so that the mixture is 15 times as heavy as water, is: **(Delhi Police & CBI 1990)**
a) 1 : 2 b) 2 : 3 c) 3 : 2 d) 19 : 135

11. If A is $\frac{1}{3}$ of B and B is $\frac{1}{2}$ of C, then A : B : C is
(Police Inspector Exam. 1988)
a) 1 : 3 : 6 b) 2 : 3 : 6 c) 3 : 2 : 6 d) 3 : 1 : 2

12. A certain amount was divided between Kavita and Reena in the ratio 4 : 3. If Reena's share was Rs. 2400, the amount is: **(SBI PO Exam 1988)**
a) Rs 5600 b) Rs 3200 c) Rs 9600 d) None of these

13. If a carton containing a dozen mirrors is dropped, which of the following cannot be the ratio of broken mirrors to unbroken mirrors?
a) 2 : 1 b) 3 : 1 c) 3 : 2 d) 7 : 5
(SBI PO Exam 1987)

14. A man spends Rs 8100 in buying tables at Rs 1200 each and chairs at Rs 300 each. The ratio of chairs to tables when the maximum number of tables is purchased, is
(SBI PO Exam 1988)
a) 1 : 4 b) 5 : 7 c) 1 : 2 d) 2 : 1

15. A sum of Rs 86700 is to be divided among A, B and C in such a manner that for every rupee that A gets, B gets 90 paise and for every rupee that B gets, C gets 110 paise. B's share is **(LIC AAO Exam 1988)**
a) Rs 26010 b) Rs 27000
c) Rs 30000 d) None of these

16. Rs 5625 is to be divided among A, B and C, so that A may receive (1/2) as much as B and C together receive and B receives (1/4) of what A and C together receive. The share of A is more than that of B by
(Excise and I. Tax Exam 1988)
a) Rs 750 b) Rs 775 c) Rs 1500 d) Rs 1600

17. If the weight of a 13 metres long iron rod be 23.4 kg., the weight of 6 metres long of such rod will be
(Bank PO Exam 1986)
a) 7.2 kg b) 12.4 kg c) 10.8 kg d) 18 kg

18. The ratio of the money with Ram and Gopal is 7 : 17 and that with Gopal and Krishan is 7 : 17. If Ram has Rs 490, Krishan has
a) Rs 2890 b) Rs 2330 c) Rs 1190 d) Rs 2680

19. The students in three classes are in the ratio 2 : 3 : 5. If 20 students are increased in each class, the ratio changes to 4 : 5 : 7. The total number of students in the three classes before the increase were **(LIC AAO Exam 1988)**
a) 10 b) 90 c) 100 d) None of these

20. One year ago the ratio between Laxman's and Gopal's salary was 3 : 4. The individual ratios between their last year's and this year's salaries are 4 : 5 and 2 : 3 respectively. At present the total of their salary is Rs 4160. The salary of Laxman now, is **(SBI Bank PO Exam 1987)**
a) Rs 1600 b) Rs 2560 c) Rs 1040 d) Rs 3120

21. The sum of the squares of three numbers is 532 and the ratio of the first to the second as also of the second to the third is 3 : 2. What is the second number?
a) 12 b) 14 c) 10 d) 8

22. If bc : ac : ab = 1 : 2 : 3, find $\frac{a}{bc} : \frac{b}{ca}$
a) 2 : 1 b) 3 : 1 c) 4 : 1 d) 1 : 4

23. The sum of two numbers is 'c' and their quotient is $\frac{p}{q}$. Find the numbers.
a) $\frac{pc}{p+q}, \frac{qc}{p+q}$ b) $\frac{qp}{p+q}, \frac{q}{(p+q)c}$
c) $\frac{qp}{p+q}, \frac{qc}{p+q}$ d) None of these

24. A bag contains rupees, fifty paise, twenty five paise and ten paise coins in the proportion 1 : 3 : 5 : 7. If the total amount is Rs 22.25, find the number of twenty-five paise coins.
a) 25 b) 5 c) 15 d) 35

25. In a school the number of boys and that of the girls are in the ratio of 2 : 3. If the number of boys is increased by 20% and that of girls is increased by 10%. What will be the new ratio of the number of boys to that of girls?
(SBI Bank PO 2001)
a) 4 : 5 b) 5 : 8
c) 8 : 11 d) Data inadequate

26. An amount of money is to be distributed among P, Q and R in the ratio of 6 : 19 : 7 respectively. If R gives Rs 200 of his share to Q the ratio among P, Q and R becomes 3 : 10 : 3 respectively. What was the total amount?

(SBI Bank PO 2000)

a) Rs 6400 b) Rs 12800
c) Rs 3200 d) Data inadequate

27. When 35 per cent of a number is added to another number, the second number increases by its 20 per cent. What is the ratio between the second number and the first number? **(BSRB Mumbai PO 1998)**
a) 4 : 7 b) 7 : 4
c) 8 : 5 d) Data inadequate

28. There is a ratio of 5 : 4 between two numbers. If 40 per cent of the first number is 12 then what would be the 50 per cent of the second number?
(Bank of Baroda PO 1999)
a) 12 b) 24
c) 18 d) Data inadequate

29. An amount of money is to be distributed among P, Q and R in the ratio of 5 : 8 : 12 respectively. If the total share of Q and R is four times that of P, what is definitely P's share? **(Bank of Baroda PO 1999)**
a) Rs 3000 b) Rs 5000
c) Rs 8000 d) Data inadequate

30. When 30 per cent of a number is added to another number the second number increases to its 140 per cent. What is the ratio between the first and the second number? **(Bank of Baroda PO 1999)**
a) 3 : 4 b) 4 : 3
c) 3 : 2 d) Data inadequate

31. If 25% of a number is subtracted from a second number the second number reduces to its five-sixths. What is the ratio between the first number and the second number? **(SBI Associates PO 1999)**
a) 2 : 3 b) 3 : 2
c) 1 : 3 d) Data inadequate

32. When 50% of one number is added to a second number, the second number increases to its four-thirds. What is the ratio between the first number and the second number? **(Guwahati PO 1999)**
a) 3 : 2 b) 3 : 4
c) 2 : 3 d) Data inadequate

33. An amount of money is to be divided among P, Q and R in the ratio of 4 : 9 : 16. If R gets 4 times more than P, what is Q's share in it? **(BSRB Calcutta PO 1999)**
a) Rs 1800 b) Rs 2700
c) Rs 3600 d) Data inadequate

34. If a : b = 2 : 5. Find (3a + 4b) : (4a + 5b).
a) 26 : 33 b) 14 : 31 c) 25 : 32 d) 33 : 26

35. A bag contains rupee, 50-paise and 25-paise coins in the ratio 5 : 6 : 7. If the total amount is Rs 390, find the number of 25-paise coins.
a) 280 b) 200 c) 240 d) 260

Answers

1. a; Let the numbers be 3x, 2x, 5x.

we have $9x^2 + 4x^2 + 25x^2 = 1862$

$\therefore\ 38x^2 = 1862\ \therefore\ x^2 = 49 = 7^2$

$\therefore$ x = 7. Hence, the required numbers are 21, 24 and 35

$\therefore$ sum of the numbers = 21 + 24 + 35 = 70.

2. (i) a; $\frac{A}{B} = \frac{3}{4}, \frac{B}{C} = \frac{8}{9}, \frac{C}{D} = \frac{15}{16}$

$\therefore\ \frac{A}{B} \times \frac{B}{C} \times \frac{C}{D} = \frac{3}{4} \times \frac{8}{9} \times \frac{15}{16} = \frac{5}{8}\quad \therefore$ A : D = 5 : 8

(ii) b; A : B = 3 : 4

$B : C = 8 : 9 = 1 : \frac{9}{8} = 4 : \frac{9}{2}$

$C : D = 15 : 16 = 1 : \frac{16}{15} = \frac{9}{2} : \frac{24}{5}$

$\therefore\ A : B : C : D = 3 : 4 : \frac{9}{2} : \frac{24}{5}$.

We put down the first ratio in its original form and change the terms of the other ratios so as to make each antecedent equal to the preceding consequent. (Also see **Rule 16**).

3. a; Let these parts be x and y. Then,

$\frac{x}{5} : \frac{y}{8} = 3 : 4$ or, $8x : 5y = 3 : 4$

$\therefore\ \frac{8x}{5y} = \frac{3}{5}$ or $\frac{x}{y} = \frac{5 \times 3}{4 \times 8} = \frac{15}{32}$.

Thus, x : y = 15 : 32.

Now, sum of ratios = 15 + 32 = 47.

$\therefore$ First part $= \frac{94 \times 15}{47} = 30$,

second part $= \frac{94 \times 32}{47} = 64$.

4. d; 4 × (1st part) = 5 × (2nd part) = 7 × (3rd part) = x (say)

Then, 1st part $= \frac{x}{4}$, 2nd part $= \frac{x}{5}$; 3rd part $= \frac{x}{7}$.

$\therefore$ Ratio of divisions $= \frac{x}{4} : \frac{x}{5} : \frac{x}{7}$

$= \frac{1}{4} : \frac{1}{5} : \frac{1}{7} = 35 : 28 : 20$.

Sum of ratios = 35 + 28 + 20 = 83.

$\therefore$ 1st part $= \frac{1162 \times 35}{83} = 490$

2nd part $= \frac{1162 \times 28}{83} = 392$

3rd part $= \frac{1162\times 20}{83} = 280$

5. c; Suppose C gets Re 1. Then, B gets Re $\frac{1}{4}$ and A gets Rs $\left(\frac{2}{3} of \frac{1}{4}\right)$ or Rs $\frac{1}{6}$.

Ratios of A, B and C's shares $= \frac{1}{6}:\frac{1}{4}:1 = 2:3:12$.

So, A's share = Rs $\frac{680\times 2}{17}$ = Rs 80.

B's share = Rs $\frac{680\times 3}{17}$ = Rs 120.

C's share = Rs $\frac{680\times 12}{17}$ = Rs 480.

6. b; 10% of m = 20% of n $\Rightarrow \frac{10}{100}m = \frac{20}{100}n$

$\therefore \frac{m}{n} = \left(\frac{20}{100}\times\frac{100}{10}\right) = \frac{2}{1}$.

7. a; $\frac{A}{B} = \frac{2}{3}, \frac{B}{C} = \frac{4}{5} \Rightarrow \frac{A}{B}\times\frac{B}{C} = \frac{2}{3}\times\frac{4}{5}$

$\Rightarrow \frac{A}{C} = \frac{8}{15} \Rightarrow \frac{C}{A} = \frac{15}{8}$.

8. c; $\frac{2}{5}A + 40 = \frac{2}{7}B + 20 = \frac{9}{17}C + 10 = x$.

$\therefore A = \frac{5}{2}(x-40) = \frac{5}{2}x - 100$;

$B = \frac{7}{2}(x-20) = \frac{7}{2}x - 70$

and $C = \frac{17}{9}(x-10) = \frac{17}{9}x - \frac{170}{9}$.

$\therefore \frac{5}{2}x - 100 + \frac{7}{2}x - 70 + \frac{17}{9}x - \frac{170}{9} = 600$

or, $\frac{71}{9}x = \frac{7100}{9} \Rightarrow x = 100$.

$\therefore$ A's share $= \left(\frac{5}{2}\times 100 - 100\right)$ = Rs 150.

9. d; Quantity of milk $= \left(729\times\frac{7}{9}\right)$ ml = 567 ml.

Quantity of water = (729 - 567) ml = 162 ml

Now, $\frac{567}{162+x} = \frac{7}{3} \Rightarrow 567\times 3 = 7(162+x)$

$\Rightarrow 1701 = 1134 + 7x \Rightarrow x = 81$.

10. b; Let x gm of water be taken
Then, gold = 19x gm and copper = 9x gm
Let 1 gm of gold be mixed with y gm of copper.

Then, $19x + 9xy = 15x(1+y) \Rightarrow y = \frac{2}{3}$.

11. a; $B = \frac{1}{2}C, A = \frac{1}{3}B, = \frac{1}{3}\left(\frac{1}{2}C\right) = \frac{1}{6}C$.

$\therefore A:B:C = \frac{1}{6}C:\frac{1}{2}C:C = \frac{1}{6}:\frac{1}{2}:1$ or 1 : 3 : 6.

12. a; Let the amount be Rs x.

Then, Reena's share = Rs $\left(x\times\frac{3}{7}\right)$

$\therefore \frac{3x}{7} = 2400$ or $x = \left(\frac{2400\times 7}{3}\right)$ = Rs 5600.

13. c; Sum of the ratios must divide 12.
Since 3 + 2 = 5 does not divide 12, so it can not be 3 : 2.

14. c; Maximum number of tables purchased at Rs 1200 per table spending within Rs 8100 is clearly 6.
Remaining amount = Rs (8100 - 1200 × 6) = Rs 900
Number of chairs for Rs 900 = (900 ÷ 300) = 3
∴ Number of chairs : Number of tables = 3 : 6 or 1 : 2.

15. b; If A gets Re 1, B gets 90 paise.
Now, if B gets Re 1, C gets 110 paise.

If B gets 90 paise, C gets $\left(\frac{110}{100}\times 90\right) = 99$ paise.

∴ A : B : C = 100 : 90 : 99.

So, B's share = Rs $\left(86700\times\frac{90}{289}\right)$ = Rs 27000.

16. a; A + B + C = 5625 and B = $\frac{1}{4}$ (A + C) i.e. A + C = 4B.

∴ 4B + B = 5625 or B = 1125.
Also, A + C = 4B = 4 × 1125 = 4500.

Also, $A = \frac{1}{2}$ (B + C) or B + C = 2A or B = 2A - C.

∴ 2A - C = 1125.
Now, solving A + C = 4500 and 2A - C = 1125, we get A = 1875 and C = 2625.
∴ A - B = (1875 - 1125) = Rs 750.

17. c; Let $\frac{13}{6} = \frac{23.4}{x}$. Then, $x = \left(\frac{23.4\times 6}{13}\right) = 10.8$ kg

18. a; Let Ram, Gopal and Krishan have rupees x, y and z respectively.

Then, $\frac{x}{y}=\frac{7}{17}$ and $\frac{y}{z}=\frac{7}{17}$

$\therefore \frac{x}{z}=\left(\frac{x}{y}\times\frac{y}{z}\right)=\left(\frac{7}{17}\times\frac{7}{17}\right)=\frac{49}{289}$.

Thus, if Ram has Rs 49, Krishan has Rs 289.

If Ram has Rs 490, Krishan has Rs $\left(\frac{289}{49}\times 490\right)$

= Rs 2890.

19. c; Let the number of students in the class be 2x, 3x and 5x respectively. Then,

(2x+20) : (3x+20) : (5x+20) : : 4 : 5 : 7.

$\frac{2x+20}{4}=\frac{3x+20}{5}=\frac{5x+20}{7}$.

Solving these equations, we get x = 10.

$\therefore$ total number of students in the class

= (2x+3x+5x) = 10x = 100.

20. a; Let the salaries of Laxman and Gopal one year before be x_1, y_1 and now it be x_2, y_2 respectively. Then,

$\frac{x_1}{y_1}=\frac{3}{4},\frac{x_1}{x_2}=\frac{4}{5},\frac{y_1}{y_2}=\frac{2}{3}$ and $x_2+y_2=4160$

Solving these equations, we get $x_2=1600$

21. a; $\frac{\text{First number}}{\text{Second number}}=\frac{3}{2}\times\frac{3}{3}=\frac{9}{6}$

and $\frac{\text{First number}}{\text{Third number}}=\frac{3}{2}\times\frac{2}{2}=\frac{6}{4}$

$\therefore$ First : Second : Third = 9 : 6 : 4

As per the question

$\therefore (9x)^2+(6x)^2+(4x)^2=532$

or, $133x^2=532$ or, x = 2

Second Number is 6x = 12

22. c; $\frac{bc}{ac}=\frac{1}{2}$ or, $a:b=2:1$

$\therefore \frac{a}{bc}:\frac{b}{ca}=\frac{a}{bc}\times\frac{ca}{b}=\frac{a^2}{b^2}=\frac{4}{1}$

23. a; Let the numbers be x, y.

$\therefore x+y=$ c (given)

and $\frac{x}{y}=\frac{p}{q}$ (given)

$\therefore \frac{x}{x+y}=\frac{p}{p+q}$

$\Rightarrow \frac{x}{c}=\frac{p}{p+q} \Rightarrow x=\frac{pc}{p+q}$

and $y=\frac{qc}{p+q}$ $\left[\text{Since } \frac{x}{y}=\frac{p}{q}\right]$

24. a; Let the number of coins be x, 3x, 5x, 7x respectively as rupees, fifty paise, twenty five paise and ten paise. Since

Number of coin × Value of coin in rupee = Amount in rupees

Now, value of 50-paise coin in rupee = $\frac{1}{2}$

Value of 25-paise coin in rupee = $\frac{1}{4}$

Value of 10-paise coin in rupee = $\frac{1}{10}$

$\therefore$

$\left[(x\times 1)+\left(3x\times\frac{1}{2}\right)+\left(5x\times\frac{1}{4}\right)+\left(7x\times\frac{1}{10}\right)\right]=22.25$

or, $\frac{89x}{20}=22.5$ or, x = 5

$\therefore$ number of rupee coins = 5 × 1 = 5

Number of 50-paise coins = 3x = 15

Number of 25-paise coins = 5x = 25

Number of 10-paise coins = 7x = 35

25. c; Let the number of boys be 2x and the number of girls be 3x.

No. of boys is increased by 20% = $2x\times\frac{120}{100}=\frac{12x}{5}$

No. of girls is increased by 10% = $3x\times\frac{110}{100}=\frac{33x}{10}$

The new ratio of the number of boys to that of girls

$=\frac{\frac{12x}{5}}{\frac{33x}{10}}=8:11$

26. a; Let the sum of P, Q, and R be 6x, 19x and 7x.

$\therefore$ total sum = 6x + 19x + 7x = 32x

From the question

6x : 19x + 200 : 7x - 200 = 3 : 10 : 3

ie 6x = 7x - 200 $\therefore$ x = 200

$\therefore$ total sum = 32 × 200 = Rs 6400

27. b; 35 per cent of x + y = $\frac{120}{100}y$

or, $\frac{35x+100y}{100}=\frac{120}{100}y\Rightarrow 35x=20y$

$\frac{y}{x} = \frac{35}{20} = 7:4$

28. a; $\frac{a}{b} = \frac{5}{4}$ $b = \frac{4}{5}a$ Given, (40% of a =) $\frac{2}{5}a = 12$

$\therefore$ a = 5 × 6 and $b = \frac{4}{5} \times 5 \times 6 = 24$

$\therefore$ 50% of $b = \frac{24}{2} = 12$

29. d; P : Q : R = 5 : 8 : 12

$\frac{\text{Total share of Q and R}}{\text{Share of P}} = \frac{8+12}{5} = \frac{20}{5} = 4$

So, we see that not new information has been given in question and P's share can't be determined.

30. b; Let the first and the second numbers be x and y respectively then
$y + 30\%\text{ of } x = 140\%\text{ of } y$ or, $y + 0.3x = 1.4y$
or, $0.3x = 0.4y$ $\therefore$ $x : y = 0.4 : 0.3 = 4 : 3$

31. a; Let the first and second number be x and y respectively

$y - x \times \frac{25}{100} = y \times \frac{5}{6}$

or, $y - \frac{x}{4} = \frac{5}{6}y$

or, $\frac{1}{6}y = \frac{x}{4}$ $\therefore$ x : y = 2 : 3

32. c; Let the numbers be y and x respectively.

$x + 50\%\text{ of } y = \frac{4x}{3}$ or, $\frac{y}{2} = \frac{4x}{3} - x$

or, $\frac{y}{2} = \frac{x}{3}$ or, $\frac{y}{x} = \frac{2}{3} = 2:3$

33. d; Here, neither the total amount nor the individual amount is given. So the share of Q cannot be determined.

34. a; Given $\frac{a}{b} = \frac{2}{5}$

Now, $\frac{3a+4b}{4a+5b} = \frac{3\left(\frac{a}{b}\right)+4}{4\left(\frac{a}{b}\right)+5}$ [Dividing numerator and denominator by b.]

$= \frac{3 \times \frac{2}{5} + 4}{4 \times \frac{2}{5} + 5} = \frac{26}{33}$

$\therefore (3a + 4b) : (4a + 5b) = 26 : 33$

35. a; Ratio of values of the coins $= \frac{5}{1} : \frac{6}{2} : \frac{7}{4} = 20 : 12 : 7$

Value of one-rupee coins = Rs $\left(390 \times \frac{20}{39}\right)$ = Rs 200

Value of 50-paise coins = Rs $\left(390 \times \frac{12}{39}\right)$ = Rs 120

Value of 25-paise coins = Rs $\left(390 \times \frac{7}{39}\right)$ = Rs 70

$\therefore$ Number of one-rupee coins = 200
Number of 50-paise coins = (120 × 2) = 240
Number of 25-paise coins = (70 × 4) = 280.

4

Partnership

Rule 1

Theorem: ***In a group of n persons invested different amount for different period then their profit ratio is*** $At_1 : Bt_2 : Ct_3 : Dt_4 : ... : Xt_n$

Here first person invested amount A for t_1 period, second persons invested amount B for t_2 period and so on.

Illustrative Examples

Ex. 1: Three partners A, B and C invest Rs 1600, Rs 1800 and Rs 2300 respectively in business. How should they divide a profit of Rs 1938?

Soln: The profit should be divided in the ratios of the capitals, i.e. in the ratio 16 : 18 : 23.

Now, $16+18+23=57$

A's share $= \frac{16}{57}$ of Rs $1938 =$ Rs 544

B's share $= \frac{18}{57}$ of Rs $1938 =$ Rs 612

C's share $= \frac{23}{57}$ of Rs $1938 =$ Rs 782

Ex. 2: A, B and C enter into partnership. A advances Rs 1200 for 4 months, B Rs 1400 for 8 months, and C Rs 1000 for 10 months. They gain Rs 585 altogether. Find the share of each.

Soln: Rs 1200 in 4 months earns as much profit as Rs 1200 × 4 or Rs 4800 in 1 month.

Rs 1400 in 8 months earns as much profit as Rs 1400 × 8 or Rs 11200 in 1 month.

Rs 1000 in 10 months earns as much profit as Rs 1000 × 10 or Rs 10,000 in 1 month.

Therefore, the profit should be divided in the ratio of 4800, 11,200 and 10,000 i.e. in the ratios of 12, 28 and 25.

Now, $12+28+25=65$

A's share $= \frac{12}{65} \times 585 =$ Rs 108

B's share $= \frac{28}{65} \times 585 =$ Rs 252

C's share $= \frac{25}{65} \times 585 =$ Rs 225

Note: In compound partnership, the ratio of profit is directly proportional to both money and time, so they are multiplied together to get the corresponding shares in the ratio of profits.

Ex. 3: A starts a business with Rs 2,000. B joins him after 3 months with Rs 4,000. C puts a sum of Rs 10,000 in the business for 2 months only. At the end of the year the business gave a profit of Rs 5600. How should the profit be divided among them?

Soln: Ratio of their profits (A's : B's : C's) $= 2\times12 : 4\times9 : 10\times2$
$= 6 : 9 : 5$

Now, $6+9+5=20$

Then A's share $= \frac{5600}{20} \times 6 =$ Rs 1680

B's share $= \frac{5600}{20} \times 9 =$ Rs 2520

C's share $= \frac{5600}{20} \times 5 =$ Rs 1400

Ex. 4: A and B enter into a speculation. A puts in Rs 50 and B puts in Rs 45. At the end of 4 months A withdraws half his capital and enters with a capital of Rs 70. At the end of 12 months, in what ratio will the profit be divided?

Soln: A's share : B's share : C's share

$$= 50\times4+\frac{50}{2}\times8 : 45\times6+\frac{45}{2}\times6 : 70\times6$$
$$= 400 : 405 : 420 = 82 : 81 : 84$$

Therefore, the profit will be divided in the ratio of 80 : 81 : 84.

Now, you must have understood both simple partnership and compound partnership. The formula for compound partnership can also be written as

$$\frac{A's\ Capital \times A's\ Time\ in\ partnership}{B's\ Capital \times B's\ Time\ in\ partnership} = \frac{A's\ Profit}{B's\ Profit}$$

The above relationship should be remembered because it is used very often in some types of question.

Exercise

1. Three friends A, B and C started a business by investing a sum of money in the ratio of 5 : 7 : 6. After 6 months C withdraws half of his capital. If the sum invested by 'A' is Rs 40,000, out of a total annual profit of Rs 33,000 C's share will be **[BSRB Hyderabad PO, 1999]**
a) Rs 9,000 b) Rs 12,000 c) Rs 11,000 d) Rs 10,000
2. Mr Shivkumar started a business, investing Rs 25000 in 1996. In 1997 he invested an additional amount of Rs 10000 and Mr Rakesh joined him with an amount of Rs 35000. In 1998, Mr Shivkumar invested another additional amount of Rs 10000 and Mr Suresh joined them with an amount of Rs 35000. What will be Rakesh's share in the profit of Rs 150000 earned at the end of three years from the start of the business in 1996?
[SBI PO Exam, 2001]
a) Rs 70000 b) Rs 50000 c) Rs 45000 d) Rs 75000
3. A starts a business by investing Rs 500, B joins him after two months by investing Rs 400 and C joins them after 6 months by investing Rs 800. If the annual profit is Rs 444, find the profit of C.
a) Rs 180 b) Rs 120 c) Rs 144 d) Rs 148
4. A, B and C enter into a partnership. A invests Rs 4000 for the whole year, B puts in Rs 6000 at the first and increasing to Rs 8000 at the end of 4 months, whilst C puts in at first Rs 8000 but withdraw Rs 2000 a the end of 9 months. Find the profit of A at the end of year, if the total profit is Rs 16950.
a) Rs 3600 b) Rs 6600 c) Rs 6750 d) Rs 6300
5. A, B and C go into business as partners and collect a profit of Rs 1000. If A's capital : B's capital = 2 : 3 and B's capital : C's capital = 2 : 5, find the share of the profit which goes of C.
a) Rs 160 b) Rs 240 c) Rs 500 d) Rs 600
6. Two partners put Rs 1000 and Rs 1500 into a business. How should a profit of Rs 336.70 be divided?
a) Rs 134.68, Rs 202.02 b) Rs 136, Rs 200.70
c) Rs 132.68, Rs 204.2 d) Rs 136, Rs 202.2
7. Three partners A, B and C invest Rs 2000, Rs 2500 and Rs 1100 respectively in a business. If the total profit is Rs 562.80, find the share of A in the profit.
a) Rs 201 b) Rs 205 c) Rs 204 d) Rs 208
8. Three men A, B and C subscribe Rs 4700 for a business. A subscribes Rs 700 more than B and B Rs 500 more than C. How much will C receive out of a profit of Rs 423?
a) Rs 135 b) Rs 198 c) Rs 90 d) Rs 190
9. A is a working and B a sleeping partner in a business. A puts in Rs 5000 and B puts in Rs 6000. A receives $12\frac{1}{2}$% of the profits for managing the business, and the rest is divided in proportion to their capitals. What does each get out of a profit of Rs 880?
a) Rs 350, Rs 420 b) Rs 460, Rs 350
c) Rs 110, Rs 420 d) Rs 460, Rs 420
10. A starts a business with Rs 1000. B joins him after 6 months with Rs 4000. C puts a sum of Rs 5000 for 4 months only. At the end of the year the business gave a profit of Rs 2800. How should the profit be divided among them?
a) Rs 600,. Rs 1200, Rs 1000
b) Rs 800, Rs 600, Rs 1400
c) Rs 1000, Rs 1200, Rs 600
d) Rs 1200, Rs 600, Rs 1000
11. A and B enter into a partnership for a year. A contributes Rs 3000 and B Rs 4000. After 8 months they admit C, who contributes Rs 4500. If B withdraws his contribution after 6 months, how would they share a profit of Rs 1000 at the end of the year?
a) Rs 250, Rs 200, Rs 550 b) Rs 150, Rs 200, Rs 650
c) Rs 375, Rs 250, Rs 375 d) Data inadequate
12. A, B and C enter into a partnership. A advances one-third of the capital for one-third of the time. B contributes one-sixth of the capital for one-third of the time C contributes the remaining capital for the whole time. How should they divide a profit of Rs 1200.
a) Rs 300, Rs 200, Rs 700 b) Rs 200, Rs 100, Rs 900
c) Rs 200, Rs 200, Rs 800 d) Rs 150, Rs 250, Rs 800
13. A and B entered into partnership with Rs 700 and Rs 600 respectively. After 3 months A withdrew $\frac{2}{7}$ of his stock but after 3 months more he put back $\frac{3}{5}$ of what he had withdrawn. The profits at the end of the year are Rs 726, how much of this should A receive?
a) Rs 633 b) Rs 336 c) Rs 663 d) Rs 366
14. A and B start a business, A puts in double of what B puts. A withdraws $\frac{1}{3}$ of his stock at the end of 3 months but at the end of 7 months puts back $\frac{1}{3}$ of what he has taken out, when B takes out $\frac{1}{4}$ of his stock. A receives Rs 300 profits at the end of the year, what does B receive?
a) Rs 108 b) Rs 272 c) Rs 192 d) Rs 208
15. A, B, C pasture in the same field. A has in it 10 oxen for 7 months, B has 12 oxen for 5 months and C has 15 oxen for 3 months. The rent is Rs 17.50. How much of the rent

should each pay?

a) Rs 7, Rs 6, Rs 4.50 b) Rs 6, Rs 8, Rs 3.50

c) Rs 7, Rs 5, Rs 5.50 d) Rs 8, Rs 5, Rs 4.50

16. Vimla and Surjeet started a shop jointly by investing Rs 9000 and Rs 10500 respectively. After 4 months Jaya joined them by investing Rs 12500 while Surjeet withdrew Rs 2000. At the end of the year there was a profit of Rs 4770. Find the share of Jaya.

a) Rs 1620 b) Rs 1650

c) Rs 1500 d) Data inadequate

(LIC 1991)

17. A and B enter into partnership investing Rs 12000 and Rs 16000 respectively. After 8 months, C also joins the business with a capital of Rs 15000. The share of C in a profit of Rs 45600 after 2 years will be:

a) Rs 12000 b) Rs 14400 c) Rs 19200 d) Rs 21200

(Central Excise 1988)

18. A and B entered into a partnership investing Rs 16000 and Rs 12000 respectively. After 3 months, A withdrew Rs 5000 while B invested Rs 5000 more. After 3 more months, C joins the business with a capital fo Rs 21000. The share of B exceeds that of C, out of a total profit of Rs 26400 after one year, by

a) Rs 1200 b) Rs 2400 c) Rs 3600 d) Rs 4800

(Central Excise, 1989)

19. Suresh invested Rs 12000 in a shop and Dinesh joined him after 4 months by investing Rs 7000. If the net profit after one year be Rs 13300, Dinesh's share in the profit is:

a) Rs 9576 b) Rs 4900 c) Rs 8400 d) None of these

(SBI PO Exam, 1987)

20. Ashok invests Rs 3000 for a year and Sunil joins him with Rs 2000 after 4 months. After the year they receive a return of Rs 2600. Sunil's share is

a) Rs 800 b) Rs 1000 c) Rs 750 d) Rs 900

(Railway Recruitment Board, 1991)

21. Jagmohan, Rooplal and Pandeyji rented a video cassette for one week at a rent of Rs 350. If they use it for 6 hours, 10 hours and 12 hours respectively, the rent to be paid by Pandeyji is

a) Rs 75 b) Rs 125 c) Rs 35 d) Rs 150

(Bank PO Exam, 1988)

22. Manoj got Rs 6000 as his share out of the total profit of Rs 9000 which he and Ramesh earned at the end of one year. If Manoj invested Rs 20000 for 6 months, whereas Ramesh invested his amount for the whole year, the amount invested by Ramesh was

a) Rs 60000 b) Rs 10000 c) Rs 40000 d) Rs 5000

(Bank PO Exam, 1991)

23. A, B and C invest Rs 2000, Rs 3000 and Rs 4000 in a business. After one year A removed his money. B and C continued the business for one more year. If the net profit after 2 years be Rs 3200, then A's share in the profit is

a) Rs 1000 b) Rs 600 c) Rs 800 d) Rs 400

(Asstt. Grade, 1987)

24. A and B enter into partnership investing Rs 12000 and Rs 16000 respectively. After 8 months, C also joins the business with a capital of Rs 15000. The share of C in a profit of Rs 45600 after 2 years will be

a) Rs 12000 b) Rs 14400 c) Rs 19200 d) Rs 21200

(Central Excise & I. Tax, 1988)

25. Kishan and Nandan started a joint firm. Kishan's investment was thrice the investment of Nandan and the period of his investment was two times the period of investment of Nandan. Nandan got Rs 4000 as profit for his investment. Their total profit if the distribution of profit is directly proportional to the period and amount, is

a) Rs 24000 b) Rs 16000 c) Rs 28000 d) Rs 20000

(Bank PO Exam, 1989)

Answers

1. a; **Hint:** Sum invested by A, B and C is in the ratio of

$5\times 12 : 7\times 12 : 6\times 6 + 3\times 6$

or, 60 : 84 : 54 or 10 : 14 : 9

$\therefore$ share of $C = \frac{9}{33}\times 33000 = \text{Rs } 9{,}000$

2. b; **Hint:** Ratio of their investments

$= 25{,}000\times 1 + 35000\times 1 + 45000\times 1 : 35000\times 2 : 35000\times 1$

$= 3 : 2 : 1$

$\therefore$ Rakesh's share $= \frac{2}{6}\times 150000 = \text{Rs } 50000.$

3. c

4. a; **Hint:** A's investment for 12 months = Rs 4000 × 12 = Rs 48000

B's investment (Rs 6000 for 4 months and Rs 8000 for 8 months) = (Rs 6000 × 4 + Rs 8000 × 8) = Rs 88000

C's investment (Rs 8000 for 9 months and Rs 6000 for 3 months) = (Rs 8000 × 9 + Rs 6000 × 3) = Rs 90,000.

Ratio of their incomes = A : B : C

= 48000 : 88000 : 90000

= 48 : 88 : 90 = 24 : 44 : 45

5. d; **Hint:** A's capital : B's capital = 2 : 3

Also, B's capital : C's capital $= 2:5 = 1:\frac{5}{2} = 3:\frac{15}{2}$

$\therefore$ A's capital : B's capital : C's capital $= 2:3:\frac{15}{2}$

$= 4:6:15$

Since 4 + 6 + 5 = 25

$\therefore$ A's share of the profit $= \frac{4}{25}\times$ Rs 1000 = Rs 160

B's share of the profit $= \frac{6}{25}\times$ Rs 1000 = Rs 240

C's share of the profit $= \frac{15}{25} \times$ Rs 1000 = Rs 600

6. a 7. a 8. c

9. d; **Hint:** The amount which A receives for managing

$= 12\frac{1}{2}\%$ of Rs 880 $= \frac{1}{8}$ of Rs 880 = Rs 110

The amount left to be divided in the ratio of 5 : 6
= Rs 880 - Rs 110 = Rs 770

Now, A's share $= \frac{5}{11}$ of Rs 770 = Rs 350

B's share $= \frac{6}{11}$ of Rs 770 = Rs 420

∴ A's total share = Rs 350 + Rs 110 = Rs 460

Note: A **working partner** is one who manages the business. A **sleeping partner** is one who provides capital but doesnot attend to the business.

10. a 11. c 12. b

13. d; **Hint:** Ratio of profits of A and B are

A's profit : B's profit $= 700\times3 + 500\times3 + 620\times6 : 600\times12$

$= 2100 + 1500 + 3720 : 7200$

$= 7320 : 7200 = 61 : 60$

∴ A's share in profit $= \frac{726}{60+61}\times61 = \frac{726}{121}\times61$

$= 6\times61 =$ Rs 366

14. a; **Hint:** A's profit : B's profit

$= 2x\times3 + \frac{4x}{3}\times4 + \frac{14x}{9}\times5 : x\times7 + \frac{3x}{4}\times5$

$= 6x + \frac{16x}{3} + \frac{70x}{9} : 7x + \frac{15x}{4} = \frac{172x}{9} : \frac{43x}{4}$

$= 16 : 9$

[We suppose that A puts Rs 2x and B puts Rs x in the business].

15. a 16. c 17. a

18. c; **Hint:** Ratio of capitals of A, B and C

$= (16000\times3 + 11000\times9) : (12000\times3 + 17000\times9) : (21000\times6) = 7 : 9 : 6$

B's share = Rs $\left(26400\times\frac{9}{22}\right)$ = Rs 10800

C's share = Rs $\left(26400\times\frac{6}{22}\right)$ = Rs 7200

∴ B's share exceeds C's share by Rs 3600

19. d 20. a

21. d; **Hint:** Ratio of rents = 6 : 10 : 12 = 3 : 5 : 6

∴ Pandey ji's share of rent = Rs $\left(350\times\frac{6}{14}\right)$ = Rs 150

22. d 23. d 24. a

25. c; **Hint:** Let Nandan's investment be Rs x and y months.
Then, Kishan's investment is Rs 3x for 2y months
∴ Ratio of their investments = xy : 6xy = 1 : 6
Nandan's share = Rs 4000
So, Kishan's share = Rs 24000
∴ Total profit = Rs 28000

Rule 2

Theorem: ***The formula for compound partnership can also be written as***

$$\frac{A's\ Capital\times A's\ Time\ in\ partnership}{B's\ Capital\times B's\ Time\ in\ partnership} = \frac{A's\ Profit}{B's\ Profit}.$$

Illustrative Examples

Ex. 1: A began a business with Rs 450 and was joined afterwards by B with Rs 300. When did B join if the profits at the end of the year were divided in the ratio 2 : 1?

Soln: Suppose B joined the business for x months.
Then using the above formula, we have

$\frac{450\times12}{300\times x} = \frac{2}{1}$ or, $300\times2x = 450\times12$

∴ $x = \frac{450\times12}{2\times300} = 9$ months.

Therefore, B joined after (12 – 3) = 3 months.

Ex. 2: A and B rent a pasture for 10 months. A puts in 100 cows for 8 months. How many can B put in for the remaining 2 months, if he pays half as much again as A?

Soln: Suppose B puts in x cows.

The ratio of A's and B's rents $= 1 : 1 + \frac{1}{2} = 1 : \frac{3}{2} = 2 : 3$

Then, $\frac{100\times8}{x\times2} = \frac{2}{3}$ or, $x = \frac{100\times8\times3}{2\times2} = 600$ cows

Ex. 3: A and B enter into a partnership with their capitals in the ratio 7 : 9. At the end of 8 months, A withdraws his capital. If they receive the profits in the ratio 8 : 9, find how long B's capital was used.

Soln: Suppose B's capital was used for x months. Following the above rule, we have, $\frac{7\times8}{9\times x} = \frac{8}{9}$

or, $x = \frac{7\times8\times9}{8\times9} = \frac{8}{9} = 7$

Therefore, B's capital was used for 7 months.

Exercise

1. A began a business with Rs 400 and was joined afterwards by B with Rs 240. When did B join if the profits at the end of the year were divided in the ratio 4 : 1?
   a) 5 months b) 8 months
   c) 6 months d) 10 months

2. A began a business with Rs 300 and was joined afterwards by B with Rs 150. When did B join if the profits at the end of the year were divided in the ratio 3 : 1?
a) 4 months b) 6 months c) 8 months d) 5 months
3. A began a business with Rs 550 and was joined afterwards by B with Rs 330. When did B join if the profits at the end of the year were divided in the ratio 10 : 3?
a) 5 months b) 6 months c) 9 months d) 3 months
4. A and B hire a meadow for 6 months. A puts in 21 cows for 4 months, how many can B put in for the remaining 2 months, if he pays $\frac{5}{7}$ of what A pays?
a) 20 b) 40 c) 35 d) 30
5. A and B rent a pasture for 12 months. A puts in 120 cows for 9 months. How many can B put in for the remaining 3 months, if he pays $\frac{1}{3}$ as much again as A?
a) 840 b) 480 c) 460 d) 640
6. A and B rent a pasture for 12 months. A puts in 350 oxen for 5 months. How many can B put in for the remaining 7 months, if he pays $\frac{2}{5}$ as much again as A?
a) 530 b) 350 c) 450 d) 600
7. A and B enter into a partnership with their capitals in the ratio 5 : 7. At the end of 6 months, A withdraws his capital. If they receive the profits in the ratio 6 : 7, find how long B's capital was used?
a) 5 months b) 7 months c) 3 months d) 9 months
8. A and B enter into a partnership with their capitals in the ratio 3 : 5. At the end of 4 months, A withdraws his capital. If they receive the profits in the ratio 4 : 5, find how long B's capital was used?
a) 2 months b) 5 months c) 4 months d) 3 months
9. A and B enter into a partnership with their capitals in the ratio 5 : 9. At the end of 8 months, A withdraws his capital. If they receive the profits in the ratio 4 : 9, find how long B's capital was used?
a) 10 months b) 9 months c) 8 months d) 4 months
10. A started a business by investing Rs 2700. After sometime B joined him by investing Rs 2025. At the end of one year, the profit was divided in the ratio 2 : 1. After how many months did B join the business?
[UDC Exam, 1984]
a) 4 months b) 6 months c) 3 months d) 2 months

Answers

1. a 2. c 3. b 4. d 5. b 6. b 7. a
8. d 9. a
10. a; **Hint:** Let 'B' joined after 'x' months, then applying the given formula, we have $\frac{2700\times12}{2025\times(12-x)}=\frac{2}{1}$.
$\therefore x=4$

Rule 3

If investments are in the ratio of a : b : c and the timing of their investments in the ratio of x : y : z then the ratio of their profits are in the ratio of ax : by : cz.

Illustrative Example

Ex.: A, B and C invested capitals in the ratio 2 : 3 : 5; the timing of their investments being in the ratio 4 : 5 : 6. In what ratio would their profit be distributed?

Soln : We should know that if the three investments be in the ratio a : b : c and the duration for their investments be in the ratio x : y : z, then the profit would be distributed in the ratio ax : by : cz.

Thus, following the same rule, the required ratio
$$=2\times4:3\times5:5\times6=8:15:30$$

Exercise

1. A, B and C invested capitals in the ratio 1 : 2 : 3; the timing of their investments being in the ratio 1 : 2 : 3. In what ratio would thier profit be distributed?
a) 3 : 2 : 1 b) 1 : 2 : 3 c) 1 : 4 : 9 d) 9 : 4 : 1
2. A, B and C invested capitals in the ratio 2 : 5 : 7; the timing of their investments being in the ratio 3 : 4 : 5. In what ratio would thier profit be distributed?
a) 2 : 10 : 15 b) 15 : 10 : 2 c) 6 : 20 : 35 d) 6 : 20 : 15
3. A, B and C invested capitals in the ratio 3 : 5 : 9; the timing of their investments being in the ratio 2 : 3 : 1. In what ratio would thier profit be distributed?
a) 6 : 15 : 8 b) 3 : 5 : 2 c) 2 : 5 : 3 d) 2 : 3 : 5
4. A, B and C invested capitals in the ratio 7 : 6 : 5; the timing of their investments being in the ratio 4 : 7 : 9. In what ratio would thier profit be distributed?
a) 9 : 14 : 15 b) 28 : 42 : 45 c) 15 : 14 : 9 d) 28 : 24 : 54
5. A, B and C invested capitals in the ratio 3 : 4 : 9; the timing of their investments being in the ratio 9 : 6 : 7. In what ratio would thier profit be distributed?
a) 9 : 8 : 21 b) 27 : 25 : 63 c) 27 : 24 : 36 d) 9 : 8 : 12

Answers

1. c 2. c 3. c 4. b 5. a

Rule 4

Theorem: ***If investments are in the ratio a : b : c and profits in the ratio p : q : r, then the ratio of time*** $=\frac{p}{a}:\frac{q}{b}:\frac{r}{c}$.

Illustrative Example

Ex.: A, B and C invested capitals in the ratio 5 : 6 : 8. At the end of the business term, they received the profits in the ratio 5 : 3 : 12. Find the ratio of time for which they contributed their capitals.

Soln: Using the above formula, we have

the required ratio $=\frac{5}{5}:\frac{3}{6}:\frac{12}{8}=1:\frac{1}{2}:\frac{3}{2}=2:1:3$

Exercise

1. A, B and C invested capitals in the ratio 6 : 9 : 10. At the end of the business term, they received the profits in the ratio 2 : 3 : 5. Find the ratio of time for which they contributed their capitals.
   a) 2 : 2 : 3 b) 12 : 27 : 50 c) 1 : 1 : 2 d) 2 : 3 : 3
2. A, B and C invested capitals in the ratio 1 : 2 : 3. At the end of the business term, they received the profits in the ratio 1 : 2 : 3. Find the ratio of time for which they contributed their capitals.
   a) 1 : 1 : 1 b) 1 : 2 : 3
   c) 1 : 4 : 9 d) Can't be possible
3. A, B and C invested capitals in the ratio 4 : 5 : 6. At the end of the business term, they received the profits in the ratio 2 : 3 : 4. Find the ratio of time for which they contributed their capitals.
   a) 6 : 5 : 8 b) 6 : 5 : 9 c) 10 : 12 : 9 d) 12 : 10 : 9
4. A, B and C invested capitals in the ratio 4 : 6 : 9. At the end of the business term, they received the profits in the ratio 2 : 3 : 5. Find the ratio of time for which they contributed their capitals.
   a) 1 : 1 : 9 b) 2 : 2 : 9 c) 10 : 10 : 9 d) 5 : 5 : 9
5. A, B and C invested capitals in the ratio 7 : 3 : 2. At the end of the business term, they received the profits in the ratio 2 : 3 : 7. Find the ratio of time for which they contributed their capitals.
   a) 49 : 41 : 14 b) 49 : 14 : 41 c) 49 : 14 : 4 d) 49 : 41 : 4

Answers

1\. a 2. a 3. d 4. c 5. c

Rule 5

Theorem: ***Three partners invest their capitals in a business. If the ratio of their periods of investments are $t_1 : t_2 : t_3$ and their profits are in the ratio of a : b : c, then the capitals will be in the ratio of $\frac{a}{t_1} : \frac{b}{t_2} : \frac{c}{t_3}$.***

Illustrative Example

Ex.: A, B and C invest their capitals in a business. If the ratio of their periods of investments are 2 : 3 : 4 and their profits are in the ratio of 4 : 9 : 8. Find the ratio in which the investments are made by A, B and C.

Soln: Applying the above rule, we have

$$\text{the required answer} = \frac{4}{2} : \frac{9}{3} : \frac{8}{4} = 2 : 3 : 2$$

Exercise

1. A, B and C invest their capitals in a business. If the ratio of their periods of investments are 1 : 2 : 3 and their profits are in the ratio of 2 : 3 : 1. Find the ratio in which the investments are made by A, B and C.
   a) 12 : 9 : 2 b) 2 : 9 : 12 c) 6 : 3 : 1 d) 1 : 3 : 6
2. A, B and C invest their capitals in a business. If the ratio of their periods of investments are 2 : 3 : 4 and their profits are in the ratio of 4 : 3 : 2. Find the ratio in which the investments are made by A, B and C.
   a) 1 : 2 : 4 b) 4 : 2 : 1 c) 3 : 2 : 1 d) 4 : 2 : 3
3. A, B and C invest their capitals in a business. If the ratio of their periods of investments are 2 : 3 : 6 and their profits are in the ratio of 4 : 5 : 6. Find the ratio in which the investments are made by A, B and C.
   a) 9 : 10 : 12 b) 4 : 5 : 6 c) 8 : 5 : 12 d) 6 : 5 : 3
4. A, B and C invest their capitals in a business. If the ratio of their periods of investments are 7 : 3 : 5 and their profits are in the ratio of 2 : 1 : 2. Find the ratio in which the investments are made by A, B and C.
   a) 30 : 35 : 42 b) 7 : 6 : 10 c) 42 : 30 : 35 d) 42 : 25 : 35

Answers

1\. a 2. b 3. d 4. a

Rule 6

Theorem: ***A, B and C are partners. A receives $\frac{x}{y}$ of the profit and B and C share the remaining profit equally. If A's income is increased by Rs 'A' when the profit rises from P% to Q%, then the capital invested by A is $\left(Rs\ \frac{100 \times A}{Q-P}\right)$ and the capital invested by B and C is Rs $\left[\frac{\left(\frac{100 \times A}{Q-P}\right)\left(1-\frac{x}{y}\right)}{2}\right]$ or $\left(\frac{A's\ Capital}{2}\right)\left(1-\frac{x}{y}\right)$.***

Illustrative Example

Ex.: A, B and C are partners. A receives $\frac{2}{5}$ of the profit and B and C share the remaining profit equally. A's income is increased by Rs 220 when the profit rises from 8% to 10%. Find the capitals invested by A, B and C.

Soln: **Detail Method:** For A's share: (10% - 8%) $\equiv$ Rs 220

$\therefore$ 100% $\equiv \frac{220}{2} \times 100$ = Rs 11000

$\therefore$ A's capital = Rs 11000

For B's & C's share: $\frac{2}{5} \equiv 11000$

$\therefore \frac{3}{5} \equiv \frac{11000}{2} \times 3$ = Rs 16500

$\therefore$ B's and C's capitals are Rs 8250 each.

Quicker Method: Applying the above rule, we have,

$$\text{A's capital} = \frac{100\times 220}{10-8} = \text{Rs } 11000$$

$$\text{B's and C's capitals} = \frac{\frac{100\times 220}{10-8}\times\left(1-\frac{2}{5}\right)}{2}$$

$$= \text{Rs } 8250 \text{ each.}$$

Exercise

1. A, B and C are partners. A receives $\frac{2}{7}$ of the profit and B and C share the remaining profit equally. A's income is increased by Rs 240 when the profit rises from 10% to 15%. Find the capitals invested by B and C.
   a) Rs 2400 b) Rs 1200 c) Rs 4800 d) Rs 6000
2. A, B and C are partners. A receives $\frac{5}{8}$ of the profit and B and C share the remaining profit equally. A's income is increased by Rs 450 when the profit rises from 4% to 9%. Find the capitals invested by B and C.
   a) Rs 3366 b) Rs 1687.5 c) Rs 3475 d) Rs 3466
3. A, B and C are partners. A receives $\frac{9}{10}$ of the profit and B and C share the remaining profit equally. A's income is increased by Rs 270 when the profit rises from 12% to 15%. Find the capitals invested by B and C.
   a) Rs 5000 b) Rs 1000 c) Rs 500 d) Rs 1500
4. A, B and C are partners. A receives $\frac{5}{9}$ of the profit and B and C share the remaining profit equally. A's income is increased by Rs 225 when the profit rises from 8% to 13%. Find the capitals invested by B and C.
   a) Rs 1000 b) Rs 2000 c) Rs 2500 d) Rs 950
5. A, B and C are partners. A receives $\frac{10}{13}$ of the profit and B and C share the remaining profit equally. A's income is increased by Rs 550 when the profit rises from 11% to 13%. Find the capitals invested by B and C.
   a) Rs 4125 b) Rs 4215 c) Rs 4251 d) Rs 4512

Answers

1. d 2. b 3. c 4. a 5. a

Rule 7

Two partners invest Rs x and Rs y respectively in a business and agree that P% of the profit should be divided equally between them and the remaining profit is to be treated as interest on capital. If one partner gets Rs X more than the other then to find the total profit made in the business we adopt the following steps.

Step I: First find the ratio of the profit ie

Rs x : Rs y = A : B (say)

Step II: Now, apply this formula to calculate the total profit.

$$\text{Total profit} = \text{Rs } X\left(\frac{100}{100-P}\right)\left(\frac{A+B}{A-B}\right)$$

Illustrative Example

Ex.: Two partners invest Rs 125,000 and Rs 85,000 respectively in a business and agree that 60% of the profit should be divided equally between them and the remaining profit is to be treated as interest on capital. If one partner gets Rs 300 more than the other, find the total profit made in the business.

Soln: Detail Method: The difference counts only due to the 40% of the profit which was distributed according to their investments.

Let the total profit be Rs x.

Then 40% of x is distributed in the ratio 125,000 : 85,000 = 25 : 17

Therefore, the share of the first partner

$$= 40\% \text{ of } x\left(\frac{25}{25+17}\right)$$

$$= 40\% \text{ of } x\left(\frac{25}{42}\right) = \frac{40x}{100}\left(\frac{25}{42}\right) = \frac{5x}{21}$$

and the share of the second partner

$$= 40\% \text{ of } x\left(\frac{17}{42}\right) = \frac{17x}{105}$$

Now, from the question,

$$\text{the difference in share} = \frac{5x}{21} - \frac{17x}{105} = 300$$

$$\text{or, } \frac{x(25-17)}{105} = 300$$

$$\therefore x = \frac{300\times 105}{8} = \text{Rs } 3937.50$$

Quicker Method: Applying the above rule, we have,

Step I: The ratio of profit = 125,000 : 85,000 = 25 : 17

Step II: total profit $= 300\left(\frac{100}{40}\right)\left(\frac{25+17}{25-17}\right)$

$= \text{Rs } 3937.50$

Exercise

1. Two partners invest Rs 24750 and Rs 16500 respectively in a business and agree that 20% of the profit should be divided equally between them and the remaining profit is to be treated as interest on capital. If one partner gets Rs 400 more than the other, find the total profit made in

the business.

a) Rs 5000 b) Rs 2500 c) Rs 3500 d) Rs 4500

2. Two partners invest Rs 26000 and Rs 16250 respectively in a business and agree that 40% of the profit should be divided equally between them and the remaining profit is to be treated as interest on capital. If one partner gets Rs 450 more than the other, find the total profit made in the business.

a) Rs 3250 b) Rs 3520 c) Rs 3230 d) Rs 3200

3. Two partners invest Rs 17000 and Rs 13000 respectively in a business and agree that 75% of the profit should be divided equally between them and the remaining profit is to be treated as interest on capital. If one partner gets Rs 532 more than the other, find the total profit made in the business.

a) Rs 16960 b) Rs 14960 c) Rs 16950 d) Rs 15960

4. Two partners invest Rs 6280 and Rs 3768 respectively in a business and agree that 30% of the profit should be divided equally between them and the remaining profit is to be treated as interest on capital. If one partner gets Rs 700 more than the other, find the total profit made in the business.

a) Rs 4000 b) Rs 3800 c) Rs 4125 d) Rs 4500

5. Two partners invest Rs 87000 and Rs 72500 respectively in a business and agree that 25% of the profit should be divided equally between them and the remaining profit is to be treated as interest on capital. If one partner gets Rs 240 more than the other, find the total profit made in the business.

a) Rs 3250 b) Rs 3520 c) Rs 3620 d) Rs 3720

Answers

1. b 2. a 3. d 4. a 5. b

Rule 8

Two partners A and B invests in the ratio x : y in a business. If P% of the total profit goes to charity and A's share is Rs X, then the total profit is given by Rs $X\left(\frac{100}{100-P}\right)\left(1+\frac{y}{x}\right)$.

Illustrative Example

Ex.: A and B invested in the ratio 3 : 2 in a business. If 5% of the total profit goes to charity and A's share is Rs 855, find the total profit.

Soln: **Detail Method:** Suppose the total profit is Rs 100.

Then Rs 5 goes to charity.

Now, Rs 95 is divided in the ratio 3 : 2.

$$\text{A's share} = \frac{95}{3+2}\times 3 = \text{Rs } 57$$

But we see that A's actual share is Rs 855.

$$\text{Actual total profit} = 855\left(\frac{100}{57}\right) = \text{Rs } 1500$$

Quicker Method: Applying the above rule, we have

$$\text{the total profit} = 855\left(\frac{100}{100-5}\right)\left(\frac{3+2}{3}\right)$$

$$= 855\left(\frac{100}{95}\right)\left(\frac{5}{3}\right) = \text{Rs } 1500.$$

Exercise

1. A and B invested in the ratio 5 : 3 in a business. If 10% of the total profit goes to charity and A's share is Rs 900, find the total profit.

a) Rs 1600 b) Rs 1400 c) Rs 1500 d) Rs 1800

2. A and B invested in the ratio 4 : 9 in a business. If 8% of the total profit goes to charity and A's share is Rs 460, find the total profit.

a) Rs 2625 b) Rs 2526 c) Rs 1526 d) Rs 1625

3. A and B invested in the ratio 5 : 7 in a business. If 7% of the total profit goes to charity and A's share is Rs 1860, find the total profit.

a) Rs 8400 b) Rs 4600 c) Rs 4850 d) Rs 4800

4. A and B invested in the ratio 5 : 4 in a business. If 4% of the total profit goes to charity and A's share is Rs 480, find the total profit.

a) Rs 900 b) Rs 800 c) Rs 850 d) Rs 950

5. A and B invested in the ratio 5 : 2 in a business. If 12% of the total profit goes to charity and A's share is Rs 770, find the total profit.

a) Rs 1325 b) Rs 1235 c) Rs 1225 d) Rs 2125

Answers

1. a 2. d 3. d 4. a 5. c

Rule 9

Formula: The ratio of investments = Ratio of Profits

Illustrative Example

Ex.: Three partners altogether invested Rs 114,000 in a business. At the end of the year, one got Rs 337.50, the second Rs 1125.00 and the third Rs 675 as profit. How much amount did each invest? What is the percentage of profit?

Soln: The ratio of investments = Ratio of profits

$$= 337.50 : 1125 : 675$$

$$= 3375 : 11250 : 6750$$

Dividing each by 1125, we have the ratio = 3 : 10 : 6.

$$\text{Shares of the partners} = \text{Rs } \frac{114000}{3+10+6}\times 3,$$

$$\text{Rs } \frac{114000}{3+10+6}\times 10 \text{ and Rs } \frac{114000}{3+10+6}\times 6$$

or, Rs 18000, Rs 60000 and Rs 36000

The required percentage of profit

$$= \frac{337.5+1125+675}{114000}\times 100 = \frac{2137.50}{1140} = 1.857\%$$

Exercise

1. Three partners A, B and C together invested Rs 375000 in a business. At the end of the year, A got Rs 52000, B got Rs 65000 and C got Rs 78000 as profit. How much amount did A invest?
a) Rs 100000 b) Rs 125000
c) Rs 150000 d) Rs 160000
2. Three partners A, B and C together invested Rs 14400 in a business. At the end of the year, A got Rs 1250, B got Rs 2500 and C got Rs 3750 as profit. How much amount did C invest?
a) Rs 2400 b) Rs 4800 c) Rs 7200 d) Rs 9600
3. Three partners A, B and C together invested Rs 36000 in a business. At the end of the year, A got Rs 4200, B got Rs 7000 and C got Rs 9800 as profit. How much amount did B invest?
a) Rs 7200 b) Rs 12000 c) Rs 16800 d) Rs 12500
4. Three partners A, B and C together invested Rs 22500 in a business. At the end of the year, A got Rs 2800, B got Rs 8400 and C got Rs 14000 as profit. How much amount did A invest?
a) Rs 2500 b) Rs 7500 c) Rs 12500 d) Rs 5000
5. Three partners A, B and C together invested Rs 21600 in a business. At the end of the year, A got Rs 700, B got Rs 800 and C got Rs 900 as profit. How much amount did B and C together invest?
a) Rs 15300 b) Rs 13500 c) Rs 14400 d) Rs 16300
6. Three partners A, B, C agree to divide the profit or losses in the ratio 1.50 : 1.75 : 2.25. If, in a particular year, they earn a profit of Rs 66000, find the share of B.
a) Rs 21000 b) Rs 27000 c) Rs 18000 d) Rs 22000
(LIC 1991)
7. X and Y invested in a business. They earned some profit and divided in the ratio 2 : 3. If X invested Rs 40, find the money invested by Y.
a) Rs 20 b) Rs 30 c) Rs 80 d) Rs 60
(Railways 1991)
8. Jayant started a business, investing Rs 6000. Six months later Madhu joined him, investing Rs 4000. If they made a profit of Rs 5200 at the end of the year, how much must be the share of Madhu?
a) Rs 1300 b) Rs 3900 c) Rs 3600 d) Rs 1200
(Bank PO 1991)
9. Dilip, Ram and Amar started a shop by investing Rs 2700, Rs 8100 and Rs 7200 respectively. At the end of one year, the profit was distributed. If Ram's share was Rs 3600, their total profit was **(Bank PO Exam 1988)**
a) Rs 10800 b) Rs 11600
c) Rs 8000 d) None of these

Answers

1. a 2. c 3. b 4. a 5. a 6. a
7. d 8. a

9. c; **Hint:** Ratio of shares of Dilip, Ram and Amar
$= 2700 : 8100 : 7200 = 3 : 9 : 8$
If Ram's share is Rs 9, total profit = Rs 20
If Ram's share is Rs 3600, total profit = Rs $\left(\frac{20}{9} \times 3600\right)$
= Rs 8000

Rule 10

Two partners A and B start a business. A puts in Rs 'A' more in the business than B, but B has invested his capital for T_2 periods while A has invested his capital for T_1 periods. If the share of A is Rs 'a' more than that of B out of the total profits of Rs 'P', then the capital contributed by B is

$$Rs\left[\frac{A}{\left\{\frac{(P+a)T_2}{(P-a)T_1}\right\}-1}\right]$$

and the capital contributed by A is given by [Rs A + Capital of B].

Illustrative Example

Ex.: A puts Rs 600 more in a business than B, but B has invested his capital for 5 months while A has invested his for 4 months. If the share of A is Rs 48 more than that of B out of the total profits of Rs 528, find the capital contributed by each?

Soln: Detail Method:

B's profit = Rs $\frac{528-48}{2}$ = Rs 240

A's profit = Rs 240 + Rs 48 = Rs 288

∴ A's profit per month = Rs $\frac{288}{4}$ = Rs 72

B's profit per month = Rs $\frac{240}{5}$ = Rs 48

Now, their capitals are proportional to their profits.
∴ A's capital : B's capital = 72 : 48 = 3 : 2
∴ If the difference between their capitals be Re 1, then A's capital is Rs 3 and B's capital is Rs 2. But the actual difference is Rs 600.
∴ A's capital = Rs 600 × 3 = Rs 1800
B's capital = Rs 600 × 2 = Rs 1200

Quicker Method: Applying the above rule, we have B's capital

$$= \frac{600}{\left\{\frac{(528+48)5}{(528-48)4}\right\}-1} = \frac{600}{\frac{576\times5}{480\times4}-1} = \frac{600}{\frac{3}{2}-1} = \text{Rs } 1200$$

A's capital = Rs 1200 + Rs 600 = Rs 1800.

Exercise

1. A puts Rs 795 more in a business than B, but B has invested his capital for 6 months while A has invested his for 5 months. If the share of A is Rs 33 more than that of B out of the total profits of Rs 432, find the total capital put in the business.
a) Rs 2000 b) Rs 2790 c) Rs 4785 d) Rs 7490
2. A puts Rs 550 more in a business than B, but B has invested his capital for 4 months while A has invested his for 3 months. If the share of A is Rs 45 more than that of B out of the total profits of Rs 235, find the capital contributed by A?
a) Rs 1120 b) Rs 550 c) Rs 570 d) Rs 1210
3. A puts Rs 375 more in a business than B, but B has invested his capital for 4 months while A has invested his for 8 months. If the share of A is Rs 75 more than that of B out of the total profits of Rs 125, find the capital contributed by B?
a) Rs 750 b) Rs 375 c) Rs 735 d) Rs 573
4. A puts Rs 768 more in a business than B, but B has invested his capital for 7 months while A has invested his for 4 months. If the share of A is Rs 42 more than that of B out of the total profits of Rs 358, find the capital contributed by B?
a) Rs 642 b) Rs 1400 c) Rs 632 d) Rs 462
5. A puts Rs 280 more in a business than B, but B has invested his capital for 3 months while A has invested his for 4 months. If the share of A is Rs 55 more than that of B out of the total profits of Rs 245, find the total capital in the business.
a) Rs 1520 b) Rs 1800 c) Rs 3320 d) Data inadequate

Answers

1. c 2. a 3. b 4. c 5. c

Miscellaneous

1. A, B and C can do a work in 20, 25 and 30 days respectively. They undertook to finish the work together for Rs 2220, then the share of A exceeds that of B by
a) Rs 120 b) Rs 180 c) Rs 300 d) Rs 600
(Central Excise, 1989)
2. A, B and C contract a work for Rs 550. Together A and B are to do $\frac{7}{11}$ of the work. The share of C should be
a) Rs 400 b) Rs 300 c) Rs 200 d) Rs $183\frac{1}{3}$
(Clerical Grade, 1991)
3. A, B and C enter into partnership by making investments in the ratio 3 : 5 : 7. After a year, C invests another Rs 337600 while A withdraws Rs 45600. The ratio of investments then changes to 24 : 59 : 167. How much did A invest initially?
a) Rs 45600 b) Rs 96000 c) Rs 141600 d) None of these
(LIC, AAO Exam 1988)
4. Three hikers A, B and C start on a trip with Rs 50 each and agree to share the expenses equally. If at the end of the trip, A has Rs 20 left with him, B Rs 30 and C Rs 40, how must they settle their accounts?
a) C must pay Rs 10 to A b) A must pay Rs 10 to B c) A must pay Rs 10 to C d) Can't be settled
(ITI 1985)
5. A and B invest Rs 3000 and Rs 4000 in a business. A receives Rs 10 per month out of the profit as a remuneration for running the business and the rest of profit is divided in proportion to the investments. If in a year 'A' totally receives Rs 390, what does B receive?
a) Rs 630 b) Rs 360 c) Rs 480 d) Rs 380
(IAS 1980)
6. A sum of money is to be divided among A, B and C in the ratio 2 : 3 : 7. If the total share of A and B together is Rs 1500 less than C, what is A's share in it?
a) Rs 1000 b) Rs 1500 c) Rs 2000 d) Data inadequate
(Bank PO Exam, 1989)

Answers

1. b; Ratio of shares = Ratio of 1 days's work = $\frac{1}{20}:\frac{1}{25}:\frac{1}{30}$

$= 15 : 12 : 10.$

$\therefore$ A's share = Rs $\left(2220\times\frac{15}{37}\right)$ = Rs 900

B's share = Rs $\left(2220\times\frac{12}{37}\right)$ = Rs 720

$\therefore$ A's share exceeds B's share = Rs 180

2. c; Do yourself.

3. c; Let the initial investments of A, B, C be Rs $3x$, Rs $5x$ and Rs $7x$ respectively. Then,

$(3x-45600):5x:(7x+337600)=24:59:167$

$\therefore \frac{3x-45600}{5x}=\frac{24}{59}\Rightarrow x=47200$

$\therefore$ A invested initially Rs (47200×3) = Rs 141600

4. a; Do yourself.

5. b; Total profit - Remuneration = Balance profit. This balance profit is divided in proportion to their investments.

$\frac{\text{Balance profit of A}}{\text{Balance profit of B}}=\frac{\text{Investment of A}}{\text{Investment of B}}$

or, $\frac{390-10\times 12}{\text{Balance profit of B}}=\frac{3000}{4000}=\frac{3}{4}$

(Since remuneration of A is Rs 10 per month)

or, Balance profit of B = $4\times\frac{270}{3}$ = Rs 360

Since B does not get any remuneration, hence B receives Rs 360 at the end of the year.

6. b; $7x-(2x+3x)$ = Rs 1500

$\therefore x$ = Rs 750 $\therefore$ A = $2x$ = Rs 1500

5 Percentage

Rule 1

To express any fraction $\frac{x}{y}$ in rate per cent, we multiply $\frac{x}{y}$ by 100 ie $\frac{x}{y} \times 100\%$.

Illustrative Example

Ex.: What percentage is equivalent to $\frac{3}{8}$?

Soln: $\frac{3}{8} \times 100 = \frac{75}{2} = 37\frac{1}{2}\%$

Exercise

1. What percentage is equivalent to $\frac{3}{4}$?

   a) 25% b) 75% c) 50% d) 125%

2. What percentage is equivalent to $\frac{5}{8}$?

   a) $45\frac{1}{2}\%$ b) $62\frac{1}{2}\%$ c) $22\frac{1}{2}\%$ d) $37\frac{1}{2}\%$

3. What percentage is equivalent to $\frac{17}{25}$?

   a) 68% b) 34% c) 85% d) 51%

4. What percentage is equivalent to $\frac{33}{35}$?

   a) 94% b) 94.18% c) 94.28% d) 94.38%

5. Give percentage equivalent to $\frac{11}{13}$.

   a) $64\frac{8}{13}\%$ b) $88\frac{8}{13}\%$ c) $88\frac{4}{13}\%$ d) $84\frac{8}{13}\%$

6. $8\frac{1}{3}\%$ expressed as a fraction is

   a) $\frac{25}{3}$ b) $\frac{3}{25}$ c) $\frac{1}{12}$ d) $\frac{1}{4}$

   [CBI Exam, 1989]

Answers

1. b 2. b 3. a 4. c 5. d
6. c; **Hint:** Also see **Rule 2**.

Rule 2

To express rate per cent x as a fraction, we divide x by 100 ie $\frac{x}{100}$ is the required fraction.

Illustrative Example

Ex.: What fraction is $12\frac{1}{2}$ per cent?

Soln: $12\frac{1}{2}\% = \frac{12\frac{1}{2}}{100} = \frac{25}{200} = \frac{1}{8}$

Exercise

1. What fraction is $22\frac{1}{2}$ per cent?

   a) $\frac{9}{40}$ b) $\frac{6}{40}$ c) $\frac{9}{20}$ d) $\frac{6}{20}$

2. What fraction is 15 per cent?

   a) $\frac{9}{20}$ b) $\frac{7}{20}$ c) $\frac{3}{20}$ d) $\frac{3}{10}$

3. What fraction is $32\frac{4}{5}$ per cent?

   a) $\frac{41}{125}$ b) $\frac{31}{125}$ c) $\frac{21}{145}$ d) $\frac{41}{75}$

4. Express as fractions in their lowest terms.

(i) 10%

a) $\frac{1}{10}$ b) $\frac{1}{5}$ c) $\frac{10}{11}$ d) $\frac{11}{10}$

(ii) $8\frac{1}{3}\%$

a) $\frac{1}{16}$ b) $\frac{3}{16}$ c) $\frac{1}{12}$ d) $\frac{1}{4}$

(iii) 100%

a) 2 b) 1 c) $\frac{1}{100}$ d) Can't be determined

Answers

1. a 2. c 3. a 4. (i) a (ii) c (iii) b

Rule 3

To find x% of an item we have the following formula: $\frac{x}{100}$ **× *Value of the item***

Illustrative Example

Ex.: Find 8% of Rs 625.

Soln: 8% of Rs 625 $= \frac{8}{100}\times 625 = \frac{2}{25}\times 625$ = Rs 50

Exercise

1. Find 4% of Rs 3125.
   a) Rs 250 b) Rs 125 c) Rs 150 d) Rs 75
2. Find 15% of Rs 600.
   a) Rs 90 b) Rs 75 c) Rs 150 d) Rs 125
3. Find $12\frac{1}{2}\%$ of Rs 1000.
   a) Rs 120 b) Rs 175 c) Rs 150 d) Rs 125
4. Find the value of
   (i) 5 per cent of Rs 1400.
   a) Rs 35 b) Rs 70 c) Rs 140 d) Rs 105
   (ii) 7 per cent of Rs 7150.
   a) Rs 500 b) Rs 800 c) Rs 500.50 d) Rs 250
   (iii) 4 per cent of 37 kg 500 g.
   a) 1 kg 500 g b) 2 kg
   c) 2 kg 500 g d) 1 kg
   (iv) 40 per cent of 4 quintals
   a) 1 quintal 50 kg b) 2 quintal 60 kg
   c) 1 quintal 60 kg d) 2 quintal 50 kg
   (v) $66\frac{2}{3}$ per cent of 2 kg 6 hg 3 dag 7 g.
   a) 1 kg 7 hg 5 dag 8 g b) 1 kg 5 hg 7 dag 8 g
   c) 1 kg 8 hg 5 dag 7 g d) 1 kg 7 hg 8 dag 5 g
5. An agent sells goods of value Rs 2764, what is his commission at $12\frac{1}{2}$ per cent?
   a) Rs 355.50 b) Rs 365.50 c) Rs 345.50 d) Rs 445.50
6. What is the duty on goods worth Rs 6200 at $2\frac{1}{2}$ per cent?
   a) Rs 125 b) Rs 135 c) Rs 145 d) Rs 155
7. A rent collector receives $1\frac{1}{2}\%$ for collecting rent. What will he receive for collecting Rs 5000?
   a) Rs 25 b) Rs 50 c) Rs 75 d) Rs 125
8. An auctioneer charges 10% for selling a piano. The sale price is Rs 3455. What is the auctioneer's commission?
   a) Rs 345.50 b) Rs 345
   c) Rs 385.50 d) Rs 385
9. A cask containing 425 litres lost 8% by leakage. How many litres were left in the cask?
   a) 34 litres b) 391 litres c) 334 litres c) 389 litres
10. At an election where there are two candidates only, the candidate who gets 62 per cent of the votes is elected by a majority of 144 votes. Find the total no. of votes recorded.
    a) 1200 b) 1800 c) 700 d) 600
11. At an election where there are two candidates ony, the candidate who gets 43 per cent of the votes is rejected by a majority of 420 votes. Find the total no. of votes recorded.
    a) 3000 b) 600 c) 1200 d) 2400

Answers

1. b 2. a 3. d
4. (i) b (ii) c (iii) a (iv) c (v) a 5. c
6. d 7. c 8. a

9. b; **Hint:** Required answer $= \frac{100-8}{100}\times 425 = 391$ litres

or Amount lost by leakage $= \frac{8}{100}\times 425 = 34$ litres

$\therefore$ required answer = 425 - 34 = 391 litres

10. d; **Hint:** Let the total no of votes be x

Now, according to the question,

62% of x – 38% of x = 144

or $\frac{62x}{100} - \frac{38x}{100} = 144$

$\therefore x = 600$.

11. a; **Hint:** Let the total no. of votes be x. Now according to the question, $\frac{57x}{100} - \frac{43x}{100} = 420$

$\therefore x = 3000$.

Rule 4

If x% of Number (N) is y, then the number (N) = $\frac{y}{x}\times100$,

Note: 1. x% means $\frac{x}{100}$

2. The number whose percentage is to be found is called the **Original Number (N)** and the number obtained after finding the percentage is called the **Result**.

Illustrative Example

Ex.: 25% of what number is 36?

Soln: Using the above formula, we have

required number $= \frac{36}{25}\times100 = 144$.

Exercise

1. 60% of what number is 30?
 a) 50 b) 25 c) 60 d) 75
2. $16\frac{2}{3}$% of what number is 75?
 a) 250 b) 550 c) 450 d) 225
3. $26\frac{2}{3}$% of what number is 164?
 a) 651 b) 615 c) 516 d) 561
4. What is the sum of money of which $3\frac{3}{4}$% is Rs 45?
 a) Rs 1500 b) Rs 1200 c) Rs 1250 d) Rs 1550
5. If 17% of a certain number of mangoes is equal to 1360, what is the number of mangoes?
 a) 8000 b) 6000 c) 8500 d) 7000
6. What is the number, $12\frac{1}{2}$% of which is 64?
 a) 510 b) 312 c) 512 d) 521
7. After spending 69% of her money, a lady has Rs 93 left. How much had she at first?
 a) Rs 300 b) Rs 400 c) Rs 75 d) Rs 150

Answers

1. a 2. c 3. b 4. b 5. a 6. c

7. a; **Hint:** $\frac{93}{100-69}\times100$ = Rs 300

Rule 5

If the x% of a number (N) is the result (R), then the value of percentage (x) = $\frac{Result\ (R)}{Original\ Number\ (N)}\times100$

Note: Here the number whose percentage is to be found is called the **Original Number (N)** and the number obtained after finding the percentage is called the **Result (R).**

Illustrative Example

Ex.: 0.625 is equal to what per cent of $1\frac{7}{28}$?

Soln: Here, $1\frac{7}{28}$ = Original Number (N) and 0.625 = Result (R)

Using the above formula,

value of percentage $= \frac{0.625}{1\frac{7}{28}}\times100 = 50\%$

Exercise

1. Rs $12\frac{1}{2}$ is what per cent of Rs $16\frac{2}{3}$?
 a) 50% b) 25% c) 75% d) 45%
2. What rate per cent is 1 quintal 25 kg of 1 metric tonne?
 a) $12\frac{1}{2}$% b) 25% c) $22\frac{1}{2}$% d) 45%
3. What rate per cent is 6 P of the Re?
 a) 5% b) 6% c) 12% d) 18%
4. Express $\frac{1}{3}$ as a percentage of $\frac{5}{9}$.
 a) 60% b) $\frac{500}{3}$ c) 30% d) 90%

Answers

1. c

2. a; **Hint:** Here Original Number (N) = 1 metric tonne = 1000 kg and Result (R) = 1 quintal 25 kg = 125 kg

$\therefore$ required percentage $= \frac{125}{1000}\times100 = 12\frac{1}{2}\%$

Note: This question is often put thus- "Express the fraction which 1 quintal 25 kg is of 1 metric tonne as a percentage".

3. b; **Hint:** Required percentage

$= \frac{6P}{1\text{Re}}\times100 = \frac{6}{100}\times100 = 6\%$

4. a; **Hint:** Required percentage $= \frac{\frac{1}{3}}{\frac{5}{9}}\times100 = 60\%$

Note: Value given after the word 'of' is the **Original Number (N)** and the other is the **Result (R)**: (**Always Remember**).

Rule 6

If x% and y% of a number (N) are x_1 and y_1 respectively, then following relationship exists $\frac{x_1}{x}=\frac{y_1}{y}=\frac{N}{100}$.

Illustrative Example

Ex.: 25% of a number is 20, what is 40% of that number? Also find the number.

Soln: Using the above relationship, we will solve this problem.

Here, $x_1=20, x=25$

$y=40, y_1=?$

From the above formula

$\frac{x_1}{x}=\frac{y_1}{y} \Rightarrow \frac{20}{25}=\frac{y_1}{40}$

$\therefore y_1=32$ and

Number (N) $=\frac{x_1}{x}\times 100$ (from the above formula)

$=\frac{20}{25}\times 100=80$

Exercise

1. $12\frac{1}{2}\%$ of a number is 20, what is 60% of that number?
   a) 32 b) 64 c) 96 d) 84
2. $16\frac{2}{3}\%$ of a number is 50, what is $26\frac{2}{3}\%$ of that number?
   a) 240 b) 80 c) 160 d) Data inadequate
3. 44% of a number is 275, what is 64% of that number?
   a) 450 b) 400 c) 375 d) 500
4. $22\frac{1}{2}\%$ of a number is 45, what per cent of that number is 90?
   a) 25% b) 65% c) 30% d) 45%
5. 35% of a number is 105, what per cent of that number is 100.
   a) 40% b) $37\frac{1}{2}\%$ c) $33\frac{1}{3}\%$ d) $33\frac{2}{3}\%$

Answers

1. c 2. b 3. b 4. d 5. c

Rule 7

If the value of absolute difference and percentage difference of two items A and B are given, then the total value =

$\frac{\textit{Value of absolute difference}}{\textit{Difference in per cent}}\times 100$

Illustrative Examples

Ex. 1: If the price of one kg of wheat is increased by $12\frac{1}{2}\%$, the increase is Rs 5. Find the original and new price of wheat per kg.

Soln: Here, Total Value = Original price = ?
(Consider original price and increased price as two items)
Value of absolute difference = Difference in price (ie increase in price) = Rs 5
Difference in per cent = per cent increase of price of one kg of wheat = $12\frac{1}{2}\%$.
Now applying the above formula, we have the

original price $=\frac{\text{Difference in price}}{\text{Difference in per cent}}\times 100$

$=\frac{5}{\frac{25}{2}}\times 100=$ Rs 40

$\therefore$ new price (ie increased price) $=40\times\frac{225}{200}=$ Rs 45

Ex. 2: Ram and Mohan appeared in an examination. If the difference of their marks is 60 and percentage difference of their marks is given as 30. Find the full marks for which examination has been held.

Soln: Applying the above formula, we have
Full marks

$=\frac{\text{Difference of their marks}}{\text{Percentage difference of their marks}}\times 100$

$=\frac{60}{30}\times 100=200$

Exercise

1. If the price of one kg of rice is increased by 25%, the increase is Rs 12. Find the new price of rice per kg.
   a) Rs 48 b) Rs 60 c) Rs 72 d) Rs 36
2. If the price of pen is increased by $7\frac{1}{2}\%$, the increase is Rs 15. Find the original price of pen.
   a) Rs 230 b) Rs 200 c) Rs 100 d) Rs 115
3. If the price of a pencil is decreased by $16\frac{2}{3}\%$ and the decrease is Rs 3, find the new price of pencil.
   a) Rs 18 b) Rs 21 c) Rs 24 d) Rs 16
4. A and B appeared in an examination. If the difference of their marks is 25 and percentage difference of their marks

is given as 20%. Find the full marks for which examination has been held.

a) 125 b) 100 c) 80 d) 120

5. A and B appeared in an examination. If the difference of their marks is 35 and percentage difference of their marks is given as 7%. Find the full marks for which examination has been held.

a) 500 b) 700 c) 350 d) 450

Answers

1. b 2. b 3. b 4. a 5. a

Rule 8

Theorem: ***If one type of liquid contains X% of milk, the other contains Y% of milk. A can is filled with x parts of the first liquid and y parts of the second liquid. Then, percentage of milk in the new mixture is*** $\frac{xX+yY}{x+y}$.

Illustrative Example

Ex.: One type of liquid contains 25% of milk, the other contains 30% of milk. A can filled with 6 parts of the first liquid and 4 parts of the second liquid. Find the percentage of milk in the new mixture.

Soln: Applying the above formula, we have the required percentage of milk in the

$$\text{new mixture} = \frac{6\times25+4\times30}{6+4} = 27\ \%$$

Note: This can also be solved by the following methods.

I. The required percentage of milk in the new mixture

$$= \frac{\text{Quantity of milk in the new mixture}}{\text{Quantity of the new mixture}}\times100$$

$$= \frac{\text{6 parts of 25\% milk + 4 parts of 30\% milk}}{\text{(6 parts + 4 parts) of the liquid}}\times100$$

$$= \frac{6\times\frac{25}{100}+4\times\frac{30}{100}}{10}\times100 = (15+12) = 27\ \%$$

II. This equation can be solved by the method of Alligation

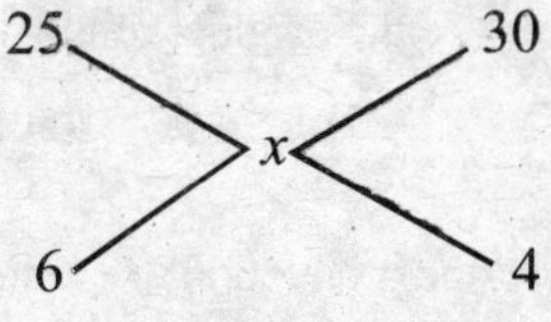

$$\frac{30-x}{x-25} = \frac{6}{4} = \frac{3}{2}$$

or, $60-2x = 3x-75$

$\therefore x = 27\%$

Exercise

1. One type of liquid contains 14% of milk, the other contains 24% of milk. A can filled with 5 parts of the first liquid and 5 parts of the second liquid. Find the percentage of milk in the new mixture.

a) 19 b) 29 c) 20 d) 21

2. One type of liquid contains 15% of milk, the other contains 5% of milk. A can filled with 7 parts of the first liquid and 8 parts of the second liquid. Find the percentage of milk in the new mixture.

a) 9% b) $9\frac{1}{3}\%$ c) $9\frac{2}{3}\%$ d) $9\frac{3}{4}\%$

3. One type of liquid contains 16% of milk, the other contains 4% of milk. A can filled with 3 parts of the first liquid and 6 parts of the second liquid. Find the percentage of milk in the new mixture.

a) 9% b) 8% c) 6% d) 10%

4. One type of liquid contains 45% of milk, the other contains 25% of milk. A can filled with 9 parts of the first liquid and 11 parts of the second liquid. Find the percentage of milk in the new mixture.

a) 34% b) 48% c) 36% d) 35%

5. One type of liquid contains 16% of milk, the other contains 26% of milk. A can filled with 5 parts of the first liquid and 7 parts of the second liquid. Find the percentage of milk in the new mixture.

a) 21% b) 22% c) 21.83% d) 23.5%

Answers

1. a 2. c 3. b 4. a 5. c

Rule 9

Original daily wage $= \frac{\text{Increased daily wage}}{100+\text{\% increase}}\times100$ ***; when increase per cent is given.***

Illustrative Example

Ex.: The daily wage is increased by 20% and a person now gets Rs 24 per day. What was his daily wage before the increase?

Soln: Applying the above formula, we have,

$$\text{required original daily wage} = \frac{24}{120}\times100 = \text{Rs } 20$$

Exercise

1. The daily wage is increased by 25% and a person now gets Rs 25 per day. What was his daily wage before the increase?

a) Rs 22 b) Rs 24 c) Rs 21 d) Rs 20

2. The daily wage is increased by 12% and a person now gets Rs 14 per day. What was his daily wage before the increase?

a) Rs $12\frac{1}{2}$ b) Rs 12 c) Rs 25 d) Rs $16\frac{1}{2}$

3. The daily wage is increased by 28% and a person now gets Rs 160 per day. What was his daily wage before the increase?
a) Rs 150 c) Rs 125 d) Rs 145 d) Rs 120
4. The daily wage is increased by 30% and a person now gets Rs 26 per day. What was his daily wage before the increase?
a) Rs 25 b) Rs 21 c) Rs 20 d) Rs 16
5. The daily wage is increased by 35% and a person now gets Rs 315 per day. What was his daily wage before the increase?
a) Rs $\frac{700}{3}$ b) Rs 10 c) Rs $\frac{100}{7}$ d) Rs $\frac{200}{9}$

Answers
1. d 2. a 3. b 4. c 5. a

Rule 10

Original daily wage $= \frac{\textit{Decreased daily wage}}{100 - \textit{\% decrease}} \times 100$ ***; when decrease per cent is given.***

Illustrative Example

Ex.: The daily wage is decreased by 15% and a person now gets Rs 17 per day. What was his daily wage before the decrease?

Soln: Applying the above formula,
the required original daily wage

$$= \frac{17}{100-15} \times 100 = \text{Rs } 20$$

Exercise
1. The daily wage is decreased by 20% and a person now gets Rs 16 per day. What was his daily wage before the decrease?
a) Rs 20 b) Rs 25 c) Rs 24 d) Rs 21
2. The daily wage is decreased by 12% and a person now gets Rs 22 per day. What was his daily wage before the decrease?
a) Rs 23 b) Rs 27 c) Rs 25 d) Rs 35
3. The daily wage is decreased by 19% and a person now gets Rs 27 per day. What was his daily wage before the decrease?
a) Rs 33 b) Rs 34 c) Rs $33\frac{2}{3}$ d) Rs $33\frac{1}{3}$
4. The daily wage is decreased by 18% and a person now gets Rs 41 per day. What was his daily wage before the decrease?
a) Rs 51 b) Rs 50 c) Rs 45 d) Rs 55
5. The daily wage is decreased by 10% and a person now gets Rs 18 per day. What was his daily wage before the decrease?
a) Rs 20 b) Rs 21 c) Rs 25 d) Rs 24

Answers
1. a 2. c 3. d 4. b 5. a

Rule 11

Ex.: Due to fall in manpower, the production in a factory decreases by 25%. By what per cent should the working hour be increased to restore the original production?

Soln: **Method I:** Decrease in production is only due to decrease in manpower. Hence, manpower is decreased by 25%.
Now, suppose that to restore the same production, working hours are increased by x%.
Production = Manpower × Working hours = M × W (say)
Now, M × W = (M - 25% of M) × (W + x% of W)

$$\text{or, } M \times W = \frac{75}{100} M \times \frac{100+x}{100} W$$

or, 100 × 100 = 75 (100 + x)

$$\text{or, } \frac{400}{3} = 100 + x \quad \therefore x = \frac{100}{3} = 33\frac{1}{3}\%$$

Method II: To make the calculations easier, suppose Manpower = 100 units and Working hours = 100 units
Suppose working hours increase by x%.
Then, $(100-25)(100+x) = 100 \times 100$

$$\text{or, } 100 + x = \frac{400}{3} \quad \therefore x = \frac{100}{3} = 33\frac{1}{3}\%$$

Direct Formula:
Required % increase in working hours

$$= \frac{25}{100-25} \times 100 = \frac{100}{3} = 33\frac{1}{3}\%$$

Exercise
1. Due to fall in manpower, the production in a factory decreases by 24%. By what per cent should the working hour be increased to restore the original production?
a) $\frac{600}{19}\%$ b) $\frac{400}{19}\%$ c) $\frac{650}{19}\%$ d) $\frac{450}{19}\%$
2. Due to fall in manpower, the production in a factory decreases by 20%. By what per cent should the working hour be increased to restore the original production?
a) 24% b) 25% c) 20% d) 35%
3. Due to fall in manpower, the production in a factory decreases by 40%. By what per cent should the working

hour be increased to restore the original production?

a) $\frac{200}{9}\%$ b) $\frac{100}{3}\%$ c) $\frac{200}{3}\%$ d) $\frac{100}{9}\%$

4. Due to fall in manpower, the production in a factory decreases by 36%. By what per cent should the working hour be increased to restore the original production?
a) 36% b) 56% c) 57% d) 56.25%
5. Due to fall in manpower, the production in a factory decreases by 30%. By what per cent should the working hour be increased to restore the original production?

a) $\frac{300}{7}\%$ b) $\frac{400}{7}\%$ c) $\frac{200}{7}\%$ d) $\frac{100}{7}\%$

Answers
1. a 2. b 3. c 4. d 5. a

Rule 12

Theorem: ***If two values are respectively x% and y% more than a third value, then the first is the*** $\frac{100+x}{100+y}\times 100\%$ ***of the second.***

Illustrative Example

Ex.: Two numbers are respectively 20% and 50% more than a third. What percentage is the first of the second?

Soln: Following the above theorem, we have

the required value $= \frac{120}{150}\times 100 = 80\%$

Exercise
1. Two numbers are respectively 25% and 20% more than a third. What percentage is the first of the second?
a) 104% b) 104.16% c) 104.26% d) 105%
2. Two numbers are respectively 20% and 35% more than a third. What percentage is the first of the second?

a) $\frac{200}{3}\%$ b) $\frac{400}{9}\%$ c) $\frac{800}{9}\%$ d) $\frac{200}{9}\%$

3. Two numbers are respectively 8% and 32% more than a third. What percentage is the first of the second?

a) $\frac{800}{11}\%$ b) $\frac{900}{11}\%$ c) $\frac{1000}{11}\%$ d) $\frac{950}{11}\%$

4. Two numbers are respectively 15% and 84% more than a third. What percentage is the first of the second?

a) $64\frac{1}{2}\%$ b) $65\frac{1}{2}\%$ c) $63\frac{1}{2}\%$ d) $62\frac{1}{2}\%$

5. Two numbers are respectively 26% and 5% more than a third. What percentage is the first of the second?
a) 120% b) 100% c) 80% d) 125%

Answers
1. b 2. c 3. b 4. d 5. a

Rule 13

Theorem: ***If two values are respectively x% and y% more than a third value, then the second is the*** $\frac{100+y}{100+x}\times 100\%$ ***of the first.***

Illustrative Example

Ex.: Two numbers are respectively 25% and 50% more than a third. What percentage is the second of the first?

Soln: Following the above theorem, we have the required value

$$= \frac{100+50}{100+25}\times 100 = \frac{150}{125}\times 100 = 120\%$$

Exercise
1. Two numbers are respectively 48% and 11% more than a third. What percentage is the second of the first?
a) 74% b) 75% c) $37\frac{1}{2}\%$ d) 80%
2. Two numbers are respectively 68% and 26% more than a third. What percentage is the second of the first?
a) 75% b) 72% c) 85% d) 78%
3. Two numbers are respectively 25% and 40% more than a third. What percentage is the second of the first?
a) 110% b) 115% c) 112% d) 122%
4. Two numbers are respectively 60% and 20% more than a third. What percentage is the second of the first?
a) 70% b) 80% c) 65% d) 75%

Answers
1. b 2. a 3. c 4. d

Rule 14

Theorem: ***If two values are respectively x% and y% less than a third value, then the second is the*** $\frac{100-y}{100-x}\times 100\%$ ***of the first.***

Illustrative Example

Ex.: Two numbers are respectively 30% and 40% less than a third number. What per cent is the second of the first?

Soln: Applying the above formula, we have the required answer

$$= \frac{100-40}{100-30}\times 100 = \frac{60}{70}\times 100 = 85\frac{5}{7}\%$$

Exercise
1. Two numbers are respectively 15% and 20% less than a third number. What per cent is the second of the first?

a) $94\frac{4}{17}\%$ b) $94\frac{2}{17}\%$ c) $94\frac{1}{17}\%$ d) 95%

2. Two numbers are respectively 25% and 30% less than a third number. What per cent is the second of the first?
a) 93% b) $93\frac{2}{3}\%$ c) $93\frac{1}{3}\%$ d) 44%
3. Two numbers are 30 and 37 per cent less than a third number. How much per cent is the second number less than the first?
a) 90% b) 80% c) 10% d) 20%
4. Two numbers are 40 and 46 per cent less than a third number. How much per cent is the second number less than the first?
a) 90% b) 10% c) 15% d) 80%

Answers

1. b 2. c

3. c; **Hint:** Second number is $\frac{100-37}{100-30}\times100=90\%$ of the first number.
$\therefore$ Second number is (100 -90 =) 10% less than the first number.

4. b

Rule 15

Theorem: ***If two values are respectively x% and y% less than a third value, then the first is the*** $\frac{100-x}{100-y}\times100\%$ ***of the second.***

Illustrative Example

Ex.: Two numbers are respectively 25% and 40% less than a third number. What per cent is the first of the second?

Soln: Using the above theorem, we have the required answer

$$=\frac{100-25}{100-40}\times100=\frac{75}{60}\times100=125\%$$

Exercise

1. Two numbers are respectively 28% and 25% less than a third number. What per cent is the first of the second?
a) 120% b) 96% c) 84% d) 108%
2. Two numbers are respectively 37% and 30% less than a third number. What per cent is the first of the second?
a) 90% b) 85% c) 95% d) 80%
3. Two numbers are respectively 32% and 20% less than a third number. What per cent is the first of the second?
a) 80% b) 75% c) 64% d) 85%
4. Two numbers are respectively 35% and 22% less than a third number. What per cent is the first number less than the second number.
a) $83\frac{1}{3}\%$ b) $16\frac{2}{3}\%$ c) $83\frac{2}{3}\%$ d) $16\frac{1}{3}\%$

Answers

1. b 2. a 3. d 4. b

Rule 16

Theorem: ***If A is x% of C and B is y% of C, then A is*** $\frac{x}{y}\times100\%$ ***of B.***

Illustrative Example

Ex.: Two numbers are respectively 20% and 25% of a third number. What percentage is the first of the second?

Soln: Following the above theorem, we have

$$\text{the required value }=\frac{20}{25}\times100=80\%$$

Note: The above relationship is very simple. When "what is the first of second" is asked, put the first as the numerator and the second as the denominator and *vice-versa*.

Exercise

1. Two numbers are respectively 15% and 20% of a third number. What percentage is the first of the second?
a) 75% b) 80% c) 70% d) 65%
2. Two numbers are respectively 7% and 28% of a third number. What percentage is the first of the second?
a) 28% b) 20% c) 30% d) 25%
3. Two numbers are respectively 10% and 16% of a third number. What percentage is the first of the second?
a) 62% b) $37\frac{1}{2}\%$ c) $62\frac{1}{2}\%$ d) $65\frac{1}{2}\%$
4. Two numbers are respectively 16% and 48% of a third number. What percentage is the first of the second?
a) $33\frac{1}{2}\%$ b) $33\frac{1}{3}\%$
c) $66\frac{1}{3}\%$ d) $33\frac{2}{3}\%$
5. Two numbers are respectively 20% and 25% of a third number. What per cent is the second of the first?
a) 120% b) 75% c) 80% d) 125%

Answers

1. a 2. d 3. c 4. b
5. d; **Hint:** *See Note*

Required answer $=\frac{25}{20}\times100=125\%$

Rule 17

Theorem: ***x% of a quantity is taken by the first, y% of the remaining is taken by the second and z% of the remaining is taken by third person. Now, if Rs A is left in the fund, then there was*** $\frac{A \times 100 \times 100 \times 100}{(100-x)(100-y)(100-z)}$ ***in the beginning.***

Illustrative Example

Ex.: A man loses $12\frac{1}{2}$% of his money and, after spending 70% of the remainder, he is left with Rs 210. How much had he at first?

Soln: **Quicker Method:**

It is a very short and fast-calculating method. The only thing is to remember the formula well.

His initial money

$$= \frac{210 \times 100 \times 100}{(100-12.5)(100-70)} = \frac{210 \times 100 \times 100}{87.5 \times 30}$$

$$= \text{Rs } 800$$

Note: As his "initial money" is definitely more than the "left money", there should not be any confusion in putting the larger value (100) in the numerator and the smaller value (100 – 12.5) in the denominator.

Exercise

1. An electrical contractor purchases a certain amount of wire, 10% of which was stolen. After using 85% of the remainder, he had 47 m 25 cm of wire left. How much wire did he purchase?
   a) 350 m b) 320 m c) 300 m d) 370 m
2. A man spends 50% of his income in board and lodging, 20% of the remainder in other personal necessities and 25% of the rest in charity, find his income, if he is left with Rs 4200.
   a) Rs 14000 b) Rs 8000
   c) Rs 12000 d) Rs 18000
3. A man loses 15% of his money and, after spending 85% of the remainder, he is left with Rs 510. How much had he at first?
   a) Rs 4500 b) Rs 3500
   c) Rs 4000 d) Rs 4200
4. A man loses 14% of his money and, after spending 25% of the remainder, he is left with Rs 1290. How much had he at first?
   a) Rs 4000 b) Rs 2000 c) Rs 2500 d) Rs 3000

Answers

1\. a; **Hint:** $\frac{4725cm \times 100 \times 100}{(100-10)(100-85)} = 350\ m$

2\. a 3. c 4. b

Rule 18

Theorem: ***If initial quantity is A and x% of the quantity is taken by the first, y% of the remaining was taken by the second and z% of the remaining is taken by third person; then is*** $\frac{A \times (100-x)(100-y)(100-z)}{100 \times 100 \times 100}$ ***left in the fund.***

Illustrative Example

Ex.: Pankaj and Chandan deposits Rs 1200 in a common fund. 20% of the initial amount is taken by Pankaj and 40% of the remaining amount is taken by Chandan. How much is left in the common fund.

Soln: Following the above theorem, we have the amount left in the fund

$$= \frac{1200 \times (100-20)(100-40)}{100 \times 100} = \text{Rs } 576$$

Note: As "left money" is definitely less than the 'initial money' there should not be any confusion in putting the smaller value (100 – 20) in the numerator and larger value (100) in the denominator.

Exercise

1. A man spends 30% of his income in board and lodging, 25% of the remainder in other personal necessities and 20% of the rest in charity. If his income is Rs 25000, find the amount left by him at the end.
   a) Rs 8500 b) Rs 9500 c) Rs 10500 d) Rs 10000
2. A man spends 15% of his income in board and lodging, 10% of the remainder in other personal necessities and 5% of the rest in charity. If his income is Rs 20000, find the amount left by him at the end.
   a) Rs 14535 b) Rs 14353 c) Rs 14533 d) Rs 15435
3. A man spends 45% of his income in board and lodging, 35% of the remainder in other personal necessities and 25% of the rest in charity. If his income is Rs 16000, find the amount left by him at the end.
   a) Rs 4920 b) Rs 4290 c) Rs 4390 d) Rs 4260
4. A man loses 25% of his money and after spending 75% of the remainder, how much is he left with if initial money is Rs 3200?
   a) Rs 800 b) Rs 400 c) Rs 900 d) Rs 600

Answers

1\. c 2. a 3. b 4. d

Rule 19

Theorem: ***x% of a quantity is added. Again, y% of the increased quantity is added. Again z% of the increased quantity is added. Now, it becomes A, then the initial amount is given by*** $\frac{A \times 100 \times 100 \times 100}{(100+x)(100+y)(100+z)}$.

Illustrative Example

Ex.: A man deposited 50% of the initial amount to his locker. And again after some time he deposited 20% of the increased amount. Now the amount becomes Rs 18,000. How much was the initial amount?

Soln: Following the above theorem, we have,

$$\text{initial amount} = \frac{18000 \times 100 \times 100}{(100+50)(100+20)} = \text{Rs } 10{,}000$$

Exercise

1. A man deposited 30% of the initial amount to his locker. And again after some time he deposited 25% of the increased amount. Now the amount becomes Rs 13,000. How much was the initial amount?
   a) Rs 8000 b) Rs 10000 c) Rs 12000 d) Rs 9000
2. A man deposited 40% of the initial amount to his locker. And again after some time he deposited 35% of the increased amount. Now the amount becomes Rs 18,900. How much was the initial amount?
   a) Rs 12000 b) Rs 10500 c) Rs 11000 d) Rs 10000
3. A man deposited 15% of the initial amount to his locker. And again after some time he deposited 45% of the increased amount. Now the amount becomes Rs 6670. How much was the initial amount?
   a) Rs 8000 b) Rs 4500 c) Rs 4000 d) Rs 7500
4. A man deposited 12% of the initial amount to his locker. And again after some time he deposited 32% of the increased amount. Now the amount becomes Rs 9240. How much was the initial amount?
   a) Rs 6250 b) Rs 6200 c) Rs 6350 d) Rs 6260
5. A man deposited 14% of the initial amount to his locker. And again after some time he deposited 45% of the increased amount. Now the amount becomes Rs 16530. How much was the initial amount?
   a) Rs 10500 b) Rs 10000 c) Rs 9500 d) Rs 9000

Answers

1. a 2. d 3. c 4. a 5. b

Rule 20

Theorem: ***If initial quantity is A and x% of the initial quantity is added. Again y% of the increased quantity is added. Again z% of the increased quantity is added, then initial quantity becomes*** $\frac{A \times (100+x)(100+y)(100+z)}{100 \times 100 \times 100}$.

Illustrative Example

Ex.: A man had Rs 4800 in his locker two years ago. In the first year, he deposited 20% of the amount in his locker. In the second year, he deposited 25% of the increased amount in his locker. Find the amount at present in his locker.

Soln: The amount is certainly more than Rs 4800. And each year, the new amount is added. So, the sum should be multiplied by

$$\frac{100+20}{100} \text{ and } \frac{100+25}{100}$$

$\therefore$ the required amount $= \frac{4800 \times 120 \times 125}{100 \times 100} = \text{Rs } 7200$

Exercise

1. A man had Rs 1200 in his locker two years ago. In the first year, he deposited 10% of the amount in his locker. In the second year, he deposited 20% of the increased amount in his locker. Find the amount at present in his locker.
   a) Rs 1584 b) Rs 1854 c) Rs 1485 d) Rs 1548
2. A man had Rs 1400 in his locker two years ago. In the first year, he deposited 30% of the amount in his locker. In the second year, he deposited 40% of the increased amount in his locker. Find the amount at present in his locker.
   a) Rs 2485 b) Rs 2584 c) Rs 2548 d) Rs 3548
3. A man had Rs 4600 in his locker two years ago. In the first year, he deposited 50% of the amount in his locker. In the second year, he deposited 15% of the increased amount in his locker. Find the amount at present in his locker.
   a) Rs 7935 b) Rs 9735 c) Rs 7953 d) Rs 7395
4. A man had Rs 600 in his locker two years ago. In the first year, he deposited 60% of the amount in his locker. In the second year, he deposited 70% of the increased amount in his locker. Find the amount at present in his locker.
   a) Rs 1362 b) Rs 1263 c) Rs 2631 d) Rs 1632

Answers

1. a 2. c 3. a 4. d

Rule 21

Population Formula I

Theorem: ***If the original population of a town is P, and the annual increase is r%, then the population in n years is given by*** $P\left(1+\frac{r}{100}\right)^n$.

Illustrative Example

Ex.: If the annual increase in the population of a town is 4% and the present number of people is 15,625, what will the population be in 3 years?

Soln: Applying the above theorem, we have

$$\text{the required population} = 15625\left(1+\frac{4}{100}\right)^3$$

$$= 15625 \times \frac{26}{25} \times \frac{26}{25} \times \frac{26}{25} = 17576$$

Exercise

1. The population of a town is 32000. It increases 15 per cent annually. What will it be in 2 years?
   a) 42320 b) 43220 c) 42520 d) 42330
2. The population of a town is 25000. It increases 2 per cent annually. What will it be in 2 years?
   a) 27010 b) 26010 c) 25010 d) 28010
3. The population of a town is 125000. It increases 6 per cent annually. What will it be in 3 years?
   a) 148877 b) 148787 c) 147788 d) 147878
4. The population of a town is 15625. It increases 8 per cent annually. What will it be in 3 years?
   a) 16983 b) 18693 c) 19683 d) 19638

Answers

1. a 2. b 3. a 4. c

Rule 22

Population Formula II

Theorem: ***If the annual increase in the population of a town be r% and the present population be P_n, then the population of the town n years ago was given as*** $\dfrac{P_n}{\left(1+\dfrac{r}{100}\right)^n}$.

Illustrative Example

Ex.: If the annual increase in the population of a town be 4% and the present population be 17576, what was it three years ago?

Soln: Following the above theorem, we have
Population 3 years ago

$$=\frac{17576}{\left(1+\frac{4}{100}\right)^3}=\frac{17576\times25\times25\times25}{26\times26\times26}=15625$$

Exercise

1. If the annual increase in the population of a town be 2% and the present population be 65025, what was it two years ago?
   a) 65200 b) 62500 c) 63500 d) 65300
2. If the annual increase in the population of a town be 4% and the present population be 16224, what was it two years ago?
   a) 15000 b) 14000 c) 15500 d) 16000
3. If the annual increase in the population of a town be 6% and the present population be 148877, what was it three years ago?
   a) 125500 b) 135000 c) 125000 d) 125600
4. If the annual increase in the population of a town be 8% and the present population be 21870, what was it two years ago?
   a) 17580 b) 15780 c) 17850 d) 18750
5. If the annual increase in the population of a town be 25% and the present population be 87500, what was it three years ago?
   a) 44400 b) 44800
   c) 48800 d) 44600
6. The income of a company increases 20% per annum. If its income is Rs 2664000 in the year 1999 what was its income in the year 1997?
   a) Rs 2220000 b) Rs 1850000
   c) Rs 2121000 d) Rs 1855000
   [BSRB Patna PO, 2001]
7. The population of a village increases by 5% annually. If the present population is 4410, what it was 2 years ago?
   a) 3410 b) 3300 c) 4000 d) 4140
   [LIC, 1991]

Answers

1. b 2. a 3. c 4. d 5. b
6. b; **Hint:** Consider income of the company as a population and apply the above rule.
7. c

Rule 23

Population Formula III

Theorem: ***If the original population of a town is P, and the annual decrease is r%, then the population in n years will be*** $P\left(1-\dfrac{r}{100}\right)^n$.

Illustrative Example

Ex.: If the annual decrease in the population of a town is 5% and the present number of people is 40,000, what will the population be in 2 years?

Soln: Following the above theorem, we have
Population in two years

$$=40000\left(1-\frac{5}{100}\right)^2=\frac{40000\times19\times19}{20\times20}=36100$$

Exercise

1. If the annual decrease in the population of a town is 4% and the present number of people is 62500, what will the population be in 2 years?
   a) 57600 b) 56700 c) 56600 d) 58600
2. If the annual decrease in the population of a town is 10% and the present number of people is 16000, what will the population be in 3 years?
   a) 12664 b) 11664 c) 11564 d) 11654
3. If the annual decrease in the population of a town is 8% and the present number of people is 68750, what will the population be in 2 years?
   a) 58920 b) 58910 c) 58290 d) 58190

4. If the annual decrease in the population of a town is 15% and the present number of people is 72000, what will the population be in 3 years?
a) 44127 b) 44217 c) 44317 d) 44227

Answers

1. a 2. b 3. d 4. b

Rule 24

Population Formula IV

Theorem: ***If the annual decrease in the population of a town be r% and the present population be P_n, then the population n years ago was*** $\dfrac{P_n}{\left(1-\dfrac{r}{100}\right)^n}$.

Illustrative Example

Ex.: If the annual decrease in the population of a town be 4% and the present population be 57600, what was it two years ago?

Soln: Applying the above formula,
Population of the town 2 years ago was

$$\frac{57600}{\left(1-\frac{4}{100}\right)^2}=\frac{57600\times25\times25}{24\times24}=62500$$

Exercise

1. If the annual decrease in the population of a town be 5% and the present population be 68590, what was it three years ago?
a) 80000 b) 60000 c) 86000 d) 65000
2. If the annual decrease in the population of a town be 10% and the present population be 21870, what was it two years ago?
a) 27036 b) 27600 c) 27000 d) 28000
3. If the annual decrease in the population of a town be 15% and the present population be 98260, what was it three years ago?
a) 160060 b) 180000 c) 150000 d) 160000
4. If the annual decrease in the population of a town be 20% and the present population be 25600, what was it two years ago?
a) 40000 b) 48000 c) 50000 d) 42000

Answers

1. a 2. c 3. d 4. a

Rule 25

Population Formula V

When the Rate of Growth is Different for Different Years

Theorem : The population of a town is P. It increases by x% during the first year, increases by y% during the second year and again increases by z% during the third year. The population after 3 years will be

$$\frac{P\times(100+x)(100+y)(100+z)}{100\times100\times100}.$$

Illustrative Example

Ex.: The population of a town is 8000. It increases by 10% during the first year and by 20% during the second year. What is the population after two years?

Soln: The required population $=\dfrac{8000\times110\times120}{100\times100}=10{,}560$.

Exercise

1. The population of a town is 7000. It increases by 5% during the first year and by 10% during the second year. What is the population after two years?
a) 8085 b) 7085 c) 9085 d) 8805
2. The population of a town is 12400. It increases by 15% during the first year, by 20% during the 2nd year, and by 25% during the third year. What is the population after three years?
a) 23190 b) 22390 c) 21390 d) 21360
3. The population of a town is 6480. It increases by 25% during the first year and by 30% during the second year. What is the population after two years?
a) 10350 b) 10530 c) 10620 d) 10360

Answers

1. a 2. c 3. b

Rule 26

Population Formula VI

When Population Increases for One Year and Then Decreases for the Next Year.

Theorem: ***The population of a town is P. It increases by x% during the first year, decreases by y% during the second year and again increases by z% during the third year. The population after 3 years will be***

$$\frac{P(100+x)(100-y)(100+z)}{100\times100\times100}.$$

Illustrative Example

Ex.: The population of a town is 10,000. It increases by 10% during the first year. During the second year, it decreases by 20% and increased by 30% during the third year. What is the population after 3 years?

Soln: The required population $=\dfrac{10000\times110\times80\times130}{100\times100\times100}$

$=11440$

Exercise

1. The population of a town is 144000. It increases by 5% during the first year. During the second year, it decreases by 10% and increased by 15% during the third year.

What is the population after 3 years?
a) 154692 b) 156492 c) 156942 d) 156462

2. The population of a town is 12500. It increases by 10% during the first year. During the second year, it decreases by 15% and increased by 20% during the third year. What is the population after 3 years?
a) 14025 b) 14625 c) 15025 d) 14035

3. The population of a town is 32000. It increases by 15% during the first year. During the second year, it decreases by 20% and increased by 25% during the third year. What is the population after 3 years?
a) 38600 b) 39800 c) 36800 d) 38900

4. The population of a town is 64000. It increases by 10% during the first year. During the second year, it decreases by 25% and increased by 5% during the third year. What is the population after 3 years?
a) 654400 b) 56440 c) 55450 d) 55440

Answers

1. b 2. a 3. c 4. d

Rule 27

Population Formula VII

Theorem: ***If during the first year, the population of town increases by x%, during the next year (ie second year) decreases by y% and again decreases by z% during the third year and the population at the end of third year is given as P. Then the population at the beginning of the first year was*** $\frac{P\times100\times100\times100}{(100+x)(100-y)(100-z)}$.

Illustrative Examples

Ex. 1: During one year, the population of a locality increases by 5% but during the next year, it decreases by 5%. If the population at the end of the second year was 7980, find the population at the beginning of the first year.

Soln : The required population $= 7980\times\left(\frac{100}{100-5}\right)\left(\frac{100}{100+5}\right)$

$$=\frac{7980\times100\times100}{95\times105}=8000$$

Note : In the above example, the population after two years is given and the population in the beginning of the first year is asked. That is why, the fractional values are inversed. Mark that point. The same thing happens to the next example.

Ex. 2: The population of a town increases at the rate of 10% during one year and it decreases at the rate of 10% during the second year. If it has 29,700 inhabitants at present, find the number of inhabitants two years ago.

Soln : The required population $=\frac{29700\times100\times100}{(100-10)\times(100+10)}$

$$=\frac{29700\times100\times100}{90\times110}=30{,}000$$

Exercise

1. During one year, the population of a locality increases by 20% but during the next year, it decreases by 15%. If the population at the end of the second year was 17340, find the population at the beginning of the first year.
a) 17000 b) 16000 c) 19000 d) 18000

2. During one year, the population of a locality increases by 5% but during the next year, it decreases by 10%. If the population at the end of the second year was 37800, find the population at the beginning of the first year.
a) 40000 b) 50000 c) 45000 d) 48000

3. The population of a town increases at the rate of 20% during one year and it decreases at the rate of 20% during the second year. If it has 57,600 inhabitants at present, find the number of inhabitants two years ago.
a) 80000 b) 65000 c) 61000 d) 60000

4. The population of a town increases at the rate of 15% during one year and it decreases at the rate of 15% during the second year. If it has 78,200 inhabitants at present, find the number of inhabitants two years ago.
a) 80000 b) 82000 c) 81000 d) 79500

5. During the first year, the population of a town increases by 20% during the second year decreases by 5% and again decreases by 10% during the third yerar and the population at the end of third year is 51300. Find the population at the beginning of the first year.
a) 50000 b) 51000 c) 49200 d) 40000

6. The population of a town increases by 12% during first year and decreases by 10% during second year. If the present population is 50400, what it was 2 years ago?
a) 40000 b) 50000 c) 42000 d) 40400
[LIC 1991]

Answers

1. a 2. a 3. d 4. a 5. a 6. b

Rule 28

Population Formula VIII

Theorem: ***The population of a town increases by x% during the first year, increases by y% during the second year and again increases by z% during the third year. If the present population of a town is P, then the population 3 years ago was*** $\frac{P\times100\times100\times100}{(100+x)(100+y)(100+z)}$.

Illustrative Example

Ex.: The population of a town increases by 10% during

the first year and by 20% during the second year. The present population of a town is 26400. Find the population of the town two years ago.

Soln: Following the above formula, we have the required population

$$= \frac{26400 \times 100 \times 100}{110 \times 120} = 20000$$

Exercise

1. The population of a town increases by 5% during the first year and by 15% during the second year. The present population of a town is 11109. Find the population of the town two years ago.
   a) 10000 b) 9820 c) 9200 d) 9300
2. The population of a town increases by 20% during the first year and by 25% during the second year. The present population of a town is 8400. Find the population of the town two years ago.
   a) 5600 b) 6500 c) 7500 d) 5700
3. The population of a town increases by 15% during the first year and by 30% during the second year. The present population of a town is 38870. Find the population of the town two years ago.
   a) 36000 b) 46000 c) 26000 d) 28000

Answers

1. c 2. a 3. c

Rule 29

Population Formula IX

Theorem: ***The population of a town is P. It decreases by x% during the first year, decreases by y% during the second year and again decreases by z% during the third year. The population after three years will be***

$$\frac{P \times (100-x)(100-y)(100-z)}{100 \times 100 \times 100}$$

Illustrative Example

Ex.: The population of a town is 8000. It decreases by 10% during the first year, 15% during the second year and 20% during the third year. What will be the population after 3 years?

Soln: Following the above theorem, we have the required population

$$= \frac{8000(100-10)(100-15)(100-20)}{100 \times 100 \times 100}$$

$$= \frac{8000 \times 90 \times 85 \times 80}{100 \times 100 \times 100} = 4896$$

Exercise

1. The population of a town is 48000. It decreases by 5% during the first year, 10% during the second year and 15% during the third year. What will be the population after 3 years?
   a) 34884 b) 44884 c) 38484 d) 34484
2. The population of a town is 64000. It decreases by 5% during the first year, 15% during the second year and 25% during the third year. What will be the population after 3 years?
   a) 37860 b) 38670 c) 38760 d) 38790
3. The population of a town is 6250. It decreases by 10% during the first year, 20% during the second year and 30% during the third year. What will be the population after 3 years?
   a) 3250 b) 3150 c) 3510 d) 3100

Answers

1. a 2. c 3. b

Rule 30

Population Formula X

Theorem: ***The population of a town decreases by x% during the first year, decreases by y% during the second year and again decreases by z% during the third year. If the present population of a town is P then the population of the town, three years ago was*** $\frac{P \times 100 \times 100 \times 100}{(100-x)(100-y)(100-z)}$.

Illustrative Example

Ex.: The population of a town decreases by 20% during the first year, decreases by 30% during the second year and again decreases by 40% during the third year. If the present population of the town is 67200 then what was the population of the town three years ago?

Soln: Following the above formula, we have the required population

$$= \frac{67200 \times 100 \times 100 \times 100}{(100-20)(100-30)(100-40)} = 200000$$

Exercise

1. The population of a town decreases by 5% during the first year, decreases by 10% during the second year and again decreases by 15% during the third year. If the present population of the town is 29070 then what was the population of the town three years ago?
   a) 40000 b) 36000 c) 40500 d) 42000
2. The population of a town decreases by 10% during the first year, decreases by 15% during the second year and again decreases by 20% during the third year. If the present population of the town is 15300 then what was the population of the town three years ago?
   a) 24000 b) 24500 c) 25000 d) 25400
3. The population of a town decreases by 15% during the first year, decreases by 20% during the second year and

again decreases by 25% during the third year. If the present population of the town is 10200 then what was the population of the town three years ago?
a) 20000 b) 30000 c) 15000 d) 19000

Answers

1. a 2. c 3. a

Rule 31

Population Formula XI

Theorem: ***The population of a town is P_1. If the males increases by x% and the females by y%, the population will be P_2 then the number of males and females are given by***

$$\left[\frac{P_2\times100-P_1(100+y)}{(x-y)}\right] \text{ and } \left[\frac{P_2\times100-P_1(100+x)}{(y-x)}\right]$$

respectively.

Illustrative Example

Ex.: The population of a town is 8000. If the males increase by 6% and the females by 10%, the population will be 8600. Find the number of females in the town.

Soln : Detail Method: Let the population of females be x.

Then 110% of x + 106% of (8000 – x) = 8600

or, $\frac{110x}{100}+\frac{106(8000-x)}{100}=8600$

or, $x(110-106)=8600\times100-8000\times106$

$\therefore x=\frac{8600\times100-8000\times106}{110-106}=\frac{12000}{4}=3000$

Quicker Method: Applying the above theorem, the required number of females

$=\frac{8600\times100-8000(100+6)}{10-6}=3000.$

Note: If we ignore the intermediate steps, we can get the population of females and males directly thus we can see that how the quicker method has been derived.

The population of females =

$\frac{8600\times100-8000(100+6)}{(10-6)}=3{,}000$

The population of males

$=\frac{8600\times100-8000(100+10)}{(6-10)}$

$=\frac{20000}{4}=5000$

Exercise

1. The population of a town is 6000. If the males increase by 5% and the females by 9%, the population will be 6500. Find the number of males in the town.
   a) 5000 b) 3000 c) 4000 d) 1000
2. The population of a town is 5000. If the males increase by 6% and the females by 14%, the population will be 5500. Find the number of females in the town.
   a) 5000 b) 2000 c) 6000 d) 2500
3. The population of a town is 8000. If the males increase by 9% and the females by 16%, the population will be 9000. Find the number of females in the town.
   a) 2000 b) 4500 c) 3000 d) 4000
4. In a certain year, the population of a certain town was 9000. If the next year the population of males increases by 5% and that of the females by 8% and the total population increases to 9600, then what was the ratio of population of males and females in that given year?
   a) 4 : 5 b) 5 : 4 c) 2 : 3 d) Data inadequate
   [Bank of Baroda PO, 1999]

Answers

1. d 2. d 3. d

4. a; **Hint:** By applying the given rule we have the no. of males = 4000 and the no. of females = 5000

$\therefore$ Required ratio $=\frac{4000}{5000}=4:5$.

Rule 32

Theorem : ***If the price of a commodity increases by r%, then the reduction in consumption so as not to increase the expenditure, is*** $\left(\frac{r}{100+r}\times100\right)\%$

Illustrative Example

Ex.: If the price of a commodity be raised by 20%, find by how much per cent must a householder reduce his consumption of that commodity so as not to increase his expenditure.

Soln: Detail Method: Present price of 1 kg of a commodity = 120 per cent of the former price of 1 kg

$=\frac{6}{5}$ of the former price of 1 kg

= former price of $\frac{6}{5}$ kg.

$\therefore$ Fromer price of 1 kg = present price of 5/6 kg

Therefore, in order that the expenditure may remain the same as before, for 1 kg consumed formerly, $\frac{5}{6}$ kg must be consumed now, that is, the consumption must be reduced by $\frac{1}{6}$ or by $\frac{100}{6}=\frac{50}{3}=16\frac{2}{3}$ per cent.

Quicker Method: From the above formula, we have

reduction in consumption $= \frac{20}{100+20} \times 100$

$= \frac{20}{120} \times 100 = \frac{50}{3} = 16\frac{2}{3}\%$

Exercise

1. If the duty on imported sugar be increased by 25 per cent. By how much per cent must a man reduce his consumption of that article so as not to increase his expenditure?
a) 20% b) 25% c) 16% d) 10%
2. If the price of a commodity be raised by 10%, find how much per cent must a householder reduce his consumption of that commodity, so as not to increase his expenditure.
a) $11\frac{1}{11}\%$ b) $9\frac{1}{11}\%$ c) $7\frac{1}{11}\%$ d) $26\frac{2}{3}\%$
3. If the price of a commodity be raised by $12\frac{1}{2}\%$, find how much per cent must a householder reduce his consumption of that commodity, so as not to increase his expenditure.
a) $9\frac{1}{9}\%$ b) $10\frac{1}{9}\%$ c) $11\frac{1}{9}\%$ d) $9\frac{1}{11}\%$
4. If the price of a commodity be raised by 15%, find how much per cent must a householder reduce his consumption of that commodity, so as not to increase his expenditure.
a) 14% b) $12\frac{1}{23}\%$ c) $23\frac{1}{13}\%$ d) $13\frac{1}{23}\%$
5. If the price of a commodity be raised by $16\frac{2}{3}\%$, find how much per cent must a householder reduce his consumption of that commodity, so as not to increase his expenditure.
a) $14\frac{1}{7}\%$ b) $13\frac{6}{7}\%$ c) $14\frac{2}{7}\%$ d) $16\frac{2}{3}\%$
6. If the price of a commodity be raised by $26\frac{2}{3}\%$, find how much per cent must a householder reduce his consumption of that commodity, so as not to increase his expenditure.
a) $20\frac{1}{19}\%$ b) $26\frac{2}{3}\%$ c) 26% d) $21\frac{1}{19}\%$
7. The price of cooking oil has increased by 25%. The percentage of reduction that a family should effect in the use of cooking oil so as not to increase the expenditure on this account is:
a) 15% b) 20% c) 25% d) 30%
[Central Excise & I. Tax, 1988]

Answers

1. a 2. b 3. c 4. d 5. c 6. d 7. b

Rule 33

Theorem: ***If the price of a commodity decreases by r%, then increase in consumption, so as not to decrease expenditure on this item is*** $\left[\frac{r}{(100-r)} \times 100\right]\%$.

Illustrative Example

Ex.: If the price of sugar falls down by 10%, by how much per cent must a householder increase its consumption, so as not to decrease expenditure in this item?

Soln: From the above formula, we have

increase in consumption $= \frac{10}{100-10} \times 100 = 11\frac{1}{9}\%$

Exercise

1. If the price of sugar falls down by 20%, by how much per cent must a householder increase its consumption, so as not to decrease expenditure in this item?
a) 25% b) 20% c) 30% d) 15%
2. If the price of tea falls down by 25%, by how much per cent must a householder increase its consumption, so as not to decrease expenditure in this item?
a) 33% b) 25% c) $33\frac{1}{3}\%$ d) $33\frac{2}{3}\%$
3. If the price of rice falls down by 5%, by how much per cent must a householder increase its consumption, so as not to decrease expenditure in this item?
a) 5% b) 20% c) $5\frac{5}{19}\%$ d) $5\frac{3}{19}\%$
4. If the price of wheat falls down by 15%, by how much per cent must a householder increase its consumption, so as not to decrease expenditure in this item?
a) $17\frac{11}{17}$ b) $17\frac{1}{17}$ c) $19\frac{1}{19}$ d) $19\frac{11}{19}$
5. If the price of salt falls down by $12\frac{1}{2}\%$, by how much per cent must a householder increase its consumption, so as not to decrease expenditure in this item?
a) 14% b) $14\frac{1}{7}\%$ c) 25% d) $14\frac{2}{7}\%$

6. If the price of coffee falls down by $16\frac{2}{3}\%$, by how much per cent must a householder increase its consumption, so as not to decrease expenditure in this item?

a) 25% b) $16\frac{2}{3}\%$ c) 20% d) 18%

7. The price of an article is cut by 10%. To restore it to the former value, the new price must be increased by

a) 10% b) $9\frac{1}{11}\%$ c) $11\frac{1}{9}\%$ d) 11%

[CPO Exam, 1990]

Answers

1. a 2. c 3. c 4. a 5. d 6. c 7. c

Rule 34

Theorem: ***If first value is r% more than the second value, then the second is*** $\left[\frac{r}{100+r}\times 100\right]\%$ ***less than the first value.***

Illustrative Example

Ex.: If A's salary is 25% more than that of B, then how much per cent is B's salary less than that of A?

Soln: The required answer $=\frac{25}{100+25}\times 100\% = 20\%$

Exercise

1. If A's salary is 20% more than that of B, then how much per cent is B's salary less than that of A?

a) $16\frac{2}{3}\%$ b) 20% c) 40% d) 10%

2. If A's salary is 10% more than that of B, then how much per cent is B's salary less than that of A?

a) $9\frac{1}{11}\%$ b) $11\frac{1}{9}\%$ c) 10% d) 20%

3. If A's salary is 5% more than that of B, then how much per cent is B's salary less than that of A?

a) 10% b) $5\frac{14}{21}\%$ c) $4\frac{16}{21}\%$ d) 5%

4. A number is 50% more than the other. Then how much per cent is the second number less than the first?

a) 50% b) 25% c) $33\frac{2}{3}\%$ d) $33\frac{1}{3}\%$

5. A number is 60% more than the other. Then how much per cent is the second number less than the first?

a) $22\frac{1}{2}\%$ b) $27\frac{1}{2}\%$ c) $37\frac{1}{2}\%$ d) 60%

6. A number is $16\frac{2}{3}\%$ more than the other. Then how much per cent is the second number less than the first?

a) $14\frac{2}{7}\%$ b) 15% c) $26\frac{2}{3}\%$ d) $14\frac{1}{7}\%$

Answers

1. a 2. a 3. c 4. d 5. c 6. a

Rule 35

Theorem: ***If the first value is r% less than the second value then, the second value is*** $\left[\frac{r}{100-r}\times 100\right]\%$ ***more than the first value.***

Illustrative Example

Ex.: If A's salary is 30% less than that of B, then how much per cent is B's salary more than that of A?

Soln: The required answer $=\frac{30}{100-30}\times 100 = 42\frac{6}{7}\%$

Exercise

1. If A's salary is 5% less than that of B, then how much per cent is B's salary more than that of A?

a) 5% b) $5\frac{5}{19}\%$ c) $5\frac{3}{19}\%$ d) 10%

2. If A's salary is 10% less than that of B, then how much per cent is B's salary more than that of A?

a) $11\frac{1}{9}\%$ b) $9\frac{1}{11}\%$ c) 5% d) 20%

3. If A's salary is 15% less than that of B, then how much per cent is B's salary more than that of A?

a) $17\frac{1}{17}$ b) $17\frac{2}{17}$ c) $17\frac{11}{17}$ d) 15%

4. If A's salary is 20% less than that of B, then how much per cent is B's salary more than that of A?

a) 21% b) $\frac{25}{2}\%$ c) 30% d) 25%

5. If A's salary is 25% less than that of B, then how much per cent is B's salary more than that of A?

a) 20% b) $33\frac{1}{3}\%$ c) 25% d) 30%

6. If A's salary is 50% less than that of B, then how much per cent is B's salary more than that of A?

a) 50% b) 75% c) 100% d) Can't be determined

Answers

1. b 2. a 3. c 4. d 5. b 6. c

Rule 36

Theorem: ***If the value of a number is first increased by x% and later decreased by x%, then net change is always a decrease which is equal to x% of x or*** $\frac{x^2}{100}$.

Illustrative Examples

Ex. 1: The salary of a worker is first increased by 10% and thereafter it was reduced by 10%. What was the change in his salary?

Soln: There is a decrease in his salary $= \frac{(10)^2}{100} = 1\%$

Ex. 2: A shopkeeper marks the price of his goods 12% higher than its original price. After that, he allows a discount of 12%. What is his percentage profit or loss?

Soln: In this case, there is always a loss. And the % value of loss $= \frac{(12)^2}{100} = 1.44\%$

Ex. 3: If the population of a town is increased by 15% in the first year and is decreased by 15% in the next year, what effect can be seen in the population of that town?

Soln: There is a decrease of $\frac{(15)^2}{100}\%$ i.e., 2.25%

Exercise

1. The salary of a worker is first increased by 5% and thereafter it was reduced by 5%. What was the change in his salary?
   a) Increase in his salary, increase % is 0.25
   b) Decrease in his salary, decrease % is 0.25
   c) Increase in his salary, increase % is 4%
   d) Decrease in his salary, decrease % is 0.5
2. The salary of a worker is first increased by 20% and thereafter it was reduced by 20%. What was the change in his salary?
   a) Decrease in his salary, decrease % is 4
   b) Decrease in his salary, decrease % is 0.4
   c) Increase in his salary, increase % is 4
   d) Decrease in his salary, decrease % is 5
3. The salary of a worker is first increased by 13% and thereafter it was reduced by 13%. What was the change in his salary?
   a) Profit, 1.69% b) Loss, 1.69%
   c) Loss, 1.09% d) Profit, 1.09%
4. The salary of a worker is first increased by 14% and thereafter it was reduced by 14%. What was the change in his salary?
   a) Loss, 1.96% b) Loss, 2.56%
   c) Profit, 1.96% d) Loss, 2.56%
5. The salary of a worker is first increased by 21% and thereafter it was reduced by 21%. What was the change in his salary?
   a) There is a decrease of 4.31%
   b) There is a increase of 4.41%
   c) There is a increase of 4.31%
   d) There is a decrease of 4.41%
6. The salary of a worker is first increased by 25% and thereafter it was reduced by 25%. What was the change in his salary?
   a) There is a decrease of 6.25%
   b) There is a decrease of 62.55%
   c) There is a increase of 6.25%
   d) There is a increase of 62.5%

Answers

1. b 2. a 3. b 4. a 5. d 6. a

Rule 37

Theorem: ***If the value is first increased by x% and then decreased by y%, then there is*** $\left(x - y - \frac{xy}{100}\right)\%$ ***increase or decrease, according to the +ve or –ve sign respectively.***

Illustrative Examples

Ex. 1: The salary of a worker was first increased by 10% and thereafter, decreased by 5%. What was the change in his salary?

Soln: Thus, in this case, $10 - 5 - \frac{10 \times 5}{100} = 4.5\%$ increase as the sign is +ve.

Ex. 2: A shopkeeper marks the prices of his goods at 20% higher than the original price. After that, he allows a discount of 10%. What profit or loss did he get?

Soln: By the theorem: $20 - 10 - \frac{20 \times 10}{100} = 8\%$

$\therefore$ he gets 8% profit as the sign obtained is +ve.

Note: If the order of increase and decrease is changed, the result remains unaffected ie if the value is first decreased by x% and then increased by y%, then there is $\left(x - y - \frac{xy}{100}\right)\%$ increase or decrease, according to the +ve or –ve sign respectively.

In other words, we may write this theorem as

% effect = % increase – % decrease – $\frac{\%\ increase \times decrease}{100}$

Illustrative Examples

Ex. 1: If the salary of a worker is first decreased by 20% and then increased by 10%. What is the percentage effect on his salary?

Soln: By Quicker Maths:

% effect = % increase – % decrease –

$$\frac{\%\ increase \times \%\ decrease}{100}$$

$$= 10 - 20 - \frac{10 \times 20}{100} = -12\%$$

∴ His salary is decreased by 12% (because the sign is -ve).

(Change of order of increase and decrease means that in the above example, firstly an increase of 10% is performed and then the decrease of 20% is performed. In both the cases, the result remains the same.)

Ex. 2: The population of a town was reduced by 12% in the year 1988. In 1989, it was increased by 15%. What is the percentage effect on the population in the beginning of 1990?

Soln: % effect = % increase – % decrease

$$- \frac{\%\ increase \times \%\ decrease}{100}$$

$$= 15 - 12 - \frac{15 \times 12}{100} = 3 - 1.8 = 1.2$$

Thus, the population is increased by 1.2%.

Exercise

1. The salary of a worker was first increased by 7% and thereafter, decreased by 5%. What was the change in his salary?
   a) increase, $\frac{33}{20}\%$ b) decrease, $\frac{33}{20}\%$
   c) increase, $\frac{23}{20}\%$ d) decrease, $\frac{23}{20}\%$
2. The salary of a worker was first increased by 10% and thereafter, decreased by 15%. What was the change in his salary?
   a) increase, 6.5% b) decrease, 6.5%
   c) increase 5.5% d) decrease 5.5%
3. The salary of a worker was first increased by 15% and thereafter, decreased by 12%. What was the change in his salary?
   a) increase, 12% b) decrease, 1.02%
   c) increase, 1.2% d) increase, 1.02%
4. A shopkeeper marks the prices of his goods at 25% higher than the original price. After that, he allows a discount of 12%. What profit or loss did he get?
   a) Profit, 15% b) Profit, 13%
   c) Profit, 10% d) Loss, 10%
5. A shopkeeper marks the prices of his goods at 30% higher than the original price. After that, he allows a discount of 20%. What profit or loss did he get?
   a) Loss, 10% b) Loss, 4%
   c) Profit, 4% d) Loss, 4%
6. A shopkeeper marks the prices of his goods at 50% higher than the original price. After that, he allows a discount of 16%. What profit or loss did he get?
   a) Loss, 26% b) Profit, 26%
   c) Profit, 34% d) Profit, 17%
7. If the salary of a worker is first decreased by 15% and then increased by 5%. What is the percentage effect on his salary?
   a) Decrease of 10% b) Increase of 10%
   c) Decrease of 10.75% d) Increase of 10.75%
8. The population of a town was reduced by 25% in the year 1988. In 1989, it was increased by 20%. What is the percentage effect on the population in the beginning of 1990?
   a) Decrease of 20% b) Increase of 10%
   c) Decrease of 10% d) Increase of 20%

Answers

1. a 2. b 3. c 4. c 5. c 6. b
7. c 8. c

Rule 38

Theorem: ***If the value is increased successively by x% and x% then the final increase is given by*** $\left[2x + \frac{x^2}{100}\right]\%$.

Illustrative Example

Ex.: A shopkeeper marks the prices of his goods at 20% higher than the original price. Due to increase in demand he again increases by 20%. What profit did he get?

Soln: Applying the above formula, we have

$$\text{the required profit} = 2 \times 20 + \frac{(20)^2}{100} = 40 + 4 = 44\%$$

Exercise

1. A shopkeeper marks the prices of his goods at 5% higher than the original price. Due to increase in demand he again increases by 5%. What profit did he get?
   a) 10% b) $10\frac{1}{4}\%$ c) 5% d) 12%
2. A shopkeeper marks the prices of his goods at 25% higher than the original price. Due to increase in demand he again increases by 25%. What profit did he get?

a) 50% b) 56% c) 56.25% d) 60%

3. A shopkeeper marks the prices of his goods at 15% higher than the original price. Due to increase in demand he again increases by 15%. What profit did he get?
a) 32.5% b) 32.25% c) 30% d) 32%
4. A shopkeeper marks the prices of his goods at 16% higher than the original price. Due to increase in demand he again increases by 16%. What profit did he get?
a) 34.56% b) 32% c) 34% d) 35%
5. A shopkeeper marks the prices of his goods at 26% higher than the original price. Due to increase in demand he again increases by 26%. What profit did he get?
a) 52% b) 58% c) 60% d) 58.76%

Answers

1. b 2. c 3. b 4. a 5. d

Rule 39

Theorem: ***If the value is increased successively by x% and y% then the final increase is given by*** $\left[x+y+\frac{xy}{100}\right]\%$.

Illustrative Example

Ex.: A shopkeeper marks the prices at 15% higher than the original price. Due to increase in demand, he further increases the price by 10%. How much % profit will he get?

Soln: By theorem: % profit $= 15+10+\frac{15\times10}{100} = 26.5\%$

Exercise

1. A shopkeeper marks the prices at 5% higher than the original price. Due to increase in demand, he further increases the price by 10%. How much % profit will he get?
a) 15% b) 15.25% c) 15.5% d) 16%
2. A shopkeeper marks the prices at 5% higher than the original price. Due to increase in demand, he further increases the price by 15%. How much % profit will he get?
a) 20% b) 20.25% c) 20.75% d) 20.5%
3. A shopkeeper marks the prices at 20% higher than the original price. Due to increase in demand, he further increases the price by 15%. How much % profit will he get?
a) 38% b) 40% c) 38.75% d) 35%
4. A shopkeeper marks the prices at 10% higher than the original price. Due to increase in demand, he further increases the price by 25%. How much % profit will he get?
a) 37% b) 35% c) 37.05% d) 37.5%
5. A shopkeeper marks the prices at 25% higher than the original price. Due to increase in demand, he further increases the price by 20%. How much % profit will he get?
a) 45% b) 48.25% c) 50% d) 50.5%

Answers

1. c 2. c 3. a 4. d 5. c

Rule 40

Theorem: If the value is decreased successively by x% and y% then the final decrease is given by $\left[x+y-\frac{xy}{100}\right]\%$.

Illustrative Example

Ex.: The population of a town is decreased by 10% and 20% in two successive years. What per cent population is decreased after two years?

Soln: Following the above theorem, we have

$$\text{per cent decrease in population} = 10+20-\frac{10\times20}{100}$$
$$= 30-2 = 28.$$

Exercise

1. The population of a town is decreased by 5% and 10% in two successive years. What per cent population is decreased after two years?
a) 15% b) 14% c) 14.5% d) 15.5%
2. The population of a town is decreased by 8% and 5% in two successive years. What per cent population is decreased after two years?
a) 13% b) 12.6% c) 12.5% d) 13%
3. The population of a town is decreased by 15% and 20% in two successive years. What per cent population is decreased after two years?
a) 32% b) 35% c) 32.5% d) 34.5%
4. The population of a town is decreased by 25% and 40% in two successive years. What per cent population is decreased after two years?
a) 65% b) 56% c) 55.5% d) 55%
5. The population of a town is decreased by 20% and 25% in two successive years. What per cent population is decreased after two years?
a) 40% b) 45% c) 35% d) 35.5%

Answers

1. c 2. b 3. a 4. d 5. a

Rule 41

Theorem: ***If the value is decreased successively by x% and x% then the final decrease is given by*** $\left[2x-\frac{x^2}{100}\right]\%$.

Illustrative Example

Ex.: The population of a town was reduced by 12% in the year 1988. In 1989, it was again reduced by 12%. What is the percentage in the population in the beginning of 1990?

Soln: Applying the above formula, we have

the required answer $= 2 \times 12 - \frac{12^2}{100}$

$= 24 - 1.44 = 22.56\%$

Exercise

1. The population of a town was reduced by 5% in the year 1988. In 1989, it was again reduced by 5%. What is the percentage in the population in the beginning of 1990?
   a) 10% b) 9.5% c) 9.75% d) 10.25%
2. The population of a town was reduced by 10% in the year 1988. In 1989, it was again reduced by 10%. What is the percentage in the population in the beginning of 1990?
   a) 20% b) 19% c) 19.5% d) 18.5%
3. The population of a town was reduced by 16% in the year 1988. In 1989, it was again reduced by 16%. What is the percentage in the population in the beginning of 1990?
   a) 34.56% b) 32% c) 29.44% d) 32.44%
4. The population of a town was reduced by 20% in the year 1988. In 1989, it was again reduced by 20%. What is the percentage in the population in the beginning of 1990?
   a) 40% b) 36% c) 44% d) 36.5%
5. The population of a town was reduced by 15% in the year 1988. In 1989, it was again reduced by 15%. What is the percentage in the population in the beginning of 1990?
   a) 27.75% b) 30% c) 27.5% d) 28%
6. The difference between a discount of 35% and two successive discounts of 20% and 20% on a certain bill was Rs 22. Find the amount of the bill.
   a) Rs 1100 b) Rs 200
   c) Rs 2200 d) Data inadequate
   [BSRB Mumbai PO, 1999]

Answers

1. c 2. b 3. c 4. b 5. a

6. c; **Hint:** Equivalent discount of two succesive discounts

$= 2 \times 20 - \frac{(20)^2}{100} = 36\%$

Now, from the question, 36% - 35% = Rs 22

$\therefore$ Amount of the bill = Rs 22 × 100 = Rs 2200

Rule 42

Theorem:

(i) ***If the one factor is decreased by x% and the other factor is increased by y%,***

(ii) ***or, if the one factor is increased by x% and the other factor is decreased by y% then the effect on the product = Increase % value – Dec. % value – $\frac{\text{Inc. \% value} \times \text{Dec. \% value}}{100}$ and the value is increased or decreased according to the +ve or –ve sign obtained.***

Note: The above written formula is the general form of both the cases.

For Case (i) it becomes: $y - x - \frac{yx}{100}$

Whereas for Case (ii) it becomes: $x - y - \frac{xy}{100}$

Thus, we see that it is more easy to remember the general formula which works in both the cases equally.

Illustrative Examples

Ex. 1: The tax on commodity is diminished by 20% and its consumption increases by 15%. Find the effect on revenue.

Soln: **Detail Method:** New Revenue = Consumption × Tax

$= (115\% \times 80\%)$ of the original

$= \left(\frac{115}{100} \times \frac{80}{100}\right)$ of the original

$= \left(\frac{115}{100} \times 80\right)\%$ of original = 92% of original

Thus, the revenue is decreased by (100 – 92) = 8%

Quicker Method: By Theorem:

Effect on revenue = Inc. % value – Dec. % value – $\frac{\text{Inc. \% value} \times \text{Dec. \% value}}{100}$

$= 15 - 20 - \frac{15 \times 20}{100} = -8\%$

Therefore, there is a decrease of 8%.

Ex. 2: If the price is increased by 10% and the sale is decreased by 5%, then what will be the effect on income?

Soln: By theorem: % effect

= Inc. % value – Dec. % value – $\frac{\text{Inc. \% value} \times \text{Dec. \% value}}{100}$

$= 10 - 5 - \frac{10 \times 5}{100} = 4.5\%$

$\therefore$ his income increases by 4.5%.

Ex. 3: If the price is decreased by 12% and sale is increased by 10% then what will be the effect on income?

Soln: $\% \text{ effect} = 10 - 12 - \frac{12 \times 10}{100} = -3.2\%$

$\therefore$ his income is decreased by 3.2%.

Exercise

1. The tax on commodity is diminished by 15% and its consumption increases by 10%. Find the effect on revenue.
   a) decrease of 6% b) decrease of 5%
   c) increase of 6.5% d) decrease of 6.5%
2. The tax on commodity is diminished by 10% and its consumption increases by 20%. Find the effect on revenue.
   a) decrease of 8% b) decrease of 10%
   c) increase of 8% d) increase of 10%
3. If the price is increased by 12% and the sale is decreased by 5%, then what will be the effect on income?
   a) income increases by 6.6%
   b) income increases by 6.4%
   c) income decreases by 6.6%
   d) income decreases by 6.4%
4. If the price is increased by 16% and the sale is decreased by 15%, then what will be the effect on income?
   a) income decreases by 1.4%
   b) income increases by 1.4%
   c) income decreases by 1.5%
   d) income increases by 1.5%
5. If the price is increased by 5% and the sale is decreased by 16%, then what will be the effect on income?
   a) income increases by 11%
   b) income decreases by 10.2%
   c) income decreases by 11.8%
   d) income increases by 11.2%

Answers

1. d 2. c 3. b 4. a 5. c

Rule 43

Theorem: ***(i) If one factor is decreased by x% and the other factor is increased by x%,***
(ii) or, if one factor is increased by x% and the other factor is decreased by x% then the product will always decrease and the effet on the product is given by $\left[\frac{x^2}{100}\right]\%$.

Illustrative Example

Ex.: Tax on commodity is diminished by 25% and consumption increases by 25%. Find the effect on revenue.

Soln: Following the above theorem,

$$\% \text{ decrease in revenue} = \left[\frac{25^2}{100}\right] = 6.25\%$$

Exercise

1. Tax on commodity is diminished by 11% and consumption increases by 11%. Find the effect on revenue.
   a) No change in the revenue
   b) 1.21 per cent decrease in revenue
   c) 1.21 per cent increase in revenue
   d) 1.31 per cent decrease in revenue
2. Tax on commodity is diminished by 17% and consumption increases by 17%. Find the effect on revenue.
   a) 2.89 per cent decrease in revenue
   b) 3.89 per cent decrease in revenue
   c) 2.79 per cent decrease in revenue
   d) 2.89 per cent increase in revenue
3. Tax on commodity is diminished by 19% and consumption increases by 19%. Find the effect on revenue.
   a) 3.61 per cent increase in revenue
   b) 3.71 per cent increase in revenue
   c) 3.61 per cent decrease in revenue
   d) 2.61 per cent decrease in revenue
4. Tax on commodity is diminished by 22% and consumption increases by 22%. Find the effect on revenue.
   a) 4.44 per cent increase in revenue
   b) 4.44 per cent decrease in revenue
   c) 4.84 per cent increase in revenue
   d) 4.84 per cent decrease in revenue
5. Tax on commodity is diminished by 29% and consumption increases by 29%. Find the effect on revenue.
   a) 8.41 per cent decrease in revenue
   b) 8.41 per cent increase in revenue
   c) 7.41 per cent increase in revenue
   d) 7.41 per cent decrease in revenue

Answers

1. b 2. a 3. c 4. d 5. a

Rule 44

Theorem: ***If one factor is decreased by x% and the other factor also decreases by y%, then the effect on the product is given by*** $\left[x + y - \frac{xy}{100}\right]\%$ ***decrease.***

Illustrative Example

Ex.: The land holding of a person is decreased by 10% due to late monsoon, the production decreases by 8%. Then what is the effect on the revenue?

Soln: Applying the above formula, we have

$$\text{the decrease in revenue} = 10 + 8 - \frac{10 \times 8}{100} = 17.2\%$$

Note: Revenue is directly proportional to (landholding × production)

Exercise

1. The land holding of a person is decreased by 12% due to late monsoon, the production decreases by 4%. Then what is the effect on the revenue?
   a) 16% b) 15% c) 15.48% d) 15.52%

2. The land holding of a person is decreased by 15% due to late monsoon, the production decreases by 12%. Then what is the effect on the revenue?
   a) 27% b) 12.18%
   c) 26.8% d) 25.2%
3. The land holding of a person is decreased by 20% due to late monsoon, the production decreases by 16%. Then what is the effect on the revenue?
   a) 36% b) 32.8%
   c) 32.2% d) 33%

Answers

1. d 2. d 3. b

Rule 45

Theorem: ***If one factor is decreased by x% and the other factor also decreases by x%, then the effect on the product is given by*** $\left[2x-\frac{x^2}{100}\right]\%$ ***decrease.***

Illustrative Example

Ex.: The number of seats in a cinema hall is decreased by 5%. The price on a ticket is also decreased by 5%. What is the effect on the revenue collected?

Soln: Applying the above formula, we have the decrease in the revenue collected

$$=10-\frac{5^2}{100}=10-\frac{1}{4}=9.75\%$$

Note: Here revenue collected is directly proportional to the product of number of seats in a cinema hall and the price on a ticket.

Exercise

1. The number of seats in a cinema hall is decreased by 8%. The price on a ticket is also decreased by 8%. What is the effect on the revenue collected?
   a) 15.36% decrease b) 16% increase
   c) 15.64% decrease d) 15.64% increase
2. The number of seats in a cinema hall is decreased by 12%. The price on a ticket is also decreased by 12%. What is the effect on the revenue collected?
   a) 24% decrease b) 24.44% decrease
   c) 24.56% decrease d) 22.56% decrease
3. The number of seats in a cinema hall is decreased by 24%. The price on a ticket is also decreased by 24%. What is the effect on the revenue collected?
   a) 42.76% decrease
   b) 42.44% decrease
   c) 48% decrease
   d) 42.34% decrease

Answers

1. a 2. d 3. b

Rule 46

Theorem: ***If one factor is increased by x% and the other increases by y% then the effect on the product is given by*** $\left[x+y+\frac{xy}{100}\right]\%$ ***increase.***

Illustrative Example

Ex.: The number of seats in a cinema hall is increased by 25%. The price on a ticket is also increased by 10%. What is the effect on the revenue collected?

Soln: Following the above formula, we have the increase in the revenue

$$=25+10+\frac{25\times10}{100}=35+2.5=37.5\%$$

Exercise

1. The number of seats in a cinema hall is increased by 30%. The price on a ticket is also increased by 5%. What is the effect on the revenue collected?
   a) 36.5% increase b) 35.5% increase
   c) 35% increase d) 36% increase
2. The number of seats in a cinema hall is increased by 25%. The price on a ticket is also increased by 20%. What is the effect on the revenue collected?
   a) 45% increase b) 49% increase
   c) 50% increase d) 52% increase
3. The number of seats in a cinema hall is increased by 16%. The price on a ticket is also increased by 12%. What is the effect on the revenue collected?
   a) 29.92% increase b) 28% increase
   c) 26.08% increase d) 28.92% increase

Answers

1. a 2. c 3. a

Rule 47

Theorem: ***If one factor is increased by x% and the other factor also increases by x% then the effect on the product is given by*** $\left[2x+\frac{x^2}{100}\right]\%$ ***increase.***

Illustrative Example

Ex.: The landholding of a person is increased by 10%. Due to early monsoon, the production increases by 10%. Then what is the effect on revenue?

Soln: Applying the above formula, we have

$$\%\text{ increase in revenue}=\left[20+\frac{100}{100}\right]=21\%$$

Exercise

1. The landholding of a person is increased by 21%. Due to early monsoon, the production increases by 21%. Then

what is the effect on revenue?
a) 46.41% increase b) 42% increase
c) 24.41% increase d) 46.59% increase

2. The landholding of a person is increased by 31%. Due to early monsoon, the production increases by 31%. Then what is the effect on revenue?
a) 71.51% increase b) 71.61% increase
c) 62.51% increase d) 63% increase

3. The landholding of a person is increased by 23%. Due to early monsoon, the production increases by 23%. Then what is the effect on revenue?
a) 46% increase b) 52.29% increase
c) 51.29% increase d) 49.29% increase

Answers

1. a 2. b 3. c

Rule 48

Theorem: ***The pass marks in an examination is x%. If a candidate who secures y marks fails by z marks, then the maximum marks, is given by*** $\frac{100(y+z)}{x}$.

Illustrative Examples

Ex. 1: A student has to secure 40% marks to get through. If he gets 40 marks and fails by 40 marks, find the maximum marks set for the examination.

Soln: By the above theorem,

$$\text{maximum marks} = \frac{100(40+40)}{40} = 200.$$

Ex. 2: In an examination, a candidate must get 80% marks to pass. If a candidate who gets 210 marks, fails by 50 marks, find the maximum marks.

Soln: By the above theorem, we have

$$\text{the maximum marks} = \frac{100(210+50)}{80} = 325.$$

Exercise

1. A student has to secure 30% marks to get through. If he gets 40 marks and fails by 20 marks, find the maximum marks set for the examination.
a) 600 b) 200 c) 100 d) 300

2. A student has to secure 15% marks to get through. If he gets 80 marks and fails by 70 marks, find the maximum marks set for the examination.
a) 100 b) 1000 c) 1500 d) 900

3. A student has to secure 16% marks to get through. If he gets 55 marks and fails by 25 marks, find the maximum marks set for the examination.
a) 400 b) 500 c) 550 d) 450

4. In an examination, a candidate must get 60% marks to pass. If a candidate who gets 120 marks, fails by 60 marks, find the maximum marks.
a) 300 b) 600 c) 250 d) 350

5. In an examination, a candidate must get 75% marks to pass. If a candidate who gets 90 marks, fails by 60 marks, find the maximum marks.
a) 300 b) 400 c) 200 d) 100

6. In an examination, a candidate must get 65% marks to pass. If a candidate who gets 645 marks, fails by 200 marks, find the maximum marks.
a) 1300 b) 1350 c) 1200 d) 1260

Answers

1. b 2. b 3. b 4. a 5. c 6. a

Rule 49

Theorem: ***A candidate scoring x% in an examination fails by 'a' marks, while another candidate who scores y% marks gets 'b' marks more than the minimum required pass marks. Then the maximum marks for that examination are M*** $= \frac{100(a+b)}{y-x}$.

Illustrative Example

Ex.: A candidate scores 25% and fails by 30 marks, while another candidate who scores 50% marks, gets 20 marks more than the minimum required marks to pass the examination. Find the maximum marks for the examination.

Soln: By the theorem, we have

$$\text{maximum marks} = \frac{100(30+20)}{50-25} = 200.$$

Note: (i) The above formula can be written as

$$\text{Maximum marks} = \frac{100(\text{Diff. of their scores})}{\text{Diff. of their \% marks}}$$

(ii) Difference of their scores = 30 + 20. Because the first candidate gets 30 less than the required pass marks, while the second candidate gets 20 more than the required pass marks.

Exercise

1. A candidate scores 35% and fails by 40 marks, while another candidate who scores 60% marks, gets 35 marks more than the minimum required marks to pass the examination. Find the maximum marks for the examination.
a) 300 b) 200 c) 350 d) 450

2. A candidate scores 46% and fails by 55 marks, while another candidate who scores 81% marks, gets 15 marks more than the minimum required marks to pass the examination. Find the maximum marks for the examination.
a) 350 b) 100 c) 150 d) 200

3. A candidate scores 26% and fails by 49 marks, while another candidate who scores 36% marks, gets 36 marks

more than the minimum required marks to pass the examination. Find the maximum marks for the examination.

a) 850 b) 750 c) 600 d) 800

4. A candidate scores 39% and fails by 58 marks, while another candidate who scores 55% marks, gets 22 marks more than the minimum required marks to pass the examination. Find the maximum marks for the examination.

a) 450 b) 650 c) 500 d) 550

5. A candidate scores 25% and fails by 45 marks, while another candidate who scores 50% marks, gets 5 marks more than the minimum required marks to pass the examination. Find the maximum marks for the examination.

a) 100 b) 150 c) 250 d) 200

6. A candidate scoring 25% in an examination fails by 30 marks while another candidate who scores 50% marks gets 20 marks more than the minimum required for a pass. Find the minimum pass percentage.

a) 20% b) 80% c) 40% d) 50%

[Hotel Management, 1991]

Answers

1. a 2. d 3. a 4. c 5. d

6. c; **Hint:** Applying the above formula,

$$\text{maxm. marks} = \frac{100 \times 50}{25} = 200$$

$\therefore$ Minimum pass marks = 25% of 200 + 30 = 80

$\therefore$ Minimum pass percentage

$$= \left(\frac{80}{200} \times 100\right)\% = 40\%$$

Rule 50

Theorem: In measuring the sides of a rectangle, one side is taken x% in excess and the other y% in deficit. The error per cent in area calculated from the measurement is $x - y - \frac{xy}{100}$ ***in excess or deficit, according to the +ve or –ve sign.***

In another form this may be written as

$$\%\ error = \%\ excess - \%\ deficit - \frac{\%\ excess \times \%\ deficit}{100}$$

Illustrative Example

Ex.: In measuring the sides of a rectangle, one side is taken 5% in excess and the other 4% in deficit. Find the error per cent in area calculated from the measurement.

Soln: By the above theorem,

$$\%\ \text{error} = 5 - 4 - \frac{5 \times 4}{100} = 1 - \frac{1}{5} = \frac{4}{5}\%\ \text{excess because}$$

sign is +ve.

Exercise

1. In measuring the sides of a rectangle, one side is taken 3% in excess and the other 5% in deficit. Find the error per cent in area calculated from the measurement.

a) 2% deficit b) 3.15% deficit

c) 2.15% deficit d) 2.15% excess

2. In measuring the sides of a rectangle, one side is taken 10% in excess and the other 4% in deficit. Find the error per cent in area calculated from the measurement.

a) $5\frac{1}{5}\%$ excess b) $5\frac{2}{5}\%$ excess

c) 6% excess d) $5\frac{3}{5}\%$ excess

3. In measuring the sides of a rectangle, one side is taken 12% in excess and the other 5% in deficit. Find the error per cent in area calculated from the measurement.

a) $7\frac{3}{5}\%$ excess b) $7\frac{2}{5}\%$ excess

c) $6\frac{2}{5}\%$ excess d) $6\frac{3}{5}\%$ excess

4. In measuring the sides of a rectangle, one side is taken 10% in excess and the other 20% in deficit. Find the error per cent in area calculated from the measurement.

a) 8% excess b) 8% deficit

c) 12% excess d) 12% deficit

Answers

1. c 2. d 3. c 4. b

Rule 51

Theorem: ***If one of the sides of a rectangle is increased by x% and the other is increased by y% then the per cent value by which area changes is given by*** $\left[x + y + \frac{xy}{100}\right]\%$ ***increase.***

Illustrative Examples

Ex.: If one of the sides of a rectangle is increased by 20% and the other is increased by 5%, find the per cent value by which the area changes.

Soln: Following the above formula,

$$\%\ \text{increase} = 20 + 5 + \frac{20 \times 5}{100} = 26\%$$

Exercise

1. If one of the sides of a rectangle is increased by 20% and the other is increased by 10%, find the per cent value by which the area changes.

a) 32% b) 30% c) 36% d) 34%

2. If one of the sides of a rectangle is increased by 25% and the other is increased by 4%, find the per cent value by

which the area changes.

a) 29% b) 30% c) 30.5% d) 30.25%

3. If one of the sides of a rectangle is increased by 13% and the other is increased by 2%, find the per cent value by which the area changes.

a) 15.36% b) 15.26% c) 16.26% d) 16.36%

4. If one of the sides of a rectangle is increased by 25% and the other is increased by 15%, find the per cent value by which the area changes.

a) 43.75% b) 40.75% c) 40% d) 43.25%

5. If one of the sides of a rectangle is increased by 30% and the other is increased by 20%, find the per cent value by which the area changes.

a) 50% b) 59% c) 56% d) 55%

Answers

1. a 2. b 3. b 4. a 5. c

Rule 52

Theorem: ***If one of the sides of a rectangle is decreased by x% and the other is decreased by y% then the per cent value by which area changes is given by*** $\left[x+y-\frac{xy}{100}\right]\%$ ***decrease.***

Illustrative Example

Ex.: In measuring sides of a rectangle, one side is taken 5% in deficit and the other 4% in deficit. Find the error per cent in area calculated from the measurement.

Soln: Applying the above formula,
percentage decrease in area

$$= 5+4-\frac{5\times 4}{100} = 9-.2 = 8.8\%$$

Exercise

1. In measuring sides of a rectangle, one side is taken 3% in deficit and the other 2% in deficit. Find the error per cent in area calculated from the measurement.

a) 5% b) 4.5% c) 4.94% d) 5.4%

2. In measuring sides of a rectangle, one side is taken 10% in deficit and the other 5% in deficit. Find the error per cent in area calculated from the measurement.

a) 14.5% b) 15% c) 15.5% d) 14%

3. In measuring sides of a rectangle, one side is taken 35% in deficit and the other 5% in deficit. Find the error per cent in area calculated from the measurement.

a) 39.5% b) 39.25% c) 40% d) 38.25%

4. In measuring sides of a rectangle, one side is taken 15% in deficit and the other 8% in deficit. Find the error per cent in area calculated from the measurement.

a) 21.8% b) 22.8% c) 23% d) 22.2%

Answers

1. c 2. a 3. d 4. a

Rule 53

Theorem: ***In an examination x% failed in English and y% failed in maths. If z% of students failed in both the subjects, the percentage of students who passed in both the subjects is*** $100-(x+y-z)$.

Illustrative Example

Ex.: In an examination, 40% of the students failed in Maths, 30% failed in English and 10% failed in both. Find the percentage of students who passed in both the subjects.

Soln: Follwoing the above theorem:
The required % = $100-(40+30-10)=40\%$

Note: We should know that the following sets complete the system, i.e.,
100% = % of students who failed in English only
+ % of students who failed in Maths only
– % of students who failed in both subjects
+ % of students who passed in both subjects.

Exercise

1. In an examination, 10% of the students failed in Maths, 20% failed in English and 5% failed in both. Find the percentage of students who passed in both the subjects.

a) 75% b) 70% c) 85% d) 80%

2. In an examination, 45% of the students failed in Maths, 15% failed in English and 30% failed in both. Find the percentage of students who passed in both the subjects.

a) 70% b) 10% c) 25% d) 75%

3. In an examination, 33% of the students failed in Maths, 24% failed in English and 17% failed in both. Find the percentage of students who passed in both the subjects.

a) 55% b) 60% c) 65% d) 70%

4. In an examination, 46% of the students failed in Maths, 29% failed in English and 25% failed in both. Find the percentage of students who passed in both the subjects.

a) 50% b) 60% c) 65% d) 40%

5. In an examination, 41% of the students failed in Maths, 29% failed in English and 10% failed in both. Find the percentage of students who passed in both the subjects.

a) 50% b) 60% c) 55% d) 40%

6. In an examination 50% of the students failed in Math, 40% failed in English and 10% failed in both. Find the percentage of students who passed in both the subjects.

a) 80% b) 20% c) 30% d) 70%

7. In an examination, 52% of the candidates failed in English, 42% failed in Mathematics and 17% failed in both.

The number of those who have passed in both the subjects is: **[CDS 1991]**

a) 23% b) 35% c) 25% d) 40%

Answers

1. a 2. a 3. b 4. a 5. d 6. b 7. a

Rule 54

Theorem: ***A man spends x% of his income. His income is increased by y% and his expenditure also increases by z%, then the percentage increase in his savings is given by***

$$\left[\frac{100y - xz}{100 - x}\right]\%.$$

Illustrative Example

Ex.: A man spends 75% of his income. His income increases by 20% and his expenditure also increases by 10%. Find the percentage increase in his savings.

Soln: **Detailed Method:** Suppose his monthly income = Rs 100

Thus, he spends Rs 75 and saves Rs 25.

His increased income = 100 + 20% of 100 = Rs 120

His increased expenditure = 75 + 10% of 75 = Rs 82.50

$\therefore$ his new savings = 120 – 82.5 = Rs 37.50

$\therefore$ % increase in his savings

$$= \frac{37.50 - 25}{25} \times 100 = 50\%$$

Quicker Method: Applying the above formula, we have

percentage increase in savings

$$= \frac{20 \times 100 - 10 \times 75}{100 - 75} = \frac{1250}{25} = 50\%$$

Exercise

1. A man spends 60% of his income. His income increases by 15% and his expenditure also increases by 5%. Find the percentage increase in his savings.
   a) 30% b) 15% c) 20% d) 25%
2. A man spends 70% of his income. His income increases by 24% and his expenditure also increases by 15%. Find the percentage increase in his savings.
   a) 35% b) 24% c) 45% d) 55%
3. A man spends 50% of his income. His income increases by 30% and his expenditure also increases by 20%. Find the percentage increase in his savings.
   a) 25% b) 50% c) 60% d) 40%
4. A man spends 80% of his income. His income increases by 40% and his expenditure also increases by 25%. Find the percentage increase in his savings.
   a) 100% b) 50% c) 80% d) 40%
5. A man spends 65% of his income. His income increases by 15% and his expenditure also increases by 14%. Find the percentage increase in his savings.
   a) 17% b) $16\frac{6}{7}\%$ c) $16\frac{5}{7}\%$ d) $17\frac{6}{7}\%$

Answers

1. a 2. c 3. d 4. a 5. b

Rule 55

Theorem: ***A solution of salt and water contains x% salt by weight. Of it 'A' kg water evaporates and the solution now contains y% of salt. The original quantity of solution is given by*** $A\left(\frac{y}{y-x}\right)$ ***kg. In other words, it may be rewritten as the original quantity of solution = Quantity of evaporated water ×*** $\left(\frac{\textit{Final \% of salt}}{\textit{\% Diff. of salt}}\right)$.

Illustrative Example

Ex.: A solution of salt and water contains 15% salt by weight. Of it 30 kg water evaporates and the solution now contains 20% of salt. Find the original quantity of solution.

Soln: **Detail Method:** Suppose there was x kg of solution initially.

Thus quantity of salt = 15% of $x = \frac{15x}{100} = \frac{3x}{20}$ kg

Now, after evaporation, only $(x - 30)$ kg of mixture contains $\frac{3x}{20}$ kg of salt.

or, 20% of(x - 30) = $\frac{3x}{20}$, or, $\frac{x-30}{5} = \frac{3x}{20}$

or, 15x = 20x – 600; $\therefore$ $x = \frac{600}{5} = 120$ kg

Quicker Method: Applying the above rule, we have,

original quantity of solution $= 30\left(\frac{20}{20-15}\right) = 120$ kg

Exercise

1. A solution of salt and water contains 5% salt by weight. Of it 20 kg water evaporates and the solution now contains 15% of salt. Find the original quantity of solution.
   a) 15 kg b) 30 kg c) 18 kg d) 24 kg
2. A solution of salt and water contains 12% salt by weight. Of it 25 kg water evaporates and the solution now contains 17% of salt. Find the original quantity of solution.
   a) 102 kg b) 85 kg c) 68 kg d) 84 kg
3. A solution of salt and water contains 17% salt by weight.

Of it 22 kg water evaporates and the solution now contains 27% of salt. Find the original quantity of solution.
a) 60 kg b) 56.4 kg c) 60.4 kg d) 59.4 kg

4. A solution of salt and water contains 13% salt by weight. Of it 42 kg water evaporates and the solution now contains 20% of salt. Find the original quantity of solution.
a) 160 kg b) 120 kg c) 125 kg d) 145 kg

5. A solution of salt and water contains 14% salt by weight. Of it 32 kg water evaporates and the solution now contains 22% of salt. Find the original quantity of solution.
a) 88 kg b) 66 kg c) 86 kg d) 68 kg

Answers

1. b 2. b 3. d 4. b 5. a

Rule 56

Theorem: ***If x% of a thing is one type, y% of the remaining thing is of second type, z% of the remaining thing is of third type and the value of remaining thing is given as 'A'. Then the total number of things is obtained by the following formula,***

Total no. of things $= A\left(\frac{100}{100-x}\right)\left(\frac{100}{100-y}\right)\left(\frac{100}{100-z}\right)$

Illustrative Example

Ex.: In a library, 20% of the books are in Hindi, 50% of the remaining are in English and 30% of the remaining are in French. The remaining 6300 books are in regional languages. What is the total number of books in the library?

Soln: **Detail Method:** Suppose there are x books in the library.

Then, the no. of books in Hindi = 20% of $x = \frac{x}{5}$

50% of the remaining, i.e. 50% of $\left(x-\frac{x}{5}\right)$ = 50% of $\frac{4x}{5} = \frac{2x}{5}$ are in English.

Now, 30% of the remaining, i.e. 30% of $\left\{x-\left(\frac{x}{5}+\frac{2x}{5}\right)\right\}$ = 30% of $\frac{2x}{5} = \frac{3x}{25}$ books are in French.

Now, $x-\left(\frac{x}{5}+\frac{2x}{5}+\frac{3x}{25}\right) = 6300$

or, $\frac{7x}{25} = 6300 \quad \therefore x = \frac{6300\times 25}{7} = 22500$

Quicker Method: Applying the above rule, we have the no. of total books

$$= 6300\left(\frac{100}{100-20}\right)\left(\frac{100}{100-50}\right)\left(\frac{100}{100-30}\right)$$

$$= \frac{6300\times 100\times 100\times 100}{80\times 50\times 70} = 22{,}500$$

Exercise

1. In a library, 5% of the books are in Hindi, 10% of the remaining are in English and 15% of the remaining are in French. The remaining 5814 books are in regional languages. What is the total number of books in the library?
a) 8000 b) 8140 c) 6000 d) 8500

2. In a library, 12% of the books are in Hindi, 15% of the remaining are in English and 18% of the remaining are in French. The remaining 15334 books are in regional languages. What is the total number of books in the library?
a) 25000 b) 26000 c) 12500 d) 13000

3. In a library, 20% of the books are in Hindi, 25% of the remaining are in English and 30% of the remaining are in French. The remaining 29400 books are in regional languages. What is the total number of books in the library?
a) 35000 b) 70000 c) 45000 d) 90000

4. In a library, 15% of the books are in Hindi, 55% of the remaining are in English and 35% of the remaining are in French. The remaining 1989 books are in regional languages. What is the total number of books in the library?
a) 8500 b) 7500 c) 7000 d) 8000

5. In a library, 8% of the books are in Hindi, 12% of the remaining are in English and 72% of the remaining are in French. The remaining 3542 books are in regional languages. What is the total number of books in the library?
a) 16525 b) 15625 c) 12655 d) 16625

Answers

1. a 2. a 3. b 4. d 5. b

Rule 57

Theorem: ***The manufacturer of an article makes a profit of x%, the wholesale dealer makes a profit of y%, and the retailer makes a profit of z%. If the retailer sold it for Rs 'A' then the manufacturing price of the article is obtained by the following formula,***

Cost of manufacturing $= A\left(\frac{100}{100+x}\right)\left(\frac{100}{100+y}\right)\left(\frac{100}{100+z}\right)$.

Illustrative Example

Ex.: The manufacturer of an article makes a profit of 25%, the wholesale dealer makes a profit of 20%, and the retailer makes a profit of 28%. Find the manufacturing price of the article if the retailer sold it for Rs 48.

Soln: By the above rule,

Cost of manufacturing

$$= 48\left(\frac{100}{100+28}\right)\left(\frac{100}{100+20}\right)\left(\frac{100}{100+25}\right)$$

$$= 48\left(\frac{100}{128}\right)\left(\frac{100}{120}\right)\left(\frac{100}{125}\right) = \text{Rs } 25.$$

Exercise

1. The manufacturer of an article makes a profit of 5%, the wholesale dealer makes a profit of 10%, and the retailer makes a profit of 15%. Find the manufacturing price of the article if the retailer sold it for Rs 5313.
   a) Rs 4000 b) Rs 4500 c) Rs 5000 d) Rs 4950
2. The manufacturer of an article makes a profit of 20%, the wholesale dealer makes a profit of 25%, and the retailer makes a profit of 30%. Find the manufacturing price of the article if the retailer sold it for Rs 39.
   a) Rs 25 b) Rs 30 c) Rs 20 d) Rs 24
3. The manufacturer of an article makes a profit of 8%, the wholesale dealer makes a profit of 12%, and the retailer makes a profit of 16%. Find the manufacturing price of the article if the retailer sold it for Rs 21924.
   a) Rs 15625 b) Rs 16525 c) Rs 15655 d) Rs 14625

Answers

1. a 2. c 3. a

Rule 58

Theorem: ***In 'A' litres of x% acidic liquid, the amount of water to be added to make y% acidic liquid is*** $\frac{A(x-y)}{y}$ ***litres.***

Note: Here, x is always greater than y.

Illustrative Example

Ex.: What quantity of water should be added to reduce 9 litres of 50% acidic liquid to 30% acidic liquid?

Soln: **Detailed Method:**

Acid in 9 litres = 50% of 9 = 4.5 litres.

Suppose x litres of water are added. Then, there are 4.5 litres of acid in (9 + x) litres of diluted liquid.

Now, according to the question,

30% of (9 + x) = 4.5

or, $\frac{3}{10}(9+x) = 4.5$

or, 27 + 3x = 45 or, 3x = 18

$\therefore x = \frac{18}{3} = 6$ litres.

Quicker Method: Applying the above rule, we have

the quantity of water to be added = $\frac{9(50-30)}{30} = 6$ litres.

Exercise

1. What quantity of water should be added to reduce 5 litres of 45% acidic liquid to 25% acidic liquid?
   a) 3 litres b) 2 litres c) 4 litres d) 4.5 litres
2. What quantity of water should be added to reduce 10 litres of 15% acidic liquid to 5% acidic liquid?
   a) 9 litres b) 20 litres c) 18 litres d) 15 litres
3. What quantity of water should be added to reduce 24 litres of 12% acidic liquid to 9% acidic liquid?
   a) 8 litres b) 6 litres c) 9 litres d) 8.5 litres
4. What quantity of water should be added to reduce 16 litres of 25% acidic liquid to 20% acidic liquid?
   a) 5 litres b) 4 litres c) 12 litres d) 8 litres
5. What quantity of water should be added to reduce 6 litres of 50% acidic liquid to 20% acidic liquid?
   a) 8 litres b) 9 litres c) 12 litres d) 9.5 litres

Answers

1. c 2. b 3. a 4. b 5. b

Rule 59

Theorem: ***In 'A' litres of x% acidic liquid, the amount of water to be taken out from the acidic liquid to make y% acidic liquid is*** $\frac{A(y-x)}{y}$ ***litres.***

Note: *Here, y is always greater than x ie acidic liquid is concentrated.*

Illustrative Example

Ex.: What quantity of water should be taken out to concentrate 15 litres of 40% acidic liquid to 60% acidic liquid.

Soln: **Detailed Method:** Acid in 15 litres = 40% of 15 = 6 litres

Suppose x litres of water are taken out.

Then, there are 6 litres of acid in (15 – x) litres of concentrated liquid.

Now, according to the question

60% of $(15 - x) = 6$

or, $\frac{3}{5}(15-x) = 6$

or, $15 - x = 10$ or, $x = 5$ litres.

Quicker Method: Following the above formula, we have

the required answer $= \frac{15(60-40)}{60} = 5$ litres

Exercise

1. What quantity of water should be taken out to concentrate 12 litres of 30% acidic liquid to 40% acidic liquid.
   a) 4 litres b) 6 litres c) 3 litres d) 8 litres
2. What quantity of water should be taken out to concentrate 21 litres of 25% acidic liquid to 35% acidic liquid.
   a) 6 litres b) 8.4 litres c) 6.4 litres d) 8 litres
3. What quantity of water should be taken out to concentrate 27 litres of 12% acidic liquid to 18% acidic liquid.
   a) 6 litres b) 12 litres c) 13.5 litres d) 9 litres
4. What quantity of water should be taken out to concentrate 29 litres of 17% acidic liquid to 29% acidic liquid.
   a) 12 litres b) 13 litres c) 12.5 litres d) 13.5 litres

Answers

1. c 2. a 3. d 4. a

Rule 60

Theorem: ***When a certain quantity of goods B is added to change the percentage of goods A in a mixture of A and B then the quantity of B to be added is***

$$\left[\frac{\text{Previous \% value of } A}{\text{Changed \% value of } A} \times \text{Mixture Quantity}\right] - \text{Mixture Quantity}$$

Illustrative Example

Ex.: In 1 kg mixture of sand and iron, 20% is iron. How much sand should be added so that the proportion of iron becomes 10%?

Soln: **Detail Method:** In 1 kg mixture, iron = 20% of 1000 gm = 200 gm and sand = 800 gm

Suppose x gm sand is added to the mixture

Then, total mixture = (1000 + x) gm

Now, % of iron $= \frac{200}{(1000+x)} \times 100 = 10$ (given)

or, 1000 + x = 2000

∴ x = 1000 gm = 1 kg

Quicker Method: Applying the above rule, we have the required quantity of sand to be added

$$= \frac{20}{10} \times 1 - 1 = 2 - 1 = 1 \text{ kg}$$

Exercise

1. In 2 kg mixture of water and milk 30% is milk. How much water should be added so that the proportion of milk becomes 15%?
   a) 4 kg b) 0.5 kg c) 2 kg d) 1 kg
2. In 3 kg mixture of water and milk 24% is milk. How much water should be added so that the proportion of milk becomes 18%?
   a) $\frac{1}{3}$ kg b) 3 kg c) 2 kg d) 1 kg
3. In 50 kg mixture of sand and cement 45% is cement. How much sand should be added so that the proportion of cement becomes 10%?
   a) 175 kg b) 225 kg c) 200 kg d) 150 kg

Answers

1. c 2. d 3. a

Rule 61

Ex.: In an examination the percentage of students qualified to the number of students appeared from school 'A' is 70%. In school 'B' the number of students appeared is 20% more than the students appeared from school 'A' and the number of students qualified from school 'B' is 50% more than the students qualified from school 'A'. What is the percentage of students qualified to the number of students appeared from school 'B'?

Soln: **Detailed Method:** Suppose 100 students appeared from school A. Then we have

	Appeared	Passed
A →	100	70
B →	120	70 + 50% of 70 = 105

Required % $= \frac{105}{120} \times 100 = 87.5\%$

Direct Formula (Quicker Method):

$$\text{Required \%} = \frac{70 \times (100+50)\%}{100 \times (100+20)\%} \times 100$$

$$= \frac{70 \times 150}{100 \times 120} \times 100 = 87.5\%$$

Exercise

1. In an examination the percentage of students qualified to the number of students appeared from school 'A' is 80%. In school 'B' the number of students appeared is 25% more than the students appeared from school 'A' and the number of students qualified from school 'B' is 40% more than the students qualified from school 'A'. What is the percentage of students qualified to the number of students appeared from school 'B'?
   a) 45% b) 90% c) 89.5% d) 89.6%
2. In an examination the percentage of students qualified to the number of students appeared from school 'A' is 60%. In school 'B' the number of students appeared is 30% more than the students appeared from school 'A' and the number of students qualified from school 'B' is 60% more than the students qualified from school 'A'. What is the percentage of students qualified to the number of students appeared from school 'B'?

a) 70% b) 75% c) $73\frac{11}{13}\%$ d) $71\frac{11}{13}\%$

3. In an examination the percentage of students qualified to the number of students appeared from school 'A' is 65%. In school 'B' the number of students appeared is 25% more than the students appeared from school 'A' and the number of students qualified from school 'B' is 40% more than the students qualified from school 'A'. What is the percentage of students qualified to the number of students appeared from school 'B'?
a) 70.8% b) 78.2% c) 72.8% d) 73%
4. In an examination the percentage of students qualified to the number of students appeared from school 'A' is 55%. In school 'B' the number of students appeared is 15% more than the students appeared from school 'A' and the number of students qualified from school 'B' is $67\frac{3}{11}\%$ more than the students qualified from school 'A'. What is the percentage of students qualified to the number of students appeared from school 'B'?
a) 80% b) 85% c) 75% d) 90%

Answers

1. d 2. c 3. c 4. a

Rule 62

Theorem: ***If the original price of a commodity is Rs X and new price of the commodity is Rs Y, then the decrease or increase in consumption so as not to increase or decrease the expenditure respectively, is*** $\left[\frac{Y-X}{Y}\times 100\right]\%$

ie $\left[\frac{\text{Difference in price}}{\text{New price}}\times 100\right]\%$.

Illustrative Example

Ex.: Wheat is now being sold at Rs 25 per kg. During last month its cost was Rs 21 per kg. Find by how much per cent a family should reduce its consumption, so as to keep the expenditure the same.

Soln: Following the above formula, we have

the required answer $= \frac{25-21}{25}\times 100 = 16\%$

Exercise

1. Rice is now being sold at Rs 20 per kg. During last month its cost was Rs 18 per kg. Find by how much per cent a family should reduce its consumption, so as to keep the expenditure the same.
a) 10% b) 20% c) 15% d) 5%
2. Tea is now being sold at Rs 16 per kg. During last month its cost was Rs 12 per kg. Find by how much per cent a family should reduce its consumption, so as to keep the expenditure the same.
a) 20% b) 28% c) 25% d) 30%
3. Sugar is now being sold at Rs 18 per kg. During last month its cost was Rs 25 per kg. Find by how much per cent a family should increase its consumption, so as to keep the expenditure the same.
a) 30% b) 29% c) 28% d) 25%
4. Wheat is now being sold at Rs 20 per kg. During last month its cost was Rs 25 per kg. Find by how much per cent a family should increase its consumption, so as to keep the expenditure the same.
a) 25% b) 20% c) 16% d) 18%
5. Tea is now being sold at Rs 30 per kg. During last month its cost was Rs 24 per kg. Find by how much per cent a family should reduce its consumption, so as to keep the expenditure the same.
a) 30% b) 20% c) 24% d) 15%

Answers

1. a 2. c 3. c 4. b 5. b

Rule 63

Theorem: ***If original price of a commodity is Rs O_r and new price of a commodity is Rs N, then keeping expenditure (E) constant, change in quantity of commodity consumed (ΔQ) is obtained by the following formula,*** $\Delta Q = \frac{E(N-O_r)}{N\times O_r}$.

Illustrative Example

Ex.: A reduction of Rs 2 per kg enables a man to purchase 4 kg more sugar for Rs 16. Find the original price of sugar.

Soln: Here, ΔQ (change in quantity consumed) = 4 kg

$O_r - N$ (change in price) = Rs 2 per kg

E (expenditure) = Rs 16

Now put the values in the above formula,

$$4 = \frac{16\times 2}{(O_r-2)\times O_r} \quad \left[\begin{array}{l}\because O_r - N = 2\\ \therefore N = O_r - 2\end{array}\right]$$

$\Rightarrow (O_r-4)(O_r+2)=0$

$\therefore$ Original price of sugar = Rs 4 per kg

Exercise

1. A reduction of 50 paise per dozen in the price of eggs means that a dozen more eggs can be bought for Rs 66. Find the original price.
a) Rs 6 b) Rs 5 c) Rs 6.5 d) Rs 8
2. A reduction of Rs 2 per kg enables a man to purchase 2 kg more tea for Rs 8. Find the original price of tea per kg.
a) Rs 4 per kg b) Rs 6 per kg
c) Rs 2 per kg d) Rs 3 per kg

3. A reduction of Rs 3 per kg enables a man to purchase 6 kg more rice for Rs 8. Find the original price of rice per kg.
a) Rs 3 per kg b) Rs 5 per kg
c) Rs 6 per kg d) Rs 4 per kg

Answers

1. a 2. a 3. d

Rule 64

Theorem: ***If original price of a commodity is Rs O_r and new price of a commodity is Rs N, keeping expenditure (E) constant, then the change in quantity of commodity consumed (ΔQ), when there is an increase or decrease of p% in the price of commodity, is obtained by the following formula, $\Delta Q = \frac{p \times E}{N \times 100}$, where $p = \frac{O_r - N}{O_r} \times 100$.***

Illustrative Example

Ex.: A reduction of 25% in the price of tea enables a person to buy 5 kg more for Rs 120. Find the original price of tea per kg.

Soln: Using the above formula,

$$N = \frac{p \times E}{\Delta Q \times 100} = \frac{25 \times 120}{5 \times 100} = 6 \text{ per kg}$$

$$\text{Here, } p = \frac{O_r - N}{O_r} \times 100$$

$$25 = \frac{O_r - 6}{O_r} \times 100 \Rightarrow 25O_r = 100Or - 600$$

$$\Rightarrow O_r = \frac{600}{75} = 8$$

∴ Original price is Rs 8 per kg

Exercise

1. A reduction of $12\frac{1}{2}$ per cent in the price of mangoes enables a purchaser to obtain 4 more for a rupee. What are the reduced price and the original price per mango?
a) Re $\frac{1}{32}$, Re $\frac{1}{28}$ b) Re $\frac{1}{32}$, Re $\frac{1}{36}$
c) Re $\frac{1}{28}$, $\frac{1}{36}$ d) None of these

2. A reduction of 20 per cent in the price of tea would enable a purchaser to obtain 4 kg more for Rs 100, what is the reduced price, and original price?
a) Rs 6.25, Rs 5 b) Rs 5, Rs 6.25
c) Rs 6, Rs 5.25 d) Rs 5.25, Rs 6

3. A reduction of $33\frac{1}{3}$ per cent in the price of oranges would enable a purchaser to obtain 8 more for a rupee. What was the price before the reduction?
a) 16 per rupee b) 24 per rupee
c) 12 per rupee d) None of these

4. A reduction of 24% in the price of tea enables a person buy 3 kg more for Rs 75. Find the original price of tea per kg.
a) Rs $6\frac{16}{19}$ b) Rs $7\frac{17}{19}$
c) Rs 6 d) Rs $6\frac{17}{19}$

Answers

1. a 2. b 3. a 4. b

Rule 65

Theorem: ***To split a number N into two parts such that one part is p% of the other. The two split parts are $\frac{100}{100+p} \times N$ and $\frac{p}{100+p} \times N$.***

Illustrative Example

Ex.: Split the number 120 into two parts such that one part is 20% of the other.

Soln: Following the above formula,

the numbers are $\frac{100}{100+20} \times 120$ and $\frac{20}{100+20} \times 120$

or, 100 and 20

Exercise

1. Split the number 150 into two parts such that one part is 25% of the other.
a) 120, 30 b) 100, 50 c) 90, 60 d) 110, 40

2. Split the number 112 into two parts such that one part is 12% of the other.
a) 84, 28 b) 80, 32 c) 100, 12 d) 102, 10

3. Split the number 280 into two parts such that one part is 40% of the other.
a) 240, 40 b) 200, 80 c) 190, 90 d) 210, 70

4. Split the number 27 into two parts such that one part is 35% of the other.
a) 18, 9 b) 21, 6 c) 24, 3 d) 20, 7

5. Split the number 31 into two parts such that one part is 24% of the other.
a) 25, 6 b) 24, 7 c) 22, 9 d) 27, 4

Answers

1. a 2. c 3. b 4. d 5. a

Rule 66

Theorem: ***If X litres of oil was poured into a tank and it was still x% empty, then the quantity of oil that must be poured into the tank in order to fill it to the brim is*** $\left(\frac{X \times x}{100 - x}\right)$ ***litres.***

Illustrative Example

Ex.: 240 litres of oil was poured into a tank and it was still 20% empty. How much oil must be poured into the tank in order to fill it to the brim?

Soln: Following the above formula, we have

the required answer $= \frac{240 \times 20}{100 - 20} = 60$ litres.

Exercise

1. 210 litres of oil was poured into a tank and it was still 30% empty. How much oil must be poured into the tank in order to fill it to the brim?
   a) 60 litres b) 90 litres c) 80 litres d) 70 litres
2. 186 litres of oil was poured into a tank and it was still 25% empty. How much oil must be poured into the tank in order to fill it to the brim?
   a) 62 litres b) 68 litres c) 43 litres d) 45 litres
3. 66 litres of oil was poured into a tank and it was still 12% empty. How much oil must be poured into the tank in order to fill it to the brim?
   a) 9 litres b) 12 litres c) 6 litres d) 8 litres
4. 1020 litres of oil was poured into a tank and it was still 15% empty. How much oil must be poured into the tank in order to fill it to the brim?
   a) 160 litres b) 90 litres c) 180 litres d) 170 litres
5. 410 litres of oil was poured into a tank and it was still 18% empty. How much oil must be poured into the tank in order to fill it to the brim?
   a) 95 litres b) 190 litres c) 90 litres d) 85 litres

Answers

1. b 2. a 3. a 4. c 5. c

Rule 67

Theorem: ***If X litres of oil was poured into a tank and it was still x% empty. Then the capacity of the tank is*** $\left(\frac{X \times 100}{100 - x}\right)$ ***litres.***

Illustrative Example

Ex.: 240 litres of oil was poured into a tank and it was still 20% empty. Find the capacity of the tank.

Soln: Applying the above formula, we have

the capacity of the tank $= \frac{240 \times 100}{100 - 20} = 300$ litres.

Exercise

1. 270 litres of oil was poured into a tank and it was still 25% empty. Find the capacity of the tank.
   a) 360 litres b) 300 litres c) 450 litres d) 350 litres
2. 170 litres of oil was poured into a tank and it was still 15% empty. Find the capacity of the tank.
   a) 260 litres b) 300 litres c) 200 litres d) 360 litres
3. 220 litres of oil was poured into a tank and it was still 12% empty. Find the capacity of the tank.
   a) 260 litres b) 350 litres c) 250 litres d) 500 litres
4. 86 litres of oil was poured into a tank and it was still 14% empty. Find the capacity of the tank.
   a) 100 litres b) 170 litres c) 150 litres d) 106 litres
5. 1260 litres of oil was poured into a tank and it was still 37% empty. Find the capacity of the tank.
   a) 2520 litres b) 2000 litres c) 2500 litres d) 2050 litres

Answers

1. a 2. c 3. c 4. a 5. b

Rule 68

Theorem: ***If a number is successively increased by x%, y% and z%, then single equivalent increase in that number will be*** $\left[(x+y+z)+\left(\frac{xy+yz+zx}{100}\right)+\frac{xyz}{100^2}\right]\%$.

Illustrative Example

Ex.: Find a single equivalent increase, if a number is successively increased by 10%, 15% and 20%.

Soln: Following the above formula, we have the required answer

$$= (10+15+20)+\frac{(10\times15+15\times20+10\times20)}{100}+\frac{(10\times15\times20)}{10000}$$

$$= 45+\frac{65}{10}+\frac{3}{10}=\frac{450+65+3}{10}=51.8\%$$

Exercise

1. Find a single equivalent increase, if a number is successively increased by 5%, 10% and 15%.
   a) 32.8% b) 31.8% c) 38.2% d) 23.8%
2. Find a single equivalent increase, if a number is successively increased by 15%, 20% and 25%.
   a) 72.2% b) 72.5% c) 72.75% d) 72%
3. Find a single equivalent increase, if a number is successively increased by 20%, 25% and 30%.
   a) 90% b) 75% c) 95% d) 85%
4. Find a single equivalent increase, if a number is successively increased by 10%, 20% and 25%.
   a) 55% b) 65% c) 70% d) 60%

Answers

1. a 2. b 3. c 4. b

Rule 69

Theorem: ***If three successive discounts of x%, y% and z% are allowed on an amount then a single discount that equivalent to the three successive discounts will be***

$$\left[x+y+z-\frac{xy+yz+zx}{100}+\frac{xyz}{100^2}\right]\%.$$

Illustrative Example

Ex.: Find a single discount equivalent to a discount series of 20%, 10% and 5%.

Soln: Applying the above rule, the equivalent successive discount

$$=\left[20+10+5-\frac{20\times10+10\times5+5\times20}{100}+\frac{20\times10\times5}{10000}\right]$$

$=31.6\%$

Exercise

1. Find a single discount equivalent to a discount series of 5%, 10% and 15%.
   a) 30% b) 27.23% c) 27.32% d) 23.72%
2. Find a single discount equivalent to a discount series of 10%, 15% and 20%.
   a) 45% b) 38.8% c) 43.8% d) 39.8%
3. Find a single discount equivalent to a discount series of 15%, 20% and 25%.
   a) 60% b) 65.5% c) 49% d) 55.6%
4. Find a single discount equivalent to a discount series of 10%, 20% and 25%.
   a) 46% b) 56% c) 55% d) 45%

Answers

1. c 2. b 3. c 4. a

Rule 70

Ex.: The price of sugar is decreased by 20% and its consumption increases by 30%. Find the new expenditure as a ratio of initial expenditure.

Soln: $\frac{\text{New expenditure}}{\text{Initial expenditure}}=\frac{(100+x)(100+y)}{(100)^2}$

Note: Put x as (+x) and y as (+y) in the case of 'increase' and x as (–x) and y as (–y) in the case of 'decrease'.
Here in the first case price of sugar decreases and in the second case consumption increases.
Hence the above formula becomes as

$$\frac{(100-x)(100+y)}{100^2}=\frac{(100-20)(100+30)}{100^2}$$

$$=\frac{80\times130}{100\times100}=\frac{26}{25}$$

Exercise

1. The price of tea is increased by 10% and its consumption also increases by 10%. Find the new expenditure as a ratio of initial expenditure.
   a) 11 : 10 b) 121 : 100
   c) 111 : 100 d) None of these
2. The price of rice is incresed by 20% and its cnsumption is decreased by 30 per cnet. Find the new expenditure as a ratio of initial expenditure.
   a) 21 : 25 b) 25 : 21 c) 7 : 8 d) 8 : 7
3. The price of sugar is decreased by 40% and its consumption is also decreased by 25%. Find the new expenditure as a ratio of initial expenditure.
   a) 9 : 20 b) 20 : 9 c) 7 : 4 d) 4 : 7
4. The price of wheat is decreased by 25% and its consumption increases by 25%. Find the new expenditure as a ratio of initial expenditure.
   a) 3 : 4 b) 5 : 4 c) 16 : 15 d) 15 : 16

Answers

1. b 2. a 3. a 4. d

Rule 71

Ex.: In an examination 30% of the students failed in Math, 25% of the students failed in English, 40% of the students failed in Hindi. If 15% of the students failed in Math and English, 20% of the students failed in English and Hindi, 25% of the students failed in Math and Hindi and 10% of the students failed in all the three subjects Math, English and Hindi, then find the percentage of students who passed in all three subjects.

Soln: We have the following formula,

$$(A\cup B\cup C)=n(A)+n(B)+n(C)-n(A\cap B)$$
$$-n(B\cap C)-n(A\cap C)+n(A\cap B\cap C)$$

∴ Total per cent of failed candiates
$=30+40+25-25-20-15+10=45\%$

∴ Total per cent of passed candidate
$=100-45=55\%$

Exercise

1. In an examination 35% of the students failed in Math, 25% of the students failed in English, 45% of the students failed in Hindi. If 10% of the students failed in Math and English, 20% of the students failed in English and Hindi, 30% of the students failed in Math and Hindi and 5% of the students failed in all the three subjects Math, English and Hindi, then find the percentage of students who passed in all three subjects.
   a) 10% b) 50% c) 80% d) 90%
2. In an examination 40% of the students failed in Math, 30% of the students failed in English, 50% of the students failed in Hindi. If 25% of the students failed in

Math and English, 15% of the students failed in English and Hindi, 22% of the students failed in Math and Hindi and 13% of the students failed in all the three subjects Math, English and Hindi, then find the percentage of students who passed in all three subjects.

a) 35% b) 29% c) 60% d) 71%

3. In an examination 20% of the students failed in Math, 15% of the students failed in English, 25% of the students failed in Hindi. If 5% of the students failed in Math and English, 10% of the students failed in English and Hindi, 15% of the students failed in Math and Hindi and 2% of the students failed in all the three subjects Math, English and Hindi, then find the percentage of students who passed in all three subjects.

a) 55% b) 65% c) 68% d) 32%

Answers

1. b 2. b 3. c

Rule 72

In a recent survey x% houses contained two or more people. Of those houses containing only one person y% were having only a male. The percentge of all houses which contain exactly one female and no males is given by

$$\left[\frac{(100-x)(100-y)}{100}\right]\%.$$

Illustrative Example

Ex.: In a recent survey 25% houses contained two or more people. Of those houses containing only one person 20% were having only a male. What is the percentage of all houses which contain exactly one female and no males?

Soln: Detail Method:

Houses containing only one person = 100 – 25 = 75%

Houses containing only a male = $75\times\frac{20}{100}=15\%$

∴ Houses containing only one female = 75–15 = 60%

Quicker Method: Applying the above theorem, we have

the required answer

$$=\frac{(100-25)(100-20)}{100}=\frac{75\times80}{100}=60\%$$

Exercise

1. In a recent survey 40% houses contained two or more people. Of those houses containing only one person 25% were having only a male. What is the percentage of all houses which contain exactly one female and no males?

a) 75 b) 40 c) 15 d) 45

[SBI Bank PO Exam, 2000]

2. In a recent survey 20% houses contained two or more people. Of those houses containing only one person 10% were having only a male. What is the percentage of all houses which contain exactly one female and no males?

a) 72% b) 27% c) 70% d) 62%

3. In a recent survey 30% houses contained two or more people. Of those houses containing only one person 15% were having only a male. What is the percentage of all houses which contain exactly one female and no males?

a) 60% b) 60.5% c) 59% d) 59.5%

4. In a recent survey 40% houses contained two or more people. Of those houses containing only one person 20% were having only a male. What is the percentage of all houses which contain exactly one female and no males?

a) 48% b) 50% c) 45% d) 56%

Answers

1. d 2. a 3. d 4. a

Rule 73

Monthly income of A is x% more than that of B. Monthly income of B is y% less than that of C. If the difference between the monthly incomes of A and C is Rs 'M', then the monthly incomes of B and C are given by Rs

$$\left[\frac{100\times(100-y)\times M}{(100+x)(100-y)-(100)^2}\right] \text{ and}$$

$$\text{Rs}\left[\frac{100^2\times M}{(100+x)(100-y)-(100)^2}\right] \text{ respectively.}$$

Illustrative Example

Ex.: Ram's monthly income is 15% more than that of Shyam. Shyam's monthly income is 10% less than that of Sohan. If the difference between the monthly incomes of Ram and Sohan is Rs 350, what is the monthly income of Shyam?

Soln: Detail Method:

Ram's monthly income

= Shyam's income + 15% of Shyam's income

= 1.15 Shyam's income

Shyam's income

= Sohan's income – 10% of Sohan's income

= 0.9 Sohan's income

∴ Ram's income = 1.15 × 0.9 Sohan's income

= 1.035 Sohan's income

Now,

Ram's income – Sohan's Income

= 1.035 Sohan's income – Sohan's income

= Rs 350 given

$\therefore$ Sohan's income $= \frac{400}{0.035} =$ Rs 10,000

$\therefore$ Shyam's income $= 0.9 \times 10,000 =$ Rs 9000

Quicker Method: Applying the above theorem, we have

$$\text{Shyam's income} = \text{Rs } \frac{100 \times (100-10) \times 350}{(100+15)(100-10)-(100)^2}$$

$$= \frac{100 \times 90 \times 350}{115 \times 90 - 10000} = \frac{9000 \times 350}{10350-10000}$$

$= $ Rs 9000

$$\text{Sohan's Income} = \text{Rs } \frac{100 \times 100 \times 350}{(100+15)(100-10)-(100)^2}$$

$= $ Rs 10,000

Exercise

1. Naresh's monthly income is 30% more than that of Raghu. Raghu's monthly income is 20% less than that of vishal. If the difference between the monthly incomes of Naresh and Vishal is Rs 800, what is the monthly income of Raghu? **[Bank of Baroda PO, 1999]**
   a) Rs 16000 b) Rs 20000
   c) Rs 12000 d) Data inadequate
2. A's monthly income is 25% more than that of B. B's monthly income is 5% less than that of C. If the difference between the monthly incomes of A and C is Rs 1875. Find B's income.
   a) Rs 9500 b) Rs 10000
   c) Rs 11375 d) None of these
3. Naresh's monthly income is 30% more than that of Raghu. Raghu's monthly income is 15% less than that of vishal. If the difference between the monthly incomes of Naresh and Vishal is Rs 1050, what is the monthly income of Raghu?
   a) Rs 9500 b) Rs 10550 c) Rs 8500 d) Rs 10000
4. Naresh's monthly income is 40% more than that of Raghu. Raghu's monthly income is 20% less than that of vishal. If the difference between the monthly incomes of Naresh and Vishal is Rs 2400, what is the monthly income of Raghu?
   a) Rs 16000 b) Rs 20000 c) Rs 30000 d) Rs 18000

Answers

1\. a 2. a 3. c 4. a

Rule 74

Weights of two persons A and B are in the ratio of a : b. A's weight increases by x% and the total weight of A and B together becomes 'W' kg, with an increase of y%, then the percentage increase in the weight of B is given by

$$\left[\left(\frac{100+y}{100}\right)\left(1+\frac{a}{b}\right)-\left\{\frac{a}{b}\left(\frac{100+x}{100}\right)+1\right\}\right] \times 100\%.$$

Illustrative Example

Ex.: Weight of two persons A and B are in the ratio of 3 : 5. A's weight increases by 20% and the total weight of A and B together becomes 80 kg, with an increase of 25%. By what per cent did the weight of B increase?

Soln: **Detail Method:**

Let the weights of A and B be $3x$ and $5x$.

Now, according to question,

$$\frac{3x \times 120}{100} + \text{B's new wt} = 80 \text{ ..... (i)}$$

$$\text{and } \frac{(3x+5x=)8x \times 125}{100} = 80 \text{ ...... (ii)}$$

From (ii) x = 8

Putting x = 8 in (i), we get

B's new wt = 80 – 28.8 = 51.2 kg

$$\% \text{ increase in B's wt} = \frac{51.2-40}{40} \times 100 = 28\%$$

Quicker Method: Applying the above rule, we have the required answer

$$= \left[\left(\frac{100+25}{100}\right)\left(1+\frac{3}{5}\right)-\left\{\frac{3}{5}\left(\frac{100+20}{100}\right)+1\right\}\right] \times 100$$

$$= \left[\frac{125}{100} \times \frac{8}{5} - \left\{\frac{3}{5} \times \frac{120}{100} + 1\right\}\right] \times 100$$

$$= \left[\frac{100-86}{50}\right] \times 100 = 28\%$$

Note: If you have to calculate percentage increase, as in the above caseyou can also use **Rule 8**.

Exercise

1. Weights of two friends Ram and Shyam are in the ratio of 4 : 5. Ram's weight increases by 10% and the total weight of Ram and Shyam together becomes 82.8 kg, with an increase of 15%. By what per cent did the weight of Shyam increase? **[Guwahati PO Exam, 1999]**
   a) 12.5% b) 17.5% c) 19% d) 21%
2. Weights of two friends Seeta and Geeta are in the ratio of 1 : 2. Seeta's weight increases by 20% and the total weight of Seeta and Geeta together becomes 60 kg, with an increase of 30%. By what per cent did the weight of Geeta increase?
   a) 35% b) 40% c) 34.5% d) 36.5%
3. Weights of two friends Rinku and Sunil are in the ratio of 3 : 7. Rinku's weight increases by 20% and the total weight of Rinku and Sunil together becomes 75 kg, with an increase of 40%. By what per cent did the weight of Sunil increase?

a) $44\frac{8}{7}\%$ b) $48\frac{4}{7}\%$
c) 49% d) None of these

4. Weights of two friends Ashok and Vinod are in the ratio of 2 : 5. Ashok's weight increases by 5% and the total weight of Ashok and Vinod together becomes 89 kg, with an increase of 15%. By what per cent did the weight of Vinod increase?
a) 19% b) 19.5% c) 16% d) 21.5%

Answers

1. c 2. a 3. b 4. a

Rule 75

A person spends x% of his monthly income on item 'A' and y% of the remaining on the item 'B'. He saves the remaining amount. If the saving amount is Rs 'S', then

(i) the monthly income of person $= Rs\left[\frac{S\times(100)^2}{(100-x)(100-y)}\right]$

(ii) the monthly amount spent on the item A

$$= Rs\ \frac{S\times x\times 100}{(100-x)(100-y)}$$

(iii) the monthly amount spent on the item B

$$= Rs\left[\frac{y\times S}{(100-y)}\right]$$

Note: Here 'S' = Saving per month.

Illustrative Example

Ex.: Mr Raju spends 20% of his monthly income on food and 25% of the remaining on room rent. He saves the remaining amount. If the saving amount is Rs 6000, find the monthly income of Raju, the amount spent on food and the amount spent on room rent.

Soln: Detail Method:

Let the monthly income of Raju be x.

Amount spent on food = 20% of $x = \frac{x}{5}$

Amount spent on room rent = 25% of $\frac{4x}{5} = \frac{x}{5}$

Remaining amount $= \frac{4x}{5} - \frac{x}{5} = \frac{3x}{5}$

According to the question, $\frac{3x}{5} = 6000$

$\therefore\ x = $ Rs 10000

$\therefore$ Monthly income of Raju = Rs 10000

Amount spent on food = Rs 2000 and amount spent on room rent = Rs 2000.

Quicker Method: Applying the above theorem, we have

Monthly income of Raju $= \frac{6000\times(100)^2}{(100-20)(100-25)}$

= Rs 10000

Amount spent on the food $= \frac{6000\times 20\times 100}{(100-20)(100-25)}$

= Rs 2000

Amount spent on the room rent $= \frac{25\times 6000}{100-25}$

= Rs 2000

Exercise

1. Mr Yadav spends 60% of his monthly salary on consumable items and 50% of the remaining on clothes and transport. He saves the remaining amount. If his savings at the end of the year were Rs 48456, how much amount per month would he have spent on clothes and transport? **[BSRB Delhi PO, 2000]**
a) Rs 4038 b) Rs 8076 c) Rs 9691.20 d) Rs 4845.60
2. Pankaj spends 15% of his monthly salary on entertainment, and 40% of the remaining on board and lodging. He saves the remaining amount. If his monthly savings are Rs 1020, find the monthly salary of Pankaj.
a) Rs 2000 b) Rs 5000 c) Rs 2200 d) Rs 5500
3. Mr Yadav spends 30% of his monthly salary on consumable items and 25% of the remaining on clothes and transport. He saves the remaining amount. If his savings at the end of the year were Rs 63000, how much amount per month would he have spent on clothes and transport?
a) Rs 1570 b)Rs 1750 c) Rs 1850 d) Rs 3000
4. Mr Yadav spends 10% of his monthly salary on consumable items and 15% of the remaining on clothes and transport. He saves the remaining amount. If his savings at the end of the year were Rs 18360, how much amount per month would he have spent on clothes and transport?
a) Rs 2700 b) Rs 720 c) Rs 270 d) None of these

Answers

1. a; **Hint:** Here savings = Rs 48456 (for a year ie 12 months)

$\therefore$ Savings per month $= \frac{48456}{12} =$ Rs 4038

Now, apply the above rule (iii), we have

the required answer $= \frac{50\times \text{Rs } 4038}{100-50} =$ Rs 4038

2. a 3. b 4. c

Rule 76

When the price of an item was increased by x%, a family reduced its consumption in such a way that the expendi-

ture on the item was only y% more than before. If 'W' kg were consumed per month before, then the new monthly consumption is given by $\left(\frac{100+y}{100+x}\right)$ ***W kg.***

Illustrative Example

Ex.: When the price of tea was increased by 25%, a family reduced its consumption in such a way that the expenditure on tea was only 20% more than before. If 25 kg were consumed per month before, find the new monthly consumption.

Soln: Applying the above theorem, we have

the required answer $= \frac{100+20}{100+25} \times 25 = 24$ kg

Note: Expenditure = Price × Consumption

Exercise

1. When the price of rice was increased by 32%, a family reduced its consumption in such a way that the expenditure on rice was only 10% more than before. If 30 kg were consumed per month before, find the new monthly consumption.
   a) 25 kg b) 24 kg
   c) 20 kg d) 18 kg
2. When the price of tea was increased by 20%, a family reduced its consumption in such a way that the expenditure on tea was only 15% more than before. If 24 kg were consumed per month before, find the new monthly consumption.
   a) 19 kg b) 18 kg c) 23 kg d) 21 kg
3. When the price of wheat was increased by 44%, a family reduced its consumption in such a way that the expenditure on wheat was only 20% more than before. If 18 kg were consumed per month before, find the new monthly consumption.
   a) 14 kg b) 15 kg c) 16 kg d) 10 kg
4. When the price of coffee was increased by 24%, a family reduced its consumption in such a way that the expenditure on coffee was only 8% more than before. If 31 kg were consumed per month before, find the new monthly consumption.
   a) 26 kg b) 25 kg c) 28 kg d) 27 kg

Answers

1. a 2. c 3. b 4. d

Rule 77

If the price of an item is increased by x% and a housewife reduced the consumption of that item by x%, then her expenditure on that item decreases by $\left(\frac{x}{10}\right)^2$ ***%. Or, in words it can be written as the following.***

Per cent Expenditue Change

$$= -\left(\frac{\text{Common increase or decrease}}{10}\right)^2 \%.$$

Note: Here -ve sign shows the decrease in expenditure ie in the above case there is always decrease in the expenditure.

Illustrative Example

Ex.: The price of sugar is increased by 20% and a housewife reduced her consumption of sugar by 20% and hence her expenditure on sugar,
a) remains unaltered b) decreases by 20%
c) decreases by 4 per cent d) increases by 4%

Soln: Applying the above rule, we have

the required answer $= -\left(\frac{20}{10}\right)^2 = -4\%$

ie expenditure decreases by 4%

$\therefore$ answer is (c)

Exercise

1. The price of coffee is increased by 15% and a housewife reduced her consumption of coffee by 15% and hence her expenditure on coffee,
   a) remains unchanged b) increases by 1%
   c) decreases by 4% d) decreases by 2.25%
2. The price of tea is increased by 10% and a housewife reduced her consumption of tea by 10% and hence her expenditure on tea,
   a) remains unaltered b) decreases by 1%
   c) decreases by 4% d) decreases by 2%
3. The price of sugar is increased by 25% and a housewife reduced her consumption of sugar by 25% and hence her expenditure on sugar,
   a) remains unaltered b) increases by 6.25%
   c) decreases by 6.25% d) decreases by 2.25%
4. The price of groundnut oil is increased by 30% and a housewife reduced her consumption of groundnut oil by 30% and hence her expenditure on groundnut oil,
   a) remains unchanged b) increases by 9%
   c) decreases by 6% d) decreases by 9%

Answers

1. d 2. b 3. c 4. d

Miscellaneous

1. In a school, a total of 110 students are studying together in two divisions A and B of Class X. The students are studying only Hindi, only Sanskrit or both Hindi and Sanskrit. The total number of students in A and B divisions are in the ratio of 5 : 6; the number of students studying only Hindi is 40% of the total number of students in the two divisions. The number of students studying both subjects in A division is 30% of the stu-

dents in that division and is equal to the number of students studying only Hindi in the same division. 36 students study both Hindi and Sanskrit. What is the total number of students studying only Sanskrit in class X?

[SBI Bank PO, 1999]

a) 44 b) 38 c) 36
d) 30 e) None of these

2. Out of a total 85 children playing badminton or table tennis or both, total number of girls in the group is 70% of the total number of boys in the group. The number of boys playing only badminton is 50% of the number of boys and the total number of boys playing badminton is 60% of the total number of boys. The number of children playing only table tennis is 40% of the total number of children and a total of 12 children play badminton and table tennis both. What is the number of girls playing only badminton? **[SBI Associates PO, 1999]**
a) 16 b) 14 c) 17
d) Data inadequate e) None of these

3. Pradip spends 40 per cent of his monthly income on food items, and 50 per cent of the remaining on clothes and conveyance. He saves one-third of the remaining amount after spending on food, clothes and conveyance. If he saves Rs 19,200 every year, what is his monthly income? **[BSRB Calcutta PO, 1999]**
a) Rs 24000 b) Rs 12000 c) 16000
d) Rs 20000 e) None of these

4. Ashok gave 40 per cent of the amount he had to Jayant. Jayant in turn gave one-fourth of what he received from Ashok to Prakash. After paying Rs 200 to the taxidriver out of the amount he got from Jayant, Prakash now has Rs 600 left with him. How much amount did Ashok have?

[BSRB Chennai PO, 2000]

a) Rs 1200 b) Rs 4000 c) Rs 8000
d) Data inadequate e) None of these

5. Rajesh solved 80 per cent of the questions in an examination correctly. If out of 41 questions solved by Rajesh 37 questions are correct and of the remaining questions out of 8 questions 5 questions have been solved by Rajesh correctly then find the total number of questions asked in the examination.

[BSRB Bangalore PO, 2000]

a) 75 b) 65 c) 60
d) Can't be determined e) None of these

6. In a class of 60 children, 30% children can speak only English, 20% Hindi and English both and the rest of the children can speak only Hindi. How many children can speak Hindi? **[BSRB Patna PO, 2001]**
a) 42 b) 36 c) 30
d) 48 e) None of these

7. The ratio of males and females in a city is 7 : 8 and the percentage of children among males and females is 25% and 20% respectively. If the number of adult females in the city is 156800 what is the total population?

[BSRB Patna PO, 2001]

a) 245000 b) 367500 c) 196000
d) 171500 e) None of these

8. The ratio of the number of students appearing for examination in the year 1998 in the states A, B and C was 3 : 5 : 6. Next year if the number of students in these states increases by 20%, 10% and 20% respectively, the ratio in states A and C would be 1 : 2. What was the number of students who appeared for the examination in the state A in 1998? **[BSRB Patna PO, 2001]**
a) 7200 b) 6000 c) 7500
d) Data inadequate e) None of these

Directions (Q. 9-13): Answer these questions on the basis of the information given below:

i) In a class of 80 students the girls and the boys are in the ratio of 3 : 5. The students can speak only Hindi or only English or both Hindi and English.
ii) The number of boys and the number of girls who can speak only Hindi is equal and each of them is 40% of the total number of girls.
iii) 10% of the girls can speak both the languages and 58% of the boys can speak only English.

[SBI Bank PO, 2001]

9. How many girls can speak only English?
a) 12 b) 29 c) 18
d) 15 e) None of these

10. In all how many boys can speak Hindi?
a) 12 b) 9 c) 24
d) Data inadequate e) None of these

11. What percentage of all the students (boys and girls together) can speak only Hindi?
a) 24 b) 40 c) 50
d) 30 e) None of these

12. In all how many students (boys and girls together) can speak both the languages?
a) 15 b) 12 c) 9
d) 29 e) None of these

13. How many boys can speak either only Hindi or only English?
a) 25 b) 38 c) 41 d) 29 e) None of these

14. Madan's salary is 25% of Ram's salary and Ram's salary is 40% of Sudin's salary. If the total salary of all the three for a month is Rs 12000, how much did Madan earn that month? **(Bank PO 1991)**
a) Rs 800 b) Rs 8000 c) Rs 600 d) Rs 850

15. ?% of 130 = 11.7 **(SBI Bank PO Exam, 1987)**
a) 90 b) 9 c) 0.9 d) 0.09

16. 40% of 70 = 4 × ? **(Bank Clerical Exam, 1990)**
a) 28 b) 280 c) 7 d) 70

17. What is 25% of 25% equal to? **(Astt. Grade 1987)**
a) 6.25 b) .625 c) .0625 d) .00625

18. 5 out of 2250 parts of earth is sulphur. What is the per-

centage of sulphur in earth? **(Hotel Management 1991)**

a) $\frac{1}{45}$ b) $\frac{2}{9}$ c) $\frac{11}{50}$ d) $\frac{2}{45}$

19. $\frac{30\% \text{ of } 80}{?} = 24$ **(Delhi Police 1989)**

a) $\frac{3}{10}$ b) 1 c) $\frac{3}{17}$ d) 2

20. 75% of a number when added to 75 becomes the number itself. The number is:
a) 150 b) 200 c) 225 d) 300
(Railway Recruitment 1991)

21. $\left(0.756 \times \frac{3}{4}\right)$ is equivalent to: **(SSC Exam 1987)**

a) 18.9% b) 37.8% c) 56.7% d) 75%

22. If 90% of A = 30% of B and B = x% of A, then the value of x is:
(Astt. Grade 1987)
a) 600 b) 800 c) 300 d) 900

23. The marked price is 10% higher than the cost price. A discount of 10% is given on the marked price. In this kind of sale, the seller **(CDS 1991)**
a) Bears no loss, no gain b) Gain 1%
c) Loses 10% d) Loses 1%

24. If x is 90% of y, what per cent of x is y?
(Astt. Grade 1990)
a) 90 b) 190 c) 101.1 d) 111.1

25. Which number is 60% less than 80?
(Astt. Grade 1990)
a) 48 b) 42 c) 32 d) 12

26. If the base of a rectangle is increased by 10% and the area is unchanged, then its corrsponding altitude must be decreased by? **(CBI Exam 1990)**

a) $11\frac{1}{9}\%$ b) $9\frac{1}{11}\%$ c) 11% d) 10%

27. The price of an article was increased by p%. Later the new price was decreased by p%. If the latest price was Re 1, the original price was: **(CPO Exam 1990)**

a) Re 1 b) Rs $\frac{1-p^2}{100}$

c) Rs $\frac{\sqrt{1-p^2}}{100}$ d) Rs $\left(\frac{10000}{10000-p^2}\right)$

28. If 10% of m is the same as 20% of n then m : n is equal to:
(CBI Exam 1990)
a) 1 : 2 b) 2 : 1 c) 5 : 1 d) 10 : 1

29. A owns a house worth Rs 10000. He sells it to B at a profit of 10% based on the worth of the house. B sells the house back to A at a loss of 10%. In this transaction, A gets **(CBI Exam 1990)**
a) a profit of Rs 2000 b) a profit of Rs 1100
c) a profit of Rs 1000 d) no profit no loss

30. p is six times as large as q. The per cent that q is less than p, is: **(CPO Exam 1990)**

a) $16\frac{2}{3}$ b) 60 c) 90 d) $83\frac{1}{3}$

31. In an examination 70% candidates passed in English and 65% in Mathematics. If 27% candidate failed in both the subjects and 248 passed the examination, the total number of candidates was: **(Bank Clerical Exam, 1991)**
a) 400 b) 348 c) 420 d) 484

Answers

1. d; 30 students

2. b; Let the number of boys = x

then $x + \frac{7x}{10} = 85 \Rightarrow x = 50$

No. of girls = 85 − 50 = 35

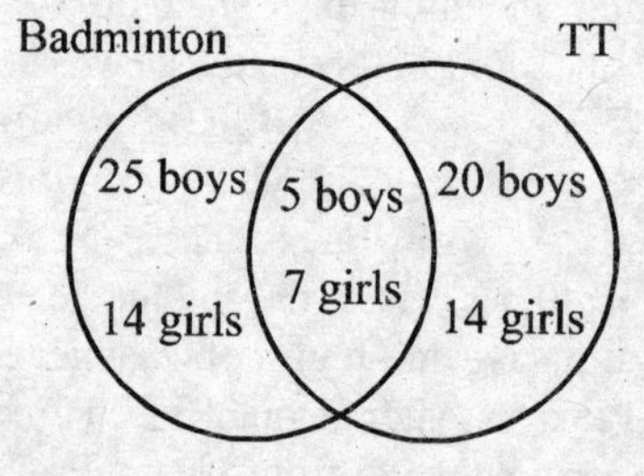

3. c; Food items = 40%

Clothes + conveyance = $\frac{1}{2}$ of 60% = 30%

$\frac{1}{3}$ of 30% = $\frac{19200}{12} \Rightarrow 10\% = 1600$

$\therefore$ 100% = Rs 16000

4. c; $J = \frac{2}{5}A \quad P = \frac{1}{4} \times \frac{2}{5}A = \frac{1}{10}A$ and $\frac{1}{10}A - 200 = 600$

$\therefore \frac{1}{10}A = 800$

A = Rs 8000

5. b; Suppose there are 8x questions apart from the 41 questions.

Then $\frac{37+5x}{41+8x} = 80\% = \frac{4}{5}$

$\Rightarrow 185 + 25x = 164 + 32x \Rightarrow 7x = 21 \Rightarrow x = 3$

$\therefore$ Total no. of questions = 41 + 8x = 65

6. a; Number of students who speak only English

$= 30\%\text{ of } 60 = 18$

Number of students who speak Hindi and English

$= 20\%\text{ of } 60 = 12$

$\therefore$ Number of students who speak only Hindi

$= (60 - 30 =)\ 30$

$\therefore$ No. of students who speak Hindi $= 30 + 12 = 42$

7\. b; Number of females $= 156800 \times \dfrac{100}{80} = 196000$

$\therefore$ Number of males $= \dfrac{7}{8} \times 196000 = 171500$

$\therefore$ Total population $= 196000 + 171500 = 367500$

8\. d; Let the number of students appearing for examination in the year 1998 in the states A, B and C be 3x, 5x and 6x respectively.

According to the question, $\dfrac{3x \times \frac{120}{100}}{6x \times \frac{120}{100}} = \dfrac{1}{2} \Rightarrow \dfrac{1}{2} = \dfrac{1}{2}$

(9-13): No. of boys in the class $= \dfrac{5}{8} \times 80 = 50$

$\therefore$ No. of girls in the class $= 80 - 50 = 30$

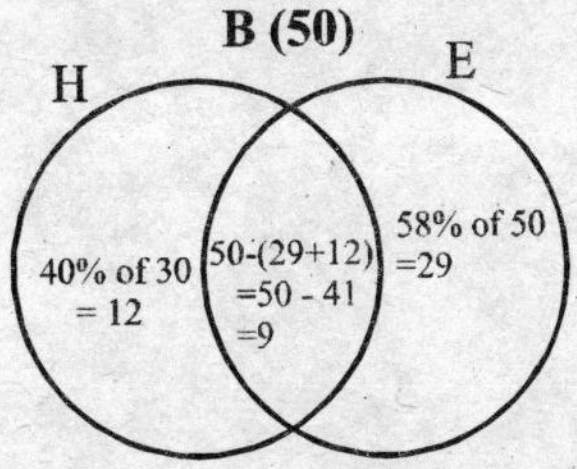

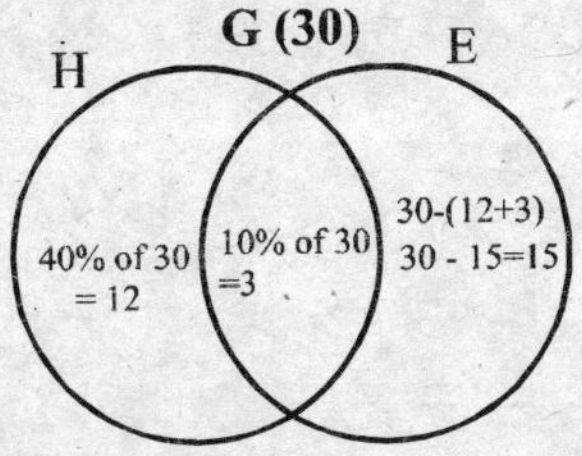

9\. d 10. e 11. d 12. b 13. c

14\. a; Let Sudin's salary = Rs x. Then,

Ram's salary $= \left(\dfrac{40}{100} x\right) = \text{Rs } \dfrac{2x}{5}$

and Madan's salary = Rs $\left(\dfrac{25}{100} \times \dfrac{2x}{5}\right) = \text{Rs}\left(\dfrac{x}{10}\right)$

$\therefore\ x + \dfrac{2x}{5} + \dfrac{x}{10} = 1200 \Rightarrow x = 8000$

So, Madan's salary = Rs $\left(\dfrac{8000}{10}\right)$ = Rs 800

15\. b; Let x% of 130 = 11.7

Then, $\dfrac{x}{100} \times 130 = 11.7 \Rightarrow x = \dfrac{11.7 \times 100}{130} = 9$

16\. c; Let 40% of 70 = 4 × x

Then, $\dfrac{40}{100} \times 70 = 4 \times x \Rightarrow x = \left(\dfrac{40}{100} \times 70 \times \dfrac{1}{4}\right) = 7$

17\. c; 25% of 25% $= \left(\dfrac{25}{100} \times \dfrac{25}{100}\right) = \dfrac{625}{10000} = 0.0625$

18\. b; Percentage of sulphur $= \left(\dfrac{5}{2250} \times 100\right)\% = \dfrac{2}{9}\%$

19\. b; $\dfrac{30\%\text{ of } 80}{x} = 24 \Rightarrow 24x = \left(\dfrac{30}{100} \times 80\right) = 24 \Rightarrow x = 1$

20\. d; 75 + (75% of x) = x

So, $75 + \left(\dfrac{75}{100}\right)x = x$

or, $75 + \dfrac{3}{4}x = x$ or, $x - \dfrac{3}{4}x = 75$

$\therefore\ \dfrac{1}{4}x = 75.$ So, $x = 75 \times 4 = 300$

21\. c;

$0.756 \times \dfrac{3}{4} = \left(\dfrac{756}{1000} \times \dfrac{3}{4}\right) = \left(\dfrac{756 \times 3}{1000 \times 4} \times 100\right)\% = 56.7\%$

22\. c; $\dfrac{90}{100} A = \dfrac{30}{100} B$ or, $3A = B$ ... (i)

Given $B = x\%$ of A or, $B = \dfrac{x}{100} \times A$ .... (ii)

From eqn (i) and (ii) we have,

$x = 300.$

23\. d; Let CP = Rs 100. Then, marked price = Rs 110

Discount = (10% of Rs 110) $= \left(\dfrac{10}{100} \text{ of Rs } 110\right)$ = Rs 11

$\therefore$ SP = Rs (110 - 11) = Rs 99

So, the seller loses 1%. [See **Rule 37**]

24\. d; $x = 90\%\text{ of } y = \dfrac{90}{100} y = \dfrac{9y}{10} \Rightarrow \dfrac{y}{x} = \dfrac{10}{9}$

Let $y = z\%$ of $x = \dfrac{z}{100} x \Rightarrow \dfrac{y}{x} = \dfrac{z}{100}$

$\therefore\ \dfrac{z}{100} = \dfrac{10}{9} \Rightarrow z = \dfrac{10 \times 100}{9} = 111.1\%$

25\. c; (80 - 60% of 80) $= \left(80 - \dfrac{60}{100} \times 80\right) = (80 - 48) = 32$

26\. b; Let length = 100 m and height = x m; Area = (100x)

New length = 110 m & let new height = (x - y% of x)

Then, $110 \times \left(x - \dfrac{y}{100} x\right) = 100 \times x$

or, $110 \times \left(1 - \dfrac{y}{100}\right) = 100$

or, $1 - \dfrac{y}{100} = \dfrac{100}{110}$ or, $\dfrac{y}{100} = 1 - \dfrac{100}{110} = \dfrac{10}{110} = \dfrac{1}{11}$

$\therefore\ y=\frac{100}{11}=9\frac{1}{11}\%$

27. d; Let original price = Rs x

$\left(\frac{100-p}{100}\right)$ of $\left(\frac{100+p}{100}\right)$ of $x=1$

$\therefore\ x=\frac{100\times100}{(100-p)(100+p)}=\frac{10000}{(10000-p^2)}$

28. b; $\frac{10}{100}m=\frac{20}{100}n\Rightarrow\frac{m}{n}=\left(\frac{20}{100}\times\frac{100}{10}\right)=\frac{2}{1}$

$\therefore\ m:n=2:1$

29. b; Price paid by B = Rs $\left(\frac{110}{100}\times10000\right)$ = Rs 11000

Price paid by A = Rs $\left(\frac{90}{100}\times11000\right)$ = Rs 9900

Thus profit made by A in two transactions
= Rs (1000 + 100) = Rs 1100

30. d; $p=6q\Rightarrow q=\frac{1}{6}p$

$\therefore$ q is less than p by $\left(p-\frac{1}{6}p\right)=\frac{5}{6}p$

$\therefore$ Required percentage = $\left(\frac{5}{6}p\times\frac{1}{p}\times100\right)\%=83\frac{1}{3}\%$

31. a; Failed in English only = (30 - 27) = 3%
Failed in Mathematics only = (35 - 27) = 8%
Failed in both the subjects = 27%
Failed in one or both of the subjects
= (3 + 8 + 27)% = 38%

$\therefore$ 62% of x = $248\Rightarrow\frac{62}{100}\times x=248$

$\therefore\ x=\frac{248\times100}{62}=400$

6 Average

Rule 1

To find the average, when the number of quantities and their sum is given. We have the following formula

$$\textit{Average} = \frac{\textit{Sum of the quantities}}{\textit{Number of quantities}}$$

Illustrative Example

Ex.: A batsman scores 35, 45 and 37 runs in first, second and third innings respectively. Find the average runs in the three innings.

Soln: Follow the above formula, we have

$$\text{average runs} = \frac{35+45+37}{3} = 39 \text{ runs.}$$

Exercise

1. Find the average of first 6 prime numbers.

   a) $5\frac{3}{5}$ b) $6\frac{5}{6}$ c) $5\frac{5}{6}$ d) $6\frac{3}{5}$

2. Find the average of first 25 multiples of 5.

   a) 65 b) 60 c) 75 d) 80

3. Find the average of first 5 prime numbers.

   [Clerical Grade Exam, 1989]

   a) 5.6 b) 5.5 c) 6.5 d) 4.6

4. Find the average of first 18 multiples of 6.

   a) 75 b) 57 c) 67 d) 76

5. If a, b, c, d, e are five consecutive odd integers, then what is their average?

   [Bank PO Exam, 1989]

   a) a + 4 b) $\frac{abcde}{5}$ c) $5(a+b+c+d+e)$ d) a + 8

6. What was the average daily expenditure of a man in 1999 who spent Rs 76535 in the first half year and Rs 88165 in the last?

   a) Rs 450 b) Rs 451.32 c) Rs 451.23 d) Rs 450.23

7. The average of first five multiples of 3 is

   [Central Excise & I. Tax, 1988]

   a) 3 b) 9 c) 12 d) 15

Answers

1. b; **Hint:** First 6 prime numbers are 2, 3, 5, 7, 11 and 13.
2. a; **Hint:** Required average $= \frac{5(1+2+3+....+25)}{25}$

   $= \frac{25\times 26}{2} \times \frac{5}{25} = 65$

   $\left[\because 1+2+...+n = \frac{n(n+1)}{2}\right]$
3. a; **Hint:** First five prime numbers are 2, 3, 5, 7, 11.
4. b
5. a; **Hint:** b = a + 2, c = a + 4, d = a + 6 and e = a + 8

   ∴ required average

   $= \frac{a+a+2+a+4+a+6+a+8}{5} = a+4$
6. c; **Hint:** Required average = Rs $\frac{76535+88165}{365}$ = Rs 451.23
7. b

Rule 2

To find the sum, when the number of quantities and their average is given. We have the following formula:
Sum of quantities = Average × Number of quantities.

Illustrative Example

Ex.: The average marks obtained by 120 candidates in a certain examination is 35. Find the total marks.

Soln: Following the above formula, we have

the total marks = 120 × 35 = 4200.

Exercise

1. The average weight of a class having 52 students is 52 kg. Find the total weight of the class.

   a) 2504 kg b) 2708 kg c) 2704 kg d) 2407 kg

2. The average scores of a batsman is 44.4 runs. If he played 125 innings so far, find the total runs made by him.

   a) 5550 runs b) 5250 runs c) 5450 runs d) 5560 runs

3. The average marks obtained by 144 candidates in a certain examination is 55. Find the total marks.

a) 7290 b) 7920 c) 7930 d) 7390

4. The population of 6 villages is 803, 900, 1100, 1023, 945 and 980. What is the population of the seventh village if the average population of the seven villages is 1000?
a) 1249 b) 1429 c) 1428 d) 1349

5. The mean temperature from the 9th to the 16th of January, both days inclusive, was 11.6° C and from the 10th to the 17th it was 12.2° C. The temperature on the 9th was 10.8° C. What was it on the 17th?
a) 15.6° C b) 4.8° C c) 9.6° C d) 15° C

6. The weights of four rowers of a boat are respectively 70 kg, 72 kg, 73 kg and 74 kg and the average weight of whole crew, including the coxswain is 70 kg. Find the weight of the coxswain.
a) 61 kg b) 68 kg c) 62 kg d) 63 kg

7. The average weight of 19 students was 25 kg. By the admission of a new student the average weight is reduced to 24.8 kg. The weight of the new student is
a) 24.8 kg b) 20.8 kg c) 20.6 kg d) 21 kg

8. The average of 6 numbers is 8. What is the 7th number so that average becomes 10? **[LIC Exam, 1991]**
a) 22 b) 18 c) 21 d) 20

9. The average of 50 numbers is 38. If two numbers namely, 45 and 55 are discarded, the average of remaining numbers is: **[CBI Exam, 1990]**
a) 36.50 b) 37.00 c) 37.50 d) 37.52

10. The average age of an adult class is 40 years. 12 new students with an average age of 32 years join the class, thereby decreasing the average of the class by 4 years. The original strength of the class was:
[Central Excise & I. Tax 1989]
a) 10 b) 11 c) 12 d) 15

11. The average expenditure of a man for the first five months is Rs 120 and for the next seven months is Rs 130. His monthly average income if he saves Rs 290 in that year, is: **[Railways 1991]**
a) Rs 160 b) Rs 170 c) Rs 150 d) 140

12. The average temperature of first 3 days is 27° and of the next 3 days is 29°. If the average of the whole week is 28.5° C, the temperature of the last day is:
[Railways 1991]
a) 31.5° b) 10.5° c) 21° d) 42°

13. A cricketer scored 180 runs in the first test and 258 runs in the second. How many runs should he score in the third test so that his average score in the three tests would be 230 runs? **[Bank PO 1991]**
a) 219 b) 242 c) 334 d) None of these

14. The average salary of 20 workers in an office is Rs 1900 per month. If the manager's salary is added, the average becomes Rs 2000 per month. The manager's annual salary (in Rs) is:
[SBI PO Exam, 1988]
a) 24000 b) 25200 c) 45600 d) None of these

Answers

1. c 2. a 3. b

4. a; **Hint:** Population of seventh village $= 7 \times 1000 - (803 + 900 + 1100 + 1023 + 945 + 980) = 1249$

5. a; Total temperature from 9th to 16th of January
$= 8 \times 11.6°C = 92.8°C$
Total temperature from 10th to 17th of January
$= 8 \times 12.2 = 97.6°C$
Now, according to the question,
Temperature on 17th of January
$= 10.8°C + 97.6°C - 92.8°C = 15.6°C.$

6. a; **Hint:** Required answer $= 5 \times 70 - (70 + 72 + 73 + 74)$
$= 61$ kg.

7. d; **Hint:** Weight of new student $= (20 \times 24.8 - 19 \times 25)$
$= 21$ kg

8. a; **Hint:** Let 7th number be x.
Sum of given 6 numbers $= (6 \times 8) = 48$
$\therefore$ Average of 7 numbers $= \frac{48 + x}{7}$.

$$\therefore \frac{48+x}{7} = 10 \text{ or, } 48 + x = 70 \text{ or } x = 22.$$

Hence, the 7th number is 22.

9. c; **Hint:** Total of 50 numbers $= 50 \times 38 = 1900.$
Average of 48 numbers
$$= \frac{1900 - (45 + 55)}{48} = \frac{1800}{48} = 37.50$$

10. c; **Hint:** $40x + 12 \times 32 = (12 + x) \times 36 \quad \therefore x = 12$

11. c

12. a; **Hint:** Temperature of the last day
$$= [28.5 \times 7 - (27 \times 3) - (29 \times 3)]^{\circ} = 31.5°$$

13. d; **Hint:** Let the runs he should score in third test be x. Then,
$$\frac{180 + 258 + x}{3} = 230 \Rightarrow x = 252.$$

14. d; **Hint:** Total monthly salary of 21 personnels
$=$ Rs $(21 \times 2000) =$ Rs 42000.
Total monthly salary of 20 personnels
$=$ Rs $(20 \times 1900) =$ Rs 38000.
Monthly salary of the manager $=$ Rs 4000.
Annual salary of the manager $=$ Rs 48000.

Rule 3

To find number of quantities, when the sum of quantities and average are given. We have the following formula:

$$\textit{Number of quantities} = \frac{\textit{Sum of quantities}}{\textit{Average}}$$

Illustrative Example

Ex.: If the sum of the 'n' number of quantities is 30 and the average is 6. Find the value of 'n'.

Soln: Applying the above formula, we have

$$n=\frac{30}{6}=5.$$

Exercise

1. If the sum of x number of quantities is 162 and the average is 9. Find the value of x.
   a) 18 b) 28 c) 19 d) 17
2. Average daily income of a rickshaw puller is Rs 45. If after x days, rickshaw puller earns Rs 315, find the value of x.
   a) 8 days b) 15 days c) 5 days d) 7 days
3. Total temperature of the month October is 775° C. If the average temperature of that month is 25° C, find of how many days is the month of October?
   a) 30 days b) 29 datys c) 31 days d) Data inadequate
4. In a coconut grove, (x + 2) trees yield 60 nuts per year, x trees yield 120 nuts per year and (x – 2) trees yield 180 nuts per year. If the average yield per year per tree be 100, find x. **[MBA 1986]**
   a) 4 b) 2 c) 8 d) 6

Answers

1. a 2. d 3. c

4. a; **Hint:** $\frac{(x+2)\times 60+x\times 120+(x-2)\times 180}{x+2+x+x-2}=100$

$\therefore x=4$

Rule 4

Theorem: ***If the average of 'm' boys is 'x' and the average of 'n' boys is 'b' then the average of all of them put together (= total average) is*** $\frac{mx+ny}{m+n}$.

Illustrative Example

Ex.: The average age of students in section A of 50 students is 14 years and the average age of students in section B of 30 students is 6 years. Find the average age of students in both sections taken together.

Soln: Following the above formula, we have

the required average $=\frac{50\times 14+30\times 6}{50+30}=11$ years.

Note: The above rule is a kind of discrete series.

In a discrete series the values of the variables are multiplied by their respective frequencies and the products so obtained are totalled. This total is divided by the number of items, which in a discrete series, is equal to the total of the frequencies. The resulting quotient is a simple arithmetic average of the series. In the form of formula it is written as

$$Average=\frac{x_1f_1+x_2f_2+...+x_nf_n}{f_1+f_2+f_3+...+f_n}$$

Where, $f_1, f_2, f_3 ..., f_n$ *are the frequencies ie no. of items and* $x_1, x_2, x_3, ..., x_n$ *are the values of each respective item.*

Ex. 1: A man bought 13 shirts of Rs 50 each, 15 pants of Rs 60 each and 12 pairs of shoes at Rs 65 a pair. Find the average value of each article.

Soln: Direct Method:

$$Average=\frac{13\times 50+15\times 60+12\times 65}{13+15+12}=Rs\ 58.25$$

Ex. 2: The average score of a cricketer in two matches is 27 and in three other matches is 32. Then find the average score in all the five matches.

Soln: Direct Method:

Average in 5 matches

$$=\frac{2\times 27+3\times 32}{2+3}=\frac{54+96}{5}=30$$

Exercise

1. A student bought 4 books for Rs 120 from one book shop and 6 books for Rs 150 from another. The average price (in rupees), he paid per book was ______.
   a) Rs 27 b) Rs 27.50 c) Rs 135 d) Rs 138
2. In a class of 100 students, the mean marks obtained in a subject is 30 and in another class of 50 students the mean marks obtained in the same subject is 60. The mean marks obtained by the students of two classes taken together is _____.
   a) 30 b) 50 c) 40 d) 45
3. A class has 20 boys and 30 girls. The average age of boys is 12 years and that of girls is 11 years what is the average age of the whole class?
   a) 11.4 years b) 11.6 years c) 11.2 years d) 12 years
4. Of 20 men 12 gain Rs 335 each and 8 men gain Rs 240 each. What is the average gain per man?
   a) Rs 297 b) Rs 290 c) Rs 279 d) Rs 397
5. If 20 chairs are bought at Rs 50 each, and 15 at Rs 45 each and 15 more at Rs 40 each. What is the average price of a chair?
   a) Rs 60 b) Rs 45 c) Rs 45.5 d) Rs 50.5
6. The average height of 30 girls out of a class of 40 is 160 cm and that of the remaining girls is 156 cm. What is the average height of the whole class?
   a) 159 cm b) 160 cm c) 159.5 cm d) 160.5 cm
7. The average expenditure of a man for the first five months is Rs 1200 and for the next seven months is Rs 1300. Find his monthly average income if he saves Rs 2900 during the year.
   a) Rs 750 b) Rs 1500 c) Rs 1750 d) Rs 500
8. A man bought 13 tins at Rs 50 each, 15 tins at Rs 60 each and 12 tins at Rs 65 each. What is the average price paid per tin?
   a) Rs 58 b) Rs 58.50 c) Rs 58.25 d) Rs 58.75

9. In a certain primary school there are fifteen boys at the age of 12, sixteen at 15, and eighteen at 14 years. Find the average age of boys.

a) $13\frac{5}{7}$ years b) $15\frac{3}{7}$ years

c) $15\frac{5}{7}$ years d) $13\frac{3}{7}$ years

10. Out of 24 girls 6 are 1 m 15 cm in height, 8 are 1 m 5 cm and the rest 1 m 11 cm. What is he average height of the girls?
a) 1 m b) 2 m c) 1 m 10 cm d) 2 m 10 cm
11. The average daily number of persons passing a certain point on Monday, Tuesday, Wednesday, and Thursday is 2163. The average daily number passing on Friday, Saturday and Sunday is 1960. What is the daily average for the whole weak?
a) 2706 b) 2067 c) 2076 d) 3076
12. The average age of the boys in a class of 20 boys is 15.6 years. What will be the average age if 5 new boys come whose average is 15.4 years?
a) 15.56 years b) 13.36 years
c) 15 years d) 15.50 years
13. The average age of 600 scholars of a school is 10.75 years. By the enrolment of 40 new scholars the average is reduced to 10.4375 years. Find the average age of the new scholars.

a) $3\frac{4}{5}$ years b) $5\frac{3}{4}$ years c) $4\frac{3}{5}$ years d) $6\frac{3}{4}$ years

14. In a certain primary school, there are 60 boys of age 12 each, 40 of age 13 each, 50 of age 14 each and 50 of age 15 each. The average age (in years) of the boys of the school is: **[Clerical Grade Exam, 1991]**
a) 13.50 b) 13 c) 13.45 d) 14
15. The average height of 30 girls out of a class of 40 is 160 cm and that of the remaining girls is 156 cm. The average height of the whole class is:
[Central Excise & I Tax Exam, 1988]
a) 158 cm b) 158.5 cm c) 159 cm d) 159.5 cm
16. The average height of 30 boys, out of a class of 50, is 160 cm. If the average height of the remaining boys is 165 cm, the average height of the whole class (in cm) is:
[Clerical Grade Exam, 1989]
a) 161 b) 162 c) 163 d) 164

Answers

1. d 2. c 3. a 4. a 5. c 6. a
7. b; **Hint:** Required average monthly income

$$= \frac{5\times1200+7\times1300+2900}{12} = \frac{18000}{12}$$

= Rs 1500

8. c 9. a 10. c
11. c; **Hint:** Required daily average

$$= \frac{4\times2163+3\times1960}{7} = 2076$$

12. a
13. b; Let the average age of the new scholars be x.
Now,

$$\frac{600\times10.75+40\times x}{600+40} = 10.4375$$

or, $6450+40x=6680$

$$\therefore x = \frac{23}{4} = 5\frac{3}{4}$$

14. c 15. c 16. b

Rule 5

Theorem: ***If the average age of 'm' boys is 'x' and the average age of 'n' boys out of them (m boys) is 'y' then the average age of the rest of the boys is*** $\frac{mx-ny}{m-n}$.

Illustrative Example

Ex.: The average of 10 quantities is 12. The average of 6 of them is 8. What is the average of remaining four numbers.

Soln: By the above theorem, we have

the required average $= \frac{10\times12-6\times8}{10-6} = 18$.

Exercise

1. A group of 20 girls has average age of 12 years. Average age of first 12 from the same group is 13 years. What is the average age of other 8 girls in the group? **[BSRB Bhopal PO 2000]**
a) 10 b) 11 c) 11.5 d) 10.5
2. A group of 30 girls has average age of 13 years. Average age of first 18 from the same group is 15 years. What is the average age of other 12 girls in the group?
a) 12 years b) 10 years c) 16 years d) 10.5 years
3. The average age of 30 students in a class is 12 years. The average age of a group of 5 of the students is 10 years and that of another group of 5 students is 14 years. Find the average age of the remaining students.
a) 14 years b) 10 years
c) 12 years d) Data inadequate
4. 30 horses are bought for Rs 150000. The average cost of 18 of them is Rs 4500. What is the average cost of others?
a) Rs 5750 b) Rs 7550 c) Rs 5760 d) Rs 4750
5. The average of 12 results is 15, and the average of the first two is 14. What is the average of the rest?
a) 15.2 b) 13.2 c) 15 d) 16

6. The average of 3 numbers is 7, that of the first two is 4, find the third number.
a) 13 b) 10 c) 12 d) 11
7. The average score of a cricketer for 10 matches is 38.9 runs. If the average for the first 6 matches is 42, find the average for the last four matches.
a) 34.52 b) 43.25 c) 34.25 d) 35

Answers

1. d 2. b

3. c; **Hint:** Required average

$$=\frac{30\times12-\{(5\times10)+(5\times14)\}}{30-(5+5)}=\frac{240}{20}=12$$

4. a; **Hint:** Required average

$$=\frac{150000(30\times5000)-18\times4500}{30-18}=\text{Rs } 5750$$

5. a

6. a; **Hint:** Required number $=\frac{3\times7-2\times4}{3-2}=13$.

7. c

Rule 6

Theorem: ***If the average of 'n' quantities is equal to 'x'. When a quantity is removed the average becomes 'y'. Then the value of the removed quantity is [n (x – y) + y].***

Illustrative Example

Ex.: The average age of 24 boys and a class teacher of a class is equal to 15 years. If class teacher left the class due to health problem the average becomes 14. Find the age of class teacher who left the class.

Soln: Following the above theorem, we have the required answer = 25 (15 – 14) + 14 = 25 + 14 = 39 years.

Exercise

1. The average age of 24 students and the class teacher is 16 years. If the class teacher's age is excluded, the average reduces by 1 year. What is the age of the class teacher? **[BSRB Mumbai PO, 1998]**
a) 50 years b) 45 years
c) 40 years d) Data inadequate
2. The total age of 26 persons are 442 years. Out of these persons one is a teacher and others are students. If the teacher's age is excluded, the average reduces by 2 years. What is the age of the teacher?
a) 50 years b) 55 years c) 60 years d) 67 years
3. The average age of 30 students and the class teacher is 20 years. If the class teacher's age is excluded, the average reduces by 1 year. What is the age of the class teacher?
a) 39 years b) 50 years c) 40 years d) 49 years
4. The average age of 13 students and the class teacher is 19 years. If the class teacher's age is excluded, the average reduces by 2 years. What is the age of the class teacher?
a) 45 years b) 40 years
c) 38 years d) 39 years
5. The average age of 15 students and the class teacher is 15 years. If the class teacher's age is excluded, the average reduces by 1 year. What is the age of the class teacher?
a) 30 years b) 31 years
c) 29 years d) 28 years

Answers

1. c

2. d; **Hint:** Here, n = 26, $x=\frac{442}{26}=17$ years and y = 17 – 2 = 15 years.
Now apply the formula and get the answer = 67 years.

3. b 4. a 5. a

Rule 7

Theorem: ***If the average of 'n' numbers is 'm' and if 'x' is added to or subtracted from each given number, the average of 'n' numbers becomes (m + x) or (m – x) respectively. In the other words average value will be increased or decreased by 'x'.***

Illustrative Examples

Ex. 1: The average of 11 numbers is 21. If 3 is added to each given number, what will be the new average?

Soln: From the above theorem, we have new average = 21 + 3 = 24

Ex. 2: The average of 6 numbers is 15. If 3 is subtracted from each given number, what will be the new average?

Soln: From the above theorem, we have new average = 15 – 3 = 12

Exercise

1. The average of 15 numbers is 25. If 5 is added to each given number, what will be the new average?
a) 20 b) 30 c) 25 d) Data inadequate
2. The average of n numbers is 4n. If n is added to each given number, what will be the new average?
a) (n + 1)4 b) 5n c) (n + 1)5 d) None of these
3. The average of 8 numbers is 14. If 2 is subtracted from each given number, what will be the new average?
a) 12 b) 10 c) 16 d) None of these
4. The average of x numbers is $3x$. If x – 1 is subtracted from each given number, what will be the new average?
a) $2x + 1$ b) $(x - 1)3$ c) $2x - 1$ d) Data inadequate

Answers

1. b 2. b 3. a 4. a

Rule 8

Theorem: ***If the average of 'n' quantities is equal to 'x' and when a new quantity is added the average becomes 'y'. Then the value of the new quantity is [n (y – x) + y]. In another words it may be written as,***

Value of new entrant = No. of old members × Increase in average + New average.

Illustrative Example

Ex.: The average age of 30 boys of a class is equal to 14 years. When the age of the class teacher is included the average becomes 15 years. Find the age of the class teacher.

Soln: **Detailed Method:** Total ages of 30 boys

$= 14 \times 30 = 420$ years

Total ages when class teacher is included

$= 15 \times 31 = 465$ years

$\therefore$ Age of class teacher $= 465 - 420 = 45$ years

Quicker Method: Applying the above theorem, we have

the required answer $= 15 + 30(15 - 14) = 45$ years

Exercise

1. The average age of 34 boys in a class is 14 years. If the teacher's age is included the average age of the boys and the teacher becomes 15 years. What is the teacher's age? **[BSRB Calcutta PO 1999]**
   a) 48 years b) 46 years c) 49 years d) 45 years
2. The average age of 24 boys in a class is 16 years. If the teacher's age is included the average age of he boys and the teacher becomes 18 years. What is the teacher's age?
   a) 64 years b) 62 years c) 66 years d) 60 years
3. The average age of 44 boys in a class is 26 years. If the teacher's age is included the average age of the boys and the teacher becomes 27 years. What is the teacher's age?
   a) 69 years b) 70 years
   c) 59 years d) Data inadequate
4. The average age of a class of 40 boys is 16.95 years, but by the admission of a new boy the average age is raised to 17 years. Find the age of the new boy.
   a) 19 years b) 18 years c) 15 years d) 19.5 years
5. The average age of 30 children in a class is 9 years. If the teacher's age be included, the average age becomes 10 years. Find the teacher's age. **[Bank PO 1991]**
   a) 40 years b) 36 years c) 42 years d) 39 years

Answers

1. c 2. c 3. a 4. a 5. a

Rule 9

Theorem: ***If the average of 'n' numbers is 'm' and if each given no. is multiplied to or divided by 'x', then the average of n numbers becomes mx or $\frac{m}{x}$ respectively.***

Illustrative Examples

Ex. 1: The average of 12 numbers is 35. If each of the numbers is multiplied by 2, find the average of new set of numbers.

Soln: Applying the above rule, we have

required answer $= 35 \times 2 = 70$

Ex. 2: The average of 12 numbers is 35. If each of the numbers is divided by 5, find the average of new set of numbers.

Soln: Required answer $= \frac{35}{5} = 7$

Exercise

1. The average of n numbers is x. If each of the numbers is multiplied by $(n - 1)$, find the average of new set of numbers.
   a) $\frac{x}{n-1}$ b) $(n-1)x$ c) nx d) $\left(\frac{n}{n-1}\right)x$
2. The average of 13 numbers is 36. If each of the numbers is multiplied by 3, find the average of new set of numbers.
   a) 108 b) 120 c) 104 d) 106
3. The average of 29 numbers is 45. If each of the numbers is divided by 9, find the average of new set of numbers.
   a) 9 b) 5 c) 6 d) 8
4. The average of 40 numbers is 405. If each of the numbers is divided by 15, find the average of new set of numbers.
   a) 27 b) 28 c) 21 d) 26
5. The average of 8 numbers is 21. If each of the numbers is multiplied by 8, find the average of new set of numbers. **[Central Excise 1989]**
   a) 168 b) 167 c) 158 d) 161

Answers

1. b 2. a 3. b 4. a 5. a

Rule 10

Theorem: ***The average weight of 'n' persons is increased by 'x' kg when some of them [$n_1, n_2, \ldots n$, where $n_1 + n_2 + \ldots < n$] who weigh [$y_1 + y_2 + \ldots$ where, $y_1 + y_2 + \ldots = y$ kg] are replaced by the same no. of persons. Then the weight of the new persons is (y + nx) ie.***

Weight of new persons = Weight of removed person + No. of persons × Increase in average.

Illustrative Examples

Ex. 1: The average weight of 4 men is increased by 3 kg when one of them who weighs 120 kg is replaced by another man. What is the weight of the new man?

Soln: **Quicker Aproach:** If the average is increased by 3 kg, then the sum of weights increases by $3 \times 4 = 12$ kg.

And this increase in weight is due to the extra weight included due to the inclusion of new person.

$\therefore$ Weight of new man = 120 + 12 = 132 kg.

Quicker Method: Weight of new person = weight of removed person + No. of persons × increase in average = $120 + 4 \times 3 = 132$ kg.

Ex. 2: The average age of 8 persons in a committee is increased by 2 years when two men aged 35 years and 45 years are substituted by two women. Find the average age of these two women.

Soln: By the direct formula, we have

the total age of two women $= 2 \times 8 + (35 + 45)$

$= 16 + 80 = 96$ years

$\therefore$ Average age of two women $= \frac{96}{2} = 48$ years.

Exercise

1. The average weight of 8 persons increases by 1.5 kg. If a person weighing 65 kg is replaced by a new person, what could be the weight of the new persons? **[BSRB Delhi PO 2000]**
   a) 76 kg b) 77 kg
   c) 76.5 kg d) Data inadequate
2. The average weight of 10 men is increased by $1\frac{1}{2}$ kg when one of the men who weighs 68 kg is replaced by a new man. Find the weight of the new man.
   a) 73 kg b) 83 kg
   c) 82.5 kg d) Data inadequate
3. The average weight of 15 men is increased by 2 kg when one of the men who weighs 48 kg is replaced by a new man. Find he weight of the new man.
   a) 88 kg b) 78 kg
   c) 77.5 kg d) Data inadequate
4. The average age of a committee of 7 trustees is the same as it was 5 years ago, a younger man having been substituted for one of them. How much younger was he than the trustee whose place he took?
   a) 30 years b) 35 years
   c) 25 years d) Data inadequate
5. The average age of 10 persons in a committee is increased by 1 year when two men aged 42 years and 38 years are substituted by two women. Find the average age of these two women.
   a) 46 years b) 45 years
   c) 42 years d) 44 years
6. The average age of 11 persons in a committee is increased by 2 years when three men aged 32 years, 33 years and 34 years are substituted by three women. Find the average age of these three women.
   a) 40 years b) $41\frac{1}{3}$ years
   c) 41 years d) $40\frac{1}{3}$ years
7. The average weight of the 8 oarsmen in a boat is increased by 1 kg when one of the crew, who weighs 60 kg is replaced by a new man. What is the weight of the new man?
   a) 78 kg b) 66 kg c) 68 kg d) 72 kg
8. The average weight of 8 persons is increased by 2.5 kg, when one of them whose weight is 56 kg is replaced by a new man. The weight of the new man is: **[Central Excise & I Tax 1988]**
   a) 66 kg b) 75 kg c) 76 kg d) 86 kg

Answers

1. b 2. b 3. b

4. b; **Hint:** The average age of committee of 7 trustees is the same as it was 5 years ago. But today committee gains $7 \times 5 = 35$ years. Hence, we can conclude that the younger man who replaced the trustee 5 years ago is 35 years younger than the trustee.
In another way it can be explained as the following. If the new member would have not been substituted, then the increased age $= (5 \times 7) = 35$ years. So the new member is 35 years younger than the trustee whose place he took.

5. b 6. d 7. c 8. c

Rule 11

Theorem: ***The average age of 'n' persons is decreased by 'x' years when some of them [$n_1, n_2 ... n$; where $n_1 + n_2 + ... \lt n$] aged [$y_1 + y_2 + ...$ where, $y_1 + y_2 + ... = y$ years] are replaced by the same number of persons. Then the age of the new persons is [$y - nx$] ie***

Age of new persons = Age of removed persons – No. of persons × Decreased in average

Illustrative Example

Ex.: In a class, there are 20 boys whose average age is decreased by 2 months, when one boy aged 18 years is replaced by a new boy. Find the age of the new boy.

Soln: Applying the above rule, we have

the required answer $= 18 - 20 \times \frac{2}{12}$

$= 18 - \frac{10}{3} = \frac{44}{3} = 14\frac{2}{3}$

= 14 years 8 months.

Exercise

1. In a class there are 24 boys whose average age is decreased by 3 months, when 1 boy aged 20 years is replaced by a new boy. Find the age of the new boy.
   a) 14 years b) 16 years c) 17 years d) 18 years

2. In a class there are 36 boys whose average age is decreased by 5 months, when 1 boy aged 30 years is replaced by a new boy. Find the age of the new boy.
a) 16 years b) 10 years c) 15 years d) 20 years
3. In a class there are 15 boys whose average age is decreased by 4 months, when 1 boy aged 23 years is replaced by a new boy. Find the age of the new boy.
a) 20 years b) 18 years c) 21 years d) 18.5 years
4. In a class there are 27 students whose average age is decreased by 4 months, when 4 students aged, 16, 17, 18 and 19 years respectively are replaced by the same number of students. Find the average age of the new students.
a) 15.25 years b) 15 years
c) 15.5 years d) 16 years
5. In a class there are 18 students whose average age is decreased by 2 months, when 3 students aged 12, 13 and 14 years respectively are replaced by the same number of students. Find the average age of the new students.
a) 18 years b) 12 years c) 16 years d) 14 years
6. In a class there are 16 students whose average age is decreased by 3 months, when 2 students aged 24 and 26 years respectively are replaced by the same number of students. Find the average age of the new students.
a) 23 years b) 21 years c) 18 years d) None of these

Answers

1. a 2. c 3. b 4. a 5. b 6. a

Rule 12

Theorem: ***The average of marks obtained by 'n' candidates in a certain examination is 'T'. If the average marks of passed candidates is 'P' and that of the failed candidates is 'F'. Then the number of candidates who passed the examination is*** $\left[\frac{n(T-F)}{P-F}\right]$.

ie Number of passed candidates

$$=\frac{\text{Total candidates (Total Average – Failed Average)}}{\text{Passed Average – Failed Average}}$$

Illustrative Example

Ex.: The average of marks obtained by 120 candidates in a certain examination is 35. If the average marks of passed candidates is 39 and that of the failed candidates is 15, what is the number of candidates who passed the examination?

Soln: **Detail Method:** Let the number of passed candidates be x.

Then total marks $= 120 \times 35 = 39x + (120 - x) \times 15$

or, $4200 = 39x + 1800 - 15x$

or, $24x = 2400$

$\therefore x = 100$

$\therefore$ number of passed candidates = 100.

Quicker Method: Applying the above formula, we have

$$\text{no. of passed candidates} = \frac{120(35-15)}{39-5} = 100$$

Exercise

1. The average of marks obtained by 90 candidates in a certain examination is 38. If the average marks of passed candidates is 40 and that of the failed candidates is 30, what is the number of candidates who passed the examination?
a) 72 b) 70 c) 75 d) 80
2. The average of marks obtained by 80 candidates in a certain examination is 32. If the average marks of passed candidates is 34 and that of the failed candidates is 18, what is the number of candidates who passed the examination?
a) 170 b) 70 c) 80 d) 75
3. The average of marks obtained by 65 candidates in a certain examination is 25. If the average marks of passed candidates is 27 and that of the failed candidates is 14, what is the number of candidates who passed the examination?
a) 55 b) 65 c) 60 d) 75
4. The average of marks obtained by 75 candidates in a certain examination is 31. If the average marks of passed candidates is 35 and that of the failed candidates is 25, what is the number of candidates who passed the examination?
a) 40 b) 46 c) 45 d) 54
5. The average of marks obtained by 77 candidates in a certain examination is 17. If the average marks of passed candidates is 19 and that of the failed candidates is 8, what is the number of candidates who passed the examination?
a) 36 b) 63 c) 40 d) 70

Answers

1. a 2. b 3. a 4. c 5. b

Rule 13

Theorem: ***The average of marks obtained by 'n' candidates in a certain examination is 'T'. If the average marks of passed candidates is 'P' and that of the failed candidates is 'F'. Then the number of candidates who failed the examination is*** $\left[\frac{n(P-T)}{P-F}\right]$.

ie, the number of failed candidates

$$=\frac{\text{Total candidates (Passed average – Total average)}}{\text{Passed average – Failed average}}$$

Illustrative Example

Ex.: The average of marks obtained by 120 candidates in a certain examination is 35. If the average marks of passed candidates is 39 and that of the failed candidates is 15, what is the number of candidates who failed the examination?

Soln: Following the above formula, we have

the no. of failed candidates $= \dfrac{120(39-35)}{39-15} = 20$.

Exercise

1. The average of marks obtained by 108 candidates in a certain examination is 20. If the average marks of passed candidates is 28 and that of the failed candidates is 16, what is the number of candidates who failed the examination?
   a) 70 b) 78 c) 81 d) 72
2. The average of marks obtained by 110 candidates in a certain examination is 15. If the average marks of passed candidates is 25 and that of the failed candidates is 14, what is the number of candidates who failed the examination?
   a) 100 b) 90 c) 105 d) 95
3. The average of marks obtained by 102 candidates in a certain examination is 18. If the average marks of passed candidates is 21 and that of the failed candidates is 15, what is the number of candidates who failed the examination?
   a) 51 b) 52 c) 61 d) 50
4. The average of marks obtained by 115 candidates in a certain examination is 36. If the average marks of passed candidates is 40 and that of the failed candidates is 17, what is the number of candidates who failed the examination?
   a) 30 b) 25 c) 20 d) 34
5. The average of marks obtained by 125 candidates in a certain examination is 29. If the average marks of passed candidates is 36 and that of the failed candidates is 11, what is the number of candidates who failed the examination?
   a) 32 b) 35 c) 30 d) 40

Answers

1. d 2. a 3. a 4. c 5. b

Rule 14

Theorem: ***If the average of n results (where n is an odd number) is 'a' and the average of first*** $\left(\dfrac{n+1}{2}\right)$ ***results is 'b' and that of last*** $\left(\dfrac{n+1}{2}\right)$ ***is 'c'. Then*** $\left(\dfrac{n+1}{2}\right)$ ***th result is*** $\left[\dfrac{n+1}{2}(b+c)-na\right]$.

Illustrative Example

Ex.: The average of 11 results is 50. If the average of first six results is 49 and that of last six is 52, find the sixth result.

Soln: Following the above formula, we have

$$\text{sixth result} = \frac{11+1}{2}(49+52)-11\times 50$$
$$= 6\times(101)-550 = 56$$

Exercise

1. The average of 17 numbers is 45. The average of first 9 of these numbers is 51 and the last 9 of these numbers is 36. What is the ninth number? **[BSRB Mumbai PO, 1998]**
   a) 14 b) 16 c) 22 d) 18
2. The average of 19 numbers is 40. The average of first 10 of these numbers is 39 and the last 10 of these numbers is 36. What is the 10th number?
   a) 12 b) 14 c) 18 d) 10
3. The average of 15 numbers is 50. The average of first 8 of these numbers is 52 and the last 8 of these numbers is 39. What is the 8th number?
   a) 32 b) 22 c) 31 d) 30
4. The average of 13 numbers is 30. The average of first 7 of these numbers is 32 and the last 7 of these numbers is 22. Find the 7th number.
   a) 18 b) 12 c) 16 d) 10
5. The average of 21 numbers is 35. The average of first 11 of these numbers is 42 and the last 11 of these numbers is 23. What is 11th number?
   a) 20 b) 22 c) 18 d) 23
6. The average of 25 results is 18; that of first 12 is 14 and of the last 12 is 17. Thirteenth result is: **[CBI Exam, 1990]**
   a) 78 b) 85 c) 28 d) 72

Answers

1. d 2. d 3. a 4. b 5. a 6. a

Rule 15

Ex.: The average of 11 results is 30, that of the first five is 25 and that of the last five is 28. Find the value of the 6th number.

Soln: **Direct Formula:**

6th number = Total of 11 results – (Total of first five + Total of last five results)

$= 11\times 30-(5\times 25+5\times 28) = 330-265 = 65.$

Exercise

1. The average of 11 numbers is 37.15, the average of the first 3 is 36.41 and of the last 7 is 37.51. The fourth number is found to be wrong. Find the average of the rest.
   a) 37.28 b) 37 c) 37.18 d) 37.08
2. The average of 12 results is 36, that of the first six is 32

and that of the last five is 35. Find the value of the 7th number.
a) 65 b) 60 c) 64 d) 56

3. The average of 15 results is 28, that of the first seven is 26 and that of the last seven is 25. Find the value of the 8th number.
a) 36 b) 66 c) 65 d) 63

4. The average of 13 results is 39, that of the first five is 38 and that of the last seven is 36. Find the value of the 6th number.
a) 64 b) 46 c) 65 d) 56

Answers

1. c; **Hint:** The fourth number $= 11 \times 37.15 - (3 \times 36.41 + 7 \times 37.51) = 36.85$

Number can not be in fraction. Hence this is wrong.

$\therefore$ Required average

$$= \frac{11 \times 37.15 - 36.85}{10} = \frac{371.8}{10} = 37.18$$

2. a 3. d 4. c

Rule 16

Theorem: ***If a batsman in his nth innings makes a score of 'x', and thereby increases his average by 'y', then the average after 'n' innings is [x – y (n – 1)].***

Illustrative Example

Ex.: A batsman in his 17th innings makes a score of 85 and thereby increases his average by 3. What is his average after 17 innings?

Soln: Following the above formula,
$85 - 3(17 - 1) = 37$.

Exercise

1. A batsman in his 16th innings makes a score of 92 and thereby increases his average by 4. What is his average after 16 innings?
a) 32 b) 30 c) 34 d) 23

2. A batsman, in his 19th innings, missed a century by 2 runs and thereby increases his average by 3. What is his average after 19 innings.
a) 54 b) 44 c) 45 d) 43

3. A batsman in his 21st innings makes a score of 88 and thereby increases his average by 2. What is his average after 21 innings?
a) 46 b) 48 c) 45 d) 44

4. A batsman in his 20th innings makes a score of 110 and thereby increases his average by 4. What is his average after 20 innings?
a) 34 b) 43 c) 36 d) 30

5. A batsman in his 44th innings makes a score of 86 and thereby increases his average by 1. What is his average after 44 innings?
a) 34 b) 43 c) 46 d) 40

Answers

1. a 2. b 3. b 4. a 5. b

Rule 17

Theorem: ***If a cricketer has completed 'n' innings and his average is 'x' runs. The number of runs, he must make in his next innings so as to raise his average to 'y' are [n (y – x) + y].***

Illustrative Example

Ex.: A cricketer has completed 10 innings and his average is 21.5 runs. How many runs must he make in his next innings so as to raise his average to 24?

Soln: Following the above theorem, we have
the required answer $= 10(24 - 21.5) + 24 = 25 + 24 = 49$.

Exercise

1. A cricketer has completed 15 innings and his average is 20 runs. How many runs must he make in his next innings so as to raise his average to 25?
a) 75 b) 50 c) 100 d) 85

2. A cricketer has completed 20 innings and his average is 44.5 runs. How many runs must he make in his next innings so as to raise his average to 45?
a) 45 b) 60 c) 40 d) 55

3. A cricketer has completed 31 innings and his average is 18 runs. How many runs must he make in his next innings so as to raise his average to 22?
a) 124 b) 146 c) 136 d) 142

4. A cricketer has completed 18 innings and his average is 26.5 runs. How many runs must he make in his next innings so as to raise his average to 27?
a) 63 b) 36 c) 45 d) 54

5. A cricketer has completed 14 innings and his average is 30 runs. How many runs must he make in his next innings so as to raise his average to 32?
a) 60 b) 55 c) 65 d) 50

Answers

1. c 2. d 3. b 4. b 5. a

Rule 18

Theorem: ***If a person travels a distance at a speed of x km/hr and the same distance at a speed of y km/hr, then the average speed during the whole journey is given by*** $\frac{2xy}{x+y}$ ***km/hr.***

or,

If half of the journey is travelled at a speed of x km/hr and the next half at a speed of y km/hr, then average speed during the whole journey is $\frac{2xy}{x+y}$ ***km/hr.***

or,

If a man goes to a certain place at a speed of x km/hr and returns to the original place at a speed of y km/hr, then the average speed during up-and-down journey is $\frac{2xy}{x+y}$ km/hr.

Note: In all the above three cases the two parts of the journey are equal, hence the last two may be considered as a special case of the first. That's why all the three lead to the same result.

Illustrative Example

Ex.: A train travels from A to B at the rate of 20 km per hour and from B to A at the rate of 30 km/hr. What is the average rate for the whole journey?

Soln: By the formula:

$$\text{Average speed} = \frac{2\times 20\times 30}{20+30} = 24 \text{ km/hr.}$$

Exercise

1. A constant distance from A to B is covered by a man at 40 km/hr. The person rides back the same disance at 30 km/hr. Find his average speed during the whole journey.
   a) 34 km/hr b) 35.29 km/hr
   c) 34.29 km/hr d) 35 km/hr
2. A man goes to a certain place at a speed of 15 km/hr and returns to the original place at a speed of 12 km/hr, find the average speed during up-and-down journey.
   a) 13 km/hr b) $13\frac{1}{3}$ km/hr
   c) $13\frac{2}{3}$ km/hr d) $11\frac{1}{3}$ km/hr
3. A man goes to a certain place at a speed of 30 km/hr and returns to the original place at a speed of 20 km/hr, find the average speed during up-and-down journey.
   a) 24 km/hr b) 25 km/hr
   c) 28 km/hr d) 23 km/hr
4. Ram travels half of a journey at the speed of 24 km/hr and the next half at a speed of 16 km/hr. What is the average speed of Ram during the whole journey?
   a) $19\frac{2}{5}$ km/hr b) 20 km/hr
   c) $19\frac{1}{5}$ km/hr d) $16\frac{1}{5}$ km/hr
5. A person travels half of a journey at the speed of 30 km/hr and the next half at a speed of 15 km/hr. What is the average speed of the person during the whole journey?
   a) 20 km/hr b) 25 km/hr c) 18 km/hr d) 24 km/hr

Answers

1. c 2. b 3. a 4. c 5. a

Rule 19

Theorem: ***If a person travels three equal distances at a speed of x km/hr, y km/hr and z km/hr respectively, then the average speed during the whole journey is $\frac{3xyz}{xy+yz+xz}$ km/hr.***

Illustrative Example

Ex.: A person divides his total route of journey into three equal parts and decides to travel the three parts with speeds of 40, 30 and 15 km/hr respectively. Find his average speed during the whole journey.

Soln: By the theorem:

$$\text{Average speed} = \frac{3\times 40\times 30\times 15}{40\times 30+30\times 15+40\times 15}$$

$$= \frac{3\times 40\times 30\times 15}{2250} = 24 \text{ km/hr.}$$

Exercise

1. A person divides his total route of journey into three equal parts and decides to travel the three parts with speeds of 20, 15 and 10 km/hr respectively. Find his average speed during the whole journey.
   a) $13\frac{11}{13}$ km/hr b) $11\frac{11}{13}$ km/hr
   c) $13\frac{3}{13}$ km/hr d) $11\frac{3}{13}$ km/hr
2. A person divides his total route of journey into three equal parts and decides to travel the three parts with speeds of 5, 10 and 15 km/hr respectively. Find his average speed during the whole journey.
   a) $8\frac{1}{11}$ km/hr b) $11\frac{1}{11}$ km/hr
   c) $8\frac{2}{11}$ km/hr d) $9\frac{2}{11}$ km/hr
3. A person divides his total route of journey into three equal parts and decides to travel the three parts with speeds of 20, 25 and 40 km/hr respectively. Find his average speed during the whole journey.
   a) $26\frac{2}{23}$ km/hr b) $26\frac{1}{23}$ km/hr
   c) $25\frac{2}{23}$ km/hr d) $25\frac{1}{23}$ km/hr
4. The average speed of a cyclist who covers first, second and third km at 20, 16 and 12 km/hr respectively (in km/hr) is ____.
   a) 16.24 km/hr b) 16 km/hr
   c) 15.66 km/hr d) 15.32 km/hr

Answers

1. a 2. c 3. a 4. d

Rule 20

Theorem: ***If a person covers A km at x km/hr and B km at y km/hr and C km at z km/hr, then the average speed in covering the whole disance is*** $\left[\dfrac{A+B+C}{\dfrac{A}{x}+\dfrac{B}{y}+\dfrac{C}{z}}\right]$ ***km/hr.***

Illustrative Example

Ex.: A person covers 12 km at 3 km/hr, 18 km at 9 km/hr and 24 km at 4 km/hr. Then find the average speed in covering the whole distance.

Soln: Applying the above formula, we have

the average speed $= \dfrac{12+18+24}{\dfrac{12}{3}+\dfrac{18}{9}+\dfrac{24}{4}} = \dfrac{54}{12} = 4.5$ km/hr.

Exercise

1. A person covers 15 km at 5 km/hr, 12 km at 6 km/hr and 16 km at 4 km/hr. Then find the average speed in covering the whole distance.

   a) $7\frac{4}{9}$ km/hr b) $9\frac{4}{7}$ km/hr

   c) $4\frac{7}{9}$ km/hr d) $7\frac{7}{9}$ km/hr

2. A person covers 9 km at 3 km/hr, 25 km at 5 km/hr and 30 km at 10 km/hr. Then find the average speed in covering the whole distance.

   a) $5\frac{9}{11}$ km/hr b) $11\frac{5}{9}$ km/hr

   c) $9\frac{5}{11}$ km/hr d) $5\frac{5}{11}$ km/hr

3. A person covers 18 km at 6 km/hr, 16 km at 8 km/hr and 30 km at 6 km/hr. Then find the average speed in covering the whole distance.

   a) 6.5 km/hr b) 6.4 km/hr c) 6.2 km/hr d) 6 km/hr

Answers

1. c 2. a 3. b

Rule 21

Theorem: ***If a person covers Ath part of the distance at x km/hr, Bth part of the distance at y km/hr and the remaining Cth part at z km/hr, then his average speed is*** $\left(\dfrac{1}{\dfrac{A}{x}+\dfrac{B}{y}+\dfrac{C}{z}}\right)$ ***km/hr.***

Note: **1.** If instead of Ath, Bth and Cth part we are given A%, B% and C% the expression changes to average speed $= \left(\dfrac{100}{\dfrac{A}{x}+\dfrac{B}{y}+\dfrac{C}{z}}\right)$ km/hr.

2. Ath Bth and Cth part or A%, B% and C% together constitute the total distance covered.

Illustrative Example

Ex.: A person runs the first $\frac{1}{5}$th of the distance at 2 km/hr, the next one half at 3 km/hr and the remaining distance at 1 km/hr. Find his average speed.

Soln: Remaining distance $= 1-\left(\dfrac{1}{5}+\dfrac{1}{2}\right) = \dfrac{3}{10}$

Now, applying the above rule,

the average speed $= \dfrac{1}{\dfrac{1/5}{2}+\dfrac{1/2}{3}+\dfrac{3/10}{1}} = \dfrac{1}{\dfrac{1}{10}+\dfrac{1}{6}+\dfrac{3}{10}} = \dfrac{30}{17} = 1\dfrac{13}{17}$ km/hr.

Exercise

1. Find the average speed of a person when he covers first one-third of the distance at 10 km/hr, next $\frac{1}{3}$rd at 8 km/hr and the last one-third at 6 km/hr.

   a) 7.66 km/hr b) 6.77 km/hr c) 6.67 km/hr d) 7.86 km/hr

2. A person runs the first $\frac{1}{4}$th of the distance of 8 km/hr, the next $\frac{3}{5}$th at 6 km/hr and the remaining distance at 10 km/hr. Find his average speed.

   a) 17 km/hr b) 17.87 km/hr

   c) 17.78 km/hr d) None of these

3. A man covers first 20% of the distance at 10 km/hr, next 50% at 5 km/hr and the remaining distance at 15 km/hr. Find his average speed.

   a) 7 km/hr b) 7.14 km/hr

   c) 7.24 km/hr d) 4.17 km/hr

4. A train covers 50% of the journey at 30 km/hr, 25% of the journey at 25 km/hr and the remaining at 20 km/hr. Find the average speed of the train during the whole journey.

   a) $25\frac{25}{47}$ km/hr b) $25\frac{5}{47}$ km/hr

   c) $25\frac{23}{47}$ km/hr d) None of these

5. On a journey across Mumbai, a taxi averages 30 km/hr for 60% of the distance, 20 km/hr for 20% of it and 10 km/hr for the remainder. The average speed for the whole journey (in km/hr) is
a) 20 km/hr b) 22.5 km/hr
c) 25 km/hr d) 24.625 km/hr

Answers
1. a 2. c 3. b 4. a 5. a

Rule 22

Theorem: ***If the average value of all the members of a group is 'x', the average value of the first part of members is 'y', the average value of the remaining part of members is 'z' and the number of the first part of members is 'n', then the number of the other part of members is*** $\left[\frac{n(x-y)}{z-x}\right]$.

Illustrative Example

Ex.: The average salary of the entire staff in a office is Rs 120 per month. The average salary of officers is Rs 460 and that of non-officers is Rs 110. If the number of officers is 15, then find the number of non-officers in the office.

Soln: **Detail Method:** Let the required number of non-officers = x

Then, $110x + 460 \times 15 = 120(15 + x)$

or, $120x - 110x = 460 \times 15 - 120 \times 15 = 15(460 - 120)$

or, $10x = 15 \times 340$; $\therefore x = 15 \times 34 = 510$.

Quicker Method: No. of non-officers = No. of officers ×

$$\left(\frac{\text{Av. salary of officers} - \text{Mean average}}{\text{Mean average} - \text{Av. salary of non-officers}}\right)$$

$$= 15\left(\frac{460-120}{120-110}\right) = 510.$$

Exercise

1. The average salary of the entire staff in a office is Rs 130 per month. The average salary of officers is Rs 540 and that of non-officers is Rs 114. If the number of officers is 16, then find the number of non-officers in the office.
a) 140 b) 410 c) 510 d) 150
2. The average salary of the entire staff in a office is Rs 220 per month. The average salary of officers is Rs 650 and that of non-officers is Rs 170. If the number of officers is 25, then find the number of non-officers in the office.
a) 215 b) 315 c) 250 d) 350
3. The average salary of the entire staff in a office is Rs 166 per month. The average salary of officers is Rs 456 and that of non-officers is Rs 142. If the number of officers is 24, then find the number of non-officers in the office.
a) 300 b) 290 c) 390 d) 310
4. The average salary of the entire staff in a office is Rs 200 per month. The average salary of officers is Rs 318 and that of non-officers is Rs 183. If the number of officers is 34, then find the number of non-officers in the office.
a) 118 b) 240 c) 246 d) 236
5. The average salary of the entire staff in a office is Rs 150 per month. The average salary of officers is Rs 450 and that of non-officers is Rs 80. If the number of officers is 14, then find the number of non-officers in the office.
a) 65 b) 55 c) 60 d) 70

Answers
1. b 2. a 3. b 4. d 5. c

Rule 23

Theorem: ***If the average value of all the members of a group is x, the average value of the first part of members is y, the average value of the remaining part of members is z and the number of the remaining part of members is n, then the number of first part of members is*** $\left[\frac{n(x-z)}{y-x}\right]$.

Illustrative Example

Ex.: The average age of all the students of a class is 18 years. The average age of boys of the class is 20 years and that of the girls is 15 years. If the number of girls in the class is 20, then find the number of boys in the class.

Soln: Following the above formula, we have

the required answer $= \frac{20(18-15)}{20-18} = 30$.

Exercise

1. The average age of all the students of a class is 16 years. The average age of boys of the class is 21 years and that of the girls is 12 years. If the number of girls in the class is 10, then find the number of boys in the elass.
a) 4 b) 8 c) 12 d) 10
2. The average age of all the students of a class is 24 years. The average age of boys of the class is 29 years and that of the girls is 20 years. If the number of girls in the class is 25, then find the number of boys in the class.
a) 30 b) 15 c) 24 d) 20
3. The average age of all the students of a class is 22 years. The average age of boys of the class is 26 years and that of the girls is 19 years. If the number of girls in the class is 16, then find the number of boys in the class.
a) 12 b) 10 c) 6 d) 8
4. The average age of all the students of a class is 25 years. The average age of boys of the class is 27 years and that of the girls is 23 years. If the number of girls in the class is 32, then find the number of boys in the class.

a) 24 b) 16 c) 32 d) 28

5. The average age of all the students of a class is 22 years. The average age of boys of the class is 24 years and that of the girls is 20 years. If the number of girls in the class is 30, then find the total number of students in the class.
a) 60 b) 30 c) 45 d) 50

Answers

1. b 2. d 3. a 4. c 5. b

Rule 24

Ex.: There were 35 students in a hostel. If the number of students increases by 7, the expenses of the mess increase by Rs 42 per day while the average expenditure per head diminishes by Re 1. Find the original expenditure of the mess.

Soln: **Detail Method:** Suppose the average expenditure was Rs x. Then total expenditure = $35x$.

When 7 more students join the mess, total expenditure = $35x + 42$

Now, the average expenditure = $\frac{35x+42}{35+7} = \frac{35x+42}{42}$

Now, we have $\frac{35x+42}{42} = x - 1$

or, $35x + 42 = 42x - 42$

or, $7x = 84 \therefore x = 12$

Thus the original expenditure of the mess = 35 × 12 = Rs 420

Direct Formula:

If decrease in average = x

increase in expenditure = y

increase in no. of students = z

and number of students (originally) = N, then

the original expenditure = $N\left[\frac{x(N+z)+y}{z}\right]$

In this case, $35\left[\frac{1(35+7)+42}{7}\right] = 35 \times 12 = \text{Rs } 420$

Exercise

1. There were 40 students in a hostel. If the number of students increases by 8, the expenses of the mess increase by Rs 48 per day while the average expenditure per head diminishes by Rs 2. Find the original expenditure of the mess.
a) Rs 620 b) Rs 720 c) Rs 750 d) Rs 820

2. There were 45 students in a hostel. If the number of students increases by 9, the expenses of the mess increase by Rs 25 per day while the average expenditure per head diminishes by Re 1. Find the original expenditure of the mess.
a) Rs 390 b) Rs 295 c) Rs 395 d) Rs 400

3. There were 42 students in a hostel. If the number of students increases by 7, the expenses of the mess increase by Rs 32.5 per day while the average expenditure per head diminishes by Rs 1.5. Find the original expenditure of the mess.
a) Rs 636 b) Rs 536 c) Rs 630 d) Rs 656

4. There were 36 students in a hostel. If the number of students increases by 4, the expenses of the mess increase by Rs 32 per day while the average expenditure per head diminishes by Re 1. Find the original expenditure of the mess.
a) Rs 640 b) Rs 648 c) Rs 650 d) Rs 658

Answers

1. b 2. c 3. a 4. b

Rule 25

The average of first 'n' natural numbers is $\frac{n+1}{2}$**.**

Illustrative Example

Ex.: Find the average of first 61 natural numbers.

Soln: Applying the above rule, we have

the required average = $\frac{61+1}{2} = 31$.

Exercise

1. Find the average of first 62 natural numbers.
a) 31 b) 31.5 c) 31.2 d) 32
2. Find the average of first 31 natural numbers.
a) 15 b) 14 c) 16 d) 17
3. Find the average of first 91 natural numbers.
a) 45 b) 47 c) 45.5 d) 46
4. Find the average of first 101 natural numbers.
a) 50.5 b) 52 c) 51 d) None of these

Answers

1. b 2. c 3. d 4. c

Rule 26

Ex.: The average weight of 50 balls is 5 gm. If the weight of the bag be included the average weight increases by 0.05 gm. What is the weight of the bag?

Soln: **Direct Formula:**

Weight of bag = Old average + Increase in average × Total no. of objects

$= 5 + 0.05 \times 51 = 5 + 2.55 = 7.55$ gm

Exercise

1. The average weight of 30 balls is 3 gm. If the weight of the bag be included the average weight increases by 0.03 gm. What is the weight of the bag?
a) 3.93 gm b) 3 gm c) 4 gm d) 4.39 gm
2. The average weight of 40 balls is 4 gm. If the weight of

the bag be included the average weight increases by 0.04 gm. What is the weight of the bag?
a) 4.64 gm b) 6.64 gm
c) 5.64 gm d) None of these

3. The average weight of 20 balls is 2 gm. If the weight of the bag be included the average weight increases by 0.02 gm. What is the weight of the bag?
a) 2.04 gm b) 2.42 gm c) 3.42 gm d) 3.04 gm

Answers

1. a 2. c 3. b

Rule 27

The average of n (where n is an odd number) consecutive numbers is always the middle number. The numbers may be

(a) merely consecutive numbers eg, 1, 2, 3, 4, 5.
Average value of these 5 consecutive numbers will be the middle number ie 3

Proof: ***Average*** $= \frac{1+2+3+4+5}{5} = \frac{15}{5} = 3.$

(b) consecutive odd numbers eg 1, 3, 5, 7, 9.
The average = middle number = 5

Proof: ***Average*** $= \frac{1+3+5+7+9}{5} = \frac{25}{5} = 5.$

(c) consecutive even numbers eg, 2, 4, 6, 8, 10.
The average = middle number = 6

Proof: Average $= \frac{2+4+6+8+10}{5} = \frac{30}{5} = 6.$

Exercise

1. Find the average of these 9 consecutive numbers.
5, 6, 7, 8, 9, 10, 11, 12, 13
2. Find the average of the consecutive numbers given below.
25, 26, 27, 28, 29, 30, 31
3. Find the average of the consecutive odd numbers given below.
33, 35, 37, 39, 41, 43, 45
4. Find the average of the consecutive odd numbers given below.
71, 73, 75, 77, 79, 81, 83, 85, 87, 89, 91
5. Find the average of the consecutive even numbers given below.
50, 52, 54, 56, 58, 60, 62, 64, 66, 68, 70, 72, 74

Answers

1. 9 2. 28 3. 39 4. 81 5. 62

Rule 28

Theorem: ***The average of 'n' (where n = even number) consecutive numbers (whether merely consecutive, consecutive odd or consecutive even) is the average of the middle two numbers.***

(a) merely consecutive numbers eg, 1, 2, 3, 4, 5, 6.
Average = average of the middle two numbers

$= \frac{3+4}{2} = 3.5.$

Proof: Average $= \frac{1+2+3+4+5+6}{6} = \frac{21}{6} = 3.5.$

(b) consecutive odd numbers eg, 1, 3, 5, 7, 9, 11

Average $= \frac{5+7}{2} = 6.$

Proof: Average $= \frac{1+3+5+7+9+11}{6} = \frac{36}{6} = 6.$

(c) *consecutive even numbers eg 2, 4, 6, 8, 10, 12.*

Average $= \frac{6+8}{2} = 7.$

Proof: Average $= \frac{2+4+6+8+10+12}{6} = \frac{72}{6} = 7.$

Note: In all the above series, there are two middle terms. Hene the required average can be calculated by the following methods.

I. In the case of **'consecutive numbers'**:
Average = Smaller middle term + 0.5 or Greater middle term − 0.5

II. In the case of **'consecutive odd'** and **'consecutive even'**:
Average = Smaller middle term + 1 or Greater middle term − 1.

Exercise

1. Find the average value of the following consecutive numbers.
(i) 5, 6, 7, 8, 9, 10, 11, 12, 13, 14
(ii) 22, 23, 24, 25, 26, 27, 28, 29, 30, 31, 32, 33
(iii) 71, 72, 73, 74, 75, 76
2. Find the average value of the following consecutive odd numbers.
(i) 9, 11, 13, 15, 17, 19
(ii) 35, 37, 39, 41, 43, 45, 47, 49
(iii) 81, 83, 85, 87, 89, 91, 93, 95
3. Find the average value of the following consecutive even numbers.
(i) 82, 84, 86, 88, 90, 92
(ii) 50, 52, 54, 56, 58, 60, 62, 64

Answers

1. (i) 9.5 (ii) 27.5
(iii) 73.5
2. (i) 14 (ii) 42
(iii) 88
3. (i) 87 (ii) 57

Rule 29

The average of odd numbers from 1 to n is $\frac{[Last\ odd\ number\ +1]}{2}$***, where n = natural odd number.***

Illustrative Example

Ex.: What is the average of odd numbers from 1 to 35?

Soln: Applying the above rule, we have

the required answer $= \frac{35+1}{2} = 18$.

Exercise

1. What is the average of odd numbers from 1 to 39?
   a) 20 b) 19 c) 18 d) 21
2. What is the average of odd numbers from 1 to 79?
   a) 30 b) 25 c) 35 d) 40
3. What is the average of odd numbers from 1 to 103?
   a) 53 b) 51 c) 52 d) 50
4. What is the average of odd numbers from 1 to 51?
   a) 27 b) 26 c) 25 d) 28

Answers

1. a 2. d 3. c 4. b

Rule 30

The average of even numbers from 1 to n is $\left[\frac{Last\ even\ number+2}{2}\right]$***; where n = natural even number.***

Illustrative Example

Ex.: What is the average of even numbers from 1 to 50?

Soln: Applying the above formula, we have

the required answer $= \frac{50+2}{2} = 26$.

Exercise

1. What is the average of even numbers from 1 to 30?
   a) 16 b) 15 c) 18 d) 17
2. What is the average of even numbers from 1 to 80?
   a) 40 b) 41 c) 42 d) 44
3. What is the average of even numbers from 1 to 42?
   a) 20 b) 24 c) 22 d) 18
4. What is the average of even numbers from 1 to 92?
   a) 45 b) 44 c) 46 d) 47

Answers

1. a 2. b 3. c 4. d

Rule 31

The average of square of natural numbers till n is $\left[\frac{(n+1)(2n+1)}{6}\right]$.

Illustrative Example

Ex.: Find the average of squares of natural numbers till 7.

Soln: Following the above formula, we have

the required answer $= \frac{(7+1)(2\times7+1)}{6} = \frac{8\times15}{6} = 20$.

Exercise

1. What is the average of square of the natural numbers from 1 to 10.
   a) 39 b) 40 c) 38.5 d) 37.5
2. What is the average of square of the natural numbers from 1 to 23.
   a) 188 b) 182 c) 180 d) 183
3. What is the average of square of the natural numbers from 1 to 35.
   a) 436 b) 426 c) 416 d) 446
4. What is the average of square of the natural numbers from 1 to 41.
   a) 580 b) 571 c) 851 d) 581

Answers

1. c 2. a 3. b 4. d

Rule 32

The average of cubes of natural numbers till n is $\left[\frac{n(n+1)^2}{4}\right]$.

Illustrative Example

Ex.: Find the average of cubes of natural numbers till 7.

Soln: Applying the above formula, we have

the required answer $= \frac{7(7+1)^2}{4} = \frac{7\times8\times8}{4} = 112$.

Exercise

1. Find the average of cubes of natural numbers from 1 to 16.
   a) 1156 b) 1516 c) 1155 d) 1165
2. Find the average of cubes of natural numbers from 1 to 15.
   a) 690 b) 960 c) 890 d) 980
3. Find the average of cubes of natural numbers from 1 to 24.
   a) 7350 b) 3570 c) 3750 d) None of these
4. Find the average of cubes of natural numbers from 1 to 27.
   a) 756 b) 9252 c) 5922 d) 5292
5. Find the average of cubes of natural numbers from 1 to 8.
   a) 162 b) 172 c) 153 d) 163

Answers

1. a 2. b 3. c 4. d 5. a

Rule 33

The average of first n consecutive even numbers is (n + 1).

Illustrative Example

Ex.: Find the average of first 50 consecutive even numbers.

Soln: Following the above rule, we have the required answer = 50 + 1 = 51.

Exercise

1. Find the average of first 52 consecutive even numbers.
   a) 52 b) 53 c) 51 d) 50
2. Find the average of first 60 consecutive even numbers.
   a) 61 b) 59 c) 62 d) 58
3. Find the average of first 14 consecutive even numbers.
   a) 14 b) 13 c) 16 d) 15
4. Find the average of first 18 consecutive even numbers.
   a) 16 b) 17 c) 19 d) 18

Answers

1. b 2. a 3. d 4. c

Rule 34

The average of first n consecutive odd numbers is 'n'.

Illustrative Example

Ex.: Find the average of first 16 consecutive odd numbers.

Soln: Applying the above rule, we have the required answer = 16.

Exercise

1. Find the average of first 17 consecutive odd numbers.
   a) 16 b) 17 c) 18 d) 15
2. Find the average of first 28 consecutive odd numbers.
   a) 28 b) 27
   c) 29 d) 26
3. Find the average of first 64 consecutive odd numbers.
   a) 64 b) 63 c) 65 d) 66
4. Find the average of first 55 consecutive odd numbers.
   a) 54 b) 56 c) 55.5 d) 55

Answers

1. b 2. a 3. a 4. d

Rule 35

The average of squares of first n consecutive even numbers

is $\left[\frac{2(n+1)(2n+1)}{3}\right]$.

Illustrative Example

Ex.: Find the average of squares of first 10 consecutive even numbers.

Soln: Following the above rule, we have the required answer

$$= \frac{2(10+1)(20+1)}{3} = \frac{2\times11\times21}{3} = 154.$$

Exercise

1. Find the average of squares of first 11 consecutive even numbers.
   a) 184 b) 148 c) 186 d) 174
2. Find the average of squares of first 14 consecutive even numbers.
   a) 280 b) 270 c) 290 d) 295
3. Find the average of squares of first 17 consecutive even numbers.
   a) 450 b) 420 c) 430 d) 410
4. Find the average of squares of first 23 consecutive even numbers.
   a) 750 b) 754 c) 725 d) 752

Answers

1. a 2. c 3. b 4. d

Rule 36

The average of squares of consecutive even numbers till n

is $\left[\frac{(n+1)(n+2)}{3}\right]$.

Ex.: Find the average of squares of consecutive even numbers till 10.

Soln: Applying the above rule, we have

$$\text{the required answer} = \frac{11\times12}{3} = 44.$$

Exercise

1. Find the average of squares of consecutive even numbers from 1 to 11.
   a) 65 b) 52 c) 44 d) 51
2. Find the average of squares of consecutive even numbers from 1 to 22.
   a) 184 b) 174 c) 182 d) 186
3. Find the average of squares of consecutive even numbers from 1 to 26.
   a) 243 b) 236 c) 252 d) 235
4. Find the average of squares of consecutive even numbers from 1 to 35.
   a) 484 b) 445
   c) 408 d) 444
5. Find the average of squares of consecutive even numbers from 1 to 44.
   a) 680 b) 960 c) 690 d) 860

Answers

1. c 2. a 3. c 4. c 5. c

Rule 37

The average of squares of consecutive odd numbers till n is

$$\left[\frac{n(n+2)}{3}\right].$$

Illustrative Example

Ex.: Find the average of squares of consecutive odd numbers till 15.

Soln: Following the above formula, we have

the required answer $= \frac{15(15+2)}{3} = 85$.

Exercise

1. Find the average of squares of consecutive odd number from 1 to 14.
   a) 70 b) 65 c) 75 d) 66
2. Find the average of squares of consecutive odd number from 1 to 20.
   a) 142 b) 136 c) 133 d) 144
3. Find the average of squares of consecutive odd number from 1 to 21.
   a) 164 b) 161 c) 144 d) 184
4. Find the average of squares of consecutive odd number from 1 to 37.
   a) 404 b) 464 c) 481 d) 444
5. Find the average of squares of consecutive odd number from 1 to 44.
   a) 645 b) 702 c) 802 d) 502

Answers

1. b 2. c 3. b 4. c 5. a

Rule 38

Theorem: ***If the average of n members is 'A' and on re-checking it is noticed that some of the numbers (ie $x_1, x_2, x_3, \ldots x_n$) are wrongly taken as $(x'_1, x'_2, x'_3, \ldots, x'_n)$ then their correct average is***

Previous average

$$+\frac{Sum\ of\ correct\ numbers - Sum\ of\ wrong\ numbers}{n}$$

or, Correct average

$$= A + \frac{(x_1 + x_2 + x_3 + \ldots x_n) - (x'_1 + x'_2 + x'_3 + \ldots + x'_n)}{n}$$

Illustrative Example

Ex.: In a calculation Mohan found that the average of 4 numbers is 25 and on rechecking Sohan noticed that a number 15 is wrongly taken as 51. Find the correct average.

Soln: Applying the above formula,

Correct average $= 25 + \frac{15-51}{4} = 25 - 9 = 16$.

Exercise

1. The average weight of a group of 20 boys was calculated to be 89.4 kg and it was later discovered that one weight was misread as 78 kg instead of the correct one of 87 kg. The correct average weight is
   a) 88.95 kg b) 89.25 kg c) 89.55 kg d) 89.85 kg
2. The average weight of a group of 15 boys was calculated to be 60 kg and it was later discovered that one weight was misread as 24 kg instead of the correct one of 42 kg. The correct average weight is
   a) 60.2 kg b) 61.2 kg c) 62 kg d) 61 kg
3. In a calculation Ram found that the average of 10 numbers is 45 and on rechecking Shyam noticed that the some numbers 18, 34, 63 is wrongly taken as 81, 43 and 36. Find the correct average.
   a) 39.5 b) 40.5 c) 45.5 d) 50.50
4. The average weight of a group of 9 boys was calculated to be 74.5 kg and it was later discovered that one weight was misread as 56 kg instead of the correct one of 65 kg. The correct average weight is
   a) 75 kg b) 73.5 kg c) 75.5 kg d) 76 kg
5. The average weight of 15 students was calculated to be 52 kg and it was later discovered that one weight was misread as 21 kg instead of the correct one of 12 kg. The correct average weight is
   a) 51.4 kg b) 50.6 kg c) 52.4 kg d) 51.6 kg

Answers

1. d 2. b 3. b 4. c 5. a

Rule 39

Geometric Mean: ***Geometric mean is useful in calculating averages of ratios such as average population growth rate, average percentage increase etc.***

Geometric mean of $x_1, x_2, x_3, \ldots, x_n$ is denoted by

$$GM = \sqrt[n]{x_1 \times x_2 \times x_3 \times \ldots \times x_n}$$

Illustrative Example

Ex.: Find the geometric mean of 4, 8, 16.

Soln: GM $= \sqrt[3]{4 \times 8 \times 16} = \sqrt[3]{512} = 8$

Exercise

1. Find the geometric mean of 2, 4 and 8.
   a) 2 b) 4 c) 3 d) 5
2. Find the geometric mean of 3, 6 and 12.
   a) 4 b) 8 c) 6 d) None of these
3. Find the geometric mean of 4, 10 and 25.
   a) 5 b) 10 c) 20 d) 15
4. Find the geometric mean of 9, 12 and 16.
   a) 8 b) 12 c) 14 d) 22

Answers

1. b 2. c 3. b 4. b

Rule 40

Harmonic Mean: ***Harmonic mean is useful for finding out average speed of a vehicle, average production per day, etc.***

Harmonic mean of $x_1, x_2, x_3, ..., x_n$ ***is denoted by***

$$HM = \frac{1}{\frac{1}{n}\left[\frac{1}{x_n}+\frac{1}{x_2}+\frac{1}{x_3}+...+\frac{1}{x_n}\right]}$$

Illustrative Example

Ex.: Find the harmonic mean of 2, 3, 4 and 5.

Soln: $HM = \frac{1}{\frac{1}{4}\left[\frac{1}{2}+\frac{1}{3}+\frac{1}{4}+\frac{1}{5}\right]}$

$$= \frac{4}{\frac{30+20+15+12}{60}} = \frac{4\times60}{77} = \frac{240}{77}$$

Exercise

1. Find the harmonic mean of 5, 6, 7 and 8.
2. Find the harmonic mean of 4, 6 and 8.
3. Find the harmonic mean of 12, 15, 18 and 21.

Answers

1. $\frac{3360}{533}$ 2. $\frac{72}{13}$ 3. $\frac{5040}{299}$

Rule 41

Average of a series having common difference 2 is $\frac{\textit{first term} + \textit{last term}}{2}$.

Illustrative Example

Ex.: Find the average of the following series 25, 27, 29, 31, 33

Soln: Applying the above formula, we have

the required average $= \frac{25+33}{2} = 29$.

Exercise

1. Find the average of the following series

(i) 22, 24, 26, 28, 30, 32, 36, 38, 40

a) 32 b) 31 c) 29 d) 33

(ii) 13, 15, 17, 19, 21, 23, 25, 27, 29, 31, 33

a) 23 b) 32 c) 24 d) 22

(iii) 85, 87, 89, 91, 93, 95

a) 91 b) 89 c) 90 d) 92

(iv) 70, 72, 74, 76, 78, 80, 82, 84, 86, 88, 90

a) 80 b) 82 c) 78 d) 84

(v) 35, 37, 39, 41, 43, 45, 47, 49, 51, 53, 55

a) 47 b) 49 c) 43 d) 45

Note: Try to solve the above questions by using the rules **27** and **28**.

Answers

i. b ii. a iii. c iv. a v. d

Rule 42

If the average of the first and the second of three numbers is 'x' more or less than the average of the second and the third of these numbers, then the differene between the first and the third of these three numbers is given by '2x'.

Note: Here only 2 numbers (ie first and second or second and third) are involved in calculating average, therefore, we multiply x by 2. If 'n' numbers are involved, for getting answer, we multiply x by n.

Illustrative Example

Ex.: The average of the first and the second of three numbers is 10 more than the average of the second and the third of these numbers. What is the difference between the first and the third of these three numbers?

Soln: **Detail Method:**

Average of the first and the second numbers

$$= \frac{\text{First} + \text{Second}}{2} \text{ and}$$

Average of the second and the third numbers

$$= \frac{\text{Second} + \text{Third}}{2}.$$

According to the question,

$$\frac{\text{First} + \text{Second}}{2} - \frac{\text{Second} + \text{Third}}{2} = 10$$

$\therefore$ First − Third = 20

Quicker Method: Applying the above rule, we can get

the required answer = 2 × 10 = 20.

Exercise

1. The average of the first and the second of three numbers is 15 more than the average of the second and the third of these numbers. What is the difference between the first and the third of these three numbers? **[SBI PO Exam, 2000]**

a) 15 b) 45

c) 60 d) 30

2. The average of the first and the second of three numbers is 12 more than the average of the second and the third of these numbers. What is the difference between the first and the third of these three numbers?
a) 24 b) 10 c) 12 d) Data inadequate
3. The average of the first and the second of three numbers is 16 more than the average of the second and the third of these numbers. What is the difference between the first and the third of these three numbers?
a) 32 b) 48 c) 61 d) 16
4. The average of the first and the second of three numbers is 13 more than the average of the second and the third of these numbers. What is the difference between the first and the third of these three numbers?
a) 25 b) 24 c) 26 d) 19
5. The average of Suresh's marks in English and History is 55. His average of marks in English and Science is 65. What is the difference between the marks which he obtained in History and Science?
[Bank of Baroda PO 1999]
a) 40 b) 60 c) 20 d) Data inadequate
6. The average marks scored by Ganesh in English, Science, Mathematics and History is less than 15 from that scored by him in English, History, Geography and Mathematics. What is the difference of marks in Science and Geography scored by him? **[BSRB Chennai PO 2000]**
a) 40 b) 50 c) 60 d) Data inadequate
7. The average temperature for Monday, Tuesday and Wednesday was 40° C. The average for Tuesday, Wednesday and Thursday was 41° C. That for Thursday being 42° C, what was the temperature on Monday?
a) 39° C b) 45° C c) 44° C d) 40° C
8. The average temperature for Monday, Tuesday and Wednesday was 40° C. The average for Tuesday, Wednesday and Thursday was 41° C and that of Thursday being 45° C. What was the temperature on Monday?
a) 48° C b) 41° C c) 46° d) 42° C
9. The mean temperature of Monday to Wednesday was 37° C and that of Tuesday to Thursday was 34° C. If the temperature on Thursday was $\left(\frac{4}{5}\right)$th that of Monday, what was the temperature on Thursday?
a) 34° b) 35.5° C c) 36° C d) 36.5° C

Answers

1. d 2. a 3. a 4. c

5. c; **Hint:** Here, we can say that the average of Suresh's marks in English and History is 10 less than the average marks in English and Science. (65 – 55 = 10) and now apply the above rule.

6. c; **Hint:** Here four subjects are involved ie English, **Science**, Mathematics and History in first group and English History, **Geography** and Mathematics in second group. We observe that English, Mathematics and History are common to both the groups. Hence the difference of marks in **Science** and **Geography** is given by 4 × 15 = 60. **[Also see note.]**

7. a; **Hint:** Average temperature for **Monday**, Tuesday and Wednesday was 1° C less (41° C – 40° C) than the average temperature of Tuesday, Wednesday and Thursday. Therefore, difference of temperature between **Thursday** and Monday is given by 3 × 1 (By the rule) = 3° C. Temperature of Thursday is 42° C given. Hence, temperature of Monday is (42 – 3 =) 39° C.

8. d

9. c; **Hint:** Monday to Wednesday = Monday, Tuesday, Wednesday
Tuesday to Thursday = Tuesday, Wednesday, Thursday
Difference of temperature between Monday and Thursday = 3 × (37 – 34) = 9° C.
According to the question,
$x - \frac{4x}{5} = 9$ [where, x = temperature of Monday]
$\therefore x = 45$
∴ Temperature of Monday = 45° C and
temperature of Thursday = $\frac{4}{5} \times 45 = 36°$ C

Rule 43

If average of 'n' consecutive odd numbers is 'x', then the difference between the smallest and the largest numbers is given by 2(n – 1).

Note: We see that the above formula is independent of x. That means, this formula always holds good irrespective of the value of x.

Illustrative Example

Ex.: If average of 7 consecutive numbers is 21, what is the difference between the smallest and the largest numbers?

Soln: Applying the above rule, we have the required answer = 2 (7 – 1) = 12.

Exercise

1. If average of 6 consecutive numbers is 48, what is the difference between the smallest and the largest numbers? **[NABARD, 1999]**
a) 10 b) 12 c) 9 d) Data inadequate
2. If average of 8 consecutive numbers is 64, what is the difference between the smallest and the largest numbers?
a) 12 b) 16 c) 14 d) 18
3. If average of 15 consecutive numbers is 32, what is the

difference between the smallest and the largest numbers?

a) 28 b) 18 c) 26 d) 24

4. If average of 18 consecutive numbers is 45, what is the difference between the smallest and the largest numbers?

a) 36 b) 34 c) 35 d) 37

Answers

1. a 2. c 3. a 4. b

Miscellaneous

1. The average attendance of a college for the first three days of a week is 325, and for first four days it is 320. How many were present on the fourth day?

a) 305 b) 350 c) 530 d) 503

2. A car runs for t_1 hours at v_1 km/hr, t_2 hours at v_2 km/hr. What is the average speed of the car for the entire journey?

a) $\frac{t_1 + t_2}{v_1 t_1 + v_2 t_2}$ km/hr b) $\frac{v_1 t_1 + v_2 t_2}{t_1 + t_2}$ km/hr

c) $\frac{v_1 t_2 + v_2 t_1}{v_1 + v_2}$ km/hr d) $\frac{v_1 + v_2}{v_1 t_1 + v_2 t_2}$ km/hr

3. A car runs x km at an average speed of v_1 km/hr and y km at an average speed of v_2 km/hr. What is the average speed of the car for the entire journey?

a) $\frac{v_1 v_2 (x+y)}{x v_2 + y v_1}$ km/hr b) $\frac{x v_2 + y v_1}{v_1 v_2 (x+y)}$ km/hr

c) $\frac{xy(v_1 + v_2)}{x v_1 + y v_2}$ km/hr d) $\frac{x v_1 + y v_2}{xy(v_1 + v_2)}$ km/hr

4. An aeroplane covers the four sides of square field at speeds of 200, 400, 600 and 800 km/hr. Then the average speed of the plane in the entire journey is

a) 600 km/hr b) 400 km/hr c) 500 km/hr d) 384 km/hr

5. The average age of the three boys is 15 years. Their ages are in the ratio 3:5:7. Then the age of the oldest is **[SBI PO Exam, 1987]**

a) 7 years b) 14 years c) 20 years d) 21 years

6. The population of a town increased by 20% during the first year, by 25% during the next year and by 44% during the third year. Find the average rate of increase during 3 years.

a) 36.87% b) 37.68% c) 38.67% d) None of these

7. An investor earns 3% return on $\frac{1}{4}$th of his capital, 5% on $\frac{2}{3}$rd and 11% on the remainder. What is the average rate of return he earns on his total capital?

a) 5% b) 10% c) 5.5% d) 10.5%

8. Out of three given numbers, the first one is twice the second and three times the third. If the average of these numbers is 88, then the difference between first and third is ______.

a) 48 b) 72 c) 96 d) 32

9. The average of 8 readings is 24.3, out of which the average of first two is 18.5 and that of next three is 21.2. If the sixth reading is 3 less than seventh and 8 less than eighth, what is the sixth reading?

a) 24.8 b) 26.5 c) 27.6 d) 29.4

10. The average age of a family of 6 members is 22 years. If the age of the youngest member be 7 years, the average age of the family at the birth of the youngest member, was **[Railways 1995]**

a) 15 years b) 17 years c) 17.5 years d) 18 years

11. The average age of a husband and wife was 23 years when they were married 5 years ago. The average age of the husband, the wife and a child who was born during the interval, is 20 years now. How old is the child now?

a) 9 months b) 1 year c) 3 years d) 4 years

12. 5 years ago, the average age of A, B, C and D was 45. With E joining them now, the average age of all the five is 49 years. How old is E?

a) 25 years b) 40 years c) 45 years d) 64 years

13. 5 years ago, the average of Ram and Shyam's ages was 20 years. Now, the average age of Ram, Shyam and Mohan is 30 years. What will be Mohan's age 10 years hence? **[LIC 1991]**

a) 45 years b) 50 years c) 49 years d) 60 years

14. The average height of 40 students is 163 cm. On a particular day, three students A, B, C were absent and the average of the remaining 37 students was found to be 162 cm. If A, B have equal heights and the height of C be 2 cm less than that of A, find the height of A. **[LIC 1991]**

a) 176 cm b) 166 cm c) 180 cm d) 186 cm

15. Out of three numbers, the first is twice the second and is half of the third. If the average of the three numbers is 56, the three numbers in order are: **[Central Excise & I. Tax 1988]**

a) 48, 96, 24 b) 48, 24, 96 c) 96, 24, 48 d) 96, 48, 24

16. Of the three numbers, second is twice the first and is also thrice the third. If the average of the three numbers is 44, the largest number is: **[Railways 1991]**

a) 24 b) 36 c) 72 d) 108

17. The sum of three numbers is 98. If the ratio between first and second be 2 : 3 and that between second and third be 5 : 8, then the second number is: **[SSC Exam 1986]**

a) 30 b) 20 c) 58 d) 48

18. The average weight of 3 men A, B and C is 84 kg. Another man D joins the group and the average now be-

comes 80 kg. If another man E, whose weight is 3 kg more than that of D, replaces A, then average weight of B, C, D and E becomes 79 kg. The weight of A is: **[Bank PO 1989]**

a) 70 kg b) 72 kg c) 75 kg d) 80 kg

19. The average age of A, B, C, D 5 years ago was 45 years. By including X, the present average of all the five is 49 years. The present age of X is: **[Bank PO, 1988]**

a) 64 years b) 48 years c) 45 years d) 40 years

20. The average age of A and B is 20 years. If C were to replace A, the average would be 19 and if C were to replace B, the average would be 21. What are the ages of A, B and C? **[MBA 1982]**

a) 22, 18, 20 b) 18, 22, 20 c) 22, 20, 18 d) 18, 20, 22

Answers

1. a; Required answer $= 320 \times 4 - 325 \times 3 = 305$.

2. b; Distance covered in t_1 hours $= t_1v_1$ km.

Distance covered in t_2 hours $= t_2v_2$ km

Total distance $= t_1v_1 + t_2v_2$

Total time $= t_1 + t_2$

$\therefore$ Average speed $= \dfrac{v_1t_1 + v_2t_2}{t_1 + t_2}$ km/hr

3. a; Time taken in the first journey $= \dfrac{x}{v_1}$ hours

Time taken in the second journey $= \dfrac{y}{v_2}$ hours

Total distance $= (x + y)$ km

Total time $= \left(\dfrac{x}{v_1} + \dfrac{x}{v_2}\right)$ hours

$\therefore$ Average speed $= \dfrac{\text{Distance}}{\text{Time}} = \left(\dfrac{x+y}{\dfrac{x}{v_1} + \dfrac{y}{v_2}}\right)$

$= \dfrac{v_1v_2(x+y)}{xv_2 + yv_1}$ km/hr

4. d; Let one side of the square be x km. Then, the total distance = 4x km.

Total time $= \dfrac{x}{200} + \dfrac{x}{400} + \dfrac{x}{600} + \dfrac{x}{800} = \dfrac{x}{96}$ hours

$\therefore$ Average speed $= \dfrac{4x \times 96}{x} = 384$ km/hr

5. d; Let their ages be 3x, 5x and 7x

$\therefore$ Total of their ages $= 3x + 5x + 7x = 3 \times 15$

or, $15x = 45 \Rightarrow x = 3$

$\therefore$ The age of the oldest $= 7x = 21$ years.

6. c; Let the initial population be 100

Population after the first year $= 100 \times 1.20 = 120$

Population after the second year $= 120 \times 1.25 = 150$

Population after the third year $= 150 \times 1.44 = 216$

Net increase $= 216 - 100 = 116$

Net per cent increase during 3 years

$= \dfrac{116}{100} \times 100 = 116\%$

Net per cent increase per year $= \dfrac{116}{3}\% = 38.67\%$

7. a; Remainder capital $= 1 - \left(\dfrac{1}{4} + \dfrac{2}{3}\right)$

$= 1 - \dfrac{3+8}{12} = 1 - \dfrac{11}{12} = \dfrac{1}{12}$

Total return

$= 3 \times \dfrac{1}{4} + 5 \times \dfrac{2}{3} + 11 \times \dfrac{1}{12} = \dfrac{9+40+11}{12} = \dfrac{60}{12} = 5$

$\therefore$ Average per cent return = 5%

8. c; Let first = x. Then, second $= \dfrac{x}{2}$ and third $= \dfrac{x}{3}$

$\therefore x + \dfrac{x}{2} + \dfrac{x}{3} = 3 \times 88$

or, $\dfrac{11x}{6} = 264$

or, $x = \dfrac{264 \times 6}{11} = 144$

$\therefore$ First – Third $= \left(144 - \dfrac{1}{3} \times 144\right) = 96$

9. c; Let 6th reading = x. Then, 7th = (x + 3) and 8th = (x + 8)

$\therefore 2 \times 18.5 + 3 \times 21.2 + x + (x+3) + (x+8) = 8 \times 24.3$

or, $37 + 63.6 + 3x + 11 = 194.4$ or, $x = 27.6$

10. d; Total present age of the family $= (6 \times 22) = 132$ years

Total age of the family 7 years ago $= (132 - 7 \times 6) = 90$ years

At that time, the number of members = 5.

$\therefore$ Average age at that time $= \left(\dfrac{90}{5}\right)$ years = 18 years.

11. d; Present total age of husband and wife

$= (2 \times 23 + 2 \times 5) = 56$ years.

Present total age of husband, wife and child

$= 3 \times 20 = 60$ years.

Present age of child $= (60-56) = 4$ years.

12\. c; 5 years ago, $(A+B+C+D) = (45 \times 4)$ years

$= 180$ years.

Now, $(A+B+C+D) = (180 + 4 \times 5)$ years $= 200$ years.

Now, $(A+B+C+D+E) = (5 \times 49)$ years $= 245$ years.

$\therefore$ Age of E now $= (245 - 200) = 45$ years.

13\. b; Total age of Ram and Shyam 5 years ago

$= (2 \times 20) = 40$ years

$\therefore$ Total age of Ram and Shyam now

$= (40 + 5 + 5) = 50$ years.

Total age of Ram, Shyam and Mohan now

$= (3 \times 30) = 90$ years.

Mohan's age now $= (90 - 50)$ years $= 40$ years.

Mohan's age 10 years hence $= (40 + 10)$ years

$= 50$ years.

14\. a; Let the heights of A, B, and C be x cm, x cm and $(x-2)$ cm

Then, $x + x + (x-2) = (163 \times 40 - 162 \times 37)$.

$\therefore$ x = 176 cm

15\. b; Let the numbers be 2x, x and 4x.

$$\text{Average} = \frac{2x+x+4x}{3} \Rightarrow \frac{7x}{3} = 56$$

$$\therefore x = \frac{3 \times 56}{7} = 24$$

Hence, the numbers in order are 48, 24 and 96.

16\. c; Let the numbers be x, 2x and $\frac{2}{3}x$.

$$\text{Average} = \frac{x + 2x + \frac{2}{3}x}{3} \Rightarrow \frac{11x}{9} = 44$$

$$\therefore x = \frac{44 \times 9}{11} = 36$$

So, the numbers are 36, 72 and 24.

Hence, the largest one is 72.

17\. a; Let the numbers be x, y and z. Then,

$$x + y + z = 98, \ \frac{x}{y} = \frac{2}{3} \text{ and } \frac{y}{z} = \frac{5}{8}$$

$$\therefore x = \frac{2y}{3} \text{ and } z = \frac{8y}{5}$$

$$\text{So, } \frac{2y}{3} + y + \frac{8y}{5} = 98 \ \text{ or, } \frac{49y}{15} = 98 \ \text{ or } y = 30$$

18\. c; Weight of $D = (80 \times 4 - 84 \times 3)$ kg $= 68$ kg

Weight of $E = (68 + 3)$ kg $= 71$ kg

$(B+C+D+E)$'s weight $= (79 \times 4)$ kg $= 316$ kg

$\therefore$ $(B+C)$'s weight $= [316 - (68 + 71)]$ kg $= 177$ kg

Hence, A's weight $= [(84 \times 3) - 177]$ kg $= 75$ kg.

19\. c; Total age of A, B, C, D 5 years ago $= (45 \times 4)$ years

$= 180$ years.

Total present age of A, B, C, D and X $= (49 \times 5)$ years

$= 245$ years

Present age of A, B, C and D $= (180 + 5 \times 4)$ years

$= 200$ years.

$\therefore$ Present age of X = 45 years.

20\. a; Say, a, b, c are the ages of A, B, and C

$\therefore$	$a+b=$		$2 \times 20 = 40$
+	$b+c=$		$2 \times 19 = 38$
+	$c+a=$		$2 \times 21 = 42$
+	$a+b+c=$		60
			[Adding all the 3 equalities]
–	$b+c$	=	38
$\therefore$	a	=	22 .
–	$a+b$	=	– 40
$\therefore$	b	=	18
and	c	=	20

$\therefore$ Age of A = 22 years

Age of B = 18 years

Age of C = 20 years

7

Problems Based on Ages

Rule 1

Theorem: ***If t years earlier the father's age was x times that of his son. At present the father's age is y times that of his son. Then the present ages of the son and the father are*** $\frac{t(x-1)}{x-y}$ ***and*** $y\left[\frac{t(x-1)}{x-y}\right]$ ***respectively.***

Illustrative Examples

Ex. 1: The age of the father 3 years ago was 7 times the age of his son. At present the father's age is five times that of his son. What are the present ages of the father and the son?

Soln: Son's age = $\frac{3\times(7-1)}{7-5}=9$ yrs

and father's age = 9 × 5 = 45 years

Ex. 2: Ten years ago A was half of B in age. If the ratio of their present ages is 3 : 4, what will be the total of their present ages?

Soln: 10 years ago A was $\frac{1}{2}$ of B's age.

At present A is $\frac{3}{4}$ of B's age.

$\therefore$ B's age = $\frac{10\left(\frac{1}{2}-1\right)}{\frac{1}{2}-\frac{3}{4}}=20$ years.

A's age = $\frac{3}{4}$ of 20 = 15 years.

$\therefore$ Total of their present ages = 20 + 15 = 35 years.

Exercise

1. A father is twice as old as his son. 20 years ago, the age of the father was 12 times the age of the son. The present age of the son is _______.
   a) 20 years b) 25 years c) 22 years d) 26 years
2. A is twice as old as B. 12 years ago, A was five times as old as B. Find the present age of A.
   a) 16 years b) 32 years c) 24 years d) 28 years
3. The age of the father 4 years ago was 8 times the age of his son. At present the father's age is 4 times that of his son. Find the present age of son.
   a) 9 years b) 7 years c) 14 years d) 18 years
4. The age of the father 8 years ago was 5 times the age of his son. At present the father's age is 3 times that of his son. Find the present age of father.
   a) 48 years b) 36 years c) 46 years d) 58 years
5. The age of the father 6 years ago was 3 times the age of his son. At present the father's age is twice that of his son. Find the present age of son.
   a) 8 years b) 16 years c) 14 years d) 12 years
6. 12 years ago, the ratio of the ages of Ram and Rahim is 2 : 3. If the ratio of their present ages is 5 : 6, what will be the total of their present ages.
   a) 46 years b) 42 years c) 44 years d) 48 years
7. The age of a man is 4 times that of his son. Five years ago, the man was nine times as old as his son was at that time. The present age of the man is:
   a) 28 years b) 32 years c) 40 years d) 23 years
8. The age of Arvind's father is 4 times his age. If 5 years ago, father's age was 7 times of the age of his son at that time, what is Arvind's father's present age?
   a) 84 years b) 70 years c) 40 years d) 35 years

(SBI PO Exam 1987)

Answers

1. c 2. a 3. b 4. a 5. d

6. c; **Hint:** 12 years ago Ram was $\frac{2}{3}$ of Rahim's age

At present Ram is $\frac{5}{6}$ of Ram's age.

Now apply the given formula to get the answer.

7. b

8. c; **Hint:** Here x = 7 and y = 4, Now apply the given formula.

Rule 2

Theorem: ***If the present age of the father is x times the age of his son. t years hence, the father's age becomes y times the age of his son. Then the present ages of the father and his son are*** $x\left[\frac{(y-1)t}{x-y}\right]$ ***and*** $\frac{(y-1)t}{x-y}$ ***years respectively.***

Illustrative Example

Ex.: At present the age of the father is five times that of the age of his son. Three years hence, the father's age would be four times that of his son. Find the present ages of the father and the son.

Soln: Applying the above rule, we have

Son's age $= \frac{(4-1)\times 3}{5-4} = 9$ yrs

and father's age = 9 × 5 = 45 years

Exercise

1. The ratio between the present ages of P and Q is 5 : 8. After four years, the ratio between their ages will be 2 : 3. What is Q's age at present.
(BSRB Mumbai PO 1998)
a) 36 years b) 20 years c) 24 years d) 32 years
2. The age of Mr Ramesh is four times the age of his son. After ten years the age of Mr Ramesh will be only twice the age of his son. Find the present age of Mr Ramesh's son.
a) 10 years b) 11 years c) 12 years d) 5 years
(BSRB Bangalore PO 2000)
3. The age of Mrs Anjali is 5 times the age of his son. After 12 years the age of Mrs Anjali will be only twice the age of his son. Find the present age of Mrs Anjali's son.
a) 4 years b) 16 years c) 12 years d) 18 years
4. The age of Mrs Rachana is 6 times the age of his son. After 6 years the age of Mrs Rachana will be only thrice the age of his son. Find the present age of Mrs Rachana's son.
a) 15 years b) 9 years c) 6 years d) 12 years
5. The age of Mr Anuj is 7 times the age of his son. After 7 years the age of Mr Anuj will be only 4 times the age of his son. Find the present age of Mr Anuj's son.
a) 7 years b) 16 years c) 12 years d) 21 years
6. A man's age is three times that of his son. In 12 years, the father's age will be double the son's age. Man's present age is:
a) 27 years b) 32 years c) 36 years d) 40 years

Answers

1. d; **Hint:** Consider Q and P as father and son to apply the given formula. From the question it is clear that Q's age is $\frac{8}{5}$ times of P ie $x = \frac{8}{5}$, and after 4 years, Q's age becomes $\frac{3}{2}$ times of the age of P. Hence t = 4 years and $y = \frac{3}{2}$. Now apply the formula and get the answer as 32 years. Hence (d) is the correct answer. Also see **Rule-8**.

2. d 3. a 4. c 5. a 6. c

Rule 3

Theorem: ***If*** t_1 ***years earlier the age of the father was x times the age of his son.*** t_2 ***years hence, the age of the father becomes y times the age of his son. Then the present ages of the son and the father are*** $\frac{t_2(y-1)+t_1(x-1)}{(x-y)}$ ***and*** $\left[\frac{\text{Son's age}}{2}(x+y)-\frac{t_1}{2}(x-1)+\frac{t_2}{2}(y-1)\right]$ ***years respectively.***

Note: *When* $t_1 = t_2 = t$, *then the formulae become the following.*

(i) present age of the son $= \left[\frac{t(x+y-2)}{(x-y)}\right]$ *years and*

(i) present age of the father

$$= \left[\frac{\text{Son's age}}{2}(x+y)-\frac{t}{2}(x-y)\right] \text{ years}$$

Illustrative Examples

Ex. 1: Three years earlier the father was 7 times as old as his son. Three years hence the father's age would be four times that of his son. What are the present ages of the father and the son?

Soln: **Quicker Method:**

Son's age $= \frac{3(4-1)+3(7-1)}{7-4} = \frac{9+18}{3} = 9$ yrs

Father's age $= \frac{9}{2}[7+4]+\frac{3}{2}(1-7)+\frac{3}{2}(4-1)$

$= \frac{99+9}{2} - 9 = 45$ years

Ex. 2: One year ago the ratio between Samir's and Ashok's age was 4 : 3. One year hence the ratio of their ages will be 5 : 4. What is the sum of their present ages in years?

Soln: One year ago Samir's age was $\frac{4}{3}$ of Ashok's age.

One year hence Samir's age will be $\frac{5}{4}$ of Ashok's age.

∴ Ashok's age (by formula (3));

$$A=\frac{1\left(\frac{4}{3}-1\right)+1\left(\frac{5}{4}-1\right)}{\frac{4}{3}-\frac{5}{4}}=\frac{\frac{1}{3}+\frac{1}{4}}{\frac{1}{12}}=7 \text{ years}$$

Now, by the first relation:

$$\frac{(S-1)}{7-1}=\frac{4}{3}$$

∴ $S=8+1=9$ years.

∴ Total ages $=A+S=9+7=16$ years.

where A = Ashok's present age and S = Samir's present age

Exercise

1. Ten years ago Mohan was thrice as old as Ram was but 10 years hence, he will be only twice as old as Ram. Find Mohan's present age.
   a) 72 years b) 70 years
   c) 30 years d) Cannot be determined
2. A father was 4 times as old as his son 8 years ago. Eight years hence, father will be twice as old as his son. Find the sum of their present ages.
   a) 56 years b) 58 years
   c) 40 years d) None of these
3. A's mother was four times as old as A, ten years ago. After ten years she will be twice as old as A. Then A's present age is _______.
   a) 30 years b) 20 years c) 24 years d) 25 years
4. A man says to his son, "seven years ago I was seven times as old as you were and three years hence 1 will be three times as old as you will be." There ages are ____ and _____ years.
   a) 60, 12 b) 52, 12 c) 42, 12 d) 50, 15
5. Sunil was three times as old as Sandeep 6 years back. Sunil will be 5/3 times as old as Sandeep 6 years hence. How old is sandeep today?
   a) 18 years b) 24 years c) 12 years d) 15 years
6. Ten years ago A was thrice as old as B was but 12 years hence, A will be only twice as old as B. Then the A's present age is ______ years.
   a) 66 b) 32 c) 76 d) 56
7. The age of a father 10 years ago was thrice the age of his son. Ten years hence, the father's age will be twice that of his son. The ratio of their present ages is:
   a) 8 : 5 b) 7 : 3 c) 5 : 2 d) 9 : 5
8. 10 years ago Chandravati's mother was 4 times older than her daughter. After 10 years, the mother will be twice older than the daughter. The present age of Chandravati is:
   (Bank PO Exam 1988)
   a) 5 years b) 10 years c) 20 years d) 30 years

Answers

1. b; **Hint:** In this problem, you have been asked to find Mohan's present age. To calculate Mohan's present age by '*Quicker Method*' you have to calculate Ram's present age first and then apply this result to the formula for Mohan's present age. This involves a complex process. Hence you may solve such problem quickly by "**traditional method**" as it has been given below. It is up to you which method you are comfortable with. If you have to find Ram's present age, you can go for '**quicker method**':
   Let Mohan's present age be x years and Ram's present age be y years.
   Thus, accordingly to the first conditions
   $(x-10)=3(y-10)$ or $x-3y=-20$
   Now, Mohan's age after 10 years $=(x+10)$ years.
   Ram's age after 10 years $=(y+10)$ years.
   ∴ $(x+10)=2(y+10)$ or $(x-2y)=10$
   Solving (i) and (ii) one gets $x=70$ and $y=30$.
   ∴ Mohan's age = 70 years. Ram's age = 30 years.

2. a 3. b 4. c 5. c 6. c

7. b; **Hint:** Present ages of father and his son are 70 and 30 years respectively.
   ∴ required ratio = 7 : 3

8. c

Rule 4

Theorem: ***If t years earlier, the father's age was x times that of his son. At present the father's age is y times that of his son. Then the sum total of the age of the father and the son is*** $\left[\left\{\frac{t(x-1)}{x-y}\right\}(y+1)\right]$ ***years.***

Illustrative Example

Ex.: Ten years ago A was half of B in age. If the ratio of their present ages is 3 : 4, what will be the total of their present ages?

Soln: 10 years ago A was $\frac{1}{2}$ of B's age

At present A is $\frac{3}{4}$ of B's age

Now, applying the above theorem,

$$\text{Total of their present ages}=\left[\frac{10\left(\frac{1}{2}-1\right)}{\frac{1}{2}-\frac{3}{4}}\right]\left(\frac{3}{4}+1\right)$$

$$=\frac{20\times 7}{4}=35 \text{ years.}$$

Exercise

1. 8 years ago Raman was thrice the age of Chandan. If the present age of Raman is twice the age of Chandan, what will be the total of their present ages?
   a) 48 years b) 42 years c) 36 years d) 54 years
2. 5 years ago Saket was 4 times the age of Alok. If the present age of Saket is thrice the age of Alok, what will be the total of their present ages?
   a) 56 years b) 60 years c) 45 years d) 65 years
3. 9 years ago Vimal was 5 times the age of Sudarshan. If the present age of Vimal is twice the age of Sudarshan, what will be the total of their present ages?
   a) 66 years b) 54 years c) 36 years d) 46 years
4. 8 years ago Kunal was 8 times the age of Kamal. If the present age of Kunal is 4 times the age of Kamal, what will be the total of their present ages?
   a) 54 years b) 60 years c) 65 years d) 70 years
5. 14 years ago Ram was 4 times the age of Pankaj. If the present age of Ram is twice the age of Pankaj, what will be the total of their present ages?
   a) 42 years b) 63 years c) 62 years d) 48 years

Answers

1. a 2. b 3. c 4. d 5. b

Rule 5

Theorem: ***If the present age of the father is x times that the age of his son. t years hence, the fathers' age becomes y times the age of his son. Then the sum of the present ages of father and his son is*** $\left[\left\{\frac{t(y-1)}{x-y}\right\}(x+1)\right]$ ***years.***

Illustrative Example

Ex.: At present the age of the father is 5 times the age of his son, three years hence the fathers' age would be four times that of his son. What is the sum of the present ages of father and his son?

Soln: Applying the above formula, we have

the required answer $= \frac{(4-1)\times 3}{5-4}\times(5+1) = 9\times 6 = 54$ years.

Exercise

1. At present the age of the father is 6 times the age of his son, 4 years hence the fathers' age would be 5 times that of his son. What is the sum of the present ages of father and his son?
   a) 116 years b) 112 years c) 114 years d) 111 years
2. At present the age of the father is 4 times the age of his son, 3 years hence the fathers' age would be thrice that of his son. What is the sum of the present ages of father and his son?
   a) 20 years b) 25 years c) 30 years d) 60 years
3. At present the age of the father is 7 times the age of his son, 4 years hence the fathers' age would be 4 times that of his son. What is the sum of the present ages of father and his son?
   a) 21 years b) 24 years c) 28 years d) 32 years
4. At present the age of the father is 3 times the age of his son, 9 years hence the fathers' age would be twice that of his son. What is the sum of the present ages of father and his son?
   a) 36 years b) 38 years c) 32 years d) 46 years
5. At present the age of the father is 7 times the age of his son, 6 years hence the fathers' age would be 5 times that of his son. What is the sum of the present ages of father and his son?
   a) 80 years b) 64 years
   c) 96 years d) None of these

Answers

1. b 2. c 3. d 4. a 5. c

Rule 6

Theorem: ***If the sum of the present ages of A and B is x years. 't' years ago, the age of A was 'y' times the age of the B. Then the present ages of A and B are as follows;***

(i) Age of B $= \frac{x+t(y-1)}{y+1}$ *(ii) Age of A* $= \frac{xy-t(y-1)}{y+1}$

Illustrative Example

Ex.: The sum of the ages of a mother and her daughter is 50 years. Also 5 years ago, the mother's age was 7 times the age of the daughter. What are the present ages of the mother and the daughter?

Soln: **Detail Method:** Let the age of the daughter be x years. Then, the age of the mother is $(50-x)$ years.

5 years ago, $7(x-5) = 50-x-5$

or, $8x = 50-5+35 = 80$

$\therefore x = 10$

Therefore, daughter's age = 10 years.

and mother's age = 40 years.

Quicker Method:

Daughter's age

$= \frac{\text{Total ages} + \text{No. of years ago (Times} - 1)}{\text{Times} + 1}$

$= \frac{50+5(7-1)}{7+1} = 10$ years

Mother's age $= \frac{50\times 7-5(7-1)}{7+1} = 40$ years

Thus, daughter's age = 10 years and mother's age = 40 years.

Exercise

1. The sum of the ages of Sweta and her mother is 63 years. Four years back her mother's age was 4 times of Sweta's age at that time. Then the present age of Sweta's mother is _______ years.
   a) 48 years b) 44 years c) 40 years d) 52 years
2. The sum of the present ages of A and B is 60 years. Also 12 years ago, the ratio of the ages of A and B is 5 : 4. Find the present age of A.
   a) 28 years b) 32 years c) 18 years d) 42 years
3. The sum of the ages of Anjali and her mother is 48 years. Six years back her mother's age was twice the age of Anjali. Find the ratio of the present ages of Anjali's mother and Anjali.
   a) 3 : 5 b) 4 : 5 c) 5 : 4 d) 5 : 3
4. The sum of the ages of Vineet and Roshan is 56 years. 4 years back Roshan's age was 3 times the age of Vineet. Find the present age of Vineet.
   a) 40 years b) 32 years c) 24 years d) 16 years
5. The sum of the ages of P and Q is 42 years. 3 years back the age of P was 5 times the age of the Q. Find the difference between the present ages of P and Q.
   a) 23 years b) 24 years c) 28 years d) 33 years
6. The sum of the ages of a father and a son is 50 years. Also, 5 years ago, the father's age was 7 times the age of the son. The present ages of the father and the son respectively, are:
   a) 35 years, 15 years b) 40 years, 10 years
   c) 38 years, 12 years d) 42 years, 8 years

Answers

1. a

2. b; **Hint:** Here A's age is $\frac{5}{4}$ times of the age of B.

3. d; **Hint:** Age of Anjali = 18 years and the age of her mother = 30 years.
   Hence the required ratio = 30 : 18 = 5 : 3.

4. d 5. b 6. b

Rule 7

Theorem: ***If the sum of the present ages of A and B is 'x' years. After 't' years, the age of A will be 'y' times that of B. Then the ages of A and B are given below.***

(i) The age of A $= \frac{xy + t(y-1)}{y+1}$ ***and***

(ii) The age of B $= \frac{x - t(y-1)}{y+1}$ **years.**

Illustrative Example

Ex.: The sum of the ages of a son and father is 56 years. After 4 years, the age of the father will be three times that of the son. What is the age of the son?

Soln: **Detail Method:** Let the age of the son be x years.
Then , the age of the father is $(56 - x)$ years.
After 4 years, $3(x+4) = 56 - x + 4$
or, $4x = 56 + 4 - 12 = 48$
$\therefore x = 12$ years.
Thus, son's age = 12 years.
Quicker Method:
Son's age

$$= \frac{\text{Total ages} - \text{No. of years after (Times} - 1)}{\text{Times} + 1}$$

$$= \frac{56 - 4(3-1)}{3+1} = \frac{48}{4} = 12 \text{ years.}$$

Exercise

1. 4 years hence, mother's age will be twice the age of her daughter. The sum of their present ages is 46 years. Find the present ages.
   a) 32 years, 14 years b) 36 years, 10 years
   c) 38 years, 8 years d) Can't be determined
2. The sum of ages of A and B is 60 years. After 10 years A will be thrice as old as B. Find the difference of their present ages.
   a) 30 years b) 40 years c) 10 years d) 50 years
3. The sum of the ages of a father and a son is 56 years. After 4 years, the age of the father will be three times that of the son. Then the age of the father is ______.
   a) 35 years b) 40 years c) 42 years d) 44 years
4. The sum of the ages of the father and his son is 41 years. After 17 years the father's age will be twice the age of his son. Then the respective ages of the father and son are ______ and ______ years.
   a) 32, 9 b) 34, 7 c) 33, 8 d) 31, 10
5. The sum of the ages of the father and his son is 66 years. After 3 years, the fathers' age will be thrice the age of his son. Then find the ratio of the present ages of father and his son.
   a) 25 : 17 b) 17 : 6 c) 17 : 5 d) 17 : 3
6. The sum of the ages of a son and father is 56 years. After four years, the age of the father will be three times that of the son. Their ages respectively are:
   (Railway Recruitment Board Exam, 1989)
   a) 12 years, 44 years b) 16 years, 48 years
   c) 16 years, 42 years d) 18 years, 36 years

Answers

1. a 2. b 3. d 4. c 5. c 6. a

Rule 8

Theorem: ***If the ratio of the ages of A and B at present is a : b. After 'T' years the ratio will become c : d. Then the present ages of A and B are as follows:***

$$\textit{Age of } A = a \times \frac{T(c-d)}{ad - bc} = a \times \frac{T(c-d)}{\textit{difference of cross products}}$$

$$\textit{Age of B} = b\times\frac{T(c-d)}{ad-bc} = b\times\frac{T(c-d)}{\textit{difference of cross products}}$$

Illustrative Example

Ex.: The ratio of the ages of the father and the son at present is 6 : 1. After 5 years the ratio will become 7 : 2. What are the present ages of the son and the father?

Soln:

	Father	:	Son
Present age =	6	:	1
After 5 years =	7	:	2

$$\text{Son's age} = 1\times\frac{5(7-2)}{6\times2-7\times1} = 5 \text{ years.}$$

$$\text{Father's age} = 6\times\frac{5(7-2)}{6\times2-7\times1} = 30 \text{ years.}$$

Exercise

1. The ratio of present ages of P and Q is 7 : 3. After four years their ages are in the ratio of 2 : 1. What is the present age of P? **(BSRB Chennai PO 2000)**
   a) 24 years b) 28 years
   c) 32 years d) Data inadequate
2. The present ages of the father and son are in the ratio 6 : 1. After 5 years, the ratio will be 13 : 3. Find the present age of the son.
   a) 60 years b) 15 years c) 30 years d) 10 years
3. The ages of A and B are in the ratio 3 : 5. After 9 years the ratio of their ages will be 3 : 4. The present age of B is ______.
   a) 9 years b) 15 years c) 20 years d) 16 years
4. The present ages of father and son are in the ratio 5 : 1. After 5 years, the ratio will be 11 : 3. Then the present age of the father is ________ years.
   a) 60 years b) 40 years c) 45 years d) 50 years
5. The ratio of the present ages of P and Q is 8 : 5. After 6 years their ages are in the ratio of 3 : 2. Find the ratio of the sum and difference of the present ages of P and Q.
   a) 13 : 3 b) 39 : 19 c) 13 : 2 d) 13 : 5
6. The ages of A and B are in the ratio 2 : 5. After 8 years their ages will be in the ratio 1 : 2. The difference of their ages is:
   a) 20 years b) 24 years c) 26 years d) 29 years
7. The ratio of the ages of father and son at present is 6 : 1. After 5 years, the ratio will become 7 : 2. The present age of the son is:
   a) 10 years b) 9 years c) 6 years d) 5 years
   (Bank PO 1991)

Answers

1. b 2. d 3. b 4. d
5. a; **Hint:** Age of P = 48 years and Age of Q = 30 years
∴ Required ratio = 48 + 30 : 48 - 30 = 78 : 18 = 13 : 3
6. b 7. d

Rule 9

Theorem: ***If the ratio of the ages of A and B at present is a : b. 'T' years earlier, the ratio was c : d. Then the present***
(i) Age of A

$$= a\times\frac{T(c-d)}{bc-ad} = a\times\frac{T(c-d)}{\textit{difference of cross products}}$$

(ii) Age of B

$$= b\times\frac{T(c-d)}{bc-ad} = b\times\frac{T(c-d)}{\textit{difference of cross products}}$$

Illustrative Example

Ex.: The ratio of the ages of the father and the son at present is 3 : 1. 4 years earlier, the ratio was 4 : 1. What are the present ages of the son and the father?

Soln: Father : Son
Present age = 3 : 1
4 years before = 4 : 1

$$\text{Son's age} = 1\times\frac{4(4-1)}{4\times1-3\times1} = 12 \text{ years.}$$

$$\text{Father's age} = 3\times\frac{4(4-1)}{4\times1-3\times1} = 36 \text{ years.}$$

Exercise

1. 6 years ago Jagunnath was twice as old as Badri. If the ratio of their present ages is 9 : 5 respectively, what is the difference between their present ages in years? **(BSRB Bhopal PO 2000)**
   a) 24 b) 30
   c) 50 d) Cannot be determined
2. 3 years ago Ambuj was thrice as old as Avinash. If the ratio of their present ages is 8 : 3 respectively, what is the difference between their present ages in years?
   a) 32 years b) 30 years c) 28 years d) 35 years
3. 8 years ago, the ratio of the ages of Nagendra and Sunil was 3 : 2. If the ratio of their present ages is 7 : 5 respectively, what is the sum of their present ages?
   a) 96 years b) 86 years c) 76 years d) 66 years
4. 8 years ago, the ratio of the ages of Rachana and Archana was 4 : 3. If the ratio of their present ages is 6 : 5 respectively, what is the ratio of the sum and difference of their present ages?
   a) 11 : 2 b) 11 : 4 c) 11 : 1 d) 11 : 3
5. The ratio of the ages of A and B at present is 4 : 3. 10 years earlier, the ratio was 3 : 2, then find the present ages of A and B.
   a) 40 years and 30 years b) 48 years and 36 years
   c) 64 years and 48 years d) 20 years and 15 years

Answers

1. a; **Hint:** Here c : d = 2 : 1 and present ages of Jagunnath and Badri is 54 and 30 years respectively.

2. b 3. a

4. c; **Hint:** Their present ages were 24 and 20 years respectively

∴ the required ratio = 24 + 20 : 24 - 20 = 44 : 4 = 11 : 1

5. a

Rule 10

Theorem: ***If the ratio of the ages of A and B at present is a : b. After 'T' years the ratio will become c : d. Then the sum of present ages of A and B is*** $\left[\frac{T(c-d)}{ad-bc}\right](a+b)$ ***or***

$$\left[\frac{T(c-d)}{\text{difference of cross products}}\times(a+b)\right].$$

Illustrative Example

Ex.: The ratio of the ages of the father and the son at present is 6 : 1. After 5 years the ratio will become 7 : 2. What is the sum of the present ages of the father and the son?

Soln: Following the above formula, we have

$$\text{the required answer} = \left[\frac{5(7-2)}{6\times2-7\times1}\right](6+1)$$

$$= 5 \times 7 = 35 \text{ years.}$$

Exercise

1. The ratio of the ages of Sweta and Rachna is 2 : 5. After 8 years, their ages will be in the ratio 1 : 2. Then the sum of their present ages is _______.
   a) 56 years b) 46 years c) 36 years d) 58 years
2. The ratio of the ages of the father and the son at present is 5 : 2. After 3 years the ratio will become 7 : 3. What is the sum of the present ages of the father and the son?
   a) 64 years b) 74 years c) 84 years d) 88 years
3. The ratio of the ages of the father and the son at present is 4 : 1. After 9 years the ratio will become 5 : 2. What is the sum of the present ages of the father and the son?
   a) 50 years b) 60 years c) 45 years d) 55 years
4. The ratio of the ages of the father and the son at present is 5 : 3. After 7 years the ratio will become 3 : 2. What is the sum of the present ages of the father and the son?
   a) 46 years b) 48 years c) 56 years d) 58 years
5. The ages of Kanchan is thrice the age of the Chandan. After 12 years the age of Kanchan will become twice the age of Chandan. Then the sum of their present ages is _______ years.
   a) 42 years b) 48 years c) 46 years d) 50 years

Answers

1. a 2. c 3. c 4. c

5. b; **Hint:** Here, a : b = 3 : 1 and c : d = 2 : 1. Now apply the given formula.

Rule 11

Theorem: ***If the product of the present ages of A and B is 'x' years and the ratio of the present ages of A and B is a : b. Then the present***

(i) Age of A = $a\times\sqrt{\frac{x}{ab}}$ ***years and (ii) Age of B =*** $b\times\sqrt{\frac{x}{ab}}$ ***years***

Illustrative Example

Ex.: If the product of the present ages of the father and his son is 900 years and the ratio of their present ages is 25 : 9. Find their present ages.

Soln: Applying the above formula, we have

Present age of the father

$$= 25\times\sqrt{\frac{900}{25\times9}} = \frac{25\times30}{15} = 50 \text{ years.}$$

Present age of his son

$$= 9\times\sqrt{\frac{900}{25\times9}} = \frac{9\times30}{15} = 18 \text{ years.}$$

Exercise

1. The ratio of A's and B's ages is 4 : 5. The product of their ages is 320 years. Then A's present age is _______.
   a) 20 years b) 16 years
   c) 24 years d) None of these
2. The ratio of A's and B's ages is 3 : 2. The product of their ages is 216 years. Then the sum of their present ages is _______ years.
   a) 18 years b) 30 years c) 36 years d) 32 years
3. The age of Jayshree is thrice the age of her younger sister. The product of their ages is 300 years. Then the Jayshree's present age is _______ years.
   a) 30 b) 20 c) 10 d) 40
4. If the product of the present ages of the father and his son is 750 years and the ratio of their present ages is 6 : 5. Find the difference between their present ages.
   a) 10 years b) 15 years c) 8 years d) 5 years
5. If the product of the present ages of the father and his son is 3600 years and the ratio of their present ages is 16 : 9. Find the difference between their present ages.
   a) 45 years b) 40 years
   c) 35 years d) None of these
6. The ratio of the father's age to the son's age is 4 : 1. The product of their ages is 196. The ratio of their ages after 5 years will be:
   a) 3 : 1 b) 10 : 3 c) 11 : 4 d) 14 : 5

Answers

1. b 2. b

3. a; **Hint:** Ratio of Jayshree and her younger sister = 3 : 1

4. d 5. c

6. c; **Hint:** Applying the given formula, we have

their present ages are 28 and 7 years respectively. Ratio of their ages after 5 years will be 28 + 5 : 7 + 5 = 33 : 12 = 11 : 4

Rule 12

Theorem: ***If the ratio of the ages of A and B at present is a : b. 'T' years earlier, the ratio was c : d, then the sum of the present ages of A and B is***

$$(a+b)\left[\frac{T(c-d)}{bc-ad}\right] \text{ or } \left[\frac{T(c-d)}{\text{difference of cross products}}\right](a+b)$$

Illustrative Example

Ex.: The ratio of the ages of the father and the son at present is 3 : 1. 4 years earlier, the ratio was 4 : 1. What is the sum of the present ages of the son and the father?

Soln: Following the above formula, we have

the required answer $= (3+1)\left[\frac{4(4-1)}{4\times1-3\times1}\right]$

$= 4 \times 12 = 48$ years.

Exercise

1. If the ratio of the ages of A and B at present is 2 : 1. 6 years earlier, the ratio was 3 : 1. What is the sum of the present ages of A and B?
   a) 24 years b) 26 years
   c) 34 years d) 36 years
2. If the ratio of the ages of A and B at present is 3 : 2. 3 years earlier, the ratio was 5 : 3. What is the sum of the present ages of A and B?
   a) 30 years b) 32 years c) 35 years d) 40 years
3. If the ratio of the ages of A and B at present is 4 : 3. 2 years earlier, the ratio was 5 : 2. What is the sum of the present ages of A and B?
   a) 37 years b) 31 years c) 42 years d) 36 years
4. If the ratio of the ages of A and B at present is 7 : 2. 7 years earlier, the ratio was 6 : 1. What is the sum of the present ages of A and B?
   a) 36 years b) 63 years c) 64 years d) 46 years
5. If the ratio of the ages of A and B at present is 5 : 3. 6 years earlier, the ratio was 2 : 1. What is the sum of the present ages of A and B?
   a) 48 years b) 46 years c) 47 years d) 49 years

Answers

1. d 2. a 3. c 4. b 5. a

Rule 13

Theorem: ***If a man's age is x% of what it was t_1 years ago, but y% of what it will be after t_2 years. Then his present age is*** $\left[\frac{xt_1+yt_2}{(x-y)}\right]$ ***years.***

Note: If $t_1 = t_2 = t$, then formula will become $\left(\frac{x+y}{x-y}\right)t$ years.

Illustrative Example

Ex.: A man's age is 125% of what it was 10 years ago, but $83\frac{1}{3}\%$ of what it will be after 10 years. What is his present age?

Soln: **Detail method:** Let the present age be x years. Then

125% of $(x-10) = x$; and $83\frac{1}{3}\%$ of $(x+10) = x$

$\therefore 125\%$ of $(x-10) = 83\frac{1}{3}\%$ of $(x+10)$

or, $\frac{5}{4}(x-10) = \frac{5}{6}(x+10)$

or, $\frac{5}{4}x - \frac{5}{6}x = \frac{50}{6} + \frac{50}{4}$

or, $\frac{5x}{12} = \frac{250}{12}$ $\therefore x = 50$ years

Quicker Method: Applying the above rule, we have the required answer

$$= \frac{125 \times \text{No. of years ago} + 83\frac{1}{3} \times \text{No. of yrs after}}{125 - 83\frac{1}{3}}$$

$$= \frac{125\times10+83\frac{1}{3}\times10}{125-83\frac{1}{3}} = \frac{10\left(\frac{375+250}{3}\right)}{\frac{375-250}{3}}$$

$$= \frac{10\times625}{125} = 50 \text{ years.}$$

Exercise

1. A man's age is 150% of what it was 10 years ago, but 75% of what it will be after 10 years. What is his present age?
   a) 25 years b) 30 years c) 35 years d) 40 years
2. A man's age is 135% of what it was 5 years ago, but 80% of what it will be after $2\frac{1}{2}$ years. What is his present age?
   a) 15 years b) 25 years
   c) 20 years d) None of these

3. A man's age is 132% of what it was 3 years ago, but 88% of what it will be after 5 years. What is his present age?
a) 11 years b) 17 years c) 29 years d) 19 years
4. A man's age is 125% of what it was 9 years ago, but 80% of what it will be after 9 years. What is his present age?
a) 26 years b) 24 years c) 41 years d) 32 years
5. A man's age is 130% of what it was 4 years ago, but 90% of what it will be after 4 years. What is his present age?
a) 9 years b) 15 years c) 12 years d) 22 years

Answers

1. b 2. d 3. d 4. c 5. d

Rule 14

The ratio of A's and B's ages is a : b. If the difference between the present ages of A and B 't' years hence is 'x' then

(i) the present age of A (younger) $\left(\dfrac{t+x}{\frac{b}{a}-1}\right)$ ***years***

(ii) the present age of B (older) $= \left[\dfrac{(t+x)\frac{b}{a}}{\frac{b}{a}-1}\right]$ ***years and***

(iii) the sum of the present ages of A and B

$$= \frac{(t+x)}{\left(\frac{b}{a}-1\right)}\left[1+\frac{b}{a}\right] \textit{ years.}$$

Note: Here A is always younger than B and a : b is younger : older. Hence $\dfrac{b}{a} = \dfrac{older}{younger}$.

Illustrative Example

Ex.: The ratio of A's and B's ages is 4 : 5. If the difference between the present age of B and the age of A 5 years hence is 3, then what is the total of present ages of A and B?

Soln: **Detail Method:**

$\dfrac{A}{B} = \dfrac{4}{5}$ or, $B = \dfrac{5}{4}A$

B − (A + 5) = 3 or, B = A + 8

$\therefore \dfrac{5}{4}A = A + 8$ or, $A\left(\dfrac{5}{4}-1\right) = 8$

∴ A = 32 years

$B = \dfrac{5}{4}\times 32 = 40$ years

∴ A + B = 40 + 32 = 72 years

Quicker Method: Applying the above theorem, we have

$$\text{the required answer} = \left[\frac{5+3}{\frac{5}{4}-1}\right]\left[1+\frac{5}{4}\right]$$

$$= 8\times 4\times\frac{9}{4} = 72 \text{ years.}$$

Exercise

1. The ratio of P's and Q's ages is 5 : 7. If the difference between the present age of Q and the age of P 6 years hence is 2 then what is the total of present ages of P and Q? **(Bank of Baroda PO 1999)**
a) 52 years b) 48 years
c) 56 years d) Data inadequate
2. The ratio of P's and Q's ages is 3 : 5. If the difference between the present age of Q and the age of P 3 years hence is 3 then what is the total of present ages of P and Q?
a) 24 years b) 27 years c) 32 years d) 20 years
3. The ratio of P's and Q's ages is 5 : 8. If the difference between the present age of Q and the age of P 9 years hence is 6 then what is the total of present ages of P and Q?
a) 56 years b) 65 years c) 45 years d) 75 years
4. The ratio of Radha's and Ruchi's ages is 9 : 4. If the difference between the present age of Radha and the age of Ruchi 5 years hence is 5 then what is the sum of the present ages of Radha and Ruchi?
a) 18 years b) 16 years c) 26 years d) 28 years
5. The ratio of the ages of Ram and Shyam is 7 : 4. If the difference between the present age of Ram and the age Shyam 2 years hence is 4 then what is the sum of the present ages of Ram and Shyam.
a) 22 years b) 18 years c) 24 years d) 32 years

Answers

1. b 2. a 3. b

4. c; **Hint:** See **Note:** Here a : b = 9 : 4 because Radha is older than Ruchi.
If you apply the given formula we get
Radha's present age = 18 years and Ruchi's present age = 8 years.

5. a

Rule 15

The difference of present ages of A and B is 'x' years. If 't' years back their ages were in the ratio a : b, then

(i) the age of $A = \left[t+\dfrac{x}{\frac{a}{b}-1}\right]+x$ ***years, (ii) the age of***

$B=\left[t+\frac{x}{\frac{a}{b}-1}\right]$ *years and (iii) the total Age of A and B*

$=\left[2\left(t+\frac{x}{\frac{a}{b}-1}\right)+x\right]$ *years.*

Note: Here A > B ie A is older than B. Hence a > b.

Illustrative Example

Ex: The difference of present ages of A and B is 12 years. If 6 years back their ages were in the ratio 3 : 2, how old are A and B?

Soln: Detail Method:

A - B = 12 years ....(i)

$\frac{A-6}{B-6}=\frac{3}{2} \Rightarrow 2A-12=3B-18$

or, 2A - 3B = -6 ....(ii)

From eqn (i) and eqn (ii), we get

A's age = 42 years and B's age = 30 years.

Quicker Method: Applying the above theorem, we have

the age of $A=\left[6+\frac{12}{\frac{3}{2}-1}\right]+12=42$ years and

the age of $B=\left[6+\frac{12}{\frac{3}{2}-1}\right]=30$ years.

Exercise

1. Rahul is younger than Radha by 10 years. If 5 years back their ages were in the ratio 1 : 2, how old is Radha?
   a) 20 years b) 15 years
   c) 25 years d) Data inadequate
2. The ages of Ram and Mohan differs by 16 years. Six years ago, Mohan's age was thrice as that of Ram's. Find the sum of their ages.
   a) 30 years b) 27 years c) 24 years d) 25 years
3. The difference between the ages of two persons is 8 years. 15 years ago, the elder one was twice as old as the younger one. Then the present age of the elder person is _______ years.
   a) 23 b) 31 c) 28 d) 24
4. The difference between the ages of two persons is 12 years. 6 years ago, the ratio of the elder and younger one is 5 : 2. Then the present age of younger one is _________ years.
   a) 16 b) 28 c) 14 d) 24
5. The difference of present ages of A and B is 10 years. If 4 years back their ages were in the ratio of 7 : 5. Find the sum of their present ages.
   a) 39 years b) 58 years c) 78 years d) 68 years

Answers

1. c; **Hint:** Here Radha - Rahul = 10 years.
   ie Radha - Rahul = A - B and
   Radha : Rahul = 2 : 1. Now by applying the given formula, we have the age of Radha $=\left[5+\frac{10}{\frac{2}{1}-1}\right]+10$
   = 25 years.
2. a; **Hint:** Here A = Mohan, B = Ram and a : b = 3 : 1
   Now apply the given rule.
3. b; **Hint:** Here a : b = 2 : 1, and apply the given **Rule - 15 (i).**
4. c 5. d

Miscellaneous

1. If the ages of P and R are added to twice the age of Q, the total becomes 59. If the ages of Q and R are added to thrice the age of P, the total becomes 68. And if the age of P is added to thrice the age of Q and thrice the age of R, the total becomes 108. What is the age of P?
   [SBI Bank PO, 1999]
   a) 15 yrs b) 19 yrs c) 17 yrs
   d) 12 yrs e) None of these
2. The product of the ages of Harish and Seema is 240. If twice the age of Seema is more than Harish's age by 4 years, what is Seema's age in years?
   [SBI Bank PO, 1999]
   a) 12 years b) 20 years c) 10 years
   d) 14 years e) Data inadequate
3. Jayesh is twice as old as Vijay and half as old as Suresh. If the sum of Suresh's and Vijay's ages is 85 years, what is Jayesh's age in years?
   [BSRB Bhopal PO, 2000]
   a) 34 b) 36 c) 68
   d) Can not be determined e) None of these
4. Present age of Rahul is 8 years less than Ritu's present age. If 3 years ago Ritu's age was x, which of the following represents Rahul's present age?
   [BSRB Delhi PO, 2000]
   a) $x+3$ b) $x-5$ c) $x-3+8$
   d) $x+3+8$ e) None of these
5. The ratio of the present ages of a son and his father is 1 : 5 and that of his mother and father is 4 : 5. After 2 years the ratio of the age of the son to that of his mother

becomes 3 : 10. What is the present age of the father?
a) 30 years b) 28 years c) 37 years
d) Data inadequate e) None of these

6. 20 years ago my age was $\frac{1}{3}$ of what it is now. What is my present age?
a) 30 years b) 25 years c) 35 years d) 40 years

7. 15 years hence, A will be twice as old as B, but five years ago A was 4 times as old as B. Find the difference of their present ages.
a) 15 years b) 45 years c) 30 years d) 25 years

8. A says to B "I am twice as old as you were when I was as old as you are". The sum of their ages is 63 years. Find the difference of their ages.
a) 27 years b) 12 years c) 9 years d) 6 years

9. A is as much younger than B as he is older than C. If the sum of B's and C's ages is 40 years. Find the age of A.
a) 20 years b) 25 yerars c) 30 years d) 27 years

10. A is twice as old as B was two years ago. If the difference in their ages be 2 years, find A's age.
a) 14 years b) 18 years c) 8 years d) 12 years

11. In ten years, A will be twice as old as B was 10 years ago. If A is now 9 years older than B. Find the present age of B.
a) 39 years b) 40 years c) 36 years d) 49 years

12. Five years ago, the total of the ages of father and son was 60 years. The ratio of their present ages is 4 : 1. Then the present age of the father is ______.
a) 48 years b) 51 years c) 56 years d) 61 years

13. Two years ago, a mother was four times as old as her daughter. 8 years hence, mother's age will exceed her daughter's age by 12 years. The ratio of the present ages of mother and daughter is ______.
a) 3 : 1 b) 4 : 1 c) 3 : 2 d) 5 : 1

14. A is 3 years younger to B. C is two years older than A. Then B's relation to C is ______.
a) two years older b) one year younger
c) one year older d) two years younger

15. If C's age is twice the average age of A, B and C. A's age is one half the average of A, B and C. If B is 5 years old, the aveage age of A, B and C is ______.
a) 10 years b) 15 years c) 12 years d) 9 years

16. A father's age is three times the sum of the ages of his two children, but 20 years hence his age will be equal to the sum of their ages. Then the father's age is ______.
a) 30 years b) 40 years c) 35 years d) 45 years

17. A father's age is four times as much as the sum of the ages of his three children but 6 years hence his age will be only double the sum of their ages. Then the age of the father is ______.
a) 30 years b) 40 years c) 60 years d) 45 years

18. The respective ages of a father and his son are 41 and 16 years. In how many years will the father be twice as old as his son?
a) 19 years b) 9 years c) 10 years d) 15 years

19. The total ages of A, B and C at present is 90 years. Ten years ago the ratio of their ages was 1 : 2 : 3. Then the present age of B is __________.
a) 30 years b) 20 years
c) 40 years d) None of these

20. The sum of the ages of a father and son is 45 years. Five years ago, the product of their ages was four times the father's age at that time, then the present ages of the father and son respectively are _______ and _______ years.
a) 39, 6 b) 35, 10 c) 36, 9 d) 40, 10

21. The ratio of the father's and son's age is 7 : 4. The product of their ages is 1008. The ratio of their ages after 6 years hence will be _______.
a) 5 : 3 b) 8 : 5 c) 7 : 4 d) 5 : 8

22. Ratio of Sujeet's age to Sameer's age is 4 : 3. Sujeet will be 26 years old after 6 years. Then the present age of Sameer is _______.
a) 21 years b) 15 years c) 24 years d) 18 years

23. If 6 years are subtracted from the present age of Randheer and the remainder is divided by 18, then the present age of his grandson Anup is obtained. If Anup is 2 years younger to Mahesh whose age is 5 years, then what is the age of Randheer? **[Bank PO 1993]**
a) 96 years b) 84 years c) 48 years d) 60 years

24. The ratio of Vimal's age and Arun's age is 3 : 5 and sum of their ages is 80 years. The ratio of their ages after 10 years will be ________. **[Bank PO 1990]**
a) 2 : 3 b) 1 : 2 c) 3 : 2 d) 3 : 5

25. Shyam is 3 times as old as his son. After 10 years, the sum of their ages will be 76 years. The respective ages of the father and the son are ________ and _______ years.
a) 42, 14 b) 39, 13
c) 45, 15 d) None of these

26. A is 20 years older than B. He is also 6 times as old as B. Then the respective ages of A and B are _______ and _______ years.
a) 24, 4 b) 42, 7 c) 30, 5 d) None of these

27. The ages of A, B and C together total 185 years. B is twice as old as A and C is 17 years older than A. Then the respective ages of A, B and C are
a) 40, 86 and 59 years b) 42, 84 and 59 years
c) 40, 80 and 65 years d) None of these

28. The ratio of Vimal's age and Arun's age is 3 : 5 and the sum of their ages is 80 years. Find the ratio of their ages.
(i) after 10 years and (ii) 10 years ago
a) 2 : 3, 2 : 1 b) 2 : 3, 1 : 2 c) 3 : 2, 1 : 2 d) 3 : 2, 2 : 1

29. In 10 years, A will be twice as old as B was 10 years ago. If A is now 9 years older than B, the present age of B is:
a) 29 years b) 39 years c) 19 years d) 49 years
[LIC Exam 1989]

30. The difference between the ages of two persons is 10 years. 15 years ago, the elder one was twice as old as the younger one. The present age of the elder person is:
a) 35 years b) 25 years c) 45 years d) 55 years

31. One year ago a father was four times as old as his son. In 6 years time his age exceeds twice his son's age by 9 years. Ratio of their ages is:
a) 13 : 4 b) 12 : 5 c) 11 : 3 d) 9 : 2

32. Ratio of Ashok's age to Pradeep's age is equal to 4 : 3. Ashok will be 26 years old after 6 years. How old is Pradeep now?
[Railway Recruitment Board Exam 1989]
a) $19\frac{1}{2}$ years b) 21 years c) 12 years d) 15 years

33. Deepak is 4 times as old as his son. Four years hence the sum of their ages will be 43 years. How old is Deepak's son now?
a) 5 years b) 7 years c) 8 years d) 10 years

34. The ratio of Mona's age to the age of her mother is 3 : 11. The difference of their ages is 24 years. The ratio of their ages after 3 years will be:
a) 1 : 3 b) 2 : 3 c) 3 : 5 d) None of these

35. Kamla got married 6 years ago. Today her age is $1\frac{1}{4}$ times her age at the time of marriage. His son's age is (1/10) times her age. The age of her son is:
a) 2 years b) 3 years c) 4 years d) 5 years
[Bank PO Exam 1988]

36. Sachin was twice as old as Ajay 10 years back. How old is Ajay today if Sachin will be 40 years old 10 years hence? **[Bank PO Exam 1991]**
a) 20 years b) 10 years c) 30 years d) 15 years

37. The sum of the ages of a father and son is 45 years. Five years ago, the product of their ages was four times the father's age at that time. The present age of the father is:
[Hotel Management, 1991]
a) 39 years b) 36 years
c) 25 years d) None of these

Answers

1. d; P + R + 2Q = 59; Q + R + 3P = 68
and P + 3(Q + R) = 108
Solving the above two equations, we get P = 12 yrs.

2. a; Let the ages of Harish and Seema be x and y respectively.
According to the question,
x.y = 240 ....(i) 2y - x = 4 ....(ii)
Solving equations (i) and (ii), we get
y = 12 years

3. a; J : V : S = 2 : 1 : 4
$\therefore$ Jayesh's age = $\frac{2}{(1+4)}\times 85 = 34$ years

4. b; Let the Rahul's present age is 'A' years.
Then Ritu's present age is **A + 8**
Now, according to the question,
A + 8 - 3 = x $\therefore$ A = **x - 5 years**
Hence, (b) is the correct answer

5. e; $\frac{S}{F}=\frac{1}{5}\Rightarrow F=5S$ $\frac{M}{F}=\frac{4}{5}\Rightarrow M=\frac{4}{5}F$
$\frac{S+2}{M+2}=\frac{3}{10}\Rightarrow 10S+20=3M+6$
$=3\times\frac{4}{5}\times 5S+6=12S+6$
$\therefore$ 2S = 14 $\Rightarrow$ S = 7 years $\therefore$ F = 5S = 35 years

6. a; Let my present age be x years.
$\therefore (x-20)=\frac{x}{3}$ or (3x - 60) = x or, 2x = 60
$\therefore$ x = My present age = 30 years.

7. c; Let A's age = x, and B's age = y years.
As per the first condition : (x + 15) = 2(y + 15)
or x - 2y = 15 ....(i)
As per the second condition : (x - 5) = 4 (y - 5) or x - 4y = -15 ....(ii)
Solving (i) and (ii), one gets, x = 45, y = 15.
$\therefore$ A's age = 45 years, B's age = 15 years

8. c; Let A's age be x years and B's age be y years.
$\therefore$ x + y = 63 ....(i)
(x - y) years ago, A was of y years age.
Now according to the question,
x = 2 [y - (x - y)] or, x = 2(y - x)
or, 3x - 4y = 0 .....(ii)
Solving equ (i) and equ (ii), we get x = 36, y = 27
$\therefore$ A's age = 36 years, B's age = 27 years.
$\therefore$ Difference of their ages = 36 - 27 = 9 years.

9. a; Given that (B - A) = (A - C) or (B + C) = 2A ....(i) and B + C = 40 ....(ii)
$\therefore$ A = 20 years.

10. c; Let B's age 2 years ago be x years.
$\therefore$ A's present age = 2x years
Also 2x - (x + 2) = 2
or, x = 4 $\therefore$ A's age = 2 × 4 = 8 years

11. a; Let B's age be x years, then A's age be (x + 9) years.
As per the given condition
(x + 9 + 10) = 2(x - 10) or, x = 39
$\therefore$ The present age of B = 39 years

12. c; Let the present age of the son be x and that of the father be 4x years.
$\therefore$ (x - 5) + (4x - 5) = 60 or 5x = 70 $\therefore$ x = 14 years
$\therefore$ Father's present age = 56 years.

13. a; Let the mother's age 2 years ago be 4x and daughter's age 2 years ago be x.
$\therefore$ (4x + 8) - (x + 8) = 12

or, 3x = 12 or x = 4
∴ Mother's present age = 4x + 2 = 18 years and daughter's present age = x + 2 = 6 years
∴ required ratio = 3 : 1

14. c; Let the age of A be (x - 3) years
∴ B's age = x years
∴ C's age = (x - 3) + 2 = x - 1
∴ B's age - C's age = x - (x - 1) = 1 year
∴ B is one year older than C.

15. a; Let the average age of A, B and C be x years
∴ total age of A, B and C = 3 × x = 3x years
Now, according to the question,

$$3x-\left(2x+\frac{x}{2}\right)=5 \quad \therefore x=10 \text{ years}$$

16. a; Let the present, age of father be x years and the present age of son be y years.
∴ x = 3y ....(i)
Also, (x + 20) = (y + 20 + 20) ....(ii)
[20 will be added twice as for 2 children]
Solving (i) and (ii), we get x = 30 years.

17. c; Let father's age be x years and the sum of ages of children be y years.
∴ x = 4y ....(i)
Also (x + 6) = 2(y + 6 + 6 + 6) ....(ii)
[6 is added thrice for three children]
Solving (i) and (ii)
x = 60 years and y = 15 years.

18. b; Suppose x years hence the father will be twice as old as his son.
x + 41 = 2(x + 16)
or, x = 41 - 32 = 9 years

19. a; Let the respective ages of A, B and C ten years ago be x, 2x and 3x years.
∴ (x + 10) + (2x + 10) + (3x + 10) = 90
or, x = 10
∴ B's present age = 2x + 10 = 30 years

20. c; Let son's age be x years
or Father's age = (45 - x) years.
∴ (x - 5) (45 - x - 5) = 4(45 - x - 5)
or, x = 9 years
∴ The son's age = 9 years
Father's age = 45 - 9 = 36 years

21. b; Let father's and son's age be 7x and 4x respectively.
$\therefore 28x^2 = 1008$ or $x^2 = 36$ or $x = 6$
∴ Father's age = 7x = 42 years
Son's age = 4x = 24 years
Father's age after 6 years hence = 48 years
Son's age 6 years hence = 30 years
Ratio = 48 : 30 or 8 : 5

22. b; Let the respective ages of Sujeet and Sameer be 4x and 3x years.
∴ 4x + 6 = 26 or x = 5
∴ Sameer's present age = 3 × 5 = 15 years

23. d; Anup's age = (5 - 2) years = 3 years
Let Randheer's age be x years.

$$\therefore \frac{x-6}{18}=3$$

or, x = 54 + 6 = 60 years

24. a; Let their ages be 3x and 5x years.
∴ 3x + 5x = 80 or x = 10
∴ Vimal's age 10 years hence = (3x + 10) = 40 years
Aruna's age 10 years hence = (5x + 10) = 60 years
Ratio = 40 : 60 or 2 : 3

25. a; Let son's present age be x years.
∴ Father's present age = 3x years.
Son's age 10 years hence = x + 10
Father's age 10 years hence = 3x + 10
As per the condition, (x + 10) + (3x + 10) = 76
or 4x = 56 ∴ x = 14
∴ Son's present age = 14 years
Father's present age = 42 years

26. a; Let the age of B be x years
According to the question,
x + 20 = 6x ∴ x = 4 years
∴ A' age = 4 × 6 = 24 years and B's age = 4 years

27. b; Let A's age be x years
B's age be 2x years
C's age = (x + 17) years
According to the question,
x + 2x + (x + 17) = 185
∴ 4x = 185 - 17 = 168
∴ x = 42, ∴ A's age = 42 years
B's age = 84 years
C's age = 42 + 17 = 59 years

28. b; Let the ages be 3x and 5x years.
∴ 3x + 5x = 80 ∴ x = 10
(i) Ratio of their ages after 10 years = 3x + 10 : 5x + 10 or 40 : 60 = 2 : 3
(ii) Ratio of their ages before 10 years = 3x - 10 : 5x - 10 or 20 : 40 = 1 : 2

29. b; Let the present ages of B and A be x yrs & (x + 9) yrs
or, (x + 9 + 10) = 2(x - 10) or x = 39.

30. a; Let the present age of the elder person be x years.
Then, the present age of another person = (x - 10) years.
(x - 15) = 2(x - 10 - 15) or x = 35
∴ The present age of the elder person is 35 years.

31. c; Let the present ages of father & son be x & y respectively.
Then, (x - 1) = 4(y - 1) or 4y - x = 3 ....(i)
And, (x + 6) - 2(y + 6) = 9 or -2y + x = 15 ....(ii)
Solving (i) & (ii) we get, x = 33 and y = 9.
∴ Ratio of their ages = 33 : 9 or, 11 : 3.

32. d; Ashok's present age = (26 - 6) years = 20 years.

$\therefore$ Pradeep's present age = $\left(\frac{3}{4}\times 20\right)$ years = 15 years.

33. b; Let the son's age be x years.
Then, $(x+4)+(4x+4)=43$
or $5x=35$ or $x=7$

34. a; Let the ages of Mona and her mother be 3x and 11x years respectively.
Then, $(11x-3x)=24$ or $x=3$.
So, their present ages are 9 years and 33 years.
Ratio of their ages after 3 years = 12 : 36 or 1 : 3.

35. b; Let son's age be x. Then, Kamla's age = 10x years.
Kamla's age at the time of marriage = (10x - 6) years.
$\therefore\ 10x=\frac{5}{4}(10x-6)$ or $40x=50x-30$ or $x=3$

36. a; Let Ajay's age 10 years back be x years.
Then, Sachin's age 10 years back = 2x years.
$\therefore\ 2x+20=40$ or $x=10$.
Ajay's present age = (10 + 10) years = 20 years.

37. b; Let father's present age = x years.
Then, son's present age = (45 - x) years.
$(x-5)(45-x-5)=4(x-5)$
or $x^2-41x+180=0$
or $(x-36)(x-5)=0$
$\therefore\ x=36$ years.

8

Profit and Loss

Rule 1

To find profit or loss when cost price and selling price are given.

*(i) When Selling Price > Cost Price, there is a **profit** and it is given by Selling Price – Cost Price.*

*(ii) When Selling Price < Cost Price, there is a **loss** and it is given by Cost Price – Selling Price.*

Illustrative Examples

Ex. 1: A man buys a toy for Rs 25 and sells it for Rs 30. Find his profit.

Soln: In this example,
Selling Price (SP) of a toy = Rs 30 and
Cost Price (CP) of a toy = Rs 25
Here, SP > CP
Now, following the above formula (i), we have
profit = Rs 30 – Rs 25 = Rs 5

Ex. 2: A boy buys a Parker pen for Rs 50 and sells it for Rs 45. Find his loss.

Soln: In the above example,
Cost Price (CP) of a Parker pen = Rs 50 and
Selling Price (SP) of a Parker pen = Rs 45
Here, SP < CP
Now, following the above formula (ii), we have
loss = Rs 50 – Rs 45 = Rs 5

Exercise

1. Given the following values, find the unknown values.
   (i) CP = Rs 500, SP = Rs 600, Profit/Loss = ?
   (ii) CP = Rs 1270, SP = Rs 1250, Profit/Loss = ?
   (iii) CP = ?, SP = Rs 2390, Profit = Rs 120.50
   (iv) CP = Rs 72, SP = ? Loss = Rs 15.60
2. Fill in the blanks in each of the following
   (i) CP = Rs 1265, SP = Rs 1253, Loss = Rs .....
   (ii) CP = Rs ..... SP = Rs 450, Profit = Rs 150
   (iii) CP = Rs 3355, SP = Rs 7355, ... = Rs ......
   (iv) CP = Rs ....., SP = Rs 2390, Loss = Rs 5.50
   (v) CP = Rs 6556, SP = Rs 6560, ... = Rs .....
3. By selling an article for Rs 96, double profit is obtained than the profit that would have been obtained by selling it for Rs 84. What is the cost price of the article?
   [BSRB Delhi PO, 2000]

Answers

1. (i) Rs 100 profit (ii) Rs 20 loss
   (iii) Rs 2269.50 (iv) Rs 56.40
2. (i) Loss = Rs 12 (ii) CP = Rs 300
   (iii) Profit = Rs 4000 (iv) CP = Rs 2395.50
   (v) Gain = Rs 4
3. a; **Hint:** Let the cost price of the article be Rs x.
   Then, $2(84 - x) = 96 - x$
   or, $168 - 2x = 96 - x$
   $\therefore x = $ Rs 72.

Rule 2

The profit or loss is generally reckoned as so much per cent on the cost.

$$\textit{Gain or Loss per cent} = \frac{\text{Loss or Gain}}{\text{CP}} \times 100$$

Illustrative Examples

Ex. 1: A man buys a toy for Rs 25 and sells it for Rs 30. Find his gain per cent.

Soln: $\% \text{ Gain} = \frac{Gain}{CP} \times 100 = \frac{5}{25} \times 100 = 20\%$

Ex. 2: A boy buys a pen for Rs 25 and sells it for Rs 20. Find his loss per cent.

Soln: $\% \text{ Loss} = \frac{\text{Loss}}{\text{CP}} \times 100 = \frac{5}{25} \times 100 = 20\%$

Exercise

1. Find the gain or loss per cent, if
   (i) CP = Rs 500 and SP = Rs 565
   (ii) CP = Rs 700 and SP = Rs 630
2. If the profit made on a packet of tea is Rs 4 and the cost price of the packet is Rs 20, then how much is the profit percentage?
   a) 20% b) 25% c) 30% d) 15%
3. A box of Alphanso mangoes was purchased by a fruit

seller for Rs 300. However, he had to sell them for Rs 255 because they began to get over ripe. What was the loss percentage?

a) 15% b) 10% c) 20% d) 18%

4. Harish bought a second-hand typewriter for Rs 1200 and spent Rs 200 on its repairs. He sold it for Rs 1680. Find his profit or loss. What was his profit or loss per cent?

a) 10% loss b) 15% loss c) 20% loss d) 20% gain

5. Karim bought 150 dozen pencils at Rs 10 a dozen. His overhead expenses were Rs 100. He sold them at Rs 1.20 each. What was his profit or loss per cent?

a) 30% profit b) 30% loss c) 35% loss d) 35% profit

6. Venkat purchased 20 dozens of toys at the rate of Rs 375 per dozen. He sold each one of them at the rate of Rs 33. What was his percentage profit? **[BSRB Chennai PO, 2000]**

a) 6.5 b) 5.6 c) 3.5 d) 4.5

7. By selling twelve notebooks, the seller earns profit equal to the selling price of two note-books, what is his percentage profit? **[BSRB Delhi PO, 2000]**

a) 20% b) 25% c) $16\frac{2}{3}$% d) Data inadequate

8. A loss of 5% was suffered by selling a plot for Rs 4,085. The cost price of the plot was: **[RRB Exam, 1991]**

a) Rs 4350 b) Rs 4259.25
c) Rs 4200 d) Rs 4300

9. The CP of an article which is sold at a loss of 25% for Rs 150, is **[RRB Exam, 1991]**

a) Rs 125 b) Rs 175 c) Rs 200 d) Rs 225

10. A man buys 10 articles for Rs 8 and sells them at the rate of Rs 1.25 per article. His gain percent is: **[CDS Exam, 1991]**

a) 50% b) $56\frac{1}{4}$% c) $19\frac{1}{2}$% d) 20%

11. Bhajan Singh purchased 120 reams of paper at Rs 80 per ream. He spent Rs 280 on transportation, paid octroi at the rate of 40 paise per ream and paid Rs 72 to the coolie. If he wants to have a gain of 8%, the selling price per ream must be: **[Bank PO Exam, 1988]**

a) Rs 86 b) Rs 89
c) Rs 90 d) Rs 87.48

12. If I purchased 11 books for Rs 10 and sold all the books at the rate of 10 books for Rs 11, the profit per cent is: **[RRB Exam, 1989]**

a) 10% b) 11% c) 21% d) 100%

13. An umbrella marked at Rs 80 is sold for Rs 68. The rate of discount is: **[RRB Exam, 1989]**

a) 12% b) 15% c) $17\frac{11}{17}$% d) 20%

Answers

1. (i) 13% gain (ii) 10% loss
2. a
3. a
4. d; **Hint:** Total CP = Rs (1200 + 200) = Rs 1400 and SP = Rs 1680

SP > CP, Hence profit and profit %

$$= \frac{1680-1400}{1400} \times 100 = 20\%$$

5. d; **Hint:** Total cost price = 150 × 10 + 100 = Rs 1600
Total selling price = 150 × 12 × 1.20 = Rs 2160
Profit = Rs 2160 – Rs 1600 = Rs 560

$$\text{Profit \%} = \frac{560}{1600} \times 100 = 35\%$$

6. b; **Hint:** Cost price of 20 dozens toys = 20 × 375 = Rs 7500
Selling price of 20 dozens toys = 20 × 33 × 12 = Rs 7920

$$\text{Profit percentage} = \frac{7920-7500}{7500} \times 100 = 5.6\%$$

7. a; **Hint:** Percentage profit

$$= \frac{2}{(12-2)} \times 100 = \frac{2}{10} \times 100 = 20\%$$

[Also see **Rule 81**]

8. d; **Hint:** $CP = Rs \left(\frac{100}{95} \times 4085\right) = Rs\ 4300$

9. c; **Hint:** $CP = Rs \left(\frac{100}{75} \times 150\right) = Rs\ 200$

10. b; **Hint:** CP of 10 articles = Rs 8
SP of 10 articles = Rs (1.25 × 10) = Rs 12.50
Profit = Rs (12.50 – 8) = Rs 4.50

$$\text{Gain \%} = \left(\frac{4.50}{8} \times 100\right) = 56\frac{1}{4}\%$$

11. c; **Hint:** Total CP of 120 reams = Rs (120 × 80 + 280 + 72 + 120 × 0.4) = Rs 10000.

$$\text{CP of 1 ream} = (10000 \div 120) = Rs \left(\frac{250}{3}\right).$$

$$\therefore \text{ SP of 1 ream} = \left[108\% \text{ of } Rs\ \frac{250}{3}\right] = Rs\ 90.$$

12. c; **Hint:** CP = Rs 10

$$SP = Rs \left(\frac{11}{10} \times 11\right) = Rs\ \frac{121}{10}$$

$$\text{Gain} = Rs \left(\frac{121}{10} - 10\right) = Rs\ \frac{21}{10}$$

$$\therefore \text{ Gain \%} = Rs \left(\frac{21}{10 \times 10} \times 100\right) = 21\%$$

13. b

Rule 3

If the profit earned by selling an article for Rs x is equal to the loss incurred when the same article is sold for Rs y, then to make z% profit the sale price of the article should be Rs $\left[\frac{(x+y)(100+z)}{200}\right]$.

Illustrative Example

Ex.: The profit earned by selling an article for Rs 600 is equal to the loss incurred when the same article is sold for Rs 400. What should be the sale price of the article for making 25 per cent profit?

Soln: **Detail Method:** Let the cost price be Rs x.

Now, according to the question,

$600 - x = x - 400$ or, $2x = 1000$ $\therefore x = \text{Rs } 500$.

Again, selling price of the article for making 25% profit

$= \frac{500 \times 125}{100} = \text{Rs } 625$

Quicker Method: Applying the above theorem, we have

the required answer

$= \frac{(600+400)(100+25)}{200} = 5 \times 125 = \text{Rs } 625.$

Exercise

1. The profit earned by selling an article for Rs 832 is equal to the loss incurred when the same article is sold for Rs 448. What should be the sale price of the article for making 50 per cent profit? **[BSRB Chennai PO, 2000]**
a) Rs 960 b) Rs 1060 c) Rs 1200 d) Rs 920
2. The profit earned by selling an article for Rs 482 is equal to the loss incurred when the same article is sold for Rs 318. What should be the sale price of the article for making 30 per cent profit?
a) Rs 560 b) Rs 520 c) Rs 540 d) Rs 580
3. The profit earned by selling an article for Rs 317 is equal to the loss incurred when the same article is sold for Rs 233. What should be the sale price of the article for making 20 per cent profit?
a) Rs 390 b) Rs 370 c) Rs 350 d) Rs 330
4. The profit earned by selling an article for Rs 515 is equal to the loss incurred when the same article is sold for Rs 475. What should be the sale price of the article for making 40% per cent profit?
a) Rs 693 b) Rs 707 c) Rs 683 d) Rs 673
5. The profit earned by selling an article for Rs 680 is equal to the loss incurred when the same article is sold for Rs 420. What should be the sale price of the article for making 60 per cent profit?
a) Rs 870 b) Rs 890 c) Rs 880 d) Rs 990

Answers

1. a 2. b 3. d 4. a 5. c

Rule 4

Theorem: ***If a reduction of x% in the price of an article enables a person to buy n kg more for Rs A, then the reduced and the original prices per kg of the article are*** $\left(\frac{Ax}{100n}\right)$ ***per kg and*** $\left(\frac{Ax}{(100-x)n}\right)$ ***per kg respectively.***

Illustrative Example

Ex.: A reduction of 20% in the price of sugar enables a person to buy 2 kg more for Rs 30. Find the reduced and the original price per kg of sugar.

Soln: Following the above theorem, we have

the reduced price of sugar $= \frac{30 \times 20}{100 \times 2}$

$= \text{Rs } 3$ per kg and

the original price of sugar $= \frac{30 \times 20}{(100-20)2} = \frac{15}{4}$

$= \text{Rs } 3\frac{3}{4}$ per kg.

Exercise

1. A reduction of 30% in the price of tea enables a person to buy 3 kg more for Rs 20. Find the original price per kg of tea.
a) Rs $2\frac{5}{7}$ b) Rs $2\frac{6}{7}$ c) Rs $2\frac{4}{7}$ d) Rs $3\frac{5}{7}$
2. A reduction of 15% in the price of coffee enables a person to buy 2 kg more for Rs 30. Find the original price per kg of coffee.
a) Rs $2\frac{11}{17}$ b) Rs $3\frac{11}{17}$ c) Rs $2\frac{10}{11}$ d) Rs $2\frac{1}{17}$
3. A reduction of 25% in the price of rice enables a person to buy 4 kg more for Rs 40. Find the reduced and the original price per kg of rice respectively.
a) Rs $\frac{10}{3}$, Rs 2.5 b) Rs 2.5, Rs $\frac{10}{3}$
c) Rs 3.5, Rs $\frac{11}{3}$ d) Can't be determined
4. A reduction of 10% in the price of salt enables a person to buy 2 kg more for Rs 18. Find the reduced and the original price per kg of salt respectively.
a) Re 1, Rs 0.9 b) Rs 0.9, Re 1
c) Rs 2, Rs 1.9 d) Rs 1.9, Rs 2
5. A reduction of 10 per cent in the price of potatoes enables me to obtain 25 kg more for Rs 225. What is the reduced price per kg? Find also the original price per kg.
a) Rs 0.9, Re 1 b) Re 1, Rs 2

c) Rs 0.8, Rs 2 d) None of these

6. A reduction of 20 per cent in the price of onions enables a purchaser to obtain 8 kg more for Rs 80. What is the reduced price per kg? What was the price per kg before reduction?
a) Rs 1.50, Rs 2 b) Rs 2, Rs 2.50
c) Rs 3, Rs 2.50 d) Re 1, Rs 2.50

7. A reduction of 25 per cent in the price of clay would enable a purchaser to obtain 25 kg more for Rs 45. What is the reduced price per kg and what was the price per kg before reduction?
a) 40 P, 60 P b) 45 P, 50 P c) 45 P, 60 P d) 45 P, 48 P

8. A reduction of $12\frac{1}{2}$ per cent in the price of oranges enables one to purchase 15 oranges more for Rs 42. Find the price per orange before and after reduction.
a) 40 P, 35 P b) 30 P, 40 P c) 45 P, 35 P d) 35 P, 45 P

9. If a reduction of 25 per cent were made in the price of sweets it would enable a purchaser to obtain 10 more than before for Rs 7.20. Find the reduced price, and the present price.
a) 18 P, 25 P b) 25 P, 30 P
c) 18 P, 24 P d) None of these

10. A reduction of 40 per cent in the price of bananas would enable a purchaser to obtain 60 more for Rs 45. What is the reduced price?
a) 50 P b) 40 P c) 60 P d) 30 P

Answers

1. b 2. a 3. b 4. b 5. a 6. b 7. c
8. a 9. c 10. d

Rule 5

Theorem: ***If a hike of x% in the price of an article forces a person to buy n kg less, then the new and the original prices per kg of the article are*** $\left[\frac{Ax}{100n} \text{ per kg}\right]$ ***and*** $\left[\frac{Ax}{(100+x)n} \text{ per kg}\right]$ ***respectively.***

Illustrative Example

Ex.: A 10% hike in the price of rice forces a person to purchase 2 kg less for Rs 110. Find the new and the original prices of the rice.

Soln: Applying the above formula, we have

$$\text{the new price} = \frac{10 \times 110}{100 \times 2} = \text{Rs } 5\frac{1}{2} \text{ per kg and}$$

$$\text{the original price} = \frac{10 \times 110}{110 \times 2} = \text{Rs 5 per kg}$$

Exercise

1. A 15% hike in the price of wheat forces a person to purchase 2 kg less for Rs 46. Find the original price of the wheat.
a) Rs 3 per kg b) Rs 2 per kg
c) 3.5 per kg d) Data inadequate

2. A 20% hike in the price of wheat forces a person to purchase 1 kg less for Rs 20. Find the original price of the wheat.
a) Rs 3 per kg b) Rs $3\frac{1}{3}$ per kg
c) Rs $3\frac{2}{3}$ per kg d) Data inadequate

3. A 25% hike in the price of tea forces a person to purchase 2 kg less for Rs 75. Find the original price of the tea.
a) Rs 7 b) Rs 8 c) Rs 7.5 d) Rs 8.5

4. A 12% hike in the price of coffee forces a person to purchase 2 kg less for Rs 56. Find the original price of the coffee.
a) Rs 2 b) Rs 2.5 c) Rs 3.5 d) Rs 3

Answers

1. a 2. b 3. c 4. d

Rule 6

Theorem: ***If a man purchases 'x' items for Rs 'y' and sells 'y' items for Rs 'x', then the profit or loss [depending upon the respective (+ve) or (–ve) sign in the final result] made by him is*** $\left[\frac{x^2 - y^2}{y^2} \times 100\right]\%$.

Illustrative Example

Ex.: If a man purchases 11 oranges for Rs 10 and sells 10 oranges for Rs 11. How much profit or loss does he make?

Soln: **Detailed Method:** Suppose that the person bought 11 × 10 = 110 oranges.

$$\text{CP of 110 oranges} = \frac{10}{11} \times 110 = \text{Rs } 100$$

$$\text{SP of 110 oranges} = \frac{11}{10} \times 110 = \text{Rs } 121$$

∴ Profit = Rs 121 – Rs 100 = Rs 21

$$\text{and \% profit} = \frac{\text{Profit}}{\text{CP}} \times 100 = \frac{21}{100} \times 100 = 21\%$$

Quicker Method: Applying the above formula, we have,

$$\frac{11^2 - 10^2}{10^2} \times 100 = 21\%.$$

Since the sign is +ve, there is a gain of 21%.

Exercise

1. If a man purchases 12 mangoes for Rs 10 and sells 10 mangoes for Rs 12. How much profit or loss does he make?
   a) 40% loss b) 40% profit c) 44% loss d) 44% profit
2. If a man purchases 15 bananas for Rs 12 and sells 12 bananas for Rs 15. How much profit or loss does he make?
   a) 56.25% profit b) 56.25% loss
   c) 56.75% profit d) 56.75% loss
3. If a man purchases 4 apples for Rs 5 and sells 5 apples for Rs 4. How much profit or loss does he make?
   a) 36% profit b) 36% loss c) 36.5% profit d) 36.5% loss
4. If a man purchases 7 oranges for Rs 8 and sells 8 oranges for Rs 7. How much profit or loss does he make?
   a) $\frac{275}{16}$% loss b) $\frac{275}{16}$% profit
   c) $\frac{375}{16}$% loss d) $\frac{375}{16}$% profit
5. A grocer buys eggs at 10 for Rs 8 and sells at 8 for Rs 10. Find his gain or loss per cent.
   a) gain per cent of $56\frac{1}{2}$ b) gain per cent of $56\frac{3}{4}$
   c) loss per cent of $56\frac{1}{4}$ d) gain per cent of $56\frac{1}{4}$

Answers

1. d 2. a 3. b 4. c 5. d

Rule 7

Theorem: ***If a man purchases 'a' items for Rs 'b' and sells 'c' items for Rs 'd', then the gain or loss [depending upon the respective (+ve) or (–ve) sign in the final result] made by him is*** $\left[\frac{ad-bc}{bc}\times 100\right]\%$.

Note: **Rule - 6** is the special case of this rule. In Rule - 6, we have,
a = d = x and b = c = y.

Illustrative Example

Ex.: A boy buys 9 oranges for Rs 16 and sells them at 11 for Rs 20. What does he gain or lose per cent?

Soln: Following the above formula, we have

$$\% \text{ profit or loss} = \frac{9\times 20-16\times 11}{16\times 11}\times 100 = 2\frac{3}{11}\%$$

Since the sign is +ve, there is a gain of $2\frac{3}{11}\%$.

Exercise

1. A person buys 8 oranges for Rs 15 and sells them at 10 for Rs 18. What does he gain or lose per cent?
   a) gain of 4% b) loss of 4% c) gain of 2% d) loss of 2%
2. A person buys 4 apples for Rs 5 and sells them at 5 for Rs 4. What does he gain or lose per cent?
   a) loss of 36% b) gain of 36%
   c) loss of 30% d) gain of 30%
3. A person buys 16 bananas for Rs 5 and sells them at 12 for Rs 5. What does he gain or lose per cent?
   a) $33\frac{1}{3}$% loss b) 33% loss
   c) $33\frac{1}{3}$% profit d) $33\frac{2}{3}$% profit
4. A person buys 12 eggs for Rs 15 and sells them at 10 for Rs 14. What does he gain or lose per cent?
   a) 10% loss b) 12% profit c) 12% loss d) 15% profit

Answers

1. b 2. a 3. c 4. b

Rule 8

Problems Based on Dishonest Dealer

$$\textbf{\% gain} = \frac{\text{Error}}{\text{True value} - \text{Error}}\times 100$$

$$\textbf{or, \% gain} = \frac{\text{True weight} - \text{False weight}}{\text{False weight}}\times 100$$

Illustrative Example

Ex.: A dishonest dealer professes to sell his goods at cost price, but he uses a weight of 960 gm for the kg weight. Find his gain per cent.

Soln: **Detailed Method:** Suppose goods cost the dealer Re 1 per kg. He sells for Re 1 what cost him Re 0.96.

$\therefore$ Gain on Re 0.96 = Re 1 – Re 0.96 = Re 0.04

$\therefore$ Gain on Rs 100 = $\frac{0.04}{0.96}\times 100$ = Rs $4\frac{1}{6}$

$\therefore$ Gain % = $4\frac{1}{6}\%$

Quicker Method: Applying the above Rule, we have

the required gain % $= \frac{40}{1000-40}\times 100 = 4\frac{1}{6}\%$

Exercise

1. A dishonest fruit vendor professes to sell his goods at cost price but he uses a weight of 900 g for the kg weight. Find his gain per cent. **[RRB Exam, 1991]**
   a) $9\frac{1}{11}\%$ b) $11\frac{1}{9}\%$ c) $11\frac{2}{9}\%$ d) $9\frac{2}{11}\%$

2. A cloth dealer professes to sell poplin at cost price, but uses a metre having a length of 92 cm only and charges for the metre. Find his gain per cent.

a) $8\frac{16}{23}\%$ b) $9\frac{16}{23}\%$

c) $7\frac{16}{23}\%$ d) Data inadequate

3. A dishonest fruit vendor professes to sell his goods at cost price but he uses a weight of 950 g for the kg weight. Find his gain per cent.

a) $5\frac{5}{19}\%$ b) $5\frac{15}{19}\%$

c) $5\frac{4}{19}\%$ d) Data inadequate

4. A dishonest fruit vendor professes to sell his goods at cost price but he uses a weight of 800 g for the kg weight. Find his gain per cent.

a) 20% b) 40% c) 25% d) 50%

5. A cloth dealer professes to sell bedsheet at cost price, but uses a metre having a length of 80 cm only and charges for the metre. Find his gain per cent.

a) 25% b) 20% c) 16% d) 30%

Answers

1. b 2. a 3. a 4. c 5. a

Rule 9

Total percentage profit $= \frac{\%\text{ profit} + \%\text{ less in wt}}{100 - \%\text{ less in wt}} \times 100$

Illustrative Example

Ex.: A grocer sells rice at a profit of 10% and uses a weight which is 20% less. Find his total percentage gain.

Soln: **Detail method:**

Suppose he bought at Rs x/kg.

Then he sells at Rs $\left(\frac{110x}{100}\right)$ per $\frac{80}{100}$ kg

or, at Rs $\frac{110x}{100} \times \frac{100}{80}$ per kg or, at Rs $\frac{11x}{8}$ per kg

Now, % profit $= \frac{\frac{11x}{8} - x}{x} \times 100 = \frac{300}{8} = 37.5\%$

Quicker Method: Applying the above rule, we have the total % profit

$= \frac{10+20}{100-20} \times 100 = \frac{30 \times 100}{80} = 37.5\%$.

Exercise

1. A grocer sells rice at a profit of 5% and uses a weight which is 20% less. Find his total percentage gain.

a) 31.25% b) 1.5% c) 31% d) Data inadequate

2. A grocer sells rice at a profit of 10% and uses a weight which is 25% less. Find his total percentage gain.

a) $36\frac{2}{3}\%$ b) $46\frac{1}{3}\%$ c) $46\frac{2}{3}\%$ d) $36\frac{1}{3}\%$

3. A grocer sells rice at a profit of 20% and uses a weight which is 20% less. Find his total percentage gain.

a) 25% b) 50% c) 75% d) 45%

4. A grocer sells rice at a profit of 25% and uses a weight which is 25% less. Find his total percentage gain.

a) 66% b) $66\frac{1}{3}\%$ c) $66\frac{3}{4}\%$ d) $66\frac{2}{3}\%$

5. A grocer sells rice at a profit of 20% and uses a weight which is 25% less. Find his total percentage gain.

a) 50% b) 55% c) 60% d) 65%

Answers

1. a 2. c 3. b 4. d 5. c

Rule 10

Theorem: ***If the shopkeeper sells his goods at x% loss on cost price but uses y gm instead of z gm, then his % profit or loss is*** $[100 - x]\frac{z}{y} - 100$ ***according as the sign is +ve or –ve.***

Illustrative Example

Ex.: A dishonest dealer sells goods at $6\frac{1}{4}\%$ loss on cost price but uses 14 gm instead of 16 gm. What is his percentage profit or loss?

Soln: Detail Method: Suppose the cost price is Rs x per kg.

Then he sells the goods for Rs $x\left(\frac{100 - \frac{25}{4}}{100}\right)$

$= \text{Rs } \frac{15x}{16}$ per kg

Now, suppose he bought y kg of goods.

Then, his total investment = Rs xy

and his total return = Rs $\frac{15x}{16} \times y\left(\frac{16}{14}\right) = \text{Rs } \frac{15}{14}xy$

$\therefore$ his % profit $= \frac{\frac{15}{14}xy - xy}{xy} \times 100 = \frac{50}{7} = 7\frac{1}{7}\%$

Quicker Method: In the above case,

$$\% \text{ profit or loss} = \left[100 - 6\frac{1}{4}\right]\left[\frac{16}{14}\right] - 100$$

$$= \frac{375}{4} \times \frac{16}{14} - 100 = \frac{1500-1400}{14} = \frac{100}{14} = \frac{50}{7} = 7\frac{1}{7}\%$$

Since the sign is +ve, there is a profit of $7\frac{1}{7}\%$.

Exercise

1. A merchant professes to lose 4% on a certain tea, but he uses a weight equal to 840 g instead of 1 kg. Find his real loss or gain per cent.

   a) $14\frac{2}{7}\%$ loss b) $16\frac{2}{7}\%$ loss

   c) $14\frac{2}{7}\%$ gain d) $16\frac{2}{7}\%$ gain

2. A merchant professes to lose 10% on a certain tea, but he uses a weight equal to 900 g instead of 1 kg. Find his real loss or gain per cent.

   a) 10% gain b) 10% loss

   c) neither gain nor loss d) Data inadequate

3. A merchant professes to lose 8% on a certain tea, but he uses a weight equal to 460 g instead of 1 kg. Find his real loss or gain per cent.

   a) 50% gain b) 100% gain c) 200% gain d) 75% gain

4. A cloth dealer professes to lose 9% on a certain garments, but he uses a metre having a length of 91 cm only and charges for the metre. Find his gain or loss per cent.

   a) 10% gain b) 9% loss

   c) 9% gain d) Neither gain nor lose

5. A dishonest fruit vendor professes to lose 20% on his goods, but he uses a weight of 720 gm for the kg weight. Find his gain or lose per cent.

   a) $11\frac{2}{9}\%$ gain b) $9\frac{1}{11}\%$ gain

   c) $11\frac{1}{9}\%$ gain d) $11\frac{1}{9}\%$ lose

6. A merchant professes to lose 6% on a certain tea, but he uses a weight equal to 900 gm instead of one kg. His real gain per cent is: **[Clerks' Grade Exam, 1991]**

   a) $5\frac{1}{9}\%$ b) 6% c) 4% d) $4\frac{4}{9}\%$

Answers

1\. c 2. c 3. b 4. d 5. c 6. d

Rule 11

Theorem: ***A dishonest dealer sells the goods at x% loss on cost price but uses y% less weight, then his percentage profit or loss is*** $\left[\frac{y-x}{100-y} \times 100\right]$ ***according as the sign is +ve or –ve.***

Illustrative Example

Ex.: A dishonest dealer sells the goods at $6\frac{1}{4}\%$ loss on cost price but uses $12\frac{1}{2}\%$ less weight. What is his percentage profit or loss?

Soln: Following the above formula we have

$$\% \text{ loss or gain} = \frac{\frac{25}{2} - \frac{25}{4}}{100 - \frac{25}{2}} \times 100 = \frac{100}{14} = 7\frac{1}{7}\%.$$

Since sign is +ve there is a profit of $7\frac{1}{7}\%$.

Exercise

1. A dishonest dealer sells the goods at 10% loss on cost price but uses 20% less weight. What is his percentage profit or loss?

   a) $12\frac{1}{2}\%$ loss b) 12.5% gain

   c) 13.5% gain d) 12% gain

2. A dishonest dealer sells the goods at 20% loss on cost price but uses 15% less weight. What is his percentage profit or loss?

   a) $5\frac{11}{17}\%$ loss b) $5\frac{15}{17}\%$ loss

   c) $5\frac{15}{17}\%$ gain d) $5\frac{11}{17}\%$ gain

3. A dishonest dealer sells the goods at 20% loss on cost price but uses 25% less weight. What is his percentage profit or loss?

   a) $6\frac{2}{3}\%$ gain b) $6\frac{1}{3}\%$ gain

   c) 6% gain d) $6\frac{2}{3}\%$ loss

4. A dishonest dealer sells the goods at 44% loss on cost price but uses 30% less weight. What is his percentage profit or loss?

   a) 20% gain b) 28% gain c) 20% loss d) 25% loss

5. A dishonest dealer sells the goods at 5% loss on cost price but uses 24% less weight. What is his percentage profit or loss?

   a) 25% loss b) 20% gain c) 20% loss d) 25% gain

Answers

1. b 2. b 3. a 4. c 5. d

Rule 12

Theorem: ***If a seller uses 'X' gm in place of one kg (1000 gm) to sell his goods and gains a profit of x% on cost price, then his actual gain or loss percentage is*** $(100+x)\left[\frac{1000}{X}\right]-100$ ***according as the sign is +ve or –ve.***

Illustrative Example

Ex.: A seller uses 840 gm in place of one kg to sell his goods. Find his actual % profit or loss, when he sells his article on 4% gain on cost price.

Soln: **Detail Method:** Selling price of 840 gm

$= \text{Rs}\,(100+4) = \text{Rs}\,104$

$\therefore$ Profit = SP – CP = 104 – 84 = Rs 20

$$\left[\because \text{CP of 840 gm} = \frac{100}{1000}\times 840 = \text{Rs } 84\right]$$

$$\therefore \%\text{ profit} = \frac{20\times 100}{84} = 23\frac{17}{21}\%$$

Quicker Method: Applying the above formula, we have

$$(100+4)\left(\frac{1000}{840}\right)-100 = 23\frac{17}{21}\%$$

Since sign is +ve, there is a profit of $23\frac{17}{21}\%$

Exercise

1. A seller uses 900 gm in place of one kg to sell his goods. Find his actual % profit or loss, when he sells his article on 20% gain on cost price.

   a) 33% gain b) $33\frac{1}{3}\%$ gain

   c) $33\frac{1}{3}\%$ loss d) $33\frac{2}{3}\%$ gain

2. A seller uses 990 gm in place of one kg to sell his goods. Find his actual % profit or loss, when he sells his article on 10% gain on cost price.

   a) $11\frac{1}{9}\%$ gain b) $9\frac{1}{11}\%$ gain

   c) $11\frac{1}{9}\%$ loss d) $9\frac{1}{11}\%$ loss

3. A seller uses 920 gm in place of one kg to sell his goods. Find his actual % profit or loss, when he sells his article on 15% gain on cost price.

   a) 20% gain b) 15% gain

   c) 25% gain d) 30% gain

4. A seller uses 1250 gm in place of one kg to sell his goods. Find his actual % profit or loss, when he sells his article on 25% gain on cost price.

   a) 10% loss b) 15% gain

   c) Neither loss nor gain d) Can't be determined

5. A seller uses 870 gm in place of one kg to sell his goods. Find his actual % profit or loss, when he sells his article on 16% gain on cost price.

   a) $33\frac{1}{3}\%$ loss b) $33\frac{2}{3}\%$ loss

   c) $33\frac{2}{3}\%$ gain d) $33\frac{1}{3}\%$ gain

Answers

1. b 2. a 3. c 4. c 5. d

Rule 13

If a dealer wants to earn x% profit on an article after offering y% discount to the customer. To arrive at label price, the marked price should be increased by $\left(\frac{x+y}{100-y}\right)\times 100$ ***per cent.***

Illustrative Example

Ex.: If a dealer wants to earn 20% profit on an article after offering 30% discount to the customer, by what percentage should he increase his marked price to arrive at the label price?

Soln: Applying the above rule, we have

$$\text{the required answer} = \frac{20+30}{100-30}\times 100 = 71\frac{3}{7}\%$$

Exercise

1. If a dealer wants to earn 5% profit on an article after offering 10% discount to the customer, by what percentage should he increase his marked price to arrive at the label price? **[NABARD, 1999]**

   a) 15 b) $16\frac{2}{3}$ c) $15\frac{1}{2}$ d) $16\frac{1}{2}$

2. If a dealer wants to earn 10% profit on an article after offering 15% discount to the customer, by what percentage should he increase his marked price to arrive at the label price?

   a) $17\frac{7}{29}\%$ b) $29\frac{7}{17}\%$ c) $29\frac{5}{17}\%$ d) $29\frac{8}{17}\%$

3. If a dealer wants to earn 15% profit on an article after offering 20% discount to the customer, by what percentage should he increase his marked price to arrive at the

label price?

a) 34.75% b) 43.25% c) 43.75% d) 34.25%

4. If a dealer wants to earn 20% profit on an article after offering 25% discount to the customer, by what percentage should he increase his marked price to arrive at the label price?

a) 60% b) 50% c) 45% d) 55%

5. If a dealer wants to earn 15% profit on an article after offering 25% discount to the customer, by what percentage should he increase his marked price to arrive at the label price?

a) 53% b) $53\frac{2}{3}\%$ c) $35\frac{1}{3}\%$ d) $53\frac{1}{3}\%$

Answers

1. b 2. b 3. c 4. a 5. d

Rule 14

Goods passing through successive hands

Theorem: ***A sells a goods to B at a profit of x% and B sells it to C at a profit of y%. If C pays Rs X for it, then the cost price for A is Rs*** $\left[\frac{100^2 X}{(100+x)(100+y)}\right]$.

Illustrative Example

Ex.: A sells a good to B at a profit of 20% and B sells it to C at a profit of 25%. If C pays Rs 225 for it, what was the cost price for A?

Soln: During both the transactions there are profits. So our calculating figures would be 120, 125 and 100. A's cost price is certainly less than C's selling price.

$$\therefore \text{ Required price} = 225\times\frac{100}{120}\times\frac{100}{125} = \text{Rs } 150$$

Note: Since we need a value which is less than the given value, so our multiplying fractions should be less than one. That is why we multiplied 225 with $\frac{100}{120}$ and $\frac{100}{125}$.

Exercise

1. A sells a horse to B at a profit of 5% and B sells it to C at a profit of 10%. If C pays Rs 2310 for it, what did it cost A?

a) Rs 2300 b) Rs 2200 c) Rs 2000 d) Rs 2050

2. A bicycle passes through the hands of three dealers each of whom gains 25%. If the third sells it for Rs 750, what did the first pay for it?

a) Rs 384 b) Rs 483 c) Rs 564 d) Rs 374

3. The manufacturer of a machine sells it to a wholesale dealer making a profit of 20 per cent on its cost, the wholesale dealer sells it to shopkeeper, making a profit of 5 per cent and the shopkeeper makes a profit of 12 per cent by selling it for Rs 141.12. Find the cost of the manufacturer.

a) Rs 100.50 b) Rs 100 c) Rs 150 d) Rs 142.51

4. A sells a good to B at a profit of 10% and B sells it to C at a profit of 15%. If C pays Rs 1265 for it, what was the cost price for A?

a) Rs 1100 b) Rs 950 c) Rs 1000 d) Rs 1250

5. A sells a good to B at a profit of 15% and B sells it to C at a profit of 20%. If C pays Rs 690 for it, what was the cost price for A?

a) Rs 600 b) Rs 500 c) Rs 630 d) Rs 580

Answers

1. c

2. a; **Hint:** Required answer $= \frac{750\times100\times100\times100}{125\times125\times125} = \text{Rs } 384$

3. b 4. c 5. b

Rule 15

Theorem: ***A sells a goods to B at a loss of x% and B sells it to C at a loss of y%. If C pays Rs X for it, then the price at which A buys is Rs*** $\left[\frac{100^2 X}{(100-x)(100-y)}\right]$.

Illustrative Example

Ex.: A sells a horse to B at a loss of 20% and B sells it to C at a loss of 25%. If C pays Rs 900 for it, at what price did A buy?

Soln: Following the above formula, we have the required answer

$$= 900\times\frac{100}{100-20}\times\frac{100}{100-25} = \frac{900\times100\times100}{80\times75}$$
$$= \text{Rs } 1500.$$

Exercise

1. A sells a horse to B at a loss of 5% and B sells it to C at a loss of 10%. If C pays Rs 855 for it, at what price did A buy?

a) Rs 955 b) Rs 1000 c) Rs 1050 d) Rs 1100

2. A sells a horse to B at a loss of 10% and B sells it to C at a loss of 20%. If C pays Rs 1440 for it, at what price did A buy?

a) Rs 2000 b) Rs 2500 c) Rs 1800 d) Rs 1840

3. A bicycle passes through the hands of three dealers each of whom loses 5%. If the third sells it for Rs 6859, what did the first pay for it?

a) Rs 6900 b) Rs 7000 c) Rs 8000 d) Rs 7950

Answers

1. b 2. a

3. c; **Hint:** Required answer $= \frac{100\times100\times100\times6859}{95\times95\times95}$

$= \text{Rs } 8000$

Rule 16

Theorem: A sells a goods to B at a profit of x% and B sells it to C at a loss of y%. If C pays Rs X for it then the price at which A buys is Rs $\left[\frac{100^2 X}{(100+x)(100-y)}\right]$.

Note: If A sells a goods to B at a loss of x% and B sells it to C at a profit of y%, then the formula for cost price will be $\frac{(100)^2 X}{(100-x)(100+y)}$

Illustrative Example

Ex.: A sells a bicycle to B at a profit of 30% and B sells it to C at a loss of 20%. If C pays Rs 520 for it, at what price did A buy?

Soln: In the whole transaction there is a gain of 30% and a loss of 20%, so our calculating figures would be 130, 80 and 100.

$$\text{B's cost of price} = 520 \times \frac{100}{80}$$

$$\text{A's cost price} = 520 \times \frac{100}{80} \times \frac{100}{130} = \text{Rs } 500.$$

Exercise

1. A sells an article to B at a profit of 5% and B sells it to C at a loss of 5%. If C pays Rs 23.94 for it, what did it cost A?
   a) Rs 24 b) Rs 25 c) Rs 26 d) Rs 30
2. A sells an article to B at a profit of 20% and B sells it to C at a loss of 20%. If C pays Rs 19.20 for it, what did it cost A?
   a) Rs 25 b) Rs 23 c) Rs 20 d) Rs 24
3. A sells an article to B at a profit of 15% and B sells it to C at a loss of 10%. If C pays Rs 207 for it, what did it cost A?
   a) Rs 197 b) Rs 199 c) Rs 201 d) Rs 200
4. A man sells a car to his friend at 10% loss. If the friend sells it for Rs 54000 and gains 20%, the original CP of the car was: **[SSC Exam, 1987]**
   a) Rs 25000 b) Rs 37500 c) Rs 50000 d) Rs 60000

Answers

1. a 2. c 3. d
4. c; **Hint:** See 'note' in the given formula,

$$\text{Cost price} = \frac{100 \times 100 \times 54000}{(100-10)(100+20)} = \text{Rs } 50000.$$

Rule 17

Theorem: ***When there are two successive profits of x% and y% , then the resultant profit per cent is given by*** $\left(x+y+\frac{xy}{100}\right)$.

Illustrative Example

Ex.: A sells a bicycle to B at a profit of 20% and B sells it to C at a profit of 25%. Find the resultant profit per cent.

Soln: Following the above formula,
The resultant profit per cent

$$= 20 + 25 + \frac{20 \times 25}{100} = 50\%$$

Exercise

1. A sells a bicycle to B at a profit of 5% and B sells it to C at a profit of 10%. Find the resultant profit per cent.
   a) 15% b) 16.2% c) 15.2% d) 15.5%
2. A sells a bicycle to B at a profit of 10% and B sells it to C at a profit of 15%. Find the resultant profit per cent.
   a) 25% b) 26.5% c) 25.6% d) 26.2%
3. A sells a bicycle to B at a profit of 15% and B sells it to C at a profit of 20%. Find the resultant profit per cent.
   a) 38% b) 35% c) 36% d) 34%
4. A sells a bicycle to B at a profit of $7\frac{1}{2}$% and B sells it to C at a profit of $17\frac{1}{2}$%. Find the resultant profit per cent.
   a) 25% b) $25\frac{5}{16}$% c) $26\frac{5}{16}$% d) $16\frac{5}{26}$%
5. A sells a bicycle to B at a profit of 25% and B sells it to C at a profit of 30%. Find the resultant profit per cent.
   a) 55% b) 62.5% c) 63.5% d) 60.5%

Answers

1. d 2. b 3. a 4. c 5. b

Rule 18

Theorem: ***When there is a profit of x% and loss of y% in a transaction, then the resultant profit or loss per cent is given by*** $\left(x-y-\frac{xy}{100}\right)$ ***according to the + ve and the -ve signs respectively.***

Illustrative Example

Ex.: A sells a bicycle to B at a profit of 30% and B sells it to C at a loss of 20%. Find the resultant profit or loss.

Soln: Applying the above formula, we have

$$\text{the resultant profit or loss} = 30 - 20 - \frac{30 \times 20}{100} = 4\%$$

profit, because sign is +ve.

Exercise

1. A sells a bicycle to B at a profit of 15% and B sells it to C at a loss of 5%. Find the resultant profit or loss.
   a) 9.75% loss b) 9.75% gain

c) 9.25% gain d) 10.25% gain

2. A sells a bicycle to B at a profit of 10% and B sells it to C at a loss of 20%. Find the resultant profit or loss.
a) 12% loss b) 12% gain c) 22% loss d) 22% gain
3. A sells a bicycle to B at a profit of 20% and B sells it to C at a loss of 15%. Find the resultant profit or loss.
a) 2% loss b) 2% gain c) 8% gain d) 8% loss
4. A sells a bicycle to B at a profit of 15% and B sells it to C at a loss of 10%. Find the resultant profit or loss.
a) 3.5% gain b) 3.5% loss c) 2.5% gain d) 4.5% gain
5. A sells a bicycle to B at a profit of 20% and B sells it to C at a loss of 5%. Find the resultant profit or loss.
a) 15% gain b) 14% gain c) 16% gain d) 13% gain

Answers

1. c 2. a 3. b 4. a 5. b

Rule 19

Theorem: ***If a person sells a goods at a profit of x%. Had he sold it for Rs X more, y% would have been gained. Then the cost price is given by Rs*** $\left[\frac{X}{y-x}\times 100\right]$***. In other words, cost***

$$price = \frac{More\ gain}{Difference\ in\ percentage\ profit}\times 100$$

Illustrative Example

Ex.: I sold a book at a profit of 12%. Had I sold it for Rs 18 more, 18% would have been gained. Find the cost price.

Soln: **Detailed Method:** Here,

118% of cost - 112% of cost = Rs 18

$\therefore$ 6% of cost = Rs 18

$\therefore$ cost $= \frac{18\times 100}{6}$ = Rs 300.

Quicker Method: Following the above formula, we have

cost $= \frac{18\times 100}{18-12}$ = Rs 300.

Exercise

1. A man sold a table at a profit at $6\frac{1}{2}$ per cent. Had he sold it for Rs 1250 more, 19 per cent would have been gained. Find the cost price.
a) Rs 10000 b) Rs 1000
c) Rs 100000 d) Can't be determined
2. I sold a pen at a profit of 15%. Had I sold it for Rs 24 more, 21% would have been gained. Find the cost price.
a) Rs 300 b) Rs 400 c) Rs 450 d) Rs 600
3. A person sold a chair at a profit of 14%. Had he sold it for Rs 15 more, 17% would have been gained. Find the cost price.
a) Rs 550 b) Rs 650 c) Rs 600 d) Rs 500
4. I sold a book at a profit of 5%. Had I sold it for Rs 17 more, 15% would have been gained. Find the cost price.
a) Rs 85 b) Rs 170 c) Rs 150 d) Rs 180

Answers

1. a 2. b 3. d 4. b

Rule 20

Theorem: ***If a person sells a goods at a loss of x%. Had he been able to sell it at a gain of y%, it would have fetched Rs X more than it did, then the cost price is given by Rs*** $\left[\frac{X}{x+y}\times 100\right]$.

Illustrative Example

Ex.: A man sold a horse at a loss of 7%. Had he been able to sell it at a gain of 9%, it would have fetched Rs 64 more than it did. What was the cost price?

Soln: **Detail Method:**

Here 109 % of cost – 93% of cost = Rs 64

$\therefore$ 16 % of cost = Rs 64

$\therefore$ cost $= \frac{64\times 100}{16}$ = Rs 400.

Quicker Method: Applying the above rule, we have

the cost price $= \frac{64\times 100}{9+7} = \frac{64\times 100}{16}$ = Rs 400

Exercise

1. A chair was sold at a loss of 10 per cent. If it was sold for Rs 84 more, there would have been a gain of 4 per cent. For how much was the chair sold?
a) Rs 600 b) Rs 640 c) Rs 540 d) Rs 500
2. A man sells a machine at 10% below cost price. Had he received Rs 1494 more than he did, he would have made a profit of $12\frac{1}{2}$%. What did the machine cost?
a) Rs 6640 b) Rs 6650 c) Rs 6460 d) Rs 6440
3. A man had a table to sell. I offered him a sum of money for the table which he refused as being 13% below the value of the table. I then offered Rs 450 more and the second offer was 5% more than the estimated value. Find the value of the table.
a) Rs 2250 b) Rs 2750 c) Rs 2400 d) Rs 2500
4. I sold a book at a loss of 8%. Had I sold it for Rs 120 more, 7% would have been gained. Find the cost price.
a) Rs 900 b) Rs 800 c) Rs 600 d) Rs 750
5. I sold a book at a loss of 5%. Had I sold it for Rs 72 more, 13% would have been gained. Find the cost price.
a) Rs 450 b) Rs 500 c) Rs 300 d) Rs 400

Answers

1. c; **Hint:** Cost price of the chair $= \frac{84}{10+4} \times 100 = \text{Rs } 600$

$\therefore$ Selling price of the chair $= \frac{600 \times 90}{100} = \text{Rs } 540.$

2. a 3. d 4. b 5. d

Rule 21

Theorem: ***When there are two successive loss of x% and y%, then the resultant loss per cent is given by*** $\left[x + y - \frac{xy}{100}\right]$.

Illustrative Example

Ex.: A sells a good to B at a loss of 20% and B sells it to C at a loss of 25%. Find the resultant loss per cent.

Soln: Applying the above formula, we have, the

required loss per cent $= 20 + 25 - \frac{20 \times 25}{100} = 40\%$.

Exercise

1. A sells a good to B at a loss of 5% and B sells it to C at a loss of 10%. Find the resultant loss per cent.
   a) 15% b) 15.5% c) 14.5% d) 14%
2. A sells a good to B at a loss of 10% and B sells it to C at a loss of 15%. Find the resultant loss per cent.
   a) 26.5% b) 23.5% c) 25.5% d) 25%
3. A sells a good to B at a loss of 5% and B sells it to C at a loss of 15%. Find the resultant loss per cent.
   a) 19.75% b) 20.75% c) 19.25% d) 20%
4. A sells a good to B at a loss of 15% and B sells it to C at a loss of 20%. Find the resultant loss per cent.
   a) 32% b) 38% c) 35% d) 32.5%
5. A sells a good to B at a loss of 10% and B sells it to C at a loss of 20%. Find the resultant loss per cent.
   a) 30% b) 32% c) 26% d) 28%

Answers

1. c 2. b 3. c 4. a 5. d

Rule 22

Theorem: ***A person sells an article at a profit of x%. If he had bought it at y% less and sold it for Rs 'X' more, he would have gained z%, then the cost price is given by Rs***

$$\left[\left(\frac{X}{z - \left(y + x + \frac{zy}{100}\right)}\right) \times 100\right]$$

or Rs $\left[\frac{X \times 100^2}{(100+z)(100-y) - 100(100+x)}\right]$.

Illustrative Example

Ex.: A person sells an article at a profit of 10%. If he had bought it at 10% less and sold it for Rs 3 more, he would have gained 25%. Find the cost price.

Soln: **Detail Method:** Let the actual cost price = Rs 100
Actual selling price at 10% profit = Rs 110
Supposed cost price at 10% less = Rs 90
Supposed selling price at 25% gain

$= \text{Rs } 90 \times \frac{125}{100} = \text{Rs } 112.5$

$\therefore$ the difference in the selling prices
$= \text{Rs } 112.5 - \text{Rs } 110 = \text{Rs } 2.5$

If the difference is Rs 2.5, the CP = Rs 100

If the difference is Rs 3, the CP $= \frac{100}{2.5} \times 3 = \text{Rs } 120.$

Quicker Method: Applying the above formula, we have

$$\text{Cost Price} = \frac{3}{25 - \left(10 + 10 + \frac{25 \times 10}{100}\right)} \times 100$$

$$= \frac{3}{2.5} \times 100 = \text{Rs } 120.$$

Exercise

1. A bookseller sells a book at a profit of 10 per cent. If he had bought it at 4 per cent less and sold it for Rs 6 more, he would have gained $18\frac{3}{4}$ per cent. What did it cost him?
   a) Rs 75 b) Rs 125 c) Rs 150 d) Rs 200
2. A man bought a chair and sold it at a gain of 10%. If he had bought it at 20% less and sold it for Rs 10 more, he would have gained 40%. Find the cost price of the chair.
   a) Rs 500 b) Rs 600 c) Rs 550 d) Rs 650
3. A man bought a chair and sold it at a gain of 20%. If he had bought it at 20% less and sold it for Rs 60 more, he would have gained 50%. Find the cost price of the chair.
   a) Rs 3200 b) Rs 2400
   c) Rs 3000 d) Can't be determined
4. A man bought a chair and sold it at a gain of 15%. If he had bought it at 15% less and sold it for Rs 160 more, he would have gained 40%. Find the cost price of the chair.
   a) Rs 4000 b) Rs 3000 c) Rs 3600 d) Rs 3200
5. A man bought a chair and sold it at a gain of 5%. If he had bought it at 5% less and sold it for Rs 45 more, he would have gained 20%. Find the cost price of the chair.
   a) Rs 400 b) Rs 500 c) Rs 450 d) Rs 550
6. A man sells an article at a profit of 10%, if he had bought it at 5% less and sold it for Rs 160 more he would have

made profit of 20%. Find the cost price of the article.
a) Rs 4000 b) Rs 8000 c) Rs 4500 d) Rs 7500

Answers

1. c 2. a

3. d; **Hint:** Required answer $= \frac{60}{50-\left(20+20+\frac{20\times 50}{100}\right)}\times 100$

$= \frac{60}{50-50}\times 100 =$ Can't be determined

$\left[\because \frac{60}{50-50} = \frac{60}{0} = \text{not defined}\right]$

4. a 5. b 6. a

Rule 23

Theorem: ***A person bought an article and sold it at a loss of x%. If he had bought it for y% less and sold it for Rs X more he would have had a profit of z%, then the cost price of the article is Rs*** $\left[\left(\frac{X}{z+\left(x-y-\frac{zy}{100}\right)}\right)\times 100\right]$.

Illustrative Example

Ex.: A person bought an article and sold it at a loss of 10%. If he had bought it for 20% less and sold it for Rs 55 more he would have had profit of 40%. Find the cost price of the article.

Soln: Following the above theorem, we have cost price

$$= \left[\frac{55}{40+\left(10-20-\frac{40\times 20}{100}\right)}\right]\times 100 = \frac{55}{22}\times 100$$

= Rs 250.

Exercise

1. A person bought an article and sold it at a loss of 5%. If he had bought it for 10% less and sold it for Rs 78 more he would have had profit of 20%. Find the cost price of the article.
a) 600 b) Rs 500 c) Rs 650 d) Rs 550
2. A person bought an article and sold it at a loss of 10%. If he had bought it for 15% less and sold it for Rs 58 more he would have had profit of 40%. Find the cost price of the article.
a) Rs 300 b) Rs 200 c) Rs 150 d) Rs 250
3. A person bought an article and sold it at a loss of 10%. If he had bought it for 20% less and sold it for Rs 270 more he would have had profit of 50%. Find the cost price of the article.
a) Rs 850 b) Rs 950 c) Rs 900 d) Rs 800
4. A person bought an article and sold it at a loss of 15%. If he had bought it for 5% less and sold it for Rs 87 more he would have had profit of 20%. Find the cost price of the article.
a) Rs 200 b) Rs 400 c) Rs 350 d) Rs 300
5. A person bought an article and sold it at a loss of 15%. If he had bought it for 20% less and sold it for Rs 114 more he would have had profit of 30%. Find the cost price of the article.
a) Rs 600 b) Rs 500 c) Rs 700 d) Rs 650
6. A person bought an article and sold it at a loss of 10%. If he had bought it for 20% less and sold it for Rs 55 more, he would have had a profit of 40%. The CP of the article is: **[Central Excise & I Tax, 1988]**
a) Rs 200 b) Rs 225
c) Rs 250 d) None of these

Answers

1. a 2. b 3. c 4. d 5. a 6. c

Rule 24

Theorem: ***A man buys an article and sells it at a profit of x%. If he had bought it at y% less and sold it for Rs X less, he would have gained z%, then the cost price is Rs*** $\left[\frac{X}{x+\left(y-z+\frac{zy}{100}\right)}\right]\times 100$.

Illustrative Example

Ex.: A man sells an article at a profit of 20%. If he bought it at 20% less and sold it for Rs 75 less, he would have gained 25%. What is the cost price?

Soln: Following the above theorem, we have Cost price

$$= \frac{75}{20+\left(20-25+\frac{25\times 20}{100}\right)}\times 100 = \frac{75}{20}\times 100$$

= Rs 375.

Exercise

1. A man sells out an article at a profit of 25%. Had he bought it at 25% less and sold it for Rs 25 less, he would have still gained 25%. Find the cost price.
a) Rs 90 b) Rs 80 c) Rs 70 d) Rs 60
2. A man sells an article at a gain of 15%. If he had bought it at 10% less and sold it for Rs 4 less, he would have gained 25%. Find the cost price.
a) Rs 160 b) Rs 260 d) Rs 610 d) Rs 180
3. A man sells an article at a profit of 10%. If he bought it at 10% less and sold it for Rs 110 less, he would have

gained 10%. What is the cost price?
a) Rs 1100 b) Rs 1050 c) Rs 1000 d) Rs 950

4. A man sells an article at a profit of 30%. If he bought it at 15% less and sold it for Rs 54.75 less, he would have gained 10%. What is the cost price?
a) Rs 1500 b) Rs 1600 c) Rs 150 d) Rs 160

5. A man sells an article at a profit of 40%. If he bought it at 30% less and sold it for Rs 140 less, he would have gained 20%. What is the cost price?
a) Rs 250 b) Rs 150 c) Rs 200 d) Rs 300

Answers

1. b 2. a 3. c 4. c 5. a

Rule 25

Theorem: ***A dealer sells an article at a loss of x%. Had he sold it for Rs X more, he would have gained y%. In order to gain z%, the selling price and the initial cost price of the article are given by Rs*** $\left[\dfrac{X(100+z)}{x+y}\right]$***, or***

$$\textit{Selling Price} = \frac{\textit{More rupees } (100 + \%\ \textit{final gain})}{\%\ \textit{gain} + \%\ \textit{loss}}$$ ***and***

Rs $\left[\dfrac{100X}{x+y}\right]$ ***respectively.***

Illustrative Example

Ex.: A dealer sold a radio at a loss of 2.5%. Had he sold it for Rs 100 more, he would have gained 7.5%. For what value should he sell it in order to gain 12½%? Find the initial cost price also.

Soln: **Detail Method:** Suppose he bought the radio for Rs x. Then selling price at 2.5% loss

$$= \text{Rs } x\left(\frac{100-2.5}{100}\right) = \frac{97.5x}{100}$$

and selling price at 7.5% gain = Rs x $\left(\dfrac{100+7.5}{100}\right)$

$$= \text{Rs } \frac{107.5x}{100}$$

From the question, $\dfrac{107.5x}{100} - \dfrac{97.5x}{100} = \text{Rs } 100$

or, 10x = 100 × 100 ∴ x = Rs 1000

Therefore to gain 12.5%, he should sell for

$$\text{Rs } 1000\left(\frac{100+12.5}{100}\right) = \text{Rs } 1125$$

Quicker Method: We have the following formula,

$$\text{Selling Price} = \frac{\text{More rupees } (100 + \%\text{ final gain})}{\%\text{ gain} + \%\text{ loss}}$$

$$= \frac{100(112.5)}{7.5+2.5} = \text{Rs } 1125.$$

$$\text{Cost Price} = \frac{100 \times 100}{7.5+2.5} = \text{Rs } 1000.$$

Exercise

1. A dealer sold a chair at a loss of 5%. Had he sold it for Rs 150 more, he would have gained 10%. For what value should he sell it in order to gain 20%? Find the initial cost price also.
a) Rs 1200, Rs 1000 b) Rs 1300, Rs 1000
c) Rs 1000, Rs 850 d) Rs 1000, Rs 950

2. A dealer sold a table at a loss of 10%. Had he sold it for Rs 75 more, he would have gained 15%. For what value should he sell it in order to gain 25%?
a) Rs 475 b) Rs 575 c) Rs 375 d) Rs 300

3. A dealer sold a pen at a loss of 15%. Had he sold it for Rs 35 more, he would have gained 20%. For what value should he sell it in order to gain 5%?
a) Rs 150 b) Rs 105 c) Rs 125 d) Rs 115

4. A dealer sold a TV set at a loss of 20%. Had he sold it for Rs 9000 more, he would have gained 25%. For what value should he sell it in order to gain 30%?
a) Rs 27000 b) Rs 36000
c) Rs 28000 d) Rs 26000

5. A dealer sold a camera at a loss of 13%. Had he sold it for Rs 6300 more, he would have gained 17%. For what value should he sell it in order to gain 32%?
a) Rs 72720 b) Rs 27270
c) Rs 72270 d) Rs 27720

Answers

1. a 2. c 3. b 4. d 5. d

Rule 26

Theorem: ***A dealer sells an article at a profit of x%. If he increases its price by Rs X, he earns a profit of y%, then the cost price and the initial selling price of the article are Rs*** $\left[\dfrac{100X}{y-x}\right]$ ***and Rs*** $\left[\dfrac{(100+x)X}{y-x}\right]$ ***respectively.***

Illustrative Example

Ex.: A hawker sells oranges at a profit of 25 per cent. If he increases the selling price of each orange by 30 paise, he earns a profit of 40%. Find the cost price and the initial selling price of each orange.

Soln: Following the above theorem,

$$\text{Cost Price} = \frac{100 \times 30}{40-25} = 200 \text{ paise}$$

$$\text{Initial Selling Price} = \frac{(100+25) \times 30}{40-25} = 250 \text{ paise.}$$

Exercise

1. A dealer sells an article at a profit of 15%. If he increases its price by Rs 120, he earns a profit of 20%. Find the cost price and the initial selling price respectively.
a) Rs 2760, Rs 2400　b) Rs 2400, Rs 2760
c) Rs 2400, Rs 2670　d) Rs 2670, Rs 2400
2. A dealer sells an article at a profit of 10%. If he increases its price by Rs 49, he earns a profit of 17%. Find the cost price and the initial selling price respectively.
a) Rs 700, Rs 770　b) Rs 770, Rs 700
c) Rs 600, Rs 660　d) Rs 660, Rs 600
3. A dealer sells an article at a profit of 12%. If he increases the its price by Rs 81, he earns a profit of 21%. Find the cost price and the initial selling price respectively.
a) Rs 1008, Rs 800　b) Rs 800, Rs 1008
c) Rs 900, Rs 1008　d) Rs 900, Rs 1000
4. A hawker sells oranges at a profit of $12\frac{1}{2}$ per cent. If he increases the selling price of each orange by 60 paise, he earns a profit of 20%. Find the cost price and the initial selling price of each orange.
a) 900 P, 700 P　b) 700 P, 900 P
c) 900 P, 800 P　d) 800 P, 900 P
5. A hawker sells oranges at a profit of 30 per cent. If he increases the selling price of each orange by 20 paise, he earns a profit of 50%. Find the cost price and the initial selling price of each orange.
a) 100 P, 130 P　b) 130 P, 100 P
c) 150 P, 100 P　d) 100 P, 150 P

Answers

1. b　2. a　3. c　4. d　5. a

Rule 27

Theorem: ***A dealer sells an article at a loss of x%. If he sells it at Rs X more he makes a profit of y%, then the cost price and the initial selling price of the article are Rs*** $\left[\frac{100X}{x+y}\right]$ ***and Rs*** $\left[\frac{(100-x)X}{x+y}\right]$ ***respectively.***

Illustrative Example

Ex.: A man had to sell wheat at a loss of 10%. If he increases the price by Rs 5 per kg, he would make a profit of 20%. Find the cost price and initial selling price per kg of wheat.

Soln: Applying the above theorem,

the required cost price $= \frac{100\times5}{10+20} = \frac{50}{3} = \text{Rs } 16\frac{2}{3}$ per kg and

the initial selling price $= \frac{(100-10)\times5}{10+20} = \text{Rs } 15$ per kg

Exercise

1. A man had to sell rice at a loss of 5%. If he increases the price by Rs 3 per kg, he would make a profit of 10%. Find the cost price and initial selling price per kg of rice.
a) Rs 20, Rs 19　b) Rs 19, Rs 20
c) Rs 18, Rs 19　d) Rs 19, Rs 18
2. A man had to sell coffee at a loss of 15%. If he increases the price by Rs 7 per kg, he would make a profit of 20%. Find the cost price and initial selling price per kg of coffee.
a) Rs 18, Rs 20　b) Rs 20, Rs 18
c) Rs 20, Rs 17　d) Rs 17, Rs 20
3. A man had to sell tea at a loss of 5%. If he increases the price by Rs 4 per kg, he would make a profit of 15%. Find the cost price and initial selling price per kg of tea.
a) Rs 19, Rs 21　b) Rs 20, Rs 19
c) Rs 19, Rs 20　d) Can't be determined
4. A man had to sell pulse at a loss of 10%. If he increases the price by Rs 5 per kg, he would make a profit of 15%. Find the cost price and initial selling price per kg of pulse.
a) Rs 20, Rs 18　b) Rs 18, Rs 20
c) Rs 15, Rs 18　d) Rs 18, Rs 15

Answers

1. a　2. c　3. b　4. a

Rule 28

Theorem: ***A dealer sells an article at a profit of x%. Had he sold it for Rs x less, he would have still gained y%. In order to gain z%, selling price and the initial cost price of the article are given by*** $\left[\frac{X(100+z)}{x-y}\right]$ ***and*** $\left[\frac{100X}{x-y}\right]$ ***respectively.***

Illustrative Example

Ex.: A dealer sold a radio at a profit of 20%. Had he sold it for Rs 100 less, he would have gained 15%. For what value should he sell it in order to gain 25%? Also find the initial cost price?

Soln: Applying the above theorem, we have the

selling price $= \frac{100\times(100+25)}{20-15} = \text{Rs } 2500$ and

initial cost price $= \frac{100\times100}{20-15} = \text{Rs } 2000.$

Exercise

1. A dealer sold a TV set at a profit of 10%. Had he sold it for Rs 200 less, he would have gained 8%. For what value should he sell it in order to gain 5%?
a) Rs 12500　b) Rs 10500
c) Rs 9500　d) Data inadequate
2. A dealer sold a camera at a profit of 12%. Had he sold it

for Rs 240 less, he would have gained 7%. For what value should he sell it in order to gain 20%?
a) Rs 5760 b) Rs 6760 c) Rs 7560 d) Rs 6750

3. A dealer sold a sofa set at a profit of 18%. Had he sold it for Rs 450 less, he would have gained 12%. For what value should he sell it in order to gain 14%?
a) Rs 5880 b) Rs 5850 c) Rs 8550 d) Rs 8850

Answers

1. b 2. a 3. c

Rule 29

Theorem: ***A dealer sells an article at a profit of x%. Had he sold it for Rs X more, he would have gained y%. In order to gain z%, selling price and the initial cost price of the article are given by Rs*** $\left[\frac{X(100+z)}{y-x}\right]$ ***and Rs*** $\left[\frac{100X}{y-x}\right]$ ***respectively.***

Illustrative Example

Ex.: A dealer sold a radio at a profit of 15%. Had he sold it for Rs 100 more, he would have gained 20%. For what value should he sell it in order to gain 25%? Also find the initial cost price.

Soln: Following the above theorem, we have

$$\text{selling price} = \frac{100(100+25)}{20-15} = \text{Rs } 2500 \text{ and}$$

$$\text{initial cost price} = \frac{100\times100}{20-15} = \text{Rs } 2000.$$

Exercise

1. A dealer sold a camera at a profit of 5%. Had he sold it for Rs 120 more, he would have gained 15%. For what value should he sell it in order to gain 10%?
a) Rs 1320 b) Rs 1330 c) Rs 1230 d) Data inadequate

2. A dealer sold a TV set at a profit of 20%. Had he sold it for Rs 525 more, he would have gained 25%. For what value should he sell it in order to gain 40%?
a) Rs 17400 b) Rs 14700 c) Rs 15700 d) Rs 17500

3. A dealer sold a pen at a profit of 12%. Had he sold it for Rs 3 more, he would have gained 18%. For what value should he sell it in order to gain 28%?
a) Rs 68 b) Rs 60 c) Rs 64 d) Rs 54

Answers

1. a 2. b 3. c

Rule 30

Theorem: ***A dealer sells an article at a loss of x%. Had he sold it for Rs X more, he would have still lost y%. In order to gain z%, the selling price and the initial cost price of the article are given by Rs*** $\left[\frac{X(100+z)}{x-y}\right]$ ***and Rs*** $\left[\frac{100X}{x-y}\right]$ ***respectively.***

Illustrative Example

Ex.: A dealer sold a radio at a loss of 20%. Had he sold it for Rs 100 more, he would have lost 15%. For what value should he sell it in order to gain 25%. Also find the initial cost price.

Soln: Applying the above theorem, we have

$$\text{selling price} = \frac{100\times(100+25)}{20-15} = \text{Rs } 2500 \text{ and}$$

$$\text{initial cost price} = \frac{100\times100}{20-15} = \text{Rs } 2000.$$

Exercise

1. A dealer sold a CD player at a loss of 25%. Had he sold it for Rs 135 more, he would have lost 20%. For what value should he sell it in order to gain 35%.
a) Rs 3645 b) Rs 3465 c) Rs 3545 d) Rs 3655

2. A dealer sold a TV set at a loss of 27%. Had he sold it for Rs 500 more, he would have lost 17%. For what value should he sell it in order to gain 40%.
a) Rs 6000 b) Rs 5000 c) Rs 7000 d) Rs 8000

3. A dealer sold a chair at a loss of 12%. Had he sold it for Rs 150 more, he would have lost 3%. For what value should he sell it in order to gain 8%.
a) Rs 1600 b) Rs 1800 c) Rs 2100 d) Rs 1500

Answers

1. a 2. c 3. b

Rule 31

Theorem: ***If an article is sold at a profit of x%. If both the cost price and selling price are Rs A less, the profit would be y% more, then the cost price is*** $\left[\frac{(x+y)\times A}{y}\right]$.

In other words, cost price

$$= \frac{[\text{Initial Profit \%} + \text{Increase in profit \%}]\times A}{\text{Increase in profit \%}}$$

Illustrative Example

Ex.: An article is sold at a profit of 20%. If both the cost price and selling price are Rs 100 less, the profit would be 4% more. Find the cost price.

Soln: **Detail Method:** Suppose the cost price of that article is Rs x.

$$\text{The selling price} = \text{Rs } x\left(\frac{120}{100}\right)$$

New cost price and selling price is Rs (x - 100) and Rs $\left[x\left(\frac{120}{100}\right)-100\right]$ respectively.

$$\text{New profit} = \text{Rs}\left[x\left(\frac{120}{100}-100\right)-(x-100)\right]$$

$$= \text{Rs}\left[x\left\{\frac{120}{100}-1\right\}\right] = \text{Rs x}\left(\frac{20}{100}\right)$$

$\therefore$ New percentage profit

$$= \frac{x\left(\frac{20}{100}\right)}{x-100}\times 100 = \frac{20x}{x-100}\%$$

We are also given that the new percentage of profit $=20+4=24\%$

or, $\frac{20x}{x-100} = 24$ or, $4x=2400$

$\therefore x=600$

Thus cost of the article = Rs 600

Quicker Method: When cost price and selling price are reduced by the same amount (say A) then Cost price

$$= \frac{[\text{Initial profit \%} + \text{Increase in profit \%}]\times A}{\text{Increase in profit \%}}$$

In this case, Cost price = Rs $\frac{(20+4)\times 100}{4}$ = Rs 600

Exercise

1. An article is sold at a profit of 10%. If both the cost price and selling price are Rs 25 less, the profit would be 5% more. Find the cost price.
   a) Rs 65 b) Rs 60 c) Rs 75 d) Rs 72
2. An article is sold at a profit of 13%. If both the cost price and selling price are Rs 36 less, the profit would be 6% more. Find the cost price.
   a) Rs 119 b) Rs 114 c) Rs 108 d) Rs 112
3. An article is sold at a profit of 14%. If both the cost price and selling price are Rs 132 less, the profit would be 12% more. Find the cost price.
   a) Rs 286 b) Rs 268 c) Rs 266 d) Rs 276
4. An article is sold at a profit of 15%. If both the cost price and selling price are Rs 56 less, the profit would be 7% more. Find the cost price.
   a) Rs 176 b) Rs 186 c) Rs 156 d) Rs 196
5. An article is sold at a profit of 14%. If both the cost price and selling price are Rs 117 less, the profit would be 9% more. Find the cost price.
   a) Rs 199 b) Rs 399
   c) Rs 299 d) Data inadequate

Answers

1. c 2. b 3. a 4. a 5. c

Rule 32

Theorem: ***A dealer sells an article at a profit of x%. If he reduces the price by Rs X, he incurs a loss of y%, then the cost price and the initial selling price of the article are Rs***

$\left[\frac{100X}{x+y}\right]$ ***and Rs*** $\left[\frac{(100+x)X}{x+y}\right]$ ***respectively.***

Illustrative Example

Ex.: A man sells wheat at a profit of 5%. If he reduces the price by Rs 4 per kg, he would make a loss of 15%. Find the cost price and the initial selling price per kg of wheat.

Soln: Applying the above theorem,

cost price $= \frac{100\times 4}{15+5}$ = Rs 20 and

selling price $= \frac{(100+5)4}{15+5}$ = Rs 21.

Exercise

1. A man sells rice at a profit of 10%. If he reduces the price by Rs 3 per kg, he would make a loss of 20%. Find the initial selling price per kg of rice.
   a) Rs 10 b) Rs 11 c) Rs 9 d) Rs 21
2. Ram sells pulse at a profit of 15%. If he reduces the price by Rs 4 per kg, he would make a loss of 25%. Find the initial selling price per kg of pulse.
   a) Rs 11 b) Rs 11.25 c) Rs 11.5 d) Rs 11.75
3. Anubhav sells tea at a profit of 20%. If he reduces the price by Rs 10 per kg, he would make a loss of 30%. Find the initial selling price per kg of tea.
   a) Rs 12 b) Rs 16 c) Rs 20 d) Rs 24
4. Shakshi sells coffee at a profit of 12%. If he reduces the price by Rs 29 per kg, he would make a loss of 17%. Find the initial selling price per kg of coffee.
   a) Rs 112 b) Rs 117 c) Rs 129 d) Rs 113
5. Shashank sells salt at a profit of 18%. If he reduces the price by Rs 8 per kg, he would make a loss of 22%. Find the initial selling price per kg of salt.
   a)l Rs 23.5 b) Rs 32.6 c) Rs 32.5 d) Rs 23.6

Answers

1. b 2. c 3. d 4. a 5. d

Rule 33

Theorem: ***A dealer sells an article at a profit of x%. If he reduces the price by Rs X, he makes a profit of y% then the cost price and the initial selling price of the article are Rs***

$\left[\frac{100X}{x-y}\right]$ ***and Rs*** $\left[\frac{(100+x)X}{x-y}\right]$ ***respectively.***

Illustrative Example

Ex.: A hawker sells oranges at a profit of 40%. If he reduces the selling price of each orange by 30 paise, he earns a profit of 25%. Find the cost price and the initial selling price of each orange.

Soln: Following the above theorem, we have

$$\text{cost price} = \frac{100 \times 30}{40-25} = 200 \text{ paise and}$$

$$\text{selling price} = \frac{(100+40)30}{40-25} = 280 \text{ paise.}$$

Exercise

1. A hawker sells mango at a profit of 35%. If he reduces the selling price of each mango by 10 paise, he earns a profit of 20%. Find the initial selling price of each mango.
a) 90 paise b) 80 paise c) 95 paise d) Re 1
2. A hawker sells apple at a profit of 40%. If he reduces the selling price of each apple by 40 paise, he earns a profit of 20%. Find the initial selling price of each apple.
a) 140 paise b) 280 paise c) 160 paise d) 260 paise
3. A hawker sells orange at a profit of 25%. If he reduces the selling price of each orange by 60 paise, he earns a profit of 5%. Find the initial selling price of each orange.
a) 375 paise b) 400 paise c) 475 paise d) 360 paise

Answers

1. a 2. b 3. a

Rule 34

Theorem: ***An article is sold at P% profit. If its cost price is lowered by Rs c and at the same time its selling price is also lowered by Rs s, and the percentage of profit increases by p%. Then the cost price of that article is***

$$= \frac{c(P+p)-100(s-c)}{p}.$$

Illustrative Example

Ex: An article is sold at 20% profit. If its cost price and selling price are less by Rs 10 and Rs 5 respectively and the percentage profit increases by 10%. Find the cost price.

Soln: Using the above formula:

$$\frac{10(20+10)-100(5-10)}{10} = \frac{800}{10} = \text{Rs } 80.$$

Exercise

1. An article is sold at 10% profit. If its cost price and selling price are less by Rs 5 and Rs 2.5 respectively and the percentage profit increases by 5%. Find the cost price.
a) Rs 40 b) Rs 45 c) Rs 60 d) Rs 65
2. An article is sold at 15% profit. If its cost price and selling price are less by Rs 10 and Rs 7.5 respectively and the percentage profit increases by 10%. Find the cost price.
a) Rs 50 b) Rs 75 c) Rs 100 d) Rs 40
3. An article is sold at 25% profit. If its cost price and selling price are less by Rs 20 and Rs 17.5 respectively and the percentage profit increases by 15%. Find the cost price.
a) Rs 75 b) Rs 80 c) Rs 70 d) Rs 65
4. An article is sold at $12\frac{1}{2}$% profit. If its cost price and selling price are less by Rs 10 and Rs 8.85 respectively and the percentage profit increases by $7\frac{1}{2}$%. Find the cost price.
a) Rs 21 b) Rs 32 c) Rs 52 d) Rs 42
5. An article is sold at 30% profit. If its cost price and selling price are less by Rs 25 and Rs 19 respectively and the percentage profit increases by 20%. Find the cost price.
a) Rs 97 b) Rs 92.5 c) Rs 98.5 d) Rs 98

Answers

1. d 2. a 3. c 4. d 5. b

Rule 35

Theorem: ***An article is sold at P% profit. If its cost price and selling price are increased by Rs c and Rs s respectively and percentage profit decreases by p%, then the cost price of that article is Rs*** $\left[\frac{c(P-p)-100(s-c)}{p}\right]$.

Illustrative Example

Ex: An article is sold at 25% profit. If its CP and SP are increased by Rs 20 and Rs 4 respectively, the percentage of profit decreases by 15%. Find the cost price.

Soln: **Quicker Method:** Applying the above formula,

$$\text{Cost Price} = \frac{20(25-15)-100(4-20)}{15} = \frac{1800}{15}$$

$$= \text{Rs } 120.$$

Exercise

1. An article is sold at 10% profit. If its CP and SP are increased by Rs 15 and Rs 5 respectively, the percentage of profit decreases by 5%. Find the cost price.
a) Rs 215 b) Rs 213 c) Rs 512 d) Rs 312
2. An article is sold at $12\frac{1}{2}$% profit. If its CP and SP are increased by Rs 10 and Rs 2 respectively, the percentage of profit decreases by $7\frac{1}{2}$%. Find the cost price.

a) Rs $116\frac{2}{3}$ b) Rs $114\frac{1}{3}$ c) Rs $113\frac{1}{3}$ d) Rs $113\frac{2}{3}$

3. An article is sold at 5% profit. If its CP and SP are increased by Rs 7.50 and Rs 2.50 respectively, the percentage of profit decreases by $2\frac{1}{2}$%. Find the cost price.

a) Rs 215 b) Rs 207.5 c) Rs 215.5 d) Rs 270.5

4. An article is sold at 25% profit. If its CP and SP are increased by Rs 16 and Rs 6 respectively, the percentage of profit decreases by 20%. Find the cost price.

a) Rs 45 b) Rs 58
c) Rs 54 d) Data inadequate

5. An article is sold at 50% profit. If its CP and SP are increased by Rs 32 and Rs 12 respectively, the percentage of profit decreases by 40%. Find the cost price.

a) Rs 58 b) Rs 60
c) Rs 68 d) Rs 54

Answers

1. a 2. c 3. b 4. c 5. a

Rule 36

Theorem: ***If a person sells an item 'A' at a profit of x% and the other item 'B' at a loss of y% but on the whole he gains Rs X. On the other hand if he sells the item 'A' at a loss of y% and the item 'B' at a profit of x% then he neither gains nor loses, then the (i) Cost Price of A is Rs*** $\left(\frac{Xx}{x^2-y^2}\times 100\right)$ ***and the (ii) Cost Price of B is Rs*** $\left(\frac{Xy}{x^2-y^2}\times 100\right)$.

Illustrative Example

Ex.: A person sells his table at a profit of $12\frac{1}{2}$% and the chair at a loss of $8\frac{1}{3}$% but on the whole he gains Rs 25. On the other hand if he sells the table at a loss of $8\frac{1}{3}$% and the chair at a profit of $12\frac{1}{2}$% then he neither gains nor loses. Find the cost price of the table and the chair.

Soln: Applying the above theorem, we have

$$\text{Cost price of the table} = \frac{25\times\frac{25}{2}}{\left(\frac{25}{2}\right)^2-\left(\frac{25}{3}\right)^2}\times 100$$

$$= \text{Rs } 360.$$

Cost price of the chair

$$= \frac{25\times\frac{25}{3}}{\left(\frac{25}{2}\right)^2-\left(\frac{25}{3}\right)^2}\times 100 = \text{Rs } 240.$$

Exercise

1. A person sells his table at a profit of 25% and the chair at a loss of 20% but on the whole he gains Rs 18. On the other hand if he sells the table at a loss of 20% and the chair at a profit of 25% then he neither gains nor loses. Find the cost price of the table and the chair.

a) Rs 200, Rs 160 b) Rs 160, Rs 200
c) Rs 250, Rs 180 d) Rs 210, Rs 170

2. A person sells his table at a profit of $12\frac{1}{2}$% and the chair at a loss of 10% but on the whole he gains Rs 9. On the other hand if he sells the table at a loss of 10% and the chair at a profit of $12\frac{1}{2}$% then he neither gains nor loses. Find the cost price of the table and the chair.

a) Rs 210, Rs 170 b) Rs 200, Rs 160
c) Rs 250, Rs 160 d) Data inadequate

3. A person sells his table at a profit of 15% and the chair at a loss of 5% but on the whole he gains Rs 16. On the other hand if he sells the table at a loss of 5% and the chair at a profit of 15% then he neither gains nor loses. Find the cost price of the table and the chair.

a) Rs 120, Rs 40 b) Rs 150, Rs 90
c) Rs 40, Rs 120 d) Data inadequate

4. A person sells his table at a profit of 30% and the chair at a loss of 10% but on the whole he gains Rs 32. On the other hand if he sells the table at a loss of 10% and the chair at a profit of 30% then he neither gains nor loses. Find the cost price of the table and the chair.

a) Rs 140, Rs 20 b) Rs 150, Rs 40
c) Rs 120, Rs 40 d) Rs 180, Rs 40

5. A person sells his table at a profit of 40% and the chair at a loss of 20% but on the whole he gains Rs 60. On the other hand if he sells the table at a loss of 20% and the chair at a profit of 40% then he neither gains nor loses. Find the cost price of the table and the chair.

a) Rs 200, Rs 100 b) Rs 100, Rs 200
c) Rs 250, Rs 150 d) Rs 150, Rs 250

Answers

1. a 2. b 3. a 4. c 5. a

Rule 37

Theorem: ***If cost price of x articles is equal to the selling price of y articles, then the profit percentage*** = $\frac{x-y}{y}\times 100\%$.

Illustrative Example

Ex.: The cost price of 10 articles is equal to the selling price of 9 articles. Find the profit per cent.

Soln: **Detail Method:** Let the cost price of 1 article be Re 1

$\therefore$ Cost of 10 articles = Rs 10

$\therefore$ Selling price of 9 articles = Rs 10

$\therefore$ Selling price of 10 articles = Rs $\frac{10\times10}{9}$ = Rs $\frac{100}{9}$

$\therefore$ gain on Rs 10 = Rs $\frac{100}{9}$ − Rs 10 = Rs $\frac{10}{9}$

$\therefore$ gain on Rs 100 = Rs $\frac{100}{9}$ = Rs $11\frac{1}{9}$

$\therefore$ profit per cent is $11\frac{1}{9}\%$.

Another Method:

To avoid much calculation we should suppose that the total investment = 10 × 9 = Rs 90

Then cost price of 1 article = $\frac{90}{10}$ = Rs 9

and selling price of 1 article = $\frac{90}{9}$ = Rs 10

$\therefore$ % profit = $\frac{10-9}{9}\times100=\frac{1}{9}\times100=11\frac{1}{9}\%$

By Direct Formula (given in theorem)

% profit = $\frac{10-9}{9}\times100=11\frac{1}{9}\%$

Exercise

1. The cost price of 24 articles is equal to the selling price of 18 articles. Find the gain per cent.
   a) $33\frac{1}{3}\%$ b) 25% c) 24% d) $33\frac{2}{3}\%$
2. I sell 16 articles for the some money as I paid for 20. What is my gain per cent?
   a) 24% b) 25% c) 30% d) 35%
3. If the selling price of $\frac{4}{5}$ of a number of books be equal to the cost price of the whole, find the profit per cent.
   a) 20% b) 25% c) $33\frac{1}{3}\%$ d) $16\frac{2}{3}\%$
4. By selling 66 metres of cloth, I gain the cost of 22 metres. Find my gain per cent.
   a) 25% b) $66\frac{2}{3}\%$ c) $33\frac{1}{3}\%$ d) 20%
5. If cost price of 12 articles is equal to the selling price of 8 articles. Find the profit per cent.
   a) 50% b) 40% c) 60% d) 30%
6. A man sells 320 mangoes at the cost price of 400 mangoes. His gain per cent is:
   [Asstt. Grade Exam. 1987]
   a) 10% b) 25% c) 15% d) 20%

Answers

1. a 2. b

3. b; **Hint:** Required answer = $\frac{1-\frac{4}{5}}{\frac{4}{5}}\times100=25\%$

4. c; **Hint:** Here, $x-y=22$ metres.

5. a 6. b

Rule 38

Theorem: ***An article is sold at x% profit. If its cost price is increased by Rs A and at the same time if its selling price is also increased by Rs B, the percentage profit becomes y%, then the cost price of the article is*** $\left[\frac{\left(A-B+\frac{Ay}{100}\right)}{(x-y)}\right]\times100$.

Note: *(x - y) is the difference in the profits.*

Illustrative Example

Ex.: An article is sold at 20% profit. If its cost price is increased by Rs 50 and at the same time if its selling price is also increased by Rs 30, the percentage of profit becomes $16\frac{2}{3}\%$. Find the cost price. [The last statement of the question may be written as "the percentage of profit decreases by $3\frac{1}{3}\%$ ie $\left(20\%-16\frac{2}{3}\%=3\frac{1}{3}\%\right)$"].

Soln: Applying the above formula, we have

$$\text{Cost Price}=\frac{50-30+\frac{50\times50}{300}}{\frac{10}{3}}\times100=\text{Rs }850.$$

Exercise

1. An article is sold at 10% profit. If its cost price is increased by Rs 25 and at the same time if its selling price

is also increased by Rs 15, the percentage of profit becomes 5%. Find the cost price.
a) Rs 325 b) Rs 525 c) Rs 225 d) Rs 255

2. An article is sold at 25% profit. If its cost price is increased by Rs 60 and at the same time if its selling price is also increased by Rs 40, the percentage of profit becomes 15%. Find the cost price.
a) Rs 390 b) Rs 290
c) Rs 190 d) Data inadequate

3. An article is sold at $12\frac{1}{2}\%$ profit. If its cost price is increased by Rs 30 and at the same time if its selling price is also increased by Rs 20, the percentage of profit becomes $7\frac{1}{2}\%$. Find the cost price.
a) Rs 225 b) Rs 245 c) Rs 345 d) Rs 249

4. An article is sold at $22\frac{1}{2}\%$ profit. If its cost price is increased by Rs 40 and at the same time if its selling price is also increased by Rs 35, the percentage of profit becomes 15%. Find the cost price.
a) Rs $146\frac{2}{3}$ b) Rs $156\frac{2}{3}$ c) Rs $146\frac{1}{3}$ d) Rs $156\frac{1}{3}$

5. An article is sold at 45% profit. If its cost price is increased by Rs 80 and at the same time if its selling price is also increased by Rs 70, the percentage of profit becomes 30%. Find the cost price.
a) Rs 246 b) Rs $226\frac{2}{3}$ c) Rs $293\frac{1}{3}$ d) Rs $226\frac{1}{3}$

Answers

1. c 2. b 3. b 4. a 5. b

Rule 39

Theorem: ***If x part is sold at m% profit, y part is sold at n% profit, z part is sold at p% profit and Rs P is earned as overall profit then the value of total consignment =*** $\frac{P \times 100}{xm + ny + pz}$, ***(Where, x + y + z = total).***

Illustrative Example

Ex.: $\frac{1}{3}$ of a commodity is sold at 15% profit, $\frac{1}{4}$ is sold at 20% profit and the rest at 24% profit. If a total profit of Rs 62 is earned, then find the value of the commodity.

Soln: **Detail Method:** Suppose the value of the commodity was Rs x. Then $\frac{x}{3}$ was sold at 15% profit, $\frac{x}{4}$ was sold at 20% profit and $x - \left(\frac{x}{3} + \frac{x}{4}\right) = \frac{5x}{12}$ was sold at 25% profit.

Now, profit $\frac{x}{3}\left(\frac{15}{100}\right) + \frac{x}{4}\left(\frac{20}{100}\right) + \frac{5x}{12}\left(\frac{24}{100}\right) = Rs\ 62$

or, $\frac{x}{20} + \frac{x}{20} + \frac{x}{10} = 62$

or, $\frac{4x}{20} = 62 \quad \therefore x = \frac{62 \times 20}{4} = \text{Rs } 310.$

Quicker Method: Applying the above rule, we have the value of commodity

$$= \frac{62 \times 100}{\frac{1}{3} \times 15 + \frac{1}{4} \times 20 + \frac{5}{12} \times 24} = \frac{62 \times 100}{5 + 5 + 10}$$

$= \text{Rs } 310.$

Exercise

1. $\frac{2}{3}$ of a commodity is sold at 30% profit, $\frac{1}{4}$ is sold at 16% profit and the rest at 12% profit. If a total profit of Rs 75 is earned, then find the value of the commodity.
a) Rs 300 b) Rs 250 c) Rs 350 d) Rs 325

2. $\frac{3}{4}$ of a commodity is sold at 16% profit, $\frac{1}{6}$ is sold at 18% profit and the rest at 60% profit. If a total profit of Rs 67 is earned, then find the value of the commodity.
a) Rs 345 b) Rs 335 c) Rs 385 d) Rs 435

3. $\frac{1}{3}$ of a commodity is sold at 24% profit, $\frac{1}{12}$ is sold at 24% profit and the rest at 60% profit. If a total profit of Rs 90 is earned, then find the value of the commodity.
a) Rs 100 b) Rs 300 c) Rs 200 d) Rs 150

4. $\frac{1}{5}$ of a commodity is sold at 25% profit, $\frac{1}{10}$ is sold at 40% profit and the rest at 30% profit. If a total profit of Rs 150 is earned, then find the value of the commodity.
a) Rs 550 b) Rs 450
c) Rs 650 d) Rs 500

5. $\frac{3}{5}$ of a commodity is sold at 15% profit, $\frac{1}{5}$ is sold at 15% profit and the rest at 15% profit. If a total profit of Rs 37.5 is earned, then find the value of the commodity.
a) Rs 250 b) Rs 200
c) Rs 350 d) Rs 300

Answers

1. a 2. b 3. c 4. d 5. a

Rule 40

Theorem: ***If x part is sold at m% profit and the rest y part is sold at n% loss and Rs P is earned as overall profit then the value of the total consignment is Rs*** $\left(\frac{P\times 100}{xm-ny}\right)$.

Illustrative Example

Ex.: Two-thirds of a consignment was sold at a profit of 6% and the rest at a loss of 3%. If there was an overall profit of Rs 540, find the value of the consignment.

Soln: **Detail Method:** Suppose the value of consignment was Rs x. Then $\frac{2}{3}$ x was sold at 6% profit, i.e., for Rs $\frac{2}{3}x\left(\frac{106}{100}\right)$.

And $\frac{1}{3}$ part is sold at Rs $\frac{x}{3}\left(\frac{97}{100}\right)$.

Now, profit $=\left\{\frac{2x}{3}\left(\frac{106}{100}\right)+\frac{x}{3}\left(\frac{97}{100}\right)\right\}-x$

$=\frac{212x+97x}{300}-x=\frac{309x-300x}{300}=\frac{9x}{300}$

Since, $\frac{9x}{300}=540$

$\therefore$ x = Rs $\frac{540\times 300}{9}$ = Rs 18,000.

Quicker Method: Applying the above rule, we have

the value of Consignment $=\frac{\text{Total Profit}\times 100}{\%\text{ profit}\times\frac{2}{3}-\%\text{ loss}\times\frac{1}{3}}$

$=\frac{540\times 100}{6\times\frac{2}{3}-3\times\frac{1}{3}}$ = Rs 18,000.

Exercise

1. 3/5th of a consignment was sold at a profit of 15% and the rest at a loss of 10%. If there was an overall profit of Rs 32, find the value of the consignment.
   a) Rs 640 b) Rs 960 c) Rs 460 d) Rs 740
2. 4/5th of a consignment was sold at a profit of 25% and the rest at a loss of 35%. If there was an overall profit of Rs 910, find the value of the consignment.
   a) Rs 6000 b) Rs 7000 c) Rs 7500 d) Rs 8000
3. 3/8th of a consignment was sold at a profit of 32% and the rest at a loss of 16%. If there was an overall profit of Rs 112, find the value of the consignment.
   a) Rs 6500 b) Rs 5700 c) Rs 5600 d) Data inadequate
4. 4/7th of a consignment was sold at a profit of 42% and the rest at a loss of 28%. If there was an overall profit of Rs 84, find the value of the consignment.
   a) Rs 650 b) Rs 750 c) Rs 600 d) Rs 700
5. 5/9th of a consignment was sold at a profit of 27% and the rest at a loss of 18%. If there was an overall profit of Rs 112, find the value of the consignment.
   a) Rs 1600 b) Rs 1400 c) Rs 1500 d) Rs 1200

Answers

1. a 2. b 3. c 4. d 5. a

Rule 41

Theorem: ***If a person bought two items for Rs P. He sold one at a loss of x% and the other at a gain of y% and he found that each item was sold at the same price, then the cost prices of two items are given as follows:***

(i) Cost price of the item sold at loss is $\frac{P(100+y)}{(100-x)+(100+y)}$

and

(ii) Cost price of the item sold at gain is $\frac{P(100-x)}{(100-x)+(100+y)}$.

Illustrative Example

Ex.: A person bought two watches for Rs 480. He sold one at a loss of 15% and the other at a gain of 19% and he found that each watch was sold at the same price. Find the cost prices of the two watches.

Soln: **Direct Formula** : CP of watch sold at loss = $\frac{480\times(100+\%\text{ profit})}{(100-15)+(100+19)}$

$=\frac{480\times 119}{204}$ = Rs 280.

$\therefore$ CP of watch sold at gain = 480 - 280 = Rs 200.

Note: CP of watch sold at gain $=\frac{480\times(100-\%\text{ loss})}{(100-15)+(100+19)}$ = Rs 200.

Exercise

1. A person bought two horses for Rs 960. He sold one at a loss of 20% and the other at a gain of 60% and he found that each horse was sold at the same price. Find the cost price of two horses.
   a) Rs 640, Rs 320 b) Rs 540, Rs 420
   c) Rs 440, Rs 520 d) Rs 650, Rs 310
2. A person bought two goats for Rs 630. He sold one at a loss of 10% and the other at a gain of 20% and he found

that each goat was sold at the same price. Find the cost price of two goats.

a) Rs 350, Rs 280 b) Rs 360, Rs 270
c) Rs 340, Rs 290 d) Rs 300, Rs 330

3. A person bought two oxen for Rs 410. He sold one at a loss of 20% and the other at a gain of 25% and he found that each ox was sold at the same price. Find the cost price of two oxen.

a) Rs 240, Rs 170 b) Rs 230, Rs 180
c) Rs 260, Rs 150 d) Rs 250, Rs 160

4. A person bought two tables for Rs 201. He sold one at a loss of 14% and the other at a gain of 15% and he found that each table was sold at the same price. Find the cost price of two tables.

a) Rs 115, Rs 86 b) Rs 114, Rs 87
c) Rs 116, Rs 85 d) Rs 119, Rs 82

5. A person bought two chairs for Rs 380. He sold one at a loss of 22% and the other at a gain of 12% and he found that each chair was sold at the same price. Find the cost price of two chairs.

a) Rs 225, Rs 155 b) Rs 226, Rs 154
c) Rs 224, Rs 156 d) Data inadequate

Answers

1. a 2. b 3. d 4. a 5. c

Rule 42

Theorem: ***A person buys certain quantity of an article for Rs A. If he sells mth part of the stock at a profit of x% and the remaining nth part at y% profit, then the per cent profit in this transaction is*** $\left[\frac{mx+ny}{m+n}\right]$

or $\left[\frac{\textit{First part} \times \textit{\% profit on first part} + \textit{Second part} \times \textit{\% profit on second part}}{\textit{Total of two parts}}\right]$.

Illustrative Example

Ex.: Nandlal purchased 20 dozen notebooks at Rs 48 per dozen. He sold 8 dozen at 10% profit and the remaining 12 dozen at 20% profit. What is his profit percentage in this transaction?

Soln: Cost price of 20 dozen notebooks = 20 × 48 = Rs 960

Selling price of 8 dozen notebooks $= \text{Rs } 8 \times 48\left(\frac{110}{100}\right)$

Selling price of 12 dozen notebooks

$$= \text{Rs } 12 \times 48\left(\frac{120}{100}\right)$$

$$\therefore \text{ total selling price} = \text{Rs }\frac{2112}{5} + \text{Rs }\frac{3456}{5}$$

$$= \text{Rs }\frac{5568}{5}$$

$$\text{Profit} = \frac{5568}{5} - 960 = \text{Rs }\frac{768}{5}$$

$$\therefore \text{ profit} = \frac{768 \times 100}{5 \times 960} = 16\%$$

Quicker Method:

Percentage profit

$$= \frac{\left(\begin{array}{l}\text{First part} \times \text{\% profit on first part} + \\ \text{second part} \times \text{\% profit on second part}\end{array}\right)}{\text{Total of two parts}}$$

Here, total = 20 dozens are sold in two parts; first part = 8 dozens and second part = 12 dozens.

$$\therefore \text{ \% profit} = \frac{8 \times 10 + 12 \times 20}{20} = \frac{320}{20} = 16\%$$

Exercise

1. A person buys 50 mangoes at Rs 2 each. He sells $\frac{1}{5}$th of the mangoes at 5% profit and the remaining mangoes at 15% profit. What is his total profit percentages in this transaction?

a) 20% b) 13% c) 16% d) 14%

2. Rachana purchased 25 dozens Jackets at Rs 5000 per dozen. He sold 12 dozens at 15% profit and the remaining 13 dozens at 25% profit. What is his profit percentage in this transaction?

a) 20% b) 25.5% c) 20.5% d) 20.2%

3. Radha purchased 13 dozens pens at Rs 70 per dozen. He sold 10 dozens at 7% profit and the remaining 3 dozens at 7% profit. What is his profit percentage in this transaction?

a) 8% b) 7% c) 10% d) 14%

4. Ramesh purchased 30 bananas at Rs 3 each. He sold 16 bananas at 15% profit and the remaining at 30% profit. What is his profit percentage in this transaction?

a) 33% b) 22% c) 17% d) 31%

Answers

1. b; **Hint:** Here $m = 50 \times \frac{1}{5} = 10$ mangoes and n = 50 – 10 = 40 mangoes

2. d 3. b 4. b

Rule 43

Theorem: ***If x part is sold at m% loss, y part is sold at n% loss, the rest part z is sold at p% loss and Rs P is overall***

loss, then the value of total consignment is $\frac{P \times 100}{xm + ny + pz}$.

Illustrative Example

Ex.: $\frac{1}{3}$ of a commodity is sold at 15% loss, $\frac{1}{4}$ is sold at 20% loss and the rest at 24% loss. If a total loss is Rs 62, then find the value of the commodity.

Soln: Applying the above theorem, we have

$$\text{Total value} = \frac{62 \times 100}{\frac{1}{3} \times 15 + \frac{1}{4} \times 20 + \frac{5}{12} \times 24} = \text{Rs } 310.$$

Exericse

1. $\frac{2}{3}$ of a commodity is sold at 15% loss, $\frac{1}{6}$ is sold at 12% loss and the rest at 18% loss. If a total loss is Rs 45, then find the value of the commodity.
a) Rs 200 b) Rs 250 c) Rs 300 d) Rs 450

2. $\frac{1}{4}$ of a commodity is sold at 16% loss, $\frac{1}{8}$ is sold at 24% loss and the rest at 32% loss. If a total loss is Rs 54, then find the value of the commodity.
a) Rs 200 b) Rs 300
c) Rs 150 d) Data inadequate

3. $\frac{1}{3}$ of a commodity is sold at 12% loss, $\frac{2}{5}$ is sold at 15% loss and the rest at 30% loss. If a total loss is Rs 72, then find the value of the commodity.
a) Rs 360 b) Rs 450
c) Rs 400 d) Data inadequate

Answers

1. c 2. a 3. c

Rule 44

Theorem: ***If x part is sold at m% loss and the other rest part y is sold at n% profit and Rs P is overall loss, then the value of total consignment is Rs*** $\left(\frac{P \times 100}{xm - yn}\right)$.

Illustrative Example

Ex.: $\frac{2}{3}$ of a consignment was sold at 6% loss and the rest at a profit of 3%. If there was an overall loss of Rs 540, find the value of the consignment.

Soln: Applying the above theorem, we have the

$$\text{total value} = \frac{540 \times 100}{6 \times \frac{2}{3} - \frac{1}{3} \times 3} = \text{Rs } 18000.$$

Exercise

1. $\frac{3}{4}$ of a consignment was sold at 8% loss and the rest at a profit of 4%. If there was an overall loss of Rs 120, find the value of the consignment.
a) Rs 2450 b) Rs 2500 c) Rs 2400 d) Rs 2600

2. $\frac{4}{5}$ of a consignment was sold at 10% loss and the rest at a profit of 5%. If there was an overall loss of Rs 840, find the value of the consignment.
a) Rs 12000 b) Rs 16000 c) Rs 14000 d) Data inadequate

3. $\frac{4}{7}$ of a consignment was sold at 21% loss and the rest at a profit of 14%. If there was an overall loss of Rs 600, find the value of the consignment.
a) Rs 60000 b) Rs 10000 c) Rs 15000 d) Rs 12000

4. $\frac{5}{8}$ of a consignment was sold at 16% loss and the rest at a profit of 8%. If there was an overall loss of Rs 140, find the value of the consignment.
a) Rs 2000 b) Rs 2500 c) Rs 1800 d) Rs 2400

5. $\frac{3}{5}$ of a consignment was sold at 15% loss and the rest at a profit of 10%. If there was an overall loss of Rs 15, find the value of the consignment.
a) Rs 200 b) Rs 250 c) Rs 350 d) Rs 300

Answers

1. c 2. a 3. b 4. a 5. d

Rule 45

Theorem: ***A person buys certain quantity of an article for Rs A. If he sells nth part of the stock at a loss of x%, then to make an overall profit of y% on the total transaction he should sell the remaining stock at the per cent profit of*** $\left[\frac{y + xn}{1 - n}\right]$.

Illustrative Example

Ex.: If goods be purchased for Rs 450, and one third be sold at a loss of 10 per cent, at what gain per cent should the remainder be sold so as to gain 20 per cent on the whole transaction?

Soln: **Detail Method:**

Cost of one third of goods = $\frac{1}{3}$ of Rs 450 = Rs 150

The selling price of one-third of goods = Rs $150 \times \frac{90}{100}$
= Rs 135

The total selling price is to be Rs $450 \times \frac{120}{100}$ or Rs 540

Hence the selling price of the remaining two-thirds of the goods must be (Rs 450 - Rs 135) or Rs 405.
But the cost of this two-thirds = Rs 300
$\therefore$ gain on Rs 300 = Rs 405 - Rs 300 = Rs 105

$\therefore$ gain on Rs 100 = $\frac{1}{3}$ of Rs 105 = Rs 35.

$\therefore$ gain % is 35.

Quicker Method: Applying the above formula, we have

the required % profit = $\frac{20+10\times\frac{1}{3}}{1-\frac{1}{3}} = 35\%$

Exercise

1. A distributor buys books for Rs 8000 from a publisher. He had to sell one-fourth at a loss of 20%. At what per cent gain he should sell the remaining stock so as to make an overall profit of 10% on the total transaction?
   a) 20% b) 25% c) 18% d) 16%
2. If goods be purchased for Rs 840, and one fourth be sold at a loss of 20 per cent, at what gain per cent should the remainder be sold so as to gain 20 per cent on the whole transaction.
   a) $33\frac{2}{3}\%$ b) 33% c) $33\frac{1}{2}\%$ d) $33\frac{1}{3}\%$
3. If goods be purchased for Rs 380, and $\frac{2}{3}$rd be sold at a loss of 15 per cent, at what gain per cent should the remainder be sold so as to gain 10 per cent on the whole transaction.
   a) 60% b) 70% c) 90% d) 40%
4. If goods be purchased for Rs 480, and $\frac{1}{5}$th be sold at a loss of 25 per cent, at what gain per cent should the remainder be sold so as to gain 25 per cent on the whole transaction.
   a) 37% b) $37\frac{1}{2}\%$ c) $32\frac{1}{2}\%$ d) Data inadequate

Answers

1. a 2. d 3. a 4. b

Rule 46

Theorem: ***A person buys certain quantity of an article for Rs A. If he sells nth part of the stock at a profit of x%, then to make an overall profit of y% on the total transaction he should sell the remaining stock at the per cent profit or loss of $\left[\frac{y-xn}{1-n}\right]$ according to the (+ve) or (–ve) sign.***

Illustrative Example

Ex.: A man buys rice for Rs 4400. He sells $\frac{1}{3}$rd of it at a profit of 5%. At what per cent gain should he sell remaining $\frac{2}{3}$rd so as to make an overall profit of 10% on the whole transaction?

Soln: Following the above theorem,

the required % profit = $\frac{10-5\times\frac{1}{3}}{1-\frac{1}{3}} = \frac{25}{2} = 12\frac{1}{2}\%$.

Exercise

1. A man buys wheat for Rs 5200. He sells $\frac{2}{3}$rd of it at a profit of 6%. At what per cent gain should he sell remaining $\frac{1}{3}$rd so as to make an overall profit of 12% on the whole transaction?
   a) 24% b) 20% c) 22% d) 26%
2. A man buys tea for Rs 8400. He sells $\frac{3}{5}$th of it at a profit of 15%. At what per cent gain should he sell remaining $\frac{2}{5}$th so as to make an overall profit of 20% on the whole transaction?
   a) 27% b) 27.5% c) 28% d) Data inadequate
3. A man buys pulse for Rs 4800. He sells $\frac{2}{5}$th of it at a profit of 25%. At what per cent gain should he sell remaining $\frac{3}{5}$th so as to make an overall profit of 19% on the whole transaction?
   a) 15% b) 18% c) 21% d) 12%
4. A man buys coffee for Rs 7200. He sells $\frac{3}{8}$th of it at a profit of 24%. At what per cent gain should he sell remaining $\frac{5}{8}$th so as to make an overall profit of 21% on

the whole transaction?

a) $18\frac{1}{5}\%$ b) $20\frac{1}{5}\%$ c) $19\frac{1}{5}\%$ d) Data inadequate

5. A man buys rice for Rs 5520. He sells $\frac{5}{7}$ th of it at a profit of 14%. At what per cent gain should he sell remaining $\frac{2}{7}$ th so as to make an overall profit of 18% on the whole transaction?

a) 28% b) 24% c) 27% d) Data inadequate

Answers

1. a 2. b 3. a 4. c 5. a

Rule 47

Theorem: ***A person buys certain quantity of an article for Rs A. If he sells nth part of the stock at a loss of x%, then to make an overall no profit no loss on the total transaction, he should sell the remaining stock at the per cent profit of***

$$\left[\frac{n}{1-n}x\right]\%.$$

Illustrative Example

Ex.: A distributor buys books for Rs 8000 from a publisher. He had to sell one-fourth at a loss of 20%. At what per cent gain he should sell the remaining stock so as to make an overall no profit no loss on the total transaction?

Soln: Applying the above formula, we have

$$\text{the required per cent profit} = \frac{20\times\frac{1}{4}}{1-\frac{1}{4}} = \frac{5\times4}{3} = 6\frac{2}{3}\%.$$

Exercise

1. A distributor buys books for Rs 6000 from a publisher. He had to sell one-third at a loss of 30%. At what per cent gain he should sell the remaining stock so as to make an overall no profit no loss on the total transaction?

a) 25% b) 12% c) 10% d) 15%

2. A distributor buys books for Rs 750 from a publisher. He had to sell $\frac{2}{5}$ th at a loss of 15%. At what per cent gain he should sell the remaining stock so as to make an overall no profit no loss on the total transaction?

a) 10% b) 9% c) 12% d) 15%

3. A distributor buys books for Rs 510 from a publisher. He had to sell $\frac{4}{9}$ th at a loss of 18%. At what per cent gain he should sell the remaining stock so as to make an overall no profit no loss on the total transaction?

a) $14\frac{3}{5}\%$ b) $16\frac{2}{3}\%$ c) $14\frac{2}{5}\%$ d) Data inadequate

4. A distributor buys books for Rs 670 from a publisher. He had to sell $\frac{3}{8}$ th at a loss of 16%. At what per cent gain he should sell the remaining stock so as to make an overall no profit no loss on the total transaction?

a) $9\frac{3}{5}\%$ b) $9\frac{2}{5}\%$ c) $9\frac{1}{5}\%$ d) $8\frac{3}{5}\%$

5. A distributor buys books for Rs 5400 from a publisher. He had to sell $\frac{5}{6}$ th at a loss of 6%. At what per cent gain he should sell the remaining stock so as to make an overall no profit no loss on the total transaction?

a) 20% b) 30% c) 32% d) 25%

Answers

1. d 2. a 3. c 4. a 5. b

Rule 48

Theorem: ***A person buys certain quantity of an article for Rs A. If he sells nth part of the stock at a profit of x%, then to make an overall no profit and no loss on the total transaction he should sell the remaining stock at the per cent loss of*** $\left[\frac{n}{1-n}x\right]\%$.

Illustrative Example

Ex.: A man buys rice for Rs 4400. He sells $\frac{1}{3}$ rd of it at a profit of 5%. At what per cent loss should he sell remaining $\frac{2}{3}$ rd so as to make no profit no loss on the whole transaction?

Soln: Following the above theorem, we have

$$\text{the required per cent loss} = \frac{5\times\frac{1}{3}}{1-\frac{1}{3}} = \frac{5}{2} = 2\frac{1}{2}\%$$

Exercise

1. A man buys coffee for Rs 4500. He sells $\frac{2}{3}$ rd of it at a profit of 6%. At what per cent loss should he sell remain-

ing $\frac{1}{3}$rd so as to make no profit no loss on the whole transaction?

a) 10% b) 6% c) 12% d) 18%

2. A man buys tea for Rs 4600. He sells $\frac{2}{5}$ th of it at a profit of 15%. At what per cent loss should he sell remaining $\frac{3}{5}$th so as to make no profit no loss on the whole transaction?

a) 10% b) 12% c) 14% d) 8%

3. A man buys wheat for Rs 4800. He sells $\frac{5}{7}$ th of it at a profit of 14%. At what per cent loss should he sell remaining $\frac{2}{7}$ th so as to make no profit no loss on the whole transaction?

a) 25% b) 30% c) 35% d) Data inadequate

4. A man buys grains for Rs 4750. He sells $\frac{3}{8}$ th of it at a profit of 16%. At what per cent loss should he sell remaining $\frac{5}{8}$ th so as to make no profit no loss on the whole transaction?

a) $9\frac{3}{5}\%$ b) $9\frac{2}{5}\%$

c) $9\frac{1}{5}\%$ d) Data inadequate

5. A man buys cereals for Rs 6000. He sells $\frac{7}{9}$ th of it at a profit of 18%. At what per cent loss should he sell remaining $\frac{2}{9}$ th so as to make no profit no loss on the whole transaction?

a) 64% b) 60% c) 65% d) 63%

Answers

1. c 2. a 3. c 4. a 5. d

Rule 49

Theorem: ***A man purchases m articles at x a rupee and n articles at y a rupee. He mixes them together and sells them at z a rupee. Then his gain or loss per cent is***

$$\left[\left(\frac{(m+n)xy}{(my+nx)z}-1\right)\times 100\right]\%$$ ***according as the +ve or –ve sign.***

Illustrative Example

Ex.: A man buys 400 oranges at 4 a rupee and 500 oranges at 2 a rupee. He mixes them together and sells them at 3 a rupee. Find his per cent loss or gain.

Soln: Applying the above formula,

$$\%\text{ profit or loss} = \left[\frac{(400+500)\times 4\times 2}{(400\times 2+500\times 4)\times 3}-1\right]\times 100\%$$

$$= -\frac{100}{7} = -14\frac{2}{7}\%$$

–ve sign shows that there is a loss in the transaction.

$\therefore$ % loss = $14\frac{2}{7}\%$.

Exercise

1. A man buys 200 bananas at 8 a rupee and 250 bananas at 4 a rupee. He mixes them together and sells them at 6 a rupee. Find his per cent loss or gain.

a) $14\frac{2}{7}\%$ loss b) $14\frac{2}{7}\%$ profit

c) $16\frac{2}{7}\%$ loss d) $16\frac{2}{7}\%$ profit

2. A man buys 300 oranges at 5 a rupee and 450 oranges at $2\frac{1}{2}$ a rupee. He mixes them together and sells them at 4 a rupee. Find his per cent loss or gain.

a) Approx. 28% loss b) Approx. 28% profit
c) Approx. 22% loss d) Approx 29% profit

3. A man buys 25 apples at 8 a rupee and 35 apples at 4 a rupee. He mixes them together and sells them at 6 a rupee. Find his per cent loss or gain.

a) 15% b) 16% c) 16.78% d) 15.78%

4. A person purchased 100 oranges at 4 a rupee and 200 oranges at 2 a rupee. He mixed them and sold at 3 oranges a rupee. Find his per cent loss or gain.

a) 20% loss b) 25% loss c) 20% profit d) 25% profit

5. A person buys 100 toffees at 10 a rupee and 200 toffees at 5 a rupee. He mixes them together and sells at 4 a rupee. Find his per cent profit.

a) 20% b) 25% c) 40% d) 50%

Answers

1. a 2. c 3. d 4. a 5. d

Rule 50

Theorem: ***A man purchases a certain number of articles at x a rupee and the same number at y a rupee. He mixes them together and sells them at z a rupee. Then his gain or loss %***

$$= \left[\frac{2xy}{z(x+y)}-1\right]\times 100$$ ***according as the sign is +ve or –ve.***

Note: This rule is the special case of **Rule - 49**. If we put m = n in the **Rule - 49** we get **Rule - 50**.

Illustrative Examples

Ex. 1: A man purchases a certain number of mangoes at 3 per rupee and the same number at 4 per rupee. He mixes them together and sells them at 3 per rupee. What is his gain or loss per cent?

Son: By the theorem:

$$\text{Profit or loss per cent} = \left[\frac{2\times3\times4}{3(3+4)}-1\right]\times100$$

$$= \left[\frac{24}{21}-1\right]\times100 = \frac{100}{7} = 14\frac{2}{7}\%$$

Since the sign is +ve, there is a gain of $14\frac{2}{7}\%$.

Ex. 2: A man purchases a certain number of toffees at 25 a rupee and the same number at 20 a rupee. He mixes them together and sells them at 45 for 2 rupees. What does he gain or lose per cent in the transaction?

Soln: By the theorem:

$x = 25,\ y = 20$ and $z = \frac{45}{2} = 22.5$

$$\therefore \%\text{ profit or loss} = \left[\frac{2xy}{z(x+y)}-1\right]\times100$$

$$= \left[\frac{2\times25\times20}{22.5(25+20)}-1\right]\times100$$

$$= \frac{1000-1012.5}{1012.5}\times100 = -\frac{12.5}{1012.5}\times100$$

$$= -\frac{100}{81} = -1\frac{19}{81}\%$$

Since the sign is –ve there is a loss of $1\frac{19}{81}\%$.

Ex. 3: Oranges are bought at 11 for a rupee and an equal number more at 9 for a rupee. If these are sold at 10 for a rupee, find the loss or gain per cent.

Soln: By the theorem:

$$\%\text{ profit or loss} = \left[\frac{2\times11\times9}{10(11+9)}-1\right]\times100$$

$$= \left[\frac{198}{200}-1\right]\times100 = \frac{-2}{200}\times100 = -1\%$$

Since the sign is –ve there is a loss of 1%.

Note: (1) From the above two eamples, we find that when $z = \frac{x+y}{2}$, there is always loss.

(2) If there is x = y = z, there is neither gain nor loss. Do you agree?

Exercise

1. A man purchases a certain number of oranges at 5 per rupee and the same number at 6 per rupee. He mixes them together and sells them at 8 per rupee. What is his gain or loss per cent?

   a) $33\frac{9}{11}\%$ loss b) $31\frac{9}{11}\%$ loss

   c) $31\frac{1}{11}\%$ loss d) $33\frac{9}{11}\%$ profit

2. A man purchases a certain number of oranges at 6 per rupee and the same number at 8 per rupee. He mixes them together and sells them at 6 per rupee. What is his gain or loss per cent?

   a) $14\frac{2}{7}\%$ gain b) $16\frac{2}{7}\%$ loss

   c) $14\frac{2}{7}\%$ loss d) $16\frac{2}{7}\%$ loss

3. A man purchases a certain number of toffees at 20 a rupee and the same number at 15 a rupee. He mixes them together and sells them at 40 for 2 rupees. What does he gain or lose per cent in the transaction?

   a) $14\frac{2}{7}\%$ gain b) $14\frac{2}{7}\%$ loss

   c) $14\frac{3}{7}\%$ gain d) $14\frac{4}{7}\%$ loss

4. A man purchases a certain number of toffees at 15 a rupee and the same number at 10 a rupee. He mixes them together and sells them at 35 for 2 rupees. What does he gain or lose per cent in the transaction?

   a) $31\frac{3}{7}\%$ gain b) $31\frac{1}{7}\%$ gain

   c) $31\frac{2}{7}\%$ loss d) $31\frac{3}{7}\%$ loss

5. Oranges are bought at 12 for a rupee and an equal number more at 8 for a rupee. If these are sold at 15 for a rupee, find the loss or gain per cent.
   a) 36% loss b) 36% profit c) 38% loss d) 38% profit

6. Oranges are bought at 16 for a rupee and an equal number more at 14 for a rupee. If these are sold at 20 for a rupee, find the loss or gain per cent.
   a) 25% loss b) 25.3% profit
   c) 25.6% loss d) 25.3% loss

7. A man buys a certain number of oranges at 20 per rupee and an equal number at 30 per rupee. He mixes them and sells them at 25 per rupee. What is his gain or loss per cent?
   a) 4% profit b) 4% loss c) 20% profit d) 20% loss

8. A man buys a certain number of oranges at 5 a rupee and an equal number at 3 a rupee. He mixes them together and sells at 4 a rupee. Find his per cent profit or loss.

a) $6\frac{1}{4}\%$ profit b) 25% profit

c) $6\frac{1}{4}\%$ loss d) Data inadequate

9. A man buys toffees at 6 a rupee and an equal number of toffees at 3 a rupee. He mixes them together and sells at 4 a rupee. Find his per cent loss or profit.

a) Neither loss nor profit b) 100% loss
c) 100% profit d) Can't be determined

Answers

1. b 2. a 3. b 4. d 5. a 6. d 7. b
8. c 9. a

Rule 51

Theorem: ***If a person bought items A and B for Rs P. He sold A at a gain of x% and the other at a gain of y% and he found that each item A and B was sold at the same price, then the***

(i) cost price of A $= \frac{(100+y)\times P}{(100+x)+(100+y)}$ ***and the***

(ii) cost price of B $= \frac{(100+x)\times P}{(100+x)+(100+y)}$.

Illustrative Example

Ex.: A person bought two horses for Rs 690. He sold one at a gain of 10% and the other at a gain of 20% and he found that each horse was sold at the same price. Find the cost prices of the two horses.

Soln: Following the above theorem, we have the

(i) cost price of the first horse

$$= \frac{(100+20)\times 690}{(100+10)+(100+20)} = \text{Rs } 360.$$

(ii) cost price of the second horse

$$= \frac{(100+10)\times 690}{(100+10)+(100+20)} = \text{Rs } 330$$

Exercise

1. A person bought two tables for Rs 860. He sold one at a gain of 5% and the other at a gain of 10% and he found that each table was sold at the same price. Find the cost prices of the two tables.

a) Rs 440, Rs 420 b) Rs 460, Rs 400
c) Rs 480, Rs 380 d) Data inadequate

2. A person bought two chairs for Rs 735. He sold one at a gain of 20% and the other at a gain of 25% and he found that each chair was sold at the same price. Find the cost prices of the two chairs.

a) Rs 375, Rs 360 b) Rs 385, Rs 350
c) Rs 370, Rs 365 d) Data inadequate

3. A person bought two goats for Rs 920. He sold one at a gain of 18% and the other at a gain of 12% and he found that each goat was sold at the same price. Find the cost prices of the two goats.

a) Rs 462, Rs 458 b) Rs 472, Rs 448
c) Rs 482, Rs 438 d) Data inadequate

4. A person bought two oxen for Rs 500. He sold one at a gain of 23% and the other at a gain of 27% and he found that each ox was sold at the same price. Find the cost prices of the two oxen.

a) Rs 264, Rs 236 b) Rs 244, Rs 266
c) Rs 254, Rs 246 d) Data inadequate

5. A person bought two sofa-sets for Rs 2800. He sold one at a gain of 35% and the other at a gain of 45% and he found that each sofa-set was sold at the same price. Find the cost prices of the two sofa-sets.

a) Rs 1430, Rs 1370 b) Rs 1440, Rs 1360
c) Rs 1460, Rs 1340 d) Rs 1450, Rs 1350

Answers

1. a 2. a 3. b 4. c 5. d

Rule 52

Theorem: ***If certain article is bought at the rate of 'A' for a rupee, then to gain x%, the article must be sold at the rate of*** $\left(\frac{100}{100+x}\right)\times A$ ***for a rupee (Remember the rule of fraction).***

Illustrative Example

Ex.: If toffees are bought at the rate of 25 for a rupee, how many must be sold for a rupee so as to gain 25%?

Soln: **Detail Method:** SP of 25 toffees = Re $1\times\frac{125}{100}$ = Rs $\frac{5}{4}$

$\therefore$ No. of toffees sold for Rs $\frac{5}{4} = 25$

$\therefore$ No. of toffees sold for Re 1 $= \frac{25\times 4}{5} = 20$.

Short-cut Method (Method of Fraction):

As there is 25% gain so our calculating figure would be 125 and 100. Now, to gain a profit the number of articles sold for one rupee must be less than the number bought for one rupee. Thus the multiplying fraction is $\frac{100}{125}$.

$\therefore$ required no. of toffees $= 25\times\frac{100}{125} = 20$.

Exercise

1. If oranges are bought at the rate of 21 for a rupee, how many must be sold for a rupee so as to gain 5%?
   a) 20 b) 18 c) 16 d) 19
2. If apples are bought at the rate of 11 for a rupee, how many must be sold for a rupee so as to gain 10%?
   a) 10 b) 12 c) 15 d) 8
3. If bananas are bought at the rate of 30 for a rupee, how many must be sold for a rupee so as to gain 20%?
   a) 24 b) 25 c) 28 d) 30
4. A man buys oranges at 6 a rupee. For how many a rupee he should sell them so as to gain 20%.
   a) 4 oranges a rupee b) 5 oranges a rupee
   c) 3 oranges a rupee d) Data inadequate

Answers

1. a 2. a 3. b 4. b

Rule 53

Theorem: ***If a man buys n items of A and m items of B for Rs P, and sells the items of A at a profit of x% and the items of B at a profit of y%, and his whole gain is Rs X, then the price he pays for***

(i) one item of A is $\left[\frac{100X - Py}{n(x-y)}\right]$ ***and***

(ii) one item of B is $\left[\frac{Px - 100X}{m(x-y)}\right]$ ***or*** $\left[\frac{100X - Px}{m(y-x)}\right]$.

Illustrative Example

Ex.: A man buys 5 horses and 7 oxen for Rs 5850. He sells the horses at a profit of 10% and oxen at a profit of 16% and his whole gain is Rs 711. What price does he pay for a horse and a ox?

Soln: Following the above theorem,

$$\text{Price of a horse} = \frac{100 \times 711 - 5850 \times 16}{5(10-16)}$$

$$= \frac{-22500}{-30} = \text{Rs } 750.$$

$$\text{Price of a ox} = \frac{5850 \times 10 - 100 \times 711}{7(10-16)}$$

$$= \frac{58500 - 71100}{-42} = \text{Rs } 300.$$

Note: 1. *If a man buys n items of A and m items of B for Rs P, and sells the items of A at a profit of x% and the items of B at a loss of y%, and his whole gain is Rs X, then the price he pays for*

(i) one item of A is $\left[\frac{100X + Py}{n(x+y)}\right]$ *and*

(ii) one item of B is $\left[\frac{Px - 100X}{m(x+y)}\right]$.

Ex.: If a man buys 10 pens and 5 pencils for Rs 500, and sells the pens at a profit of 10% and the pencils at a loss of 15%, and his whole gain is Rs 25. What price does he pay for a pen and a pencil?

Soln: Following the above theorem, we have the

$$\text{price of a pen} = \frac{100 \times 25 + 500 \times 15}{10(10+15)} = \text{Rs } 40 \text{ and}$$

$$\text{the price of a pencil} = \frac{500 \times 10 - 100 \times 25}{5(10+15)} = \text{Rs } 20.$$

2. *If a man buys n items of A and m items of B for Rs P, and sells the items of A at a loss of x% and the items of B at a loss of y%, and his whole loss is Rs X, then the price he pays for*

(i) one item of A is $\left[\frac{100X - Py}{n(x-y)}\right]$ *and*

(ii) one item of B is $\left[\frac{Px - 100X}{m(x-y)}\right]$ *or* $\left[\frac{100X - Px}{m(y-x)}\right]$

Ex.: A man buys 5 horses and 7 oxen for Rs 5850. He sells the horses at a loss of 10% and oxen at a loss of 16% and his whole loss is Rs 711. What price does he pay for a horse and a ox?

Soln: Following the above theorem, we have the

$$\text{price of a horse} = \frac{100 + 711 - 5850 \times 16}{5(10-16)} = \text{Rs } 750 \text{ and}$$

$$\text{price of a ox} = \frac{5850 \times 10 - 100 \times 711}{7(10-16)} = \text{Rs } 300.$$

3. *If a man buys n items of A and m items of B for Rs P, and sells the items of A at a loss of x% and the items of B at a profit of y%, and his whole loss is Rs X, then the price he pays for one item of A is* $\left[\frac{100X + Py}{n(x+y)}\right]$ *and one item of B is* $\left[\frac{Px - 100X}{m(x+y)}\right]$.

Ex.: If a man buys 10 pens and 5 pencils for Rs 500, and sells the pens at a loss of 10% and the pencils at a profit of 15% and his whole loss is Rs 25. What price does he pay for a pen and a pencil?

Soln: Following the above theorem, we have the

$$\text{price of a pen} = \frac{100 \times 25 + 500 \times 15}{10(10+15)} = \text{Rs } 40 \text{ and the}$$

price of a pencil $= \dfrac{500 \times 10 - 100 \times 25}{5(10+15)} =$ Rs 20.

Exercise

1. A man buys 20 pens and 16 books for Rs 360. He sells pens at a profit of 40% and books with a gain of 25%. If his overall gain is Rs 120, the CP of the pen is ____.
   a) Rs 13 b) Rs 12 c) Rs 15 d) Rs 10
2. 6 horses and 8 cows cost me Rs 37500. I sell them for Rs 44625 making a profit of 25 per cent on the horses and 10 per cent on the cows. What is the average cost of each horse and of each cow.
   a) Rs 3750, Rs 1875 b) Rs 3650, Rs 1975
   c) Rs 4750, Rs 2875 d) Data inadequate
3. A man buys5 horses and 7 oxen for Rs 58500. He sells the horses at a profit of 10 per cent and oxen at a profit of 16 per cent and his whole gain is Rs 7110. What price does he pay for a horse?
   a) Rs 7500 b) Rs 3000
   c) Rs 8500 d) Rs 5700
4. A man buys 3 tables and 12 chairs for Rs 2400. He sells the tables at a profit of 20% and chairs at a profit of 10% and makes a total profit of Rs 300. At what price did he buy tables and chairs?
   a) Rs 600, Rs 1800 b) Rs 200, Rs 350
   c) Rs 800, Rs 1400 d) Can't be determined
5. Raman buys 5 pens and 30 pencils for Rs 1000. He sells the pens at a profit of 15% and pencils at a profit of 10% and makes a total profit of Rs 120. Find the cost of a pen and of a pencil.
   a) Rs 70, Rs 30 b) Rs 85, Rs 15
   c) Rs 80, Rs 20 d) Data inadequate

Answers

1. d

2. a; **Hint:** Here 'X' = gain = Rs 44625 - Rs 37500 = Rs 7125
Now apply the above rule and get the answer.

3. a

4. a; **Hint:** By the formula we calculate price of one table and one chair ie Rs 200 and Rs 150 respectively.
$\therefore$ required answer = the price of 3 tables
$= 3 \times 200 =$ Rs 600
and the price of 12 chairs $= 12 \times 150 =$ Rs 1800.

5. c

Rule 54

Ex.: A person bought some oranges at the rate of 5 per rupee. He bought the same number of oranges at the rate of 4 per rupee. He mixes both the types and sells at 9 for rupees 2. In this business he bears a loss of Rs 3. Find out how many oranges he bought in all?

Soln: **Detail Method:** Suppose he bought x oranges of each quality.

Then his total investment $= \dfrac{x}{5} + \dfrac{x}{4} =$ Rs $\dfrac{9x}{20}$

Total selling price = Rs $\dfrac{2x \times 2}{9} =$ Rs $\dfrac{4x}{9}$

$\therefore$ total loss $= \dfrac{9x}{20} - \dfrac{4x}{9} = \dfrac{81x - 80x}{180} = \dfrac{x}{180}$

then Rs $\dfrac{x}{180} =$ Rs 3 $\therefore x = 180 \times 3 = 540$

Therefore he bought $2 \times 540 = 1080$ oranges in total.

Quicker Method: In the above question:
If x oranges/rupee and y oranges/rupee are mixed in same numbers and sold at z oranges/rupee then

Number of total oranges bought $= \dfrac{\text{loss rupees} \times 2xyz}{z(x+y) - 2xy}$

$$= \frac{3 \times 2 \times 5 \times 4 \times 4.5}{4.5(5+4) - 2 \times 5 \times 4} = \frac{120 \times 4.5}{40.5 - 40} = 1080 \text{ oranges.}$$

Exercise

1. A person bought some oranges at the rate of 6 per rupee. He bought the same number of oranges at the rate of 4 per rupee. He mixes both the types and sells at 10 for rupees 2. In this business he bears a loss of Rs 4. Find out how many oranges he bought in all?
   a) 480 oranges b) 840 oranges
   c) 490 oranges d) Data inadequate
2. A person bought some oranges at the rate of 3 per rupee. He bought the same number of oranges at the rate of 2 per rupee. He mixes both the types and sells at 5 for rupees 2. In this business he bears a loss of Rs 2. Find out how many oranges he bought in all?
   a) 240 oranges b) 120 oranges
   c) 180 oranges d) Cann't possible
3. A person bought some oranges at the rate of 6 per rupee. He bought the same number of oranges at the rate of 5 per rupee. He mixes both the types and sells at 10 for rupees 2. In this business he bears a loss of Rs 4. Find out how many oranges he bought in all.
   a) 240 oranges b) 680 oranges
   c) 480 oranges d) Cann't possible
4. A person bought some oranges at the rate of 12 per rupee. He bought the same number of oranges at the rate of 8 per rupee. He mixes both the types and sells at 20 for rupees 2. In this business he bears a loss of Rs 8. Find out how many oranges he bought in all?
   a) 1680 b) 1820 c) 1920 d) 1290

Answers

1. a 2. b

3. d; **Hint:** Applying the given formula, we get no. of oranges = -240, that is not possible.

4. c

Rule 55

Theorem: ***If a tradesman marks his goods at x% above his cost price and allows purchasers a discount of y% for cash, then there is*** $\left(x-y-\frac{xy}{100}\right)$ ***% profit or loss according to +ve or –ve sign respectively.***

Note: When $x = y$, then formula becomes $-\frac{x^2}{100}$. –ve sign indicates that there will be always loss.

Illustrative Examples

Ex. 1: A tradesman marks his goods at 25% above his cost price and allows purchasers a discount of $12\frac{1}{2}$% for cash. What profit % does he make?

Soln: **Detail Method:** Let the cost price = Rs 100
Marked price = Rs 125

Discount = $12\frac{1}{2}$% of Rs 125 = Rs $15\frac{5}{8}$

$\therefore$ reduced price = Rs 125 – Rs $15\frac{5}{8}$ = Rs $109\frac{3}{8}$

$\therefore$ gain per cent = $109\frac{3}{8}-100=9\frac{3}{8}$%

Quicker Method: Applying the above theorem, we have

$x = 25\%,\ y = 12\frac{1}{2}\%$

$\therefore \left(x-y-\frac{xy}{100}\right)\%$

$=\left(25-12\frac{1}{2}-\frac{25\times25}{200}\right)\% = 9\frac{3}{8}\%$ profit

Note: Thus we see that if
x = marked percentage above CP
y = discount in per cent
z = profit in per cent
Then there exists a relationship;

$$z = x-y-\frac{xy}{100}$$

Ex. 2: A trader allows a discount of 10% for cash payment. How much % above cost price must he mark his goods to make a profit of 10%?

Soln: If we use the relationship discussed above, we have

$$10 = x-10-\frac{10x}{100}$$

$$10x-100-x=100$$

$$\therefore x=\frac{200}{9}=22\frac{2}{9}\%$$

Exercise

1. A shopkeeper labels the price of articles 20% above the cost price. If he allows Rs 31.20 off on a bill of Rs 312; find his profit per cent on the article. **[BSRB Mumbai PO, 1999]**

   a) 8 b) $12\frac{1}{3}$ c) $11\frac{2}{3}$ d) $8\frac{1}{3}$

2. A trader allows a discount of 5 per cent for cash payment. How much per cent above the cost price must he mark his goods to make a profit of 10 per cent?

   a) $15\frac{11}{19}$% b) $15\frac{15}{19}$% c) $19\frac{15}{19}$% d) $15\frac{14}{19}$%

3. A tradesman marks his goods 30 per cent above cost price but makes a reduction of $6\frac{1}{4}$ per cent on the marked price for ready money, find his gain per cent.

   a) $26\frac{1}{10}$% b) $27\frac{1}{8}$% c) $21\frac{7}{8}$% d) $22\frac{7}{8}$%

4. A tradesman's charges are 20 per cent over cost price. If he allows his customers 10 per cent off their bills for cash payment, what is his gain per cent? **[RRB Exam 1999]**

   a) 4% b) 8% c) 10% d) 12%

5. A tradesman marks his goods at 30 per cent above the cost price. If he allows a discount of 3 paise in a Re off the marked price, what percentage of profit does he make?

   a) 25% b) 27% c) 28% d) 26%

6. How much per cent above the cost price should a shopkeeper mark his goods so that after allowing a discount of 10%, he should gain 26%.

   a) 16% b) 40% c) 30% d) 36%

7. The marked price is 10% higher than C.P. A discount of 10% is given on marked price. In this kind of sale the seller. **[RRB Exam 1991]**

   a) bears no loss, no gain b) gains
   c) loses d) loses 1%

Answers

1. a; **Hint:** Here, $x = 20\%,\ y=\frac{31.20}{312}\times100=10\%$

2. b; **Hint:** $x-5-\frac{5x}{100}=10$

   $\therefore x=\frac{300}{19}=15\frac{15}{19}\%$

3. c 4. b

5. d; **Hint:** Here x = 30% and y = discount = 3 paise in 1 Re

ie $\frac{3}{100}\times 100 = 3\%$. Now apply the given formula and get the answer.

6. b; **Hint:** $x-10-\frac{x\times 10}{100}=26$

$\therefore x = 40\%$

7. d; **Hint:** See 'Note" given in the formula

$\text{loss }\% = \frac{(10)^2}{100} \doteq 1\%$

Rule 56

Theorem: ***If a man buys two items A and B for Rs P, and sells one item A so as to lose x% and the other item B so as to gain y%, and on the whole he neither gains nor loses,***

then (i) the cost of the item A is $\left(\frac{Py}{x+y}\right)$ ***and***

(ii) the cost of the item B is $\left(\frac{Px}{x+y}\right)$.

Illustrative Example

Ex.: A man buys two horses for Rs 1350. He sells one so as to lose 6% and the other so as to gain 7.5%. On the whole he neither gains nor loses. What does each horse cost?

Soln: **Detail Method:** Loss on one hors e= gain on the other

$\therefore$ 6% of the cost of first horse = 7.5% of the cost of the second.

$$\therefore \frac{\text{Cost of first horse}}{\text{Cost of second horse}} = \frac{7.5\%}{6\%} = \frac{15}{12} = \frac{5}{4}$$

Dividing Rs 1350 in the ratio of 5 : 4,

Cost of first horse = Rs 750

Cost of the second = Rs 600

Quicker Method: The above rule may be written as given below.

Cost of first horse

$$= \frac{\text{CP of both} \times \%\text{ loss or gain on 2nd}}{\%\text{ loss or gain on 1st} + \%\text{ loss or gain on 2nd}}$$

Cost of second horse

$$= \frac{\text{CP of both} \times \%\text{ loss or gain on 1st}}{\%\text{ loss or gain on 1st} + \%\text{ loss or gain on 2nd}}$$

Now, applying the formula, we have the

$$\text{cost of 1st horse} = \frac{1350\times 7.5}{6+7.5} = \text{Rs 750 and the}$$

$$\text{cost of 2nd hose} = \frac{1350\times 6}{6+7.5} = \text{Rs 600.}$$

Exercise

1. A man buys two chairs for Rs 1250. He sells one so as to lose 5% and the other so as to gain 20%. On the whole he neither gains nor loses. What does each chair cost?
   a) Rs 1000, Rs 250 b) Rs 1100, Rs 150
   c) Rs 900, Rs 350 d) Rs 950, Rs 300
2. A man buys two tables for Rs 1750. He sells one so as to lose 10% and the other so as to gain 15%. On the whole he neither gains nor loses. What does each table cost?
   a) Rs 1000, Rs 750 b) Rs 1050, Rs 700
   c) Rs 950, Rs 800 d) Can't be determined
3. A man buys two oxen for Rs 3000. He sells one so as to lose 5% and the other so as to gain 25%. On the whole he neither gains nor loses. What does each ox cost?
   a) Rs 2500, Rs 500 b) Rs 2100, Rs 900
   c) Rs 2400, Rs 600 d) Rs 2700, Rs 300
4. A man buys two goats for Rs 440. He sells one so as to lose 12% and the other so as to gain 28%. On the whole he neither gains nor loses. What does each goat cost?
   a) Rs 306, Rs 134 b) Rs 300, Rs 140
   c) Rs 308, Rs 132 d) Can't be determined
5. A man buys two horses for Rs 1550. He sells one so as to lose 23% and the other so as to gain 27%. On the whole he neither gains nor loses. What does each horse cost?
   a) Rs 807, Rs 743 b) Rs 817, Rs 733
   c) Rs 827, Rs 723 d) Rs 837, Rs 713

Answers

1. a 2. b 3. a 4. c 5. d

Rule 57

Theorem: ***By selling a certain item at the rate of 'X' items a rupee, a man loses x%. If he wants to gain y%, then the number of items should be sold for a rupee is*** $\left[\left(\frac{100-x}{100+y}\right)X\right]$.

Illustrative Example

Ex.: By selling oranges at 32 a rupee, a man loses 40%. How many for a rupee should he sell in order to gain 20%?

Soln: Applying the above formula, we have,

$$\text{the required answer} = 32\left(\frac{100-40}{100+20}\right) = \frac{32\times 60}{120} = 16$$

Exercise

1. By selling bananas at 21 a rupee, a man loses 30%. How many for a rupee should he sell in order to gain 5%?
   a) 15 b) 12 c) 14 d) 16
2. By selling apples at 46 a rupee, a man loses 15%. How many for a rupee should he sell in order to gain 15%?
   a) 17 b) 34 c) 27 d) 24

3. By selling mangos at 44 a rupee, a man loses 20%. How many for a rupee should he sell in order to gain 10%?
a) 32 b) 30 c) 33 d) 28
4. By selling oranges at 35 a rupee, a man loses 25%. How many for a rupee should he sell in order to gain 25%?
a) 35 b) 21 c) 25 d) 30
5. By selling 12 oranges for one rupee a man loses 20%. How many for a rupee should he sell to get a gain of 20%? **[CDS Exam 1989]**
a) 5 b) 8 c) 10 d) 15

Answers

1. c 2. b 3. a 4. b 5. b

Rule 58

Theorem: ***By selling an article for Rs A, a dealer makes a profit of x%. If he wants to make profit of y%, then he should increase his selling price by Rs*** $\left[\left(\frac{y-x}{100+x}\right)A\right]$ ***and the selling price is given by Rs*** $\left[\left(\frac{100+y}{100+x}\right)\times A\right]$.

Illustrative Example

Ex.: By selling an article for Rs 19.50 a dealer makes a profit of 30%. By how much should he increase his selling price so as to make a profit of 40%. Find the selling price also.

Soln: Applying the above formula, we have

$$\text{increase in selling price} = \left(\frac{40-30}{100+30}\right)19.50 = \text{Rs } 1.5$$

and

$$\text{the selling price} = \left(\frac{100+40}{100+30}\right)19.50 = \text{Rs } 21.$$

Exercise

1. By selling wheat at Rs 2.34 a kg a grocer gained 17%. At what price should he have sold it to gain 20%?
a) Rs 1.40 b) Rs 2.40 c) Rs 3.40 d) Rs 3.50
2. A machine is sold for Rs 5060 at a gain of 10%. What would have been gained or lost per cent if it had been sold for Rs 4370?
a) 5% loss b) 5% gain c) 22% gain d) 22% loss
3. A time piece is sold for Rs 64 at a gain of 28%. What would be gained or lost per cent by selling it for Rs 47.50 P?
a) 4% profit b) 5% loss c) 4% loss d) 5% profit
4. By selling 12 kg of onions for Rs 63, I gain 5%. What do I gain or lose per cent by selling 50 kg of the same onions for Rs 247.50?
a) 20% loss b) 1% loss c) 5% loss d) 1% profit
5. By selling an article for Rs 77 a person gained $\frac{1}{10}$ of his outlay. What would he have gained or lost per cent by selling it for Rs 63?
a) 10% profit b) 20% loss
c) 10% loss d) Data inadequate

Answers

1. b
2. a; **Hint:** From the given formula we have to find the value y as given below.

$$\left(\frac{100+y}{100+10}\right)\times 5060 = 4370 \therefore y = -5\%$$

Here, -ve sign shows that there is a loss.

3. b
4. b; **Hint:** $\left(\frac{100+y}{100+5}\right)\frac{63}{12} = \frac{247.50}{50}$

$\therefore$ y = -1% [-ve sign shows loss]
$\therefore$ required answer = 1% loss

5. c; **Hint:** $\left(\frac{100+y}{100+10}\right)77 = 63$

$[x = \frac{1}{10} \text{ of outlay} = \frac{1}{10}\times 100 = 10\%]$

$\therefore$ y = -10% $\therefore$ there is a loss of 10%.

Rule 59

Theorem: ***By selling an article for Rs A, a dealer makes a loss of x%. If he wants to make a profit of y%, then he should increase his selling price by Rs*** $\left[\left(\frac{x+y}{100-x}\right)A\right]$ ***and the selling price is given by Rs*** $\left[\left(\frac{100+y}{100-x}\right)A\right]$.

Illustrative Examples

Ex. 1: By selling an article for Rs 160 a dealer makes a loss of 20%. By how much should he increase his selling price so as to make a profit of 25%. Also find the selling price.

Soln: Following the above theorem, we have

$$\text{increase in selling price} = \frac{25+20}{100-20}\times 160 = \text{Rs } 90 \text{ and}$$

$$\text{the selling price} = \frac{100+25}{100-20}\times 160 = \text{Rs } 250.$$

Ex. 2: A man bought a certain quantity of rice at the rate of Rs 150 per quintal. 10% of the rice was spoiled. At what price should he sell the remaining to gain 20% of his outlay?

Soln: "10% of the rice is spoiled" may be considered as if he bought the rice at 10% loss. ie if he sells the rice at Rs 150 per quintal, 10% loss will occur. Now, applying the above formula, we have the selling price

$$= \text{Rs } 150\left(\frac{100+20}{100-10}\right) = \text{Rs } 150 \times \frac{120}{90} = \text{Rs } 200.$$

Exercise

1. By selling salt at Rs 55.80 per quintal a dealer lost 7 per cent. At what price should he have sold it to gain 7 per cent?
a) Rs 64 b) Rs 64.02 c) Rs 64.20 d) Rs 64.25
2. By selling a carriage for Rs 570, I would lose 5%. At what price must I sell it to gain 5%?
a) Rs 630 b) Rs 840 c) Rs 420 d) Rs 730
3. By selling a book for Rs 14 a book seller would lose $12\frac{1}{2}\%$. At what price should he sell it to gain $12\frac{1}{2}\%$.
a) Rs 16 b) Rs 28 c) Rs 24 d) Rs 18
4. By selling clay at Rs 38 a kilogram, a dealer lost 5%. At what price should he have sold it to gain 30%?
a) Rs 52 b) Rs 78 c) Rs 54 d) Rs 64
5. If $6\frac{1}{2}$ per cent is lost by selling an article for Rs 9.35, for what price should it be sold to gain 13%?
a) Rs 11 b) Rs 12.30 c) Rs 11.30 d) Rs 13.30
6. A horse is sold for Rs 1230 at a loss of 18%. What would have been gained or lost per cent if it had been sold for Rs 1600?
a) $6\frac{2}{3}\%$ loss b) $6\frac{2}{3}\%$ gain
c) $6\frac{1}{3}\%$ gain d) Can't be determined
7. If 13% is lost by selling goods for Rs 295.80, what would be gained or lost per cent by selling them for Rs 323?
a) 5% loss b) 5% gain c) 3% loss d) 10% gain
8. If by selling an article for Rs 60, a person loses $\frac{1}{7}$ of his outlay. What would he have gained or lost per cent by selling it for Rs 77?
a) 20% gain b) 20% loss c) 10% gain d) 10% loss
9. On selling an article for Rs 240, a trader loses 4%. In order to gain 10%, he must sell that article for **[NDA Exam, 1990]**
a) Rs 264.00 b) Rs 273.20 c) Rs 275.00 d) Rs 280.00

Answers

1. c 2. a 3. d 4. a 5. c

6. b; **Hint:** $\left(\frac{100+y}{100-18}\right)\times 1230 = 1600$

$\therefore y = \frac{20}{3} = 6\frac{2}{3}\%$ gain

7. a

8. c; **Hint:** $\left(\frac{100+y}{100-\frac{100}{7}}\right)\times 60 = 77$

$\left[x = \frac{1}{7} \text{ of outlay} = \frac{1}{7}\times 100 = \frac{100}{7}\%\right]$

$\therefore$ y = 10% $\therefore$ required answer = 10% gain.

9. c

Rule 60

Theorem: ***A dealer buys an item at x% discount on its original price. If he sells it at a y% increase on the original price, then the per cent profit he gets is*** $\left(\frac{y+x}{100-x}\right)\times 100$.

Illustrative Example

Ex.: A dealer bought a horse at 20% discount on its original price. He sold it at 40% increase on the original price. What percentage profit did he get?

Soln: Following the above theorem, we have

dealer's profit per cent $= \frac{40+20}{100-20}\times 100 = 75\%$.

Exercise

1. A dealer bought a horse at 10% discount on its original price. He sold it at 20% increase on the original price. What percentage profit did he get?
a) $33\frac{1}{3}\%$ b) $9\frac{1}{11}\%$ c) $11\frac{4}{9}\%$ d) $9\frac{7}{11}\%$
2. A dealer bought a horse at 5% discount on its original price. He sold it at 10% increase on the original price. What percentage profit did he get?
a) $5\frac{5}{9}\%$ b) $5\frac{15}{19}\%$ c) $7\frac{5}{19}\%$ d) $5\frac{7}{19}\%$
3. A dealer bought a horse at 25% discount on its original price. He sold it at 45% increase on the original price. What percentage profit did he get?
a) $26\frac{1}{3}\%$ b) $16\frac{2}{3}\%$ c) $93\frac{1}{3}\%$ d) Data inadequate

Answers

1. a 2. b 3. c

Rule 61

Theorem: ***If a trader buys an article at x% discount on its original price and sells it at y% increase on the price he buys it, then the percentage of profit he makes on the original price is*** $\left[y-x-\frac{xy}{100}\right]$.

Illustrative Example

Ex.: A trader bought a car at 20% discount on its original price. He sold it at 40% increase on the price he bought it. What percentage of profit did he make on the original price?

Soln: Applying the above theorem, we have

$$\%\text{ profit} = 40-20-\frac{40\times 20}{100}=12\%.$$

Exercise

1. A trader bought a car at 15% discount on its original price. He sold it at 30% increase on the price he bought it. What percentage of profit did he make on the original price?
   a) 11.5% b) 10.5% c) 11% d) 10%
2. A trader bought a car at 10% discount on its original price. He sold it at 20% increase on the price he bought it. What percentage of profit did he make on the original price?
   a) 8% b) 6% c) 10% d) 12%
3. A trader bought a car at 5% discount on its original price. He sold it at 10% increase on the price he bought it. What percentage of profit did he make on the original price?
   a) 5.5% b) 6.5% c) 4.5% d) 3.5%

Answers

1. b 2. a 3. c

Rule 62

Ex.: A horse worth Rs 9000 is sold by A to B at 10% loss. B sells the horse back to A at 10% gain. Who gains and who loses? Find also the values.

Soln: **Detail Method:**

A sells to B for Rs $9000\left(\frac{90}{100}\right)$ = Rs 8100

Again, B sells to A for Rs $8100\left(\frac{110}{100}\right)$ = Rs 8910

Thus, A loses Rs (8910 - 8100) = Rs 810.

In this whole transaction, A's investment is only Rs 9000 (the cost of the horse) because the horse returned to his hand.

$\therefore$ A's% loss $=\frac{810}{9000}\times 100=9\%$

B gains Rs 810 (the same as A loses) and his investment in this transaction is Rs 8100.

$\therefore$ B's % gain $=\frac{810}{8100}\times 100=10\%$

Quicker Maths (direct formula): In such case, the first buyer bears loss and his % of loss is given by $\frac{\%\text{ gain }(100-\%\text{ loss})}{100}$

In this case, A's loss% $=\frac{10(100-10)}{100}=9\%$

and loss amount $=9000\left(\frac{9}{100}\right)$ = Rs 810.

Exercise

1. A horse worth Rs 16000 is sold by A to B at 5% loss. B sells the horse back to A at 5% gain. Find the value of loss amount.
   a) Rs 860 b) Rs 760 c) Rs 1260 d) Rs 960
2. A horse worth Rs 12000 is sold by A to B at 15% loss. B sells the horse back to A at 15% gain. Find the value of loss amount.
   a) Rs 1530 b) Rs 1630 c) Rs 1350 d) Rs 1550
3. A horse worth Rs 8000 is sold by A to B at 20% loss. B sells the horse back to A at 20% gain. Find the value of loss amount.
   a) Rs 1380 b) Rs 1480 c) Rs 1180 d) Rs 1280
4. A horse worth Rs 5000 is sold by A to B at 25% loss. B sells the horse back to A at 25% gain. Find the value of loss amount.
   a) Rs 937.5 b) Rs 938
   c) Rs 938.5 d) Can't be determined

Answers

1. b 2. a 3. d 4. a

Rule 63

Theorem: ***A person marks his goods x% above the cost price but allows y% discount for cash payment. If he sells the article for Rs X, then the cost price is*** $X\left(\frac{100}{100+x}\right)\left(\frac{100}{100-y}\right)$.

Illustrative Example

Ex.: Satish marks his goods 25% above the cost price but allows 12.5% discount for cash payment. If he sells the article for Rs 875, find his cost price.

Soln: Applying the above formula, we have the

$$\text{cost price} = 875\left(\frac{100}{100+25}\right)\left(\frac{100}{100-12.5}\right) = \text{Rs } 800.$$

Exercise

1. Ramesh marks his goods 20% above the cost price but allows 20% discount for cash payment. If he sells the article for Rs 960, find his cost price.
   a) Rs 800 b) Rs 900
   c) Rs 1000 d) Data inadequate
2. Rakesh marks his goods 10% above the cost price but allows 10% discount for cash payment. If he sells the article for Rs 1980, find his cost price.
   a) Rs 3000 b) Rs 2400 c) Rs 2500 d) Rs 2000
3. Ritesh marks his goods 15% above the cost price but allows 25% discount for cash payment. If he sells the article for Rs 3450, find his cost price.
   a) Rs 4000 b) Rs 4200 c) Rs 4300 d) Rs 4125
4. Rishav marks his goods 30% above the cost price but allows 30% discount for cash payment. If he sells the article for Rs 2730, find his cost price.
   a) Rs 3500 b) Rs 2800 c) Rs 2950 d) Rs 3000

Answers

1. c 2. d 3. a 4. d

Rule 64

Ex.: A milkman buys some milk. If he sells it at Rs 5 a litre, he loses Rs 200, but when he sells it at Rs 6 a litre, he gains Rs 150. How much milk did he purchase?

Soln: **Detail Method:** Difference in selling price = Rs 6/litre – Rs 5/litre = Re 1/litre.

If he increases the SP by Re 1/litre, he gets Rs 200 + Rs 150 = Rs 350 more.

$$\therefore \text{ he purchased } \frac{\text{Rs } 350}{\text{Re } 1/\text{litre}} = 350 \text{ litres milk.}$$

Quicker Maths (direct formula)

Quantity of milk

$$= \frac{\text{Difference of amount}}{\text{Difference of rate}} = \frac{150-(-200)}{6-5}$$

$$= \frac{350}{1} = 350 \text{ litres.}$$

Exercise

1. A milkman buys some milk. If he sells it at Rs 10 a litre, he loses Rs 400, but when he sells it at Rs 12 a litre, he gains Rs 300. How much milk did he purchase?
   a) 360 litres b) 700 litres c) 350 litres d) 175 litres
2. A milkman buys some milk. If he sells it at Rs 2 a litre, he loses Rs 125, but when he sells it at Rs 7 a litre, he gains Rs 125. How much milk did he purchase?
   a) 100 litres b) 50 litres c) 75 litres d) Data inadequate
3. A milkman buys some milk. If he sells it at Rs 8 a litre, he loses Rs 233, but when he sells it at Rs 9 a litre, he gains Rs 317. How much milk did he purchase?
   a) 550 litres b) 450 litres c) 600 litres d) 560 litres

Answers

1. c 2. b 3. a

Rule 65

Ex.: A fruit merchant makes a profit of 25% by selling mangoes at a certain price. If he charges Re 1 more on each mango, he would gain 50%. Find what price per mango did he sell at first. Also find the cost price per mango.

Soln: **Detail Method:** Suppose the cost price of a mango be Rs x.

$$\text{Then, first selling price} = \text{Rs } x\left(\frac{100+25}{100}\right) = \text{Rs } \frac{5x}{4}$$

If he charges Re 1 more and gets 50% profit then there exists a relationship.

$$\frac{5x}{4} + 1 = x\left(\frac{100+50}{100}\right) = \frac{3x}{2}$$

$$\text{or, } \frac{3x}{2} - \frac{5x}{4} = 1$$

$\therefore x = \text{Rs } 4$

$\therefore$ Cost price/mango = Rs 4

$$\text{and first selling price} = 4\left(\frac{125}{100}\right) = \text{Rs } 5.$$

Quicker Maths (direct formula):

$$\text{Cost price} = \frac{100 \times \text{More charge}}{\%\text{ Difference in profit}} \text{ and}$$

$$\text{Selling price} = \frac{\text{More charge }(100 + \%\text{ first profit})}{\%\text{ Difference in profit}}$$

Thus in this case

$$CP = \frac{100 \times 1}{50-25} = \text{Rs } 4,\ SP = \frac{1 \times 125}{50-25} = \text{Rs } 5.$$

Exercise

1. A fruit merchant makes a profit of 15% by selling oranges at a certain price. If he charges Rs 2 more on each orange, he would gain 35%. Find what price per orange did he sell at first. Also find the cost price per orange.
   a) Rs 11.5, Rs 10 b) Rs 10.5, Rs 11
   c) Rs 12, Rs 11.5 d) Rs 11.5, Rs 9
2. A fruit merchant makes a profit of 5% by selling apples at a certain price. If he charges Rs 2 more on each apple, he would gain 30%. Find what price per apple did he sell

at first. Also find the cost price per apple.

a) Rs 8.4, Rs 6 b) Rs 8, Rs 6.4
c) Rs 8.4, Rs 8 d) None of these

3. A fruit merchant makes a profit of 10% by selling bananas at a certain price. If he charges Re 1 more on each banana, he would gain 20%. Find what price per banana did he sell at first. Also find the cost price per mango.

a) Rs 24, Rs 20 b) Rs 12, Rs 10
c) Rs 24, Rs 12 d) Rs 20, Rs 10

4. A fruit merchant makes a profit of 10% by selling mangoes at a certain price. If he charges Re 1 more on each mango, he would gain 25%. Find what price per mango did he sell at first. Also find the cost price per mango.

a) Rs $8\frac{1}{3}$, Rs $6\frac{2}{3}$ b) Rs $8\frac{2}{3}$, Rs $6\frac{1}{3}$

c) Rs $6\frac{2}{3}$, Rs $5\frac{1}{3}$ d) Can't be determined

Answers

1\. a 2. c 3. b 4. a

Rule 66

Theorem: The rule of column: ***x kg of milk costs as much as y kg of rice, z kg of rice costs as much as p kg of pulse, w kg of pulse costs as much as t kg of wheat, u kg of wheat costs as much as v kg of edible oil. If n kg of milk costs Rs A, then the cost of m kg of edible oil is Rs*** $\left(\frac{A\times x\times z\times w\times u\times m}{n\times y\times p\times t\times v}\right)$.

Note: 1. *x kg of milk costs as much as y kg of rice, z kg of rice costs as much as p kg of pulse, w kg of pulse costs as much as t kg of wheat, u kg of wheat costs as much as v kg of edible oil. If k kg of edible oil costs Rs B, then the cost of n kg of milk is Rs* $\left(\frac{B\times v\times t\times p\times y\times n}{k\times u\times w\times z\times x}\right)$.

Ex.: 9 kg of rice costs as much as 4 kg of sugar, 14 kg of sugar costs as much as 1.5 kg of tea, 2 kg of tea costs as much as 5 kg of coffee. Find the cost of 2.5 kg of rice, if 11 kg of coffee costs Rs 462.

Soln: Applying the above theorem, the required answer

$$=\frac{462\times2.5\times4\times1.5\times5}{11\times2\times14\times9}=\text{Rs }12.50.$$

2. For detail please consult "*Magical Book on Quicker Maths*"

Illustrative Example

Ex.: 9 kg of rice costs as much as 4 kg of sugar, 14 kg of sugar costs as much as 1.5 kg of tea, 2 kg of tea costs as much as 5 kg of coffee, find the cost of 11 kg of coffee, if 2.5 kg of rice costs Rs 12.50.

Soln: Applying the above formula, we have

$$\text{the required answer}=\frac{12.5\times9\times14\times2\times11}{2.5\times4\times1.5\times5}=\text{Rs }462.$$

Exercise

1. 5 chairs cost as much as 12 stools, 7 stools as much as 2 tables, 3 tables as much as 2 sofas, if the cost of 5 sofas be Rs 8750, find that of a chair.

a) Rs 800 b) Rs 875
c) Rs 950 d) Can't be determined

2. If 2 horses are worth 3 camels, and 9 camels are worth 10 bicycles and 100 bicycles are worth 3 motor cars, what is the price of a horse, if a motor car costs Rs 192000?

a) Rs 960 b) Rs 9600 c) Rs 320 d) Rs 3200

3. If 3 cups cost as much as 2 plates and 9 plates as much as 2 kettles and one kettle as much as 3 dishes, what is the price of a cup, if a dish costs Rs 13.50?

a) Rs 6 b) Rs 3 c) Rs 9 d) Rs 12

4. A gets Rs 3 as often as B gets Rs 4, B gets Rs 5 as often as C gets Rs 6, C gets Rs 8 as often as D gets Rs 15, if A gets Rs 3.25, what will D get?

a) Rs 9.50 b) Rs 9.25
c) Rs 9.75 d) Can't be determined

5. 10 kg of rice costs as much as 20 kg of wheat, 25 kg of wheat costs as much as 2 kg of tea, 5 kg of tea costs as much as 25 kg of sugar. Find the cost of 6 kg of sugar if 4 kg of rice costs Rs 32.

a) Rs 60 b) Rs 50 c) Rs 65 d) Rs 45

6. 20 kg of potato costs as much as 5 kg of tomato, 12 kg of tomato costs as much as 30 kg of onion, 15 kg of onion costs as much as 18 kg of cabbage. If 10 kg of cabbage costs Rs 50. Find the cost of 24 kg of potato.

a) Rs 90 b) Rs 72 c) Rs 108 d) Rs 96

Answers

1\. a; **Hint: See Note:**

x rupees = 1 chair
5 chairs = 12 stools
7 stools = 2 tables
3 tables = 2 sofas
5 sofas = Rs 8750

$$\therefore\ x=\frac{1\times12\times2\times2\times875}{5\times7\times3\times5}=800$$

$\therefore$ the price of a chair si Rs 800.

2\. b; **Hint:** Here cost of one motor car is given as Rs 192000
$\therefore$ cost of 3 motor cars is as Rs 192000 × 3.

$$\therefore\ \text{required answer}=\frac{1\times3\times10\times3\times192000}{2\times9\times100}=\text{Rs }9600$$

3\. a 4. c

5\. a; **Hint:** Cost of 6 kg of sugar $=\frac{10\times25\times5\times6\times32}{20\times2\times25\times4}=\text{Rs }60$

6. a; **Hint:** Cost of 24 kg of potato $= \frac{5\times30\times18\times50\times24}{20\times12\times15\times10}$

$= $ Rs 90.

Rule 67

Theorem: ***A person saves x% of his income. If his income increases by y% and he decides to save z% of his increased income, then the per cent by which his saving is increased is given by*** $\left[z\left(\frac{100+y}{x}\right)-100\right]$.

Note: *If he sticks to his previous saving habit of x% then by the direct formula,*

$$\% \text{ increase in saving} = x\left(\frac{100+y}{x}\right)-100$$

= y%, which is same as the percentage increase in income.

Illustrative Example

Ex.: A person saves 10% of his income. If his income increases by 20% and he decides to save 15% of his income, by what per cent has his saving increased?

Soln: Applying the above rule, we have the

$$\% \text{ increase in saving} = \frac{(100+20)15-10\times100}{10} = 80\%$$

Exercise

1. A person saves 20% of his income. If his income increases by 40% and he decides to save 30% of his income, by what per cent has his saving increased?
   a) 90% b) 100% c) 160% d) 110%
2. A person saves 25% of his income. If his income increases by 15% and he decides to save 24% of his income, by what per cent has his saving increased?
   a) 10% b) 15% c) 10.4% d) 10.5%
3. A person saves 15% of his income. If his income increases by 25% and he decides to save 20% of his income, by what per cent has his saving increased?
   a) $66\frac{2}{3}\%$ b) $66\frac{1}{3}\%$ c) $33\frac{1}{3}\%$ d) $33\frac{2}{3}\%$

Answers

1. d 2. c 3. a

Rule 68

Theorem: ***When each of the two commodities is sold at the same price Rs A, and a profit of P% is made on the first and a loss of L% is made on the second, then the percentage gain or loss is*** $\frac{100(P-L)-2PL}{(100+P)+(100-L)}$ ***according to the +ve or –ve sign.***

Illustrative Examples

Ex. 1: Each of the two cars is sold at the same price. A profit of 10% is made on the first and a loss of 7% is made on the second. What is the combined loss or gain?

Soln: By the theorem, we have

$\frac{100(10-7)-2\times10\times7}{200+10-7} = \frac{160}{203}\%$ gain as the sign is +ve.

Ex. 2: Each of the two horses is sold for Rs 720. The first one is sold at 25% profit and the other one at 25% loss. What is the % loss or gain in this deal?

Soln: **Detail Method:** Total selling price of two horses $= 2 \times 720 =$ Rs 1,440

The CP of first horse $= 720 \times \frac{100}{125} =$ Rs 576

The CP of second horse $= 720 \times \frac{100}{75} =$ Rs 960

Total CP of two horses $= 576 + 960 =$ Rs 1,536

Therefore, loss = Rs 1,536 – Rs 1,440 = Rs 96

$\therefore$ % loss $= \frac{96\times100}{1536} = 6.25\%$

Direct Formula: (See theorem; note):

In this type of question where SP is given and profit and loss percentage are same, there is always loss and the

$$\% \text{ loss} = \frac{(25)^2}{100} = \frac{625}{100} = 6.25\%$$

[The above example is a special case when percentage values of loss and gain are the same.]

Note: 1. In the special case when P = L we have

$$\frac{100\times0-2P^2}{200} = -\frac{P^2}{100}$$

Since the sign is –ve, there is always loss and the value is given as $\frac{(\%\text{ value})^2}{100}$.

2. *When each of the two commodities is sold at the same price Rs A, and a profit of P% is made on the first and a profit of L% is made on the second, then the percentage gain is* $\frac{100(P+L)+2PL}{(100+P)+(100+L)}$.

Ex.: A man sells two articles, each for the same price Rs 500. He earns 25% profit on the first and 15% profit on the second. Find his overall per cent profit.

Soln: Applying the above theorem, we have

% profit

$$= \frac{100(25+15)+2\times25\times15}{(100+25)+(100+15)} = \frac{4750}{240} = 19.79\%$$

3. When each of the two commodities is sold at the same price Rs A and a profit of P% is made on the first and a profit of P% is made on the second, then the percentage gain is [P%].

Ex.: A man sells two articles, each for the same price Rs 300. He earns 20% profit on the first and 20% profit on the second. Find is overall per cent profit.

Soln: Following the above theorem, we have the required per cent profit = 20%

Exercise

1. A man sells two goats at Rs 120 each and by doing so he gains 15% on one goat and loses 15% on the other. What is the % loss or gain in this deal?
   a) 22.5% gain b) 2.25% gain
   c) 2.25% loss d) 2.25% gain
2. A person sells two horses at Rs 240 each and by doing so he gains 27% on one goat and loses 27% on the other. What is the % loss or gain in this deal?
   a) 7.29% gain b) 7.29% loss c) 8.29% loss d) 6.29% loss
3. A man sells two articles, each for the same price Rs 890. He earns 20% profit on the first and 14% profit on the second. Find his overall per cent profit.
   a) $\frac{160}{203}\%$ gain b) $\frac{320}{203}\%$ loss
   c) $\frac{20}{103}\%$ gain d) $\frac{40}{103}\%$ gain
4. A man sells two articles, each for the same price Rs 640. He earns 20% profit on the first and 10% profit on the second. Find his overall per cent profit.
   a) 14.78% b) 14.08% c) 14.58% d) None of these
5. A man sells two articles, each for the same price Rs 550. He earns 25% profit on the first and 35% profit on the second. Find his overall per cent profit.
   a) 29.8% b) 29% c) 30.6% d) 30.8%
6. A man sells two houses at the rate of Rs 1.995 lakh each. On one he gains 5% and on the other he loses 5%. His gain or loss per cent in the whole transaction is:
   [Central Excise & I. Tax 1988]
   a) 0.25% loss b) 0.25% gain
   c) 2.5% loss d) neither loss nor gain

Answers

1. c 2. b 3. c 4. a 5. a
6. a; **Hint:** See **'note'** in the given rule.

Rule 69

Theorem: ***A person calculates his percentage on the selling price whereas the other person calculates his profit on the cost price. They find that the difference of their profits is Rs A. If the selling price of both of them are the same, and one gets x% profit and the other gets y% profit then selling price*** $= \text{Rs}\frac{A\times100\times(100+y)}{(100)^2-(100+y)(100-x)}$

or $\frac{\text{Difference in profit}\times100\times(100+y)}{(100)^2-(100+y)(100-x)}$

Illustrative Example

Ex.: Rakesh calculates his profit percentage on the selling price whereas Ramesh calculates his profit on the cost price. They find that the difference of their profits is Rs 150. If the selling price of both of them are the same, and Rakesh gets 25% profit and Ramesh gets 20% profit, then find their selling price.

Soln: Applying the above formula, we have the

$$\text{selling price} = \frac{150\times100\times(100+20)}{(100)^2-(100+20)(100-25)}$$

$$= \frac{150\times100\times120}{100\times100-120\times75} = \text{Rs}\,1800.$$

Note: *A person calculates his profit percentage on the selling price where the other person calculates his profit on the cost price. They find that the difference of their profits is Rs A. If the selling price of both of them are the same and both of them gets x% profit, then the selling price* $= \text{Rs}\frac{A\times100\times(100+x)}{x^2}$ *or*

$$\frac{\text{Difference in profit}\times100\times(100+x)}{x^2}.$$

Ex.: Rakesh calculates his profit percentage on the selling price whereas Ramesh calculates his on the cost price. They find that the difference of their profits is Rs 100. If the selling price of both of them are the same and both of them get 25% profit, find their selling price.

Soln: **Detail Method:** Suppose the selling price for both of them is Rs x.

Now, cost price of Rakesh $= x\left(\frac{100-25}{100}\right) = \frac{3}{4}x$.

and cost price of Ramesh $= x\left(\frac{100}{100+25}\right) = \frac{4}{5}x$.

Rakesh's profit $= x - \frac{3}{4}x = \frac{x}{4}$

Ramesh's profit $= x - \frac{4}{5}x = \frac{x}{5}$

Now, difference of their profits

$$= \frac{x}{4} - \frac{x}{5} = \text{Rs}\,100\ \text{(given)}$$

or, $\frac{x}{20} = 100$

$\therefore$ x = Rs 2000

Thus selling price = Rs 2000

Quicker Method: Applying the above rule, we have

$$\text{the selling price} = \frac{\text{Diff. in profit} \times 100 \times (100 + 25)}{(25)^2}$$

$$= \frac{100 \times 100 \times 125}{25 \times 25} = \text{Rs } 2000.$$

Exercise

1. Sunil calculates his profit percentage on the selling price whereas Sujeet calculates his profit on the cost price. They find that the difference of their profits is Rs 900. If the selling price of both of them are the same, and Sunil gets 50% profit and Sujeet gets 40% profit, then find their selling price.
   a) Rs 4200 b) Rs 4500 c) Rs 4000 d) Rs 4800
2. Ajay calculates his profit percentage on the selling price whereas Vijay calculates his profit on the cost price. They find that the difference of their profits is Rs 130. If the selling price of both of them are the same, and Ajay gets 15% profit and Vijay gets 10% profit, then find their selling price.
   a) Rs 220 b) Rs 2200 c) Rs 2300 d) Rs 230
3. Pankaj calculates his profit percentage on the selling price whereas Chandan calculates his profit on the cost price. They find that the difference of their profits is Rs 135. If the selling price of both of them are the same, and Pankaj gets 30% profit and Chandan gets 25% profit, then find their selling price.
   a) Rs 1250 b) Rs 1150 c) Rs 1450 d) Rs 1350
4. Vineet calculates his profit percentage on the selling price whereas Roshan calculates his profit on the cost price. They find that the difference of their profits is Rs 275. If the selling price of both of them are the same, and Vineet gets 25% profit and Roshan gets 15% profit, then find their selling price.
   a) Rs 2350 b) Rs 2300
   c) Rs 2100 d) Rs 2250
5. Miheer calculates his profit percentage on the selling price whereas Satya calculates his profit on the cost price. They find that the difference of their profits is Rs 110. If the selling price of both of them are the same, and Miheer gets 10% profit and Satya gets 5% profit, then find their selling price.
   a) Rs 2200 b) Rs 2250
   c) Rs 1750 d) Rs 2100

Answers

1. a 2. b 3. d 4. b 5. d

Rule 70

Theorem: ***A man sells two items for Rs A. The cost price of the first is equal to the selling price of the second. If the first is sold at x% loss and the second at y% gain, then the total gain or loss is Rs*** $\left[\frac{(100-x)-\left(\frac{100^2}{100+y}\right)}{100+(100-x)}\right] \times A$ ***according as the +ve or –ve sign.***

Illustrative Example

Ex.: A man sells two horses for Rs 1710. The cost price of the first is equal to the selling price of the second. If the first is sold at 10% loss and the second at 25% gain, what is his total gain or loss (in rupees)?

Soln: **Detail Method:** We suppose that the cost price of the first horse is Rs 100. Then we arrange our values in a tabular form:

	1st horse	2nd horse	Total
CP	100	$100\left(\frac{100}{125}\right) = 80$	180
SP	$100\left(\frac{90}{100}\right) = 90$	100	190

$\therefore$ CP : SP = 180 : 190 = 18 : 19

$\therefore$ Profit $= \frac{19-18}{19} \times 1710 = \text{Rs } 90$

Quicker Method: Applying the above rule, we have

$$\text{profit} = \frac{(100-10)-100\left(\frac{100}{100+25}\right)}{100+(100-10)} \times 1710$$

$$= \frac{90-80}{190} \times 1710 = \text{Rs } 90.$$

Exercise

1. A man sells two horses for Rs 1850. The cost price of the first is equal to the selling price of the second. If the first is sold at 15% loss and the second at 25% gain, what is his total gain or loss (in rupees)?
   a) Rs 50 loss b) Rs 500 loss
   c) Rs 50 gain d) Rs 500 gain
2. A man sells two horses for Rs 1950. The cost price of the first is equal to the selling price of the second. If the first is sold at 5% loss and the second at 25% gain, what is his total gain or loss (in rupees)?
   a) Rs 150 loss b) Rs 150 gain
   c) Rs 250 gain d) None of these

3. A man sells two horses for Rs 1475. The cost price of the first is equal to the selling price of the second. If the first is sold at 20% loss and the second at 25% gain, what is his total gain or loss (in rupees)?
a) Rs 60 loss b) Rs 60 gain
c) Rs 80 gain d) Neither gain nor loss
4. A man sells two horses for Rs 11900. The cost price of the first is equal to the selling price of the second. If the first is sold at 30% loss and the second at 25% gain, what is his total gain or loss (in rupees)?
a) Rs 600 loss b) Rs 700 loss
c) Rs 750 gain d) Rs 700 gain
5. A man sells two horses for Rs 1760. The cost price of the first is equal to the selling price of the second. If the first is sold at 24% loss and the second at 25% gain, what is his total gain or loss (in rupees)?
a) Rs 40 gain b) Rs 40 loss
c) Rs 60 loss d) Neither loss nor gain

Answers

1. c 2. b 3. d 4. b 5. b

Rule 71

Theorem: ***If a dealer sells an item for Rs A, making a profit of x%, and he sells another item at a loss of y%, and on the whole he makes neither profit nor loss, then the cost of the second table is Rs*** $\left[A\left(\frac{100}{100+x}\right)\frac{x}{y}\right]$.

Illustrative Example

Ex.: A dealer sells a table for Rs 400, making a profit of 25%. He sells another table at a loss of 10%, and on the whole he makes neither profit nor loss. What did the second table cost him?

Soln: **Detail Method:** Profit on the first table

$$= 400\left(\frac{25}{125}\right) = \text{Rs } 80$$

He loses Rs 80 on the second table (since there is neither profit nor loss)

$\therefore$ Cost price of second table $= \frac{80}{10}\times 100 = \text{Rs } 800$.

Quicker Method: Applying the above rule, we have cost price of the second table

$$= 400\left(\frac{100}{125}\right)\left(\frac{25}{10}\right) = \text{Rs } 800.$$

Exercise

1. A dealer sells a chair for Rs 600, making a profit of 20%. He sells another chair at a loss of 5%, and on the whole he makes neither profit nor loss. What did the second chair cost him?
a) Rs 4000 b) Rs 2000 c) Rs 4500 d) Rs 2500
2. A dealer sells a horse for Rs 460, making a profit of 15%. He sells another horse at a loss of 5%, and on the whole he makes neither profit nor loss. What did the second horse cost him?
a) Rs 1200 b) Rs 1250
c) Rs 1100 d) None of these
3. A dealer sells an ox for Rs 1260, making a profit of 20%. He sells another ox at a loss of 10%, and on the whole he makes neither profit nor loss. What did the second ox cost him?
a) Rs 2000 b) Rs 2200
c) Rs 2100 d) Data inadequate
4. A dealer sells a goat for Rs 260, making a profit of 30%. He sells another goat at a loss of 20%, and on the whole he makes neither profit nor loss. What did the second goat cost him?
a) Rs 300 b) Rs 360 c) Rs 350 d) Rs 290
5. A dealer sells a table for Rs 405, making a profit of 35%. He sells another table at a loss of 30%, and on the whole he makes neither profit nor loss. What did the second table cost him?
a) Rs 360 b) Rs 350 c) Rs 340 d) Rs 300

Answers

1. b 2. a 3. c 4. a 5. b

Rule 72

Theorem: If a discount of $x_1\%$ is given on the marked price of an article, the shopkeeper gets a profit of P%. If he offers a discount of $x_2\%$ on the same article, then his per cent profit is given by $(100+P)\left[\frac{100-x_2}{100-x_1}\right]-100$.

In other words, required % profit

$$= (100+\%\text{ profit})\left[\frac{100-\%2nd\text{ discount}}{100-\%1st\text{ discount}}\right]-100$$

Illustrative Example

Ex.: If a discount of 10% is given on the marked price of an article, the shopkeeper gets a profit of 20%. Find his % profit if he offers a discount of 20% on the same article.

Soln: **Detail Method:**
Suppose the marked price = Rs 100
Then selling price at 10% discount = Rs (100 – 10)
= Rs 90
Since he gets 20% profit, his cost price

$$= 90\left(\frac{100}{120}\right) = \text{Rs } 75$$

Now, at 20% discount, the selling price = Rs (100 – 20)
= Rs 80

Thus his % profit

$$= \frac{80-75}{75} \times 100 = \frac{500}{75} = \frac{20}{3} = 6\frac{2}{3}\%$$

Quicker Method: Applying the above rule

the required % profit $= (100+20)\left[\frac{100-20}{100-10}\right] - 100$

$$= 120\left(\frac{80}{90}\right) - 100 = \frac{320}{3} - 100 = \frac{20}{3} = 6\frac{2}{3}\%.$$

Exercise

1. If a discount of 15% is given on the marked price of an article, the shopkeeper gets a profit of 30%. Find his % profit if he offers a discount of 30% on the same article.

   a) $7\frac{1}{17}\%$ b) $7\frac{2}{17}\%$ c) $17\frac{3}{7}\%$ d) $17\frac{1}{7}\%$

2. If a discount of $12\frac{1}{2}\%$ is given on the marked price of an article, the shopkeeper gets a profit of 25%. Find his % profit if he offers a discount of 25% on the same article.

   a) $7\frac{2}{7}\%$ b) $7\frac{5}{7}\%$ c) $7\frac{1}{7}\%$ d) Data inadequate

3. If a discount of 5% is given on the marked price of an article, the shopkeeper gets a profit of 10%. Find his % profit if he offers a discount of 10% on the same article.

   a) $4\frac{5}{19}\%$ b) $4\frac{2}{19}\%$ c) $4\frac{3}{19}\%$ d) $4\frac{4}{19}\%$

4. If a discount of 15% is given on the marked price of an article, the shopkeeper gets a profit of 25%. Find his % profit if he offers a discount of 20% on the same article.

   a) $\frac{100}{7}\%$ b) $\frac{100}{17}\%$ c) $\frac{300}{7}\%$ d) $\frac{300}{17}\%$

5. If a discount of 20% is given on the marked price of an article, the shopkeeper gets a profit of 30%. Find his % profit if he offers a discount of 25% on the same article.

   a) $21\frac{1}{8}\%$ b) $11\frac{7}{8}\%$ c) $21\frac{7}{8}\%$ d) $11\frac{1}{8}\%$

Answers

1. a 2. c 3. d 4. d 5. c

Rule 73

Ex.: A farmer sold a cow and a calf for Rs 760 and got a profit of 10% on the cow and 25% on the calf. If he sells the cow and the calf for Rs 767.50 and gets a profit of 25% on the cow and 10% on the calf, find the individual cost price of the cow and the calf.

Soln: Quicker Method: (i) For cost of cow:

	Cow	Calf
(1)	110%+	125% = 760
(2)	125%+	110% = 767.5

$$\text{Cost of cow} = \frac{125\% \text{ of } 767.5 - 110\% \text{ of } 760}{(125\%)^2 - (110\%)^2}$$

$$= \frac{\frac{5}{4}\times 767.5 - \frac{11}{10}\times 760}{(1.25)^2 - (1.1)^2}$$

$$= \frac{959.375 - 836}{(1.25+1.1)(1.25-1.1)} = \frac{123.375}{2.35\times 0.15} = \text{Rs } 350.$$

(ii) For cost of calf:

	Cow		Calf
(1) SP	110% + 125%	=	760
(2) SP	125% + 110%	=	767.5

$$\text{Cost of calf} = \frac{125\% \text{ of } 760 - 110\% \text{ of } 767.5}{(125\%)^2 - (110)^2}$$

$$= \frac{950 - 844.25}{2.35\times 0.15} = \text{Rs } 300.$$

Exercise

1. A farmer sold a horse and a mule for Rs 1520 and got a profit of 20% on the horse and 50% on the mule. If he sells the horse and the mule for Rs 1535 and gets a profit of 50% on the horse and 20% on the mule, find the individual cost price of the horse and the mule.

   a) Rs 590.74, Rs 540.74 b) Rs 590.84, Rs 540.84
   c) Rs 580.74, Rs 550.74 d) Can't be determined

2. A farmer sold a cow and a ox for Rs 800 and got a profit of 20% on the cow and 25% on the ox. If he sells the cow and the ox for Rs 820 and gets a profit of 25% on the cow and 20% on the ox, find the individual cost price of the cow and the ox.

   a) Rs 530.6, Rs 130.6 (Approx)
   b) Rs 531.5, Rs 135.5 (Approx)
   c) Rs 515.6, Rs 115.6 (Approx)
   d) Cannot be determined

Answers

1. a 2. a

Rule 74

Theorem: ***An article is sold at a certain price. If there is a loss of x% when the article is sold at $\frac{1}{n}$ of the previous selling price, then the percentage profit is*** $[n(100-x)-100]$ ***or*** $[n(100-\%\ loss)-100]$.

Illustrative Example

Ex.: What will be the percentage profit after selling an article at a certain price if there is a loss of 12½% when the article is sold at half of the previous selling price?

Soln: Detail Method: Suppose the previous selling price = Rs x

Now, the later selling price = Rs $\frac{x}{2}$

There is a loss of $12\frac{1}{2}\%$ when selling price = $\frac{x}{2}$

$\therefore$ cost price = $\frac{x}{2}\left(\frac{100}{100-12.5}\right)=\frac{100x}{175}=\frac{4x}{7}$

Now, when selling price is Rs x, % profit

$$=\frac{x-\frac{4x}{7}}{\frac{4x}{7}}\times100=\frac{7x-4x}{4x}\times100=\frac{3}{4}\times100=75\%$$

Quicker Method:

Required % profit = 100 – 2 × % loss [$\because$ Here, n = 2

$\therefore$ $n(100-x)-100=200-2x-100$

$=100-2x$ or, 100 – 2 × % loss.]

$=100-2\times12\frac{1}{2}=100-25$

$=75\%$.

Exercise

1. What will be the percentage profit after selling an article at a certain price if there is a loss of 40% when the article is sold at 1/3 rd of the previous selling price?
   a) 20% b) 80% c) 75% d) 60%
2. What will be the percentage profit after selling an article at a certain price if there is a loss of 30% when the article is sold at half of the previous selling price?
   a) 40% b) 30% c) 50% d) 35%
3. What will be the percentage profit after selling an article at a certain price if there is a loss of 35% when the article is sold at half of the previous selling price?
   a) 30% b) 25% c) 35% d) 20%
4. What will be the percentage profit after selling an article at a certain price if there is a loss of 25% when the article is sold at half of the previous selling price?
   a) 45% b) 60% c) 55% d) 50%
5. What will be the percentage profit after selling an article at a certain price if there is a loss of 60% when the article is sold at $\frac{1}{4}$ th of the previous selling price?
   a) 65% b) 80% c) 60% d) Can't be determined

Answers

1. b 2. a 3. a 4. d 5. c

Rule 75

Ex.: 5% more is gained by selling a cow for Rs 1010 than by selling it for Rs 1000. Find the cost price of the cow.

Soln: Detail Method: Suppose the cost price = Rs x

Then $\frac{100-x}{x}\times100+5=\frac{1010-x}{x}\times100$

or, $\frac{100}{x}[1010-x-1000+x]=5$

or, $\frac{100}{x}(10)=5$

$\therefore$ x = Rs 200

Quicker Method:

5% of cost price = Rs (1010 – 1000) = Rs 10

$\therefore$ CP = $\frac{10\times100}{5}$ = Rs 200.

Direct Formula:

Cost price = $\frac{100\times\text{Diff. in SP}}{\%\text{ diff. in profit or loss}}=\frac{100\times10}{5}$

= Rs 200.

Exercise

1. 10% more is gained by selling an ox for Rs 750 than by selling it for Rs 730. Find the cost price of the ox.
   a) Rs 250 b) Rs 300 c) Rs 200 d) Rs 150
2. 12% more is gained by selling a goat for Rs 248 than by selling it for Rs 224. Find the cost price of the goat.
   a) Rs 200 b) Rs 100 c) Rs 150 d) Rs 250
3. 15% more is gained by selling a camel for Rs 1260 than by selling it for Rs 1215. Find the cost price of the camel.
   a) Rs 350 b) Rs 300 c) Rs 250 d) Rs 325
4. 16% more is gained by selling a cow for Rs 760 than by selling it for Rs 720. Find the cost price of the cow.
   a) Rs 250 b) Rs 300
   c) Rs 200 d) Not possible
5. 20% more is gained by selling a horse for Rs 800 than by selling it for Rs 880. Find the cost price of the horse.
   a) Rs 400 b) Rs 300
   c) Rs 500 d) Not possible

Answers

1. c 2. a 3. b 4. a

5. d; **Hint:** 20% more can not be gained by selling a horse for Rs 800 than Rs 880. Hence, it is not possible.

Rule 76

Theorem: ***A shopkeeper uses a false scale (or weight) for selling and purchasing an article. If, on purchasing, he***

deceives the seller by x% and on selling, he deceives the customer by y%, then the overall per cent gain in the whole transaction is $\left[\frac{(100+x)(100+y)}{100}-100\right]\%$.

Illustrative Example

Ex.: A shopkeeper, on purchasing, deceives the seller by 20% and on selling, deceives the customer by 30%. What is the overall per cent gain in the whole transaction?

Soln: Applying the above theorem, we have

$$\% \text{ gain} = \frac{(100+20)(100+30)}{100}-100=56\%.$$

Exercise

1. A shopkeeper, on purchasing, deceives the seller by 10% and on selling, deceives the customer by 5%. What is the overall per cent gain in the whole transaction?
   a) 15.5% b) 15%
   c) 25.5% d) Neither lose nor gain
2. A shopkeeper, on purchasing, deceives the seller by 15% and on selling, deceives the customer by 10%. What is the overall per cent gain in the whole transaction?
   a) 26.25% b) 26.6% c) 26.5% d) 26%
3. A shopkeeper, on purchasing, deceives the seller by 10% and on selling, deceives the customer by 20%. What is the overall per cent gain in the whole transaction?
   a) 32% b) 33% c) 31% d) 26%
4. A shopkeeper, on purchasing, deceives the seller by 20% and on selling, deceives the customer by 25%. What is the overall per cent gain in the whole transaction?
   a) 25% b) 40% c) 60% d) 50%
5. A shopkeeper, on purchasing, deceives the seller by 25% and on selling, deceives the customer by 30%. What is the overall per cent gain in the whole transaction?
   a) 62.2% b) 62.5% c) 72.5% d) 63.2%

Answers

1. a 2. c 3. a 4. d 5. b

Rule 77

Theorem: ***A shopkeeper increases the original selling price by n times. If initial gain per cent is x%, then the new gain % is*** $[(100+x)n-100]$.

Illustrative Example

Ex.: A shopkeeper increases the original selling price of a pen by 3 times. Initial gain per cent is given as 5%. Find out the new gain per cent.

Soln: Applying the above formula, we have
new gain per cent = (100 + 5) 3 – 100 = 215%.

Exercise

1. A shopkeeper increases the original selling price of a pen by 2 times. Initial gain per cent is given as 5%. Find out the new gain per cent.
   a) 10% b) 110% c) 170% d) 140%
2. A shopkeeper increases the original selling price of a pen by 2 times. Initial gain per cent is given as 15%. Find out the new gain per cent.
   a) 130% b) 30% c) 140% d) 40%
3. A shopkeeper increases the original selling price of a pen by 3 times. Initial gain per cent is given as $2\frac{1}{2}\%$. Find out the new gain per cent.
   a) 192% b) 207.5% c) 193.5% d) 194.5%
4. A shopkeeper increases the original selling price of a pen by 2 times. Initial gain per cent is given as 20%. Find out the new gain per cent.
   a) 120% b) 20% c) 140% d) 40%

Answers

1. b 2. a 3. b 4. c

Rule 78

Theorem: ***If a person sells*** n_1 ***part of the property at*** x_1 ***% gain,*** n_2 ***part of the property at*** x_2 ***% gain and the remaining part at*** x_3 ***% loss, then the over all gain or loss per cent is given by*** $n_1x_1+n_2x_2+n_3(-x_3)$ ***according as the +ve or –ve sign. Here,*** $n_1+n_2+n_3$ ***= the whole part.***

Note: This formula is applicable to any number of parts of the property, incurring different % gain and % loss on each part and when overall % gain or % loss is to be found out. Put +ve for profit and -ve sign for loss in the formula.

Illustrative Example

Ex.: A person sells $\frac{1}{4}$th part of his property at 20% gain, $\frac{2}{3}$rd part of the property at 21% gain and the remaining part at 12% loss what is the overall % gain or % loss incurred by him.

Soln: Applying the above formula, we have

$$\% \text{ gain or } \% \text{ loss} = \frac{1}{4}\times 20+\frac{2}{3}\times 21+\frac{1}{12}\times(-12)$$
$$=5+14-1=18\%.$$

–ve sign indicates that there is a loss in transaction.
∴ % loss = 18%

Exercise

1. A merchant buys 1260 kg of corn, $\frac{1}{4}$ of which he sells at

a gain of 5 per cent, $\frac{1}{3}$ at a gain of 8 per cent, and the remainder at a gain of 12 per cent. If he had sold the whole at a gain of 10 per cent, he would have gained Rs 27.30 more. Find the cost price per kg.
a) Rs 2 b) Rs 2.5 c) Rs 3 d) Rs 3.5

2. A merchant buys 4000 kg of wheat, one-fifth of which he sells at a gain of 5 per cent, one-fourth at a gain of 10%, one-half at a gain of 12 per cent, and the remainder at a gain of 16 per cent. If he had sold the whole at a gain of 11 per cent, he would have made Rs 72.80 more. What was the cost price of the corn per kg?
a) Rs 2 b) Rs 2.60 c) Rs 2.50 d) None of these

3. A person sells $\frac{1}{3}$rd part of his property at 15% gain, $\frac{1}{6}$th part of the property at 12% gain and the remaining part at 10% loss what is the overall % gain or % loss incurred by him.
a) 2% loss b) 12% gain c) 2% gain d) 2.5% gain

4. A person sells $\frac{1}{5}$th part of his property at 25% gain, $\frac{2}{5}$th part of the property at 25% gain and the remaining part at 25% loss what is the overall % gain or % loss incurred by him.
a) 5% loss b) 5% gain
c) 25% gain d) None of these

Answers

1. a; **Hint:** % profit $= 5\times\frac{1}{4}+8\times\frac{1}{3}+12\times\frac{5}{12}=\frac{107}{12}$

From the question,

$$10\%-\frac{107}{12}\% \equiv \text{Rs } 27.30$$

or, $\frac{13}{12}\% \equiv \text{Rs } 27.30$

$\therefore$ $100\% = \frac{2730\times 12}{13} = \text{Rs } 2520$ = cost price

$\therefore$ cost price of corn per kg $= \frac{2520}{1260} = \text{Rs } 2$.

2. b; **Hint:** % profit $= \frac{1}{5}\times 5+\frac{1}{4}\times 10+\frac{1}{2}\times 12+\frac{1}{20}\times 16$

$$=\frac{103}{10}\%$$

From the question, we have

$$11\%-\frac{103}{10}\% \equiv \text{Rs } 72.80$$

or, $\frac{7}{10}\% \equiv \text{Rs } 72.80$

$\therefore$ $100\% \equiv$ Rs 10400 = cost price

$\therefore$ cost price of the corn per kg $= \frac{10400}{4000} = \text{Rs } 2.60$.

3. c 4. b

Rule 79

Theorem: ***If a merchant, by selling his goods, has a gain of x% of the selling price, then his real gain per cent on the cost price is*** $\left[\frac{x}{100-x}\times 100\right]\%$.

Note: Real profit per cent is always calculated on cost price and real profit per cent is always more than the % profit on selling price.

Illustrative Example

Ex.: If a merchant estimates his profit as 20% of the selling price, what is his real profit per cent?

Soln: Applying the above formula, we have

the real gain per cent $= \left[\frac{20}{100-20}\times 100\right]\% = 25\%$.

Exercise

1. If a merchant estimates his profit as 25% of the selling price, what is his real profit per cent?
a) 20% b) $33\frac{1}{3}\%$ c) $16\frac{2}{3}\%$ d) 25%

2. If a merchant estimates his profit as 15% of the selling price, what is his real profit per cent?
a) $17\frac{11}{17}\%$ b) $11\frac{11}{17}\%$ c) $17\frac{1}{17}\%$ d) $7\frac{11}{17}\%$

3. If a merchant estimates his profit as 5% of the selling price, what is his real profit per cent?
a) $5\frac{4}{19}\%$ b) $6\frac{5}{19}\%$ c) $5\frac{5}{19}\%$ d) $19\frac{5}{19}\%$

4. If a merchant estimates his profit as 10% of the selling price, what is his real profit per cent?
a) $11\frac{1}{3}\%$ b) $11\frac{2}{9}\%$ c) $9\frac{1}{11}\%$ d) $11\frac{1}{9}\%$

5. If a merchant estimates his profit as 30% of the selling price, what is his real profit per cent?
a) $42\frac{6}{7}\%$ b) $43\frac{6}{7}\%$ c) $42\frac{5}{7}\%$ d) $43\frac{4}{7}\%$

6. A man purchased a watch for Rs 400 and sold it at a gain of 20% of the selling price. The selling price of the watch is: **[Clerks' Grade Exam 1990]**
a) Rs 300 b) Rs 320 c) Rs 440 d) Rs 500

Answers

1. b 2. a 3. c 4. d 5. a

6. d; Real gain per cent $= \frac{20}{100-20} \times 100 = 25\%$

Selling price $= 400\left(\frac{125}{100}\right) = \text{Rs } 500$.

Rule 80

Theorem: ***If a merchant, by selling his goods, has a loss of x%, of the selling price, then his real loss per cent on the cost price is*** $\left[\frac{x}{100+x} \times 100\right]\%$.

Illustrative Example

Ex.: If a merchant estimates his loss as 10% of the selling price, what is his real loss per cent?

Soln: Applying the above theorem, we have the

real loss per cent $= \frac{10}{100+10} \times 100 = 9\frac{1}{11}\%$.

Note: Real loss % (ie per cent loss on cost price) is always less than % loss on selling price.

Exercise

1. If a merchant estimates his loss as 5% of the selling price, what is his real loss per cent?
a) $4\frac{15}{21}\%$ b) $4\frac{16}{21}\%$ c) $3\frac{16}{21}\%$ d) None of these
2. If a merchant estimates his loss as 15% of the selling price, what is his real loss per cent?
a) $13\frac{1}{23}\%$ b) $23\frac{1}{13}\%$ c) $13\frac{2}{23}\%$ d) $23\frac{1}{23}\%$
3. If a merchant estimates his loss as 20% of the selling price, what is his real loss per cent?
a) $14\frac{2}{3}\%$ b) $26\frac{2}{3}\%$ c) $16\frac{2}{3}\%$ d) $9\frac{1}{11}\%$
4. If a merchant estimates his loss as 25% of the selling price, what is his real loss per cent?
a) 30% b) 24% c) 25% d) 20%
5. If a merchant estimates his loss as 30% of the selling price, what is his real loss per cent?
a) $23\frac{1}{13}\%$ b) $23\frac{2}{13}\%$ c) $13\frac{1}{23}\%$ d) None of these

Answers

1. b 2. a 3. c 4. d 5. a

Rule 81

Theorem: ***If a merchant, by selling N articles gains the selling price of n articles, then the gain per cent is*** $\left[\frac{n}{N-n} \times 100\right]\%$.

Illustrative Example

Ex.: By selling 66 metres of cloth, I gain the selling price of 22 metres. Find the gain per cent.

Soln: Following the above formula, we have

gain % $= \frac{22}{66-22} \times 100 = 50\%$.

Exercise

1. By selling 15 metres of cloth, I gain the selling price of 7 metres. Find the gain per cent.
a) 77.5% b) 87.5% c) 37.5% d) None of these
2. By selling 75 metres of cloth, I gain the selling price of 25 metres. Find the gain per cent.
a) $33\frac{1}{3}\%$ b) 50% c) 25% d) 45%
3. By selling 25 metres of cloth, I gain the selling price of 5 metres. Find the gain per cent.
a) 20% b) 25% c) 24% d) 16%

Answers

1. b 2. b 3. b

Rule 82

Theorem: ***If a merchant, by selling N articles, loses the selling price of n articles, then the loss per cent is*** $\left[\frac{n}{N+n} \times 100\right]\%$.

Illustrative Example

Ex.: By selling 66 metres of cloth, I lose the selling price of 22 metres. Find the loss per cent.

Soln: Applying the above theorem, we have

loss % $= \frac{22}{66+22} \times 100 = 25\%$.

Exercise

1. By selling 75 metres of cloth, I lose the selling price of 25 metres. Find the loss per cent.
a) 20% b) $33\frac{1}{3}\%$ c) 25% d) None of these
2. By selling 85 metres of cloth, I lose the selling price of 15 metres. Find the loss per cent.
a) 12% b) 15% c) 18% d) 20%
3. By selling 60 metres of cloth, I lose the selling price of 30 metres. Find the loss per cent.
a) $33\frac{1}{3}\%$ b) $16\frac{2}{3}\%$ c) $33\frac{2}{3}\%$ d) 50%

4. By selling 48 metres of cloth, I lose the selling price of 12 metres. Find the loss per cent.
a) 24% b) 20% c) 25% d) Can't be determined

Answers

1. c 2. b 3. a 4. b

Rule 83

Theorem: ***If a merchant, by selling N articles, gains or loses the cost price of n articles, the gain or loss per cent is given by the*** $\left[\frac{n}{N}\times 100\right]\%$.

Illustrative Example

Ex.: By selling 66 metres of cloth a person gains the cost price of 22 metres. Find the gain per cent.

Soln: Applying the above formula, we have

$$\% \text{ gain} = \frac{22}{66}\times 100 = 33\frac{1}{3}\%.$$

Exercise

1. By selling 48 metres of cloth a person gains the cost price of 12 metres. Find the gain per cent.
a) 25% b) 20% c) 28% d) 30%
2. By selling 75 metres of cloth a person gains the cost price of 15 metres. Find the gain per cent.
a) 20% b) 25% c) 24% d) 30%
3. By selling 96 metres of cloth a person gains the cost price of 16 metres. Find the gain per cent.
a) $14\frac{2}{3}\%$ b) $16\frac{2}{3}\%$ c) $26\frac{2}{3}\%$ d) 20%
4. By selling 100 metres of cloth a person gains the cost price of 32.5 metres. Find the gain per cent.
a) 32.6% b) 30% c) 32.5% d) None of these

Answers

1. a 2. a 3. b 4. c

Rule 84

Theorem: ***A businessman marks an article at Rs A and allows x% discount (on the marked price). He gains y%. If the cost price of the article is Rs B, then the selling price of the article can be calculatd from the equation given below***

$$\frac{A(100-x)}{100} = \frac{B(100+y)}{100} = \textit{selling price.}$$

Note: Remember discount is given on marked price, and gain is calculated on the cost price.

Illustrative Example

Ex.: A discount of 16% on the marked price of a book enables a man to buy a pen which costs Rs 80. How much did he pay for the book?

Soln: Applying the above formula, we have

$$\frac{A(100-x)}{100} = \text{Selling Price}$$

(Marked Price – Selling Price = Customer's saving on marked price due to discount) = Rs 80, given)

$$A - \frac{A(100-x)}{100} = \text{Rs } 80$$

or, A = Rs 500

$$\therefore \text{ Selling Price} = \frac{500(100-16)}{100} = \text{Rs } 420.$$

Exercise

1. How much per cent above the cost price should a shopkeeper mark his goods so as to earn a profit of 26% after allowing a discount of 10% on the marked price.
a) 40% b) 30% c) 20% d) 50%
2. A shopkeeper's price is 50% above the cost price. If he allows his customer a discount of 30%, what profit does he make? **[IT Inspectors' Exam, 1991]**
a) 5% b) 10% c) 8% d) 4%
3. A man's price is 20% above the cost price. He allows his customers a discount and makes a profit of 8%. Find the rate of discount. **[RBI Exam 1989]**
a) 20% b) 15% c) 10% d) 25%
4. How much per cent above the cost price should a shopkeeper mark his goods so as to earn a profit of 36% after allowing a discount of 15% on the marked price?
a) 60% b) 50% c) 40% d) 65%

Answers

1. a; **Hint:** From the given formula, we have
A(100 - x) = C(100 + y)
or, A(100 - 10) = C(100 + 26)
or, $A = \frac{126}{90}C = 1.4C = (1+0.4)C$
ie A is + 0.4 more than C
∴ Marked price A is 40% above the cost price.

2. a 3. c 4. a

Rule 85

If a person buys an article with x per cent discount on the marked price and sells the article with y per cent profit on the marked price, then his per cent profit on the price he buys the article is given by $\left(\frac{x+y}{100-x}\right)\times 100$ ***per cent.***

Illustrative Example

Ex.: Raman bought on article with 20 per cent discount on the labelled price. He sold the article with 30 per cent profit on the labelled price. What was his per cent profit on the price he bought?

Soln: **Detail Method:**
Let the labelled price of the article be Rs x.

$$\text{Cost price} = x\left(\frac{100-20}{100}\right) = \text{Rs}\,\frac{4x}{5}$$

$$\text{Selling price} = x\left(\frac{100+30}{100}\right) = \text{Rs}\,\frac{13}{10}x$$

$$\text{Profit} = \frac{13x}{10} - \frac{4x}{5} = \frac{13x - 8x}{10} = \frac{x}{2}$$

$$\%\ \text{profit} = \frac{\frac{x}{2}}{\frac{4x}{5}} \times 100 = \frac{5}{8} \times 100 = \frac{125}{2} = 62.5\%$$

Quicker Method: Applying the above theorem, we have

$$\text{the required per cent profit} = \left(\frac{20+30}{100-20}\right) \times 100$$

$$= \frac{5}{8} \times 100 = 62.5\%$$

Exercise

1. Jeevan bought an article with 30 per cent discount on the labelled price. He sold the article with 12 per cent profit on the labelled price. What was his per cent profit on the price he bought? **[Bank of Baroda PO 1999]**
   a) 40 b) 50 c) 60 d) Data inadequate
2. Anjali bought an article with $12\frac{1}{2}$ per cent discount on the labelled price. He sold the article with $17\frac{1}{2}$ per cent profit on the labelled price. What was his per cent profit on the price he bought?
   a) 35% b) $34\frac{1}{7}\%$ c) $34\frac{2}{7}\%$ d) $35\frac{2}{7}\%$
3. Deepti bought an article with 15 per cent discount on the labelled price. He sold the article with 10 per cent profit on the labelled price. What was his per cent profit on the price he bought?
   a) $28\frac{7}{17}\%$ b) $29\frac{7}{17}\%$
   c) $29\frac{5}{17}\%$ d) Data inadequate
4. Naval bought an article with 25 per cent discount on the labelled price. He sold the article with 5 per cent profit on the labelled price. What was his per cent profit on the price he bought?
   a) 40% b) 45% c) 35% d) 30%
5. Vinay bought an article with 20 per cent discount on the labelled price. He sold the article with 12 per cent profit on the labelled price. What was his per cent profit on the price he bought?
   a) 35% b) 25% c) 20% d) 40%

Answers

1. c 2. c 3. b 4. a 5. d

Rule 86

A person sells articles at Rs A each after giving x% discount on marked price. Had he not given the discount, he would have earned a profit of y% on the cost price. Then the cost price of each article is given by Rs

$$\left[\frac{100^2 A}{(100-x)(100+y)}\right].$$

Illustrative Example

Ex.: A shopkeeper sold certain articles at Rs 425 each after giving 15% discount on labelled price. Had he not given the discount, he would have earned a profit of 25% on the cost price. What was the cost price of each article?

Soln: Detail Method:

$$\text{Labelled price of the article} = \frac{425 \times 100}{85} = \text{Rs } 500$$

Let the cost price be Rs x

Now, according to the question,

$$\frac{500-x}{x} \times 100 = 25$$

$$\text{or } x(100+25) = 50000$$

$$\therefore\ x = \frac{50000}{125} = \text{Rs } 400$$

$\therefore$ cost price of each article = Rs 400

Exercise

1. A shopkeeper sold sarees at Rs 266 each after giving 5% discount on labelled price. Had he not given the discount, he would have earned a profit of 12% on the cost price. What was the cost price of each saree? **[SBI Associates PO 1999]**
   a) Rs 280 b) Rs 260 c) Rs 240 d) Rs 250
2. A shopkeeper sold tables at Rs 2139 each after giving 7% discount on labelled price. Had he not given the discount, he would have earned a profit of 15% on the cost price. What was the cost price of each table?
   a) Rs 2500 b) Rs 2100 c) Rs 2000 d) Rs 1900
3. A shopkeeper sold chairs at Rs 528 each after giving 12% discount on labelled price. Had he not given the discount, he would have earned a profit of 20% on the cost price. What was the cost price of each chair?

a) Rs 500 b) Rs 520 c) Rs 480 d) Rs 490

4. A shopkeeper sold almirahs at Rs 166 each after giving 17% discount on labelled price. Had he not given the discount, he would have earned a profit of 25% on the cost price. What was the cost price of each almirah?
a) Rs 165 b) Rs 155 c) Rs 160 d) Rs 164
5. A shopkeeper sold beds at Rs 1134 each after giving 19% discount on labelled price. Had he not given the discount, he would have earned a profit of 40% on the cost price. What was the cost price of each bed?
a) Rs 1100 b) Rs 1000 c) Rs 1050 d) Rs 900

Answers

1. d 2. c 3. a 4. c 5. b

Rule 87

A certain company declares x per cent discount for wholesale buyers. If a person buys articles from the company for Rs A after getting discount. He fixed up the selling price of the articles in such a way that he earned a profit y% on original company price. Then the total selling price is given by Rs $A\left(\frac{100+y}{100-x}\right)$.

Illustrative Example

Ex: A garment company declared 15% discount for wholesale buyers. Mr Sushil bought garments from the company for Rs 8500 after getting discount. He fixed up selling price of garments in such a way that he earned a profit of 10% on original company price. What is the total selling price?

Soln: Detail Method:

$$\text{Original company price} = \frac{8500\times100}{100-15} = \text{Rs } 10000.$$

Let the total selling price be Rs x.
Now, according to the question,

$$\frac{x-10000}{10000}\times100 = 10$$

or, 100x - 1000000 = 100000
or, x = Rs 11000.
∴ total selling price = Rs 11000.

Quicker Method: Applying the above rule, we have

$$\text{the total selling price} = \frac{100+10}{100-15}\times8500 = \text{Rs } 11000.$$

Exercise

1. A garment company declared 15% discount for wholesale buyers. Mr Sachdev bought garments from the company for Rs 25000 after getting discount. He fixed up the selling price of garments in such a way that he earned a profit of 8% on original company price. What is the approximate total selling price? **[Guwahati PO 1999]**
a) Rs 28000 b) Rs 29000 c) Rs 31000 d) Rs 28500
2. A garment company declared 12% discount for wholesale buyers. Mr Mohan bought garments from the company for Rs 8800 after getting discount. He fixed up the selling price of garments in such a way that he earned a profit of 4% on original company price. What is the approximate total selling price?
a) Rs 10400 b) Rs 14000
c) Rs 10800 d) Data inadequate
3. A garment company declared 17% discount for wholesale buyers. Mr Sameer bought garments from the company for Rs 1660 after getting discount. He fixed up the selling price of garments in such a way that he earned a profit of 7% on original company price. What is the approximate total selling price?
a) Rs 2130 b) Rs 2140 c) Rs 2410 d) Rs 2310
4. A garment company declared 14% discount for wholesale buyers. Mr Sujeet bought garments from the company for Rs 860 after getting discount. He fixed up the selling price of garments in such a way that he earned a profit of 6% on original company price. What is the approximate total selling price?
a) Rs 1060 b) Rs 1160
c) Rs 960 d) Can't be determined

Answers

1. c 2. a 3. b 4. a

Rule 88

A shopkeeper sold an article for Rs A after giving x% discount on the labelled price and made y% profit on the cost price. Had he not given the discount, the percentage profit would have been $\left[\frac{x+y}{100-x}\times100\right]$ ***per cent.***

Illustrative Example

Ex: A shopkeeper sold an article for Rs 400 after giving 20% discount on the labelled price and made 30% profit on the cost price. What would have been the percentage profit, had he not given the discount?

Soln: Detail Method:

$$\text{Labelled price} = \frac{400\times100}{80} = \text{Rs } 500$$

$$\text{Cost price} = \frac{400\times100}{130} = \text{Rs } \frac{4000}{13}$$

Now, according to the question,

$$\text{Profit \%} = \frac{500-\frac{4000}{13}}{\frac{4000}{13}}\times100$$

$$= \frac{250}{4} = \frac{125}{2} = 62.5\%$$

Quicker Method: Applying the above rule, we have

$$\text{the required per cent profit} = \frac{20+30}{100-20} \times 100$$

$$= \frac{50}{80} \times 100 = \frac{125}{2} = 62.5\%$$

Exercise

1. A shopkeeper sold an article for Rs 720 after giving 10% discount on the labelled price and made 20% profit on the cost price. What would have been the percentage profit, had he not given the discount? **[Guwahati PO 1999]**

   a) 25% b) 23% c) $33\frac{1}{3}\%$ d) 28%

2. A shopkeeper sold an article for Rs 750 after giving 20% discount on the labelled price and made 40% profit on the cost price. What would have been the percentage profit, had he not given the discount?

   a) 75% b) 85% c) 60% d) 70%

3. A shopkeeper sold an article for Rs 880 after giving 12% discount on the labelled price and made 32% profit on the cost price. What would have been the percentage profit, had he not given the discount?

   a) 25% b) 35% c) 40% d) 50%

4. A shopkeeper sold an article for Rs 860 after giving 14% discount on the labelled price and made 29% profit on the cost price. What would have been the percentage profit, had he not given the discount?

   a) 50% b) 60% c) 45% d) 55%

5. A shopkeeper sold an article for Rs 210 after giving 16% discount on the labelled price and made 5% profit on the cost price. What would have been the percentage profit, had he not given the discount?

   a) 50% b) 25% c) 30% d) 20%

6. A shopkeeper sells a TV set for Rs 16560 at 10% discount on its marked price and earns 15% profit. If no discount is offered, then what will be his present per cent profit? **[BSRB Patna PO 2001]**

   a) $27\frac{7}{9}$ b) $22\frac{7}{9}$ c) $25\frac{7}{9}$ d) Data inadequate

Answers

1. c 2. a 3. d 4. a 5. b 6. a

Rule 89

Theorem: If an item is bought at the rate of X items for a rupee, then the number of items sold for a rupee in order to gain x% is $\left[X\left(\frac{100}{100+x}\right)\right]$.

Illustrative Example

Ex: If oranges are bought at the rate of 30 for a rupee, how many must be sold for a rupee in order to gain 25%?

Soln: Applying he above formula, we have

$$\text{the required number of oranges} = 30\left(\frac{100}{100+25}\right) = 24.$$

Exercise

1. If bananas are bought at the rate of 24 for a rupee, how many must be sold for a rupee in order to gain 20%?

   a) 20 b) 18 c) 22 d) 16

2. If apples are bought at the rate of 39 for a rupee, how many must be sold for a rupee in order to gain 30%?

   a) 33 b) 36 c) 30 d) None of these

3. If mangoes are bought at the rate of 56 for a rupee, how many must be sold for a rupee in order to gain 40%?

   a) 44 b) 42 c) 43 d) 40

4. If oranges are bought at the rate of 27 for a rupee, how many must be sold for a rupee in order to gain 35%?

   a) 26 b) 25 c) 20 d) 24

5. If bananas are bought at the rate of 46 for a rupee, how many must be sold for a rupee in order to gain 15%?

   a) 40 b) 30 c) 35 d) 45

Answers

1. a 2. c 3. d 4. c 5. a

Miscellaneous

1. An article when sold for Rs 200 fetches 25 per cent profit. What would be the percentage profit/loss if 6 such articles are sold for Rs 1056? **[BSRB Calcutta PO, 1999]**

   a) 10 per cent loss b) 10 per cent profit
   c) 5 per cent loss d) 5 per cent profit
   e) None of these

2. A shopkeeper gave an additional 20 per cent concession on the reduced price after giving 30 per cent standard concession on an article. If Arun bought that article for Rs 1120, what was the original price? **[BSRB Calcutta PO, 1999]**

   a) Rs 3000 b) Rs 4000 c) Rs 2400
   d) Rs 2000 e) None of these

3. A shopkeeper bought 150 calculators at the rate of Rs 250 per calculator. He spent Rs 2500 on transportation and packing. If the marked price of calculator is Rs 320 per calculator and the shopkeeper gives a discount of 5% on the marked price then what will be the percentage profit gained by the shopkeeper? **[BSRB Hyderabad PO, 1999]**

   a) 20% b) 14% c) 15%
   d) 16% e) None of these

4. An article when sold for Rs 960 fetches 20% profit. What would be the per cent profit or loss if 5 such articles are

sold for Rs 825 each?

[BSRB Bhopal PO, 2000]

a) 3.125% profit b) 3.125% loss
c) Neither profit nor loss d) 16.5% profit
e) None of these

5. Two chairs and three tables cost Rs 1025 and 3 chairs and two tables cost Rs 1100. What is the difference between the cost of one table and that of one chair?

[BSRB Bhopal PO, 2000]

a) Rs 75 b) Rs 35 c) Rs 125
d) Cannot be determined e) None of these

6. What per cent of selling price would be 34% of cost price if gross profit is 26% of the selling price?

[BSRB Bangalore PO, 2000]

a) 17.16 b) 74.00 c) 25.16
d) 88.40 e) None of these

7. A sells a horse to B for Rs 4860, thereby losing 19 per cent, B sells it to C at a price which would have given A 17 per cent profit. Find B's gain.

[SBI Bank PO, 1998]

a) Rs 2160 b) Rs 2610 c) Rs 1260 d) Rs 2260

8. Profit after selling a commodity for Rs 425 is same as loss after selling it for Rs 355. The cost of the commodity is: **[Bank PO 1989]**

a) Rs 385 b) Rs 390 c) Rs 395 d) Rs 400

9. The cost price of an article, which on being sold at a gain of 12% yields Rs 6 more than when it is sold at a loss of 12%, is **[CBI Exam 1990]**

a) Rs 30 b) Rs 25 c) Rs 20 d) Rs 24

10. Alok bought 25 kg of rice at the rate of Rs 6.00 per kg and 35 kg of rice at the rate of Rs 7.00 per kg. He mixed the two and sold the mixture at the rate of Rs 6.75 per kg. What was his gain or loss in this transaction?

[PO Exam 1990]

a) Rs 16.00 gain b) Rs 16.00 loss
c) Rs 20.00 gain d) None of these

11. When the price of pressure cooker was increased by 15%, its sale fell down by 15%. The effect on the money receipt was: **[SBI PO Exam 1987]**

a) no effect b) 15% decrease
c) 7.5% increase d) 2.25% decrease

12. A boy buys oranges at Rs 2 for 3 oranges and sells them at a rupee each. To make a profit of Rs 10, he must sell:

[CDS Exam 1991]

a) 10 oranges b) 20 oranges
c) 30 oranges d) 40 oranges

13. Subhash purchased a tape recorder at $\frac{9}{10}$th of its selling price and sold it at 8% more than its selling price. His gain is: **[SBI PO Exam 1987]**

a) 9% b) 10% c) 18% d) 20%

14. At what price must Kantilal sell a mixture of 80 kg sugar at Rs 6.75 per kg with 120 kg at Rs 8 per kg to gain 20%?

[SBI PO Exam 1987]

a) Rs 7.50 per kg b) Rs 9 per kg
c) Rs 8.20 per kg d) Rs 8.85 per kg

15. The loss incurred on selling an article for Rs 270 is as much as the profit made after selling it at 10% profit. The CP of the article is: **[Bank PO Exam 1989]**

a) Rs 90 b) Rs 110 c) Rs 363 d) Rs 300

16. An item costing Rs 200 is being sold at 10% loss. If the price is further reduced by 5%, the selling price will be:

[NDA Exam 1987]

a) 179 b) Rs 175 c) Rs 171 d) Rs 170

17. A retailer purchases a sewing machine at a discount of 15% and sells it for Rs. 1955. In the bargain he makes a profit of 15%. How much is the discount which he got from the whole sale? **[LIC AAO Exam 1988]**

a) Rs 270 b) Rs 290
c) Rs 300 d) None of these

18. A discount series of 10%, 20% and 40% is equal to a single discount of: **[Central Excise & I Tax 1989]**

a) 50% b) 56.80% c) 70% d) 70.28%

19. While selling a watch a shopkeeper gives a discount of 5%. If he gives a discount of 7%, he earns Rs 15 less as profit. What is the marked price of the watch?

[LIC AAO Exam 1988]

a) Rs 697.50 b) Rs 712.50
c) Rs 787.50 d) None of these

20. A shopkeeper earns a profit of 12% after selling a book at 10% discount on the printed price. The ratio of the cost price and printed price of the book is:

[GIC Exam 1988]

a) 45:56 b) 50:61
c) 99:125 d) None of these

21. Tarun bought a TV with 20% discount on the labelled price. Had he bought it with 25% discount he would have saved Rs 500. At what price did he buy the TV?

[PO Exam 1990]

a) Rs 5000 b) Rs 10000
c) Rs 12000 d) None of these

22. A reduction of 20% in the price of mangoes enables a person to purchase 12 more for Rs 15. The price of 16 mangoes before reduction was:

[RRB Exam 1989]

a) Rs 5 b) Rs 6 c) Rs 7 d) Rs 9

23. There would be 10% loss if rice is sold at Rs 5.40 per kg. At what price per kg should it be sold to earn a profit of 20%? **[SBI PO Exam 1988]**

a) Rs 7.20 b) Rs 7.02 c) Rs 6.48 d) Rs 6

Answers

1. b; $CP = \frac{200}{125} \times 100 = \text{Rs } 160$

$\therefore$ CP of 6 articles $= 6 \times 160 = 960$

$\therefore$ profit = 1056 - 960 = 96

Percentage profit = $\frac{96}{960} \times 100 = 10\%$

2. d; Original price = $1120 \times \frac{100}{70} \times \frac{100}{80}$ = Rs 2000

3. b; CP of 150 calculators = 150 × 250 = Rs 37500
$\therefore$ total CP = 37500 + 2500 = Rs 40000
Marked price of 150 calculators = 150 × 320 = Rs 48000

Selling price after discount = $48000 \times \frac{95}{100}$ = Rs 45600

$\therefore$ percentage profit = $\frac{45600 - 40000}{40000} \times 100 = 14\%$

4. a; Cost price of the article = $960 \times \frac{100}{120}$ = Rs 800
$\therefore$ Cost price of 5 articles = Rs 800 × 5 = Rs 4000
$\therefore$ Selling price of 5 articles = 825 × 5 = Rs 4125

$\therefore$ Gain % = $\frac{4125 - 4000}{4000} \times 100 = 3.125\%$

5. a; Let the cost of each table and chair be Rs x and y respectively.
$\therefore$ 2y + 3x = 1025 and
3y + 2x = 1100
Solving the above two equations, we get
x = Rs 175 and y = Rs 250
$\therefore$ Diff. between the cost of one table and one chair
= Rs (250 - 175) = Rs 75

6. c; Let the selling price of the article of Rs 100.
$\therefore$ Profit = Rs 26
$\therefore$ Cost price of the article = 100 - 26 = Rs 74

$\therefore$ Reqd. % = $\frac{34 \times 74}{100} = 25.16\%$

7. a; Cost of the horse paid by $A = 4860 \times \frac{100}{81}$ = Rs 6000.

Cost of the horse paid by $C = 6000 \times \frac{117}{100}$ = Rs 7020

Gain of B = Rs 7020 - Rs 4860 = Rs 2160.

8. b; Let CP = Rs x Then,
425 - x = x - 355 ⇒ 2x = 780 ⇒ x = 390.

9. b; Let the CP be Rs x

Then, SP when gain is $12\% = \left(\frac{12x}{100} + x\right) = \frac{112x}{100}$.

$\therefore \frac{112x}{100} - \frac{88x}{100} = 6$.

or, $\frac{24x}{100} = 6$ or, $x = \frac{600}{24}$ = Rs 25

10. d; CP of total 60 kg of rice = Rs (6 × 25 + 7 × 35) = Rs 395.
SP of total 60 kg of rice = Rs (6.75 × 60) = Rs 405.
Gain = Rs (405 - 395) = Rs 10.

11. d; Let the original cost of each cooker be Re 1 and let the number sold originally be 100.
Total sale proceed = Rs (100 × 1) = Rs 100.
New rate = (115% of Re 1) = Rs 1.15.
Number sold now = 85.
$\therefore$ Sale proceed now = Rs (1.15 × 85) = Rs 97.75.
So, there is a decrease of 2.25% in the money receipt.

12. c; Suppose he sells x oranges

Then, CP of x oranges = Rs $\frac{2}{3}$x.

SP of x oranges = Rs x

Profit on x oranges = Rs $\left(x - \frac{2}{3}x\right)$ = Rs $\frac{x}{3}$

$\therefore \frac{x}{3} = 10 \Rightarrow x = 30$.

13. d; Let the SP be Rs x.

Then, CP paid by Subhash = Rs $\frac{9x}{10}$.

SP received by Subhash = (108% of Rs x) = Rs $\frac{27x}{25}$.

$\therefore$ Gain = Rs $\left(\frac{27x}{25} - \frac{9x}{10}\right)$ = Rs $\left(\frac{9x}{50}\right)$.

Hence, gain % = $\left(\frac{9x}{50} \times \frac{10}{9x} \times 100\right)\% = 20\%$.

14. b; Total CP of 200 kg sugar = Rs (80 × 6.75 + 120 × 8) = Rs 1500

CP of 1 kg = Rs $\left(\frac{1500}{200}\right)$ = Rs 7.50.

Gain required = 20%.
$\therefore$ SP of 1 kg = (120% of Rs 7.50)

= Rs $\left(\frac{120}{100} \times 7.50\right)$ = Rs 9 per kg.

15. d; Let CP be Rs x. Then,

x - 270 = 10% of $x = \frac{x}{10}$ or x = 300.

16. c; SP = 90% of Rs 200 = Rs 180.
Further, SP = (95% of Rs 180) = Rs 171

17. c; Let the marked price be Rs x.
Discount availed by the retailer = 15% of Rs x.
$\therefore$ CP of the machine by the retailer

$= (x - 15\% \text{ of } x) = \text{Rs } \frac{17x}{20}.$

So, 15% of $\frac{17x}{20} = 1955 - \frac{17x}{20}$

$\therefore \frac{51x}{400} + \frac{17x}{20} = 1955$ or $x = 2000.$

Discount received by retailer = (15% of Rs 2000) = Rs 300.

18. b; Let original price = Rs 100.
Price after 1st discount = Rs 90.

Price after 2nd discount = Rs $\left(\frac{80}{100} \times 90\right)$ = Rs 72.

Price after 3rd discount = Rs $\left(\frac{60}{100} \times 72\right)$ = Rs 43.20

$\therefore$ Single discount = (100 - 43.20) = 56.8%.

19. d; Let the marked price be Rs x.
Then, (7% of x) - 15 = (5% of x)

or, $\frac{7x}{100} - \frac{5x}{100} = 15$ or $x = 750.$

20. a; Let the printed price of the book be Rs 100.
After a discount of 10%, SP = Rs 90.
Profit earned = 12%.

$\therefore$ CP of the book = Rs $\left(\frac{100}{112} \times 90\right)$ = Rs $\frac{1125}{14}$.

Hence, (CP) : (Printed price) = $\frac{1125}{14} : 100$ or 45 : 56.

21. d; Let SP of TV (by trader) = Rs 100.
If SP is Rs 80, MP = Rs 100.

If SP is Rs 100, MP = Rs $\left(\frac{100}{80} \times 100\right)$ = Rs 125.

Now, if discount is 25%, then SP = (75% of Rs 125)

= Rs $\frac{375}{4}$.

Diff. between two SP = Rs $\left(100 - \frac{375}{4}\right)$ = Rs $\frac{25}{4}$.

If diff. is Rs $\frac{25}{4}$, SP = Rs 100

If diff. is Rs 500, SP = Rs $\left(100 \times \frac{4}{25} \times 500\right)$ = Rs 8000.

22. a; Suppose the price of 1 mango be x paise.

Number of mangoes for Rs 15 $= \frac{1500}{x}$

New price of one mango = (80% of x) = $\frac{4x}{5}$ paise.

Number of mangoes for Rs 15 $= \left(\frac{1500 \times 5}{4x}\right)$.

$\therefore \frac{7500}{4x} - \frac{1500}{x} = 12$ or $x = 31.25$

$\therefore$ Cost of 16 mangoes before reduction

= Rs $\left(\frac{31.25 \times 16}{100}\right)$ = Rs 5.

23. a; Let CP per kg be Rs x. Then,
$x - 10\%$ of $x = 5.40$ or $x = 6.$
$\therefore$ SP = Rs [6 + 20% of 6] = Rs 7.20

9

Simple Interest

Rule 1

(i) To find simple interest

Simple Interest (SI) = $\frac{P \times t \times r}{100}$;

where, P = Principal, t = number of years, r = rate per cent

Illustrative Example

Ex.: Find the simple interest on Rs 400 for 5 years at 6 per cent.

Soln: SI = $\frac{400 \times 5 \times 6}{100}$ = Rs 120.

(ii) To find principal

$P = \frac{100I}{tr}$;

Where, P = Principal, I = Interest, t = Number of years, r = Rate per cent

Illustrative Example

Ex.: What sum of money will produce Rs 143 interest in $3\frac{1}{4}$ years at $2\frac{1}{2}$ per cent simple interest?

Soln: Applying the above formula, we have

$P = \frac{100 \times 143}{3\frac{1}{4} \times 2\frac{1}{2}} = Rs \frac{100 \times 143 \times 4 \times 2}{13 \times 5}$ = Rs 1760.

(iii) To find rate per cent

$r = \frac{100I}{Pt}$; Where I = Interest, P = Principal, t = No. of years

Illustrative Example

Ex.: A sum of Rs 468.75 was lent out at a simple interest and at the end of 1 year 8 months the total amount was Rs 500. Find the rate of interest per cent per annum.

Soln: I = [A – P]; where A = Amount, P = Principal
= 500 – 468.75 = Rs 31.25

Now, applying the above formula, we have

$r = \frac{100 \times 31.25}{468.75 \times \frac{5}{3}} = 100 \times \frac{3125}{46875} \times \frac{3}{5} = 4\ \%.$

(iv) To find time

$t = \frac{100I}{Pr}$

Where, I = Interest, P = Principal, r = Rate per cent

Illustrative Example

Ex.: In what time will Rs 8500 amount to Rs 15767.50 at $4\frac{1}{2}$ per cent per annum?

Soln: Here, Interest = Rs 15767.50 – Rs 8500 = Rs 7267.50

$\therefore t = \frac{7267.50 \times 100}{8500 \times 4.5} = 19$ years.

Exercise

1. A sum of Rs 800 is lent for one year at the rate of 18% per annum. Find the interest.
   a) Rs 144 b) Rs 140 c) Rs 150 d) Rs 154
2. A sum of Rs 4000 is lent for 5 years at the rate of 15% per annum. Find the interest.
   a) Rs 3000 b) Rs 2000 c) Rs 1000 d) None of these
3. Anita borrowed Rs 400 from her friend at the rate of 12% per annum for $2\frac{1}{2}$ years. Find the interest and the amount paid by her.
   a) Rs 140, Rs 540 b) Rs 130, Rs 530
   c) Rs 125, Rs 525 d) Rs 120, Rs 520
4. A farmer borrowed Rs 2400 at 12% interest per annum. At the end of $2\frac{1}{2}$ years, he cleared his account by paying Rs 1200 and a cow. Find the cost of the cow.

a) Rs 1820 b) Rs 1720
c) Rs 1920 d) None of these

5. Find the interest on Rs 1200 at 6% per annum for 146 days.
a) Rs 28.80 b) Rs 30 c) Rs 25.50 d) Rs 28

6. What sum of money will produce Rs 150 interest in 2 years at 5 per cent simple interest?
a) Rs 1400 b) Rs 1050
c) Rs 1500 d) Data inadequate

7. A sum of Rs 400 was lent out at a simple interest and at the end of $6\frac{1}{4}$ years the total amount was Rs 625. Find the rate of interest per cent per annum.
a) 9% b) 8% c) 5% d) 6%

8. In what time will Rs 500 amount to Rs 625 at 5 per cent per annum?
a) 6 years b) 5 years c) 10 years d) None of these

9. At what rate per cent will Rs 425 amount to Rs 476 in 3 years?
a) 4% b) 5% c) 6% d) 2%

10. In how many years will Rs 300 amount to Rs 405 at 5 per cent?
a) 7 years b) 3 years c) 5 years d) 9 years

11. The simple interest on Rs 500 for 6 years at 5% per annum is **[Clerical Grade, 1991]**
a) Rs 250 b) Rs 150 c) Rs 140 d) Rs 120

12. A man will get Rs 87 as simple interest on Rs 725 at 4% per annum in: **[Clerical Grade, 1991]**
a) 3 years b) $3\frac{1}{2}$ years c) 4 years d) 5 years

13. Interest on a certain sum of money for $2\frac{1}{3}$ years at $3\frac{3}{4}$% per annum is Rs 210. The sum is **[Railways, 1989]**
a) Rs 2800 b) Rs 1580 c) Rs 2400 d) None of these

14. The simple interest at x% for x years will be Rs x on a sum of
a) Rs x b) Rs $100x$ c) Rs $\left(\frac{100}{x}\right)$ d) Rs $\left(\frac{100}{x^2}\right)$

15. A borrowed Rs 5000 from B at simple interest. After 4 years, B received Rs 1000 more than the amount given to B on loan. The rate of interest was: **[BSRB Exam, 1991]**
a) 5% b) 25% c) 20% d) 4%

16. If 1 Re produces Rs 9 as interest in 60 years at simple interest, the rate per cent is: **[Clerical Grade, 1991]**
a) 30 b) 15 c) $13\frac{1}{3}$ d) 9

17. The sum of money that will give Re 1 as interest per day at 5% per annum simple interest is: **[Clerical Grade, 1991]**
a) Rs 3650 b) Rs 36500 c) Rs 730 d) Rs 7300

Answers

1. a 2. a 3. d

4. c; **Hint:** $SI = \frac{2400 \times 12 \times 5}{200} = \text{Rs } 720$

Amount = Rs 2400 + Rs 720 = Rs 3120
$\therefore$ Cost of cow = Rs 3120 – Rs 1200 = Rs 1920

5. a; **Note:** 73, 146, 219 and 292 days are respectively $\frac{1}{5}$, $\frac{2}{5}$, $\frac{3}{5}$ and $\frac{4}{5}$ of a year.

6. c 7. c 8. b 9. a 10. a 11. b
12. a 13. c

14. c; **Hint:** Principal = Rs $\left(\frac{100 \times x}{x \times x}\right)$ = Rs $\left(\frac{100}{x}\right)$

15. a; **Hint:** Rate % = $\left(\frac{100 \times 1000}{5000 \times 4}\right) = 5\%$

16. b; **Hint:** Rate = $\left(\frac{100 \times 9}{1 \times 60}\right) = 15\%$

17. d; **Hint:** Sum = Rs $\left(\frac{365 \times 100}{1 \times 5}\right)$ = Rs 7300.

Rule 2

When time (t) changes from t_1 to t_2,

$$SI_1 - SI_2 = \frac{P \times r \times (t_1 - t_2)}{100};$$

Where, SI = Simple Interest, P = Principal, r = Rate per cent per annum.

Illustrative Example

Ex.: If the simple interest on Rs 2000 increases by Rs 40, when the time increases by 4 years. Find the rate per cent per annum.

Soln: Applying the above formula, we have

$$40 = \frac{2000 \times r \times 4}{100}$$

$$\therefore r = \frac{40 \times 100}{2000 \times 4} = 0.5\%.$$

Note: Here, $SI_1 - SI_2$ = Rs 40 and $t_1 - t_2 = 4$ years.

Exercise

1. If the simple interest on Rs 1000 increases by Rs 20, when the time increases by 2 years. Find the rate per cent per annum.
   a) 0.5% b) 0.25% c) 1% d) None of these
2. If the simple interest on Rs 625 increases by Rs 25, when the time increases by 2 years. Find the rate per cent per annum.
   a) 2% b) 3% c) 1% d) 0.5%
3. If the simple interest on Rs 1500 increases by Rs 30, when the time increases by 8 years. Find the rate per cent per annum.
   a) 0.5% b) 0.25% c) 0.75% d) 1.25%
4. If the simple interest on Rs 170 increases by Rs 17, when the time increases by 5 years. Find the rate per cent per annum.
   a) 2% b) 1% c) 2.5% d) 1.5%

Answers

1. c 2. a 3. b 4. a

Rule 3

Theorem: ***If a person deposits Rs x_1 in a bank at r_1% per annum and Rs x_2 in another bank at r_2% per annum, then the rate of interest for the whole sum is*** $\left[\frac{x_1r_1 + x_2r_2}{x_1 + x_2}\right]$.

Illustrative Example

Ex.: A man deposits Rs 2000 in a bank at 4% per annum and Rs 3000 in UTI at 14% per annum. Find the rate of interest for the whole sum.

Soln: Following the above formula, we have the rate of interest for the whole sum

$$= \frac{(2000\times 4)+(3000\times 14)}{2000+3000} = \frac{50000}{5000} = 10\%.$$

Exercise

1. Parameshwaran invested an amount of Rs 12000 at the simple interest rate of 10 per cent per annum and another amount at the simple interest rate of 20 per cent per annum. The total interest earned at the end of one year on the total amount invested became 14 per cent per annum. find the total amount invested.
   [SBI Associates PO Exam, 1999]
   a) Rs 22000 b) Rs 25000 c) Rs 20000 d) Rs 24000
2. Randhir invested an amount of Rs 6000 at the simple interest rate of 5 per cent per annum and another amount at the simple interest rate of 10 per cent per annum. The total interest earned at the end of one year on the total amount invested became 7 per cent per annum. find the total amount invested.
   a) Rs 8000 b) Rs 4000 c) Rs 10000 d) Rs 15000
3. Raju invested an amount of Rs 3000 at the simple interest rate of $2\frac{1}{2}$ per cent per annum and another amount at the simple interest rate of 5 per cent per annum. The total interest earned at the end of one year on the total amount invested became $3\frac{1}{2}$ per cent per annum. find the total amount invested.
   a) Rs 5000 b) Rs 8000 c) Rs 6000 d) Rs 2000
4. A man deposits Rs 1350 in a bank at 5% per annum and Rs 1150 in another bank at 6% per annum. Find the rate of interest for the whole sum.
   a) 5.40% b) 6.40% c) 5.46% d) 11%
5. A man deposits Rs 4000 in a bank at 15% per annum and Rs 6000 in another bank at 16% per annum. Find the rate of interest for the whole sum.
   a) 15.86% b) 31% c) 14.6% d) 15.6%

Answers

1. c; **Hint:** Applying the rule we have,

$$14 = \frac{12000\times 10 + x_2\times 20}{12000 + x_2}$$

or, x_2 = Rs 8000

∴ Total amount = Rs 12000 + Rs 8000 = Rs 20000.

2. c 3. a 4. c 5. d

Rule 4

Theorem: ***If the simple interest on a sum of money is $1/n$ of the principal, and the number of years is equal to the rate per cent per annum, then the rate per cent is*** $\left[\sqrt{100\times\frac{1}{n}}\right]$%.

Note: *1.* Time is also given by the same formula

ie $\left[\sqrt{100\times\frac{1}{n}}\right]$ years.

2. Also see **Rule 35.**

Illustrative Example

Ex.: The simple interest on a sum of money is $\frac{1}{9}$ of the principal, and the number of years is equal to the rate per cent per annum. Find the rate per cent.
[Clerical Grade, 1991]

Soln: Applying the above formula, we have

$$\text{rate} = \sqrt{100\times\frac{1}{9}} = \frac{10}{3} = 3\frac{1}{3}\%.$$

Exercise

1. The simple interest on a sum of money is $\frac{1}{4}$ of the prin-

cipal, and the number of years is equal to the rate per cent per annum. Find the rate per cent.

a) 6% b) 4% c) 5% d) 10%

2. The simple interest on a sum of money is $\frac{1}{16}$ of the principal, and the number of years is equal to the rate per cent per annum. Find the rate per cent.

a) $2\frac{1}{2}\%$ b) $3\frac{1}{3}\%$ c) $4\frac{1}{2}\%$ d) None of these

3. The simple interest on a sum of money is $\frac{1}{25}$ of the principal, and the number of years is equal to the rate per cent per annum. Find the rate per cent.

a) 3% b) 4% c) 2% d) 2.5%

4. The simple interest on a sum of money is equal to the principal and the number of years is equal to the rate per cent per annum. Find the rate per cent.

a) 25% b) 100% c) 10% d) Can't be possible

5. The simple interest on a sum of money is $\frac{4}{9}$ of the principal, and the number of years is equal to the rate per cent per annum. Find the rate per cent.

a) $6\frac{2}{3}\%$ b) $5\frac{3}{5}\%$ c) $7\frac{2}{3}\%$ d) $6\frac{1}{3}\%$

Answers

1. c 2. a 3. c 4. c 5. a

Rule 5

When Principal (P) changes from P_1 to P_2, then $SI_1 - SI_2$ is given by $\frac{(P_1 - P_2) \times r \times t}{100}$.

Illustrative Example

Ex.: If the simple interest on Rs 1200 be more than the interest on Rs 1,000 by Rs 30 in 3 years, find the rate per cent per annum.

Soln: Applying the above formula, we have,

$$30 = \frac{(1200 - 1000) \times r \times 3}{100}$$

$$\text{or, } r = \frac{100 \times 30}{3 \times 200} = 5\%.$$

Exercise

1. If the simple interest on Rs 1350 be more than the interest on Rs 1250 by Rs 20 in 2 years, find the rate per cent per annum.

a) 5% b) 10% c) 6% d) 8%

2. If the simple interest on Rs 2764 be more than the interest on Rs 2464 by Rs 15 in 5 years, find the rate per cent per annum.

a) 2% b) 2.5% c) 4% d) 1%

3. If the simple interest on Rs 1888 be more than the interest on Rs 1763 by Rs 25 in 4 years, find the rate per cent per annum.

a) 5% b) 3% c) 8% d) 10%

Answers

1. b 2. d 3. a

Rule 6

When Rate (r) changes from R_1 to R_2, $SI_1 - SI_2$ is given by $\frac{P \times (r_1 - r_2) \times t}{100}$.

Illustrative Example

Ex.: If simple interest on Rs 2000 increases by Rs 40, when the rate % increases by 2% per annum. Find the time.

Soln: Applying the above formula, we have,

$$40 = \frac{2000 \times 2 \times t}{100}$$

$$\text{or, } t = \frac{100 \times 40}{2000 \times 2} = 1 \text{ year.}$$

Exercise

1. If simple interest on Rs 300 increases by Rs 15, when the rate % increases by 4% per annum. Find the time.

a) $1\frac{1}{4}$ years b) 1 year c) $2\frac{1}{4}$ years d) 2 years

2. If simple interest on Rs 1250 increases by Rs 50, when the rate % increases by 4% per annum. Find the time.

a) 1 year b) 1.5 years
c) 2 years d) 1.25 years

3. If simple interest on Rs 375 increases by Rs 75, when the rate % increases by 5% per annum. Find the time.

a) 2 years b) 8 years
c) 4 years d) None of these

4. The difference between the interests received from two different banks on Rs 1000 for 2 years is Rs 20. Thus the difference between their rates is ____ per cent.

a) 2 b) 1 c) 1.5 d) 2.5

Answers

1. a 2. a 3. c

4. b; **Hint:** $\frac{1000 \times (r_1 - r_2) \times 2}{100} = 20$

$$\therefore r_1 - r_2 = \frac{2000}{1000 \times 2} = 1 \text{ per cent}$$

Rule 7

Theorem: The annual payment that will discharge a debt of Rs A due in t years at the rate of interest r% per annum

is $\left[\dfrac{100A}{100t+\dfrac{rt(t-1)}{2}}\right]$.

Illustrative Example

Ex.: What annual payment will discharge a debt of Rs 770 due in 5 years, the rate of interest being 5% per annum?

Soln: Detail Method: Let the annual payment be P rupees.

The amount of Rs P in 4 years at 5%

$$=\frac{100P+4\times5P}{100}=\frac{120P}{100}$$

The amount of Rs P in 3 years at 5% $=\dfrac{115P}{100}$

The amount of Rs P in 2 years at 5% $=\dfrac{110P}{100}$

The amount of Rs P in 1 year at 5% $=\dfrac{105P}{100}$

These four amounts together with the last annual payment of Rs P will discharge the debt of Rs 770.

$$\therefore\ \frac{120P}{100}+\frac{115P}{100}+\frac{110P}{100}+\frac{105P}{100}+P=700$$

$$\therefore\ \frac{550P}{100}=770\quad\therefore\ P=\frac{770\times100}{550}=140$$

Hence annual payment = Rs 140

Quicker Method: Using the above theorem, we have

$$\text{annual payment}=\frac{100\times770}{5\times100+\dfrac{5(4)(5)}{2}}=\frac{770\times100}{550}$$

= Rs 140.

Exercise

1. What annual payment will discharge a debt of Rs 19350 due 4 years hence at the rate of 5% simple interest?
a) Rs 4500 b) Rs 5400 c) Rs 4000 d) None of these
2. What annual payment will discharge a debt of Rs 1540 due 7 years hence at the rate of 10% simple interest?
a) Rs $170\frac{3}{13}$ b) Rs $168\frac{3}{13}$ c) Rs $169\frac{3}{23}$ d) Rs $169\frac{3}{13}$
3. What annual payment will discharge a debt of Rs 47250 due 3 years hence at the rate of 5% simple interest?
a) Rs 15500 b) Rs 16000 c) Rs 15000 d) Rs 14000
4. What annual payment will discharge a debt of Rs 5600 due 5 years hence at the rate of 6% simple interest?
a) Rs 500 b) Rs 560 c) Rs 1000 d) None of these
5. What annual payment will discharge a debt of Rs 9270 due 3 years hence at the rate of 3% simple interest?
a) Rs 3000 b) Rs 2000 c) Rs 2500 d) Rs 3500
6. What annual instalment will discharge a debt of Rs 4,200 due in 5 years at 10% simple interest? **[AAO 1982]**
a) Rs 700 per year b) Rs 350 per year
c) Rs 750 per year d) Rs 650 per year

Answers

1. a 2. d 3. c 4. c 5. a 6. a

Rule 8

Ex.: The rate of interest for the first 2 years is 3% per annum, for the next 3 years is 8% per annum and for the period beyond 5 years 10% per annum. If a man gets Rs 1520 as a simple interest for 6 years, how much money did he deposit?

Soln: Detail Method: Let his deposit = Rs 100

Interest for first 2 years = Rs 6

Interest for next 3 years = Rs 24

Interest for the last year = Rs 10

Total interest = Rs 40

When interest is Rs 40, deposited amount is Rs 100

$\therefore$ when interest is Rs 1520, deposited amount

$$=\frac{100}{40}\times1520=\text{Rs }3800$$

Direct Formula: Principal $=\dfrac{\text{Interest}\times100}{t_1r_1+t_2r_2+t_3r_3+....}$

$$=\frac{1520\times100}{2\times3+3\times8+1\times10}=\frac{1520\times100}{40}$$

= Rs 3800.

Note: Here $t_3=6-5=1$ year.

Exercise

1. Arun borrowed a sum of money from Jayant at the rate of 8% per annum simple interest for the first four years, 10% per annum for the next 6 years and 12% per annum for the period beyond 10 years. If he pays a total of Rs 12160 as interest only at the end of 15 years, how much money did he borrow? **[BSRB Mumbai PO, 1998]**
a) Rs 8000 b) Rs 10000 c) Rs 12000 d) Rs 9000
2. Ashok borrowed some money at the rate of 6 per cent per annum for the first two years, at the rate of 9 per cent per annum for the next three years and at the rate of 14% per cent per annum for the period beyond five years. If he pays a total interest of Rs 11400 at the end of 9 years how much money did he borrow? **[Bank of Baroda PO, 1999]**
a) Rs 16,000 b) Rs 14,000 c) Rs 18,000 d) Rs 12,000

3. Nelson borrowed some money at the rate of 6 per cent per annum for the first three years, 9 per cent per annum for the next five years and 13 per cent per annum for the period beyond eight years. If the total interest paid by him at the end of eleven years is Rs 8160, how much money did he borrow? **[BSRB Chennai PO 2000]**
a) Rs 12000 b) Rs 10000
c) Rs 8000 d) Data inadequate
4. Manish borrowed some money at the rate of 7 per cent per annum for the first three years, 9 per cent per annum for the next six years and 10 per cent per annum for the period beyond nine years. If the total interest paid by him at the end of fifteen years is Rs 4050, how much money did he borrow?
a) Rs 2800 b) Rs 3600 c) Rs 3000 d) Rs 3500
5. Manoranjan borrowed some money at the rate of 5 per cent per annum for the first two years, 4 per cent per annum for the next four years and 3 per cent per annum for the period beyond six years. If the total interest paid by him at the end of eight years is Rs 3840, how much money did he borrow?
a) Rs 12000 b) Rs 1200 c) Rs 8000 d) Rs 6000

Answers

1. a 2. d 3. c 4. c 5. a

Rule 9

Theorem: ***If a sum of money becomes 'x' times in 't' years at SI, the rate of interest is given by*** $\frac{100(x-1)}{t}\%$.

Illustrative Examples

Ex. 1: A sum of money doubles itself in 10 years at simple interest. What is the rate of interest?

Soln: **Detail Method:** Let the sum be Rs 100
After 10 years it becomes Rs 200
$\therefore$ Interest = 200 – 100 = 100

Then, rate $= \frac{100I}{Pt} = \frac{100\times100}{100\times10} = 10\%$

Direct Formula:

Using the above formula: rate $= \frac{100(2-1)}{10} = 10\%$.

Ex. 2: A sum of money trebles itself in 20 years at SI. Find the rate of interest.

Soln: Rate $= \frac{100(3-1)}{20} = 10\%$.

Note: *If a sum of money becomes 'x' times at the rate of r% simple interest, then the time (t) is given by* $\frac{100(x-1)}{r}$ *years.*

Ex.: In what time does a sum of money become four times at the simple interest rate of 5% per annum?

Soln: Using the above formula, we have

$$t = \frac{100(4-1)}{5} = 60 \text{ years.}$$

Exercise

1. A sum of money becomes 6 times in 20 years at SI. Find the rate of interest.
a) 20% b) 15% c) 16% d) 25%
2. A sum of money becomes 4 times in 12 years at SI. Find the rate of interest.
a) 25% b) 24% c) 14% d) 15%
3. A sum of money becomes 5 times in 20 years at SI. Find the rate of interest.
a) 20% b) 16% c) 25% d) 10%
4. A sum of money becomes 3 times in 10 years at SI. Find the rate of interest.
a) 16% b) 18% c) 20% d) 25%
5. In what time does a sum of money become four times at the simple interest rate of 10% per annum?
a) 30 years b) 25 years c) 35 years d) 40 years
6. In what time does a sum of money become twice at the simple interest rate of 5% per annum?
a) 25 years b) 24 years c) 20 years d) 16 years
7. In what time does a sum of money become thrice at the simple interest rate of 8% per annum?
a) 30 years b) 15 years c) 20 years d) 25 years

Answers

1. d 2. a 3. a 4. c 5. a 6. c 7. d

Rule 10

Theorem: ***A certain sum is invested for certain time. It amounts to Rs A_1 at r_1 % per annum. But when invested at r_2 % per annum, it amounts to Rs A_2, then the time is given by*** $\left[\frac{A_1 - A_2}{A_2 r_1 - A_1 r_2}\right]\times100$ ***years.***

Illustrative Example

Ex.: A certain sum is invested for certain time. It amounts to Rs 80 at 5% per annum. But when invested at 2% per annum, it amounts to Rs 40. Find the time.

Soln: Applying the above formula, we have,

$$\text{time} = \frac{80-40}{40\times5-80\times2}\times100 = 100 \text{ years.}$$

Exercise

1. A certain sum is invested for certain time. It amounts to Rs 400 at 10% per annum. But when invested at 4% per annum, it amounts to Rs 200. Find the time.
a) 100 years b) 75 years c) 50 years d) 60 years
2. A certain sum is invested for certain time. It amounts to Rs 150 at 5% per annum. But when invested at 3% per

annum, it amounts to Rs 100. Find the time.
a) 120 years b) 100 years c) 80 years d) 60 years

3. A certain sum is invested for certain time. It amounts to Rs 450 at 7% per annum. But when invested at 5% per annum, it amounts to Rs 350. Find the time.
a) 50 years b) 60 years c) 45 years d) 40 years
4. A certain sum is invested for certain time. It amounts to Rs 60 at 6% per annum. But when invested at 3% per annum, it amounts to Rs 30. Find the time.
a) 15 years b) 20 years c) 5 years d) Not possible
5. A certain sum is invested for certain time. It amounts to Rs 500 at 8% per annum. But when invested at 3% per annum, it amounts to Rs 200. Find the time.
a) 100 years b) 200 years c) 50 years d) 300 years

Answers

1. c 2. b 3. a
4. d; **Hint:** Applying the given rule,

required time $= \frac{60-30}{30\times 6-60\times 3} = \frac{30}{0}$ cannot be defined.

5. d

Rule 11

Theorem: ***A certain sum is invested for certain time. It amounts to Rs A_1 at r_1% per annum. But when invested at r_2% per annum, it amounts to Rs A_2, then the sum is given by Rs $\left[\frac{A_2 r_1 - A_1 r_2}{r_1 - r_2}\right]$.***

Illustrative Example

Ex.: A certain sum is invested for certain time. It amounts to Rs 80 at 5% per annum. But when invested at 2% per annum, it amounts to Rs 40. Find the sum.

Soln: Applying the above formula, we have

$$\text{sum} = \frac{40\times 5-80\times 2}{5-2} = \text{Rs } \frac{40}{3} = \text{Rs } 13\frac{1}{3}$$

Exercise

1. A certain sum is invested for certain time. It amounts to Rs 400 at 10% per annum. But when invested at 4% per annum, it amounts to Rs 200. Find the sum.
a) Rs $\frac{200}{3}$ b) Rs 100 c) Rs $\frac{400}{3}$ d) None of these
2. A certain sum is invested for certain time. It amounts to Rs 150 at 5% per annum. But when invested at 3% per annum, it amounts to Rs 100. Find the sum.
a) Rs 50 b) Rs 25 c) Rs 30 d) Rs 60
3. A certain sum is invested for certain time. It amounts to Rs 450 at 7% per annum. But when invested at 5% per annum, it amounts to Rs 350. Find the sum.
a) Rs 150 b) Rs 250 c) Rs 100 d) Rs 200
4. A certain sum is invested for certain time. It amounts to Rs 500 at 8% per annum. But when invested at 3% per annum, it amounts to Rs 200. Find the sum.
a) Rs 20 b) Rs 50 c) Rs 25 d) Rs 35

Answers

1. a 2. b 3. c 4. a

Rule 12

Theorem: ***A sum was put at SI at a certain rate for t years. Had it been put at x% higher rate, it would have fetched Rs 'A' more, then the sum is Rs $\left[\frac{A\times 100}{t\times x}\right]$ or $\frac{\text{More Interest}\times 100}{\text{Time}\times\text{More Rate}}$.***

Illustrative Example

Ex.: A sum was put at SI at a certain rate for 2 years. Had it been put at 3% higher rate, it would have fetched Rs 300 more. Find the sum.

Soln: **Detail Method:** Let the sum be Rs x and the original rate be y% per annum. Then, new rate = (y + 3)% per annum.

$$\therefore \frac{x(y+3)\times 2}{100} - \frac{x(y)\times 2}{100} = 300$$

$xy + 3x - xy = 15{,}000$
or, $x = 5000$
Thus, the sum = Rs 5000.

Quicker Method: Applying the above formula, we have,

$$\text{sum} = \frac{300\times 100}{2\times 3} = \text{Rs } 5000.$$

Exercise

1. A sum was put at SI at a certain rate for 3 years. Had it been put at 4% higher rate, it would have fetched Rs 600 more. Find the sum.
a) Rs 5000 b) Rs 4000 c) Rs 6000 d) Rs 3000
2. A sum was put at SI at a certain rate for 5 years. Had it been put at 5% higher rate, it would have fetched Rs 500 more. Find the sum.
a) Rs 2500 b) Rs 2000 c) Rs 1500 d) Rs 1800
3. A sum was put at SI at a certain rate for 6 years. Had it been put at 4% higher rate, it would have fetched Rs 960 more. Find the sum.
a) Rs 3000 b) Rs 3500 c) Rs 4000 d) Rs 4500
4. A sum was put at SI at a certain rate for 4 years. Had it been put at 4% higher rate, it would have fetched Rs 160 more. Find the sum.
a) Rs 1500 b) Rs 800 c) Rs 1200 d) Rs 1000
5. A sum was put at SI at a certain rate for 2 years. Had it

been put at 5% higher rate, it would have fetched Rs 125 more. Find the sum.
a) Rs 1250 b) Rs 1000 c) Rs 750 d) Rs 1500

Answers

1. a 2. b 3. c 4. d 5. a

Rule 13

Theorem: ***If a certain sum of money amounts to Rs A_1 in t_1 years and to Rs A_2 in t_2 years, then the sum is given by*** $\left(\frac{A_2t_1 - A_1t_2}{t_1 - t_2}\right)$.

Illustrative Example

Ex.: A certain sum of money amounts to Rs 756 in 2 years and to Rs 873 in 3.5 years. Find the sum and the rate of interest.

Soln: Applying the above theorem, we have

the required answer $= \frac{873 \times 2 - 756 \times 3.5}{2 - 3.5}$

$= \frac{-900}{-1.5} =$ Rs 600

$\therefore$ sum = Rs 600

SI = 756 – 600 = Rs 156

rate $= \frac{100 \times 156}{600 \times 2} = 13\%$ per annum.

(Also see **Rule 31**)

Exercise

1. A certain sum of money amounts to Rs 550 in 3 years and to Rs 650 in 4 years. Find the sum.
a) Rs 250 b) Rs 300 c) Rs 150 d) Rs 350
2. A certain sum of money amounts to Rs 758 in 4 years and to Rs 875 in 6 years. Find the sum.
a) Rs 534 b) Rs 524 c) Rs 624 d) Rs 434
3. A certain sum of money amounts to Rs 1125 in 5 years and to Rs 1200 in 8 years. Find the sum.
a) Rs 900 b) Rs 500 c) Rs 1000 d) Rs 800
4. A certain sum of money amounts to Rs 625 in 4 years and to Rs 680 in 5 years. Find the sum.
a) Rs 505 b) Rs 305 c) Rs 405 d) Rs 504
5. A certain sum of money at simple interest amounts to Rs 379.50 in 3 years and to Rs 453.75 in 7½ years. Find the sum and the rate of interest.
a) Rs 330, 5% b) Rs 370, 5%
c) Rs 330, 4% d) Rs 370, 4%
6. A sum of money lent out at simple interest amounts to Rs 460 in 3 years while in 5 years it amounts to Rs 500. Find the sum and the rate of interest.
a) Rs 400, 4% b) Rs 300, 5%
c) Rs 400, 5% d) None of these
7. A sum of money lent out at simple interest amounts to Rs 560 in 3 years while in 5 years it amounts to Rs 600. Find the sum and the rate of interest.
a) Rs 400, 5% b) Rs 500, 4%
c) Rs 300, 5% d) Rs 400, 4%

Answers

1. a 2. b 3. c 4. c 5. a 6. c 7. b

Rule 14

Theorem: ***A sum was put at SI at a certain rate for t years. Had it been put at x% lower rate, it would have fetched Rs A less, then the sum is Rs*** $\left[\frac{A \times 100}{t \times x}\right]$ ***or*** $\frac{\text{Less Interest} \times 100}{\text{Time} \times \text{Less rate}}$.

Ex.: A sum was put at SI at a certain rate for 2 years. Had it been put at 3% lower rate, it would have fetched Rs 300 less. Find the sum.

Soln: **Quicker Method:** Applying the above formula, we have

Sum $= \frac{300 \times 100}{2 \times 3} =$ Rs 5000.

Exercise

1. A sum was put at SI at a certain rate for 4 years. Had it been put at 5% lower rate, it would have fetched Rs 100 less. Find the sum.
a) Rs 500 b) Rs 5000 c) Rs 400 d) Rs 4000
2. A sum was put at SI at a certain rate for 5 years. Had it been put at 2% lower rate, it would have fetched Rs 150 less. Find the sum.
a) Rs 1000 b) Rs 1500 c) Rs 1800 d) Rs 2000
3. A sum was put at SI at a certain rate for 3 years. Had it been put at 4% lower rate, it would have fetched Rs 600 less. Find the sum.
a) Rs 500 b) Rs 4000 c) Rs 5000 d) Rs 6000
4. A sum was put at SI at a certain rate for 2 years. Had it been put at 7% lower rate, it would have fetched Rs 280 less. Find the sum.
a) Rs 1500 b) Rs 1800 c) Rs 2800 d) Rs 2000
5. A sum was put at SI at a certain rate for 4 years. Had it been put at 6% lower rate, it would have fetched Rs 720 less. Find the sum.
a) Rs 3000 b) Rs 4000 c) Rs 3500 d) Rs 2400

Answers

1. a 2. b 3. c 4. d 5. a

Rule 15

Theorem: ***Rs X is divided into two parts such that if one part be invested at r_1% and the other at r_2%, the annual interest from both the investments is Rs A. Then the first***

part is given by $\left[\frac{100A - r_2 X}{r_1 - r_2}\right]$.

Illustrative Example

Ex.: Rs 4000 is divided into two parts such that if one part be invested at 3% and the other at 5%, the annual interest from both the investments is Rs 144. Find each part.

Soln: Detail Method:

Let the amount lent at 3% rate be Rs x,
then 3% of x + 5% of $(4000 - x) = 144$
or, $3x + 5 \times 4000 - 5x = 14400$
or, $2x = 5600 \therefore x = 2800$
Thus, the two amounts are Rs 2800 and Rs (4000 – 2800) or Rs 1200

Quicker Method: Applying the above theorem, we have,

$$\text{the first part} = \frac{100 \times 144 - 5 \times 4000}{3-5}$$

$$= \frac{-5600}{-2} = \text{Rs } 2800 \text{ and}$$

the second part = Rs 4000 – Rs 2800 = Rs 1200.

Exercise

1. Anish borrowed Rs 15000 at the rate of 12% and an other amount at the rate of 15% for two years. The total interest paid by him was Rs 9000. How much did he borrow? **[BSRB Hyderabad PO, 1999]**
   a) Rs 32000 b) Rs 33000 c) Rs 30000 d) Rs 63000
2. Aniket deposited two parts of a sum of Rs 25000 in different banks at the rates of 15% per annum and 18% per annum respectively. In one year he got Rs 4050 as the total interest. What was the amount deposited at the rate of 18% per annum? **[BSRB Patna PO, 2001]**
   a) Rs 9000 b) Rs 18000
   c) Rs 15000 d) None of these
3. A man had Rs 2000, part of which he lent at 5 per cent and the rest at 4 per cent. The whole annual interest was Rs 92. How much did he lend at 5 per cent?
   a) Rs 1200 b) Rs 800 c) Rs 700 d) Rs 1300
4. Sudarshan had Rs 1500, part of which he lent at 3 per cent and the rest at 2 per cent. The whole annual interest was Rs 32. How much did he lend at 2 per cent?
   a) Rs 200 b) Rs 1300 c) Rs 300 d) Rs 1200
5. Saket deposited two parts of a sum of Rs 12500 in different banks at the rates of 15% per annum and 18% per annum respectively. In one year he got Rs 2025 as the total interest. What was the amount deposited at the rate of 15% per annum?
   a) Rs 5000 b) Rs 6500 c) Rs 7500 d) Rs 8000
6. A milk man borrowed Rs 2,500 from two money lenders. For one loan, he paid 5% per annum and for the other, he paid 7% per annum. The total interest paid for two years was Rs 265. How much did he borrow at each rate? **[MBA '1986]**
   a) Rs 2125, Rs 375 b) Rs 2000, Rs 500
   c) Rs 1875, Rs 625 d) None of these

Answers

1. b; **Hint:** Let the amount borrowed second time be Rs x.

$\therefore$ X = Rs $(15000 + x)$, A = Rs $\frac{9000}{2}$ = Rs 4500

[Since total interest is given for 2 years.]

$r_1 = 12\%$ and $r_2 = 15\%$

Now, applying the above formula, we have

$$\frac{100 \times 4500 - 15(15000 + x)}{12 - 15} = 15000$$

or, x = Rs 18000

$\therefore$ Total amount borrowed = Rs 15000 + Rs 18000 = Rs 33000

2. d; **Hint:** Applying the formula, we have the amount deposited at the rate of 15% per annum

$$= \frac{100 \times 4050 - 18 \times 25000}{15 - 18}$$

= Rs 15000

$\therefore$ Amount deposited at the rate of 18% per annum
= Rs 25000 – Rs 15000
= Rs 10000.

3. a 4. b 5. c 6. a

Rule 16

Theorem: ***At a certain rate of simple interest Rs X amounted to Rs A in t_1 years. If the rate of interest be decreased by r%, then after t_2 years the new interest is given by Rs***

$$\left[\left(\frac{A - X}{t_1}\right) - \left(\frac{rX}{100}\right)\right] t_2.$$

Illustrative Example

Ex.: At a certain rate of simple interest Rs 800 amounted to Rs 920 in 3 years. If the rate of interest be decreased by 3%, what will be the amount after 3 years.

Soln: Detail Method:

$$\text{First rate of interest} = \frac{120 \times 100}{800 \times 3} = 5\%$$

New rate = 5 – 3 = 2%

$$\therefore \text{New Interest} = \frac{800 \times 3 \times 2}{100} = \text{Rs } 48$$

$\therefore$ New amount = 800 + 48 = Rs 848

Quicker Method: Applying the above formula,

New Interest = $\left[\frac{920-800}{3}-\frac{3\times800}{100}\right]\times3$

$= (40 - 24)3 = \text{Rs } 48.$

$\therefore$ New Amount = 800 + 48 = Rs 848.

Exercise

1. At a certain rate of simple interest Rs 400 amounted to Rs 460 in 3 years. If the rate of interest be decreased by 3%, what will be the amount after 3 years?
a) Rs 424 b) Rs 484 c) Rs 242 d) Rs 848
2. At a certain rate of simple interest Rs 900 amounted to Rs 1260 in 4 years. If the rate of interest be decreased by 2%, what will be the amount after 4 years?
a) Rs 1338 b) Rs 1188 c) Rs 1378 d) Rs 1128
3. At a certain rate of simple interest Rs 1600 amounted to Rs 1840 in 5 years. If the rate of interest be decreased by 3%, what will be the amount after 5 years?
a) Rs 1720 b) Rs 1680 c) Rs 1600 d) Not possible
4. At a certain rate of simple interest Rs 1220 amounted to Rs 1320 in 2 years. If the rate of interest be decreased by 3%, what will be the amount after 5 years?
a) Rs 1387 b) Rs 1287 c) Rs 1278 d) Rs 1388
5. At a certain rate of simple interest Rs 1230 amounted to Rs 1550 in 5 years. If the rate of interest be decreased by 4%, what will be the amount after 6 years?
a) Rs 1320 b) Rs 1320.8 c) Rs 1318.8 d) Rs 1638.8

Answers

1. a 2. b

3. c; **Hint:** New interest = $\left[\frac{1840-1600}{5}-\frac{3\times1600}{100}\right]5$

$= (48 - 48)5 = 0$

$\therefore$ New amount = Rs 1600 + 0 = Rs 1600

4. b 5. c

Rule 17

Theorem: ***At a certain rate of simple interest Rs X amounted to Rs A in t_1 years. If the rate of interest be increased by r% then after t_2 years the new interest is given by Rs***

$$\left[\left(\frac{A-X}{t_1}\right)+\left(\frac{rX}{100}\right)\right]t_2.$$

Illustrative Example

Ex.: At a certain rate of simple interest Rs 800 amounted to Rs 920 in 3 years. If the rate of interest be increased by 3%, what will be the amount after 3 years?

Soln: Detail Method:

First Rate of interest = $\frac{120\times100}{800\times3}=5\%$

New rate = 5 + 3 = 8%

$\therefore$ New Interest = $\frac{800\times3\times8}{100}$ = Rs 192

$\therefore$ New Amount = 800 + 192 = Rs 992.

Quicker Method: Applying the above formula, we have

The New Interest = $\left[\frac{920-800}{3}+\frac{3\times800}{100}\right]\times3$

$= (40 + 24)3 = 64 \times 3 = \text{Rs } 192.$

$\therefore$ New Amount = 800 + 192 = Rs 992.

Exercise

1. At a simple interest Rs 800 becomes Rs 956 in three years. If the interest rate is increased by 3%, how much would Rs 800 become in three years?
a) Rs 1020.80 b) Rs 1004
c) Rs 1028 d) Data inadequate
2. At a simple interest Rs 900 becomes Rs 1060 in 4 years. If the interest rate is increased by 2%, how much would Rs 900 become in 5 years?
a) Rs 1206 b) Rs 1206.50
c) Rs 1190 d) Rs 1260.70
3. At a simple interest Rs 850 becomes Rs 1250 in 5 years. If the interest rate is increased by 5%, how much would Rs 850 become in 4 years?
a) Rs 1340 b) Rs 1430
c) Rs 1340.50 d) None of these
4. At a simple interest Rs 1240 becomes Rs 1340 in 4 years. If the interest rate is increased by 2%, how much would Rs 1240 become in 3 years?
a) Rs 1839.40 b) Rs 1389
c) Rs 1380.40 d) Rs 1389.40
5. At a simple interest Rs 1300 becomes Rs 1550 in 5 years. If the interest rate is increased by 3%, how much would Rs 1300 become in 2 years?
a) Rs 1478 b) Rs 1748 c) Rs 1578 d) Rs 1874
6. At a simple interest Rs 1400 becomes Rs 1727 in 3 years. If the interest rate is increased by 4%, how much would Rs 1400 become in 6 years?
a) Rs 2490 b) Rs 2390 c) Rs 2890 d) Rs 2590
7. Rs 1,200 amounts to Rs 1,632 in 4 years at a certain rate of simple interest. If the rate of interest is increased by 1%, it would amount to how much?
[Bank PO, 1991]
a) Rs 1635 b) Rs 1644 c) Rs 1670 d) Rs 1680

Answers

1. c; **Hint:** Applying the above theorem, we have,

the new interest = $\left[\frac{956-800}{3}+\frac{3\times800}{100}\right]\times3$ = Rs 228

$\therefore$ required answer = Rs 800 + Rs 228 = Rs 1028

2. c 3. a 4. d 5. a 6. b 7. d

Rule 18

Theorem: ***The simple interest on a sum of money will be Rs x after 't' years. If in the next 't' years principal becomes n times, then the total interest at the end of the '2t'th year is given by Rs [(n + 1)x].***

Illustrative Example

Ex.: The simple interest on a sum of money will be Rs 300 after 5 years. In the next 5 years principal is trebled, what will be the total interest at the end of the 10th year?

Soln: Detail Method:

$SI = \frac{P \times r \times t}{100}$; Here SI = Rs 300, t = 5 years given.

$300 = \frac{P \times r \times 5}{100} \Rightarrow 5\text{Pr} = 300 \times 100$ .... (i)

Now, we calculate SI for next 5 years.

$SI = \frac{3P \times r \times 5}{100}$ [$\because$ In the next 5 years Principal (P) is trebled.]

From equation (i)

$SI = \frac{3 \times 300 \times 100}{100}$ = Rs 900.

$\therefore$ Total SI at the end of 10th year

= Rs 300 + Rs 900 = Rs 1200.

Quicker Method: Applying the above theorem, we have,

the required Simple Interest = (3 + 1) 300 = Rs 1200.

Exercise

1. The simple interest on a sum of money will be Rs 250 after 6 years. In the next 6 years principal is doubled, what will be the total interest at the end of the 12th year?
   a) Rs 850 b) Rs 750 c) Rs 650 d) None of these
2. The simple interest on a sum of money will be Rs 400 after 3 years. In the next 3 years principal becomes 4 times, what will be the total interest at the end of the 6th year?
   a) Rs 2000 b) Rs 2500 c) Rs 1800 d) Rs 1600
3. The simple interest on a sum of money will be Rs 150 after 4 years. In the next 4 years principal becomes 5 times, what will be the total interest at the end of the 8th year?
   a) Rs 950 b) Rs 850 c) Rs 900 d) Rs 860
4. The simple interest on a sum of money will be Rs 190 after 7 years. In the next 7 years principal becomes 3 times, what will be the total interest at the end of the 14th year?
   a) Rs 760 b) Rs 850 c) Rs 750 d) Rs 780

Answers

1. b 2. a 3. c 4. a

Rule 19

Theorem: ***The simple interest on a sum of money will be Rs x after t_1 years. If in the next t_2 years principal becomes n times, then the total interest at the end of $(t_1 + t_2)$th year is given by Rs*** $x\left[1+\left(\frac{t_2}{t_1}\right)n\right]$.

Note: **Rule - 18** is a special case of this rule.
In **rule - 18**, $t_1 = t_2 = t$, hence we have the formula like [x (1 + n)].

Illustrative Example

Ex.: The simple interest on a sum of money will be Rs 300 after 4 years. In the next 6 years principal becomes 4 times, what will be the total interest at the end of the 10th year?

Soln: Detail Method:

Simple Interest for 4 years = $\frac{P \times r \times 4}{100}$ = Rs 300

$\Rightarrow$ 4Pr = 300 × 100 .... (i)

Simple interest for the next 6 years = $\frac{4P \times r \times 6}{100}$

[$\because$ P becomes four times in the next 6 years]

From equation (i)

$SI = \frac{300 \times 100 \times 6}{100}$ = Rs 1800.

$\therefore$ Total simple interest at the end of 10th year

= Rs 300 + Rs 1800 = Rs 2100.

Quicker Method: Applying the above formula, we have

the total simple interest at the end of 10th year

$= 300\left[1+\left(\frac{6}{4}\right)\times 4\right] = 300 \times 7$

= Rs 2100.

Exercise

1. The simple interest on a sum of money will be Rs 450 after 6 years. In the next 8 years principal becomes 3 times, what will be the total interest at the end of the 14th year?
   a) Rs 2250 b) Rs 1250
   c) Rs 2050 d) None of these
2. The simple interest on a sum of money will be Rs 235 after 5 years. In the next 7 years principal becomes 5 times, what will be the total interest at the end of the 12th year?
   a) Rs 1980 b) Rs 1080 c) Rs 1880 d) Rs 1860
3. The simple interest on a sum of money will be Rs 165

after 2 years. In the next 4 years principal becomes 2 times, what will be the total interest at the end of the 6th year?

a) Rs 835 b) Rs 825 c) Rs 625 d) None of these

4. The simple interest on a sum of money will be Rs 225 after 3 years. In the next 5 years principal becomes 3 times, what will be the total interest at the end of the 8th year?

a) Rs 1250 b) Rs 1330 c) Rs 1360 d) Rs 1350

5. The simple interest on a sum of money will be Rs 625 after 4 years. In the next 12 years principal becomes 8 times, what will be the total interest at the end of the 16th year?

a) Rs 16525 b) Rs 16625 c) Rs 15625 d) Rs 15525

Answers

1. a 2. c 3. b 4. d 5. c

Rule 20

Theorem: ***A sum of Rs X is lent out in n parts in such a way that the interest on first part at r_1% for t_1 years, the interest on second part at r_2% for t_2 years the interest on third part at r_3% for t_3 years, and so on, are equal, the ratio in which the sum was divided in n parts is given by***

$$\frac{1}{r_1t_1}:\frac{1}{r_2t_2}:\frac{1}{r_3t_3}:....\frac{1}{r_nt_n}.$$

Note: Compare this rule with **Rule - 24**. In this rule **interest** on different parts of the sum are equal. Whereas in **Rule - 24 amounts** of different parts of the sum become equal after different time periods.

Illustrative Example

Ex.: A sum of Rs 2600 is lent out in two parts in such a way that the interest on one part at 10% for 5 years is equal to that on another part at 9% for 6 years. Find the two sums.

Soln: Detail Method:

$$\text{Each Interest} = \frac{\text{1st Part}\times5\times10}{100} = \frac{\text{2nd Part}\times6\times9}{100}$$

$$\text{or, } \frac{\text{1st Part}}{\text{2nd Part}} = \frac{6\times9}{5\times10} = \frac{27}{25} = 27:25$$

$$\therefore \text{1st Part} = \frac{2600}{27+25}\times27 = \text{Rs } 1350$$

and 2nd Part = 2600 – 1350 = Rs 1250

Quicker Method: If we use the above theorem,

$$S_1:S_2 = \frac{1}{50}:\frac{1}{54} = 54:50 = 27:25$$

$$\therefore \text{1st Part} = \frac{2600}{27+25}\times27 = \text{Rs 1350 and}$$

2nd Part = 2600 – 1350 = Rs 1250.

Exercise

1. A sum of Rs 1440 is lent out in three parts in such a way that the interests on first part at 2% for 3 years, second part at 3% for 4 years and third part at 4% for 5 years are equal. Then find the difference between the largest and the smallest sum.

a) Rs 460 b) Rs 560 c) Rs 400 d) Rs 200

2. A sum of Rs 2560 is lent out in two parts in such a way that the interest on one part at 4% for 5 years is equal to that on another part at 3% for 4 years. Find the two sums.

a) Rs 960, Rs 1600 b) Rs 560, Rs 2000
c) Rs 860, Rs 1700 d) Rs 900, Rs 1660

3. A sum of Rs 1085 is lent out in two parts in such a way that the interest on one part at 9% for 3 years is equal to that on another part at 18% for 2 years. Find the two sums.

a) Rs 620, Rs 465 b) Rs 600, Rs 485
c) Rs 520, Rs 565 d) Rs 630, Rs 655

4. A sum of Rs 1521 is lent out in two parts in such a way that the interest on one part at 10% for 5 years is equal to that on another part at 8% for 10 years. Find the two sums.

a) Rs 926, Rs 595 b) Rs 906, Rs 615
c) Rs 916, Rs 605 d) Rs 936, Rs 585

5. A sum of Rs 2525 is lent out in two parts in such a way that the interest on one part at 7% for 4 years is equal to that on another part at 6% for 7 years. Find the difference between the two parts of the sum.

a) Rs 505 b) Rs 1515 c) Rs 1010 d) None of these

Answers

1. b; **Hint:** Required ratio $= \frac{1}{2\times3}:\frac{1}{3\times4}:\frac{1}{4\times5} = 10:5:3$

1st part = Rs 800, second part = Rs 400 and third part = Rs 240

$\therefore$ required answer = Rs 800 – Rs 240 = Rs 560

2. a 3. a 4. d

5. a; **Hint:** Two parts will be Rs 1515 and Rs 1010

$\therefore$ difference = Rs 1515 – Rs 1010 = Rs 505

Rule 21

Theorem: ***If a sum of money becomes 'n' times at the simple interest rate of r% per annum, then it will become 'm' times at the simple interest rate of $\left[\left(\frac{m-1}{n-1}\right)\times r\right]$ per cent.***

Illustrative Example

Ex.: A sum of money becomes four times at the simple interest rate of 5% per annum. At what rate per cent will it become sevenfold?

Soln: Applying the above formula, we have

$$\text{the required rate} = \frac{7-1}{4-1} \times 5 = 10 \text{ per cent.}$$

Exercise

1. A sum of money becomes three times at the simple interest rate of 4% per annum. At what rate per cent will it become six fold?
   a) 10% b) 12% c) 8% d) None of these
2. A sum of money becomes eight times at the simple interest rate of 7% per annum. At what rate per cent will it become four fold?
   a) 3% b) 3.5% c) 4% d) 2%
3. A sum of money becomes five times at the simple interest rate of 8% per annum. At what rate per cent will it become seven fold?
   a) 6% b) 10% c) 12% d) 14%
4. A sum of money becomes two times at the simple interest rate of 2% per annum. At what rate per cent will it become five fold?
   a) 10% b) 8% c) 6% d) 9%
5. A sum of money becomes six times at the simple interest rate of 5% per annum. At what rate per cent will it become twelve fold?
   a) 10% b) 12% c) 9% d) 11%

Answers

1. a 2. a 3. c 4. b 5. d

Rule 22

Theorem: ***A certain sum of money amounted to Rs A_1 at r_1% in a time in which Rs P amounted to Rs A_2 at r_2%. If the rate of interest is simple, then the sum is given by Rs***

$$\left[\frac{A_1}{\left(1-\frac{r_1}{r_2}\right)+\left(\frac{A_2 r_1}{Pr_2}\right)}\right].$$

Note: If $r_1\% = r_2\%$, the formula becomes $\left[\frac{A_1}{A_2} \times P\right]$.

Illustrative Example

Ex.: A certain sum of money amounted to Rs 575 at 5% in a time in which Rs 750 amounted to Rs 840 at 4%. If the rate of interest is simple, find the sum.

Soln: Detail Method:

Interest = Rs 840 – Rs 750 = Rs 90

$$\therefore \text{ Time} = \frac{90 \times 100}{750 \times 4} = 3 \text{ years.}$$

Now, by the formula,

$$\text{Sum} = \frac{100 \times \text{Amount}}{100 + \text{rt}} = \frac{100 \times 575}{100 + 3 \times 5} = \text{Rs } 500.$$

Quicker Method: Applying the above formula, we have

$$\text{Sum} = \frac{575}{\left(1-\frac{5}{4}\right)+\frac{840 \times 5}{750 \times 4}} = \frac{575}{-\frac{1}{4}+\frac{7}{5}}$$

$$= \frac{575}{\frac{28-5}{20}} = \frac{575 \times 20}{23} = \text{Rs } 500.$$

Exercise

1. A certain sum of money amounted to Rs 710 at 7% in a time in which Rs 700 amounted to Rs 910 at 5%. If the rate of interest is simple, find the sum.
   a) Rs 500 b) Rs 450 c) Rs 650 d) Rs 600
2. A certain sum of money amounted to Rs 810 at 4% in a time in which Rs 450 amounted to Rs 720 at 3%. If the rate of interest is simple, find the sum.
   a) Rs 500 b) Rs 450 c) Rs 600 d) Rs 475
3. A certain sum of money amounted to Rs 825 at 7% in a time in which Rs 560 amounted to Rs 960 at 6%. If the rate of interest is simple, find the sum.
   a) Rs 550 b) Rs 475 c) Rs 650 d) Rs 450
4. A certain sum of money amounted to Rs 765 at 8% in a time in which Rs 640 amounted to Rs 750 at 5%. If the rate of interest is simple, find the sum.
   a) Rs 650 b) Rs 600 c) Rs 700 d) Rs 450
5. A certain sum of money amounted to Rs 1020 at 9% in a time in which Rs 720 amounted to Rs 880 at 4%. If the rate of interest is simple, find the sum.
   a) Rs 680 b) Rs 780 c) Rs 700 d) Rs 580

Answers

1. a 2. b 3. d 4. b 5. a

Rule 23

Theorem: ***A certain sum of money amounts to Rs A_1 in t years at r% per annum, then the time in which it will amount to Rs A_2 at the same rate of interest is given by***

$$\left[\frac{A_2}{A_1}\left(t+\frac{100}{r}\right)-\frac{100}{r}\right] \textit{ years.}$$

Illustrative Example

Ex.: A certain sum of money amounts to Rs 2613 in 6 years at 5% per annum. In how many years will it amount to Rs 3015 at the same rate?

Soln: Detail Method:

$$\text{Principal} = \frac{100 \times \text{Amount}}{100 + rt} = \frac{100 \times 2613}{100 + 30} = \text{Rs } 2010$$

[See **Rule - 26**].

Again,

$$2010 = \frac{100 \times 3015}{100 \times 5t} \Rightarrow t = \frac{1}{5}\left[\frac{100 \times 3015 - 100 \times 2010}{2010}\right]$$

$$= \frac{100 \times 1005}{2010 \times 5} = 10 \text{ years.}$$

Quicker Method: Applying the above theorem,

the required time $= \frac{3015}{2613}\left(6 + \frac{100}{5}\right) - \frac{100}{5}$

$$= \frac{3015}{2613} \times 26 - 20 = 30 - 20 = 10 \text{ years.}$$

Exercise

1. A certain sum of money amounts to Rs 4800 in 5 years at 4% per annum. In how many years will it amount to Rs 5120 at the same rate?
   a) 6 years b) 7 years c) 8 years d) 9 years
2. A certain sum of money amounts to Rs 456 in 2½ years at 8% per annum. In how many years will it amount to Rs 608 at the same rate?
   a) 6½ years b) 12½ years c) 7½ years d) 6¼ years
3. A certain sum of money amounts to Rs 5000 in 5 years at 10% per annum. In how many years will it amount to Rs 6000 at the same rate?
   a) 8 years b) 6 years c) 10 years d) 9 years
4. A certain sum of money amounts to Rs 1572 in 4 years at 5% per annum. In how many years will it amount to Rs 1703 at the same rate?
   a) 5 years b) 6 years c) 7 years d) 8 years
5. A certain sum of money amounts to Rs 1400 in 8 years at 5% per annum. In how many years will it amount to Rs 1500 at the same rate?
   a) 10 years b) 8 years
   c) 12 years d) None of these

Answers

1. b 2. c 3. a 4. b 5. a

Rule 24

Theorem: ***When different amounts mature to the same amount at simple rate of interest, the ratio of the amounts invested are in inverse ratio of (100 + time × rate). That is, the ratio in which the amounts are invested is***

$$\frac{1}{100 + r_1t_1} : \frac{1}{100 + r_2t_2} : \frac{1}{100 + r_3t_3} : .... : \frac{1}{100 + r_nt_n}.$$

Note: Also see **Rule - 20**. Try to understand the difference between these two rules.

Illustrative Example

Ex.: A man invests an amount of Rs 15,860 in the names of his three sons A, B, and C in such a way that they get the same amount after 2, 3, and 4 years respectively. If the rate of simple interest is 5% then find the ratio in which the amount was invested for A, B and C?

Soln: **Quicker Method:** Following the above theorem, the required ratio

$$= \frac{1}{100 + 2 \times 5} : \frac{1}{100 + 3 \times 5} : \frac{1}{100 + 4 \times 5}$$

$$= \frac{1}{110} : \frac{1}{115} : \frac{1}{120}$$

$$= 276 : 264 : 253.$$

Exercise

1. Rs 7930 is so divided into three parts such that their amounts after 2, 3 and 4 years respectively are equal, the simple interest being at the rate of 5% per annum. Find the difference between the greatest and the smallest parts of the sum.
   a) Rs 250 b) Rs 230 c) Rs 280 d) Rs 330
2. Rs 8829 is divided into three parts in such a way that their amounts at 4% per annum simple interest after 5, 6 and 8 years are equal. Find each part of the sum.
   a) Rs 3069, Rs 2970, Rs 2790
   b) Rs 3609, Rs 2970, Rs 2790
   c) Rs 3089, Rs 2970, Rs 2790
   d) Rs 3069, Rs 2960, Rs 2760
3. Rs 7914 is divided into three parts in such a way that the first part at 3% per annum after 8 years, the second part at 4% per annum after 5 years and third part at 6% per annum after 2 years give equal amounts. Find each part.
   a) Rs 2520, Rs 2604, Rs 2790
   b) Rs 2620, Rs 2504, Rs 2790
   c) Rs 2520, Rs 2704, Rs 2690
   d) None of these
4. Divide Rs 2379 into three parts so that their amounts after 2, 3 and 4 years respectively may be equal, the rate of interest being 5 per cent per annum.
   a) Rs 838, Rs 782, Rs 759
   b) Rs 828, Rs 792, Rs 759
   c) Rs 828, Rs 782, Rs 769
   d) None of these
5. A sum of Rs 18750 is left by will by a father to be divided between two sons of 12 and 14 years of age, so that when they attain maturity at 18, the amount (principal + interest) received by each at 5 per cent simple interest will be the same. Find the sum allotted at present to each son.
   a) Rs 9000, Rs 9750 b) Rs 8000, Rs 1750
   c) Rs 9500, Rs 9250 d) None of these

Answers

1. b 2. a 3. a 4. b

5. a; **Hint:** Required ratio

$$= \frac{1}{100+(18-12)\times 5} : \frac{1}{100+(18-14)\times 5}$$

$$= \frac{1}{130} : \frac{1}{120} = 12:13$$

Sum allotted to first son $= \frac{18750}{12+13} \times 12$ = Rs 9000

Sum allotted to second son $= \frac{18750}{12+13} \times 13$ = Rs 9750

Rule 25

Theorem: ***There is a direct relationship between the principal and the amount and is given by sum*** $= \frac{100\times Amount}{100+rt}$.

Illustrative Example

Ex.: A certain sum of money amounts to Rs 2613 in 6 years at 5% per annum. Find the sum.

Soln: Applying the above formula, we have

the sum $= \frac{100\times 2613}{100+30}$ = Rs 2010.

Exercise

1. What principal will amount to Rs 274.05 in 2 years 6 months at 3½% per annum simple interest?
   a) Rs 252 b) Rs 250 c) Rs 248 d) Rs 270
2. What principal will amount to Rs 560 in 3 years at 4 per cent per annum simple interest?
   a) Rs 540 b) Rs 500 c) Rs 550 d) Rs 560
3. Find the sum of money that will amount to Rs 5105 in 6½ years at 4¼ per cent per annum simple interest.
   a) Rs 3600 b) Rs 4500 c) Rs 4000 d) Rs 4400
4. Find to the nearest rupee, what principal amount to Rs 3456.50 at 5½ per cent in 4 years?
   a) Rs 2832 b) Rs 2633 c) Rs 2732 d) Rs 2833
5. What principal will amount to Rs 6133.75 in 3 years 9 months at 2½ per cent?
   a) Rs 5806 b) Rs 5608 c) Rs 5506 d) Rs 5508

Answers

1. a 2. b 3. c

4. d; **Hint:** Principal $= \frac{100\times 3456.5}{100+22}$ = Rs 2833.2 (nearly)

5. b

Rule 26

Theorem: ***A person lent a certain sum of money at r% simple interest and in 't' years the interest amounted to Rs A less than the sum lent, then the sum lent is given by Rs*** $\left[\frac{100A}{100-rt}\right]$.

Illustrative Example

Ex.: A person lent a certain sum of money at 4% simple interest, and in 8 years the interest amounted to Rs 340 less than the sum lent. Find the sum lent.

Soln: **Detail Method:** Let the sum be Rs x.

$\therefore$ Interest $= \frac{x\times 8\times 4}{100} = \frac{32x}{100}$

From the question,

$$\frac{32x}{100} = x - 340$$

$$\Rightarrow 32x = 100x - 34000$$

$$\Rightarrow x = \frac{34000}{68} = Rs\ 500.$$

Quicker Method: Applying the above theorem, we have

Sum $= \frac{100\times 340}{100-8\times 4} = \frac{340\times 100}{68}$ = Rs 500.

Exercise

1. A person lent a certain sum of money at 5% simple interest, and in 6 years the interest amounted to Rs 350 less than the sum lent. Find the sum lent.
   a) Rs 600 b) Rs 500 c) Rs 650 d) Rs 450
2. A person lent a certain sum of money at 4% simple interest, and in 3 years the interest amounted to Rs 880 less than the sum lent. Find the sum lent.
   a) Rs 1000 b) Rs 1250 c) Rs 1100 d) Rs 1200
3. A person lent a certain sum of money at 7½% simple interest, and in 12½ years the interest amounted to Rs 625 less than the sum lent. Find the sum lent.
   a) Rs 5000 b) Rs 10,000 d) Rs 1100 d) Rs 1200
4. A person lent a certain sum of money at 4% simple interest, and in 4 years the interest amounted to Rs 336 less than the sum lent. Find the sum lent.
   a) Rs 400 b) Rs 450 c) Rs 500 d) Rs 560
5. A person lent a certain sum of money at 4% simple interest, and in 5 years the interest amounted to Rs 520 less than the sum lent. Find the sum lent.
   a) Rs 600 b) Rs 650 c) Rs 700 d) Rs 750

Answers

1. b 2. a 3. b 4. a 5. b

Rule 27

Theorem: ***If a sum of money becomes 'n' times in 't' years at a simple interest, then the time in which it will amount to 'm' times itself is given by*** $\left(\frac{m-1}{n-1}\right)$ ***t years.***

Illustrative Example

Ex.: A sum of money doubles itself in 4 years at a simple interest. In how many years will it amount to 8 times itself?

Soln: Applying the above formula, we have,

the required time = $\left(\frac{8-1}{2-1}\right) \times 4 = 28$ years.

Exercise

1. A sum of money becomes 4 times itself in 3 years at a simple interest. In how many years will it amount to 7 times itself?
   a) 6 years b) 12 years c) 8 years d) None of these
2. A sum of money becomes 3 times itself in 3 years at a simple interest. In how many years will it amount to 5 times itself?
   a) 8 years b) 6 years c) 7 years d) 5.5 years
3. A sum of money becomes 5 times itself in 5 years at a simple interest. In how many years will it amount to 7 times itself?
   a) 7 years b) 6 years 6 months
   c) 6 years d) 7 years, 6 months
4. A sum of money becomes 5 times itself in 4 years at a simple interest. In how many years will it amount to 9 times itself?
   a) 8 years b) 10 years c) 9 years d) 8½ years
5. A sum of money doubles itself in 20 years, in how many years would it treble itself?
   a) 30 years b) 40 years c) 50 years d) 60 years
6. A sum of money doubles itself in 5 years. It will become 4 times of itself in
   [Clerical Grade Exam, 1991]
   a) 10 years b) 12 years c) 15 years d) 20 years

Answers

1\. a 2. b 3. d 4. a 5. b 6. c

Rule 28

Theorem: ***If the simple interest on Rs P_1 is less than the interest on Rs P_2 at r% simple interest by Rs A, then the time is given by $\left[\frac{A \times 100}{r(P_2 - P_1)}\right]$ years.***

Illustrative Example

Ex.: The simple interest on Rs 1650 will be less than the interest on Rs 1800 at 4% simple interest by Rs 30. Find the time.

Soln: Applying the above formula, we have,

$$\text{Time} = \frac{30 \times 100}{4(1800 - 1650)} = \frac{30 \times 100}{4 \times 150} = 5 \text{ years.}$$

Exercise

1. The simple interest on Rs 825 will be less than the interest on Rs 900 at 2% simple interest by Rs 15. Find the time.
   a) 5 years b) 8 years c) 10 years d) 12 years
2. The simple interest on Rs 601 will be less than the interest on Rs 726 at 5% simple interest by Rs 25. Find the time.
   a) 3 years b) 4 years c) 5 years d) 6 years
3. The simple interest on Rs 750 will be less than the interest on Rs 845 at 10% simple interest by Rs 57. Find the time.
   a) 5 years b) 4 years c) 6 years d) 7 years
4. The simple interest on Rs 750 will be less than the interest on Rs 900 at 8% simple interest by Rs 60. Find the time.
   a) 3 years b) 4 years c) 9 years d) 5 years
5. The simple interest on Rs 1250 will be less than the interest on Rs 1400 at 3% simple interest by Rs 45. Find the time.
   a) 10 years b) 9 years c) 8 years d) 6 years

Answers

1\. c 2. b 3. c 4. d 5. a

Rule 29

Theorem: ***Two equal amounts of money are deposited at r_1% and r_2% for t_1 and t_2 years respectively. If the difference between their interests is I_d then the sum = $\frac{I_d \times 100}{r_1 t_1 - r_2 t_2}$.***

Illustrative Example

Ex.: Two equal amounts of money are deposited in two banks each at 15% per annum for 3.5 years and 5 years respectively. If the difference between their interests in Rs 144, find each sum.

Soln: **Detail Method:** Let the sum be Rs x, then

$$\frac{x \times 15 \times 5}{100} - \frac{x \times 15 \times 7}{200} = 144$$

or, $150x - 105x = 144 \times 200$

$$\therefore x = \frac{144 \times 200}{45} = \text{Rs } 640$$

Quicker Method: Applying the above formula, we have,

$$\text{Sum} = \frac{144 \times 100}{15 \times 5 - 15 \times 3.5} = \frac{144 \times 100}{22.5} = \text{Rs } 640.$$

Exercise

1. The simple interest on a certain sum of money at 4% per annum for 4 years is Rs 80 more than the interest on the same sum for 3 years at 5% per annum. Find the sum.
   a) Rs 8000 b) Rs 8500 c) Rs 9000 d) Rs 4500
2. The simple interest in 14 months on a certain sum at the rate of 6 per cent per annum is Rs 250 more than the

interest on the same sum at the rate of 8 per cent in 8 months. How much amount was borrowed?

[BSRB Bangalore PO, 2000]

a) Rs 15000 b) Rs 25000 c) Rs 7500 d) Rs 14500

3. The difference in simple interests on a certain sum of money for 3 years and 5 years at 18% per annum is Rs 2160. Then the sum is ______.

a) Rs 6500 b) Rs 4500 c) Rs 6000 d) Rs 7500

4. The difference in simple interests on a certain sum at 5% for 4 years and 6% for 3 years is Rs 20. Find the sum.

a) Rs 1000 b) Rs 1200 c) Rs 800 d) Rs 900

5. The difference in simple interests on a certain sum at 4% per annum for 3 years and at 5% per annum for 2 years is Rs 50. Find the sum.

a) Rs 5000 b) Rs 4000 c) Rs 3000 d) Rs 2500

6. The difference in simple interests earned on a certain sum of money at 6% per annum at the end of 2 years and at the end of 4 years is Rs 1200. What is the sum?

a) Rs 5000 b) Rs 7000 c) Rs 10000 d) Rs 9000

Answers

1. a; **Hint:** Here Id = Rs 80, $r_1 = 4\%$, $t_1 = 4$ years

$r_2 = 5\%$ and $t_2 = 3$ years.

2. a; **Hint:** Sum $= \dfrac{250\times100}{\frac{14}{12}\times6-\frac{8}{12}\times8} =$ Rs 15000

3. c 4. a 5. d 6. c

Rule 30

Theorem: ***If the difference between the interest received from two different banks on Rs X for t years is Rs I_d, then the difference between their rates is given by*** $\left[\dfrac{I_d\times100}{X\times t}\right]$ ***per cent.***

Illustrative Example

Ex.: The difference between the interest received from two different banks on Rs 500 for 2 years is Rs 2.5. Find the differene between their rates.

Soln: **Detail Method:**

$$I_1 = \frac{500\times2\times r_1}{100} = 10r_1 \qquad I_2 = \frac{500\times2\times r_2}{100} = 10r_2$$

$$I_1 - I_2 = 10r_2 = 2.5$$

$$\text{or, } r_1 - r_2 = \frac{2.5}{10} = 0.25\%$$

Quicker Method: Applying the above theorem, we have,

the difference between their rates

$$= \frac{2.5\times100}{500\times2} = 0.25\%.$$

Exercise

1. The difference between the interest received from two different banks on Rs 450 for 4 years is Rs 18. Find the differene between their rates.

a) 2% b) 1% c) 1.5% d) 3%

2. The difference between the interest received from two different banks on Rs 150 for 5 years is Rs 15. Find the differene between their rates.

a) 2% b) 1% c) 2.5% d) 1.5%

3. The difference between the interest received from two different banks on Rs 200 for 3 years is Rs 60. Find the differene between their rates.

a) 5% b) 7% c) 10% d) 9%

4. The difference between the interest received from two different banks on Rs 600 for 6 years is Rs 72. Find the differene between their rates.

a) 3% b) 3.5% c) 4% d) 2%

5. The difference between the interest received from two different banks on Rs 750 for 2 years is Rs 90. Find the differene between their rates.

a) 4% b) 6% c) 8% d) None of these

Answers

1. b 2. a 3. c 4. d 5. b

Rule 31

Theorem: ***If a sum amounts to Rs A_1 in t_1 years and Rs A_2 in t_2 years at simple rate of interest, then rate per annum*** $= \dfrac{100[A_2 - A_1]}{(A_1t_2 - A_2t_1)}$.

Illustrative Example

Ex.: A sum of money at simple interest amounts to Rs 600 in 4 years and Rs 650 in 6 years. Find the rate of interest per annum.

Soln: **Detail Method:** Suppose the rate of interest = r% and the sum = Rs A

$$\text{Now, } A + \frac{A\times r\times4}{100} = 600 \quad \text{or, } A + \frac{Ar}{25} = 600$$

$$\text{or, } A\left[1+\frac{r}{25}\right] = 600 \quad ...\ (1)$$

$$\text{And } A + \frac{A\times r\times6}{100} = 650$$

$$\text{or, } A\left[1+\frac{3r}{50}\right] = 650 \quad ....\ (2)$$

Dividing (1) by (2), we have

$$\frac{1+\frac{r}{25}}{1+\frac{3r}{50}}=\frac{600}{650}; \quad \text{or, } \frac{(25+r)\times 2}{50+3r}=\frac{12}{13}$$

or, (50 + 2r) × 13 = (50 + 3r) × 12

or, 650 + 26r = 600 + 36r; or, 10r = 50

∴ r = 5%.

Quicker Method: Following the above theorem, we have

$$r=\frac{100[650-600]}{6\times 600-4\times 650}=\frac{100\times 50}{1000}=5\%.$$

Note: Also see **Rule - 13**.

Exercise

1. A sum of money lent out at simple interest amounts to Rs 720 in 2 years and to Rs 1020 in 7 years. Find the rate per cent per annum.
   a) 10% b) 12% c) 5% d) 15%
2. The amount of a certain sum with simple interest for 20 years is Rs 586.40 and with simple interest for 10 years more is Rs 696.35. Find the rate per cent per annum at which interest is reckoned.
   a) 3% b) 4% c) 5% d) 8%
3. A certain sum of money at simple interest amounts to Rs 379.50 in 3 years and to Rs 453.75 in 7½ years. Find the rate of interest.
   a) 6% b) 4% c) 5% d) 8%
4. A sum of money lent out at simple interest amounts to Rs 460 in 3 years while in 5 years it amounts to Rs 500. Find the rate of interest.
   a) 6% b) 5% c) 4% d) 3%
5. A sum of money lent out at simple interest amount to Rs 560 in 3 years while in 5 years it amounts to Rs 600. Find the rate of interest.
   a) 3% b) 5% c) 4% d) 6%
6. A certain sum of money at SI amounts to Rs 1012 in 2½ years and to Rs 1067.20 in 4 years. The rate of interest per annum is **[Central Excise, 1989]**
   a) 2.5% b) 3% c) 4% d) 5%
7. A certain sum of money at simple interest amounts to Rs 1260 in 2 years and to Rs 1350 in 5 years. The rate per cent per annum is **[Central Excise, 1988]**
   a) 2.5% b) 3.75% c) 5% d) 7.5%

Answers

1. a

2. a; **Hint:** Here $t_2 = 20 + 10 = 30$ years

$$\therefore \text{ Required rate \%} = \frac{(696.35-586.40)100}{586.40\times 30-696.35\times 20}=3\%$$

3. c 4. b 5. c 6. c 7. a

Rule 32

Theorem: ***If a person borrows Rs X from a bank at simple interest and after t_1 years he paid Rs x_1 to the bank and at the end of t_2 years from the date of borrowing he paid Rs x_2 to the bank to settle the account, then the rate of interest is given by*** $\left[\frac{x_1+x_2-X}{x_1t_1+t_2(X-x_1)}\right]\times 100$ ***%.***

Illustrative Example

Ex.: Ramesh borrows Rs 7000 from a bank at SI. After three years he paid Rs 3000 to the bank and at the end of 5 years from the date of borrowing he paid Rs 5450 to the bank to settle the account. Find the rate of interest.

Soln: Detail Method: Any sum that is paid back to the bank before the last instalment is deducted from the *principal* and not from the *interest*. Thus,

Total interest = Interest on Rs 7000 for 3 years + Interest on (Rs 7000 – Rs 3000) = Rs 4000 for 2 years.

or, (5450 + 3000 – 7000)

$$=\frac{7000\times 3\times r}{100}+\frac{4000\times 2\times r}{100}$$

or, 1450 = 210r + 80r

$$\therefore r=\frac{1450}{290}=5\%.$$

Quicker Method: Following the above theorem, we have

$$\text{the rate of interest }=\left[\frac{3000+5450-7000}{(3000\times 3)+(5\times 4000)}\right]\times 100$$

$$=\frac{1450\times 100}{29000}=\frac{1450}{290}=5\%.$$

Exercise

1. Rakesh borrows Rs 3500 from a bank at SI. After three years he paid Rs 1500 to the bank and at the end of 5 years from the date of borrowing he paid Rs 2725 to the bank to settle the account. Find the rate of interest.
   a) 10% b) 5% c) 2.5% d) None of these
2. Sudhir borrows Rs 6000 from a bank at SI. After 4 years he paid Rs 2500 to the bank and at the end of 5 years from the date of borrowing he paid Rs 4560 to the bank to settle the account. Find the rate of interest.
   a) 3% b) 3.5% c) 3.85% d) None of these
3. Uday borrows Rs 5000 from a bank at SI. After 5 years he paid Rs 1700 to the bank and at the end of 5 years from the date of borrowing he paid Rs 3550 to the bank to settle the account. Find the rate of interest.
   a) 1% b) 1.5% c) 2% d) 10%
4. Binod borrows Rs 5500 from a bank at SI. After 2 years

he paid Rs 2410 to the bank and at the end of 2 years from the date of borrowing he paid Rs 3690 to the bank to settle the account. Find the rate of interest.

a) $3\frac{1}{3}\%$ b) $2\frac{2}{3}\%$ c) $6\frac{1}{2}\%$ d) $5\frac{5}{11}\%$

Answers

1. b 2. c 3. a 4. d

Rule 33

Theorem: ***Some amount out of Rs P was lent at r_1 % per annum and the remaining at r_2% per annum. If the total simple interest from both the fractions in t years was Rs 'A', then the sum lent at r_1% per annum was given by Rs*** $\left[\frac{100A - Pr_2t}{(r_1 - r_2)t}\right]$.

Illustrative Example

Ex.: Some amount out of Rs 7000 was lent at 6% per annum and the remaining at 4% per annum. If the total simple interest from both the fractions in 5 years was Rs 1600, find the sum lent at 6% per annum.

Soln: Detail Method: Suppose Rs x was lent at 6% per annum.

Thus, $\frac{x \times 6 \times 5}{100} + \frac{(7000 - x) \times 4 \times 5}{100} = 1600$

or, $\frac{3x}{10} + \frac{7000 - x}{5} = 1600$

or, $\frac{3x + 14000 - 2x}{10} = 1600$

$\therefore x = 16000 - 14000 =$ Rs 2000.

Quicker Method: Applying the above formula, we have

the required answer $= \frac{100 \times 1600 - 7000 \times 4 \times 5}{(6-4)5}$

= Rs 2000.

Exercise

1. Some amount out of Rs 8000 was lent at 7% per annum and the remaining at 5% per annum. If the total simple interest from both the fractions in 6 years was Rs 2600, find the sum lent at 7% per annum.

   a) Rs $\frac{2000}{3}$ b) Rs 1500 c) Rs $\frac{5000}{3}$ d) Rs $\frac{4000}{3}$

2. Some amount out of Rs 9000 was lent at 8% per annum and the remaining at 6% per annum. If the total simple interest from both the fractions in 7 years was Rs 3900, find the sum lent at 8% per annum.

   a) Rs $\frac{5000}{7}$ b) Rs $\frac{6000}{7}$ c) Rs 1600 d) None of these

3. Some amount out of Rs 950 was lent at 6% per annum and the remaining at 4% per annum. If the total simple interest from both the fractions in 5 years was Rs 200, find the sum lent at 6% per annum.

   a) Rs 700 b) Rs 100 c) Rs 250 d) Rs 450

Answers

1. c 2. b 3. b

Rule 34

Theorem: ***A person invested $\frac{1}{n_1}$ of his capital at x_1%, $\frac{1}{n_2}$ at x_2% and the remainder $\frac{1}{n_3}$ at x_3%. If his annual income is Rs A, the capital is given by Rs*** $\left[\frac{A \times 100}{\frac{x_1}{n_1} + \frac{x_2}{n_2} + \frac{x_3}{n_3}}\right]$.

Illustrative Example

Ex.: A man invested $\frac{1}{3}$ of his capital at 7%, $\frac{1}{4}$ at 8% and the remainder at 10%. If his annual income is Rs 561, find the capital.

Soln: Detail Method:

Let the capital be Rs x.

From the question,

$\frac{1}{3}x$ of 7% + $\frac{1}{4}x$ of 8% + $\left[1 - \left(\frac{1}{3} + \frac{1}{4}\right)\right]x$ of 10%

= Rs 561

or, $\frac{\frac{1}{3}x \times 7}{100} + \frac{\frac{1}{4}x \times 8}{100} + \frac{\frac{5}{12}x \times 10}{100} =$ Rs 561

or, $\frac{7}{3}x + 2x + \frac{25}{6}x = 561 \times 100$

or, $x = \frac{561 \times 100 \times 6}{51} =$ Rs 6600.

Quicker Method: Applying the above theorem, we have

Capital $= \frac{561 \times 100}{\frac{7}{3} + 2 + \frac{25}{6}} =$ Rs 6600.

Exercise

1. $\frac{1}{3}$ of my capital is invested at 4 per cent, $\frac{1}{4}$ at 3 per cent and the remainder at 5 per cent. If my annual in-

come is Rs 50, find the capital.
a) Rs 1100 b) Rs 1300 c) Rs 1200 d) Rs 1600

2. $\frac{2}{3}$ of my capital is invested at 3 per cent, $\frac{1}{6}$ at 6 per cent and the remainder at 12 per cent. If my annual income is Rs 25, find the capital.
a) Rs 500 b) Rs 600 c) Rs 650 d) Rs 450

3. $\frac{1}{4}$ of my capital is invested at 4 per cent, $\frac{1}{8}$ at 6 per cent and the remainder at 8 per cent. If my annual income is Rs 270, find the capital.
a) Rs 5000 b) Rs 4400 c) Rs 4000 d) Rs 3800

4. $\frac{3}{4}$ of my capital is invested at 8 per cent, $\frac{1}{8}$ at 8 per cent and the remainder at 8 per cent. If my annual income is Rs 480, find the capital.
a) Rs 6000 b) Rs 6600 c) Rs 4800 d) Rs 5600

5. $\frac{1}{5}$ of my capital is invested at 5 per cent, $\frac{1}{5}$ at 5 per cent and the remainder at 5 per cent. If my annual income is Rs 50, find the capital.
a) Rs 800 b) Rs 1000 c) Rs 1200 d) Rs 1500

6. Out of a certain sum, $\frac{1}{3}$rd is invested at 3%, $\frac{1}{6}$th at 6% and the rest at 8%. If the simple interest for 2 years from all these investments amounts to Rs 600, find the original sum. **[BSRB Bank PO Exam, 1988]**
a) Rs 5000 b) Rs 6000 c) Rs 5200 d) Rs 5500

Answers

1. c 2. a 3. c 4. a 5. b

6. a; **Hint:** Here A = annual income = Rs $\frac{600}{2}$ = Rs 300
= simple interest per annum.
Now, apply the given rule and get the answer.

Rule 35

Theorem: ***If the simple interest on certain sum 'P' is 'I' and the number of years is equal to the rate per cent per annum, then the rate per cent or time is given by*** $\left[\sqrt{\frac{100 \times I}{P}}\right]$.

Note: Compare this rule with **Rule - 4**.

Illustrative Example

Ex.: The simple interest on certain sum Rs 160 is Rs 40, and the number of years is equal to the rate per cent per annum. Find the rate per cent.

Soln: Applying the above formula, we have

$$\text{rate} = \sqrt{\frac{100 \times 40}{160}} = 5\%.$$

Exercise

1. The simple interest on certain sum Rs 625 is Rs 100, and the number of years is equal to the rate per cent per annum. Find the rate per cent.
a) 5% b) 4% c) 3% d) 4.5%
2. The simple interest on certain sum Rs 225 is Rs 4, and the number of years is equal to the rate per cent per annum. Find the rate per cent.
a) 1% b) $2\frac{1}{3}\%$ c) $1\frac{1}{3}\%$ d) None of these
3. The simple interest on certain sum Rs 100 is Rs 9, and the number of years is equal to the rate per cent per annum. Find the rate per cent.
a) 3% b) 4% c) 1% d) 2%
4. The simple interest on certain sum Rs 729 is Rs 144, and the number of years is equal to the rate per cent per annum. Find the rate per cent.
a) $4\frac{4}{9}\%$ b) $4\frac{1}{3}\%$ c) $3\frac{1}{3}\%$ d) $9\frac{1}{4}\%$
5. The simple interest on certain sum Rs 900 is Rs 100, and the number of years is equal to the rate per cent per annum. Find the rate per cent.
a) $3\frac{2}{3}\%$ b) 3% c) $3\frac{1}{3}\%$ d) None of these

Answers

1. b 2. c 3. a 4. a 5. c

Rule 36

Theorem: ***To find the amount (Principal + Simple Interest).***
If a borrower has to pay at r% simple interest per annum for t years and simple interest is given as Rs I, then the following formula is used to find the amount.

$$A\ (\text{Amount}) = I\ (\text{Simple Interest}) \left[1 + \frac{100}{rt}\right].$$

Illustrative Example

Ex.: A certain sum of money is borrowed by a person at 4% simple interest for 5 years. If he has to pay Rs 160 as interest, find the total amount he has to pay.

Soln: Applying the above formula, we have

$$\text{Amount} = 160\left(1 + \frac{100}{20}\right) = \text{Rs } 960.$$

Exercise

1. A certain sum of money is borrowed by a person at 5% simple interest for 5 years. If he has to pay Rs 100 as

interest, find the total amount he has to pay.
a) Rs 400 b) Rs 500 c) Rs 550 d) None of these

2. A certain sum of money is borrowed by a person at 3% simple interest for 4 years. If he has to pay Rs 120 as interest, find the total amount he has to pay.
a) Rs 1020 b) Rs 820 c) Rs 1120 d) Rs 1220

3. A certain sum of money is borrowed by a person at 5% simple interest for 3 years. If he has to pay Rs 150 as interest, find the total amount he has to pay.
a) Rs 1150 b) Rs 1250 c) Rs 1050 d) Rs 1350

4. A certain sum of money is borrowed by a person at 3% simple interest for 6 years. If he has to pay Rs 144 as interest, find the total amount he has to pay.
a) Rs 1044 b) Rs 844 c) Rs 1144 d) Rs 944

Answers

1. b 2. c 3. a 4. d

Rule 37

Theorem: ***If a person lends Rs P_1 for t_1 years and Rs P_2 for t_2 years received altogether Rs I as a simple interest, the rate per cent per annum is*** $\left(\frac{100I}{P_1t_1+P_2t_2}\right)$***%.***

Illustrative Example

Ex.: A person lends Rs 600 for 5 years and Rs 750 for 2 years, received altogether Rs 450 as a simple interest. Find the rate per annum.

Soln: Applying he above formula, we have

$$r=\frac{100\times 450}{(600\times 5)+(750\times 2)}=10\%.$$

Exercise

1. A lent Rs 600 to B for 2 years, and Rs 150 to C for 4 years and received altogether from both Rs 90 as interest. Find the rate of interest, simple interest being calculated.
a) 4% b) 5% c) 6% d) 10%

2. A man lends Rs 350 for 6 years and Rs 540 for 2½ years to two persons and receives in all Rs 115 as interest. Fin the rate per cent, if simple interest be reckoned.
a) $3\frac{1}{3}\%$ b) $2\frac{1}{3}\%$ c) $3\frac{2}{3}\%$ d) $2\frac{2}{3}\%$

3. The simple interest on Rs 400 for 5 years together with that on Rs 600 for 4 years came to Rs 132, the rate being the same in both the cases. Find the rate per cent of interest.
a) 1% b) 5% c) 4% d) 3%

4. The simple interest on Rs 300 for 6 years together with that on Rs 500 for 5 years came to Rs 430, the rate being the same in both the cases. Find the rate per cent of interest.
a) 10% b) 8% c) 7% d) 5%

5. The simple interest on Rs 150 for 3 years together with that on Rs 250 for 6 years came to Rs 97.50, the rate being the same in both the cases. Find the rate per cent of interest.
a) 10% b) 5% c) 6% d) 8%

6. A lent Rs 600 to B for 2 years and Rs 150 to C for 4 years and received altogether from both Rs 90 as simple interest. The rate of interest is: **[Railways, 1988]**
a) 12% b) 10% c) 5% d) 4%

Answers

1. b 2. a 3. d 4. a 5. b 6. c

Rule 38

Theorem: ***On Rs P_1 invested at a simple interest rate r_1 per cent per annum, Rs I_1 is obtained as interest in certain years. In order to earn Rs I_2 as interest on Rs P_2 in the same number of years, the rate of simple interest is given by*** $\left[\frac{I_2}{I_1}\times\frac{P_1}{P_2}\times r_1\right]$ ***per cent.***

Illustrative Example

Ex.: On Rs 2500 invested at a simple interest rate 4 per cent per annum, Rs 500 is obtained as interest in certain years. In order to earn Rs 2000 as interest on Rs 4000, in the same number of years, what should be the rate of simple interest?

Soln: **Detail Method:**

$$\text{Time}=\frac{500\times 100}{2500\times 4}=5 \text{ years}$$

$$\text{Rate}=\frac{2000\times 100}{5\times 4000}=10\%$$

Quicker Method: Applying the above rule, we have

$$\text{the required answer}=\frac{2500}{4000}\times\frac{2000}{500}\times 4=10\%$$

Exercise

1. On Rs 3000 invested at a simple interest rate 6 per cent per annum, Rs 900 is obtained as interest in certain years. In order to earn Rs 1600 as interest on Rs 4000 in the same number of years, what should be the rate of simple interest? **[BSRB Mumbai PO, 1999]**
a) 7 per cent b) 8 per cent
c) 9 per cent d) Data inadequate

2. On Rs 1250 invested at a simple interest rate 2 per cent per annum, Rs 250 is obtained as interest in certain years. In order to earn Rs 1000 as interest on Rs 2000 in the same number of years, what should be the rate of simple

interest?

a) 3% b) 4% c) 5% d) None of these

3. On Rs 1500 invested at a simple interest rate 3 per cent per annum, Rs 450 is obtained as interest in certain years. In order to earn Rs 800 as interest on Rs 2000 in the same number of years, what should be the rate of simple interest?

a) 4% b) 6% c) 5% d) 8%

4. On Rs 3500 invested at a simple interest rate 7 per cent per annum, Rs 500 is obtained as interest in certain years. In order to earn Rs 800 as interest on Rs 4900 in the same number of years, what should be the rate of simple interest?

a) 9% b) 10% c) 12% d) 8%

5. On Rs 4500 invested at a simple interest rate 12 per cent per annum, Rs 1200 is obtained as interest in certain years. In order to earn Rs 1800 as interest on Rs 5400 in the same number of years, what should be the rate of simple interest?

a) 14% b) 15% c) 18% d) 16%

Answers

1. b 2. c 3. a 4. d 5. b

Rule 39

Theorem: ***The time in which a sum of money becomes n times itself at r% per annum simple interest is given by***

$$\left(\frac{n-1}{r}\right)\times 100 \textit{ years.}$$

Illustrative Example

Ex.: In what time will a sum of money double itself at 5% per annum simple interest being charged?

Soln: **Detail Method**

Suppose the capital is Rs 100, in order to double itself the interest on it should also amount to Rs 100.

Now, the interest on Rs 100 for 1 year = Rs 5

$\therefore$ the required no. of years $= \frac{100}{5} = 20$ years.

Quicker Method: Applying the above theorem, we have

the required answer $= \frac{2-1}{5}\times 100 = 20$ years

Exercise

1. At what rate per cent simple interest will a sum of money double itself in 20 years?

[Bank PO Exam, 1990]

a) 4% b) 3% c) 5% d) None of these

2. At what rate per cent simple interest will a sum of money treble itself in 25 years?

a) 4% b) 5% c) 8% d) 6%

3. In how many years will a sum of money treble itself at $3\frac{1}{8}$%?

a) 64 years b) 32 years c) 54 years d) 60 years

4. At what rate of interest per annum will a sum double itself in 8 years?

[Bank PO Exam, 1986]

a) 12½% b) 5% c) 6% d) 10½%

Answers

1. c; **Hint:** $\left(\frac{2-1}{r}\right)\times 100 = 20 \quad \therefore r = 5\%$

2. c 3. a 4. a

Miscellaneous

1. If the rate of interest rises from 6½ to 8%, a man's annual income increases by Rs 4050. Find the capital

a) Rs 270000 b) Rs 370000
c) Rs 300000 d) None of these

2. If Rs 5600 amount to Rs 6678 in 3½ years, what will Rs 9400 amount to in $5\frac{1}{4}$ years, at the same rate per cent per annum simple interest?

a) Rs 9400 b) Rs 12114.25
c) Rs 12115 d) None of these

3. A man derives his income from the investment of Rs 4150 at a certain rate of interest and Rs 3500 at 1 per cent higher. His whole income for 4 years is Rs 1211. Find the rates of interest.

a) 3½%, 4½% b) 2½%, 3½%
c) 4½%, 5½% d) None of these

4. The simple interest on a sum of money will be Rs 600 after 10 years. If the principal is trebled after 5 years, the total interest at the end of 10 years will be:

[Bank PO 1987]

a) Rs 600 b) Rs 900
c) Rs 1200 d) Data inadequate

5. If x is the simple interest on y and y is the simple interest on z, the rate % and the time being the same in both cases, what is the relation between x, y and z?

[Bank PO, 1989]

a) $x^2 = yz$ b) $y^2 = xz$ c) $z^2 = xy$ d) xyz = 1

Answers

1. a; Due to the rise in the rate of interest, annual income increases by Rs $\left(8-6\frac{1}{2}\right)$ = Rs $1\frac{1}{2}$, when the capital is Rs 100.

Thus the required capital $= \dfrac{100\times 2\times 4050}{3} =$ Rs 270000

2. b; We first find the rate per cent

Interest on Rs 5600 = Rs 6678 – Rs 5600 = Rs 1078

$\therefore$ rate % = $\dfrac{100\times 1078}{5600\times 3\frac{1}{2}} = \dfrac{100\times 1078\times 2}{5600\times 7} = 5\dfrac{1}{2}$

Interest on Rs 9400 = Rs $\dfrac{9400\times 21\times 11}{100\times 4\times 2}$

= Rs $\dfrac{10857}{4}$ = Rs 2714.25

$\therefore$ the required amount = Rs 9400 + Rs 2714.25
= Rs 12114.25

3. a; Income for 1 year = Rs $\dfrac{1211}{4}$

Since the rate of interest for Rs 3500 is 1% higher, therefore, if we subtract 1% on Rs 3500 from Rs $\dfrac{1211}{4}$ the remainder will be 1 year's interest on (Rs 4150 + Rs 3500) at the lower rate of interest.

Interest on (Rs 4150 + Rs 3500)

= Rs $\dfrac{1211}{4}$ – 1% on Rs 3500

$\therefore$ Interest on Rs 7650 = Rs $\dfrac{1211}{4}$ – Rs 35 = Rs $\dfrac{1071}{4}$

$\therefore$ Interest on Rs 100 = Rs $\dfrac{1071}{4}\times\dfrac{100}{7650}$ = Rs $3\dfrac{1}{2}$

$\therefore$ The lower rate is $3\frac{1}{2}$% and the higher rate is $4\frac{1}{2}$%.

4. c; Let the sum be Rs x.

SI = Rs 600, Time = 10 years

$\therefore$ Rate = $\left(\dfrac{600\times 100}{x\times 10}\right)\% = \left(\dfrac{6000}{x}\right)\%$ per annum.

SI for first 5 years = Rs $\left(\dfrac{x\times 5\times 6000}{100\times x}\right)$ = Rs 300.

SI for last 5 years = Rs $\left(\dfrac{3x\times 5\times 6000}{100\times x}\right)$ = Rs 900

Hence, total interest at the end of 10 years = Rs 1200

5. b; $x = \dfrac{y\times r\times t}{100}$ and $y = \dfrac{z\times r\times t}{100}$

or, $\dfrac{x}{y} = \dfrac{y}{z}$ or, $y^2 = xz$

$\therefore$ Correct answer is (b).

10

Compound Interest

Rule 1

If Principal = Rs P
Time = t years
Rate = r per cent per annum and
Interest is compounded annually, then,

$$\textbf{Amount} = P\left[1+\frac{r}{100}\right]^t$$

Illustrative Example

Ex.: Rs 7500 is borrowed at CI at the rate of 4% per annum. What will be the amount to be paid after 2 years?

Soln: Applying the above formula,

$$\text{Amount} = 7500\left[1+\frac{4}{100}\right]^2 = \frac{7500\times 26\times 26}{25\times 25}$$

$$= \text{Rs } 8112.$$

Exercise

1. Raviraj invested an amount of Rs 10,000 at compound interest rate of 10 per cent per annum for a period of three years. How much amount will Raviraj get after 3 years? **[SBI Associates PO, 1999]**
   a) Rs 12340 b) Rs 13210 c) Rs 13320 d) Rs 13310
2. Seema invested an amount of Rs 16000 for two years at compound interest and received an amount of Rs 17640 on maturity. What is the rate of interest? **[SBI Associates PO, 1999]**
   a) 8 pcpa b) 5 pcpa c) 4 pcpa d) None of these
3. Amit Kumar invested an amount of Rs 15000 at compound interest rate of 10 pcpa for a period of 2 years. What amount will he receive at the end of two years? **[Guwahati PO, 1999]**
   a) Rs 18000 b) Rs 18500 c) Rs 17000 d) Rs 18150
4. Find the amount at compound interest of Rs 625 in 2 years at 4 per cent.
   a) Rs 676 b) Rs 756 c) Rs 767 d) Rs 675
5. Find the amount at compound interest of Rs 2700 in 3 years at $3\frac{1}{3}$ per cent
   a) Rs 2979.10 b) Rs 2997.10
   c) Rs 2797.10 d) None of these
6. What is the amount at compound interest for 2 years at 2 per cent?
   a) 2.0404 times the principal
   b) 1.0404 times the principal
   c) 1.404 times the principal
   d) Data inadequate

Answers

1. d; **Hint:** Amount $= 10000\left(1+\frac{10}{100}\right)^3$

$$= 10000\times\frac{11}{10}\times\frac{11}{10}\times\frac{11}{10} = \text{Rs } 13310$$

2. b; **Hint:** $\frac{A}{P} = \left(1+\frac{r}{100}\right)^t$ or $\frac{17640}{16000} = \left(1+\frac{r}{100}\right)^2$

or, $\sqrt{\frac{11025}{10000}} = 1+\frac{r}{100}$ or, $\frac{5\times 7\times 3}{100} - 1 = \frac{r}{100}$

or, $\frac{r}{100} = \frac{105-100}{100}$ or, r = 5%

3. d 4. a 5. a 6. b

Rule 2

When interest is compounded half-yearly

$$\textbf{\textit{Amount}} = P\left[1+\frac{\frac{r}{2}}{100}\right]^{2t} = P\left[1+\frac{r}{200}\right]^{2t}$$

Illustrative Example

Ex.: Rs 7500 is borrowed at CI at the rate of 4% per annum. What will be the amount to be paid after 1 year, if interest is compounded half-yearly?

Soln: Applying the above formula,

$$\text{Amount} = 7500\left[1+\frac{4}{200}\right]^2$$

$$= \frac{7500\times102\times102}{100\times100} = \text{Rs } 7803.$$

Exercise

1. Find the amount of Rs 1000 in 1 year at 5 per cnet compound interest payable half-yearly.
   a) Rs 1050 (Approx) b) Rs 950 (Approx)
   c) Rs 1125 (Approx) d) Rs 1025 (Approx)
2. Find the amount of Rs 6400 in 1 year 6 months at 5 per cent compound interest, interest being calculated half yearly.
   a) Rs 6882.10 b) Rs 6892.10
   c) Rs 6982.10 d) None of these
3. Find the compound interest on Rs 350 for 1 year at 4 per cent per annum, the interest being payable half yearly.
   a) Rs 364.14 b) Rs 365.15 c) Rs 14.14 d) Rs 15.15

Answers

1. a 2. b

3. c; **Hint:** $\text{Amount} = 350\times\frac{204}{200}\times\frac{204}{200} = \text{Rs } 364.14$

$\therefore$ Compound interest = Rs 364.14 - Rs 350 = Rs 14.14

Rule 3

When interest is compounded quarterly

$$\textit{Amount} = P\left[1+\frac{\frac{r}{4}}{100}\right]^{4t} = P\left[1+\frac{r}{400}\right]^{4t}$$

Illustrative Example

Ex.: Rs 7500 is borrowed at CI at the rate of 4% per annum. What will be the amount to be paid after 6 months, if interest is compounded quarterly?

Soln: Applying the above formula,

$$\text{Amount} = 7500\left[1+\frac{4}{400}\right]^{4\times\frac{1}{2}} = \frac{7500\times101\times101}{100\times100}$$

$$= \text{Rs } 7650.75.$$

Exercise

1. Find the compound interest on Rs 2000 for $1\frac{1}{4}$ years at 10 per cent per annum, the interest being payable quarterly.
   a) Rs 2262.81 b) Rs 262.81
   c) Rs 262.18 d) None of these
2. Find the compound interest on Rs 10000 in 9 months at 4 per cent interest payable quarterly.
   a) Rs 303 (Approx) b) Rs 313 (Approx)
   c) Rs 203 (Approx) d) None of these
3. Find the compound interest on Rs 8000 in 3 months at 5 per cent interest payable quarterly.
   a) Rs 250 b) Rs 200 c) Rs 150 d) Rs 100

Answers

1. b; **Hint:** $\text{Amount} = 2000\left(1+\frac{10}{400}\right)^{4\times\frac{5}{4}}$

$$= 2000\left(\frac{41}{40}\right)^5 = \text{Rs } 2262.81$$

$\therefore$ Compound interest = 2262.81 - 2000 = Rs 262.81

2. a 3. d

Rule 4

Let Principal = Rs P
Time = t years
Rate = r per cent per annum

$$\textbf{CI (Compound Interest)} = A - P = P\left[\left(1+\frac{r}{100\times n}\right)^{n\times t} - 1\right]$$

Where, n = 1, when the interest is compounded yearly
n = 2, when the interest is compounded half-yearly
n = 4, when the interest is compounded quarterly
n = 12, when the interest is compounded monthly
When, interest is compounded yearly, n = 1

$$\therefore \textbf{CI} = P\left[\left(1+\frac{r}{100}\right)^t - 1\right]$$

Note: Rule 1, 2 and 3 are special cases of this general rule.

Illustrative Example

Ex.: If the interest is compounded annually, find the compound interest on Rs 2000 for 3 years at 10% per annum.

Soln: Applying the above formula,

$$\text{CI} = 2000\left[\left(1+\frac{10}{100}\right)^3 - 1\right] = 2000\left[\left(\frac{11}{10}\right)^3 - 1\right]$$

$$= \text{Rs } 662.$$

Exercise

1. The compound interest on any sum at the rate of 5% for two years is Rs 512.50. Find the sum.
   [BSRB Hyderabad PO, 1999]
   a) Rs 5200 b) Rs 4800 c) Rs 5000 d) Rs 5500
2. Find the amount on Rs 60,000, if the interest is com-

pounded half-yearly at 4 per cent per annum for $1\frac{1}{2}$ years. **[NABARD, 1999]**

a) Rs 63672.48 b) Rs 62424.00
c) Rs 67491.84 d) Rs 64896.00

3. What will be the compound interest acquired on a sum of Rs 12000 for 3 years at the rate of 10% per annum? **[BSRB Bhopal PO, 2000]**
a) Rs 2652 b) Rs 3972 c) Rs 3960 d) Rs 3852
4. Find the compound interest on Rs 4000 for 3 years at 5 per cent per annum.
a) Rs 630 b) Rs 620.50
c) Rs 630.50 d) None of these
5. Find the compound interest on Rs 1600 for 3 years at $3\frac{3}{4}$ per cent per annum.
a) Rs 186.83 b) Rs 168.83 c) Rs 186.38 d) Rs 168.38
6. What principal will amount to Rs 1352 in 2 years at 4 per cent compound interest?
a) Rs 1520 b) Rs 1260 c) Rs 1250 d) Rs 1220
7. On what principal will the compound interest for 3 years at 5 per cent amount to Rs 63.05?
a) Rs 400 b) Rs 500 c) Rs 450 d) Rs 550
8. A property decreases in value every year at the rate of $6\frac{1}{4}$ per cent on its value at the beginning of that year. If its value at the end of 3 years be Rs 21093.75, what was it worth at the beginning of these three years?
a) Rs 25600 b) Rs 26500 c) Rs 25500 d) Rs 25800
9. One man offers Rs 80000 for an estate and another Rs 84270 to be paid in 3 year's time, allowing 6 per cent compound interest. Which is the better offer.
a) first b) second c) both d) None of these
10. A merchant commences with a certain capital and gains annually at the rate of 25 per cent. At the end of 3 years he is worth Rs 10,000. What was his original capital.
a) Rs 5120 b) Rs 5220 c) Rs 5210 d) Rs 5130
11. What sum put out for $1\frac{1}{2}$ years at 4 per cent compound interest, payable half-yearly will amount to Rs 6632.55?
a) Rs 6250 b) Rs 6520 c) Rs 6350 d) None of these
12. Rs 800 at 5% per annum compound interest amount to Rs 882 in **[Clerical Grade 1991]**
a) 4 years b) 3 years c) 2 years d) 1 year

Answers

1. c; **Hint:** Let the sum be Rs x

$$512.50 = x\left[\left(1+\frac{5}{100}\right)^2 - 1\right] = x\left(\frac{441-400}{400}\right)$$

$$\therefore\ x = \frac{512.50\times 400}{41} = \text{Rs } 5000$$

2. a; **Hint:** Here $r = \frac{4}{2} = 2\%$ and $t = \frac{3}{2}\times 2 = 3$ years

$$\therefore\ \text{Amount} = 60000\left(1+\frac{2}{100}\right)^3 = 60000\times\frac{51}{50}\times\frac{51}{50}\times\frac{51}{50}$$

$$= \text{Rs } 63672.48$$

3. b; **Hint:** Compound interest $= 12000\left(1+\frac{10}{100}\right)^3 - 12000$

$$= 15972 - 12000 = \text{Rs } 3972$$

4. c 5. a

6. c; **Hint:** Let Rs P be the principal

Then $P\left(1+\frac{4}{100}\right)^2 = 1352$

$$\therefore\ P\times\frac{26}{25}\times\frac{26}{25} = 1352$$

$$\therefore\ P = \frac{1352\times 25\times 25}{26\times 26} = 1250$$

$\therefore$ the required principal is Rs 1250

7. a; **Hint:** Let Rs P be the principal

Then Rs $P\left\{\left(1+\frac{5}{100}\right)^3 - 1\right\} = \text{Rs } 63.05$

$$\therefore\ P\left(\frac{21}{20}\times\frac{21}{20}\times\frac{21}{20} - 1\right) = \frac{6305}{100}$$

$\therefore\ P = \text{Rs } 400$

$\therefore$ the required principal is Rs 400

8. a; **Hint:** Here $P\left(1-\frac{6\frac{1}{4}}{100}\right)^3 = 21093\frac{3}{4}$

$$\therefore\ P\left(\frac{15}{16}\right)^3 = \frac{84375}{4}$$

$$\therefore\ P = \frac{84375\times 16\times 16\times 16}{4\times 15\times 15\times 15} = \text{Rs } 25600$$

$\therefore$ Original value of the property = Rs 25600

9. a; **Hint:** Rs 80000 after 3 years will become

$80000\times\frac{106}{100}\times\frac{106}{100}\times\frac{106}{100} = \text{Rs } 95281.28$, which is greater than Rs 84270. Hence, first one is a better offer.

10. a; **Hint:** Original capital $=\frac{10000\times100\times100\times100}{125\times125\times125}=80$

$\times 64 = \text{Rs } 5120$

11. a; **Hint:** Sum $=\frac{6632.55\times100\times100\times100}{102\times102\times102}=\text{Rs } 6250$

$$\left[\because 6632.55=\text{Sum}\left(1+\frac{\frac{4}{2}}{100}\right)^{2\times\frac{3}{2}}\right]$$

12. c; **Hint:** Let time be t years

$$882=800\left(1+\frac{5}{100}\right)^t \Rightarrow \frac{882}{800}=\left(\frac{21}{20}\right)^t$$

$$\Rightarrow\left(\frac{21}{20}\right)^2=\left(\frac{21}{20}\right)^t \Rightarrow t=2\text{ years}$$

Rule 5

When rate of interest is r_1%, r_2% and r_3% for 1st year, 2nd year, and 3rd year respectively, then Amount = P

$$\left[1+\frac{r_1}{100}\right]\times\left[1+\frac{r_2}{100}\right]\times\left[1+\frac{r_3}{100}\right].$$

Illustrative Example

Ex.: Rs 7500 is borrowed at CI at the rate of 2% for the first year, 4% for the second year and 5% for the third year. Find the amount to be paid after 3 years.

Soln: Applying the above theorem,

$$\text{Amount}=7500\left[1+\frac{2}{100}\right]\left[1+\frac{4}{100}\right]\left[1+\frac{5}{100}\right]$$

$$=\frac{7500\times102\times104\times105}{100\times100\times100}=\text{Rs } 8353.8.$$

Exercise

1. Rs 50000 is borrowed at CI at the rate of 1% for the first year, 2% for the second year and 3% for the third year. Find the amount to be paid after 3 years.
   a) Rs 50355.3 b) Rs 53055.3 c) Rs 53505.3 d) Rs 53053.5
2. Rs 125000 is borrowed at CI at the rate of 2% for the first year, 3% for the second year and 4% for the third year. Find the amount to be paid after 3 years.
   a) Rs 135678 b) Rs 136587 c) Rs 163578 d) Rs 136578
3. Rs 25000 is borrowed at CI at the rate of 3% for the first year, 4% for the second year and 5% for the third year. Find the amount to be paid after 3 years.
   a) Rs 28119 b) Rs 29118 c) Rs 28129 d) Rs 28117
4. What sum will amount to Rs 15916.59 in 3 years at compound interest, the interest for 1st, 2nd and 3rd years being 3, 2 and 1 per cent respectively.
   a) Rs 15900 b) Rs 15000 c) Rs 16000 d) None of these
5. What sum of money will amount to Rs 699.66 in 2 years, reckoning compound interest for 1 year at 4 per cent and for the other at $3\frac{1}{2}$ per cent per annum.
   a) Rs 560 b) Rs 650 c) Rs 670 d) Rs 580

Answers

1. b 2. d 3. a

4. b; **Hint:** $15916.59=P\left(\frac{103}{100}\times\frac{102}{100}\times\frac{101}{100}\right)$

$$\therefore P=\frac{1591659\times100\times100}{103\times102\times101}=\text{Rs } 15000$$

5. b; **Hint:** Sum $=\frac{699.66\times100\times200}{104\times207}=\text{Rs } 650$

Rule 6

Let, Amount = Rs A
Principal = Rs P
Time = t years

Rate of interest (r) = n × 100 $\left[\left(\frac{A}{P}\right)^{\frac{1}{t}\times n}-1\right]$ **% per annum.**

Where, n = 1, when the interest is compounded yearly
n = 2, when the interest is compounded half-yearly
n = 4, when the interest is compounded quarterly
n = 12, when the interest is compounded monthly
If the interest is compounded yearly, then n = 1

$$\therefore r=100\left[\left(\frac{A}{P}\right)^{1/t}-1\right]\%\text{ per annum.}$$

Illustrative Examples

Ex. 1: At what rate per cent per annum will Rs 1000 amount to Rs 1331 in 3 years? The interest is compounded yearly.

Soln: Applying the above formula,

$$r=100\left[\left(\frac{1331}{1000}\right)^{1/3}-1\right]=100\left[\left(\frac{11}{10}\right)^{3\times\frac{1}{3}}-1\right]$$

$$=10\%\text{ per annum}$$

Ex. 2: At what rate per cent compounded yearly will Rs 80,000 amount to Rs 88,200 in 2 years?

Soln: Detail Method: We have $80{,}000\left(1+\frac{r}{100}\right)^2$ = Rs 88200

or, $\left(1+\frac{r}{100}\right)^2=\frac{88200}{80000}=\frac{441}{400}=\left(\frac{21}{20}\right)^2$

or, $1+\frac{r}{100}=\frac{21}{20}$

$\therefore$ r = 5%

Quicker Method: Applying the above formula,

$$r=100\left[\left(\frac{88200}{80000}\right)^{1/2}-1\right]=100\left[\frac{21}{20}-1\right]$$

= 5% per annum.

Exercise

1. At what rate per cent compound interest, will Rs 400 amount to Rs 441 in 2 years?
   a) 4% b) 5% c) 6% d) 3%
2. At what rate per cent compound interest will Rs 625 amount to Rs 676 in 2 years?
   a) 3% b) Rs 2% c) 4% d) 5%
3. At what rate per cent will the compound interest on Rs 2500 amount to Rs 477.54 in 3 years?
   a) 6% b) 4% c) 5% d) None of these

Answers

1. b 2. c

3. a; **Hint:** rate % = $100\left[\left(\frac{2977.54}{2500}\right)^{\frac{1}{3}}-1\right]$

$\because$ Here compound interest = Rs 477.54 (given)

$\therefore$ Amount (A) = Rs 2500 + Rs 477.54 = Rs 2977.54

$=100\times\left[\frac{53}{50}-1\right]=6\%$

Rule 7

Let Principal = P
Rate = r% per annum
Amount = A and

Time is given in the form of fraction like $2\frac{1}{2}$ years.

$$\textbf{Amount (A)} = P\left[1+\frac{r}{100}\right]^2\times\left[1+\frac{\frac{r}{2}}{100}\right]$$

Illustrative Example

Ex.: Find the compound interest on Rs 2000 at 5% per annum, compounded yearly for $2\frac{1}{2}$ years.

Soln: We know that, compound interest = Amount – Principal

CI = A – P

Now, using the above formula,

$$A=P\left[1+\frac{r}{100}\right]^2\times\left[1+\frac{r}{200}\right]$$

$$=2000\left[\frac{105}{100}\right]^2\times\left[\frac{205}{200}\right]$$

$$=\frac{2000\times105\times105\times205}{100\times100\times200}=\text{Rs }2260.12$$

$\therefore$ CI = 2260.12 – 2000 = Rs 260.12.

Exercise

1. Find the amount of Rs 800 at compound interest in $2\frac{1}{2}$ years at 5 per cent.
   a) Rs 904.05 b) Rs 904.50 c) Rs 904 d) Rs 905.04
2. On what sum will the compound interest for $2\frac{1}{2}$ years at 10% amount to Rs 6352.50?
   a) Rs 4900 b) Rs 5500 c) Rs 5000 d) Rs 5800
3. Find the amount of Rs 4000 for $2\frac{1}{2}$ years at 6% compound interest.
   a) Rs 4629.23 b) Rs 4692.32
   c) Rs 4639.32 d) None of these

Answers

1. a
2. c; **Hint:** Sum = $\frac{6352.50\times100\times100\times100}{110\times110\times105}$ = Rs 5000
3. a

Rule 8

Theorem: ***A sum of money, placed at compound interest, becomes n times in t years and m times in x years. We calculate the value of x from the equation given below*** $\left(n^{1/t}=m^{1/x}\right)$.

Illustrative Examples

Ex. 1: A sum of money placed at compound interest doubles itself in 4 years. In how many years will it amount to eight times itself?

Soln: Detail Method:

We have $P\left(1+\frac{r}{100}\right)^4=2P \quad \therefore\left(1+\frac{r}{100}\right)^4=2$

Cubing both sides, we get

$$\left(1+\frac{r}{100}\right)^{12}=2^3=8$$

Or, $P\left(1+\frac{r}{100}\right)^{12}=8P$

Hence, the required time is 12 years.

Quicker Method- I:

x becomes 2x in 4 years

2x becomes 4x in next 4 years

4x becomes 8x in yet another 4 years

Thus, x becomes 8x in $4+4+4=12$ years

Quicker Method - II

Applying the above formula, we have

$$2^{1/4}=8^{1/x}\Rightarrow 2^{1/4}=2^{3/x}\Rightarrow\frac{1}{4}=\frac{3}{x}$$

$\therefore x=12$ years.

Quicker Method - III: Remember the following conclusion:

If a sum becomes n times in t years at CI then it will be (n)^m times in mt years. Thus, if a sum becomes 2 times in 4 years it will be $(2)^3$ times in $3\times4=12$ years.

Ex. 2: A sum of money at compound interest amounts to thrice itself in three years. In how many years will it be 9 times itself?

Soln: Detail Method:

Suppose the sum = Rs x

Then, we have

$$3x=x\left(1+\frac{r}{100}\right)^3 \text{ or, } 3=\left(1+\frac{r}{100}\right)^3$$

Squaring both sides

$$(3)^2=\left\{\left(1+\frac{r}{100}\right)^3\right\}^2 \text{ or, } 9=\left(1+\frac{r}{100}\right)^6$$

Now multiply both sides by x; then $9x=x\left(1+\frac{r}{100}\right)^6$

$\therefore$ the sum x will be 9 times in 6 years.

Quicker Method: Remember the following conclusion:

If a sum becomes n times in t years at CI then it will be $(n)^m$ times in mt years.

Thus, if a sum becomes 3 times in 3 years it will be $(3)^2$ times in $2\times3=6$ years.

Exercise

1. A sum of money placed at compound interest doubles itself in 5 years. In how many years will it amount to 8 times itself?
   a) 18 years b) 15 years c) 16 years d) 12 years
2. A sum of money placed at compound interest doubles itself in 6 years. In how many years will it amount to 16 times itself?
   a) 24 years b) 26 years c) 22 years d) 20 years
3. A sum of money placed at compound interest thrice itself in 4 years. In how many years will it amount to 27 times itself?
   a) 12 years b) 15 years c) 14 years d) 10 years
4. If a sum of money at compound interest amounts to thrice itself in 3 years, then in how many years will it be 9 times itself? **[Bank PO Exam, 1981]**
   a) 12 year b) 6 years c) 9 years d) 15 years

Answers

1. b 2. a 3. a 4. c

Rule 9

Theorem: ***If a certain sum becomes 'm' times in 't' years, the rate of compound interest r is equal to 100*** $\left[(m)^{1/t}-1\right]$.

Illustrative Example

Ex.: At what rate per cent compound interest does a sum of money become nine-fold in 2 years?

Soln: Detail Method: Let the sum be Rs x and the rate of compound interest be r% per annum, then

$$9x=x\left(1+\frac{r}{100}\right)^2 \text{ or, } 9=\left(1+\frac{r}{100}\right)^2$$

$$\text{or, } 3=1+\frac{r}{100}\text{; or, }\frac{r}{100}=2\quad\therefore r=200\%$$

Quicker Method: Applying the above rule, we have $r=100\left[(9)^{1/2}-1\right]=100(3-1)=200\%$.

Exercise

1. At what rate per cent compound interest, does a sum of money become $\frac{9}{4}$ times itself in 2 years?
   a) 50% b) 100% c) 25% d) 40%
2. At what rate per cent will the compound interest, does a sum of money become four fold in 2 years? **[Bank PO Exam, 1977]**
   a) 150% b) 100% c) 200% d) 75%
3. At what rate per cent will the compound interest, does a sum of money become 27 times in 3 years?
   a) 100% b) 150% c) 75% d) 200%
4. At what rate per cent will the compound interest, does a sum of money become 16 times in 4 years?
   a) 100% b) 150% c) 50% d) 75%

Answers

1. a 2. b 3. d 4. a

Rule 10

Theorem: ***If the CI on a certain sum for t years at r% be Rs C, then the SI is given by the following formula, Simple***

$$\textit{Interest (SI)} = \frac{rt}{100\left[\left(1+\frac{r}{100}\right)^t - 1\right]} \times \textit{Compound Interest}$$

(CI).

Note: *When t = 2*

$$SI = \frac{2r}{100\left[1+\frac{r^2}{100^2}+\frac{2r}{100}-1\right]} \times CI$$

$$= \frac{2r \times CI \times 100}{r^2+200r} = \frac{200r}{r(r+200)} \times CI$$

$$\therefore \quad SI = \left[\frac{200}{r+200} \times CI\right]$$

Illustrative Example

Ex.: If the CI on a certain sum for 2 years at 3% be Rs 101.50, what would be the SI?

Soln: **Detail Method:**

$$\text{CI on 1 rupee} = \left(1+\frac{3}{100}\right)^2 - 1 = \left(\frac{103}{100}\right)^2 - 1$$

$$= \text{Re}\ \frac{609}{10000}$$

$$\text{SI on 1 rupee} = \text{Re}\ \frac{2\times 3}{100} = \text{Re}\ \frac{6}{100}$$

$$\therefore \frac{\text{SI}}{\text{CI}} = \frac{6}{100} \times \frac{10000}{609} = \frac{200}{203}$$

$$\therefore\ \text{SI} = \frac{200}{203}\ \text{of CI} = \frac{200}{203} \times 101.5 = \text{Rs } 100$$

Quicker Method I: Applying the above formula, we have,

$$\text{Simple Interest} = \frac{rt}{100\left[\left(1+\frac{r}{100}\right)^t - 1\right]} \times \text{Compound Interest}$$

$$= \frac{3\times 2}{100\left[\left(1+\frac{3}{100}\right)^2 - 1\right]} \times 101.50 = \frac{6}{100\left[\left(\frac{103}{100}\right)^2 - 1\right]} \times 101.50$$

$$= \frac{6}{100} \times \frac{10000}{609} \times 101.50 = \text{Rs } 100$$

Quicker Method II: Here, t = 2

$$\therefore\ \text{SI} = \frac{200}{203} \times 101.50 = \text{Rs } 100 \quad \text{[See \textbf{Note}]}$$

Exercise

1. If the CI on a certain sum for 2 years at 5% be Rs 410, what would be the SI?
   a) Rs 200 b) Rs 300 c) Rs 350 d) Rs 400
2. If the CI on a certain sum for 2 years at 4% be Rs 510, what would be the SI?
   a) Rs 500 b) Rs 505 c) Rs 400 d) 475
3. If the CI on a certain sum for 2 years at 6% be Rs 25.75, what would be the SI?
   a) Rs 25 b) Rs 24 c) Rs 20 d) Rs 15
4. If the compound interest on a certain sum for 3 years at 5 per cent be Rs 50.44, what would be the simple interest?
   a) Rs 49 b) Rs 48 c) Rs 44 d) Rs 40
5. The compound interest on a sum of money for 3 years at 5 per cent is Rs 1324.05. What is the simple interest?
   a) Rs 1260 b) Rs 1560 c) Rs 1160 d) Rs 1360
6. The compound interest on a certain sum of money for 2 years at 10% per annum is Rs 420. The simple interest on the same sum at the same rate and same time will be: **[Clerical Grade Exam, 1991]**
   a) Rs 350 b) Rs 375 c) Rs 380 d) Rs 400

Answers

1. d 2. a 3. a 4. b 5. a 6. d

Rule 11

Theorem: ***If the SI on a certain sum for 2 years at r% be Rs 'S' then the CI is given by the following formula, CI =*** $\left[\frac{r+200}{200} \times S\right]$.

Illustrative Example

Ex.: The simple interest on a certain sum of money for 2 years at 10% per annum is Rs 400. Find the compound interest at the same rate and for the same time.

Soln: Using the above formula,

$$\text{CI} = \frac{210}{200} \times 400 = \text{Rs } 420.$$

Exercise

1. The simple interest on a certain sum of money for 2 years at 5% per annum is Rs 100. Find the compound interest at the same rate and for the same time.
   a) Rs 102.50 b) Rs 103 c) Rs 103.50 d) Rs 102.25
2. The simple interest on a certain sum of money for 2 years at 6% per annum is Rs 300. Find the compound interest at the same rate and for the same time.
   a) Rs 310 b) Rs 308 c) Rs 307 d) Rs 309
3. The simple interest on a certain sum of money for 2 years

at 4% per annum is Rs 350. Find the compound interest at the same rate and for the same time.

a) Rs 387 b) Rs 367 c) Rs 357 d) None of these

4. The simple interest on a certain sum of money for 2 years at 20% per annum is Rs 200. Find the compound interest at the same rate and for the same time.

a) Rs 320 b) Rs 220 c) Rs 210 d) Rs 225

5. SI on a sum at 4% per annum for 2 years is Rs 80. The CI on the same sum for the same period is:

[Asstt. Grade 1987]

a) Rs 1081.60 b) Rs 81.60 c) Rs 160 d) None of these

Answers

1. a 2. d 3. c 4. b 5. b

Rule 12

Theorem: ***If the compound interest on a certain sum for 2 years is Rs 'C' and simple interest is Rs 'S', then the rate of interest per annum is*** $\left[\frac{2\times(C-S)}{S}\times100\right]\%$ ***per or*** $\frac{2\times Diff.\times100}{SI}$.

Illustrative Example

Ex.: The compound interest on a certain sum for 2 years is Rs 40.80 and simple interest is Rs 40.00. Find the rate of interest per annum and the sum.

Soln: **Detail Method:** A little reflection will show that the difference between the simple and compound interests for 2 years is the interest on the first year's interest.

First year's SI = Rs $\frac{40}{2}$ = Rs 20

CI – SI = Rs 40.8 – Rs 40 = Re 0.80

Interest on Rs 20 for 1 year = Re 0.80

$\therefore$ Interest on Rs 100 for 1 year = Rs $\frac{80\times100}{100\times20}$

= Rs 4

$\therefore$ rate = 4%

Now, principal P is given by

$$P=\frac{100\times I}{tr}=\frac{100\times40}{2\times4}=\text{Rs }500$$

Quicker Method: Applying the above rule, we have

$$\text{the rate}=\frac{2\times0.8}{40}\times100=4\%$$

$$\text{and sum}=\frac{40\times100}{4\times2}=\text{Rs }500.$$

Exercise

1. The compound interest on a certain sum for 2 years is Rs 50.50 and simple interest is Rs 50. Find the rate of interest per annum and the sum.

a) 4%, Rs 1000 b) 2%, Rs 1150
c) 2%, Rs 1250 d) 4%, Rs 1250

2. The compound interest on a certain sum for 2 years is Rs 105 and simple interest is Rs 100. Find the rate of interest per annum and the sum.

a) 10%, Rs 500 b) 10%, Rs 1000
c) 20%, Rs 1000 d) None of these

3. The compound interest on a certain sum for 2 years is Rs 60.60 and simple interest is Rs 60. Find the rate of interest per annum and the sum.

a) 2%, Rs 1600 b) 2%, Rs 1400
c) 3%, Rs 1500 d) 2%, Rs 1500

Answers

1. c 2. a 3. d

Rule 13

Theorem: ***If on a certain sum of money, the simple interest for 2 years at the rate of r% per annum is Rs X, then the difference in compound interest and simple interest is given by Rs*** $\left(\frac{Xr}{200}\right)$.

Note: This formula is applicable only for **2 years.**

Illustrative Example

Ex.: On a certain sum of money, the simple interest for 2 years is Rs 50 at the rate of 5% per annum. Find the difference in CI and SI.

Soln: Applying the above formula, we have

$$\text{difference in CI and SI}=\frac{50\times5}{200}=\text{Rs }1.25.$$

Exercise

1. On a certain sum of money, the simple interest for 2 years is Rs 140 at the rate of 4% per annum. Find the difference in CI and SI.

a) Rs 3 b) Rs 1.5 c) Rs 2.8 d) Rs 1.8

2. On a certain sum of money, the simple interest for 2 years is Rs 160 at the rate of 5% per annum. Find the difference in CI and SI.

a) Rs 4 b) Rs 5 c) Rs 6 d) Rs 8

3. On a certain sum of money, the simple interest for 2 years is Rs 150 at the rate of 3% per annum. Find the difference in CI and SI.

a) Rs 5 b) Rs 4.5 c) Rs 2.5 d) Rs 2.25

4. On a certain sum of money, the simple interest for 2 years is Rs 200 at the rate of 7% per annum. Find the difference in CI and SI.

a) Rs 7 b) Rs 6 c) Rs 3.5 d) None of these

Answers

1. c 2. a 3. d 4. a

Rule 14

Theorem: ***When difference between the compound interest and simple interest on a certain sum of money for 2 years at r% rate is Rs x, then the sum is given by***

$$Sum = \frac{\text{Difference} \times 100 \times 100}{\text{Rate} \times \text{Rate}} = \frac{x(100)^2}{r^2} = x\left(\frac{100}{r}\right)^2.$$

Illustrative Example

Ex.: The difference between the compound interest and the simple interest on a certain sum of money at 5% per annum for 2 years is Rs 1.50. Find the sum.

Soln: Using the above theorem:

$$\text{Sum} = 1.5\left(\frac{100}{5}\right)^2 = 1.5 \times 400 = \text{Rs } 600.$$

Exercise

1. The difference between the compound interest and the simple interest on a certain sum of money at 4% per annum for 2 years is Rs 2. Find the sum.
   a) Rs 1260 b) Rs 1225 c) Rs 1250 d) Rs 1230
2. The difference between the compound interest and the simple interest on a certain sum of money at 5% per annum for 2 years is Rs 3. Find the sum.
   a) Rs 600 b) Rs 1200
   c)Rs 1400 d) Data inadequate
3. The difference between the compound interest and the simple interest on a certain sum of money at 8% per annum for 2 years is Rs 4. Find the sum.
   a) Rs 625 b) Rs 1260
   c) Rs 312.5 d) None of these
4. The difference between the compound interest and the simple interest on a certain sum of money at 10% per annum for 2 years is Rs 2.50. Find the sum.
   a) Rs 350 b) Rs 275 c) Rs 250 d) Rs 325
5. The difference between the compound interest and the simple interest on a certain sum of money at 4% per annum for 2 years is Rs 1.40. Find the sum.
   a) Rs 875 b) Rs 857 c) Rs 785 d) None of these
6. The difference between the compound interest and simple interest on a certain sum at 5% for 2 years is Rs 1.50. The sum is **[Bank PO 1987]**
   a) Rs 600 b) Rs 500 c) Rs 400 d) Rs 300

Answers

1. c 2. b 3. a 4. c 5. a 6. a

Rule 15

Theorem: ***On a certain sum of money, the difference between compound interest and simple interest for 2 years at r% rate is given by*** *sum* $\left(\frac{r}{100}\right)^2$.

Illustrative Example

Ex.: Find the difference between the compound interest and the simple interest for the sum Rs 1500 at 10% per annum for 2 years.

Soln: Using the above theorem, we have

$$\text{Difference} = \text{Sum}\left(\frac{r}{100}\right)^2 = 1500\left(\frac{10}{100}\right)^2 = \text{Rs } 15.$$

Exercise

1. Find the difference between the compound interest and the simple interest for the sum Rs 1250 at 4% per annum for 2 years.
   a) Rs 2 b) Rs 2.5 c) Rs 1.5 d) Re 1
2. Find the difference between the compound interest and the simple interest for the sum Rs 1500 at 5% per annum for 2 years.
   a) Rs 3.25 b) Rs 7.5 c) Rs 3.75 d) None of these
3. Find the difference between the compound interest and the simple interest for the sum Rs 625 at 8% per annum for 2 years.
   a) Rs 3 b) Rs 4 c) Rs 4.5 d) Rs 1.5
4. Find the difference between the compound interest and the simple interest for the sum Rs 2500 at 6% per annum for 2 years.
   a) Rs 9 b) Rs 8 c) Rs 7.5 d) Rs 6

Answers

1. a 2. c 3. b 4. a

Rule 16

Theorem: ***If the difference between CI and SI on a certain sum for 3 years at r% is Rs x, the sum will be*** $\frac{\text{Difference} \times (100)^3}{r^2(300+r)}$.

Illustrative Example

Ex.: If the difference between CI and SI on a certain sum of money for 3 years at 5% per annum is Rs 122, find the sum.

Soln: By the above theorem:

$$\text{Sum} = \frac{122 \times 100 \times 100 \times 100}{5^2(300+5)} = \text{Rs } 16{,}000.$$

Exercise

1. On what sum will the difference between the simple and compound interests for 3 years at 5 per cent per annum amount to Rs 12.20?
   a) Rs 1600 b) Rs 800 c) Rs 1200 d) Rs 1500
2. On what sum will the difference between the simple and compound interests for 3 years at 4 per cent per annum amount to Rs 3.04?
   a) Rs 1250 b) Rs 625 c) Rs 650 d) Rs 675
3. On what sum will the difference between the simple and

compound interests for 3 years at 6 per cent per annum amount to Rs 13.77?
a) Rs 1250 b) Rs 1320 c) Rs 1520 d) Rs 1150
4. On what sum will the difference between the simple and compound interests for 3 years at 3 per cent per annum amount to Rs 27.27?
a) Rs 5000 b) Rs 10000 c) Rs 8000 d) None of these

Answers

1. a 2. b 3. a 4. b

Rule 17

Theorem: ***On a certain sum of money, the difference between compound interest and simple interest for 3 years at r% per annum is given by difference =*** $\frac{Sum \times r^2(300+r)}{(100)^3}$.

Illustrative Example

Ex.: Find the difference between CI and SI on Rs 8000 for 3 years at 2.5% pa.

Soln: Using the above theorem,

$$\text{Difference} = \frac{\text{Sum} \times r^2(300+r)}{(100)^3}$$

$$= \frac{8000 \times 2.5 \times 2.5(300+2.5)}{100 \times 100 \times 100}$$

$$= \frac{8 \times 25 \times 25 \times 3025}{100 \times 100 \times 100} = \frac{121}{8} = \text{Rs } 15.125.$$

Exercise

1. Find the difference between the simple and compound interest on Rs 500 for 3 years at 4 per cent.
a) Rs 2.432 b) Rs 3.432 c) Rs 2.342 d) Rs 2.423
2. What is the difference between the simple and compound interest for 3 years at 5 per cent?
a) 0.0007625 times the principal
b) 0.07625 times the principal
c) 0.007625 times the principal
d) Data inadequate
3. Find the difference between the simple and compound interest on Rs 10000 for 3 years at 3 per cent.
a) Rs 27.8 b) Rs 27.27 c) Rs 37.27 d) Rs 37.8
4. Find the difference between the simple and compound interest on Rs 8000 for 3 years at 5 per cent.
a) Rs 61 b) Rs 60 c) Rs 51 d) Rs 59

Answers

1. a
2. c; **Hint:** Let P be the principal
Difference between CI and SI

$$= \frac{P \times 5 \times 5 \times (300+5)}{100 \times 100 \times 100} - 0.007625\,P$$

3. b 4. a

Rule 18

Theorem: ***If an amount of money grows upto Rs A_1 in n years and upto Rs A_2 in (n + 1) years on compound interest, then the rate per cent is given by*** $\left[\frac{(A_2 - A_1)100}{A_1}\right]$

or $\frac{\text{Difference of amount after n years and (n + 1) years} \times 100}{\text{Amount after n years}}$.

Illustrative Examples

Ex. 1: An amount of money grows upto Rs 4840 in 2 years and upto Rs 5324 in 3 years on compound interest. Find the rate per cent.

Soln: **Detail Method:** We have,
P + CI of 3 years = Rs 5324 .... (1)
P + CI of 2 years = Rs 4840 ... (2)
Subtracting (2) from (1), we get
CI of 3rd year = 5324 – 4840 = Rs 484
Thus, the CI calculated in the third year which is Rs 484 is basically the amount of interest on the amount generated after 2 years which is Rs 4840.

$$\therefore r = \frac{484 \times 100}{4840 \times 1} = 10\%$$

Quicker Method (Direct Formula):

$$\text{Rate} = \frac{\text{Difference of amount after n years and (n+1) years} \times 100}{\text{Amount after n years}}$$

In this case, n = 2

$$\therefore \text{Rate} = \frac{\text{Difference of amount after 2 years and 3 years} \times 100}{\text{Amount after 2 years}}$$

$$= \frac{(5324 - 4840)}{4840} \times 100 = \frac{484 \times 100}{4840} = 10\%.$$

Note: The above generalised formula can be used for any positive value of n. See in the following example.

Ex. 2: A certain amount of money at compound interest grows upto Rs 51168 in 15 years and upto Rs 51701 in 16 years. Find the rate per cent per annum.

Soln: Using the above formula:

$$\text{Rate} = \frac{(51701 - 51168) \times 100}{51168} = \frac{533 \times 100}{51168}$$

$$= \frac{100}{96} = \frac{25}{24} = 1\frac{1}{24}\%.$$

Exercise

1. A certain amount of money at compound interest grows upto Rs 6560 in 3 years and upto Rs 7216 in 4 years. Find the rate per cent per annum.
a) 10% b) 5% c) 8% d) 6%

2. A certain amount of money at compound interest grows upto Rs 7520 in 15 years and upto Rs 7896 in 16 years. Find the rate per cent per annum.
a) 10% b) 8% c) 5% d) None of these
3. A certain amount of money at compound interest grows upto Rs 3840 in 4 years and upto Rs 3936 in 5 years. Find the rate per cent per annum.
a) 2.05% b) 2.5% c) 2% d) 3.5%
4. A certain amount of money at compound interest grows upto Rs 4950 in 19 years and upto Rs 5049 in 20 years. Find the rate per cent per annum.
a) 2% b) 2.5% c) 1% d) 1.5%
5. A certain amount of money at compound interest grows upto Rs 12960 in 2 years and upto Rs 13176 in 3 years. Find the rate per cent per annum.
a) $1\frac{1}{3}\%$ b) $2\frac{1}{3}\%$ c) $2\frac{2}{3}\%$ d) $1\frac{2}{3}\%$

Answers
1. a 2. c 3. b 4. a 5. d

Rule 19

Theorem: ***If an amount of money grows upto Rs A_1 in n years and upto Rs A_2 in (n + 1) years on compound interest, then the sum is given by Rs*** $\left[A_1\left(\frac{A_1}{A_2}\right)^n\right]$.

Illustrative Example
Ex.: An amount of money grows upto Rs 4840 in 2 years and up to Rs 5324 in 3 years on compound interest. Find the sum.

Soln: Applying the above theorem,

$$\text{Sum} = 4840\times\left(\frac{4840}{5324}\right)^2 = \text{Rs } 4000.$$

Exercise
1. What sum of money at compound interest will amount to Rs 650 at the end of the first year and Rs 676 at the end of the second year?
a) Rs 625 b) Rs 630 c) Rs 620 d) None of these
2. What sum of money at compound interest will amount to Rs 480 at the end of the first year and Rs 576 at the end of the second year?
a) Rs 420 b) Rs 450 c) Rs 400 d) Rs 375
3. An amount of money grows upto Rs 2750 in 2 years and upto Rs 3125 in 3 years on compound interest. Find the sum.
a) Rs 2129.6 b) Rs 1229.6
c) Rs 2219.6 d) Data inadequate
4. An amount of money grows upto Rs 1200 in 2 years and upto Rs 1440 in 3 years on compound interest. Find the sum.
a) Rs $\frac{2200}{3}$ b) Rs 800
c) Rs $\frac{2500}{3}$ d) None of these

Answers
1. a 2. c 3. a 4. c

Rule 20

Theorem: ***To find the ratio of Compound Interest (CI) to Simple Interest (SI) on a certain sum at r% for 2 years, we use the following formula*** $\frac{CI}{SI} = \frac{r}{200}+1$.

Illustrative Example
Ex.: Find the ratio of CI to SI on a certain sum at 4% per annum for 2 years.

Soln: Applying the above formula,

$$\frac{CI}{SI} = \frac{4}{200}+1 = \frac{204}{200} = \frac{51}{50}$$

Exercise
1. Find the ratio of CI to SI on a certain sum at 5% per annum for 2 years.
a) 41 : 40 b) 42 : 41 c) 43 : 40 d) 41 : 35
2. Find the ratio of CI to SI on a certain sum at 8% per annum for 2 years.
a) 27 : 26 b) 26 : 25 c) 26 : 21 d) 25 : 24
3. Find the ratio of CI to SI on a certain sum at 45% per annum for 2 years.
a) 49 : 47 b) 49 : 42 c) 49 : 40 d) None of these
4. Find the ratio of CI to SI on a certain sum at 15% per annum for 2 years.
a) 53 : 40 b) 53 : 50 c) 43 : 40 d) 50 : 43
5. Find the ratio of CI to SI on a certain sum at 10% per annum for 2 years.
a) 7 : 5 b) 21 : 20 c) 8 : 5 d) 20 : 19

Answers
1. a 2. b 3. c 4. c 5. b

Rule 21

Theorem: ***If a sum 'A' becomes 'B' in t_1 years at compound rate of interest, then after t_2 years the sum becomes*** $\frac{(B)^{t_2/t_1}}{(A)^{t_2/t_1 - 1}}$ ***rupees.***

Illustrative Example
Ex.: Rs 4800 becomes Rs 6000 in 4 years at a certain rate of compound interest. What will be the sum after 12 years?

Soln: **Detail Method:** We have:

$$4800\left(1+\frac{r}{100}\right)^4 = 6000$$

or, $\left(1+\frac{r}{100}\right)^4 = \frac{6000}{4800} = \frac{5}{4}$

Now, $\left(1+\frac{r}{100}\right)^{4\times3} = \left(\frac{5}{4}\right)^3 = \frac{125}{64}$

or, $\left(1+\frac{r}{100}\right)^{12} = \frac{125\times75}{64\times75} = \frac{9375}{4800}$

or, $4800\left(1+\frac{r}{100}\right)^{12} = 9375$

The above equation shows that Rs 4800 becomes Rs 9375 after 12 years.

Quicker Method: Applying the above rule, we have

the required amount $= \frac{(6000)^{12/4}}{(4800)^{12/4-1}} = \frac{(6000)^3}{(4800)^2}$

$= \text{Rs } 9375$

Exercise

1. Rs 2400 becomes Rs 3000 in 3 years at a certain rate of compound interest. What will be the sum after 6 years?
   a) Rs 4750 b) Rs 3750
   c) Rs 3570 d) None of these
2. Rs 1200 becomes Rs 1500 in 2 years at a certain rate of compound interest. What will be the sum after 6 years?
   a) Rs 2433.25 b) Rs 2334.75 c) Rs 2343.75 d) Rs 2343.25
3. Rs 9600 becomes Rs 12000 in 6 years at a certain rate of compound interest. What will be the sum after 12 years?
   a) Rs 15000 b) Rs 14000 c) Rs 16000 d) Rs 18000
4. Rs 1600 becomes Rs 2000 in 2 years at a certain rate of compound interest. What will be the sum after 4 years?
   a) Rs 2500 b) Rs 2400 c) Rs 2200 d) None of these

Answers

1. b 2. c 3. a 4. a

Rule 22

Theorem: ***If the compound rate of interest for the first t_1 years is r_1 %, for the next t_2 years is r_2 %, for the next t_3 years is r_3 %, ... and the last t_n years is r_n %, then compound interest on Rs x for $(t_1+t_2+t_3+...t_n)$ years is***

$$\left[x\left(1+\frac{r_1}{100}\right)^{t_1}\left(1+\frac{r_2}{100}\right)^{t_2}....\left(1+\frac{r_n}{100}\right)^{t_n}\right]-x.$$

Illustrative Examples

Ex. 1: Find the compound interest on Rs 18,750 in 2 years, the rate of interest being 4% for the first year and 8% for the second year.

Soln: After first year the amount $= 18750\left(1+\frac{4}{100}\right)$

$= 18750\left(\frac{104}{100}\right)$

After 2nd year the amount $= 18750\left(\frac{104}{100}\right)\left(\frac{108}{100}\right)$

$= 18750\left(\frac{26}{25}\right)\left(\frac{27}{25}\right)$

$= \text{Rs } 21060$

$\therefore$ CI $= 21060 - 18750 = \text{Rs } 2310$.

Ex. 2: Find the compound interest on Rs 10000 for 3 years if the rate of interest is 4% for the first year, 5% for the second year and 6% for the third year.

Soln: The compound interest on Rs x in 't' years if the rate of interest is r_1% for the first year, r_2% for the second year ... and r_t% for the *tth* year is given by

$$x\left(1+\frac{r_1}{100}\right)\left(1+\frac{r_2}{100}\right)...\left(1+\frac{r_t}{100}\right)-x$$

Note: Here, $t_1 = t_2 = t_3 = 1$

In this case

$$\text{CI} = 10000\left(1+\frac{4}{100}\right)\left(1+\frac{5}{100}\right)\left(1+\frac{6}{100}\right)-10000$$

$$= 10000\left(\frac{26}{25}\right)\left(\frac{21}{20}\right)\left(\frac{53}{50}\right)-10000$$

$= 11575.20 - 10000 = \text{Rs } 1575.2$.

Exercise

1. Find the compound interest on Rs 9375 in 2 years, the rate of interest being 2% for the first year and 4% for the second year.
   a) Rs 570 b) Rs 1140 c) Rs 1155 d) Rs 670
2. Find the compound interest on Rs 8000 in 2 years, the rate of interest being 5% for the first year and 10% for the second year.
   a) Rs 1340 b) Rs 1420 c) Rs 1240 d) None of these
3. Find the compound interest on Rs 3200 in 2 years, the rate of interest being $7\frac{1}{2}$% for the first year and $12\frac{1}{2}$% for the second year.
   a) Rs 620 b) Rs 670 c) Rs 770 d) Rs 760

4. Find the compound interest on Rs 50000 for 3 years if the rate of interest is 5% for the first year, 6% for the second year and 10% for the third year.
a) Rs 10632 b) Rs 16032 c) Rs 10362 d) Rs 13062
5. Find the compound interest on Rs 80000 for 3 years if the rate of interest is 5% for the first year, 4% for the second year and 5% for the third year.
a) Rs 17128 b) Rs 11728 c) Rs 11278 d) Rs 11738

Answers
1. a 2. c 3. b 4. a 5. b

Rule 23

Theorem: ***Certain sum of money at compound interest will amount to Rs A in $(t_1 + t_2 + t_3 + ... + t_n)$ years. If the rate of interest for the first t_1 years is r_1 %, for the next t_2 years is r_2 %, for the next t_3 years is r_3 % ... and the last t_n years is r_n %, then the sum is given by***

$$A\left(\frac{100}{100+r_1}\right)\left(\frac{100}{100+r_2}\right)\left(\frac{100}{100+r_3}\right)...\left(\frac{100}{100+r_n}\right).$$

Where $t_1 = t_2 = t_3 = ... = t_n = 1$

Illustrative Example

Ex.: What sum of money at compound interest will amount to Rs 2249.52 in 3 years, if the rate of interest is 3% for the first year, 4% for the second year and 5% for the third year?

Soln: **Detail Method:** The general formula for such question is

$$A = P\left(1+\frac{r_1}{100}\right)\left(1+\frac{r_2}{100}\right)\left(1+\frac{r_3}{100}\right)...$$

Where, A = Amount, P = Principal, and r_1, r_2, r_3 are the rates of interest for different years.
In the above case,

$$2249.52 = P\left(1+\frac{3}{100}\right)\left(1+\frac{4}{100}\right)\left(1+\frac{5}{100}\right)$$

or, 2249.52 = P(1.03)(1.04)(1.05)

$$\therefore P = \frac{2249.59}{1.03\times1.04\times1.05} = \text{Rs } 2000.$$

Quicker Method: Applying the above rule, we have

$$\text{principal} = 2249.52\left(\frac{100}{103}\right)\left(\frac{100}{104}\right)\left(\frac{100}{105}\right) = \text{Rs } 2000.$$

Exercise

1. What sum of money at compound interest will amount to Rs 5305.53 in 3 years, if the rate of interest is 1% for the first year, 2% for the second year and 3% for the third year?
a) Rs 5000 b) Rs 5200 c) Rs 5130 d) Rs 4910
2. What sum of money at compound interest will amount to Rs 1365.78 in 3 years, if the rate of interest is 2% for the first year, 3% for the second year and 4% for the third year?
a) Rs 1360 b) Rs 1250 c) Rs 1160 d) Rs 1240
3. What sum of money at compound interest will amount to Rs 562.38 in 3 years, if the rate of interest is 3% for the first year, 4% for the second year and 5% for the third year?
a) Rs 400 b) Rs 450 c) Rs 500 d) Rs 520
4. What sum of money at compound interest will amount to Rs 2893.8 in 3 years, if the rate of interest is 4% for the first year, 5% for the second year and 6% for the third year?
a) Rs 2500 b) Rs 2400 c) Rs 2200 d) None of these

Answers
1. a 2. b 3. c 4. a

Rule 24

Theorem: ***If a man borrows Rs P at r% compound interest and pays back Rs A at the end of each year, then at the end of the nth year he should pay***

$$Rs\ P\left[1+\frac{r}{100}\right]^n - A\left[\left(1+\frac{r}{100}\right)^{n-1}+\left(1+\frac{r}{100}\right)^{n-2}+...+\left(1+\frac{r}{100}\right)^1\right]$$

Illustrative Example

Ex.: A man borrows Rs 3000 at 10% compound rate of interest. At the end of each year he pays back Rs 1000. How much amount should he pay at the end of the third year to clear all his dues?

Soln: Using the above formula, the required answer

$$= 3000\left[1+\frac{10}{100}\right]^3 - 1000\left[\left(1+\frac{10}{100}\right)^2+\left(1+\frac{10}{100}\right)^1\right]$$

$$= 3000\left(\frac{11}{10}\times\frac{11}{10}\times\frac{11}{10}\right) - 1000\left[\left(\frac{11}{10}\right)^2+\frac{11}{10}\right]$$

$$= 3993 - \left\{1000\times\frac{121}{100}+1000\times\frac{11}{10}\right\}$$

= 3993 − 1210 − 1100 = Rs 1683.

Exercise

1. A man borrows Rs 1500 at 5% compound rate of interest. At the end of each year he pays back Rs 500. How much amount should he pay at the end of the third year to clear all his dues?

a) Rs $680\frac{3}{16}$ b) Rs $600\frac{3}{16}$

c) Rs $660\frac{3}{16}$ d) None of these

2. A man borrows Rs 4000 at 20% compound rate of interest. At the end of each year he pays back Rs 1500. How much amount should he pay at the end of the third year to clear all his dues?
a) Rs 2592 b) Rs 2852 c) Rs 2952 d) Rs 2953
3. A man borrows Rs 3000 at 30% compound rate of interest. At the end of each year he pays back Rs 1000. How much amount should he pay at the end of the third year to clear all his dues?
a) Rs 3602 b) Rs 3601 c) Rs 3603 d) Rs 3604

Answers

1. c 2. c 3. b

Rule 25

Theorem: ***If a sum of money say Rs x is divided among n parts in such a manner that when placed at compound interest, amount obtained in each case remains equal while the rate of interest on each part is $r_1, r_2, r_3, ..., r_n$ respectively and time period for each part is $t_1, t_2, t_3, ..., t_n$ respectively, then the divided parts of the sum will in the ratio of***

$$\frac{1}{\left(1+\frac{r_1}{100}\right)^{t_1}}:\frac{1}{\left(1+\frac{r_2}{100}\right)^{t_2}}:\frac{1}{\left(1+\frac{r_3}{100}\right)^{t_3}}:...:\frac{1}{\left(1+\frac{r_n}{100}\right)^{t_n}}.$$

Illustrative Example

Ex.: Divide Rs 3903 between A and B, so that A's share at the end of 7 years may equal B's share at the end of 9 years, compound interest being at 4%.

Soln: Applying the above theorem,

$$\text{A's share : B's share} = \frac{1}{\left(1+\frac{4}{100}\right)^7}:\frac{1}{\left(1+\frac{4}{100}\right)^9}$$

$$= 1:\frac{1}{\left(1+\frac{4}{100}\right)^2} = 1:\frac{625}{626}$$

$$= 626 : 625$$

Dividing Rs 3903 in the ratio 676 : 625

$$\text{A's present share} = \frac{676}{678+625}\times 3903 = \text{Rs } 2028.$$

B's present share = Rs 3903 – Rs 2028 = Rs 1875.

Exercise

1. Divide Rs 2708 between A and B, so that A's share at the end of 6 years may equal B's share at the end of 8 years, compound interest being at 8%.
a) Rs 1458, Rs 1250 b) Rs 1448, Rs 1260
c) Rs 1438, Rs 1270 d) Rs 1468, Rs 1240
2. Divide Rs 1105 between A and B, so that A's share at the end of 5 years may equal B's share at the end of 7 years, compound interest being at 10%.
a) Rs 505, Rs 600 b) Rs 605, Rs 500
c) Rs 705, Rs 400 d) Rs 625, Rs 480
3. Divide Rs 6100 between A and B, so that A's share at the end of 3 years may equal B's share at the end of 5 years, compound interest being at 20%.
a) Rs 3600, Rs 2500 b) Rs 3500, Rs 2600
c) Rs 3400, Rs 2700 d) Rs 3450, Rs 2650

Answers

1. a 2. b 3. a

Miscellaneous

1. The difference between the simple and the compound interest compounded every six months at the rate of 10 per cent per annum at the end of two years is Rs 124.05. What is the sum? **[SBI PO Exam, 2000]**
a) Rs 10000 b) Rs 6000 c) Rs 12000 d) Rs 8000
2. A person invested a certain amount at simple interest at the rate of 6 per cent per annum earning Rs 900 as an interest at the end of three years. Had the interest been compounded every year, how much more interest would he have earned on the same amount with the same interest rate after three years? **[NABARD, 1999]**
a) Rs 38.13 b) Rs 25.33 c) Rs 55.08 d) Rs 35.30
3. Find the effective annual rate of 5 per cent per annum compound interest paid helf yearly.
a) 1.025% b) 6.0625% c) 5.0625% d) None of these
4. Find the effective annual rate of 4 per cent per annum compound interest paid quarterly.
a) 4.0604% b) 4.604% c) 5.0605% d) 5.605%
5. In what time will Rs 390625 amount to Rs 456976 at 4 per cent compound interest?
a) 2 years b) 4 years c) 3 years d) 5 years
6. Find the least number of complete years in which a sum of money put out at 20 per cent compound interest will be more than doubled.
a) 2 years b) 3 years
c) 4 years d) Data inadequate
7. In what time will Rs 6250 amount to Rs 6632.55 at 4% compound interest payable half-yearly?
a) 3 years b) $\frac{3}{2}$ years c) 1 years d) $\frac{5}{2}$ years
8. Find what is that first year in which a sum of money will

become more than double in amount if put out at compound interest at the rate of 10% per annum.
a) 6th year b) 7th year
c) 8th year d) Data inadequate

9. A sum of money put out at compound interest amounts in 2 years to Rs 578.40 and in 3 years to Rs 614.55. Find the rate of interest.
a) 6% b) $6\frac{1}{4}\%$ c) $6\frac{1}{2}\%$ d) $6\frac{3}{4}\%$

10. Divide Rs 3903 between A and B, so that A's share at the end of 7 years may equal to B's share at the end of 9 years, compound interest being at 4 per cent.
a) Rs 2028, Rs 1875 b) Rs 2018, Rs 1885
c) Rs 2008, Rs 1895 d) Rs 2038, Rs 1865

11. Vijay obtains a loan of Rs 64,000 against his fixed deposits. If the rate of interest be 2.5 paise per rupee per annum, calculate the compound interest payable after 3 years.
a) Rs 4921 b) Rs 5020 c) Rs 4821 d) Rs 4920

12. A certain sum is interested at compound. The interest accrued in the first two years is Rs 272 and that in the first three years is Rs 434. Find the rate per cent.
a) $12\frac{1}{2}\%$ b) $7\frac{1}{2}\%$ c) $17\frac{1}{2}\%$ d) 25%

Answers

1. d; Let the sum be Rs x.

Then $\left[x\left(1+\frac{5}{100}\right)^4-x\right]-\left[\frac{x\times10\times2}{100}\right]=124.05$

Solving the above eqn, we get x = Rs 8000

2. c; Certain sum for the person $=\frac{900\times100}{18}=$ Rs 5000

$\therefore$ interest on Rs 5000 by CI

$=5000\left(1+\frac{6}{100}\right)^3-5000=$ Rs 955.08

$\therefore$ More interest = Rs (955.08 - 900) = Rs 55.08.

3. c; The amount of Rs 100 in one year at compound interest at 5% per annum payable half-yearly

$=\text{Rs }100\left(1+\frac{\frac{5}{2}}{100}\right)^2=\text{Rs }100(1.025)^2=\text{Rs }105.0625$

Thus the *nominal rate* of 5% payable half-yearly has the same effect as the rate 5.0625 per cent would have if payable yearly.
Hence 5.0625 per cent is calle the **effective annual rate** of 5% per annum payable half-yearly

4. a; Effective annual rate

$=100\left(1+\frac{\frac{4}{4}}{100}\right)^4-100$

$=104.0604-100=4.0604\%$

5. b; $\because P\left(1+\frac{r}{100}\right)^t=A$ (See **Rule - 1**)

$\therefore 390625\left(1+\frac{4}{100}\right)^t=456976$

$\therefore \left(1+\frac{4}{100}\right)^t=\frac{456976}{390625}=\left(\frac{26}{25}\right)^4$

$\therefore \left(\frac{26}{25}\right)^t=\left(\frac{26}{25}\right)^4 \quad \therefore t=4$

$\therefore$ the required time is 4 years.

6. c; Here $P\left(1+\frac{20}{100}\right)^t>2P$

or, $\left(\frac{6}{5}\right)^t>2$

By trial $\frac{6}{5}\times\frac{6}{5}\times\frac{6}{5}\times\frac{6}{5}>2$

$\therefore$ the required time is 4 years.

7. b; $6250\left(1+\frac{\frac{4}{2}}{100}\right)^{2t}=6632.55$

or, $\left(1+\frac{2}{100}\right)^{2t}=\frac{663255}{625000}=\frac{132651}{12500}=\left(\frac{51}{50}\right)^3$

or, $\left(\frac{51}{50}\right)^{2t}=\left(\frac{51}{50}\right)^3$ or, 2t = 3

$\therefore t=\frac{3}{2}$ years

8. c; Here $P\left(1+\frac{10}{100}\right)^t>2P$

or, $\left(\frac{11}{10}\right)^t>2$

By trial $\frac{11\times11\times11\times11\times11\times11\times11\times11}{10\times10\times10\times10\times10\times10\times10\times10}>2$

Hence, the first year in which a sum of money will become more than double in amount is 8th year.

9. b; Clearly, the difference between Rs 578.40 and Rs 614.55 is the interest on Rs 578.40 for 1 year.

$\therefore$ interest on Rs 578.40 for 1 year = Rs 614.55 - Rs 578.40 = Rs 36.15

$\therefore$ interest on Rs 100 for 1 year = Rs 36.15 × $\frac{100}{578.40}$

$= \text{Rs } \frac{3615}{57840}\times\frac{100}{1} = \text{Rs } 6\frac{1}{4}$

$\therefore$ the required rate is $6\frac{1}{4}$ per cent.

10. a; We have at once

$$(\text{A's present share})\left(1+\frac{4}{100}\right)^7$$

$$= (\text{B's present share})\left(1+\frac{4}{100}\right)^9$$

$$\therefore \frac{\text{A's present share}}{\text{B's present share}} = \left(1+\frac{4}{100}\right)^2 = \left(\frac{26}{25}\right)^2 = \frac{676}{625}$$

Dividing Rs 3903 in the ratio 676 : 625

$\therefore$ A's present share = $\frac{676}{676+625}$ of Rs 3903

= Rs 2028

$\therefore$ B's present share = Rs 3903 - Rs 2028 = Rs 1875

11. a; P = Rs 64000

r = 2.5 paise per rupee per annum (given)

= 0.025 rupee per rupee per annum.

= 0.025 × 100 rupee per hundred rupee per annum.

= 0.025 ×100 per cent per annum

= 2.5 per cent per annum

t = 3 yrs

$$\text{CI} = 64000\left[\left(\frac{1+0.025\times100}{100}\right)^2-1\right]$$

$$= 64000\left[(1+0.025)^2-1\right] = \text{Rs } 4921$$

$\therefore$ The compound interest payable is Rs 4921

Note: Remember that x paise per rupee per annum = Rs x per cent per annum.

12. a; Amount $A = P\left[1+\frac{r}{100}\right]^t$;

$$\text{CI} = P\left[\left(1+\frac{r}{100}\right)^t-1\right]$$

Putting $\left(1+\frac{r}{100}\right) = q$ in the above equation,

$\therefore$ for two years, t = 2, then, $P\left[q^2-1\right] = 272$ .... (i)

For three years $P\left[q^3-1\right] = 434$ .... (ii)

Dividing (ii) by (i)

$$\frac{(q^2+q+1)(q-1)}{(q+1)(q-1)} = \frac{434}{272}$$

or, $\frac{q^2+q+1}{q+1} = \frac{217}{136} \Rightarrow \frac{q^2}{q+1} = \frac{81}{136}$

or, $136q^2 + 81q + 81 = 0$

or, $q = \frac{9}{8}$

$\therefore 1+\frac{r}{100} = \frac{9}{8} \Rightarrow r = \frac{1}{8}\times100\% = 12\frac{1}{2}\%$.

11 Problems Based on Instalment

Introduction

1. Instalment Purchase Schemes: To increase their sales, different business organisations offer different schemes to the buyers to enable them to buy costly articles even if their income is very low. One such scheme is the instalment scheme. The cost of the article bought is paid in **instalments**.

2. In instalment scheme the buyer has to pay more because in addition to the selling price, the buyer has to pay the interest also on it.

3. Cash Price: The amount for which the article can be purchased on full payment is called **cash price.**

4. Cash Down Payment: While purchasing an article under instalment scheme, some payment has to be made initially. It is called **cash down payment**. The remaining amount is paid in equal monthly, quarterly or annual instalements as may be decided at the time of purchase.

Case-I: Based on Simple Interest

When the period over which the instalment scheme is operative is **less than a year**. The payment is made in a specified number of equal monthly instalments. Naturally we calculate the **simple interest** in such cases.

Type I: ***Monthly instalment being given, to find the rate of interest.***

Illustrative Examples

Ex. 1: A coat is sold for Rs 60 or Rs 20 cash down payment and Rs 8 per month for 6 months. Determine the rate of interest.

Soln: Cash price of the coat = Rs 60
Cash down payment = Rs 20
Amount paid in 6 instalments = Rs (8 × 6) = Rs 48
Total amount paid under instalment plan
= Rs 20 + Rs 48 = Rs 68
Interest charged = Rs 68 - Rs 60 = Rs 8
Now we find the principals for each of the six months.
Principal for the first month = Rs 60 - Rs 20 = Rs 40
Principal for the second month = Rs 40 - Rs 8 = Rs 32
Principal for the third month = Rs 32 - Rs 8 = Rs 24
Principal for the fourth month = Rs 24 - Rs 8 = Rs 16
Principal for the fifth month = Rs 16 - Rs 8 = Rs 8
Principal for the sixth month = Rs 8 - Rs 8 = Rs 0
∴ Total Principal = Rs 120

Therefore, interest on Rs 120 for 1 month or $\frac{1}{12}$ year is Rs 8.

$$\text{Rate }\% = \frac{SI \times 100}{P \times T} = \frac{8 \times 100}{120 \times \frac{1}{12}}$$

$$= \frac{8 \times 100 \times 12}{120 \times 1} = 80\%$$

Hence the rate of interest is 80%.

Ex 2: A television is marked at Rs 3575 cash or Rs 1500 as cash down payment and Rs 420 a month for 5 months. Find the rate of interest for this instalment plan.

Soln: Cash price = Rs 3575
Cash down payment = Rs 1500
Amount paid in 5 monthly instalments
= Rs (420 × 5) = Rs 2100
∴ Total amount paid under instalment plan
= Rs 1500 + Rs 2100 = Rs 3600
∴ Interest charged = Rs 3600 – Rs 3575 = Rs 25
The principal for each month is as under:
Principal for the 1st month = Rs 3575 – Rs 1500
= Rs 2075
Principal for the 2nd month = Rs 2075 – Rs 420
= Rs 1655
Principal for the 3rd month = Rs 1655 - Rs 420
= Rs 1235
Principal for the 4th month = Rs 1235 - Rs 420
= Rs 815
Principal for the 5th month = Rs 815 - Rs 420
= Rs 395
Total Principal = Rs 6175
The final (sixth) instalment of Rs 420 consists of the amount of Rs 395 and the interest of Rs 25 (Rs 395 + Rs 25 = Rs 420).

Thus, the interest on Rs 6175 for 1 month or $\frac{1}{12}$ year is Rs 25.

$$\therefore I = \frac{P \times T \times R}{100}$$

$$\Rightarrow \text{R (Rate)} = \frac{I \times 100}{P \times T} = \frac{25 \times 100}{6175 \times \frac{1}{12}}$$

$$= \frac{25 \times 100 \times 12}{6175 \times 1} = \frac{1200}{247} = 4.86\%$$

Hence the rate of interest is 4.86%.

Ex 3: A house is sold for Rs 30000 cash or Rs 17500 as cash down payment and instalments of Rs 1600 per month for 8 month. Determine the rate of interest correct to one decimal place, under the instalment plan.

Soln: Cash price = Rs 30000
Cash down payment = Rs 17500
Total amount paid in 8 monthly instalments
= Rs (1600 × 8) = Rs 12800
Total amount paid under instalment plan
= Rs 17500 + Rs 12800 = Rs 30300
Interest charged = Rs 30300 - Rs 30000 = Rs 300
Principal for 1st month = Rs 30000 - Rs 17500
= Rs 12500
Principal for 2nd month = Rs 12500 - Rs 1600
= Rs 10900
Principal for 3rd month = Rs 10900 - Rs 1600
= Rs 9300
Principal for 4th month = Rs 9300 - Rs 1600
= Rs 7700
Principal for 5th month = Rs 7700 - Rs 1600
= Rs 6100
Principal for 6th month = Rs 6100 - Rs 1600
= Rs 4500
Principal for 7th month = Rs 4500 - Rs 1600
= Rs 2900
Principal for 8th month = Rs 2900 - Rs 1600
= Rs 1300
Total Principal = Rs 55200
The last instalment of Rs 1600 includes Rs 1300 plus Rs 300 interest.

Time = 1 month = $\frac{1}{12}$ year, Interest = Rs 300

$$\therefore I = \frac{P \times T \times R}{100} \Rightarrow \text{R (Rate)} = \frac{I \times 100}{P \times T}$$

$$= \frac{300 \times 100}{55200 \times \frac{1}{12}} = \frac{300 \times 100 \times 12}{55200} = \frac{150}{23} = 6.52$$

Hence the rate of interest is 6.5%

Ex 4: A radio is available for Rs 950 cash or Rs 200 cash down payment and 10 monthly instalments of Rs 80 each. Find the rate of interest charged.

Soln: Cash price = Rs 950
Cash down payment = Rs 200
∴ Amount paid in 10 monthly instalments = Rs (80 × 10) = Rs 800
∴ Total amount paid under instalment plan = Rs 800 + Rs 200 = Rs 1000
∴ Interest charged = Rs 1000 - Rs 950 = Rs 50
Principal for 1st month = Rs 950 - Rs 200 = Rs 750
Principal for 2nd month = Rs 750 - Rs 80 = Rs 670
Principal for 3rd month = Rs 670 - Rs 80 = Rs 590
Principal for 4th month = Rs 590 - Rs 80 = Rs 510
Principal for 5th month = Rs 510 - Rs 80 = Rs 430
Principal for 6th month = Rs 430 - Rs 80 = Rs 350
Principal for 7th month = Rs 350 - Rs 80 = Rs 270
Principal for 8th month = Rs 270 - Rs 80 = Rs 190
Principal for 9th month = Rs 190 - Rs 80 = Rs 110
Principal for 10th month = Rs 110 - Rs 80 = Rs 30
Total Principal for 1 month = Rs 3900
The last instalment of Rs 80 comprises Rs 30 plus Rs 50 interest.

Time = 1 month = $\frac{1}{12}$ year, I = Rs 50

$$\therefore I = \frac{P \times T \times R}{100} \Rightarrow \text{R (Rate)} = \frac{I \times 100}{P \times T}$$

$$\Rightarrow Rate = \frac{50 \times 100}{3900 \times \frac{1}{12}} = \frac{50 \times 100 \times 12}{3900} = \frac{200}{13} = 15\frac{5}{13}$$

Hence the rate of interest is $15\frac{5}{13}\%$.

Type 2: Rate of interest being given, to find the monthly instalment.

Illustrative Examples

Ex. 1: A pocket transistor is sold for Rs 125 cash or for Rs 26 as cash down payment followed by 4 equal monthly instalments. If the rate of interest charged is 25% per annum, determine the monthly instalment.

Soln: Cash price of the transistor = Rs 125
Cash down payment = Rs 26
Balance of price due = Rs 125 - Rs 26 = Rs 99
Rate of interest charged = 25% p.a.
Interest on Rs 99 for 4 months

$$= \text{Rs}\ \frac{99 \times \frac{4}{12} \times 25}{100} = Rs\frac{33}{4} = Rs\ 8.25$$

$$\left(\because I = \frac{P \times T \times R}{100}\right)$$

$\therefore$ Amount due = Rs 99 + Rs 8.25 = Rs 107.25 ....(i)

Let monthly instalment be x rupees

$\therefore$ At the end of 4th month

1st instalment of Rs x will amount to

$$\text{Rs}\left(x + \frac{x \times \frac{3}{12} \times 25}{100}\right) = Rs\left(x + \frac{x}{16}\right) = Rs\frac{17x}{16}$$

$$\left[\because A = P + I = P + \frac{P \times T \times R}{100}\right]$$

2nd instalment of Rs x will amount to

$$Rs\left(x + \frac{x \times \frac{2}{12} \times 25}{100}\right) = Rs\left(x + \frac{x}{24}\right) = Rs\frac{25x}{24}$$

3rd instalment of Rs x will amount to

$$Rs\left(x + \frac{x \times \frac{1}{12} \times 25}{100}\right) = Rs\left(x + \frac{x}{48}\right) = Rs\frac{49x}{48}$$

4th instalment of Rs x will amount to

$$Rs\left(x + \frac{x \times 0 \times 25}{100}\right) = Rs\ x$$

$\therefore$ Total amount of 4 instalments at the end of 4th month $= Rs\left(\frac{17x}{16} + \frac{25x}{24} + \frac{49x}{48} + x\right)$

$$= Rs\frac{51x + 50x + 49x + 48x}{48}$$

$$= Rs\frac{198x}{48} = Rs\frac{33}{8}x \text{ ....(ii)}$$

From (i) and (ii), we get

$$\frac{33}{8}x = 107.25 \Rightarrow x = \frac{107.25 \times 28}{33} = \frac{858}{33} = 26$$

Hence the monthly instalment is Rs 26.

Ex. 2: A ceiling fan is marked at Rs 485 cash or Rs 105 cash down payment followed by 3 equal monthly instalments. If the rate of interest charged under this instalment plan is 16% per annum, find the monthly instalment.

Soln: Cash price of the ceiling fan = Rs 485

Cash down payment = Rs 105

$\therefore$ Balance to be paid in 3 equal monthly instalments = Rs 485 - Rs 105 = Rs 380

Rate of interest = 16% p.a.

Interest on Rs 380 for 3 months

$$= Rs\frac{380 \times \frac{3}{12} \times 16}{100} = Rs\ 15.20$$

$$\left[\because I = \frac{P \times T \times R}{100}\right]$$

$\therefore$ Amount due = Rs 380 + Rs 15.20 = Rs 395.20 ... (i)

Let the monthly instalment be Rs x.

1st instalment of Rs x will amount to

$$Rs\left(x + \frac{x \times \frac{2}{12} \times 16}{100}\right) = Rs\left(x + \frac{2x}{75}\right) = Rs\frac{77x}{75}$$

2nd instalment of Rs x will amount to

$$Rs\left(x + \frac{x \times \frac{1}{12} \times 16}{100}\right) = Rs\left(x + \frac{2x}{75}\right) = Rs\frac{76x}{75}$$

3rd instalment of Rs x will amount to

$$Rs\left(x + \frac{x \times 0 \times 16}{100}\right) = Rs\ x$$

$\therefore$ Total amount payable under instalment plan

$$= Rs\left(\frac{77x}{75} + \frac{76x}{75} + x\right) = Rs\frac{77x + 76x + 75x}{75}$$

$$= Rs\frac{228}{75}x = Rs\frac{76}{25}x$$

From (i) and (ii), we get

$$\frac{76}{25}x = 395.20 \Rightarrow x = \frac{395.20 \times 25}{76} = \frac{9880}{76} = 130$$

Hence the monthly instalment = Rs 130.

Exercise

1. An electric iron is sold for Rs 110 cash or for Rs 50 cash down payment followed by Rs 62 after a month. Find the rate of interest charged under instalment plan.
2. A bicycle is sold for Rs 400 cash or Rs 160 cash down payment followed by 2 monthly instalments of Rs 130 each. Find the rate of interest.
3. A pressure cooker is available on Rs 180 cash or for Rs 70 cash down payment followed by Rs 60 a month for 2 months. Find the rate of interest charged under instalment plan.

4. A television set is priced at Rs 2400 cash or Rs 1200 cash down payment followed by 6 monthly instalments of Rs 225 each. What rate of interest will the dealer charge under instalment plan?
5. A mixi is marked at Rs 1000 cash or Rs 250 cash down payment followed by Rs 200 a month for 4 months. Find the rate of interest for this instalment plan.
6. A room cooler is marked at Rs 2000 cash or Rs 400 cash down payment followed by Rs 300 per month for 6 months. Determine the rate of interest charged under this instalment plan.
7. A watch is sold either for Rs 180 cash or for Rs 40 cash down payment followed by Rs 30 a month for 5 months. Determine the rate of interest.
8. Determine the interest rate charged under each of the following instalment plans.

Article	Cash price	Cash down payment	Each instalment	No. of monthly instalments
(i) TV	2575	1000	300	6
(ii) Refrigerator	3580	1500	440	5
(iii) Typewriter	3600	1200	280	10
(iv) Tape recorder	1600	300	175	8

9. An article is sold for Rs 100 each or for Rs 10 as cash down payment followed by 5 equal monthly instalments. If the rate of interest charged under instalment plan be 48% per annum, determine the monthly instalment.

Answers

1. Cash price of the electric iron = Rs 110
Cash down payment = Rs 50
Balance to be paid after 1 month
= Rs 110 – Rs 50 = Rs 60
Monthly instalment = Rs 62
∴ Interest = Rs 62 – Rs 60 = Rs 2
Thus Rs 2 is charged as interest on Rs 60 for 1 month or $\frac{1}{12}$ year

$$\therefore I = \frac{P \times T \times R}{100} \Rightarrow \text{R (Rate)} = \frac{I \times 100}{P \times T}$$

$$\Rightarrow R = \frac{2 \times 100}{60 \times \frac{1}{12}} = \frac{2 \times 100 \times 12}{60} = 40$$

Hence the rate of interest is 40%.

2. Cash price of bicycle = Rs 400
Cash down payment = Rs 160
Balance to be paid = Rs 400 – Rs 160 = Rs 240
Monthly instalment = Rs 130
∴ Amount paid in 2 equal monthly instalments
= Rs (130 × 2) = Rs 260
∴ Interest charged = Rs 260 – Rs 240 = Rs 20
Principal for 1st month = Rs 240
Principal for 2nd month = Rs 240 – Rs 130 = Rs 110
Total Principal = Rs 350
Thus, Rs 20 is the interest on Rs 350 for 1 month or $\frac{1}{12}$ year.

$$\therefore I = \frac{P \times T \times R}{100} \Rightarrow R = \frac{I \times 100}{P \times T}$$

$$\Rightarrow R = \frac{20 \times 100}{350 \times \frac{1}{12}} = \frac{20 \times 100 \times 12}{350} = \frac{480}{7} = 68\frac{4}{7}$$

Hence the rate of interest is $68\frac{4}{7}$% pa

3. Cash price of the pressure cooker = Rs 180
Cash down payment = Rs 70
Balance to be paid = Rs 180 – Rs 70 = Rs 110
Monthly instalment = Rs 60
∴ Amount paid in 2 equal monthly instalments = Rs (60 × 2) = Rs 120
∴ Interest charged = Rs 120 – Rs 110 = Rs 10
Principal for 1st month = Rs 110
Principal for 2nd month = Rs 110 – Rs 60 = Rs 50
Total Principal for 1 month = Rs 160
Thus, Rs 10 is the interest on Rs 160 for 1 month or $\frac{1}{12}$ year.

$$\therefore R = \frac{I \times 100}{P \times T} = \frac{10 \times 100}{160 \times \frac{1}{12}} = \frac{10 \times 100 \times 12}{160} = 75$$

Hence the rate of interest is 75% per annum.

4. Cash price of the television set = Rs 2400
Cash down payment = Rs 1200
Balance to be paid = Rs 2400 – Rs 1200 = Rs 1200
Monthly instalment = Rs 225
∴ Amount paid in 6 equal monthly instalments = Rs (225 × 6) = Rs 1350
Interest charged = Rs 1350 – Rs 1200 = Rs 150
Principal for 1st month = Rs 1200
Principal for 2nd month = Rs 1200 – Rs 225 = Rs 975
Principal for 3rd month = Rs 975 – Rs 225 = Rs 750
Principal for 4th month = Rs 750 – Rs 225 = Rs 525
Principal for 5th month = Rs 525 – Rs 225 = Rs 300
Principal for 6th month = Rs 300 – Rs 225 = Rs 75
Total principal for 1 month = Rs 3825
Thus, Rs 150 is the interest on Rs 3825 for 1 month or $\frac{1}{12}$ year

$$\therefore\ R = \frac{I \times 100}{P \times T} = \text{Rs } \frac{150 \times 100}{3825 \times \frac{1}{12}}$$

$$= \frac{150 \times 100 \times 12}{3825} = \frac{800}{17} = 47\frac{1}{17}$$

Hence the rate of interest is $47\frac{1}{17}\%$ pa

5. Cash price of the mixi = Rs 1000
Cash down payment = Rs 250
Balance to be paid = Rs 1000 – Rs 250 = Rs 750
Monthly instalment = Rs 200
∴ Amount paid in 4 equal monthly instalments = Rs (200×4) = Rs 800
∴ Interest charged = Rs 800 – Rs 750 = Rs 50
Principal for 1st month = Rs 750
Principal for 2nd month = Rs 750 – Rs 200 = Rs 550
Principal for 3rd month = Rs 550 – Rs 200 = Rs 350
Principal for 4th month = Rs 350 – Rs 200 = Rs 150
Total Principal for 1 month = Rs 1800
Thus Rs 50 is the interest on Rs 1800 for 1 month or $\frac{1}{12}$ year

$$\therefore\ R = \frac{I \times 100}{P \times T} = \frac{50 \times 100}{1800 \times \frac{1}{12}}$$

$$= \frac{50 \times 100 \times 12}{1800} = \frac{100}{3} = 13\frac{1}{3}$$

Hence the rate of interest is $33\frac{1}{3}\%$ pa

6. Cash price of the room cooler = Rs 2000
Cash down payment = Rs 400
Balance to be paid = Rs 2000 – Rs 400 = Rs 1600
Monthly instalment = Rs 300
∴ Amount paid in 6 equal monthly instalments = Rs (300×6) = 1800
∴ Interest charged = Rs 1800 – Rs 1600 = Rs 200
Principal for 1st month = Rs 1600
Principal for 2nd month = Rs 1600 – Rs 300 = Rs 1300
Principal for 3rd month = Rs 1300 – Rs 300 = Rs 1000
Principal for 4th month = Rs 1000 – Rs 300 = Rs 700
Principal for 5th month = Rs 700 – Rs 300 = Rs 400
Principal for 6th month = Rs 400 – Rs 300 = Rs 100
Total principal for 1 month = Rs 5100
Thus Rs 200 is the interest on Rs 5100 for 1 month or $\frac{1}{12}$ year

$$\therefore\ R = \frac{I \times 100}{P \times T} = \frac{200 \times 100}{5100 \times \frac{1}{12}}$$

$$= \frac{200 \times 100 \times 12}{5100} = \frac{800}{17} = 47\frac{1}{17}$$

Hence, the rate of interest is $47\frac{1}{7}\%$ pa

7. Cash price of the watch = Rs 180
Cash down payment = Rs 40
∴ Balance to be paid = Rs 180 – Rs 40 = Rs 140
Monthly instalment = Rs 30
∴ Amount paid in 5 equal monthly instalments = Rs (30×5) = Rs 150
∴ Interest charged = Rs 150 – Rs 140 = Rs 10
Principal for 1st month = Rs 140
Principal for 2nd month = Rs 140 – Rs 30 = Rs 110
Principal for 3rd month = Rs 110 – Rs 30 = Rs 80
Principal for 4th month = Rs 80 – Rs 30 = Rs 50
Principal for 5th month = Rs 50 – Rs 30 = Rs 20
Total principal for 1 month = Rs 400
Thus Rs 10 is the interest for 1 month or $\frac{1}{12}$ year on Rs 400

$$\therefore\ R = \frac{I \times 100}{P \times T} = \frac{10 \times 100}{400 \times \frac{1}{12}} = \frac{10 \times 100 \times 12}{400} = 30$$

Hence, the rate of interest is 30% pa

8. (i) Cash price of TV = Rs 2575
Cash down payment = Rs 1000
∴ Balance to be paid = Rs 2575 – Rs 1000 = Rs 1575
Monthly instalment = Rs 300
∴ Amount paid in 6 equal monthly instalments = Rs (300×6) = Rs 1800
∴ Interest charged = Rs 1800 – Rs 1575 = Rs 225
Principal for 1st month = Rs 1575
Principal for 2nd month = Rs 1575 – Rs 300 = Rs 1275
Principal for 3rd month = Rs 1275 – Rs 300 = Rs 975
Principal for 4th month = Rs 975 – Rs 300 = Rs 675
Principal for 5th month = Rs 675 – Rs 300 = Rs 375
Principal for 6th month = Rs 375 – Rs 300 = Rs 75
Total principal for 1 month = Rs 4950
Thus Rs 225 is the interest on Rs 4950 for 1 month or $\frac{1}{12}$ year

$$\therefore\ R = \frac{I \times 100}{P \times T} = \frac{225 \times 100}{4950 \times \frac{1}{12}}$$

$$= \frac{225\times100\times12}{4950\times1} = \frac{600}{11} = 54.55$$

Hence the rate of interest is 54.55% pa

(ii) Cash price of the refrigerator = Rs 3580
Cash down payment = Rs 1500
∴ Balance to be paid = Rs 2080
Monthly instalment = Rs 440
∴ Amount paid in 5 equal monthly instalments = Rs (440 × 5) = Rs 2200
∴ Interest charged = Rs 2200 – Rs 2080 = Rs 120
Principal for 1st month = Rs 2080
Principal for 2nd month = Rs 2080 – Rs 440 = Rs 1640
Principal for 3rd month = Rs 1640 – Rs 440 = Rs 1200
Principal for 4th month = Rs 1200 – Rs 440 = Rs 760
Principal for 5th month = Rs 760 – Rs 440 = Rs 320
Total principal for 1 month = Rs 6000
Thus, Rs 120 is the interest on Rs 6000 for 1 month or $\frac{1}{12}$ year

$$\therefore R = \frac{I\times100}{P\times T} = \frac{120\times100}{6000\times\frac{1}{12}} = \frac{120\times100\times12}{600\times1} = 24$$

Hence the rate of interest is 24% pa

(iii) Cash price of the typewriter = Rs 3600
Cash down payment = Rs 1200
∴ Balance to be paid = Rs 3600 – Rs 1200 = Rs 2400
Monthly instalment = Rs 280
∴ Amount paid in 10 equal monthly instalments = Rs (280 × 10) = Rs 2800
∴ Interest charged = Rs 2800 – Rs 2400 = Rs 400
Principal for 1st month = Rs 2400
Principal for 2nd month = Rs 2400 – Rs 280 = Rs 2120
Principal for 3rd month = Rs 2120 – Rs 280 = Rs 1840
Principal for 4th month = Rs 1840 – Rs 280 = Rs 1560
Principal for 5th month = Rs 1560 – Rs 280 = Rs 1280
Principal for 6th month = Rs 1280 – Rs 280 = Rs 1000
Principal for 7th month = Rs 1000 – Rs 280 = Rs 720
Principal for 8th month = Rs 720 – Rs 280 = Rs 440
Principal for 9th month = Rs 440 – Rs 280 = Rs 160
Principal for 10th month = Rs 160 – Rs 280 = – Rs 120
Ignore the negative principal
∴ Total principal for 1 month = Rs 11520
Thus, Rs 400 is the interest on Rs 11520 for 1 month or $\frac{1}{12}$ year

$$\therefore R = \frac{I\times100}{P\times T} = \frac{400\times100}{11520\times\frac{1}{12}}$$

$$= \frac{400\times100\times12}{11520\times1} = \frac{125}{3} = 41.67$$

Hence, the rate of interest is 41.67% pa

(iv) Cash price of the tape recorder = Rs 1600
Cash down payment = Rs 300
Balance to be paid = Rs 1600 – Rs 300 = Rs 1300
Monthly instalment = Rs 175
∴ Amount paid in 8 equal monthly instalments = Rs (175 × 8) = Rs 1400
∴ Interest charged = Rs 1400 – Rs 1300 = Rs 100
Principal for 1st month = Rs 1300
Principal for 2nd month = Rs 1300 – Rs 175 = Rs 1125
Principal for 3rd month = Rs 1125 – Rs 175 = Rs 950
Principal for 4th month = Rs 950 – Rs 175 = Rs 775
Principal for 5th month = Rs 775 – Rs 175 = Rs 600
Principal for 6th month = Rs 600 – Rs 175 = Rs 425
Principal for 7th month = Rs 425 – Rs 175 = Rs 250
Principal for 8th month = Rs 250 – Rs 175 = Rs 75
Total principal for 1 month = Rs 5500
Thus, Rs 100 is the interest on Rs 5500 for 1 month or $\frac{1}{12}$ year

$$\therefore R = \frac{I\times100}{P\times T} = \frac{100\times100}{5500\times\frac{1}{12}}$$

$$= \frac{100\times100\times12}{5500\times1} = \frac{240}{11} = 21.81$$

Hence, the rate of interest charged is 21.8% pa

9. Cash price of the article = Rs 100
Cash down payment = Rs 10
∴ Balance to be paid = Rs 100 – Rs 10 = Rs 90
Rate of interest charged = 48% pa

Interest on Rs 90 for 5 months or $\frac{5}{12}$ year is

$$= \text{Rs}\ \frac{90\times\frac{5}{12}\times48}{100} = \text{Rs}\ \frac{90\times5\times48}{12\times100} = \text{Rs}\ 18$$

Amount due = Rs 90 + Rs 18 = Rs 108 ....(i)
Let the monthly instalment be x rupees
∴ At the end of 5th month:
1st instalment of Rs x will amount to

$$\text{Rs}\left(x + \frac{x\times\frac{4}{12}\times48}{100}\right)$$

$$A = P + I = P + \frac{P\times T\times R}{100}$$

$$= \text{Rs}\left(x + \frac{4x}{25}\right) = \text{Rs}\ \frac{29x}{25}$$

2nd instalment of Rs x will amount to

$$\text{Rs}\left(x+\frac{x\times\frac{3}{12}\times 48}{100}\right)=\text{Rs}\left(x+\frac{3x}{25}\right)=\text{Rs}\frac{28x}{25}$$

3rd instalment of Rs x will amount to

$$\text{Rs}\left(x+\frac{x\times\frac{2}{12}\times 48}{100}\right)=\text{Rs}\left(x+\frac{2x}{25}\right)=\text{Rs}\frac{27x}{25}$$

4th instalment of Rs x will amount to

$$\text{Rs}\left(x+\frac{x\times\frac{1}{12}\times 48}{100}\right)=\text{Rs}\left(x+\frac{x}{25}\right)=\text{Rs}\frac{26x}{25}$$

5th instalment of Rs x will amount to

$$\text{Rs}\left(x+\frac{x+0\times 48}{100}\right)=\text{Rs}\,x$$

$\therefore$ Total amount of 5 instalments at the end of 5 months

$$=\text{Rs}\left(\frac{29x}{25}+\frac{28x}{25}+\frac{27x}{25}+\frac{26x}{25}+x\right)$$

$$=\text{Rs}\ \frac{29x+28x+27x+28x+25x}{25}$$

$$=\text{Rs}\ \frac{135x}{25}=\text{Rs}\ \frac{27x}{5} \quad ....(ii)$$

From (i) and (ii), we get

$$\frac{27x}{5}=108 \Rightarrow x=\frac{108\times 5}{27}=20$$

Each instalment = Rs 20

Type 3: To find the annual payment to discharge a debt if the rate per cent is given.

Theorem: ***The annual payment that will discharge a debt of Rs A due in t years at the rate of interest r% per annum is***

$$\frac{100A}{100t+\frac{rt(t-1)}{2}}.$$

Illustrative Example

Ex.: What annual payment will discharge a debt of Rs 770 due in 5 years, the rate of interest being 5% per annum?

Soln: **Detail Method:** Let the annual payment be P rupees. The amount of Rs P in 4 years at 5%

$$=\frac{100\text{P}+4\times 5\text{P}}{100}=\frac{120\text{P}}{100}$$

The amount of Rs P in 3 years at 5% $=\frac{115\text{P}}{100}$

The amount of Rs P in 2 years at 5% $=\frac{110\text{P}}{100}$

The amount of Rs P in 1 year at 5% $=\frac{105\text{P}}{100}$

These four amounts together with the last annual payment of Rs P will discharge the debt of Rs 770.

$$\therefore\ \frac{120\text{P}}{100}+\frac{115\text{P}}{100}+\frac{110\text{P}}{100}+\frac{105\text{P}}{100}+\text{P}=770$$

$$\therefore\ \frac{550\text{P}}{100}=770$$

$$\therefore\ \text{P}=\frac{770\times 100}{550}=140$$

Hence, annual payment = Rs 140

Quicker Method: Applying the above theorem, we have,

$$\text{Annual payment}=\frac{100\times 770}{100\times 5+\frac{5\times 5(5-1)}{2}}=Rs140.$$

Exercise

1. What annual instalment will discharge a debt of Rs 2210 due in 4 years at 7% simple interest?
   a) Rs 450 b) Rs 500 c) Rs 550 d) Rs 575
2. What quarterly payment will discharge a debt of Rs 2120 in one year at 16% per annum simple interest?
   a) Rs 1000 b) Rs 400 c) Rs 850 d) Rs 500
3. What annual payment will discharge a debt of Rs 19350 due 4 years hence at the rate of 5% simple interest?
   a) Rs 4600 b) Rs 3500 c) Rs 4500 d) Rs 4550
4. Find the annual instalment that will discharge a debt of Rs 12900 due in 4 years at 5% per annum simple interest.
   a) Rs 3500 b) Rs 2500 c) Rs 3000 d) Rs 3200
5. Find the annual instalment that will discharge a debt of Rs 5400 due in 5 years at 4% per annum simple interest.
   a) Rs 1200 b) Rs 1000 c) Rs 800 d) Rs 1050
6. What quarterly payment will discharge a debt of Rs 2280 due in two years at 16% per annum simple interest?
   a) Rs 500 b) Rs 450 c) Rs 550 d) Rs 250
   (Bank PO Exam 1989)
7. What annual payment will discharge a debt of Rs 580 due in 5 years, the rate being 8% per annum?
   a) Rs 166.40 b) Rs 120 c) Rs 100 d) Rs 65.60

Answers

1. b; **Hint:** Required annual payment

$$=\frac{100\times 2210}{100\times 4+\frac{7\times 4(4-1)}{2}}=\frac{100\times 2210}{442}=\text{Rs }500.$$

2. d; **Hint:** Here instalment is quarterly. Hence from the question, we have, t = 4 and $r = \frac{16}{4} = 4\%$

Now applying the given rule,

$$\text{required answer} = \frac{100 \times 2120}{100 \times 4 + \frac{4 \times 4(4-1)}{2}} = \frac{100 \times 2120}{424}$$

$= \text{Rs } 500.$

3. c; **Hint:** Required answer

$$= \frac{100 \times 19350}{100 \times 4 + \frac{5 \times 4 \times 3}{2}} = \frac{100 \times 19350}{430} = \text{Rs } 4500.$$

4. c; **Hint:** Required answer

$$= \frac{100 \times 12900}{100 \times 4 + \frac{5 \times 4(4-1)}{2}} = \frac{100 \times 12900}{430} = \text{Rs } 3000$$

5. b; **Hint:** Required answer

$$= \frac{100 \times 5400}{100 \times 5 + \frac{4 \times 5 \times (5-1)}{2}} = \frac{100 \times 5400}{540} = \text{Rs } 1000.$$

6. d; **Hint:** Here, t = 8 and r = 4% [Since, payment is quarterly for 2 years]

Now, applying the given rule, we have the required answer

$$= \frac{100 \times 2280}{100 \times 8 + \frac{8 \times 7 \times 4}{2}} = \frac{2280 \times 100}{912} = \text{Rs } 250.$$

7. c; **Hint:** Required answer

$$= \frac{100 \times 580}{100 \times 5 + \frac{8 \times 5 \times (5-1)}{2}} = \frac{100 \times 580}{500 \times 580} = \text{Rs } 100$$

Case - 2: Based on Compound Interest

The problems of money lending in which the payment is made in instalments and the range normally **is in years**. In such cases **compound interest** computations are used.

Type I: To find each instalment when the instalments are equal.

Theorem: ***A sum of Rs P is to be paid back in n equal annual instalments. If the interest is compounded annually at R% per annum, then the value of each instalment is given by***

$$\left[\frac{P}{\left(\frac{100}{100+R}\right) + \left(\frac{100}{100+R}\right)^2 + \ldots + \left(\frac{100}{100+R}\right)^n}\right]$$

Illustrative Example

Ex: A sum of Rs 3310 is to be paid back in 3 equal annual instalments. How much is each instalment if the interest is compounded annually at 10% per annum.

Soln: First Method: Let each equal annual instalment be Re 1.

∴ The 1st instalment is paid after a year

∴ Principal of the 1st instalment

$$= \text{Re}1 \times \frac{10}{11} = \text{Re}\frac{10}{11} \quad \left[\begin{array}{l} A = P\left(1 + \frac{r}{100}\right)^r \\ I = P\left(1 + \frac{10}{100}\right)^1 = P \times \frac{11}{10} \\ \Rightarrow P = I \times \frac{10}{11} \end{array}\right]$$

Similarly, the principal of 2nd instalment

$$= \text{Re}1 \times \left(\frac{10}{11}\right)^2 = \text{Re}\frac{100}{121}$$

The principal of 3rd instalment

$$= \text{Re}1 \times \left(\frac{10}{11}\right)^3 = \text{Re}\frac{1000}{1331}$$

∴ Total of the three principals

$$= Rs\left(\frac{10}{11} + \frac{100}{121} + \frac{1000}{1331}\right)$$

$$= Rs\frac{1210 + 1100 + 1000}{1331} = Rs\frac{3310}{1331}$$

When the principal is Rs $\frac{3310}{1331}$,

each instalment = Re 1

When the principal is Re 1 each instalment

$$= \text{Re}1 \times \frac{1331}{3310}$$

When the principal is Rs 3310 each instalment

$$= \text{Rs } \frac{1331}{3310} \times 3310 = \text{Rs } 1331$$

∴ Each instalment = Rs 1331

Second Method: Let each equal annual instalment be Rs x and P_1, P_2, P_3, be respectively the principals for the three instalments.

The first instalment is paid after a year

∴ Principal (P_1) of the first instalment

$$= x \times \frac{10}{11} = \frac{10x}{11}$$

Similarly $P_2 = x\left(\frac{10}{11}\right)^2$

$P_3 = x\left(\frac{10}{11}\right)^3$

Now $P_1 + P_2 + P_3 = 3310$

i.e. $\frac{10x}{11} + x\left(\frac{10}{11}\right)^2 + x\left(\frac{10}{11}\right)^3 = 3310$

$\Rightarrow x\left(\frac{10}{11}\right)\left[1 + \frac{10}{11} + \left(\frac{10}{11}\right)^2\right] = 3310$

$\Rightarrow x\left(\frac{10}{11}\right)\left(1 + \frac{10}{11} + \frac{100}{121}\right) = 3310$

$\Rightarrow x\left(\frac{10}{11}\right)\left(\frac{121+110+100}{121}\right) = 3310$

$\Rightarrow x\left(\frac{10}{11}\right)\left(\frac{331}{121}\right) = 3310$

$\Rightarrow x = 3310 \times \frac{11}{10} \times \frac{121}{331} = 1331$

Hence the required annual instalment = Rs 1331

Quicker Method: Applying the above theorem, we have the

$$\text{required annual instalment} = \frac{3310}{\frac{10}{11} + \frac{100}{121} + \frac{1000}{1331}}$$

$$= \frac{3310 \times 1331}{3310} = Rs\ 1331.$$

Type II: To find the Principal when each instalment is given.

Theorem: *A man borrows some money on compound interest and returns it in t years in n equal instalments. If the rate of interest is R% and the yearly instalment is Rs X, then the amount borrowed is given by*

$$Rs\ X\left[\left(\frac{100}{100+R}\right) + \left(\frac{100}{100+R}\right)^2 + \ldots + \left(\frac{100}{100+R}\right)^t\right]$$

Note: 1. To find the total interest charged we use the following formula,

$$Rs\ X\left[n - \left\{\left(\frac{100}{100+R}\right) + \left(\frac{100}{100+R}\right)^2 + \ldots + \left(\frac{100}{100+R}\right)^t\right\}\right]$$

2. To calculate the principal and interest charged with each instalment following formula is used.

Principal:

For the 1st year = $RsX\left(\frac{100}{100+R}\right)$

For the 2nd year = $RsX\left(\frac{100}{100+R}\right)^2$

" "

" "

For the tth year = $Rs\ X\left(\frac{100}{100+R}\right)^t$

Interest Charged:

Interest in 1st instalment = $Rs\ X\left(1 - \frac{100}{100+R}\right)$

Interest in 2nd instalment = $Rs\ X\left[1 - \left(\frac{100}{100+R}\right)^2\right]$

" " "

" " "

Interest in nth instalment = $Rs\ X\left[1 - \left(\frac{100}{100+R}\right)^n\right]$

Illustrative Example

Ex: A man borrowed some money and paid back in 3 equal annual instalments of Rs 2160 each. What sum did he borrow, if the rate of interest charged by the money lender was 20% per annum compounded annually? Find also the total interest charged. Also calculate the principal and interest charged with each instalment.

Soln: Detail Method

Amount of each annual instalment = Rs 2160

Rate of interest = 20% p.a. Number of instalments = 3

Principal for the 1st year = Rs $\frac{2160}{1 + \frac{20}{100}}$

$= \text{Rs } 2160 \times \frac{5}{6} = \text{Rs } 1800$

$\left[\begin{array}{l} \because A = P\left(1 + \frac{r}{100}\right)^t \\ 2160 = P\left(1 + \frac{20}{100}\right) \end{array}\right]$

Principal for the 2nd year

$= Rs\ 2160 \times \left(\frac{5}{6}\right)^2 = Rs\ 2160 \times \frac{25}{36} = Rs\ 1500$

Principal for the 3rd year

$= Rs\ 2160 \times \left(\frac{5}{6}\right)^3 = Rs\ 2160 \times \frac{125}{216} = Rs\ 1250$

Amount borrowed = Sum of the principals for all the three years = Rs (1800 + 1500 + 1250) = Rs 4550

Total interest charged

= Total amount of the three instalments – Amount borrowed

= Rs (2160 × 3 - 4550) = Rs (6480 - 4550) = Rs 1930

Interest in 1st instalment

= 1st instalment - Principal for 1st instalment

= Rs (2160 - 1800) = Rs 360

Interest in 2nd instalment

= 2nd instalment - Principal for 2nd instalment

= Rs (2160 - 1500) = Rs 660

Interest in 3rd instalment

= 3rd instalment - Principal for 3rd instalment

= Rs (2160 - 1250) = Rs 910

Quicker Method: Applying the above theorem, we have

(i) Amount borrowed

$$= 2160\left[\frac{100}{100+20}+\left(\frac{100}{100+20}\right)^2+\left(\frac{100}{100+20}\right)^3\right]$$

$$= 2160\left[\frac{5}{6}+\frac{25}{36}+\frac{125}{216}\right]$$

= 1800 + 1500 + 1250 = Rs 4550

(ii) Total interest charged $= 2160\left[3-\left\{\frac{5}{6}+\frac{25}{36}+\frac{125}{216}\right\}\right]$

= Rs (6480 - 4550) = Rs 1930

(iii) Interests - in

1st instalment = $2160\left(1-\frac{5}{6}\right) = Rs\ 360$

2nd instalment = $2160\left[1-\frac{25}{36}\right] = Rs\ 660$

3rd instalment = $2160\left[1-\frac{125}{216}\right] = Rs\ 910$

(iv) Principal for the

1st year = $Rs\ 2160\times\frac{5}{6} = Rs\ 1800$

2nd year = $Rs\ 2160\times\frac{25}{36} = Rs\ 1500$

3rd year = $Rs\ 2160\times\frac{125}{216} = Rs\ 1250$

Type III: To find each instalment when the instalments are not equal.

Theorem: *A sum of Rs P is to be paid back in n annual instalments. If the interest is compounded annually on the balance at R% and is to be included in each instalment, then the value of each instalment is given by the following,*

Instalment at the end of 1st year = Rs $\frac{P}{n}\left[1+\frac{Rn}{100}\right]$

Instalment at the end of 2nd year = Rs $\frac{P}{n}\left[1+\frac{R(n-1)}{100}\right]$

Instalment at the end of 3rd year = Rs $\frac{P}{n}\left[1+\frac{R(n-2)}{100}\right]$

Instalment at the end of 4th year = Rs $\frac{P}{n}\left[1+\frac{R(n-3)}{100}\right]$

" " "

" " "

Instalment at the end of nth year = Rs $\frac{P}{n}\left[1+\frac{R}{100}\right]$

Note: Number of instalments = no. of years.

Now, we can alternatively write the above theorem as follows.

Value of instalment at the end of required year

$$= \frac{\text{Sum which is to be paid back}}{\text{No. of instalments}} \times \left[1+\frac{\text{(Rate per cent) (no. of instalments - one less the year after which instalment is payable)}}{100}\right]$$

Illustrative Examples

Ex.1: A sum of Rs 7500 is to be paid back in 3 annual instalments. How much is each instalment, if the interest is compounded annually on the balance at 4% and is to be included in each instalment.

Soln: The loan is to be paid in 3 annual instalments.

∴ Each instalment will be of Rs (7500 ÷ 3) or Rs 2500 together with the interest on the balance for 1 year.

Amount payable at the end of 1st year

= Rs 2500 + 4% of Rs 7500

= Rs 2500 + Rs $\frac{4}{100}\times 7500$

= Rs 2500 + Rs 300 = Rs 2800

Balance at the end of first year

= Rs 7500 – Rs 2500 = Rs 5000

∴ Amount payable at the end of 2nd year

= Rs 2500 + 4% of Rs 5000

= Rs 2500 + Rs $\frac{4}{100}\times 5000$

= Rs 2500 + Rs 200 = Rs 2700

Balance at the end of 2nd year = Rs 5000 – Rs 2500 = Rs 2500

∴ Amount payable at the end of 3rd year

$= \text{Rs } 2500 + 4\% \text{ of Rs } 2500$

$= \text{Rs } 2500 + \text{Rs } \frac{4}{100} \times 2500$

$= \text{Rs } 2500 + \text{Rs } 100 = \text{Rs } 2600$

Hence the three instalments are Rs 2800, Rs 2700 and Rs 2600

Quicker Method: Applying the above theorem, we have

Amount payable at the end of 1st year

$= \frac{7500}{3}\left[1 + \frac{4 \times (3-0)}{100}\right]$ **[see note]**

$= \text{Rs } 2800$

Amount payable at the end of 2nd year

$= \frac{7500}{3}\left[1 + \frac{4 \times (3-1)}{100}\right] = \text{Rs } 2700$

Amount payable at the end of 3rd year

$= \frac{7500}{3}\left[1 + \frac{4(3-2)}{25}\right]$

$= 2500 + 100 = \text{Rs } 2600$

Ex 2: A loan of Rs 12000 is to be paid back in 6 annual instalments. How much is each instalment if the interest is compounded annually on the balance at 5% and is included in each instalment?

Soln: **Detail Method:**

The loan is to be paid in 6 annual instalments.

∴ Each instalment will be of Rs 12000 ÷ 6 or Rs 2000 together with interest on the balance for 1 year.

Amount payable at the end of 1st year

$= \text{Rs } 2000 + 5\% \text{ of Rs } 12000$

$= \text{Rs } 2000 + \text{Rs } \frac{5}{100} \times 12000$

$= \text{Rs } 2000 + \text{Rs } 600 = \text{Rs } 2600$

Balance at the end of first year

$= \text{Rs } 12000 - \text{Rs } 2000 = \text{Rs } 10000$

Amount payable at the end of 2nd year

$= \text{Rs } 2000 + 5\% \text{ of Rs } 10000$

$= \text{Rs } 2000 + \text{Rs } \frac{5}{100} \times 10000$

$= \text{Rs } 2000 + \text{Rs } 500 = \text{Rs } 2500$

Balance at the end of 2nd year

$= \text{Rs } 10000 - \text{Rs } 2000 = \text{Rs } 8000$

Amount payable at the end of 3rd year

$= \text{Rs } 2000 + 5\% \text{ of Rs } 8000$

$= \text{Rs } 2000 + \text{Rs } \frac{5}{100} \times 8000$

$= \text{Rs } 2000 + \text{Rs } 400 = \text{Rs } 2400$

Balance at the end of 3rd year

$= \text{Rs } 8000 - \text{Rs } 2000 = \text{Rs } 6000$

Amount payable at the end of 4th year

$= \text{Rs } 2000 + 5\% \text{ of Rs } 6000$

$= \text{Rs } 2000 + \text{Rs } \frac{5}{100} \times 6000$

$= \text{Rs } 2000 + \text{Rs } 300 = \text{Rs } 2300$

Balance at the end of 4th year

$= \text{Rs } 6000 - \text{Rs } 2000 = \text{Rs } 4000$

Amount payable at the end of 5th year

$= \text{Rs } 2000 + 5\% \text{ of Rs } 4000$

$= \text{Rs } 2000 + \text{Rs } \frac{5}{100} \times 4000$

$= \text{Rs } 2000 + \text{Rs } 200 = \text{Rs } 2200$

Balance at the end of 5th year

$= \text{Rs } 4000 - \text{Rs } 2000 = \text{Rs } 2000$

Amount payable at the end of 6th year

$= \text{Rs } 2000 + 5\% \text{ of Rs } 2000$

$= \text{Rs } 2000 + \text{Rs } \frac{5}{100} \times 2000$

$= \text{Rs } 2000 + \text{Rs } 100 = \text{Rs } 2100$

Hence the six instalments are Rs 2600, Rs 2500, Rs 2400, Rs 2300, Rs 2200, Rs 2100.

Quicker Method: Applying the given rule we have,

Amount payable at the end of 1st year

$= \frac{12000}{6}\left[1 + \frac{5 \times (6-0)}{100}\right]$

$= 2000 + \frac{2000 \times 3}{10} = Rs\ 2600$

Amount payable at the end of 2nd year

$= \frac{12000}{6}\left[1 + \frac{5 \times (6-1)}{100}\right]$

$= 2000 + 500 = \text{Rs } 2500$

Similarly we can find the remaining instalments as Rs 2400, Rs 2300, Rs 2200 and Rs 2100.

Type IV: To find cash price when different instalments are given.

Theorem: ***A person buys an item on the terms that he is required to Rs P cash down payment followed by Rs x at the end of first year, Rs y at the end of second year and Rs z at the end of third year. Interest is charged at the rate of R% per annum, then the***

(i) Cash price of the item is given by

$$\text{Rs } P + \frac{100}{100+R}\left[x + y\left(\frac{100}{100+R}\right) + z\left(\frac{100}{100+R}\right)^2\right] \text{ and}$$

(ii) The total interest charged is given by

Rs [P + x + y + z - Cash Price]

Illustrative Example

Ex: Subash purchased a refrigerator on the terms that he is required to pay Rs 1500 cash down payment followed by Rs 1020 at the end of first year, Rs 1003 at the end of second year and Rs 990 at the end of third year. Interest is charged at the rate of 10% per annum. Calculate the cash price and the total interest charged.

Soln: Detail Method: Let the cash price of the refrigerator be Rs x

Cash down payment = Rs 1500

$\therefore$ Remaining amount = Rs (x - 1500)

Let P_1, P_2, P_3 be the principals of the three annual instaments. Then

$$1020 = P_1\left(1+\frac{10}{100}\right)^1 = P_1\left(\frac{11}{10}\right)$$

$$\left[\because A = P\left(1+\frac{r}{100}\right)^t\right]$$

$$\Rightarrow P_1 = Rs\ 1020\times\left(\frac{10}{11}\right)$$

Similarly $1003 = P_2\left(1+\frac{10}{100}\right)^2 = P_2\left(\frac{11}{10}\right)^2$

$$\Rightarrow P_2 = 1003\left(\frac{10}{11}\right)^2$$

and $990 = P_3\left(1+\frac{10}{100}\right)^3 = P_3\left(\frac{11}{10}\right)^3$

$$\Rightarrow P_3 = 990\left(\frac{10}{11}\right)^3$$

$$\therefore P_1+P_2+P_3 = Rs\left[1020\times\frac{10}{11}+1003\left(\frac{10}{11}\right)^2+990\left(\frac{10}{11}\right)^3\right]$$

$$= Rs\frac{10}{11}\left(1020+1003\times\frac{10}{11}+990\times\frac{100}{121}\right)$$

$$= Rs\frac{10}{11}\left(\frac{123420+110330+99000}{121}\right)$$

$$= Rs\frac{10}{11}\times\frac{332750}{121} = Rs\ 2500$$

$\therefore$ x - 1500 = 2500 $\Rightarrow$ x = 2500 + 1500 = 4000

$\therefore$ Cash price of the refrigerator = Rs 4000

Total sum paid = Rs (1500 + 1020 + 1003 + 990)

= Rs 4513

$\therefore$ Total interest charged = Rs (4513 - 4000) = Rs 513

Quicker Method: Applying the above theorem, we have

cash price of the refrigerator

$$= 1500+\frac{10}{11}\left[1020+1003\times\frac{10}{11}+990\times\frac{100}{121}\right]$$

$$= 1500+\frac{10}{11}\left[\frac{123420+110330+99000}{121}\right]$$

= 1500 + 2500 = Rs 4000

Total interest charged

= 1500 + 1020 + 1003 + 990 - 4000 = Rs 513

Exercise

1. The price of a tape recorder is Rs 1500. A customer purchased it by paying a cash sum of Rs 300 and the balance with due interest in three half yearly equal instalments. If the dealer charges interest at the rate of 5% per annum compounded half yearly, find the value of each instalment.
2. A dealer offers a refrigerator for Rs 3000 on cash payment. A customer agrees to pay Rs 1000 cash down and the balance with due interest in three equal annual instalments. If the dealer charges an interest of $12\frac{1}{2}\%$ pa compounded annually, what should be the annount of each instalment?
3. A man borrows some money on compound interest and returns it in two years in two equal instalments. If the rate of interest is 5% and yearly instalment is Rs 441; find the amount borrowed.
4. A sum of money is to be paid back in 3 annual instalments of Rs 2800, Rs 2700 and Rs 2600 payable at the end of 1st year, 2nd year and 3rd year respectively. If rate of interest be 4% pa, calculate the principal and the interest charged.
5. One can purchase a flat from a house building society for Rs 55000 cash or on the terms that he should pay Rs 4275 as cash down payment and the rest in three equal yearly instalments. The society charges interest at the rate of 165 per annum compounded half yearly. If the flat is purchased under instalment plan, find the value of each instalment.
6. A sum of Rs 5600 is paid back in yearly instalments. How much is each instalment, if the interest is compounded annually on the balance at 8% per annum and is to be included in each instalment?
7. A sum of Rs 6000 is paid back in 3 annual instalments. How much is each instalment if the interest is compounded annually on the balance at 10% per annum and is to be included in each instalment?
8. A sum of Rs 8400 is to be returned in three annual instalments. What is the annual instalment, if the rate of interest is $9\frac{1}{2}\%$ per annum compounded annually on

the balance and is to be included in each instalment?

9. The price of a tape-recorder is Rs 1561. A customer purchased it by paying a cash of Rs 300 and balance with due interest in 3 half yearly equal instalments. If the dealer charges interest at the rate of 10% per annum compounded half yearly, find the value of each instalment.
10. A loan of Rs 2550 is to be paid back in two equal half yearly instalments. How much is each instalment, if the interest is compounded half yearly at 8% per annum?
11. A sum of Rs 2600 is to be paid back in 2 equal annual instalments. What is the annual instalment, if the rate of interest is 8% per annum compounded annually?
12. A man borrows Rs 816 and agrees to return it in two equal annual instalments. What is the annual instalment, if the rate of interest is 12.5% per annum compounded annually?
13. Govind borrowed money from a money lender and agreed to pay back in 3 equal annual instalments of Rs 665.50 each. What sum did he borrow, if the rate of interest charged by the money lender was 10% per annum compounded annually?
14. A man takes loan on compound interest and returns it in two equal annual instalments. If the rate of interest is 16% per annum and the yearly instalment is Rs 1682, find the principal and the interest charged with each instalment.
15. A man borrowed some money and paid back in 3 equal annual instalments of Rs 2160 each. What sum did he borrow if the rate of interest charged by the money lender was 20% per annum compounded annually? Find also the total interest charged.
16. Kusum borrowed money and returned it in 3 equal quarterly instalments of Rs 1630.50 each. What sum did she borrow if the rate of interest charged by the money lender was 20% per annum compounded quarterly? Find also the total interest charged.
17. Naresh took loan from a bank and the payment was made in 3 annual instalments of Rs 2600, Rs 2400 and Rs 2200 payable at the end of first, second and third year respectively. Interest was charged at 10% per annum. Calculate the amount of loan taken and the interest paid by him.
18. A person borrows Rs 5407.50 and agrees to pay the loan back with compound interest at the rate of $13\frac{1}{2}\%$ per annum in 3 equal half yearly instalments. Find the amount of each instalment, if the interest is compounded half yearly.
19. Sanjay bought a gas stove on instalment basis. He has to pay Rs 500 cash down payment and Rs 810 at the end of first year, Rs 520 at the end of second year and Rs 460 at the end of third year. Interest is charged at the rate of 15% per annum. Calculate the total cash price of the gas stove.
20. A dealer advertises that a cassette recorder is sold at Rs 450 cash down followed by two yearly instalments of Rs 680 and Rs 590 at the end of first year and second year respectively. If the interest charged is 18% per annum compounded annually, find the cash price of the cassette recorder.
21. A colour TV set purchased under instalment purchase system. Cash down payment is Rs 2000 and 3 annual instalments of Rs 1800, Rs 1560 and Rs 1430 are payable at the end of first year, second year and third year respectively. If the rate of interest is 10% per annum respectively, find the cash price of the TV set and the total interest charged under instalment plan.
22. A sewing machine is available at Rs 240 cash down payment followed by 3 annual instalments of Rs 380, Rs 240 and Rs 200 payable at the end of first year, second year and third year respectively. If the rate of interest is 25% per annum compound interest, find the cash price and total interest being charged under the instalment plan.

Answers

1. Price of the tape recorder = Rs 1500
 Paid cash = Rs 300
 Balance to be paid = Rs (1500 – 300) = Rs 1200
 Let each half-yearly instalment = Rs x
 r = 5% pa = (5/2)% half-yearly, Amount = Rs 1200
 ∴ Using the formula

$$A = P\left(1+\frac{r}{100}\right)^n$$

$$x = P\left(1+\frac{5/2}{100}\right)^1 = P\left(\frac{205}{200}\right) = P\left(\frac{41}{40}\right)$$

∴ Principal (P) included in the first instalment

$$= x \times \frac{40}{41} = \frac{40}{41}x$$

Similarly, Principal included in the second instalment

$$= \left(\frac{40}{41}\right)^2 x$$

Principal included in the third instalment $= \left(\frac{40}{41}\right)^3 x$

$$\therefore \frac{40}{41}x + \left(\frac{40}{41}\right)^2 x + \left(\frac{40}{41}\right)^3 x = 1200$$

$$\Rightarrow \frac{40}{41}x\left[1+\frac{40}{41}+\left(\frac{40}{41}\right)^2\right] = 1200$$

$$\Rightarrow \frac{40}{41}x\left(1+\frac{40}{41}+\frac{1600}{1681}\right)=1200$$

$$\Rightarrow \frac{40}{41}x\left(\frac{1681+1640+1600}{1681}\right)=1200$$

$$\Rightarrow \frac{4921}{1681}x=1200\times\frac{41}{40}=1230$$

$$\Rightarrow x=1230\times\frac{1681}{4921}=\text{Rs } 20.20$$

2. Cash price = Rs 3000
Cash down payment = Rs 1000
Balance = Rs (3000 - 1000) = Rs 2000
Let the amount of each instalment = Rs x

Rate = $12\frac{1}{2}$ % pa = (25/2)% pa

Principal included in the first instalment is given by

$$x=P\left(1+\frac{25/2}{100}\right)$$

$$A=P\left(1+\frac{r}{100}\right)^t=P\left(\frac{225}{200}\right)=P\left(\frac{9}{8}\right)$$

$$\Rightarrow P=x\times\frac{8}{9}$$

Similarly, principal included in the 2nd instalment =

$$x\times\left(\frac{8}{9}\right)^2$$

Principal included in the 3rd instalment = $x\times\left(\frac{8}{9}\right)^3$

$$\therefore\ x\times\frac{8}{9}+x\left(\frac{8}{9}\right)^2+x\left(\frac{8}{9}\right)^3=2000$$

$$\Rightarrow \frac{8}{9}x\left(1+\frac{8}{9}+\frac{64}{81}\right)=2000$$

$$\Rightarrow \frac{8}{9}x\left(\frac{81+72+64}{81}\right)=2000$$

$$\Rightarrow \frac{8}{9}\times\frac{217x}{81}=2000$$

$$\Rightarrow x=\frac{2000\times81\times9}{8\times217}=\text{Rs } 839.86$$

$\therefore$ Amount of each instalment = Rs 839.86

3. The man paid Rs 441 as the amount at the end of first year and another Rs 441, as the amount at the end of second year.

$\therefore$ Principal of the first year

$$=441\div\left(1+\frac{5}{100}\right)=441\div\frac{105}{100}$$

$$=441\div\frac{21}{20}=441\times\frac{20}{21}=\text{Rs } 420$$

Principal for the second year = $441\div\left(1+\frac{5}{100}\right)^2$

$$=441\times\frac{20}{21}\times\frac{20}{21}=\text{Rs } 400$$

$\therefore$ Total principal = Rs 420 + Rs 400 = Rs 820

4. Let the principals for the three annual instalments be P_1, P_2, P_3. Then

$$2800=P_1\left(1+\frac{4}{100}\right)^1 \Rightarrow P_1=\text{Rs } 2800\times\frac{25}{26}$$

$$\left[\because A=P\left(1+\frac{r}{100}\right)^t\right]$$

$$2700=P_2\left(1+\frac{4}{100}\right)^2 \Rightarrow P_2=\text{Rs } 2700\times\left(\frac{25}{26}\right)^2$$

$$2600=P_3\left(1+\frac{4}{100}\right)^3 \Rightarrow P_3=\text{Rs } 2600\times\left(\frac{25}{26}\right)^3$$

Total sum borrowed = $P_1+P_2+P_3$

$$=\text{Rs}\left[2800\left(\frac{25}{26}\right)+2700\left(\frac{25}{26}\right)^2+2600\left(\frac{25}{26}\right)^3\right]$$

$$=\text{Rs } \frac{25}{26}\times100\left[28+27\times\frac{25}{26}+26\times\frac{25}{26}\times\frac{25}{26}\right]$$

$$=\text{Rs } \frac{2500}{26}\left(28+\frac{675}{26}+\frac{625}{26}\right)$$

$$=\text{Rs } \frac{2500}{26}\left(\frac{728+675+625}{26}\right)=\text{Rs } \frac{2500}{26}\times\frac{2028}{26}$$

= Rs 2500 × 3 = Rs 7500

Total money paid = Rs (2800 + 2700 + 2600) = Rs 8100

$\therefore$ Total interest charged = Rs 8100 – Rs 7500 = Rs 600

5. Cash price of the flat = Rs 55000
In the instalment plan, cash down payment = Rs 4275
$\therefore$ Present value of the price to be paid in instalments = Rs 55000 – Rs 4275 = Rs 50725

Let each instalment be Rs x

Rate = 16% pa = 8% half-yearly

$\therefore$ Principal for the 1st instalment at the end of 1st

$$\text{half yearly} = \text{Rs } \frac{x}{1+\frac{8}{100}}$$

$$\left[\because A = P\left(1+\frac{r}{100}\right)^t \Rightarrow x = P\left(1+\frac{8}{100}\right)^1\right]$$

$$= \text{Rs } \frac{100}{108}x = \text{Rs}\left(\frac{25}{27}\right)x$$

Similarly, Principal for the 2nd instalment

$$= \text{Rs}\left(\frac{25}{27}\right)^2 x$$

$$\text{Principal for the 3rd instalment} = \text{Rs}\left(\frac{25}{27}\right)^3 x$$

Total principal for the three instalments

$$= \text{Rs}\left[\frac{25}{27}x + \left(\frac{25}{27}\right)^2 x + \left(\frac{25}{27}\right)^3 x\right]$$

It should be equal to Rs 50725

$$\therefore \frac{25}{27}x\left[1+\frac{25}{27}+\left(\frac{25}{27}\right)^2\right] = 50725$$

$$\Rightarrow \frac{25}{27}x\left(1+\frac{25}{27}+\frac{625}{729}\right) = 50725$$

$$\Rightarrow \frac{25}{27}x\left(\frac{729+675+625}{729}\right) = 50725$$

$$\Rightarrow \frac{25}{27}x \times \frac{2029}{729} = 50725$$

$$\Rightarrow x = \frac{50725 \times 27 \times 729}{25 \times 2029} = 19683$$

$\therefore$ Each instalment = Rs 19683

6. The sum is to be paid back in 4 annual instalments.

$\therefore$ Each instalment will be of Rs (5600 ÷ 4)

ie Rs 1400 together with interest on the balance for one year

$\therefore$ Amount payable at the end of 1st year

$$= \text{Rs } 1400 + 8\% \text{ of Rs } 5600$$

$$= \text{Rs } 1400 + \text{Rs}\left(5600 \times \frac{8}{100}\right)$$

$$= \text{Rs } 1400 + \text{Rs } 448 = \text{Rs } 1848$$

Balance at the end of 1st year

$$= \text{Rs } (5600 - 1400) = \text{Rs } 4200$$

Amount payable at the end of 2nd year

$$= \text{Rs } 1400 + 8\% \text{ of Rs } 4200$$

$$= \text{Rs } 1400 + \text{Rs}\left(\frac{8}{100} \times 4200\right)$$

$$= \text{Rs } 1400 + \text{Rs } 336 = \text{Rs } 1736$$

Balance at the end of 2nd year

$$= \text{Rs } (4200 - 1400) = \text{Rs } 2800$$

Amount payable at the end of 3rd year

$$= \text{Rs } 1400 + 8\% \text{ of Rs } 2800$$

$$= \text{Rs } 1400 + \text{Rs}\left(\frac{8}{100} \times 2800\right)$$

$$= \text{Rs } 1400 + \text{Rs } 224 = \text{Rs } 1624$$

Balance at the end of 3rd year

$$= \text{Rs } (2800 - 1400) = \text{Rs } 1400$$

$\therefore$ Amount payable at the end of 4th year

$$= \text{Rs } 1400 + 8\% \text{ of Rs } 1400$$

$$= \text{Rs } 1400 + \text{Rs}\left(\frac{8}{100} \times 1400\right)$$

$$= \text{Rs } 1400 + \text{Rs } 112 = \text{Rs } 1512$$

Hence the four instalments are

Rs 1848, Rs 1736, Rs 1624 and Rs 1512

7. The sum is to be paid back in 3 annual instalments

$\therefore$ Each instalment will be of Rs (6000 ÷ 3), ie Rs 2000 together with interest on the balance for one year

$\therefore$ Amount payable at the end of 1st year

$$= \text{Rs } 2000 + 10\% \text{ of Rs } 6000$$

$$= \text{Rs } 2000 + \text{Rs}\left(\frac{10}{100} \times 6000\right)$$

$$= \text{Rs } 2000 + \text{Rs } 600 = \text{Rs } 2600$$

Balance at the end of 1st year

$$= \text{Rs } (6000 - 2000) = \text{Rs } 4000$$

$\therefore$ Amount payable at the end of 2nd year

$$= \text{Rs } 2000 + 10\% \text{ of Rs } 4000$$

$$= \text{Rs } 2000 + \text{Rs}\left(\frac{10}{100} \times 4000\right)$$

$$= \text{Rs } 2000 + \text{Rs } 400 = \text{Rs } 2400$$

Balance at the end of 2nd year

$$= \text{Rs } 4000 - \text{Rs } 2000 = \text{Rs } 2000$$

$\therefore$ Amount payable at the end of 3rd year

$$= \text{Rs } 2000 + 10\% \text{ of Rs } 2000$$

$$= \text{Rs } 2000 + \text{Rs}\left(\frac{10}{100} \times 2000\right)$$

$$= \text{Rs } 2000 + \text{Rs } 200 = \text{Rs } 2200$$

Hence the three instalments are: Rs 2600, Rs 2400 and Rs 2200.

8. The sum is to be returned in 3 annual instalments.

$\therefore$ Each instalment will be of Rs (8400 ÷ 3), ie Rs 2800 together with interest on the balance for one year.

$\therefore$ Amount payable at the end of 1st year

$$= \text{Rs } 2800 + \frac{19}{2}\% \text{ of Rs } 8400$$

$$= \text{Rs } 2800 + \text{Rs}\left(\frac{19}{2} \times \frac{1}{100} \times 8400\right)$$

$$= \text{Rs } 2800 + \text{Rs } 798 = \text{Rs } 3598$$

Balance at the end of 1st year

$$= \text{Rs } 8400 - \text{Rs } 2800 = \text{Rs } 5600$$

$\therefore$ Amount payable at the end of 2nd year

$$= \text{Rs } 2800 + \frac{19}{2}\% \text{ of Rs } 5600$$

$$= \text{Rs } 2800 + \text{Rs}\left(\frac{19}{200} \times 5600\right)$$

$$= \text{Rs } 2800 + \text{Rs } 532 = \text{Rs } 3332$$

Balance at the end of 2nd year

$$= \text{Rs } 5600 - \text{Rs } 2800 = \text{Rs } 2800$$

$\therefore$ Amount payable at the end of 3rd year

$$= \text{Rs } 2800 + \frac{19}{2}\% \text{ of Rs } 2800$$

$$= \text{Rs } 2800 + \text{Rs}\left(\frac{19}{200} \times 2800\right)$$

$$= \text{Rs } 2800 + \text{Rs } 266 = \text{Rs } 3066$$

Hence, the three instalments are: Rs 3598, Rs 3332 and Rs 3066

9. Cash price of the tape recorder = Rs 1561

Cash down payment = Rs 300

$\therefore$ Price to be paid in instalment has its present value

$$= \text{Rs } 1561 - \text{Rs } 300 = \text{Rs } 1261$$

Let each instalment be Rs x

Rate = 10% pa = 5% half yearly

$\therefore$ Principal (P) included in the 1st instalment

$$= \text{Rs } \frac{x}{\left(1 + \frac{5}{100}\right)^1}$$

$$\left[\because A = P\left(1 + \frac{r}{100}\right)^t x = P\left(1 + \frac{5}{100}\right)^1\right]$$

$$= \text{Rs}\left(x \div \frac{105}{100}\right) = \text{Rs}\left(x \times \frac{100}{105}\right)$$

$$= \text{Rs}\left(\frac{20}{21}\right)x$$

Similarly, principals for the next two instalments are

$$P_2 = \text{Rs}\left(\frac{20}{21}\right)^2 x \text{ and } P_3 = \text{Rs}\left(\frac{20}{21}\right)^3 x$$

$$P_1 + P_3 + P_3 = 1261$$

$$\text{ie, } \left(\frac{20}{21}\right)x + \left(\frac{20}{21}\right)^2 x + \left(\frac{20}{21}\right)^3 x = 1261$$

$$\Rightarrow \left(\frac{20}{21}\right)x\left[1 + \frac{20}{21} + \left(\frac{20}{21}\right)^2\right] = 1261$$

$$\Rightarrow \frac{20}{21}x\left(1 + \frac{20}{21} + \frac{400}{441}\right) = 1261$$

$$\Rightarrow \frac{20}{21}x\left(\frac{441 + 420 + 400}{441}\right) = 1261$$

$$\Rightarrow \frac{20}{21}x \times \frac{1261}{441} = 1261 \Rightarrow x = \frac{1261 \times 21 \times 441}{20 \times 1261}$$

$$\Rightarrow x = \frac{9261}{20} = 463.05$$

$\therefore$ Each instalment = Rs 463.05

10. Let each instalment be Rs x

Rate = 8% pa = 4% half yearly

$\therefore$ Principal (P_1) included in the 1st instalment

$$= \text{Rs } \frac{x}{\left(1 + \frac{4}{100}\right)^1} = \text{Rs}\left(x \div \frac{26}{25}\right) = \text{Rs}\left(\frac{25}{26}\right)x$$

Similarly, principal (P_2) for the next instalment

$$= \text{Rs}\left(\frac{25}{26}\right)^2 x$$

$$P_1 + P_2 = 2550$$

$$\text{ie } \left(\frac{25}{26}\right)x + \left(\frac{25}{26}\right)^2 x = 2550$$

$$\Rightarrow \left(\frac{25}{26}\right)x\left(1 + \frac{25}{26}\right) = 2550$$

$$\Rightarrow \left(\frac{25}{26}\right)x\left(\frac{26 + 25}{26}\right) = 2550$$

$$\Rightarrow \frac{25}{26} \times x \times \frac{51}{26} = 2550$$

$\Rightarrow x = 2550 \times \frac{26}{25} \times \frac{26}{51} = 1352$

$\therefore$ Each instalment = Rs 1352

11. Let each instalment be Rs x
Rate = 8% pa
$\therefore$ Principal (P_1) included in the 1st instalment

$$= \text{Rs} \frac{x}{\left(1+\frac{8}{100}\right)} = \text{Rs}\left(x \div \frac{108}{100}\right)$$

$$= \text{Rs}\left(x \times \frac{100}{108}\right) = \text{Rs}\left(\frac{25}{27}\right)x$$

Similarly, principal (P_2) included in the 2nd instalment

$$= \text{Rs}\left(\frac{25}{27}\right)^2 x$$

$\therefore P_1 + P_2 = 2600$

ie $\left(\frac{25}{27}\right)x + \left(\frac{25}{27}\right)^2 x = 2600$

On solving this equation, you will obtain x = 1458
$\therefore$ Each annual instalment = Rs 1458

12. Let each annual instalment be Rs x
Rate = 12.5% pa
$\therefore$ Principal (P_1) included in the 1st instalment

$$= \text{Rs} \frac{x}{\left(1+\frac{12.5}{100}\right)^1} = \text{Rs}\left(x \div \frac{112.5}{100}\right)^1$$

$$= \text{Rs}\left(x \times \frac{100}{112.5}\right)^1 = \text{Rs}\left(x \times \frac{1000}{1125}\right)^1 = \text{Rs}\left(\frac{8}{9}\right)^1 x$$

Similarly principal (P_2) included in the 2nd instalment = Rs $\left(\frac{8}{9}\right)^2 x$.

$\therefore P_1 + P_2 = 816$

ie $\left(\frac{8}{9}\right)x + \left(\frac{8}{9}\right)^2 x = 816$

On solving this equations you will get x = 486
$\therefore$ Each annual instalment = Rs 486

13. Govind paid Rs 665.50 as the amount at the end of 1st year, 2nd year and 3rd year.
$\therefore$ Principal (P_1) for the first year

$$= 665.50 \div \left(1+\frac{10}{100}\right)^1$$

$$\left[\because A = P\left(1+\frac{r}{100}\right)^t \Rightarrow 665.50 = P\left(1+\frac{10}{100}\right)^1\right]$$

$$= 665.50 \div \left(\frac{11}{10}\right)^1 = 665.50 \times \frac{10}{11} = \text{Rs } 605$$

Principal (P_2) for the second year

$$= 665.50 \div \left(1+\frac{10}{100}\right)^2 = 665.50 \div \left(\frac{11}{10}\right)^2$$

$$= 665.50 \times \frac{10}{11} \times \frac{10}{11} = \text{Rs } 550$$

Principal (P_3) for the third year

$$= 665.50 \div \left(1+\frac{10}{100}\right)^3 = 665.50 \times \left(\frac{11}{10}\right)^3$$

$$= 665.50 \times \frac{10}{11} \times \frac{10}{11} \times \frac{10}{11} = \text{Rs } 500$$

$\therefore$ Total principal = $P_1 + P_2 + P_3$
= Rs (605 + 550 + 500) = Rs 1655

14. The yearly instalment paid at the end of 1st year and 2nd year = Rs 1682
$\therefore$ Principal (P_1) for the first year

$$= 1682 \div \left(1+\frac{16}{100}\right)^1$$

$$\left[\because A = P\left(1+\frac{r}{100}\right)^t \Rightarrow 1682 = P_1\left(1+\frac{16}{100}\right)^1\right]$$

$$= \text{Rs}\left(1682 \div \frac{116}{100}\right)$$

$$= \text{Rs}\left(1682 \times \frac{100}{116}\right) = \text{Rs } 1682\left(\frac{25}{29}\right)$$

= Rs 1450

Principal (P_2) for the 2nd year

$$= 1682 \div \left(1+\frac{16}{100}\right)^2 = \text{Rs } 1682 \times \left(\frac{25}{29}\right)^2$$

$$= \text{Rs } 1682 \times \frac{25}{29} \times \frac{25}{29} = \text{Rs } 1250$$

$\therefore$ Total principal = $P_1 + P_2$

$= \text{Rs } 1450 + \text{Rs } 1250 = \text{Rs } 2700$

Total amount paid = Rs (1682 × 2) = Rs 3364

$\therefore$ Total interest = Rs 3364 – Rs 2700 = Rs 664

Interest charged with first instalment

$$= \text{Rs } 2700 \times \frac{16}{100} = \text{Rs } 432$$

Interest charged with second instalment

$= \text{Rs } 664 - \text{Rs } 432 = \text{Rs } 232$

15. See the solution of **Q. No. 12.**

16. The quarterly instalment paid at the end of 1st, 2nd and 3rd quarter = Rs 4630.50

$\therefore$ Principal (P_1) for the first quarter

$$= 4630.50 \div \left(1 + \frac{5}{100}\right)^1$$

$$= \text{Rs}\left(4630.50 \div \frac{105}{100}\right)$$

$$= \text{Rs}\left(4630.50 \times \frac{100}{105}\right)$$

$$= \text{Rs}\left(4630.50 \times \frac{20}{21}\right) = 4410$$

Similarly, principal (P_2) for the 2nd quarter

$$= \text{Rs } 4630.50\left(\frac{20}{21}\right)^2$$

$$= \text{Rs } 4630.50 \times \frac{20}{21} \times \frac{20}{21} = \text{Rs } 4200$$

Principal (P_3) for the 3rd quarter

$$= \text{Rs } 4630.50\left(\frac{20}{21}\right)^3$$

$$= \text{Rs } 4630.50 \times \frac{20}{21} \times \frac{20}{21} \times \frac{20}{21} = \text{Rs } 4000$$

Total principal = $P_1 + P_2 + P_3$

$= \text{Rs } (4410 + 4200 + 4000) = \text{Rs } 12610$

Total amount paid = Rs (4630.50 × 3) = Rs 13891.50

$\therefore$ Total interest = Rs 13891.50 – Rs 12610

= Rs 1281.50

17. Instalment paid at the end of 1st year = Rs 2600

$\therefore$ Principal of 1st instalment

$$= \text{Rs } 2600 \div \left(1 + \frac{10}{100}\right)^1$$

$$= \text{Rs}\left(2600 \div \frac{110}{100}\right)$$

$$= \text{Rs}\left(2600 \times \frac{100}{110}\right)$$

$$= \text{Rs}\left(2600 \times \frac{10}{11}\right) = \text{Rs } \frac{26000}{11}$$

2nd instalment paid at the end of 2nd year = Rs 2400

$$\therefore \text{ Principal of 2nd instalment} = \text{Rs } 2400 \times \left(\frac{10}{11}\right)^2$$

$$= \text{Rs } 2400 \times \frac{10}{11} \times \frac{10}{11} = \text{Rs } \frac{240000}{121}$$

3rd instalment paid at the end of 3rd year = Rs 2200

$\therefore$ Principal of 3rd instalment

$$= \text{Rs } 2200 \times \left(\frac{10}{11}\right)^3$$

$$= \text{Rs } 2200 \times \frac{10}{11} \times \frac{10}{11} \times \frac{10}{11}$$

$$= \text{Rs } \frac{200000}{121}$$

$$\therefore \text{ Total principal} = \text{Rs}\left(\frac{26000}{11} + \frac{240000}{121} + \frac{200000}{121}\right)$$

$$= \text{Rs } \frac{286000 + 240000 + 200000}{121}$$

$$= \text{Rs } \frac{7260000}{121} = \text{Rs } 6000$$

$\therefore$ The amount of loan taken = Rs 6000

Total amount paid = Rs (2600 + 2400 + 2200) = Rs 7200

Total interest paid = Rs (7200 – 6000) = Rs 1200

18. Let each instalment be of Rs x

$$\text{Rate} = 13\frac{1}{3}\% \text{ pa} = \frac{40}{3}\% \text{ pa}$$

$$= \frac{40}{6}\% = \frac{20}{3}\% \text{ half-yearly}$$

$\therefore$ Principal (P_1) for amount x at the end of 1st six

$$\text{months} = \text{Rs } x \div \left(1 + \frac{20/3}{100}\right)^1$$

$$\left[\because P = A \div \left(1 + \frac{r}{100}\right)^t\right]$$

$$= \text{Rs}\left(x \div \frac{320}{300}\right) = \text{Rs } x\left(\frac{300}{320}\right) = \text{Rs } x\left(\frac{15}{16}\right)$$

Similarly, the principal for amount x at the end of second six months = Rs $x\left(\frac{15}{16}\right)^2$.

Principal (P_3) for amount x at the end of third six months = Rs $x\left(\frac{15}{16}\right)^3$

$$P_1 + P_2 + P_3 = 5407.50$$

ie $x\left(\frac{15}{16}\right) + x\left(\frac{15}{16}\right)^2 + x\left(\frac{15}{16}\right)^3 = 5407.50$

$$\Rightarrow x\left(\frac{15}{16}\right)\left[1 + \frac{15}{16} + \left(\frac{15}{16}\right)^2\right] = 5407.50$$

$$\Rightarrow x\left(\frac{15}{16}\right)\left[\frac{256 + 240 + 225}{256}\right] = 5407.50$$

$$\Rightarrow x\left(\frac{15}{16}\right)\left(\frac{721}{256}\right) = 5407.50$$

$$\therefore x = 5407.50 \times \frac{16}{15} \times \frac{256}{721} = 2048$$

$\therefore$ Each instalment = Rs 2048

19. Let the cash price of the gas stove be Rs x
Cash down payment = Rs 500
$\therefore$ Remaining amount = Rs (x – 500)
1st instalment paid at the end of first year = Rs 810

$\therefore$ Principal of 1st instalment = Rs $810 \div \left(1 + \frac{15}{100}\right)^1$

$$= \text{Rs}\left(810 \div \frac{115}{100}\right)$$

$$\left[\because A = P\left(1 + \frac{r}{100}\right)^t \quad \therefore P = A \div \left(1 + \frac{r}{100}\right)^t\right]$$

$$= \text{Rs}\left(810 \times \frac{100}{115}\right) = \text{Rs } 810\left(\frac{20}{23}\right) = \text{Rs } \frac{16200}{23}$$

2nd instalment = Rs 520

$\therefore$ Principal of 2nd instalment = Rs $520\left(\frac{20}{23}\right)^2$

$$= \text{Rs } 520 \times \frac{20}{23} \times \frac{20}{23} = \text{Rs } \frac{208000}{529}$$

3rd instalment = Rs 460
$\therefore$ Principal of 3rd instalment

$$= \text{Rs } 460\left(\frac{20}{23}\right)^3 = \text{Rs } 460 \times \frac{20}{23} \times \frac{20}{23} \times \frac{20}{23}$$

$$= \text{Rs } \frac{160000}{529}$$

$\therefore$ Total principal = Rs $\left(\frac{16200}{23} + \frac{208000}{529} + \frac{160000}{529}\right)$

$$= \text{Rs } \frac{372600 + 208000 + 160000}{529} = \text{Rs } \frac{740600}{529}$$

$$\therefore x - 500 = \frac{740600}{529} \Rightarrow x = 500 + \frac{740600}{529}$$

$$\Rightarrow x = \frac{264500 + 740600}{529} = \frac{1005100}{529} = 1900$$

$\therefore$ Cash price of the gas stove = Rs 1900

20. Let the cash price of the cassette recorder be Rs x
Cash down payment = Rs 450
$\therefore$ Remaining amount = Rs (x – 450)
Rate = 18% pa
1st instalment paid at the end of first year = Rs 680

$\therefore$ Principal of first instalment = Rs $680 \div \left(1 + \frac{18}{100}\right)^1$

$$\left[\because A = P\left(1 + \frac{r}{100}\right)^t \quad \therefore P = A \div \left(1 + \frac{r}{100}\right)^t\right]$$

$$= \text{Rs } 680 \div \frac{118}{100} = \text{Rs } 680\left(\frac{100}{118}\right)$$

$$= \text{Rs } 680\left(\frac{50}{59}\right) = \text{Rs } \frac{34000}{59}$$

2nd instalment at the end of 2nd year = Rs 590

$\therefore$ Principal of second instalment = Rs $590\left(\frac{50}{59}\right)^2$

$$= \text{Rs } 590 \times \frac{50}{59} \times \frac{50}{59} = \text{Rs } \frac{25000}{59}$$

Total principal = Rs $\left(\frac{34000}{59} + \frac{25000}{59}\right)$

$$= \text{Rs } \frac{59000}{59} = \text{Rs } 1000$$

$$\therefore x - 450 = 1000 \Rightarrow x = 1000 + 450 = 1450$$

Hence, the cash price of the casette recorder is Rs 1450

21. Let the cash price of the colour TV be Rs x

Cash down payment = Rs 2000

$\therefore$ Remaining amount = Rs (x – 2000)

Rate = 10% pa

1st instalment paid at the end of first year = Rs 1800

$\therefore$ Principal of first instalment = Rs $1800 \div \left(1+\frac{10}{100}\right)^1$

$$\left[\because A = P\left(1+\frac{r}{100}\right)^t \quad \therefore P = A \div \left(1+\frac{r}{100}\right)^t\right]$$

$$= \text{Rs}\left(1800 \div \frac{110}{100}\right)$$

$$= \text{Rs}\left(1800 \times \frac{100}{110}\right) = \text{Rs } 1800\left(\frac{10}{11}\right)$$

$$= \text{Rs } \frac{18000}{11}$$

2nd instalment paid at the end of 2nd year = Rs 1560

$\therefore$ Principal of 2nd instalment = Rs $1560\left(\frac{10}{11}\right)^2$

$$= \text{Rs } 1560 \times \frac{10}{11} \times \frac{10}{11} = \text{Rs } \frac{156000}{121}$$

3rd instalment paid at the end of third year = Rs 1430

$\therefore$ Principal of 3rd instalment = Rs $1430\left(\frac{10}{11}\right)^3$

$$= \text{Rs } 1430 \times \frac{10}{11} \times \frac{10}{11} \times \frac{10}{11} = \text{Rs } \frac{130000}{121}$$

$\therefore$ Total principal of the three instalments

$$= \text{Rs}\left(\frac{18000}{11} + \frac{156000}{21} + \frac{130000}{121}\right)$$

$$= \text{Rs } \frac{198000+156000+130000}{121}$$

$$= \text{Rs } \frac{484000}{121} = \text{Rs } 4000$$

$x - 2000 = 4000 \Rightarrow x = 2000 + 4000 = 6000$

$\therefore$ The cash price of the colour TV = Rs 6000

Amount paid in instalment plan

= Rs(2000 + 1800 + 1560 + 1430) = Rs 6790

$\therefore$ Total interest paid = Rs 6790 – Rs 6000 = Rs 790.

22. Let the cash price of the sewing machine be Rs x

Cash down payment = Rs 240

$\therefore$ Remaining amount = Rs $(x - 240)$

1st instalment at the end of first year = Rs 380

$\therefore$ Principal of 1st instalment = $380 \div \left(1+\frac{25}{100}\right)^1$

$$\left[\because A = P\left(1+\frac{r}{100}\right)^t \quad \therefore P = A \div \left(1+\frac{r}{100}\right)^t\right]$$

$$= \text{Rs}\left(380 \div \frac{125}{100}\right)$$

$$= \text{Rs } 380\left(\frac{100}{125}\right) = \text{Rs } 380\left(\frac{4}{5}\right) = \text{Rs } 304$$

2nd instalment at the end of 2nd year = Rs 240

$\therefore$ Principal of 2nd instalment

$$= \text{Rs } 240\left(\frac{4}{5}\right)^2 = \text{Rs } 240 \times \frac{16}{25} = \text{Rs } \frac{768}{5}$$

3rd instalment at the end of 3rd year = Rs 200

$\therefore$ Principal of 3rd instalment

$$= \text{Rs } 200\left(\frac{4}{5}\right)^3 = \text{Rs } 200 \times \frac{4}{5} \times \frac{4}{5} \times \frac{4}{5}$$

$$= \text{Rs } \frac{512}{4}$$

The total principal of the three instalments

$$= \text{Rs}\left(304 + \frac{768}{5} + \frac{512}{5}\right)$$

$$= \text{Rs}\left(\frac{1520+768+512}{5}\right)$$

$$= \text{Rs } \frac{2800}{5} = \text{Rs } 560$$

$\therefore x - 240 = 560 \Rightarrow x = 560 + 240 = 800$

$\therefore$ Cash price of the sewing machine = Rs 800

Total amount paid = Rs(240 + 380 + 240 + 200)

= Rs 1060

$\therefore$ Total interest paid = Rs 1060 – Rs 800 = Rs 260

Case 3: Instalments to pay off debt

Suppose borrower is to pay a sum of Rs P due after T years, but he wants to pay through equal instalments at an agreed interval (may be quarterly, monthly, half-yearly or yearly) to discharge off his debt in T years. Hence, through equal instalment payments, the debt of Rs P due in T years is cleared in T years.

Then,

$$A = nx + \frac{xr}{100m}\left[\frac{n(n-1)}{2}\right];$$

Where A = amount due after T years
n = number of instalments to be actually paid to discharge the debt,
r = rate per cent per annum charged on simple interest
m = number of instalments per year,
= 1, if yearly instalment is paid.
= 4, if quarterly instalment is paid.
= 2, if half yearly instalment is paid,
= 12, if monthly instalment is paid,
x = amount of each instalment in Rs.
Note: Also see **Type 3** of **Case I**.

Illustrative Example

Ex: A man borrows a sum of money at 16% per annum simple interest, promising to pay Rs 8480 after a year from the date of borrowing. If he wants to discharge the debt by paying four equal quarterly instalments, how much should be each instalment.

Soln: Applying the above formula, we have

$$8480 = 4x + \frac{x \times 16}{100 \times 4} \times \frac{4 \times 3}{2} \left[\because m = 4 \left(quarterly\right)\right]$$

$$\Rightarrow \frac{106}{25} x = 8480 \Rightarrow x = Rs\ 2000.$$

Exercise

1. A sum of Rs 2 is lent to be paid back in 3 equal monthly instalments of Re 1 each. Find the rate per cent.
   a) 400% b) 140% c) 340% d) 40%
2. A money lender lends out Rs 9 on the condition that the loan is payable in 10 months in 10 equal instalments of Re 1 each. Find the rate per cent per annum.
   a) $26\frac{2}{3}\%$ b) $22\frac{2}{3}\%$ c) $16\frac{2}{3}\%$ d) $33\frac{1}{3}\%$
3. A sum of Rs 10 is lent out to be returned in 11 monthly instalments of Re 1 each, interest being simple. The rate of interest is ________.
   a) 10% b) 11% c) $9\frac{1}{11}\%$ d) $21\frac{9}{11}\%$

Answers

1. a; **Hint:** Applying the given rule, we have
A = Rs 2, x = Re 1, n = 3, m = 12

$$\text{or, } 2 = 3 \times 1 + \frac{r \times 1}{100 \times 12} \times \frac{3 \times 2}{2}$$

or, r = 400% ∴ the rate per cent per annum = 400%

2. a; **Hint:** Applying the given rule, we have,

$$9 = 10 \times 1 + \frac{1 \times r}{100 \times 12} \times \frac{10 \times (10-1)}{2}$$

$$\text{or, } \frac{15r}{400} = -1 \text{ [we omit the -ve sign]}$$

$$\therefore r = \frac{400}{15} = \frac{80}{3} = 26\frac{2}{3}\%$$

3. d; **Hint:** $10 = 11 \times 1 + \frac{1 \times r}{100 \times 12} \times \frac{11 \times (11-1)}{2}$

$$\text{or, } r = \frac{240}{11} = 21\frac{9}{11}\%.$$

12

Alligation

Rule 1

Theorem: ***The proportion in which rice at Rs x per kg must be mixed with rice at Rs y per kg, so that the mixture be worth Rs z a kg, is given by*** $\left(\frac{y-z}{z-x}\right)$.

Illustrative Example

Ex.: In what proportion must rice at Rs 3.10 per kg be mixed with rice at Rs 3.60 per kg, so that the mixture be worth Rs 3.25 a kg?

Soln: **Detail Method:** Let the required ratio be x : y.

As per the question,

$310x + 360y = 325(x+y)$

or, $310x + 360y = 325x + 325y$

or, $325x - 310x = 360y - 325y$

or, $15x = 35y \therefore \frac{x}{y} = \frac{35}{15} = 7:3$.

Alligation Method:

CP of 1 kg cheaper rice (310 paise)   CP of 1 kg dearer rice (360 paise)

Mean Price (325 paise)

35   15

$$\frac{\text{Quantity of cheaper}}{\text{Quantity of dearer}} = \frac{\text{CP of dearer} - \text{Mean Price}}{\text{Mean Price} - \text{CP of cheaper}}$$

$$= \frac{360-325}{325-310} = \frac{35}{15} = 7:3.$$

$\therefore$ they must be mixed in the ratio of 7 : 3.

Note: This result can be obtained directly by applying the above theorem.

Exercise

1. In what proportion must wheat at Rs 3.20 per kg be mixed with wheat at Rs 3.70 per kg, so that the mixture be worth Rs 3.35 a kg?
   a) 9:5 b) 7:5 c) 7:3 d) 3:1
2. In what proportion must tea at Rs 14 per kg be mixed with tea at Rs 18 per kg, so that the mixture be worth Rs 17 a kg?
   a) 1:1 b) 1:3 c) 2:3 d) 3:1
3. In what proportion must coffee at Rs 21 per kg be mixed with coffee at Rs 28 per kg, so that the mixture be worth Rs 25 a kg?
   a) 4:3 b) 4:5 c) 5:4 d) 3:4
4. In what proportion must cotton at Rs 24.50 per kg be mixed with cotton at Rs 30.50 per kg, so that the mixture be worth Rs 26 a kg?
   a) 3:1 b) 1:3 c) 3:2 d) 2:3
5. In what proportion must sugar at Rs 16.60 a kg be mixed with sugar at Rs 16.45 a kg so that the mixture may be worth Rs 16.54 a kg?
   a) 2:1 b) 2:3 c) 3:2 d) 4:1
6. In what proportion must tea at Rs 47.50 per kg be mixed with tea at Rs 50.50 per kg to produce a mixture worth Rs 48.50 per kg?
   a) 2:1 b) 1:2 c) 4:1 d) 3:2
7. In what proportion must a brewer mix beer at Rs 11 a litre with bear at Rs 6 a litre, so that the mixture may be worth Rs 8 a litre?
   a) 2:1 b) 1:2 c) 3:2 d) 2:3
8. How must a grocer mix teas at Rs 6 a kg and Rs 6.50 a kg so that the mixture may be worth Rs 6.20 a kg.
   a) 2:3 b) 3:2 c) 3:1 d) 1:3
9. In what ratio should gold at Rs 15 per gm be mixed with gold at Rs 10 per gm so that the resulting mixture be worth Rs 13 per gm.
   a) 3:2 b) 3:1 c) 1:1 d) 2:3
10. In what ratio must a grocer mix sugar at 72 paise per kg with sugar at 48 paise per kg so that by selling the mixture at 63 paise per kg he may gain $\frac{1}{6}$ of his outlay?
    a) 1:3 b) 3:1 c) 2:3 d) 3:2
11. Sugar at Rs 15 per kg is mixed with sugar at Rs 20 per kg in the ratio 2:3. Find the price per kg of the mixture.
    a) Rs 18 b) Rs 16 c) Rs 17 d) Rs 19
12. A grocer buys black tea at Rs 5.25 per kg and green tea at Rs 7.50 per kg. How must he mix them so that by

selling the mixture at Rs 7 per kg he may gain $\frac{1}{6}$ of his outlay.
a) 1:2 b) 1:3 c) 2:1 d) 3:1

13. In what proportion should water and wine at Rs 22.50 a litre be mixed to reduce the price to Rs 18 a litre?
a) 1:4 b) 4:1 c) 2:3 d) 3:2

14. Currants at Rs 50 per kg are mixed with currants at Rs 90 per kg to make a mixture of 17 kg worth Rs 70 per kg, how many kilograms of each are taken?
a) 8 kg, 9 kg b) $8\frac{1}{2}$ kg of each
c) 7 kg, 10 kg d) None of these

15. A person bought 60 quintals of rice of two different sorts for Rs 4642.50. The better sort costs Rs 80 per quintal and the worse Rs 75.50 per quintal. How many quintals were there of each sort?
a) 25 quintals, 35 quintals b) 20 quintals, 40 quintals
c) 32 quintals, 28 quintals d) None of these

16. A man has whisky worth Rs 22 a litre and another lot worth Rs 18 a litre. Equal quantities of these are mixed with water to obtain a mixture of 50 litres worth Rs 16 a litre. Find how much water the mixture contains?
a) 5 litres b) 10 litres c) 15 litres d) 20 litres

Answers

1. c 2. b 3. d 4. a 5. b 6. a 7. c
8. b 9. d

10. a; **Hint:** $\left(1+\frac{1}{6}=\right)1\frac{1}{6}$ of the cost price of a kg of the mixture = 63P

$\therefore$ cost price of a kg of the mixture $=\frac{63}{1\frac{1}{6}}=54P$

Now, applying the given formula, we have

the required answer $=\frac{54-48}{72-54}=1:3$.

11. a; **Hint:** $\frac{20-2}{Z-15}=\frac{2}{3}$ $\therefore$ Z = Rs 18 per kg

12. c; **Hint:** See Q. No. 10.

13. a; **Hint:** Required proportion $=\frac{20.50-18}{18-0}$

[$\because$ Water worths Rs 0 a litre]

$=\frac{4.50}{18}=1:4$.

14. b; **Hint:** Required ratio $=\frac{90-70}{70-50}=1:1$

Hence, $\frac{17}{1+1}=8\frac{1}{2}$ kg of each are taken.

15. a; **Hint:** Per quintal cost of two different sorts of rice

$=\frac{4642.50}{60}=$ Rs 77.375 per quintal

Proportion $=\frac{75.50-77.375}{77.375-80}=\frac{1.875}{2.625}=5:7$

The quantity of better sort $=\frac{60}{12}\times5=25$ quintals and

the quantity of worse sort $=\frac{60}{12}\times7=35$ quintals.

16. b; **Hint:** Two lots of whisky having equal quantities are mixed.

Let the price of mixture of whisky be Rs x per litre.

$\therefore \frac{18-x}{x-22}=1$ $\therefore$ x = Rs 20 a litres.

Now this mixture is mixed with water and worth Rs 16 a litre.

Hence, the proportion of water to mixture

$=\frac{20-16}{16-0}=1:4$.

$\therefore$ quantity of water $=\frac{50}{1+4}\times1=10$ litres.

Rule 2

Theorem: ***The quantity of salt at Rs x per kg that a man must mix with n kg of salt at Rs y per kg, so that he may, on selling the mixture at Rs z per kg, gain p% on the outlay is given by*** $n\times\left[\frac{100z-y(100+p)}{x(100+p)-100z}\right]$ ***kg.***

Note: If we suppose that the quantity of salt at Rs x be m, then we have.

$$\frac{m}{n}=\frac{100z-y(100+p)}{x(100+p)-100z}$$

Illustrative Example

Ex.: How many kg of salt at 42 P per kg must a man mix with 25 kg of salt at 24 P per kg, so that he may, on selling the mixture at 40 per kg, gain 25% on the outlay?

Soln: **Detail Method:**

Let the required amount of salt be x kg.

According to the question,

$42\times x+24\times25\,(x+25)\times40\times\frac{100}{125}$

$$\left[\begin{array}{l}\because \text{Selling price of the mixture} = 40 \text{ per kg given} \\ \therefore \text{Cost price of the mixture} = 40\times\frac{100}{125}\times(x+25)\end{array}\right]$$

or, $42x+24\times25=32x+32\times25$

or, $10x=25\times8$ $\therefore$ $x=20$ kg.

Method of Alligation:

Cost price of mixture = $40+\frac{100}{125}$ P = 32 P per kg ....

By the rule of fraction

42   24

       32

8   10

Ratio = 4 : 5

Thus for every 5 kg of salt at 24 P, 4 kg of salt at 42 P is used.

$\therefore$ the required no. of kg = $25\times\frac{4}{5}=20$.

Quicker Method: Applying the above theorem,

Required answer = $\left[\frac{100\times40-24\times(100+25)}{42\times(100+25)-100\times40}\right]\times25$

$=\left[\frac{4000-3000}{5250-4000}\right]\times25=\left[\frac{1000}{1250}\right]\times25=20$ kg.

Exercise

1. Jaydeep purchased 25 kg of rice at the rate of Rs 16.50 per kg and 35 kg of rice at the rate of Rs 25.50 per kg. He mixed the two and sold the mixture. Approximately, at what price per kg did he sell the mixture to make 25 per cent profit? **(BSRB Mumbai PO 1998)**
   a) Rs 26.50 b) Rs 27.50 c) Rs 28.50 d) Rs 30.00
2. Jagtap purchases 30 kg of wheat at the rate of Rs 11.50 per kg and 20 kg of wheat at the rate of Rs 14.25 per kg. He mixed the two and sold the mixture. Approximately at what price per kg should he sell the mixture to make 30 per cent profit?
   a) Rs 16.30 b) Rs 18.20 c) Rs 15.60 d) Rs 14.80
   (BSRB Calcutta PO 1999)
3. Prabhu purchased 30 kg of rice at the rate of Rs 17.50 per kg and another 30 kg rice at a certain rate. He mixed the two and sold the entire quantity at the rate of Rs 18.60 per kg and made 20 per cent overall profit. All what price per kg did he purchase the lot of another 30 kg rice?
   a) Rs 14.50 b) Rs 12.50 c) Rs 15.50 d) Rs 13.50
   (BSRB Chennai PO 2000)
4. A grocer purchased 20 kg of rice at the rate of Rs 15 per kg and 30 kg of rice at the rate of Rs 13 per kg. At what price per kg should he sell the mixture to earn $33\frac{1}{3}\%$ profit on the cost price?
   a) Rs 28.00 b) Rs 20.00 c) Rs 18.40 d) Rs 17.40
   (BSRB Delhi PO 2000)
5. A grocer buys two kinds of barley at Re 1.50 P and 95 paise per kilogram respectively. In what proportion should these be mixed so that by selling the mixture at Re 1.60 P per kilogram, 25% may be gained?
   a) 3:1 b) 3:2 c) 4:1 d) 2:3
6. In what proportion must a grocer mix one kind of wheat at Rs 4.50 per kg with another at Rs 4 per kg in order that by selling the mixture at Rs 5.20 per kg he may make a profit of 20 per cent?
   a) 3:1 b) 4:1 c) 3:2 d) 2:1
7. How many kg of salt costing 40 P per kg must be mixed with 16 kg of salt costing 55 P per kg so that 25 per cent may be gained by selling the mixture at 60 P per kg?
   a) 14 kg b) 16 kg c) 12 kg d) 15 kg
8. What weight of wheat worth Rs 4.20 per kg should be mixed with 60 kg of sugar worth Rs 2.70 per kg so that when the mixture is sold at Rs 3.30 per kg, there may be neither gain nor loss.
   a) 50 kg b) 45 kg c) 55 kg d) 40 kg
9. Kantilal mixes 80 kg. of sugar worth of Rs. 6.75 per kg. with 120 kg. worth of Rs. 8 per kg. At what rate shall he sell the mixture to gain 20%?
   a) Rs 7.50 b) Rs. 9 c) Rs. 8.20 d) Rs. 8.85
   (SBI PO Exam 1987)

Answers

1\. a; **Hint:** $35\left[\frac{100\times z-24.50(100+25)}{16.50(100+25)-100\times z}\right]=25$

or, $\frac{100z-3062.5}{2062.5-100z}=\frac{5}{7}$

or, 700z - 21437.5 = 10312.5 - 500z

or, 1200z = 31750

$\therefore$ z = Rs 26.458 per kg $\approx$ Rs 26.50 per kg.

2\. a; **Hint:** $20\left[\frac{100\times z-14.25(100+30)}{11.50(100+30)-100\times z}\right]=30$

or, $\frac{100z-1852.5}{1495-100z}=\frac{3}{2}$ or 500z = 8190

$\therefore$ $z=\frac{18190}{500}$ = Rs 16.38 $\approx$ Rs 16.30

3\. d; **Hint:** $30\left[\frac{100\times18.60-y\times120}{17.50\times120-100\times18.60}\right]=30$

or, 1860 - 120y = 2100 - 1860 = 240

or, 120y = 1620 $\therefore$ $y=\frac{1620}{120}$ = Rs 13.50

4. c; **Hint:** $\left[\dfrac{100\times z-13\left(100+\dfrac{100}{3}\right)}{15\times\left(100+\dfrac{100}{3}\right)-100\times z}\right]30=20$

$\therefore$ z = Rs 18.40

5. b; **Hint:** See **Note.**

$$\text{Required proportion}=\frac{100\times1.60-0.95(100+25)}{1.50(100+25)-100\times1.60}$$

$$=\frac{160-118.75}{187.5-160}=\frac{4125}{2750}=\frac{3}{2}$$

$=3:2$

6. d 7. a

8. d; **Hint:** Put the value of p = 0 in the given rule.

$\therefore$ Required answer

$$=\frac{100\times3.30-2.70(100+0)}{4.20\times(100+0)-100\times3.30}\times60$$

$$=\frac{60\times60}{90}=40\text{ kg.}$$

9. b; **Hint:** $120\left[\dfrac{100z-8(100+20)}{675(100+20)-100z}\right]=80$

$\therefore$ z = Rs 9 per kg.

Rule 3

Theorem: ***A mixture of a certain quantity of milk with 'l' litres of water is worth Rs x per litre. If pure milk be worth Rs y per litre, then the quantity of milk is given by*** $l\left(\dfrac{x}{y-x}\right)$ ***litres.***

Illustrative Example

Ex.: A mixture of a certain quantity of milk with 16 litres of water is worth 90 P per litre. If pure milk be worth 108 P per litre how much milk is there in the mixture?

Soln: **Detail Method:** Let the quantity of milk be x litres.

$(x+16)\,90=x\times108+16\times0$ [$\because$ the price of water is 0 P)

or, $90x+16\times90=108x$

or, $18x=16\times90$ $\therefore x=80$ litres.

$\therefore$ The quantity of milk = 80 litres.

Alligation Method: The mean value is 90 P and the price of water is 0 P.

milk   water
108   0
     90
90 − 0   108 − 90

By the Alligation Rule, milk and water are in the ratio of 5 : 1.

$\therefore$ quantity of milk in the mixture = 5 × 16 = 80 litres.

Quicker Method: Applying the above theorem,
Quantity of milk in the mixture

$$=16\left(\frac{90}{108-90}\right)=16\times5=80\text{ litres.}$$

Exercise

1. A mixture of a certain quantity of milk with 25 litres of water is worth Rs 2 per litre. If pure milk be worth Rs 12 per litre how much milk is there in the mixture?
   a) 5 litres b) 7 litres c) 6 litres d) 4 litres
2. A mixture of a certain quantity of milk with 16 litres of water is worth Rs 3 per litre. If pure milk be worth Rs 7 per litre how much milk is there in the mixture?
   a) 10 litres b) 12 litres
   c) 14 litres d) None of these
3. A mixture of a certain quantity of milk with 32 litres of water is worth Rs 1.50 per litre. If pure milk be worth Rs 4.50 per litre how much milk is there in the mixture?
   a) 18 litres b) 14 litres c) 16 litres d) 20 litres

Answers

1. a 2. b 3. c

Rule 4

Theorem: ***The proportion in which water must be mixed with spirit to gain or to lose x% by selling it at cost price is given by*** $\left(\dfrac{x}{100}\right)$.

Illustrative Example

Ex.: In what proportion must water be mixed with spirit to gain $16\frac{2}{3}$% by selling it at cost price?

Soln: **Detail Method:** Let the required proportion of water to spirit be a : b and the cost price of spirit be Rs x per litre.

As per the question,
Selling price of the mixture = Rs x per litre.
Cost price of the mixture

$$=x\times\frac{100}{100+\frac{50}{3}}=\text{Rs }\frac{6x}{7}\text{ per litre.}$$

Now, assume that the cost price of water = Rs 0 per litre.

$$\therefore (a\times0+b\times x)=(a+b)\frac{6x}{7}$$

or, $bx=(a+b)\dfrac{6}{7}x$ or, $b\left(1-\dfrac{6}{7}\right)=\dfrac{6a}{7}$

or, $\frac{b}{7}=\frac{6a}{7}$ or, $\frac{a}{b}=\frac{1}{6}$

$\therefore$ required ratio = 1 : 6

Alligation Method: Let CP of sprit be Re 1 per litre.

Then SP of 1 litre of mixture = Re 1. Gain = $16\frac{2}{3}\%$

CP of 1 litre of mixture = Rs $\left(\frac{100\times3\times1}{350}\right)$ = Re $\left(\frac{6}{7}\right)$

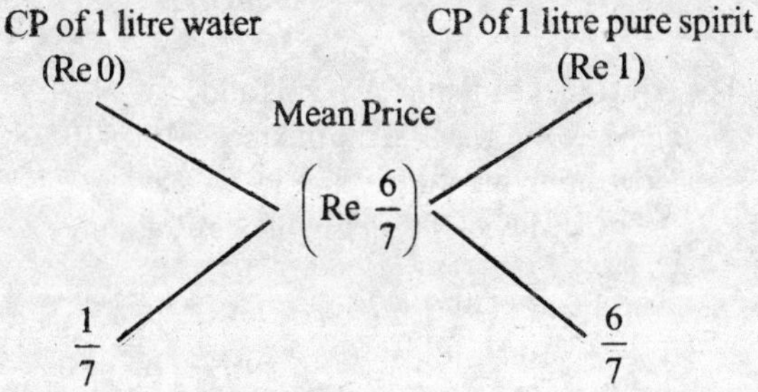

$\frac{\text{Quantity of water}}{\text{Quantity of spirit}}=\frac{\frac{1}{7}}{\frac{6}{7}}=\frac{1}{6}$

or Ratio of water and spirit = 1 : 6.

Quicker Method: Applying the above theorem,

the required proportion = $\frac{50}{300}=1:6$.

Exercise

1. In what proportion must water be mixed with spirit to gain $26\frac{2}{3}\%$ by selling it at cost price?
   a) 4:15 b) 2:7 c) 1:11 d) 15:4
2. In what proportion must water be mixed with spirit to gain $33\frac{1}{3}\%$ by selling it at cost price?
   a) 3:1 b) 1:2 c) 1:3 d) 2:3
3. In what proportion must water be mixed with spirit to gain 16% by selling it at cost price?
   a) 4:25 b) 2:9 c) 1:6 d) 25:4
4. In what proportion must water be mixed with spirit to gain 25% by selling it at cost price?
   a) 4:1 b) 3:4 c) 4:3 d) 1:4

Answers

1. a 2. c 3. a 4. d

Rule 5

Theorem: ***n gm of sugar solution has x% sugar in it. The quantity of sugar should be added to make it y% in the solution is given by*** $n\left(\frac{y-x}{100-y}\right)$ ***gm. or Quantity of sugar added*** = $\frac{\text{Solution}(\text{required \% value} - \text{present \% value})}{(100-\text{required \% value})}$.

Illustrative Example

Ex.: 300 gm of sugar solution has 40% sugar in it. How much sugar should be added to make it 50% in the solution?

Soln: **Alligation Method:** The existing solution has 40% sugar. And sugar is to be mixed; so the other solution has 100% sugar. So by alligation method:

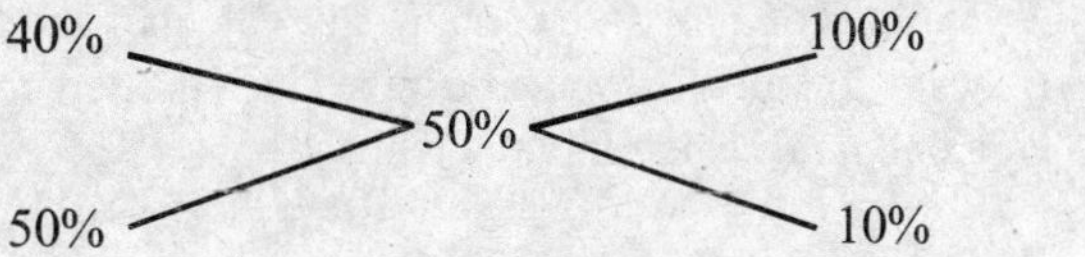

$\therefore$ The two mixtures should be added in the ratio 5 : 1.

Therefore, required sugar = $\frac{300}{5}\times1=60$ gm.

Quicker Method: Applying the above theorem, we have

quantity of sugar added = $\frac{300(50-40)}{100-50}=60$ gm.

Exercise

1. A mixture of 40 litres of milk and water contains 10% water. How much water must be added to make 20% water in the new mixture?
   a) 5 litres b) 6 litres c) 8 litres d) 10 litres
2. A petrol pump owner mixed leaded and unleaded petrol in such a way that the mixture contains 10% unleaded petrol. What quantity of leaded petrol should be added to 1 litre mixture so that the percentage of unleaded petrol becomes 5%.
   a) 1000 ml b) 900 ml c) 1900 ml d) 1800 ml
   (SBI Associates PO - 1999)
3. 150 gm of sugar solution has 20% sugar in it. How much sugar should be added to make it 25% in the solution?
   a) 10 gm b) 45 gm c) 35 gm d) 40 gm
4. A petrol pump owner mixed leaded and unleaded petrol in such a way that the mixture contains 20% unleaded petrol. What quantity of leaded petrol should be added to 2 litres mixture so that the percentage of unleaded petrol becomes 10%.
   a) 1000 ml b) 2000 ml c) 1500 ml d) None of these
5. A 40 litres mixture of milk and water contains 10 per cent of water. How much water must be added to make the water 28% in the new mixture?
   a) 10 litres b) 14 litres c) 8 litres d) 12 litres
6. In a mixture of wheat and barley the wheat is 60%. To 400 quintals of the mixture a quantity of barley is added and then the wheat is $53\frac{1}{3}\%$ of the resulting mixture. How

many quintals of barley are added?

a) $\frac{400}{7}$ quintals b) 50 quintals

c) $46\frac{2}{3}$ quintals d) $53\frac{2}{3}$ quintals

7. 50 gm of an alloy of gold and silver contains 80% gold (by weight). Find the quantity of gold that is to be mixed up with this alloy so that it may contain 95% gold.
a) 130 gm b) 140 gm c) 145 gm d) 150 gm

8 15 litres of a mixture contains 20% alcohol and the rest water. If 3 litres of water be mixed in it, the percentage of alcohol in the new mixture will be:

a) 17 b) $16\frac{2}{3}$ c) $18\frac{1}{2}$ d) 15

(Clerical Grade 1991)

9. 729 ml of a mixture contains milk and water in the ratio 7:2. How much more water is to be added to get a new mixture containing milk and water in ratio 7:3?
a) 600 ml b) 710 ml c) 520 ml d) None of these

(Railways, 1991)

10. In a mixture of 60 litres, the ratio of milk and water is 2 : 1. If the ratio of the milk and water is to be 1 : 2, then the amount of water to be further added is:
a) 20 litres b) 30 litres c) 40 litres d) 60 litres

(NDA Exam 1990)

11. A mixture of 66 litres of milk and water are in the ratio 5 : 1, and water is added to make the ratio 5 : 3. Find the quantity of water added.
a) 20 litres b) 18 litres c) 22 litres d) 24 litres

(LIC Exam 1988)

Answers

1. a; **Hint:** Required amount of water $= \left(\frac{20-10}{100-20}\right)\times 40$

$$= \frac{400}{80} = 5 \text{ litres.}$$

2. a; **Hint:** Here we have to find the quantity of leaded petrol Hence, we have to make certain changes in the given data.
% of leaded petrol in the mixture = 100 - 10 = 90%.
After addition of leaded petrol (that has to be calculated) percentage of leaded petrol becomes (100 - 5 =) 95%.
Now, applying the given theorem, we have

the required answer $= \left(\frac{95-90}{100-95}\right) 1000 \text{ ml} = 1000 \text{ ml}$

3. a 4. b 5. a

6. b; **Hint:** Here barley is added. Hence $y = 100 - 53\frac{1}{3} = 46\frac{2}{3}$, $x = 100 - 60 = 40\%$. Now, applying the given rule, we have

the required answer $= \left[\frac{46\frac{2}{3}-40}{100-46\frac{2}{3}}\right]\times 400$

$= 50$ quintals.

7. d

8. b; **Hint:** In the mixture, water is added.
Hence, % of water in the mixture = 100 - 20 = 80%
Now, applying the given rule, we have
the percentage of water in the new mixture

$$= 15\left(\frac{y-80}{100-y}\right) = 3$$

$$\therefore\ y = \frac{500}{6}\%$$

∴ required answer ie % of alcohol in the new mixture =

$$100 - \frac{500}{6} = \frac{100}{6} = \frac{50}{3} = 16\frac{2}{3}\%$$

9. d; **Hint:** Percentage of water in first mixture

$$= \frac{2}{2+7}\times 100 = \frac{200}{9}\%$$

percentage of water in the second mixture

$$= \frac{3}{7+3}\times 100 = 30\%$$

Now applying the given rule,

required answer $= \left[\frac{30-\frac{200}{9}}{100-30}\right] 729 = 81$ ml

10. d; **Hint:** $60\left[\frac{\frac{200}{3}-\frac{100}{3}}{100-\frac{200}{3}}\right] = 60$ litres.

11. c

Rule 6

Theorem: ***There are 'N' students in a class. Rs X are distributed among them so that each boy gets Rs x and each girl gets Rs y. Then the ratio of boys to the girls is given by*** $\left[\frac{X-Ny}{Nx-X}\right]$ ***and the no. of boys and the no. of girls are***

$\left(\frac{X-Ny}{x-y}\right)$ ***and*** $\left(\frac{Nx-X}{x-y}\right)$ ***respectively.***

Illustrative Example

Ex.: There are 65 students in a class, 39 rupees are distributed among them so that each boy gets 80 P and each girl gets 30 P. Find the number of boys and girls in that class.

Soln: **Detail Method:** Let the ratio of boys to the girls in the class be a : b.

As per the question,

No. of boys $= \frac{65\times a}{a+b}$ and the no. of girls $= \frac{65\times b}{a+b}$

or, $\frac{65\times a}{a+b}\times 80+\frac{65\times b}{a+b}\times 30=3900$

or, $(5200-3900)a=(3900-1950)b$

or, $\frac{a}{b}=\frac{1950}{1300}=\frac{3}{2}$

$\therefore$ a : b = 3 : 2

$\therefore$ the no. of boys $= \frac{65\times 3}{5}=39$ and

the no. of girls $= \frac{65\times 2}{5}=26$.

Alligation Method: Here alligation is applicable for "money per boy or girl."

Mean value of money per student $= \frac{3900}{65} = 60$ P

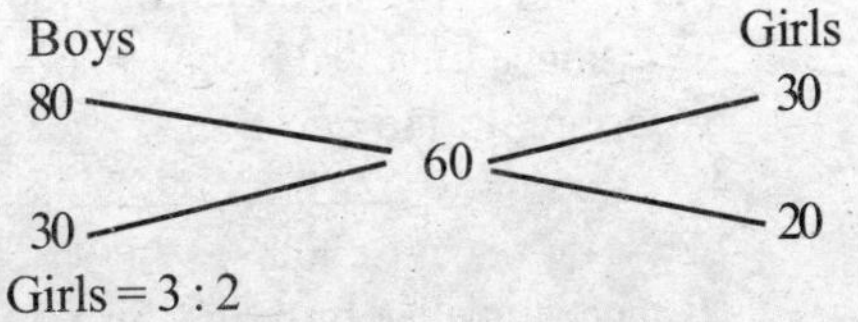

$\therefore$ Boys : Girls = 3 : 2

$\therefore$ Number of boys $= \frac{65}{3+2}\times 3=39$

and number of girls = 65 − 39 = 26.

Quicker Method: Applying the above theorem, we have

N = 65 $\qquad$ X = Rs 39 = 3900 P

x = 80 P $\qquad$ y = 30 P

No. of boys $= \frac{3900-65\times 30}{80-30}=\frac{1950}{50}=39$.

No. of girls $= \frac{65\times 80-3900}{80-30}=\frac{1300}{50}=26$.

Exercise

1. There are 60 students in a class, 120 rupees are distributed among them so that each boy gets Rs 2.50 and each girl gets 50 P. Find the ratio of boys to the girls.
   a) 3 : 5 $\quad$ b) 1 : 2 $\quad$ c) 3 : 4 $\quad$ d) 5 : 3
2. There are 75 students in a class, 48 rupees are distributed among them so that each boy gets Re 1 and each girl gets 40 P. Find the number of boys and girls in that class.
   a) 30, 45 $\quad$ b) 40, 35 $\quad$ c) 25, 50 $\quad$ d) 35, 40
3. There are 50 students in a class, 32 rupees are distributed among them so that each boy gets Re 1 and each girl gets 50 P. Find the number of girls and boys in that class.
   a) 14 girls, 36 boys $\quad$ b) 36 girls, 14 boys
   c) 20 girls, 30 boys $\quad$ d) 30 girls, 20 boys

Answers

1. a $\quad$ 2. a $\quad$ 3. b

Rule 7

Theorem: ***A person has a liquid of Rs x per litre. The ratio in which water should be mixed in that liquid, so that after selling the mixture at Rs y per litre he may get a profit of P%, is given by*** $\left[\frac{y}{(x-y)+\left(\frac{P}{100}\right)x}\right]$.

Illustrative Example

Ex.: A person has a chemical of Rs 25 per litre. In what ratio should water be mixed in that chemical so that after selling the mixture at Rs 20/litre he may get a profit of 25%.

Soln: **Detail Method:** Let the ratio of chemical to water in the mixture be a : b.

Cost price of the chemical is Rs 25 per litre

$\therefore$ cost price of a litre of the chemical = Rs 25

Assume that the cost price of water be Rs 0 per litre

Now, according to the question,

Selling price of the mixture = Rs 20 per litre

$\therefore$ Selling price of (a + b) litres of the mixture = Rs (a + b) 20

Cost price of (a + b) litres of the mixture

$= (a+b)\times 20\times\frac{100}{125}$

(By the rule of fraction)

= Rs (a + b) 16

or, $25\times a+0\times b=(a+b)\,16$

or, $9a = 16b$

or, $\frac{a}{b}=\frac{16}{9}$

$\therefore$ a : b = 16 : 9

$\therefore$ Required ratio = 16 : 9.

Alligation Method: In this question the alligation method is applicable on prices, so we should get the average price of mixture.

SP of mixture = Rs 20/ litre, profit = 25%

$\therefore$ average price = $20 \times \frac{100}{125}$ = Rs 16/ litre

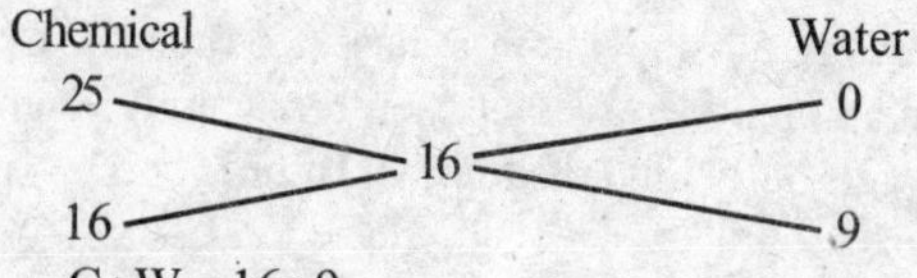

$\therefore$ C : W = 16 : 9

Quicker Method: Applying the above theorem, we have the

$$\text{reqd ratio} = \frac{20}{(25-20)+\frac{25}{100}\times 25} = \frac{80}{45} = \frac{16}{9} = 16:9.$$

Exercise

1. A man buys milk at Rs 7.50 a litre and after adding water sells it at Rs 9 a litre thereby making profit of $33\frac{1}{3}\%$. Find the proportion of water he has added.
   a) 9:1 b) 7:1 c) 9:2 d) 3:1
2. A man buys milk at Rs 5 a litre and mixes it with water. By selling the mixture at Rs 4 a litre he gains $12\frac{1}{2}$ per cent on his outlay. How much water did each litre of the mixture contain?
   a) $\frac{32}{45}$ litre b) $\frac{13}{45}$ litre c) $\frac{32}{13}$ litre d) None of these
3. A milk seller pays Rs 500 per kilolitre for his milk. He adds water to it and sells the mixture at 56 P a litre, thereby making altogether 40% profit. Find the proportion of water to milk which his customers receive.
   a) 1:4 b) 2:3 c) 1:5 d) 4:1
4. A person has a chemical of Rs 50 per litre. In what ratio should water be mixed in that chemical so that after selling the mixture at Rs 40 per litre he may get a profit of 50%.
   a) 8:7 b) 9:8 c) 10:7 d) 4:3

Answers

1. a

2. b; **Hint:** $\text{Required ratio} = \frac{4}{(5-4)+\left(\frac{25}{200}\right)5} = \frac{32}{13} = 32:13$

The quantity of water that the each litre of tho mixture contains = $\frac{13}{32+13} \times 1 = \frac{13}{45}$ litre

3. a; **Hint:** Here, $x = \frac{500}{1000} = 50P$, y = 56P, P = 40%

$$\text{Ratio of milk to water} = \frac{56}{(50-56)+\frac{40}{100}\times 50} = \frac{4}{1} = 4:1$$

$\therefore$ required answer (ie ratio of water to milk) = 1:4.

4. a

Rule 8

Theorem: ***A person travels D km in T hours in two stages. In the first part of the journey, he travels by bus at the speed of x km/hr. In the second part of the journey, he travels by train at the speed of y km/hr. Then the distance travelled by bus is $\left[\frac{yT-D}{y-x}\right]x$ km and the distance travelled by train is $\left[\frac{D-xT}{y-x}\right]y$ km.***

Illustrative Example

Ex.: A person travels 285 km in 6 hours in two stages. In the first part of the journey, he travels by bus at the speed of 40 km per hr. In the second part of the journey, he travels by train at the speed of 55 km per hr. How much distance did he travel by train?

Soln: **Detail Method:** Let the person travels for x hours by the train.

$\therefore$ Time for which he travels by bus = $(6-x)$ hours.

The distance travelled by train = $55 \times x$ km and the distance travelled by bus = $(6-x)$ 40 km

According to the question,

$40(6-x)+55x = 285$

$15x = 45 \quad \therefore x = 3$ hours.

Distance travelled by train = $55 \times 3 = 165$ km.

Alligation Method: In this question, the alligation method is applicable for the speed.

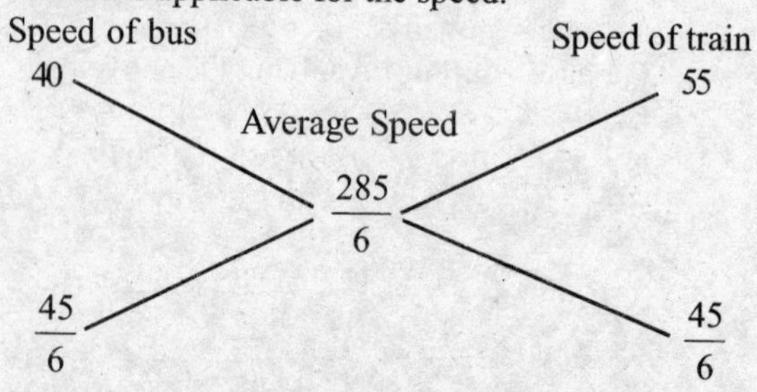

$\therefore$ time spent in bus : time spent in train

$$= \frac{45}{6} : \frac{45}{6} = 1:1$$

$\therefore$ distance travelled by train = $55 \times 3 = 165$ km.

Quicker Method: Applying the above theorem, we have

$$\text{reqd distance} = \frac{285-40\times 6}{55-40}\times 55 = 3\times 55 = 165 \text{ km.}$$

Exercise

1. A person travels 255 km in 7 hours in two stages. In the first part of the journey, he travels by bus at the speed of 30 km per hr. In the second part of the journey, he travels by train at the speed of 45 km per hr. How much distance did he travel by bus?
   a) 120 km b) 135 km c) 145 km d) 125 km
2. A person travels 245 km in 6 hours in two stages. In the first part of the journey, he travels by bus at the speed of 30 km per hr. In the second part of the journey, he travels by train at the speed of 50 km per hr. How much distance did he travel by train?
   a) 162.5 km b) 82.5 km c) 164 km d) 83 km
3. A person travels 490 km in 6 hours in two stages. In the first part of the journey, he travels by bus at the speed of 60 km per hr. In the second part of the journey, he travels by train at the speed of 100 km per hr. What is the ratio between distances travelled by bus and train?
   a) 65:33 b) 5:3 c) 3:5 d) 33:65
4. A man travels a distance of 200 km in 4 hours, partly by bus at 40 km/hr and the rest by train at 75 km/hr. Find the distance covered in each part?

   a) $85\frac{5}{7}$ km, $114\frac{2}{7}$ km b) $114\frac{2}{7}$ km, $85\frac{5}{7}$ km

   c) $84\frac{2}{7}$ km, $115\frac{5}{7}$ km d) $115\frac{5}{7}$ km, $84\frac{2}{7}$ km

Answers

1. a 2. a 3. d 4. b

Rule 9

Theorem: ***A trader has N kg of certain item, part of which he sells at x% profit and the rest at y% profit. He gains P% on the whole. The quantity of item sold at x% profit is*** $\left[\left(\frac{y-P}{y-x}\right)N\right]$ ***kg and the quantity of item sold at y% profit is given by*** $\left[\left(\frac{P-x}{y-x}\right)N\right]$ ***kg.***

Illustrative Example

Ex: A trader has 50 kg of pulses, part of which he sells at 8% profit and the rest at 18% profit. He gains 14% on the whole. What is the quantity sold at 18% profit?

Ex.: **Detail Method:**
Let the quantity sold at 18% profit be x kg. Then the quantity sold at 8% profit will be $(50-x)$ kg.

For a matter of convenience suppose that the price of pulse is 1 rupee per kg.
Then price of x kg pulse = Rs x and price of $(50-x)$ kg pulse = Rs $(50-x)$
Now we get an equation,
18% of $x + 8\%$ of $(50-x) = 14\%$ of 50
$\Rightarrow 18x + 8(50-x) = 14\times 50$
$\Rightarrow 10x = 300 \quad \therefore x = 30$

By Alligation Method:

I Part **II Part**
8% profit 18% profit
14%
(mean profit)
4% 6%

$= 4:6 = 2:3$
Therefore the quantity sold at 18% profit

$$= \frac{50}{2+3}\times 3 = 30 \text{ kg.}$$

Quicker Method: Applying the above theorem, we have the required quantity

$$= \left(\frac{14-8}{18-8}\right)\times 50 = \frac{6}{10}\times 50 = 30 \text{ kg.}$$

Exercise

1. A trader has 25 kg of rice, part of which he sells at 4% profit and the rest at 9% profit. He gains 7% on the whole. What is the quantity sold at 9% profit?
   a) 15 kg b) 10 kg c) 18 kg d) 12 kg
2. A trader has 100 kg of wheat, part of which he sells at 16% profit and the rest at 36% profit. He gains 28% on the whole. What is the quantity sold at 36% profit?
   a) 50 kg b) 60 kg c) 45 kg d) 65 kg
3. A trader has 40 kg of pulses, part of which he sells at 10% profit and the rest at 20% profit. He gains 16% on the whole. What is the quantity sold at 20% profit?
   a) 28 kg b) 30 kg c) 24 kg d) 26 kg

Answers

1. a 2. b 3. c

Rule 10

Theorem: ***A trader has N kg of a certain item, a part of which he sells at x% profit and the rest of y% loss. He gains P% on the whole. Then the quantity sold at x% profit is given by*** $\left[\left(\frac{P+y}{x+y}\right)N\right]$ ***kg and the quantity sold at y% loss is given by*** $\left[\left(\frac{x-P}{x+y}\right)N\right]$ ***kg.***

Illustrative Example

Ex.: A trader has 50 kg of rice, a part of which he sells at

10% profit and the rest at 5% loss. He gains 7% on the whole. What is the quantity sold at 10% gain and 5% loss?

Soln: **Detail Method:** Let the quantity sold at 10% profit be x kg.

Then the quantity sold at 5% loss will be $(50-x)$ kg.

For a matter of convenience suppose that the price of rice is 1 rupee per kg.

Then price of x kg rice = Rs x and price of (50 – x) kg rice = Rs $(50-x)$

Now we get an equation,

10% profit of x + 5% loss of (50 – x) = 7% gain of 50

or, 10% of x – 5% of (50 – x) = 7% of 50

or, 10x – 250 + 5x = 350

$\therefore x = 40$ kg and $(50-x) = 50 - 40 = 10$ kg.

Therefore, the quantity sold at 10% profit = 40 kg and the quantity sold at 5% loss = 50 – 40 = 10 kg.

Alligation Method:

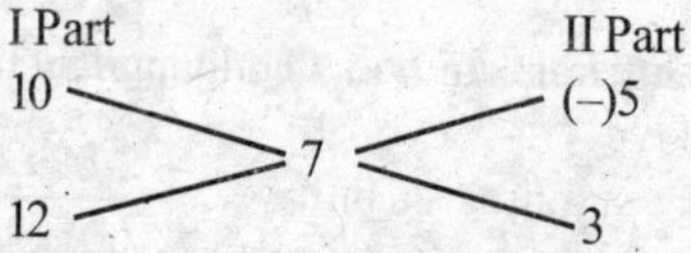

$\therefore$ Ratio of quantities sold at 10% profit and 5% loss = 12 : 3 = 4 : 1

Therefore, the quantity sold at 10% profit

$$= \frac{50}{4+1} \times 4 = 40 \text{ kg}$$

and the quantity sold at 5% loss = 50 – 40 = 10 kg.

Note: Whenever there is loss, take the negative value. Here, difference between 7 and (–5) = 7 – (–5) = 7 + 5 = 12. Never take the difference that counts negative value.

Quicker Method: Applying the above theorem, we have

$$\text{Quantity sold at 10\% profit} = \left(\frac{7+5}{10+5}\right)50$$

$$= \frac{12}{15} \times 50 = 40 \text{ kg.}$$

$$\text{Quantity sold at 5\% loss} = \left(\frac{10-7}{10+5}\right) \times 50 = 10 \text{ kg.}$$

Exercise

1. A trader has 90 kg of pulse, a part of which he sells at 20% profit and the rest at 10% loss. He gains 14% on the whole. What is the quantity sold at 20% gain and 10% loss?
   a) 72 kg, 18 kg b) 70 kg, 20 kg
   c) 62 kg, 28 kg d) None of these
2. A trader has 45 kg of wheat, a part of which he sells at 30% profit and the rest at 15% loss. He gains 21% on the whole. What is the quantity sold at 30% gain and 15% loss?
   a) 26 kg, 19 kg b) 36 kg, 9 kg
   c) 35 kg, 10 kg d) None of these
3. A trader has 40 kg of tea, a part of which he sells at 12% profit and the rest at 8% loss. He gains 9% on the whole. What is the quantity sold at 12% gain and 8% loss?
   a) 30 kg, 10 kg b) 32 kg, 8 kg
   c) 33 kg, 7 kg d) 34 kg, 6 kg

Answers

1. a 2. b 3. d

Rule 11

Theorem: ***A trader has N kg of a certain item, a part of which he sells at x% profit and the rest at y% loss. On the whole his loss is P%. Then the quantity sold at x% profit is*** $\left(\frac{y-P}{x+y}\right)$ ***N kg and the quantity sold at y% loss is given by*** $\left(\frac{x+P}{x+y}\right)N$ ***kg.***

Illustrative Example

Ex.: A trader has 50 kg of rice, a part of which he sells at 14% profit and the rest at 6% loss. On the whole his loss is 4%. What is the quantity sold at 14% profit and that at 6% loss?

Soln: **Detail Method:** Let the quantity sold at 14% profit be x kg. Then the quantity sold at 6% loss will be $(50-x)$ kg.

For a matter of convenience suppose that the price of rice is 1 rupee per kg.

Then price of x kg rice = Rs x and price of $(50-x)$ kg rice = Rs $(50-x)$

Now, we have

14% profit of x + 6% loss of $(50-x)$ = 4% loss of 50

or, 14% of x – 6% of $(50-x)$ = –4% of 50

or, $14x - 300 + 6x = -200$

or, $20x = 100$

$\therefore x = 5$ kg

and $50 - x = 50 - 5 = 45$ kg.

Therefore, the quantity sold at 14% profit = 5 kg and the quantity sold at 6% loss = 45 kg.

Alligation Method:

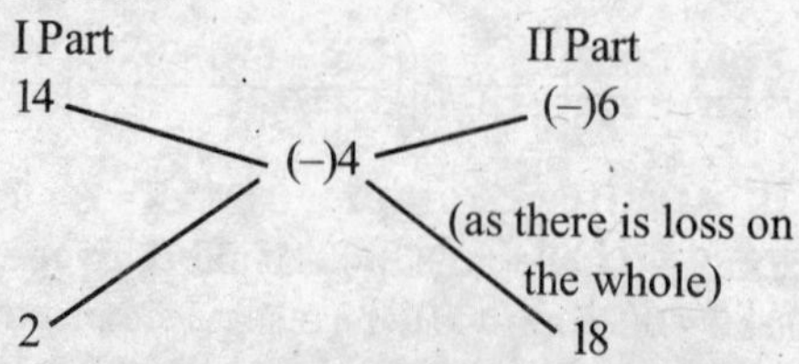

$\therefore$ ratio of quantities sold at 14% profit and 6% loss = 2 : 18 = 1 : 9

$$\therefore \text{quantity sold at 14\% profit} = \frac{50}{1+9} \times 1 = 5 \text{ kg}$$

and sold at 6% loss = 50 – 5 = 45 kg.

Note: Numbers in the third line should always be +ve. That is why (–) 6 – (–)4 = –2 is not taken under consideration.

Quicker Method: Applying the above theorem,
Quantity sold at 14% profit

$$=\left(\frac{6-4}{14+6}\right)\times 50 = 5 \text{ kg and}$$

the quantity sold at 6% loss $=\left(\frac{14+4}{14+6}\right)\times 50 = 45$ kg.

Exercise

1. A trader has 40 kg of rice, a part of which he sells at 28% profit and the rest at 12% loss. On the whole his loss is 8%. What is the quantity sold at 28% profit and that at 12% loss?
   a) 4 kg, 36 kg   b) 10 kg, 30 kg
   c) 8 kg, 32 kg   d) None of these
2. A trader has 48 kg of rice, a part of which he sells at 16% profit and the rest at 8% loss. On the whole his loss is 6%. What is the quantity sold at 16% profit and that at 8% loss?
   a) 42 kg, 6 kg   b) 44 kg, 4 kg
   c) 4 kg, 44 kg   d) 6 kg, 42 kg
3. A trader has 44 kg of rice, a part of which he sells at 26% profit and the rest at 18% loss. On the whole his loss is 16%. What is the quantity sold at 26% profit and that at 18% loss?
   a) 2 kg, 42 kg   b) 4 kg, 40 kg
   c) 42 kg, 2 kg   d) 40 kg, 4 kg

Answers

1. a   2. c   3. a

Rule 12

Theorem: ***A person's expenditure and savings are in the ratio a : b. His income increases by x%. His expenditure also increases by y%. His percentage increase in saving is given by*** $\left[\left(\frac{a}{b}+1\right)x-\frac{a}{b}y\right]$ **%.**

Illustrative Example

Ex.: Mira's expenditure and saving are in the ratio 3 : 2. Her income increases by 10%. Her expenditure also increases by 12%. By how many % does her saving increase?

Soln: **Detail Method:** Let the Mira's expenditure and saving be Rs $3x$ and Rs $2x$

Mira's income $= 3x + 2x = 5x$

Increased income $= 5x\times\frac{110}{100}=\frac{11}{2}x$

Increased expenditure $= 3x\times\frac{112}{100}=\frac{84}{25}x$

Increased saving $=\frac{11}{2}x-\frac{84}{25}x=\frac{107x}{50}$

Increase in saving $=\frac{107x}{50}-2x=\frac{7x}{50}$

% increase in saving $=\frac{7x}{50\times 2x}\times 100 = 7\%$.

Alligation Method:

Expenditure   Saving
12   x
(% increase in exp)   (% increase in saving)

10
(% increase in income)

3   2 (given)

We get two values of x, 7 and 13. But to get a viable answer, we must keep in mind that the central value (10) must lie between x and 12. Thus the value of x should be 7 and not 13.

∴ required % increase = 7%

Quicker Method: Applying the above theorem, we have
the required percentage increase in saving

$$=\left(\frac{3}{2}+1\right)\times 10-\frac{3}{2}\times 12$$

$$=25-18=7\%.$$

Exercise

1. Ritu's expenditure and saving are in the ratio 5 : 2. Her income increases by 12%. Her expenditure also increases by 14%. By how many % does her saving increase?
   a) 14%   b) 7%   c) 8%   d) 9%
2. Sita's expenditure and saving are in the ratio 5 : 3. Her income increases by 15%. Her expenditure also increases by 9%. By how many % does her saving increase?
   a) 20%   b) 30%   c) 25%   d) 24%
3. Ranju's expenditure and saving are in the ratio 4 : 5. Her income increases by 25%. Her expenditure also increases by 35%. By how many % does her saving increase?
   a) 15%   b) 16%   c) 18%   d) 17%

Answers

1. b   2. c   3. d

Rule 13

Theorem: ***A vessel of L litres is filled with liquid A and B. x% of A and y% of B is taken out of the vessel. It is found that the vessel is vacated by z%. Then the initial quantity of***

liquid A and B is given by $\left(\frac{z-y}{x-y}\right)L$ ***litres and*** $\left(\frac{x-z}{x-y}\right)L$ ***litres respectively.***

Illustrative Example

Ex.: A vessel of 80 litres is filled with milk and water. 70% of milk and 30% of water is taken out of the vessel. It is found that the vessel is vacated by 55%. Find the initial quantity of milk and water.

Soln: **Detail Method:** Let the initial quantity of milk be x litres. Therefore, initial quantity of water = (80 – x) litres.

According to the question,

70% of $x + 30\%$ of $(80 - x) = 55\%$ of 80

or, $70x + 2400 - 30x = 4400$

or, $40x = 2000$

$\therefore x = 50$ litres.

Initial quantity of water = (80 – 50) = 30 litres.

Alligation Method: Here the % values of milk and water that is taken from the vessel should be taken into consideration.

Milk 70%   Water 30%

55%

25%   15%

$\Rightarrow 5:3$

Ratio of milk to water = 5 : 3

$\therefore$ quantity of milk $= \frac{80}{5+3} \times 5 = 50$ litres.

and quantity of water $= \frac{80}{5+3} \times 3 = 30$ litres.

Quicker Method: Applying the above theorem,

Initial quantity of milk

$= \left(\frac{55-30}{70-30}\right) \times 80 = \frac{25}{40} \times 80 = 50$ litres.

Initial quantity of water

$= \left(\frac{70-55}{70-30}\right) \times 80 = \frac{15}{40} \times 80 = 30$ litres.

Exercise

1. A vessel of 120 litres is filled with milk and water. 80% of milk and 40% of water is taken out of the vessel. It is found that the vessel is vacated by 65%. What is the ratio of milk to water?
   a) 5:3 b) 6:5
   c) 3:5 d) 4:3
2. A vessel of 40 litres is filled with milk and water. 75% of milk and 35% of water is taken out of the vessel. It is found that the vessel is vacated by 60%. Find the initial quantity of milk and water.
   a) 25 litres, 15 litres b) 30 litres, 10 litres
   c) 22 litres, 18 litres d) None of these
3. A vessel of 80 litres is filled with milk and water. 65% of milk and 25% of water is taken out of the vessel. It is found that the vessel is vacated by 50%. Find the initial quantity of milk and water.
   a) 45 litres, 35 litres b) 50 litres, 30 litres
   c) 55 litres, 25 litres d) None of these

Answers

1. a 2. a 3. b

Rule 14

Theorem: ***In a group, there are some 4-legged creatures and some 2-legged creatures. If heads are counted, there are x and if leggs are counted there are y, then the no. of 4-legged creatures are given by*** $\left(\frac{y-2x}{2}\right)$ ***or*** $\left(\frac{Total\ legs - 2 \times Total\ heads}{2}\right)$ ***and the no. of 2-legged creatures are given by*** $\left(\frac{4x-y}{2}\right)$ ***or*** $\left(\frac{4 \times Total\ heads - Total\ legs}{2}\right)$.

Illustrative Example

Ex.: In a zoo, there are rabbits and pigeons. If heads are counted, there are 200 and if legs are counted, there are 580. How many pigeons are there?

Soln: **Detail Method:** Let the no. of rabbits be R and the pigeons be P.

According to the question,

$R + P = 200$ ..... (i) and

$4R + 2P = 580$ .... (ii)

[$\because$ Rabbits are 4-legged creatures and pigeons are 2-legged creatures.]

From solving eqn (i) and (ii) we get

$R = 90$, and $P = 110$

$\therefore$ No. of rabbits = 90 and

No. of pigeons = 110.

Alligation Method: Rule of Alligation is applicable on number of legs per head.

Average number of legs per head $= \frac{580}{200} = \frac{29}{10}$

Rabbit 4   Pigeons 2

$\frac{29}{10}$

$\frac{9}{10}$ $\frac{11}{10}$

$\therefore$ Rabbit : Pigeons = 9 : 11

$\therefore$ Number of pigeons $= \frac{200}{9+11} \times 11 = 110$

Quicker Method: Applying the above theorem, we have
the number of pigeons (2-legged)

$$= \frac{4 \times 200 - 580}{2} = 110.$$

Exercise

1. In a courtyard there are many chickens and goats. If heads are counted, it comes to 100 but when legs are counted, it comes to 320. Find the number of chickens and goats in the courtyard.
 a) 40, 60 b) 60, 40
 c) 45, 55 d) 55, 45
2. In a zoo, there are rabbits and pigeons. If heads are counted, there are 100 and if legs are counted, there are 290. How many rabbits are there?
 a) 55 b) 45 c) 40 d) 50
3. In a zoo, there are rabbits and pigeons. If heads are counted, there are 50 and if legs are counted, there are 140. How many pigeons are there?
 a) 20 b) 25 c) 30 d) 35

Answers

1. a 2. b 3. c

Rule 15

Theorem: ***A jar contains a mixture of two liquids A and B in the ratio a : b. When L litres of the mixture is taken out and P litres of liquid B is poured into the jar, the ratio becomes x : y. Then the amount of liquid A, contained in the jar, is given by***

$$\left[\frac{L\left(\frac{y}{x}-\frac{b}{a}\right)+P\left(1+\frac{b}{a}\right)}{\left(\frac{a}{b}-\frac{b}{a}\times\frac{x}{y}\right)+\left(1-\frac{x}{y}\right)}\times\left(\frac{x}{y}\times\frac{a}{b}\right)\right] \textit{ litres}$$

and the amount of liquid B in the jar is given by

$$\left[\frac{L\left(\frac{y}{x}-\frac{b}{a}\right)+P\left(1+\frac{b}{a}\right)}{\left(\frac{a}{b}-\frac{b}{a}\times\frac{x}{y}\right)+\left(1-\frac{x}{y}\right)}\times\left(\frac{x}{y}\right)\right] \textit{ litres.}$$

Illustrative Example

Ex.: A jar contains a mixture of two liquids A and B in the ratio 4 : 1. When 10 litres of the mixture is taken out and 10 litres of liquid B is poured into the jar, the ratio becomes 2 : 3. How many litres of liquid A was contained in the jar?

Soln: Method I:

In original mixture, % of liquid B $= \frac{1}{4+1} \times 100 = 20\%$

In the resultant mixture, % of liquid B

$$= \frac{3}{2+3} \times 100 = 60\%$$

Replacement is made by the liquid B, so the % of B in second mixture = 100%

Then by the method of Alligation:

20% 100%
60%
40% 40%

$\therefore$ Ratio in which first and second mixtures should be added is 1 : 1. What docs it imply? It simply implies that the reduced quantity of the first mixture and the quantity of mixture B which is to be added are the same.

$\therefore$ Total mixture = 10 + 10 = 20 litres.

and liquid A $= \frac{20}{5} \times 4 = 16$ litres.

Method II

The above method is explained through percentage. Now, method II will be explained through fraction.

Fraction of B in original mixture $= \frac{1}{5}$

Fraction of B in second mixture (liquid B) = 1

Fraction of B in resulting mixture $= \frac{3}{5}$

So,

$\frac{1}{5}$ 1
$\frac{3}{5}$
$\frac{2}{5}$ $\frac{2}{5}$

Thus, we see that the original mixture and liquid B are mixed in the same ratio. That is, if 10 litres of liquid B is added then after taking out 10 litres of mixture from the jar, there should have been 10 litres of mixture left. So, the quantity of mixture in the jar = 10 + 10 = 20 litres

and quantity of A in the jar $= \frac{20}{5} \times 4 = 16$ litres.

Method III

This method is different from the Method of Alligation. Let the quantity of mixture in the jar be $5x$ litres. Then

$$4x-10\left(\frac{4}{4+1}\right):x-10\left(\frac{1}{4+1}\right)+10=2:3 \text{ .... (*)}$$

or, $4x-8:x-2+10=2:3$

or, $\frac{4x-8}{x+8}=\frac{2}{3}$ $\therefore x=4$

Then quantity of A in the mixture = $4x = 4 \times 4 = 16$ litres.

Note: (*): Liquid A in original mixture $= 4x$

Liquid A taken out with 10 litres of mixture $= 10\times\frac{4}{4+1}$ litres.

$\therefore$ Remaining quantity of A in the mixture

$$= 4x-10\left(\frac{4}{5}\right)$$

Liquid B in original mixture $= x$

Liquid B taken out with 10 litres of mixture

$$= 10\left(\frac{1}{5}\right) \text{ litres}$$

Liquid B added = 10 litres.

$\therefore$ Total quantity of liquid B $= x-10\left(\frac{1}{5}\right)+10$

And the ratio of the two should be 2 : 3.

Quicker Method: Applying the above theorem, we have

amount of liquid A contained in the jar

$$= \frac{10\left(\frac{3}{2}-\frac{1}{4}\right)+10\left(1+\frac{1}{4}\right)}{\left(4-\frac{1}{4}\times\frac{2}{3}\right)+\left(1-\frac{2}{3}\right)}\times\left(\frac{2}{3}\times\frac{4}{1}\right)$$

$$= \frac{\frac{25}{2}+\frac{25}{2}}{\frac{23}{6}+\frac{1}{3}}\times\frac{2}{3}\times\frac{4}{1}=8\times 2=16 \text{ litres.}$$

Exercise

1. A jar contains a mixture of two liquids A and B in the ratio 3 : 1. When 15 litres of the mixture is taken out and 9 litres of liquid B is poured into the jar, the ratio becomes 3 : 4. How many litres of liquid A was contained in the jar?
   a) 27 litres b) 24 litres c) 30 litres d) 21 litres
2. A vessel contains mixture of liquids A and B in the ratio 3 : 2. When 20 litres of the mixture is taken out and replaced by 20 litres of liquid B, the ratio changes to 1 : 4. How many litres of liquid A was there initially present in the vessel?
   a) 14 litres b) 20 litres c) 18 litres d) 30 litres
3. A can contains a mixture of two liquids in proportion 7:5. When 9 litres of mixture are drawn off and the can is filled with B, the proportion of A and B becomes 7:9. How many litres of liquid A was contained by the can initially?
   a) 21 litres b) 18 litres c) 24 litres d) None of these
   (Railways 1991)

Answers

1. a 2. c 3. a

Rule 16

Theorem: ***L litres of a mixture contains two liquids A and B in the ratio a : b. The amount of liquid B, that is added to get a new mixture containing liquid A and B in the ratio x : y, is given by*** $\left[\left(\frac{\frac{y}{x}}{1+\frac{b}{a}}\right)-\left(\frac{1}{1+\frac{a}{b}}\right)\right]L$ ***litres.***

Illustrative Example

Ex.: 729 litres of a mixture contains milk and water in the ratio 7 : 2. How much water is to be added to get a new mixture containing milk and water in the ratio 7 : 3?

Soln: **Detail Method:** Let the amount of water be x litres.

The original mixture contains $\left(\frac{729\times7}{7+2}\right)$ litres of milk and $\left(\frac{729\times2}{7+2}\right)$ litres of water.

Now, from the question,

x litres of water is added. Therefore $\frac{\frac{729\times7}{9}}{\frac{729\times2}{9}+x}=\frac{7}{3}$

or, $729\times7\times3=729\times2\times7+9\times7x$

or, $63x=7\times729$

$\therefore x=\frac{7\times729}{63}=81$ litres.

Alligation Method:

To solve this question by the method of alligation, we can use either of the two, **percentage** or **fractional** value.

Percentage value $\Rightarrow$ change the ratio into percentage.

% of water in the original mixture

$$=\frac{2}{7+2}\times100=\frac{200}{9}\%$$

% of water in the resulting mixture $= \frac{3}{7+3} \times 100 = 30\%$

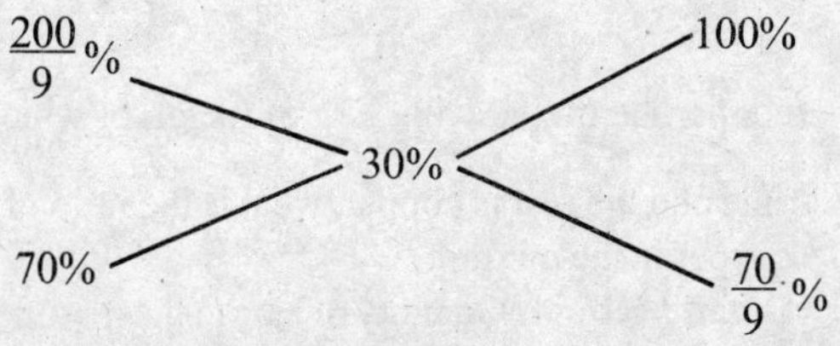

Therefore, the ratio in which the mixture and water are to be added is $1 : \frac{1}{9}$ or 9 : 1

Then quantity of water to be added $= \frac{729}{9} \times 1$

$= 81$ litres.

Fractional value ⇒ Change the ratio into fraction.

Fraction of water in the original mixture $= \frac{2}{9}$

Fraction of water in the resulting mixture $= \frac{3}{10}$.

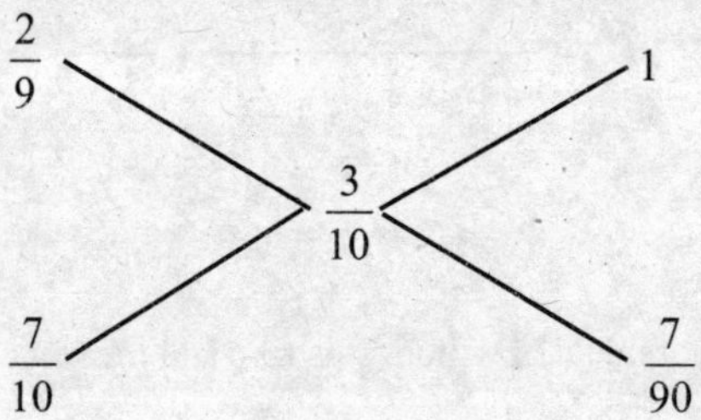

Therefore, the ratio in which the mixture and water are to be added is $\frac{7}{10} : \frac{7}{90} = 1 : \frac{1}{9} = 9 : 1$

Then quantity of water to be added to the mixture =

$\frac{729}{9} = 81$ litres.

Quicker Method: Applying the above theorem,

$$\text{required amount of water} = \left[\left(\frac{\frac{3}{7}}{1+\frac{2}{7}}\right) - \left(\frac{1}{1+\frac{7}{2}}\right)\right] \times 729$$

$$= \left[\frac{3}{7} \times \frac{7}{9} - \frac{2}{9}\right] \times 729$$

$$= \left(\frac{1}{3} - \frac{2}{9}\right) \times 729 = \frac{729}{9} = 81 \text{ litres.}$$

Exercise

1. 56 litres of a mixture contains milk and water in the ratio 5 : 2. How much water is to be added to get a new mixture containing milk and water in the ratio 5 : 3?
   a) 9 litres b) 6 litres c) 7 litres d) 8 litres
2. 36 litres of a mixture contains milk and water in the ratio 2 : 1. How much water is to be added to get a new mixture containing milk and water in the ratio 1 : 1?
   a) 12 litres b) 16 litres c) 8 litres d) 15 litres
3. 25 litres of a mixture contains milk and water in the ratio 3 : 2. How much water is to be added to get a new mixture containing milk and water in the ratio 3 : 4?
   a) 12 litres b) 8 litres c) 10 litres d) 14 litres

Answers

1\. d 2. a 3. c

Rule 17

Theorem: ***If x glasses of equal size are filled with a mixture of spirit and water. The ratio of spirit and water in each glass are as follows:*** $a_1 : b_1, a_2 : b_2, \ldots a_x : b_x$. ***If the contents of all the x glasses are emptied into a single vessel, then proportion of spirit and water in it is given by***

$$\left(\frac{a_1}{a_1+b_1} + \frac{a_2}{a_2+b_2} + \ldots + \frac{a_x}{a_x+b_x}\right) : \left(\frac{b_1}{a_1+b_1} + \frac{b_2}{a_2+b_2} + \ldots + \frac{b_x}{a_x+b_x}\right)$$

Illustrative Example

Ex.: In three vessels each of 10 litres capacity, mixture of milk and water is filled. The ratios of milk and water are 2 : 1, 3 : 1 and 3 : 2 in the three respective vessels. If all the three vessels are emptied into a single large vessel, find the proportion of milk and water in the mixture.

Soln: By the above theorem the required ratio is

$$\left(\frac{2}{2+1} + \frac{3}{3+1} + \frac{3}{3+2}\right) : \left(\frac{1}{2+1} + \frac{1}{3+1} + \frac{2}{3+2}\right)$$

$$= \left(\frac{2}{3} + \frac{3}{4} + \frac{3}{5}\right) : \left(\frac{1}{3} + \frac{1}{4} + \frac{2}{5}\right)$$

$$= \frac{40+45+36}{3 \times 4 \times 5} : \frac{20+15+24}{3 \times 4 \times 5}$$

$$= 121 : 59$$

Note: This question can also be solved without using the theorem. For convenience in calculation, you will have to suppose the capacity of the vessels to be the LCM of (2 + 1), (3 + 1) and (3 + 2), i.e. 60 litres. Because it hardly matters whether the capacity of each vessel is 10 litres or 60 litres or 1000 litres. The only thing is that they should have equal quantity of mixture.

Exercise

1. Three equal glasses are filled with a mixture of spirit and

water. The proportion of spirit to water in each glass is as follows. In the first glass as 2:3, in the second as 3:4, in the third as 4:5. The contents of the three glasses are emptied into a single vessel. What is the proportion of spirit and water in it?
a) 401:544 b) 501:445 c) 544:401 d) 455:401

2. Three equal glasses are filled with mixtures of milk and water. The proportion of milk and water in each glass is as follows. In the first glass as 3:1, in the second glass as 5:3 and in the third as 9:7. The contents of the three glasses are emptied into a single vessel. What is the proportion of milk and water in it?
a) 31:17 b) 17:31 c) 15:31 d) 31:15

3. Four vessels of equal sizes contain mixture of spirit and water. The concentration of spirit in 4 vessels are 60%, 70%, 75% and 80% respectively. If all the four mixtures are mixed, find in the resultant mixture the ratio of spirit to water.
a) 57:13 b) 23:57 c) 57:23 d) None of these

4. Two equal glasses filled with mixtures of alcohol and water in the proportions of 2 : 1 and 1 : 1 respectively were emptied into a third glass. What is the proportion of alcohol and water in the third glass?
a) 5:7 b) 7:5 c) 3:5 d) 5:3
(Bank PO Exam, 1990)

Answers

1. a 2. a
3. c; **Hint:** Ratio of spirit to water in the different vessels

$$\Rightarrow \frac{60}{40}=3:2,\ \frac{75}{25}=3:1\ \frac{70}{30}=7:3,\ \frac{80}{20}=4:1$$

Now applying the given rule, we have the required ratio

$$=\left[\frac{3}{5}+\frac{7}{10}+\frac{3}{4}+\frac{4}{5}\right]:\left[\frac{2}{5}+\frac{3}{10}+\frac{1}{4}+\frac{1}{5}\right]$$

$$=\frac{12+14+15+16}{20}:\frac{6+6+5+4}{20}=57:23$$

4. b

Rule 18

Theorem: ***If M kg of mixture, of which $\frac{a}{b}$ is A and the rest is B, be mixed with N kg of metal, of which $\frac{x}{y}$ is A and the rest is B, then the ratio of A to B in the mixture is given by***

$$\left[\frac{M\frac{a}{b}+N\frac{x}{y}}{M\left(1-\frac{a}{b}\right)+N\left(1-\frac{x}{y}\right)}\right].$$

Illustrative Example

Ex.: If 2 kg of metal, of which $\frac{1}{3}$ is zinc and the rest of copper, be mixed with 3 kg of metal, of which $\frac{1}{4}$ is zinc and the rest is copper, what is the ratio of zinc to copper in the mixture?

Soln: **Detail Method:** Quantity of zinc in the mixture

$$=2\left(\frac{1}{3}\right)+3\left(\frac{1}{4}\right)=\frac{2}{3}+\frac{3}{4}=\frac{8+9}{12}=\frac{17}{12}$$

Quantity of copper in the metal

$$=3+2-\frac{17}{12}=5-\frac{17}{12}=\frac{43}{12}$$

$$\therefore \text{ratio}=\frac{17}{12}:\frac{43}{12}=17:43$$

Quicker Method: Applying the above theorem, we have
the required ratio

$$=\frac{2\times\frac{1}{3}+3\times\frac{1}{4}}{2\left(1-\frac{1}{3}\right)+3\left(1-\frac{1}{4}\right)}=\frac{\frac{2}{3}+\frac{3}{4}}{\frac{4}{3}+\frac{9}{4}}=17:43.$$

Exercise

1. If 1 kg of metal of which $\frac{1}{3}$ is zinc and the rest copper be mixed with 2 kg of metal of which $\frac{1}{4}$ is zinc and the rest copper, what is the ratio of zinc to copper in the mixture?
a) 5:13 b) 6:13 c) 13:5 d) 13:6

2. If 4 kg of metal of which $\frac{1}{4}$ is zinc and the rest copper be mixed with 5 kg of metal of which $\frac{1}{5}$ is zinc and the rest copper, what is the ratio of zinc to copper in the mixture?
a) 2:7 b) 3:7 c) 4:7 d) 5:7

3. If 5 kg of metal of which $\frac{3}{5}$ is zinc and the rest copper be mixed with 3 kg of metal of which $\frac{2}{3}$ is zinc and the rest copper, what is the ratio of zinc to copper in the mixture?
a) 3:5 b) 5:3 c) 5:2 d) 2:5

Answers

1. a 2. a 3. b

Rule 19

Theorem: ***If x glasses of different sizes, say S_1, S_2, S_3,... S_x, are filled with a mixture of spirit and water. The ratio of spirit and water in each glass are as follows, $a_1 : b_1$, $a_2 : b_2$, $a_3 : b_3$,...., $a_x : b_x$. If the contents of all the glasses are emptied into a single vessel, then proportion of spirit and water in it is given by***

$$\left[\frac{a_1S_1}{a_1+b_1}+\frac{a_2S_2}{a_2+b_2}+\frac{a_3S_3}{a_3+b_3}+...+\frac{a_xS_x}{a_x+b_x}\right]:$$

$$\left[\frac{b_1S_1}{a_1+b_1}+\frac{b_2S_2}{a_2+b_2}+\frac{b_3S_3}{a_3+b_3}+...+\frac{b_xS_x}{a_x+b_x}\right]$$

Note: ***Rule 17*** is the special case of this rule.

Illustrative Example

Ex.: Three glasses of sizes 3 litres, 4 litres and 5 litres contain mixture of spirit and water in the ratio 2 : 3, 3 : 7 and 4 : 11 respectively. The contents of all the three glasses are poured into single vessel. Find the ratio of spirit to water in the resultant mixture.

Soln: Applying the above theorem,

Spirit : Water

$$=\left[\frac{2\times3}{2+3}+\frac{3\times4}{3+7}+\frac{4\times5}{4+11}\right]:\left[\frac{3\times3}{2+3}+\frac{7\times4}{3+7}+\frac{11\times5}{4+11}\right]$$

$$=\left[\frac{6}{5}+\frac{12}{10}+\frac{20}{15}\right]:\left[\frac{9}{5}+\frac{28}{10}+\frac{55}{15}\right]$$

$$=\frac{56}{15}:\frac{124}{15}=56:124$$

or, Spirit : Water = 14 : 31.

Exercise

1. Two casks of 48 and 42 litres are filled with mixtures of wine and water, the proportions in the two casks being respectively 13:7 and 18:17. If the contents of the two casks be mixed, and 20 litres of water added to the whole what will be the proportion of wine to water in the result?

   a) 13:12 b) 12:13 c) 21:31 d) 31:21

2. Three glasses of capacity 2 litres, 5 litres and 9 litres contain mixture of milk and water with milk concentrations 90%, 80% and 70% respectively. The contents of three glasses are emptied into a large vessel. Find the milk concentration and ratio of milk to water in the resultant mixture.

   a) 121:39 b) 131:49 c) 39:121 d) 49:131

3. Four glasses of sizes 3 litres, 4 litres, 6 litres and 7 litres contain mixture of milk and water in the ratio 2:1, 5:3, 6:3 and 9:5 respectively. Find the ratio milk to water if the contents of all the four glasses are poured into one large vessel.

   a) 13:6 b) 13:7 c) 11:7 d) 7:13

4. Three vessels of sizes 3 litres, 4 litres and 5 litres contain mixture of milk and water. The concentration of milk in the three vessels are 60%, 75% and 80% respectively. If all the three mixtures are mixed, what is the ratio of milk to water in the resultant mixture.

   a) 11:4 b) 12:5 c) 4:11 d) 5:12

Answers

1. b; **Hint:** Ratio of wine to water, when 20 litres of water are not added

$$=\left[\frac{13\times48}{20}+\frac{18\times42}{35}\right]:\left[\frac{7\times48}{20}+\frac{17\times42}{35}\right]$$

$=264:186=44:31$

Now, 20 litres of water are added,

$$\text{quantity of wine}=\frac{48+42}{44+31}\times44=\frac{264}{5}\text{ litres}$$

$$\text{and quantity of water}=20+\left(\frac{48+42}{44+31}\times31\right)$$

$$=\frac{186}{5}+20=\frac{286}{5}$$

$$\therefore\text{ required ratio}=\frac{264}{5}:\frac{286}{5}=12:13$$

2. a; **Hint:** Ratios are $\frac{90}{10}=9:1$, $\frac{80}{20}=4:1$, $\frac{70}{30}=7:3$

3. b 4. a

Rule 20

Theorem: ***A man mixes M_1 litres of milk at Rs x per litre with M_2 litres at Rs y per litre. Amount of water, that should be added to make the average value of the mixture Rs z per litre, is given by $\left[\frac{M_1(x-z)+M_2(y-z)}{z}\right]$ litres.***

Illustrative Example

Ex.: A man mixes 5 kilolitres of milk at Rs 600 per kilolitre with 6 kilolitres at Rs 540 per kilolitre. How many kilolitres of water should be added to make the average value of the mixture Rs 480 per kilolitre?

Soln: **Detail Method:** According to the question,

Cost of 5 kilolitres of milk = 600 × 5 = Rs 3000

Cost of 6 kilolitres of milk = 540 × 6 = Rs 3240

Now, we suppose that x kilolitres of water is added.

Total amount of the mixture = 5 + 6 + x = (11 + x) kilolitres.

Total cost of the mixture = Rs 3000 + Rs 3240 = R 6240

From the question,

$$\frac{6240}{11+x} = 480 \quad \text{or, } 11+x = \frac{6240}{480} = 13$$

$\therefore x = 2$ kilolitres.

Alligation Method: This question should be solved by the method of alligation.

Cost of milk when two qualities are mixed

$$= \frac{5\times 600 + 6\times 540}{5+6} = \text{Rs } \frac{6240}{11} \text{ per kilolitre.}$$

Cost of water = Rs 0/ kilolitre.

So, First mixture (milk) Second mixture (water)

$\frac{6240}{11}$ 0

480

480 $\frac{960}{11}$

$\therefore$ Ratio of milk and water $= 480 : \frac{960}{11}$

$$= 1 : \frac{2}{11} = 11 : 2$$

Which implies that 11 kilolitres of milk should be mixed with 2 kilolitres of water. Thus 2 kilolitres of water should be added.

Quicker Method: Applying the above theorem, we have

the required amount of water

$$= \frac{5\times(600-480) + 6\times(540-480)}{480}$$

$$= \frac{5\times 120 + 6\times 60}{480} = \frac{960}{480} = 2 \text{ kilolitres.}$$

Exercise

1. A man mixes 5 kilolitres of milk at Rs 6000 per kilolitre with 6 kilolitres at Rs 5400, and with sufficient water to make the average value Rs 4800 per kilolitre. How many kilolitres of water has he added?

   a) 2 kilolitres b) 4 kilolitres

   c) 3 kilolitres d) 1.5 kilolitres

2. A man mixes 6 kilolitres of milk at Rs 650 per kilolitre with 7 kilolitres at Rs 600, and with sufficient water to make the average value Rs 540 per kilolitre. How many kilolitres of water has he added?

   a) 3 kilolitres b) 2 kilolitres

   c) 4 kilolitres d) None of these

3. A man mixes 6 kilolitres of milk at Rs 325 per kilolitre with 9 kilolitres at Rs 300, and with sufficient water to make the average value Rs 270 per kilolitre. How many kilolitres of water has he added?

   a) $2\frac{2}{9}$ kilolitres b) $2\frac{1}{9}$ kilolitres

   c) $3\frac{1}{3}$ kilolitres d) $6\frac{2}{3}$ kilolitres

Answers

1. a 2. b 3. a

Rule 21

Ex.: In an alloy, zinc and copper are in the ratio 1 : 2. In the second alloy the same elements are in the ratio 2 : 3. In what ratio should these two alloys be mixed to form a new alloy in which the two elements are in ratio 5 : 8?

Soln: **Detail Method:** Let them be mixed in the ratio x : y

Then, in 1st alloy, Zinc $= \frac{x}{3}$ and Copper $= \frac{2x}{3}$

2nd alloy, Zinc $= \frac{2y}{5}$ and Copper $= \frac{3y}{5}$

Now, we have $\frac{x}{3} + \frac{2y}{5} : \frac{2x}{3} + \frac{3y}{5} = 5 : 8$

or, $\frac{5x+6y}{10x+9y} = \frac{5}{8}$ or, $40x + 48y = 50x + 45y$

or, $10x = 3y$ $\therefore \frac{x}{y} = \frac{3}{10}$

Thus, the required ratio = 3 : 10.

By Method of Alligation:

You must know that we can apply this rule over the fractional value of either zinc or copper. Let us consider the fractional value of zinc.

1st alloy 2nd alloy

$\frac{1}{3}$ $\frac{2}{5}$

$\frac{5}{13}$ (mixture)

$\frac{1}{65}$ $\frac{2}{39}$

Therefore, they should be mixed in the ratio

$\frac{1}{65} : \frac{2}{39}$ or, $\frac{1}{65} \times \frac{39}{2} = \frac{3}{10}$ or 3 : 10

Note: Now, we try to solve it by taking fractional value of Copper.

1st alloy 2nd alloy

$\frac{2}{3}$ $\frac{3}{5}$

$\frac{8}{13}$

$\frac{1}{65}$ $\frac{2}{39}$

Therefore, they should be mixed in the ratio

$$\frac{1}{65}:\frac{2}{39} \text{ or, } \frac{1}{65}\times\frac{39}{2}=\frac{3}{10} \text{ or, } 3:10$$

Exercise

1. In an alloy, zinc and copper are in the ratio 1 : 3. In the second alloy the same elements are in the ratio 3 : 4. In what ratio should these two alloys be mixed to form a new alloy in which the two elements are in ratio 5 : 4?
   a) 7 : 11 b) 4 : 11 c) 5 : 11 d) None of these
2. In an alloy, zinc and copper are in the ratio 2 : 3. In the second alloy the same elements are in the ratio 4 : 5. In what ratio should these two alloys be mixed to form a new alloy in which the two elements are in ratio 6 : 5?
   a) 5 : 36 b) 25 : 36 c) 35 : 36 d) None of these
3. In an alloy, zinc and copper are in the ratio 3 : 4. In the second alloy the same elements are in the ratio 4 : 5. In what ratio should these two alloys be mixed to form a new alloy in which the two elements are in ratio 7 : 3?
   a) 161 : 181 b) 171 : 181 c) 161 : 171 d) 151 : 161
4. A jar full of whisky contains 40% of alcohol. A part of this whisky is replaced by another containing 19% alcohol and now the percentage of alcohol was found to be 26. The quantity of whisky replaced is:
   a) $\frac{2}{5}$ b) $\frac{1}{3}$ c) $\frac{2}{3}$ d) $\frac{3}{5}$

(Hotel Management, 1991)

Answers

1\. a 2. b 3. c

4\. c; **Hint:** Ratio of alcohol to whisky in the Jar = 40:60 = 2:3.
Ratio of alcohol to whisky in another jar = 19:81.
Ratio of alcohol to whisky in the new mixture = 26:74 = 13:37
Now, applying the given **alligation method**, we have

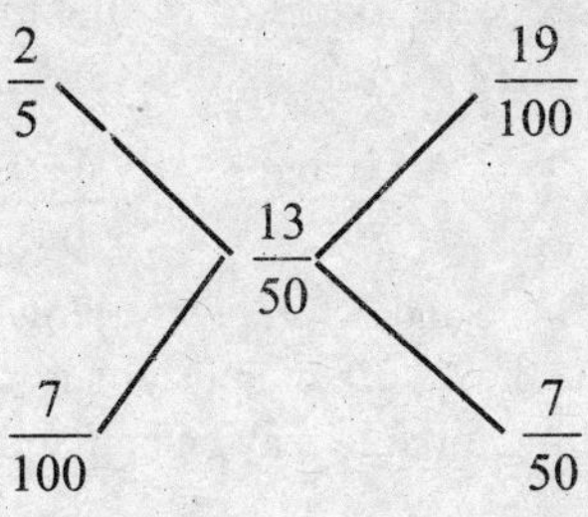

$\therefore$ ratio of alcohol to whisky in the replaced mixture

$$=\frac{7}{100}:\frac{7}{50}=1:2$$

$$\therefore \text{ quantity of whisky replaced } = \frac{2}{1+2}=\frac{2}{3}.$$

Rule 22

Theorem: ***If a person buys n kg of an item at the rate of Rs P per kg. If he sells m kg at a profit of x%, then the rate per kg, at which he should sell the remaining to get a profit of y% on the total deal, is given by Rs P*** $\left[1+\frac{ny-mx}{(n-m)100}\right]$.

Illustrative Example

Ex.: Jayshree purchased 150 kg of wheat at the rate of Rs 7 per kg. She sold 50 kg at a profit of 10%. At what rate per kg should she sell the remaining to get a profit of 20% on the total deal?

Soln: Detail Method:

Selling price of 150 kg wheat at 20% profit

$$=150\times7\left(\frac{120}{100}\right)=\text{Rs } 1260$$

Selling price of 50 kg wheat at 10% profit

$$=50\times7\left(\frac{110}{100}\right)=\text{Rs } 385$$

$\therefore$ Selling price per kg of remaining 100 kg wheat

$$=\frac{1260-385}{100}=\text{Rs } 8.75$$

By Method of Alligation: Selling price per kg at 10% profit = Rs 7.70
Selling price per kg at 20% profit = Rs 8.40
Now, the two lots are in ratio = 1 : 2

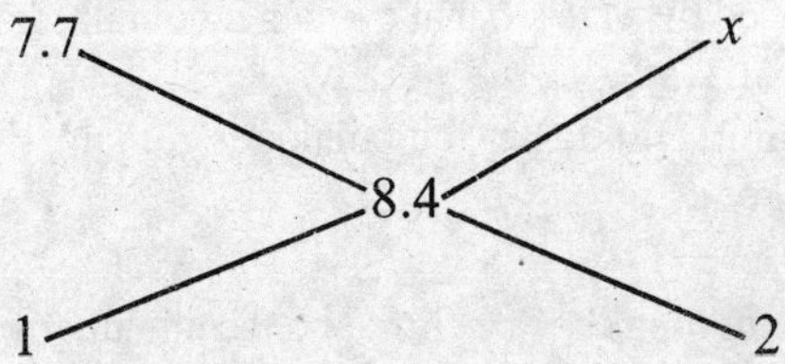

$$\Rightarrow \frac{8.4-7.7}{x-8.4}=\frac{2}{1} \therefore x-8.4=\frac{0.7}{2}=0.35 \quad \therefore x=8.75$$

$\therefore$ Selling price per kg of remaining 100 kg = Rs 8.75

Quicker Method: Applying the above theorem, we have

$$\text{the required answer} = \left[\frac{150\times20-50\times10}{(150-50)\times100}+1\right]\times7$$

$$=\left[\frac{3000-500}{100\times100}+1\right]\times7$$

$$=\frac{5}{4}\times7=\frac{35}{4}=\text{Rs } 8.75 \text{ per kg.}$$

Exercise

1. Sugandha purchased 160 kg of rice at the rate of Rs 25 per kg. She sold 60 kg at a profit of 20%. At what rate per kg should she sell the remaining to get a profit of 30% on the total deal?

a) Rs 17 b) Rs 24 c) Rs 31 d) Rs 34

2. Sunanda purchased 80 kg of wheat at the rate of Rs 10 per kg. She sold 30 kg at a profit of 10%. At what rate per kg should she sell the remaining to get a profit of 15% on the total deal?

a) Rs 11.8 b) Rs 10.8 c) Rs 11 d) Rs 10.75

3. Mala purchased 75 kg of pulses at the rate of Rs 8 per kg. She sold 25 kg at a profit of 5%. At what rate per kg should she sell the remaining to get a profit of 10% on the total deal?

a) Rs 8.25 b) Rs 9.50 c) Rs 9 d) Rs 9.75

Answers

1. d 2. a 3. c

Rule 23

Theorem: ***A container contains x part milk and y part water. From this container, 'a' part of the mixture is taken out and replaced by water. Now, half of the container contains milk and another half contains water. The value of 'a' is given by*** $\frac{1}{2}\left(\frac{x-y}{x}\right)$ ***part.***

Illustrative Example

Ex.: A container contains 7 part milk and 3 part water. How many parts of mixture should be taken out and replaced by water so that container contains half milk and half water.

Soln: **Detail Method:** Let the container contain 1 litre of mixture

Amount of milk = $\frac{7}{10}$ litre and the amount of water

$= \frac{3}{10}$ litre.

Now, let us suppose that x part of the mixture is taken out. In the container amount of milk $= \left(\frac{7}{10} - \frac{7x}{10}\right)$ litres and the amount of water $\left(\frac{3}{10} - \frac{3x}{10}\right)$ litres.

If container is replaced by x part of water, then the amount of water in the container becomes $\left(\frac{3}{10} - \frac{3x}{10} + x\right)$ litres.

As per the question

$$\frac{\frac{7}{10} - \frac{7x}{10}}{\frac{3}{10} - \frac{3x}{10} + x} = \frac{\frac{1}{2}}{\frac{1}{2}}$$

or, $\frac{7}{10}(1-x) = \frac{3}{10}(1-x) + x$

or, $(1-x)\frac{2}{5} = x$ or, $\frac{7}{5}x = \frac{2}{5}$ $\therefore x = \frac{2}{7}$

$\therefore \frac{2}{7}$ part of the mixture is taken out.

Alligation Method: Let us suppose that initially container contains x litres of the mixture, then

Milk : Water $= \frac{7x}{10} : \frac{3x}{10} = 7:3$

Now, applying the alligation method,

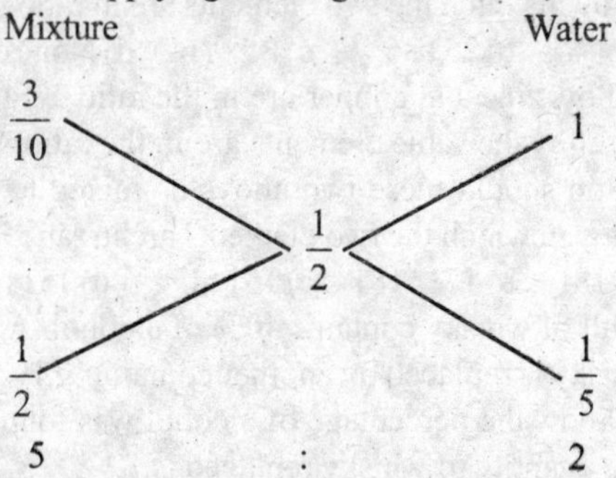

Now, according to the question,

taken out mixture = replaced water $= \frac{2}{7}$ part.

Quicker Method: Applying the above theorem,

the required answer $= \frac{1}{2}\left(\frac{7-3}{7}\right) = \frac{2}{7}$ part.

Exercise

1. A vessel is filled with a liquid, 3 parts of which are water and 5 parts syrup. How much of the mixture must be drawn off and replaced with water so that the mixture may be half water and half syrup?

a) $\frac{1}{5}$ b) $\frac{1}{7}$ c) $\frac{4}{5}$ d) $\frac{3}{10}$

2. A cask contains 3 parts ale and 1 part porter. How much of the mixture must be drawn off and porter substituted in order that the resulting mixture may be half and half?

a) $\frac{1}{3}$ b) $\frac{1}{2}$ c) $\frac{1}{5}$ d) $\frac{2}{3}$

3. A container contains 8 parts milk and 4 parts water. How many parts of mixture should be taken out and replaced by water so that container contains half milk and half water.

a) $\frac{1}{5}$ parts b) $\frac{1}{8}$ parts c) $\frac{1}{4}$ parts d) $\frac{1}{2}$ parts

4. A container contains 9 parts milk and 6 parts water. How many parts of mixture should be taken out and replaced by water so that container contains half milk and half water.

a) $\frac{1}{3}$ b) $\frac{1}{6}$ c) $\frac{1}{9}$ d) $\frac{1}{2}$

5. A container contains 4 parts milk and 1 parts water. How many parts of mixture should be taken out and replaced by water so that container contains half milk and half water.

a) $\frac{3}{4}$ b) $\frac{1}{2}$ c) $\frac{5}{8}$ d) $\frac{3}{8}$

Answers

1. a 2. a 3. c 4. b 5. d

Rule 24

Theorem: ***There is a cask which contains 'L' litres of milk as cost price Rs x / litre. The amount of water, which should be added to the cask so that the cost of milk reduces to Rs y / litre, is given by*** $L\left(\frac{x-y}{y}\right)$ ***litres.***

Illustrative Example

Ex.: How much water must be added to a cask which contains 40 litres of milk at cost price Rs 3.5/litre so that the cost of milk reduces to Rs 2/litre?

Soln: **Detail Method:** Let the x litres of water be added to the cask.

Cost price of 40 litres of milk = 40 × 3.5 = Rs 140.

According to the question,

$\frac{140}{40+x} = 2$ or, $2x = 140 - 80 = 60$

$\therefore x = 30$ litres.

Alligation Method: We will apply the alligation on price of milk, water and mixture.

Milk Water
3.5 0
Mean
2
2 1.5

$\therefore$ ratio of milk and water should be 2 : 1.5 = 4 : 3.

$\therefore$ added water $= \frac{40}{4} \times 3 = 30$ litres.

Quicker Method: Applying the above theorem, we have required amount of water

$$= 40\left(\frac{3.5-2}{2}\right) = 40 \times \frac{3}{4} = 30 \text{ litres.}$$

Exercise

1. How much water must be added to 14 kilolitres of milk worth Rs 5.40 a litre so that the value of the mixture may be Rs 4.20 a litre?

a) 4 kilolitres b) 8 kilolitres
c) 6 kilolitres d) 5 kilolitres

2. How much water should be added to 60 litres of milk at $1\frac{1}{2}$ litres for Rs 10 so as to have a mixture worth Rs $5\frac{1}{3}$ per litre?

a) 16 litres b) 15 litres
c) 18 litres d) 20 litres

3. How much water must be added to a cask containing $40\frac{1}{2}$ litres of spirit worth Rs 15.68 a litre to reduce the price to Rs 12.96 a litre?

a) $7\frac{2}{3}$ litres b) $8\frac{1}{3}$ litres
c) $8\frac{1}{2}$ litres d) None of these

4. How much chicory at Rs 24 a kg should be added to 15 kg of tea at Rs 60 a kg, as to make the mixture worth Rs 39 a kg?

a) 21 kg b) 20 kg
c) 27 kg d) 18 kg

5. How many bananas at 5 for Re 1.20 should be mixed with 300 bananas at 6 for Rs 2.10 so that they should all be worth Rs 3.60 a dozen?

a) 350 b) 280 c) 320 d) 250

Answers

1. a

2. b; **Hint:** Here $x = \frac{10 \times 2}{3} = \text{Rs } \frac{20}{3}$ a litre

Now, applying the given rule, we have

the required answer $= \left(\frac{\frac{20}{3} - \frac{16}{3}}{\frac{16}{3}}\right) \times 60 = 15$ litres

3. c

4. a; **Hint:** By alligation Method:

Tea Chicori
60 24
39
15 21

$\therefore$ ratio of tea and chicori = 5:7.

$\therefore$ added chicori $= \frac{15}{5} \times 7 = 21$ kg.

5. d; **Hint:**

Bananas at 6 Bananas at 5

$\frac{210}{6}=35$ $\frac{120}{5}=24$

$\frac{360}{12}=30$

6 5

$\therefore$ Required answer $=\frac{300}{6}\times 5=250$.

Rule 25

Theorem: ***One type of liquid contains x% of A, the other contains y% of A. A can is filled with n parts of the first liquid and m parts of the second liquid. Then the percentage of liquid A in the new mixture is given by*** $\left[\frac{nx+my}{(n+m)}\right]$ ***per cent.***

Illustrative Example

Ex.: One type of liquid contains 25% of milk, the other contains 30% of milk. A can is filled with 6 parts of the first liquid and 4 parts of the second liquid. Find the percentage of milk in the new mixture.

Soln: Detail Method: The reqd. percentage of milk in the new mixture

$$=\frac{\text{Quantity of milk in the new mixture}}{\text{Quantity of the new mixture}}\times 100$$

$$=\frac{6\text{ parts of }25\%\text{ milk}+4\text{ parts of }30\%\text{ milk}}{(6\text{ parts}+4\text{ parts})\text{of the liquid}}\times 100$$

$$=\frac{6\times\frac{25}{100}+4\times\frac{30}{100}}{10}\times 100=(15+12)=27$$

Alligation Method: This equation can be solved by the method of Alligation.

25 30

x

6 4

$$\frac{30-x}{x-25}=\frac{6}{4}=\frac{3}{2}$$

or, $60-2x=3x-75$

or, $5x=60+75$

$\therefore x=27\%$

Quicker Method: Applying the above theorem, we have the

required answer $=\frac{6\times 25+4\times 30}{6+4}=\frac{270}{10}=27\%$.

Exercise

1. A solution of sugar syrup has 15% sugar. Another solution has 5% sugar. How many litres of the second solution must be added to 20 litres of the first solution to make a solution of 10% sugar?
   a) 10 b) 5 c) 15 d) 20
   (NABARD-1999)

2. One liquid contains $22\frac{1}{2}$ per cent of water, another 27 per cent. A glass is filled with 5 parts of one liquid and 7 parts of the other. What percentage of water in the glass?
   a) $25\frac{1}{8}\%$ b) 25.75% c) 25.25% d) 25%

3. One type of liquid contains 15% of milk, the other contains 20% of milk. A can is filled with 4 parts of the first liquid and 11 parts of the second liquid. Find the percentage of milk in the new mixture.
   a) $18\frac{1}{3}\%$ b) 18% c) $18\frac{2}{3}\%$ d) $18\frac{3}{8}\%$

4. One type of liquid contains $12\frac{1}{2}\%$ of milk, the other contains 15% of milk. A can is filled with 8 parts of the first liquid and 12 parts of the second liquid. Find the percentage of milk in the new mixture.
   a) 12% b) 13% c) 15% d) 14%

5. One type of liquid contains $3\frac{1}{3}\%$ of milk, the other contains $5\frac{1}{5}\%$ of milk. A can is filled with 3 parts of the first liquid and 5 parts of the second liquid. Find the percentage of milk in the new mixture.
   a) $4\frac{1}{2}\%$ b) $7\frac{1}{2}\%$ c) $4\frac{1}{3}\%$ d) $7\frac{1}{3}\%$

Answers

1. d; **Hint:** $\frac{15\times 20+5\times m}{20+m}=10$ $\therefore$ m = 20 litres.

2. a 3. c 4. d 5. a

Rule 26

Theorem: ***Weights of two friends A and B are in the ratio of a : b. A's weight increases by x% and the total weight of A and B together becomes w kg, with an increase of y%.***

Then the weight of A $=\left[\frac{a\times 100\times w}{(a+b)(100+y)}\right]$ ***kg.***

Weight of B $=\left[\frac{b\times 100\times w}{(a+b)(100+y)}\right]$ ***kg.***

Total weight = Weight of A + Weight of B = $\left(\frac{100w}{100+y}\right)$ ***kg***

and the per cent by which weight of B increases = $\left[\frac{y(a+b)-ax}{b}\right]\%$.

Illustrative Example

Ex.: Weights of two friends Ram and Shyam are in the ratio of 4 : 5. Ram's weight increases by 10% and the total weight of Ram and Shyam together becomes 82.8 kg, with an increase of 15%. By what per cent did the weight of Shyam increase?

Soln: **Detail Method:** Let the weights of Ram and Shyam be $4x$ and $5x$.

Now, according to question,

$$\frac{4x\times100}{100} + \text{Shyam's new wt} = 82.8 \quad .... \text{(i)}$$

and $$\frac{(4x+5x=)9x\times115}{100} = 82.8 \quad .... \text{(ii)}$$

From (ii), $x = 8$

Putting in (i), we get

Shyam's new wt = (82.8 – 35.2 =) 47.6

$$\%\text{ increase in Shyam's wt} = \left(\frac{47.6-40}{40}\times100\right) = 19\%$$

Alligation Method:

Ram — Shyam

10% — x%

15%

4 — 5 (given)

By the rule of alligation

$$\frac{x-15}{15-10} = \frac{4}{5}$$

or, $x - 15 = 4 \quad \therefore x = 19$

Quicker Method: Applying the above theorem,

% increase in Shyam's weight

$$= \frac{15\times9-4\times10}{5} = \frac{95}{5} = 19\%.$$

Exercise

1. Weights of two friends Naval and Keval are in the ratio of 3 : 4. Naval's weight increases by 16% and the total weight of Naval and Keval together becomes 83 kg, with an increase of 20%. By what per cent did the weight of Keval increase?
   a) 23% b) 32% c) 24% d) 28%
2. Weights of two friends Pran and Prem are in the ratio of 2 : 3. Pran's weight increases by $6\frac{1}{2}\%$ and the total weight of Pran and Prem together becomes 41 kg, with an increase of 8%. By what per cent did the weight of Prem increase?
   a) 10% b) 12% c) 9% d) None of these
3. Weights of two friends Sudhir and Sudhesh are in the ratio of 4 : 1. Sudhir's weight increases by 12% and the total weight of Sudhir and Sudhesh together becomes 50 kg, with an increase of 25%. By what per cent did the weight of Sudhesh increase?
   a) 77% b) 75% c) 74% d) 70%

Answers

1. a 2. c 3. a

Rule 27

Theorem: ***Suppose a container contains M units of mixture of A and B. From this, R unit of mixture is taken out and replaced by an equal amount of ingredient B only. This process (of taking out and replacing it) is repeated n times, then after n operations,***

$$\frac{\text{Amount of A left}}{\text{Amount of A originally present}} = \left(1-\frac{R}{M}\right)^n$$ ***and the amount of B left = M – amount of A left.***

Illustrative Examples

Ex. 1: An 8-litres cylinder contains a mixture of oxygen and nitrogen, the volume of oxygen being 16% of total volume. A few litres of the mixture is released and an equal amount of nitrogen is added. Then the same amount of the mixture as before is released and replaced by nitrogen for the second time. As a result, the oxygen content becomes 9% of the total volume. How many litres of mixture is released each time?

Soln: The cylinder originally contains a mixture of oxygen and nitrogen. An equal amount of released mixture is replaced by an equal amount of nitrogen. So, applying the above formula,

$$\frac{\text{Amount of A (oxygen) left}}{\text{Amount of A (oxygen) originally present}} = \left(1-\frac{R}{M}\right)^n$$

Where, total volume of mixture = Volume of cylinder = M = 8 litres.

Released amount of mixture = R litres

Number of operations done (n) = 2

$$\therefore \frac{0.09\times8}{0.16\times8} = \left(1-\frac{R}{8}\right)^2 \Rightarrow R = 2$$

$\therefore$ 2 litres of mixture is released each time.

Ex. 2: A dishonest hair dresser uses a mixture having 5 parts pure after-shave lotion and 3 parts pure water. After taking out some portion of the mixture, he adds equal amount of pure water to the remaining portion of mixture such that the amount of after-shave lotion and water become equal. Find the part of mixture taken out.

Soln: The hair dresser originally uses mixture. Equal part of the mixture is replaced by an equal part of water. So, using the above theorem,

$$\frac{\text{Amount of A (after-shave lotion) left}}{\text{Amount of A (after-shave lotion) originally present}} = \left[1-\frac{R}{M}\right]^n$$

Where, original amount of mixture = 1 litre (suppose)

$$\frac{\frac{1}{2}}{\frac{5}{8}} = \left[1-\frac{R}{1}\right]^1, \text{(Since n = 1)}$$

$$\Rightarrow R = \frac{1}{5}$$

$\therefore \frac{1}{5}$ part of the mixture has been taken out.

Note: Also see **Rule 23**.

Exercise

1. An 12-litres cylinder contains a mixture of oxygen and nitrogen, the volume of oxygen being 40% of total volume. A few litres of the mixture is released and an equal amount of nitrogen is added. Then the same amount of the mixture as before is released and replaced by nitrogen for the second time. As a result, the oxygen content becomes 10% of the total volume. How many litres of mixture is released each time?
   a) 3 litres b) 9 litres
   c) 6 litres d) None of these
2. An 25-litres cylinder contains a mixture of oxygen and nitrogen, the volume of oxygen being 25% of total volume. A few litres of the mixture is released and an equal amount of nitrogen is added. Then the same amount of the mixture as before is released and replaced by nitrogen for the second time. As a result, the oxygen content becomes 9% of the total volume. How many litres of mixture is released each time?
   a) 15 litres b) 10 litres c) 14 litres d) 18 litres
3. An 50-litres cylinder contains a mixture of oxygen and nitrogen, the volume of oxygen being 25% of total volume. A few litres of the mixture is released and an equal amount of nitrogen is added. Then the same amount of the mixture as before is released and replaced by nitrogen for the second time. As a result, the oxygen content becomes 16% of the total volume. How many litres of mixture is released each time?
   a) 24 litres b) 10 litres
   c) 28 litres d) 20 litres

Answers

1. c 2. b 3. b

Rule 28

Theorem: ***Amount of liquid left after n operations, when the container originally contains x units of liquid from which y units is taken out each time is*** $x\left(\frac{x-y}{x}\right)^n$ ***units. Or, we can alternately write as***

$$\frac{\textit{Amount of the liquid left}}{\textit{Amount of the liquid originally present}} = \left(1-\frac{y}{x}\right)^n.$$

Illustrative Example

Ex.: A container contained 80 kg of milk. From this container 8 kg of milk was taken out and replaced by water. This process was further repeated two times. How much milk is now contained by the container.

Soln: Applying the above theorem, we have

$$\text{the amount of milk left} = \left[\frac{80-8}{80}\right]^3 \times 80 \text{ kg} = 58.32 \text{ kg}$$

Note: Consider a container containing only ingredient 'A' of x_0 unit. From this, x_r unit is taken out and replaced by an equal amount of ingredient B. This process is repeated n times, then, after n operations.

$$\frac{\text{Amount of A left}}{\text{Amount of B left}} = \frac{\left(1-\frac{x_r}{x_0}\right)^n}{1-\left(1-\frac{x_r}{x_0}\right)^n}$$

Ex.: A bottle is full of dettol. One third of it is taken out and then an equal amount of water is poured into the bottle to fill it. This operation is done four times. Find the final ratio of dettol and water in the bottle.

Soln: The bottle originally contains dettol only.
Let the bottle contain 1 litre of dettol originally.
So, applying the above formula,

$$\frac{\text{Amount of A (dettol) left}}{\text{Amount of B (water) left}} = \frac{\left(1-\frac{x_r}{x_0}\right)^n}{1-\left(1-\frac{x_r}{x_0}\right)^n}$$

$$\frac{\text{Dettol}}{\text{Water}} = \frac{\left[1-\frac{\frac{1}{3}}{1}\right]^4}{1-\left[1-\frac{\frac{1}{3}}{1}\right]^4} = \frac{\left(\frac{2}{3}\right)^4}{1-\left(\frac{2}{4}\right)^4} = \frac{16}{65}.$$

$\therefore$ Finally, the bottle contains dettol and water in the ratio 16 : 65.

Exercise

1. From a cask of wine, containing 64 litres, 8 litres are drawn out and the cask is filled up with water. If the same process is repeated a second, then a third time, what will be the number of litres of wine left in the cask?
a) $42\frac{1}{8}$ kg b) $42\frac{3}{8}$ kg c) $48\frac{7}{8}$ kg d) $42\frac{7}{8}$ kg

2. From a vessel filled with alcohol, $\frac{1}{5}$ of its contents is removed, and the vessel is then filled up with water. If this be done 5 times in succession, what proportion of the alcohol originally contained in the vessel will have been removed from it?
a) $\frac{1024}{3125}$ b) $\frac{2101}{3125}$ c) $\frac{1024}{2101}$ d) None of these

3. From a cask full of spirits one-hundredth part is drawn and the cask filled with water. From the mixture one-hundredth part is drawn and the cask again filled with water, and a similar operation is again performed. Find the ratio of the quantity of wine left in the cask to the original quantity after the third operation.
a) 970299 : 1000000 b) 29701 : 1000000
c) 970399 : 1000000 d) 971099 : 1000000

4. From a cask of wine containing 25 litres, 5 litres are with drawn and the cask is filled with water. The process is repeated a second and then a third time. Find the ratio of wine to water in the resulting mixture.
a) 64 : 61 b) 61 : 64 c) 51 : 54 d) 46 : 61

5. A vessel contains 125 litres of wine. 25 litres of wine was taken out of the vessel and replaced by water. Then 25 litres of mixture was withdrawn and again replaced by water. The operation was repeated for third time. How much wine is now left in the vessel?
a) 54 litres b) 25 litres c) 64 litres d) None of these

6. From a cask of wine, containing 64 litres, 8 litres are drawn out and the cask is filled up with water. If the same process is repeated a second, then a third time, what will be the proportion of wine to water in the resulting mixture?
a) 343 : 169 b) 343 : 512 c) 169 : 343 d) 512 : 343

7. A vessel contains 24 litres of milk. 4 litres are withdrawn and replaced by water. The process is repeated a second time. Find the ratio of milk to water in the resulting mixture?
a) 25 : 36 b) 36 : 11 c) 11 : 25 d) 25 : 11

Answers

1. d; **Hint:** Required answer

$$= \left(1-\frac{8}{64}\right)^3 \times 64 = \left(\frac{7}{8}\right)^3 \times 64 = 42\frac{7}{8} \text{ kg}$$

2. b; **Hint:** The alcohol now contained in the vessel

$$= \left(1-\frac{1}{5}\right)^5 = \left(\frac{4}{5}\right)^5 = \frac{1024}{3125}$$

$$\therefore \text{Required answer} = 1-\frac{1024}{3125} = \frac{2101}{3125}.$$

3. a

4. a; **Hint:** Quantity of a wine left in the cask

$$= \left(1-\frac{1}{5}\right)^3 = \left(\frac{4}{5}\right)^3 = \frac{64}{125}$$

$$\text{Quantity of water left in the cask} = 1-\frac{64}{125} = \frac{61}{125}$$

$$\therefore \text{required ratio} = \frac{64}{125}\Big/\frac{61}{125} = \frac{64}{61} = 64 : 61.$$

5. c; **Hint:** Amount of wine left $= 125\left(1-\frac{25}{125}\right)^3$

$$= 125 \times \frac{64}{125} = 64 \text{ litres}$$

6. a; **Hint:** Required proportion

$$= \frac{\left(1-\frac{1}{8}\right)^3}{1-\left(1-\frac{1}{8}\right)^3} = \frac{\frac{343}{512}}{1-\frac{343}{512}} = 343 : 169.$$

7. d; **Hint:** See **Note** of the given rule.

Rule 29

Theorem: ***'L' litres are drawn from a caskfull of water and it is then filled with milk. After n operations, if the quantity of water now left in the cask is to that of milk in it as a : b,***

then the capacity of cask is given by $\left[\frac{L}{1-\left(\frac{a}{a+b}\right)^{1/n}}\right]$ ***litres.***

Illustrative Example

Ex.: Nine litres are drawn from a cask full of water and it is then filled with milk. Nine litres of mixture are drawn and the cask is again filled with milk. The quantity of water now left in the cask is to that of the milk in it as 16 : 9. How much does the cask hold?

Soln: Here no. of operations are 2

$\therefore$ n = 2

Applying the above theorem, we have

the capacity of the cask

$$= \frac{9}{1-\left(\frac{16}{25}\right)^{1/2}} = 9 \times 5 = 45 \text{ litres.}$$

Exercise

1. Eight litres are drawn off from a vessel full of water and substituted by pure milk. Again eight litres of the mixture are drawn off and substituted by pure milk. If the vessel now contains water and milk in the ratio 9 : 40, find the capacity of the vessel.
   a) 14 litres b) 24 litres
   c) 16 litres d) 12 litres
2. Ten litres of wine are drawn from a vessel full of wine. It is then filled up with water. Ten litres of the mixture are drawn and the vessel is again filled up with water. The ratio of the quantity of wine now left in the vessel is to that of the water in it as 144:25. Find the capacity of the vessel.
   a) 135 litres b) 120 litres
   c) 130 litres d) None of these
3. 19 litres are drawn from a vessel full of spirit and it is filled with water. Then 19 litres of the mixture are drawn and the vessel is again filled with water. The ratio of the spirit to water now present in the vessel is 81:19. What is the full capacity of the vessel?
   a) 190 litres b) 180 litres
   c) 170 litres d) 195 litres
4. 6 litres are drawn from a cask full of wine and it is then filled with water. 6 litres of the mixture are drawn and the cask is again filled with water. The quantity of wine now left in the cask is to that of the water in it as 121:23. How much does the cask hold?
   a) 54 litres b) 62 litres c) 70 litres d) 72 litres

Answers

1. a 2. c 3. a 4. d

Miscellaneous

1. There are two vessels of equal capacity, one full of milk, and the second one-third full of water. The second vessel is then filled up out of the first, the contents of the second are then poured back into the first till it is full and then again the contents of the first are poured back into the second till it is full. What is the proportion of milk in the second vessel?
   a) $\frac{20}{37}$ b) $\frac{20}{27}$ c) $\frac{37}{20}$ d) $\frac{27}{20}$
2. Three lumps of gold, weighing respectively 6, 5, 4 g and of 15, 14, $12\frac{1}{2}$ carats fineness are mixed together, what is the fineness of the resulting compound?
   a) 14 carats b) 16 carats c) 12 carats d) 18 carats
3. In what ratio must a person mix three kinds of wheat costing him Rs 1.20, Rs 1.44 and Rs 1.74 per kg, so that the mixture may be worth Rs 1.41 per kg?
   a) 11 : 77 : 7 b) 7 : 11 : 77
   c) 11 : 7 : 77 d) None of these
4. Fresh fruit contains 72% water and dry fruit contains 20% water. How much dry fruit from 100 kg of fresh fruit can be obtained?
   a) 32 kg b) 33 kg c) 30 kg d) 35 kg
   (MBA 1991)
5. In two alloys, copper and zinc are related in the ratios of 4:1 and 1:3. 10 kg of 1st alloy 16 kg of 2nd alloy and some of pure copper are melted together. An alloy was obtained in which the ratio of copper to zinc was 3:2. Find the weight of the new alloy.
   a) 34 kg b) 35 kg c) 36 kg d) 30 kg
   (MBA 1984)

Answers

1. b; Let M be the vessel containing milk and W the vessel containing water.

	First Vessel	**Second Vessel**
1st operation	1M	$\frac{1}{3}W$
2nd operation	$\frac{1}{3}M$	$\frac{1}{3}W+\frac{2}{3}M$
3rd operation	$\frac{1}{3}M+\frac{2}{3}\left(\frac{1}{3}W+\frac{2}{3}M\right)$	$\frac{1}{3}\left(\frac{1}{3}W+\frac{2}{3}M\right)$
4th operation	$\frac{1}{3}\left[\frac{1}{3}M+\frac{2}{3}\left(\frac{1}{3}W+\frac{2}{3}M\right)\right]$	$\left[\frac{1}{3}\left(\frac{1}{3}W+\frac{2}{3}M\right)+\frac{2}{3}\left[\frac{1}{3}M+\frac{2}{3}\left(\frac{1}{3}W+\frac{2}{3}M\right)\right]\right]$

Simplifying the quantity on the right hand side, we get the proportions of water and milk in the second vessel.

$$\left[\frac{1}{9}W+\frac{2}{9}M+\frac{2}{3}\left\{\frac{1}{3}M+\frac{2}{9}W+\frac{4}{9}M\right\}\right]$$

$$= \frac{1}{9}W+\frac{2}{9}M+\frac{2}{9}M+\frac{4}{27}W+\frac{8}{27}M$$

$$\therefore \text{ proportion of milk} = \frac{2}{9}M+\frac{2}{9}M+\frac{8}{27}M = \frac{20}{27}M$$

$\therefore \frac{20}{27}$ of the second vessel is milk.

2. a; Fineness of the compound

$$= \frac{6\times15+5\times14+4\times12\frac{1}{2}}{6+5\times4} \text{ carats}$$

$$= \frac{210}{15} \text{ or 14 carats.}$$

3. a; **Step I.** Mix wheats of first and third kind to get a mixture worth Rs 1.41 per kg.?

C.P. of 1 kg wheat of 1st kind 120 P — C.P. of 1 kg wheat of 3rd type 174 P

Mean price 141 P

33 — 21

By alligation rule:

$$\frac{(\text{Quantity of 1st kind of wheat})}{(\text{Quantity of 3rd kind of wheat})} = \frac{33}{21} = \frac{11}{7}$$

i.e., they must be mixed in the ratio 11 : 7.

Step II. Mix wheats of 1st and 2nd kind to obtain a mixture worth of Rs 1.41 per kg.

C.P. of 1 kg wheat of 1st kind 120 P — C.P. of 1 kg wheat of 2nd kind 144 P

Mean Price 141 P

3 — 21

∴ By alligation rule:

$$\frac{(\text{Quantity of 1st kind of wheat})}{(\text{Quantity of 2nd kind of wheat})} = \frac{3}{21} = \frac{1}{7}.$$

i.e., they must be mixed in the ratio 1 : 7.

Thus, $\frac{(\text{Quantity of 2nd kind of wheat})}{(\text{Quantity of 3rd kind of wheat})}$

$$= \frac{(\text{Quantity of 2nd kind of wheat})}{(\text{Quantity of 1st kind of wheat})} \times \frac{(\text{Quantity of 1st kind of wheat})}{(\text{Quantity of 3rd kind of wheat})}$$

$$= \left(\frac{7}{1} \times \frac{11}{7}\right) = \left(\frac{11}{1}\right)$$

∴ Quantities of wheat of (1st kind : 2nd kind : 3rd kind)

$$= 1:7:\frac{7}{11} = 11:77:7.$$

4. d; We are concerned with solid part of the fruit (pure portion). Assume x kg of dry fruit is obtained.

∴ Solid part in fresh fruit = Solid part in dry fruit

or, $0.28 \times 100 = 0.8 \times x$

or, $x = 35$ kg.

∴ 35 kg of dry fruit can be obtained from 100 kg fresh fruit.

5. b; Here two alloys are mixed to form a third alloy, hence quantity of only one of the ingredients in each of the alloy will be considered. [Refer to **Rule 21**]

Here, pure copper is also added, hence, quantity of copper in all the three alloy will be considered.

Let the amount of pure copper = x kg.

∴ pure copper + copper in 1st alloy + copper in 2nd alloy = copper in 3rd alloy

or, $x + \frac{4}{5}\times10 + \frac{1}{4}\times16 = \frac{3}{5}(10+16+x)$

or, $12 + x = \frac{3}{5}(26+x)$

or, $x = 9$ kg

∴ Weight of new alloy = 10 + 16 + 9 = 35 kg

Note: In place of pure copper, if pure zinc were added then quantity of zinc in all the three alloys have to be considered for finding the weight of the new alloy.

13 Time and Work

Rule 1

Theorem: ***If M_1 persons can do W_1 works in D_1 days and M_2 persons can do W_2 works in D_2 days then we have a very general formula in the relationship of $M_1D_1W_2 = M_2D_2W_1$.***

Illustrative Examples

Ex. 1: 16 men can do a piece of work in 10 days. How many men are needed to complete the work in 40 days?

Soln: **Detail Method:** To do a work in 10 days, 16 men are needed. or, to do the work in 1 day, 16 × 10 men are needed. So to do the work in 40 days, $\frac{16 \times 10}{40} = 4$ men are needed.

Quicker Method: $M_1D_1W_2 = M_2D_2W_1$

$M_1 = 16, D_1 = 10, W_1 = 1$ and

$M_2 = ?, D_2 = 40, W_2 = 1$

Thus, from $M_1D_1W_2 = M_2D_2W_1$

$16 \times 10 = M_2 \times 40$

or, $M_2 = \frac{16 \times 10}{40} = 4$ men.

Ex. 2: 40 men can cut 60 trees in 8 hours. If 8 men leave the job, how many trees will be cut in 12 hours?

Soln: **Detail Method:**

40 men - working 8 hrs - cut 60 trees

or, 1 men - working 1 hr - cuts $\frac{60}{40 \times 8}$ trees

Thus, 32 men - working 12 hrs - cut $\frac{60 \times 32 \times 12}{40 \times 8} = 72$ trees.

Quicker Method:

$M_1 = 40$, $D_1 = 8$ (as days and hrs both denote time)

$W_1 = 60$ (cutting of trees is taken as work)

$M_2 = 40 - 8 = 32$, $D_2 = 12, W_2 = ?$

Putting the values in the formula,

$M_1D_1W_2 = M_2D_2W_1$

We have, $40 \times 8 \times W_2 = 32 \times 12 \times 60$

or, $W_2 = \frac{32 \times 12 \times 60}{40 \times 8} = 72$ trees.

Ex. 3: A can do a piece of work in 5 days. How many days will he take to complete 3 works of the same type?

Soln: **Quicker Method:**

$M_1D_1W_2 = M_2D_2W_1$

As 'A' is the only person to do the work in both the cases, so $M_1 = M_2 = 1$ (Useless to carry it)

$D_1 = 5$ days, $W_1 = 1$, $D_2 = ?$ and $W_2 = 3$

Putting the values in the formula we have,

$5 \times 3 = D_2 \times 1$ or, $D_2 = 15$ days.

Exercise

1. 8 men can do a piece of work in 5 days. How many men are needed to complete the work in 10 days?
   a) 8 men b) 4 men c) 2 men d) None of these
2. 15 men can do a piece of work in 6 days. How many men are needed to complete the work in 3 days?
   a) 30 men b) 25 men c) 35 men d) 40 men
3. 20 men can cut 30 trees in 4 hours. If 4 men leave the job, how many trees will be cut in 6 hours?
   a) 30 trees b) 36 trees c) 40 trees d) None of these
4. 10 men can cut 15 trees in 2 hours. If 2 men leave the job, how many trees will be cut in 3 hours?
   a) 15 trees b) 20 trees c) 16 trees d) 18 trees
5. A can do a piece of work in 6 days. How many days will he take to complete 2 works of the same type?
   a) 12 days b) 10 days c) 6 days d) 3 days

Answers

1. c 2. a 3. b 4. d 5. a

Rule 2

Theorem: ***If M_1 persons can do W_1 works in D_1 days working T_1 hours a day and M_2 persons can do W_2 works in D_2 days working T_2 hours a day then we have a very general formula in the relationship of $M_1D_1T_1W_2 = M_2D_2T_2W_1$.***

Illustrative Example

Ex: 5 men can prepare 10 toys in 6 days working 6 hours a day. Then in how many days can 12 men prepare 16 toys working 8 hrs a day?

Soln: By using the above theorem

$5 \times 6 \times 6 \times 16 = 12 \times D_2 \times 8 \times 10$

$\therefore D_2 = \frac{5 \times 6 \times 6 \times 16}{12 \times 8 \times 10} = 3 \text{ days}$

Note: Number of toys is considered as work in the above example.

Exercise

1. The work done by a woman in 8 hours is equal to the work done by a man in 6 hours and by a boy in 12 hours. If working 6 hours per day 9 men can complete a work in 6 days then in how many days can 12 men, 12 women and 12 boys together finish the same work working 8 hours per day?

   a) $1\frac{1}{2}$ days b) $3\frac{2}{3}$ days c) 3 days d) $1\frac{1}{2}$ days

   (BSRB Patna PO-2001)

2. 10 men can prepare 20 toys in 3 days working 12 hours a day. Then in how many days can 24 men prepare 32 toys working 4 hrs a day?

   a) 2 days b) 3 days c) 4 days d) 6 days

3. 20 men can prepare 40 toys in 24 days working 18 hours a day. Then in how many days can 36 men prepare 48 toys working 16 hrs a day?

   a) 16 days b) 12 days c) 21 days d) 18 days

Answers

1. d; **Hint:** 8 Women = 6 Men = 12 Boys

   $12M + 12W + 12B = 12M + 9M + 6M = 27M$

   Now, applying the above formula, we have

   $9 \times 6 \times 6 = 27 \times 8 \times D_2$

   $\therefore D_2 = \frac{9 \times 6 \times 6}{27 \times 8} = 1\frac{1}{2}$ days

2. a 3. d

Rule 3

Theorem: ***If A and B can do a piece of work in x days, B and C in y days, C and A in z days, then (A + B + C) working together will do the same work in $\left[\frac{2xyz}{xy + yz + xz}\right]$ days.***

Let $\left(\frac{2xyz}{xy + yz + xz}\right)$ be 'r' then

'A' alone will do the same work in $\left(\frac{yr}{y - r}\right)$ days or $\left(\frac{2xyz}{xy + yz - zx}\right)$ days,

'B' alone will do the same work in $\left(\frac{zr}{z - r}\right)$ days or $\left(\frac{2xyz}{yz + zx - xy}\right)$ days and

'C' alone will do the same work in $\left(\frac{xr}{x - r}\right)$ days or $\left(\frac{2xyz}{xz + xy - yz}\right)$ days.

Illustrative Example

Ex: A and B can do a piece of work in 12 days, B and C in 15 days, C and A in 20 days. How long would each take separately to do the same work?

Soln: Using the above theorem,

$r = \frac{2 \times 12 \times 15 \times 20}{12 \times 15 + 12 \times 20 + 15 \times 20} = 10$ days.

Now, A can do the work in $\frac{10 \times 15}{15 - 10} = 30$ days.

B can do the work in $\frac{10 \times 20}{20 - 10} = 20$ days.

C can do the work in $\frac{10 \times 12}{12 - 10} = 60$ days.

Exercise

1. A and B can finish a piece of work in 30 days, B and C in 40 days while C and A in 60 days. How long will they take to finish it together?

   a) $26\frac{2}{3}$ days b) $16\frac{2}{3}$ days c) 25 days d) 24 days

2. A and B can do a piece of work in 10 days, B and C in 15 days and C and A in 20 days. They all work at it for 6 days, and then A leaves, and B and C go on together for 4 days more. If B then leaves, how long will C take to complete the work?

   a) 20 days b) 25 days c) 10 days d) 15 days

3. A and B can do a piece of work in 6 days, A and C in $5\frac{1}{2}$ days, B and C in 4 days. In what time could each do it?

a) $20\frac{4}{13}, 8\frac{16}{31}, 7\frac{19}{35}$ b) $8\frac{16}{31}, 20\frac{4}{13}, 7\frac{19}{35}$

c) $7\frac{19}{35}, 8\frac{16}{31}, 20\frac{4}{13}$ d) None of these

4. A and B can mow a field in $3\frac{1}{2}$ days, A and C in 4 days, B and C in 5 days. In what time could they mow it, all working together?

a) $3\frac{75}{104}$ b) $2\frac{74}{103}$ c) $3\frac{74}{103}$ d) $2\frac{47}{103}$

Answers

1. a

2. c; **Hint:** A, B and C together can do the work in

$$\frac{2\times10\times15\times20}{10\times15+20\times10+15\times20}=\frac{120}{30}\text{ days}$$

work done by all in 6 days $=\frac{13}{20}$

work done by B and C in 4 days $=\frac{4}{15}$

Remaining work $=1-\left(\frac{13}{20}+\frac{4}{15}\right)=\frac{1}{12}$, which is to be done by C.

Now, from the question,

C alone can do the whole work in $\frac{\frac{120}{13}\times10}{10-\frac{120}{13}}=120$ days

[See Rule-6]

$\therefore \frac{1}{12}$ of the work is done by C in $\frac{120}{12}=10$ days.

3. a; **Hint:** Here x = 6, y = 4 and z = $\frac{11}{2}$. Now apply the given rule.

4. b

Rule 4

Theorem: ***If A can do a piece of work in x days and B can do it in y days then A and B working together will do the same work in*** $\left(\frac{xy}{x+y}\right)$ ***days.***

Illustrative Example

Ex.: A can do a piece of work in 5 days, and B can do it in 6 days. How long will they take if both work together?

Soln: Detail Method: 'A' can do $\frac{1}{5}$ work in 1 day.

'B' can do $\frac{1}{6}$ work in 1 day.

Thus 'A' and 'B' can do $\left(\frac{1}{5}+\frac{1}{6}\right)$ work in 1 day.

$\therefore$ 'A' and 'B' can do the work in $\frac{1}{\frac{1}{5}+\frac{1}{6}}$ days

$=\frac{30}{11}=2\frac{8}{11}$ days.

Quicker Method: Applying the above theorem,

A + B can do the work in $\frac{5\times6}{5+6}$ days

$=\frac{30}{11}=2\frac{8}{11}$ days.

Exercise

1. 10 men can complete a piece of work in 15 days and 15 women and complete the same work in 12 days. If all the 10 men and 15 women work together, in how many days will the work get completed?

a) 6 b) $7\frac{2}{3}$ c) $6\frac{2}{3}$ d) $6\frac{1}{3}$

(SBI Associates PO-1999)

2. A can do a piece of work in 20 days and B can do it in 30 days. How long would they take to do it working together?

a) 12 days b) 10 days c) 15 days d) 16 days

3. A can do a piece of work in 6 days. B takes 8 days. C takes as long as A and B would take working together. How long will it take B and C to complete the work together?

a) $2\frac{1}{5}$ day b) $2\frac{2}{5}$ days c) 6 days d) $4\frac{2}{5}$ days

4. A does $\frac{7}{10}$ of a piece of work in 15 days. He does the remainder with the assistance of B in 4 days. In what time could A and B together do it?

a) $13\frac{1}{3}$ days b) 12 days

c) $12\frac{2}{3}$ days d) None of these

5. A can do a piece of work in 16 days, B in 10 days. A and B work at it together for 6 days and then C finishes it in 3 days, in how many days could C have done it alone?
a) 40 days b) 80 days c) 90 days d) 120 days

6. A can do a piece of work in 4 hours, B and C can do it in 3 hours. A and C can do it in 2 hours. How long would B alone take to do it?
a) 14 hours b) 12 hours c) 10 hours d) 16 hours

7. A can do a piece of work in 30 days while B can do it in 40 days. A and B working together can do it in
a) 70 days b) $42\frac{3}{4}$ days c) $27\frac{1}{7}$ days d) $17\frac{1}{7}$ days
(Railways 1989)

8. A can do (1/3) of a work in 5 days and B can do (2/5) of the work in 10 days. In how many days both A and B together can do the work?
a) $7\frac{3}{4}$ days b) $9\frac{3}{8}$ days c) $8\frac{4}{5}$ days d) 10 days
(Railways 1991)

Answers

1. c; **Hint:** x = 15 days, y = 12 days, Now apply the above rule.

2. a

3. b; **Hint:** C completes the work in $\left(\frac{6\times8}{6+8}=\frac{24}{7}\right)$ days

$\therefore$ B and C together complete the work in

$$\left(\frac{\frac{24}{7}\times8}{\frac{24}{7}+8}=\frac{12}{5}=2\frac{2}{5}\right)\text{ days.}$$

4. a; **Hint:** $\left(1-\frac{7}{10}=\frac{3}{10}\right)$ work is done by A and B in 4 days

$\therefore$ The whole work is done by A and B in $\frac{4\times10}{3}$

$$=\frac{40}{3}=13\frac{1}{3}\text{ days.}$$

5. d; **Hint:** A and B together can do the whole work in $\frac{16\times10}{16+10}=\frac{80}{13}$ days.

$\therefore$ In 6 days A and B together can do $\frac{13}{80}\times6=\frac{39}{40}$ work.

Now, remaining work $\left(1-\frac{39}{40}=\frac{1}{40}\right)$ is done by C in 3 days

$\therefore$ The whole work is done by C in 40 × 3 = 120 days.

6. b; **Hint:** A, B and C together can do a piece of work in $\left(\frac{4\times3}{4+3}=\frac{12}{7}\right)$ hours.

B alone take to complete a piece of work

$$\left(\frac{\frac{12}{7}\times2}{2-\frac{12}{7}}=12\right)\text{ hours.}$$

[See Rule-6].

7. d

8. b; **Hint:** A can do the whole work in (5 × 3 = 15) days.

B can do the whole work in $\left(\frac{10\times5}{2}=25\right)$ days.

A and B together can do the work in $\frac{25\times15}{25+15}$

$$=\frac{75}{8}=9\frac{3}{8}\text{ days.}$$

Rule 5

Theorem: ***If A, B and C can do a work in x, y and z days respectively then all of them working together can finish the work in*** $\left(\frac{xyz}{xy+yz+xz}\right)$ ***days.***

Illustrative Example

Ex.: A can do a piece of work in 5 days, and B can do it in 6 days. If C, who can do the work in 12 days, joins them, how long will they take to complete the work?

Soln: By the theorem:

A, B and C can do the work in

$$\frac{5\times6\times12}{5\times6+6\times12+5\times12}=\frac{360}{162}=2\frac{2}{9}\text{ days.}$$

Exercise

1. A can do a piece of work in 5 days, B in 4 days and A, B and C together in 2 days. In what time would C do it alone?
a) 25 days b) 12 days c) 15 days d) 20 days

2. A takes half as long to do a piece of work as B takes, and if C does it in the same time as A and B together, and if all three working together would take 7 days, how long would each take separately?

a) 21 days, 42 days, 14 days

b) 20 days, 40 days, $\frac{40}{3}$ days

c) 15 days, 45 days, $\frac{45}{4}$ days

d) None of these

3. Five men can do a piece of work in 2 hours, which 7 women could do in 3 hours, or 9 children in 4 hours. How long would 1 man, 1 woman and 1 child together take to do the work?

a) $\frac{1260}{221}$ b) $\frac{1270}{231}$ c) $\frac{1221}{260}$ d) None of these

4. A takes twice as much time as B and thrice as much time as C to finish a piece of work, working together they can finish the work in 2 days, find the time each will take to finish the work.

a) 12, 6, 4 b) 18, 9, 6

c) 24, 12, 8 d) None of these

5. If A does a piece of work in 4 days, which B can do in 5, and C can do in 6, in what time will they do it, all working together?

a) 2 days b) $\frac{50}{27}$ days

c) $\frac{60}{37}$ days d) None of these

6. A, B and C can do a piece of work in 6, 12 and 24 days respectively. They altogether will complete the work in

a) $3\frac{3}{7}$ days b) $\frac{7}{24}$ days c) $4\frac{4}{5}$ days d) $\frac{5}{24}$ days

(Clerical Grade Exam, 1991)

Answers

1. d; **Hint:** Applying the given rule, we have

the required answer $= \frac{5\times4\times z}{5\times4+5\times2+4\times z} = 2$

or, $20z = 40 + 18z$

or, $2z = 40 \therefore z = 20$ days.

2. a; **Hint:** Let A takes x days to complete the work.

$\therefore$ B takes 2x days

From the question, C takes to complete the work $=$

$\frac{2x\times x}{2x+x} = \frac{2}{3}x$ days

(See **Rule 4**)

Now, applying the given rule, we have

$$\frac{x\times2x\times\frac{2}{3}x}{x\times2x+2x\times\frac{2}{3}x+x\times\frac{2x}{3}} = 7$$

or, $\frac{4/3\,x^3}{4x^2} = 7$

or, $\frac{x}{3} = 7 \therefore x = 21$ days

Hence, A completes the work in 21 days, B in $(21\times2 = 42)$ days and C in $\left(\frac{2}{3}\times21 = 14\right)$ days.

3. a; **Hint:** 1 men can do a piece of work in $(2\times5 = 10)$ days

1 woman can do the same work in $(7\times3 = 21)$ days.

1 child can do the same work in $(9\times4 = 36)$ days.

Now, applying the given rule, we have

the required answer $= \frac{10\times21\times36}{10\times21+21\times36+10\times36}$

$= \frac{7560}{1326} = \frac{1260}{221}$ days.

4. a; **Hint:** Let A takes x days to complete the work.

B takes $\frac{x}{2}$ and C takes $\frac{x}{3}$ days.

Now, applying the given rule, we have

$$\frac{x\times\frac{x}{2}\times\frac{x}{3}}{x\times\frac{x}{2}+\frac{x}{2}\times\frac{x}{3}+x\times\frac{x}{3}} = 2$$

or, $\frac{\frac{x^3}{6}}{x^2} = 2 \qquad \therefore x = 12$ days.

$\therefore$ A takes 12 days, B takes $\left(\frac{12}{2} = 6\right)$ days and C takes $\left(\frac{12}{3} = 4\right)$ days to complete the work.

5. c 6. a

Rule 6

Theorem: ***If A and B together can do a piece of work in x days and A alone can do it in y days, then B alone can do the work in $\frac{xy}{y-x}$ days.***

Illustrative Example

Ex.: A and B together can do a piece of work in 6 days and A alone can do it in 9 days. In how many days can B alone do it?

Soln: **Detail Method:** A and B can do $\frac{1}{6}$ of the work in 1 day.

A alone can do $\frac{1}{9}$ of the work in 1 day.

$\therefore$ B alone can do $\left(\frac{1}{6}-\frac{1}{9}\right)=\frac{1}{18}$ of the work in 1 day.

$\therefore$ B alone can do the whole work in 18 days.

Quicker Method: By the theorem:

B alone can do the whole work in $\frac{6\times 9}{9-6}=\frac{54}{3}=18$ days.

Exercise

1. A, B and C can do a piece of work in 6, 12 and 24 days respectively. In what time will they altogether do it?

   a) $3\frac{4}{7}$ days b) $3\frac{2}{7}$ days c) $3\frac{3}{7}$ days d) $2\frac{3}{7}$ days

2. A and B working together could mow a field in 28 days and with the help of C they could have mowed it in 21 days. How long would C take by himself?

   a) 86 days b) 48 days c) 84 days d) None of these

3. B can do a piece of work in 6 hours, B and C can do it in 4 hours and A, B and C in $2\frac{2}{3}$ hours. In how many hours can A and B do it?

   a) $3\frac{3}{7}$ hours b) $3\frac{5}{7}$ hours c) $3\frac{2}{7}$ hours d) None of these

4. A and B together can do a piece of work in $4\frac{4}{5}$ days, B and C together can do it in 8 days, and A, B and C together in 4 days. How long would A and C together take to do it? In what time would B do it alone?

   a) 8 days, 12 days b) 4 days, 8 days c) 6 days, 12 days d) 8 days, 16 days

5. A does $\frac{2}{5}$ of a piece of work in 9 days, he then calls in B, and they finish the work in 6 days. How long would B take to do the whole work by himself?

   a) 18 days b) 16 days c) 12 days d) 21 days

6. A and B together can do a piece of work in 6 days, B alone could do it in 16 days. If B stops after 3 days, how long afterwards will A have finished the work?

   a) $7\frac{4}{5}$ days b) $5\frac{4}{5}$ days c) $4\frac{4}{5}$ days d) $4\frac{1}{5}$ days

7. A and B can reap a field in 30 days, working together. After 11 days, however, B is called off and A finishes it by himself in 38 days more. In what time could each alone do the whole?

   a) 60 days each b) 15 days, 30 days c) 20 days, 60 days d) None of these

8. A, B and C together can do a piece of work in 6 days, which B alone can do in 16 days and B and C together can do in 10 days, in how many days can A and B together do it?

   a) $\frac{240}{31}$ b) $\frac{242}{31}$ c) $\frac{241}{31}$ d) None of these

9. A and B together can do a piece of work in 8 days, B alone can do it in 12 days, supposing B alone works at it for 4 days, in how many more days could A alone finish it?

   a) 18 days b) 24 days c) 16 days d) 20 days

10. A and B can together do a piece of work in 15 days. B alone can do it in 20 days. A alone can do it in:

    a) 30 days b) 40 days c) 45 days d) 60 days

    (Railways 1991)

11. A and B finish a job in 12 days while A, B and C can finish it in 8 days. C alone will finish the job in:

    a) 20 days b) 14 days c) 24 days d) 16 days

    (Hotel Management 1991)

Answers

1. c

2. c; **Hint:** Required answer $=\frac{28\times 21}{28-21}=84$ days.

3. a; **Hint:** Applying the given rule, we have

   C can do a the whole work in $\left(\frac{6\times 4}{6-4}=12\right)$ days.

   Now from the question,

   $$\text{required answer}=\frac{12\times\frac{8}{3}}{12-\frac{8}{3}}=\frac{96}{28}=\frac{24}{7}=3\frac{3}{7}\text{ days.}$$

4. c; **Hint:** C alone can do the work in $\frac{\frac{24}{5}\times 4}{\frac{24}{5}-4}=24$ days.

   A alone can do the work in $\left(\frac{8\times 4}{8-4}=8\right)$ days.

   Now applying the **Rule-4**, we have

   A and C together can do the work in $\left(\frac{24\times 8}{24+8}=6\right)$ days

B alone can do the work in $\left(\frac{6\times4}{6-4}=12\right)$ days.

5. a; **Hint:** A takes $\left(\frac{5}{2}\times9=\frac{45}{2}\right)$ days to complete the whole work.

A and B do $\left(1-\frac{2}{5}=\frac{3}{5}\right)$ work in 6 days.

$\therefore$ A and B do the whole work in $\left(\frac{6\times5}{3}=10\right)$ days.

Now applying the given rule, we have

the required answer $=\dfrac{\frac{45}{2}\times10}{\frac{45}{2}-10}=\dfrac{45\times10}{25}=18$ days.

6. c; **Hint:** A alone can do the whole work in $\left(\frac{16\times6}{16-10}=\frac{48}{5}\right)$ days.

Work done by A and B together in 3 days $=\frac{3}{6}=\frac{1}{2}$.

Remaining work $\left(1-\frac{1}{2}=\frac{1}{2}\right)$ is done by A in $\frac{1}{2}\times\frac{48}{5}$

$=\frac{24}{5}=4\frac{4}{5}$ days.

7. a; **Hint:** Work done in 11 days by A and B $=\frac{11}{30}$

Remaining work $\left(1-\frac{11}{30}=\frac{19}{30}\right)$ is done by A in 38 days

$\therefore$ the whole work is done by A in $\frac{30\times38}{19}=60$ days.

B alone can do the whole work in $\left(\frac{30\times60}{60-30}=60\right)$ days.

8. a; **Hint:** A alone can do a piece of work in $\left(\frac{10\times6}{10-6}=15\right)$ days.

A and B together can do a piece of work in $\frac{15\times16}{15+16}=\frac{240}{31}$ days.

[See **Rule-4**].

9. c; **Hint:** A alone can do the work in $\left(\frac{12\times8}{12-8}=24\right)$ days.

Work done by B in 4 days $=\frac{4}{12}=\frac{1}{3}$.

Remaining work $\left(1-\frac{1}{3}=\frac{2}{3}\right)$ is done by A in

$\frac{2}{3}\times24=16$ days.

10. d

11. c; **Hint:** C alone will finish the job in $\frac{12\times8}{12-8}=24$ days.

Rule 7

Theorem: ***To do a certain work B would take n times as long as A and C together and C m times as long as A and B together. If the three men together complete the work in x days, then if the number of days taken by B = (no. of days taken by A + B + C) × (n + 1) and the number of days taken by C = (no. of days taken by A + B + C) × (m + 1).***

Illustrative Example

Ex.: To do a certain work B would take three times as long as A and C together and C twice as long as A and B together. The three men together complete the work in 10 days. How long would each take separately?

Soln: **Detail Method:** By the question

3 times B's daily work = (A + C)'s daily work.

Add B's daily work to both sides.

$\therefore$ 4 times B's daily work = (A + B + C)'s daily work

$=\frac{1}{10}$

$\therefore$ B's daily work $=\frac{1}{40}$

Also, 2 times C's daily work = (A + B)'s daily work.

Add C's daily work to both sides.

$\therefore$ 3 times C's daily work = (A + B + C)'s daily work

$=\frac{1}{10}$

$\therefore$ C's daily work $=\frac{1}{30}$

Now A's daily work $=\frac{1}{10}-\left(\frac{1}{40}+\frac{1}{30}\right)=\frac{1}{24}$

$\therefore$ A, B and C can do the work in 24, 40 and 30 days respectively.

Quicker Method:
Number of days taken by B
= (Number of days taken by A + B + C) × (3 + 1)
= 10 (3 + 1) = 40 days
Similarly,
Number of days taken by C = 10 (2 + 1) = 30 days

Number of days taken by A = $\dfrac{1}{\dfrac{1}{10}-\left(\dfrac{1}{40}+\dfrac{1}{30}\right)} = 24$ days.

Exercise

1. A, B and C together can finish a piece of work in 12 days, A and C together work twice as much as B, A and B together work thrice as much as C. In what time could each do it separately?
   a) $28\frac{4}{5}$, 42, 48 b) $28\frac{4}{5}$, 36, 48
   c) 28, $36\frac{4}{5}$, 48 d) None of these
2. To do a certain work B would take 4 times as long as A and C together and C 3 times as long as A and B together. The three men together complete the work in 5 days. How long would take B and C to complete the work?
   a) $9\frac{1}{11}$ days b) $11\frac{1}{9}$ days c) $26\frac{2}{3}$ days d) $28\frac{3}{5}$ days
3. To do a certain work B would take 2 times as long as A and C together and C 3 times as long as A and B together. The three men together complete the work in 16 days. How long would take B and C to complete the work?
   a) $27\frac{2}{7}$ days b) 27 days c) $27\frac{3}{7}$ days d) $27\frac{4}{7}$ days

Answers

1. b; **Hint:** B completes the work in [12(2 + 1) = 36] days and
C completes the same work in [12(3 + 1) = 48] days
B and C together complete the work in

$$\left(\frac{36\times48}{36+48}=\frac{144}{7}\right) \text{ days.}$$

Now, from the question,
A takes to complete the same work

$$= \frac{12\times\frac{144}{7}}{\frac{144}{7}-12} = \frac{144}{5} = 28\frac{4}{5} \text{ days.}$$

2. b; **Hint:** B alone can do the work in (4 + 1) × 5 = 25 days
C alone can do the work in (3 + 1) × 5 = 20 days
B and C together can do the work in $\dfrac{25\times20}{25+20} = 11\dfrac{1}{9}$ days

3. c

Rule 8

Theorem: ***If x_1 men or y_1 women can reap a field in 'D' days, then x_2 men and y_2 women take to reap it*** $\left[\dfrac{D(x_1y_1)}{x_2y_1+x_1y_2}\right]$ *days*.

Illustrative Example

Ex.: If 3 men or 4 women can reap a field in 43 days, how long will 7 men and 5 women take to reap it?

Soln: First Method:

3 men reap $\frac{1}{43}$ of the field in 1 day.

∴ 1 man reaps $\frac{1}{43\times3}$ of the field in 1 day.

4 women reap $\frac{1}{43}$ of the field in 1 day.

∴ 1 woman reaps $\frac{1}{43\times4}$ of the field in 1 day.

∴ 7 men and 5 women reap $\left(\frac{7}{43\times3}+\frac{5}{43\times4}\right)=\frac{1}{12}$ of the field in 1 day.

∴ 7 men and 5 women will reap the whole field in 12 days.

Second Method:

3 men = 4 women

∴ 1 man = $\frac{4}{3}$ women

∴ 7 men = $\frac{28}{3}$ women

∴ 7 men + 5 women = $\frac{28}{3}+5=\frac{43}{3}$ women

Now, the question becomes:
If 4 women can reap a field in 43 days, how long will $\frac{43}{3}$ women take to reap it?

The "basic-formula" gives

$$4\times 43=\frac{43}{3}\times D_2$$

$$\text{or, } D_2=\frac{4\times 43\times 3}{43}=12 \text{ days.}$$

Quicker Method:

$$\text{Required number of days}=\frac{1}{\left[\frac{7}{43\times 3}+\frac{5}{43\times 4}\right]}$$

$$=\frac{43\times 3\times 4}{7\times 4+5\times 3}=12 \text{ days}$$

Note:

1. The above formula is very easy to remember.
If we divide the question in two parts and call the first part as OR-part and the second part as AND-part then

$$\frac{7}{43\times 3}=\frac{\text{Number of men in AND - part}}{\text{Number of days}\times\text{Number of men in OR - part}}$$

Similarly, you can look for the second part in denominator.

2. Second step of the quicker method $\left(\text{ie } \frac{43\times 3\times 4}{7\times 4+5\times 3}\right)$ is the form of formula given in the **Rule 8**.

Exercise

1. If 3 men or 5 women can reap a field in 43 days, how long will 5 men and 6 women take to reap it?
a) 15 days b) 25 days c) 18 days d) 12 days
2. If 2 men or 4 women can reap a field in 44 days, how long will 3 men and 5 women take to reap $\frac{3}{4}$th of the field?
a) 10 days b) 8 days c) 12 days d) None of these
3. If 6 men or 10 women can reap a field in 86 days, how long will 10 men and 12 women take to reap it?
a) 30 days b) 35 days c) 25 days d) 40 days
4. If 4 men or 6 boys can finish a piece of work in 20 days, in how many days can 6 men and 11 boys finish it?
a) 8 days b) 6 days c) 7 days d) 9 days
(LIC Exam, 1991)
5. 10 men can finish a piece of work in 10 days whereas it takes 12 women to finish it in 10 days. If 15 men and 6 women undertake to complete the work how many days will they take to complete it?
a) 11 b) 5 c) 4 d) 2
(Bank PO Exam, 1991)

Answers

1. a 2. c 3. a 4. b

5. b; **Hint:** 10 men and 12 women can finish a piece of work in the same no. of days ie 10 days. Hence we can say that 10 men or 12 women finish the work in 10 days. Now we can apply the given formula.

Rule 9

Theorem: ***If a_1 men and b_1 boys can do a piece of work in x days and a_2 men and b_2 boys can do it in y days, then the following relationship is obtained:***

$$1\ man=\left[\frac{yb_2-xb_1}{xa_1-ya_2}\right]boys$$

Illustrative Example

Ex: If 12 men and 16 boys can do a piece of work in 5 days and 13 men and 24 boys can do it in 4 days, how long will 7 men and 10 boys take to do it?

Soln: **Detail Method:**

12 men and 16 boys can do the work in 5 days .... (1)
13 men and 24 boys can do the work in 4 days .... (2)
Now it is easy to see that if the no. of workers be multiplied by any number, the time must be divided by the same number (derived from: *more workers less time*). Hence multiplying the no. of workers in (1) and (2) by 5 and 4 respectively, we get

5 (12 men + 16 boys) can do the work in $\frac{5}{5}=1$ day

4 (13 men + 24 boys) can do the work in $\frac{4}{4}=1$ day

or, 5 (12 m + 16 b) = 4(13 m + 24 b)
or, 60 m + 80 b = 52 m + 96 b ... (*)
or, 60 m – 52 m = = 96 b – 80 b
or, 8 m = 16 b
$\therefore$ 1 man = 2 boys.
Thus, 12 men + 16 boys = 24 boys + 16 boys = 40 boys
and 7 men + 10 boys = 14 boys + 10 boys = 24 boys
The question now becomes:
"If 40 boys can do a piece of work in 5 days how long will 24 boys take to do it?"
Now, by "basic formula", we have
$40\times 5=24\times D_2$ ..... (*) (*)

$$\text{or, } D_2=\frac{40\times 5}{24}=8\frac{1}{3} \text{ days.}$$

Note: During practice session (*) should be your first step to be written down. Further calculations should be done mentally. Once you get that 1 man = 2 boys, your next step should be (*) (*). This way you can get the result within seconds.

Quicker Method: Applying the above theorem,

$$1 \text{ men}=\frac{24\times 4-16\times 5}{12\times 5-4\times 13}=\frac{16}{8}=2 \text{ boys.}$$

Thus, 12 men + 16 boys = 24 boys + 16 boys = 40 boys
and 7 men + 10 boys = 14 boys + 10 boys = 24 boys
Now, by basic formula, we have

$40 \times 5 = 24 \times D_2$

or $D_2 = \frac{40 \times 5}{24} = 8\frac{1}{3} days$.

Exercise

1. If 12 men and 16 boys can do a piece of work in 5 days and 13 men and 24 boys can do it in 4 days, compare the daily work done by a man with that done by a boy.
   a) 1 : 2 b) 2 : 1 c) 1 : 3 d) 3 : 1
2. If 30 men and 14 boys can reap a field in 21 days, in how many days will 20 men and 4 boys reap it, supposing that 3 men can do as much as 5 boys?
   a) 36 days b) 30 days c) 42 days d) 45 days
3. If 5 men and 2 boys working together can do 4 times as much work per hour as a man and a boy together, compare the work of a man with that of a boy.
   a) 2 : 1 b) 3 : 1 c) 4 : 1 d) Data inadequate
4. If I must hire 2 men and 3 boys for 6 days to do the same piece of work as 11 men and 5 boys could do in $1\frac{1}{2}$ days, compare the work of a boy with that of a man.
   a) 7 : 3 b) 3 : 7 c) 2 : 5 d) 5 : 2
5. 8 children and 12 men complete a certain piece of work in 9 days. Each child takes twice the time by a man to finish the work. In how many days will 12 men finish the same work?
   a) 8 b) 9 c) 12 d) 15

(Bank PO Exam, 1988)

Answers

1. b; **Hint:** Applying the above theorem,

$$\frac{\text{a man's work}}{\text{a boy's work}} = \frac{24 \times 4 - 16 \times 5}{12 \times 5 - 4 \times 13} = \frac{16}{8} = 2:1$$

2. a; **Hint:** Here relationship between men and boys is given.

$3 \text{ men} = 5 \text{ boys} \therefore 1 \text{ man} = \frac{5}{3}$ boys.

Now, $30M + 14B = \frac{30 \times 5}{3} + 14 = 64$ boys and

$20M + 4B = \frac{112}{3}$ boys.

From the formula, $64 \times 21 = \frac{112}{3} \times D_2$

$\therefore D_2 = 36$ days.

3. a; **Hint:** Let 5 men and 2 boys can do the work in x days. Hence a man and a boy together can do the same work in 4x days.
Now, applying the given rule, we have

the required answer $= \frac{4x \times 1 - x \times 2}{x \times 5 - 4x \times 1} = 2:1$.

Note: Also see **Rule-15**

4. b; **Hint:** Applying the given rule, we have

$$\frac{\text{Man}}{\text{Boy}} = \frac{\frac{3}{2} \times 5 - 6 \times 3}{6 \times 2 - \frac{3}{2} \times 11} = \frac{7}{3}$$

$\therefore$ Boy : Man = 3 : 7.

5. c; **Hint:** 2 children = 1 man
$\therefore$ 8 children + 12 men = 16 men
From the question,
Since 16 men can complete a certain piece of work in 9 days

$\therefore$ 12 men finish the work in $\left(\frac{16 \times 9}{12} = 12\right)$ days.

Rule 10

Theorem: ***A certain number of men can do a work in 'D' days. If there were 'x' men less it could be finished in 'd' days more, then the number of men originally are*** $\left[\frac{x(D+d)}{d}\right]$.

Illustrative Example

Ex: A certain number of men can do a work in 60 days. If there were 8 men less it could be finished in 10 days more. How many men are there?

Soln: Using the above formula, we have

the original number of men $= \frac{8(60+10)}{10} = 56$ *men*.

Exercise

1. A certain number of men can do a work in 45 days. If there were 4 men less it could be finished in 15 days more. How many men are there?
   a) 28 men b) 16 men c) 24 men d) 20 men
2. A certain number of men can do a work in 30 days. If there were 6 men less it could be finished in 20 days more. How many men are there?
   a) 15 men b) 12 men
   c) 18 men d) 20 men
3. A certain number of men can do a work in 60 days. If there were 6 men less it could be finished in 12 days more. How many men are there?

a) 30 men b) 32 men c) 28 men d) 31 men

4. A certain number of men can do a work in 70 days. If there were 2 men less it could be finished in 10 days more. How many men are there?
a) 15 men b) 17 men c) 16 men d) 12 men

Answers

1. b 2. a 3. d 4. c

Rule 11

Theorem: ***If A is 'n' times as fast (or slow) as B, and is therefore able to finish a work in 'D' days less (or more) than B, then the time in which they can do it working together is given by*** $\left(\frac{Dn}{n^2-1}\right)$ *days*.

Illustrative Example

Ex: A is thrice as fast as B, and is therefore able to finish a work in 60 days less than B. Find the time in which they can do it working together.

Soln: **Detail Method:**
A is thrice as fast as B, means that if A does a work in 1 day then B does it in 3 days.
Hence, if the difference be 2 days, then A does the work in 1 day and B in 3 days. But the difference is 60 days. Therefore, A does the work in 30 days and B in 90 days.
Now A and B together will do the work in

$$\frac{30\times 90}{30+90}\textit{days} = \frac{45}{2} = 22.5 \textit{ days}.$$

Quicker Method: Applying the above theorem, we have

the required answer $= \frac{60\times 3}{3^2-1} = \frac{60\times 3}{8} = \frac{45}{2}$
$= 22.5$ days.

Exercise

1. A is twice as fast as B, and is therefore able to finish a work in 30 days less than B. Find the time in which they can do it working together.
a) 18 days b) 20 days c) 24 days d) 22 days
2. A is 4 times as fast as B, and is therefore able to finish a work in 45 days less than B. Find the time in which they can do it working together.
a) 12 days b) 16 days c) 8 days d) 20 days
3. A is thrice as fast as B, and is therefore able to finish a work in 40 days less than B. Find the time in which they can do it working together.
a) 16 days b) 10 days c) 15 days d) None of these
4. A is thrice as good a workman as B and is therefore able to finish a work in 80 days less than B. Find the time in which they can do it working together.
a) 30 days b) 20 days c) 24 days d) 25 days

(MBA, 1983)

Answers

1. b 2. a 3. c 4. a

Rule 12

Theorem: ***If a person can finish a work in*** d_1 ***days at*** h_1 ***hours a day and another person can finish the same work in*** d_2 ***days at*** h_2 ***hours a day, then the no. of days in which they can finish the works working together 'h' hours a day is*** $\left[\frac{(h_1d_1)(h_2d_2)}{(h_1d_1)+(h_2d_2)}\right]\frac{1}{h}$ *days*.

Illustrative Example

Ex: I can finish a work in 15 days at 8 hrs a day. You can finish it in $6\frac{2}{3}$ days at 9 hrs a day. Find in how many days we can finish it working together 10 hrs a day.

Soln: **Detail Method:** First suppose each of us works for only one hour a day.
Then I can finish the work in $15 \times 8 = 120$ days
and you can finish the work in $\frac{20}{3}\times 9 = 60$ days
Now, we together can finish the work in $\frac{120\times 60}{120+60} = 40$ *days*.
But here we are given that we do the work 10 hrs a day. Then clearly we can finish the work in 4 days.
Quicker Method: Applying the above formula, we have

the required answer $= \left[\frac{15\times 8\times \frac{20}{3}\times 9}{15\times 8+\frac{20}{3}\times 9}\right]\frac{1}{10} = 4$ days.

Exercise

1. I can finish a work in 10 days at 4 hrs a day. You can finish it in 15 days at 5 hrs a day. Find in how many days we can finish it working together 10 hrs a day.
a) $\frac{50}{23}$ days b) $\frac{70}{23}$ days c) $\frac{60}{23}$ days d) $\frac{40}{23}$ days
2. I can finish a work in 16 days at 5 hrs a day. You can finish it in 12 days at 4 hrs a day. Find in how many days we can finish it working together 6 hrs a day.
a) 5 days b) 4 days c) 6 days d) None of these
3. I can finish a work in 14 days at 6 hrs a day. You can finish it in 8 days at 2 hrs a day. Find in how many days we can finish it working together 4 hrs a day.

a) $3\frac{8}{25}$ days b) $9\frac{3}{25}$ days

c) $3\frac{9}{25}$ days d) $4\frac{9}{25}$ days

Answers

1. c 2. a 3. c

Rule 13

Theorem: ***If A can do a work in x days, B takes y days to complete it and C takes as long as A and B would take working together, then B and C together take to complete the work*** $= \left[\frac{xy}{2x+y}\right]$ ***days*. A and C together take to complete the work** $= \left[\frac{xy}{x+2y}\right]$ ***days* and A, B and C together take to complete the work** $= \left[\frac{xy}{2(x+y)}\right]$ ***days*.**

Illustrative Eample

Ex: A can do a work in 6 days. B takes 8 days to complete it. C takes as long as A and B would take working together. How long will it take B and C, A and C, and A, B and C to complete the work together?

Soln: Using the above formula, we have,

(B + C) together take to complete the work $= \frac{6\times 8}{12+8}$

$$= \frac{48}{20} = \frac{12}{5} = 2\frac{2}{5} \textit{days}.$$

(A + C) together take to complete the work $= \frac{6\times 8}{6+16}$

$$= \frac{48}{22} = 2\frac{2}{11} \textit{days}.$$

(A + B + C) together take to complete the work

$$= \frac{6\times 8}{2(6+8)} = \frac{48}{28} = \frac{12}{7} = 1\frac{5}{7} \textit{days}.$$

Exercise

1. A can do a work in 3 days. B takes 4 days to complete it. C takes as long as A and B would take working together. How long will it take B and C to complete the work together?

   a) $\frac{5}{6}$ days b) $\frac{6}{5}$ days c) $\frac{4}{3}$ days d) $\frac{3}{4}$ days

2. A can do a work in 4 days. B takes 5 days to complete it. C takes as long as A and B would take working together. How long will it take B and C to complete the work together?

   a) $\frac{20}{13}$ b) $\frac{25}{14}$ c) $\frac{22}{13}$ d) $\frac{20}{7}$

3. A can do a work in 6 days. B takes 7 days to complete it. C takes as long as A and B would take working together. How long will it take A and C to complete the work together?

   a) $2\frac{1}{5}$ days b) 2 days c) $2\frac{1}{10}$ days d) $3\frac{1}{10}$ days

4. A can do a work in 8 days. B takes 6 days to complete it. C takes as long as A and B would take working together. How long will it take A and C to complete the work together?

   a) $2\frac{1}{5}$ days b) $3\frac{2}{5}$ days c) $2\frac{3}{5}$ days d) $2\frac{2}{5}$ days

5. A can do a work in 10 days. B takes 15 days to complete it. C takes as long as A and B would take working together. How long will it take A, B and C to complete the work together?

   a) 6 days b) 3 days c) 4 days d) 8 days

6. A can do a work in 20 days. B takes 5 days to complete it. C takes as long as A and B would take working together. How long will it take A, B and C to complete the work together?

   a) 2 days b) 4 days c) 3 days d) 6 days

Answers

1. b 2. a 3. c 4. d 5. b 6. a

Rule 14

Theorem: ***A is n times as good a workman as B. If together, they finish the work in x days, then A and B separately can do the same work in*** $\left(\frac{n+1}{n}\right)x$ ***days and (n + 1)x days respectively.***

Illustrative Example

Ex: A is twice as good a workman as B. Together, they finish the work in 14 days. In how many days can it be done by each separately?

Soln: **Detail Method:**

Let B finish the work in $2x$ days. Since A is twice as active as B therefore, A finishes the work in x days.

(A + B) finish the work in $\frac{2x^2}{3x} = 14$

or $x = 21$

$\therefore$ A finishes the work in 21 days and B finishes the work in $21 \times 2 = 42$ days.

Quicker Method I: Using the above theorem:

A finishes the work in $\frac{(2+1)\times 14}{2} = 21\ days$.

B finishes the work in $(2+1)\ 14 = 42\ days$.

Quicker Method II: Twice + One time = Thrice active person does the work in 14 days. Then one-time active person (B) will do it in 14 × 3 = 42 days and twice active person (A) will do it in $\frac{14}{2} = 21$ days.

Note: Efficient person takes less time. In other words we may say that "*Efficiency (E) is indirectly proportional to number of days (D) taken to complete a work*". Then mathematically

$E \alpha \frac{1}{D}$ or, $E = \frac{K}{D}$, whee K is a constant

or, ED = constant

or, $E_1D_1 = E_2D_2 = E_3D_3 = E_4D_4 = ...E_nD_n$, And we see in the above case: $E_1D_1 = E_2D_2 = E_3D_3$ or, 3 × 14 = 2 × 21 = 1 × 4^2.

Thus, our answer verifies the above statement.

Exercise

1. A and B together can do a piece of work in 7 days. If A does twice as much work as B in a given time, find how long A alone would take to do the work?
   a) 21 days b) 20 days c) 10 days d) $10\frac{1}{2}$ days
2. A and B together can do a piece of work in 8 days. If A does twice as much work as B in a given time, find how long A alone would take to do the work?
   a) 10 days b) 12 days c) 14 days d) 16 days
3. A and B together can do a piece of work in 9 days. If A does thrice as much work as B in a given time, find how long A alone would take to do the work?
   a) 12 days b) 14 days c) 16 days d) 18 days
4. A and B together can do a piece of work in 6 days. If A does twice as much work as B in a given time, find how long A alone would take to do the work?
   a) 16 days b) 9 days c) 18 days d) 21 days
5. A and B together can do a piece of work in 3 days. If A does thrice as much work as B in a given time, find how long A alone would take to do the work?
   a) 4 days b) 10 days c) 14 days d) 12 days

Answers

1. d 2. b 3. a 4. b 5. a

Rule 15

Theorem: ***If a_1 men and b_1 boys working together can do n times as much work per hour as a_2 men and b_2 boys together, then the comparison of the work of a man with that of a boy is given by*** $\frac{Man}{Boy} = \frac{nb_2 - b_1}{a_1 - na_2}$.

Illustrative Example

Ex: 5 men and 2 boys working together can do 4 times as much work per hour as a man and a boy together. Compare the work of a man with that of a boy.

Soln: Applying the above formula, we have

$$\frac{Man}{Boy} = \frac{4\times 1 - 2}{5 - 4\times 1} = \frac{2}{1} \quad [\text{Here, } a_2 = 1 \text{ and } b_2 = 1]$$

That is, a man is twice as efficient as a boy.

Exercise

1. 6 men and 3 boys working together can do 5 times as much work per hour as a man and a boy together. Compare the work of a man with that of a boy.
   a) 2 : 1 b) 3 : 1
   c) 3 : 2 d) 4 : 1
2. 8 men and 4 boys working together can do 6 times as much work per hour as a man and a boy together. Compare the work of a man with that of a boy.
   a) 2 : 1 b) 3 : 1 c) 1 : 1 d) 1 : 2

Answers

1. a 2. c

Rule 16

Theorem: ***If A and B can do a work in x and y days respectively, they began the work together, but A left after some time and B finished the remaining work in z days; then the no. of days after which A left is given by*** $\left(\frac{xy}{x+y}\right)\left(\frac{y-z}{y}\right)$***days***.

Illustrative Example

Ex: A and B can do a work in 45 and 40 days respectively. They began the work together, but A left after some time and B finished the remaining work in 23 days. After how many days did A leave?

Soln: **Detail Method:** B works alone for 23 days.

$\therefore$ Work done by B in 23 days = $\frac{23}{40}$ work

$\therefore$ A + B do together $1 - \frac{23}{40} = \frac{17}{40}$ work

Now, A + B do 1 work in $\frac{40\times 45}{40+45} = \frac{40\times 45}{85}$ days.

$\therefore$ A + B do $\frac{17}{40}$ work in $\frac{40\times 45}{85}\times\frac{17}{40} = 9$ days.

Quicker Method: Applying the above formula,

the required answer $= \left(\frac{40\times45}{40+45}\right)\left(\frac{40-23}{40}\right) = 9$ days.

Exercise

1. A and B can do a work in 40 and 35 days respectively. They began the work together, but A left after some time and B finished the remaining work in 10 days. After how many days did A leave?

   a) $13\frac{2}{3}$ days b) 13 days c) $13\frac{1}{3}$ days d) 14 days

2. A and B can do a work in 35 and 25 days respectively. They began the work together, but A left after some time and B finished the remaining work in 15 days. After how many days did A leave?

   a) 6 days b) 5 days c) $5\frac{5}{6}$ days d) $6\frac{5}{6}$ days

3. A and B can do a work in 20 and 15 days respectively. They began the work together, but A left after some time and B finished the remaining work in 8 days. After how many days did A leave?

   a) 4 days b) 5 days c) 3 days d) 6 days

4. A and B can together finish a work in 30 days. They worked for it for 20 days and then B left. The remaining work was done by A alone in 20 more days. A alone can finish the work in:

   a) 54 days b) 60 days c) 48 days d) 50 days

   (Central Excise 1988)

Answers

1. c 2. c 3. a

4. b; **Hint:** In the given formula, we have

$$\frac{xy}{x+y} = 30 \text{ days.}$$

Now, from the question,

B left the work, ie y = time taken by A to complete the whole work and z = 20 days.

Now, applying the given formula, we have

$$30\times\left(\frac{y-20}{y}\right) = 20 \quad \therefore \ y = 60 \text{ days.}$$

Rule 17

Theorem: ***If x_1 men or x_2 women or x_3 boys can do a work in 'D' days, then the no. of days in which 1 man, 1 woman and 1 boy do the same work is given by the following formula, number of required days =***

$$\left[\frac{D\times x_1\times x_2\times x_3}{x_1x_2+x_1x_3+x_2x_3}\right] \textit{days.}$$

Illustrative Example

Ex: 1 man or 2 women or 3 boys can do a work in 44 days. Then in how many days will 1 man, 1 woman and 1 boy do the work?

Soln: Applying the above formula, we have

the no. of required days $= \frac{44\times1\times2\times3}{1\times2+2\times3+1\times3}$

$$= \frac{44\times1\times2\times3}{2+6+3} = 24 \text{ days.}$$

Exercise

1. 2 men or 3 women or 4 boys can do a work in 52 days. Then in how many days will 1 man, 1 woman and 1 boy do the work?

   a) 24 days b) 42 days c) 36 days d) 48 days

2. 3 men or 4 women or 5 boys can do a work in 47 days. Then in how many days will 1 man, 1 woman and 1 boy do the work?

   a) 40 days b) 50 days c) 60 days d) 45 days

3. 1 man or 3 women or 4 boys can do a work in 38 days. Then in how many days will 1 man, 1 woman and 1 boy do the work?

   a) 24 days b) 12 days c) 18 days d) 36 days

4. 1 man or 2 women or 4 boys can do a work in 56 days. Then in how many days will 1 man, 1 woman and 1 boy do the work?

   a) 24 days b) 28 days c) 20 days d) 32 days

Answers

1. d 2. c 3. a 4. d

Rule 18

Theorem: ***A group of men decided to do a work in x days, but 'n' of them became absent. If the rest of the group did the work in 'y' days, then the original number of men is given by $\left(\frac{ny}{y-x}\right)$ men.***

Illustrative Example

Ex: A group of men decided to do a work in 10 days, but five of them became absent. If the rest of the group did the work in 12 days, find the original number of men.

Soln: Applying the above formula, we have

the required answer $= \frac{5\times12}{12-10} = 30$ men.

Exercise

1. A group of men decided to do a work in 13 days, but 6 of them became absent. If the rest of the group did the work in 15 days, find the original number of men.

a) 30 men b) 35 men c) 40 men d) 45 men

2. A group of men decided to do a work in 12 men, but 8 of them became absent. If the rest of the group did the work in 20 days, find the original number of men.
a) 18 men b) 20 men c) 22 men d) 24 men

3. A group of men decided to do a work in 15 days, but 2 of them became absent. If the rest of the group did the work in 25 days, find the original number of men.
a) 5 men b) 4 men c) 7 men d) 6 men

Answers

1. d 2. b 3. a

Rule 19

Theorem: ***A certain number of men can do a work in 'D' days. If there were 'x' men more it could be finished in 'd' days less, then the number of men originally are*** $\left[\dfrac{x(D-d)}{d}\right]$

or

$$\frac{\text{No. of more workers} \times \text{Number of days taken by the second group}}{\text{No. of less days}}$$

Illustrative Example

Ex.: A certain number of men can do a work in 60 days. If there were 8 men more it could be finished in 10 days less. How many men are there?

Soln: Applying the above rule, we have original number of workers

$$= \frac{\text{No. of more workers} \times \text{No. of days taken by the second group}}{\text{No. of less days}}$$

$$= \frac{8 \times (60-10)}{10} = \frac{8 \times 50}{10} = 40 \text{ men.}$$

Exercise

1. A certain number of men can do a work in 50 days. If there were 3 men more it could be finished in 5 days less. How many men are there?
a) 36 men b) 18 men c) 27 men d) 30 men

2. A certain number of men can do a work in 75 days. If there were 6 men more it could be finished in 15 days less. How many men are there?
a) 20 men b) 24 men
c) 28 men d) 32 men

3. A certain number of men can do a work in 35 days. If there were 10 men more it could be finished in 10 days less. How many men are there?
a) 25 men b) 20 men c) 15 men d) 30 men

Answers

1. c 2. b 3. a

Rule 20

Theorem: ***A builder decided to build a farmhouse in 'D' days. He employed 'x' men in the beginning and 'y' more men after 'd' days and completed the construction in stipulated time. If he had not employed the additional men, then the men in the beginning would have finished it in*** $\left[\dfrac{D(x+y)-yd}{x}\right]$ ***days and it would have been*** $\left[\dfrac{y(D-d)}{x}\right]$ ***days behind the schedule.***

Illustrative Example

Ex.: A builder decided to build a farmhouse in 40 days. He employed 100 men in the beginning and 100 more after 35 days and completed the construction in stipulated time. If he had not employed the additional men, how many days behind schedule would it have been finished?

Soln: **Detail Method:** Let 100 men only complete the work in **x** days.

Work done by 100 men in 35 days + Work done by 200 men in

(40 – 35 =) 5 days = 1.

or, $\dfrac{35}{x} + \dfrac{200 \times 5}{100x} = 1$

or, $\dfrac{45}{x} = 1 \therefore x = 45$ days

Therefore, if additional men were not employed, the work would have lasted 45 – 40 = 5 days behind schedule time.

Quicker Approach:

200 men do the rest of the work in 40 – 35 = 5 days.

$\therefore$ 100 men can do the rest of the work in $\dfrac{5 \times 200}{100} = 10$ days.

$\therefore$ required number of days = 10 – 5 = 5 days.

Quicker Method: Applying the above theorem, we have

the required number of days $= \dfrac{100(40-35)}{100} = 5$ days.

Exercise

1. A builder decided to build a farmhouse in 45 days. He employed 150 men in the beginning and 120 more after 30 days and completed the construction in stipulated time. If he had not employed the additional men, how many days behind schedule would it have been finished?
a) 12 days b) 10 days c) 15 days d) 8 days

2. A builder decided to build a farmhouse in 50 days. He employed 50 men in the beginning and 50 more after 40 days and completed the construction in stipulated time. If he had not employed the additional men, in how many

days would it have been finished by the men in the beginning?

a) 80 days b) 60 days c) 40 days d) 75 days

3. A builder decided to build a farmhouse in 60 days. He employed 150 men in the beginning and 130 more after 45 days and completed the construction in stipulated time. If he had not employed the additional men, how many days behind schedule would it have been finished?

a) 10 days b) 23 days c) 13 days d) 15 days

4. A builder decided to build a farmhouse in 20 days. He employed 40 men in the beginning and 20 more after 10 days and completed the construction in stipulated time. If he had not employed the additional men, in how many days would it have been finished by the men in the beginning?

a) 50 days b) 60 days c) 40 days d) 5 days

Answers

1. a 2. b 3. c 4. a

Rule 21

Theorem: ***A, B and C can do a work in x, y and z days respectively. They all begin together. If A continues to work till it is finished, C leaves after working d_2 days and B d_1 days before its completion, then the time in which work is finished, is given by*** $\left[\dfrac{x(yz+d_1z+d_2y)}{xy+xz+yz}\right]$ ***days.***

Illustrative Example

Ex: A, B and C can do a work in 8, 16, 24 days respectively. They all begin together. A continues to work till it is finished, C leaving off 2 days and B one day before its completion. In what time is the work finished?

Soln: **Detail Method:**

Let the work be finished in x days.

Then, A's x day's work + B's $(x-1)$ day's work + C's $(x-2)$ day's work = 1

or, $\dfrac{x}{8}+\dfrac{x-1}{16}+\dfrac{x-2}{24}=1 \quad \therefore x=5$ days.

Quicker Method: Applying the above formula, we have

the required answer $=\dfrac{8[(16\times24)+(1\times24)+(2\times16)]}{(8\times16)+(16\times24)+(8\times24)}$

$=\dfrac{3072+448}{704}=\dfrac{3520}{404}=5$ days.

Exercise

1. A, B and C can do a piece of work in 16, 32 and 48 days respectively, they start working together but C leaves after working 4 days and B, 2 days before the completion of work. Find in how many days the work was finished?

a) 5 days b) 8 days c) 10 days d) 12 days

2. A, B and C can do a piece of work in 10, 12 and 15 days respectively, they start working together but C leaves after working 3 days and B, 4 days before the completion of work. Find in how many days the work was finished?

a) $6\frac{2}{15}$ days b) $5\frac{1}{3}$ days c) $7\frac{2}{15}$ days d) $6\frac{2}{5}$ days

3. A, B and C can do a piece of work in 5, 8 and 10 days respectively, they start working together but C leaves after working 2 days and B, 1 days before the completion of work. Find in how many days the work was finished?

a) 3 days b) $3\frac{1}{17}$ ays c) $3\frac{2}{7}$ days d) $2\frac{11}{13}$ days

Answers

1. c 2. a 3. d

Rule 22

Theorem: ***There is a sufficient food for 'M' men for 'D' days. If after 'd' days 'm' men leave the place, then the rest of the food will last for the rest of the men for*** $\left[\dfrac{D-d}{M-m}\times M\right]$ ***days.***

Illustrative Example

Ex: There is a sufficient food for 400 men for 31 days. After 28 days, 280 men leave the place. For how many days will the rest of the food last for the rest of the men?

Soln: **Detail Method:** The rest of the food will last for $(31-28)=3$ days if no body leaves the place.

Thus the rest of the food will last for $3\left[\dfrac{400}{120}\right]$ days for the 120 men left.

$\therefore$ Ans $=3\left(\dfrac{400}{120}\right)=10$ days.

Note: For less persons the food will last longer, therefore, 3 is multiplied by $\dfrac{400}{120}$, a more than one fraction.

Quicker Method: Using the above formula, we have

The required answer $=\dfrac{31-28}{400-280}\times400=10$ days.

Exercise

1. There is a sufficient food for 200 men for 36 days. After 33 days, 140 men leave the place. For how many days

will the rest of the food last for the rest of the men?

a) 5 days b) 10 days c) 18 days d) 15 days

2. There is a sufficient food for 116 men for 25 days. After 21 days, 100 men leave the place. For how many days will the rest of the food last for the rest of the men?

a) 19 days b) 24 days c) 29 days d) 15 days

3. There is a sufficient food for 300 men for 32 days. After 29 days, 210 men leave the place. For how many days will the rest of the food last for the rest of the men?

a) 12 days b) 14 days c) 15 days d) 10 days

4. There is a sufficient food for 150 men for 15 days. After 10 days, 75 men leave the place. For how many days will the rest of the food last for the rest of the men?

a) 10 days b) 8 days c) 5 days d) 15 days

Answers

1. b 2. c 3. d 4. a

Rule 23

Theorem: ***A takes as much time as B and C together take to finish a job. If A and B working together finish the job in x days. C alone can do the same job in y days, then B alone can do the same work in*** $\left(\frac{2xy}{y-x}\right)$ ***days and A alone can do the same work in*** $\left(\frac{2xy}{y+x}\right)$ ***days.***

Illustrative Example

Ex: A takes as much time as B and C together take to finish a job. A and B working together finish the job in 10 days. C alone can do the same job in 15 days. In how many days can B alone do the same work?

Soln: **Quicker Method I:** Using the above theorem, B alone can do the same work in $\frac{2\times15\times10}{15-10}=60$ days

Quicker Method II:

(A + B) + (C) can do in $\frac{15\times10}{15+10}=6$ days.

Since A's days = (B + C)'s days.

B + C can do in 6 × 2 = 12 days.

$\therefore$ B [B = {B + C} − C] can do in $\frac{15\times12}{15-12}=60$ days.

Exercise

1. A can do a certain work in the same time in which B and C together can do it. If A and B together could do it in 10 days, and C alone in 50 days, in what time could B alone do it?

a) 25 days b) 30 days c) 24 days d) 20 days

2. A can do a certain work in the same time in which B and C together can do it. If A and B together could do it in 15 days, and C alone in 30 days, in what time could B alone do it?

a) 40 days b) 60 days c) 45 days d) 35 days

3. A can do a certain work in the same time in which B and C together can do it. If A and B together could do it in 12 days, and C alone in 24 days, in what time could B alone do it?

a) 36 days b) 40 days c) 44 days d) 48 days

4. A can do a certain work in the same time in which B and C together can do it. If A and B together could do it in 10 days, and C alone in 15 days, in how many days can A alone do the same work?

a) 12 days b) 60 days c) 24 days d) 48 days

Answers

1. a 2. b 3. d 4. a

Rule 24

Theorem: ***A team of x persons is supposed to do a work in 'D' days. After 'd_1' days, 'y' more persons were employed and the work was finished 'd_2' days earlier, then the number of days it would have been delayed if 'x' more persons were not employed is given by*** $\left[\frac{y\{D-(d_1+d_2)\}-d_2x}{x}\right]$ ***days and the number of days in which the work would have been finished is given by*** $\left[\frac{(x+y)(D-d_2)-d_1y}{x}\right]$ ***days***

Illustrative Example

Ex: A team of 30 men is supposed to do a work in 38 days. After 25 days, 5 more men were employed and the work finished one day earlier. How many days would it have been delayed if 5 more men were not employed?

Soln: **Quicker Approach:**

35 men do the rest of the job in 12 days (12 = 38 − 25 − 1)

$\therefore$ 30 men can do the rest of the job in $\frac{12\times35}{30}=14$ days.

Thus the work would have been finished in 25 + 14 = 39 days that is, (39 − 38) = 1 day after the scheduled time.

Quicker Method: Applying the above formula, we have

the required answer = $\frac{5\{38-(25+1)\}-1\times30}{30}$

$=\frac{5\times12-30}{30}=1$ *day*.

Exercise

1. A team of 40 men is supposed to do a work in 48 days. After 35 days, 15 more men were employed and the work

finished 2 days earlier. How many days would it have been delayed if 15 more men were not employed?

a) 2 days b) $2\frac{1}{8}$ days c) $1\frac{1}{8}$ days d) 1 day

2. A team of 25 men is supposed to do a work in 44 days. After 18 days, 2 more men were employed and the work finished 1 day earlier. How many days would it have been delayed if 2 more men were not employed?
a) 1 day b) 2 days c) 1.5 days d) None of these
3. A team of 20 men is supposed to do a work in 30 days. After 12 days, 5 more men were employed and the work finished 2 days earlier. In how many days would it have been finished if 5 more men were not employed?
a) 30 days b) 28 days c) 32 days d) 34 days
4. A team of 27 men is supposed to do a work in 36 days. After 30 days, 9 more men were employed and the work finished 3 days earlier. In how many days would it have been finished if 9 more men were not employed?
a) 35 days b) 28 days c) 34 days d) 39 days

Answers

1. b 2. a 3. c 4. c

Rule 25

Theorem: ***A, B and C can do a work in x days, y days and z days respectively. They started the work together but after d_1 days A left. If B left the work d_2 days before the completion of the work, then the whole work will be completed in***

$$\frac{z}{x}\left[\frac{y(x-d_1)+d_2x}{y+z}\right] days.$$

Illustrative Example

Ex: A, B and C can do a work in 16 days, $12\frac{4}{5}$ days and 32 days respectively. They started the work together but after 4 days A left. B left the work 3 days before the completion of the work. In how many days was the work completed?

Soln: **Detail Method:**
Suppose the work is completed in x days,
A's 4 day's work + B's (x - 3) day's work + C's x day's work = 1

or, $\frac{4}{16}+\frac{(x-3)5}{64}+\frac{x}{32}=1$

or, $\frac{16+5x-15+2x}{64}=1$

or, $7x+1=64$ $\therefore x=9$ days.

Quicker Method: Applying the above formula, we have

the required answer $= \frac{32}{16}\left[\frac{\frac{64}{5}(16-4)+(3\times16)}{\frac{64}{5}+32}\right]$

$=2\times4.5=9$ days.

Exercise

1. A, B and C can do a piece of work in 12, 18 and 24 days respectively, they work at it together, A stops the work after 4 days and B is called off 2 days before the work is done. In what time was the work finished?
a) 12 days b) 14 days c) 16 days d) 8 days
2. A, B and C can do a piece of work in 6, 9 and 12 days respectively, they work at it together, A stops the work after 2 days and B is called off 1 day before the work is done. In what time was the work finished?
a) 4 days b) 6 days c) 7 days d) 3 days
3. A, B and C can do a piece of work in 18, 27 and 12 days respectively, they work at it together, A stops the work after 6 days and B is called off 3 days before the work is done. In what time was the work finished?
a) 6 days b) 8 days c) 10 days d) $6\frac{6}{13}$ days
4. A, B and C can do a piece of work in 24, 36 and 48 days respectively, they work at it together, A stops the work after 8 days and B is called off 4 days before the work is done. In what time was the work finished?
a) 10 days b) 8 days c) 16 days d) 14 days

Answers

1. d 2. a 3. d 4. c

Rule 26

Theorem: ***A started a work and left after working a_1 days. Then B was called and he finished the work in b_1 days. Had A left the work after working for a_2 days, B would have finished the remaining work in b_2 days. Then, each of them ie A and B, working alone finish the whole work in*** $\left[\frac{b_2a_1-b_1a_2}{b_2-b_1}\right]$ ***days and*** $\left[\frac{a_2b_1-a_1b_2}{a_2-a_1}\right]$ ***days respectively.***

Illustrative Example

Ex: A started a work and left after working for 2 days. Then B was called and he finished the work in 9 days. Had A left the work after working for 3 days, B would have finished the remaining work in 6 days. In how many days can each of them, working alone, finish the whole work?

Soln: **Detailed Method:** Suppose A and B do the work in **x** and y days respectively. Now, work done by A in 2 days + work done by B in 9 days = 1 or, $\frac{2}{x}+\frac{9}{y}=1$ Similarly, $\frac{3}{x}+\frac{6}{y}=1$

To solve the above equation put $\frac{1}{x}$ = a and $\frac{1}{y}=b$.

Thus

2a + 9b = 1 ..... (1) and 3a + 6b = 1 .... (2)

Performing (2) × 3 – (1) × 3 we have

$5a=1 \therefore a=\frac{1}{5}$ or, $x=\frac{1}{a}=5$ days.

and $y=\frac{1}{b}=15$ days.

Quicker Method: In such case: (Using the above theorem)

A will finish the work in $\frac{3\times9-2\times6}{9-6}=\frac{15}{3}=5$ days.

For B, we should use the above result.

B does $1-\frac{2}{5}=\frac{3}{5}$ work in 9 days.

$\therefore$ B does 1 work in $9\times\frac{5}{3}=15$ days.

Exercise

1. A started a work and left after working for 1 day. Then B was called and he finished the work in $4\frac{1}{2}$ days. Had A left the work after working for $1\frac{1}{2}$ days, B would have finished the remaining work in 3 days. In how many days can each of them, working alone, finish the whole work?
   a) 5 days, 15 days b) 2.5 days, 7.5 days
   c) 3.5 days, 8.5 days d) None of these
2. A started a work and left after working for 3 days. Then B was called and he finished the work in $13\frac{1}{2}$ days. Had A left the work after working for $4\frac{1}{2}$ days, B would have finished the remaining work in 9 days. In how many days can each of them, working alone, finish the whole work?
   a) 7.5 days, 22.5 days b) 7 days, 9 days
   c) 5 days, 15 days d) 23.5 days, 8.5 days
3. A started a work and left after working for 4 days. Then B was called and he finished the work in 18 days. Had A left the work after working for 6 days, B would have finished the remaining work in 12 days. In how many days can each of them, working alone, finish the whole work?
   a) 5 days, 20 days b) 10 days, 30 days
   c) 15 days, 30 days d) 5 days, 30 days

Answers

1. b 2. a 3. b

Rule 27

Theorem: ***A can do a work in x days and B can do the same work in y days. If they work together for 'd' days and A goes away, then the number of days in which B finishes the work is given by*** $y-\left(1+\frac{y}{x}\right)d$ ***days.***

Illustrative Example

Ex: A can do a work in 25 days and B can do the same work in 20 days. They work together for 5 days and then A goes away. In how many days will B finish the work?

Soln: Detail Method

A + B can do the work in 5 days $=5\left[\frac{1}{25}+\frac{1}{20}\right]$

$=\frac{5\times45}{25\times20}=\frac{9}{20}$

Rest of the work $=1-\frac{9}{20}=\frac{11}{20}$

B will do the rest of the work in $=1-\frac{9}{20}=\frac{11}{20}$ days.

Quicker Method: Applying the above theorem, we have

the required answer $=20-\left(1+\frac{20}{25}\right)\times5$

$=20-9=11$ days.

Exercise

1. A can do a piece of work in $6\frac{2}{3}$ days and B in 5 days. They work together for 2 days and then A leaves B to finish the work alone. How long will B take to finish it?
   a) $1\frac{1}{2}$ b) 3 days c) 2 days d) 1 day
2. A can do a piece of work in 50 days and B in 40 days. They work together for 10 days and then A leaves B to finish the work alone. How long will B take to finish it?
   a) 11 days b) 18 days c) 22 days d) 26 days
3. A can do a piece of work in 20 days and B in 15 days. They work together for 6 days and then A leaves B to

finish the work alone. How long will B take to finish it?

a) 3 days b) 4 days c) $3\frac{1}{2}$ days d) $4\frac{1}{2}$ days

4. A can do a piece of work in $12\frac{1}{2}$ days and B in 10 days. They work together for $2\frac{1}{2}$ days and then A leaves B to finish the work alone. How long will B take to finish it?

a) $\frac{11}{3}$ days b) $\frac{11}{5}$ days c) 6 days d) $\frac{11}{2}$ days

Answers

1. a 2. c 3. d 4. d

Rule 28

Theorem: ***If A can complete $\frac{a}{b}$ part of a work in x days, then the $\frac{c}{d}$ part of the work will be done in y days. We can calculate the value of y from the given equation $\frac{x}{a/b}=\frac{y}{c/d}$.***

Note: $\frac{\text{No. of days worked}}{\text{Part of work done}}$ = constant for a person

Illustrative Example

Ex: A can do $\frac{3}{4}$ of a work in 12 days. In how many days can he finish $\frac{1}{8}$ of the work?

(SBI PO Exam 1987)

Soln: Using the above theorem, we have

$$\frac{12}{3/4}=\frac{y}{1/8} \text{ or } y=\frac{12}{3}\times 4\times\frac{1}{8}=2 \text{ days.}$$

Exercise

1. Ram can do $\frac{2}{3}$ of a work in 16 days. In how many days can he finish $\frac{1}{12}$ of the work?

a) 1 day b) 3 days c) 2 days d) $2\frac{1}{2}$ days

2. Vinay can do $\frac{1}{4}$ of a work in 5 days. In how many days can he finish $\frac{1}{5}$ of the work?

a) 20 days b) 5 days
c) 4 days d) Data inadequate

3. Sudhir can do $\frac{4}{5}$ of a work in 8 days. In how many days can he finish $\frac{1}{10}$ of the work?

a) 1 day b) 2 days c) 3 days d) None of these

Answers

1. c 2. c 3. a

Rule 29

Ex.: 38 men, working 6 hours a day can do a piece of work in 12 days. Find the number of days in which 57 men working 8 hrs a day can do twice the work. Assume that 2 men of the first group do as much work in 1 hour as 3 men of the second group do in $1\frac{1}{2}$ hr.

Soln: Detailed Method:

2 × 1 men of first group = 3 × 1.5 men of second group

or, 2 men of first group = 4.5 men of second group

$\therefore$ 38 men of first group $=\frac{4.5}{2}\times 38=19\times 4.5$

$\because$ (19 × 4.5) men do 1 work, working 6 hrs/day in 12 days.

$\therefore$ 1 man does 1 work working 1 hr/day in (12 × 19 × 4.5 × 6) days.

$\therefore$ 57 men do 2 work working 8 hrs/day in

$$=\frac{12\times 19\times 4.5\times 6}{57\times 8}\times 2=27 \text{ days.}$$

Quicker Method:

Ratio of efficiency of persons in first group to the second group

$$=E_1:E_2=(3\times 1.5):2\times 1=4.5:2 \text{ ..... (*)}$$

Now, use the formula:

$$M_1D_1T_1E_1W_2=M_2D_2T_2E_2W_1 \text{ ..... (*)(*)}$$

$$\therefore D_2=\frac{38\times 12\times 6\times 4.5\times 2}{57\times 8\times 2\times 1}=27 \text{ days.}$$

Note: (*) Less number of persons from the first group do the same work in less number of days, so they are more efficient.

(*)(*) M represents the number of men.
D represents the number of days.
T represents the number of working hours.
E represents the efficiency.
W represents the work and
the suffix represents the respective groups.

Exercise

1. 40 men, working 8 hours a day can do a piece of work in 15 days. Find the number of days in which 60 men working 4 hrs a day can do twice the work. Assume that 3 men of the first group do as much work in 2 hour as 4 men of the second group do in 3 hrs.
   a) 60 days b) 40 days c) 80 days d) None of these
2. 30 men, working 4 hours a day can do a piece of work in 10 days. Find the number of days in which 45 men working 8 hrs a day can do twice the work. Assume that 2 men of the first group do as much work in 2 hour as 4 men of the second group do in 1 hr.
   a) $6\frac{1}{3}$ days b) $6\frac{2}{3}$ days c) $5\frac{3}{6}$ days d) $3\frac{1}{6}$ days

Answers

1. c; **Hint:** 3 × 2 men of first group = 4 × 3 men of second group
   ∴ Ratio of efficiency of persons in first group to the second group = $E_1 : E_2 = 2:1$. Now apply the given formula.
2. b

Rule 30

Theorem: ***If A working alone takes 'x' days more than A and B, and B working alone takes 'y' days more than A and B together then the number of days taken by A and B working together is given by $\left[\sqrt{xy}\right]$ days.***

Illustrative Example

Ex: A alone would take 14 hours more to complete the job than if both A and B would together. If B worked alone, he took $3\frac{1}{2}$ hours more to complete the job than A and B worked together. What time, would they take if both A and B worked together?

Soln: Applying the above theorem, we have

the required answer = $\sqrt{\dfrac{14\times 7}{2}} = 7$ hours.

Exercise

1. A alone would take 8 hours more to complete the job than if both A and B would together. If B worked alone, he took $4\frac{1}{2}$ hours more to complete the job than A and B worked together. What time, would they take if both A and B worked together?
   a) 6 hours b) 5 hours c) 7 hours d) 8 hours
   (Income Tax and Excise Exam, 1985)
2. A alone would take 16 hours more to complete the job than if both A and B would together. If B worked alone, he took 4 hours more to complete the job than A and B worked together. What time, would they take if both A and B worked together?
   a) 5 hours b) 8 hours c) 9 hours d) None of these
3. A alone would take 27 hours more to complete the job than if both A and B would together. If B worked alone, he took 3 hours more to complete the job than A and B worked together. What time, would they take if both A and B worked together?
   a) 8 hours b) 10 hours c) 9 hours d) 6 hours

Answers

1. a 2. b 3. c

Rule 31

Theorem: ***If A, B and C can do a job alone in x days, y days and z days respectively.***
∴ alone time for A = x days
alone time for B = y days
alone time for C = z days
Now consider the following cases,

Case I: To find the amount of work done by A, B and C separately.
Using the formula,

$$\text{Amount of work} = \frac{\text{Number of days actually worked}}{\text{alone time}}$$

and assuming that A, B and C have worked for d_1 days, d_2 days and d_3 days respectively, then

amount of work by $A = \dfrac{d_1}{x}$, amount of work by $B = \dfrac{d_2}{y}$ and amount of work by $C = \dfrac{d_3}{z}$.

Case II: If the job is complete, then add the amount of work done by A, B and C and equate it to 1.

ie $\dfrac{d_1}{x}+\dfrac{d_2}{y}+\dfrac{d_3}{z}=1$, if the job is half complete the following equation is obtained,

$$\frac{d_1}{x}+\frac{d_2}{y}+\frac{d_3}{z}=\frac{1}{2}$$

Illustrative Example

Ex: A man, a woman or a boy can do a job in 20 days, 30 days or 60 days respectively. How many boys must assist 2 men and 8 women to do the work in 2 days.
(MBA 1992)

Soln: Let the required number of boys be x.
Now, using the above theorem
(2 men's work for 2 days) + women's work for

2 days) + (x boy's work for 2 days) = 1

or, $\left(2\times2\times\frac{1}{20}\right)+\left(8\times2\times\frac{1}{30}\right)+\left(x\times2\times\frac{1}{60}\right)=1$

$\therefore x = 8$ boys.

Exercise

1. A and B together can do a piece of work in 12 days which B and C together can do in 16 days. After A has been working at it for 5 days, and B for 7 days. C finishes it in 13 days. In how many days could each do the work by himself?
   a) 16, 48 and 26 days respectively
   b) 16, 48 and 24 days respectively
   c) 26, 48 and 24 days respectively
   d) 16, 46 and 24 days respectively
2. A can do a job in 20 days, B in 30 days and C in 60 days. If A is helped on every 3rd day by B and C, then in how many days, the job is finished? **[ITI 1989]**
   a) 20 days b) 15 days c) 18 days d) 24 days
3. A can do a job in 12 days, B in 15 days. They work together for 2 days. Then B leaves and A alone continues the work. After 1 day C joins A and work is completed in 5 more days. In how many days can C do it alone?
   a) 15 days b) 20 days c) 25 days d) 30 days
4. A and B can do a job in 15 days and 10 days respectively. They began the work together but A leaves after some days and B finished the remaining job in 5 days. After how many days did A leave?
   a) 2 days b) 4 days c) 3 days d) 6 days
5. A and B can do a job in 16 days and 12 days respectively. 4 days before finishing the job, A joins B. B has started the work alone. Find how many days B has worked alone? **[Bank PO 1989]**
   a) 8 days b) 10 days c) 4 days d) 5 days
6. A man, a woman or a boy can do a job in 20 days, 30 days or 60 days respectively. How many boys must assist 2 men and 8 women to do the work in 2 days? **[MBA 1992]**
   a) 8 boys b) 10 boys c) 12 boys d) 16 boys
7. A can do a job in 3 days less time than B. A works at it alone for 4 days and then B takes over and completes it. If altogether 14 days were required to finish the job, how many days would each of them take alone to finish it?
   a) 13 days, 16 days b) 12 days, 15 days
   c) 15 days, 12 days d) 15 days, 18 days
8. A can do a piece of work in 24 days, while B alone can do it in 16 days. With the help of C they finish the work in 8 days. Find in how many days alone C can do the work? **[MBA 1988]**
   a) 48 days b) 36 days c) 40 days d) 50 days

Answers

1. b; **Hint:** Let the whole work be 1

A and B in 1 day do $\frac{1}{12}$.

B and C in 1 day do $\frac{1}{16}$.

A's 5 day's work + B's 7 day's work + C's 13 day's work = 1

Or, A's 5 day's work + B's 5 day's work + B's 2 day's work + C's 2 day's work + C's 11 day's work = 1

$\therefore \frac{5}{12}+\frac{2}{16}+C's$ 11 day's work = 1

$\therefore$ C's 11 day's work $=1-\left(\frac{5}{12}+\frac{2}{16}\right)=\frac{11}{24}$

$\therefore$ C's 1 day's work $=\frac{1}{24}$

$\therefore$ B's 1 day's work $=\frac{1}{16}-\frac{1}{24}=\frac{1}{48}$

$\therefore$ A's day's work $=\frac{1}{12}-\frac{1}{48}=\frac{1}{16}$

$\therefore$ A, B and C can do the work in 16, 48 and 24 days respectively.

2. b; **Hint:** Since A is helped by B and C on every 3rd day, A works for 3 days while B and C work for 1 day

$\frac{1}{20}\times3+\frac{1}{30}\times1+\frac{1}{60}+1=\frac{1}{5}$ [$\because$ B and C help only on 3rd day]

$\therefore$ Total time for the job = 3 × 5 = 15 days.

3. c; **Hint:** Let C do it alone in x days

A's amount of work +B's amount of work +C's amount of work = 1

or, $(2+1+5)\frac{1}{12}+\left(2\times\frac{1}{15}\right)+\left(5\times\frac{1}{x}\right)=1$

or, $\frac{5}{x}=1-\left(\frac{8}{12}+\frac{2}{15}\right)$ or $\frac{5}{x}=\frac{1}{5}$ $\therefore x=25$

$\therefore$ C can do it alone in 25 days.

4. c; **Hint:** In this problem, total time for the work is not known and also it is not to be found out. Hence total time for the work is not to be considered.

If A leaves after x days ie A works for x days and B works for x + 5 days,

then applying the given rule, we have

$$\frac{\text{No. of days A worked}}{\text{Alone time for A}}+\frac{\text{No. of days B worked}}{\text{Alone time for B}}=1$$

or, $\frac{x}{15}+\frac{x+5}{10}=1$ or, x = 3

∴ **A leaves** after 3 days.

5. d; **Hint:** If B works alone for x days; A's amount of work + B's amount of work = 1

or, $\frac{4}{16}+\frac{(x+4)}{12}=1 \quad \therefore x=5$

6. a; **Hint:** Using the given rule we have (2 men's work) + (8 women's work) + (x boy's work) = 1

or, $\left(2\times2\times\frac{1}{20}\right)+\left(8\times2\times\frac{1}{30}\right)+\left(x\times2\times\frac{1}{60}\right)=1$

or, $\frac{1}{5}+\frac{8}{15}+\frac{x}{30}=1$

or, $\frac{6+16+x}{30}=1 \quad \therefore x=8$ boys.

7. b; **Hint:** Let A alone takes x days to finish the work and hence B alone takes (x + 3) days.
Now, using the given rule, we have
A's amount of work + B's amount of work = 1

or, $\frac{4}{x}+\frac{10}{x+3}=1 \quad \therefore x=12$

∴ A alone takes 12 days and B alone takes (12 + 3 = 15) days to complete the work.

8. a; **Hint:** $\frac{8}{24}+\frac{8}{16}+\frac{8}{x}=1 \qquad \therefore x=48$ days.

Rule 32

Theorem: ***Two persons A and B can finish a job alone in x and y days respectively. If they start working on alternate days, then to find the total job completion time, following steps are taken.***

Note: This formula is applicable only when x and y are integers.

Case: If A starts the work

Step I: First calculate the value of p; where p = nearest integer value to be considered $=\left(\frac{xy}{x+y}\right)$

Step II:

(a) When, $x-y=\pm2, \pm4$ then, apply the following formula.

T (Total job completion time) $=\frac{xy+p(x-y)}{x}$

(b) When, $x-y=\pm1, \pm3$, then apply the following formula,

T (Total job completion time) $=\frac{xy-p(x-y)}{y}$

Illustrative Examples

Ex. 1: A and B working alone can finish a job in 5 days and 7 days respectively. They work at it alternately for a day. If A starts the work, find in how many days the job will be finished?

Soln: Applying the above theorem:

Step I: $p=\frac{xy}{x+y}=\frac{5\times7}{5+7}=\frac{35}{12}\approx3$ (nearest integer value)

Step II: $x-y=5-7=-2$, Here, formula (a) will be applied

∴ Total time to finish the job if A starts the work

$$=\frac{xy+p(x-y)}{x}=\frac{5\times7+3(5-7)}{5}$$

$$=\frac{29}{5}=5\frac{4}{5} \text{ days.}$$

Ex. 2: A and B working separately can do a work in 9 and 12 days respectively. A starts the work and they work on alternate days. In how many days will the work be completed?

Soln: Applying the above theorem,

Step I: $p=\frac{12\times9}{12+9}=\frac{108}{21}\approx5$ (nearest integer value)

Step II: $x-y=9-12=-3$, Here formula (b) will be applied.

∴ Total time to finish the job if A starts the work

$$=\frac{xy-p(x-y)}{y}=\frac{(9\times12)-5(9-12)}{12}$$

$$=\frac{108+15}{12}=\frac{41}{4}=10\frac{1}{4} \text{ days}$$

Now we try to solve the above examples by Detail Method.

Ex. 1: **Detail Method:**

In the first day A does $\frac{1}{5}$ of the work

In the second day B does $\frac{1}{7}$ of the work

∴ in the first 2 days $\left(\frac{1}{5}+\frac{1}{7}=\frac{12}{35}\right)$ of the work

∴ in 4 days $\frac{12}{35}\times2=\frac{24}{35}$ of the work

Now, $\left(1-\frac{24}{35}\right)=\frac{11}{35}$ of the work remains to be done.

In fifth day A does $\frac{1}{5}$ of the work

∴ B will finish the work $\left(\frac{11}{35}-\frac{1}{5}\right)=\frac{4}{35}$ of the work in $\frac{4}{35} \div \frac{1}{7}$ or $\frac{4}{5}$ days.

$\therefore$ the total time required $= \left(4+1+\frac{4}{5}\right) = 5\frac{4}{5}$ days.

Ex. 2: Detail Method: (A + B)'s day's work $= \frac{1}{9} + \frac{1}{12} = \frac{7}{36}$

We see that $5 \times \frac{7}{36} = \frac{35}{36}$ (just less than 1) ie (A + B) work for 5 pairs of days ie for 10 days.

Now rest of the work $\left(1 - \frac{35}{36}\right) = \frac{1}{36}$ is to be done by A

A can do $\frac{1}{36}$ work in $9 \times \frac{1}{36} = \frac{1}{4}$ day

$\therefore$ Total days $= 10 + \frac{1}{4} = 10\frac{1}{4}$ days.

Note: *Two persons A and B can finish a job alone in x and y days respectively. If they start working on alternate days, then to find the total job completion time, following steps are taken.*

Case: If B starts the work

Step I: First calculate the value of p;
where p = nearest integer value to be considered

$$= \left(\frac{xy}{x+y}\right)$$

Step II:

a) when, $x - y = \pm 2, \pm 4$, then apply the following formula,

T (Total job completion time) $= \frac{xy - p(x-y)}{y}$

b) when $x - y = \pm 1, \pm 3$, then apply the following formula,

T (Total job completion time) $= \frac{xy + p(x-y)}{x}$

Illustrative Examples

Ex. 1: A and B working alone can finish a job in 5 days and 7 days respectively. They work at it alternately for a day. If B starts the work, find in how many days the job will be finished?

Soln: Apply the above theorem:

Step I: $p = \frac{xy}{x+y} = \frac{35}{12} \approx 3$ (Nearest integer value)

Step II: x - y = 5 - 7 = -2; Here, formula (a) will be applied.

$\therefore$ Total time to finish the job if B starts the work

$$= \frac{xy - p(x-y)}{y} = \frac{(5\times 7) - 3(5-7)}{7}$$

$$= \frac{35 - 3(-2)}{7} = \frac{41}{7} = 5\frac{6}{7} \text{ days.}$$

Ex. 2: A and B working separately can do a work in 9 and 12 days respectively. B starts the work and they work on alternate days. In how many days will the work be completed?

Soln: Applying the above theorem, we have

Step I: $p = \frac{12 \times 9}{12 + 9} = \frac{108}{21} \approx 5$ (nearest integer value)

Step II: $x - y = 9 - 12 = -3$, Here formula (b) will be applied.

$\therefore$ total time to finish the job is B starts the work

$$= \frac{xy + p(x-y)}{x}$$

$$= \frac{9 \times 12 + 5(9-12)}{9} = \frac{108 - 15}{9} = \frac{31}{3} = 10\frac{1}{3} \text{ days.}$$

Exercise

1. Two women, Ganga and Jamuna, working separately can mow a field in 8 and 12 hours respectively. If they work for an hour alternately, Ganga beginning at 9 am, when will the mowing be finished?

   a) $6\frac{1}{2}$ pm b) $8\frac{1}{2}$ pm c) $3\frac{1}{2}$ pm d) None of these

2. Ram and Mohan can do a job alone in 10 and 8 days respectively. On 1st January Ram starts the job and then they work on alternate days. When will the work be finished?

   a) 8th January b) 10th January
   c) 9th January d) None of these

3. A and B working separately can plough a field in 6 and 10 hours respectively. At 8 AM A starts the work and they work in stretches of one hour alternately, when will the ploughing be completed?

   a) 3 : 20 PM b) 2.20 PM
   c) 12.30 PM d) None of these

Answers

1. a; **Hint:** $p = \frac{8 \times 12}{8 + 12} \approx 5$, Here $x - y = 8 - 12 = -4$

   Hence apply the formula (a).

   $\therefore$ reqd answer $= \frac{96 - 20}{8} = \frac{76}{8} = \frac{19}{2} = 9\frac{1}{2}$ hours.

   Ganga started at 9 am hence she completes the work

at 9 am $+9\frac{1}{2}$ hrs $=6\frac{1}{2}$ pm.

2. c; **Hint:** $p=\frac{8\times10}{18}\approx5$, here x - y = 10 - 8 = +2, hence apply the formula (a).

$\therefore$ required answer $=\frac{80+10}{10}=9$ days.

Since Ram starts on 1st January

$\therefore$ work will be completed on 9th day ie on 9th of January.

3. a; **Hint:** $p=\frac{6\times10}{16}\approx4$

Here x - y = 6 - 10 = 4, hence formula (a) will be applied.

$$\therefore \text{required time}=\frac{6\times10+4(6-10)}{6}$$

$$=\frac{22}{3}=7\frac{1}{3}\text{ hours.}$$

$\therefore$ required answer = 8 am + $7\frac{1}{3}$ hours = 3:20 pm.

Rule 33

Theorem: ***If A, B and C together can do a work in x days, A alone can do the work in 'a' days and B alone can do the work in 'b' days, then C will do the same work in*** $\left[\frac{x\,ab}{ab-x(a+b)}\right]$ ***days.***

Illustrative Example

Ex: A, B and C together can do a work in 6 days. A alone can do the work in 18 days and B alone can do the same work in 27 days. Find in what time C alone can do that work?

Soln: Applying the above formula, we have

$$\text{the required answer }=\frac{6\times18\times27}{18\times27-6(18+27)}$$

$$=13\frac{1}{2}\text{ days.}$$

Exericse

1. A, B and C together can do a work in 2 days. A alone can do the work in 6 days and B alone can do the same work in 9 days. Find in what time C alone can do that work?

   a) $4\frac{1}{2}$ days b) $6\frac{3}{4}$ days c) 9 days d) None of these

2. A, B and C together can do a work in 8 days. A alone can do the work in 24 days and B alone can do the same work in 36 days. Find in what time C alone can do that work?

   a) 9 days b) 15 days c) 18 days d) 24 days

3. A, B and C together can do a work in 4 days. A alone can do the work in 12 days and B alone can do the same work in 18 days. Find in what time C alone can do that work?

   a) 8 days b) 27 days c) 9 days d) 18 days

4. A, B and C together can do a work in 12 days. A alone can do the work in 36 days and B alone can do the same work in 54 days. Find in what time C alone can do that work?

   a) 9 days b) 18 days c) 24 days d) 27 days

Answers

1. a 2. c 3. c 4. d

Rule 34

Theorem: ***If A and B can do a work in x and y days respectively and A leaves the work after doing for 'a' days, then B does the remaining work in*** $\left[\frac{(x-a)y}{x}\right]$ ***days.***

Illustrative Example

Ex: A can complete a work in 25 days and B can do the same work in 10 days. If A after doing 4 days, leaves the work, find in how many days B will do the remaining work?

Soln: Applying the above formula, we have the required answer

$$=\frac{(25-4)\times10}{25}=\frac{21\times10}{25}=\frac{42}{5}=8\frac{2}{5}\text{ days.}$$

Exercise

1. A can complete a work in 20 days and B can do the same work in 25 days. If A after doing 5 days, leaves the work, find in how many days B will do the remaining work?

   a) $18\frac{3}{4}$ days b) $8\frac{3}{4}$ days

   c) $17\frac{3}{4}$ days d) None of these

2. A can complete a work in 35 days and B can do the same work in 28 days. If A after doing 10 days, leaves the work, find in how many days B will do the remaining work?

   a) 25 days b) 20 days c) 27 days d) 24 days

3. A can complete a work in 24 days and B can do the same work in 18 days. If A after doing 4 days, leaves the work, find in how many days B will do the remaining work?

   a) 10 days b) 12 days c) 15 days d) 16 days

Answers

1. a 2. b 3. c

Rule 35

Theorem: ***If A and B can do a work in x and y days respectively, and B leaves the work after doing for 'a' days, then A does the remaining work in*** $\left[\frac{(y-a)x}{y}\right]$ ***days.***

Illustrative Example

Ex: A can do a work in 15 days and B alone can do that work in 25 days. If B after doing 5 days leaves the job, find in how many days A will do the remaining work.

Soln: Applying the above formula, we have

the required answer $= \frac{(25-5)\times 15}{25}$

$= \frac{20\times 15}{25} = 12$ days

Exercise

1. A and B working together can do a piece of work in 6 days, B alone could do it in 8 days. Supposing B works at it for 5 days, in how many days A alone could finish the remaining work?
   a) 9 days b) 8 days c) 6 days d) 12 days
2. A and B working together can do a piece of work in 10 days, B alone could do it in 20 days. Supposing B works at it for 4 days, in how many days A alone could finish the remaining work?
   a) 9 days b) 12 days c) 16 days d) 10 days
3. A and B working together can do a piece of work in 30 days, B alone could do it in 50 days. Supposing B works at it for 10 days, in how many days A alone could finish the remaining work?
   a) 12 days b) 60 days c) 16 days d) 18 days
4. A and B working together can do a piece of work in $7\frac{1}{2}$ days, B alone could do it in $12\frac{1}{2}$ days. Supposing B works at it for $2\frac{1}{2}$ days, in how many days A alone could finish the remaining work?
   a) 5 days b) 8 days c) 7 days d) 15 days
5. A can complete a job in 9 days. B in 10 days and C in 15 days. B and C start the work and are forced to leave after 2 days. The time taken to complete the remaining work is: **[NDA Exam 1987]**
   a) 13 days b) 10 days c) 9 days d) 6 days

Answers

1. a; **Hint:** First apply the **Rule-6**, and find the no. of days in which A alone could do the whole work ie $\frac{6\times 8}{8-6} = 24$ days.
   Now, applying the given rule, we have
   the required answer $= \frac{(8-5)24}{8} = 9$ days.

2. c 3. b 4. d

5. d; **Hint:** B and C together can do the work in $\frac{15\times 10}{15+10} = 6$ days. (See **Rule-5**)
   Here, y = 6 days, and x = 9 days.
   Now applying the given rule, we have
   the required answer $= \frac{(6-2)\times 9}{6} = 6$ days.

Rule 36

Theorem: ***A and B can do a piece of work in x and y days respectively and both of them starts the work together. If B leaves the work 'a' days before the completion of work, then the total time, in which the whole work is completed,*** $= \left[\frac{(y+a)x}{x+y}\right]$ ***days.***

Illustrative Example

Ex: A and B can do a piece of work in 15 days and 25 days. Both starts the work together for some time, but B leaves the job 7 days before the work is completed. Find the time in which work is finished.

Soln: Applying the above formula, we have

the required answer $= \frac{(25+7)15}{25+15} = 12$ days.

Exercise

1. A and B can do a piece of work in 20 days and 30 days. Both starts the work together for some time, but B leaves the job 5 days before the work is completed. Find the time in which work is finished.
   a) 7 days b) 12 days c) 14 days d) 16 days
2. A and B can do a piece of work in 25 days and 35 days. Both starts the work together for some time, but B leaves the job 7 days before the work is completed. Find the time in which work is finished.
   a) 17 days b) $17\frac{1}{2}$ days c) 18 days d) 20 days
3. A and B can do a piece of work in 30 days and 45 days. Both starts the work together for some time, but B leaves the job 15 days before the work is completed. Find the time in which work is finished.
   a) 24 days b) 28 days c) 20 days d) 16 days
4. A and B can do a piece of work in 16 days and 24 days. Both starts the work together for some time, but B leaves the job 6 days before the work is completed. Find the time in which work is finished.
   a) 18 days b) 14 days c) 12 days d) 8 days
5. A and B can do a piece of work in 17 days and 33 days.

Both starts the work together for some time, but B leaves the job 7 days before the work is completed. Find the time in which work is finished.

a) $3\frac{5}{13}$ days b) $5\frac{3}{13}$ days

c) $13\frac{3}{5}$ days d) None of these

Answers

1. c 2. b 3. a 4. c 5. c

Rule 37

Theorem: ***A and B can do a piece of work in x and y days respectively and both of them starts the work together. If A leaves the work 'a' days before the completion of the work, then the total time in which the whole work is completed***

$= \frac{(x+a)y}{(x+y)}$ ***days.***

Illustrative Example

Ex: A and B can do a piece of work in 10 days and 20 days respectively. Both starts the work together but A leaves the work 5 days before its completion time. Find the time in which work is finished.

Soln: Applying the above formula, we have

the required answer $= \frac{(10+5)20}{10+20} = 10$ days.

Exercise

1. A can do a piece of work in 14 days and B in 21 days. They begin together. But 3 days before the completion of the work, A leaves off. In how many days is the work completed?

   a) 10 days b) 5 days c) $5\frac{1}{5}$ days d) $10\frac{1}{5}$ days

2. A can do a piece of work in 15 days and B in 25 days. They begin together. But 5 days before the completion of the work, A leaves off. In how many days is the work completed?

   a) $12\frac{1}{2}$ days b) $13\frac{1}{2}$ days c) $11\frac{3}{4}$ days d) 25 days

3. A can do a piece of work in 20 days and B in 40 days. They begin together. But 10 days before the completion of the work, A leaves off. In how many days is the work completed?

   a) 10 days b) 15 days c) 20 days d) 25 days

4. A can do a piece of work in 5 days and B in 10 days. They begin together. But $2\frac{1}{2}$ days before the completion of the work, A leaves off. In how many days is the work completed?

   a) 2 days b) 4 days c) 5 days d) 8 days

Answers

1. d 2. a 3. c 4. c

Rule 38

Theorem: ***A can do a piece of work in x days. If A does the work only for 'a' days and the remaining work is done by B in 'b' days, the B alone can do the work in*** $\left(\frac{xb}{x-a}\right)$ ***days.***

Illustrative Example

Ex: A can do a piece of work in 12 days. A does the work for 2 days only and leaves the job. B does the remaining work in 5 days. In how many days B alone can do the work?

Soln: Applying the above formula, we have

the required answer $= \frac{12\times5}{12-2} = 6$ days.

Exercise

1. A can do a piece of work in 15 days. A does the work for 3 days only and leaves the job. B does the remaining work in 8 days. In how many days B alone can do the work?

   a) 12 days b) 10 days c) 15 days d) 8 days

2. A can do a piece of work in 25 days. A does the work for 5 days only and leaves the job. B does the remaining work in 4 days. In how many days B alone can do the work?

   a) 5 days b) 15 days
   c) 9 days d) None of these

3. A can do a piece of work in 23 days. A does the work for 11 days only and leaves the job. B does the remaining work in 9 days. In how many days B alone can do the work?

   a) 17 days b) 18 days c) $17\frac{1}{4}$ days d) $17\frac{3}{4}$ days

4. A can do a piece of work in 22 days. A does the work for 12 days only and leaves the job. B does the remaining work in 5 days. In how many days B alone can do the work?

   a) 11 days b) 10 days c) 12 days d) 14 days

5. A can do a piece of work in 80 days. He works at it for 10 days and then B alone finishes the work in 42 days. The two together could complete the work in:

   a) 24 days b) 25 days c) 30 days d) 35 days

   (Clerical Grade Exam, 1991)

Answers

1. b 2. a 3. c 4. a

5. c; **Hint:** B alone can do the work in $\left(\frac{80 \times 42}{80-10} = 48\right)$ days.

The two together could complete the work in $\left(\frac{48 \times 80}{80+48} = 30\right)$ days. (See **Rule-4**)

Rule 39

Theorem: ***A and B can do a piece of work in x and y days respectively. Both starts the work together. But due to some problems A leaves the work after some time, and B does the remaining work in 'a' days, then the time after which A leaves the work is given by*** $\left[\frac{(y-a)x}{x+y}\right]$ ***days.***

Illustrative Example

Ex: A and B can do a piece of work in 45 days and 40 days respectively. They start the work together but after some days, A leaves the job. B alone does the remaining work in 23 days. Find after how many days does A leave the job?

Soln: Using the above theorem, we have

the required answer $= \frac{(40-23)45}{45+40} = 9$ days.

Exercise

1. A and B can do a piece of work in 20 days and 25 days respectively. They start the work together but after some days, A leaves the job. B alone does the remaining work in 10 days. Find after how many days does A leave the job?

   a) $6\frac{2}{3}$ days b) $6\frac{1}{3}$ days c) 6 days d) $5\frac{2}{3}$ days

2. A and B can do a piece of work in 25 days and 30 days respectively. They start the work together but after some days, A leaves the job. B alone does the remaining work in 8 days. Find after how many days does A leave the job?

   a) 12 days b) 8 days c) 10 days d) 16 days

3. A and B can do a piece of work in 14 days and 21 days respectively. They start the work together but after some days, A leaves the job. B alone does the remaining work in 6 days. Find after how many days does A leave the job?

   a) 7 days b) 6 days c) 8 days d) 9 days

4. A and B can do a piece of work in 22 days and 23 days respectively. They start the work together but after some days, A leaves the job. B alone does the remaining work in 8 days. Find after how many days does A leave the job?

   a) 6 days b) $6\frac{2}{3}$ days c) $7\frac{2}{3}$ days d) $7\frac{1}{3}$ days

Answers

1. a 2. c 3. b 4. d

Rule 40

Theorem: ***A completes a work in 'x' days. B completes the same work in 'y' days. A started working alone and after 'a' days B joined him. Then the time in which, they will take together to complete the remaining work is given by*** $\left[\frac{(x-a)y}{x+y}\right]$.

Illustrative Example

Ex: Ram completes a work in 10 days. Shyam completes the same work in 15 days. Ram starts working alone and after 5 days B joins him. How many days will they now take together to complete the remaining work?

Soln: Applying the above rule, we have

the required answer $= \frac{(10-5)15}{25} = 3$ days

Exercise

1. A completes a work in 12 days. B completes the same work in 15 days. A started working alone and after 3 days B joined him. How many days will they now take together to complete the remaining work?

   a) 5 b) 8 c) 6 d) 4

   (BSRB Calcutta PO 1999)

2. A completes a work in 20 days. B completes the same work in 25 days. A started working alone and after 2 days B joined him. How many days will they now take together to complete the remaining work?

   a) 12 days b) 10 days c) 8 days d) 16 days

3. A completes a work in 12 days. B completes the same work in 13 days. A started working alone and after 7 days B joined him. How many days will they now take together to complete the remaining work?

   a) $1\frac{3}{5}$ days b) $3\frac{3}{5}$ days c) 2 days d) $2\frac{3}{5}$ days

4. A completes a work in 21 days. B completes the same work in 24 days. A started working alone and after 6 days B joined him. How many days will they now take together to complete the remaining work?

   a) 6 days b) 8 days c) 10 days d) 12 days

5. A completes a work in 15 days. B completes the same work in 20 days. A started working alone and after 1 day B joined him. How many days will they now take together to complete the remaining work?

   a) 8 days b) 7 days c) 6 days d) None of these

6. Mahesh and Umesh can complete a work in 10 and 15 days respectively. Umesh starts the work and after 5 days Mahesh also joins him. In all, the work would be completed in:
   a) 9 days b) 7 days
   c) 11 days d) None of these
   (Clercial Grade 1991)

Answers

1. a 2. b 3. d 4. b 5. a

6. a; **Hint:** Here A = Umesh, B = Mahesh

$\therefore$ x = 15 days and y = 10 days

Now, applying the given rule, we have the time taken by A and B together to complete the remaining work

$$= \frac{(15-5)10}{10+15} = 4 \text{ days.}$$

$\therefore$ total time consumed to complete the work

$= 5 + 4 = 9$ days.

Miscellaneous

1. Twenty-four men can complete a work in sixteen days. Thirty-two women can complete the same work in twenty-four days. Sixteen men and sixteen women started working and worked for twelve days. How many more men are to be added to complete the remaining work in 2 days?
   [Bank of Baroda PO, 1999]
   a) 48 b) 24 c) 36 d) None of these
2. 25 men and 15 women can complete a piece of work in 12 days. All of them start working together and after working for 8 days the women stopped working. 25 men completed the remaining work in 6 days. How many days will it take for completing the entire job if only 15 women are put on the job? **[Guwahati PO, 1999]**
   a) 60 days b) 88 days
   c) 94 days d) None of these
3. 10 men and 15 women finish a work in 6 days. One man alone finishes that work in 100 days. In how many days will a woman finish the work?
   [BSRB Hyderabad PO, 1999]
   a) 125 days b) 150 days c) 90 days d) 225 days
4. A can do a piece of work in 12 days, B can do the same work in 8 days, and C can do the same job in $\frac{4}{5}$ th time required by both A and B. A and B work together for 3 days, then C completes the job. How many complete days did C work? **[NABARD, 1999]**
   a) 8 b) 6 c) 3 d) None of these
5. 12 men take 18 days to complete a job whereas 12 women in 18 days can complete $\frac{3}{4}$ of the same job. How many days will 10 men and 8 women together take to complete the same job? **[BSRB Delhi PO, 2000]**
   a) 6 b) $13\frac{1}{2}$ c) 12 d) None of these
6. If 5 men and 3 boys can reap 23 hectares in 4 days and if 3 men and 2 boys can reap 7 hectares in 2 days, how many boys must assist 7 men in order that they may reap 45 hectares in 6 days?
   a) 2 boys b) 6 boys c) 4 boys d) 5 boys
7. 25 men can reap a field in 20 days. When should 15 men leave the work, if the whole field is to be reaped in $37\frac{1}{2}$ days after they leave the work?
   a) 6 days b) 4 days
   c) 5 days d) None of these
8. A can copy 75 pages in 25 hours, A and B together can copy 135 pages in 27 hours. In what time can B copy 42 pages?
   a) 21 hrs b) 5 hrs 36 secs
   c) 18 hrs d) 24 hrs
9. 15 men would finish a piece of work in 210 days. But at the end of every 10 days, 15 additional men are employed. In how many days will it be finished?
   a) 30 days b) 70 days c) 35 days d) 60 days
10. A piece of work was to be completed in 40 days, a number of men employed upon it did only half the work in 24 days, 16 more men were then set on, and the work was completed in the specified time, how many men were employed at first?
    a) 16 men b) 32 men c) 24 men d) 48 men
11. Ramesh can finish a job in 20 days. He worked for 10 days alone and completed the remaining job working with Dinesh, in 2 days. How many days would both Dinesh and Ramesh together take to complete the entire job?
    a) 4 b) 5 c) 10 d) 12
    [BSRB Bank PO Exam, 1991]
12. A can do a piece of work in 12 days. B is 60% more efficient than A. The number of days, it takes B to do the same piece of work, is:
    a) $7\frac{1}{2}$ b) $6\frac{1}{4}$ c) 8 d) 6
    [CBI Exam, 1991]
13. 12 men can complete a work within 9 days. After 3 days they started the work, 6 men joined them to replace 2 men. How many days will they take to complete the remaining work?
    a) 2 b) 3 c) 4 d) $4\frac{1}{2}$
    [BSRB Bank PO Exam, 1991]

14. A can do a piece of work in 5 hours, B in 9 hours and C in 15 hours. If C could work with them for 1 hour only, the time taken by A and B together to complete the work is:

a) 2 hours b) 3 hours c) $3\frac{1}{2}$ hours d) 4 hours

[Clerical Grade, 1991]

15. A does half as much work as B in three-fourth of the time. If together they take 18 days to complete a work, how much time shall B take to do it?

a) 40 days b) 35 days
c) 30 days d) None of these

[LIC AAO Exam, 1988]

16. Two workers A and B working together completed a job in 5 days. If A worked twice as efficiently as he actually did and B worked $\frac{1}{3}$ as efficiently as he actually did, the work would have completed in 3 days. Find the time for A to complete the job alone.

(MBA, 1982)

a) $6\frac{1}{2}$ days b) $6\frac{1}{4}$ days c) $6\frac{3}{4}$ days d) $12\frac{1}{2}$ days

17. Mohan can mow his lawn in x hours. After 2 hours it begins to rain. The unmoved part of the lawn is _______.

a) $\frac{2}{x}$ b) $\frac{2-x}{2}$ c) $\frac{x}{2}$ d) $\frac{x-2}{x}$

[ITI, 1988]

18. If factory A turns out x cars an hour and factory B turns out y cars every 2 hours, the number of cars which both factories turn out in 8 hours is ______.

a) $8(x+y)$ b) $8x+\frac{y}{2}$ c) $16(x+y)$ d) $(2x+y)4$

[MBA, 1985]

Answers

1. b; 24 men complete the work in 16 days

$\therefore$ 16 men complete $\left(\frac{16}{24}\times\frac{12}{16}=\right)\frac{1}{2}$ part of work in 12 days

32 women complete the work in 24 days

$\therefore$ 16 women complete $\frac{16}{32}\times\frac{14}{24}=\frac{7}{24}$ part of work in $(12+2=)$ 14 days

So, the remaining part of the work which is done by sixteen men + sixteen women and the reqd additional no. of men in 2 days

$$=1-\left(\frac{1}{2}+\frac{7}{24}\right)=\frac{1}{2}-\frac{7}{24}=\frac{5}{24} \text{ (part)}$$

Now, in 2 days $\frac{5}{24}$ part of the work is done by

$$24\times\frac{16}{2}\times\frac{5}{24}=40 \text{ men}$$

2. d; 25 men and 15 women can complete, a piece of work in 12 days.

$\therefore$ work done by them in 8 days $=\frac{8}{12}=\frac{2}{3}$.

Remaining work is completed by 25 men in 6 days.

$\therefore$ Time taken by 25 men to complete the whole work

$$=\frac{3\times6}{1}=18 \text{ days.}$$

From the question,
Time taken by 25 men to complete the whole work

$$=\frac{36}{3-2}=36 \text{ days}$$

$\left[\because \frac{1}{12}-\frac{1}{18}=\frac{1}{36}\right.$ work is completed by 15 women in one day.]

3. d; One man alone finishes the work in 100 days.
$\therefore$ 10 men finish the work in 10 days.
From the question,

15 women finish in one day, $\frac{1}{6}-\frac{1}{10}=\frac{1}{15}$ work

$\therefore$ 15 women finish the whole work in 15 days.
$\therefore$ 1 women finishes the whole work in 15×15 $=225$ days.

4. d

5. b; $12\text{ M}\times18=12\text{ W}\times18\times\frac{4}{3}$ $\therefore \text{W}=\frac{3}{4}\text{ M}$

$10\text{ M}+8\text{ W}=10\text{ M}+8\times\frac{3}{4}\text{ M}=16\text{ M}$

$\therefore$ 16 men can complete the same work in

$\frac{12\times18}{16}=\frac{27}{2}=13\frac{1}{2}$ days

6. a; 5 men + 3 boys can reap 23 hectares in 4 days .....(i)
3 men + 2 boys can reap 7 hectars in 2 days .....(ii)
$\therefore$ from (1).
14 (5 men + 3 boys) can reap 23 × 14 hectares in 4 days ....(iii)
Now, from (2)
23 (3 men + 2 boys) can reap 7 × 2 × 23 hectares in 4 days ....(iv)
$\therefore$ 14(5 men + 3 boys) = 23 (3 men + 2 boys)
$\therefore$ 70 men + 42 boys = 69 men + 46 boys

1 men = 4 boys
Now, 5 men + 3 boys = 23 boys
∴ 23 boys can reap 23 hectares in 4 days
∴ 1 boy can reap 1 hectare in 4 days
∴ 4 boys can reap 1 hectare in 1 day
∴ 4 × 45 boys can reap 45 hectares in 1 day

∴ $\frac{4\times 45}{6}$ boys can reap 45 hectares in 6 days

∴ 30 boys can reap 45 hectares in 6 days
But 30 boys = 28 boys + 2 boys = 7 men + 2 boys
Hence 2 boys must assist 7 men.

7. c; 25 men can reap the field in 20 days.

∴ 10 men can reap the field in $\frac{20\times 25}{10}$ or 50 days.

when 15 men leave the work, 10 men remain and

these can reap in $37\frac{1}{2}$ days $\frac{37\frac{1}{2}}{50}$ or $\frac{3}{4}$ of the field.

Hence all men must work till $\left(1-\frac{3}{4}\right)$ or $\frac{1}{4}$ of the field

is reaped.

Now 25 men reap $\frac{1}{4}$ of the field in $\frac{20}{4}$ or 5 days.

8. a; In 25 hours A can copy 75 pages

In 1 hour A can copy $\frac{75}{25}$ = 3 pages

In 27 hours A and B can copy 135 pages

In 1 hour A and B can copy $\frac{135}{27} = 5$ pages

∴ In 1 hour B can copy (5 - 3 = 2) pages
∴ B can copy 42 pages in 21 hours.

9. d; 10 days' work by 15 men = $\frac{10}{210} = \frac{1}{21}$

At the end of every 10 days 15 additoonal men are employed ie for the next 10 days we have
15 + 15 = 30 men.

∴ Next 10 day's work by 30 men = $\frac{2}{21}$

Hence, in 20 days only $\left(\frac{1}{21}+\frac{2}{21}=\frac{3}{21}\right)$ work is completed.

To complete the whole work we have to reach the value of $\left(\frac{21}{21}\right)$ work

Now,

$\left(\frac{1}{21}+\frac{2}{21}+\frac{3}{21}+.....\frac{6}{21}\right)=\frac{21}{21}=1$

Hence total time to complete the whole work = 10 + 10 + 10 + 10 + 10 + 10 = 60 days.

10. b; Let x men are employed at first.

x men do $\frac{1}{2}$ of the work in 24 days

∴ 1 man do the whole work in 24 × 2 × x = 48x days.
Now, from the question,

(x + 16) men do the remaining work $\left(1-\frac{1}{2}=\frac{1}{2}\right)$ in (40 - 24 = 16) days.

∴ 1 man do the whole work in 16 × 2 (x + 16) days.
or, 48x = 32(x + 16) ∴ x = 32 men.

11. a; Ramesh alone finished $\frac{1}{2}$ of the work in 10 days.

Remaining $\frac{1}{2}$ of the job was finished by Ramesh and Dinesh together in 2 days.
Therefore, they both together can finish the complete job in 4 days.

12. a; A's 1 day's work = $\frac{1}{12}$

B's 1 day's work = $\frac{1}{12}$ + 60% of $\frac{1}{12}=\frac{2}{15}$.

∴ B can do the work in $\frac{15}{2}$ ie $7\frac{1}{2}$ days

13. d; 12 men can complete $\frac{1}{3}$ of the work in 3 days and the remaining $\frac{2}{3}$ of the work in 6 days.

1 man can complete $\frac{2}{3}$ of the work in $(12\times 6) = 72$ days.

∴ 12 - 2 + 6 = 16 men can complete $\frac{2}{3}$ of the work in $\frac{72}{16}=4\frac{1}{2}$ days.

14. a; $\left(\frac{1}{5}+\frac{1}{9}+\frac{1}{15}\right)$ ie. $\frac{17}{45}$ work is finished in 1 hour.

∴ Remaining work = $1-\frac{17}{45}=\frac{28}{45}$.

(A + B)'s 1 hour's work = $\frac{1}{5}+\frac{1}{9}=\frac{14}{45}$.

$\frac{14}{45}$ work is done by A and B in 1 hour.

$\frac{28}{45}$ work will be done by A and B in $\left(\frac{45}{14}\times\frac{28}{45}\right)=2$ hours.

15. c; Suppose B takes x days to do the work.

$\therefore$ A takes $\left(2\times\frac{3}{4}x\right)$ ie $\frac{3x}{2}$ days to do it.

Now, (A + B)'s 1 day's work = $\frac{1}{18}$.

$\therefore \frac{1}{x}+\frac{2}{3x}=\frac{1}{18}$ or $x=30$.

16. b; Efficiency is proportional to work done per day and work done per day × number of days worked = amount of work done
Considering efficiency of A and B initially as 1
Suppose A alone can do the work in x days and B alone can do the same work in y days.

Then, $\frac{5}{x}+\frac{5}{y}$ = total work done = 1

Since efficiency of A and B are 2 and $\frac{1}{3}$ respectively,

$\therefore \frac{1}{x}\times 2\times 3+\frac{1}{y}\times\frac{1}{3}\times 3=1$

or, $\frac{6}{x}+\frac{1}{y}=1$ ....(i) and $\frac{1}{x}+\frac{1}{y}=\frac{1}{5}$ ....(ii)

Now, subtracting eqn (i) from eqn (ii) we have $x=\frac{25}{4}=6\frac{1}{4}$ days.

17. d; Mohan mows the whole lawn in x hours.

$\therefore$ Mohan mows, in 2 hours, $\frac{2}{x}$ of the lawn.

$\therefore$ Unmowed part = $1-\frac{2}{x}=\frac{x-2}{x}$ part.

18. d; Factory A turns out x cars in one hour. Factory B turns out $\frac{y}{2}$ cars in one hour.

In one hour both the factories A and B can turn out $\left(x+\frac{y}{2}\right)$ cars.

$\therefore$ in 8 hours both factories turn out $8\left(x+\frac{y}{2}\right)$ cars ie $4(2x+y)$ cars.

14

Work and Wages

Rule 1

Theorem: ***A can do a work in x days and B can do the same work in y days. If the contract for the work is Rs X, and both of them work togehter, then the share of A and B is given by***

$$Rs\left(\frac{X}{x+y}\times y\right) \text{ and } \left(\frac{X}{x+y}\times x\right) \text{ respectively.}$$

Illustrative Example

Ex: A can do a work in 6 days and B can do the same work in 5 days. The contract for the work is Rs 220. How much shall B get if both of them work together?

Soln: **Method I:** A's 1 day's work $=\frac{1}{6}$

B's 1 day's work $=\frac{1}{5}$

$\therefore$ ratio of their wages $=\frac{1}{6}:\frac{1}{5}=5:6$

$\therefore$ B's share $=\frac{220}{5+6}\times 6 = Rs\ 120$.

Method II: As wages are distributed in inverse proportion of number of days, their share should be in the ratio 5:6

$\therefore$ B's share $=\frac{220}{11}\times 6$ Rs 120.

Method III: Applying the above theorem, we have

B's share $=\frac{X}{x+y}\times x=\frac{220}{5+6}+6$ = Rs 120.

Exercise

1. Two men A and B, working together complete a piece of work which it would have taken them respectively 12 and 18 days to complete if they worked separately. They received in payment Rs 1492.50, find their shares.
   a) Rs 895.50, Rs 597 b) Rs 895, Rs 597.50
   c) Rs 885.50, Rs 607 d) Rs 885, Rs 607.50
2. A can do a work in 10 days while B can do it in 12 days. Both of them work together to do the work. If the total amount paid for the work is Rs 440, how much is A's share in it?
   a) Rs 200 b) Rs 250 c) Rs 240 d) Rs 260
3. Ram can do a certain work in 15 days while Chandan can do it in 25 days. Both work together and finish the work. In what ratio should the total earnings be divided between them?
   a) 3:5 b) 2:5 c) 5:2 d) 5:3
4. A, B and C can do a work in 4, 6 and 10 days respectively. They finish the work together and earn Rs 310. What is the share of each?
   a) Rs 150, Rs 100, Rs 60 b) Rs 140, Rs 110, Rs 60
   c) Rs 160, Rs 90, Rs 60 d) Rs 150, Rs 110, Rs 50

Answers

1. a 2. c

3. d; **Hint:** Ram's wage: Chandan's wage

$$=\left(\frac{X}{x+y}\right)\times y:\left(\frac{X}{x+y}\right)\times x$$

or, y:x [i.e. the ratio in which the wages are divided is in inverse proportion to the time taken by them to do the work alone] = 25 : 15 = 5:3

Note: This rule can be extended to more than two workers also. See the **Q.No. 4**.

4. a; **Hint:** A's share : B's share : C's share

$$=\frac{1}{4}:\frac{1}{6}:\frac{1}{10}=15:10:6$$

A's share $=\left(\frac{15}{15+10+6}\right)\times 310$ = Rs 150

B's share $=\left(\frac{10}{15+10+6}\right)\times 310$ = Rs 100

C's share $=\left(\frac{6}{15+10+6}\right)\times 310$ = Rs 60

Note: (Also see **Rule-2**)

Rule 2

Theorem: ***A, B and C can do a work in x, y and z days respectively. If doing that work together they get an amount of Rs X, then the share of A = Rs*** $\left(\frac{Xyz}{xy+xz+yz}\right)$,

Share of B = Rs $\left(\frac{Xxz}{xy+xz+yz}\right)$ ***and Share of C = Rs*** $\left(\frac{Xxy}{xy+xz+yz}\right)$

and ratio of their shares is given by A:B:C = yz:xz:xy

Illustrative Example

Ex: A, B and C can do a work in 6, 8 and 12 days respectively. Doing that work together they get an amount of Rs 1350. What is the share of B in that amount?

Soln: Detail Method:

A's one day's work $= \frac{1}{6}$

B's one day's work $= \frac{1}{8}$

C's one day's work $= \frac{1}{12}$

A's share : B's share : C's share $= \frac{1}{6}:\frac{1}{8}:\frac{1}{12}$

Multiplying each into ratio by the LCM of their denominators, the ratios become 4:3:2.

$\therefore$ B's share $= \frac{1350}{9}\times 3 =$ Rs 450.

Quicker Method I: Applying the above theorem, we have:

Share of B $= \frac{1350\times 6\times 12}{48+96+72} = \frac{1350}{216}\times 72 =$ Rs 450.

Quicker Method II:

A's share : B's share : C's share

= B's time × C's time : A's time × C's time : A's time × B's time

$= 96:72:48 = 4:3:2$

$\therefore$ B's share $= \frac{1350}{9}\times 3 =$ Rs 450.

Exercise

1. A, B and C together do a piece of work for Rs 374.50. A working alone could do it in 5 days, B working alone could do it in 6 days and C working alone could do it in 7 days. How should the money be divided among them?
   a) Rs 147, Rs 122.50, Rs 105
   b) Rs 157, Rs 122.50, Rs 95
   c) Rs 147, Rs 112.50, Rs 115
   d) Rs 137, Rs 132.50, Rs 105
2. A, B and C contract to do a work for Rs 4200. A can do the work in 6 days, B in 10 days and C in 12 days. If they work together to do the work, what is the share of C?
   a) Rs 2000 b) Rs 1200 c) Rs 1000 d) Rs 1500
3. A, B and C contract to do a work for Rs 6500. A can do the work in 10 days, B in 15 days and C in 20 days. If they work together to do the work, what is the share of B?
   a) Rs 200 b) Rs 3000 c) Rs 1500 d) Rs 2500

Answers

1. a 2. c 3. a

Rule 3

Theorem: ***A person A can do a work in x days. With the help of a another person B, he can do the same work in y days. If they get Rs X for that work, then the share of A and B is given by Rs*** $\left[\frac{Xy}{x}\right]$ ***and Rs*** $\left[\frac{X(x-y)}{x}\right]$ ***respectively.***

Illustrative Example

Ex: A man can do a work in 10 days. With the help of a boy he can do the same work in 6 days. If they get Rs 50 for that work, what is the share of that boy?

Soln: Detail Method:

The boy can do the work in $\frac{10\times 6}{10-6} = 15$ days

(Recall the theorem)

Man's share : Boy's share = 15 : 10 = 3 : 2

Man's share $= \frac{50}{5}\times 3 = Rs\ 30$.

Boy's share $= \frac{50}{5}\times 2 = Rs\ 20$.

Quicker Method: Applying the above theorem

Man's share $= \frac{50\times 6}{10} =$ Rs 30.

Boy's share $= \frac{50\times(10-6)}{10} =$ Rs 20.

Exercise

1. Ram can do a work in 20 days. Ram and Shyam together do the same work in 15 days. If they are paid Rs 400 for that work, what is the share of each.
   a) Rs 300, Rs 100 b) Rs 200, Rs 200
   c) Rs 250, Rs 150 d) Rs 350, Rs 50
2. Suresh can do a work in 15 days. Suresh and Ramesh

together do the same work in 10 days. If they are paid Rs 1500 for the work, how should the money be divided between them?
a) Rs 1000, Rs 500 b) Rs 700, Rs 800
c) Rs 1200, Rs 300 d) None of these

3. Sohan can do a work in 25 days. Sohan and Mohan together do the same work in 20 days. If they are paid Rs 625 for the work, how much money should Mohan get?
a) Rs 400 b) Rs 200 c) Rs 100 d) Rs 125

Answers

1. a 2. a 3. d

Rule 4

Theorem: ***A and B undertake to do a work for Rs X. A can do it alone in x days and B in y days. If with the assistance of a boy they finish the work in 'd' days, then the share that A gets is Rs*** $\left(\frac{dX}{x}\right)$***, B gets is Rs*** $\left(\frac{dX}{y}\right)$ ***and the Boy gets is*** $Rs\ X\left[1-\left(\frac{x+y}{xy}\right)d\right]$

And the ratio of shares is given by A : B : Boy = dy : dx : xy - d (x + y)

Illustrative Example

Ex: A and B undertake to do a work for Rs 56. A can do it alone in 7 days and B in 8 days. If with the assistance of a boy they finish the work in 3 days, then the boy gets Rs _____.

Soln: **Detail Method:**

A's 3 days' work + B's 3 days' work + Boy's 3 day's work = 1

Or, $\frac{3}{7}+\frac{3}{8}+$ Boy's 3 day's work = 1

Or, Boy's 3 day's work = $1-\left(\frac{3}{7}+\frac{3}{8}\right)=\frac{11}{56}$

Ratio of shares = $\frac{3}{7}:\frac{3}{8}:\frac{11}{56}=\frac{3\times56}{7}:\frac{3\times56}{8}:\frac{11\times56}{56}$

$=24:21:11$

$\therefore$ Boy's share = $\frac{56}{24+21+11}\times11=$ Rs 11.

Quicker Method: Applying the above formula, we have

Boy's share = $56\left[1-\left(\frac{7+8}{7\times8}\right)3\right]=Rs\ 11$.

Exercise

1. Two men undertake to do a piece of work for Rs 200. One alone could do it in 6 days, the other in 8 days. With the assistance of a boy they finish it in 3 days. Find the share of the boy.
a) Rs 25 b) Rs 100 c) Rs 75 d) Rs 50

2. A and B contract to do a work together for Rs 300. A alone can do it in 8 days and B alone in 12 days. But with the help of C they finish it in 4 days. Find the share of C.
a) Rs 30 b) Rs 60 c) Rs 100 d) Rs 50

3. Sita and Gita undertake to do a work together for Rs 600. Sita alone can do it in 15 days and Gita alone in 20 days. But with the assistance of Rita they finish it in 5 days. Find the share of Rita.
a) Rs 150 b) Rs 200 c) Rs 250 d) Rs 300

Answers

1. a 2. d 3. c

Rule 5

Theorem: ***A, B and C contract a work for Rs X. If together, A and B are supposed to do*** $\frac{x}{y}$ ***of the work, then the share of C is given by*** $Rs\left[X\left(1-\frac{x}{y}\right)\right]$.

Illustrative Example

Ex: A, B and C contract a work for Rs 550. Together A and B are supposed to do $\frac{7}{11}$ of the work. How much does C get?

Soln: **Detail Method:** (A + B) did $\frac{7}{11}$ work and C did $\left(1-\frac{7}{11}\right)=\frac{4}{11}$ work

$\therefore$ (A + B)'s share : C's share = $\frac{7}{11}:\frac{4}{11}=7:4$

$\therefore$ C's share = $\frac{550}{11}\times4=Rs\ 200$.

Quicker Method: Applying the above theorem, we have

C's share = $550\left(1-\frac{7}{11}\right)=Rs\ 200$.

Exercise

1. A, B and C undertake to do a work for Rs 660. A and B together do $\frac{8}{11}$ of the work and rest is done by C alone. How much should C get?
a) Rs 200 b) Rs 160 c) Rs 180 d) Rs 190

2. A, B and C undertake to do a work for Rs 707. A and B

together do $\frac{5}{7}$ of the work and rest is done by C alone. How much should C get?

a) Rs 202 b) Rs 200 c) Rs 102 d) Rs 150

3. A, B and C undertake to do a work for Rs 480. A and B together do $\frac{1}{4}$ of the work and rest is done by C alone. How much should C get?

a) Rs 360 b) Rs 120 c) Rs 240 d) Rs 180

Answers

1. c 2. a 3. a

Rule 6

Theorem: ***Wages for x women amount to Rs X_1 in d_1 days. If the daily wage of a man is n times that of a woman, then the number of men that receive Rs X_2 for the work of d_2 days is*** $\left(\frac{X_2}{X_1}\right)\times\left(\frac{d_1}{d_2}\right)\times\left(\frac{x}{n}\right)$.

Ex: Wages for 45 women amount to Rs 15525 in 48 days. How many men must work 16 days to receive Rs 5750, the daily wages of a man being double those of a woman?

Soln: Detail Method:

Wage of a woman for a day $= \frac{15525}{45\times 48} = \text{Rs } \frac{115}{16}$

Thus, wage of a man for a day $= 2\times\frac{115}{16} = \text{Rs } \frac{115}{8}$

Now, number of men

$$= \frac{\text{Total wage}}{\text{No. of days} \times 1 \text{ man's 1 day's wage}}$$

$$= \frac{5750\times 8}{16\times 115} = 25 \text{ men.}$$

Note: We should remember the relationship: Total wage = One person's one day's wage × No. of persons × No. of days

Quicker Method: Applying the above theorem, we have

the no. of men $= \frac{5750}{15525}\times\frac{48}{16}\times\frac{45}{2} = 25$ men.

Exercise

1. If the wages of 45 women amount to Rs 46575 in 48 days, how many men must work 16 days to receive Rs 17250, the daily wages of a man being double than those of a woman?

a) 20 men b) 25 men c) 30 men d) 15 men

2. Wages of 20 boys for 15 days is Rs 9000. If the daily wage of a man is one and half times that of a boys, how many men must work for 30 days to earn Rs 13500?

a) 12 men b) 20 men c) 16 men d) 10 men

3. Wages of 10 women for 5 days is Rs 1250. The daily wage of a man is twice that of a woman. How many men must work for 8 days to earn Rs 1600?

a) 5 men b) 8 men c) 4 men d) 6 men

Answers

1. b 2. d 3. c

Rule 7

Theorem: ***x_1 men and y_1 boys can earn Rs X_1 in d_1 days, if x_2 men and y_2 boys can earn Rs X_2 in d_2 days, then the following relationship is obtained***

$$\frac{men}{boys} = \left[\frac{X_1y_2d_2 - X_2y_1d_1}{X_2x_1d_1 - X_1x_2d_2}\right].$$

Illustrative Example

Ex: 3 men and 4 boys can earn Rs 756 in 7 days. 11 men and 13 boys can earn Rs 3008 in 8 days. In what time will 7 men with 9 boys earn Rs 2480?

Soln: (3m + 4b) in 1 day earn Rs $\frac{756}{7}$ = Rs 108 ....(1)

(11m + 13b) in 1 day earn Rs $\frac{3008}{8}$ = Rs 376 ....(2)

From (1), we see that to earn Re 1 in 1 day there should be $\frac{3m+4b}{108}$ persons. Similarly, from (2), to earn Re 1 in 1 day there should be $\frac{11m+13b}{376}$ persons.

And also; $\frac{3m+4b}{108} = \frac{11m+13b}{376}$ ....(*)

or, m(3 × 376 - 11 × 108) = b(108 × 13 - 4 × 376) ....(**)

$\therefore \frac{m}{b} = \frac{100}{60} = \frac{5}{3}$

Now, from (1)

(3m + 4b) in 1 day earn Rs 108

or, $3m + 4\times\frac{3}{5}m$ in 1 day earn Rs 108

or, $\frac{27m}{5}$ in 1 day earn Rs 108

$\therefore$ 1 m in 1 day earns Rs $\frac{108\times 5}{27}$ = Rs 20

Thus, we get that a man earns Rs 20 daily and a boy earns

Rs $20\times\frac{3}{5}$ = Rs 12 daily.

$\because$ 7m + 9b earn Rs (7 × 20 + 9 × 12) = Rs 248 in 1 day.
$\therefore$ 7m + 9b earn Rs 2480 in 10 days.

Note: (*) Since both the LHS and the RHS denote the same quantity: "Number of persons earning Re 1 in 1 day".
(**) We can arrive at this step directly be using cross-multiplication-division rule. Arrange the given information as follows:

Men	Boys		Earning		Days
3	4	×	756	÷	7
11	13	×	3008	÷	8

Now,

$$\text{Men}\left(\frac{3\times3008}{8}-\frac{11\times756}{7}\right)$$

$$=\text{Boys}\left(\frac{13\times756}{7}-\frac{4\times3008}{8}\right)$$

or, m(3 × 376 - 11 × 108) = b(108 × 13 - 4 × 376)

or, $\frac{m}{b}=\frac{5}{3}$

Quicker Method: Applying the above theorem, we have

$$\frac{men}{boys}=\frac{756\times13\times8-3008\times4\times7}{3008\times3\times7-756\times11\times8}=\frac{-5600}{-3360}=\frac{5}{3}$$

Now, 3m + 4b in 1 day earn Rs $\frac{756}{7}$ = Rs 108.

or, $3m\times4\times\frac{3}{5}m$ in 1 day earn Rs 108

or, $\frac{27m}{5}$ in 1 day earn Rs 108.

$\therefore$ 1m in 1 day earns Rs $\frac{108\times5}{27}$ = Rs 20.

Thus, we get that a man earns Rs 20 daily and a boy earns Rs $20\times\frac{3}{5}$ = Rs 12 daily

$\because$ 7m + 9b earn Rs (7 × 20 + 9 × 12) = Rs 248 in 1 day
$\therefore$ 7m + 9b earn Rs 2480 in 10 days.

Exercise

1. If 5 men with 7 boys can earn Rs 3825 in 6 days and 2 men with 3 boys can earn Rs 1050 in 4 days, in what time will 7 men with 6 boys earn Rs 22500?
a) 15 days b) 20 days c) 25 days d) 30 days
2. If 3 men with 4 boys can earn Rs 2100 in 7 days and 11 men with 13 boys can earn Rs 8300 in 8 days, in what time will 7 men with 9 boys earn Rs 11000?
a) 16 days b) 18 days c) 14 days d) 20 days
3. If 12 men with 13 boys can earn Rs 4893.75 in 3 days and 5 men with 6 boys can erarn Rs 3562.50 in 5 days, in what time will 3 men with 4 boys earn Rs 210?
a) 8 days b) 7 days c) 10 days d) 9 days
4. 5 men and 5 women earn Rs 660 in 3 days. 10 men and 20 women earn Rs 3500 in 5 days. In how many days can 6 men and 4 women earn Rs 1300?
a) 5 days b) 10 days c) 6 days d) 12 days
5. 4 men and 6 boys earn Rs 1600 in 5 days. 3 men and 7 boys earn Rs 1740 in 6 days, in what time will 7 men and 6 boys earn Rs 3760?
a) 6 days b) 8 days c) 10 days d) 12 days

Answers

1. d; **Hint:**

$$\frac{Men}{Boys}=\frac{3825\times3\times4-1050\times7\times6}{1050\times6\times5-3825\times2\times4}=\frac{1800}{900}=2:1$$

1 men = 2 boys.
(5m + 7b) can earn in 6 days *Rs 3825.*

5m + 7b can earn in 1 day $Rs\frac{3825}{6}Rs=637.5$

$5m+\frac{7}{2}m$ can earn in 1 day *Rs 637.5*

$\therefore$ *1m* can earn in 1 day $\frac{637.5\times2}{17}=Rs\ 75$

and *1 b* can earn in 1 day $Rs\frac{75}{2}$

(7m + 6b) can earn in 1 day = $7\times75+6\times\frac{75}{2}=Rs\ 750$

$\therefore$ *Rs 22500* can be earned by

$$(7m+6b)\text{ in}\left(\frac{22500}{750}=30\right)\text{days}$$

2. a 3. b 4. a 5. b

Rule 8

Theorem: ***If A, B and C together earn Rs x_1 in d_1 days, A and C together earn Rs x_2 in d_2 days and B and C together earn Rs x_3 in d_3 days, then the daily earning of A is Rs $\left[\frac{x_1d_3-x_3d_1}{d_1.d_3}\right]$, B is Rs $\left[\frac{x_1d_2-x_2d_1}{d_1.d_2}\right]$ and C is Rs $\left[\frac{x_3}{d_3}+\frac{x_2}{d_2}-\frac{x_1}{d_1}\right]$.***

Illustrative Example

Ex: A, B and C together earn Rs 1350 in 9 days. A and C together earn Rs 470 in 5 days. B and C together earn Rs 760 in 10 days. Find the daily earning of C.

Soln: **Detail Method:** Daily earning of A + B + C

$$= \frac{Rs\,1350}{9} = Rs\,150 \;....(1)$$

Daily earning of A + C = $\frac{Rs\,470}{5} = Rs\,94$ ....(2)

Daily earning of B + C = $\frac{Rs\,760}{10} = Rs\,76$ ....(3)

From (1) and (2) daily earning of B

= 150 - 94 = Rs 56 ....(4)

From (3) and (4) daily earning of C = 76 - 56 = Rs 20

Quicker Method: Applying the above theorem, we have

daily earning of $C = \frac{760}{10} + \frac{470}{5} - \frac{1350}{9}$

$= 76 + 94 - 150 =$ Rs 20.

Exercise

1. A, B and C together earn Rs 2700 in 18 days. A and C together earn Rs 940 in 10 days. B and C together earn Rs 1520 in 20 days. Find the daily earning of C.
   a) Rs 20 b) Rs 40 c) Rs 10 d) None of these
2. A, B and C together earn Rs 640 in 8 days. A and C together earn Rs 250 in 5 days. B and C together earn Rs 420 in 6 days. Find the daily earning of C.
   a) Rs 60 b) Rs 50 c) Rs 80 d) Rs 40
3. A, B and C together earn Rs 1500 in 10 days. A and C together earn Rs 800 in 8 days. B and C together earn Rs 900 in 9 days. Find the daily earning of B.
   a) Rs 50 b) Rs 60 c) Rs 40 d) Rs 30
4. A, B and C together earn Rs 750 in 5 days. A and C together earn Rs 400 in 4 days. B and C together earn Rs 450 in $4\frac{1}{2}$ days. Find the daily earning of A.
   a) Rs 40 b) Rs 60 c) Rs 55 d) Rs50

Answers

1. a 2. a 3. a 4. d

15

Pipes and Cisterns

Rule 1

Theorem: ***If a pipe can fill a tank in x hours, then the part filled in 1 hour*** $= \frac{1}{x}$.

Illustrative Example

Ex: A pipe can fill a tank in 10 hours. Find the part of tank filled in one hour.

Soln: Applying the above theorem, we have

the part filled in 1 hour $= \frac{1}{10}$.

Exercise

1. A pipe can fill a cistern in 25 hours. Find the part of tank filled in 5 hours.

   a) $\frac{1}{25}$ b) $\frac{1}{5}$ c) $\frac{1}{10}$ d) Data inadequate

2. A pipe can fill cistern in 33 minutes. Find the time in which $\frac{1}{11}$ part of the cistern will be filled.

   a) 3 minutes b) 2 minutes
   c) 11 minutes d) None of these

Answers

1\. b 2. a

Rule 2

Theorem: ***If a pipe can empty a tank in y hours, then the part of the full tank emptied in 1 hour*** $= \frac{1}{y}$.

Illustrative Example

Ex: A pipe can empty a tank in 12 hours. Find the part of the tank emptied in one hour.

Soln: Applying the above theorem, we have

the part emptied in 1 hour $= \frac{1}{12}$.

Exercise

1. A pipe can empty a tank in 14 hours. Find the part of the tank emptied in 7 hours.

   a) $\frac{1}{4}$ b) $\frac{1}{7}$ c) $\frac{1}{2}$ d) None of these

2. A pipe can empty a cistern in 27 hours. Find the time in which $\frac{2}{3}$ part of the cistern will be emptied.

   a) 9 hours b) 12 hours c) 15 hours d) 18 hours

Answers

1\. c 2. d

Rule 3

Theorem: ***If a pipe can fill a tank in x hours and another pipe can empty the full tank in y hours, then the net part filled in 1 hour, when both the pipes are opened*** $= \left(\frac{1}{x} - \frac{1}{y}\right)$.

∴ time (T) taken to fill the tank, when both the pipes are opened $= \frac{xy}{y-x}$

Note: If T is (+ve), then cistern gets filled up and if T is (-ve), then cistern gets emptied

Illustrative Example

Ex: A pipe can fill a tank in 10 hours and another pipe can empty it in 12 hours. If both the pipes are opened, find the time in which tank is filled.

Soln: Applying the above theorem, we have

the required time $= \frac{10 \times 12}{12-10} = 60$ hrs.

Exercise

1. A water tank is $\frac{2}{5}$th full. Pipe A can fill the tank in 10 minutes and the pipe B can empty it in 6 minutes. If both the pipes are open, how long will it take to empty or fill the tank completely?

a) 6 minutes to empty b) 6 minutes to fill
c) 9 minutes to empty d) 9 minutes to fill

(BSRB Mumbai PO-1999)

2. Pipe A fills the cistern in half an hour and pipe B in 40 minutes, but owing to a crack in the bottom of the cistern it is found that pipe A now takes 40 minutes to fill the cistern. How long will B take now to fill it and how long will the crack take to empty it?
a) The leak empties in 1 hour and B fills in 2 hours.
b) B fills in an hour and the leak empties in 2 hours
c) B fills in an hour and the leak empties in an hour
d) Data inadequate
3. A cistern which could be filled in 9 hours takes one our more to be filled owing to a leak in its bottom. If the cistern is full, in what time will the leak empty it?
a) 80 hours b) 85 hours c) 90 hours d) 95 hours
4. A tap can fill a cistern in 8 hours and another can empty it in 16 hours. If both the taps are opened simultaneously, the time (in hours) to fill the tank is:
a) 8 b) 10 c) 16 d) 24

(Clerical Grade Exam, 1991)

5. A tap can fill a tank in 25 minutes and another can empty it in 50 minutes. If the tank is already half full and both the taps are opened together, the
a) tank is emptied in 20 minutes
b) tank is filled up in 25 minutes
c) tank is filled up in 20 minutes
d) tank is emptied in 25 minutes

(SBI Bank PO Exam, 1985)

Answers

1. a; **Hint:** Time taken to fill or empty the whole tank

$$= \frac{6\times10}{6-10} = -15 \text{ minutes}$$

-ve sign shows that the tank will be emptied.

$\therefore \frac{2}{5}$ th full of the tank will be emptied in $\frac{15\times2}{5}=6$ minutes.

2. b; **Hint:** Let the leak empties it in x hours.
From the given rule, we have

$$\frac{x\times30}{x-30}=40 \therefore x=120 \text{ minutes}=2 \text{ hours.}$$

Now, from the question, applying the rule we have, time taken by B to fill the tank when crack in the bottom develps $=\frac{120\times40}{120-40}=60 \text{ minutes}=1 \text{ hour}$

3. c; **Hint:** Let the leak empty the full cistern in x hours.
Now applying the given rule,

$$\frac{9\times x}{x-9}=9+1=10$$

or, $x=90$ hours.

4. c

5. b; **Hint:** $T=\frac{1}{2}\left(\frac{25\times50}{50-25}\right)=+25$ minutes.

+ve sign shows that tank is filled in 25 minutes.

Rule 4

Theorem: ***If a tap can fill a x_1 part of the cistern in t_1 minutes and x_2 part in t_2 minutes, then following expression is obtained $\frac{t_1}{x_1}=\frac{t_2}{x_2}$.***

Illustrative Example

Ex: A fill pipe can fill $\frac{1}{4}$ of cistern in 16 minutes. In how many minutes, it can fill $\frac{3}{4}$ of the cistern.

Soln: Applying the above theorem, we have

$$\frac{16}{1/4}=\frac{t_2}{3/4}\Rightarrow t_2=\frac{16\times4}{4}\times3=48 \text{ minutes.}$$

Note: If a tap can empty x_1 part of the cistern in t_1 minutes and x_2 part in t_2 minutes, then following expression is obtained $\frac{t_1}{x_1}=\frac{t_2}{x_2}$.

Ex: An empty pipe can empty $\frac{2}{3}$ of cistern in 8 minutes. In 9 minutes what part of cistern will be emptied?

Soln: Applying the above theorem,

$$\frac{8}{2/3}=\frac{9}{x_2}$$

or, $x_2=\frac{9}{12}=\frac{3}{4}$ part of the cistern will be emptied.

Exercise

1. A fill pipe can fill $\frac{3}{4}$ of cistern in 27 minutes. In how many minutes, it can fill $\frac{2}{3}$ of the cistern.
a) 36 minutes b) 24 minutes c) 28 minutes d) 21 minutes
2. A fill pipe can fill $\frac{3}{5}$ of cistern in 21 minutes. In how many minutes, it can fill $\frac{3}{7}$ of the cistern.
a) 12 minutes b) 18 minutes
c) 15 minutes d) None of these

3. An empty pipe can empty $\frac{1}{3}$ of cistern in 3 minutes. In $7\frac{1}{2}$ minutes what part of cistern will be emptied?

a) $\frac{3}{5}$ b) $\frac{5}{6}$ c) $\frac{1}{2}$ d) $\frac{2}{3}$

Answers

1. b 2. c 3. b

Rule 5

Theorem: ***A tap A can empty a cistern in x hours and the other tap B can empty is in y hours. If both emptying taps are opened together, then the time taken to empty the full cistern is given by*** $\left(\frac{xy}{x+y}\right)$ ***hrs.***

Illustrative Example

Ex: A pipe can empty a tank in 10 hrs and another pipe can empty it in 5 hours. If both the pipes are opened simultaneously, find the time in which a full tank is emptied.

Soln: Applying the above theorem, we have

the required time $=\frac{10\times5}{10+5}=\frac{50}{15}=\frac{10}{3}=3\frac{1}{3}$ hrs.

Exercise

1. A pipe can empty a tank in 5 hrs and another pipe can empty it in 15 hours. If both the pipes are opened simultaneously, find the time in which a full tank is emptied.

a) 7 hrs b) $\frac{15}{2}$ hrs

c) $\frac{15}{4}$ hrs d) None of these

2. A pipe can empty a tank in 15 hrs and another pipe can empty it in 10 hours. If both the pipes are opened simultaneously, find the time in which a full tank is emptied.

a) 8 hrs b) 6 hrs c) 4 hrs d) 5 hrs

3. A pipe can empty a tank in 12 minutes and another pipe can empty it in 16 minutes. If both the pipes are opened simultaneously, find the time in which a full tank is emptied.

a) 6 minutes b) $6\frac{1}{7}$ minutes

c) $6\frac{2}{7}$ minutes d) None of these

Answers

1. c 2. b

3. d; **Hint:** Required answer $=\frac{12\times16}{12+16}=\frac{48}{7}=6\frac{6}{7}$ minutes

Rule 6

Theorem: ***If a pipe can fill a tank in x hours and another can fill the same tank in y hours, then the net part filled in 1 hr, when both the pipes are opened*** $=\left(\frac{1}{x}+\frac{1}{y}\right)$

$\therefore$ ***time taken to fill the tank*** $=\left[\frac{xy}{x+y}\right]$ ***hours.***

Illustrative Example

Ex.: Two pipes A and B can fill a tank in 36 hours and 45 hours respectively. If both the pipes are opened simultaneously, how much time will be taken to fill the tank?

Soln: **Detail Method:** Part filled by A alone in 1 hour $=\frac{1}{36}$

Part filled by B alone in 1 hour $=\frac{1}{45}$

$\therefore$ Part filled by (A + B) in 1 hour

$=\left(\frac{1}{36}+\frac{1}{45}\right)=\frac{9}{180}=\frac{1}{20}$

Hence, both the pipes together will fill the tank in 20 hours.

Quicker Method: Applying the above theorem:

Time taken $=\frac{36\times45}{36+45}=20$ hrs.

Exercise

1. Two pipes A and B can fill a tank in 30 minutes and 18 minutes respectively. If both the pipes are opened simultaneously, how much time will be taken to fill the tank?

a) $\frac{45}{2}$ minutes b) $\frac{45}{4}$ minutes

c) $\frac{45}{7}$ minutes d) $\frac{45}{8}$ minutes

2. Two pipes A and B can fill a tank in 30 minutes and 15 minutes respectively. If both the pipes are opened simultaneously, how much time will be taken to fill the tank?

a) 10 minutes b) 12 minutes

c) 8 minutes d) 9 minutes

3. Two pipes can fill a cistern in 9 hours and 12 hours respectively. In how much time will they fill the cistern when opened together?

a) $5\frac{1}{7}$ hours b) $5\frac{2}{7}$ hours

c) $5\frac{3}{7}$ hours d) 5 hours

Answers

1. b 2. a 3. a

Rule 7

Theorem: ***A pipe can fill a tank in x hours. Due to a leak in the bottom it is filled in y hours. If the tank is full, the time taken by leak to empty the tank*** $= \left(\frac{xy}{y-x}\right)$ ***hrs.***

Illustrative Example

Ex.: A pipe can fill a tank in 15 hours. Due to a leak in the bottom, it is filled in 20 hours. If the tank is full, how much time will the leak take to empty it?

Soln: **Detail Method:** Work done by the leak in 1 hour

$$= \left(\frac{1}{15}-\frac{1}{20}\right)=\frac{1}{60}$$

$\therefore$ the leak will empty the full tank in 60 hrs.

Quicker Method: Applying the above theorem, we have

$$\text{required time} = \frac{15\times 20}{20-15} = 60 \text{ hrs.}$$

Exercise

1. There is a leak in the bottom of a cistern. When the cistern is thoroughly repaired, it would be filled in $3\frac{1}{2}$ hours. It now takes half an hour longer. If the cistern is full, how long would the leak take to empty the cistern?
a) 28 hours b) 27 hours c) 32 hours d) 24 hours.
2. There is a leak in the bottom of a cistern. When the cistern is thoroughly repaired, it would be filled in 12 minutes. It now takes 18 minutes longer. If the cistern is full, how long would the leak take to empty the cistern?
a) 20 minutes b) 24 minutes
c) 26 minutes d) 30 minutes
3. There is a leak in the bottom of a cistern. When the cistern is thoroughly repaired, it would be filled in 8 hours. It now takes 12 hours. If the cistern is full, how long would the leak take to empty the cistern?
a) 20 hours b) 24 hours c) 28 hours d) 32 hours
4. There is a leak in the bottom of a cistern. When the cistern is thoroughly repaired, it would be filled in 4 minutes. It now takes 16 minutes. If the cistern is full, how long would the leak take to empty the cistern?
a) $5\frac{2}{3}$ minutes b) $4\frac{2}{3}$ minutes
c) $5\frac{1}{3}$ minutes d) None of these
5. A cistern is normally filled in 8 hrs but takes 2 hrs longer to fill because of a leak in its bottom. If the cistern is full, the leak will empty it in: **[Railways 1991]**
a) 16 hrs b) 40 hrs c) 25 hrs d) 20 hrs

Answers

1. a; **Hint:** Here $x = 3.5$ hours and $y = 3.5 + 0.5 = 4$ hours. Now apply the given rule.

2. a 3. b 4. c

5. b; **Hint:** Here x = 8 hrs and y = 8 + 2 = 10 hrs
Now, applying the given rule, we have

$$\text{the required answer} = \frac{8\times 10}{10-8} = 40 \text{ hrs.}$$

Rule 8

Theorem: ***If a pipe fills a tank in x hours and another fills the same tank in y hours, but a third one empties the full tank in z hours, and all of them are opened together, the net part filled in 1 hour*** $= \left[\frac{1}{x}+\frac{1}{y}-\frac{1}{z}\right]$

$\therefore$ ***time taken to fill the tank*** $= \left[\frac{xyz}{yz+xz-xy}\right]$ ***hours.***

Illustrative Example

Ex.: Pipe A can fill a tank in 20 hours while pipe B alone can fill it in 30 hours and pipe C can empty the full tank in 40 hours. If all the pipes are opened together, how much time will be needed to make the tank full?

Soln: **Detail Method:** Net part filled in 1 hour

$$= \left(\frac{1}{20}+\frac{1}{30}-\frac{1}{40}\right)=\frac{7}{120}$$

$\therefore$ The tank will be full in $\frac{120}{7}$ i.e. $17\frac{1}{7}$ hours.

Quicker Method: Applying the above theorem, we have

time taken to fill the tank

$$= \frac{20\times 30\times 40}{30\times 40+20\times 40-20\times 30} = \frac{120}{7} = 17\frac{1}{7} \text{ hrs.}$$

Exercise

1. Top A can fill a water tank in 25 minutes, tap B can fill the same tank in 40 minutes and tap C can empty the tank in 30 minutes. If all the three taps are opened together, in how many minutes will the tank be completely filled up or emptied? **(BSRB Patna PO, 2001)**
a) $3\frac{2}{13}$ b) $15\frac{5}{13}$ c) $8\frac{2}{13}$ d) $31\frac{11}{19}$
2. A cistern can be filled by two pipes, A and B in 12 minutes and 14 minutes respectively and can be emptied by

a third pipe C in 8 minutes. If all the taps be turned on at the same moment, what part of the cistern will remain unfilled at the end of 7 minutes?

a) $\frac{5}{24}$ b) $\frac{19}{24}$ c) $\frac{7}{24}$ d) $\frac{17}{24}$

3. A cistern has 3 pipes, A, B and C. A and B can fill it in 2 and 3 hours respectively. C is a waste pipe. If all the 3 pipes be opened at once, $\frac{7}{24}$ of the cistern will be filled up in 30 minutes. In what time can C empty the full cistern?

a) 3 hours b) 4 hours c) 5 hours d) 6 hours

Answers

1. d; **Hint:** required answer $= \frac{25\times40\times30}{40\times30+25\times30-25\times40}$

$= \frac{600}{19} = 31\frac{11}{19}$ minutes

2. b; **Hint:** Time taken to fill the whole tank

$= \frac{12\times14\times8}{14\times8+12\times8-12\times14} = \frac{168}{5}$ minutes

$\therefore$ in 7 minutes $\frac{5}{168}\times7 = \frac{5}{24}$ part of the tank will be filled.

$\therefore$ required answer $= 1-\frac{5}{24} = \frac{19}{24}$ part.

3. b; **Hint:** $\because$ $\frac{7}{24}$ of the cistern will be filled up in $\frac{1}{2}$ hr.

$\therefore$ The whole of the cistern will be filled up in $\left(\frac{1}{2}\times\frac{24}{7} = \frac{12}{7}\right)$ hrs.

Let the pipe C be empty the whole cistern in x hours. Now, applying the given rule we have,

$$\frac{2\times3\times x}{3\times x+2\times x-2\times3} = \frac{12}{7}$$

or, $42x = 60x - 72$ $\therefore$ $x = 4$ hours.

Rule 9

Theorem: ***Two pipes A and B can fill a cistern in x hrs and y hrs respectively. There is also an outlet C. If all the three pipes are opened together, the tank is full in T hrs, then the time taken by C to empty the full tank is*** $\left[\frac{xyT}{yT+xT-xy}\right]$ ***hrs.***

Illustrative Example

Ex: Two pipes A and B can fill a cistern in 1 hour and 75 minutes respectively. There is also an outlet C. If all the three pipes are opened together, the tank is full in 50 minutes. How much time will be taken by C to empty the full tank?

Soln: **Detail Method:**

Work done by C in 1 min

$$= \left(\frac{1}{60}+\frac{1}{75}-\frac{1}{50}\right) = \frac{3}{300} = \frac{1}{100}$$

$\therefore$ C can empty the full tank in 100 minutes.

Quicker Method: Applying the above theorem, we have

the required time $= \frac{60\times75\times50}{75\times50+60\times50-60\times75}$

= 100 minutes.

Exercise

1. Two pipes A and B can fill a cistern in 12 minutes and 15 minutes respectively. There is also an outlet C. If all the three pipes are opened together, the tank is full in 10 minutes. How much time will be taken by C to empty the full tank?

a) 10 min b) 20 min c) 15 min d) Data inadequate

2. Two pipes A and B can fill a cistern in 36 minutes and 45 minutes respectively. There is also an outlet C. If all the three pipes are opened together, the tank is full in 30 minutes. How much time will be taken by C to empty the full tank?

a) 1 hour b) $\frac{3}{4}$ hour c) $\frac{1}{2}$ hour d) $1\frac{1}{4}$ hours

3. Two pipes A and B can fill a cistern in 18 hours and $22\frac{1}{2}$ hours respectively. There is also an outlet C. If all the three pipes are opened together, the tank is full in 15 hours. How much time will be taken by C to empty the full tank?

a) 60 hours b) 45 hours c) 30 hours d) 42 hours

4. Two pipes A and B can fill a cistern in 24 minutes and 30 minutes respectively. There is also an outlet C. If all the three pipes are opened together, the tank is full in 20 minutes. How much time will be taken by C to empty the full tank?

a) 30 min b) 40 min c) 45 min d) 1 hour

Answers

1. b 2. a 3. c 4. b

Rule 10

Theorem: ***A cistern is filled by three pipes whose diameters are x cm, y cm and z cm respectively (where, $x < y < z$). Three pipes are running together. If the largest pipe alone will fill it in P minutes and the amount of water flowing in by each pipe being proportional to the square of its diameter, then the time in which the cistern will be filled by the three pipes***

is $\left[\dfrac{Pz^2}{x^2+y^2+z^2}\right]$ *minutes.*

Illustrative Example

Ex: In what time would a cistern be filled by three pipes whose diameters are 1 cm, $1\frac{1}{3}$ cm, 2 cm, running together, when the largest alone fill it in 61 minutes, the amount of water flowing in by each pipe being proportional to the square of its diameter?

Soln: Detail Method: In 1 minute the pipe of 2 cm diameter fills $\frac{1}{61}$ of the cistern.

In 1 minute the pipe of 1 cm diameter fills $\frac{1}{61}\times\frac{1}{4}$ of the cistern --(*)

In 1 minute the pipe of $1\frac{1}{3}$ cm diameter fills $\frac{1}{61}\times\frac{4}{9}$ of the cistern. --(**)

$\therefore$ In 1 minute $\left(\frac{1}{61}+\frac{1}{61\times4}+\frac{4}{61\times9}\right)=\frac{1}{36}$ of the cistern is filled.

$\therefore$ the whole is filled in 36 minutes.

Note: (*) We are given that amount of water flowing is proportional to the square of the diameter of the pipe.

Since 2 cm diameter fills $\frac{1}{61}$ of the cistern,

1 cm diameter fills $\frac{1}{61}\left(\frac{1}{2}\right)^2=\frac{1}{61}\times\frac{1}{4}$ of the cistern.

(**) $1\frac{1}{3}=\frac{4}{3}$ cm diameter fills $\frac{1}{61}\times\frac{1}{4}\left(\frac{4}{3}\right)^2=\frac{1}{61}\times\frac{4}{9}$ of the cistern.

Quicker Method: Applying the above theorem, we have

the required time $=\dfrac{61\times(2)^2}{(1)^2+\left(\frac{4}{3}\right)^2+(2)^2}=\dfrac{61\times4}{1+\frac{16}{9}+4}$

$=\dfrac{61\times4\times9}{9+16+36}=36$ minutes.

Exercise

1. In what time would a cistern be filled by three pipes whose diameters are 1 cm, 2 cm, 3 cm, running together, when the largest alone fill it in 42 minutes, the amount of water flowing in by each pipe being proportional to the square of its diameter?
   a) 27 minutes b) 36 minutes
   c) 18 minutes d) 24 minutes
2. In what time would a cistern be filled by three pipes whose diameters are 2 cm, 3 cm, 4 cm, running together, when the largest alone fill it in 58 minutes, the amount of water flowing in by each pipe being proportional to the square of its diameter?
   a) 23 minutes b) 32 minutes
   c) 36 minutes d) 28 minutes
3. In what time would a cistern be filled by three pipes whose diameters are 1 cm, 3 cm, 4 cm, running together, when the largest alone fill it in 26 minutes, the amount of water flowing in by each pipe being proportional to the square of its diameter?
   a) 20 minutes b) 24 minutes
   c) 16 minutes d) 12 minutes
4. In what time would a cistern be filled by three pipes whose diameters are 1 cm, 2 cm, 4 cm, running together, when the largest alone fill it in $1\frac{1}{20}$ hours, the amount of water flowing in by each pipe being proportional to the square of its diameter?
   a) 38 minutes b) 42 minutes
   c) 44 minutes d) 48 minutes

Answers

1. a 2. b 3. c 4. d

Rule 11

Theorem: ***Two pipes A and B can fill a tank in x minutes and y minutes respectively. If both the pipes are opened simultaneously, then the time after which pipe B should be closed, so that the tank is full in t minutes, is*** $\left[y\left(1-\frac{t}{x}\right)\right]$ ***minutes.***

Illustrative Example

Ex: Two pipes A and B can fill a tank in 24 minutes and 32 minutes respectively. If both the pipes are opened simultaneously, after how much time should B be closed so that the tank is full in 18 minutes?

Soln: Detail Method: Let B be closed after x minutes. Then, part filled by (A + B) in x min. + part filled by A in (18 - x) min. = 1.

$\therefore x\left(\frac{1}{24}+\frac{1}{32}\right)+(18-x)\times\frac{1}{24}=1$

or, $\frac{7x}{96}+\frac{18-x}{24}=1$

or, $7x+4(18-x)=96$

or, $3x = 24 \quad \therefore x = 8$

So, B should be closed after 8 min.

Quicker Method: Applying the above theorem,

Pipe B should be closed after $\left(1-\frac{18}{24}\right)\times 32 = 8$ min.

Exercise

1. Two pipes A and B can fill a tank in 12 minutes and 16 minutes respectively. If both the pipes are opened simultaneously, after how much time should B be closed so that the tank is full in 9 minutes?
a) 8 min b) 6 min c) 4 min d) 10 min
2. Two pipes A and B can fill a tank in 6 hours and 8 hours respectively. If both the pipes are opened simultaneously, after how much time should B be closed so that the tank is full in $4\frac{1}{2}$ hours?
a) 1 hour b) 2 hours c) 3 hours d) $2\frac{1}{2}$ hours
3. Two pipes A and B can fill a tank in 36 minutes and 48 minutes respectively. If both the pipes are opened simultaneously, after how much time should B be closed so that the tank is full in 27 minutes?
a) 10 min b) 12 min c) 14 min d) 16 min
4. Two pipes A and B can fill a tank in 18 minutes and 24 minutes respectively. If both the pipes are opened simultaneously, after how much time should B be closed so that the tank is full in $13\frac{1}{2}$ minutes?
a) 9 min b) 6 min c) 8 min d) 10 min
5. Two pipes A and B can fill a cistern in 20 minutes and 25 minutes respectively. Both are opened together, but at the end of 5 minutes, B is turned off. How much longer will the cistern take to fill?
a) 16 minutes b) 18 minutes
c) 11 minutes d) None of these

Answers

1. c 2. b 3. b 4. b

5. a; **Hint:** $25\left(1-\frac{t}{20}\right) = 5 \quad \therefore$ t = 16 minutes.

Rule 12

Theorem: ***Two pipes P and Q will fill a cistern in x hours and y hours respectively. If both pipes are opened together, then the time after which the first pipe must be turned off, so that the cistern may be just filled in t hours, is*** $\left[x\left(1-\frac{t}{y}\right)\right]$ ***hours.***

Illustrative Example

Ex: Two pipes P and Q would fill a cistern in 24 hours and 32 hours respectively. If both pipes are opened together, find when the first pipe must be turned off so that the cistern may be just filled in 16 hours.

Soln: **Detail Method:** Suppose the first pipe was closed after x hrs.

Then, first's x hrs' supply + second's 16 hrs' supply = 1

or, $\frac{x}{24}+\frac{16}{32}=1 \quad \therefore \frac{x}{24}=1-\frac{1}{2}=\frac{1}{2}$

$\therefore x = 12$ hrs.

Quicker Method: Applying the above theorem, we have

the first pipe should work for $\left(1-\frac{16}{32}\right)\times 24$ hrs

= 12 hrs.

Exercise

1. Two pipes, P and Q can fill a cistern in 12 and 15 minutes respectively. Both are opened together, but at the end of 3 minutes the first is turned off. How much longer will the cistern take to fill?
a) $8\frac{1}{4}$ minutes b) $11\frac{1}{4}$ minutes
c) $7\frac{3}{4}$ minutes d) $8\frac{3}{4}$ minutes
2. Two pipes P and Q would fill a cistern in 12 and 16 minutes respectively. Both pipes being opened, find when the first pipe must be turned off so that the cistern may be just filled in 8 minutes
a) 15 minutes b) 8 minutes
c) 6 minutes d) 10 minutes
3. A cistern can be filled by two pipes in 30 and 40 minutes respectively. Both the pipes were opened at once, but after some time the first was shut up, and the cistern was filled in 10 minutes more. How long after the pipes had been opened was the first pipe shut up?
a) $\frac{90}{11}$ minutes b) $\frac{90}{7}$ minutes
c) $\frac{90}{13}$ minutes d) $\frac{45}{2}$ minutes

Answers

1. a; **Hint:** $12\left(1-\frac{t}{15}\right)=3 \quad \therefore t=\frac{45}{4}=11\frac{1}{4}$ minutes

$\therefore$ required answer $= 11\frac{1}{4}-3 = 8\frac{1}{4}$ minutes

2. c

3. b; **Hint:** Let the first pipe be shut up after x minutes.

Now, applying the above rule, we have

$$30\left(1-\frac{x+10}{40}\right)=x \quad \text{[Here t = (x + 10) minutes]}$$

or, $x=\frac{90}{7}$ minutes.

Rule 13

Theorem: ***If two pipes A and B function simultaneously, the reservoir is filled in x hours and pipe A fills the reservoir y hours faster than the other, then the time taken by the faster pipe A to fill the reservoir is*** $\left[\frac{\sqrt{y^2+4x^2}-(y-2x)}{2}\right]$ ***hours.***

Illustrative Example

Ex: If two pipes function simultaneously, the reservoir is filled in 12 hrs. One pipe fills the reservoir 10 hours faster than the other. How many hours does the faster pipe take to fill the reservoir?

Soln: **Detail Method:** Let the faster pipe fills the tank in x hrs.

Then the slower pipe fills the tank in x + 10 hrs.

When both of then are opened, the reservoir will be filled in

$$\frac{x(x+10)}{x+(x+10)}=12$$

or, $x^2-14x-120=0$

$\therefore$ x = 20, -6

But x can't be -ve, hence the faster pipe will fill the reservoir in 20 hrs.

Quicker Method: Applying the above theorem, we have

the time taken by the faster pipe

$$=\frac{\sqrt{(10)^2+4(12)^2}-(10-2\times 12)}{2}$$

$$=\frac{\sqrt{100+576}+14}{2}=\frac{40}{2}=20 \text{ hours.}$$

Exercise

1. If two pipes function simultaneously, the reservoir is filled in 6 hrs. One pipe fills the reservoir 5 hours faster than the other. How many hours does the faster pipe take to fill the reservoir?
   a) 20 hours b) 10 hours c) 15 hours d) 12 hours
2. If two pipes function simultaneously, the reservoir is filled in 18 hrs. One pipe fills the reservoir 15 hours faster than the other. How many hours does the faster pipe take to fill the reservoir?
   a) 60 hours b) 30 hours c) 40 hours d) 45 hours
3. If two pipes function simultaneously, the reservoir is filled in 9 hrs. One pipe fills the reservoir $7\frac{1}{2}$ hours faster than the other. How many hours does the faster pipe take to fill the reservoir?
   a) 15 hours b) 20 hours c) 25 hours d) 30 hours
4. If two pipes function simultaneously, the reservoir is filled in 24 minutes. One pipe fills the reservoir 20 minutes faster than the other. How many hours does the faster pipe take to fill the reservoir?
   a) 60 min b) 45 min c) 40 min d) 30 min

Answers

1. b 2. b 3. a 4. c

Rule 14

Theorem: ***Three pipes A, B and C can fill a cistern in x hours. If after working together for t hours, C is closed and A and B fill the cistern in y hours, then the time in which the cistern can be filled by the pipe C is*** $\left[\frac{xy}{y-x+t}\right]$ ***hours.***

Illustrative Example

Ex: Three pipes A, B and C can fill a cistern in 6 hrs. After working together for 2 hours, C is closed and A and B fill the cistern in 8 hrs. Then find the time in which the cistern can be filled by pipe C.

Soln: **Detail Method:** A + B + C can fill in 1 hr $=\frac{1}{6}$ of cistern.

A + B + C can fill in 2 hrs $=\frac{2}{6}=\frac{1}{3}$ of cistern.

Unfilled part $=\left(1-\frac{1}{3}=\right)\frac{2}{3}$ is filled by A + B in 8 hrs.

$\therefore$ (A + B) can fill the cistern in $\frac{8\times 3}{2}=12$ hrs.

And we have (A + B + C) can fill the cistern in 6 hrs.

$\therefore$ C = (A + B + C) - (A + B) can fill the cistern in $\frac{12\times 6}{12-6}=12$ hrs.

Quicker Method: Applying the above theorem, we have

the required time $=\frac{6\times 8}{8-6+2}=12$ hrs.

Exercise

1. Three taps A, B and C can fill a cistern in 10, 15 and 20 minutes respectively. They are all turned on at once, but after 3 minutes C is turned off. How many minutes longer will A and B take to fill the cistern?
   a) 2 min b) 2 min 6 sec
   c) 1 min 6 sec d) 3 min 8 sec
2. Three taps, A, B and C can fill a cistern in 10 min, 12 min and 15 min respectively. They are all turned on at once, but after $1\frac{1}{2}$ min B and C are turned off. How many minutes longer will A take then to fill the cistern?
   a) $6\frac{1}{4}$ min b) $7\frac{1}{4}$ min c) $6\frac{3}{4}$ min d) $8\frac{3}{4}$ min
3. Three pipes A, B and C can fill a cistern in 12 hrs. After working together for 4 hours, C is closed and A and B fill the cistern in 16 hrs. Then find the time in which the cistern can be filled by pipe C.
   a) 12 hrs b) 16 hrs c) 20 hrs d) 24 hrs
4. Three pipes A, B and C can fill a cistern in 36 minutes. After working together for 12 minutes, C is closed and A and B fill the cistern in 48 minutes. Then find the time in which the cistern can be filled by pipe C.
   a) 72 minutes b) 60 minutes
   c) 48 minutes d) 64 minutes
5. Three pipes A, B and C can fill a cistern in 18 minutes. After working together for 6 minutes, C is closed and A and B fill the cistern in 24 minutes. Then find the time in which the cistern can be filled by pipe C.
   a) 30 minutes b) 24 minutes
   c) 36 minutes d) 45 minutes

Answers

1\. b; **Hint:** $x = \frac{10\times15\times20}{10\times15+10\times20+15\times20} = \frac{60}{13}$ minutes.

Now, applying the given rule, we have

$$\frac{\frac{60}{13}\times y}{y-\frac{60}{13}+3} = 20 \text{ or, } y = \frac{21}{10} = 2 \text{ min 6 seconds.}$$

2\. a; **Hint:** $x = \frac{10\times12\times15}{10\times12+12\times15+10\times15} = 4$ min

B and C are turned off after $1\frac{1}{2}$ min

$\therefore$ B and C together can fill a cistern in

$\left(\frac{12\times15}{12+15} = \frac{20}{3}\right)$ min

Now, applying the given rule, we have

$$\frac{4+y}{y-4+\frac{3}{2}} = \frac{20}{3}$$

$$\therefore\ y = \frac{25}{4} = 6\frac{1}{4} \text{ minutes.}$$

3\. d 4. a 5. c

Rule 15

Theorem: ***A pipe can fill a tank in x units of time and another pipe in y units of time, but a third pipe can empty it in z units of time. If the first two pipes are kept open for t units of time in the beginning and then the third pipe is also opened, the time in which the cistern is emptied*** **is given by**

$$\left[\frac{zt}{\left(\frac{xy}{x+y}\right)-z}\right]$$ **units of time.**

Illustrative Example

Ex. A pipe can fill a tank in 12 minutes and another pipe in 15 minutes, but a third pipe can empty it in 6 minutes. The first two pipes are kept open for 5 minutes in the beginning and then the third pipe is also opened. In what time is the cistern emptied?

Soln: Detail Method:

Cistern filled in 5 minutes $= 5\left(\frac{1}{12}+\frac{1}{15}\right) = \frac{3}{4}$

Net work done by 3 pipes in 1 minute

$$= \left(\frac{1}{12}+\frac{1}{15}\right) - \frac{1}{6} = -\frac{1}{60}$$

-ve sign shows that $\frac{1}{60}$ part is emptied in 1 minutes.

$\therefore\ \frac{3}{4}$ part is emptied in $60\times\frac{3}{4} = 45$ minutes.

Quicker Method: Applying the above theorem, we have

the required time $= \frac{6\times5}{\frac{12\times15}{15+12}-6} = \frac{6\times5\times3}{2} = 45$ minutes.

Exercise

1. A pipe can fill a tank in 10 minutes and another pipe in 15 minutes, but a third pipe can empty it in 5 minutes. The first two pipes are kept open for 4 minutes in the beginning and then the third pipe is also opened. In what time

is the cistern emptied?

a) 25 minutes b) 20 minutes
c) 24 minutes d) 28 minutes

2. A pipe can fill a tank in 12 minutes and another pipe in 18 minutes, but a third pipe can empty it in 7 minutes. The first two pipes are kept open for 1 minute in the beginning and then the third pipe is also opened. In what time is the cistern emptied?

a) 35 minutes b) 28 minutes
c) 45 minutes d) 38 minutes

3. A pipe can fill a tank in 24 minutes and another pipe in 30 minutes, but a third pipe can empty it in 12 minutes. The first two pipes are kept open for 10 minutes in the beginning and then the third pipe is also opened. In what time is the cistern emptied?

a) 45 minutes b) 60 minutes
c) 75 minutes d) 90 minutes

4. A pipe can fill a tank in 20 minutes and another pipe in 30 minutes, but a third pipe can empty it in 10 minutes. The first two pipes are kept open for 8 minutes in the beginning and then the third pipe is also opened. In what time is the cistern emptied?

a) 35 minutes b) 38 minutes
c) 40 minutes d) 30 minutes

Answers

1. b 2. a 3. d 4. c

Rule 16

Theorem: ***A, B and C are three pipes connected to a tank. A and B together fill the tank in x hours. B and C together fill the tank in y hours. A and C together fill the tank in z hours.***

(i) ***Time taken by A to fill the tank*** $=\left(\frac{2xyz}{xy+yz-xz}\right)$ ***hrs,***

(ii) ***Time taken by B to fill the tank*** $=\left(\frac{2xyz}{yz+xz-xy}\right)$ ***hrs,***

(iii) ***Time taken by C to fill the tank*** $=\left(\frac{2xyz}{xz+xy-yz}\right)$ ***hrs***

Illustrative Example

Ex: A, B and C are three pipes connected to a tank. A and B together fill the tank in 6 hrs. B and C together fill the tank in 10 hrs. A and C together fill the tank in $7\frac{1}{2}$ hrs. In how much time will A, B and C fill the tank separately?

Soln: Detail Method: A + B fill in 6 hrs.

B + C fill in 10 hrs,

A + C fill in $7\frac{1}{2}=\frac{15}{2}$ hrs

$\therefore$ 2(A + B + C) fill in $\frac{6\times10\times\frac{15}{2}}{6\times10+6\times\frac{15}{2}+10\times\frac{15}{2}}$

$=\frac{6\times5\times15}{180}=\frac{5}{2}$ hrs

$\therefore$ A + B + C fill the tank in 5 hrs.

Now, A[= (A + B + C) − (B + C)] fills in $=\frac{10\times5}{10-5}=10$ hrs.

Similarly, B fills in $\frac{\frac{15}{2}\times5}{\frac{15}{2}-5}=15$ hrs.

and C fills in $\frac{5\times6}{6-5}=30$ hrs.

Quicker Method: Applying the above theorem, Time taken by A to fill the tank

$$=\frac{2\times6\times10\times\frac{15}{2}}{6\times10+10\times\frac{15}{2}-6\times\frac{15}{2}}=\frac{900}{90}=10 \text{ hrs.}$$

Time taken by B to fill the tank

$$=\frac{2\times6\times10\times\frac{15}{2}}{10\times\frac{15}{2}+6\times\frac{15}{2}-6\times10}=\frac{900}{60}=15 \text{ hrs.}$$

Time taken by C to fill the tank

$$=\frac{2\times6\times10\times\frac{15}{2}}{6\times\frac{15}{2}+6\times10-10\times\frac{15}{2}}=\frac{900}{30}=30 \text{ hrs.}$$

Exercise

1. Three pipes A, B and C are connected to a tank. A and B together can fill the tank in 60 minutes, B and C together in 40 minutes and C and A together in 30 minutes. In how much time will each pipe fill the tank separately?

a) 80 min, 240 min, 48 min b) 40 min, 120 min, 24 min
c) 60 min, 250 min, 64 min d) None of these

2. Three pipes A, B and C are connected to a tank. A and B together can fill the tank in 6 hours, B and C together in 4 hours and C and A together in 3 hours. In how much time will each pipe fill the tank separately?

a) 4 hrs, 12 hrs, $2\frac{2}{5}$ hrs b) 8 hrs, 24 hrs, $4\frac{4}{5}$ hrs

c) 8 hrs, 12 hrs, $4\frac{4}{5}$ hrs d) 4 hrs, 24 hrs, $4\frac{4}{5}$ hrs

3. Three pipes A, B and C are connected to a tank. A and B together can fill the tank in 12 hrs, B and C together in 20 hrs and C and A together in 15 hrs. In how much time will each pipe fill the tank separately?
a) 10 hrs, 15 hrs, 30 hrs b) 20 hrs, 15 hrs, 60 hrs
c) 20 hrs, 30 hrs, 60 hrs d) 20 hrs, 30 hrs, 45 hrs

4. Three pipes A, B and C are connected to a tank. A and B together can fill the tank in 12 hrs, B and C together in 8 hrs and C and A together in 6 hrs. In how much time will each pipe fill the tank separately?
a) 16 hrs, 48 hrs, $9\frac{3}{5}$ hrs b) 16 hrs, 24 hrs, $9\frac{3}{5}$ hrs
c) 8 hrs, 48 hrs, $9\frac{3}{5}$ hrs d) 16 hrs, 48 hrs, $8\frac{4}{5}$ hrs

Answers

1. a 2. b 3. c 4. a

Rule 17

Theorem: ***Two pipes can separately fill a tank in x hrs and y hrs respectively. If both the pipes are opened to fill the tank but when the tank is n part full a leak develops in the tank through which m part of the total water supplied by both the pipes leak out, then the total time to fill the tank is***

$$\left(\frac{xy}{x+y}\right)\left(\frac{1-mn}{1-m}\right) \text{ hrs.}$$

Ilustrative Example

Ex: Two pipes can separately fill a tank in 20 hrs and 30 hrs respectively. Both the pipes are opened to fill the tank but when the tank is $\frac{1}{3}$ full a leak develops in the tank through which $\frac{1}{3}$ of the water supplied by both the pipes leak out. What is the total time taken to fill the tank?

Soln: Detail Method: Time taken by the two pipes to fill the tank $= \frac{20\times30}{20+30}$ hrs = 12 hrs.

$\therefore \frac{1}{3}$ of tank is filled in $\frac{12}{3} = 4$ hrs.

Now, $\frac{1}{3}$ of the supplied water leaks out

$\Rightarrow$ the filler pipes are only $1-\frac{1}{3}=\frac{2}{3}$ as efficient as earlier.

$\Rightarrow$ the work of (12 -4 =) 8 hrs will be completed now in

$8\div\frac{2}{3}=\frac{8\times3}{2}=12$ hrs $\therefore$ total time = 4 + 12 = 16 hrs

OR

Since $\frac{1}{3}$ of supplied water leaks out, the leakage empties the tank in 12 × 3 = 36 hrs. Now, time taken to fill the tank by the two pipes and the leakage

$$=\frac{36\times12}{36-12}=18 \text{ hrs.}$$

$\therefore$ time taken by the two pipes and the leakage to fill $\frac{2}{3}$ of the tank $=18\times\frac{2}{3}=12$ hrs.

$\therefore$ total time = 4 hrs + 12 hrs = 16 hrs.

Quicker Method: Applying the above theorem,

$$\text{the total time} = \frac{20\times30}{20+30}\left(\frac{1-\frac{1}{3}\times\frac{1}{3}}{1-1/3}\right)$$

$$=\frac{600}{50}\times\frac{4}{3}=16 \text{ hrs.}$$

Exercise

1. Two pipes can separately fill a tank in 10 hrs and 15 hrs respectively. Both the pipes are opened to fill the tank but when the tank is $\frac{1}{6}$ full a leak develops in the tank through which $\frac{1}{6}$ of the water supplied by both the pipes leak out. What is the total time taken to fill the tank?
a) 8 hrs b) 5 hrs c) 6 hrs d) 9 hrs

2. Two pipes can separately fill a tank in 30 hrs and 45 hrs respectively. Both the pipes are opened to fill the tank but when the tank is $\frac{2}{3}$ full a leak develops in the tank through which $\frac{2}{3}$ of the water supplied by both the pipes leak out. What is the total time taken to fill the tank?
a) 25 hrs b) 30 hrs c) 35 hrs d) None of these

3. Two pipes can separately fill a tank in 20 hrs and 30 hrs respectively. Both the pipes are opened to fill the tank but when the tank is $\frac{3}{4}$ full a leak develops in the tank

through which $\frac{3}{4}$ of the water supplied by both the pipes leak out. What is the total time taken to fill the tank?
a) 20 hrs b) 18 hrs c) 21 hrs d) 27 hrs

Answers
1. c 2. b 3. c

Rule 18

Theorem: ***A cistern is normally filled in x hrs but takes t hrs longer to fill because of a leak in its bottom. If the cistern is full, the leak will empty it in*** $\left[\frac{x\times(x+t)}{t}\right]$ ***hrs.***

Illustrative Example

Ex.: A cistern is normally filled in 8 hrs but takes two hrs longer to fill because of a leak in its bottom. If the cistern is full, the leak will empty it in _____ hrs.

Soln: Detailed Method: Suppose the leak can empty the tank in x hrs.

Then part of cistern filled in 1 hr = $\frac{1}{8}-\frac{1}{x}=\frac{x-8}{8x}$

$\therefore$ Cistern will be filled in $\frac{8x}{x-8}$ hrs

or, $8x=10x-80$ $\therefore x=40$ hrs.

Quicker Approach: The filler takes 2 hrs more

$\Rightarrow$ the leak empties in 10 hrs what the filler fills in 2 hrs.

$\Rightarrow$ the leak empties in 10 hrs = $\frac{2}{8}=\frac{1}{4}$ of cistern

$\Rightarrow$ the leak empties the full cistern in $4\times10=40$ hrs.

Quicker Method:

The leak will empty in $\frac{8\times(8+2)}{2}=40$ hrs.

Exercise

1. A cistern is normally filled in 4 hrs but takes 1 hr longer to fill because of a leak in its bottom. If the cistern is full, the leak will empty it in ____ hr.
a) 10 hrs b) 20 hrs c) 15 hrs d) 12 hrs
2. A cistern is normally filled in 6 hrs but takes 3 hrs longer to fill because of a leak in its bottom. If the cistern is full, the leak will empty it in ____ hrs.
a) 24 hrs b) 18 hrs c) 30 hrs d) 21 hrs
3. A cistern is normally filled in 5 hrs but takes 1 hr longer to fill because of a leak in its bottom. If the cistern is full, the leak will empty it in ____ hr.
a) 10 hrs b) 12 hrs c) 30 hrs d) 18 hrs
4. A cistern is normally filled in 10 hrs but takes 4 hrs longer to fill because of a leak in its bottom. If the cistern is full, the leak will empty it in _____ hr.
a) 70 hrs b) 60 hrs c) 50 hrs d) 35 hrs

Answers
1. b 2. b 3. c 4. d

Rule 19

Theorem: ***If three taps are opened together, a tank is filled in t hours. One of the taps can fill it in x hours and another in y hours. The third tap fills or empties the tank in***

$$\left[\frac{t}{1-t\left(\frac{x+y}{xy}\right)}\right] \text{ hours.}$$

Nature of the third tap — whether it is filler or waste pipe — depends upon the (+ve) or (-ve) sign of the above expression.

Illustrative Example

Ex: If three taps are opened together, a tank is filled in 12 hrs. One of the taps can fill it in 10 hrs and another in 15 hrs. How does the third tap work?

Soln: Detail Method: We have to find the nature of the third tap — whether it is a filler or a waste pipe.
Let it be a filler pipe which fills in x hrs.

Then, $\frac{10\times15\times x}{10\times15+10x+15x}=12$

or, $150x=150\times12+25x\times12$

or, $-150x=1800$

$\therefore x=-12$

–ve sign shows that the third pipe is a waste pipe which vacates the tank in 12 hrs.

Quicker Method: Applying the above theorem, we have

$$\frac{12}{1-12\left(\frac{10+15}{10\times15}\right)}=-12 \text{ hrs.}$$

$\therefore$ –ve sign shows that the third pipe is a waste pipe which vacates the tank in 12 hrs.

Exercise

1. If three taps are opened together, a tank is filled in 6 hrs. One of the taps can fill it in 5 hrs and another in $7\frac{1}{2}$ hrs. How does the third tap work?
a) 6 hours, fill pipe b) 8 hours, waste pipe
c) 6 hours, waste pipe d) 8 hours, fill pipe
2. If three taps are opened together, a tank is filled in 18 hrs. One of the taps can fill it in 15 hrs and another in $22\frac{1}{2}$

hrs. How does the third tap work?

a) 16 hrs, fill pipe b) 18 hrs, fill pipe
c) 18 hrs, waste pipe d) 16 hrs, waste pipe

3. If three taps are opened together, a tank is filled in 24 hrs. One of the taps can fill it in 20 hrs and another in 30 hrs. How does the third tap work?
a) 24 hrs, waste pipe b) 20 hrs, waste pipe
c) 20 hrs, fill pipe d) 24 hrs, fill pipe

4. If three taps are opened together, a tank is filled in 36 hrs. One of the taps can fill it in 30 hrs and another in 45 hrs. How does the third tap work?
a) 36 hrs, waste pipe b) 30 hrs, waste pipe
c) 36 hrs, fill pipe d) 24 hrs, waste pipe

Answers

1. c 2. c 3. a 4. a

Rule 20

Theorem: ***Two pipes can fill a cistern in x and y minutes respectively. A waste pipe carries of w litres of water per minute from the cistern. If all three pipes are opened together and a full cistern gets emptied in z minutes, then the capacity of the cistern is*** $\left[\frac{w(xyz)}{xy+xz+yz}\right]$ **litres.**

Illustrative Example

Ex: Two pipes A and B can separately fill in 15 and 10 minutes respectively and a waste pipe C can carry off 7 litres per minute. If all the pipes are opened when the cistern is full, it is emptied in 2 hours. How many litres does the cistern hold?

Soln: Applying the above theorem, we have the capacity of cistern

$$=\frac{7\times15\times10\times120}{15\times10+10\times120+15\times120}=40 \text{ litres.}$$

Exercise

1. Two pipes A and B can separately fill in $7\frac{1}{2}$ and 5 minutes respectively and a waste pipe C can carry off 14 litres per minute. If all the pipes are opened when the cistern is full, it is emptied in 1 hour. How many litres does the cistern hold?
a) 40 litres b) 30 litres c) 35 litres d) 45 litres

2. Two pipes A and B can separately fill in 30 and 20 minutes respectively and a waste pipe C can carry off 6 litres per minute. If all the pipes are opened when the cistern is full, it is emptied in 60 minutes. How many litres does the cistern hold?
a) 10 litres b) 30 litres c) 60 litres d) 45 litres

3. Two pipes A and B can separately fill in 15 and 10 minutes respectively and a waste pipe C can carry off 3 litres per minute. If all the pipes are opened when the cistern is full, it is emptied in 30 minutes. How many litres does the cistern hold?
a) 15 litres b) 30 litres c) 25 litres d) 45 litres

4. Two pipes A and B can separately fill in 45 and 30 minutes respectively and a waste pipe C can carry off 9 litres per minute. If all the pipes are opened when the cistern is full, it is emptied in $1\frac{1}{2}$ hours. How many litres does the cistern hold?
a) 225 litres b) 135 litres c) 125 litres d) None of these

5. Two pipes A and B can separately fill a cistern in $7\frac{1}{2}$ and 5 minutes respectively and a waste pipe C can carry off 14 litres per minute. If all the pipes are opened when the cistern is full, it is emptied in 1 hour. How many litres does it hold?
a) 40 litres b) 35 litres c) 45 litres d) 60 litres

Answers

1. a 2. c 3. a 4. b 5. a

Rule 21

Theorem: ***To find out the capacity (C) of the cistern in litres, if n number of filling pipes, each capable of filling a cistern alone in x minutes, and m number of emptying pipes, each capable of emptying a cistern alone in y minutes, are opened together and as a result w is the rate at which the tank fills per minute, the following formula is used,*** $C=\left(\frac{wxy}{ny-mx}\right)$ ***litres.***

Illustrative Example

Ex: There are 5 filling pipes, each capable of filling a cistern alone in 12 minutes, and 3 emptying pipes each capable of emptying a cistern alone in 16 minutes. All pipes are opened together and as a result, tank fills 11 litres of water per minute. Find the capacity of the tank.

Soln: Applying the above theorem, we have the capacity of the cistern

$$=\frac{11\times12\times16}{16\times5-3\times12}=\frac{11\times12\times16}{44}=48 \text{ litres.}$$

Exercise

1. There are 10 filling pipes each capable of filling a cistern alone in 6 minutes and 6 emptying pipes each capable of emptying a cistern alone in 8 minutes. All pipes are opened together and as a result, tank fills 22 litres of water per minute. Find the capacity of the tank.
a) 48 litres b) 36 litres
c) 24 litres d) 16 litres

2. There are 6 filling pipes each capable of filling a cistern

alone in 16 minutes and 4 emptying pipes each capable of emptying a cistern alone in 20 minutes. All pipes are opened together and as a result, tank fills 14 litres of water per minute. Find the capacity of the tank.
a) 60 litres b) 80 litres c) 75 litres d) 45 litres

3. There are 3 filling pipes each capable of filling a cistern alone in 8 minutes and 2 emptying pipes each capable of emptying a cistern alone in 10 minutes. All pipes are opened together and as a result, tank fills 7 litres of water per minute. Find the capacity of the tank.
a) 20 litres b) 25 litres c) 40 litres d) 30 litres

4. There are 12 filling pipes each capable of filling a cistern alone in 32 minutes and 8 emptying pipes each capable of emptying a cistern alone in 40 minutes. All pipes are opened together and as a result, tank fills 28 litres of water per minute. Find the capacity of the tank.
a) 160 litres b) 120 litres c) 100 litres d) 80 litres

Answers

1. c 2. b 3. c 4. a

Rule 22

Theorem: ***Two pipes A and B can fill a cistern in x and y hours respectively. If opened together they take t hours extra to fill the cistern due to a leak, then the time in which the leak alone empties the full cistern is***

$$\frac{xy}{x+y}\left[1+\frac{xy}{(x+y)t}\right] \textit{ hours.}$$

Illustrative Example

Ex: Two pipes can fill a cistern in 14 and 16 hours respectively. The pipes are opened simultaneously and it is found that due to leakage in the bottom, $\frac{8}{15}$ hrs extra are taken for the cistern to be filled up. If the cistern is full, in what time would the leak empty it?

Soln: Applying the above theorem, we have time taken by leak to empty the cistern

$$=\frac{14\times16}{14+16}\left[1+\frac{14\times16}{(14+16)\frac{8}{15}}\right]$$

$$=\frac{112}{15}\left[1+\frac{112}{15\times\frac{8}{15}}\right]=\frac{112}{15}\times15=112\text{ hrs.}$$

Exercise

1. Two pipes can fill a cistern in 7 and 8 hours respectively. The pipes are opened simultaneously and it is found that due to leakage in the bottom, 16 minutes extra are taken for the cistern to be filled up. If the cistern is full, in what time would the leak empty it?
a) 112 hrs b) 56 hrs c) 84 hrs d) 98 hrs

2. Two pipes can fill a cistern in 8 and 10 hours respectively. The pipes are opened simultaneously and it is found that due to leakage in the bottom, $\frac{2}{9}$ hrs extra are taken for the cistern to be filled up. If the cistern is full, in what time would the leak empty it?
a) 90 hrs b) $93\frac{1}{3}$ hrs
c) 93 hrs d) $90\frac{1}{3}$ hrs

3. Two pipes can fill a cistern in 10 and 15 hours respectively. The pipes are opened simultaneously and it is found that due to leakage in the bottom, 2 hrs extra are taken for the cistern to be filled up. If the cistern is full, in what time would the leak empty it?
a) 20 hrs b) 21 hrs c) 24 hrs d) 28 hrs

4. Two pipes can fill a cistern in 30 and 15 hours respectively. The pipes are opened simultaneously and it is found that due to leakage in the bottom, 5 hrs extra are taken for the cistern to be filled up. If the cistern is full, in what time would the leak empty it?
a) 60 hrs b) 45 hrs c) 35 hrs d) 30 hrs

Answers

1. b 2. b 3. c 4. d

Rule 23

Theorem: ***A cistern has a leak which would empty it in x hours. If a tap is turned on which admits water at the rate of w litres per hour into the cistern, and the cistern is now emptied in y hours, then the capacity of the cistern is***

$$w\times\left(\frac{xy}{y-x}\right) \textit{ litres.}$$

Illustrative Example

Ex: A cistern has a leak which would empty it in 8 hours. A tap is turned on which admits 6 litres a minute into the cistern, and it is now emptied in 12 hours. How many litres does the cistern hold?

Soln: Applying the above theorem, we have
w = 6 litres per minute = 6 × 60 litres per hour
x = 8 hours and y = 12 hours

$$\therefore \text{ capacity of cistern}=\frac{12\times8}{12-8}\times6\times60=8640\text{ litres.}$$

Exercise

1. A cistern has a leak which would empty it in 4 hours. A tap is turned on which admits 3 litres a minute into the cistern, and it is now emptied in 6 hours. How many litres does the cistern hold?
a) 360 litres b) 1080 litres

c) 2160 litres d) None of these

2. A cistern has a leak which would empty it in 10 hours. A tap is turned on which admits 2 litres per hr into the cistern, and it is now emptied in 15 hours. How many litres does the cistern hold?
a) 50 litres b) 60 litres c) 45 litres d) 360 litres
3. A cistern has a leak which would empty it in 8 hours. A tap is turned on which admits 4 litres per minute into the cistern, and it is now emptied in 16 hours. How many litres does the cistern hold?
a) 3840 litres b) 2840 litres
c) 3880 litres d) None of these

Answers

1. c
2. b; **Hint:** Here w = 2 litres per hour

$\therefore$ required answer $= \frac{15 \times 10}{15-10} \times 2 = 60$ litres

3. a

Rule 24

Theorem: ***One filling pipe A is n times faster than the other filling pipe B. If B can fill a cistern in x hours, then the time in which the cistern will be full, if both the filling pipes are opened together, is $\left(\frac{x}{n+1}\right)$ hours.***

Note: Value of the slower filling pipe is given.

Illustrative Example

Ex: One filling pipe A is 5 times faster than second filling pipe B. If B can fill a cistern in 18 minutes, then find the time when the cistern will be full if both fill pipes are opened together.

Soln: Applying the above theorem,
x = 18 minutes [Filling pipe B is slower than the filling pipe A]
n = 5

$\therefore$ the required time $= \frac{18}{5+1} = 3$ minutes.

Exercise

1. One filling pipe A is 3 times faster than second filling pipe B. If B can fill a cistern in 16 hours, then find the time when the cistern will be full if both fill pipes are opened together.
a) 5 hrs b) 6 hrs c) 4 hrs d) Data inadequate
2. One filling pipe A is 5 times faster than second filling pipe B. If B can fill a cistern in 36 minutes, then find the time when the cistern will be full if both fill pipes are opened together.
a) 6 minutes b) 8 minutes c) 4 minutes d) 12 minutes
3. One filling pipe A is 7 times faster than second filling pipe B. If B can fill a cistern in 56 minutes, then find the time when the cistern will be full if both fill pipes are opened together.
a) 6 minutes b) 5 minutes c) 9 minutes d) 7 minutes
4. One filling pipe A is 10 times faster than second filling pipe B. If B can fill a cistern in 55 minutes, then find the time when the cistern will be full if both fill pipes are opened together.
a) 5 minutes b) 4 minutes
c) 7 minutes d) None of these

Answers

1. c 2. a 3. d 4. a

Rule 25

Theorem: ***One filling pipe A is n times faster than the other filling pipe B. If A can fill a cistern in x hours, then the time in which the cistern will be full, if both the filling pipes are opened together, is $\left(\frac{n}{n+1}\right)x$ hours.***

Note: *Value of the faster filling pipe is given.*

Illustrative Example

Ex: One filling pipe A is 5 times faster than second filling pipe B. If A can fill a cistern in 18 minutes, then find the time when the cistern will be full if both fill pipes are opened together.

Soln: Applying the above theorem,
n = 5
x = 18 minutes [Here, filling pipe A is faster than the filling pipe B.]

$\therefore$ the required time $= \left(\frac{5}{5+1}\right)18 = 15$ minutes.

Exercise

1. One filling pipe A is 4 times faster than second filling pipe B. If A can fill a cistern in 15 minutes, then find the time when the cistern will be full if both fill pipes are opened together.
a) 10 minutes b) 12 minutes
c) 15 minutes d) 14 minutes
2. One filling pipe A is 3 times faster than second filling pipe B. If A can fill a cistern in 12 minutes, then find the time when the cistern will be full if both fill pipes are opened together.
a) 9 minutes b) 10 minutes
c) 12 minutes d) None of these
3. One filling pipe A is 6 times faster than second filling pipe B. If A can fill a cistern in 28 minutes, then find the time when the cistern will be full if both fill pipes are opened together.
a) 20 minutes b) 24 minutes
c) 18 minutes d) 12 minutes

4. One filling pipe A is 9 times faster than second filling pipe B. If A can fill a cistern in 30 minutes, then find the time when the cistern will be full if both fill pipes are opened together.
   a) 28 minutes b) 25 minutes
   c) 24 minutes d) 27 minutes

Answers

1. b 2. a 3. b 4. d

Rule 26

Theorem: ***If one filling pipe A is n times faster and takes x minutes less time than the other filling pipe B, then the time, they will take to fill a cistern, if both the pipes are opened together, is $\left[\frac{nx}{(n^2-1)}\right]$ minutes. A will fill the cistern in $\left(\frac{x}{n-1}\right)$ minutes and B will take to fill the cistern $\left(\frac{nx}{n-1}\right)$ minutes.***

Note: Here, A is the faster filling pipe and B is the slower one.

Illustrative Example

Ex: One fill pipe A is 4 times faster than second fill pipe B and takes 30 minutes less than the fill pipe B. When will the cistern be full if both fill pipes are opened together?

Soln: Applying the above theorem, we have

the required time $= \frac{4\times 30}{(4^2-1)} = 8$ minutes.

Exercise

1. One fill pipe A is 3 times faster than second fill pipe B and takes 24 minutes less than the fill pipe B. When will the cistern be full if both fill pipes are opened together?
   a) 14 min b) 9 min c) 18 min d) Data inadequate
2. One fill pipe A is 4 times faster than second fill pipe B and takes 15 minutes less than the fill pipe B. When will the cistern be full if both fill pipes are opened together?
   a) 4 min b) 6 min c) 9 min d) 12 min
3. One fill pipe A is 5 times faster than second fill pipe B and takes 48 minutes less than the fill pipe B. When will the cistern be full if both fill pipes are opened together?
   a) 12 min b) 8 min c) 10 min d) 15 min

Answers

1. b 2. a 3. c

Rule 27

Theorem: ***If one filling pipe A is n times slower and takes x minutes more time than the other filling pipe B, then the time, they will take to fill a cistern, if both the pipes are opened together is $\left[\frac{nx}{(n^2-1)}\right]$ minutes. A will fill the cistern in $\left(\frac{nx}{n-1}\right)$ minutes and B will fill the cistern in $\left(\frac{x}{n-1}\right)$ minutes.***

Note: Here A is the slower filling pipe and B is the faster one.

Illustrative Example

Ex: One fill pipe A is 4 times slower than the second fill pipe B and takes 30 minutes more time than the fill pipe B. When will the cistern be full if both fill pipes are opened together? Also find, how much time will A and B separately take to fill the cistern?

Soln: Following the above formula, we have

the required time $= \frac{4\times 30}{(4^2-1)} = 8$ minutes

A will fill the cistern in $\frac{4\times 30}{4-1} = 40$ minutes and

B will fill the cistern in $\frac{30}{4-1} = 10$ minutes.

Exercise

1. One fill pipe A is 2 times slower than the second fill pipe B and takes 9 minutes more time than the fill pipe B. Find how much time will A take to fill the cistern?
   a) 6 minutes b) 10 minutes c) 15 minutes
   d) 8 minutes
2. One fill pipe A is 3 times slower than the second fill pipe B and takes 16 minutes more time than the fill pipe B. Find how much time will B take to fill the cistern?
   a) 6 minutes b) 21 minutes c) 14 minutes
   d) Data inadequate

Answers

1. a 2. a

Rule 28

Theorem: ***'P' pipes are fitted to a water tank. Some of these are filling pipes and the other emptying pipes. Each filling pipe can fill the tank in 'x' hours and each waste pipe can empty the tank in 'y' hours. On pening all the pipes, an empty tank is filled in 'T' hours. Then the number of filling pipes is $\left(\frac{y+PT}{x+y}\times\frac{x}{T}\right)$ and the number of waste pipes is $\left(\frac{PT-x}{x+y}\times\frac{y}{T}\right)$.***

Illustrative Example

Ex: 6 pipes are fitted to a water tank.Some of these are filling pipes and the other emptying pipes. Each filling pipe can fill the tank in 9 hours and each waste pipe can empty the tank in 6 hours. On opening all the pipes, an empty tank is filled in 9 hours. Find the number of filling pipes.

Soln: Following the above theorem, we have

P = 6, T = 9 hours, x = 9 hours and y = 6 hours.

$\therefore$ the required no. of filling pipes $= \frac{6+6\times 9}{9+6}\times\frac{9}{9}$

$= \frac{60}{15} = 4$

$\therefore$ the no. of waste pipes = 6 - 4 = 2.

Check the answer from the above formula,

no. of waste pipes $= \frac{6\times 9-9}{9+6}\times\frac{6}{9} = \frac{45}{15}\times\frac{6}{9} = 2.$

Exercise

1. 8 taps are fitted to a water tank. Some of them are water taps to fill the tank and the remaining are outlet taps used to empty the tank. Each water tap can fill the tank in 12 hours and each outlet tap can empty it in 36 hours. On opening all the taps, the tank is filled in 3 hours. Find the number of water taps.
   a) 5 b) 4 c) 3 d) 2
2. 16 taps are fitted to a water tank. Some of them are water taps to fill the tank and the remaining are outlet taps used to empty the tank. Each water tap can fill the tank in 6 hours and each outlet tap can empty it in 18 hours. On opening all the taps, the tank is filled in $1\frac{1}{2}$ hours. Find the number of empty taps.
   a) 7 b) 9 c) 6 d) 8
3. 9 taps are fitted to a water tank. Some of them are water taps to fill the tank and the remaining are outlet taps used to empty the tank. Each water tap can fill the tank in 9 hours and each outlet tap can empty it in 9 hours. On opening all the taps, the tank is filled in 9 hours. Find the number of water taps.
   a) 4 b) 5 c) 6 d) Can't be determined

Answers

1. a 2. b 3. b

Rule 29

Theorem: ***Two filling pipes A and B opened together can fill a cistern in t minutes. If the first filling pipe A alone takes x minutes more or less than t and the second fill pipe B alone takes y minutes more or less than t minutes, then t is given by*** $\left[t=\sqrt{xy}\right]$ ***minutes.***

Illustrative Example

Ex: One fill pipe A takes $4\frac{1}{2}$ minutes more to fill the cistern than two fill pipes A and B opened together to fill it. Second fill pipe B takes 8 minutes more to fill cistern than two fill pipes A and B opened together to fill it. When will the cistern be full if both pipes are opened simultaneously?

Soln: Applying the above theorem, we have

the required time $= \sqrt{\frac{9}{2}\times 8} = 6$ minutes.

Exercise

1. One fill pipe A takes $2\frac{1}{2}$ minutes more to fill the cistern than two fill pipes A and B opened together to fill it. Second fill pipe B takes 10 minutes more to fill cistern than two fill pipes A and B opened together to fill it. When will the cistern be full if both pipes are opened simultaneously?
   a) 6 minutes b) 5 minutes
   c) 4 minutes d) Data inadequate
2. One fill pipe A takes 3 minutes more to fill the cistern than two fill pipes A and B opened together to fill it. Second fill pipe B takes $21\frac{1}{3}$ minutes more to fill cistern than two fill pipes A and B opened together to fill it. When will the cistern be full if both pipes are opened simultaneously.
   a) 7 minutes b) 16 minutes
   c) 8 minutes d) 10 minutes
3. One fill pipe A takes 4 minutes more to fill the cistern than two fill pipes A and B opened together to fill it. Second fill pipe B takes 9 minutes more to fill cistern than two fill pipes A and B opened together to fill it. When will the cistern be full if both pipes are opened simultaneously.
   a) 4 minutes b) 6 minutes
   c) 5 minutes d) None of these

Answers

1. b 2. c 3. b

Rule 30

A General Method to solve the Problems on Pipes and Cisterns

Theorem: ***If first fill pipe can fill a cistern in*** x_1 ***minutes alone, second fill pipe can fill the same alone in*** x_2 ***minutes, and similarly, first empty pipe can empty the full cistern alone in*** y_1 ***minutes, second empty pipe can empty the full cistern alone in*** y_2 ***minutes, then***

the alone filling time for first fill pipe = x_1 ***minutes,***

alone filling time for second fill pipe = x_2 ***mintues***

alone emptying time for first empty pipe = y_1 ***minutes and***

alone emptying time for second empty pipe = y_2 ***mintues.***

Now, if first fill pipe and second fill pipe are opened for t_1 ***minutes and*** t_2 ***minutes respectively. First empty pipe and second empty pipe are opened for*** t_3 ***minutes and*** t_4 ***minutes respectively, then***

Step I

To find the amount of work (filling or emptying) done by each pipe (fill or empty), we use the following formula,

Amount of work (filling or emptying) done

$$= \frac{\text{No. of minutes opened}}{\text{Alone time (empty or fill)}}$$

Note: Put (-ve) sign for '**emptying work**'.

Step II Add the amount of work done by each pipe and equate it to the part of cistern filled.

(i) $\frac{t_1}{x_1}+\frac{t_2}{x_2}-\frac{t_3}{y_1}-\frac{t_4}{y_2}=1$, if cistern is filled completely

(ii) $\frac{t_1}{x_1}+\frac{t_2}{x_2}-\frac{t_3}{y_1}-\frac{t_4}{y_2}=\frac{1}{2}$, if cistern is made half full.

(iii) $\frac{t_1}{x_1}+\frac{t_2}{x_2}-\frac{t_3}{y_1}-\frac{t_4}{y_2}=0$, if full cistern is emptied completely and so on.

Step III: Find the value of unknown

Illustrative Examples

Ex 1: Two fill pipes A and B can fill a cistern in 12 and 16 minutes respectively. Both fill pipes are opened together, but 4 minutes before the cistern is full, one pipe A is closed. How much time will the cistern take to fill.

Soln: Let the cistern be filled in x minutes.

$\therefore$ Pipe B is opened for x minutes and pipe A is opened for (x - 4) minutes.

Using the above method

$$\frac{x-4}{12}+\frac{x}{16}=1 \text{ or, } \frac{4x-16+3x}{48}=1$$

$$\therefore\ x=\frac{64}{7}=9\frac{1}{7} \text{ minutes}$$

$\therefore$ The cistern is filled in $9\frac{1}{7}$ minutes.

Ex 2: Two fill taps A and B can separately fill a cistern in 45 and 40 minutes respectively. They started to fill a cistern together but fill tap A is turned off after few minutes and fill tap B fills the rest part of cistern in 23 minutes. After how many minutes, was tap A turned off?

Soln: Here, total filling time is not given and you don't need to calculate also.

Let the tap A be turned off after x minutes

$\therefore$ tap B is opened for (x + 23) minutes.

Now, use the above method

$$\frac{x}{45}+\frac{x+23}{40}=1$$

$$\text{or, } \frac{8x+9x+207}{360}=1$$

or, 17x = 153 $\therefore$ x = 9 minutes.

Ex 3: A cistern can be filled by two pipes filling separately in 12 and 16 minutes respectively. Both pipes are opened together for a certain time but being clogged, only $\frac{7}{8}$ of full quantity water flows through the former and only $\frac{5}{6}$ through the latter pipe. The obstructions, however, being suddenly removed, the cistern is filled in 3 minutes from that moment. How long was it before the full flow began?

Soln: Let both pipes remain clogged for x minutes and hence full flow began after x minutes only.

$\therefore$ part of cistern filled in x minutes + part of cistern filled in 3 minutes = cistern filled

Now use the above method,

$$\text{or, } \left[\left(\frac{7}{8}\times\frac{x}{12}\right)+\left(\frac{5}{6}\times\frac{x}{16}\right)\right]+\left[\frac{3}{12}+\frac{3}{16}\right]=1$$

$$\text{or, } \frac{12x}{96}+\frac{7}{16}=1 \quad \therefore x=4.5 \text{ minutes.}$$

Exercise

1. Two fill pipes A and B can fill a cistern in 6 and 8 minutes respectively. Both fill pipes are opened together, but 2 minutes before the cistern is full, one pipe A is closed. How much time will the cistern take to fill.

   a) $4\frac{4}{7}$ min b) $4\frac{5}{7}$ min c) $6\frac{4}{7}$ min d) $6\frac{5}{7}$ min

2. Two fill pipes A and B can fill a cistern in 18 and 24 minutes respectively. Both fill pipes are opened together, but 6 minutes before the cistern is full, one pipe A is closed. How much time will the cistern take to fill.

   a) $12\frac{4}{7}$ min b) $12\frac{5}{7}$ min c) $13\frac{5}{7}$ min d) None of these

3. Two fill taps A and B can separately fill a cistern in 10 and 20 minutes respectively. They started to fill a cistern

together but fill tap A is turned off after few minutes and fill tap B fills the rest part of cistern in 8 minutes. After how many minutes, was tap A turned off?
a) 3 min b) 4 min c) 5 min d) 2 min

4. Two fill taps A and B can separately fill a cistern in 15 and 30 minutes respectively. They started to fill a cistern together but fill tap A is turned off after few minutes and fill tap B fills the rest part of cistern in 9 minutes. After how many minutes, was tap A turned off?
a) 6 min b) 8 min
c) 7 min d) Data inadequate

5. A cistern can be filled by two pipes filling separately in 15 and 25 minutes respectively. Both pipes are opened together for a certain time but being clogged, only $\frac{5}{6}$ of full quantity water flows through the former and only $\frac{5}{8}$ through the latter pipe. The obstructions, however, being suddenly removed, the cistern is filled in 5 minutes from that moment. How long was it before the full flow began?
a) $\frac{161}{29}$ minutes b) $\frac{168}{29}$ minutes
c) $\frac{148}{29}$ minutes d) None of these

6. A cistern can be filled by one of two pipes in 30 minutes and by the other in 36 minutes. Both pipes are opened together for a certain time but being particularly clogged, only $\frac{5}{6}$ of the full quantity of water flows through the former and only $\frac{9}{10}$ through the latter. The obstructions, however, being suddenly removed, the cistern is filled in $15\frac{1}{2}$ minutes from that moment. How long was it before the full flow of water began?
a) 1 minute b) 2 minute
c) minute d) $1\frac{1}{2}$ minute

Answers

1. a 2. c 3. b 4. c 5. b

6. a; **Hint:** Net filling in last $15\frac{1}{2}$ minutes

$$=\frac{31}{2}\left(\frac{1}{30}+\frac{1}{36}\right)=\frac{341}{360}$$

Now, suppose they remained clogged for x minutes.

Net filling in these x minutes

$$=\left(\frac{x}{30}\times\frac{5}{6}+\frac{x}{36}\times\frac{9}{10}\right)=\frac{19x}{360}.$$

$$\text{Remaining part}=\left(1-\frac{19x}{360}\right)=\left(\frac{360-19x}{360}\right).$$

$$\therefore \frac{360-19x}{360}=\frac{341}{360} \text{ or } x=1.$$

Hence, the pipes remained clogged for 1 minute.

Rule 31

Theorem: ***Three pipes A, B and C are attached to a cistern. A can fill it in x minutes and B in y minutes. C is a waste pipe for emptying it. After opening both the pipes A and B, a man leaves the cistern and returns when the cistern should have been just full. Finding, however, that the waste pipe had been left open, he closes it and the cistern now gets filled in t minutes. The time in which the pipe C, if opened alone, empty the full cistern is*** $\left(\frac{xy}{x+y}\right)^2\frac{1}{t}$ ***minutes.***

Illustrative Example

Ex: Three pipes A, B and C are attached to a cistern. A can fill it in 10 minutes and B in 15 minutes. C is a waste pipe for emptying it. After opening both the pipes A and B, a man leaves the cistern and returns when the cistern should have been just full. Finding, however, that the waste pipe had been left open, he closes it and the cistern now gets filled in 2 minutes. In how much time the pipe C, if opened alone, empty the full cistern.

Soln: **Detail Method:**
Let the pipe C alone empty the cistern in x minutes.

A and B together can fill the cistern in $\frac{10\times15}{10+15}=6$ minutes

$\because$ waste pipe C had been left open for 6 minutes

$\therefore$ in 6 minutes $\frac{6}{x}$ part of the cistern will be emptied by the waste pipe C.

Now, $\frac{6}{x}$ part of the cistern would be filled by A and B together in 2 minutes.

$\therefore$ Cistern will be full in $\frac{x}{3}$ minutes.

From the question, we have

$\frac{x}{3} = 6 \therefore x = 18$ minutes.

Quicker Method: Applying the above theorem,
Here, x = 10 minutes
y = 15 minutes
t = 2 minutes

$\therefore$ Empty time for waste pipe $C = \left(\frac{10 \times 15}{10+15}\right)^2 \times \frac{1}{2}$

$= \frac{6 \times 6}{2} = 18$ min.

Exercise

1. A bath can be filled by the cold water pipe in 10 minutes and by the hot water pipe in 15 minutes. A person leaves the bathroom after turning on both pipes simultaneously and returns at the moment when the bath should be full. Finding, however, that the waste pipe has been open, he now closes it. In 4 minutes more the bath is full. In what time would the waste pipe empty it?
   a) 9 min b) 8 min c) 12 min d) 6 min
2. A bath can be filled by the cold water pipe in 15 minutes and by the hot water pipe in 30 minutes. A person leaves the bathroom after turning on both pipes simultaneously and returns at the moment when the bath should be full. Finding, however, that the waste pipe has been open, he now closes it. In 5 minutes more the bath is full. In what time would the waste pipe empty it?
   a) 25 min b) 20 min c) 30 min d) None of these
3. A bath can be filled by the cold water pipe in 20 minutes and by the hot water pipe in 30 minutes. A person leaves the bathroom after turning on both pipes simultaneously and returns at the moment when the bath should be full. Finding, however, that the waste pipe has been open, he now closes it. In 6 minutes more the bath is full. In what time would the waste pipe empty it?
   a) 16 min b) 29 min c) 24 min d) 27 min
4. A bath can be filled by the cold water pipe in 30 minutes and by the hot water pipe in 60 minutes. A person leaves the bathroom after turning on both pipes simultaneously and returns at the moment when the bath should be full. Finding, however, that the waste pipe has been open, he now closes it. In 8 minutes more the bath is full. In what time would the waste pipe empty it?
   a) 25 min b) 30 min c) 40 min d) 50 min
5. Three pipes A, B and C are attached to a cistern, A can fill it in 20 minutes and B in 30 minutes. C is a waste pipe meant for emptying it. After opening both the pipes A and B, a man leaves the cistern and returns when the cistern should have been just full. Finding however that the waste pipe had been left open, he closes it and the cistern now gets filled in 3 minutes. In how much time would the waste pipe C, if opened alone, empty the full cistern?
   a) 48 min b) 24 min c) 36 min d) 42 min

Answers

1. a 2. b 3. c 4. d 5. a

Rule 32

Theorem: ***If a pipe A fills a cistern in x hours and suddenly a leak develops through which every hour n part of the water filled by the pipe A leaks out, then the time in which tank is full*** $= \left(\frac{x}{1-n}\right)$ ***hours.***

Illustrative Example

Ex: A pipe A can fill a tank in 12 hours. Due to development of a whole in the bottom of the tank $\frac{1}{3}$rd of the water filled by the pipe A leaks out. Find the time when the tank will be full.

Soln: Detail Method:

In 1 hour $\frac{1}{12}$ part of the tank is filled.

Due to the leak, every hour $\left(\frac{1}{3} \times \frac{1}{12} = \frac{1}{36}\right)$ part of the water leaks out.

$\therefore$ the whole tank will be emptied in 36 hours.

$\therefore$ time to fill the tank $= \frac{12 \times 36}{36-12} = 18$ hours.

Quicker Method: Applying the above rule, we have

the required answer $= \frac{12}{1-\frac{1}{3}} = \frac{12 \times 3}{2} = 18$ hours.

Exercise

1. A pipe A can fill a tank in 18 minutes. Due to development of a whole in the bottom of the tank $\frac{1}{4}$ of the water filled by the pipe A leaks out. Find the time when the tank will be full.
   a) 24 min b) 20 min c) 27 min d) Data inadequate
2. A pipe A can fill a tank in 27 minutes. Due to development of a whole in the bottom of the tank $\frac{1}{10}$ of the water filled by the pipe A leaks out. Find the time when the tank will be full.
   a) 32 min b) 34 min c) 36 min d) 30 min
3. A pipe A can fill a tank in 28 minutes. Due to develop-

ment of a whole in the bottom of the tank $\frac{1}{8}$ of the water filled by the pipe A leaks out. Find the time when the tank will be full.

a) 30 min b) 32 min c) 35 min d) 34 min

4. A pipe A can fill a tank in 16 minutes. Due to development of a whole in the bottom of the tank $\frac{1}{5}$ of the water filled by the pipe A leaks out. Find the time when the tank will be full.

a) 18 min b) 24 min c) 20 min d) Data inadequate

Answers

1. a 2. d 3. b 4. c

Rule 33

Theorem: ***If pipes A, B and C can fill a cistern in x, y and z hours respectively. If pipe C is closed t hours before the cistern is full, then the time in which tank is filled =***

$$\left(\frac{xy(t+z)}{xy+yz+xz}\right) \textit{hrs.}$$

Illustrative Example

Ex: Three pipes A, B and C can fill a tank in 12 minutes, 16 minutes and 24 minutes respectively. The pipe C is closed 3 minutes before the tank is filled. In what time will the tank be full?

Soln: **Detail Method:**

Let the tank be full in x minutes.

Now, $\frac{x}{12}+\frac{x}{16}+\frac{x-3}{24}=1$

or, $\frac{4x+3x+2x-6}{48}=1$

or, $9x=54 \therefore x=6$ minutes

Quicker Method: Applying the above theorem,

Here, $t=3$ minutes

$x=12$ minutes

$y=16$ minutes

$z=24$ minutes

$\therefore$ the required time

$$=\frac{12\times16\times27}{16\times12+16\times24+12\times24}=6 \text{ minutes.}$$

Exercise

1. Three pipes A, B and C can fill a tank in 6 minutes, 8 minutes and 12 minutes respectively. The pipe C is closed 6 minutes before the tank in filled. In what time will the tank be full?

a) 6 min b) 4 min c) 5 min d) Data inadequate

2. Three pipes A, B and C can fill a tank in 3 minutes, 4 minutes and 6 minutes respectively. The pipe C is closed 3 minutes before the tank is filled. In what time will the tank be full?

a) 3 min b) 1 min c) 2 min d) 4 min

3. Three pipes A, B and C can fill a tank in 15 minutes, 20 minutes and 30 minutes respectively. The pipe C is closed 6 minutes before the tank is filled. In what time will the tank be full?

a) 5 min b) 8 min c) 6 min d) 12 min

Answers

1. b 2. c 3. b

Miscellaneous

1. Two pipes A and B can fill a tank in 15 hours and 20 hours respectively while a third pipe C can empty the full tank in 25 hours. All the three pipes are opened in the beginning. After 10 hours C is closed. Find, in how much time will the tank be full?

a) 12 hrs b) 8 hrs c) 10 hrs d) 14 hrs

2. Three pipes A, B and C can fill a cistern in 10 hours, 12 hours and 15 hours respectively. First A was opened. After 1 hour, B was opened and after 2 hours from the start of A, C was also opened. Find the time in which the cistern is just full.

a) 2 hrs b) 4 hrs c) 2 hrs 52 min d) 4 hrs 52 min

3. A, B, C are pipes attached to a cistern. A and B can fill it in 20 and 30 minutes respectively, while C can empty it in 15 minutes. If A, B, C be kept open successively for 1 minute each, how soon will the cistern be filled?

a) 167 min b) 160 min c) 166 min d) 164 min

Answers

1. a; Tank filled in 10 hours $=10\left(\frac{1}{15}+\frac{1}{20}-\frac{1}{25}\right)=\frac{23}{30}$.

Remaining part $=\left(1-\frac{23}{30}\right)=\frac{7}{30}$.

Work done by (A + B) in 1 hour $=\left(\frac{1}{15}+\frac{1}{20}\right)=\frac{7}{60}$.

Now, $\frac{7}{60}$ part is filled by (A + B) in 1 hour

$\therefore \frac{7}{30}$ part will be filled by (A + B) in $\left(\frac{60}{7}\times\frac{7}{30}\right)$ hrs

= 2 hours

$\therefore$ Total time in which the tank is full $=(10+2)$

= 12 hours

2. d; [(A's 1 hour work) + (A + B)'s 1 hour work]

$$=\frac{1}{10}+\left(\frac{1}{10}+\frac{1}{12}\right)=\frac{17}{60}.$$

Remaining part $= \left(1 - \frac{17}{60}\right) = \frac{43}{60}$.

Now, (A+B+C)'s 1 hour work $= \left(\frac{1}{10} + \frac{1}{12} + \frac{1}{15}\right) = \frac{1}{4}$.

$\frac{1}{4}$ part is filled by 3 pipes in 1 hour.

$\frac{43}{60}$ part will be filled by them in $\left(4 \times \frac{43}{60}\right)$ hrs

= 2 hours 52 min.

∴ Total time taken to fill the cistern = 4 hours 52 min.

3. a; Work done in 3 minutes $= \left(\frac{1}{20} + \frac{1}{30} - \frac{1}{15}\right) = \frac{1}{60}$

Clearly, $\frac{55}{60}$ part of cistern is filled in 3 × 55 or 165 min.

Remaining part $= \left(1 - \frac{55}{60}\right) = \frac{5}{60} = \frac{1}{12}$.

Now, $\frac{1}{20}$ part is filled by A in 1 min.

and $\left(\frac{1}{12} - \frac{1}{20}\right)$ ie $\frac{1}{30}$ part is filled by B in 1 min.

∴ required time = (3 × 55 + 1 + 1) min = 167 min.

16 Time and Distance

Rule 1

(i) Distance = Speed × Time

(ii) Time $= \frac{\text{Distance}}{\text{Speed}}$

(iii) Speed $= \frac{\text{Distance}}{\text{Time}}$

(iv) x km/hr $= \left(x \times \frac{5}{18}\right)$ m/sec

(v) x m/sec $= \left(x \times \frac{18}{5}\right)$ km/hr

Illustrative Examples

Ex. 1: Find the distance covered by a man walking for 10 minutes at a speed of 6 km/hr.

Soln: Applying the above formula (i), we have

$$\text{Distance} = 6 \text{ km/hr} \times \frac{10}{60} \text{ hr} \left[\because 10 \text{ minutes} = \frac{10}{60} \text{ hr}\right]$$

$$= 1 \text{ km}$$

Ex. 2: Find the time taken to cover a distance of 125 km by a train moving at a speed of 50 km/hr.

Soln: Applying the above formula (ii), we have,

$$\text{Time} = \frac{125}{50} \text{ hours} = 2.5 \text{ hours}$$

Ex. 3: A train covers a distance of 1250 km in 25 hours. Find the speed of the train.

Soln: Applying the above formula (iii),

$$\text{Speed} = \frac{1250}{25} = 50 \text{ km/hr}$$

Ex. 4: Express a speed of 18 km/hr in metres per second.

Soln: Applying the above formula (iv), we have

$$18 \text{ km/hr} = \left(18 \times \frac{5}{18}\right) \text{ m/sec} = 5 \text{ m/sec.}$$

Ex. 5: Express 10 m/sec in km/hr.

Soln: Applying the above formula (v), we have,

$$10 \text{ m/sec} = \left(10 \times \frac{18}{5}\right) \text{ km/hr} = 36 \text{ km/hr.}$$

Exercise

1. A train runs at the rate of 45 km an hour. What is its speed in metres per second?
   a) $12\frac{1}{2}$ m/sec b) 25 m/sec
   c) $10\frac{1}{2}$ m/sec d) None of these
2. A motor car takes 50 seconds to travel 500 metres. What is its speed in km per hour?
   a) 32 km/hr b) 36 km/hr c) 34 km/hr d) 38 km/hr
3. How many km per hour does a man walk who passes through a street 600 m long in 5 minutes?
   a) 3.6 km/hr b) 7.2 km/hr
   c) 8 km/hr d) None of these
4. Compare the rates of two trains, one travelling at 45 km an hour and the other at 10 m a second.
   a) 4 : 5 b) 3 : 5 c) 5 : 4 d) 5 : 3
5. The distance of the sun from the earth is one-hundred forty-three million four hundred thousand km and light travels from the former to the latter in seven minutes and fifty-eight seconds. Find the velocity of light per second.
   a) 300000 km/sec b) 30000 km/sec
   c) 3×10^6 km/sec d) Can't be determined
6. The wheel of an engine $4\frac{2}{7}$ metres in circumference makes seven revolutions in 4 seconds. Find the speed of the train in km per hour.
   a) 18 km/hr b) 24 km/hr c) 36 km/hr d) 27 km/hr
7. Sound travels 330 metres a second. How many kilometres is a thunder-cloud distant when the sound follows the flash after 10 seconds?

a) 3.3 km b) 3 km c) 3.2 km d) 3.5 km

8. What is the length of the bridge which a man riding 15 km an hour can cross in 5 minutes?

a) 1 km b) 2 km c) $2\frac{1}{2}$ km d) $1\frac{1}{4}$ km

9. A man crosses a street 600 m long in 5 minutes. His speed in kilometres per hour is **[Clerical Grade Exam, 1991]**

a) 7.2 b) $\frac{5}{36}$ c) 3.6 d) 10

10. If a man covers $10\frac{1}{5}$ km in 3 hours, the distance covered by him in 5 hours is **[Asst. Grade Exam, 1987]**

a) 18 km b) 15 km c) 16 km d) 17 km

11. A train travels 92.4 km/hr. How many metres will it travel in 10 minutes? **[Bank PO Exam, 1991]**

a) 15400 b) 1540 c) 154 d) 15.40

12. A train covers a distance in 50 minutes if it runs at a speed of 48 km per hour on an average. The speed at which the train must run to reduce the time of journey to 40 minutes, will be **[Central Excise and I Tax Exam, 1988]**

a) 50 km/hr b) 55 km/hr c) 60 km/hr d) 70 km/hr

Answers

1. a; **Hint:** Use the given rule (iv)

ie required answer $= \frac{45\times5}{18} = \frac{25}{2} = 12\frac{1}{2}$ m per sec

2. b; **Hint:** Speed of the motor car in m/sec $= \frac{500}{50} = 10$ m/sec

10 m/sec $= \frac{10\times18}{5} = 36$ km/hr (Use rule (v))

3. b; **Hint:** Man's speed in m/sec $= \frac{600}{5\times60} = 2$ m/sec

2 m/sec $= \frac{18\times2}{5} = 7.2$ km/hr

4. c; **Hint:** Speed of the first train = 45 km/hr

Speed of the second train = 10 m/sec $= 10\times\frac{18}{5} = 36$ km/hr

$\therefore$ ratio of the speeds of the trains = 45 : 36 = 5 : 4

5. a; **Hint:** Speed $= \frac{\text{Distance}}{\text{Time}}$

Here, distance between the sun and the earth = 143400000 km and

time = 7 × 60 + 58 = 478 seconds

$\therefore$ speed $= \frac{143400000}{478} = 300000$ km/sec

6. d; **Hint:** Distance travelled by the train in 4 seconds

$= \frac{30}{7}\times7 = 30$ metres.

$\therefore$ speed of the train $= \frac{30}{4} = \frac{15}{2}$ m/sec

(using the given rule (iii))

$\frac{15}{2}$ m/sec $= \frac{15}{2}\times\frac{18}{5} = 27$ km/hr

(Using the given rule (v))

7. a; **Hint:** Distance = Speed × Time $= \frac{330\times10}{1000} = 3.3$ km

8. d; **Hint:** Required answer $= 15\times\frac{5}{60} = \frac{5}{4} = 1\frac{1}{4}$ km

(Using the given rule (i))

9. a; **Hint:** Required answer $= \frac{600}{5\times60}\times\frac{18}{5} = 7.2$ km/hr

10. d; **Hint:** Distance covered in 5 hours $= \left(\frac{51}{5\times3}\times5\right)$ km

= 17 km

11. a; **Hint:** Speed of train $= \left(92.4\times\frac{5}{18}\right)$ m/sec $= \frac{77}{3}$ m/sec

Distance covered in (10 × 60) sec $= \left(\frac{77}{3}\times10\times60\right)$ metres = 15400 metres

12. c; **Hint:** Distance = (speed × time) $= \left(48\times\frac{50}{60}\right) = 40$ km

Speed $= \frac{\text{Distance}}{\text{Time}} = \left(\frac{40\times60}{40}\right)$ km/hr = 60 km/hr

Rule 2

Theorem: ***If a certain distance is covered at x km/hr and the same distance is covered at y km/hr then the average speed during the whole journey is*** $\frac{2xy}{x+y}$ ***km/hr.***

Illustrative Example

Ex.: A man covers a certain distance by car driving at 70 km/hr and he returns back to the starting point riding on a scooter at 55 km/hr. Find his average speed for the whole journey.

Soln: Applying the above theorem, we have,

$$\text{Average speed} = \frac{2\times70\times55}{70+55}\text{ km/hr} = 61.5\text{ km/hr.}$$

Exercise

1. A man covers a certain distance by car driving at 40 km/hr and he returns back to the starting point riding on a scooter at 10 km/hr. Find his average speed for the whole journey.
   a) 8 km/hr b) 16 km/hr
   c) 12 km/hr d) None of these
2. A man covers a certain distance by car driving at 30 km/hr and he returns back to the starting point riding on a scooter at 20 km/hr. Find his average speed for the whole journey.
   a) 24 km/hr b) 27 km/hr c) 15 km/hr d) 12 km/hr
3. A man covers a certain distance by car driving at 15 km/hr and he returns back to the starting point riding on a scooter at 35 km/hr. Find his average speed for the whole journey.
   a) 20 km/hr b) 18 km/hr c) 21 km/hr d) 24 km/hr
4. A man covers a certain distance by car driving at 40 km/hr and he returns back to the starting point riding on a scooter at 20 km/hr. Find his average speed for the whole journey.
   a) $16\frac{2}{3}$ km/hr b) 26 km/hr
   c) $36\frac{2}{3}$ km/hr d) $26\frac{2}{3}$ km/hr
5. A train goes from a station A to another station B at a speed of 64 km/hr but returns to A at a slower speed. If its average speed for the trip is 56 km/hr, the return speed of the train is nearly. **[Hotel Management, 1991]**
   a) 48 km/hr b) 50 km/hr c) 52 km/hr d) 47.4 km/hr
6. On a tour a man travels at the rate of 64 km an hour for the first 160 km, then travels the next 160 km at the rate of 80 km an hour. The average speed in km per hour for the first 320 km of the tour, is **[SBI PO Exam, 1988]**
   a) 35.55 b) 71.11 c) 36 d) 72

Answers

1. b 2. a 3. c 4. d

5. b; **Hint:** Let the return speed of the train be y km/hr
Then, applying the given rule, we have,

$$\frac{2\times64\times y}{(64+y)} = 56$$

or, $2 \times 64 \times y = 56(64 + y)$
or, $y = 49.8 = 50$ km/hr (nearly)

6. b; **Hint:** $\therefore$ Average speed $= \left(\frac{2\times64\times80}{64+80}\right)$ km/hr $= 71.11$ km/hr

Rule 3

Theorem: ***A person is walking at a speed of x km/hr. After every kilometre, if he takes rest for t hours, then the time he will take to cover a distance of y km is $\left[\frac{y}{x}+(y-1)t\right]$ hr. Or, in other words, required time***

$$= \frac{\textit{Distance to be covered}}{\textit{Speed}} + \textit{Number of rest} \times \textit{Time for each rest}$$

Illustrative Example

Ex.: A man is walking at a speed of 12 km per hour. After every km he takes rest for 12 minutes. How much time will he take to cover a distance of 36 km.

Soln: **Detail Method:**

Rest time = Number of rest × Time for each rest

$$= 35 \times \frac{12}{60} = 7 \text{ hr}$$

[$\because$ To cover 36 km, the man has to take rest 35 times, as he takes rest every hour]

$$\text{Total time to cover 36 km} = \frac{36}{12} + 7 = 10 \text{ hours.}$$

Quicker Method: Applying the above theorem,
Here, x = 12 km/hr
y = 36 km

$$t = 12\text{ minutes} = \frac{12}{60} = \frac{1}{5}\text{ hr.}$$

$$\text{Total time} = \frac{36}{12} + (36-1)\frac{1}{5} = 3 + 7 = 10\text{ hours.}$$

Exercise

1. A man is walking at a speed of 6 km per hour. After every km he takes rest for 6 minutes. How much time will he take to cover a distance of 18 km.
   a) 5 hrs b) 5 hrs 42 min
   c) 4 hrs 42 min d) None of these
2. A man is walking at a speed of 18 km per hour. After every km he takes rest for 18 minutes. How much time will he take to cover a distance of 54 km.
   a) 18 hrs b) 18 hrs 54 min
   c) 16 hrs 54 min d) None of these
3. A man is walking at a speed of 9 km per hour. After every km he takes rest for 9 minutes. How much time will he take to cover a distance of 27 km.
   a) 6 hrs b) 6 hrs 45 min
   c) 6 hrs 54 min d) None of these
4. A man rides at the rate of 18 km an hour, but stops 6 minutes to change hours at the end of every 7th kilometre. How long will he take to go a distance of 90 kilometres?

a) 6 hrs b) 6 hrs 12 min
c) 6 hrs 15 min d) None of these

5. A man rides at the rate of 20 kilometres an hour. But stops 10 minutes to change horses at the end of every 25th kilometre. How long will he take to go a distance of 175 kilometres?
a) 9 hrs 15 min b) 9 hrs 45 min
c) 10 hrs d) 8 hrs 45 min

Answers

1. c 2. b 3. c

4. b; **Hint:** Here instead of every kilometre, every 7th kilometre is given. Hence, we should first calculate

no. of rest ie $\frac{90}{7} \approx 12$

Now applying the given rule, we have

the required answer $= \frac{90}{18} + 12 \times \frac{6}{60}$
$= 5$ hrs $+ 1$ hr 12 min
$= 6$ hrs 12 min

5. b; **Hint:** No. of rest $= \frac{175}{25} - 1 = 6$

$\therefore$ required answer $= \frac{175}{20} + 6 \times \frac{10}{60} = 9$ hrs 45 min

Rule 4

Theorem: ***A person covers a certain distance between two points. Having an average speed of x km/hr, he is late by x_1 hours. However, with a speed of y km/hr he reaches his destination y_1 hours earlier. The distance between the two points is given by $\left[\frac{xy(x_1 + y_1)}{(y - x)}\right]$ km. Or,***

Required distance

$= \frac{\textit{Product of two speeds}}{\textit{Difference of two speeds}} \times$ *Difference between arrival times*

(See note)

Illustrative Example

Ex.: A man covers a certain distance between his house and office on scooter. Having an average speed of 30 km/hr, he is late by 10 min. However, with a speed of 40 km/hr, he reaches his office 5 min earlier. Find the distance between his house and office.

Soln: Detail Method Let the distance be x km.

Time taken to cover x km at 30 km/hr $= \frac{x}{30}$ hrs.

Time taken to cover x km at 40 km/hr $= \frac{x}{40}$ hrs.

Difference between the time taken = 15 min $= \frac{1}{4}$ hr.

$\therefore \frac{x}{30} - \frac{x}{40} = \frac{1}{4}$ or, $4x - 3x = 30$ or, $x = 30$

Hence, the required distance is 30 km.

Quicker Method: Applying the above theorem, we have

the required distance $= \frac{30 \times 40}{40 - 30} \times \frac{10 + 5}{60} = 30$ km.

Note: 10 minutes late and 5 minutes earlier make a difference of $10 + 5 = 15$ minutes. As the other units are in km/hr, the difference in time should also be changed into hours.

$\therefore$ 15 minutes $= \frac{15}{60} = \frac{1}{4}$ hr.

Exercise

1. A man covers a certain distance between his house and office on scooter. Having an average speed of 15 km/hr, he is late by 5 min. However, with a speed of 20 km/hr, he reaches his office $2\frac{1}{2}$ min earlier. Find the distance between his house and office.
a) 15 km b) $15\frac{1}{2}$ km c) 7 km d) $7\frac{1}{2}$ km

2. A man covers a certain distance between his house and office on scooter. Having an average speed of 60 km/hr, he is late by 20 min. However, with a speed of 80 km/hr, he reaches his office 10 min earlier. Find the distance between his house and office.
a) 120 km b) 90 km c) 80 km d) 60 km

3. A man covers a certain distance between his house and office on scooter. Having an average speed of 45 km/hr, he is late by 15 min. However, with a speed of 60 km/hr, he reaches his office $7\frac{1}{2}$ min earlier. Find the distance between his house and office.
a) $45\frac{1}{2}$ km b) $67\frac{1}{2}$ km c) 45 km d) $37\frac{1}{2}$ km

4. Ram travels at the rate of 3 km/hr and he reaches 15 minutes late. If he travels at the rate of 4 km/hr, he reaches 15 minutes earlier. The distance Ram has to travel is **[CDS Exam, 1989]**
a) 1 km b) 6 km c) 7 km d) 12 km

Answers

1. d 2. a 3. b 4. b

Rule 5

Theorem: ***A person walking at a speed of x km/hr reaches his destination x_1 hrs late. Next time he increases his speed by y km/hr, but still he is late by y_1 hrs. The distance of his destination from his house is given by*** $\left[(x_1-y_1)(x+y)\frac{x}{y}\right]$ ***km.***

Illustrative Example

Ex.: A boy walking at a speed of 10 km/hr reaches his school 15 minutes late. Next time he increases his speed 2 km/hr, but still he is late by 5 minutes. Find the distance of his school from his house.

Soln: Applying the above theorem, we have the required distance

$$=\left(\frac{15-5}{60}\right)(10+2)\times\frac{10}{2}=10\text{ km.}$$

Exercise

1. A boy walking at a speed of 15 km/hr reaches his school 20 minutes late. Next time he increases his speed 5 km/hr, but still he is late by 10 minutes. Find the distance of his school from his house.
   a) 15 km b) 10 km c) 18 km d) 20 km
2. A boy walking at a speed of 45 km/hr reaches his school 10 minutes late. Next time he increases his speed 15 km/hr, but still he is late by 5 minutes. Find the distance of his school from his house.
   a) 10 km b) 12 km c) 25 km d) 15 km
3. A boy walking at a speed of 30 km/hr reaches his school 40 minutes late. Next time he increases his speed 10 km/hr, but still he is late by 20 minutes. Find the distance of his school from his house.
   a) 30 km b) 25 km
   c) 40 km d) None of these
4. A man walking with a speed of 5 km/hr reaches his target 5 minutes late. If he walks at a speed of 6 km/hr, he still reaches 3 minutes late. Find the distance of his target from his house.
   a) 3 km b) 2 km c) $1\frac{1}{2}$ km d) 1 km

Answers

1. b 2. d 3. c

4. d; **Hint:** Here y = 6 km/hr – 5 km/hr = 1 km/hr
Now applying the given rule we have

the required distance $=\left(\frac{5-3}{60}\right)\times(5+1)\frac{5}{1}=1$ km

Rule 6

Theorem: ***A person goes to a destination at a speed of x km/hr and returns to his place at a speed of y km/hr. If he takes T hours in all, the distance between his place and destination is*** $\left(\frac{xy}{x+y}\times T\right)$ ***km. In other words,***

Required distance

$$=\textit{Total time taken}\times\frac{\textit{Product of the two speeds}}{\textit{Addition of the two speeds}}$$

Illustrative Example

Ex.: A boy goes to school at a speed of 3 km/hr and returns to the village at a speed of 2 km/hr. If he takes 5 hrs in all, what is the distance between the village and the school?

Soln: **Detail Method:** Let the required distance be x km.

Then time taken during the first journey $=\frac{x}{3}$ hr.

and time taken during the second journey $=\frac{x}{2}$ hr.

$$\therefore \frac{x}{3}+\frac{x}{2}=5\Rightarrow\frac{2x+3x}{6}=5\Rightarrow 5x=30$$

$\therefore x=6$

$\therefore$ required distance = 6 km

Quicker Method: Applying the above rule, we have

the required distance $=5\times\frac{3\times2}{3+2}=6$ km.

Exercise

1. A man walks to a town at the rate of $5\frac{1}{2}$ kilometres an hour and rides back at the rate of 10 kilometres an hour, how far has he walked, the whole time occupied having been 6 hours 12 minutes?
   a) 31 km b) 29 km c) 22 km d) 17 km
2. A man walks to a town at the rate of 4 kilometres an hour and rides back at the rate of 5 kilometres an hour, how far has he walked, the whole time occupied having been 9 hours?
   a) 20 km b) 25 km c) 18 km d) 15 km
3. A man walks to a town at the rate of 5 kilometres an hour and rides back at the rate of 7 kilometres an hour, how far has he walked, the whole time occupied having been 6 hours?
   a) 17 km b) 35 km c) 17.5 km d) 35.2 km
4. A man walks to a town at the rate of 8 kilometres an hour and rides back at the rate of 6 kilometres an hour, how far

has he walked, the whole time occupied having been 7 hours?

a) 14 km b) 18 km c) 22 km d) 24 km

5. A boy goes to school with the speed of 3 km an hour and returns with a speed of 2 km/hr. If he takes 5 hours in all, the distance in km between the village and the school is **[NDA 1990]**

a) 6 b) 7 c) 8 d) 9

6. A car completes a certain journey in 8 hours. It covers half the distance at 40 km/hr and the rest at 60 km/hr. The length of the journey in kilometres is **[Clerical Grade, 1991]**

a) 192 b) 384 c) 400 d) 420

Answers

1. c; **Hint:** $x = \frac{11}{2}$ km/hr, y = 10 km/hr, $T = 6 + \frac{12}{60} = \frac{31}{5}$ hours

Now applying the given rule, we have

$$\text{the required distance} = \frac{\frac{11}{2} \times 10}{\frac{11}{2} + 10} \times \frac{31}{5} = 22 \text{ km}$$

2. a 3. c 4. d 5. a

6. b; **Hint:** $\frac{1}{2}$ of the journey $= \frac{40 \times 60}{40 + 60} \times 8 = 192$ km

∴ The total length of the journey = 192 × 2 = 384 km

Rule 7

Theorem: ***If a person does a journey in T hours and the first half at S_1 km/hr and the second half at S_2 km/hr, then the***

$$distance = \frac{2 \times \text{Time} \times S_1 \times S_2}{S_1 + S_2}$$

Where, S_1 = Speed during first half and

S_2 = Speed during second half of journey

Illustrative Example

Ex.: A motor car does a journey in 10 hrs, the first half at 21 km/hr and the second half at 24 km/hr. Find the distance.

Soln: Detail Method: Let the distance be x km.

Then $\frac{x}{2}$ km is travelled at a speed of 21 km/hr and $\frac{x}{2}$ km at a speed of 24 km/hr.

Then, time taken to travel the whole journey

$$= \frac{x}{2 \times 21} + \frac{x}{2 \times 24} = 10 \text{ hrs}$$

$$\text{So, } x = \frac{2 \times 10 \times 21 \times 24}{21 + 24} = 224 \text{ km.}$$

Quicker Method: Applying the above theorem, we have

$$\text{Distance} = \frac{2 \times 10 \times 21 \times 24}{21 + 24} = 224 \text{ km.}$$

Note: Remember that half of the journey means half of the distance and not the time.

Exercise

1. A motor car does a journey in 12 hrs, the first half at 15 km/hr and the second half at 30 km/hr. Find the distance.

a) 190 km b) 280 km c) 240 km d) 210 km

2. A motor car does a journey in 6 hrs, the first half at 10 km/hr and the second half at 20 km/hr. Find the distance.

a) 90 km b) 80 km c) 60 km d) None of these

3. A motor car does a journey in 27 hrs, the first half at 14 km/hr and the second half at 13 km/hr. Find the distance.

a) 264 km b) 351 km c) 251 km d) 364 km

4. A motor car does a journey in 9 hrs, the first half at 12 km/hr and the second half at 15 km/hr. Find the distance.

a) 120 km b) 100 km c) 124 km d) 96 km

Answers

1. c 2. b 3. d 4. a

Rule 8

Theorem: ***The distance between two stations, A and B is D km. A train starts from A and moves towards B at an average speed of x km/hr. If an another train starts from B, t hours earlier than the train at A, and moves towards A at an average speed of y km/hr, then the distance from A, where the two trains will meet, is $\left[(D - ty)\left(\frac{x}{x+y}\right)\right]$ km.***

Illustrative Example

Ex.: The distance between two stations, Delhi and Amritsar, is 450 km. A train starts at 4 pm from Delhi and moves towards Amritsar at an average speed of 60 km/hr. Another train starts from Amritsar at 3.20 pm and moves towards Delhi at an average speed of 80 km/hr. How for from Delhi will the two trains meet and at what time?

Soln: Detail Method:

Suppose the trains meet at a distance of x km from Delhi. Let the trains from Delhi and Amritsar be A and B respectively. Then,

[Time taken by B to cover (450 – x) km]

– [Time taken by A to cover x km] $= \frac{40}{60}$ ..... (see note)

$$\frac{450 - x}{80} - \frac{x}{60} = \frac{40}{60}$$

$\therefore 3(450-x)-4x=160 \Rightarrow 7x=1190 \Rightarrow x=170.$

Thus, the trains meet at a distance of 170 km from Delhi.

Time taken by A to cover 170 km $=\left(\frac{170}{60}\right)$ hrs

$= 2$ hrs 50 min.

So, the trains meet at 6.50 p.m.

Note: RHS = 4 : 00 p.m. -- 3.20 p.m. = 40 minutes $=\frac{40}{60}$ hr

LHS comes from the fact that the train from Amritsar took 40 minutes more to travel up to the meeting point because it had started its journey at 3.20 p.m. whereas the train from Delhi had started its journey at 4 p.m. and the meeting time is the same for both the trains.

Quicker Method: Applying the above theorem, we have

the distance from Delhi at which two trains meet

$$=\left(450-\frac{40}{60}\times 80\right)\left(\frac{60}{60+80}\right)$$

$$=\frac{1190}{7}=170 \text{ km.}$$

Time taken by A to cover 170 km $=\left(\frac{170}{60}\right)$ hrs

$= 2$ hrs 50 min.

So, the trains meet at 6.50 p.m.

Exercise

1. At what distance from Delhi will a train, which leaves Delhi for Amritsar at 2.45 pm and goes at the rate of 50 km an hour, meet a train which leaves Amritsar for Delhi at 1.35 pm and goes at the rate of 60 km per hour, the distance between the two towns being 510 km?
   a) 150 km b) 170 km c) 200 km d) 210 km
2. The distance between two stations A and B is 900 km. Atrain starts from A and moves towards B at an average speed of 30 km/hr. Another train starts from B, 20 minutes earlier than the train at A, and moves towards A at an average speed of 40 km/hr. How far from A will the two trains meet?
   a) 380 km b) 320 km c) 240 km d) None of these
3. The distance between two stations A and B is 450 km. Atrain starts from A and moves towards B at an average speed of 15 km/hr. Another train starts from B, 20 minutes earlier than the train at A, and moves towards A at an average speed of 20 km/hr. How far from A will the two trains meet?
   a) 180 km b) 320 km c) 190 km d) 260 km
4. Two stations A and B are 110 km apart on a straight line. A train starts from A and travels towards B at 40 km/hr. Another train, starting from B, 2 hrs earlier, travels towards A at 50 km/hr. Find the distance from station A at which two trains meet?
   a) 5 km b) $5\frac{4}{9}$ km c) 4 km d) $4\frac{4}{9}$ km

Answers

1. c 2. a 3. c 4. d

Rule 9

Theorem: ***The distance between two stations A and B, is D km. A train starts from A and moves towards B at an average speed of x km/hr. If an another train starts from B, t hours later than the train at A, and moves towards A at an average speed of y km/hr, then the distance from A, where the two trains will meet, is*** $\left[(D+ty)\left(\frac{x}{x+y}\right)\right]$ ***km.***

Illustrative Example

Ex.: The distance between two stations, Delhi and Amritsar, is 450 km. A train starts at 3.10 pm from Delhi and moves towards Amritsar at an average speed of 20 km/hr. Another train starts from Amritsar at 4 pm and moves towards Delhi at an average speed of 60 km/hr. How far from Delhi will the two trains meet and at what times.

Soln: **Detail Method:**

Suppose the trains meet at a distance of x km from Delhi. Let the trains from Delhi and Amritsar be A and B respectively. Then,

[Time taken by A to cover x km] – [Time taken by B to cover (450 – x) km] $=\frac{50}{60}$ .... (see note)

$$\frac{x}{20}-\frac{450-x}{60}=\frac{50}{60}$$

$$3x-450+x=50$$

or, $4x=500 \therefore x=\frac{500}{4}=125$ km.

Thus, the trains meet at a distance of 125 km from Delhi.

Time taken by A to cover 125 km $=\frac{125}{20}=6$ hrs 15 minutes.

So the trains meet at 9.25 p.m.

Note: RHS = 4 p.m. – 3.10 p.m. = 50 minutes or, $\frac{50}{60}$ hrs.

LHS comes from the fact that the train from Delhi took 50 minutes more to travel up to the meeting point because it had started its journey at 3.10 p.m. whereas the train from Amritsar had started its journey at 4 p.m. and the meeting time is the same for both the trains.

Quicker Method: Applying the above theorem, we have

distance from Delhi at which two trains meet

$$= \left(450 + \frac{50}{60} \times 60\right)\left(\frac{20}{20+60}\right) = \frac{500}{4} = 125 \text{ km}$$

Time taken by A to cover 125 km

$$= \frac{125}{20} = 6 \text{ hrs } 15 \text{ min.}$$

So the trains meet at 9.25 pm.

Exercise

1. A train which travels at the uniform rate of 10 metres per second leaves Patna for Kanpur at 7 am. At what distance from Patna will it meet a train which leaves Kanpur for Patna at 7.20 am and travels one-third faster than it does, the distance from Patna to Kanpur being 68 km?
   a) 28 km b) 42 km c) 36 km d) 40 km
2. A train going 50 km an hour leaves Calcutta for Allahabad (900 km) at 9 pm. Another train going 70 km an hour leaves Allahabad for Calcutta at the same time, when and where will they pass each other?
   a) 375 km from Calcutta, 4.30 am
   b) 525 km from Calcutta, 4.30 pm
   c) 525 km from Allahabad, 4.20 am
   d) None of these
3. A starts from Allahabad to Kanpur and walks at the rate of 12 km an hour. B starts from Kanpur 2 hours later and walks towards Allahabad at the rate of 8 kilometres an hour, if they meet in 9 hours after B started, find the distance from Allahabad to Kanpur.
   a) 204 km b) 104 km c) 140 km d) 240 km
4. The distance between Delhi and Patna is 1000 km. A train leaves Delhi for Patna at 5 pm at 150 km/hr. Another train leaves Patna for Delhi at 6.30 pm at 100 km/hr. How far from Delhi will the two trains meet?
   a) 690 km b) 310 km c) 590 km d) 410 km
5. The distance between two stations A and B is 220 km. A train leaves A towards B at an average speed of 80 km per hr. After half an hour, another train leaves B towards A at an average speed of 100 km/hr. Find the distance from A of the point where the two trains meet.
   a) 180 km b) 120 km c) 160 km d) 80 km

Answers

1. c; **Hint:** $t = 7.20 \text{ am} - 7 \text{ am} = 20 \text{ minutes} = \frac{1}{3}$ hr

$$x = 10 \text{ m/sec} = 10 \times \frac{18}{5} = 36 \text{ km/hr}$$

$$y = 36 + 36 \times \frac{1}{3} = 48 \text{ km/hr}$$

Now, applying the given rule we have the required answer

$$= \left(68 + \frac{1}{3} \times 48\right)\left(\frac{36}{48+36}\right) = 36 \text{ km}$$

2. a; **Hint:** Here $t = 0$, and applying the given rule, we get

$$\frac{50}{5+70} \times 900 = 375 \text{ km from Calcutta}$$

$$\text{time} = \frac{375}{50} = 7\frac{1}{2} \text{ hrs} = 9 \text{ pm} + 7\frac{1}{2} \text{ hrs} = 4.30 \text{ am}$$

Note: This problem can also be solved by **Rule 9**. Try to solve by that rule also.

3. a; **Hint:** The distance from B where the two meet $= 9 \times 8 = 72$ km

$\therefore$ the distance from A where they meet $= D - 72$ km

Let the D be the distance between Allahabad and Kanpur.

$$(D + 2 \times 8)\left(\frac{12}{20}\right) = D - 72$$

or, $3D + 48 = 5D - 360$

or, $2D = 408$ $\therefore$ $D = 204$ km

4. a 5. b

Rule 10

Theorem: ***If the new speed of a person is $\frac{a}{b}$ of the usual speed, then the change in time taken to cover the same distance is $\left(\frac{b}{a} - 1\right)$ × usual time or, usual time is given by***

$$\left[\frac{\textit{Change in time}}{\left(\frac{b}{a} - 1\right)}\right] \textit{hrs.}$$

Note: A person improves his timing or becomes late depends on the –ve or +ve sign respectively of the above expression.

Illustrative Examples

Ex. 1: Walking $\frac{3}{4}$ of his usual speed, a person is 10 min late to his office. Find his usual time to cover the distance.

Soln: **Detail Method:** Let the usual time be x min.

Time taken at $\frac{3}{4}$ of the usual speed $= \frac{4x}{3}$ min

[Note: If the speed of a body is changed in the ratio a : b, then the ratio of the time taken changes in the ratio b : a.]

$\therefore \frac{4}{3}x - x = 10 \Rightarrow \frac{x}{3} = 10 \Rightarrow x = 30$ min.

Quicker Method: Applying the above theorem, we have

Usual time $= \frac{\text{Change in time}}{\left(\frac{4}{3}-1\right)} = \frac{10}{\frac{4}{3}-1} = \frac{10}{\frac{1}{3}} = 30$ minutes.

Here sign is (+ve) and hence in the question late time has been given.

Ex.2: Running $\frac{4}{3}$ of his usual speed, a person improves his timing by 10 minutes. Find his usual time to cover the distance.

Soln: **Direct Formula:** Applying the above theorem,

Usual time $= \frac{\text{Change in time}}{\frac{3}{4}-1} = \frac{10}{-\frac{1}{4}} = -40$ minutes

Here (–ve) sign shows that the person improves his timing. Hence in the question improved time has been given.

$\therefore$ correct answer = 40 minutes.

Exercise

1. Walking $\frac{2}{3}$ of his usual speed, a person is 15 min late to his office. Find his usual time to cover the distance.
   a) 30 minutes b) 25 minutes
   c) 15 minutes d) None of these
2. Walking $\frac{4}{5}$ of his usual speed, a person is 6 min late to his office. Find his usual time to cover the distance.
   a) 12 minutes b) 18 minutes
   c) 24 minutes d) Data inadequate
3. Walking $\frac{1}{3}$ of his usual speed, a person is 12 min late to his office. Find his usual time to cover the distance.
   a) 36 minutes b) 18 minutes
   c) 6 minutes d) Can't be determined
4. Walking $\frac{3}{5}$ of his usual speed, a person is 18 min late to his office. Find his usual time to cover the distance.
   a) 27 minutes b) 25 minutes
   c) 24 minutes d) 20 minutes
5. By walking at $\frac{3}{4}$ of his usual speed, a man reaches office 20 minutes later than usual. His usual time is **[Railways, 1991]**
   a) 30 minutes b) 60 minutes
   c) 75 minutes d) $1\frac{1}{2}$ hours

Answers

1. a 2. c 3. c 4. a 5. b

Rule 11

Theorem: ***If a train travelling x km an hour leaves a place and t hours later another train travelling y km an hour, where y > x, in the same direction, then they will be together after travelling $\left[\frac{t(xy)}{y-x}\right]$ km from the starting place.***

Illustrative Example

Ex.: A train travelling 25 km an hour leaves Delhi at 9 a.m. and another train travelling 35 km an hour starts at 2 p.m. in the same direction. How many km from Delhi will they be together?

Soln: **Detail Method I:**

Let the required distance be x km

From the question,

$\frac{x}{25} - \frac{x}{35} = 2 \text{ p.m.} - 9 \text{ a.m.} = 5$ hours.

or, $\frac{x(35-25)}{35 \times 25} = 5$

$\therefore x = \frac{35 \times 25 \times 5}{10} = 437\frac{1}{2}$ km.

Detail Method II:

The first train has a start of 25 × 5 km and the second train gains (35 – 25) or 10 km per hour.

$\therefore$ the second train will gain 25 × 5 km in $\frac{25 \times 5}{10}$ or $12\frac{1}{2}$ hours.

$\therefore$ the required distance from Delhi $= 12\frac{1}{2} \times 35$ km

$= 437\frac{1}{2}$ km.

Quicker Method: Applying the above theorem, we have

$$\text{the required distance} = \frac{25\times35\times(2\text{ p.m.}-9\text{ a.m.})}{35-25}$$

$$= \frac{25\times35\times5}{10} = 437\frac{1}{2}\text{ km.}$$

Exercise

1. A train leaves Calcutta at 7.30 am and travels 40 km an hour, another train leaves Calcutta at noon and travels 64 km an hour, when and where will the second train overtake the first?
   a) 480 km, 7.30 pm b) 480 km, 2.30 pm
   c) 840 km, 7.30 pm d) 480 km, 6.30 pm

2. A man starts at 7 am and travels at the rate of $4\frac{3}{4}$ km an hour. At 8.15 am a coach starts from the same place and follows the man travelling at the rate of $6\frac{1}{2}$ km an hour, at what o'clock will the coach overtake the man?
   a) 12.50 pm b) $11.38\frac{4}{7}$ am
   c) $12.53\frac{4}{7}$ pm d) None of these

3. A starts from Delhi to Alwar (24 km) at 6 am walking 4 kilometres an hour. B starts from Delhi an hour later and reaches Alwar one hour before A, where did they meet?
   a) 12 km from Delhi
   b) 10 km from Alwar
   c) 10 km from Delhi
   d) Data inadequate

Answers

1. a; **Hint:** t = 7.30 am – 12 noon = $4\frac{1}{2} = \frac{9}{2}$ hrs

Now applying the given rule, we have

$$\frac{64\times40}{60-40}\times\frac{9}{2} = 480\text{ km}$$

$$\text{Required time} = \frac{480}{64} = \frac{15}{2} = 7\frac{1}{2}\text{ hrs}$$

$\therefore$ required answer = 12 noon + $7\frac{1}{2}$ hrs = 7.30 pm.

2. b; **Hint:** Here t = 7 am – 8.15 am = $1\frac{1}{4}$ hrs

$\therefore$ Applying the given rule we have the

$$\text{required distance} = \left(\frac{\frac{19}{4}\times\frac{13}{2}}{\frac{19}{4}-\frac{13}{12}}\right)\frac{5}{4} = \frac{19\times13\times5}{4\times2\times7}\text{ km}$$

$$\text{or, time} = \frac{19\times13\times5}{4\times2\times7}\times\frac{4}{19} = \frac{65}{14}\text{ hrs}$$

$\therefore$ required answer = 7 am + $4\frac{9}{14}$ hrs

$$= 11\text{ am} + \frac{9\times60}{14}\text{ min} = 11.38\frac{4}{7}\text{ am.}$$

3. a; **Hint:** Here, $x = 4$ km/hr

Time taken by A to cover the distance of 24 km = $\frac{24}{4} = 6$ hrs

Time taken by B to cover the same distance = 6 – 2 = 4 hrs (according to the question)

$\therefore y = \frac{24}{4} = 6$ km/hr and $t = 1$ hr

Now applying the given rule, we have

$$\text{the required answer} = \frac{6\times4}{6-4}\times1 = 12\text{ km from Delhi.}$$

Rule 12

Theorem: ***If two persons A and B start from a place walking at x km/hr and y km/hr respectively, at the end of t hours, when they are moving in same direction and x < y, they will be (y - x) t km apart.***

Illustrative Example

Ex.: Two men A and B start from a place P walking at 3 km and 3.5 km an hour respectively. How many km will they be apart at the end of 3 hours, if they walk in same direction?

Soln: Applying the above theorem, we have the required distance = (3.5 – 3) 3 = 1.5 km.

Exercise

1. One man takes 100 steps a minute, each 5 decimetres long, another walks 4 km an hour, if they start together, how soon will one of them be 75 metres ahead to the other?
   a) 3 min 30 sec b) 2 min 40 sec
   c) 4 min 20 sec d) 4 min 30 sec

2. Two men A and B start from a place P walking at 4 km and 5 km an hour respectively. How many km will they be apart at the end of 4 hours, if they walk in same direction?
   a) 3 km b) 4 km c) 2 km d) 4.5 km

3. Two men A and B start from a place P walking at 6 km and 8 km an hour respectively. How many km will they be apart at the end of 2 hours, if they walk in same direction?
   a) 2 km b) 8 km c) 6 km d) 4 km

Answers

1. d; **Hint:** 100 steps = 5 × 100 = 500 dm = 50 m

Speed of the first man = 50 m per minute = $\frac{50}{60} = \frac{5}{6}$ m/sec

$\therefore \frac{5}{6}$ m/sec = $\frac{5}{6} \times \frac{18}{5} = 3$ km/hr

Now, applying the given rule, we have

$(4-3)t = \frac{75}{1000} \times 60$

$\therefore t = \frac{9}{2} = 4$ min 30 sec

2. b 3. d

Rule 13

Theorem: ***If two persons A and B start from a place walking at x km/hr and y km/hr respectively, at the end of t hours, when they are moving in opposite directions, they will be (x + y)t km apart.***

Illustrative Example

Ex.: Two men A and B start from a place P walking at 3 km and 3.5 km an hour respectively. How many km will they be apart at the end of 3 hrs, if they walk in the opposite directions?

Soln: Applying the above theorem, we have the required distance = (3.5 + 3) × 3 = 19.5 km.

Exercise

1. Two men A and B start from a place P walking at 2 km and 5 km an hour respectively. How many km will they be apart at the end of 2 hrs, if they walk in the opposite directions?
   a) 7 km b) 8 km c) 12 km d) 14 km
2. Two men A and B start from a place P walking at 3.5 km and 4.5 km an hour respectively. How many km will they be apart at the end of 3 hrs, if they walk in the opposite directions?
   a) 21 km b) 24 km
   c) 18 km d) Data inadequate
3. Two men A and B start from a place P walking at 5.5 km and 6.5 km an hour respectively. How many km will they be apart at the end of 5 hrs, if they walk in the opposite directions?
   a) 60 km b) 50 km c) 45 km d) 30 km

Answers

1. d 2. b 3. a

Rule 14

Theorem: ***Two men A and B walk from P to Q, a distance of 'D' km, at the speed of 'a' km/hr and 'b' km/hr respectively. If B reaches Q, returns immediately and meets A at R, then the distance travelled by A (or distance from P to R) is $2D\left(\frac{a}{a+b}\right)$ km and the distance travelled by B (or PQ + QR) is $2D\left(\frac{b}{a+b}\right)$ km.***

Illustrative Example

Ex.: Two men A and B walk from P to Q, a distance of 21 km, at 3 and 4 km an hour respectively. B reaches Q, returns immediately and meets A at R. Find the distance from P to R.

Soln: **Detail Method:** When B meets A at R, B has walked the distance PQ + QR and A the distance PR. That is, both of them have together walked twice the distance from P to Q, i.e. 42 km.

Now the rates of A and B are 3 : 4 and they have walked 42 km.

Hence the distance PR travelled by A

$= \frac{3}{7}$ of 42 km = 18 km

Quicker Method: When the ratio of speeds of A and B is a : b, then in this case:

Distance travelled by A = 2 × Distance of two points $\left(\frac{a}{a+b}\right)$

and distance travelled by B = 2 × Distance of two points $\left(\frac{b}{a+b}\right)$

Thus, distance travelled by A (PR)

$= 2 \times 21\left(\frac{3}{3+4}\right) = 18$ km.

Exercise

1. Two men A and B walk from P to Q, a distance of 18 km, at 4 and 5 km an hour respectively. B reaches Q, returns immediately and meets A at R. Find the distance from P to R.
   a) 15 km b) 16 km
   c) 12 km d) Can't be determined
2. Two men A and B walk from P to Q, a distance of 22 km, at 5 and 6 km an hour respectively. B reaches Q, returns immediately and meets A at R. Find the distance from P to R.
   a) 16 km b) 18 km c) 20 km d) 15 km

3. Two men A and B walk from P to Q, a distance of 30 km, at 7 and 8 km an hour respectively. B reaches Q, returns immediately and meets A at R. Find the total distance travelled by B.
a) 35 km b) 36 km c) 32 km d) 33 km
4. Two men A and B walk from P to Q, a distance of 24 km, at 11 and 13 km an hour respectively. B reaches Q, returns immediately and meets A at R. Find the total distance travelled by B.
a) 26 km b) 28 km c) 30 km d) 32 km

Answers

1. b 2. c 3. c 4. a

Rule 15

Theorem: ***If two persons A and B start at the same time in opposite directions from two points and after passing each other they complete the journeys in 'a' and 'b' hrs respectively then A's speed : B's speed =*** $\sqrt{b}:\sqrt{a}$.

Illustrative Example

Ex.: A man sets out to cycle from Delhi to Rohtak, and at the same time another man starts from Rohtak to cycle to Delhi. After passing each other they complete their journeys in $3\frac{1}{3}$ and $4\frac{4}{5}$ hours respectively. At what rate does the second man cycle if the first cycle at 8 km per hour?

Soln: If two persons (or train) A and B start at the same time in opposite direction from two points, and arrive at the point a and b hrs respectively after having met, then

A's rate : B's rate = $\sqrt{b}:\sqrt{a}$ (from the theorem)

Thus in the above case

$$\frac{\text{1st man's rate}}{\text{2nd man's rate}}=\frac{\sqrt{4\frac{4}{5}}}{\sqrt{3\frac{1}{3}}}=\frac{6}{5}$$

$\therefore$ 2nd man's rate $=\frac{5}{6}\times 8=6\frac{2}{3}$ km/hr.

Exercise

1. A man sets out to cycle from Delhi to Rohtak, and at the same time another man starts from Rohtak to cycle to Delhi. After passing each other they complete their journeys in 4 and 9 hours respectively. At what rate does the second man cycle if the first cycle at 9 km per hour?
a) 4 km b) 6 km c) 8 km d) Data inadequate
2. A man sets out to cycle from Delhi to Rohtak, and at the same time another man starts from Rohtak to cycle to Delhi. After passing each other they complete their journeys in 4 and 16 hours respectively. At what rate does the second man cycle if the first cycle at 18 km per hour?
a) 8 km b) 9 km c) 12 km d) 6 km
3. A man sets out to cycle from Delhi to Rohtak, and at the same time another man starts from Rohtak to cycle to Delhi. After passing each other they complete their journeys in 16 and 25 hours respectively. At what rate does the second man cycle if the first cycle at 25 km per hour?
a) 21 km b) 18 km c) $12\frac{1}{2}$ km d) 20 km
4. A man sets out to cycle from Delhi to Rohtak, and at the same time another man starts from Rohtak to cycle to Delhi. After passing each other they complete their journeys in 9 and 16 hours respectively. At what rate does the second man cycle if the first cycle at 16 km per hour?
a) 12 km b) 15 km c) 8 km d) 6 km

Answers

1. b 2. b 3. d 4. a

Rule 16

Report of guns

Theorem: ***If two guns were fired from the same place at an interval of t_1 minutes but a person in a train approaching the place hears the second report t_2 minutes after the first, then the speed of the train, supposing that sound travels at 330 metres per second, is*** $1188\left(\frac{t_1-t_2}{t_2}\right)$ ***km/hr.***

Illustrative Example

Ex.: Two guns were fired from the same place at an interval of 13 minutes but a person in a train approaching the place hears the second report 12 minutes 30 seconds after the first. Find the speed of the train, supposing that sound travels at 330 metres per second.

Soln: Detail Method: It is easy to see that the distance travelled by the train in 12 min. 30 seconds could be travelled by sound in (13 min – 12 min 30 seconds) = 30 seconds

$\therefore$ the train travels 330 × 30 metres in $12\frac{1}{2}$ min.

$\therefore$ the speed of the train per hour $=\frac{330\times 30\times 2\times 60}{25\times 1000}$

$=\frac{1188}{25}$ or $47\frac{13}{25}$ km.

Quicker Method: Applying the above theorem, we have

$$\text{speed of the train per hour} = 1188\left(\frac{13-\frac{25}{2}}{\frac{25}{2}}\right)$$

$$= \frac{1188}{25} \text{ or } 47\frac{13}{25} \text{ km.}$$

Exercise

1. Two bullets were fired at a place at an interval of 12 minutes. A person approaching the firing point in his car hears the two sounds at an interval of 11 minutes 40 seconds. The speed of sound is 330 m/sec. What is the speed of the car?

   a) $\frac{1188}{35}$ km/hr b) $\frac{594}{35}$ km/hr

   c) $\frac{1881}{35}$ km/hr d) Data inadequate

2. Two bullets were fired at a place at an interval of 34 minutes. A person approaching the firing point in his car hears the two sounds at an interval of 33 minutes. The speed of sound is 330 m/sec. What is the speed of the car?

   a) 72 km b) 36 km
   c) 45 km d) 51 km

3. Two bullets were fired at a place at an interval of 28 minutes 30 seconds. A person approaching the firing point in his car hears the two sounds at an interval of 27 minutes. The speed of sound is 330 m/sec. What is the speed of the car?

   a) 44 km b) 66 km c) 64 km d) 54 km

4. Two bullets were fired at a place at an interval of 38 minutes. A person approaching the firing point in his car hears the two sounds at an interval of 36 minutes. The speed of sound is 330 m/sec. What is the speed of the car?

   a) 66 km b) 49 km c) 99 km d) 98 km

Answers

1. a 2. b 3. b 4. a

Rule 17

Theorem: ***If two runners A and B cover the same distance at the rate of x km/hr and y km/hr respectively, then the distance travelled, when A takes t hours longer than the B,***

is $\left[\frac{xy}{y-x}\times t\right]$ ***km. or***

Distance =

$$\frac{\textit{Multiplication of speeds}}{\textit{Difference of Speeds}} \times \textit{Difference in time to cover the distance}$$

(See Note)

Illustrative Example

Ex.: Two runners cover the same distance at the rate of 15 km and 16 km per hour respectively. Find the distance travelled when one takes 16 minutes longer than the other.

Soln: **Detail Method:** Let the distance be x km.

Time taken by the first runner = $\frac{x}{15}$ hrs

Time taken by the second runner = $\frac{x}{16}$ hrs

Now, $\frac{x}{15}-\frac{x}{16}=\frac{16}{60}$

or, $\frac{x(16-15)}{15\times16}=\frac{16}{60}$

$\therefore x=\frac{16}{60}\times15\times16=64$ km.

Quicker Method: Applying the above theorem,

Distance $=\frac{15\times16}{16-15}\times\frac{16}{60}=64$ km.

Note: The above theorem may be written as, "If two runners A and B cover the same distance at the rate of x km/hr and y km/hr respectively, then the distance travelled, when B takes t hours less than the A, is $\left[\frac{xy}{y-x}\times t\right]$ km."

See the following example,

Ex.: Two cars run to a place at the speeds of 45 km/hr and 60 km/hr respectively. If the second car takes 5 hrs less than the first for the journey, find the length of the journey.

Soln: Distance $=\frac{45\times60}{60-45}\times5=900$ km.

Exercise

1. Two men start together to walk a certain distance, one at $3\frac{3}{4}$ km an hour, and the other at 3 km an hour. The former arrives half an hour before the latter. Find the distance.

   a) $7\frac{1}{2}$ km b) 15 km c) $15\frac{1}{2}$ d) 7 km

2. Two bicyclists do the same journey by travelling respectively at the rates of 9 and 10 km an hour. Find the length of the journey when one takes 32 minutes longer than the other.

a) 38 km b) 36 km c) 48 km d) 24 km

3. A man walks from A to B and back in a certain time at the rate of $3\frac{1}{2}$ km per hour. But if he had walked from A to B at the rate of 3 km an hour and back from B to A at the rate of 4 km an hour, he would have taken 5 minutes longer. Find the distance between A and B.
a) 14 km b) 12 km c) 6 km d) 7 km

4. A car starts from P for Q travelling 20 kilometres an hour. $1\frac{1}{2}$ hours later another car starts from P and travelling at the rate of 30 kilometres an hour reaches Q $2\frac{1}{2}$ hrs before the first car. Find the distance from P to Q.
a) 90 km b) 150 km c) 240 km d) 270 km

Answers

1. a 2. c

3. d; Man's speed in the first case = $3\frac{1}{2}$ km/hr

Man's average speed in the second case

$= \frac{2\times3\times4}{3+4} = \frac{24}{7}$ km/hr (See **Rule 2**)

Now, applying the given rule,

we have $2(AB) = \frac{\frac{7}{2}\times\frac{24}{7}}{\frac{7}{2}-\frac{24}{7}}\times\frac{5}{60} = 14$ km

$\therefore AB = \frac{A}{2} = 7$

4. c; **Hint:** Here $t = 2\frac{1}{2}+1\frac{1}{2} = 4$ hours

Now applying the given rule we have

the required distance $= \frac{30\times20}{30-20}\times4 = 240$ km

Rule 18

Carriage driving in a fog

Theorem: ***A carriage driving in a fog passed a person who was walking at the rate of x km/hr in the same direction. If he could see the carriage for t hours and it was visible to him upto a distance of 'd' km, then the distance travelled by the carriage in t hours is $(xt+d)$ metres and speed of the carriage is $\left(x+\frac{d}{t}\right)$ km/hr.***

Illustrative Example

Ex.: A carriage driving in a fog passed a man who was walking at the rate of 3 km an hour in the same direction. He could see the carriage for 4 minutes and it was visible to him upto a distance of 100 m. What was the speed of the carriage?

Soln: **Detail Method:** The distance travelled by the man in 4 minutes $= \frac{3\times1000}{60}\times4 = 200$ metres.

$\therefore$ distance travelled by the carriage in 4 minutes $= (200+100) = 300$ metres.

$\therefore$ speed of the carriage $= \frac{300}{4}\times\frac{60}{1000}$ km per hour

$= 4\frac{1}{2}$ km per hour

Quicker Method: Applying the above theorem, we have

$x = 3$ km/hr

$t = 4$ minutes $= \frac{4}{60} = \frac{1}{15}$ hrs

$d = 100$ metres $= \frac{100}{1000} = \frac{1}{10}$ km

Now, applying the above rule, we have

the required answer $= 3 + \frac{\frac{1}{10}}{\frac{1}{15}} = 3+\frac{3}{2} = 4\frac{1}{2}$ km/hr

Exercise

1. A carriage driving in a fog passed a man who was walking at the rate of 6 km an hour in the same direction. He could see the carriage for 8 minutes and it was visible to him upto a distance of 200 m. What was the speed of the carriage?
a) 9 km/hr b) $7\frac{1}{2}$ km/hr c) 7 km/hr d) $8\frac{1}{2}$ km/hr

2. A carriage driving in a fog passed a man who was walking at the rate of 5 km an hour in the same direction. He could see the carriage for 6 minutes and it was visible to him upto a distance of 120 m. What was the speed of the carriage?
a) $6\frac{1}{5}$ km/hr b) $5\frac{1}{5}$ km/hr
c) $6\frac{3}{5}$ km/hr d) None of these

3. A carriage driving in a fog passed a man who was walk-

ing at the rate of 6 km an hour in the same direction. He could see the carriage for 12 minutes and it was visible to him upto a distance of 150 m. What was the speed of the carriage?

a) $7\frac{3}{4}$ km/hr b) $6\frac{3}{4}$ km/hr c) $5\frac{3}{4}$ km/hr d) 8 km/hr

4. A carriage driving in a fog passed a man who was walking at the rate of 8 km an hour in the same direction. He could see the carriage for 15 minutes and it was visible to him upto a distance of 500 m. What was the speed of the carriage?

a) 10 km/hr b) $8\frac{1}{2}$ km/hr
c) 12 km/hr d) None of these

Answers

1. b 2. a 3. b 4. a

Rule 19

Theorem: ***A person takes x hours to walk to a certain place and ride back. However, he could have gained t hours, if he had covered both ways by riding, then the time taken by him to walk both ways is (x + t) hours.***

or,

Both ways walking = One way walking and one way riding time + gain in time

Illustrative Example

Ex.: A man takes 8 hours to walk to a certain place and ride back. However, he could have gained 2 hrs, if he had covered both ways by riding. How long would he have taken to walk both ways?

Soln: **Detail Method:**

Walking time + Riding time = 8 hrs ...... (1)
2 Riding time = 8 – 2 = 6 hrs ..... (2)
2 × (1) – (2) gives the result
2 × walking time = 2 × 8 – 6 = 10 hrs.
∴ both ways walking will take 10 hrs.

Quicker Approach: Two ways riding saves a time of 2 hrs. It simply means that one way riding takes 2 hrs less than one way walking. It further means that one way walking takes 2 hrs more than one way riding. Thus, both way walking will take 8 + 2 = 10 hrs.

Quicker Method: Applying the above theorem, we have

Both ways walking = One way walking and one way riding time + Gain in time = 8 + 2 = 10 hrs.

Exercise

1. A man takes 6 hrs 30 min in walking to a certain place and riding back. He would have gained 2 hrs 10 min by riding both ways. How long would he take to walk both ways?

a) 8 hrs b) 8 hrs 20 min
c) 4 hrs 20 min d) 8 hrs 40 min

2. A man takes 5 hrs 42 min in walking to a certain place and riding back. He would have gained 1 hr 18 min by riding both ways. How long would he take to walk both ways?

a) 6 hours b) 6 hours 50 min
c) 7 hours d) Data inadequate

3. A man takes 7 hours in walking to a certain place and riding back. He would have gained 3 hours by riding both ways. How long would he take to walk both ways?

a) 10 hours b) 12 hours c) 8 hours d) 9 hours

4. A man takes 8 hrs 32 min in walking to a certain place and riding back. He would have gained 2 hrs 14 min by riding both ways. How long would he take to walk both ways?

a) 10 hrs b) 10 hrs 36 min
c) 10 hrs 46 min d) 10 hrs 40 min

Answers

1. d 2. c 3. a 4. c

Rule 20

Theorem: ***A man takes x hours to walk to a certain place and ride back. However, if he walks both ways he needs t hours more, then the time taken by him to ride both ways is (x – t) hours.***

Illustrative Example

Ex.: A man takes 12 hrs to walk to a certain place and ride back. However, if he walks both ways he needs 3 hours more. How long would he have taken to ride both ways?

Soln: Applying the above theorem, we have
required time = 12 – 3 = 9 hrs.

Exercise

1. I walk a certain distance and ride back by taking $6\frac{1}{4}$ hours altogether. I could walk both ways in $7\frac{3}{4}$ hours. How long would it take me to ride both ways?

a) $4\frac{1}{4}$ hrs b) $5\frac{3}{4}$ hrs c) $4\frac{3}{4}$ hrs d) $4\frac{1}{2}$ hrs

2. A man takes $5\frac{2}{3}$ hrs to walk to a certain place and ride back. However, if he walks both ways he needs $1\frac{3}{4}$ hours more. How long would he have taken to ride both ways?

a) $3\frac{1}{12}$ hrs b) $3\frac{11}{12}$ hrs c) $4\frac{11}{12}$ hrs d) $4\frac{1}{12}$ hrs

3. A man takes 8 hrs to walk to a certain place and ride back. However, if he walks both ways he needs 1 hour more. How long would he have taken to ride both ways?
a) 9 hrs b) 7 hrs c) 6 hrs d) None of these

Answers

1. c; **Hint:** Here $t = 7\frac{3}{4} - 6\frac{1}{4} = \frac{3}{2}$ hrs

Applying the given rule, we have

the required answer $= 6\frac{1}{4} - \frac{3}{2} = 4\frac{3}{4}$ hrs

2. b 3. b

Rule 21

Theorem: ***A person A leaves a point P and reaches Q in x hours. If another person B leaves the point Q, t hours later than A and reaches the point P in y hours, then the time in which A meets to B is*** $(y+t)\left(\frac{x}{x+y}\right)$ ***hrs.***

Illustrative Example

Ex.: A man leaves a point P and reaches the point Q in 4 hours. Another man leaves the point Q, 2 hours later and reaches the point P in 4 hours. Find the time in which first man meets to the second man.

Soln: **Detail Method:** Let the distance PQ = A km.
And they meet x hrs after the first man starts.

Average speed of first man $= \frac{A}{4}$ km/hr

Average speed of second man $= \frac{A}{4}$ km/hr

Distance travelled by first man $= \frac{Ax}{4}$ km

They meet x hrs after the first man starts. The second man, as he starts 2 hrs late, meets after $(x-2)$ hrs from his start. Therefore, the distance travelled by the second man $= \frac{A(x-2)}{4}$ km

Now, $\frac{Ax}{4} + \frac{A(x-2)}{4}$ km = A

or, $2x - 2 = 4$

$\therefore x = 3$ hrs

Quicker Method: Applying the above theorem, we have

$x = 4$ hrs, $y = 4$ hrs and $t = 2$ hrs

$\therefore$ the required time $= (4+2)\left(\frac{4}{4+4}\right) = 3$ hrs.

Note: The above example may be written as "A man leaves a point P at 6 a.m. and reaches the point Q at 10 a.m. Another man leaves the point Q at 8 a.m. and reaches the point P at 12 noon. At what time do they meet?"

Soln: Applying the above theorem, we have

x = 10 am − 6 am = 4 hrs

y = 12 noon − 8 am = 4 hrs

First man starts at 6 am and second man starts at 8 am. Therefore, second man starts (8 am − 6 am = 2 hrs) late than the first.

$\therefore$ t = 2 hrs.

$\therefore$ the required time $= (4+2)\left(\frac{4}{4+4}\right) = 3$ hours

Hence, they meet at 6 am + 3 am = 9 am.

Exercise

1. A man leaves a point P and reaches the point Q in 3 hours. Another man leaves the point Q, 1 hour later and reaches the point P in 3 hours. Find the time in which first man meets to the second man.
a) 2 hrs b) 3 hrs
c) $1\frac{1}{2}$ hrs d) Data inadequate
2. A man leaves a point P and reaches the point Q in 5 hours. Another man leaves the point Q, 2 hours later and reaches the point P in 7 hours. Find the time in which first man meets to the second man.
a) 4 hrs b) $4\frac{1}{4}$ hrs c) $3\frac{3}{4}$ hrs d) $3\frac{1}{4}$ hrs
3. A man leaves a point P and reaches the point Q in 6 hours. Another man leaves the point Q, 4 hours later and reaches the point P in 6 hours. Find the time in which first man meets to the second man.
a) 6 hrs b) 5 hrs c) 4 hrs d) 3 hrs
4. A man leaves a point P at 8 am and reaches the point Q at 12 noon. Another man leaves the point Q at 9 am and reaches the point P at 1 pm. At what time do they meet?
a) 10.30 am b) 11.30 am
c) 10 am d) They will never meet

Answers

1. a 2. c 3. b 4. a

Rule 22

Theorem: ***A person A leaves a point P and reaches Q in x hours. If another person B leaves the point Q, t hours earlier than A and reaches the point P in y hours, then the time in which A meets to B is*** $(y-t)\left(\frac{x}{x+y}\right)$ ***hrs.***

Illustrative Example

Ex.: A man leaves a point P and reaches the point Q in 4 hours. Another man leaves the point Q. 2 hours earlier and reaches the point P in 4 hours. Find the time in which the first man meets to the second man.

Soln: **Detail Method:** Let the distance PQ = A km

And they meet x hrs after the first man starts.

Average speed of first man $= \frac{A}{4}$ km/hr

Average speed of second man $= \frac{A}{4}$ km/hr

Distance travelled by first man $= \frac{Ax}{4}$ km

They meet x hrs after the first man starts. The second man, as he starts 2 hrs late, meets after $(x+2)$ hrs from his start. Therefore,

the distance travelled by the second man

$$= \frac{A(x+2)}{4} \text{ km}$$

Now, $\frac{Ax}{4} + \frac{A(x+2)}{4}$ km = A

$2x + 2 = 4 \quad \therefore x = 1$ hr

Quicker Method: Applying the above theorem, we have

$x = 4$ hrs, $y = 4$ hrs and $t = 2$ hrs

$\therefore$ the required time $= (4-2)\left(\frac{4}{4+4}\right) = 1$ hr

Note: The above example may be written as "A man leaves a point P at 8 am and reaches the point Q at 12 noon. Another man leaves the point Q at 6 am and reaches the point P at 10 am. At what time do they meet?"

Soln: Applying the above theorem, we have

$x =$ 12 noon – 8 a.m. = 4 hrs

$y =$ 10 a.m. – 6 a.m. = 4 hrs

First man starts at 8 a.m. and second man starts at 6 a.m. Therefore, second man starts (8 a.m. – 6 a.m. = 2 hrs) earlier than the first.

$\therefore t = 2$ hrs

$\therefore$ the required time $= (4-2)\left(\frac{4}{4+4}\right) = 1$ hr

Hence, they meet at 8 a.m. + 1 hour = 9 a.m.

Exercise

1. A man leaves a point P and reaches the point Q in 3 hours. Another man leaves the point Q. 1 hour earlier and reaches the point P in 3 hours. Find the time in which the first man meets to the second man.
   a) 1 hr b) 2 hrs
   c) 30 min d) Can't be determined
2. A man leaves a point P and reaches the point Q in 6 hours. Another man leaves the point Q. 2 hours earlier and reaches the point P in 6 hours. Find the time in which the first man meets to the second man.
   a) 1 hr b) 3 hrs c) 2 hrs d) $1\frac{1}{2}$ hrs
3. A man leaves a point P and reaches the point Q in 5 hours. Another man leaves the point Q. 3 hours earlier and reaches the point P in 7 hours. Find the time in which the first man meets to the second man.
   a) 1 hr b) $1\frac{2}{3}$ hrs
   c) $2\frac{1}{3}$ hrs d) None of these

Answers

1. a 2. c 3. b

Rule 23

Theorem: Speed and time taken are inversely proportional.

Therefore, $S_1T_1 = S_2T_2 = S_3T_3$ ....

Where, S_1, S_2, S_3 ***... are the speeds***

and T_1, T_2, T_3 ***.... are the time taken to travel the same distance.***

Note: Also see **Rule 32.**

Illustrative Example

Ex.: A person covers a distance in 40 minutes if he runs at a speed of 45 km per hour on an average. Find the speed at which he must run to reduce the time of journey to 30 minutes.

Soln: Required speed $= 45 \times 40 = S_2 \times 30$

$$\therefore S_2 = \frac{45 \times 40}{30} = 60 \text{ km/hr.}$$

Exercise

1. A person covers a distance in 8 minutes if he runs at a speed of 9 km per hour on an average. Find the speed at which he must run to reduce the time of journey to 6 minutes.
   a) 12 km/hr b) 10 km/hr c) 9 km/hr d) None of these
2. A person covers a distance in 24 minutes if he runs at a speed of 27 km per hour on an average. Find the speed at which he must run to reduce the time of journey to 18 minutes.
   a) 27 km/hr b) 36 km/hr c) 45 km/hr d) 48 km/hr
3. A person covers a distance in 12 minutes if he runs at a speed of $13\frac{1}{2}$ km per hour on an average. Find the speed at which he must run to reduce the time of journey to 9

minutes.

a) 16 km/hr b) 21 km/hr c) 24 km/hr d) 18 km/hr

4. A person covers a distance in 16 minutes if he runs at a speed of 18 km per hour on an average. Find the speed at which he must run to reduce the time of journey to 12 minutes.

a) 16 km/hr b) 18 km/hr
c) 24 km/hr d) None of these

Answers

1. a 2. b 3. d 4. c

Rule 24

Theorem: ***Without any stoppage if a person travels a certain distance at an average speed of x km/hr, and with stoppages he covers the same distance at an average speed of y km/hr, then he stops*** $\left[60\left(\frac{x-y}{y}\right)\right]$ ***minutes per hour.***

Or

Time of rest per hour = $\left(\frac{\text{Difference of speed}}{\text{Speed without stoppage}}\right)\times 60$ ***minutes.***

Illustrative Example

Ex.: Without any stoppage a person travels a certain distance at an average speed of 80 km/hr, and with stoppages he covers the same distance at an average speed of 60 km/hr. How many minutes per hour does he stop?

Soln: **Detail Method:** Let the total distance be x km.

Time taken at the speed of 80 km/hr = $\frac{x}{80}$ hrs.

Time taken at the speed of 60 km/hr = $\frac{x}{60}$ hrs.

$\therefore$ he rested for $\left(\frac{x}{60}-\frac{x}{80}\right)$ hrs = $\frac{20x}{60\times 80}=\frac{x}{240}$ hrs

$\therefore$ his rest per hour = $\frac{x}{240}\div\frac{x}{60}=\frac{x}{240}\times\frac{60}{x}=\frac{1}{4}$ hrs

= 15 minutes.

Quicker Method: Applying the above theorem,
Time of rest per hour

$$=\left(\frac{\text{Difference of speed}}{\text{Speed without stoppage}}\right)\times 60 \text{ minutes}$$

$$=\left(\frac{80-60}{80}\right)\times 60=\frac{1}{4}\times 60 \text{ hr} = 15 \text{ minutes.}$$

Exercise

1. Without any stoppage a person travels a certain distance at an average speed of 40 km/hr, and with stoppages he covers the same distance at an average speed of 30 km/hr. How many minutes per hour does he stop?

a) 7 min b) $7\frac{1}{2}$ min c) 15 min d) None of these

2. Without any stoppage a person travels a certain distance at an average speed of 42 km/hr, and with stoppages he covers the same distance at an average speed of 28 km/hr. How many minutes per hour does he stop?

a) 15 minutes b) 14 minutes c) 28 minutes d) 20 minutes

3. Without any stoppage a person travels a certain distance at an average speed of 27 km/hr, and with stoppages he covers the same distance at an average speed of 15 km/hr. How many minutes per hour does he stop?

a) 26 min 40 sec b) 30 min
c) 26 min d) 16 min 40 sec

4. Without any stoppage a person travels a certain distance at an average speed of 15 km/hr, and with stoppages he covers the same distance at an average speed of 12 km/hr. How many minutes per hour does he stop?

a) 15 min b) 12 min c) 16 min d) 18 min

Answers

1. c 2. d 3. a 4. b

Rule 25

Theorem: ***A person has to cover a distance of x km in t hours. If he covers nth part of the journey in mth of the total time, then his speed should be*** $\left(\frac{x}{t}\times\frac{1-n}{1-m}\right)$ ***km/hr to cover the remaining distance in the remaining time.***

Illustrative Example

Ex.: A person has to cover a distance of 80 km in 10 hrs. If he covers half of the journey in $\frac{3}{5}$th time, what should be his speed to cover the remaining distance in the time left?

Soln: **Detail Method:**

Left distance = $80\left(1-\frac{1}{2}\right)=40$ km

Time left = $10\left(1-\frac{3}{5}\right)=4$ hrs

$\therefore$ required speed = $\frac{40}{4}=10$ km/hr

Quicker Method: Applying the above theorem, we have

the required speed $= \dfrac{\left(1-\frac{1}{2}\right)80}{\left(1-\frac{3}{5}\right)10} = 10$ km/hr.

Exercise

1. A person has to cover a distance of 15 km in 3 hrs. If he covers $\frac{1}{3}$ of the journey in $\frac{2}{3}$rd time, what should be his speed to cover the remaining distance in the time left?
   a) 12 km/hr b) 10 km/hr c) 5 km/hr d) 15 km/hr
2. A person has to cover a distance of 60 km in 15 hrs. If he covers $\frac{1}{4}$ of the journey in $\frac{2}{5}$th time, what should be his speed to cover the remaining distance in the time left?
   a) 5 km/hr b) 9 km/hr c) 6 km/hr d) 12 km/hr
3. A person has to cover a distance of 40 km in 5 hrs. If he covers $\frac{1}{5}$ of the journey in $\frac{3}{5}$th time, what should be his speed to cover the remaining distance in the time left?
   a) 18 km/hr b) 15 km/hr c) 16 km/hr d) 14 km/hr
4. Laxman has to cover a distance of 6 km in 45 min. If he covers one half of the distance in $\frac{2}{3}$rd time, what should be his speed in km/hr to cover the remaining distance in the remaining time? **[Bank PO Exam, 1991]**
   a) 12 b) 16 c) 3 d) 8

Answers

1. b 2. a 3. c 4. a

Rule 26

Theorem: ***A man rode out a certain distance by train at the rate of x km/hr and walked back at the rate of y km/hr. If the whole journey took 't' hours, then the distance he rode is $\left(\frac{xy}{x+y}\right)t$ km.***

Illustrative Example

Ex.: A man rode out a certain distance by train at the rate of 25 km an hour and walked back at the rate of 4 km per hour. The whole journey took 5 hours and 48 minutes. What distance did he ride?

Soln: Detail Method: Let the distance be x km.

Then time spent in journey by train $= \frac{x}{25}$ hrs.

And time spent in journey by walking $= \frac{x}{4}$ hrs.

Therefore, $\frac{x}{25}+\frac{x}{4} = 5$ hrs 48 minutes.

or, $\frac{29x}{100} = 5\frac{48}{60} = \frac{29}{5}$ $\therefore x = \frac{100}{5} = 20$ km

Quicker Method: Applying the above theorem, the required distance

$$= 5\frac{48}{60} \times \frac{25\times 4}{25+4} = \frac{29}{5} \times \frac{25\times 4}{29} = 20 \text{ km}$$

Exercise

1. A man rode out a certain distance by train at the rate of 30 km an hour and walked back at the rate of 5 km per hour. The whole journey took 7 hrs. What distance did he ride?
   a) 30 km b) 25 km c) 28 km d) 35 km
2. A man rode out a certain distance by train at the rate of 13 km an hour and walked back at the rate of 12 km per hour. The whole journey took 5 hours. What distance did he ride?
   a) 32 km b) 26 km c) 31.2 km d) 26.5 km
3. A man rode out a certain distance by train at the rate of 15 km an hour and walked back at the rate of 12 km per hour. The whole journey took 9 hours. What distance did he ride?
   a) 48 km b) 60 km c) 36 km d) None of these

Answers

1. a 2. c 3. b

Rule 27

Theorem: ***A man travels D km in x hours, partly by air and partly by train. If he had travelled all the way by air, he would have saved $\frac{a}{b}$ of the time he was in train and would have arrived at his destination y hours early, then the distance he travelled by air is $D\left[\frac{x-\frac{y}{a/b}}{x-y}\right]$ km.***

Illustrative Example

Ex.: A man travels 360 km in 4 hrs, partly by air and partly by train. If he had travelled all the way by air, he would have saved $\frac{4}{5}$ of the time he was in train and would have arrived at his destination 2 hours early. Find the distance he travelled by air and train.

Soln: **Detail Method:** $\frac{4}{5}$ of total time in train = 2 hours.

$\therefore$ Total time in train $= \frac{2\times5}{4} = \frac{5}{2}$ hrs.

$\therefore$ Total time spent in air $= 4 - \frac{5}{2} = \frac{3}{2}$ hrs.

By the given hypothesis, if 360 km is covered by air, then time taken is (4 – 2 =) 2 hrs.

$\therefore$ when $\frac{3}{2}$ hrs is spent in air, distance covered

$$= \frac{360}{2} \times \frac{3}{2} = 270 \text{ km.}$$

$\therefore$ Distance covered by train = 360 – 270 = 90 km.

Quicker Method: Applying the above theorem,

$$\text{Distance covered by air} = 360\left[\frac{4 - \frac{5\times2}{4}}{4-2}\right]$$

$$= 360 \times \frac{3}{4} = 270 \text{ km.}$$

$\therefore$ distance covered by train = 360 – 270 = 90 km.

Exercise

1. A man travels 480 km in 6 hrs, partly by air and partly by train. If he had travelled all the way by air, he would have saved $\frac{3}{4}$ of the time he was in train and would have arrived at his destination 3 hours early. Find the distance he travelled by train.
   a) 320 km b) 160 km c) 260 km d) 220 km
2. A man travels 120 km in 5 hrs, partly by air and partly by train. If he had travelled all the way by air, he would have saved $\frac{2}{3}$ of the time he was in train and would have arrived at his destination 2 hours early. Find the distance he travelled by air.
   a) 80 km b) 40 km c) 85 km d) 90 km
3. A man travels 400 km in 8 hrs, partly by air and partly by train. If he had travelled all the way by air, he would have saved $\frac{4}{5}$ of the time he was in train and would have arrived at his destination 4 hours early. Find the ratio between distances travelled by the air and the train.
   a) 1 : 3 b) 3 : 1 c) 2 : 1 d) 1 : 2

Answers

1. b 2. a

3. b; **Hint:** Distance travelled by the air $= 400\left(\frac{8 - \frac{4}{4/5}}{8-4}\right)$

$= 300$ km

$\therefore$ distance travelled by the train = 400 – 300 = 100 km

$\therefore$ required ratio = 3 : 1.

Rule 28

Theorem: ***One aeroplane starts t hours later than the scheduled time from a place D km away from its destination. To reach the destination at the schedued time the pilot has to increase the speed by 'p' km/hr. Then the plane takes*** $\left[\frac{\sqrt{t^2 + \frac{4Dt}{p}} + t}{2}\right]$ ***hours in the normal case. And the normal speed of the aeroplane is*** $\left(\frac{2D}{\sqrt{t^2 + \frac{4Dt}{p}} + t}\right)$ ***km/hr.***

Note: Normal case indicates the original case in which speed of the aeroplane has not been changed.

Illustrative Example

Ex.: One aeroplane started 30 minutes later than the scheduled time from a place 1500 km away from its destination. To reach the destination at the scheduled time the pilot had to increase the speed by 250 km/hr. What was the speed of the aeroplane per hour during the journey?

Soln: **Detail Method:**

Let it take x hrs in second case.

Then speed $= \frac{1500}{x} = \frac{1500}{x + \frac{1}{2}} + 250$

or, $\frac{1500\left(x + \frac{1}{2}\right) - 1500x}{x\left(x + \frac{1}{2}\right)} = 250$

or, $750 = 250x\left(x + \frac{1}{2}\right)$ or, $x^2 + \frac{x}{2} - 3 = 0$

or, $2x^2 + x - 6 = 0$

or, $2x^2+4x-3x-6=0$

or, $x(2x-3)+2(2x-3)=0$

or, $(x+2)(2x-3)=0$

$\therefore x=-2, \frac{3}{2}$

Therefore, the plane takes $\frac{3}{2}$ hrs in second case, ie

$\frac{3}{2}+\frac{1}{2}=2$ hrs in normal case. Thus, normal speed =

$\frac{1500}{2}=750$ km/hr.

Quicker Method: Applying the above theorem, Normal speed of the aeroplane

$$=\frac{2\times1500}{\sqrt{\frac{1}{4}+\frac{4\times1500\times1}{2\times250}}+\frac{1}{2}}=\frac{2\times1500}{\sqrt{\frac{49}{4}}+\frac{1}{2}}=750 \text{ km/hr.}$$

Exercise

1. One aeroplane started 1 hour later than the scheduled time from a place 3000 km away from its destination. To reach the destination at the scheduled time the pilot had to increase the speed by 500 km/hr. What was the speed of the aeroplane per hour during the journey?
   a) 1500 km/hr b) 1000 km/hr
   c) 850 km/hr d) None of these
2. One aeroplane started $1\frac{1}{2}$ hrs later than the scheduled time from a place 2400 km away from its destination. To reach the destination at the scheduled time the pilot had to increase the speed by 800 km/hr. What was the speed of the aeroplane per hour during the journey?
   a) 1600 km/hr b) 800 km/hr
   c) 1200 km/hr d) 1550 km/hr
3. One aeroplane started 30 minutes later than the scheduled time from a place 1800 km away from its destination. To reach the destination at the scheduled time the pilot had to increase the speed by 300 km/hr. What was the speed of the aeroplane per hour during the journey?
   a) 600 km/hr b) 800 km/hr c) 1200 km/hr d) 900 km/hr

Answers

1. b 2. b 3. d

Rule 29

Theorem: ***A train leaves the station t hours before the scheduled time. The driver decreases its speed by p km/hr. At the next station D km away, the train reached on time. Then the train takes*** $\left(\frac{\sqrt{t^2+\frac{4Dt}{p}}-t}{2}\right)$ ***hours in the normal case (original case) and the normal speed (original speed) of the train is*** $\left(\frac{2D}{\sqrt{t^2+\frac{4Dt}{p}}-t}\right)$ ***km/hr.***

Illustrative Example

Ex.: A train leaves the station 1 hour before the scheduled time. The driver decreases its speed by 4 km/hr. At the next station 120 km away, the train reached on time. Find the original speed of the train.

Soln: **Detail Method:** Let it takes x hours in second case.

Then speed = $\frac{120}{x}=\frac{120}{x-1}-4$

or, $\frac{120}{x-1}-\frac{120}{x}=4$ or, $\frac{120(x-x+1)}{(x-1)x}=4$

or, $120=4x^2-4x$ or, $4x^2-4x-120=0$

or, $x^2-x-30=0$

or, $x^2-6x+5x-30=0$

or, $x(x-6)+5(x-6)=0$

or, $x=-5, 6$ $\therefore x=6$

Therefore, the train takes 6 hours in second case, ie $(6-1=5)$ hours in original case.

$\therefore$ Original speed = $\frac{120}{5}=24$ km/hr

Quicker Method: Applying the above theorem,

Original speed of the train = $\frac{2\times120}{\sqrt{1^2+\frac{4\times120\times1}{4}}-1}$

$=\frac{2\times120}{11-1}=24$ km/hr.

Exercise

1. A train leaves the station 1 hour before the scheduled time. The driver decreases its speed by 50 km/hr. At the next station 300 km away, the train reached on time. Find the original speed of the train.
   a) 100 km/hr b) 150 km/hr

c) 200 km/hr d) None of these

2. A train leaves the station $1\frac{1}{2}$ hours before the scheduled time. The driver decreases its speed by 80 km/hr. At the next station 240 km away, the train reached on time. Find the original speed of the train.
a) 180 km/hr b) 160 km/hr c) 200 km/hr d) 120 km/hr

3. A train leaves the station 30 minutes before the scheduled time. The driver decreases its speed by 30 km/hr. At the next station 180 km away, the train reached on time. Find the original speed of the train.
a) 140 km/hr b) 125 km/hr c) 120 km/hr d) 100 km/hr

Answers

1. b 2. b 3. c

Rule 30

Theorem: ***When a person travels equal distance at speed V_1 and V_2 km/hr, his average speed is x km/hr. But when he travels at these speeds for equal times his average speed is y km/hr, then the values of V_1 and V_2 are $\left(y+\sqrt{y(y-x)}\right)$ km/hr and $\left(y-\sqrt{y(y-x)}\right)$ km/hr respectively. And the difference of the two speeds is given by $2\sqrt{y(y-x)}$ km/hr.***

Illustrative Example

Ex.: When a man travels equal distance at speeds V_1 and V_2 km/hr, his average speed is 4 km/hr. But when he travels at these speeds for equal times his average speed is 4.5 km/hr. Find the difference of the two speeds and also find the values of V_1 and V_2.

Soln: **Detail Method:** Suppose the equal distance = D km
Then time taken with V_1 and V_2 speeds are

$\frac{D}{V_1}$ hrs and $\frac{D}{V_2}$ hrs respectively.

$$\therefore \text{ average speed} = \frac{\text{Total distance}}{\text{Total time}} = \frac{2D}{\frac{D}{V_1}+\frac{D}{V_2}}$$

$$= \frac{2V_1V_2}{V_1+V_2} = 4 \text{ km/hr}$$

In second case,

$$\text{average speed} = \frac{V_1+V_2}{2} = 4.5 \text{ km/hr}$$

That is; $V_1+V_2=9$ and $V_1V_2=18$

Now, $(V_1-V_2)^2 = (V_1+V_2)^2 - 4V_1V_2$
$= 81-72=9$

$\therefore V_1-V_2 = 3$ km/hr and $V_1+V_2=9$ km/hr

$\therefore V_1=6$ km/hr and $V_2=3$ km/hr

Quicker Method: Applying the above theorem,

$$V_1-V_2 = 2\sqrt{(4.5)(4.5-4)} = 3 \text{ km/hr}$$

$$V_1 = y+\sqrt{y(y-x)}$$

$$= 4.5+\frac{3}{2} = 6 \text{ km/hr and}$$

$$V_2 = y-\sqrt{y(y-x)}$$

$$= 4.5-\frac{3}{2} = 3 \text{ km/hr.}$$

Exercise

1. When a man travels equal distance at speeds V_1 and V_2 km/hr, his average speed is 8 km/hr. But when he travels at these speeds for equal times his average speed is 9 km/hr. Find the difference of the two speeds.
a) 6 km/hr b) 5 km/hr c) 4 km/hr d) 8 km/hr

2. When a man travels equal distance at speeds V_1 and V_2 km/hr, his average speed is 6 km/hr. But when he travels at these speeds for equal times his average speed is 8 km/hr. Find the difference of the two speeds.
a) 7 km/hr b) 6 km/hr c) 8 km/hr d) 10 km/hr

3. When a man travels equal distance at speeds V_1 and V_2 km/hr, his average speed is 5 km/hr. But when he travels at these speeds for equal times his average speed is 9 km/hr. Find the difference of the two speeds.
a) 10 km/hr b) 12 km/hr c) 14 km/hr d) 8 km/hr

Answers

1. a 2. c 3. b

Rule 31

Theorem: ***A person travels for T_1 hours at the speed of R_1 km/hr and for T_2 hours at the speed of R_2 km/hr. At the end of it, if he finds that he has covered 'f' of the total distance, then his average speed, to cover the remaining distance in T hours, should be $\left(\frac{(R_1T_1+R_2T_2)\left(\frac{1}{f}-1\right)}{T}\right)$ km/hr.***

Illustrative Example

Ex.: A person travels for 3 hrs at the speed of 40 km/hr and for 4.5 hrs at the speed of 60 km/hr. At the end of it, he finds that he has covered $\frac{3}{5}$ of the total distance. At what average speed should he travel to cover the remaining distance in 4 hrs?

Soln: **Detail Method:** Total distance covered in $(3+4.5)$ hrs.
$=3\times 40+4.5\times 60=390$ km.

Now, since $\frac{3}{5}$ of the distance = 390 km

$\therefore \frac{2}{5}$ of the distance $= 390\times\frac{5}{3}\times\frac{2}{5}=260$ km.

$\therefore$ average speed for the remaining distance =

$\frac{260}{4}=65$ km/hr.

Quicker Method: Applying the above theorem, we have
the average speed for the remaining distance

$$=\frac{(R_1T_1+R_2T_2)\left(\frac{1}{f}-1\right)}{T}$$

$$=\frac{(40\times 3+60\times 4.5)\left(\frac{5}{3}-1\right)}{4}$$

$$=\frac{390\times 2}{4\times 2}=65 \text{ km/hr.}$$

Exercise

1. A person travels for 4 hrs at the speed of 30 km/hr and for 6 hrs at the speed of 40 km/hr. At the end of it, he finds that he has covered $\frac{3}{4}$ of the total distance. At what average speed should he travel to cover the remaining distance in 3 hrs?
   a) 40 km/hr b) 35 km/hr c) 45 km/hr d) 60 km/hr
2. A person travels for 2 hrs at the speed of 15 km/hr and for 3 hrs at the speed of 20 km/hr. At the end of it, he finds that he has covered $\frac{2}{3}$ of the total distance. At what average speed should he travel to cover the remaining distance in 5 hrs?
   a) 8 km/hr b) 9 km/hr c) 6 km d) None of these
3. A person travels for 3 hrs at the speed of 12 km/hr and for 4 hrs at the speed of 15 km/hr. At the end of it, he finds that he has covered $\frac{4}{5}$ of the total distance. At what average speed should he travel to cover the remaining distance in 6 hrs?
   a) 6 km/hr b) 8 km/hr c) 4 km/hr d) 5 km/hr

Answers

1. a 2. b 3. c

Rule 32

Theorem: ***If a person A walking at the rate of S_1 km/hr, takes t_1 hours to cover a distance and another person B, walking at the rate of S_2 km/hr, takes t_2 hours to cover the same distance, then $S_1t_1=S_2t_2$ = Distance or, $S_1t_1=S_2t_2$ = constant.***

Thus we see that both speed and time are inversely proportional to each other. That is, if the speed increases to 4 times, the time will decrease to $\frac{1}{4}$ times.

Application of the above theorem,

Case I: If A takes 8 hours to cover a distance and B is four times faster than A, then what time will B take to cover the same distance?

We have, $S_1t_1=S_2t_2$

$\Rightarrow S_1t_1=4S_1t_2$

$\Rightarrow t_2=\frac{t_1}{4}=\frac{8}{4}=2$ hrs.

Case II: If A takes 8 hours to cover a distance and he is 4 times faster than B, then what time will B take to cover the same distance?

We have,

$S_1t_1=S_2t_2 \Rightarrow 4S_2\times 8=S_2\times t_2$

$\therefore t_2=32$ hrs.

Case III: If B is 20% faster than A, then what time will he take to travel the distance which A travels in 20 minutes?

We have, $S_1t_1=S_2t_2$

$S_1\times 20=\frac{120}{100}S_1\times t_2$

$\therefore t_2=\frac{20\times 100}{120}=\frac{50}{3}=16\frac{2}{3}$ min

Case IV: If B takes 30% less time than A, to cover the same distance. What should be the speed of B if A walks at a rate of 7 km/hr?

Again, $7\times t_1=S_2\left(\frac{100-30}{100}\right)t_1$

$\therefore S_2=\frac{7\times 100}{70}=10$ km/hr.

Exercise

1. If A takes 6 hours to cover a distance and B is 2 times faster than A, then what time will B take to cover the

same distance?

a) 2 hrs b) 3 hrs c) $\frac{3}{2}$ hrs d) 4 hrs

2. If A takes 4 hours to cover a distance and B is four times faster than A, then what time will B take to cover the same distance?

a) 2 hrs b) $2\frac{1}{2}$ hrs c) 1 hr d) $1\frac{1}{2}$ hrs

3. If A takes 3 hours to cover a distance and he is 3 times faster than B, then what time will B take to cover the same distance?

a) 9 hrs b) 6 hrs c) 8 hrs d) None of these

4. If A takes 4 hours to cover a distance and he is 2 times faster than B, then what time will B take to cover the same distance?

a) 6 hrs b) 8 hrs c) 9 hrs d) None of these

5. If B is 25% faster than A, then what time will he take to travel the distance which A travels in 25 minutes?

a) 25 minutes b) 20 minutes
c) 30 minutes d) None of these

6. If B is 30% faster than A, then what time will he take to travel the distance which A travels in 26 minutes?

a) 15 minutes b) 20 minutes
c) 25 minutes d) None of these

7. If B takes 10% less time than A, to cover the same distance. What should be the speed of B if A walks at a rate of 9 km/hr?

a) 10 km/hr b) 15 km/hr c) 20 km/hr d) 5 km/hr

8. If B takes 40% less time than A, to cover the same distance. What should be the speed of B if A walks at a rate of 15 km/hr?

a) 15 km/hr b) 20 km/hr
c) 25 km/hr d) None of these

Answers

1. b 2. c 3. a 4. b 5. b 6. b 7. a
8. c

Rule 33

Theorem: ***A person travels x_1 km by steamer, x_2 km by train and x_3 km by horse. It took T hours. If the rate of train is 'n' times that of the horse and 'm' times that of the steamer, then the rate of the train is $\left(\frac{mx_1 + x_2 + nx_3}{T}\right)$ km/hr, rate of the horse is $\frac{1}{n}\left(\frac{mx_1 + x_2 + nx_3}{T}\right)$ km/hr and the rate of the steamer is $\frac{1}{m}\left(\frac{mx_1 + x_2 + nx_3}{T}\right)$ km/hr.***

Illustrative Example

Ex.: A person travelled 120 km by steamer, 450 km by train and 60 km by horse. It took 13 hours 30 minutes. If the rate of the train is 3 times that of the horse and 1.5 times that of the steamer, find the rate of the train per hour.

Soln: **Detail Method:** Suppose the speed of horse = x km/hr. Then speed of the train = $3x$ km/hr and speed of the steamer = $2x$ km/hr

Now, $\frac{120}{2x} + \frac{450}{3x} + \frac{60}{x} = 13.5$ hours

(Since 13 hrs 30 minutes = 13.5 hrs)

or, $\frac{360 + 900 + 360}{6x} = 13.5$

$\therefore x = \frac{1620}{6 \times 13.5} = 20$

$\therefore$ Speed of train = $3x = 3 \times 20 = 60$ km/hr

Quicker Method: Applying the above theorem, we have

Speed of the train = $\frac{1.5 \times 120 + 450 + 3 \times 60}{13.5}$

= 60 km/hr.

Speed of the horse = $\frac{1}{3} \times 60 = 20$ km/hr.

Speed of the steamer = $\frac{1}{3/2} \times 60 = \frac{2}{3} \times 60 = 40$ km/hr.

Exercise

1. A person travelled 60 km by steamer, 225 km by train and 30 km by horse. It took 15 hours. If the rate of the train is 3 times that of the horse and 2 times that of the steamer. Find the speed of the train per hour.

a) $\frac{29}{2}$ km/hr b) $\frac{29}{3}$ km/hr
c) 29 km/hr d) None of these

2. A person travelled 50 km by steamer, 60 km by train and 60 km by horse. It took 15 hours. If the rate of the train is 4 times that of the horse and 3 times that of the steamer. Find the rate of the steamer.

a) 10 km/hr b) 30 km/hr c) 15 km/hr d) 18 km/hr

3. A person travelled 25 km by steamer, 40 km by train and 30 km by horse. It took 7 hours. If the rate of the train is 4 times that of the horse and 2 times that of the steamer. Find the rate of the horse.

a) 15 km/hr b) $7\frac{1}{2}$ km/hr c) 30 km/hr d) 16 km/hr

Answers

1. c 2. a 3. b

Rule 34

Theorem: ***A person covers a certain distance on scooter. Had he moved x_1 km/hr faster, he would have taken t_1 hours less. If he had moved x_2 km/hr slower, he would have taken t_2 hours more, then the original speed (S) is***

$$\left[\frac{x_1x_2(t_1+t_2)}{t_2x_1-t_1x_2}\right]$$ ***km/hr and the distance is given by***

$$\left[\frac{t_1\times S(S+x_1)}{x_1}\right]$$ ***km.***

Illustrative Example

Ex.: A man covers a certain distance on scooter. Had he moved 3 km/hr faster, he would have taken 40 minutes less. If he had moved 2 km/hr slower, he would have taken 40 minutes more. Find the distance (in km) and original speed.

Soln: **Detail Method:** Suppose the distance is D km and the initial speed is x km/hr.

Then, we have $\frac{D}{x+3}=\frac{D}{x}-\frac{40}{60}$ and $\frac{D}{x-2}=\frac{D}{x}+\frac{40}{60}$

or, $\frac{D}{x}-\frac{D}{x+3}=\frac{2}{3}$ or, $\frac{3D}{x(x+3)}=\frac{2}{3}$ .... (1)

and $\frac{D}{x-2}-\frac{D}{x}=\frac{2}{3}$ or, $\frac{2D}{x(x-2)}=\frac{2}{3}$ .... (2)

From (1) and (2) we have $\frac{3D}{x(x+3)}=\frac{2D}{x(x-2)}$

or, $3(x-2)=2(x+3)$ or, $3x-6=2x+6$

$\therefore x=12$ km/hr

Now, if we put this value in (1) we get

$$D=\frac{2}{3}\times\frac{12\times15}{3}=40 \text{ km.}$$

Quicker Method I: Applying the above theorem, we have the required speed

$$=\frac{(3\times2)\left(\frac{2}{3}+\frac{2}{3}\right)}{\frac{2}{3}\times3-\frac{2}{3}\times2}=\frac{8\times3}{2}=12 \text{ km/hr and}$$

$$\text{the required distance}=\frac{12\times\frac{2}{3}(12+3)}{3}=40 \text{ km.}$$

Quicker Method II: In the above question when time reduced in arrival (40 minutes) is equal to the time increased in arrival (40 minutes) then

Speed

$$=\frac{2\times(\text{Increase in speed}\times\text{Decrease in speed})}{\text{Difference in increase and decrease in speeds}}$$

$$=\frac{2\times(3\times2)}{(3-2)}=12 \text{ km/hr}$$

$$\text{Now, Distance}=\frac{(12+3)\times(12-2)}{(12+3)-(12-2)}\times \text{Diff. between arrival time}$$

$$=\frac{15\times10}{5}\times\frac{40+40}{60}=40 \text{ km.}$$

Note: 1. 40 minutes late and 40 minutes earlier make a difference of $40+40=80$ minutes $=\frac{80}{60}$ hrs.

2. This question is the special case of the above theorem.

Here, $(t_1=t_2)$, put this into the formula and get the above method.

Exercise

1. A man covers a certain distance on scooter. Had he moved 4 km/hr faster, he would have taken 30 minutes less. If he had moved 3 km/hr slower, he would have taken 30 minutes more. Find the original speed.
   a) 24 km/hr b) 20 km/hr c) 28 km/hr d) 18 km/hr
2. A man covers a certain distance on scooter. Had he moved 6 km/hr faster, he would have taken 30 minutes less. If he had moved 4 km/hr slower, he would have taken 1 hr 30 minutes more. Find the original speed.
   a) 7 km/hr b) $6\frac{5}{7}$ km/hr
   c) $6\frac{6}{7}$ km/hr d) None of these
3. A man covers a certain distance on scooter. Had he moved 8 km/hr faster, he would have taken 30 minutes less. If he had moved 6 km/hr slower, he would have taken 15 minutes more. Find the distance (in km) and original speed.
   a) 36 km/hr b) 24 km/hr c) 18 km/hr d) 30 km/hr

Answers

1. a 2. c 3. a

Rule 35

Theorem: ***A thief is spotted by a policeman from a distance of d km. When the policeman starts the chase, the thief also***

starts running. Assuming the speed of the thief x kilometres an hour, and that of the policeman y kilometres an hour, then the thief will run before he is overtaken $= d\left(\frac{x}{y-x}\right)$ ***km.***

Or,

The distance covered by the thief before he gets caught

$$= \frac{\textit{Lead of distance}}{\textit{Relative speed}} \times \textit{Speed of thief}$$

Illustrative Example

Ex.: A thief is spotted by a policeman from a distance of 200 metres. When the policeman starts the chase, the thief also starts running. Assuming the speed of the thief 10 kilometres an hour, and that of the policeman 12 kilometres an hour, how far will have the thief run before he is overtaken?

Soln: **Detail Method:** Relative speed = 12 – 10 = 2 km/hr

$\therefore$ the thief will be caught after $= \frac{0.2}{2} = \frac{1}{10}$ hr.

$\therefore$ distance covered by the thief before he gets caught

$$= 10 \times \frac{1}{10} = 1 \text{ km}$$

Quicker Method: Applying the above theorem, we have

the required distance $= \frac{0.2 \text{ km}}{2 \text{ km/hr}} \times 10$ km/hr = 1 km.

Exercise

1. A policeman goes after a thief who has 100 metres' start. If the policeman runs a kilometre in 8 minutes, and the thief a kilometre in 10 minutes, how far will the thief have gone before he is overtaken?
   a) 350 metres b) 400 metres
   c) 450 metres d) 460 metres
2. A thief is spotted by a policeman from a distance of 400 metres. When the policeman starts the chase, the thief also starts running. Assuming the speed of the thief 8 kilometres an hour, and that of the policeman 10 kilometres an hour, how far will have the thief run before he is overtaken?
   a) 2 km b) 1 km c) $1\frac{2}{5}$ km d) $1\frac{3}{5}$
3. A thief is spotted by a policeman from a distance of 150 metres. When the policeman starts the chase, the thief also starts running. Assuming the speed of the thief 12 kilometres an hour, and that of the policeman 18 kilometres an hour, how far will have the thief run before he is overtaken?
   a) 150 metres b) 200 metres
   c) 300 metres d) 1 km
4. A thief is spotted by a policeman from a distance of 300 metres. When the policeman starts the chase, the thief also starts running. Assuming the speed of the thief 12 kilometres an hour, and that of the policeman 15 kilometres an hour, how far will have the thief run before he is overtaken?
   a) 1 km b) 1.2 km c) 1.5 km d) 2 km

Answers

1. b; **Hint:** d = 100 metres $= \frac{100}{1000} = \frac{1}{10}$ km

x = 1 kilometre in 10 minutes = 6 km/hr

y = 1 kilometre in 8 minutes $= 7\frac{1}{2}$ km/hr

Now applying the given rule, we have

$$\left(\frac{6}{\frac{15}{2}-6}\right) \times \frac{100}{1000} = \frac{4}{10} \text{ km} = 400 \text{ metres}$$

2. d 3. c 4. b

Rule 36

To find the average speed

Theorem: ***If a moving body travels*** $x_1, x_2, x_3, \ldots x_n$***, metres, moving with different speeds*** $S_1, S_2, S_3, \ldots, S_n$ ***metres per second in time*** $T_1, T_2, T_3, \ldots, T_n$ ***seconds respectively, then it is necessary to calculate the average speed of the body throughout the journey. If the average speed is denoted by*** V_a***, then,***

$$V_a = \frac{\textit{Total distance travelled}}{\textit{Total time taken}}$$

$$\textit{Case I} = \frac{x_1 + x_2 + x_3 + \ldots + x_n}{T_1 + T_2 + T_3 + \ldots + T_n}$$

$$\textit{Case II} = \frac{S_1T_1 + S_2T_2 + S_3T_3 + \ldots + S_nT_n}{T_1 + T_2 + T_3 + \ldots + T_n}$$

$$\textit{Case III} = \frac{x_1 + x_2 + x_3 + \ldots + x_n}{\frac{x_1}{S_1} + \frac{x_2}{S_2} + \frac{x_3}{S_3} + \ldots + \frac{x_n}{S_n}}$$

Illustrative Examples

Ex. 1: A train travels 225 km in 3.5 hours and 370 km in 5 hours. Find the average speed of train.

Soln: Applying the above theorem (case-I)

Average speed $= \frac{225 + 370}{3.5 + 5} = 70$ km/hr

Ex. 2: A car during its journey travels 30 minutes at a speed of 40 km/hr, another 45 minutes at a speed of 60 km/hr and 2 hours at a speed of 70 km/hour. Find its average speed of the car.

Soln: Applying the above theorem (case-II)

Average speed

$$=\frac{\left(\frac{30}{60}\times 40\right)+\left(\frac{45}{60}\times 60\right)+(2\times 70)}{\frac{30}{60}+\frac{45}{60}+2}=63\text{ km/hr.}$$

Ex. 3: A man walks 3 km at a speed of 3 km/hr, runs 4 km at a speed of 4 km/hr and goes by bus another 16 km. Speed of the bus is 16 km/hr. If the speed of the bus is considered as the speed of the man, find the average speed of the man.

Soln: Applying the above theorem (case-III)

$$\text{Average speed}=\frac{3+4+16}{\frac{3}{3}+\frac{4}{4}+\frac{16}{16}}=\frac{23}{3}=7\frac{2}{3}\text{ km/hr.}$$

Exercise

1. A train travels 150 km in 5 hours and 250 km in 3 hours. Find the average speed of train.
   a) 50 km/hr b) 60 km/hr c) 100 km/hr d) 80 km/hr
2. A train travels 220 km in 6 hours and 340 km in 2 hours. Find the average speed of train.
   a) 80 km/hr b) 70 km/hr c) 50 km/hr d) 35 km/hr
3. A car during its journey travels 2 hrs at a speed of 25 km/hr, another 4 hrs at a speed of 30 km/hr and 4 hours at a speed of 35 km/hour. Find its average speed of the car.
   a) 62 km/hr b) 31 km/hr c) 29 km/hr d) 58 km/hr
4. A car during its journey travels 40 minutes at a speed of 30 km/hr, another 50 minutes at a speed of 60 km/hr and 1 hour at a speed of 30 km/hr. Find its average speed of the car.
   a) 40 km/hr b) 35 km/hr
   c) 45 km/hr d) None of these
5. A man walks 6 km at a speed of $1\frac{1}{2}$ km/hr, runs 8 km at a speed of 2 km/hr and goes by bus another 32km. Speed of the bus is 8 km/hr. If the speed of the bus is considered as the speed of the man, find the average speed of the man.
   a) $15\frac{1}{3}$ km/hr b) $7\frac{2}{3}$ km/hr
   c) $3\frac{5}{6}$ km/hr d) None of these

Answers

1. a 2. b 3. b 4. a 5. c

Rule 37

Theorem: ***A train does a journey without stopping in T hours. If it had travelled x km an hour faster, it would have done the journey in t hours, then the origninal speed is*** $\left(\frac{t}{T-t}\times x\right)$ ***km/hr and the length of the journey is given by*** $\left(\frac{Tt}{T-t}\right)x$ ***km.***

Illustrative Example

Ex.: A train does a journey without stopping in 8 hours. If it had travelled 5 km an hour faster, it would have done the journey in 6 hours 40 minutes. What is its original speed?

Soln: Applying the above theorem, we have, the required original speed

$$=\frac{6\frac{2}{3}}{8-6\frac{2}{3}}\times 5=\frac{\frac{20}{3}}{\frac{4}{3}}=25\text{ km/hr.}$$

Exercise

1. A car finishes a journey in 10 hours at the speed of 80 km/hr. If the same distance is to be covered in eight hours how much more speed does the car have to gain? **[BSRB Delhi PO, 2000]**
   a) 8 km/hr b) 10 km/hr c) 20 km/hr d) 16 km/hr
2. A train does a journey without stopping in 8 hours. If it had travelled 6 km an hour faster, it would have done the journey in 6 hours. What is its original speed?
   a) 20 km/hr b) 15 km/hr c) 21 km/hr d) 18 km/hr
3. A train does a journey without stopping in 5 hours. If it had travelled 3 km an hour faster, it would have done the journey in 2 hours. What is its original speed?
   a) 10 km/hr b) 5 km/hr
   c) 2 km/hr d) None of these
4. A train does a journey without stopping in 9 hours. If it had travelled 8 km an hour faster, it would have done the journey in 6 hours. What is its original speed?
   a) 27 km/hr b) 16 km/hr c) 24 km/hr d) 18 km/hr

Answers

1. c; **Hint:** $\frac{8}{10-8}\times x=80 \quad \therefore x=20$ km/hr

2. d 3. c 4. b

Rule 38

Theorem: ***If a car travels a distance of D km in T hours partly at a speed of x km/hr and partly at y km/hr, then the distance travelled at a speed of x km/hr is*** $\left[(D-Ty)\left(\frac{x}{x-y}\right)\right]$ ***km.***

Illustrative Example

Ex.: A car travels a distance of 170 km in 2 hours partly at a speed of 100 km/hr and partly at 50 km/hr. Find the distance travelled at a speed of 100 km/hr.

Soln: **Detail Method:** Let the distance travelled at the speed of 100 km/hr = x km.

The distance travelled at the speed of 50 km/hr = $(170-x)$ km

or, $\frac{170}{2}=\frac{170}{\frac{x}{100}+\frac{170-x}{50}}$

or, $\frac{x}{100}+\frac{170-x}{50}=2$

$\therefore x=140$ km

$\therefore$ car travels 140 km at the speed of 100 km/hr.

Quicker Method: Applying the above theorem, we have

the required distance

$$=(170-2\times 50)\left(\frac{100}{100-50}\right)=70\times 2=140 \text{ km.}$$

Exercise

1. A car travels a distance of 85 km in 2 hours partly at a speed of 50 km/hr and partly at 25 km/hr. Find the distance travelled at a speed of 50 km/hr.
   a) 60 km b) 15 km c) 70 km d) 25 km
2. A car travels a distance of 80 km in 3 hours partly at a speed of 45 km/hr and partly at 20 km/hr. Find the distance travelled at a speed of 45 km/hr.
   a) 36 km b) 44 km c) 46 km d) 34 km
3. A car travels a distance of 160 km in 3 hours partly at a speed of 90 km/hr and partly at 40 km/hr. Find the distance travelled at a speed of 90 km/hr.
   a) 54 km b) 64 km c) 81 km d) 72 km

Answers

1. c 2. a 3. d

Rule 39

Theorem: ***A person covers a certain distance in T hours. He covers $\frac{a}{b}$ of it at x km/hr and the remaining at y km/hr. The total distance is given by $\left[\frac{T}{\frac{a/b}{x}+\frac{1-a/b}{y}}\right]$ km.***

Illustrative Example

Ex.: A long distance is covered by Rakesh in 56 minutes. He covers $\frac{2}{3}$ of it at 4 km/hr and the remaining at 5 km/hr. Find the total distance.

Soln: Applying the above theorem, we have

the required distance $=\frac{\frac{56}{60}}{\frac{2/3}{4}+\frac{1-2/3}{5}}$

$=\frac{56}{60}\times\frac{30}{7}=4$ km.

Exercise

1. A long distance is covered by Rakesh in 5 hours. He covers $\frac{3}{4}$ of it at 12 km/hr and the remaining at 16 km/hr. Find the total distance.
   a) 64 km b) 128 km c) 32 km d) 45 km
2. A long distance is covered by Rakesh in 3 hours. He covers $\frac{2}{5}$ of it at 5 km/hr and the remaining at 15 km/hr. Find the total distance.
   a) 30 km b) 35 km c) 25 km d) 25 km
3. A long distance is covered by Rakesh in 6 hours. He covers $\frac{1}{2}$ of it at 4 km/hr and the remaining at 2 km/hr. Find the total distance.
   a) 16 km b) 12 km c) 14 km d) 18 km

Answers

1. a 2. c 3. a

Rule 40

Theorem: ***A thief goes away with a car at a speed of x km/hr. If the theft has been discovered after 't' hours and the owner sets off in another car at y km/hr, then the owner will overtake the thief from the start in $\left(\frac{x}{y-x}\times t\right)$ hours.***

Illustrative Example

Ex.: A thief goes away with a Maruti car at a speed of 40 km/hr. The theft has been discovered after half an hour and the owner sets off in another car at 50 km/hr. When will the owner overtake the thief from the start?

Soln: **Detail Method:** Distance to be covered by the thief and by the car owner is the same.

Let after 't' hours owner catches the thief,

$S_1T_1 = S_2T_2$ = Disance

$t \times 50 = \left(t + \frac{1}{2}\right)40 \quad \therefore t = 2$ hours.

Quicker Method: Applying the above theorem, we have

the required time $= \frac{40}{50-40} \times \frac{1}{2} = 2$ hours.

Exercise

1. A thief steals a motor car at 1 PM and drives it at 45 km an hour. The theft is discovered at 2 PM and the owner sets off in another car at 54 km an hour. When will he overtake the thief?
   a) 5 PM b) 3 PM
   c) 7 PM d) Can't be determined
2. A thief goes away with a Maruti car at a speed of 20 km/hr. The theft has been discovered after 15 minutes and the owner sets off in another car at 30 km/hr. When will the owner overtake the thief from the start?
   a) 30 minutes b) 20 minutes
   c) 45 minutes d) None of these
3. A thief goes away with a Maruti car at a speed of 15 km/hr. The theft has been discovered after 1 hour and the owner sets off in another car at 25 km/hr. When will the owner overtake the thief from the start?
   a) 1 hr b) 2 hrs c) $1\frac{1}{2}$ hrs d) $2\frac{1}{2}$ hrs

Answers

1. c; **Hint:** t = 2 PM – 1 PM = 1 hour
   Applying the given rule, we have
   required answer $= \frac{45}{54-45} \times 1 = 5$ hours
   ie 2PM + 5 hrs = 7 PM
2. a 3. c

Rule 41

Theorem: ***A car travels a certain distance 'D' km at uniform speed. If the speed of the car is 'x' km/hr more, it takes 't' hours less to cover the same distance, then the original speed of the car is*** $\left(\frac{\sqrt{(xt)^2 + (4Dxt)} - xt}{2t}\right)$ ***km/hr.***

Illustrative Example

Ex.: A car travels a distance of 840 km at uniform speed. If the speed of the car is 10 km/hr more, it takes 2 hours less to cover the same distance. Find the original speed of the car.

Soln: Detail Method: Let the original speed be x km/hr.

From the question, we have

$$\frac{840}{x+10} = \frac{840}{x} - 2$$

or, $\frac{840}{x} - \frac{840}{x+10} = 2$

or, $\frac{840(x+10) - 840x}{x(x+10)} = 2$

or, $8400 = 2x^2 + 20x$

or, $x^2 + 10x - 4200 = 0$

or, $x = +60, -70$

$\therefore$ x = 60 km/hr [We take only +ve value of x]

Quicker Method: Applying the above theorem, we have

the original speed

$$= \frac{\sqrt{(2\times10)^2 + (4\times840\times10\times2)} - (2\times10)}{2\times2}$$

$$= \frac{\sqrt{400+67200} - 20}{4} = \frac{260-20}{4} = 60 \text{ km/hr.}$$

Exercise

1. A car travels a distance of 84 km at uniform speed. If the speed of the car is 1 km/hr more, it takes 2 hours less to cover the same distance. Find the original speed of the car.
   a) 6 km/hr b) 8 km/hr c) 7 km/hr d) None of these
2. A car travels a distance of 350 km at uniform speed. If the speed of the car is 20 km/hr more, it takes 2 hours less to cover the same distance. Find the original speed of the car.
   a) 25 km/hr b) 50 km/hr c) 75 km/hr d) 35 km/hr
3. A car travels a distance of 35 km at uniform speed. If the speed of the car is 2 km/hr more, it takes 2 hours less to cover the same distance. Find the original speed of the car.
   a) 5 km/hr b) 4 km/hr c) 7 km/hr d) None of these

Answers

1. a 2. b 3. a

Rule 42

Theorem: ***A car travels a certain distance 'D' km at uniform speed. If the speed of the car is x km/hr less, it takes 't' hours more to cover the same distance, then original speed of the car is*** $\left(\frac{\sqrt{(xt)^2 + 4Dxt} + xt}{2t}\right)$ ***km/hr.***

Illustrative Example

Ex.: A car travels a distance of 840 km at uniform speed. If the speed of the car is 10 km/hr less, it takes 2 hours more to cover the same distance. Find the original speed of the car.

Soln: **Detail method:** Let the original speed be x km/hr

From the question, we have

$$\frac{840}{x-10} = \frac{840}{x} + 2$$

or, $\frac{840}{x-10} - \frac{840}{x} = 2$

or, $\frac{840x - 840x + (840 \times 10)}{(x-10)x} = 2$

or, $8400 = 2x^2 - 20x$

or, $x^2 - 10x - 4200 = 0$

or, $x^2 - 70x + 60x - 4200 = 0$

or, $x(x-70) + 60(x-70) = 0$

or, $x = 70$ and -60

$\therefore x = 70$ km/hr

[$\because$ we take only +ve value of x]

Quicker Method: Applying the above theorem,

Original speed

$$= \frac{\sqrt{(2\times10)^2 + (4\times840\times10\times2)} + (2\times10)}{2\times2}$$

$$= \frac{260+20}{4} = \frac{280}{4} = 70 \text{ km/hr.}$$

Exercise

1. A car travels a distance of 84 km at uniform speed. If the speed of the car is 1 km/hr less, it takes 2 hours more to cover the same distance. Find the original speed of the car.
   a) 8 km/hr b) 7 km/hr
   c) 6 km/hr d) None of these
2. A car travels a distance of 35 km at uniform speed. If the speed of the car is 2 km/hr less, it takes 2 hours more to cover the same distance. Find the original speed of the car.
   a) 5 km/hr b) 7 km/hr c) 8 km/hr d) 6 km/hr
3. A car travels a distance of 350 km at uniform speed. If the speed of the car is 20 km/hr less, it takes 2 hours more to cover the same distance. Find the original speed of the car.
   a) 50 km/hr b) 60 km/hr c) 40 km/hr d) 70 km/hr

Answers

1\. b 2. b 3. d

Rule 43

Theorem: ***The ratio between the speeds of A and B is x : y. If A takes T hours more than B to cover a distance, then the actual time taken by A and B is*** $\left(\frac{T}{1-\frac{x}{y}}\right)$ ***hours and*** $\frac{x}{y}\left(\frac{T}{1-\frac{x}{y}}\right)$ ***hours respectively.***

Illustrative Example

Ex.: The ratio between the speeds of Ram and Shyam is 6 : 7. If Ram takes 30 minutes more than Shyam to cover a distance, then find the actual time taken by Ram and Shyam.

Soln: **Detail Method:** Let the speed of Ram and Shyam be S_1 km/hr and S_2 km/hr.

Distance $= S_1T_1 = S_2T_2$ [where, T_1 hrs and T_2 hrs are the time taken by Ram and Shyam respectively to cover the distance]

$\frac{S_1}{S_2} = \frac{T_2}{T_1}$ or, $\frac{6}{7} = \frac{T_2}{T_1}$ or, $\frac{7-6}{7} = \frac{T_1 - T_2}{T_1}$

or, $\frac{1}{7} = \frac{\frac{30}{60}}{T_1}$ or, $T_1 = \frac{7}{2}$ hours

$\therefore T_2 = \frac{7}{2} \times \frac{6}{7} = 3$ hours

$\therefore$ Actual time taken by Ram $= \frac{7}{2}$ hours and Shyam = 3 hours.

Quicker Method: Applying the above theorem, we have

Time taken by Ram $= \frac{\frac{1}{2}}{1-\frac{6}{7}} = \frac{7}{2}$ hours

Time taken by Shyam $= \left(\frac{\frac{1}{2}}{1-\frac{6}{7}}\right)\frac{6}{7} = \frac{7}{2} \times \frac{6}{7} = 3$ hours

Exercise

1. The ratio between the speeds of Sita and Rita is 5 : 6. If Sita takes 1 hour more than Rita to cover a distance, then find the actual time taken by Sita and Rita.
   a) 6 hrs, 5 hrs b) 6 hrs, 4 hrs

c) 5 hrs, 4 hrs   d) None of these

2. The ratio between the speeds of Ramesh and Suresh is 3 : 4. If Ramesh takes 30 minutes more than Suresh to cover a distance, then find the actual time taken by Ramesh and Suresh.

a) 2 hrs, 1 hr   b) $2\frac{1}{2}$ hrs, $1\frac{1}{2}$ hrs

c) 2 hrs, $1\frac{1}{2}$ hrs   d) $2\frac{1}{2}$ hrs, 1 hr

3. The ratio between the speeds of Sudhesh and Vivek is 2 : 5. If Sudhesh takes 3 hrs more than Vivek to cover a distance, then find the actual time taken by Sudhesh and Vivek.

a) 5 hrs, 3 hrs   b) 5 hrs, 2 hrs
c) 4 hrs, 3 hrs   d) None of these

Answers

1. a   2. c   3. b

Rule 44

Theorem: ***If a person covers $\frac{m}{n}$th part of the total journey at speed S_1 km/hr and the remaining journey at speed S_2 km/hr then his average speed for the total journey is***

$$\left[\frac{S_1S_2}{\left(1-\frac{m}{n}\right)S_1+\frac{m}{n}S_2}\right] \textit{km/hr.}$$

Illustrative Example

Ex.: A man covers one-third of his journey at 15 km/hr and the remaining journey at 30 km/hr. If the total journey is of 175 km, what is his average speed for the whole journey?

Soln: Applying the above theorem,
Average speed

$$=\frac{15\times30}{\left(1-\frac{1}{3}\right)\times15+\frac{1}{3}\times30}=\frac{15\times30}{10+10}=\frac{45}{2}=22\frac{1}{2}\text{ km/hr}$$

Exercise

1. A man covers $\frac{2}{3}$rd of his journey at 30 km/hr and the remaining journey at 60 km/hr. If the total journey is of 270 km, what is his average speed for the whole journey?

a) 36 km/hr   b) 37 km/hr   c) 45 km/hr   d) 42 km/hr

2. A man covers $\frac{3}{4}$th of his journey at 12 km/hr and the remaining journey at 16 km/hr. If the total journey is of 32 km, what is his average speed for the whole journey?

a) $12\frac{1}{5}$ km/hr   b) $12\frac{3}{5}$ km/hr

c) 15 km/hr   d) $12\frac{4}{5}$ km/hr

3. A man covers $\frac{2}{5}$th of his journey at 25 km/hr and the remaining journey at 15 km/hr. If the total journey is of 50 km, what is his average speed for the whole journey?

a) 18 km/hr   b) 20 km/hr   c) $17\frac{6}{7}$ km/hr   d) $16\frac{2}{3}$ km/hr

Answers

1. a   2. d   3. c

Rule 45

Theorem: ***To plant a pillar at every certain distance***

Case I: When the system is open, ie road, path etc.

If pillars are to be planted at every d km distance on the road or the path of l km length, then the total no. of pillars are $\left[\frac{l}{d}+1\right]$. In other words

In other words,
Total no. of pillars

$$=\frac{\textit{Total length of the road}}{\textit{Distance between two adjacent pillars}}+1$$

Illustrative Example

Ex.: A road is of 900 km length. A contractor wants to plant some pillars on the road at every 10 km of distance. Find the total no. of pillars that the contractor has to plant.

Soln: Applying the Case-I of the above theorem, we have,

the total no. of pillars $=\frac{900}{10}+1=91$.

Case II: When the system is closed ie, circle, square, rectangle etc.
If pillars are to be planted at every d km distance on a rectangle, or a circle, or a square etc, then the total no. of pillars are given by

$$\left(\frac{\text{Perimeter}}{\text{Distance between two adjacent pillars}}\right)$$

Illustrative Example

Ex.: A person wants to plant trees at the edge of a circular field of radius 7 km at every 11 km of the distance. Find the no. of trees he will have to buy for this purpose.

Soln: Applying the aboe theorem,

$$\text{Required no. of trees} = \frac{\frac{22}{7}\times 2\times 7}{11} = 4.$$

[Since perimeter of the circular field

$$= 2\pi r = \left(2\times\frac{22}{7}\times 7\right)\text{ km}]$$

Exercise

1. Supposing that telegraph poles on a railroad are 50 metres apart, how many will be passed by a train in 4 hours if the speed of the train is 45 km an hour.
   a) 3601 b) 3600 c) 360 d) 361
2. A railway passenger counts the telegraph posts on the line as he passes them. If they are 50 metres apart and the train is going at the speed of 48 km per hour, how many will he pass per minute?
   a) 15 b) 18 c) 16 d) 20
3. A road is of 800 km length. A contractor wants to plant some pillars on the road at every 16 km of distance. Find the total no. of pillars that the contractor has to plant.
   a) 50 b) 49 c) 51 d) Data inadequate
4. A person wants to plant trees at the edge of a circular field of radius 21 km at every 6 km of the distance. Find the no. of trees he will have to buy for this purpose.
   a) 21 b) 22 c) 23 d) Data inadequate

Answers

1. b; **Hint:** In the given condition system will be closed. Hence the required no. of telegraph poles

$$= \frac{45\times 4\times 100}{50} = 3600$$

2. c; **Hint:** Distance covered by the train in one minute

$$= \frac{48\times 1000}{60} = 800 \text{ metres.}$$

$\therefore$ No. of telegraph posts passed by the train per minute

$$= \frac{800}{50} = 16$$

3. c 4. b

Rule 46

Dog and Hare

Theorem: ***A hare sees a dog d km away from her and scuds off in the opposite direction at a speed of x km/hr. If t hours later the dog perceives her and gives chase at a speed of y km/hr, then the dog will overtake hare in $\left(\frac{xt+d}{y-x}\right)$ hours and at $x\left(\frac{xt+d}{y-x}\right)$ km distance from the spot whence the hare took flight.***

Illustrative Example

Ex.: A hare sees a dog 100 metres away from her and scuds off in the opposite direction at a speed of 12 km an hour. A minute later the dog perceives her and gives chase at a speed of 16 km per hour. How soon will the dog overtake the hare, and at what distance from the spot whence the hare took flight?

Soln: **Detail Method:** Suppose the hare at H sees the dog at D.

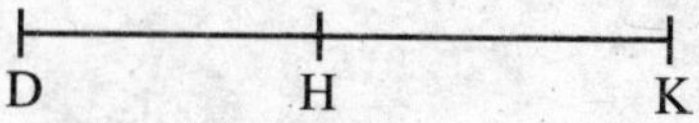

$\therefore$ DH = 100 metres

Let K be the position of the hare where the dog sees her.

$\therefore$ HK = the distance gone by the hare in 1 min

$$= \frac{12\times 1000}{60}\times 1 = 200 \text{ m}$$

$\therefore$ DK = 100 m + 200 m = 300 m

The hare thus has a start of 300 m

Now, the dog gains (16 – 12) or 4 km in an hour.

$\therefore$ the dog will gain 300 m in $\frac{60\times 300}{4\times 1000}$ or $4\frac{1}{2}$ min.

Again the distance gone by the hare in $4\frac{1}{2}$ min

$$= \frac{12\times 1000}{60}\times 4\frac{1}{2}\ 900\text{ m}$$

Quicker Method: Applying he above theorem, we have

$x = 12$ km/hr $d = 100\text{ m} = \frac{100}{1000} = \frac{1}{10}$ km

$y = 16$ km/hr and $t = 1\text{ min} = \frac{1}{60}$ hr.

$$\therefore \text{the required time} = \frac{12\times\frac{1}{60}+\frac{1}{10}}{16-12} = \frac{3}{40}\text{ hrs}$$

$$= \frac{3}{40}\times 60 = 4\frac{1}{2}\text{ min and}$$

$$\text{the required distance} = \frac{3}{40}\times 12 = \frac{9}{10}\text{ km}$$

$$= \frac{9}{10}\times 1000 = 900 \text{ metres.}$$

Exercise

1. A hare sees a dog 50 metres away from her and scuds off in the opposite direction at a speed of 6 km/hr. Two minutes later the dog perceives her and gives chase at a speed of 8 km per hour. How soon will the dog overtake the hare, and at what distance from the spot whence the hare took flight?

a) $7\frac{1}{2}$ min, 750 metres b) 15 min, 1500 metress

c) 7 min, 350 metres d) None of these

2. A hare sees a dog 200 metres away from her and scuds off in the opposite direction at a speed of 24 km/hr. Two minutes later the dog perceives her and gives chase at a speed of 32 km per hour. How soon will the dog overtake the hare, and at what distance from the spot whence the hare took flight?

a) 8 min, 2 km b) $7\frac{1}{2}$ min, 2 km

c) $7\frac{1}{2}$ min, 3 km d) $7\frac{1}{2}$ min, 1 km

Answers

1. a 2. c

Rule 47

Theorem: ***A hare, pursued by a grey-hound, is 'L' of her own leaps ahead of him. While the hare takes 'h' leaps the grey-hound takes 'g' leaps. If in one leap the hare goes d_1 m and the grey-hound d_2 m, then the no. of leaps in which the grey-hound will overtake the hare are*** $\left[\dfrac{L}{\dfrac{d_2}{d_1}-\dfrac{h}{g}}\right]$.

Illustrative Example

Ex.: A hare, pursued by a grey-hound, is 50 of her own leaps ahead of him. While the hare takes 4 leaps the grey-hound takes 3 leaps. In one leap the hare goes $1\frac{3}{4}$ metres and the grey-hound $2\frac{3}{4}$ metres. In how many leaps will the grey-hound overtake the hare?

Soln: Detail Method:

50 leaps of the hare $= 50 \times 1\frac{3}{4}$ m $= \frac{175}{2}$ m

$\therefore$ the grey-hound should gain $\frac{175}{2}$ m over the hare.

Now the grey-hound takes 3 leaps whilst the hare takes 4 leaps.

$\therefore$ the grey-hound takes 1 leap whilst the hare takes $\frac{4}{3}$ leaps.

$\therefore$ the grey-hound goes $2\frac{3}{4}$ m whilst the hare goes $\frac{4}{3}\times 1\frac{3}{4}$ m

$\therefore$ the grey-hound gains $\left(2\frac{3}{4}-\frac{4}{3}\times 1\frac{3}{4}\right)$

or $\frac{5}{12}$ m in one leap.

$\therefore$ the grey-hound gains $\frac{175}{2}$ m in $\left(\frac{175}{2}\times\frac{12}{5}\right)$

or 210 leaps.

Quicker Method: Applying the above theorem, we have

the required no. of leaps

$$=\frac{50}{\frac{11}{4}\times\frac{4}{7}-\frac{4}{3}}=\frac{50\times 21}{5}=210 \text{ leaps.}$$

Exercise

1. A hare, pursued by a grey-hound, is 30 metres before him at starting, whilst the hare takes 4 leaps the grey-hound takes 3. In one leap the hare goes $1\frac{1}{2}$ metres and the grey-houind $2\frac{1}{2}$ metres. How far will the hare have gone when she is caught by the hound?

a) 120 m b) 150 m c) 80 m d) Data inadequate

2. A hare, pursued by a grey-hound, is 60 of her own leaps ahead of him. While the hare takes 6 leaps the grey-hound takes 3 leaps. In one leap the hare goes 4 metres and the grey-hound 12 metres. In how many leaps will the grey-hound overtake the hare?

a) 30 b) 60 c) 40 d) 45

3. A hare, pursued by a grey-hound, is 20 of her own leaps ahead of him. While the hare takes 5 leaps the grey-hound takes 4 leaps. In one leap the hare goes $2\frac{3}{4}$ metres and the grey-hound $3\frac{3}{4}$ metres. In how many leaps will the grey-hound overtake the hare?

a) 176 b) 186 c) 276 d) None of these

Answers

1. a; **Hint:** Here, $L = 30 \times \frac{2}{3} = 20$ leaps

Now applying the given rule, we have

$$\frac{20}{\frac{\frac{5}{2}}{\frac{3}{2}} - \frac{4}{3}} = \frac{20}{\frac{5}{3} - \frac{4}{3}} = 60 \text{ leaps.}$$

ie grey-hound will take 60 leaps to overtake the hare.
Now, grey-hound's 3 leaps = hare's 4 leaps

$\therefore$ grey-hound's 60 leaps = hare's $\frac{4}{3} \times 60 = 80$ leaps.

$\therefore$ Hare's 80 leaps $= 80 \times \frac{3}{2} = 120$ metres

2. b 3. a

Rule 48

Theorem: ***A hare, pursued by a grey-hound, is 'L' of her own leaps ahead of him. While the hare takes 'h' leaps, the grey-hound takes 'g' leaps. If h_1 leaps of hare is equal to g_1 leaps of grey-hound, then the number of leaps in which the grey-hound will overtake the hare are*** $\left(\frac{L}{\frac{h_1}{g_1} \times \frac{h}{g}}\right)$.

Illustrative Example

Ex.: A hare, pursued by a grey-hound is 40 of her own leaps ahead of him. While the hare takes 8 leaps, the grey-hound takes 6 leaps. 4 leaps of grey-hound is equal to 6 leaps of hare. In how many leaps will the grey-hound overtake the hare?

Soln: Detail Method:

Since grey-hound is overtaking hare, therefore, we have to express leaps of grey-hound in terms of the leaps of hare.

$\because$ 4 leaps of grey-hound is equal to 6 leaps of hare.

$\therefore$ 1 leap of grey-hound is equal to $\frac{6}{4}$ leaps of hare.

$\therefore$ 6 leaps of grey-hound is equal to $\frac{6}{4} \times 6$ leaps of hare = 9 leaps of hare.

In equal time, when hare takes 8 leaps, then grey-hound takes 9 leaps equivalent to hare.

$\therefore$ No. of leaps that grey-hound takes to overtake hare $= \frac{40}{9-8} \times 6 = 240$ leaps.

Quicker Method:

Here, $L = 40, h = 8, g = 6, h_1 = 6$ and $g_1 = 4$

Now, applying the above rule, we have

The required answer $= \frac{40}{\frac{6}{4} - \frac{8}{6}} = 240$ leaps.

Exercise

1. A hare, pursued by a grey-hound is 20 of her own leaps ahead of him. While the hare takes 4 leaps, the grey-hound takes 3 leaps. 2 leaps of grey-hound is equal to 3 leaps of hare. In how many leaps will the grey-hound overtake the hare?
   a) 360 leaps b) 90 leaps
   c) 120 leaps d) 270 leaps
2. A hare, pursued by a grey-hound is 25 of her own leaps ahead of him. While the hare takes 6 leaps, the grey-hound takes 5 leaps. 4 leaps of grey-hound is equal to 5 leaps of hare. In how many leaps will the grey-hound overtake the hare?
   a) 625 leaps b) 652 leaps
   c) 265 leaps d) 500 leaps
3. A hare, pursued by a grey-hound is 22 of her own leaps ahead of him. While the hare takes 9 leaps, the grey-hound takes 4 leaps. 2 leaps of grey-hound is equal to 5 leaps of hare. In how many leaps will the grey-hound overtake the hare?
   a) 200 leaps b) 110 leaps c) 88 leaps d) 210 leaps

Answers

1. c 2. d 3. c

Rule 49

Problems on monkey

Theorem: ***A monkey trys to ascend a pole. If he ascends in first minute and slips down in the second minute, then the multiple of ascend and descend is***

$$\left[\frac{\textit{Length of the pole} - \textit{Distance of ascent}}{\textit{Distance of ascent} - \textit{Distance of slip down}}\right].$$

Now consider the following cases.

Case I: If the result is a whole number, then this whole number will be multiple. And the required answer will be $(2 \times \text{multiple} + 1)$.

Illustrative Example

Ex.: A monkey tries to ascend a greased pole 14 metres high. He ascends 2 metres in first minute and slips down 1 metre in the alternate minute. If he continues to ascend in this fashion, how long does he take to reach the top?

Soln: **Detail Method:** In every 2 minutes he is able to ascend 2 – 1 = 1 metre. This way he ascends upto 12 metres because when he reaches at the top, he does not slip down. Thus, upto 12 metres he takes 12 × 2 = 24 minutes and for the last 2 metres he takes 1 minute. Therefore, he takes 24 + 1 = 25 minutes to reach the top. That is, in 26th minute he reaches the top.

Quicker Method: Applying the above theorem, we have

$$\text{Multiple} = \frac{14-2}{2-1} = 12$$

Required answer = 2 × 12 + 1 = 25 minutes.

Case II: If the result is not the whole number ie it is a fraction, then the multiple will be the whole number of that fraction and the required answer will be

$$\left(2\times\text{Multiple} + \frac{\text{Length of the pole} - \left(\begin{array}{l}\text{Distance of ascent}\\ - \text{Distance of slip down}\end{array}\right)\times\text{Multiple}}{\text{Distance of ascent}}\right)$$

Illustrative Example

Ex.: A monkey tries to ascend a greased pole 92 metres high. He ascends 10 metres in first minute and slips down 1 metre in the alternate minute. If he continues to ascend in this fashion, how long does he take to reach the top?

Soln: **Detail Method:**

In every 2 minutes he is able to ascend 10 – 1 = 9 metres.

This way he ascends up to 9 × 10 = 90 metres.

Thus upto 90 metres he takes 10 × 2 = 20 minutes and for the remaining distance 92 – 90 = 2 metres, he takes

$\frac{2}{10} = \frac{1}{5}$ min = 12 seconds.

∴ total time = 20 minutes 12 seconds.

Quicker Method: Applying the above theorem,

$$\text{Multiple} = \frac{92-10}{10-1} = \frac{82}{9} = 9\frac{1}{9}$$

Multiple is in fraction, hence multiple will be the whole number of $9\frac{1}{9} = 9 + 1 = 10$.

$$\therefore \text{the required answer} = 2\times 10 + \frac{92-(10-1)\times 10}{10}$$

$$= 20 + \frac{1}{5} = 20 \text{ min } 12 \text{ sec.}$$

Exercise

1. A monkey climbing up a greased pole, ascends 10 m and slips down 3 m in alternate minutes. If the pole is 63 m high, how long will it take him to reach the top?
   a) 16 min 42 sec b) 16 min 40 sec
   c) 18 min 42 sec d) None of these
2. A monkey tries to ascend a greased pole 91 metres high. He ascends 10 metres in first minute and slips down 1 metre in the alternate minute. If he continues to ascend in this fashion, how long does he take to reach the top?
   a) 19 minutes b) 18 minutes
   c) 20 minutes d) None of these
3. A monkey tries to ascend a greased pole 58 metres high. He ascends 12 metres in first minute and slips down 4 metres in the alternate minute. If he continues to ascend in this fashion, how long does he take to reach the top?
   a) 13 minutes 30 seconds b) 12 minutes 50 seconds
   c) 12 minutes 40 seconds d) 12 minutes 5 seconds
4. If a snail, on an average, creeps 31 cm up a pole during 12 hours in the night, and slip down 16 cm during the 12 hours in the day. How many hours will he be in getting to the top of a pole 4.2 m high?
   a) $312\frac{30}{31}$ hrs b) $624\frac{30}{31}$ hrs
   c) $635\frac{19}{31}$ hrs d) $655\frac{30}{31}$ hrs

Answers

1. a; **Hint:** $\text{Multiple} = \frac{63-10}{10-3} = 7\frac{4}{7} \approx 8$

$$\therefore \text{required answer} = 2\times 8 + \frac{63-(10-3)\times 8}{10}$$

[See **case II**]

$$= 16 + \frac{7}{10} = 16 \text{ hrs } 42 \text{ min}$$

2. a 3. b

4. c; **Hint:** Length of the pole = 420 cm. Now in 24 hrs, the snail creeps up (31 – 16) cm or 15 cm, therefore in (24 × 26) hrs, the snail creeps up (15 × 26) cm or 390 cm. Therefore, he has (420 – 390) cm or 30 cm more to get up. And he goes 31 cm in 12 hrs and therefore over 30 cm in $\frac{12\times 30}{31}$ hrs. Therefore, he reaches on top in (24 × 26 + $\frac{12\times 30}{31} = 635\frac{19}{31}$) hrs [The number of days (26) has been so determined that (420 cm – 15 cm × 26) may be equal to 31 cm or just less than 31 cm].

This kind of questions can also be solved by the given

rule (**Quicker Method**). See the following steps given below.

Step I: Multiple $= \frac{4.2-0.31}{0.31-0.16} = \frac{389}{15} = 25\frac{14}{15} \approx 26$

Here, **Case II:** of the given rule will be applied.

Step II: Required answer

$$= 12\left[2\times 26 + \times\frac{4.2-(0.31-0.16)\times 26}{0.31}\right] = 24\times 26 + \frac{12\times 30}{31}$$

$$= 635\frac{19}{31} \text{ hrs}$$

Note: Here difference between the periods of ascend and descend is 12 hrs, hence we multiply the given formula by 12. We hope, now you can appreciate this formula.

Miscellaneous

1. Two cars A and B are running towards each other from two different places 88 km apart. If the ratio of the speeds of the cars A and B is 5:6 and the speed of the car B is 90 km per hour, after how long will the two meet each other? **[BSRB Patna PO, 2001]**

   a) $26\frac{2}{3}$ minutes b) 24 minutes
   c) 32 minutes d) 36 minutes

2. Two boys begin together to write out a booklet containing 817 lines. The first boy starts with first line, writing at the rate of 200 lines an hour, and the second boy starts with the last line, then writes line 816, and so on, backwards proceeding at the rate of 150 lines an hour. At what line will they meet?

   a) 466th b) 465th c) 467th d) 468th

3. A starts from P to walk to Q, a distance of 51.75 kilometres at the rate of 3.75 km an hour. An hour later B starts from Q for P and walks at the rate of 4.25 km an hour. When and where will A meet B?

   a) 26.25 km from Q b) 25.50 km from Q
   c) 25.30 km from P d) Can't be determined

4. A train leaves Delhi at 5 am and reaches Kanpur at 10 am. Another train leaves Kanpur at 7 am and reaches Delhi at 2 pm. At what time do the two trains meet?

   a) 8.45 am b) 3.45 pm
   c) 6.45 am d) Data inadequate

5. Suresh travelled 1200 km by air which formed (2/5) of his trip. The part of his trip which was one third of the whole trip, he travelled by car. The rest of the journey was performed by train. The distance travelled by train was **[Hotel Management Exam, 1991]**

   a) 1600 km b) 800 km c) 480 km d) 1800 km

6. A certain distance is covered at a certain speed. If half this distance is covered in double the time, the ratio of the two speeds is **[Bank PO Exam, 1986]**

   a) 4 : 1 b) 1 : 4 c) 2 : 1 d) 1 : 2

7. If a boy takes as much time in running 10 metres as a car takes in covering 25 metres; the distance covered by the boy during the time the car covers 1 km, is **[Asst Grade Exam, 1987]**

   a) 400 metres b) 40 metres c) 4 metres d) 2.5 metres

8. Suresh started cycling along the boundaries of a square field from corner point A. After half an hour, he reached the corner point C, diagonally opposite to A. If his speed was 8 km per hour, the area of the field in square km is **[Bank PO Exam, 1988]**

   a) 64 b) 8 c) 4 d) Can't be determined

9. Two trains start at the same time from Aligarh and Delhi and proceed towards each other at the rate of 16 km and 21 km per hour respectively. When they meet, it is found that one train has travelled 60 km more than the other. The distance between two stations is **[Bank PO Exam, 1988]**

   a) 445 km b) 444 km c) 440 km d) 450 km

Answers

1. c; **Hint:** Speed of the car A $= \frac{5}{6}\times 90 = 75$ km/hr

   $\therefore$ Reqd time $= \frac{88}{90\times 75}\times 60 = 32$ minutes

2. c; **Hint:** Let the two meet at the xth line.
   From the question,

   $\frac{x}{200} = \frac{817-x}{150}$ $\therefore x = 466.85$

   ie at the 467th line, they will meet.

3. b; **Hint:** A has already gone 3.75 km when B starts. Of the remaining 48 km, A walks 3.75 km and B walks 4.25 km in one hour ie they together pass over 3.75 + 4.25 or 8 km in one hour. Therefore, 48 km are passed over in $\frac{48}{8}$ or 6 hours. Therefore A meets B in 6 hours after B started. And therefore they meet at a distance of 4.25 × 6 or 25.5 km from Q.

4. a; **Hint:** Let the distance between Delhi and Kanpur be x km

   Suppose the train leaving from Delhi is A and the train leaving from Kanpur be B.

   A's speed $= \frac{x}{10\,am - 5\,am} = \frac{x}{5}$ km/hr

   B's speed $= \frac{x}{2\,pm - 7\,am} = \frac{x}{7}$ km/hr

   Since B starts two hours later than A, the distance already covered by A at the time of start of B $= \frac{2x}{5}$ km

Remaining distance $= x - \frac{2x}{5} = \frac{3x}{5}$ km

Relative speed of approach of two trains $= \frac{x}{5} + \frac{x}{7} = \frac{12x}{35}$ km/hr

Time taken to cover the remaining distance by both trains

$$= \frac{\frac{3}{5}x}{\frac{12x}{35}} = \frac{3}{5} \times \frac{35}{12} = \frac{7}{4} \text{ hrs} = 1\frac{3}{4} \text{ hrs} = 1 \text{ hr } 45 \text{ min}$$

$\therefore$ the two trains will meet at 7 am + 1 hr 45 min = 8.45 am

5. b; **Hint:** Let the total distance be x km. Then,

$\frac{2}{5}x = 1200$ or $x = 3000$ km

Distance covered by car $= \left(3000 \times \frac{1}{3}\right)$ km $= 1000$ km

$\therefore$ Distance travelled by train $= [3000 - (1000 + 1200)]$ km $= 800$ km

6. a; **Hint:** Let x km be covered in y hours. Then,

1st speed $= (x/y)$ km per hour

2nd speed $= \left(\frac{\frac{1}{2}x}{2y}\right) = \left(\frac{x}{4y}\right)$ km per hour

Ratio of speeds $= \frac{x}{y} : \frac{x}{4y} = 1 : \frac{1}{4} = 4 : 1$

7. a; **Hint:** $25 : 10 :: 1000 : x$

or, $x = \frac{10 \times 1000}{25} = 400$ metres

8. c; **Hint:** Distance covered in half an hour = 4 km
2 × (side of the square) = 4 km
or side of the square = 2 km
$\therefore$ Area of the square field = 4 sq km

9. b; **Hint:** Suppose they meet after x hours. Then,
$21x - 16x = 60$ or, $x = 12$
Now, distance between the stations
$= (16 \times 12 + 21 \times 12)$ km $= 444$ km

Note: See **Rule 11** of the chapter "Trains".

17

Trains

Rule 1

Theorem: ***When train passes a telegraph post or a stationary man it should travel the length equal to the length of the train.***

Illustrative Example

Ex: How many seconds will a train 100 metres long running at the rate of 36 km an hour take to pass a certain telegraph post?

Soln: In passing the post the train must travel its own length.

Now, $36 \text{ km/hr} = 36 \times \frac{5}{18} = 10$ m/sec

$\therefore$ required time $= \frac{100}{10} = 10$ seconds

Exercise

1. How long will a train 130 m long travelling at 40 km an hour, take to pass a kilometre stone?
   a) 12 sec b) 11.7 sec c) 11.2 sec d) 11 sec
2. A train travelling at 30 km an hour took $13\frac{1}{2}$ secs in passing a certain point. Find the length of the train.
   a) 113 m b) 112 m c) 112.5 m d) None of these
3. A train 110 m in length runs through a station at the rate of 36 km per hour. How long will it take to pass a given point?
   a) 11 sec b) 12 sec c) 13 sec d) 15 sec
4. A train 135 m long is running with a speed of 54 km per hour. In what time will it pass a telegraph post?
   a) 9 sec b) 12 sec c) 8 sec d) 6 sec
5. A train 550 metres long is running with a speed of 55 km per hour. In what time will it pass a signal post?
   a) 30 sec b) 24 sec c) 42 sec d) 36 sec
6. A train 160 metres long passes a standing man in 18 seconds. What is the speed of the train?
   a) 28 km/hr b) 36 km/hr c) 32 km/hr d) None of these
7. A train 120 m long, crosses a pole in 10 seconds. The speed of the train is:
   [Clerical Grade Exam, 1989]
   a) 40 km/hr b) 43.2 km/hr
   c) 45 km/hr d) None of these
8. A train crosses a platform in 60 seconds at a speed of 45 km/hr. How much time will it take to cross an electric pole if the length of the platform is 100 metres?
   [Bank PO Exam, 1990]
   a) 8 seconds b) 1 minute
   c) 52 seconds d) None of these
9. A train 110 metres long travels at 60 km/hr. How long does it take to cross a telegraph post?
   a) 6 sec b) 5.6 sec c) 6.6 sec d) 6.8 sec

Answers

1. b

2. c; **Hint:** Speed = 30 km/hr = $30 \times \frac{5}{18} = \frac{25}{3}$ m/sec

   Length of the train $= \frac{25}{3} \times \frac{27}{2} = \frac{225}{2} = 112\frac{1}{2}$ m

3. a; **Hint:** Speed of the train = 36 km/hr $= \frac{36 \times 5}{18} = 10$ m/sec

   $\therefore$ required time $= \frac{110}{10} = 11$ sec

4. a
5. d

6. c; **Hint:** Speed of the train $= \frac{160}{18} \times \frac{18}{5} = 32$ km/hr

7. b

8. c; **Hint:** Distance covered by the train in crossing the platform $= \left(\frac{45 \times 60}{3600}\right)$ km $= \frac{3}{4}$ km = 750 m

   $\therefore$ Length of train = (750 – 100) m = 650 metres

   $\therefore$ Time taken to cross the pole $= \left(650 \div \frac{750}{60}\right)$ sec

   $= \left(650 \times \frac{60}{750}\right)$ sec = 52 sec

9. c

Rule 2

Theorem: ***When a train passes a platform or crosses a bridge it should travel the length equal to the sum of the lengths of train and platform or bridge both.***

Illustrative Example

Ex.: How long does a train 110 metres long running at the rate of 36 km/hr take to cross a bridge 132 metres in length?

Soln: In crossing the bridge the train must travel its own length plus the length of the bridge.

Now, 36 km/hr = $36 \times \frac{5}{18} = 10$ m/sec

$\therefore$ required time = $\frac{242}{10} = 24.2$ seconds.

Exercise

1. How long will a train 60 m long travelling at 40 km an hour, take to pass through a station whose platform is 90 m long?
   a) 14 sec b) 13.5 sec c) 14.5 sec d) None of these
2. Find the length of a bridge, which a train 130 long, travelling at 45 km an hour, can cross in 30 secs.
   a) 240 m b) 235 m c) 250 m d) 245 m
3. A column of men, extending 250 m in length takes one hour to march through a street at the rate of 50 paces a minute, each pace being 75 cm. Find the length of the street.
   a) 2 km b) 1 km c) 1.5 km d) 2.5 km
4. It is noticed that exactly half a minute elapses between the time when the engine of a train 50 m long enters a tunnel 500 m long and the time when the last carriage of the train leaves the tunnel. Find at how many km per hour the train is travelling?
   a) 66 km/hr b) 55 km/hr c) 64 km/hr d) None of these
5. A train 540 m long is running with a speed of 72 km/hr. In what time will it pass a tunnel 160 m long?
   a) 40 sec b) 30 sec c) 35 sec d) 42 sec
6. A train 200 m long is running with a speed of 72 km/hr. In what time will it pass a platform 160 m long?
   a) 18 sec b) 21 sec c) 15 sec d) 20 sec
7. A train 240 m long passes a bridge 120 m long in 24 sec. Find the speed with which the train is moving.
   a) 45 km/hr b) 54 km/hr c) 36 km/hr d) 42 km/hr
8. A train 150 m long passes a telegraph post in 12 seconds. Find, in what time, it will pass a bridge 250 m long?
   a) 32 sec b) 36 sec c) 25 sec d) 24 sec
9. A train 280 m long is moving at a speed of 60 km/hr. The time taken by the train to cross a platform 220 m long is **[RRB Exam, 1989]**
   a) 20 sec b) 25 sec c) 30 sec d) 35 sec
10. A train 50 m long passes a platform 100 m long in 10 seconds. The speed of the train in m/sec is **[Railways, 1989]**
    a) 150 b) 50 c) 10 d) 15
11. A train 700 m long, is running at the speed of 72 km/hr. If it crosses a tunnel in 1 minute, then the length of the tunnel (in metres) is **[NDA Exam, 1990]**
    a) 700 b) 600 c) 550 d) 500
12. A train 110 m long travels at 60 km/hr. How long does it take to cross a platform 240 metres long?
    a) 21 sec b) 20 sec c) 18 sec d) 24 sec
13. A train with 90 km/hr crosses a bridge in 36 seconds. Another train 100 m shorter crosses the same bridge at 45 km/hr. Find the time taken by the second train to cross the bridge.
    a) 64 sec b) 60 sec c) 72 sec d) 1 hr

Answers

1. b

2. d; **Hint:** Speed of the train = 45 km/hr

$$= 45 \times \frac{5}{18} = \frac{25}{2} \text{ m/sec}$$

Let the length of the bridge be x m

$$\text{or, } \frac{x+130}{\frac{25}{2}} = 30$$

or, $x + 130 = 15 \times 25$ $\therefore x = 375 - 130 = 245$ m

3. a; **Hint:** Speed of the train = $\frac{50 \times 75}{100 \times 60}$ m/sec = $\frac{5}{8}$ m/sec

Let length of the street be x m.

$$\frac{x+250}{\frac{5}{8}} = \frac{60 \times 60 \times 5}{8} = 2250 \text{ m}$$

$\therefore x = 2000$ m = 2 km

4. a; **Hint:** Speed of the train = $\frac{50+500}{30} = \frac{55}{3}$ m/sec

$$= \frac{55}{3} \times \frac{18}{5} = 66 \text{ km/hr}$$

5. c; **Hint:** Speed of the train = 72 km/hr

$$= \left(72 \times \frac{5}{18}\right) \text{ m/sec} = 20 \text{ m/sec}$$

Sum of the lengths of the train and tunnel = (540 + 160) m = 700 m

$\therefore$ Time taken to pass the tunnel
= Time taken to cover 700 m at 20 m/sec

$$= \left(\frac{700}{20}\right) \text{ sec} = 35 \text{ sec}$$

6. a

7. b; **Hint:** Speed $= \left(\frac{240+120}{24}\right)$ m/sec = 54 km/hr

8. a; **Hint:** Speed of the train $= \left(\frac{150}{12}\right)$ m/sec

Time taken to cross the bridge

$$= \left[(150+250)\times\frac{12}{150}\right] = 32 \text{ sec}$$

9. c 10. d

11. d; **Hint:** Let the length of tunnel be x metres.

Speed of train $= \left(72\times\frac{5}{18}\right)$ m/sec = 20 m/sec

Time taken by the train to cover $(700+x)$ m

$$= \left(\frac{700+x}{20}\right) \text{ sec}$$

$$\therefore \frac{700+x}{20} = 60 \Rightarrow x = 500$$

12. a

13. a; **Hint:** Speed of the first train = 90 km/hr

$$= \frac{90\times5}{18} = 25 \text{ m/sec}$$

Let the length of the bridge be x m and that of the train be y m.

$\therefore x+y = 25\times36 = 900$ .... (i)

Again,

Speed of the second train = 45 km/hr

$$= \frac{45\times5}{18} = \frac{25}{2} \text{ m/sec}$$

$\therefore$ time taken by the second train to cross the bridge

$$= \frac{(x-100)+y}{\frac{25}{2}} = \frac{(x+y-100)2}{25}$$

Now, putting the value of $(x+y)$ from eqn (i) we have

$$\frac{(900-100)2}{25} = \frac{800\times2}{25} = 64 \text{ sec}$$

[**Note:** Also see **Rule 36**]

Rule 3

Theorem: ***When two trains are moving in opposite directions their speeds should be added to find the relative speed.***

Illustrative Example

Ex.: Two trains 121 metres and 99 metres in length respectively are running in opposite directions, one at the rate of 40 km and the other at the rate of 32 km an hour. In what time will they be completely clear of each other from the moment they meet?

Soln: As the two trains are moving in opposite directions their relative speed = 40 + 32 = 72 km/hr, i.e. they are approaching each other at 72 km/hr or 20 m/sec.

$\therefore$ the required time

$$= \frac{\text{Total length}}{\text{Relative speed}} = \frac{121+99}{20} = 11 \text{ secs.}$$

Exercise

1. A train 110 metres in length travels at 60 km/hr. In what time will it pass a man who is walking against the train at 6 km a hour?
   a) $7\frac{1}{3}$ seconds b) 6 seconds
   c) $6\frac{2}{3}$ seconds d) Data inadequate
2. Two trains 70 m and 80 m long respectively, run at the rates of 68 and 40 km an hour respectively on parallel rails in opposite directions. How long do they take to pass each other?
   a) 5 seconds b) 10 seconds
   c) 12 seconds d) 6 seconds
3. Two trains 132 metres and 108 metres in length are running towards each other on parallel lines, one at the rate of 32 km/hr and another at 40 km/hr. In what time will they be clear of each other from the moment they meet?
   a) 12 sec b) 9 sec
   c) 15 sec d) Data inadequate
4. A train 100 metres long takes $7\frac{1}{5}$ seconds to cross a man walking at the rate of 5 km/hr in a direction opposite to that of the train. Find the speed of the train.
   a) 54 km/hr b) 45 km/hr c) 42 km/hr d) 36 km/hr
5. A train 135 metres long is running with a speed of 49 km per hour. In what time will it pass a man who is walking at 5 km/hr in the direction opposite to that of the train?
   a) 9 sec b) 12 sec c) 15 sec d) 18 sec
6. Two trains 127 metres and 113 metres in length respectively are running in opposite directions, one at the rate of 46 km/hr and another at the rate of 26 km per hour. In what time will they be clear of each other from the moment they meet?
   a) 17 sec b) 12 sec
   c) 14 sec d) None of these
7. A train 270 metres long is moving at a speed of 25 km/hr. It will cross a man coming from the opposite direction at a speed of 2 km/hr in

a) 36 seconds b) 32 seconds
c) 28 seconds d) 24 seconds

8. A train 150 metres long crosses a man walking at a speed of 6 km/hr in the opposite direction in 6 seconds. The speed of the train is
a) 96 km/hr b) 84 km/hr c) 106 km/hr d) 66 km/hr

9. A train of length 150 metres, takes 10 seconds to pass over another train 100 metres long coming from the opposite direction. If the speed of the first train be 30 km/hr, the speed of the second train is **[Railways, 1991]**
a) 54 km/hr b) 60 km/hr c) 72 km/hr d) 36 km/hr

10. A train 300 metres long is running at a speed of 18 km/hr. How many seconds will it take to cross a 200 m long train running in the opposite direction at a speed of 12 km/hr? **[Bank PO Exam, 1989]**
a) 60 b) $7\frac{1}{5}$ c) 12 d) 20

11. Two trains are running in opposite directions with the same speed. If the length of each train is 135 metres and they cross each other in 18 seconds, the speed of each train is **[Railways, 1991]**
a) 104 km/hr b) 27 km/hr
c) 54 km/hr d) None of these

12. A train 110 metres long travels at 60 km/hr. How long does it take to cross a man running at 6 km/hr in the opposite direction?
a) 6 sec b) 9 sec c) 5 sec d) 8 sec

13. A train 110 metres long travels at 60 km/hr. How long does it take to cross another train 170 metres long, running at 80 km/hr in opposite direction?
a) 7 sec b) 8.2 sec c) 6.2 sec d) 7.2 sec

14. A train 100 metres long takes $3\frac{3}{5}$ seconds to cross a man walking at the rate of 6 km/hr in a direction opposite to that of the train. Find the speed of the train.
a) 94 km/hr b) 90 km/hr c) 64 km/hr d) 60 km/hr

15. Two trains start simultaneously from stations P and Q, 500 km apart and meet after 5 hours. If the difference in their speeds be 20 km/hr, find their speeds.
a) 60 km/hr, 40 km/hr b) 30 km/hr, 70 km/hr
c) 65 km/hr, 35 km/hr d) 45 km/hr, 55 km/hr

Answers

1. b; **Hint:** Relative speed = 60 + 6 = 66 km/hr

$$= \frac{66\times5}{18} = \frac{55}{3} \text{ m/sec}$$

$$\therefore \text{ required time} = \frac{110\times3}{55} = 6 \text{ sec}$$

2. a

3. a; **Hint:** Relative speed = (32 + 40) km/hr = 72 km/hr

$$= \left(72\times\frac{5}{18}\right) \text{ m/sec} = 20 \text{ m/sec}$$

Sum of lengths of the trains = (132 + 108) m = 240 m
Time taken by the trains in passing each other

$$= \left(\frac{240}{20}\right) \text{ sec} = 12 \text{ sec}$$

4. b; **Hint:** Let the speed of the train be x km/hr

$$\therefore \text{ relative speed} = (x+5) \text{ km/hr} = (x+5)\times\frac{5}{18} \text{ m/sec}$$

Now, $\frac{(x+5)5}{18}\times\frac{36}{5} = \text{m}$

or, $2x + 10 = 100$ $\therefore x = 45$

$\therefore$ required speed of the train = 45 km/hr

5. a; **Hint:** Relative speed = (49 + 5) km/hr = 15 m/sec
$\therefore$ Time taken by the train to pass the man

$$= \left(\frac{135}{15}\right) \text{ sec} = 9 \text{ sec}$$

6. b; **Hint:** Relative speed = (46 + 26) km/hr = 20 m/sec
Time taken by the trains in passing each other
= Time taken to cover (127 + 113) m at 20 m/sec

$$= \frac{240}{20} = 12 \text{ sec}$$

7. a

8. b; **Hint:** In 6 seconds, the man walking at the rate of 6 km/hr, covers 10 metres.
So, the train has to move actually (150 – 10) i.e. 140 metres in 6 seconds to cross the man.

Hence, speed of the train = $\frac{140}{6}$ m/sec

$$= \left(\frac{140}{3}\times\frac{18}{5}\right) \text{ km/hr} = 84 \text{ km/hr}$$

9. b; **Hint:** Let the speed of second train be x km/hr
Then, relative speed = $(30 + x)$ km/hr
$\therefore$ Time taken to cover (150 + 100) metres at $(30 + x)$ km/hr = 10 seconds

$$\therefore \frac{250}{(30+x)\times\frac{5}{18}} = 10 \Rightarrow \frac{250\times18}{150+5x} = 10 \Rightarrow x = 60 \text{ km/hr}$$

10. a; **Hint:** Relative speed = (18 + 12) km/hr = 30 km/hr

$$= \left(30\times\frac{5}{18}\right) \text{ m/sec} = \frac{25}{3} \text{ m/sec}$$

Time taken to cover (300 + 200) m at $\frac{25}{3}$ m/sec

$$= \left(\frac{500 \times 3}{25}\right) \text{ sec} = 60 \text{ sec}$$

11. b; **Hint:** Let the speed of each train be x m/sec

Then, $\frac{135+135}{x+x} = 18$

$\therefore x = \frac{15}{2}$ m/sec $= \frac{15}{2} \times \frac{18}{5} = 27$ km/hr

12. a 13. d

14. a; **Hint:** Let the speed of the train be x km/hr
Now, according to the question
Relative speed $= (x+6)$ km/hr

$$= \frac{(x+6) \times 5}{18} \text{ m/sec}$$

Again, $\frac{18 \times 100}{5(x+6)} = \frac{18}{5}$

$\therefore x = 100 - 6 = 94$ km/hr

15. a; **Hint:** Let the speeds of the two trains be x and y km/hr

$\therefore$ relative speed $= (x + \text{y})$ km/hr $= \frac{500}{5} = 100$ km/hr and
$x - \text{y} = 20$ km/hr (given)
After solving the above equations, we get
$x = 60$ km/hr and y $= 40$ km/hr

Rule 4

Theorem: ***When two trains are moving in the same direction the relative speed will be the difference of their speeds.***

Illustrative Example

Ex.: Two trains 121 metres and 99 metres in length respectively are running in the same direction, one at the rate of 40 km and the other at the rate of 32 km an hour. In what time will they be completely clear of each other from the moment they meet?

Soln: Relative speed $= 40 - 32 = 8$ km/hr $= \frac{20}{9}$ m/sec

Total length $= 121 + 99 = 220$ m

$\therefore$ reqd time $= \frac{\text{Total length}}{\text{Relative speed}} = \frac{220}{20} \times 9 = 99$ sec.

Exercise

1. How many seconds will a train 60 m in length travelling at the rate of 42 km an hour, take in passing another train 84 m long proceeding in the same direction at the rate of 30 km an hour?
a) 43 sec b) $43\frac{1}{5}$ sec c) 43.5 sec d) $43\frac{4}{5}$ sec
2. A passenger train going 35 km an hour and 1213 dm long overtakes a goods train on a parallel line of rails. The goods train is going 26 km an hour, and is 1712 dm long. How long does the passenger train takes in passing the other?
a) 117 sec b) 116 sec c) 114 sec d) 112 sec
3. A train 110 metres in length travels at 60 km per hour. In what time will it pass a man who is walking in the same direction at 6 km an hour?
a) 6 sec b) $6\frac{1}{3}$ sec c) $7\frac{1}{3}$ sec d) $6\frac{2}{3}$ sec
4. Two trains 70 m and 80 m long respectively, run at the rates of 68 km and 40 km an hour respectively on parallel rails in the same direction. How long do they take to pass each other?
a) $19\frac{2}{7}$ seconds b) $19\frac{1}{7}$ seconds
c) $9\frac{2}{7}$ seconds d) 5 seconds
5. A train 100 m long is running with a speed of 70 km per hour. In what time will it pass a man who is running at 10 km per hour in the same direction in which the train is going?
a) 6 sec b) 12 sec c) 9 sec d) 16 sec
6. A train 550 metres long is running with a speed of 55 km per hour. In what time will it pass a man who is walking at 10 km per hour in the same direction in which the train is going?
a) 42 sec b) 33 sec c) 40 sec d) 44 sec
7. A train 100 metres long meets a man going in opposite direction at 5 km per hour and passes him in $7\frac{1}{5}$ seconds. At what rate is the train going?
a) 45 km/hr b) 54 km/hr c) 42 km/hr d) 36 km/hr
8. Two trains of length 110 metres and 90 metres are running on parallel lines in the same direction with a speed of 35 km per hour and 40 km per hour respectively. In what time will they pass each other?
a) 144 sec b) 140 sec c) 134 sec d) 154 sec
9. A train 110 metres long travels at 60 km/hr. How long does it take to cross a man running at 6 km/hr in the same direction?
a) 7.33 sec b) 8 sec c) 7 sec d) 8.33 sec
10. A train 110 metres long travels at 60 km/hr. How long does it take to cross another train 170 metres long, running at 54 km/hr in the same direction?
a) 2 min 40 sec b) 2 min 48 sec
c) 3 min 48 sec d) 3 min 40 sec

Answers

1. b; **Hint:** Relative speed $= 42 - 30 = 12$ km/hr

$$= \frac{12 \times 5}{18} = \frac{10}{3} \text{ m/sec}$$

Total length = 60 m + 84 m = 144 m

$\therefore$ Required time = $\frac{144 \times 3}{10} = 43\frac{1}{5}$ sec

2. a; **Hint:** Relative speed = 35 – 26 = 9 km/hr

$= \frac{9 \times 5}{18} = \frac{5}{2}$ m/sec

$\therefore$ Required time = $\frac{121.3 + 171.2}{\frac{5}{2}} = \frac{292.5 \times 2}{5} = 117$ sec

3. c; **Hint:** Relative speed = 60 – 6 = 54 km/hr

$= \frac{54 \times 5}{48} = 15$ m/sec

$\therefore$ required time = $\frac{110}{15} = 7\frac{1}{3}$ seconds.

4. a 5. a

6. d; **Hint:** Relative speed = (55 – 10) km/hr = $\left(\frac{25}{2}\right)$ m/sec

$\therefore$ Time taken by the train to pass the man

$= \left(550 \times \frac{2}{25}\right)$ sec = 44 sec

7. a; **Hint:** Let the speed of the train be x m/sec

Speed of man = $\left(5 \times \frac{5}{18}\right)$ m/sec = $\left(\frac{25}{18}\right)$ m/sec

Relative speed = $\left(x + \frac{25}{18}\right)$ m/sec = $\left(\frac{18x + 25}{18}\right)$ m/sec

$\therefore \frac{100}{\left(\frac{18x + 25}{18}\right)} = \frac{36}{5} \Rightarrow x = \left(\frac{225}{18}\right)$ m/sec

$= \left(\frac{225}{18} \times \frac{18}{5}\right)$ km/hr = 45 km/hr

8. a; **Hint:** Relative speed = (40 – 35) km/hr = $\left(\frac{25}{18}\right)$ m/sec

Time taken by the trains in passing each other

= Time taken to cover (110 + 90) m at $\left(\frac{25}{18}\right)$ m/sec

$= \frac{200 \times 18}{25} = 144$ sec

9. a 10. b

Rule 5

Theorem: ***Two trains are moving in the same direction at x km/hr and y km/hr (where x > y). If the faster train crosses a man in the slower train in 't' seconds, then the length of the faster train is given by*** $\left[\frac{5}{18}(x - y)t\right]$ ***metres.***

Illustrative Example

Ex.: Two trains are moving in the same direction at 50 km/hr and 30 km/hr. The faster train crosses a man in the slower train in 18 seconds. Find the length of the faster train.

Soln: Applying the above theorem, we have

the required length = $\frac{5}{18}(50 - 30) \times 18 = 100$ m.

Exercise

1. Two trains travel in the same direction at 56 km and 29 km an hour and the faster train passes a man in the slower train in 16 seconds. Find the length of the faster train.
   a) 100 m b) 120 m
   c) 124 m d) Data inadequate
2. Two trains are moving in the same direction at 25 km/hr and 15 km/hr. The faster train crosses a man in the slower train in 9 seconds. Find the length of the faster train.
   a) 50 m b) 75 m c) 25 m d) 30 m
3. Two trains are moving in the same direction at 33 km/hr and 24 km/hr. The faster train crosses a man in the slower train in 12 seconds. Find the length of the faster train.
   a) 30 m b) 40 m c) 20 m d) 50 m
4. Two trains are moving in the same direction at 50 km/hr and 32 km/hr. The faster train crosses a man in the slower train in 10 seconds. Find the length of the faster train.
   a) 45 m b) 60 m c) 75 m d) 50 m

Answers

1. b 2. c 3. a 4. d

Rule 6

Theorem: ***A train running at x km/hr takes*** t_1 ***seconds to pass a platform. Next it takes*** t_2 ***seconds to pass a man walking at y km/hr in the opposite direction, then the length of the train is*** $\left[\frac{5}{18}(x + y)t_2\right]$ ***metres and that of the platform is*** $\frac{5}{18}\left[x(t_1 - t_2) - yt_2\right]$ ***metres.***

Illustrative Example

Ex.: A train running at 25 km/hr takes 18 seconds to pass a platform. Next, it takes 12 seconds to pass a man walking at 5 km/hr in the opposite direction. Find the length of the train and that of the platform.

Soln: Detail Method:

Speed of the train relative to man = 25 + 5 = 30 km/hr

$$= 30 \times \frac{5}{18} = \frac{25}{3} \text{ m/sec}$$

Distance travelled in 12 seconds at this speed

$$= \frac{25}{3} \times 12 = 100 \text{ m.}$$

∴ Length of the train = 100 m

Speed of train = 25 km/hr = $25 \times \frac{5}{18} = \frac{125}{18}$ m/sec

Distance travelled in 18 secs at this speed

$$= \frac{125}{18} \times 18 = 125 \text{ m.}$$

∴ length of train + length of platform = 125 m

∴ length of platform = 125 – 100 = 25 m

Quicker Method: Applying the above theorem, we have,

the length of the train $= \frac{5}{18}(25+5)\times 12 = 100$ m and

the length of the platform $= \frac{5}{18}[25(18-12)-5\times 12]$

$$= \frac{5}{18} \times 90 = 25 \text{ m.}$$

Exercise

1. A train running at 36 km/hr takes 12 seconds to pass a platform. Next it takes 6 seconds to pass a man running at the rate of 9 km/hr in the opposite direction. Find the length of the train and length of the platform.
   a) 75 m, 45 m b) 70 m, 50 m
   c) 65 m, 35 m d) Data inadequate
2. A train running at 24 km/hr takes 15 seconds to pass a platform. Next, it takes 10 seconds to pass a man walking at 12 km/hr in the opposite direction. Find the length of the train.
   a) 50 m b) 100 m c) 75 m d) 120 m
3. A train running at 20 km/hr takes 12 seconds to pass a platform. Next, it takes 6 seconds to pass a man walking at 4 km/hr in the opposite direction. Find the length of the platform.
   a) 40 m b) 26 m c) $40\frac{2}{3}$ m d) $26\frac{2}{3}$ m

Answers

1. a 2. b 3. d

Rule 7

Theorem: ***Two trains are moving in opposite directions at x km/hr and y km/hr (where x > y), if the faster train crosses a man in the slower train in t seconds, then the length of the faster train is given by*** $\left[\frac{5}{18}(x+y)t\right]$ ***metres.***

Illustrative Example

Ex.: Two trains are moving in the opposite direction at 50 km and 40 km/hr. The faster train crosses a man in the slower train in 3 seconds. Find the length of the faster train.

Soln: Applying the above theorem, we have the required length of the train

$$= \frac{5}{18}(50+40)\times 30 = 75 \text{ m}$$

Exercise

1. A man sitting in the train which is travelling at the rate of 50 km per hour observes that it takes 9 seconds for a goods train travelling in the opposite direction to pass him. If the goods train is 187.5 metres long, find its speed.
   a) 25 km/hr b) 40 km/hr c) 35 km/hr d) 36 km/hr
2. Two trains are moving in the opposite direction at 24 km and 12 km/hr. The faster train crosses a man in the slower train in 2 seconds. Find the length of the faster train.
   a) 25 m b) 30 m c) 20 m d) Data inadequate
3. Two trains are moving in the opposite direction at 12 km and 6 km/hr. The faster train crosses a man in the slower train in 10 seconds. Find the length of the faster train.
   a) 50 m b) 75 m c) 40 m d) 150 m
4. Two trains are moving in the opposite direction at 30 km and 24 km/hr. The faster train crosses a man in the slower train in 6 seconds. Find the length of the faster train.
   a) 80 m b) 100 m c) 110 m d) 90 m

Answers

1. a; **Hint:** Let the speed of the goods train be y km/hr
   Now, applying the given rule, we have
   $$\frac{5}{18}(50+y)9 = 187.5$$
   ∴ y = 25 km/hr

2. c 3. a 4. d

Rule 8

Theorem: ***A train running at x km/hr takes t_1 seconds to pass a platform. Next it takes t_2 seconds to pass a man walking at y km/hr in the same direction, then the length of the train is*** $\left[\frac{5}{18}(x-y)t_2\right]$ ***metres and that of the platform is*** $\frac{5}{18}[x(t_1-t_2)+yt_2]$ ***metres.***

Illustrative Example

Ex.: A train running at 25 km/hr takes 18 seconds to pass a platform. Next, it takes 9 seconds to pass a man walking at 5 km/hr in the same direction. Find the length of the train and that of the platform.

Soln: **Detail Method:** Speed of the train relative to man

$25-5=20$ km/hr $=20\times\frac{5}{18}=\frac{50}{9}$ m/sec

Distance travelled in 9 seconds at this speed

$=\frac{50}{9}\times 9=50$ m

$\therefore$ Length of the train = 50 m

Speed of the train = 25 km/hr = $25\times\frac{5}{18}=\frac{125}{18}$ m/sec

Distance travelled in 18 seconds at this speed

$=\frac{125}{18}\times 18=125$ m

$\therefore$ length of train + length of platform = 125 m

$\therefore$ length of platform = 125 − 50 = 75 m

Quicker Method: Applying the above theorem, we have

the length of the train $=\frac{5}{18}(25-5)\times 9=50$ m

the length of the platform $=\frac{5}{18}[25(18-9)+5\times 9]$

$=\frac{5}{18}\times 270=15\times 5=75$ m.

Exercise

1. A train running at 35 km per hour takes 18 seconds to pass a platform. Next, it takes 12 seconds to pass a man walking at the rate of 5 km/hr in the same direction. Find the length of the train and that of the platform.
   a) 50 m, 75 m b) 100 m, 75 m
   c) 75 m, 25 m d) None of these
2. A train running at 25 km/hr takes 18 seconds to pass a platform. Next it takes 10 seconds to pass a man walking at the rate of 7 km/hr in the same direction. Find the length of the platform and the length of the train.
   a) 25 m, 50 m b) 45 m, 85 m c) 50 m, 75 m d) 50 m, 80 m
3. A train running at 30 km/hr takes 20 seconds to pass a platform. Next, it takes 8 seconds to pass a man walking at 12 km/hr in the same direction. Find the length of the train.
   a) 40 m b) 50 m c) 44 m d) 42 m
4. A train running at 36 km/hr takes 24 seconds to pass a platform. Next, it takes 6 seconds to pass a man walking at 18 km/hr in the same direction. Find the length of the platform.
   a) 200 m b) 190 m c) 210 m d) 100 m

Answers

1. b 2. c 3. a 4. c

Rule 9

Theorem: ***L metres long train crosses a bridge of length L_1 metres in T seconds. Time taken by the train to cross a platform of L_2 metres is given by $\left[\frac{L+L_2}{L+L_1}\right]T$ seconds.***

Illustrative Example

Ex.: 150 metres long train crosses a platform of length 250 metres in 30 seconds. Find the time for train to cross a bridge of 130 metres.

Soln: **Detail Method:**

According to the question,

speed of the train $=\frac{150+250}{30}=\frac{40}{3}$ m/sec

In the second case train has to cover (150 + 130) metres.

$\therefore$ required time $=\frac{280\times 3}{40}=21$ seconds.

Quicker Method: Applying the above theorem, we have the

reqd time $=\frac{(130+150)\times 30}{150+250}=\frac{280\times 30}{400}=21$ secs.

Exercise

1. A train speeds past a pole in 15 seconds and speeds past a platform of 100 metres in 30 seconds. Its length (in metres) is **[Bank PO Exam, 1989]**
   a) 200 b) 100 c) 50 d) Data inadequate
2. 75 metres long train crosses a platform of length 125 metres in 20 seconds. Find the time for train to cross a bridge of 65 metres.
   a) 12 sec b) 18 sec c) 14 sec d) $\frac{21}{2}$ sec
3. 250 metres long train crosses a platform of length 350 metres in 50 seconds. Find the time for train to cross a bridge of 230 metres.
   a) 45 sec b) 50 sec c) 40 sec d) 54 sec
4. 125 metres long train crosses a platform of length 175 metres in 20 seconds. Find the time for train to cross a bridge of 115 metres.
   a) 18 sec b) 15 sec c) 16 sec d) 12 sec

Answers

1. b; **Hint:** Here $L_2=0$ because in place of bridge, pole has been given in the question.

$\therefore$ required answer $= \left(\frac{L+0}{L+100}\right)30 = 15$

$\therefore$ L = 100 m

2. c 3. b 4. c

Rule 10

Theorem: ***If L metres long train crosses a bridge or a platform of length L_1 metres in T seconds, then the time taken by train to cross a pole is given by $\left(\frac{L \times T}{L + L_1}\right)$ seconds.***

Illustrative Example

Ex.: 120 metres long train crosses a tunnel of length 80 metres in 20 seconds. Find the time for train to cross a man standing on a platform of length 130 metres.

Soln: Detail Method:

Speed of the train $= \frac{120+80}{20} = 10$ m/sec

Time taken by train to cross the man $= \frac{120}{10} = 12$ secs.

Note: Here length of the platform is given to confuse the student. Because it is of no use in solving the above example.

Quicker Method: Applying the above theorem, we have

the required time $= \frac{120 \times 20}{120+80} = 12$ seconds.

Exercise

1. 60 metres long train crosses a tunnel of length 40 metres in 10 seconds. Find the time for train to cross a man standing on a platform of length 65 metres.
   a) 6 sec b) 8 sec c) 5 sec d) 4 sec
2. 240 metres long train crosses a tunnel of length 160 metres in 40 seconds. Find the time for train to cross a man standing on a platform of length 260 metres.
   a) 12 sec b) 18 sec c) 24 sec d) 30 sec
3. 100 metres long train crosses a tunnel of length 60 metres in 16 seconds. Find the time for train to cross a man standing on a platform of length 65 metres.
   a) 12 sec b) 11 sec c) 8 sec d) 10 sec

Answers

1. a 2. c 3. d

Rule 11

Theorem: ***Two trains start at the same time from A and B and proceed towards each other at the rate of x km/hr and y km/hr respectively. When they meet it is found that one train has travelled d km more than the other. Then the distance between A and B is $\left(\frac{x+y}{x-y}\right)d$ km. or***

Distance = Difference in distance $\times \frac{\text{Sum of speeds}}{\text{Diff. in speeds}}$.

Illustrative Example

Ex.: Two trains start at the same time from Hyderabad and Delhi and proceed towards each other at the rate of 80 km and 95 km per hour respectively. When they meet, it is found that one train has travelled 180 km more than the other. Find the distance between Delhi and Hyderabad.

Soln: Detail Method-I: Faster train moves 95 – 80 = 15 km more in 1 hr.

$\therefore$ faster train moves 180 km more in $\frac{1}{15} \times 180 = 12$ hrs.

Since, they are moving in opposite directions, they cover a distance of 80 + 95 = 175 km in 1 hr.

$\therefore$ in 12 hrs they cover a distance = 175 × 12 = 2100 km

$\therefore$ distance = 2100 km

Detail Method II:

Let the train started from Delhi, covers x km of the distance and the train started from Hyderabad covers y km of the distance when they meet.

From the question,

$x - y = 180$ km ..... (i)

and $\frac{95}{x} = \frac{80}{y}$ or, $x = \frac{19}{16}y$ .... (ii)

Put the value of x of equation (ii) into equation (i)

$\frac{19}{16}y - y = 180$

$\therefore$ y = 960 km

or, $x - y = 180$ km

$\therefore$ $x = 960 + 180 = 1140$ km

$\therefore$ the distance between Hyderabad and Delhi

= 1140 + 960 = 2100 km

Quicker Method: Applying the above theorem, we have

Distance = Difference in distance $\times \frac{\text{Sum of speed}}{\text{Diff. in speed}}$

$= 180 \times \frac{175}{15} = 2100$ km.

Exercise

1. Two trains start at the same time from Patna and Gaya and proceed towards each other at the rate of 60 km and 40 km per hour respectively. When they meet, it is found that one train has travelled 20 km more than the other. Find the distance between Gaya and Patna.
   a) 100 km b) 80 km c) 120 km d) 90 km
2. Two trains start at the same time from Ranchi and Rourkela and proceed towards each other at the rate of 85 km and 63 km per hour respectively. When they meet, it is found that one train has travelled 11 km more than the other. Find the distance between Rourkela and Ranchi.
   a) 84 km b) 148 km c) 74 km d) None of these
3. Two trains start at the same time from Allahabad and Kanpur and proceed towards each other at the rate of 73 km and 47 km per hour respectively. When they meet, it is found that one train has travelled 13 km more than the other. Find the distance between Kanpur and Allahabad.
   a) 70 km b) 60 km c) 75 km d) 65 km
4. Two trains start at the same time from Lucknow and Jaunpur and proceed towards each other at the rate of 75 km and 65 km per hour respectively. When they meet, it is found that one train has travelled 10 km more than the other. Find the distance between Jaunpur and Lucknow.
   a) 75 km b) 65 km c) 80 km d) 140 km

Answers

1. a 2. c 3. b 4. d

Rule 12

Theorem: ***Two stations A and B are D km apart on a straight line. A train starts from A towards B at x km/hr. t hours later another train starts from B towards A at y km/hr. The time after which the train starting from A will meet the train starting from B is*** $\left(\frac{D+ty}{x+y}\right)$ ***hours.***

Illustrative Example

Ex.: Two stations A and B are 110 km apart on a straight line. A train starts from A towards B at the rate of 40 km/hr. 2 hours later another train starts from B and travels towards A at the rate of 50 km/hr. When will the first train meet to the second train?

Soln: **Detail Method:** Let the first train meet the second x hrs after it starts, then

$40x+(x-2)\times 50=110$ .... (see note)

Distance covered by the first train = $40x$ km

The second train starts 2 hrs after the first starts its journey, so the distance covered by the second train $=50(x-2)$

$\therefore$ Total distance = 110 km = $40x+50(x-2)$

Quicker Method: Applying the above theorem, we have

the required time $=\frac{110+2\times 50}{40+50}=\frac{210}{90}=\frac{7}{3}=2\frac{1}{3}$ hrs.

Note: The above example may be put as "Two stations A and B are 110 km apart on a straight line. One train starts from A at 8 am and travels towards B at 40 km per hour. Another train starts from B at 10 am and travels towards A at 50 km per hour. At what time will they meet?

Soln: $8\ am+\frac{110+(10\ am-8\ am)\times 50}{40+50}$

$=8\text{ am}+\frac{210}{90}=10.20$ am.

Exercise

1. Two stations A and B are 110 kms apart on a straight line. One train starts from A at 7 AM and travels towards B at 20 km/hr speed. Another train starts from B at 8 AM and travels towards A at 25 km/hr speed. At what time will they meet? **[Railways, 1989]**
   a) 9 AM b) 10 AM
   c) 11 AM d) None of these
2. Two stations A and B are 60 km apart on a straight line. A train starts from A towards B at the rate of 20 km/hr. 3 hours later another train starts from B and travels towards A at the rate of 25 km/hr. When will the first train meet to the second train?
   a) 1 hr b) 2 hrs c) 3 hrs d) 4 hrs
3. Two stations A and B are 145 km apart on a straight line. A train starts from A towards B at the rate of 25 km/hr. 1 hours later another train starts from B and travels towards A at the rate of 35 km/hr. When will the first train meet to the second train?
   a) 3 hrs b) 2 hrs c) 1½ hrs d) 2½ hrs
4. A train starts from station A at 9 am and travels at 50 km/hr towards station B, 210 km away. Another train starts from station B at 11 am and travels at 60 km/hr towards station A. At what time will they meet and at what distance from A?
   a) 12 noon, 150 km b) 11 am, 100 km
   c) 10 am, 50 km d) 1 pm, 200 km

Answers

1. b 2. c 3. a

4. a; **Hint:** Required answer $=9\text{ am}+\frac{210+(11am-9am)60}{50+60}$

$=9$ am + 3 hours = 12 noon

and the required distance from A

= Distance travelled by the first train = 3 × 50 = 150 km

Rule 13

Theorem: ***A train passes by a stationary man standing on the platform or a pole in t_1 seconds and passes by the platform completely in t_2 seconds. If the length of the platform is 'p' metres, then the length of the train is $\left(\frac{t_1 \times p}{t_2 - t_1}\right)$ metres and the speed of the train is $\left(\frac{p}{t_2 - t_1}\right)$ m/sec. Or Length of the train $= \frac{\text{Length of the platform}}{\text{Difference in time}} \times \left(\begin{array}{l}\text{Time taken to cross a}\\ \text{stationary pole or man}\end{array}\right)$ and speed of the train $= \left(\frac{\text{Length of the platform}}{\text{Difference in time}}\right)$.***

Illustrative Example

Ex.: A train passes by a stationary man standing on the platform in 7 seconds and passes by the platform completely in 28 seconds. If the length of the platform is 330 metres, what is the length of the train?

Soln: **Detail Method:** Let the length of the train be x m.

Then speed of the train $= \frac{x}{7}$ m per sec.

And also the speed of the train $= \frac{x+330}{28}$ m per sec.

Both the speeds should be equal, ie, $\frac{x}{7} = \frac{x+330}{28}$

or, $28x - 7x = 7 \times 330$

$\therefore x = \frac{7 \times 330}{21} = 110$ m

Quicker Approach: The train covers its length in 7 seconds and covers its length plus length of platform in 28 seconds. That is, it covers the length of the platform in $28 - 7 = 21$ seconds.

Now, since it covers 330 m in 21 seconds.

$\therefore$ Distance covered in 7 seconds $= \frac{330}{21} \times 7 = 110$ m

Thus we get a **direct formula as:**

Length of train

$= \frac{\text{Length of platform}}{\text{Difference in time}} \times$ Time taken to cross a stationary pole or man

$= \frac{330}{21} \times 7 = 110$ m.

Exercise

1. A train passes a pole in 15 seconds and passes a platform 100 m long in 25 seconds. Find its length.
   a) 150 m b) 100 m
   c) 250 m d) Data inadequate
2. A man standing on a railway platform noticed that a train took 21 seconds to completely pass through the station, which was 88 metres long and that it took 9 seconds in passing him. Find the length of the train, and the rate of the train in kilometres per hour.
   a) 198 m, 26 km/hr b) 66 m, 26.4 km/hr
   c) 132 m, 25 km/hr d) Data inadequate
3. A man is standing on a railway bridge which is 50 metres long. He finds that a train crosses the bridge in $4\frac{1}{2}$ seconds but himself in 2 seconds. Find the length of the train and its speed.
   a) 40 m, 72 km/hr b) 30 m, 92 km/hr
   c) 42 m, 75 km/hr d) 4 m, 70 km/hr
4. A railway train travelling at a uniform speed, clears a platform 200 metres long in 10 seconds and passes a telegraph post in 6 seconds. Find the length of the train and its speed.
   a) 300 m, 180 km/hr b) 200 m, 180 km/hr
   c) 300 m, 50 km/hr d) 200 m, 50 km/hr
5. A train passes a station platform in 36 seconds and a man standing on the platform in 20 seconds. If the speed of the train is 54 km/hr, find the length of the platform.
   a) 240 m b) 120 m c) 160 m d) 140 m
6. The train crosses a man standing on a platform 150 metres long in 10 seconds and crosses the platform completely in 22 seconds. Find the length and speed of the train.
   a) 125 m, 12 m/sec b) 120 m, 25 m/sec
   c) 125 m, 12.5 m/sec d) Data inadequate

Answers

1. a

2. b; **Hint:** Length of the train $= \frac{88 \times 9}{11 - 9} = 66$ m

Speed of the train $= \frac{88}{21-9} \times \frac{18}{5} = \frac{132}{5} = 26.4$ km/hr

3. a 4. a

5. a; **Hint:** Speed of the train = 54 km/hr $= \frac{54 \times 5}{18} = 15$ m/sec

Now, applying the given theorem, we have

$15 = \frac{p}{36 - 20}$ $\therefore$ p = 15 × 16 = 240 m

6. c

Rule 14

Theorem: ***Two stations A and B are D km apart on a straight line. A train starts from A and travels towards B at x km/hr. Another train, starting from B 't' hours earlier, trvels to-***

wards A at y km/hr. The time after which the train starting from A will meet the train starting from B is $\left(\frac{D-ty}{x+y}\right)$ hours.

Illustrative Example

Ex.: Two stations A and B are 110 km apart on a straight line. A train starts from A and travels towards B at 40 km/hr. Another train, starting from B 2 hours earlier, travels towards A at 50 km/hr. When will the first train meet to the second train?

Soln: **Detail Method:** Let the first train meet the second train x hours after it starts, then

$40x+(x+2)\times 50=110$

[Distance covered by the first train = $40x$ km

The second train starts 2 hrs earlier than the first starts its journey, so the distance covered by the second train $=50(x+2)$

$\therefore$ Total distance $=110$ km $=40x+50(x+2)$]

or, $40x+50x+100=110$

$90x=10$

$\therefore x=\frac{1}{9}$ hrs $=\frac{60}{9}=\frac{20}{3}=6\frac{2}{3}$ minutes.

Quicker Method: Applying the above theorem, we have

the required time $=\frac{110-50\times 2}{50+40}=\frac{1}{9}$ hrs $=6\frac{2}{3}$ minutes.

Note: Above example may be put as the following "Two stations A and B are 110 km apart on a straight line. A train starts from A at 10 am and travels towards B at 40 km per hour. Another train starting from B at 8 am travels towards A at 50 km per hour. At what time will they meet?

Soln: They will meet at

$8\text{ am}+\frac{110-(10\text{ am}-8\text{ am})\times 50}{40+50}=$ 8.6 a.m. (approx).

Exercise

1. Two stations A and B are 95 km apart on a straight line. A train starts from A and travels towards B at 35 km/hr. Another train, starting from B 30 minutes earlier, travels towards A at 40 km/hr. When will the first train meet to the second train?

   a) 2 hrs b) 1 hr c) $1\frac{1}{2}$ hrs d) $2\frac{1}{2}$ hrs

2. Two stations A and B are 145 km apart on a straight line. A train starts from A and travels towards B at 15 km/hr. Another train, starting from B 1 hr earlier, travels towards A at 25 km/hr. When will the first train meet to the second train?

   a) 3 hrs b) 2 hrs c) $2\frac{1}{2}$ hrs d) $1\frac{1}{2}$

3. Two stations A and B are 126 km apart on a straight line. A train starts from A and travels towards B at 23 km/hr. Another train, starting from B 30 minutes earlier, travels towards A at 32 km/hr. When will the first train meet to the second train?

   a) 1 hr b) $1\frac{1}{2}$ hrs c) 2 hrs d) $2\frac{1}{2}$ hrs

Answers

1. b 2. a 3. c

Rule 15

Theorem: ***If a train l metres long moving at a speed of x km/hr crosses another train in t seconds, then time taken by it to cover its own length is given by $\frac{18}{5}\left(\frac{l}{x}\right)$ seconds.***

Case - I: **If $\frac{18}{5}\left(\frac{l}{x}\right)<t$, trains are moving in the same direction.**

Case - II: **If $\frac{18}{5}\left(\frac{l}{x}\right)>t$, trains are moving in opposite direction.**

Illustrative Example

Ex.: A train 105 metres long moving at a speed of 54 km per hour crosses another train in 6 seconds. Then which of the following is true?

(a) Trains are moving in the same direction.
(b) Trains are moving in the opposite direction.
(c) The other train is not moving.
(d) Data inadequate

Soln: Applying the above theorem, first

find the value of $\frac{18}{5}\times\frac{l}{x}=\frac{18}{5}\times\frac{105}{54}=7$ seconds

7 seconds > 6 seconds

Hence trains are moving in the opposite direction ie (b) is the correct answer.

Exercise

1. A train 100 metres long moving at a speed of 36 km per hour crosses another train in 8 seconds. Then which of the following is true?

   a) Trains are moving in the same direction.
   b) Trains are moving in the opposite direction.
   c) The other train is not moving.
   d) Data inadequate

2. A train 125 metres long moving at a speed of 45 km per

hour crosses another train in 12 seconds. Then which of the following is true?
a) Trains are moving in the same direction.
b) Trains are moving in the opposite direction.
c) The other train is not moving.
d) Data inadequate

Answers

1. b 2. a

Rule 16

Theorem: ***Two trains of the same length but with different speeds pass a static pole in t_1 seconds and t_2 seconds respectively. They are moving in the same direction. The time they will take to cross each other is given by $\left(\frac{2t_1t_2}{t_2-t_1}\right)$ seconds.***

Illustrative Example

Ex.: Two trains of the same length but with different speeds pass a static pole in 4 seconds and 5 seconds respectively. In what time will they cross each other when they are moving in the same direction?

Soln: **Detail Method:** Let the length of the trains be x m.

The speeds of the two trains $= \frac{x}{4}$ m/s & $\frac{x}{5}$ m/s.

Total distance to be travelled = $2x$ m.

Relative speed when they are moving in the same direction $= \frac{x}{4} - \frac{x}{5} = \frac{x}{20}$ m/sec.

$\therefore$ required time $= 2x \div \frac{x}{20} = 40$ seconds.

Quicker Method: When they are moving in the same direction:

$$\text{Time} = \frac{2(4\times5)}{5-4} = 40 \text{ seconds.}$$

Exercise

1. Two trains of the same length but with different speeds pass a static pole in 5 seconds and 6 seconds respectively. In what time will they cross each other when they are moving in the same direction.
a) 1 hr b) 50 sec c) 1 min d) 60 min
2. Two trains of the same length but with different speeds pass a static pole in 6 seconds and 9 seconds respectively. In what time will they cross each other when they are moving in the same direction.
a) 36 sec b) 30 sec c) 40 sec d) 42 sec
3. Two trains of the same length but with different speeds pass a static pole in 4 seconds and 6 seconds respectively. In what time will they cross each other when they are moving in the same direction.
a) 12 sec b) 24 sec c) 28 sec d) 16 sec
4. Two trains of the same length but with different speeds pass a static pole in 8 seconds and 12 seconds respectively. In what time will they cross each other when they are moving in the same direction.
a) 58 sec b) 38 sec c) 46 sec d) 48 sec

Answers

1. c 2. a 3. b 4. d

Rule 17

Theorem: ***Two trains of the same length but with different speeds pass a static pole in t_1 seconds and t_2 seconds respectively. They are moving in the opposite directions. The time they will take to cross each other is given by $\left(\frac{2t_1t_2}{t_1+t_2}\right)$ seconds.***

Illustrative Example

Ex.: Two trains of the same length but with different speeds pass a static pole in 4 seconds and 5 seconds respectively. In what time will they cross each other when they are moving in the opposite direction.

Soln: **Detail Method:** Relative speed when they are moving in opposite directions $= \frac{x}{4} + \frac{x}{5} = \frac{9x}{20}$ m/sec.

$\therefore$ required time $= 2x \div \frac{9x}{20} = \frac{40}{9} = 4\frac{4}{9}$ seconds.

Quicker Method: When they are moving in opposite directions:

$$\text{Time} = \frac{2(4\times5)}{5+4} = \frac{40}{9} = 4\frac{4}{9} \text{ seconds.}$$

Exercise

1. Two trains of the same length but with different speeds pass a static pole in 4 seconds and 8 seconds respectively. In what time will they cross each other when they are moving in the opposite direction.
a) $\frac{16}{3}$ sec b) 5 sec c) 6 sec d) $\frac{14}{3}$ sec
2. Two trains of the same length but with different speeds pass a static pole in 10 seconds and 15 seconds respectively. In what time will they cross each other when they are moving in the opposite direction.
a) 13 sec b) 11 sec c) 12.5 sec d) 12 sec
3. Two trains of the same length but with different speeds pass a static pole in 12 seconds and 18 seconds respectively. In what time will they cross each other when they are moving in the opposite direction.

a) 14 sec b) 13.4 sec c) 14.4 sec d) 15 sec

Answers

1. a 2. d 3. c

Rule 18

Theorem: ***Two trains of the length l_1 and l_2 m respectively with different speeds pass a static pole in t_1 seconds and t_2 seconds respectively. When they are moving in the same direction, they will cross each other in*** $\left[\frac{(l_1+l_2)t_1t_2}{t_2l_1-t_1l_2}\right]$ ***seconds.***

Illustrative Example

Ex.: Two trains of the length 100 m and 125 m respectively with different speeds pass a static pole in 4 seconds and 7 seconds respectively. In what time will they cross each other when they are moving in the same direction?

Soln: **Detail Method:**

Length of the trains are 100 m and 125 m

The speeds of the trains = $\frac{100}{4}$ m/sec and $\frac{125}{7}$ m/sec

Total distance to be travelled = 100 + 125 = 225 m

Relative speed when they are moving in the same direction $=\frac{100}{4}-\frac{125}{7}=\frac{175-125}{7}=\frac{50}{7}$ m/sec.

$\therefore$ required time $=\frac{225\times7}{50}=31.5$ seconds.

Quicker Method: Applying the above theorem, we have the required time

$$=\frac{(100+125)\times4\times7}{7\times100-4\times125}=\frac{225\times4\times7}{200}=\frac{63}{2}=31.5 \text{ sec.}$$

Exercise

1. Two trains of the length 200 m and 250 m respectively with different speeds pass a static pole in 8 seconds and 14 seconds respectively. In what time will they cross each other when they are moving in the same direction?
   a) 63 sec b) 64 sec c) 72 sec d) 81 sec
2. Two trains of the length 150 m and 350 m respectively with different speeds pass a static pole in 4 seconds and 10 seconds respectively. In what time will they cross each other when they are moving in the same direction?
   a) 150 sec b) 75 sec c) 200 sec d) None of these
3. Two trains of the length 125 m and 150 m respectively with different speeds pass a static pole in 6 seconds and 8 seconds respectively. In what time will they cross each other when they are moving in the same direction?
   a) 130 sec b) 128 sec c) 132 sec d) 136 sec

Answers

1. a 2. c 3. c

Rule 19

Theorem: ***Two trains of the length l_1 m and l_2 m respectively with different speeds pass a static pole in t_1 seconds and t_2 seconds respectively. When they are moving in the opposite direction they will cross each other in*** $\left[\frac{(l_1+l_2)t_1t_2}{t_2l_1+t_1l_2}\right]$ ***seconds.***

Illustrative Example

Ex.: Two trains of the length 100 m and 125 m respectively with different speeds pass a static pole in 4 seconds and 7 seconds respectively. In what time will they cross each other when they are moving in the opposite direction?

Soln: **Detail Method:**

Lengths of the trains are 100 m and 125 m

Speeds of the trains = $\frac{100}{4}$ = 25 m/sec and $\frac{125}{7}$ m/sec

Total distance to be travelled = 100 + 125 = 225 m

Relative speed when they are moving in the opposite direction $=25+\frac{125}{7}=\frac{300}{7}$ m/sec

$\therefore$ required time $=\frac{225\times7}{300}=5.25$ seconds.

Quicker Method: Applying the above theorem, we have

$$\text{the required time}=\frac{(100+125)\times4\times7}{(7\times100)+(4\times125)}=\frac{225\times4\times7}{1200}$$

= 5.25 seconds.

Exercise

1. Two trains of the length 100 m and 150 m respectively with different speeds pass a static pole in 1 min and 3 min respectively. In what time will they cross each other when they are moving in the opposite direction?
   a) $\frac{5}{3}$ sec b) 100 sec c) 120 sec d) 50 sec
2. Two trains of the length 200 m and 250 m respectively with different speeds pass a static pole in 4 min and 6 min respectively. In what time will they cross each other

when they are moving in the opposite direction?

a) 150 min b) 4 min 48 sec

c) $4\frac{10}{11}$ min d) None of these

3. Two trains of the length 150 m and 350 m respectively with different speeds pass a static pole in 5 min and 9 min respectively. In what time will they cross each other when they are moving in the opposite direction?

a) $\frac{225}{31}$ min b) $\frac{235}{31}$ min

c) $\frac{225}{21}$ min d) None of these

Answers

1. b 2. c 3. a

Rule 20

Theorem: ***Two trains of the length l_1 m and l_2 m with the same speed pass a static pole in t_1 and t_2 seconds respectively.***

Case I: When they are moving in the same direction, they cannot pass each other.

Case II: When they are moving in the opposite direction, they will cross each other in $\left(\frac{t_1+t_2}{2}\right)$ seconds ie average of the two times.

Illustrative Example

Ex.: Two trains of the length 100 m and 125 m respectively with the same speeds pass a static pole in 4 seconds and 5 seconds respectively. In what time will they cross each other when they are moving in

(i) the same direction

(ii) opposite directions.

Soln: (i) Applying the case-I of the above theorem, they cannot pass each other.

Proof: Length of the trains = 100 m and 125 m

Speed of the two trains = $\frac{100}{4}$ and $\frac{125}{5}$

= 25 m/sec = same.

Because they are moving with the same speed relative speed will be zero. Hence they will never cross each other.

(ii) Detail Method:

Length of the trains = 100 m and 125 m

Speed of the two trains = $\frac{100}{4}$ and $\frac{125}{5}$ = 25 m/sec

Relative speed = 25 + 25 = 50 m/sec

Required time = $\frac{100+125}{50} = \frac{9}{2} = 4.5$ seconds.

Quicker Method: Applying the case-II of the above theorem, we have

the required time = $\frac{4+5}{2} = 4.5$ seconds.

Exercise

1. Two trains of the length 200 m and 100 m respectively with the same speeds pass a static pole in 6 seconds and 5 seconds respectively. In what time will they cross each other when they are moving in opposite direction.
a) 4.5 sec b) 5.5 sec c) 6 sec d) 6.5 sec
2. Two trains of the length 150 m and 275 m respectively with the same speeds pass a static pole in 7 seconds and 9 seconds respectively. In what time will they cross each other when they are moving in opposite direction.
a) 8 sec b) 6 sec c) 9 sec d) 7.5 sec
3. Two trains of the length 125 m and 175 m respectively with the same speeds pass a static pole in 8 seconds and 10 seconds respectively. In what time will they cross each other when they are moving in opposite direction.
a) 8 sec b) 10 sec c) 9 sec d) 9.5 sec

Answers

1. b 2. a 3. c

Rule 21

Theorem: ***Two trains of length l_1 m and l_2 m respectively run on parallel lines of rails. When running in the same direction the faster train passes the slower one in t_1 seconds, but when they are running in opposite directions with the same speeds as earlier, they pass each other in t_2 seconds. Then the speed of each train is given as the following.***

Speed of the faster train = $\left(\frac{l_1+l_2}{2}\right)\left(\frac{t_1+t_2}{t_1t_2}\right)$ ***m/sec and the***

speed of the slower train = $\left(\frac{l_1+l_2}{2}\right)\left(\frac{t_1-t_2}{t_1t_2}\right)$ ***m/sec.***

Thus a general formula for the speed is given as Average length of two trains

$$\times\left[\frac{1}{\text{Opposite direction's time}} \pm \frac{1}{\text{Same direction's time}}\right]$$

Illustrative Example

Ex.: Two trains of length 100 m and 80 m respectively run on parallel lines of rails. When running in the same direction the faster train passes the slower one in 18 seconds, but when they are running in opposite directions with the same speeds as earlier, they pass each other in 9 seconds. Find the speed of each train.

Soln: **Detail Method:** Let the speeds of the trains be x m/s and y m/s.
When they are moving in the same direction,
the relative speed = $(x-y)$ m/s

$$\therefore\ x-y=\frac{100+80}{18}=10$$

Similarly, $x+y=\frac{100+80}{9}=20$

Solving the two equations
$x=15$ m/s and $y=5$ m/s

Quicker Method: Applying the above theorem, we have

the speed of the faster train $=\frac{100+80}{2}\left(\frac{18+9}{18\times 9}\right)$

$=90\left(\frac{27}{18\times 9}\right)=15$ m/s and

the speed of the slower train

$=\frac{100+80}{2}\left(\frac{18-9}{18\times 9}\right)=5$ m/s.

Exercise

1. Two trains, each 80 m long, pass each other on parallel lines. If they are going in the same direction, the faster one takes one minute to pass the other completely. If they are going in different directions, they completely pass each other in 3 seconds. Find the rate of each trains in m per second.
   a) 28 m/sec, $25\frac{1}{3}$ m/sec b) $28\frac{1}{3}$ m/sec, 25 m/sec
   c) 26 m/sec, 24 m/sec d) 28 m/sec, 25 m/sec
2. Two trains 200 metres and 175 metres long are running on parallel lines. They take $7\frac{1}{2}$ seconds when running in opposite directions and $37\frac{1}{2}$ seconds when running in the same direction, to pass each other. Find their speeds in km per hour.
   a) 108 km/hr, 72 km/hr b) 98 km/hr, 82 km/hr
   c) 100 km/hr, 75 km/hr d) None of these
3. Two trains, 130 and 110 metres long, while going in the same direction, the faster train takes one minute to pass the other completely. If they are moving in opposite direction, they pass each other completely in 3 seconds. Find the speed of trains.
   a) 24 m/sec, 19 m/sec b) 42 m/sec, 38 m/sec
   c) 40 m/sec, 36 m/sec d) Data inadequate

Answers

1\. a 2. a 3. b

Rule 22

Theorem: ***Two trains can run at the speed of S_1 km/hr and S_2 km/hr respectively on parallel tracks. When they are running in opposite directions they pass each other in T_1 seconds. When they are running in the same direction, a person sitting in the faster train observes that he passes the other train in T_2 seconds.***

Then the length of the faster train = $[T_1(S_1+S_2)-T_2(S_1-S_2)]\frac{5}{18}$ metres and the length of the slower train = $T_2(S_1-S_2)\frac{5}{18}$ metres.

Illustrative Example

Ex.: Two trains can run at the speed of 54 km/hr and 36 km/hr respectively on parallel tracks. When they are running in opposite directions they pass each other in 10 seconds. When they are running in the same direction, a person sitting in the faster train observes that he passes the other train in 30 seconds. Find the length of the trains.

Soln: **Detail Method:** Speeds of trains in metres per seconds is 15 m/s and 10 m/s respectively. Let the length of faster & slower trains be x m and y m respectively.
When they are running in opposite directions:
Relative speed = 15 + 10 = 25 m/s
Total length = $(x+y)$ m

$\therefore$ time to cross each other $\frac{x+y}{25}=10$

$\therefore x+y=250$ .... (1)

In the second case, man passes the length of the slower train (y) with a speed of (15 – 10) m/s = 5 m/s

Then time $=\frac{y}{5}=30$

$\therefore$ y = 150 m
$\therefore$ length of slower train = 150 m
And from (1), $x=100$ m
$\therefore$ Length of faster train = 100 m

Quicker Method: Applying the above theorem, we have the
the length of the faster train

$=[10(54+36)-30(54-36)]\frac{5}{18}$

$=360\times\frac{5}{18}=100$ m and

the length of the slower train $=\frac{5}{18}\times 30[54-36]=150$ m.

Exercise

1. Two trains running at the rates of 45 and 36 km an hour respectively, on parallel rails in opposite directions, are observed to pass each other in 8 seconds, and when they are running in the same direction at the same rate as before, a person sitting in the faster train observes that he passes the other in 30 seconds. Find the length of the trains.
   a) 105 m, 75 m b) 50 m, 25 m
   c) 120 m, 90 m d) 100 m, 75 m
2. A certain passenger train travels at the rate of 42 km/hr, and a goods train whose length is half more than the passenger train, travels at 33 km/hr. When the two are travelling in the ame direction it takes 50 seconds for the passenger train to clear the goods train. How long does it take the two trains to pass when they are travelling in opposite directions?
   a) 6 sec b) 18 sec c) 21 sec d) 12 sec
3. Two trains run on parallel tracks at 90 km/hr and 72 km/hr respectively. When they are running in the opposite directions they cross each other in 5 seconds. When they are running in the same direction at speeds same as before a passenger sitting in the faster train sees the other train passing him in 25 seconds. Find the length of each train.
   a) 225 m, 100 m b) 125 m, 200 m
   c) 125 m, 100 m d) Data inadequate

Answers

1. a

2. a; **Hint:** $\frac{T_1(42+33)-50(42-33)}{50(42-33)}=\frac{2}{3}$

$\therefore T_1=\frac{50\times 27}{3\times 75}=6$ sec

3. c

Rule 23

Theorem: ***A train overtakes two persons who are walking in the same direction as the train is moving, at the rate of M_1 km/hr and M_2 km/hr and passes them completely in T_1 seconds and T_2 seconds respectively. Speed of the train is $\frac{5}{18}\left[\frac{M_2T_2-M_1T_1}{T_2-T_1}\right]$ m/sec or $\left(\frac{M_2T_2-M_1T_1}{T_2-T_1}\right)$ km/hr and the length of the train is given by $\left[\frac{(M_2-M_1)T_1T_2}{T_2-T_1}\right]\frac{5}{18}$ metres***

or $\frac{5}{18}\left(\frac{\text{Difference in speed of two men}\times T_1\times T_2}{T_2-T_1}\right)$ metres.

Illustrative Example

Ex.: A train overtakes two persons who are walking in the same direction as the train is moving, at the rate of 2 km/hr and 4 km/hr and passes them completely in 9 and 10 seconds respectively. Find the speed and the length of the train.

Soln: **Detail Method:** Speeds of two men are:

$2\text{ km/hr}=2\times\frac{5}{18}=\frac{5}{9}$ m/s and $4\text{ km/hr}=4\times\frac{5}{18}=\frac{10}{9}$ m/s

Let the speed of the train be x m/s. Then relative speeds are $\left(x-\frac{5}{9}\right)$ m/s and $\left(x-\frac{10}{9}\right)$ m/s

Now, length of train = Relative Speed × Time taken to pass a man

$\therefore$ length of train $=\left(x-\frac{5}{9}\right)\times 9=\left(x-\frac{10}{9}\right)\times 10$

$\therefore x=\frac{100}{9}-\frac{45}{9}=\frac{55}{9}$ m/s

$\therefore$ speed of the train $=\frac{55}{9}\times\frac{18}{5}=22$ km/hr and

length of the train $=\left(x-\frac{5}{9}\right)9=\left(\frac{55}{9}-\frac{5}{9}\right)9=50$ m.

Quicker Method: Applying the above theorem, we have
the length of the train

$=\left[\frac{\text{Diff. in Speed of two men}\times T_1\times T_2}{(T_2-T_1)}\right]\frac{5}{18}$

where T_1 and T_2 are times taken by the train to pass the two men, all in the same direction.

Thus in this case $=\left[\frac{(4-2)\times 9\times 10}{10-9}\times\frac{5}{18}\right]=50$ m

Speed of the train $=\left(\frac{M_2T_2-M_1T_1}{T_2-T_1}\right)$ km/hr

$=\frac{4\times 10-2\times 9}{10-9}=22$ km/hr.

Exercise

1. A train 75 metres long over took a person who was walking at the rate of 6 km an hour, and passes him in $7\frac{1}{2}$ seconds. Subsequently it overtook a second person,

and passed him in $6\frac{3}{4}$ seconds. At what rate was the second person travelling?

a) 1 km/hr b) 4 km/hr c) 2 km/hr d) 5 km/hr

2. A train 75 metres long overtook a man who was walking at the rate of 6 km/hr and passed him in 18 seconds. Again, the train overtook a second person in 15 seconds. At what rate was the second person travelling?

a) 3 km/hr b) 5 km/hr c) 8 km/hr d) 9 km/hr

3. A train overtakes two persons walking in the same direction at 5 km/hr and 8 km/hr respectively and completely passes them in 10 seconds and 12 seconds respectively. Find the length of the train.

a) 100 m b) 125 m c) 75 m d) 50 m

4. A train overtakes two persons walking in the same direction at 3 km/hr and 5 km/hr respectively and completely passes them in 10 seconds and 11 seconds respectively. Find the speed and length of the train.

a) 25 km/hr, $\frac{560}{9}$ m b) 24 km/hr, $\frac{550}{9}$ m

c) 25 km/hr, $\frac{550}{9}$ m d) None of these

Answers

1. c; **Hint:** Speed of the second person will be less than the first. Because train takes more time to overtake first person than the second.

$$\text{or, } \frac{(6-x)\frac{15\times 27}{2\times 4}}{\frac{15}{2}-\frac{27}{4}}\times\frac{5}{18}=75 \quad \therefore x=2 \text{ km/hr}$$

2. a; **Hint:** Since the train takes lesser time to cross the second person than the first. Hence, speed of the second person will be lesser than the first. Let the speed of the second person be x km/hr.

Now, applying the given rule, we have

$$\left[\frac{(6-x)18\times 15}{18-15}\right]\frac{5}{18}=75$$

or, $6-x=3 \therefore x=3$ km/hr

3. d 4. c

Rule 24

Theorem: ***A train passes two persons who are walking in the direction opposite which the train is moving, at the rate of M_1 m/sec and M_2 m/sec in T_1 seconds and T_2 seconds respectively. Then the speed of the train is $\left(\frac{M_2T_2-M_1T_1}{T_1-T_2}\right)$ m/sec or $\frac{18}{5}\left[\frac{M_2T_2-M_1T_1}{T_1-T_2}\right]$ km/hr and the length of the train is given by $\left[\frac{(M_2-M_1)T_1T_2}{T_1-T_2}\right]$ m.***

Illustrative Example

Ex.: A train passes two persons who are walking in the direction opposite which the train is moving, at the rate of 5 m/s and 10 m/s in 6 seconds and 5 seconds respectively. Find the length of the train and speed of the train.

Soln: **Detail Method:** Let the speed of the train be x m/s.

Length of the train = $(x+5)6=(x+10)5$

or, $(x+5)6=(x+10)5$

$\therefore x=20$ m/s

and length of the train = $(20+5)\times 6=150$ m.

Quicker Method: Applying the above theorem, we have

the speed of the train = $\frac{5\times 10-6\times 5}{6-5}=20$ m/sec and

the length of the train = $\frac{(10-5)\times 6\times 5}{6-5}=150$ m.

Exercise

1. If a train overtakes two persons who are walking in the same direction in which the train is going at the rate of 2 km/hr and 4 km/hr and passes them completely in 9 and 10 seconds. Find the length of train and its speed in km per hour.

a) 50 m, 22 km/hr b) 52 m, 25 km/hr

c) 100 m, 44 km/hr d) Data inadequate

2. A train passes two men walking in the direction opposite to the train at 7 m/sec and at 12 m/sec in 5 and 4 seconds respectively. Find the length of the train.

a) 100 m b) 120 m c) 75 m d) 125 m

3. A train passes two men walking in the direction opposite to the train at 3 m/sec and 5 m/sec in 6 seconds and 5 seconds respectively. Find the length of the train.

a) 75 m b) 80 m c) 65 m d) 60 m

Answers

1. a; **Hint:** $M_1=2\text{ km/hr}=2\times\frac{5}{18}=\frac{5}{9}$ m/sec

$M_2=4\text{ km/hr}=4\times\frac{5}{18}=\frac{10}{9}$ m/sec

$\therefore$ Speed of the train = $\frac{\frac{10}{9}\times 10-\frac{5}{9}\times 9}{10-9}=\frac{55}{9}$ m/sec

$$= \frac{55}{9} \times \frac{18}{5} = 22 \text{ km/hr}$$

$$\therefore \text{ length of the train} = \frac{\left(\frac{10}{9} - \frac{5}{9}\right) 10 \times 9}{10 - 9} = 50 \text{ m}$$

2. a   3. d

Rule 25

Theorem: ***A train L_1 metres in length passes a pole in T_1 seconds and another train of the length L_2 metres travelling in opposite direction in T_2 seconds. Then the speed of the second train is*** $\left[L_1\left(\frac{T_1 - T_2}{T_1.T_2}\right) + \frac{L_2}{T_2}\right]$ **m/sec or**

$\left[L_1\left(\frac{T_1 - T_2}{T_1.T_2}\right) + \frac{L_2}{T_2}\right]\frac{18}{5}$ **km/hr.**

Illustrative Example

Ex.: A train 100 metres in length passes a pole in 10 seconds and another train of the same length travelling in opposite direction in 8 seconds. Find the speed of the second train.

Soln: Detail Method:

$$\text{Speed of the first train} = \frac{100}{10} = 10 \text{ m/s}$$

$$\text{Relative speed in second case} = \frac{100 + 100}{8} = 25 \text{ m/.s}$$

$\therefore$ Speed of the second train = 25 – 10 = 15 m/s

$$\text{or, } 15 \times \frac{18}{5} = 54 \text{ km/hr}$$

Quicker Method: Applying the above theorem, we have

$$\text{the speed of the train} = \left[100\left(\frac{10-8}{10\times8}\right) + \frac{100}{8}\right]\frac{18}{5}$$

$$= \frac{120}{8} \times \frac{18}{5} = 54 \text{ km/hr.}$$

Exercise

1. A train 120 m in length passes a pole in 12 seconds and another train of the length 100 m travelling in opposite directions in 10 seconds. Find the speed of the train in km per hour.
   a) 43.2 km/hr b) 43 km/hr
   c) 44 km/hr d) 43.5 km/hr
2. A train 150 metres in length passes a pole in 15 seconds and another train of the 200 m length travelling in opposite direction in 10 seconds. Find the speed of the second train.
   a) 108 km/hr b) 90 km/hr
   c) 72 km/hr d) 81 km/hr
3. A train 180 metres in length passes a pole in 20 seconds and another train of the 108 m length travelling in opposite direction in 18 seconds. Find the speed of the second train.
   a) 21.6 km/hr b) 25.2 km/hr
   c) 21.2 km/hr d) 25.6 km/hr

Answers

1. a   2. b   3. b

Rule 26

Theorem: ***If a train crosses L_1 m and L_2 m long bridge or platform or tunnel in T_1 seconds and T_2 seconds respectively, then the length of the train is*** $\left(\frac{L_1T_2 - L_2T_1}{T_1 - T_2}\right)$ ***m and the speed of the train is*** $\left(\frac{L_1 - L_2}{T_1 - T_2}\right)$ ***m/sec.***

Illustrative Example

Ex.: A train crosses 210 metres and 122 metres long bridge in 25 seconds and 17 seconds respectively. Find the length and speed of the train.

Soln: Detail Method: Let the length of the train L m and speed of the train be x m/sec.

According to the question equating the speed, ie

$$\frac{210 + L}{25} = \frac{122 + L}{17} = x$$

or, (210 + L) 17 = 25(122 + L)

or, 25L – 17L = 210 × 17 – 122 × 25

$$\text{or, L} = \frac{210\times17 - 122\times25}{25-17} = \frac{520}{8} = 65 \text{ m}$$

$\therefore$ Length of the train = 65 metres

$$\therefore \text{ speed } (x) = \frac{210 + 65}{25} = \frac{275}{25} = 11 \text{ m/sec}$$

Quicker Method: Applying the above theorem, we have

the length of the train

$$= \frac{210\times17 - 122\times25}{25-17} = \frac{520}{8} = 65 \text{ m and}$$

$$\text{the speed of the train} = \frac{210 - 122}{25 - 17} = \frac{88}{8} = 11 \text{ m/sec.}$$

Exercise

1. A train takes 18 seconds to pass completely through a station 162 metres long and 15 seconds to pass completely through another station 120 metres long. Find the length of the train and its speed in km/hr.
   a) 90 metres, 50 km/hr b) 90 metres, 50.4 km/hr
   c) 92 metres, 50 km/hr d) 92 metres, 50.4 km/hr
2. A train crosses 105 metres and 61 metres long bridge in $12\frac{1}{2}$ seconds and $8\frac{1}{2}$ seconds respectively. Find the length and speed of the train.
   a) 65 m, 11 m/sec b) 32.5 m, 5.5 m/sec
   c) 32.5 m, 11 m/sec d) Data inadequate
3. A train crosses 420 metres and 244 metres long bridge in 50 seconds and 34 seconds respectively. Find the length and speed of the train.
   a) 130 m, 44 m/sec b) 130 m, 22 m/sec
   c) 130 m, 11 m/sec d) 65 m, 11 m/sec

Answers

1. b; **Hint:** Length of the train $= \frac{162 \times 15 - 120 \times 18}{18 - 15} = 90$ metres

Speed of the train $= \frac{162 - 120}{3} = 14$ m/sec

$= \frac{14 \times 18}{5} = 50.4$ km/hr

2. c 3. c

Rule 27

Theorem: ***A train L_1 metres in length passes a pole in T_1 seconds and another train of the length L_2 metres travelling in the same direction in T_2 seconds. The speed of the second train is $\frac{18}{5}\left[L_1\left(\frac{T_2 - T_1}{T_1 T_2}\right) - \frac{L_2}{T_2}\right]$ km/hr or $\left[L_1\left(\frac{T_2 - T_1}{T_1 . T_2}\right) - \frac{L_2}{T_2}\right]$ m/s.***

Illustrative Example

Ex.: A train 96 metres in length passes a pole in 6 seconds and another train of the length 144 metres travelling in the same direction in 24 seconds. Find the speed of the second train.

Soln: Detail Method:

Speed of the first train $= \frac{96}{6} = 16$ m/s

Relative speed in second case $= \frac{96 + 144}{24} = 10$ m/s

$\therefore$ speed of the second train

= speed of the first train – relative speed. [See **Note**]

$\therefore$ speed of the second train = 16 m/s – 10 m/s

= 6 m/s

$= 6 \times \frac{18}{5} = \frac{108}{5} = 21\frac{3}{5}$ km/hr

Quicker Method: Applying the above theorem, we have

the speed of the second train

$$= \frac{18}{5}\left[96\left(\frac{24 - 6}{24 \times 6}\right) - \frac{144}{24}\right]$$

$$= \frac{18}{5} \times \left[\frac{96 \times 18}{24 \times 6} - 6\right] = 6 \times \frac{18}{5} = \frac{108}{5} = 21\frac{3}{5} \text{ km/hr.}$$

[**Note:** In this case speed of the first train will be always more than the speed of the second train, because, first train crosses the second train travelling in the same direction.

$\therefore$ Relative speed = First speed – Second speed

or, Second speed = First speed – Relative speed]

Exercise

1. A train 192 metres in length passes a pole in 12 seconds and another train of the length 288 metres travelling in the same direction in 48 seconds. Find the speed of the second train.
   a) 42.6 km/hr b) 32.4 km/hr c) 32 km/hr d) 21.6 km/hr
2. A train 72 metres in length passes a pole in 8 seconds and another train of the length 108 metres travelling in the same direction in 36 seconds. Find the speed of the second train.
   a) 14.4 km/hr b) 24 km/hr c) 15 km/hr d) 15.4 km/hr
3. A train 144 metres in length passes a pole in 16 seconds and another train of the length 216 metres travelling in the same direction in 1 min 12 seconds. Find the speed of the second train.
   a) 14.4 km/hr b) 28.4 km/hr
   c) 104.4 km/hr d) Data inadequate

Answers

1. d 2. a 3. a

Rule 28

Theorem: ***A goods train and a passenger train are running on parallel tracks in the same or in the opposite direction. The driver of the goods train observes that the passenger train coming from behind overtakes and crosses his train completely in T_1 seconds. Whereas a passenger on the passenger train marks that he crosses the goods train in T_2***

seconds. If the speeds of the trains be in the ratio of a : b, then the ratio of their lengths is given by $\left(\dfrac{T_2}{T_1 - T_2}\right)$.

Illustrative Examples

Ex. 1: A goods train and a passenger train are running on parallel tracks in the same direction. The driver of the goods train observes that the passenger train coming from behind overtakes and crosses his train completely in 60 seconds. Whereas a passenger on the passenger train marks that he crosses the goods train in 40 seconds. If the speeds of the trains be in the ratio of 1 : 2, find the ratio of their lengths.

Soln: **Detail Method:** Suppose the speeds of the two trains are x m/s and $2x$ m/s respectively. Also, suppose that the lengths of the two trains are A m and B m respectively.

Then, $\dfrac{A+B}{2x-x} = 60$ ..... (1)

and $\dfrac{A}{2x-x} = 40$ ..... (2)

Dividing (1) by (2) we have

$$\frac{A+B}{A} = \frac{60}{40}$$

or, $\dfrac{B}{A} + 1 = \dfrac{3}{2}$ or, $\dfrac{B}{A} = \dfrac{1}{2}$

$\therefore A : B = 2 : 1$

Quicker Method: Applying the above theorem, we have

the required ratio $= \dfrac{40}{60-40} = 40 : 20 = 2 : 1$

Ex. 2: A goods train and a passenger train are running on parallel tracks in the opposite directions. The passenger train crosses the goods train completely in 40 seconds. Whereas a passenger on the passenger train marks that he crosses the goods train in 25 seconds. If the speeds of the trains be in the ratio of 4 : 5, find the ratio of their lengths.

Soln: **Detail Method:** Suppose the speeds of the two trains are $4x$ m/sec and $5x$ m/sec respectively. Also suppose that the lengths of the two trains are A m and B m respectively.

Then, $\dfrac{A+B}{4x+5x} = 40$ ..... (1) and

$\dfrac{A}{4x+5x} = 25$ .... (2)

Dividing (1) by (2), we have

$$\frac{A+B}{A} = \frac{40}{25}$$

or, $\dfrac{B}{A} + 1 = \dfrac{8}{5}$ or, $\dfrac{B}{A} = \dfrac{3}{5}$

$\therefore$ A : B = 5 : 3

Quicker Method: Applying the above theorem, we have

the required ratio $= \dfrac{25}{40-25} = \dfrac{5}{3} = 5 : 3$.

Exercise

1. A goods train and a passenger train are running on parallel tracks in the same direction. The driver of the goods train observes that the passenger train coming from behind overtakes and crosses his train completely in 30 seconds. Whereas a passenger on the passenger train marks that he crosses the goods train in 20 seconds. If the speeds of the trains be in the ratio of 1 : 2, find the ratio of their lengths.
   a) 3 : 2 b) 3 : 1 c) 2 : 1 d) 4 : 1
2. A goods train and a passenger train are running on parallel tracks in the same direction. The driver of the goods train observes that the passenger train coming from behind overtakes and crosses his train completely in 40 seconds. Whereas a passenger on the passenger train marks that he crosses the goods train in 30 seconds. If the speeds of the trains be in the ratio of 1 : 2, find the ratio of their lengths.
   a) 3 : 1 b) 1 : 2 c) 1 : 3 d) 1 : 2
3. A goods train and a passenger train are running on parallel tracks in the same direction. The driver of the goods train observes that the passenger train coming from behind overtakes and crosses his train completely in 50 seconds. Whereas a passenger on the passenger train marks that he crosses the goods train in 25 seconds. If the speeds of the trains be in the ratio of 1 : 2, find the ratio of their lengths.
   a) 2 : 1 b) 1 : 2
   c) 3 : 2 d) both trains will be of same length
4. A goods train and a passenger train are running on parallel tracks in the same direction. The driver of the goods train observes that the passenger train coming from behind overtakes and crosses his train completely in 15 seconds. Whereas a passenger on the passenger train marks that he crosses the goods train in 12 seconds. If the speeds of the trains be in the ratio of 1 : 2, find the ratio of their lengths.
   a) 4 : 1 b) 3 : 1 c) 1 : 4 d) 2 : 3

Answers

1. c 2. a 3. d 4. a

Rule 29

Theorem: ***A train after travelling d_1 km meets with an accident and then proceeds at $\frac{x}{y}$ of its former speed and arrives at its destination t_1 hours late. Had the accident occurred d_2 km further, it would have reached the destination only t_2 hours late. The speed of the train is $\left[\frac{d_2\left(1-\frac{x}{y}\right)}{\frac{x}{y}(t_1-t_2)}\right]$ km/hr and the distance which train travels is $\left(d_1+\frac{d_2t_1}{t_1-t_2}\right)$ km.***

Illustrative Example

Ex.: A train after travelling 50 km meets with an accident and then proceeds at $\frac{3}{4}$ of its former speed and arrives at its destination 35 minutes late. Had the accident occurred 24 km further, it would have reached the destination only 25 minutes late. Find the speed of the train and the distance which the train travels.

Soln: **Detail Method:** Let the speed of the train be x km/hr and the distance D km.

From the question we have,

$$\frac{50}{x}+\frac{(D-50)4}{3x}=\frac{D}{x}+\frac{35}{60}$$

or, $\frac{150+4D-200}{3x}=\frac{12D+7x}{12x}$

or, $4D-7x=200$ .... (i) and

$$\frac{74}{x}+\frac{(D-74)4}{3x}=\frac{D}{x}+\frac{25}{60}$$

or, $\frac{222+4D-296}{3x}=\frac{12D+5x}{12x}$

$4D-5x=296$ .... (ii)

Now, subtracting equation (i) from the equation (ii), we have

$2x=96$

$\therefore$ x = 48 km/hr

$\therefore$ Speed of the train = 48 km/hr

To find the distance, put the value of x in equation (ii)

4D – 5x = 296

or, 4D – 5 × 48 = 296

or, D = $\frac{536}{4}=134$ km

$\therefore$ Distance which the train travels = 134 km.

Quicker Method: Applying the above theorem, we have

$$\text{speed of train} = \frac{24\left(1-\frac{3}{4}\right)}{\frac{3}{4}\left(\frac{35-25}{60}\right)}=\frac{6\times4\times6}{3}=48 \text{ km/hr}$$

$$\text{and distance} = 50+\frac{24\times35}{10}=50+84=134 \text{ km}$$

Note: A train after travelling d_1 km meets with an accident and then proceeds at $\frac{x}{y}$ of its former speed and arrives at its destination t_1 hours late. Had the accident occurred d_2 km behind, it would have reached the destination only t_2 hours late. The speed of the train is given by $\left[\frac{\left(1-\frac{x}{y}\right)d_2}{\frac{x}{y}(t_2-t_1)}\right]$ km/hr and the distance is given by $\left[d_1+\frac{d_2t_1}{t_2-t_1}\right]$ km.

Ex.: A train after travelling 50 km meets with an accident and then proceeds at $\frac{3}{4}$ of its former speed and arrives at its destination 25 minutes late. Had the accident occurred 24 km behind, it would have reached the destination only 35 minutes late. Find the speed of the train and the distance travelled by the train.

Soln: **Detail Method:** Let the distance be D km and speed be the x km/hr

From the question, we have

$$\frac{50}{x}+\frac{(D-50)4}{3x}=\frac{D}{x}+\frac{25}{60}=\frac{D}{x}+\frac{5}{12}$$

or, $\frac{150+4D-200}{3x}=\frac{12D+5x}{12x}$

or, $\frac{4D-50}{3x}=\frac{12D+5x}{12x}$

or, $16D-200=12D+5x$

$\therefore$ 4D – 5x = 200 ... (i) and

$$\frac{50-24}{x}+\frac{(D-26)4}{3x}=\frac{D}{x}+\frac{35}{60}=\frac{D}{x}+\frac{7}{12}=\frac{12D+7x}{12x}$$

or, $\frac{26}{x}+\frac{4D-104}{3x}=\frac{12D+7x}{12x}$

or, $\frac{78+4D-104}{3x}=\frac{12D+7x}{12x}$

or, $\frac{4D-26}{3x}=\frac{12D+7x}{12x}$

or, $4D-7x=104$ .... (ii)

Now, subtracting equation (ii) from equation (i) we have

$2x=96$

$\therefore x=48$ km/hr

Put the value of x in equation (i) and find the distance (D)

or, $4D-5\times 48=200$

or, $4D=200+240=440$

$\therefore D=\frac{440}{4}=110$ km.

Quicker Method: Applying the above theorem, we have

$$\text{the speed of the train}=\frac{\left(1-\frac{3}{4}\right)24}{\frac{3}{4}\left(\frac{35-25}{60}\right)}=\frac{6\times 4\times 6}{3}=48$$

km/hr and

$$\text{the distance}=50+\frac{24\times 25}{10}=50+60=110 \text{ km.}$$

Exercise

1. A train running from A to B meets with an accident 50 kilometres from A, after which it moves with $\frac{3}{5}$th of its original speed, and arrives at B 3 hours late. Had the accident happened 50 kilometres further on, it would have been only 2 hours late. Find the original speed of the train and the distance from A to B.

a) 30 km/hr, 200 km b) $33\frac{1}{3}$ km/hr, $200\frac{1}{2}$ km

c) $33\frac{1}{3}$ km/hr, 200 km d) $30\frac{1}{3}$ km/hr, 200 km

2. A train after travelling 54 km meets with an accident and then proceeds at $\frac{1}{4}$ of its former speed and arrives at its destination 45 minutes late. Had the accident occurred 36 km further, it would have reached the destination only 15 minutes late. Find the speed of the train and the distance which the train travels.

a) 216 km/hr, 108 km b) 108 km/hr, 108 km

c) 206 km/hr, 180 km d) Data inadequate

3. A train after travelling 60 km meets with an accident and then proceeds at $\frac{2}{3}$ of its former speed and arrives at its destination 40 minutes late. Had the accident occurred 30 km further, it would have reached the destination only 20 minutes late. Find the speed of the train and the distance which the train travels.

a) 45 km/hr, 90 km b) 60 km/hr, 120 km

c) 45 km/hr, 120 km d) None of these

Answers

1. c 2. a 3. c

Rule 30

Theorem: ***A train after travelling d_1 km meets with an accident and then proceeds at $\frac{x}{y}$ of its former speed and arrives at the terminus t_1 hours late. Had the accident happened d_2 km further on, it would have arrives t_2 minutes sooner. The speed of the train is $\left[\frac{\left(1-\frac{x}{y}\right)d_2}{\left(\frac{x}{y}\right)t_2}\right]$ km/hr and the distance is $\left[\frac{d_1t_2+d_2t_1}{t_2}\right]$ km.***

Illustrative Example

Ex.: A train after travelling 50 km meets with an accident and then proceeds at $\frac{4}{5}$ of its former rate and arrives at the terminus 45 minutes late. Had the accident happened 20 km further on, it would have arrived 12 minutes sooner. Find the rate of the train and the distance.

Soln: **Detail Method - I:** Let the distance be D km and the speed of the train be x km/hr.

From the question we have,

$$\frac{50}{x}+\frac{(D-50)5}{4x}=\frac{D}{x}+\frac{45}{60}=\frac{D}{x}+\frac{3}{4}$$

or, $\frac{200+5D-250}{4x}=\frac{4D+3x}{4x}$

or, $5D-50=4D+3x$

or, $D-3x=50$ ..... (i) and

$$\frac{50+20}{x}+\frac{\{D-(50+20)\}5}{4x}=\frac{D}{x}+\frac{45-12}{60}=\frac{D}{x}+\frac{11}{20}$$

or, $\frac{70}{x}+\frac{5D-350}{4x}=\frac{20D+11x}{20x}$

or, $\frac{280+5D-350}{4x}=\frac{20D+11x}{20x}$

or, $25D-350=20D+11x$

or, $5D-11x=350$ .... (ii)

Now, 5 × equation (i) – equation (ii), we have

$4x=100$

$\therefore x=\frac{100}{4}=25$ km/hr

$\therefore$ speed of the train = 25 km/hr

Put the value of x in equation (i) to find the value of distance (D).

$D-3x=50$

or, $D-3\times 25=50$

$\therefore$ D = 125 km

$\therefore$ distance which train travels = 125 km.

Detail Method - II: Let A be the starting place, B the terminus, C and D the places where the accidents take place.

By travelling at $\frac{4}{5}$ of its original rate, the train would take $\frac{5}{4}$ of its usual time ie $\frac{1}{4}$ of its original time more.

$\therefore$ $\frac{1}{4}$ of the usual time taken to travel the distance CB = 45 min .... (1)

and $\frac{1}{4}$ of the usual time taken to travel the distance DB = (45 – 12) = 33 min .... (2)

Subtracting (2) from (1)

$\frac{1}{4}$ of the usual time taken to travel the distance CD = 12 min

$\therefore$ Usual time taken to travel 20 km = 48 min.

$\therefore$ Speed of the train per hour $=\frac{20}{48}\times 60=25$ km

From (1) we have,

time taken to travel CB = 45 × 4 min = 3 hours

$\therefore$ the distance CB = 25 × 3 or 75 km

Hence the distance AB = the distance (AC + CB) = (50 + 75) or 125 km.

Quicker Method: Applying the above theorem, we have

the required speed $=\frac{\left(1-\frac{4}{5}\right)\times 20}{\frac{4}{5}\times\frac{12}{60}}=\frac{\frac{1}{5}\times 20}{4\times\frac{1}{5}\times\frac{1}{5}}=5\times 5$

= 25 km/hr and

the required distance

$$=\frac{50\times 12+20\times 45}{12}=\frac{600+900}{12}=\frac{1500}{12}=125\text{ km.}$$

Note: The above example can be solved by applying the previous theorem (ie **Rule 29**),

Here, $t_2=45-12=33$ min

Now, speed of the train

$$=\frac{20\left(1-\frac{4}{5}\right)}{\frac{4}{5}\left(\frac{45-33}{60}\right)}=\frac{4}{\frac{4}{5}\times\frac{1}{5}}=25\text{ km/hr}$$

$$\text{Distance}=50+\frac{20\times 45}{12}=50+75=125\text{ km.}$$

Exercise

1. A train after travelling 100 km meets with an accident and then proceeds at $\frac{3}{5}$ of its former rate and arrives at the terminus 48 minutes late. Had the accident happened 30 km further on, it would have arrived 24 minutes sooner. Find the rate of the train and the distance.
a) 50 km/hr, 160 km b) 45 km/hr, 150 km
c) 50 km/hr, 150 km d) 50 km/hr, 120 km
2. A train after travelling 90 km meets with an accident and then proceeds at $\frac{2}{3}$ of its former rate and arrives at the terminus 40 minutes late. Had the accident happened 45 km further on, it would have arrived 20 minutes sooner. Find the rate of the train and the distance.
a) 67.5 km/hr, 160 km b) 67.5 km/hr, 180 km
c) 68 km/hr, 180 km d) 72 km/hr, 180 km
3. A train after travelling 54 km meets with an accident and

then proceeds at $\frac{3}{4}$ of its former rate and arrives at the terminus 45 minutes late. Had the accident happened 36 km further on, it would have arrived 15 minutes sooner. Find the rate of the train and the distance.

a) 48 km/hr, 189 km   b) 48 km/hr, 180 km
c) 42 km/hr, 190 km   d) None of these

Answers

1. a   2. b   3. a

Rule 31

Theorem: ***A train meets with an accident 't_1' hours after starting, which detains it for 't' hours, after which it proceeds at $\frac{x}{y}$ of its original speed. It arrives at the destination 't_2' hours late. Had the accident taken place 'd' km farther along the railway line, the train would have arrived only 't_3' hours late. The original speed of the train is given by*** $\left[\frac{d\left(1-\frac{x}{y}\right)}{\frac{x}{y}(t_2-t_3)}\right]$ ***km/hr and the length of the trip is given by*** $\left(\frac{d}{t_2-t_3}\right)\left[t_2+t_1\left(\frac{y}{x}-1\right)\right]$ ***km.***

Illustrative Example

Ex.: A train meets with an accident 3 hours after starting, which detains it for 1 hour, after which it proceeds at 75% of its original speed. It arrives at the destination 4 hours late. Had the accident taken place 150 km farther along the railway line, the train would have arrived only $3\frac{1}{2}$ hours late. Find the length of the trip and the original speed of the train.

Soln: **Detail Method:** Let the distance be d km and the initial speed be x km/hr.

As the accident taken place after 3 hours

Distance of accident site = $3x$ km

Distance left = (d - $3x$) km

Total time taken, if no accident happened = $\frac{d}{x}$ hrs

Given is

$$3+1+\frac{d-3x}{x\times\frac{75}{100}}=\frac{d}{x}+4 \Rightarrow 4+\frac{4d-12x}{3x}=\frac{d}{x}+4$$

or, $\frac{4d-12x}{3x}=\frac{d}{x}$ or, $4d-12x=3d$

or, $d=12x$

In 2nd case distance of accident site = $(3x+150)$ km

Distance left = $d-(3x+150)$

$$\therefore \frac{3x+150}{x}+1+\frac{d-(3x+150)}{x\times\frac{75}{100}}=\frac{d}{x}+\frac{7}{2}$$

$$\text{or, } \frac{3x+150}{x}+\frac{4d-12x-600}{3x}=\frac{12x}{x}+\frac{5}{2}$$

$$\left[\because \frac{d}{x}=12 \text{ or, } d=12x\right]$$

$$\text{or, } \frac{3x+150}{x}+\frac{36x-600}{3x}=\frac{29}{2}$$

$$\text{or, } \frac{15x-50}{x}=\frac{29}{2} \quad \text{or, } x=100$$

$\therefore$ original speed of the train = 100 km/hr

$\because$ distance (d) = $12x$

$\therefore$ distance = $12\times100=1200$ km

Quicker Method: Applying the above theorem, we have,

the original speed = $\frac{150\left(1-\frac{3}{4}\right)}{\frac{3}{4}\left(4-\frac{7}{2}\right)}=100$ km/hr and

the required distance = $\left[\frac{150}{\left(4-\frac{7}{2}\right)}\right]\times\left[4+3\left(\frac{4}{3}-1\right)\right]$

$=300\times4=1200$ km.

Note: 1. Change the percentage into fraction. In the above example $75\%=\frac{75}{100}=\frac{3}{4}$.

2. In the above formula, detention time (t) has been cancelled.

Exercise

1. A train meets with an accident 4 hours after starting, which detains it for 2 hours, after which it proceeds at 60% of its original speed. It arrives at the destination 5 hours late. Had the accident taken place 180 km farther along the railway line, the train would have arrived only 4 hours late. Find the length of the trip and the original speed of the train.

a) 120 km/hr, 1830 km b) 120 km/hr, 1380 km
c) 100 km/hr, 1530 km d) 100 km/hr, 1280 km

2. A train meets with an accident 5 hours after starting, which detains it for 1 hour, after which it proceeds at 40% of its original speed. It arrives at the destination 6 hours late. Had the accident taken place 200 km farther along the railway line, the train would have arrived only 5 hours late. Find the length of the trip and the original speed of the train.
a) 300 km/hr, 1700 km b) 200 km/hr, 1650 km
c) 350 km/hr, 1750 km d) 250 km/hr, 1700 km

Answers

1. b 2. a

Rule 32

Theorem: ***A train covers a distance between station A and B in T_1 hours. If the speed is reduced by x km/hr, it will cover the same distance in T_2 hours, then the distance between the two stations A and B is $\left(\frac{xT_1T_2}{T_2-T_1}\right)$ km and the speed of the train is given by $\left(\frac{xT_2}{T_2-T_1}\right)$ km/hr.***

Illustrative Example

Ex.: A train covers a distance between stations A and B in 45 minutes. If the speed is reduced by 5 km/hr, it will cover the same distance in 48 minutes. What is the distance between the two stations A and B (in km)? Also, find the speed of the train.

Soln: **Detail Method:** Suppose the distance is x km and the speed of the train is y km/hr.
Thus we have two relationships:

(1) $\frac{x}{y}=\frac{45}{60}=\frac{3}{4}\Rightarrow x=\frac{3}{4}y$

(2) $\frac{x}{y-5}=\frac{48}{60}=\frac{4}{5}\Rightarrow x=\frac{4}{5}(y-5)$

From (1) and (2)

$\frac{3}{4}y=\frac{4}{5}(y-5)$ or, $y\left(\frac{4}{5}-\frac{3}{4}\right)=4$

or, $y=\frac{4\times 20}{16-15}=80$ km/hr

Therefore speed = 80 km/hr and distance

$x=\frac{3}{4}\times 80=60$ km

Quicker Method: Applying the above theorem, we have

the speed of the train = $\frac{48}{48-45}\times 5=80$ km/hr and

the distance = $5\left[\frac{45\times 48}{48-45}\right]\frac{1}{60}=60$ km.

Note: **(1)** In the formula for distance we have used $\frac{1}{60}$ to change minutes into hours.

(2) We don't need to remember the formula for distance. Once we find the speed, we may use the first information to find the distance.

Thus, distance = $80\times\frac{45}{60}=60$ km

Exercise

1. A train covers a distance between stations A and B in 2 hours. If the speed is reduced by 6 km/hr, it will cover the same distance in 3 hours. What is the distance between the two stations A and B (in km)? Also, find the speed of the train.
a) 36 km, 18 km/hr b) 42 km, 21 km/hr
c) 18 km, 9 km/hr d) 28 km, 14 km/hr

2. A train covers a distance between stations A and B in 1 hour. If the speed is reduced by 4 km/hr, it will cover the same distance in 3 hours. What is the distance between the two stations A and B (in km)? Also, find the speed of the train.
a) 8 km, 5 km/hr b) 6 km, 6 km/hr
c) 9 km, 6 km/hr d) Data inadequate

3. A train covers a distance between stations A and B in 3 hours. If the speed is reduced by 8 km/hr, it will cover the same distance in 6 hours. What is the distance between the two stations A and B (in km)? Also, find the speed of the train.
a) 36 km, 12 km/hr b) 40 km, 15 km/hr
c) 48 km, 16 km/hr d) Data inadequate

Answers

1. a 2. b 3. c

Rule 33

Theorem: ***Two places P and Q are D km apart. A train leaves P for Q and at the same time another train leaves Q for P. Both the trains meet T hours after they start moving. If the train travelling from P to Q travels x km/hr faster or slower than the other train, then the speed of the faster train is $\left(\frac{D+Tx}{2T}\right)$ km/hr and the slower train is $\left(\frac{D-Tx}{2T}\right)$ km/hr.***

Illustrative Examples

Ex. 1: Two places P and Q are 162 km apart. A train leaves P for Q and at the same time another train leaves Q for P. Both the trains meet 6 hrs after they start moving. If the train travelling from P to Q travels 8 km/hr faster than the other train, find the speed of the two trains.

Soln: **Detail Method:** Suppose the speeds of the two trains are p km/hr and q km/hr respectively. Thus

$p+q=\frac{162}{6}=27$ .... (i) and $p-q=8$ .... (ii)

(i) + (ii) implies that

$2p=35 \quad \therefore p=17.5$ km/hr

and (i) – (ii) implies that

$2q=19 \quad \therefore q=9.5$ km/hr

Quicker Method: Applying the above theorem, we have,

speeds of the trains $=\frac{162+6\times 8}{2\times 6}$ and $\frac{162-6\times 8}{2\times 6}$

= 17.5 km/hr and 9.5 km/hr.

Ex. 2: Two places P and Q are 162 km apart. A train leaves P for Q and at the same time another train leaves Q for P. Both the trains meet 6 hrs after they start moving. If the train travelling from P to Q travels 8 km/hr slower than the other train, find the speed of the two trains.

Soln: Applying the above theorem, we have the speed of the slower train

$=\frac{162-6\times 8}{2\times 6}=9.5$ km/hr and

the speed of the faster train

$=\frac{162+6\times 8}{2\times 6}=17.5$ km/hr.

Exercise

1. Two places P and Q are 160 km apart. A train leaves P for Q and at the same time another train leaves Q for P. Both the trains meet 5 hrs after they start moving. If the train travelling from P to Q travels 6 km/hr faster than the other train, find the speed of the two trains.
   a) 19 km/hr, 16 km/hr b) 13 km/hr, 9 km/hr
   c) 19 km/hr, 13 km/hr d) Can't be determined
2. Two places P and Q are 92 km apart. A train leaves P for Q and at the same time another train leaves Q for P. Both the trains meet 4 hrs after they start moving. If the train travelling from P to Q travels 7 km/hr faster than the other train, find the speed of the two trains.
   a) 15 km/hr, 8 km/hr b) 12 km/hr, 8 km/hr
   c) 12 km/hr, 9 km/hr d) 15 km/hr, 9 km/hr
3. Two places P and Q are 132 km apart. A train leaves P for Q and at the same time another train leaves Q for P. Both the trains meet 6 hrs after they start moving. If the train travelling from P to Q travels 7 km/hr slower than the other train, find the speed of the two trains.
   a) 7 km/hr, 15 km/hr b) 8 km/hr, 16 km/hr
   c) 8 km/hr, 15 km/hr d) 7 km/hr, 16 km/hr
4. Two places P and Q are 150 km apart. A train leaves P for Q and at the same time another train leaves Q for P. Both the trains meet 2 hrs after they start moving. If the train travelling from P to Q travels 5 km/hr slower than the other train, find the speed of the two trains.
   a) $21\frac{1}{2}$ km/hr, 16 km/hr b) $42\frac{1}{2}$ km/hr 35 km/hr
   c) 42 km/hr, 36 km/hr d) Can't be determined

Answers

1. c 2. a 3. a 4. b

Rule 34

Theorem: ***Two trains A and B start from P and Q towards Q and P respectively. After passing each other they take T_1 hours and T_2 hours to reach Q and P respectively. if the train from P is moving x km/hr, then the speed of the other train is $\left(x\sqrt{\frac{T_1}{T_2}}\right)$ km/hr.***

Or

Speed of the first train

$\times\sqrt{\frac{\text{Time taken by first train after meeting}}{\text{Time taken by second train after meeting}}}$

Illustrative Example

Ex.: Two trains A and B start from Delhi and Patna towards Patna and Delhi respectively. After passing each other they take 4 hours 48 minutes and 3 hours and 20 minutes to reach Patna and Delhi respectively. If the train from Delhi is moving at 45 km/hr then find the speed of the other train.

Soln: **Detail Method:**

A	*	B
Delhi (45 km)	M	x km/hr Patna

Suppose the speed of train B is x km/hr and they meet at M. Now,

distance MB = 45 × (4 hrs + 48 minutes)

$=45\times\left(4\frac{4}{5}\right)=45\times\frac{24}{5}=216$ km

And the distance AM = x × (3 hrs + 20 minutes)

$=x\left(3\frac{1}{3}\right)=\frac{10x}{3}$ km

Now, the time to reach the train from Patna to M = the time to reach the train from Delhi to M.

or, $\frac{MB}{x}=\frac{AM}{45}$ or, $\frac{216}{x}=\frac{10x}{3+45}$

or, $10x^2 = 216\times3\times45$ or, $x^2 = 2916$

$\therefore x = 54$ km/hr.

Quicker Method: Applying the above theorem, we have

the speed of the other train

$$= 45\sqrt{\frac{4\frac{4}{5}}{3\frac{1}{3}}} = 45\sqrt{\frac{24}{5}\times\frac{3}{10}} = 45\sqrt{\frac{36}{25}}$$

$$= 45\times\frac{6}{5} = 54 \text{ km/hr.}$$

Exercise

1. Two trains A and B start from Lucknow and Delhi towards Delhi and Lucknow respectively. After passing each other they take 4 hours and 9 hours to reach Delhi and Lucknow respectively. If the train from Lucknow is moving at 60 km/hr then find the speed of the other train.
   a) 40 km/hr b) 30 km/hr c) 35 km/hr d) 50 km/hr
2. Two trains A and B start from Delhi and Chandigarh towards Chandigarh and Delhi respectively. After passing each other they take $3\frac{1}{5}$ hours and 5 hours to reach Chandigarh and Delhi respectively. If the train from Delhi is moving at 35 km/hr then find the speed of the other train.
   a) 26 km/hr b) 42 km/hr c) 28 km/hr d) 32 km/hr
3. Two trains A and B start from Kanpur and Patna towards Patna and Kanpur respectively. After passing each other they take 4 hours and 1 hour to reach Patna and Kanpur respectively. If the train from Kanpur is moving at 30 km/hr then find the speed of the other train.
   a) 15 km/hr b) 60 km/hr
   c) 45 km/hr d) Can't be determined

Answers

1. a 2. c 3. b

Rule 35

Theorem: ***The speed of two trains are in the ratio x : y. They are moving in the opposite directions on paralle tracks. The first train crosses a telegraph pole in 'a' seconds whereas the second train crosses a telegraph pole in 'b' seconds. Time taken by the trains to cross each other completely is given by*** $\left(\frac{ax+by}{x+y}\right)$ ***seconds.***

Illustrative Examples

Ex. 1: The speeds of two trains are in the ratio 3 : 4. They are going in opposite directions along parallel tracks. If each takes 3 seconds to cross a telegraph post, find the time taken by the trains to cross each other completely?

Soln: **Detail Method:** Since both the trains cross a telegraph pole in equal time, the ratio of their speeds should be equal to the ratio of their lengths. That is, the lengths of the two trains are in the ratio of 3 : 4.

Suppose the lengths of the two trains be $3x$ and $4x$ metres respectively. Since each of them takes 3 seconds to cross a telegraph pole, speed of the first train

$= \frac{3x}{3} = x$ m/s and speed of the second train

$= \frac{4x}{3}$ m/s

Since they are moving in opposite directions their relative speed $= x + \frac{4x}{3} = \frac{7x}{3}$ m/s

Sum of their lengths $= 3x + 4x = 7x$ m

$\therefore$ time taken to cross each other $= \frac{7x}{\left(\frac{7x}{3}\right)} = 3$ secs.

Quicker Method: Applying the above theorem,

Here x : y = 3 : 4 and a = b = 3 seconds

Thus, time to cross each other

$$= \frac{3\times3+3\times4}{3+4} = \frac{21}{7} = 3 \text{ seconds.}$$

Note: As in the above example, if a = b then the general formula becomes:

Required time to cross each other

$$= \frac{a(x+y)}{(x+y)} = a \text{ secs.}$$

Ex. 2: The speeds of two trains are in the ratio of 7 : 9. They are moving on the opposite directions on parallel tracks. The first train crosses a telegraph pole in 4 seconds whereas the second train crosses the pole in 6 seconds. Find the time taken by the trains to cross each other completely.

Soln: **Detail Method:** Suppose the speeds are $7x$ m/s and $9x$ m/s

Then length of first train $= 7x \times 4 = 28x$ metres and the length of second train $= 9x \times 6 = 54x$ metres

Time to cross each other $= \frac{\textit{Sum of lengths}}{\textit{Sum of speeds}}$

$$= \frac{28x+54x}{7x+9x} = \frac{82}{16} = \frac{41}{8} = 5\frac{1}{8} \text{ seconds.}$$

Quicker Method: Using the above theorem, we have

$$\text{the required time} = \frac{7\times4+9\times6}{7+9} = \frac{82}{16} = 5\frac{1}{8} \text{ seconds.}$$

Exercise

1. Two trains are moving in the opposite directions on parallel tracks at the speeds of 64 km/hr and 96 km/hr respectively. The first train passes a telegraph post in 5 seconds whereas the second train passes the post in 6 seconds. Find the time taken by the trains to cross each other completely.

   a) $\frac{18}{5}$ sec b) $\frac{28}{5}$ sec c) 6 sec d) None of these

2. The speeds of two trains are in the ratio of 3 : 5. They are moving on the opposite directions on parallel tracks. The first train crosses a telegraph pole in 3 seconds whereas the second train crosses the pole in 5 seconds. Find the time taken by the trains to cross each other completely.

   a) $\frac{17}{4}$ sec b) 4 sec c) $\frac{15}{4}$ sec d) $4\frac{1}{2}$ sec

3. The speeds of two trains are in the ratio of 2 : 3. They are moving on the opposite directions on parallel tracks. The first train crosses a telegraph pole in 10 seconds whereas the second train crosses the pole in 15 seconds. Find the time taken by the trains to cross each other completely.

   a) 23 sec b) 14 sec c) 13 sec d) 16 sec

4. The speeds of two trains are in the ratio 5 : 9. They are going in opposite directions along parallel tracks. If each takes 5 seconds to cross a telegraph post, find the time taken by the trains to cross each other completely?

   a) 3 sec b) 5 sec c) 6 sec d) 9 sec

Answers

1. b; **Hint:** $x : y = 64 : 96 = 2 : 3$

   Now applying the given rule, we have

   $$\text{the required answer} = \frac{2\times5+3\times6}{2+3} = \frac{28}{5} = 5\frac{3}{5} \text{ sec}$$

2. a 3. c 4. b

Rule 36

Theorem: ***A train with x km/hr crosses a bridge in T seconds. Another train L metres shorter crosses the same bridge at y km/hr. Time taken by the second train to cross the bridge is given by*** $\left[\frac{x}{y}T - \frac{L}{y\times\frac{5}{18}}\right]$ ***seconds.***

Illustrative Example

Ex.: A train with 90 km/hr crosses a bridge in 36 seconds. Another train 100 metres shorter crosses the same bridge at 45 km/hr. Find the time taken by the second train to cross the bridge.

Soln: Detail Method:

Speed of the first train $= 90\times\frac{5}{18} = 25$ m/sec

Let the length of the bridge be x m and length of the train be y m.

or, $x + y = 25\times36$ metres ... (i)

Speed of the second train $= 45\times\frac{5}{18} = \frac{25}{2}$ m/sec

$x + (y-100) = \frac{25}{2}\times t$ ... (ii) [where t = required time]

Now, putting equation (i) into equation (ii), we have,

$$(25\times36) - 100 = \frac{25t}{2}$$

$$\text{or, } t = \frac{(900-100)2}{25} \quad \therefore t = \frac{800\times2}{25} = 64 \text{ seconds.}$$

Quicker Method: Applying the above theorem, we have the required time

$$= \frac{90}{45}\times36 - \frac{100}{\frac{45\times5}{18}} = 72 - 8 = 64 \text{ secs.}$$

Exercise

1. A train with 60 km/hr crosses a bridge in 25 seconds. Another train 120 metres shorter crosses the same bridge at 30 km/hr. Find the time taken by the second train to cross the bridge.

   a) 35 sec b) 34.6 sec c) 35.2 sec d) 35.6 sec

2. A train with 36 km/hr crosses a bridge in 18 seconds. Another train 90metres shorter crosses the same bridge at 27 km/hr. Find the time taken by the second train to cross the bridge.

   a) 20 sec b) 18 sec c) 16 sec d) 12 sec

3. A train with 72 km/hr crosses a bridge in 36 seconds. Another train 180 metres shorter crosses the same bridge at 54 km/hr. Find the time taken by the second train to cross the bridge.

   a) 24 sec b) 32 sec
   c) 36 sec d) Can't be determined

Answers

1. d 2. d 3. c

Miscellaneous

1. Train 'A' leaves Mumbai Central for Lucknow at 11 am, running at the speed of 60 km/hr. Train 'B' leaves Mumbai Central for Lucknow by the same route at 2 pm on the same day, running at the speed of 72 km/hr. At what time

will the two trains meet each other?

[BSRB Patna PO, 2001]

a) 2 am on the next day b) 5 am on the next day
c) 5 pm on the next day d) None of these

2. A passenger train leaves Calcutta at 4 PM and travels at the rate of 30 kilometres an hour. The mail train leaves Calcutta at 9 PM and travels, on a parallel line of rails, at the rate of 45 km an hour, when will the second train overtake the first?
a) 10 hrs after the first train start
b) 12 hrs after the second train starts
c) 10 hrs after the second train starts
d) 12 hrs after the first train starts

Answers

1. b; Distance covered by train A before the train B leaves Mumbai Central $= 60 \times 3 = 180$ km

$\therefore$ Time taken to cross each other $= \frac{180}{12} = 15$ hrs

$\therefore$ required time $= 2$ pm $+ 15$ hours
$= 5$ am on the next day.

2. c; **Hint:** The first train has started 5 hrs before the second. Therefore, $(30 \times 5 = 150)$ km away when the second train starts. Therefore the second train has to gain 150 km on the first, at the rate of 15 ie $(45 - 30 = 15)$ km an hour.
Second train gains 15 km in 1 hour on the first.
$\therefore$ Second train gains 150 km in 10 hours on the first.
$\therefore$ the time required is 10 hours after the second train starts.
$\therefore$ the second overtakes the first at a distance of $(45 \times 10) = 450$ km from Calcutta.

18

Streams

Rule 1

Theorem: ***If the speed of the boat (or the swimmer) is x km/hr and if the speed of the stream is y km/hr then, while upstream the effective speed of the boat = (x – y) km/hr.***

Illustrative Example

Ex.: Speed of a man is 8 km/hr in still water. If the rate of current is 3 km/hr, find the effective speed of the man upstream.

Soln: Applying the above theorem, we have effective speed of the man upstream

$= (8 - 3 = 5)$ km/hr

Note: Normally by speed of the boat or swimmer we mean the speed of the boat (or swimmer) in still water.

Exercise

1. The speed of a boat in still water is 2 km/hr. If its speed upstream be 1 km/hr, then speed of the stream is **[Asst. Grade Exam, 1997]**
   a) 2 km/hr b) 3 km/hr
   c) 1 km/hr d) None of these
2. A boat goes 14 km upstream in 56 minutes. The speed of stream is 2 km/hr. The speed of boat in still water is
   a) 6 km/hr b) 15 km/hr c) 14 km/hr d) 17 km/hr
3. Speed of a man is 7 km/hr in still water. If the rate of current is 2 km/hr, find the effective speed of the man upstream.
   a) 9 km/hr b) 5 km/hr
   c) 14 km/hr d) Data inadequate
4. Speed of a man is 9 km/hr in still water. If the rate of current is 5 km/hr, find the effective speed of the man upstream.
   a) 14 km/hr b) 12 km/hr c) 4 km/hr d) 5 km/hr

Answers

1. c; **Hint:** $2 - y = 1 \quad \therefore y = 2 - 1 = 1$ km/hr

2. d; **Hint:** Rate upstream $= \frac{14 \times 60}{56} = 15$ km/hr

$(x - 2) = 15 \qquad \therefore x = 17$ km/hr

3. b 4. c

Rule 2

Theorem: ***If the speed of the boat (or the swimmer) is x km/hr and if the speed of the stream is y km/hr then, while downstream the effective speed of the boat = (x + y) km/hr.***

Illustrative Example

Ex.: Speed of a swimmer is 8 km/hr in still water. If the rate of stream is 3 km/hr, find the effective speed of the swimmer downstream.

Soln: Applying the above theorem, we have effective speed of the swimmer downstream

$= (8 + 3) = 11$ km/hr.

Exercise

1. The speed of a boat in still water is 10 km/hr. If its speed downstream be 13 km/hr, then speed of the stream is:
   a) 1.5 km/hr b) 3 km/hr c) 11.5 km/hr d) 5.75 km/hr
2. The rowing speed of man in still water is 20 km/hr. Going downstream, he moves at the rate of 25 km/hr. The rate of stream is
   a) 45 km/hr b) 2.5 km/hr c) 12.5 km/hr d) 5 km/hr
3. If a man goes upstream at 6 km/hr and the rate of stream is 2 km/hr, then the man's speed in still water is
   a) 4 km/hr b) 8 km/hr c) 2 km/hr d) 12 km/hr
4. A boat goes 12 kms upstream in 48 minutes. The speed of stream is 2 km/hr. The speed of boat in still water is
   a) 13 km/hr b) 2.25 km/hr c) 17 km/hr d) 15 km/hr

Answers

1. b; **Hint:** $10 + y = 13 \quad \therefore y = 13 - 10 = 3$ km/hr

2. d 3. b

4. c; **Hint:** $\frac{12 \times 60}{48} = x - 2 \qquad \therefore x = 15 + 2 = 17$ km/hr

Rule 3

Theorem: ***If x km per hour be the man's rate in still water, then $x = \frac{1}{2}$ (man's rate with current + his rate against current) ie "A man's rate in still water is half the sum of his rates with and against the current."***

Illustrative Example

Ex.: A man can row upstream at 10 km/hr and downstream at 16 km/hr. Find the man's rate in still water.

Soln: **Detail Method:** Let the speed of the man in still water be x km/hr and speed of the stream be y km/r.
According to the question,
$x+y=16$ .... (i) and $x-y=10$ ..... (ii)
Adding eqn (i) with eqn (ii), we have
$2x=26 \therefore x=13$ km/hr
$\therefore$ Speed of the man in still water = 13 km/hr.

Quicker Method: Applying the above theorem, we have

Rate of man in still water $=\frac{1}{2}(16+10)=13$ km/hr.

Exercise

1. A man can row downstream at the rate of 14 km/hr and upstream at 5 km/hr. Find man's rate in still water.
a) 9.5 km/hr b) 8 km/hr c) 8.5 km/hr d) 9 km/hr
2. A man can row downstream at the rate of 16 km/hr and upstream at 11 km/hr. Find man's rate in still water.
a) 14 km/hr b) 13.5 km/hr c) 14.5 km/hr d) 15.5 km/hr
3. The speed of a boat in still water is 12 km per hour. Going downstream it moves at the rate of 19 km per hour. The speed of the boat against the stream is _____ km/hr.
a) 5 km/hr b) 3 km/hr
c) 8 km/hr d) Data inadequate
4. A man can row 15 km downstream in 3 hours and 5 km upstream in $2\frac{1}{2}$ hours. His speed in still water is _____ km/hr.
a) 4 km/hr b) 4.5 km/hr
c) 3.5 km/hr d) Data inadequate
5. A man can row with the stream at 10 km/hr and against the stream at 5 km/hr. Man's rate in still water is
a) 5 km/hr b) 2.5 km/hr c) 7.5 km/hr d) 15 km/hr
6. A boat goes 40 km upstream in 8 hours and a distance of 36 km downstream in 6 hours. The speed of the boat in standing water is
a) 6.5 km/hr b) 6 km/hr c) 5.5 km/hr d) 5 km/hr
7. If a man rows at the rate of 5 km/hr in still water and his rate against the current is 3.5 km/hr, then the man's rate along the current is
a) 8.5 km/hr b) 6.5 km/hr c) 6 km/hr d) 4.25 km/hr
8. A man can row 44 km downstream in 4 hours. If the man's rowing rate in still water is 8 km/hr, then find in what time will he cover 25 km upstream?
a) 5 hours b) 6 hours c) 4.5 hours d) 4 hours
9. A man can row his boat with the stream at 6 km/hr and against the stream at 4 km/hr. The man's rate is
a) 1 km/hr b) 5 km/hr c) 8 km/hr d) 6 km/hr
10. A man rows 40 km upstream in 8 hours and a distance of 36 km downstream in 6 hours, then the speed of man in still water is
a) 0.5 km/hr b) 5.5 km/hr c) 6 km/hr d) 5 km/hr
11. If a man's downstream rate is 10 km/hr, and the rate of stream is 1.5 km/hr, then the man's upstream rate is
a) 13 km/hr b) 10 km/hr c) 3 km/hr d) 7 km/hr
12. If a man rows at 8 km/hr in still water and his upstream rate is 5 km/hr, then the man's rate along the current (downstream) is
a) 21 km/hr b) 8 km/hr c) 16 km/hr d) 11 km/hr

Answers

1. a 2. b

3. a; **Hint:** $(x+19)\frac{1}{2}=12 \therefore x=24-19=5$ km/hr

4. c 5. c 6. c

7. b; **Hint:** $(x+3.5)\frac{1}{2}=5 \therefore x=10-3.5=6.5$ km/hr

8. a; **Hint:** Man's rate in still water $=\frac{1}{2}$ [man's rate with current + his rate against current]

or, $8=\frac{1}{2}\left[\frac{44}{4}+\frac{25}{t}\right]$

$\therefore$ t = 5 hours.

9. b 10. b

11. d; **Hint:** Rate of stream $=\frac{1}{2}$ (downstream rate − upstream rate)

or, $1.5=\frac{1}{2}(10-x) \therefore x=7$ km/hr

12. d; **Hint:** Man's rate in still water $=\frac{1}{2}$ (downstream rate + upstream rate)

or, $8=\frac{1}{2}(5+x)=11$ km/hr

Rule 4

Theorem: ***If x km per hour be the rate of the current, then y $=\frac{1}{2}$ (man's rate with current − his rate against current) ie "The rate of the current is half the difference between the rate of the man with and against the current."***

Illustrative Example

Ex.: A man can row upstream at 10 km/hr and downstream at 16 km/hr. Find the rate of the current.

Soln: Detail Method:

Let the speed of the man in still water be x km/hr and the rate of the current be y km/hr

According to the question,

Effective speed of man downstream

$= x + y = 16$ km/hr .... (i)

Effective speed of man upstream

$= x - y = 10$ km/hr .... (ii)

Subtracting eqn (ii) from eqn (i), we have

$2y - 6$ km/hr

or, $y = 3$ km/hr

∴ Speed of the current = 3 km/hr

Quicker Method: Applying the above theorem,

Rate of current $= \frac{1}{2}(16 - 10) = 3$ km/hr

Exercise

1. A man rows upstream 16 km and downstream 27 km taking 5 hours each time. What is the velocity of current?
   a) 2 km/hr b) 2.1 km/hr
   c) 1.1 km/hr d) None of these
2. A boat moves downstream at the rate of one km in $7\frac{1}{2}$ minutes and upstream at the rate of 5 km an hour. What is the velocity of current?
   a) 1.3 km/hr b) 1.2 km/hr c) 1.6 km/hr d) 1.5 km/hr
3. A person rows a kilometre down the stream in 10 minutes and upstream in 30 minutes. Find the velocity of the stream.
   a) 1 km/hr b) 2 km/hr c) 3 km/hr d) 4 km/hr
4. A man can row three quarters of a km against the stream in 11 minutes 15 seconds and return in 7 minutes 30 seconds. Find the speed of the man in still water and also the speed of the stream
   a) 5 km/hr, 2 km/hr b) 5 km/hr, 1 km/hr
   c) 6 km/hr, 2 km/hr d) 4 km/hr, 1 km/hr
5. A boat's man goes 48 km downstream in 8 hours and returns back in 12 hours. Find the speed of the boat in still water and the rate of the stream.
   a) 5 km/hr, 1 km/hr b) 10 km/hr, 2 km/hr
   c) 6 km/hr, 1.5 km/hr d) None of these
6. A boat moves with a speed of 11 km per hour and along the stream and 7 km per hour against the stream. The rate of the stream is ______ km/hr.
   a) 1 km/hr b) 1.5 km/hr c) 2 km/hr d) 2.5 km/hr
7. A man rows upstream 11 km and downstream 26 km taking 5 hours each time. The velocity of the current is ______ km/hr.
   a) 1 km/hr b) 1.3 km/hr c) 1.5 km/hr d) 2.5 km/hr
8. A boat moves downstream at the rate of 1 km in 6 minutes and upstream at the rate of 1 km in 10 minutes. The speed of the current is
   a) 2 km/hr b) 1 km/hr c) 1.5 km/hr d) 2.5 km/hr
9. The speed of a boat downstream is 15 km/hr and the speed of the stream is 1.5 km/hr. The speed of the boat upstream is
   a) 13.5 km/hr b) 16.5 km/hr c) 12 km/hr d) 8.25 km/hr
10. If a man's rate with the current is 12 km/hr and the rate of current is 1.5 km/hr, then the man's rate against the current is
    a) 9 km/hr b) 6.75 km/hr c) 5.25 km/hr d) 7.5 km/hr
11. A man can swim downstream at 8 km/hr and upstream at 2 km/hr. Find man's rate in still water and the speed of current.
    a) 5 km, 2 km/hr b) 5 km, 1 km/hr
    c) 6 km, 3 km/hr d) 5 km, 3 km/hr
12. A man rows upstream 20 km and downstream 30 km taking 5 hours each. Find the speed of current.
    a) 2 km/hr b) 1 km/hr
    c) 1.5 km/hr d) None of these
13. A boat man can row 2 km against the stream in 20 minutes and return in 15 minutes. Find the rate of rowing in still water and the speed of current.
    a) 7 km/hr, 2 km/hr b) 6 km/hr, 2 km/hr
    c) 7 km/hr, 1 km/hr d) 7.5 km/hr, 1.5 km/hr
14. A boat moves downstream at the rate of 12 km/hr and upstream at 4 km/hr. Find the speed of the boat in still water and also the speed of current.
    a) 8 km/hr, 4 km/hr b) 4 km/hr, 2 km/hr
    c) 6 km/hr, 3 km/hr d) 3 km/hr, 1.5 km/hr
15. A man can row downstream at the rate of $2\frac{7}{9}$ metres per second and upstream at the rate of 5 km/hr. Find the man's rowing rate in still water and speed of current.
    a) 8 km/hr, 3 km/hr b) 7.5 km/hr, 3 km/hr
    c) 7.5 km/hr, 2.5 km/hr d) None of these
16. A boatman can row $1\frac{1}{2}$ km against the stream in $22\frac{1}{2}$ minutes and return in 15 minutes. Find the rate of current.
    a) 1 km/hr b) 2 km/hr c) 1.5 km/hr d) 1.3 km/hr
17. A man can row 30 km downstream in 2 hours and 15 km upstream in 5 hours. Find the man's rowing rate in still water and speed of current.
    a) 9 km/hr, 6 km/hr b) 8 km/hr, 5 km/hr
    c) 9 km/hr, 5 km/hr d) Data inadequate
18. A man can row 60 km downstream in 6 hours. If the speed of the current is 3 km/hr, then find in what time will he be able to cover 16 km upstream?
    a) 4.5 hours b) 4 hours c) 5 hours d) 5.5 hours
19. A person rows 2 km downstream in 20 minutes and upstream in one hour. Find the velocity of the stream.
    a) 2.1 km/hr b) 3.1 km/hr c) 2 km/hr d) 1.5 km/hr
20. A boatman rows 64 km downstream in 8 hours and returns back in 16 hours. Find the speed of the boat in still

water and the rate of the stream.
a) 6 km/hr, 2 km/hr b) 4 km/hr, 1 km/hr
c) 5 km/hr, 1.5 km/hr d) Data inadequate

21. A boat goes 100 km downstream in 10 hours, and 75 km upstream in 15 hours. The speed of the stream is
a) 7 km/hr b) 5 km/hr c) 3 km/hr d) $2\frac{1}{2}$ km/hr

22. A man rows 40 km upstream in 8 hours and a distance of 36 km downstream in 6 hours, then speed of the stream is
a) 0.5 km/hr b) 5.5 km/hr c) 6 km/hr d) 5 km/hr

23. A man can row three quarters of a km against the stream in $11\frac{1}{4}$ minutes and return in $7\frac{1}{2}$ minutes. Find the speed of the man in still water. What is the speed of the stream?
a) 5 km/hr, 1 km/hr b) 6 km/hr, 2 km/hr
c) 4 km/hr, 1 km/hr d) None of these

24. A man rows upstream 13 km and downstream 28 km taking 5 hours each time. What is the velocity of the currrent?
a) 1.5 km/hr b) 3 km/hr
c) 2.5 km/hr d) Data inadequate

25. A boat is rowed down a river at 10 km/hr and up the river at $4\frac{3}{4}$ km/hr. Find the velocity of the river.
a) $2\frac{1}{2}$ km/hr b) $2\frac{3}{8}$ km/hr
c) $2\frac{3}{4}$ km/hr d) $2\frac{5}{8}$ km/hr

26. In 3 hours a boat can be rowed 9 km upstream or 18 km downstream. Find the speed of the boat in still water and the rate at which the stream is running.
a) 4.5 km/hr, 1.5 km/hr b) 5 km/hr, 3 km/hr
c) 6 km/hr, 4 km/hr d) Data inadequate

Answers

1. c; **Hint:** Man's rate upstream = $\frac{16}{5}$ km/hr

Man's rate downstream = $\frac{27}{5}$ km/hr

Velocity of the current = $\frac{1}{2}\left(\frac{27}{5}-\frac{16}{5}\right)$ km/hr = 1.1 km/hr

2. d; **Hint:** Rate downstream = $\left(\frac{2}{15}\times 60\right)$ km/hr = 8 km/hr

Rate upstream = 5 km/hr

Velocity of the current = $\frac{1}{2}(8-5)$ km/hr = 1.5 km/hr

3. b; **Hint:** Rate downstream = $\frac{1}{10}\times 60$ 6 km/hr

Rate upstream = $\frac{1}{30}\times 60 = 2$ km/hr

Velocity of the stream = $\frac{1}{2}(6-2) = 2$ km/hr

4. b; **Hint:** Rate upstream = $\frac{3}{4}\times\frac{4}{45}\times 60 = 4$ km/hr

Rate downstream = $\frac{3}{4}\times\frac{2}{15}\times 60 = 6$ km/hr

Speed of the man in still water $(6+4)\frac{1}{2} = 5$ km/hr

(See **Rule 3**)

Speed of the stream = $(6-4)\frac{1}{2} = 1$ km/hr

5. a; **Hint:** Rate downstream = $\frac{48}{8} = 6$ km/hr

Rate upstream = $\frac{48}{12} = 4$ km/hr

6. c 7. c 8. a

9. c; **Hint:** $(15-y)\frac{1}{2} = 1.5$ $\therefore$ y = 15 – 3 = 12 km/hr

10. a 11. d 12. b 13. c 14. a 15. c 16. a 17. a

18. b; **Hint:** $3 = \frac{1}{2}\left[\frac{60}{6}-\frac{16}{t}\right]$ $\therefore$ t = 4 hours

19. c 20. a 21. d 22. a

23. a; **Hint:** The boat travels with stream at $\frac{3}{4}\times\frac{60}{7\frac{1}{2}} = 6$ km/hr

The boat travels against the stream at

$\frac{3}{4}\times\frac{60}{11\times\frac{1}{4}} = 4$ km/hr

$\therefore$ speed of man in still water = $\frac{1}{2}(6+4) = 5$ km/hr and

speed of stream = $\frac{1}{2}(6-4) = 1$ km/hr

24. a 25. d 26. a

Rule 5

Theorem: ***If x km be the rate of stream and a man takes n times as long to row up as to row down the river, then the rate of the man in still water is given by x $\left(\frac{n+1}{n-1}\right)$ km/hr.***

Illustrative Example

Ex.: A man takes twice as long to row up as to row down the river. If the rate of river is 4 km/hr, find the rate of the man in still water.

Soln: **Detail Method:**

Let rate of man in still water be x km/hr

Then,

$x+4=2(x-4)$

or, $x=12$ km/hr

Quicker Method: Applying the above theorem, we have

the speed of the man in still water $=4\times\left(\frac{2+1}{2-1}\right)$

$=4\times3=12$ km/hr.

Exercise

1. A man can row 4.5 km/hr in still water and he finds that it takes him twice as long to row up as to row down the river. Find the rate of stream.
   a) 2 km/hr b) 1.5 km/hr c) 2.5 km/hr d) 1.75 km/hr
2. A man can row 6 km/hr in still water. It takes him twice as long to row up as to row down the river. Find the rate of the stream.
   a) 2 km/hr b) 3 km/hr c) 1.5 km/hr d) 1 km/hr
3. A man can row at the rate of 3.5 km/hr in still water. If the time taken to row a certain distance upstream is $2\frac{1}{2}$ times as much as to row the same distance downstream, find the speed of the current.
   a) 2.5 km/hr b) 1.5 km/hr c) 3 km/hr d) 1.25 km/hr
4. A man can row 4 km/hr in still water and he finds that it takes him twice as long to row up as to row down the river. Find the rate of stream.
   a) 1.3 km/hr b) 2 km/hr c) 1 km/hr d) 1.5 km/hr
5. A man can row at the rate of 4 km/hr in still water. If the time taken to row a certain distance upstream is 3 times as much as to row the same distance downstream, find the speed of the current.
   a) 3 km/hr b) 1.5 km/hr c) 1 km/hr d) 2 km/hr
6. A person can row $7\frac{1}{2}$ km an hour in still water and he finds that it takes him twice as long to row up as to row down the river. Find the rate of the stream.
   a) 2.5 km/hr b) 2 km/hr c) 3 km/hr d) 1.5 km/hr
7. A man swimming in a stream which flows 1.5 km/hr finds that in a given time he can swim twice as far with the stream as he can against it, at what rate does he swim?
   a) 4 km/hr b) 4.5 km/hr c) 5 km/hr d) 3.5 km/hr

Answers

1. b; **Hint:** $x\left(\frac{2+1}{2-1}\right)=4.5$

$\therefore x=\frac{4.5}{3}=1.5$ km/hr

2. a 3. b 4. a 5. d 6. a 7. b

Rule 6

Theorem: ***If the speed of the boat in still water is x km/hr and the rate of current is y km/hr, then the distance travelled downstream in 'T' hours is (x + y)T km i.e. Distance travelled downstream = Downstream Rate × Time. And the distance travelled upstream in 'T' hours is (x – y)T km ie Distance travelled upstream = Upstream Rate × Time.***

Illustrative Example

Ex.: The speed of a boat in still water is 8 km/hr and the rate of current is 4 km/hr. Find the distance travelled downstream and upstream in 5 minutes.

Soln: Applying the above theorem,

Distance travelled downstream $=(8+4)\times\frac{5}{60}=1$ km.

Distance travelled upstream $=(8-4)\times\frac{5}{60}=\frac{1}{3}$ km.

Exercise

1. The speed of a boat in still water is 15 km/hr and the rate of current is 3 km/hr. The distance travelled downstream in 12 minutes is
   a) 3.6 km b) 2.4 km c) 1.2 km d) 1.8 km
2. Speed of a boat in standing water is 7 km/hr and the speed of the stream is 1.5 km/hr. A distance of 7.7 km, going upstream is covered in
   a) 1 hr 15 min b) 1 hr 12 min c) 1 hr 24 min d) 2 hr 6 min
3. The speed of a boat in still water is 15 km/hr and the rate of current is 13 km/hr. Find the distance travelled downstream in 15 minutes.
   a) 7 km b) 8 km c) 7.5 km d) 7.6 km
4. A man can row upstream 32 km in 4 hours. If the speed of current is 2 km/hr, find how much he can go downstream in 6 hours.
   a) 70 km b) 72 km c) 64 km d) 81 km
5. A man can row upstream 36 km in 6 hours. If the speed of a man in still water is 8 km/hr, find how much he can go downstream in 10 hours.
   a) 150 km b) 80 km c) 90 km d) 100 km

6. The speed of a boat in still water is 5 km/hr and the rate of current is 1 km/hr. Find the distance travelled in 20 minutes in (i) downstream, (ii) upstream.
a) 2 km, 1.33 km b) 3 km, 1.33 km
c) 1.5 km, 1 km d) Data inadequate
7. The speed of a boat in still water is 4 km/hr and the speed of current is 2 km/hr. If the time taken to reach a certain distance upstream is 9 hours, find the time it will take to go to same distance downstream.
a) 2 hrs b) 2.5 hrs c) 3.5 hrs d) 3 hrs
8. A person can swim in still water at 4 km/hr. If the speed of water is 2 km/hr, how many hours will the man take to swim back against the current for 6 km.
[UTI Exam, 1990]
a) 3 b) 4 c) $4\frac{1}{2}$ d) Data inadequate
9. A man can row in still water at 7 km/hr and the rate of stream is 3.5 km/hr. A distance of 10.5 km in going upstream is covered in
a) 1.5 hrs b) 1 hr c) 3 hrs d) 15 hrs
10. The speed of a boat in still water is 15 km/hr and the rate of stream is 5 km/hr the distance travelled downstream in 24 minutes is
a) 4 km b) 8 km c) 6 km d) 16 km

Answers

1. a; **Hint:** Required distance $= (15+3)\frac{12}{60} = \frac{18}{5} = 3.6$ km

2. c; **Hint:** $(7-1.5)T = 7.7 \therefore T = \frac{7.7}{5.5} = 1$ hr 24 minutes

3. a

4. b; **Hint:** Downstream rate $= \frac{32}{4} = 8$ km/hr
Speed of man in still water $= 8+2 = 10$ km/hr
Now, the required distance $= (10+2)6 = 72$ km

5. d; **Hint:** Speed of current $= 8 - \frac{36}{6} = 2$ km/hr
$\therefore$ required distance $= (8+2)10 = 100$ km

6. a

7. d; **Hint:** Distance $= (4-2)9 = 18$ km
$\therefore$ Required time $= \frac{18}{4+2} = 3$ hrs

8. a 9. c 10. b

Rule 7

Theorem: ***A man can row x km/hr in still waters. If in a stream which is flowing at y km/hr, it takes him z hrs to row to a place and back, the distance between the two places is*** $\frac{z(x^2-y^2)}{2x}$.

OR

$$\textit{Distance} = \left[\frac{\text{Total Time} \times \{(\text{Speed in still water})^2 - (\text{Speed of current})^2\}}{2 \times \text{Speed in still water}}\right] \textit{km}$$

Note: *The speed of a boat in still water is x km/hr and the speed of the stream is y km/hr. A man rows to a place at a distance of D km and comes back to the starting point then the total time taken by him is* $\left(\frac{2Dx}{x^2-y^2}\right)$ *hours OR Total time* $= \left(\frac{2 \times \text{Distance} \times \text{Speed in still water}}{(\text{Speed in still water})^2 - (\text{Speed in current})^2}\right)$ *hrs.*

Illustrative Examples

Ex. 1: The speed of a boat in still water is 6 km/hr and the speed of the stream is 1.5 km/hr. A man rows to a place at a distance of 22.5 km and comes back to the starting point. Find the total time taken by him.

Soln: Detail Method:
Boat's upstream speed $= 6 - 1.5 = 4.5$ km/hr
Boat's downstream speed $= 6 + 1.5 = 7.5$ km/hr

$\therefore$ Total time $= \frac{22.5}{4.5} + \frac{22.5}{7.5} = 5 + 3 = 8$ hrs.

Quicker Method: Applying the above theorem, we have

the total time $= \frac{2 \times 22.5 \times 6}{6^2 - (1.5)^2} = 8$ hours.

Ex. 2: A man can row 6 km/hr in still water. When the river is running at 1.2 km/hr, it takes him 1 hour to row to a place and back. How far is the place?

Soln: Detail Method: Man's rate downstream $= (6+1.2)$ km/hr $= 7.2$ km/hr
Man's rate upstream $= (6-1.2)$ km/hr $= 4.8$ km/hr
Let the required distance be x km. Then

$\frac{x}{7.2} + \frac{x}{4.8} = 1$ or, $4.8x + 7.2x = 7.2 \times 4.8$

$\Rightarrow x = \frac{7.2 \times 4.8}{12} = 2.88$ km.

Quicker Method: Applying the above theorem, we have

the required distance $= \frac{1 \times [6^2 - (1.2)^2]}{2 \times 6}$

$= \frac{36 - 1.44}{12} = 3 - 0.12 = 2.88$ km

Exercise

1. A man rows 8 km/hr in still water. If the river is running at 2 km/hr, it takes 32 minutes to row to a place and back.

How far is the place?
a) 1.5 km b) 2.5 km c) 2 km d) 3 km

2. A man can row 5 km/hr in still water. If the river is running at 1.5 km/hr, it takes him 1 hour to row to a place and back. How far is the place?
a) 2 km b) 2.5 km c) 2.275 km d) 2.175 km

3. A man can row 8 km per hour in still water. If the river is running at 2 km an hour, it takes him 48 minutes to row to a place and back, how far is the place?
a) 5 km b) 4 km c) 2 km d) 3 km

4. A man can row 5 km per hour in still water. If the river is running at 1 km an hour, it takes him 75 minutes to row to a place and back. The place is at a distance of ___ km from the starting point.
a) 3 km b) 4 km c) 5 km d) 2 km

5. A man can row 5 km/hr in still water. If the river is running at 1 km/hr, it takes him 1 hour to row to a place and back. How far is the place?
a) 2.5 km b) 2.4 km c) 3 km d) 3.6 km

6. A boat travels upstream from B to A and downstream from A to B in 3 hours. If the speed of the boat in still water is 9 km/hr and the speed of the current is 3 km/hr, the distance between A and B (in km) is

[BSRB Bank PO Exam, 1990]

a) 4 b) 6 c) 8 d) 12

7. A man can row 6 km/hr in the still water. If the river is running at 2 km/hr, it takes him 3 hours to row to a place and back. How far is the place?
a) 8 km b) 12 km c) 9 km d) 6 km

8. A man can row 4 km/hr in still water if the river is running at 2 km/hr, it takes 6 hours to row to a place and back. How far is the place?
a) 6 km b) 8 km c) 9 km d) 9.5 km

9. A boat's crew rowed down a stream from A to B and up again in $7\frac{1}{2}$ hours. If the stream flows at 3 km/hr and the crew can row in still water at 5 km/hr, find the distance from A to B.
a) 16 km b) 8 km c) 14 km d) 12 km

10. A boat's crew rowed down a stream from A to B and up again in 9 hours. If the stream flows at $2\frac{1}{2}$ km/hr and the crew can row in still water at $4\frac{1}{2}$ km/hr, find the distance from A to B.
a) 14 km b) 13 km c) 12 km d) 16 km

11. A man can row $7\frac{1}{2}$ km/hr in still water. If in a river running at $1\frac{1}{2}$ km/hr, it takes him 50 minutes to row to a place and back, how far off is the place?
a) 2 km b) 3 km c) 4 km d) 6 km

Answers

1. c; **Hint:** $\frac{2.D.8}{8^2-2^2}=\frac{32}{60}$ $\therefore$ D = 2 km

2. c 3. d 4. a 5. b 6. d 7. a 8. c
9. d 10. a 11. b

Rule 8

Theorem: ***If in a stream running at x km/hr, a motorboat goes D km upstream and back again to the starting point in 'T' hours, then the speed of the motorboat in still water is***

$$\left[\frac{D+\sqrt{D^2+(Tx)^2}}{T}\right] \textit{km/hr.}$$

Illustrative Example

Ex.: In a stream running at 2 km/hr, a motorboat goes 10 km upstream and back again to the starting point in 55 minutes. Find the speed of the motorboat in still water.

Soln: **Detail Method:** Let the speed of the motorboat in still water be x km/hr.

$$\frac{10}{x+2}+\frac{10}{x-2}=\frac{55}{60}$$

or, $240x = 11x^2 - 44$

or, $11x^2 - 240x - 44 = 0$

$\therefore (x-22)(11x+2)=0$

So, x = 22 km/hr (neglect the –ve value)

$\therefore$ speed of the motorboat in still water = 22 km/hr.

Quicker Method: Applying the above theorem, we have

the speed of the motorboat in still water

$$=\frac{10+\sqrt{(10)^2+\left(\frac{11}{12}\times 2\right)^2}}{\frac{11}{12}}=\frac{\frac{60+61}{6}}{\frac{11}{12}}$$

$$=\frac{121}{11}\times\frac{12}{6}=22 \text{ km/hr.}$$

Exercise

1. The current of a stream runs at the rate of 4 km an hour. A boat goes 6 km and back to its starting point in 2 hours. Find the speed of the boat in still water.
a) 8 km/hr b) 9 km/hr c) 6 km/hr d) 4 km/hr

2. A motor boat can travel at 10 km/hr in still water. It travelled 91 km downstream in a river and then returned,

taking altogether 20 hours. Find the rate of flow of river.
a) 6 km/hr b) 2 km/hr c) 3 km/hr d) 4 km/hr

3. The rate of flow of river water is 4 km/hr. A boat goes 6 km and back to the starting point in 2 hours. Find the speed of the boat in still water.
a) 6 km/hr b) 8 km/hr c) 9 km/hr d) 10 km/hr

4. In a stream running at 2 km/hr, a motorboat goes 12 km upstream and back again to the starting point in $2\frac{1}{2}$ hours. Find the speed of the motorboat in still water.
a) 15 km/hr b) 12 km/hr
c) 10 km/hr d) None of these

Answers

1. a

2. c; **Hint:** $\frac{91+\sqrt{91^2+(20x)^2}}{20}=10$

or, $400x^2 = 11881-8281=3600$

or, $x^2=9$ $\therefore x=3$ km/hr

3. b 4. c

Rule 9

Theorem: ***A man can row x_1 km upstream and y_1 km downstream in T_1 hours. Also, he can row x_2 km upstream and y_2 km downstream in T_2 hours. Then, the rate of the current and speed of the man in still water is calculated by the use of multiple cross-multiplication method as given below.***

Step I: Arrange the given figures in the following form

Upstream	Downstream	Time
x_1	y_1	T_1
x_2	y_2	T_2

Upstream speed of man $=\left(\frac{x_1y_2-x_2y_1}{y_2T_1-y_1T_2}\right)$ km/hr

Downstream speed of man $=\left(\frac{x_1y_2-x_2y_1}{x_1T_2-x_2T_1}\right)$ km/hr

Step II: Now to calculate the speed of man and current, use the following formula,

Speed of man $=\frac{1}{2}$ [upstream speed of man + downstream speed of man] **[Remember Rule 3]**

and speed of stream $=\frac{1}{2}$ [Downstream speed of man – Upstream speed of man] **[Remember Rule 4]**

Note: 1. How do the denominators of the above two formulae differ? For upstream speed we use the figures of downstream speed and time and for downstream speed we use the figures of upstream speed and time.
2. Numerators remain the same in both formulae.

Illustrative Example

Ex.: A man can row 30 km upstream and 44 km downstream in 10 hrs. Also, he can row 40 km upstream and 55 km downstream in 13 hrs. Find the rate of the current and the speed of the man in still water.

Soln: **Detail Method:**
Let, upstream rate = x km/hr and downstream rate = y km/hr

Then, $\frac{30}{x}+\frac{44}{y}=10$ and $\frac{40}{x}+\frac{55}{y}=13$

or, $30u+44v=10$
$40u+55v=13$

Where $u=\frac{1}{x}$ and $v=\frac{1}{y}$

Solving, we get $u=\frac{1}{5}$ and $v=\frac{1}{11}$

$\therefore x=5$ and $y=11$

$\therefore$ rate in still water $=\frac{5+11}{2}=8$ km/hr.

Rate of current $=\frac{11-5}{2}=3$ km/hr.

Quicker Method: (By use of multiple cross-multiplication)
Step I: Arrange the given figures in the following form:

Upstream	Downstream	Time
30	44	10
40	55	13

Upstream speed of man

$$=\frac{30\times55-40\times44}{55\times10-44\times13}=\frac{-110}{-22}=5 \text{ km/hr.}$$

Downstream speed of man

$$=\frac{30\times55-40\times44}{30\times13-40\times10}=\frac{-110}{-10}=11 \text{ km/hr.}$$

Step II:

$\therefore$ Speed of man $=\frac{5+11}{2}=8$ km/hr.

and speed of stream $=\frac{11-5}{2}=3$ km/hr.

Exercise

1. A man can row 15 km upstream and 22 km downstream in 5 hrs. Also, he can row 22 km upstream and 27.5 km

downstream in $6\frac{1}{2}$ hrs. Find the rate of the current and the speed of the man in still water.

a) 11 km/hr, 5 km/hr b) 8 km/hr, 3 km/hr
c) 5 km/hr, 2 km/hr d) None of these

2. A man can row 45 km upstream and 66 km downstream in 15 hrs. Also, he can row 66 km upstream and 82.5 km downstream in $19\frac{1}{2}$ hrs. Find the rate of the current and the speed of the man in still water.

a) 8 km/hr, 3 km/hr b) 11 km/hr, 3 km/hr
c) 11 km/hr, 8 km/hr d) Data inadequate

3. A man can row 60 km upstream and 88 km downstream in 20 hrs. Also, he can row 80 km upstream and 110 km downstream in 26 hrs. Find the rate of the current and the speed of the man in still water.

a) 12 km/hr, 4 km/hr b) 16 km/hr, 6 km/hr
c) 8 km/hr, 3 km/hr d) None of these

Answers

1. b 2. a 3. c

Rule 10

Theorem: ***A man rows a certain distance downstream in x hours and returns the same distance in y hrs. If the stream flows at the rate of z km/hr then the speed of the man in still water is given by*** $\frac{z(x+y)}{y-x}$ ***km/hr. Or, Speed in still water =***

$$\left[\frac{\text{Rate of stream (Sum of upstream and downstream time)}}{\text{Difference of upstream and downstream time}}\right]$$

km/hr.

Illustrative Example

Ex.: Ramesh can row a certain distance downstream in 6 hours and return the same distance in 9 hours. If the stream flows at the rate of 3 km per hour find the speed of Ramesh in still water.

Soln: **Detail Method:** Let the speed of Ramesh in still water be x km/hr. Then his upstream speed = $(x-3)$ km/hr and downstream speed = $(x+3)$ km/hr.

Now, we are given that up and down journey are equal, therefore,

$(x+3)6 = (x-3)9$

or, $6x+18 = 9x-27$

or, $3x = 45$

$\therefore x = 15$ km/hr

Quicker Method: By the above theorem, we have

Ramesh's speed in still water $= \frac{3(9+6)}{9-6} = 15$ km/hr.

Exercise

1. Ramesh can row a certain distance downstream in 6 hours and return the same distance in 9 hours. If the speed of Ramesh in still water is 12 km/hr,find the speed of the stream.

a) 2.4 km/hr b) 2 km/hr
c) 3 km/hr d) Data inadequate

2. A can row a certain distance down a stream in 6 hours and return the same distance in 9 hours. If the stream flows at the rate of $2\frac{1}{4}$ km/hr, find how far he can row in an hour in still water?

a) $11\frac{3}{4}$ km/hr b) 12 km/hr
c) $11\frac{1}{4}$ km/hr d) Data inadequate

3. Ajay can row a certain distance downstream in 5 hours and return the same distance in 7 hours. If the stream flows at the rate of 2 km per hour find the speed of Ajay in still water.

a) 12 km/hr b) 10 km/hr c) 18 km/hr d) 16 km/hr

4. Rohit can row a certain distance downstream in 8 hours and return the same distance in 12 hours. If the stream flows at the rate of 5 km per hour find the speed of Rohit in still water.

a) 20 km/hr b) 30 km/hr c) 15 km/hr d) 25 km/hr

Answers

1. a; **Hint:** $z\left(\frac{9+6}{9-6}\right) = 12 \quad \therefore z = 2.4$ km/hr

2. c 3. a 4. d

Rule 11

Theorem: ***If a man can row at a speed of x km/hr in still water to a certain upstream point and back to the starting point in a river which flows at y km/hr, then the averge speed for total journey (up + down) is given by*** $\left[\frac{(x+y)(x-y)}{x}\right]$ ***km/hr.***

OR

Average speed for the total journey

$$= \left[\frac{\text{Upstream rate} \times \text{Downstream rate}}{\text{Speed in still water}}\right].$$

Illustrative Example

Ex.: A man can row at a speed of 5 km/hr in still water to a certain upstream point and back to the starting point in a river which flows at 2 km/hr. Find his average speed for total journey.

Soln: **Detail Method:** Let the distance be x km.

$$\text{Average speed} = \frac{\text{Total Distance}}{\text{Total Time}}$$

$$= \frac{2x}{\frac{x}{(5+2)}+\frac{x}{(5-2)}} = \frac{2x}{\frac{x}{7}+\frac{x}{3}} = \frac{2x\times 21}{10x}$$

$$= \frac{21}{5} = 4\frac{1}{5} \text{ km/hr.}$$

Quicker Method: Applying the above theorem, we have

$$\text{average speed} = \frac{(5+2)(5-2)}{5} = \frac{21}{5} = 4\frac{1}{5} \text{ km/hr.}$$

Exercise

1. A man can row at a speed of 4.5 km/hr in still water to a certain upstream point and back to the starting point in a river which flows at 1.5 km/hr. Find his average speed for total journey.
   a) 4 km/hr b) 6 km/hr c) 4.5 km/hr d) 5 km/hr
2. A man row at a speed of 8 km/hr in still water to a certain distance upstream and back to the starting point in a river which flows at 4 km/hr. Find his average speed for total journey.
   a) 8 km/hr b) 6 km/hr c) 4 km/hr d) 10 km/hr
3. A man can row at a speed of 10 km/hr in still water to a certain upstream point and back to the starting point in a river which flows at 4 km/hr. Find his average speed for total journey.
   a) 2 km/hr b) 3 km/hr
   c) 1.5 km/hr d) Data inadequate
4. A man can row at a speed of 15 km/hr in still water to a certain upstream point and back to the starting point in a river which flows at 3 km/hr. Find his average speed for total journey.
   a) 9 km/hr b) 6 km/hr
   c) 3 km/hr d) 2 km/hr

Answers

1. a 2. b 3. a 4. d

Rule 12

Theorem: ***A man can row x km/hr in still water. If the river is running at y km/hr, it takes T hours more in upstream than to go downstream for the same distance, then the distance is given by*** $\left[\frac{(x^2-y^2)T}{2y}\right]$ ***km.***

Illustrative Example

Ex.: A man can row 6 km/hr in still water. If the river is running at 2 km/hr, it takes 3 hours more in upstream than to go downstream for the same distance. How far is the place?

Soln: **Detail Method:** Let the distance of the place be x km. According to the question,

$$\frac{x}{6-2}-\frac{x}{6+2}=3$$

$$\text{or, } \frac{x}{4}-\frac{x}{8}=3$$

$$\therefore x = 8\times 3 = 24 \text{ km}$$

Quicker Method: Applying the above theorem, we have the required distance

$$= \frac{(6^2-2^2)3}{2\times 2} = \frac{32\times 3}{4} = 8\times 3 = 24 \text{ km.}$$

Exercise

1. A man can row 5 km/hr in still water. If the river is running at 1 km/hr, it takes 2 hours more in upstream than to go downstream for the same distance. How far is the place?
   a) 24 km b) 20 km c) 18 km d) 16 km
2. A man can row 7 km/hr in still water. If the river is running at 3 km/hr, it takes 6 hours more in upstream than to go downstream for the same distance. How far is the place?
   a) 48 km b) 36 km c) 42 km d) 40 km
3. A man can row 8 km/hr in still water. If the river is running at 4 km/hr, it takes 1 hour more in upstream than to go downstream for the same distance. How far is the place?
   a) 16 km b) 12 km c) 8 km d) 6 km
4. A man can row 9 km/hr in still water. If the river is running at 3 km/hr, it takes 3 hours more in upstream than to go downstream for the same distance. How far is the place?
   a) 30 km b) 36 km d) 24 km d) None of these

Answers

1. a 2. d 3. d 4. b

Miscellaneous

1. A boat takes 3 hours to travel from place M to N downstream and back from N to M upstream. If the speed of he boat in still water is 4 km, what is the distance between the two places? **[BSRB Delhi PO, 2000]**
   a) 8 km b) 12 km
   c) 6 km d) Data inadequate
2. A man rows to a place 48 km distant and back in 14 hours. He finds that he can row 4 km with the stream in

the same time as 3 km against the stream. Find the rate of the stream?

a) 1 km/hr b) 2 km/hr c) 1.5 km/hr d) 2.5 km/hr

3. P, Q and R are the three towns on a river which flows uniformly. Q is equidistant from P and R. I row from P to Q and back in 10 hours and I can row from P to R in 4 hours. Compare the speed of my boat in still water with that of the river.

a) 5 : 3 b) 4 : 3 c) 6 : 5 d) 7 : 3

Answers

1. d; Let the distance between M and N and the speed of current in still water be d km and x km/hr respectively.

According to the question, $\frac{d}{4+x}+\frac{d}{4-x}=3$

In the above equation we have only one equation but two variables. Hence can't be determined,

2. a; Suppose that the man takes x hours to cover 4 km downstream and x hours to cover 3 km upstream.

Then, $\frac{48x}{4}+\frac{48x}{3}=14$ or $x=\frac{1}{2}$

$\therefore$ Rate upstream = 6 km/hr and rate downstream = 8 km/hr

$\therefore$ Rate of the stream = $(8-6)\frac{1}{2}$ = 1 km/hr

[See **Rule 4**]

3. a; 

I can row from P to R in 4 hours

$\therefore$ I can row from P to Q in 2 hours

But 1 can row from P to Q and back in 10 hours.

$\therefore$ I can row from Q to P in (10 – 2 =) 8 hours

Hence in rowing with the current I take 2 hours and in rowing against the current I take 8 hours, the distance being same in both the cases.

Now, distance being the same the 'down rate' and the 'up rate' are inversely proportional to the times.

$\therefore$ down rate : up rate = 8 : 2 = 4 : 1

$\therefore$ speed of boat in still water : speed of river

$=(4+1):(4-1)=5:3.$

19 Elementary Mensuration - I

Triangle

Rule 1

To find the area of a triangle if its base and height are given.

Area of a triangle = $\frac{1}{2}$ × *Base* × *Height*

Illustrative Example

Ex.: The base of a triangular field is 880 metres and its height 550 metres. Find the area of the field. Also calculate the charges for supplying water to the field at the rate of Rs 24.25 per sq hectometre.

Soln: Area of the field = $\frac{\text{Base}\times\text{Height}}{2}$

$= \frac{880\times 550}{2}$ sq metres $= \frac{440\times 550}{100\times 100}$ sq hectometres.

= 24.20 sq hectometres.

Cost of supplying water to 1 sq hectometre = Rs 24.25

∴ cost of supplying water to the whole field

= Rs 24.20 × 24.25 = Rs 586.85

Exercise

1. Find the area of a triangle in which base is 1.5 m and height is 75 cm.
   a) 5625 sq cm b) 5265 sq cm
   c) 5635 sq cm d) 5525 sq cm
2. Find the area of a triangle whose one angle is 90°, the hypotenuse is 9 metres and the base is 6.5 metres.
   a) 20 sqm b) 20.5 sq m c) 20.15 sq m d) 21 sq m
3. The base of a triangular field is three times its altitude. If the cost of cultivating the field at Rs 24.60 per hectare is Rs 332. 10, find its base and height.
   a) 250 m, 650 m b) 300 m, 900 m
   c) 350, 850 m d) None of these
4. A lawn is in the form of a triangle having its base and height in the ratio 2 : 3. The area of the lawn is $\frac{1}{12}$ hectare. Find the base and the height of the lawn.
   a) 55 m, 34 m b) 50 m, $33\frac{1}{3}$ m
   c) 50 m, 35 m d) Data inadequate
5. The base of a triangular field is three times its height. If the cost of cultivating the field at Rs 36.72 per hectare is Rs 495.72, find its base and height.
   a) 900 m, 300 m b) 600 m, 300 m
   c) 900 m, 600 m d) Can't be determined
6. Find the area of a triangle in which base is 36.8 cm and height is 7.5 cm.
   a) 128 sq cm b) 148 sq cm c) 130 sq cm d) 138 sq cm
7. If the area of a triangle with base x is equal to the area of a square with side x, then the altitude of the triangle is:
   a) $\frac{x}{2}$ b) x c) 2x d) 3x
   [I tax & Central Excise 1988]
8. If the area of a triangle is 150 sq m and base : height is 3 : 4, find its height and base.
   a) 20 m, 15 m b) 30 m, 10 m
   c) 60 m, 5 m d) Data inadequate
   [GIC Exam 1983]
9. The base of a triangular field is three times its height. If the cost of cultivating the field at Rs 1505.52 per hectare is Rs 20324.52 find its base and height.
   a) 900 m, 300 m b) 300 m, 100 m
   c) 600 m, 200 m d) Data inadequate
10. The base of a triangular field is 880 metres, and its height 550 metres. Find the area of the field. Also calculate the charges for supplying water to the field at the rate of Rs 242.50 per sq hectometre.
    a) Rs 5688.50 b) Rs 5868.50
    c) Rs 6858.50 d) None of these

Answers

1. a

2. c; **Hint:** Height of the triangle = $\sqrt{9^2-(6.5)^2} = \sqrt{38.75}m$

$= 6.2$ m

$\therefore$ Area of the triangle $= \frac{1}{2} \times Base \times Height$

$= \left(\frac{1}{2} \times 6.5 \times 6.2\right)$ sq m $= 20.15$ sq m

3. b; **Hint:** Area of the field $= \frac{332.10}{24.60} = 13.5$ hectares

$= (13.5 \times 10000) = 135000$ sq m

Let, altitude be x metres. Then, base = 3x metres.

Area $= \left(\frac{1}{2} \times base \times altitude\right) = \left(\frac{1}{2} \times x \times 3x\right) = \frac{3x^2}{2}$ sq m

$\therefore \frac{3x^2}{2} = 135000$ or, $x = \sqrt{\frac{135000 \times 2}{3}} = 300$

Hence, altitude = 300 metres and base = 900 metres (Also see **Rule - 68**)

4. b; **Hint:** Let the base be 2x metres and height 3x metres.

Then $\frac{1}{2}(2x \times 3x) = \left(\frac{1}{12} \times 10000\right)$

[$\because$ 1 hectare = 10000 sq m]

or, $x = \sqrt{\frac{10000 \times 2}{6 \times 12}} = \frac{100}{6} = \frac{50}{3}$

$\therefore$ Base $= \frac{2 \times 50}{3} = 33\frac{1}{3}$ metres,

Height $= \frac{3 \times 50}{3} = 50$ metrees. [See **Rule - 68**]

5. a; **Hint:** Area $= \frac{495.72}{36.72} \times 10000 = 135000$ sq m

$\frac{1}{2} \times 3x \times x = 135000$

$\therefore$ x = 300 ie height = 300 m

and base = 300 × 3 = 900 m [See **Rule 68**]

7. c; **Hint:** $\frac{1}{2} \times x \times h = x^2$ $\therefore$ h = 2x

8. a; **Hint:** $\frac{1}{2} \times 3x \times 4x = 150$ or, x = 5

$\therefore$ base = 3 × 5 = 15 m and height = 4x = 4 × 5 = 20 m

Rule 2

If a, b, c are the lengths of the sides of a triangle and $S = \frac{1}{2}$ ***(a + b + c) then, Area of the triangle***

$= \sqrt{s(s-a)(s-b)(s-c)}$

Illustrative Example

Ex.: Find the area of a triangle whose sides are 50 metres, 78 metres, 112 metres respectively and also find the perpendicular from the opposite angle on the side 112 metres.

Soln: Here a = 50 metres, b = 78 metres, c = 112 metres.

$\therefore s = \frac{1}{2}(50 + 78 + 112)$

$= \frac{1}{2} \times 240$ metres = 120 metres.

$\therefore$ s − a = (120 − 50) metres = 70 metres

s − b = (120 − 78) metres = 42 metres

s − c = (120 − 112) metres = 8 metres

$\therefore$ area $= \sqrt{120 \times 70 \times 42 \times 8} = 1680$ sq metres

Perpendicular $= \frac{2\ \text{Area}}{\text{Base}} = \frac{1680 \times 2}{112}$ metres

= 30 metres. [See **Rule - 1**]

Exercise

1. Find the area of a triangle with two sides equal, each being 5.1 metres and the third side 4.6 metres.
   a) 10 sq m b) 10.5 sq m c) 10.46 sq m d) 11.46 sq m
2. Find the area of a triangle in which a = 25 cm, b = 17 cm and c = 12 cm.
   a) 90 sq cm b) 80 sq cm c) 85 sq cm d) 75 sq cm
3. If the sides of a triangle are doubled, its area
   a) remains same b) becomes doubles
   c) becomes 3 times d) becomes 4 times
   [Railway Recruitment Board Exam, 1991)
4. The sides of a triangular field are 949, 1095, 1022 metres. It is let at Rs 10000 per hectare. Find the rent of the field.
   a) Rs 447636 b) Rs 446736
   c) Rs 447663 d) Data inadequate
5. The sides of a triangular field are 165 metres, 143 metres and 154 metres, find its area.
   a) 10164 sq m b) 10146 sq m
   c) 10614 sq m d) None of these
6. Two sides of a triangular field are 85 metres and 154 metres respectively and its perimeter is 324 metres. Find
   (i) the area of the field
   a) 2882 sq m b) 2782 sq m
   c) 2772 sq m d) 2672 sq m
   (ii) the perpendicular from the opposite angle on the side 154 metres
   a) 36 metres b) 18 metres
   c) 45 metres d) 27 metres
   (iii) the cost of levelling the field at the rate of Rs 5 per sq

m
a) Rs 12860 b) Rs 13760
c) Rs 13860 d) Rs 13960

7. The sides of a triangle are 51, 52, 53 cm, find the perpendicular from the opposite angle on the side of 52 cm. Also find the areas of the two triangles into which the original triangle is divided.
a) 45 cm, 560 sq cm, 640 sq cm
b) 45 cm, 540 sq cm, 630 sq cm
c) 48 cm, 540 sq cm, 630 sq cm
d) 48 cm, 530 sq cm, 640 sq cm

Answers

1. c 2. a

3. d; **Hint:** Let the original sides be a, b, c then

$$s = \frac{1}{2}(a+b+c)$$

Area of this triangle = $\sqrt{s(s-a)(s-b)(s-c)}$

For new triangle, the sides are 2a, 2b, 2c & S = 2S.

$\therefore$ Area of new triangle = $\sqrt{S(S-2a)(S-2b)(S-2c)}$

$= \sqrt{2s(2s-2a)(2s-2b)(2s-2c)}$

$= \sqrt{16s(s-a)(s-b)(s-c)}$

$= 4\sqrt{s(s-a)(s-b)(s-c)} = 4 \times$ (area of original triangle).

4. a; **Hint:** $S = \frac{1}{2}(949+1095+1022) = 1533$ m

Area = $\sqrt{1533 \times 584 \times 438 \times 511}$

$= \sqrt{511 \times 3 \times 146 \times 2 \times 3 \times 146 \times 511}$

$= 511 \times 3 \times 146 \times 2$ sq m

$\therefore$ required rent = $\frac{10000 \times 511 \times 3 \times 146 \times 2}{10000}$ = Rs 447636

5. a

6. **Hint:** The third side of the triangle = 324 - (154 + 85) = 85 metres

Now find the area by applying the given rule.

(i) c; Area = 2772 sq m

(ii) a; perpendicular distance = $\frac{2 \times 2772}{154}$ = 36 metres

(iii) c; the required cost = 2772 × 5 = Rs 13860

7. b; **Hint:** $S = \frac{51+52+53}{2} = 78$ cm

Area = $\sqrt{78(78-51)(78-52)(78-53)}$

$= \sqrt{78 \times 27 \times 26 \times 25} = 1170$ sq cm

$\frac{1}{2} \times h \times 52 = 1170 \quad \therefore h = \frac{1170 \times 2}{52} = 45$

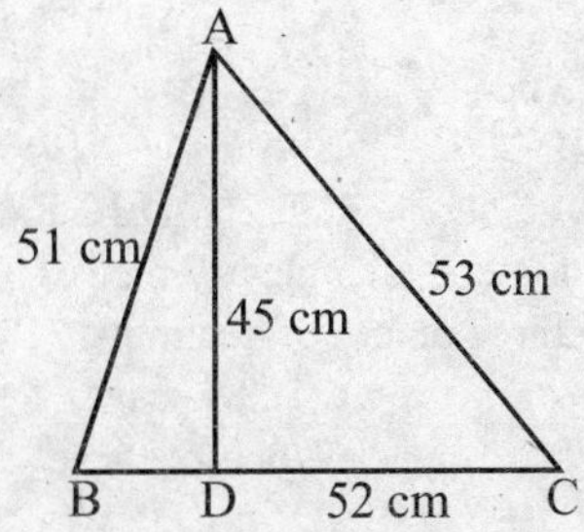

BD = $\sqrt{51^2 - 45^2} = 24$ cm

Δ ABD = $\frac{1}{2} \times 24 \times 45 = 540$ sq cm

DC = 52 - 24 = 28 cm

Δ ADC = $\frac{1}{2} \times 28 \times 45 = 630$ sq cm

Rule 3

Area of an equilateral triangle = $\frac{\sqrt{3}}{4} \times (\text{side})^2$ and perimeter of an equilateral triangle = 3 × side.

Illustrative Example

Ex.: Find the area of an equilateral triangle each of whose sides measures 8 cm. Also find perimeter of the equilateral triangle.

Soln: Applying the above formula,

Area of an equilateral triangle = $\frac{\sqrt{3}}{4} \times (8)^2$

$= \frac{\sqrt{3}}{4} \times 8 \times 8 = 16\sqrt{3}$ sq cm

Perimeter of an equilateral triangle
= 3 × side = 3 × 8 = 24 cm

Exercise

1. Find the area of an equilateral triangle each of whose sides measures 12 cm.
a) $36\sqrt{3}$ sq cm b) $18\sqrt{3}$ sq cm
c) $24\sqrt{3}$ sq cm d) $30\sqrt{3}$ sq cm

2. Find the area of a triangle in which each side measures 8 cm.
a) $2\sqrt{3}$ sq cm b) $8\sqrt{3}$ sq cm
c) $16\sqrt{3}$ sq cm d) $12\sqrt{3}$ sq cm

3. Each side of an equilateral triangle is increased by 1.5%. The percentage increase in its area is:

a) 1.5% b) 3% c) 4.5% d) 5.7%

(Railway Recruitment Board Exam, 1991)

4. If the perimeter of an equilateral triangle is 12 metres, find its area.

a) $4\sqrt{3}m^2$ b) $16\sqrt{3}m^2$ c) $8\sqrt{3}m^2$ d) $6m^2$

(LIC Exam 1986)

5. The side of an equilateral triangle is 7 metres. Calculate its area correct to three places of decimals.

a) 21.218 sq m b) 21.281 sq m
c) 21.128 sq m d) None of these

Answers

1. a 2. c

3. a; Let original length of each side = a

Then, area = $\frac{\sqrt{3}}{4}a^2 = A$

New area

$$= \frac{\sqrt{3}}{4}\left[\left(\frac{101.5}{100}a\right)^2\right] = \frac{\sqrt{3}}{4}\left(\frac{20.3}{20}\right)a^2 = \left(\frac{20.3}{20}\right)A$$

Increase in area = $\left(\frac{0.3}{20}A \times \frac{I}{A} \times 100\right)\% = 1.5\%$

4. a; **Hint:** 3 × side = 12 m ∴ side = $\frac{12}{3}$ = 4 m

Area = $\frac{\sqrt{3}}{4} \times 4 \times 4 = 4\sqrt{3}$ sq m

5. a

Rule 4

(i) Area of an isosceles triangle = $\frac{b}{4}\sqrt{4a^2 - b^2}$

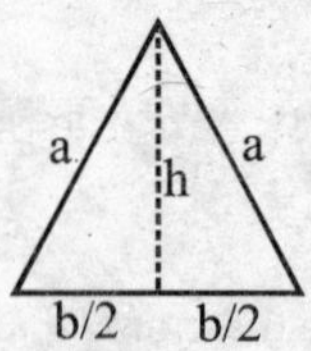

(ii) Height (h) = $\sqrt{a^2 - \left(\frac{b}{2}\right)^2} = \frac{1}{2}\sqrt{4a^2 - b^2}$

(iii) Perimeter = (2a + b)

Illustrative Example

Ex.: The base and the other side of an isosceles triangle is 10 cm and 13 cm respectively. Find its area and perimeter.

Soln: Applying the above formula,

Area = $\frac{10}{4}\sqrt{4 \times (13)^2 - (10)^2}$

$= \frac{10}{4}\sqrt{676 - 100} = \frac{10}{4} \times 24 = 60$ cm^2

Perimeter = 2 × 13 + 10 = 36 cm.

Exercise

1. The perimeter of an isosceles triangle is equal to 14 cm; the lateral side is to the base in the ratio 5 to 4. The area, in cm^2, of the triangle is:

a) $\frac{1}{2}\sqrt{21}$ b) $\sqrt{21}$ c) $\frac{3}{2}\sqrt{21}$ d) $2\sqrt{21}$

(CDS Exam 1989)

2. A plot of land is in the shape of a right angled isosceles triangle. The length of hypotenuse is $50\sqrt{2}$ m. The cost of fencing is Rs 3 per metre. The cost of fencing the plot will be:

a) less than Rs 300 b) less than Rs 400
c) more than Rs 500 d) more than Rs 600

(CDS Exam 1991)

3. In an isosceles right-angled triangle, the length of one leg is 10 metres. Find its area and its perimeter.

a) 50 sq m, 34.15 m b) 50 sq m, 44.14 m
c) 50 sq m, 34.41 m d) Data inadequate

Answers

1. d; **Hint:** Let lateral side = $5x$ & base = $4x$

Then, $5x + 5x + 4x = 14 \Rightarrow x = 1$

∴ The sides are 5 cm, 5 cm, 4 cm

Now, $h^2 = 5^2 - 2^2 \Rightarrow h = \sqrt{21}$

∴ Area = $\left(\frac{1}{2} \times 4 \times \sqrt{21}\right)cm^2 = 2\sqrt{21}cm^2$

2. c; **Hint:** Let each of the equal sides be a metres long.

Then, $a^2 + a^2 = (50\sqrt{2})^2 = 5000$

$\Rightarrow a^2 = 2500 \Rightarrow a = 50$

∴ Perimeter of the triangle = $(50 + 50 + 50\sqrt{2})$

= 100 + 50 × 1.4146 = 170.73 m

∴ Cost of fencing = Rs (170.73 × 3) = Rs 512.19

3. a; **Hint:** Area = $\frac{1}{2} \times 10 \times 10$ = 50 sq m

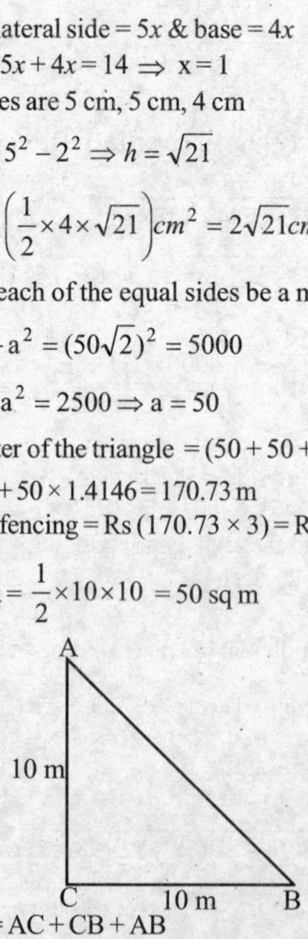

Perimeter = AC + CB + AB

$[AB = \sqrt{10^2 + 10^2} = \sqrt{200} = 10\sqrt{2}]$

$= 10 + 10 + 10\sqrt{2} = 20 + 14.14 = 34.14$ m

Rule 5

Theorem: ***The perimeter of an isosceles triangle is given as P cm. Now consider the following cases.***

Case I: If the base of the isosceles triangle is given by 'b' cm, then the length of the equal sides is $\left(\frac{P-b}{2}\right)$ ***cm.***

Case II: If the length of equal sides is given by 'a' cm, then the length of the base is (P – 2a) cm.

Illustrative Examples

Ex. 1: The perimeter of an isosceles triangle is 120 cm. If the base is 60 cm, find the length of equal sides.

Soln: Applying the above formula, (case - I)

Length of equal sides $= \frac{120-60}{2} = 30$ cm

Ex. 2: The perimeter of an isosceles triangle is 100 cm. If the length of the equal sides is given by 32 cm, find the length of the base.

Soln: Applying the above formula (case - II),
Length of the base = 100 – 2 × 32 = 36 cm

Exercise

1. The perimeter of an isosceles triangle is 60 cm. If the base is 30 cm, find the length of equal sides.
   a) 30 cm b) 15 cm c) 12 cm d) 20 cm
2. The perimeter of an isosceles triangle is 45 cm. If the base is 25 cm, find the length of equal sides.
   a) 20 cm b) 10 cm c) 8 cm d) 15 cm
3. The perimeter of an isosceles triangle is 32 cm. If the base is 18 cm, find the length of equal sides.
   a) 7 cm b) 9 cm c) 14 cm d) 8 cm
4. The perimeter of an isosceles triangle is 120 cm. If the length of the equal sides is given by 50 cm, find the length of the base.
   a) 25 cm b) 20 cm c) 15 cm d) 30 cm
5. The perimeter of an isosceles triangle is 50 cm. If the length of the equal sides is given by 12 cm, find the length of the base.
   a) 26 cm b) 24 cm c) 36 cm d) 16 cm

Answers

1. b 2. b 3. a 4. b 5. a

Rule 6

Theorem: ***To find the height of the equilateral triangle when the length of its side is given.***

Height of the equilateral triangle $= \frac{\sqrt{3}}{2} \times$ ***side.***

Illustrative Example

Ex.: Length of the side of an equilateral triangle is $4\sqrt{3}$ cm. Find its height.

Soln: Applying the above rule, we have height

$= \frac{\sqrt{3}}{2} \times 4\sqrt{3} = 6$ cm.

Exercise

1. Height of an equilateral triangle is 6 cm. Find its side.
   a) 4 cm b) $3\sqrt{3}$ cm c) $4\sqrt{3}$ cm d) $5\sqrt{3}$ cm
2. Length of the side of an equilateral triangle is $\frac{2}{\sqrt{3}}$ cm. Find its height.
   a) 1 m b) 1.5 m c) $\frac{1}{\sqrt{3}}$ m d) 0.5 m
3. Length of the side of an equilateral triangle is $3\sqrt{3}$ cm. Find its height.
   a) 4.5 m b) 4 m c) 5 m d) 5.5 m

Answers

1. c; **Hint:** $6 \text{ cm} = \frac{\sqrt{3}}{2} \times side \quad \therefore \text{ Side} = \frac{6 \times 2}{\sqrt{3}} = 4\sqrt{3}$ cm

2. a 3. a

Rule 7

Theorem: ***To find the area of an equilateral triangle if its height is given.***

Area of the equilateral triangle $= \frac{(Height)^2}{\sqrt{3}}$

Illustrative Example

Ex.: Height of an equilateral triangle is 6 cm. Find its area.

Soln: Detail Method: Let the base of an equilateral triangle be x and the height be h.

Now, $\frac{\sqrt{3}}{4}x^2 = \frac{1}{2} \times x \times h \left(\frac{1}{2} \times base \times height\right)$

or, $x = \frac{2 \times h}{\sqrt{3}} = \frac{2 \times 6}{\sqrt{3}} = \frac{12}{\sqrt{3}}$

Area $= \frac{\sqrt{3}}{4} \times x^3 = \frac{\sqrt{3}}{4} \times \frac{12 \times 12}{3} = 12\sqrt{3}$ sq cm.

Quicker Method: Applying the above theorem, we have

$$\text{Area} = \frac{6\times 6}{\sqrt{3}} = \frac{6\times 2\times\sqrt{3}\times\sqrt{3}}{\sqrt{3}} = 12\sqrt{3} \text{ sq cm.}$$

Exercise

1. Height of an equilateral triangle is 12 cm. Find its area.
   a) $48\sqrt{3}$ sq cm b) $36\sqrt{3}$ sq cm
   c) $12\sqrt{3}$ sq cm d) Data inadequate
2. Height of an equilateral triangle is 9 cm. Find its area.
   a) $27\sqrt{3}$ sq cm b) $36\sqrt{3}$ sq cm
   c) $54\sqrt{3}$ sq cm d) Data inadequate
3. Area of an equilateral triangle is $75\sqrt{3}$ sq cm. Find its height.
   a) 12 cm b) 25 cm c) 15 cm d) Data inadequate

Answers

1. a 2. a 3. c

Rule 8

Theorem: ***The perimeter of a square is equal to the perimeter of an equilateral triangle. If the diagonal of the square is 'd' units, then***

(i) the side of the square = $\frac{d}{\sqrt{2}}$ *units,*

(ii) the side of the equilateral triangle = $\frac{4d}{3\sqrt{2}}$ *units,*

(iii) the area of the square = $\frac{d^2}{2}$ *square units, and*

(iv) the area of the equilateral triangle = $\frac{2}{3\sqrt{3}}\times d^2$

sq units.

Illustrative Example

Ex.: Perimeter of a square and an equilateral triangle is equal. If the diagonal of the square is $12\sqrt{2}$ cm, then find the area of the equilateral triangle.

Soln: Detail Method: Diagonal of the square = $12\sqrt{2}$ cm. or, side × $\sqrt{2}$ = $12\sqrt{2}$

∴ Side of the square = 12 cm

Perimeter of an equilateral triangle = Perimeter of the square

or, 3 × side of the triangle = 4 × 12

∴ side of the triangle = 16

$$\therefore \text{ area} = \frac{\sqrt{3}}{4}\times 16\times 16 = 16\sqrt{3} \text{ sq cm.}$$

Quicker Method: Applying the above theorem, we have the

$$\text{area of the equilateral triangle} = \frac{2}{3\sqrt{3}}\times 12\times 12\times 2$$

$$= 64\sqrt{3} \text{ sq cm.}$$

Exercise

1. Perimeter of a square and an equilateral triangle is equal. If the diagonal of the square is $15\sqrt{2}$ cm, then find the
   (i) side of the square.
   a) 12 cm b) 15 cm c) $12\sqrt{2}$ cm d) $3\sqrt{2}$ cm
   (ii) side of the equilateral triangle.
   a) 20 cm b) 12 cm c) 24 cm d) 10 cm
   (iii) area of the square.
   a) 144 sq cm b) 225 sq cm
   c) 288 sq cm d) Data inadequate
   (iv) area of the equilateral triangle
   a) 100 sq cm b) $100\sqrt{3}$ sq cm
   c) $50\sqrt{3}$ sq cm d) $100\sqrt{2}$ sq cm
2. Perimeter of a square and an equilateral triangle is equal. If the diagonal of the square is $18\sqrt{2}$ cm, then find the
   (i) side of the square.
   a) 36 cm b) $9\sqrt{2}$ cm c) 18 cm d) Data inadequate
   (ii) side of the equilateral triangle
   a) 24 cm b) 18 cm c) 28 cm d) 32 cm
   (iii) area of the square.
   a) 324 sq cm b) 320 sq cm
   c) 648 sq cm d) Data inadequate
   (iv) area of the equilateral triangle
   a) $144\sqrt{3}$ sq cm b) $144\sqrt{2}$ sq cm
   c) $288\sqrt{2}$ sq cm d) Data inadequate

Answers

1. (i) b (ii) a (iii) b (iv) b
2. (i) c (ii) a (iii) a (iv) a

Rule 9

Rectangle

(i) To find the area of a rectangle if its length and breadth are given. Area of a rectangle = Length × Breadth

Illustrative Example

Ex.: Find the area of a rectangular field of length 12 m and width 10 m.

Soln: Applying the above formula, we have
Area = 12 × 10 = 120 sq m.

(ii) To find the breadth of a rectangle, if area and length of

the rectangle are given. Breadgh $= \dfrac{Area}{Length}$

Illustrative Example

Ex.: Area of a rectangular field of length 12 m is 120 sq m. Find the breadth of the field.

Soln: Applying the above formula, we have Breadth

$= \dfrac{120}{12} = 10$ cm.

(iii) To find the length of a rectangle, if area and breadth of the rectangle given. Length $= \dfrac{Area}{Breadth}$

Illustrative Example

Ex.: Area of a rectangular field of breadth 10 cm is 120 sq m. Find the length of the field.

Soln: Applying the above formula, we have Length

$= \dfrac{120}{10} = 12$ cm

Exercise

1. Find the area and perimeter of a rectangular plot whose length is 24.5 metres and breadth is 16.8 metres.
   a) 411.6 sq m, 82.6 m b) 412 sq m, 83 m
   c) 416.1 sq m, 86.2 m d) None of these
2. Calculate the area of a rectangular field whose length is 13.5 m and breadth is 8 m.
   a) 180 sq m b) 108 sq m
   c) 140 sq m d) None of these
3. The length and breadth of a rectangular field are in the ratio 5 : 3. If the cost of cultivating the field at 25 paise per square metre is Rs 6000, find the dimensions of the field.
   a) 250 m by 100 m b) 50 m by 30 m
   c) 200 m by 120 m d) Can't be determined
4. A room 15 m long requires 7500 tiles, each 15 cm by 12 cm, to cover the entire floor. Find the breadth of the room.
   a) 10 m b) 12 m c) 6 m d) 9 m
5. A lawn in the form of a rectangle is half as long again as it is broad. The area of the lawn is $\frac{2}{3}$ hectares. The length of the lawn is:
   a) 100 m b) $33\frac{1}{3}$ m c) $66\frac{2}{3}$ m d) $\left(\frac{100}{\sqrt{3}}\right)$ m
6. The width of a rectangular hall is $\frac{3}{4}$ of its length. If the area of the hall is 300 sq m, then the difference between its length and width is:
   a) 3 m b) 4 m c) 5 m d) 15 m
   [Bank PO Exam - 1990]
7. A room 8 m × 6 m is to be carpeted by a carpet 2 m wide. The length of carpet required is
   a) 12 m b) 36 m c) 24 m d) 48 m
   [Railway Recruitment, 1990]
8. The length of a plot of land is 4 times its breadth. A playground measuring 1200 sq m occupies one-third of the total area of the plot. What is the length of the plot, in metres?
   a) 90 b) 80 c) 60 d) None of these
   [Bank PO Exam - 1990]
9. Length of a room is 6 m longer than its breadth. If the area of the room is 72 sq m, its breadth will be:
   a) 12 m b) 6 m c) 8 m d) 10 m
10. If only the length of a rectangular plot is reduced to $\frac{2}{3}$rd of its original length, the ratio of original area to reduced area is:
    a) 2 : 3 b) 3 : 2 c) 1 : 2 d) None of these
    [Railway Recruitment 1991[
11. The cost of carpeting a room 15 m long with a carpet 75 cm wide at 30 paise per metre is Rs 36. The breadth of the room is:
    a) 8 m b) 12 m c) 9 m d) 6 m
12. Calculate the area of a rectangle 23 metres 7 decimetres long and 14 metres 4 decimetres 8 centimetres wide.
    a) 343 sq m b) 363 sq m c) 334 sq m d) 365 sq m
13. The sides of a rectangular field of 726 sq metres are in the ratio 3 : 2. Find the sides.
    a) 33 m, 22 m b) 30 m, 20 m
    c) 45 m, 30 m d) Can't be determined
14. The length of a room is 3 times its breadth and its breadth is 5 m 5dm. Find the area of its floor.
    a) 90.75 sq m b) 81.12 sq m
    c) 80.75 sq m d) 90.25 sq m

Answers

1. a; **Hint:** Area = (24.5) × (16.8) sq m = 411.6 sq m
   Perimeter = 2 × (24.5 + 16.8) m = 82.6 m (See **Rule - 11**)
2. b
3. c; **Hint:** Area $= \dfrac{6000 \times 100}{25} = 24000$ sq m
   Let the length be 5x and breadth be 3x
   or, 5x × 3x = 24000 ∴ x = $\sqrt{1600} = 40$
   ∴ Length = 5 × 40 = 200 m and breadth = 3 × 40 = 120 m
4. d; **Hint:** Area of 1 tile $= \left(\dfrac{15}{100} \times \dfrac{12}{100}\right)$ sq m

Area of the floor $= \left(7500 \times \frac{15}{100} \times \frac{12}{100}\right) = 135$ sq m

$\therefore$ Breadth of the room $= \frac{Area}{Length} = \left(\frac{135}{15}\right)$ m = 9 m

5. a; **Hint:** Let breadth = x metres. Then, length = $\frac{3}{2}$ x metres.

$\therefore\ x \times \frac{3}{2} x = \frac{2}{3} \times 10000$

or, $x^2 = \left(\frac{4}{9} \times 10000\right)$ So, x = $\frac{200}{3}$

$\therefore$ Length $= \left(\frac{3}{2} \times \frac{200}{3}\right) = 100$ metres.

6. c; **Hint:** Let length = x metres. Then breadth = $\frac{3}{4}$ x metres

$\therefore\ x \times \frac{3}{4} x = 300 \Rightarrow x^2 = \frac{300 \times 4}{3} = 400$ or, x = 20

$\therefore\ (length - breadth) = \left(20 - \frac{3}{4} \times 20\right) = 5$ metres.

7. c; **Hint:** Length of the carpet $= \left(\frac{8 \times 6}{2}\right)$ m = 24 m

(Also see **Rule - 53**)

8. d; **Hint:** $x \times 4x = 3600 \Rightarrow x^2 = 900 \Rightarrow x = 30$

$\therefore$ Length = (4 × 30) m = 120 m

9. b; **Hint:** $x(x-6) = 72 \Rightarrow x^2 - 6x - 72 = 0$

$\Rightarrow$ (x - 12) (x + 6) = 0 $\Rightarrow$ x = 12 (neglecting x = -6)

$\therefore$ Breadth = (12 - 6) m = 6 m

10. b; **Hint:** Let length = x & breadth = y

New length = $\frac{2}{3}$ x & breadth = y

$\therefore \frac{Original\ area}{Reduced\ area} = \frac{xy}{\frac{2}{3}xy} = \frac{3}{2}$

11. d; **Hint:** Length of carpet $= \frac{3600}{30} = 120$ m

Area of the carpet $= \left(120 \times \frac{75}{100}\right) m^2 = 90 m^2$

$\therefore$ Area of the room = $90 m^2$

So, breadth of the room $= \left(\frac{90}{15}\right)$ m = 6 m

12. a; **Hint:** Length = 23.70 metres [since 10 decimetres = 1 m]

Breadth = 14.48 metres

$\therefore$ Area = 23.70 × 14.48 = 343.176 ≈ 343 sq m

13. a; **Hint:** 3x × 2x = 726

or, x = 11

$\therefore$ sides = 3 × 11 = 33 m and 2x = 2 × 11 = 22 m

14. a

Rule 10

To find the length of diagonal of a rectangle, if length and breadth are given.

$$(Diagonal)^2 = (Length)^2 + (Breadth)^2$$

$$\textbf{Or, Diagonal} = \sqrt{(Length)^2 + (Breadth)^2}$$

Illustrative Example

Ex.: Find the length of diagonal of a rectangle of length 8 cm and breadth 6 cm.

Soln: Applying the above theorem, we have

Diagonal = $\sqrt{(length)^2 + (breadth)^2}$

$= \sqrt{(8)^2 + (6)^2} = \sqrt{64 + 36} = \sqrt{100} = 10$ cm.

Exercise

1. Two roads XY and YZ of 15 metres and 20 metres length respectively are perpendicular to each other. What is the distance between X and Z by the shortest route?
a) 35 metres b) 30 metres c) 24 metres d) 25 metres
(SBI Associates PO - 1999)

2. The length of the longest rod which can be laid across of floor of a rectangular room 12 m in length and 5 m in breadth will be:
a) 17 m b) 7 m c) 2.4 m d) 13 m

3. The legs of a right triangle are in the ratio of 1 : 2 and its area is 36. The hypotenuse of the triangle is:
a) 3 b) $\sqrt{5}$ c) $\sqrt{3}$ d) $6\sqrt{5}$
(Railway Recruitment 1991)

4. Find the diagonal of a rectangle whose sides are 12 metres and 5 metres
a) 13 metres b) 14 metres
c) 16 metres d) Can't be determined

5. A ladder is placed so as to reach a window 63 m high. The ladder is then turned over to the opposite side of the street and is found to reach a point 56 m high. If the ladder is 65 m long, find the width of the street.
a) 49 m b) 45 m c) 40m d) 59 m

Answers

1. d; **Hint:**

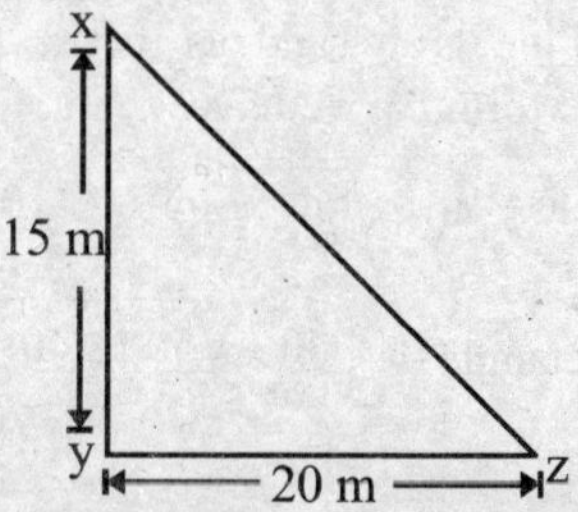

$XZ = \sqrt{15^2 + 20^2} = \sqrt{625} = 25\,m$

2. d

3. d; **Hint:** $\frac{1}{2} \times x \times 2x = 36 \Rightarrow x = 6$

$\therefore$ Hypotenuse $= \sqrt{6^2 + (12)^2} = \sqrt{180} = 6\sqrt{5}$

4. a

5. a; **Hint:**

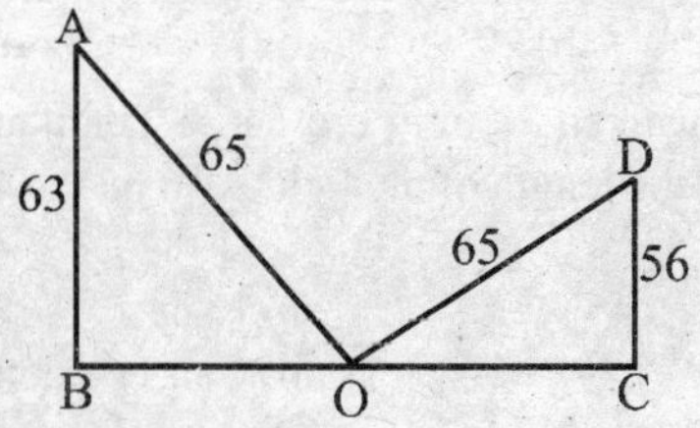

$OB = \sqrt{65^2 - 63^2} = 16\,m \qquad OC = \sqrt{65^2 - 56^2} = 33\,m$

$\therefore$ width of the street = OB + OC = 16 + 33 = 49 m

Rule 11

To find the perimeter of a rectangle if length and breadth are given.

Perimeter = 2(length + breadth)

Illustrative Example

Ex.: Find the perimeter of a rectangle of length 8 cm and breadth 6 cm.

Soln: Applying the above theorem, we have

Perimeter $= 2(8 + 6) = 28$ cm

Exercise

1. The length of a rectangular plot is 20 metres more than its breadth. If the cost of fencing the plot at the rate of Rs 26.50 per metre is Rs 5300, what is the length of the plot (in metres)?
   a) 40 b) 120 c) 50 d) None of these
   (Bank of Baroda PO 1999)
2. The length and breadth of a playground are 36 m and 21 m respectively. Flagstaffs are required to be fixed on all along the boundary at a distance 3 m apart. The number of flagstaffs will be:
   a) 37 b) 38 c) 39 d) 40
   (I Tax & Central Excise 1989)
3. A rectangle is having 15 cm as its length and 150 sq cm as its area. Its area is increased to $1\frac{1}{3}$ times the original area by increasing only its length. Its new perimeter is:
   a) 50 cm b) 60 cm c) 70 cm d) 80 cm
   (Bank PO Exam 1989)
4. The length and breadth of a rectangular piece of land are in the ratio of 5 : 3. The owner spent Rs 3000 for surrounding it from all the sides at the rate of Rs 7.50 per metre. The difference between length breadth is:
   a) 50 m b) 100 m c) 200 m d) 150 m
   (BSRB Bank PO Exam 1991)
5. The sides of a rectangular park are in the ratio 3 : 2 and its area is 3750 m^2. The cost of fencing it at 50 paise per metre is:
   a) Rs 312.50 b) Rs 375 c) Rs 187.50 d) Rs 125
6. The length of the rectangular floor is twice its width. If the length of a diagonal is $9\sqrt{5}$ m, then perimeter of the rectangle is:
   a) 27 m b) 54 m c) 81 m d) 162 m
7. The area of a rectangular field is 27000 sq m and the ratio between its length and breadth is 6 : 5. Find the cost of the wire required to go four time round the field at Rs 740 per km of length of the wire.
   a) Rs 1953.60 b) Rs 448.40
   c) Rs 1963.50 d) Data inadequate
8. The perimeter of a rectangle is 640 metres and the length is to the breadth as 5 : 3. Find its area.
   a) 2400 sq m b) 24000 sq m
   c) 24 hectare d) Can't be determined
9. The length of a rectangular field is twice its breadth. If the rent of the field at Rs 3500 a hectare is Rs 28000, find the cost of surrounding it with a fencing at Rs 5 per metre.
   a) Rs 6000 b) Rs 7000 c) Rs 6500 d) Rs 8000

Answers

1. d; **Hint:** $(l + b)2 = \frac{5300}{26.50} = 200\,m$

$\therefore$ l + b = 100 m .... (i) and l - b = 20 m (given) .... (ii)

From eqn (i) and (ii), we have

$l = \frac{120}{2} = 60\,m$

2. b; **Hint:** Petimeter $= 2\,(36 + 21) = 114$ m

$\therefore$ required no of flagstaffs $= \frac{114}{3} = 38$

3. b; **Hint:** Original length = 15 cm & breadth $= \frac{150}{15} = 10$ cm

New area $= \left(150 \times \frac{4}{3}\right) m^2 = 200 m^2$

$\therefore$ New length $= \frac{New\ area}{Original\ breadth} = \frac{200}{10} = 20$ cm

New perimeter = 2 (20 + 10) = 60 cm

4. a; **Hint:** Perimeter of the field $= \frac{3000}{7.50} = 400$ m

$\therefore 2(5x + 3x) = 400 \Rightarrow x = 25$

So, length = 125 m & breadth = 75 m

Difference between length & breadth

= (125 - 75) m = 50 m

5. d; **Hint:** $3x \times 2x = 3750 \Rightarrow x^2 = 625 \Rightarrow x = 25$

$\therefore$ Length = 75 m & breadth = 50 m

Perimeter = [2 × (75 + 50)] m = 250 m

$\therefore$ Cost of fencing = Rs $\left(250 \times \frac{1}{2}\right)$ = Rs 125

6. b; **Hint:** Let breadth = x metres and length = 2x metres

Then, $x^2 + (2x)^2 = (9\sqrt{5})^2$

$\Rightarrow 5x^2 = 405 \Rightarrow x = \sqrt{81} = 9$ [See **Rule - 10**]

$\therefore$ Perimeter = 2 (18 + 9) = 54 m

7. a; **Hint:** 6x × 5x = 27000 or, x = 30

$\therefore$ length = 180 m, breadth = 150 m

Length of wire required to go round the field four times = [4× 2 (180 + 150)] = 2.64 km

$\therefore$ required cost = Rs (2.64 × 740) = Rs 1953.60

8. b; **Hint:** 2 (5x + 3x) = 640 or, x = 40

$\therefore$ length = 200 m and breadth = 120 m

$\therefore$ area = 200 × 120 = 24000 sq m = 2.4 hectare

9. a; **Hint:** Area of the rectangular field $= \frac{28000}{3500}$

= 8 hectare = 80000 sq m

2x × x = 80000 [See **Rule - 9**]

$\therefore x = \sqrt{40000} = 200$ m and length = 400 m

Perimeter = 2 (400 + 200) = 1200 m

$\therefore$ Cost of fencing the rectangular field = 1200 × 5 = Rs 6000

Rule 12

To find the area and perimeter of a rectangle, if its one side and one diagonal are given.

(i) ***Area of rectangle*** $= \left(l \times \sqrt{d^2 - l^2}\right)$ ***sq units***

(ii) ***Perimeter of rectangle*** $= 2\,(l + \sqrt{d^2 - l^2})$ ***units.***

Illustrative Example

Ex.: One side and the diagonal of a rectangle are 40 m and 50 m respectively. Find its area and perimeter.

Soln: Applying the above formula, we have

Area of the rectangle $= 40 \times \sqrt{50^2 - 40^2} = 1200$ m².

Perimeter of the rectangle $= 2\left(40 + \sqrt{50^2 - 40^2}\right)$

= 140 metres.

Exercise

1. Find the area of a rectangle whose one side is 3 metres and the diagonal is 5 metres.
   a) 12 sq m b) 8 sq m c) 16 sq m d) 14 sq m
2. Calculate the area of a rectangular field whose one side is 12 m and the diagonal is 13 m.
   a) 70 sq m b) 60 sq m c) 45 sq m d) 75 sq m
3. One side of a rectangular field is 4 metres and its diagonal is 5 metres. The area of the field is:
   a) 12 m^2 b) 20 m^2 c) 15 m^2 d) $4\sqrt{5}\ m^2$
4. A man walked 20 m to cross a rectangular field diagonally. If the length of the field is 16 m, the breadth of the field is:
   a) 4 m b) 16 m
   c) 12 m d) Can't be determined
   (Railway Recruitment 1991)

Answers

1. a 2. b 3. a

4. c; **Hint:** Area $= 16 \times \sqrt{20^2 - 16^2} = 16 \times 12$

$\therefore$ breadth $= \frac{16 \times 12}{16} = 12$ m

Rule 13

To find the area of a rectangle when its perimeter and diagonal are given.

Area of a rectangle $= \left[\frac{(Perimeter)^2}{8} - \frac{(Diagonal)^2}{2}\right]$ ***sq units.***

Illustrative Example

Ex.: If the perimeter and diagonal of a rectangle are 14 cm and 5 cm respectively. Find its area.

Soln: **Detail Method:** Let the length and width of the rectangle be x and y cm respectively.

2(x + y) = Perimeter = 14 cm $\therefore$ x + y = 7 cm ..... (i)

$\sqrt{x^2 + y^2}$ = diagonal = 5 cm

$\therefore x^2 + y^2 = 25$ cm .... (ii)

Now, squaring equ (i)

$(x+y)^2 = 49$

$\Rightarrow x^2 + y^2 + 2xy = 49 \Rightarrow 25 + 2xy = 49$

$\therefore$ xy = area of the rectangle

$\therefore xy = \frac{49-25}{2} = \frac{24}{2} = 12$ sq cm.

Quicker Method: Applying the above theorem, we have

required area $= \frac{14\times14}{8} - \frac{5\times5}{2} = \frac{49-25}{2} = 12$ sq cm.

Exercise

1. A rectangular carpet has an area of 120 m^2 and a perimeter of 46 m. The length of its diagonal is:
a) 15 m b) 16 m c) 17 m d) 20 m
(Railway Recruitment 1991)
2. If the perimeter and diagonal of a rectangle are 16 cm and 4 cm respectively. Find its area.
a) 32 sq cm b) 26 sq cm c) 24 sq cm d) Data inadequate
3. If the perimeter and diagonal of a rectangle are 24 cm and 6 cm respectively. Find its area.
a) 72 sq cm b) 54 sq cm
c) 45 sq cm d) Data inadequate
4. A rectangular carpet has an area of 96 sq m and a diagonal of 8 m. Find the perimeter of the carpet.
a) 32 m b) 16 m c) 24 m d) 28 m

Answers

1. c; **Hint:** $120 = \frac{46\times46}{8} - \frac{(Diagonal)^2}{2}$

Or, $46\times46 - 4(Diagonal)^2 = 120\times8$

Or, $4\times(Diagonal)^2 = 1156$

$\therefore$ diagonal $= \sqrt{289} = 17$ m

2. c 3. b

4. a; **Hint:** $96 = \frac{(Perimeter)^2}{8} - \frac{8\times8}{2}$

Or, $(Perimeter)^2 = (96+32)\,8 = 1024$

$\therefore$ Perimeter $= \sqrt{1024} = 32$ m

Rule 14

To find length and breadth of a rectangle if its area and perimeter are given.

(i) Length of the rectangle

$$= \left(\sqrt{\frac{(Perimeter)^2}{16} - Area} + \frac{Perimeter}{4}\right) \textit{ units.}$$

(ii) Breadth of the rectangle

$$= \left(\frac{Perimeter}{4} - \sqrt{\frac{(Perimeter)^2}{16} - Area}\right) \textit{ units.}$$

Illustrative Example

Ex.: If the area and perimeter of a rectangle are 240 cm^2 and 68 cm respectively, find its length and breadth.

Soln: Detail Method: Let the length and breadth of the rectangle be x and y cm.

Area of the rectangle $= xy = 240$ cm^2

Perimeter of the rectangle $= (x+y)\,2 = 68$ cm

$\therefore x + y = 34$ cm ..... (i)

$(x-y)^2 = (x+y)^2 - 4xy$

$= (34)^2 - 4\times240 = 1156 - 960 = 196$

$\therefore x - y = \sqrt{196} = 14$ .... (ii)

By adding equ (i) and equ (ii), we have $2x = 48$

$\therefore x = 24$ cm

$2y = 20 \quad \therefore y = 10$ cm

Quicker Method: Applying the above theorem, we have the

length of the rectangle $= \sqrt{\frac{68\times68}{16} - 240} + \frac{68}{4}$

$= \sqrt{289-240} + 17 = 7 + 17 = 24$ cm

Breadth of the rectangle $= \frac{68}{4} - \sqrt{\frac{68\times68}{16} - 240}$

$= 17 - \sqrt{49} = 17 - 7 = 10$ cm.

Exercise

1. When the length of a rectangular plot is increased by four times its perimeter becomes 480 metres and area 12800 sq m. What was its original length (in metre)?
a) 160 b) 40
c) 20 d) Can't be determined
(BSRB Bhopal PO - 2000)
2. Calculate the area of a rectangular field whose length is 66 m and perimeter is 242 m.
a) 3630 sq m b) 3360 sq m
c) 3560 sq m d) None of these
3. The cost of fencing a rectangular field at Rs 3.50 per metre is Rs 595. If the length of the field be 60 metres,

find the cost of levelling it at 50 paise per square metre.

a) Rs 700 b) Rs 860
c) Rs 750 d) Data inadequate

4. The perimeter of a rectangle is 82 m and its area is 400 sq m. The breadth of the rectangle is:

a) 14 m b) 16 m c) 18 m d) 12 m

Answers

1. b; **Hint:** 4× original length of the rectangular plot

$$= \sqrt{\frac{(480)^2}{16} - 12800} + \frac{480}{4} = 40 + 120 = 160$$

$$\therefore \text{ Original length} = \frac{460}{4} = 40 \text{ metres}$$

2. a; **Hint:** $66 = \sqrt{\frac{(242)^2}{16} - Area} + \frac{242}{4}$

or, $\sqrt{\frac{(242)^2}{16} - Area} = 66 - 60.5 = 5.5$

$\therefore$ Area $= \frac{242 \times 242}{16} - (5.5)(5.5)$

$= 3660.25 - 30.25 = 3630$ sq m

3. c; **Hint:** Perimeter of the rectangular field $= \frac{595}{3.50} = 170$ m

Now, applying the given rule,

$60 = \sqrt{\frac{(170)^2}{16} - Area} + \frac{170}{4}$

or, $\sqrt{\frac{170 \times 170}{16} - Area} = 60 - \frac{170}{4} = 17.5$

$\therefore$ Area $= \frac{170 \times 170}{16} - (17.5 \times 17.5)$

$= 1806.25 - 306.25 = 1500$ sq m

$\therefore$ Cost of levelling $= \frac{1500 \times 50}{100} =$ Rs 750

4. b; **Hint:** Required breadth $= \frac{82}{4} - \sqrt{\frac{82 \times 82}{16} - 400}$

$= 20.5 - 4.5 = 16$ m

Rule 15

To find the difference in length and width of a rectangle when perimeter and area are given.

Difference in length and width of a rectangle

$$= \left(\sqrt{\left(\frac{Perimeter}{2}\right)^2 - 4 \times Area}\right) \textbf{ units.}$$

Illustrative Example

Ex.: Perimeter and area of a rectangle are 82 cm and 400 sq cm. Find the difference in length and width.

Soln: Detail Method: Let the length be l cm and width be b cm

As per the question, $(l+b)2 = 82$ cm

or, $l + b = 41$ cm .... (i) and $l \times b = 400$ sq cm ... (ii)

$(l-b)^2 = (l+b)^2 - 4lb = (41)^2 - 4 \times 400$

$\therefore l - b = \sqrt{1681 - 1600} = \sqrt{81} = 9$ cm

Quicker Method: Applying the above rule, we have the required answer

$$= \sqrt{\left(\frac{82}{2}\right)^2 - 4 \times 400} = \sqrt{(41)^2 - 1600} = \sqrt{81} = 9 \text{ cm.}$$

Exercise

1. If the width of a rectangle is 2 m less than its length, and its perimeter is 32 m, the area of the rectangle is:

a) 224 m^2 b) 108 m^2 c) 99 m^2 d) 63 m^2

2. A man drives 4 km distance to go around a rectangular park. If the area of the rectangle is 0.75 sq km, the difference between the length and the breadth of the rectangle is:

a) 10.25 km b) 0.5 km c) 1 km d) 2.75 km

3. The breadth of a rectangular tennis court is 7 metres less than its length and its perimeter 138 metres. Find its area.

a) 1178 sq m b) 1187 sq m c) 1168 sq m d) 1278 sq m

Answers

1. d; **Hint:** $2 = \sqrt{\left(\frac{32}{2}\right)^2 - 4 \times Area}$

$4 = 256 - 4 \times$ Area $\therefore$ area $= \frac{256 - 4}{4} = \frac{252}{4} = 63$ sq m

2. c; **Hint:** Perimeter = 4 km, Area = 0.75 sq km

By applying the given rule find the difference between length and breadth.

3. a; **Hint:** $7 = \sqrt{\left(\frac{138}{2}\right)^2 - 4 \times Area}$

or, $(69)^2 - 4 \times Area = 49$

$$\text{or, Area} = \frac{4761-49}{4} = 1178 \text{ sq m}$$

Rule 16

Theorem: ***If length of a rectangle is increased by 'x' units and due to this increase, area of the rectangle also increases by 'y' sq units, then width is given by*** $\left(\frac{y}{x}\right)$ ***units.***

Illustrative Example

Ex.: If increasing the length of a rectangular field by 5 metres, area also increases by 30 sq metres, then find the value of its width.

Soln: **Detail Method:** Let the length and breadth of the rectangular field be *l* m and b m respectively.

In first case area = lb sq m

In second case area = (l + 5) b = lb + 30

or, lb + 5b = lb + 30 $\quad\therefore$ b = 6 metres.

Quicker Method: Applying the above theorem, we have

$$\text{the width of the rectangular field} = \frac{30}{5} = 6 \text{ m.}$$

Exercise

1. The area of a rectangular courtyard is 100 sq metres. Had the length of the courtyard been longer by 2 metres, the area would have been increased by 10 sq metres. Find the length and breadth of the courtyard.
   a) 20 m, 5 m $\quad$ b) 25 m, 4 m
   c) 30 m, $30\frac{1}{3}$ m $\quad$ d) Data inadequate
2. If increasing the length of a rectangular field by 4 metres, area also increases by 16 sq metres, then find the value of its width.
   a) 4 m $\quad$ b) 8 m $\quad$ c) 6 m $\quad$ d) Data inadequate
3. If increasing the length of a rectangular field by 8 metres, area also increases by 32 sq metres, then find the value of its width.
   a) 4 m $\quad$ b) 6 m $\quad$ c) 9 m $\quad$ d) 12 m
4. If increasing the length of a rectangular field by 9 metres, area also increases by 54 sq metres, then find the value of its width.
   a) 6 m $\quad$ b) 8 m $\quad$ c) 9 m $\quad$ d) 15 m

Answers

1. a; **Hint:** Width $= \frac{10}{2} = 5$ m

$\therefore$ length $= \frac{100}{5} = 20$ m $\quad$ [See **Rule - 9**]

2. a $\quad$ 3. a $\quad$ 4. a

Rule 17

Theorem: ***There is a rectangle of area 'A' sq unit. If the sum of its diagonal and length is n times of its breadth, then the length and breadth of the rectangle are*** $\sqrt{\frac{A(n^2-1)}{2n}}$ ***and*** $\sqrt{\frac{2An}{n^2-1}}$ ***units respectively.***

Illustrative Example

Ex.: There is a rectangular field of area 60 sq cm. Sum of its diagonal and length is 5 times of its breadth. Find the breadth of the rectangular field.

Soln: **Detail Method:** Let the length and breadth of the rectangular field be x cm and y cm respectively.

As per the question,

$xy = 60$ sq cm. ...... (i) and $\sqrt{x^2+y^2}+x = 5y$ .... (ii)

or, $x^2 + y^2 = (5y - x)^2$

or, $x^2 + y^2 = 25y^2 + x^2 - 10xy$

or, $24y^2 = 10 \times 60$ $\quad$ or, $y^2 = 5 \times 5$

$\therefore$ y = 5 cm and x $= \frac{60}{5} = 12$ cm

$\therefore$ Length and breadth of the rectangular field are 12 cm and 5 cm respectively.

Quicker Method: Applying the above theorem, we have the

length of the rectangular field

$$= \sqrt{\frac{60 \times (5^2-1)}{2 \times 5}} = \sqrt{\frac{60 \times 24}{10}} = 12 \text{ cm and}$$

the breadth of the rectangular field

$$= \sqrt{\frac{2 \times 60 \times 5}{5^2 - 1}} = 5 \text{ cm.}$$

Exercise

1. There is a rectangular field of area 48 sq cm. Sum of its diagonal and length is 3 times of its breadth. Find the length and the breadth of the rectangle.
   a) 8 cm, 6 cm $\quad$ b) 12 cm, 4 cm
   c) 16 cm, 3 cm $\quad$ d) Data inadequate
2. There is a rectangular field of area 120 sq cm. Sum of its diagonal and length is 4 times of its breadth. Find the perimeter of the rectangle.
   a) 46 m $\quad$ b) 15 m
   c) 8 m $\quad$ d) Data inadequate
3. There is a rectangular field of area 420 sq cm. Sum of its

diagonal and length is 6 times of its breadth. Find the diagonal of the rectangle.
a) 35 cm b) 37 cm c) 33 cm d) 32 cm

Answers

1. a
2. a; **Hint:** First find the length and breadh.
Length = 15 m and breadth = 8 m
$\therefore$ perimeter = (15 + 8) 2 = 46 m
3. b; **Hint:** First find the length and breadth of the rectangle.
Length = 35 cm, breadth = 12 cm
$\therefore$ diagonal $= \sqrt{35^2 + 12^2} = \sqrt{1369} = 37$ m

Rule 18

Theorem: ***There is a rectangle. Its length is 'x' units more than its breadth. If its length is increased by 'y' units and its breadth is decreased by 'z' units, the area of the rectangle is unchanged. Length and breadth of the rectangle are***

$\left[\frac{(x+z)y}{y-z}\right]$ **and** $\left[\frac{(x+y)z}{y-z}\right]$ **units respectively.**

Illustrative Example

Ex.: Length of a rectangular blackboard is 8 cm more than that of its breadth. If its length is increased by 7 cm and its breadth is decreased by 4 cm, its area remains unchanged. Find the length and breadth of the rectangular blackboard.

Soln: **Detail Method:** Let the breadth of the blackboard be x cm, then
length = (x + 8) cm
As per the question,
$(x+8+7)(x-4) = (x+8)x$
or, $(x+15)(x-4) = (x+8)x$ or, $3x = 60$
$\therefore x = 20$
$\therefore$ Breadth = 20 cm and Length = 20 + 8 = 28 cm

Quicker Method: Applying the above theorem, we have

the length of the blackboard $= \frac{(8+4)7}{7-4} = 28$ cm and

the breadth of the blackboard $= \frac{(8+7)4}{7-4} = 4 \times 5$

$= 20$ cm.

Exercise

1. Length of a rectangular blackboard is 16 cm more than that of its breadth. If its length is increased by 14 cm and its breadth is decreased by 8 cm, its area remains unchanged. Find the length and breadth of the rectangular blackboard.
a) 28 cm, 20 cm b) 56 cm, 40 cm
c) 26 cm, 10 cm d) Data inadequate
2. Length of a rectangular blackboard is 12 cm more than that of its breadth. If its length is increased by 13 cm and its breadth is decreased by 8 cm, its area remains unchanged. Find the length and breadth of the rectangular blackboard.
a) 52 cm, 40 cm b) 48 cm, 42 cm
c) 26 cm, 20 cm d) Data inadequate
3. Length of a rectangular blackboard is 15 cm more than that of its breadth. If its length is increased by 9 cm and its breadth is decreased by 6 cm, its area remains unchanged. Find the length and breadth of the rectangular blackboard.
a) 60 cm, 40 cm b) 63 cm, 48 cm
c) 64 cm, 48 cm d) Data inadequate
4. Length of a rectangular blackboard is 20 cm more than that of its breadth. If its length is increased by 15 cm and its breadth is decreased by 10 cm, its area remains unchanged. Find the perimeter of the black board.
a) 150 cm b) 280 cm d) 270 cm d) 160 cm
5. Length of a rectangular blackboard is 10 cm more than that of its breadth. If its length is increased by 8 cm and its breadth is decreased by 5 cm, its area remains unchanged. Find the area of the black board.
a) 1200 sq cm b) 1250 sq cm
c) 1320 sq cm d) Data inadequate

Answers

1. b 2. a 3. b 4. b 5. a

Rule 19

Theorem: ***Length of a rectangle is increased by 'a' units and breadth is decreased by 'b' units, area of the rectangle remains unchanged. If length be decreased by 'c' units and breadth by increased by 'd' units, in this case also area of the rectangle remains unchanged. Length and breadth of the rectangle are given by*** $ac\left(\frac{d+b}{ad-bc}\right)$ ***and*** $bd\left(\frac{a+c}{ad-bc}\right)$ ***units respectively.***

Illustrative Example

Ex.: Length of a rectangular field is increased by 7 metres and breadth is decreased by 3 metres, area of the field remains unchanged. If length be decreased by 7 metres and breadth be increased by 5 metres, again area remains unchanged. Find the length and breadth of the rectangular field.

Soln: **Detail Method:** Let the length and breadth of the field be x m and y m respectively.
As per the question,
In the first case,
$(x+7)(y-3) = xy$ or, $xy + 7y - 3x - 21 = xy$

or, $3x-7y=-21$ .... (i)

In the second case,

$(x-7)(y+5)=xy$ or, $xy-7y+5x-35=xy$

or, $5x-7y=35$ .... (ii)

From equ (i) and equ (ii), we get $x=28$ m and $y=15$ m

$\therefore$ Length and breadth of the rectangular field are 28 metres and 15 metres respectively.

Quicker Method: Applying the above theorem, we have

$$\text{Length} = \left(\frac{7\times5+7\times3}{35-21}\right)\times7=\frac{56}{14}\times7=28 \text{ metres}$$

$$\text{Breadth} = \left(\frac{21+21}{35-21}\right)\times5=3\times5=15 \text{ metres.}$$

Exercise

1. Length of a rectangular field is increased by 14 metres and breadth is decreased by 6 metres, area of the field remains unchanged. If length be decreased by 14 metres and breadth be increased by 10 metres, again area remains unchanged. Find the perimeter of the rectangle.
   a) 172 m b) 192 m
   c) 162 m d) Data inadequate
2. Length of a rectangular field is increased by 8 metres and breadth is decreased by 4 metres, area of the field remains unchanged. If length be decreased by 6 metres and breadth be increased by 5 metres, again area remains unchanged. Find the area of the rectangle.
   a) 283.5 sq m b) 284 sq m
   c) 285 sq m d) Data inadequate
3. Length of a rectangular field is increased by 21 metres and breadth is decreased by 9 metres, area of the field remains unchanged. If length be decreased by 21 metres and breadth be increased by 15 metres, again area remains unchanged. Find the length of diagonal of the rectangle.
   a) 90 m b) 64 m
   c) 95.3 m (approx) d) 64.8 m (approx)

Answers

1. a 2. a

3. c; **Hint:** First find the length and breadth of the rectangle.
Length = 84 m and breadth = 45 m.

$\therefore$ diagonal $=\sqrt{84^2+45^2}=\sqrt{9081}\approx 95.3$ m

Square

Rule 20

To find the area of a square if length of one of the sides is given.

Area of a square $= (side)^2$

Illustrative Example

Ex.: Find the area of a square whose length of the side is 5 cm.

Soln: Applying the above formula, we have

Area of a square $=(5)^2=25$ sq cm

Exercise

1. The length of a rectangular plot is 144 m and its area is same as that of a square plot with one of its sides being 84 m. The width of the plot is:
   a) 7 m b) 49 m c) 14 m d) Data inadequate
2. The length of a rectangular hall is 16 metres. If it can be partitioned into two equal square rooms, what is the length of the partition?
   a) 16 m b) 8 m c) 4 m d) Data inadequate
   [UTI Exam 1990]
3. Find the area of a square whose side is 75 metres.
   a) 5625 sq m b) 5265 sq m c) 5635 sq m d) 5675 sq m
4. The side of a square field is 89 metres. By how many square metres does its area fall short of a hectare?
   a) 2179 sq m b) 2099 sq m c) 2079 sq m d) 7921 sq m
5. Two carpets are made at the same price per sq m. One of them is 25.6 m long and 8.1 m broad, and costs Rs 14400, the other which is square costs Rs 28900. What is the length of each side of the square carpet?
   a) 20.4 m b) 21 m
   c) 21.5 m d) Data inadequate
6. Find, to the nearest cm, the length of the side of a square piece of ground whose area is $\frac{1}{10}$ of a hectare.
   a) 31 m 62 cm b) 31 m 52 cm
   c) 30 m 62 cm d) 32 m 62 cm
7. Find the side of a square whose area is $68\frac{1}{16}$ sq m.
   a) $7\frac{1}{4}$ m b) $8\frac{3}{4}$ m c) $8\frac{1}{4}$ m d) $7\frac{3}{4}$ m
8. Find the sides of two squares, which contain together 12.25 hectares, the sides of the squares being in the ratio of 3 : 4.
   a) 210 m, 280 m b) 90 m, 120 m
   c) 150 m, 200 m d) 180 m, 240 m

Answers

1. b; **Hint:** Required answer $=\frac{(84)^2}{144}=7\times7=49$ m

2. b; **Hint:** Let the length of partition be x m

or, $16\times x=x^2+x^2$ $\therefore x=8$ m

3. a
4. c; **Hint:** 1 hectare = 10000 sq m
5. a; **Hint:** Rate of the carpet per sq m $= \frac{14400}{8.1 \times 25.6} = $ Rs 69.4

Area of the square $= \frac{28900}{69.4} = 416.42$ sq m

$\therefore$ length of the square $= \sqrt{416.42} = 20.4$ m

6. a; **Hint:** 1 hectare = 10000 sq m

$\frac{1}{10}$ hectare = 1000 sq m

$\therefore$ length of the side $= \sqrt{1000} = 31.62$ m = 31 m 62 cm

7. c
8. a; **Hint:** $(3x)^2 + (4x)^2 = 12.25 \times 10000 = 122500$ or, $x = 70$

$\therefore$ sides are $3x = 3 \times 70 = 210$ m and $4x = 4 \times 70 = 280$ m.

Rule 21

To find the area of a square if length of the diagonal is given.

$$\textbf{\textit{Area of the square}} = \frac{1}{2}(diagonal)^2$$

Illustrative Example

Ex.: Find the area of a square whose length of diagonal is 6 cm.

Soln: Applying the above formula, we have

Area of a square $= \frac{1}{2} \times (6)^2 = 18$ sq cm.

Exercise

1. Area of a square field is 18 sq m find its length of diagonal.
   a) 5 m b) 6 m c) 4 m d) Data inadequate
2. What would be the length of the diagonal of a square plot whose area is equal to the area of a rectangular plot of 45 m length and 40 m width
   a) 42.5 m b) 60 m
   c) 4800 m d) Data inadequate
   (Bank of Baroda PO - 1999)
3. Find the area of a square whose diagonal is 2.9 metres long.
   a) 4.5 sq m b) 5 sq m
   c) 4.205 sq m d) Can't be determined
4. Find the area of a square field, the length of whose diagonal is 36 metres.
   a) 648 sq m b) 678 sq m c) 684 sq m d) 668 sq m
5. Find the length of the diagonal of a square of area 200 square centimetres.
   a) 30 cm b) 25 cm c) 20 cm d) 24 cm
6. The area of a square field is 8 hectares. How long would a man take to cross it diagonally by walking at the rate of 4 km per hour.
   a) 5 min b) 6 min c) 4 min d) 8 min
7. The diagonal of a square is 3.2 m. Its area is:
   a) 10.24 m^2 b) 2.56 m^2 c) 3.41 m^2 d) 5.12 m^2
8. Area of a square field is $\frac{1}{2}$ hectare. The diagonal of the square is:
   a) 50 m b) 100 m c) 250 m d) $50\sqrt{2}$ m
9. A square field of 2 sq km is to be divided into two equal parts by fence which coincides with a diagonal. Find the length of the fence.
   a) 2 km b) 3 km c) 1 km d) 1.5 km
10. What is the area of a square whose diagonal is 15 metres?
    a) 225 sq m b) 112.5 sq m c) 115 sq m d) 125 sq m
11. The area of a square 11370.32 sq metres. Find the length of its diagonal.
    a) 158.8 m b) 148.8 m c) 150.6 m d) 150.8 m

Answers

1. b; **Hint:** $\frac{1}{2} \times (diagonal)^2 = 18$ $\therefore$ diagonal = 6 m
2. b; **Hint:** Area of the square plot
$= 45 \times 40 = 1800$ sq m [See **Rule - 9**]

$\frac{1}{2} \times (diagonal)^2 = 1800$

$\therefore$ diagonal $= \sqrt{1800 \times 2} = 60$ m

3. c 4. a
5. c; **Hint:** $\frac{1}{2} \times (diagonal)^2 = 200$

$\therefore$ diagonal $= \sqrt{400} = 20$ cm

6. b; **Hint:** Area = 8 hectares = $8 \times 10000 = 80000$ sq m

$\frac{1}{2} \times (diagonal)^2 = 80000$

$\therefore$ diagonal $= \sqrt{160000} = 400$ m

$\therefore$ required time $= \frac{60 \times 400}{4000} = 6$ min.

7. d
8. b; **Hint:** $\frac{1}{2} \times (\text{diagonal})^2 = \frac{1}{2} \times 10000$

$\Rightarrow$ diagonal $= \sqrt{10000} = 100$ m.

9. a; **Hint:** Area $= 2 \times 100 \times 100 \times 100 = 2000000$ sq m

Diagonal $= \sqrt{2 \times 2000000} = 2000$ m = 2 km.

10. b 11. d

Rule 22

(i) To find perimeter of a square if its length of side is given.
Perimeter of a square = 4 × side

Illustrative Example

Ex.: Find the perimeter of a square field of length of one of its sides 4 m.

Soln: Applying the above formula, we have perimeter of the square field = 4 × 4 = 16 m.

(ii) To find the length of the side of a square if perimeter of the square is given.

$$\textit{Length of the side of a square} = \frac{\textit{Perimeter}}{4}$$

Illustrative Example

Ex.: Perimeter of a square field is 16 m. Find the length of its sides.

Soln: Applying the above formula, we have

length of sides = $\frac{16}{4}$ = 4 m.

Exercise

1. Find the perimeter of a square field of length of one of its sides 5 m.
   a) 25 m b) 10 m c) 20 m d) Data inadequate
2. Find the perimeter of a square field of length of one of its sides 6 m.
   a) 24 m b) 12 m c) 30 m d) None of these
3. Perimeter of a square field is 28 m. Find the length of its sides.
   a) 7 m b) 6 m c) 8 m d) 5 m
4. Perimeter of a square field is 24 m. Find the length of its sides.
   a) 6 m b) 5 m c) 9 m d) None of these

Answers

1. c 2. a 3. a 4. a

Rule 23

To find the diagonal of a square whose sides are given.

Length of the diagonal of a square = $\sqrt{2}\times$ ***side***

Illustrative Example

Ex.: Find the diagonal of a square field whose side is of 10 m length.

Soln: Applying the above formula, we have

Length of the diagonal = $\sqrt{2} \times 10 = 10\sqrt{2}$ m.

Exercise

1. Find the diagonal of a square field whose side is of 5 m length.
   a) $5\sqrt{2}$ m b) $10\sqrt{2}$ m c) 10 m d) None of these
2. Find the diagonal of a square field whose side is of 4 m length.
   a) $4\sqrt{2}$ m b) $16\sqrt{2}$ m c) 16 m d) Data inadequate
3. Find the length of sides of a square field whose diagonal is $12\sqrt{2}$ m.
   a) 12 m b) 10 m c) 24 m d) Data inadequate

Answers

1. a 2. a

3. a; **Hint:** $12\sqrt{2} = \sqrt{2} \times$ side ∴ side = 12 m.

Rule 24

To find the diagonal and the perimeter of a square if its area is given.

(i) Length of diagonal of a square = $\sqrt{2 \times area}$

(ii) Perimeter of a square = $\sqrt{16 \times area}$

Illustrative Example

Ex.: Find the length of the diagonal and the perimeter of a square plot if its area is 400 square metres.

Soln: Applying the above formulae, we have

(i) length of diagonal of the square

$= \sqrt{2} \times \sqrt{400} = 20\sqrt{2}$ metres.

(ii) perimeter of the square

$= \sqrt{16 \times 400} = 4 \times 20 = 80$ metres.

Exercise

1. In order to fence a square Manish fixed 48 poles. If the distance between two poles is 5 metres then what will be the area of the square so formed?
   a) 2600 cm^2 b) 2500 cm^2
   c) 3025 cm^2 d) None of these
   [BSRB Bangalore PO 2000]
2. The cost of cultivating a square field at the rate of Rs 160 per hectare is Rs 1440. Find the cost of putting a fence around it at the rate of 75 paise per metre.
   a) Rs 900 b) Rs 850
   c) Rs 950 d) Rs 940
3. If the ratio of areas of two squares is 9 : 1, the ratio of their perimeters is:
   a) 9 : 1 b) 3 : 4 c) 3 : 1 d) 1 : 3
   (Asstt. Grade 1990)
4. How long will a man take to walk round the boundary of a sq field containing 9 hectares at the rate of 6 km an hour?
   a) 12 min b) 10 min
   c) 24 min d) Can't be determined
5. The perimeter of a square field is 400 m. What is its area?
   a) 1 hectare b) 0.845 hectare
   c) 1.2 hectare d) Can't be determined

6. The area of a square field is 6.25 hectares. How long will it take a man to walk round the outside of it at the rate of $2\frac{1}{2}$ km/hr? (1 hectare = 10000 sq metres)
a) 32 min b) 24 min c) 28 min d) 20 min

7. Find in km the length of the wire required to go 10 times round a square field of $6\frac{1}{4}$ hectares.
a) 8 km b) 5 km c) 10 km d) 16 km

8. Calculate the cost of surrounding with a fence a square field of 16 hectares at Rs 20 per metre.
a) Rs 30000 b) Rs 24000 c) Rs 32000 d) Rs 36000

9. The cost of levelling and turfing a square cricket field at Rs 16000 per hectare is Rs 262440. Find the cost of surrounding it with a railing costing Rs 25 per metre.
a) Rs 45000 b) Rs 40500 c) Rs 42500 d) Rs 48500

10. How long will it take to run round a square field containing 1681 sq m at the rate of 4 km an hour?
a) 3 min b) 2.45 min c) 3.46 min d) 2.46 min

Answers

1. d; **Hint:** Perimeter $= 48 \times 5 = 240$ metres

$$\text{Area} = \left(\frac{Perimeter}{4}\right)^2 = \left(\frac{240}{4}\right)^2$$

$= 60 \times 60 = 3600$ sq m [See **Rule - 24 (ii)**]

2. a; **Hint:** Area $= \left(\frac{Total\ cost}{Rate/hectare}\right) = \frac{1440}{160}$

$= 9$ hectares $= 90000$ sq m

Perimeter $= \sqrt{16 \times 90000} = 1200$ m

$\therefore$ Cost of fencing $= \frac{1200 \times 75}{100} =$ Rs 900

3. c; **Hint:** Required ratio $= \sqrt{\frac{9}{1}} = 3:1$

[$\because$ Perimeter $\alpha \sqrt{Area}$, See **Rule - 24 (ii)**]

4. a; **Hint:** Area $= 9 \times 10{,}000$ sq m

$\therefore$ perimeter $= \sqrt{16 \times 9 \times 10000} = 4 \times 3 \times 100 = 1200$ m

$\therefore$ required time $= \frac{60 \times 1200}{6 \times 1000} = 12$ min

5. a; **Hint:** $\sqrt{16 \times Area} = 400$

$\therefore$ Area $= \frac{400 \times 400}{16} = 10000$ sq m $= 1$ hectare

6. b; **Hint:** Area $= 6.25$ hectare $= 6.25 \times 10000 = 62500$ sq m

Perimeter $= \sqrt{16 \times 62500} = 4 \times 250 = 1000$ m

$\therefore$ required time $= \frac{60}{2500} \times 1000 = 24$ min

7. c; **Hint:** Area $= \frac{25}{4} \times 10000 = 62500$ sq m

Perimeter $= \sqrt{16 \times 625000} = 1000$ m $= 1$ km

$\therefore$ required length of the wire $= 10 \times 1 = 10$ km

8. c; **Hint:** Perimeter $= \sqrt{16 \times 16 \times 10000} = 1600$ m

$\therefore$ required cost $= 1600 \times 20 =$ Rs 32000

9. b; **Hint:** Area $= \frac{262440}{16000} = 16.4025$ hectare $= 164025$ sq m

Perimeter $= \sqrt{16 \times 164025} = 1620$ m

$\therefore$ required cost $= 1620 \times 25 =$ Rs 40500

10. d; **Hint:** Perimeter $= \sqrt{16 \times 1681} = 164$ m

$\therefore$ time required $= \frac{60 \times 164}{4 \times 1000} = 2.46$ min

Rule 25

To find the perimeter of a square if its diagonal is given.

Perimeter of the square $= \left(2\sqrt{2} \times Diagonal\right)$

Illustrative Example

Ex.: The diagonal of a square is 10 cm. Find its perimeter and area.

Soln: Applying the above formulae, we have

Perimeter $= 2\sqrt{2} \times 10 = 20\sqrt{2}$ cm

Area $= \frac{(10)^2}{2} = 50$ sq cm. (See **Rule - 21**)

Exercise

1. The diagonal of a square is 5 cm. Find its perimeter.
a) $5\sqrt{2}$ cm b) $6\sqrt{2}$ cm c) $10\sqrt{2}$ cm d) $15\sqrt{2}$ cm

2. The diagonal of a square is 6 cm. Find its perimeter.
a) $12\sqrt{2}$ cm b) $6\sqrt{2}$ cm c) 24 cm d) Data inadequate

3. Perimeter of a square is $24\sqrt{2}$ cm. Find its diagonal.
a) 12 cm b) $12\sqrt{2}$ cm c) $6\sqrt{2}$ cm d) 8 cm

Answers

1. c 2. a

3. a; **Hint:** $24\sqrt{2} = 2\sqrt{2} \times$ Diagonal

$\therefore$ Diagonal $= \frac{24\sqrt{2}}{2\sqrt{2}} = 12$ cm.

Rule 26

Theorem: ***If the diagonal of a square becomes x times, then the area of the square becomes x^2 times.***

Illustrative Example

Ex.: The diagonal of a square increases to its thrice. How many times will area of the new square become?

Soln: Detail Method: Let the diagonal of the original square be x m.

Original square, Diagonal = x, Area = $\frac{x^2}{2}$

New square, Diagonal = $3x$, Area = $\frac{(3x)^2}{2}$

Area of the new square $= \frac{\left(\frac{3x}{2}\right)^2}{\frac{x^2}{2}}$ = 9 times of the original square.

Quicker Method: Applying the above theorem, we have

the required answer = $(3)^2 = 9$ times.

Exercise

1. The ratio of areas of two squares, one having double its diagonal then the other is
   a) 2 : 1 b) 3 : 1 c) 3 : 2 d) 4 : 1
2. The diagonal of a square increases to its twice. How many times will area of the new square become?
   a) 2 times b) 6 times c) 4 times d) Data inadequate
3. The diagonal of a square increases to its 4 times. How many times will area of the new square become?
   a) 16 times b) 2 times c) 8 times d) None of these

Answers

1. d; **Hint:** *Diagonal* α $(Area)^2$ 2. c 3. a

Rule 27

Theorem: ***If the ratio of the areas of square A and square B is a : b, then***

(i) the ratio of their sides = $\sqrt{a}:\sqrt{b}$,

(ii) the ratio of their perimeters = $\sqrt{a}:\sqrt{b}$ ***and***

(iii) the ratio of their diagonals = $\sqrt{a}:\sqrt{b}$.

Illustrative Example

Ex.: Ratio of the areas of the two squares is 16 : 9. Find
(i) the ratio of their sides,
(ii) the ratio of their perimeters and
(iii) the ratio of their diagonals.

Soln: Detail Method:
Let the sides of the first square be x and the second square be y.

Ratio of sides = $\frac{x^2}{y^2} = \frac{16}{9}$ $\therefore \frac{x}{y} = \sqrt{\frac{16}{9}} = 4:3$

Ratio of perimeters = $4x : 4y = x : y = 4 : 3$

Ratio of diagonals = $x\sqrt{2} : y\sqrt{2} = x : y = 4 : 3$

Quicker Method: Applying the above theorem, we have
the required ratios = 4 : 3.

Exercise

1. Ratio of the areas of the two squares is 25 : 9. Find the ratio of their perimeters.
   a) 5 : 9 b) 9 : 5 c) 5 : 3 d) 3 : 5
2. Ratio of the areas of the two squares is 9 : 4. Find the ratio of their diagonals.
   a) 3 : 2 b) 3 : 1 c) 9 : 2 d) 9 : 1

Answers

1. c 2. a

Rule 28

Theorem: ***If the perimeter of a square is equal to the perimeter of a circle, then the side of the square is*** $\left(\pi\times\frac{r}{2}\right)$ ***and radius of the circle is*** $\left(\frac{2x}{\pi}\right)$. ***Where, x is the side of the square and r is the radius of the circle.***

Illustrative Examples

Ex: 1. There is a square of side 22 cm. Find the radius of the circle whose perimeter equals the perimeter of the square.

Soln: Applying the above theorem, we have the

radius of the circle = $\frac{2\times 22}{\frac{22}{7}} = 14$ cm.

Ex: 2. There is a circle of radius 7 cm. Find the side of the square whose perimeter equals the perimeter of the circle.

Soln: Applying the above theorem, we have

the side of the square = $\frac{22}{7}\times\frac{7}{2} = 11$ cm.

Exercise

1. There is a square of side 44 cm. Find the radius of the circle whose perimeter equals the perimeter of the square.

a) 7 cm b) 14 cm c) 28 cm d) Data inadequate

2. There is a square of side 11 cm. Find the radius of the circle whose perimeter equals the perimeter of the square.
a) 7 cm b) 21 cm c) 12 cm d) 9 cm

3. There is a circle is radius 21 cm. Find the side of the square whose perimeter equals the perimeter of the circle.
a) 11 cm b) 22 cm c) 33 cm d) Data inadequate

Answers

1. c 2. a 3. c

Rule 29

Theorem: ***If the side of a square is increased by 'x' units and its area becomes 'y' square units, the side of the square is given by*** $\left(\frac{y}{x}\right)$ ***units, its area is given by*** $\left(\frac{y^2}{x^2}\right)$ ***sq units and its perimeter is given by*** $4\left(\frac{y}{x}\right)$ ***units.***

Note: If the side of a square is increased by 'x' units and its area increases by 'y' units then the side of the square is given by $\frac{1}{2}\left[\frac{y}{x}-x\right]$ units.

Illustrative Example

Ex.: Length of a square is increased by 8 cm. Its area becomes 208 sq cm. Find its perimeter.

Soln: **Detail Method:** Let the side of the square be x cm.

As per the question, $(x+8)x = x^2 + 208$

or, $x^2 + 8x = x^2 + 208$

$\therefore x = 26$ cm

$\therefore$ Perimeter $= 4x = 4 \times 26 = 104$ cm

Quicker Method: Applying the above theorem, we have

the required answer $= 4 \times \frac{208}{8} = 104$ cm

Exercise

1. If the side of a square be increased by 4 cm, the area increases by 60 sq cms. The side of the square is:
a) 12 cm b) 13 cm
c) 14 cm d) None of these

2. Length of a square is increased by 4 cm. Its area becomes 44 sq cm. Find the area of the square.
a) 11 sq cm b) 44 sq cm
c) 121 sq cm d) Data inadequate

3. Length of a square is increased by 9 cm. Its area becomes 135 sq cm. Find its perimeter.
a) 60 cm b) 30 cm
c) 45 cm d) None of these

Answers

1. d; **Hint:** Required answer $= \left(\frac{60}{4} - 4\right)\frac{1}{2} = 5.5$ cm [**See Note**]

2. c 3. a

Rule 30

Theorem: ***A square room is surrounded by a verandah (on the outside of the square room) of width 'd' metres. If the area of the verandah is 'A' sq metres, then the area of the room is*** $\left(\frac{A-4d^2}{4d}\right)^2$ ***sq metres and obviously side of the square room is given by*** $\left(\frac{A-4d^2}{4d}\right)$ ***metres.***

Illustrative Example

Ex.: A square room is surrounded by a verandah of width 2 metres. Area of the verandah is 64 sq metres. Find the area of the room.

Soln: **Detail Method:** Let the side of the room ABCD be x metres.

Area of the room ABCD $= x \times x = x^2$ sq m.

Width of the path = 2 metres (given)

Sides of the figure A′B′C′D′ $= x + 2 + 2$
$= (x+4)$ metres.

Area of the figure A′B′C′D′ $= (x+4)^2$ sq m.

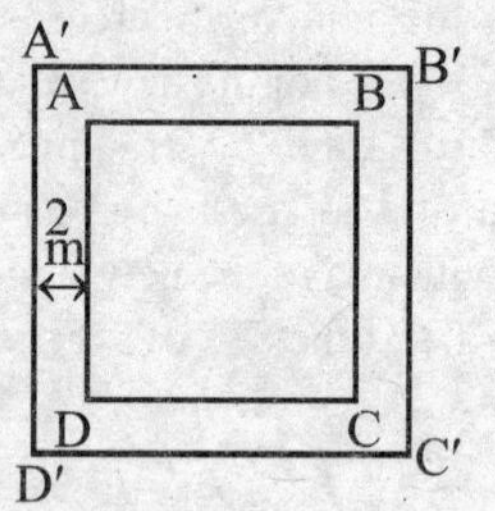

As per the question,

Area of the path = 64 sq metres

or, $(x+4)^2 - x^2 = 64$

or, $x^2 + 16 + 8x - x^2 = 64$ or, $8x = 48$

$\therefore x = 6$ metres.

Area $= x^2 = 6 \times 6 = 36$ sq metres.

Quicker Method: Applying the above theorem, we have

the area of the room $= \left(\frac{64 - 4\times 2^2}{4\times 2}\right)^2$

$$= \left(\frac{64-16}{8}\right)^2 = (6)^2 = 36 \text{ sq metres.}$$

Exercise

1. A square room is surrounded by a verandah of width 3 metres. Area of the verandah is 96 sq metres. Find the area of the room.
   a) 36 sq m b) 25 sq m
   c) 49 sq m d) Data inadequate
2. A square room is surrounded by a verandah of width 4 metres. Area of the verandah is 160 sq metres. Find the area of the room.
   a) 42 sq m b) 49 sq m c) 36 sq m d) None of these
3. A square room is surrounded by a verandah of width 2 metres. Area of the verandah is 72 sq metres. Find the area of the room.
   a) 49 sq m b) 64 sq m c) 81 sq m d) 36 sq m
4. A square room is surrounded by a verandah of width 1 metre. Area of the verandah is 24 sq metres. Find the area of the room.
   a) 25 sq m b) 16 sq m c) 36 sq m d) 30.25 sq m
5. A path 2m wide running all round a square garden has an area of 9680 sq m. Find the area of the part of the garden enclosed by the path.
   a) $(1208)^2$ sq m b) $(1028)^2$ sq m
   c) $(2208)^2$ sq m d) $(1308)^2$ sq m

Answers

1. b 2. c 3. a 4. a 5. a

Rule 31

Theorem: ***If a square room has a verandah of area 'A' sq metres and width 'd' metres all round it on its inside, then the area of the room is*** $\left(\frac{A+4d^2}{4d}\right)^2$ ***sq metres and obviously side of the square room is given as*** $\left(\frac{A+4d^2}{4d}\right)$ ***metres.***

Illustrative Example

Ex.: A square room has a verandah of area 64 sq metres and width 2 metres all round it on its inside. Find the area of the room.

Soln: **Detail Method:** Let the side of the room ABCD be x metres.

Area of the room ABCD = x^2 sq metres
Width of the path = 2 metres (given)
Sides of the figure A′B′C′D′
= x - (2 + 2) = (x - 4) metres

Area of the figure A′B′C′D′ = $(x-4)^2$ sq m.

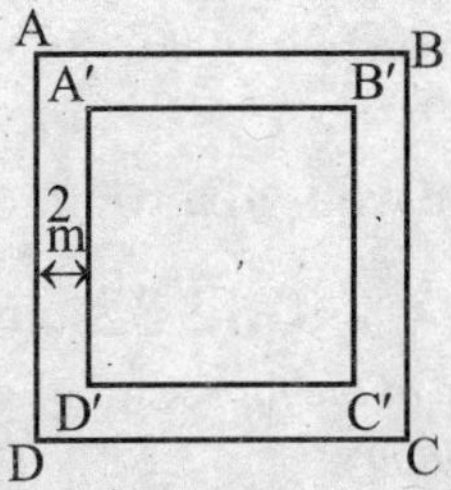

As per the question,
Area of the path = 64 sq metres

or, $x^2 - (x-4)^2 = 64$

or, $x^2 - x^2 - 16 + 8x = 64$ or, $8x = 80$

$\therefore x = 10$ metres.

Area of the room = 10 × 10 = 100 sq metres.

Quicker Method: Applying the above theorem,
Area of the room

$$= \left(\frac{64+4\times 2^2}{4\times 2}\right) = \left(\frac{80}{8}\right)^2 = 10\times 10 = 100 \text{ sq m.}$$

Exercise

1. A square field contains 2.89 hectares. It has to be fenced all-round and a path 10 m wide has to be laid out all-round close to the fence inside. If the cost of fencing is Rs 50 per m and the cost of preparing the path is Rs 10 per sq metre. Find the total expenses.
   a) Rs 64000 b) Rs 34000 c) Rs 94000 d) Rs 98000
2. A square room has a verandah of area 96 sq metres and width 3 metres all round it on its inside. Find the area of the room.
   a) 121 sq m b) 132 sq m
   c) 25 sq m d) Data inadequate
3. A square room has a verandah of area 160 sq metres and width 4 metres all round it on its inside. Find the area of the room.
   a) 196 sq m b) 169 sq m c) 256 sq m d) 36 sq m
4. A square room has a verandah of area 24 sq metres and width 1 metre all round it on its inside. Find the area of the room.
   a) 49 sq m b) 25 sq m
   c) 64 sq m d) Data inadequate

Answers

1. d; **Hint:** Area of the square = 2.89 hectares = 28900 sq m

Perimeter = $\sqrt{16\times 28900} = 680$ m [See **Rule - 24**]

$\therefore$ Cost of fencing the square field = 680 × 50 = Rs 34000
Now applying the given rule we have

$$\frac{A+4\times 10^2}{4\times 10} = \sqrt{28900} = 170$$

Or, A = 6400 sq m = Area of the path

∴ Cost in preparing the path = 6400 × 10 = Rs 64000

∴ total expenses = Rs 34000 + Rs 64000 = Rs 98000

2. a 3. a 4. a

Area of the four walls of a room

Rule 32

(i) To find the area of the four walls of a room, if its length, breadth and height are given. Area of the four walls of a room = 2 × (Length + Breadth) × Height.

Illustrative Example

Ex.: A room is 8 metres long, 6 metres broad and 3 metres high. Find the area of the four walls of the room.

Soln: Applying the above formula, we have

Area of the four walls of a room = 2 × (8 + 6) × 3 = 84 sq m.

(ii) To find the height of a room, if area of four walls of the room and its length and breadth are given.

$$Height = \left[\frac{Area\ of\ four\ walls\ of\ the\ room}{2(Length + Breadth)}\right] metres.$$

Illustrative Example

Ex.: Area of a hall, whose length is 16 metres and breadth is half of its length, is 576 sq metres. Find the height of the room.

Soln: Applying the above formula, we have

$$\text{the height of the room} = \frac{576}{2(16+8)} = 12 \text{ metres.}$$

Exercise

1. A room is 13 metres long, 9 metres broad and 10 metres high. Find the cost of carpeting the room with a carpet 75 cm board at the rate of Rs 2.40 per metre. What will be the cost of painting the four walls of the room at Rs 4.65 per sq metre, it being given that the doors and windows occupy 40 sq metres?
   a) Rs 375.50, Rs 1850 b) Rs 374.40, Rs 1860
   c) Rs 376, Rs 1875 d) Rs 374.04, Rs 1806
2. The cost of papering the walls of a room 12 metres long at the rate of 45 paise per square metre is Rs 113.40 and the cost of matting the floor at the rate of 35 paise per square metre is Rs 37.80. Find the height of the room.
   a) 9 m b) 8 m c) 6 m d) 12 m
3. The length and breadth of a room are in the ratio 4 : 3 and its height is 5.5 metres. The cost of decorating its walls at Rs 6.60 per square metre is Rs 5082. Find the length and breadth of the room.
   a) 40 m, 30 m b) 50 m, 40 m c) 30 m, 25 m d) 40 m, 20 m
4. Area of four walls of a room is 77 m^2. The length and breadth of the room are 7.5 m and 3.5 m respectively. The height of the room is:
   a) 7.7 m b) 3.5 m c) 6.77 m d) 5.4 m
5. Area of four walls of a room is 168 m^2. The breadth and height of the room are 8 m and 6 m respectively. The length of the room is:
   a) 14 m b) 12 m c) 3.5 m d) 6 m
6. The cost of papering four walls of a room is Rs 48. Each one of length, breadth and height of another room is double that of the room. The cost of papering the walls of this new room is:
   a) Rs 96 b) Rs 192 c) Rs 384 d) Rs 288
7. A hall, whose length is 16 metres and breadth twice its height, takes 168 metres of paper 2 metres wide for its four walls. Find the area of the floor.
   a) 192 sq m b) 196 sq m c) 129 sq m d) 190 sq m
8. Find the cost of painting the walls of a room of 5 metres long, 4 metres broad and 4 metres high at Rs 8.50 per sq metre.
   a) Rs 610 b) Rs 216 c) Rs 512 d) Rs 612
9. The cost of painting the walls of a room $7\frac{5}{6}$ metres long, 4 metres wide at Rs 16.20 per sq metre is Rs 1940.40. How high is the room?
   a) $4\frac{3}{4}$ m b) $4\frac{1}{4}$ m c) $4\frac{2}{3}$ m d) $4\frac{1}{3}$ m
10. How many metres of wall paper 2 metres wide will be required for a room 8.3 metres long, 4.2 metres wide and 4 metres high?
    a) 40 m b) 50 m c) 45 m d) 75 m
11. The area of the four walls of a room is 5940 sq dm and the length is twice the breadth, the height being 33 dm. Find the area of the ceiling.
    a) 18 sq m b) 1.8 sq m c) 16 sq m d) 1.6 sq m
12. A rectangular room is 6 m wide and 3 m high. If the area of its walls is 81 sq m, find the length.
    a) 6.5 m b) 5 m c) 6 m d) 7.5 m
13. A room is 10.5 metres long and 6.25 metres broad. The cost of papering the walls with paper 1.5 m wide at Rs 24 per metre is Rs 2680. Find the height of the room.
    a) 5 metres b) 6 metres c) 8 metres d) 10 metres
14. The length of room is $1\frac{1}{2}$ times its breadth. The cost of carpeting it at Rs 150 per sq metres is Rs 14400 and the cost of white washing the four walls at Rs 5 per sq metre is Rs 625. Find the length, breadth and height of the room.
    a) 12 m, 8 m, $3\frac{1}{8}$ m b) $12\frac{1}{2}$ m, $8\frac{1}{8}$ m, $3\frac{1}{8}$ m

c) $12\frac{3}{4}$ m, $8\frac{2}{3}$ m, $3\frac{2}{4}$ m d) Data inadequte

15. The length of a room is 6.5 metres. The cost of painting the walls at Rs 56 per sq metre is Rs 4928 and the cost of carpeting the room at Rs 224 per sq metre is Rs 6552. Find the height and width of the room.
a) 4 m, 5 m b) 4.5 m, 5.5 m
c) 4 m, 4.5 m d) Data inadequate
16. The length of a room is double the breadth. The cost of colouring the ceiling at Rs 25 per sq m is Rs 5000 and the cost of painting the four walls at Rs 240 per sq m is Rs 64800. Find the height of the room.
a) 4 m b) 4.5 m c) 3.5 m d) 5 m
17. Two square rooms, one a metre longer each way than the other, are of equal height, and cost respectively Rs 33600 and Rs 35280 to paper the walls at Rs 70 per sq m. Find the height.
a) 6 m b) 8 m c) 5 m d) 4 m

Answers

1. b; **Hint:** Cost of carpet $= \left(2.40 \times \frac{13 \times 9 \times 100}{75}\right) =$ Rs 374.40

[See **Rule - 54**]
Area of four walls = [2 (13 + 9) × 10] = 440 sq m
Area to be painted = Rs (440 - 40) = 400 sq m
Cost of painting = Rs (400 × 4.65) = Rs 1860

2. c; **Hint:** Area of floor

$= \left(\frac{\textit{Total cost}}{\textit{Rate per sqm}}\right) = \left(\frac{3780}{35}\right) = 108\ sqm$

$\therefore$ Breadth of the room $= \frac{\textit{Area of floor}}{\textit{Length of the room}} = \left(\frac{108}{12}\right)$

= 9 metres
Now, area of four walls

$= \left(\frac{\textit{Total cost of Papering}}{\textit{Rate per sq metre}}\right) = \left(\frac{11340}{35}\right) = 252$ sq m

Let the height of room be h metres.
Then, 2 × (12 + 9) × h = 252

$\therefore$ Height, h $= \frac{252}{2 \times 21} = 6$ metres.

3. a; **Hint:** Area of the four walls $= \left(\frac{5082}{6.6}\right) = 770$ sq m

Now, 2 (4x + 3x) × 5.5 = 770 or, x = 10
$\therefore$ length = 4x = 4 × 10 = 40 m and breadth = 3x = 3 × 10 = 30 m.

4. b 5. d

6. b; **Hint:** Cost of papering $[2(l+b)\times h]m^2 =$ Rs 48

$\therefore$ cost of papering $[2(2l+2b)\times 2h]m^2$

ie $4[2(l+b)\times h] = 4 \times 48 =$ Rs 192

7. a; **Hint:** $2(16+2h)h = 168 \times 2$ or, $(8+h)h = 84$

or, $h^2 + 8h - 84 = 0$
By solving the above equation we get h= 6 and -14
$\therefore$ height = 6 m and breadth = 12 m (neglecting negative value of h)
$\therefore$ area of the floor = 16 × 12 = 192 sq m

8. d

9. c; **Hint:** Area of the four walls $= \frac{1940.40}{16.20}$ sq m

$2h\left(\frac{47}{6}+5\right) = \frac{19404}{162} = \frac{1078}{9}$ sq m

$h = \frac{1078}{9 \times 2\left(\frac{47}{6}+5\right)} = \frac{14}{3} = 4\frac{2}{3}$ m.

10. b; **Hint:** Required answer $= \frac{2 \times 4(8.3+4.2)}{2} = 50$ m.

11. a; **Hint:** 2 × 33 (x + 20) = 5940
$\therefore$ x = 30 dm = 3 m and length = 2x = 2 × 30 = 60 dm = 6 m
$\therefore$ area of the ceiling = 6 × 3 = 18 sq m

12. d; **Hint:** 2 × 3 (x + 6) = 81 $\therefore$ x $= \frac{15}{2} = 7.5$ m

13. a; **Hint:** Length of the paper $= \frac{2680}{24}$ m

Area of the paper $= \frac{2680}{24} \times \frac{15}{10} =$ sq m

Now, as per the question $2 \times h(10.5+6.25) = \frac{2680}{24} \times \frac{15}{10}$

$\therefore$ h $= \frac{167.5}{16.75 \times 2} = 5$ m.

14. a; **Hint:** Area of the room $= \frac{14400}{150} = 96$ sq m

Or, $\frac{3}{2}x \times x = 96$ $\therefore$ x = 8 m = breadth

$\therefore$ length $= \frac{3}{2} \times 8 = 12$ m

Area of the four walls $= \frac{625}{5}$ 125 sq m

Or, 2 × h (12 + 8) = 125 $\therefore$ h $= \frac{25}{8} = 3\frac{1}{8}$ m.

15. c; **Hint:** Area of floor of the room $= \frac{6552}{224}$

Now, $6.5 \times x = \frac{6552}{224}$ $\therefore$ x = 4.5 m

Area of the four walls of the room $= \frac{4928}{56} = 88$ sq m

Or, $2 \times h(6.5 + 4.5) = 88$ $\therefore$ h = 4 m.

16. b

17. a; **Hint:** Let the side of one square be x m and the other be (x + 1) m.

Now, as per the question, $2h(x+x) = \frac{33600}{70} = 480$

$\therefore$ hx = 120 .... (i) and

$2(x+1) \times 2h = \frac{35280}{70} = 504$

$\therefore$ xh + h = 126 .... (ii)

Putting the value of xh from equ (i) into the equ (ii)

$\therefore$ h = 126 - 120 = 6 m

Parallelogram

Rule 33

Theorem: ***To find the area of a parallelogram if its Base and Height are given.***

Area of a parallelogram = Base × Height.

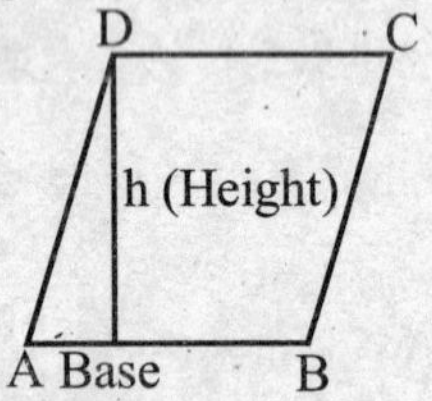

Illustrative Example

Ex.: One side of a parallelogram is 17 cm. The perpendicular distance between this and the opposite side is 13 cm. Find the area of the parallelogram.

Soln: Here, b = 17 cm and h = 13 cm

Now, applying the above formula,

Area of parallelogram = Base × Height = 17 × 13 = 221 cm^2.

Exercise

1. Find the area of a parallelogram whose base is 35 metres and altitude 18 metres.
   a) 630 sq m b) 650 sq m c) 730 sq m d) 660 sq m
2. The area of a parallelogram is 338 sq m. If its altitude is twice the corresponding base, determine the base and the altitude.
   a) 13 m, 26 m b) 14 m, 28 m c) 15 m, 30 m d) 12 m, 24 m
3. One side of a parallelogram is 14 cm. Its distance from the opposite side is 16 cm. The area of the parallelogram is:
   a) 112 cm^2 b) 224 cm^2 c) 56 π cm^2 d) 210 cm^2
4. Find the areas of the following parallelograms
   (i) Base 26 metres, height 8 metres
   a) 208 sq m b) 206 sq m
   c) 200 sq m d) 205 sq m
   (ii) Base 54 metres, height 22 metres
   a) 1288 sq m b) 1388 sq m
   c) 1188 sq m d) 1088 sq m

Answrs

1. a
2. a; **Hint:** $338 = x \times 2x$ $\therefore$ x = 13 and 2x = 26
   $\therefore$ base = 13 metres and the altitude = 26 metres
3. b 4. (i) a (ii) c

Rule 34

Theorem: ***To find the area of a parallelogram, if the lengths of the two adjacent sides and the length of the diagonal connecting the ends of the two sides are given. (see the figure).***

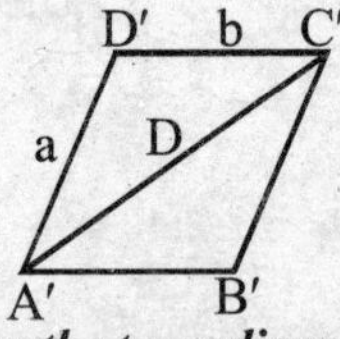

Where, 'a' and 'b' are the two adjacent sides and 'D' is the diagonal connecting the ends of the two sides.

Area of a parallelogram $= 2\sqrt{s(s-a)(s-b)(s-D)}$ ***and*** $S = \frac{a+b+D}{2}$

Illustrative Example

Ex.: The two adjacent sides of a parallelogram are 5 cm and 4 cm respectively, and if the respective diagonal is 7 cm then find the area of the parallelogram?

Soln: Required area $= 2\sqrt{s(s-a)(s-b)(s-D)}$

Where $S = \frac{a+b+D}{2} = \frac{5+4+7}{2} = 8$

$= 2\sqrt{8(8-5)(8-4)(8-7)}$

$= 2\sqrt{8 \times 3 \times 4} = 8\sqrt{6} = 19.6$ sq cm.

Exercise

1. Find the area of a parallelogram; if its two adjacent sides are 12 cm and 14 cm and if the diagonal connecting the ends is 18 cm.
   a) 176.49 sq cm b) 167.49 sq cm
   c) 167.94 sq cm d) None of these
2. Find the area of a parallelogram whose two adjacent sides are 130 metres and 140 metres and one of the diagonals is 150 metres long.
   a) 16800 sq m b) 17800 sq m

c) 18600 sq m d) Can't be determined

3. Find the area of a parallelogram whose two adjacent sides are 130 metres and 140 metres and one of the diagonals is 150 metres long. Find also the cost of gravelling it at the rate of Rs 10 per square metre.
   a) 15800 sq m, Rs 158000 b) 16800 sq m, Rs 168000
   c) 14800 sq m, Rs 148000 d) None of these

Answers

1. c 2. a 3. b

Rule 35

Theorem: In a parallelogram, the sum of the squares of the diagonals = 2 × (the sum of the squares of the two adjacent sides)

or, $D_1^2 + D_2^2 = 2(a^2 + b^2)$

Where, D_1 ***and*** D_2 ***are the diagonals and a and b are the adjacent sides.***

Illustrative Example

Ex.: In a parallelogram, the lengths of adjacent sides are 12 cm and 14 cm respectively. If the length of one diagonal is 16 cm, find the length of the other diagonal.

Soln: In a parallelogram, the sum of the squares of the diagonals = 2 × (the sum of the squares of the two adjacent sides)

or, $D_1^2 + D_2^2 = 2(a^2 + b^2)$

or, $16^2 + x^2 = 2(12^2 + 14^2)$

or, $256 + x^2 = 2(144 + 196)$

or, $x^2 = 680 - 256 = 424$ $\therefore x = \sqrt{424} = 20.6$ cm

Exercise

1. A parallelogram, the lengths of whose sides are 11 cm and 13 cm has one diagonal 20 cm long. Find the length of another diagonal.
   a) 15 cm b) 18 cm
   c) 20 cm d) Can't be determined
2. A parallelogram, the lengths of whose sides are 11 cm and 8 cm has one diagonal 10 cm long, find the length of the other diagonal.
   a) 17.78 cm (approx) b) 18.68 cm (approx)
   c) 17.87 cm (approx) d) Data inadequate
3. In a parallelogram, the lengths of adjacent sides are 24 cm and 28 cm respectively. If the length of one diagonal is 32 cm, find the length of the other diagonal.
   a) 41.2 m (approx) b) 31 m (approx)
   c) 43.2 m (approx) d) None of these

Answers

1. c 2. a 3. a

Rule 36

Theorem: ***To find the sides of a parallelogram if the distance between its opposite sides and the area of the parallelogram is given.***

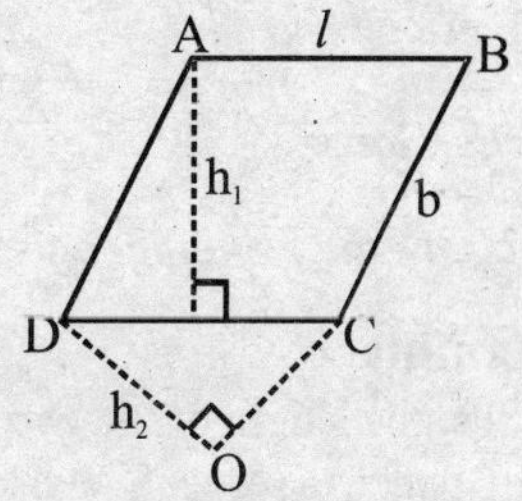

Here, ABCD is a parallelogram, h_1 ***and*** h_2 ***are the distance between opposite sides, 'l' and 'b' are the sides of the parallelogram. 'A' is area of the parallelogram.***

$A = lh_1 = bh_2$

$\therefore l = \frac{A}{h_1}$ ***and*** $b = \frac{A}{h_2}$

Illustrative Example

Ex.: A parallelogram has an area of 160 cm². If the distance between its opposite sides are 10 cm and 16 cm. Find the sides of the parallelogram.

Soln: Applying the above formula, we have

Length of the parallelogram $= \frac{160}{10} = 16$ cm.

Breadth of the parallelogram $= \frac{160}{16} = 10$ cm.

Exercise

1. A parallelogram has an area of 150 cm². If the distance between its opposite sides are 15 cm and 25 cm. Find the sides of the parallelogram.
   a) 10 cm, 6 cm
   b) 12 cm, 8 cm
   c) 8 cm, 4 cm
   d) Data indequate
2. A parallelogram has an area of 144 cm². If the distance between its opposite sides are 12 cm and 16 cm. Find the sides of the parallelogram.
   a) 12 cm, 9 cm b) 10 cm, 6 cm
   c) 14 cm, 10 cm d) None of these
3. A parallelogram has an area of 196 cm². If the distance between its opposite sides are 7 cm and 14 cm. Find the sides of the parallelogram.
   a) 28 cm, 14 cm
   b) 14 cm, 7 cm
   c) 28 cm, 21 cm
   d) Data inadequate

Answers

1. a 2. a 3. a

Rhombus

Rule 37

Theorem: To find perimeter of a rhombus if the length of the two diagonals are given.

Perimeter of the rhombus $= \left(2\sqrt{\left(d_1^{\,2}+d_2^{\,2}\right)}\right)$ *units.*

Where, d_1 *and* d_2 *are the two diagonals.*

Illustrative Example

Ex.: In a rhombus, the length of the two diagonals are 40 metres and 30 metres respectively. Find its perimeter.

Soln: Applying the above formula, we have perimeter of the rhombus

$$= 2\sqrt{(40)^2+(30)^2} = 2\sqrt{2500} = 50\times 2 = 100 \text{ m.}$$

Exercise

1. In a rhombus, the length of the two diagonals are 3 metres and 4 metres respectively. Find its perimeter.
   a) 14 m b) 10 m c) 5 m d) 7 m
2. In a rhombus, the length of the two diagonals are 12 metres and 16 metres respectively. Find its perimeter.
   a) 20 m b) 40 m c) 25 m d) 45 m
3. In a rhombus, the length of the two diagonals are 18 metres and 24 metres respectively. Find its perimeter.
   a) 30 m b) 45 m c) 60 m d) 55 m

Answers

1. b 2. b 3. c

Rule 38

Theorem: ***To find area of a rhombus if the side and the height are given.***

Area of the rhombus = (side × height) sq units.

Illustrative Example

Ex.: The side and the height of a rhombus are 14 cm and 30 cm respectively. Find its area.

Soln: Applying the above formula, we have area of the rhombus = 14 cm × 30 cm = 420 sq cm.

Exercise

1. If a square and a rhombus stand on the same base, then the ratio of areas of square and rhombus is:
   a) greater than 1 b) equal to 1
   c) equal to $\frac{1}{4}$ d) equal to $\frac{1}{2}$
2. The side and the height of a rhombus are 15 cm and 25 cm respectively. Find its area.
   a) 325 sq cm b) 375 sq cm
   c) 345 sq cm d) None of these
3. The side and the height of a rhombus are 20 cm and 30 cm respectively. Find its area.
   a) 900 sq cm b) 600 sq cm
   c) 625 sq cm d) Data inadequate
4. The side and the height of a rhombus are 12 cm and 18 cm respectively. Find its area.
   a) 216 sq cm b) 261 sq cm
   c) 316 sq cm d) Data inadequate

Answers

1. a; **Hint:** $\dfrac{\text{Area of square}}{\text{Area of rhombus}} = \dfrac{\text{base}\times\text{base}}{\text{base}\times\text{height}}$

$= \dfrac{a\times a}{a\times h} = \dfrac{a}{h} > 1$, since $a > h$

2. b 3. b 4. a

Rule 39

Theorem: ***To find the side and one of the diagonals of a rhombus if area and one of its diagonals are given.***

(i) Diagonal of the rhombus $(d_2) = \dfrac{2A}{d_1}$

(ii) Side of the rhombus $= \dfrac{1}{2}\left(\sqrt{d_1^{\,2}+\dfrac{4A^2}{d_1^{\,2}}}\right)$

Where, A *= area of the rhombus*

d_1 *= length of the one diagonal*

d_2 *= length of the other diagonal.*

Illustrative Example

Ex.: A rhombus of area 24 sq cm has one of its diagonals of 6 cm. Find the other diagonal and side of the rhombus.

Soln: **Detail Method:** Area = 24 sq cm

Length of the diagonal = d_1 = 6 cm

Area = $\frac{1}{2}$ (Product of its diagonals) $= \frac{1}{2}\times d_1\times d_2$

$\therefore d_2 = \dfrac{2\,Area}{d_1} = \dfrac{24\times 2}{6} = 8$ cm

Side $= \frac{1}{2}\sqrt{d_1^{\,2}+d_2^{\,2}} = \frac{1}{2}\sqrt{8^2+6^2} = 5$ cm

Quicker Method: Applying the above formula, we have

(i) Diagonal of the rhombus $= \dfrac{2\times 24}{6} = 8$ cm

(ii) Side of the rhombus $= \dfrac{1}{2}\left(\sqrt{36+\dfrac{4\times 24\times 24}{6\times 6}}\right)$

$= \dfrac{10}{2} = 5$ cm

Exercise

1. A rhombus of area 6 sq m has one of its diagonals of 3 m. Find the other diagonal and side of the rhombus.
   a) 4 m, 5 m b) 6 m, 8 m
   c) 4 m, 2.5 m d) Data inadequate
2. A rhombus of area 96 sq m has one of its diagonals of 12 m. Find the other diagonal and side of the rhombus.
   a) 16 m, 10 m b) 8 m, 20 m
   c) 16 m, 20 m d) 8 m, 10 m
3. A rhombus of area 216 sq cm has one of its diagonals of 24 cm. Find the other diagonal and side of the rhombus.
   a) 18 cm, 30 cm b) 18 cm, 15 cm
   c) 9 cm, 15 cm d) Data inadequate

Answers

1. c 2. a 3. b

Rule 40

Theorem: ***If one of the diagonals of a rhombus of side 'x' units measures 'd' units, then the area of the rhombus is given by*** $d\left(\sqrt{x^2-\left(\frac{d}{2}\right)^2}\right)$ ***sq units and the length of the other diagonal is*** $2\times\left(\sqrt{x^2-\left(\frac{d}{2}\right)^2}\right)$ ***units.***

Illustrative Example

Ex.: One of the diagonals of a rhombus of side 5 cm measures 8 cm. Find the area of the rhombus.

Soln: **Detail Method:** We know that the diagonals of a rhombus bisect each other at right angle.

From the figure we can write for right angled triangle,

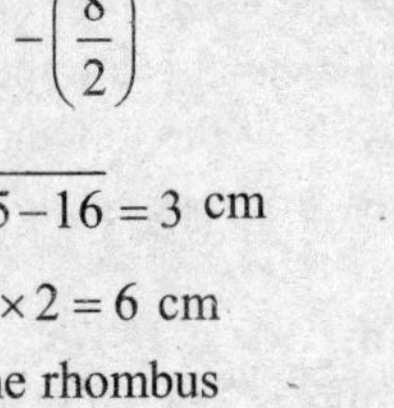

$$\frac{d_2}{2}=\sqrt{5^2-\left(\frac{8}{2}\right)^2}$$

$$=\sqrt{25-16}=3 \text{ cm}$$

$\therefore\ d_2=3\times 2=6$ cm

Area of the rhombus

$$=\frac{1}{2}\times d_1\times d_2=\frac{1}{2}\times 8\times 6=24 \text{ cm}$$

Quicker Method: Applying the above theorem,

$$\text{Area}=8\left(\sqrt{5^2-\left(\frac{8}{2}\right)^2}\right)=8\times\sqrt{9}=8\times 3=24 \text{ sq cm}$$

Note: Expression of the above theorem can be written as follows,

(i) Area of the rhombus

$$=d\left[\sqrt{\left(\frac{Perimeter}{4}\right)^2-\left(\frac{d}{2}\right)^2}\right] \text{ sq units.}$$

(ii) Length of the other diagonal

$$=2\left[\sqrt{\left(\frac{Perimeter}{4}\right)^2-\left(\frac{d}{2}\right)^2}\right] \text{ units.}$$

"If perimeter and one of the diagonals of a rhombus are given, then the area and the length of the other diagonal can be calculated."

Exercise

1. Find the area of a rhombus one side of which measures 20 cm and one of whose diagonals is 24 cm.
   a) 384 sq cm b) 348 sq cm
   c) 484 sq cm d) Can't be determined
2. One side of a rhombus is 10 cm and one of its diagonals is 12 cm. The area of the rhombus is:
   a) 120 cm^2 b) 60 cm^2
   c) 80 cm^2 d) 96 cm^2

Answers

1. a; **Hint:** Required answer $=24\left[\sqrt{20^2-\left(\frac{24}{2}\right)^2}\right]=16\times 24$

$$=384 \text{ sq cm}$$

2. d

Rule 41

To find the area of a rhombus if its diagonals are given.

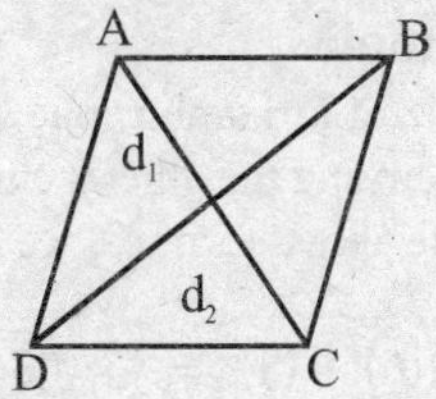

Area of a rhombus

$$=\frac{1}{2}\times D_1\times D_2=\frac{1}{2}(\text{Product of diagonals})$$

Illustrative Example

Ex.: Find the area of a rhombus one of whose diagonals measures 8 cm and the other 10 cm.

Soln: Area = $\frac{1}{2}$ (product of diagonals) = $\frac{8\times10}{2}$ = 40 sq cm.

Exercise

1. If the perimeter of a rhombus is 4a and lengths of the diagonals are x and y, then its area is:
   a) a (x + y) b) x^2+y^2 c) xy d) $\frac{1}{2}$xy
   (NDA 1990)
2. In a rhombus whose area is 144 sq cm one of its diagonals is twice as long as the other. The lengths of its diagonals are:
   a) 24 cm, 48 cm b) 12 cm, 24 cm
   c) $6\sqrt{2}$ cm, $12\sqrt{2}$ cm d) 6 cm, 12 cm
   (CDS 1989)
3. Find the area of a rhombus one of whose diagonals measures 6 cm and the other 12 cm.
   a) 36 sq cm b) 24 sq cm
   c) 20 sq cm d) None of these
4. Find the area of a rhombus one of whose diagonals measures 8 cm and the other 18 cm.
   a) 42 sq cm b) 72 sq cm
   c) 52 sq cm d) Data inadequate

Answers

1. d;
2. b; **Hint:** Let its diagonals be x cm and 2x cm. Then
   $\frac{1}{2}\times x\times 2x=144 \Rightarrow x^2=144$ or, x = 12
   ∴ Lengths of diagonals are 12 cm, 24 cm
3. a 4. b

Rule 42

To find the sides of the rhombus if its two diagonals are given.

Side of rhombus = $\frac{1}{2}\times\sqrt{D_1^2+D_2^2}$ ***; Where,*** D_1 ***and*** D_2 ***are the two diagonals.***

Illustrative Example

Ex.: Find the side of a rhombus one of whose diagonals measures 6 cm and the other 8 cm.

Soln: Applying the above formula, we have

$$\text{side}=\frac{1}{2}\times\sqrt{(6)^2+(8)^2}=\frac{1}{2}\times\sqrt{(36+64)}=5 \text{ cm.}$$

Exercise

1. Find the side of a rhombus one of whose diagonals measures 4 cm and the other 3 cm.
   a) 6 cm b) 3.5 cm c) 2.5 cm d) 5 cm
2. Find the side of a rhombus one of whose diagonals measures 12 cm and the other 16 cm.
   a) 10 cm b) 20 cm c) 12 cm d) 15 cm
3. Find the side of a rhombus one of whose diagonals measures 18 cm and the other 24 cm.
   a) 20 cm b) 30 cm c) 15 cm d) Data inadequate

Answers

1. c 2. a 3. c

Trapezium

Rule 43

To find the area of a trapezium, when length of parallel sides and the perpendicular distance between them is given.

Area of a trapezium = $\frac{1}{2}$ ***(sum of parallel sides × perpendicular distance between the parallel sides) =*** $\frac{1}{2}$ ***(a + b)h; where a and b are the parallel sides of the trapezium and h is the perpendicular distance between the sides a and b.***

Illustrative Example

Ex.: A trapezium has the perpendicular distance between the two parallel sides 60 m. If the lengths of the parallel sides be 40 m and 130 m, then find the area of the trapezium.

Soln: Applying the above formula,

Area of a trapezium = $\frac{1}{2}(130+40)60=85\times60=5100$ sq m.

Exercise

1. The cost of ploughing trapezoid field at the rate of Rs 1.35 per square metre is Rs 421.20. The difference between the parallel sides is 8 metres and the perpendicular distance between them is 24 metres. Find the length of parallel sides.
   a) 17 m, 9 m b) 28 m, 20 m
   c) 34 m, 26 m d) Can't be determined
2. The two parallel sides of a trapezium are 1 m and 2 m respectively. The perpendicular distance between them is 6 m. The area of trapezium is:
   a) 9 m^2 b) 12 m^2 c) 6 m^2 d) 18 m^2
3. The cross section of a canal is a trapezium in shape. If the canal is 10 m wide at the top and 6 m wide at the bottom and the area of cross section is 640 m^2, the length of canal is:
   a) 40 m b) 80 m c) 160 m d) 384 m
4. The area of a trapezium is 384 sq cm. If its parallel sides are in the ratio 3 : 5 and the perpendicular distance between them be 12 cm, the smaller of parallel sides is:
   a) 16 cm b) 24 cm c) 32 cm c) 40 cm

Answers

1. a; **Hint:** Let the length of parallel sides be x and y.

$$\text{Area} = \frac{421.20}{1.35} = 312 \text{ sq m}$$

Now, applying the given rule,

$$312 = \frac{1}{2}(x+y)\times 24 \qquad \therefore\ x+y=26 \text{ .... (i)}$$

and x - y = 8 (given) .... (ii)

By solving equ (i) and equ (ii), we get

x = 17 m and y = 9 m.

2. a

3. b; **Hint:** let the length of canal be x m.

$$\text{Then, } \frac{1}{2}(10+6)\times x = 640 \Rightarrow x = \frac{640\times 2}{16} = 80 \text{ m.}$$

4. b; **Hint:** $\frac{1}{2}(3x+5x)\times 12 = 384 \Rightarrow x = \frac{384\times 2}{8\times 12} = 8$

$\therefore$ Smaller side = 24 cm.

Rule 44

Theorem: ***To find the area of a trapezium, when the lengths of parallel sides and non-parallel sides are given.***

Area of a trapezium $= \frac{a+b}{k}\sqrt{s(s-k)(s-c)(s-d)}$

where, k = (a – b) ie the difference between the parallel sides and c and d are the two non-parallel sides of the trapezium. And $s = \frac{k+c+d}{2}$

Illustrative Example

Ex.: In a trapezium, parallel sides are 60 and 90 cms respectively and non-parallel sides are 40 and 50 cms respectively. Find its area.

Soln: k = difference between the parallel sides

= 90 – 60 = 30 cm

Let c be 40 cm then d = 50 cm

$$\text{Now, } s = \frac{k+c+d}{2} = \frac{30+40+50}{2} = \frac{120}{2} = 60 \text{ cm}$$

$$\therefore \text{Area} = \frac{a+b}{k}\sqrt{s(s-k)(s-c)(s-d)}$$

$$= \frac{60+90}{30}\sqrt{60(60-30)(60-40)(60-50)}$$

$$= 5\sqrt{60\times 30\times 20\times 10} = 5\times 600 = 3000 \text{ sq cm.}$$

Exercise

1. A field is in the form of a trapezium whose parallel sides are 120 metres and 75 metres and the non-parallel sides are 105 metres and 72 metres. Find the cost of ploughing the field at the rate of 60 paise per square metres.
   a) Rs 3404 b) Rs 3440 c) Rs 3574 d) Rs 3414
2. The two parallel sides of a trapezium measure 58 metres and 42 metres respectively. The other two sides are equal, each being 17 metres. Find its area.
   a) 570 sq m b) 750 sq m
   c) 740 sq m d) 760 sq m
3. Find the area of a trapezium whose parallel sides are 11 metres and 25 metres long and the non-parallel sides are 15 metres and 13 metres long respectively.
   a) 216 sq m b) 316 sq m
   c) 215 sq m d) 206 sq m

Answers

1. a; **Hint:** k = 120 - 75 = 45

$$s = \frac{45+105+72}{2} = 111$$

$$\text{Area} = \frac{120+75}{45}\sqrt{111(111+45)(111-105)(111-72)}$$

= 5673.66

$$\therefore \text{ the cost of ploughing the field} = \frac{5673.66\times 60}{100}$$

= Rs 3404

2. b 3. a

Rule 45

To find the perpendicular distance between the two parallel sides of the trapezium.

Perpendicular distance $= \frac{2}{k}\sqrt{s(s-k)(s-c)(s-d)}$

where, k = (a – b) ie the difference between the parallel sides and c and d are the two non-parallel sides of the trapezium.

And $s = \frac{k+c+d}{2}$

Illustrative Example

Ex.: In a trapezium parallel sides are 60 and 90 cm respectively and non-parallel sides are 40 and 50 cm respectively. Find the perpendicular distance between the two parallel sides of the trapezium.

Soln: $h = \frac{2}{k}\sqrt{s(s-k)(s-c)(s-d)}$ [k = (90 - 60) = 30 and s $= \frac{30+40+50}{2} = 60$]

$$= \frac{2}{30}\times\sqrt{60\times 30\times 20\times 10} = \frac{1}{15}\times 600 = 40 \text{ cm}$$

Exercise

1. In a trapezium parallel sides are 30 and 45 cm respectively and non-parallel sides are 20 and 25 cm respectively. Find the perpendicular distance between the two parallel sides of the trapezium.
   a) 40 cm b) 45 cm c) 20 cm d) 25 cm
2. In a trapezium parallel sides are 50 and 80 cm respectively and non-parallel sides are 30 and 40 cm respectively. Find the perpendicular distance between the two parallel sides of the trapezium.
   a) $\frac{40\sqrt{5}}{3}$ cm b) $\frac{40\sqrt{3}}{3}$ cm
   c) $8\sqrt{3}$ cm d) Data inadequate
3. In a trapezium parallel sides are 25 and 40 cm respectively and non-parallel sides are 15 and 20 cm respectively. Find the perpendicular distance between the two parallel sides of the trapezium.
   a) $\frac{30\sqrt{3}}{7}$ cm b) $\frac{20\sqrt{3}}{5}$ cm
   c) $\frac{20\sqrt{5}}{3}$ cm d) Data inadequate

Answers

1. c 2. a 3. c

Rule 46

Theorem: ***To find the height of the trapezium if its area and parallel sides are given.***

Height $= \left(\frac{2A}{a+b}\right)$ ***units.***

Where, A = Area of the trapezium, a and b are the length of parallel sides of the trapezium.

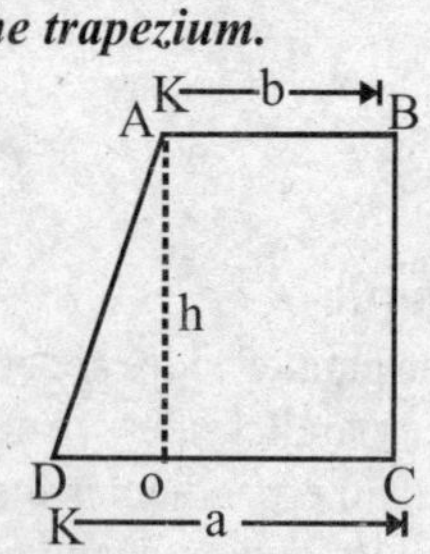

ABCD is a trapezium.

Illustrative Example

Ex.: The two parallel sides of a trapezium of area 800 sq cm measure 25 cm and 55 cm. What is the height of the trapezium.

Soln: Applying the above formula,

$$\text{Height} = \frac{2\times800}{25+55} = \frac{2\times800}{80} = 20 \text{ cm.}$$

Exercise

1. The two parallel sides of a trapezium of area 400 sq cm measure 15 cm and 35 cm. What is the height of the trapezium.
   a) 15 cm b) 25 cm c) 16 cm d) 24 cm
2. The height of a trapezium is 20 cm. Area and one of its parallel sides are 250 sq cm and 16 cm respectively. Find the other parallel side.
   a) 9cm b) 8 cm
   c) 12 cm d) Data inadequate
3. The two parallel sides of a trapezium of area 150 sq cm measure 12 cm and 18 cm. What is the height of the trapezium.
   a) 5 cm b) 10 cm c) 15 cm d) 21 cm

Answers

1. c 2. a; **Hint:** $20 = \frac{2\times250}{x+16}$ $\therefore$ x = 9 cm. 3. b

Circle

Rule 47

Theorem: ***To find the circumference of a circle when radius is given.***

Circumference of a circle = $2\pi r$ ***or,*** πd ***[∵ Diameter (d) = 2r]***

Note: ***To find the radius of a circle when perimeter or circumference is given.***

(i) Radius of a circle $= \frac{\textit{Perimeter or circumference}}{2\pi}$ ***and***

(ii) Diameter $= \frac{\textit{Perimeter}}{\pi}$ $[\because \textit{Diameter} = 2r]$

Illustrative Examples

Ex. 1: Find the perimeter or circumference of a circle of radius 7 cm.

Soln: Applying the above formula,

$$\text{Circumference} = 2\pi r = 2\times\frac{22}{7}\times7 = 44 \text{ cm}$$

Ex. 2: Find the radius of a circular field whose circumference measures $5\frac{1}{2}$ km.

$$\left[\textit{Take } \pi = \frac{22}{7}\right]$$

Soln: $r = \frac{C}{2\pi}$

$\therefore$ required radius

$$= \frac{\frac{11}{2}\times1000}{2\pi} \text{ m} = \frac{\frac{11}{2}\times1000\times7}{2\times22} \text{ m} = 875 \text{ m.}$$

Exercise

1. The perimeter of a semi circle of 56 cm diameter will be:
   a) 144 cm b) 232 cm c) 154 cm d) 116 cm
   [Bank PO Exam 1989]
2. Find the circumference of a circle whose radius is 42 metres.
   a) 264 metres b) 624 metres c) 426 metres d) 264 metres
3. The difference between the circumference and diameter of a circle is 210 metres. Find the radius of the circle.
   a) 49 metres b) 52 metres c) 39 metres d) 45 metres
4. A circular wire of radius 42 cm is cut and bent in the form of a rectangle whose sides are in ratio 6 : 5. The smaller side of the rectangle is:
   a) 30 cm b) 60 cm c) 72 cm d) 132 cm
   [I. Tax and Central Excise 1989]
5. The difference between the circumference and the radius of a circle is 74 metres. Find the diameter of the circle.
   a) 28 m b) 14 m c) 7 m d) 35 m
6. Find the cost of fencing a circular field of 560 metres radius at Rs 332 per 10 metres.
   a) Rs 118684 b) Rs 118584
   c) Rs 116864 d) Rs 116854

Answers

1. a; **Hint:** Perimeter $= (\pi r + 2r) = \left(\frac{22}{7}\times 28 + 56\right) = 144$ cm

2. a

3. a; **Hint:** $2\pi r - 2r = 210$

$$r(\pi - 1) = 105 \text{ or, } r\left(\frac{22}{7} - 1\right) = 105$$

$$\therefore r = \frac{105\times 7}{15} = 49 \text{ m}$$

4. b; **Hint:** $2(6x + 5x) = 2\times\frac{22}{7}\times 42$ or, $x = 12$

$\therefore$ smaller side $= 5x = 5\times 12 = 60$ cm

5. a; **Hint:** $2\pi r - r = 74$

$$\text{or, } r(2\pi - 1) = 74 \text{ or, } r = \frac{74\times 7}{37} = 14 \text{ m}$$

$\therefore$ diameter $= 14\times 2 = 28$ m

6. c; **Hint:** Required cost $= \frac{332}{10}\times 2\times\frac{22}{7}\times 560 =$ Rs 116864

Rule 48

Theorem: To find the area of a circle if radius is given.

$$\textit{Area of a circle} = \pi r^2 = \pi\left(\frac{Diameter}{2}\right)^2$$

$\left[\because Diameter = 2\times radius\right]$

Note: ***To find the radius of a circle when its area is given.***

(i) Radius of the circle $= \sqrt{\frac{Area}{\pi}}$ ***and***

(ii) Diameter of the circle $= 2\left(\sqrt{\frac{Area}{\pi}}\right)$

Illustrative Examples

Ex. 1: Find the area of a circular field of radius 7 m.

Soln: Applying the above formula, we have
Area of the circular field

$$= \pi r^2 = \frac{22}{7}\times 7\times 7 = 154 \text{ sq m.}$$

Ex. 2: Find the radius of a circular table whose surface area is 154 cm^2.

Soln: Applying the above theorem, we have

$$\text{radius} = \sqrt{\frac{154}{22/7}} = \sqrt{7\times 7} = 7 \text{ cm.}$$

Exercise

1. Find the circumference and the area of a circle of diameter 98 cm.
   a) 308 cm, 7546 sq cm b) 380 cm, 7456 sq cm
   c) 380 cm, 7645 sq cm d) 308 cm, 7545 sq cm
2. The length of a rope by which a buffalo must be tethered in order that she may be able to graze an area of 9856 sq m, is:
   a) 56 m b) 64 m c) 88 m d) 168 m
   [I. Tax and Central Excise 1989]
3. Find the area of a circle whose radius is $3\frac{1}{2}$ km.
   a) 38.5 sq km b) 83.5 sq km
   c) 36.5 sq km d) None of these
4. If the radius of one circle is twelve times the radius of another, how many times does the area of the greater contain the area of the smaller?
   a) 12 times b) 64 times
   c) 144 times d) Data inadequate
5. The radius of a circle is 2 metres. What is the radius of another circle whose area is 9 times that of the first?
   a) 18 m b) 12 m c) 6 m d) 9 m
6. Find the area of a circle whose diameter is 200 cm [Take $\pi = 3.1416$]
   a) 31416 sq cm b) 31516 sq cm
   c) 31216 sq cm d) 31816 sq cm
7. The area of a triangular plate of which the base and the altitude are 33 cm and 14 cm respectively is to be reduced to one third by drilling a circular hole through it. Calculate the diameter of the hole.
   a) 7 cm b) 14 cm c) 6 cm d) 12 cm

Answers

1. a; **Hint:** Diameter of the circle = 98 cm

$\therefore$ Its radius = 49 cm

$\therefore$ Circumference = $2\pi r = \left(2\times\frac{22}{7}\times 49\right) = 3.08$ metres

(See **Rule - 47**)

And, area = $\pi r^2 = \left(\frac{22}{7}\times 49\times 49\right) = 7546$ sq cm

2. a; **Hint:** $\pi r^2 = 9856 \Rightarrow r^2 = \left(9856\times\frac{7}{22}\right) = (448\times 7)$

$\Rightarrow$ r = 56 m.

3. a; **Hint:** Area = $\frac{22}{7}\times\frac{7}{2}\times\frac{7}{2} = \frac{77}{2} = 38.5$ sq km

4. c; **Hint:** $\frac{\pi(12r)^2}{\pi r^2} = \frac{144}{1} = 144$ times

5. c; **Hint:** $\frac{9}{1} = \frac{\pi r^2}{\pi(2)^2}$ or, $r^2 = 36$

$\therefore$ r = 6 m

6. a

7. b; **Hint:** Area of plate = $\left(\frac{1}{2}\times 33\times 14\right) = 231$ sq cm

Reduced area = $\left(\frac{1}{3}\times 231\right) = 77$ sq cm

Area of hole = (231 - 77) = 154 sq cm

$\pi r^2 = 154$

or, $r^2 = \frac{154}{22}\times 7 = 7\times 7$ $\therefore$ $r = \sqrt{7\times 7} = 7$ cm

$\therefore$ Diameter = 2r = 2 × 7 = 14 cm

Rule 49

Theorem: ***To find the area of circle if its perimeter or circumference is given.***

$$\textit{Area of a circle} = \frac{(circumference)^2}{4\pi}$$

Illustrative Example

Ex.: The circumference of a circular garden is 1012 m. Find the area.

Soln: Applying the above formula, we have

$$\text{Area} = \frac{(1012)^2}{4\times\frac{22}{7}} = 81466 \text{ sq m.}$$

Exercise

1. The area of a circle is 38.5 square metres. Find its circumference.
   a) 22 metres b) 20 metres c) 18 metres d) 24 metres
2. The circumference of a circle is 6.6 metres. Find its area.
   a) 3.465 sq m b) 4.365 sq m c) 3.565 sq m d) 3.466 sq m
3. The circumference of a circle is 352 m. Its area is:
   a) 9856 m^2 b) 8956 m^2 c) 6589 m^2 d) 5986 m^2
   [NDA Exam 1990]
4. Find the area of a circle whose circumference is $6\frac{2}{7}$ km.
   a) π sq km b) $\frac{\pi}{2}$ sq km c) 3π sq km d) 2π sq km

Answers

1. a; **Hint:** $38.5 = \frac{(circumference)^2}{4\pi}$

$\therefore$ circumference = $\sqrt{\frac{38.5\times 4\times 22}{7}} = \sqrt{484} = 22$ metres

2. a 3. a 4. a

Rule 50

Theorem: ***To find arc of a sector, when θ (angle subtended by the arc at the centre of a circle of which arc is a part) and circumference (or perimeter) is given.***

(i) Arc of a sector = $\left(\frac{\theta}{360}\right)$ × ***circumference***

(ii) Circumference = $\frac{360\times \textit{Arc of sector}}{\theta}$

(iii) Area of a sector = $\frac{\theta}{360}\times\pi r^2$ ***[If only radius (r) of the circle is given]***

= $\frac{1}{2}$ × ***radius × length of arc***

Illustrative Example

Ex.: Length of a metal wire is 60 cm. Metal wire is bent and made an arc as a part of perimeter of a circle. If this arc subtends an angle of 60° at the centre, then find the perimeter of the circle.

Soln: Applying the above formula, we have

$$\text{Circumference} = \frac{360\times 60}{60} = 360 \text{ cm.}$$

Exercise

1. In a circle of radius 28 cm, an arc subtends an angle of

72° at the centre. Find the length of the arc and the area of the sector so formed.
a) 35.2 cm, 492.8 sq cm b) 36 cm, 493 sq cm
c) 35.4 cm, 492.8 sq cm d) None of these

2. The length of an arc subtending an angle of 72° is 22 cm. Find the radius of the circle.
a) 17 cm b) 17.5 cm c) 18.5 cm d) 27.5 cm

3. The radius of a circle is 35 cm. Find the area of a sector enclosed by two radii and an arc 44 cm in length.
a) 770 sq cm b) 670 sq cm
c) 780 sq cm d) Can't be determined

4. If a piece of wire 20 cm long is bent into an arc of a circle subtending an angle of 60° at the centre, then the radius of the circle (in cm) is:
a) $\frac{\pi}{120}$ b) $\frac{\pi}{60}$ c) $\frac{120}{\pi}$ d) $\frac{60}{\pi}$
[NDA Exam 1990]

5. Find the area of sectors of a circle whose radius is 6 metres
(i) when the angle at the centre is 42°.
a) 13.2 sq m b) 14 sq m
c) 25 sq m d) 12 sq m
(ii) when the length of the arc is 11 metres.
a) 33 sq m b) 34 sq m
c) 32 sq m d) Data inadequate

6. Taking $\pi = \frac{22}{7}$, find the area of the sector when
(i) angle = 90° and radius = 21 cm
a) $346\frac{1}{2}$ sq m b) $348\frac{1}{2}$ sq m
c) $347\frac{1}{2}$ sq m d) Data inadequate
(ii) arc = $9\frac{3}{8}$ metres and radius = $9\frac{3}{5}$ metres
a) 48 sq m b) 45 sq m
c) 44 sq m d) 55 sq m

7. From a circular piece of cardboard of radius 3 metres two sectors of 40° have been cut off. Find the area of the remaining portion.
a) 22 sq m b) 44 sq m c) 28 sq m d) 18 sq m

8. The area of a sector is one-twelfth that of the complete circle. Find the angle of the sector.
a) 45° b) 60° c) 90° d) 30°

9. (i) A horse is placed inside a rectangular enclosure 40 metres by 36 metres and is tethered to one corner by a rope 14 metres long. Over what area can it graze?
a) 154 sq m b) 124 sq m
c) 164 sq m d) Data inadequate
(ii) If the horse is outside the enclosure and is tethered to the corner by the same rope, over what area can it graze?
a) 254 sq m b) 462 sq m
c) 616 sq m d) Data inadequate

10. Find the length of the arcs cut off from a circle of radius 7 cm by a chord 7 cm long.
a) $7\frac{1}{3}$ m, $36\frac{2}{3}$ m b) $7\frac{2}{3}$ m, $36\frac{1}{3}$ m
c) $8\frac{1}{3}$ m, $32\frac{2}{3}$ m d) None of these

11. The radius of a circle of centre O is 5 cm. Two radii OA and OB are drawn at right angles to each other. Find the areas of the two segments made by the chord BA.
a) $7\frac{1}{7}$ sq m, $71\frac{3}{7}$ sq m b) $6\frac{2}{7}$ sq m, $72\frac{4}{7}$ sq m
c) $5\frac{2}{7}$ sq m, $75\frac{3}{7}$ sq m d) Data inadequate

Answers

1. a; **Hint:** Length of arc = $\frac{2\pi r \times \theta}{360} = \left(2\times\frac{22}{7}\times 28\times\frac{72}{360}\right)$
$= 35.2$ cm
Area of the sector = $\frac{\pi r^2 \times \theta}{360} = \left(\frac{22}{7}\times 28\times 28\times\frac{72}{360}\right)$
$= 492.8$ sq cm

2. b; **Hint:** $22 = \frac{72}{360}\times 2\times\frac{22}{7}\times r$ $\therefore$ r = 17.5 cm

3. a; **Hint:** $44 = \left(2\times\frac{22}{7}\times 35\times\frac{\theta}{360}\right)$ or, $\theta = 72°$.
Area of sector = $\left(\frac{22}{7}\times\frac{72}{360}\times 35\times 35\right)$ = 770 sq cm

4. d; **Hint:** $2\pi r = \frac{360\times 20}{60}$ $\therefore r = \frac{360\times 20}{60} = \frac{60}{\pi}$ cm

5. (i) a; **Hint:** Required area = $\frac{42}{360}\times\frac{22}{7}\times 6\times 6 = \frac{66}{5} = 13\frac{1}{5}$
= 13.2 sq m.
(ii) a; **Hint:** Required area = $\frac{1}{2}\times 6\times 11 = 33$ sq m

6. (i) a (ii) b

7. a; **Hint:** Required area = $\pi 3^2 - \frac{2\times 40}{360}\times\pi 3^2$

$$= 9\pi\left(1-\frac{2}{9}\right) = \frac{9\times22\times7}{7\times9} = 22 \text{ sq m.}$$

8. d; **Hint:** $\frac{\pi r^2}{12} = \frac{\theta}{360}\times\pi r^2 \quad \therefore\ \theta = \frac{360}{12} = 30^\circ.$

9. (i) a; **Hint:**

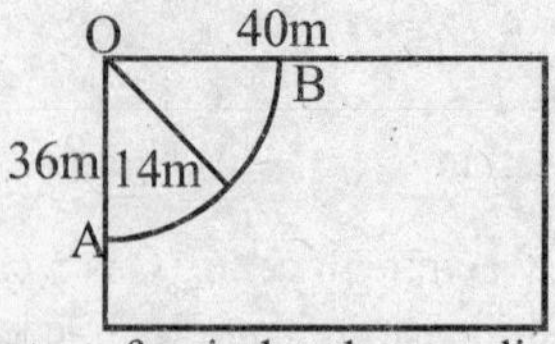

OAB is a sector of a circle whose radius is 14 metres.

$\therefore$ Area of the sector OAB $= \frac{90}{360}\times\pi\times14\times14$

$$= \frac{1}{4}\times\frac{22}{7}\times14\times14 = 154 \text{ sq m}$$

(ii) b; **Hint:** Required area $= \frac{22}{7}\times14\times14-154$

$= 616 - 154 = 462$ sq m.

10. a; **Hint:** Two arcs will be cut off, one smaller and the other bigger. (See the figure given below)

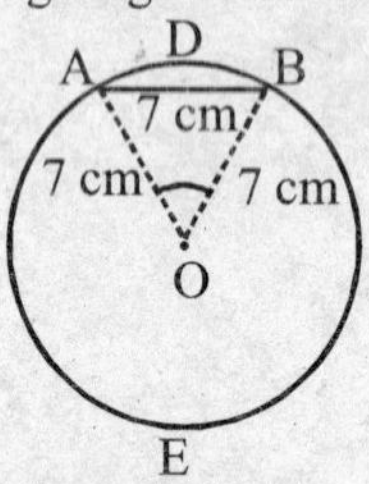

Δ OAB is an equilateral triangle

$\therefore\ \angle O = 60^\circ$

$\therefore$ length of the arc ADB

$$= \frac{60}{360}\times2\times\frac{22}{7}\times7 = \frac{22}{3} = 7\frac{1}{3} \text{ m and}$$

length of the arc AEB

$$= 2\times\frac{22}{7}\times7-\frac{22}{3} = \frac{110}{3} = 36\frac{2}{3} \text{ m.}$$

11. a; **Hint:**

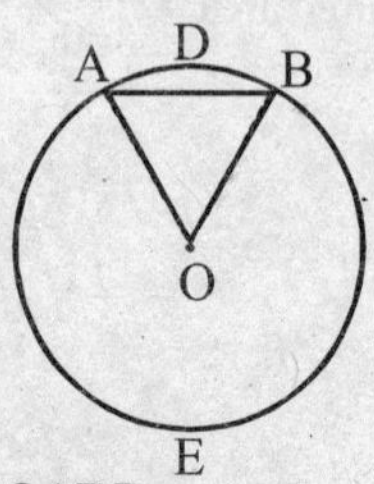

Area of the sector OADB

$$= \frac{90}{360}\times\frac{22}{7}\times5\times5 = \frac{275}{14} \text{ sq m}$$

Area of the ΔOAB $= \frac{1}{2}\times5\times5 = \frac{25}{2}$ sq m

$\therefore$ Area of the segment ADB

$$= \frac{275}{14}-\frac{25}{2} = \frac{50}{7} = 7\frac{1}{7} \text{ sq m and}$$

area of the segment AEB

$$= \frac{22}{7}\times5\times5-\frac{50}{7} = \frac{500}{7} = 71\frac{3}{7} \text{ sq m.}$$

Rule 51

Theorem: There are two concentric circles of radii R and r respectively. Now consider the following cases.

Case I: If larger circle makes 'n' revolutions to cover a certain distance, then the smaller circle makes $\left(\frac{R}{r}\right)$ n revolutions to cover the same distance.

Case II: If smaller circle makes n revolutions to cover a certain distance, then the larger circle makes $\left(\frac{r}{R}\right)n$ revolutions to cover the same distance.

Illustrative Example

Ex.: There are two concentric circles of radii 8 cm and 3 cm respectively. If larger circle makes 120 revolutions to cover a certain distance, then find the number of revolutions made by smaller circle to cover the same distance.

Soln: Applying the above theoerm, we have

the required no. of revolutions $= \frac{8}{3}\times120 = 320$ revolutions.

Exercise

1. There are two concentric circles of radii 15 cm and 5 cm respectively. If larger circle makes 100 revolutions to cover a certain distance, then find the number of revolutions made by smaller circle to cover the same distance.
   a) 300 revolutions b) 250 revolutions
   c) 125 revolutions d) Data inadequate
2. There are two concentric circles of radii 12 cm and 4 cm respectively. If larger circle makes 70 revolutions to cover a certain distance, then find the number of revolutions made by smaller circle to cover the same distance.
   a) 210 b) 120 c) 240 d) 225
3. There are two concentric circles of radii 10 cm and 6 cm respectively. If smaller circle makes 50 revolutions to cover a certain distance, then find the number of revolutions made by larger circle to cover the same distance.
   a) 15 b) 30 c) 16 d) 25

Answers

1. a 2. a 3. b

Rule 52

Theorem: ***There are two concentric circles. Radius of the larger circle is R. If the area of the smaller circle is $\frac{x}{y}$ of the area of the region (shaded portion) between two circles, then***

(i) the radius of the smaller circle = $Rs\sqrt{\frac{1}{\left(1+\frac{y}{x}\right)}}$,

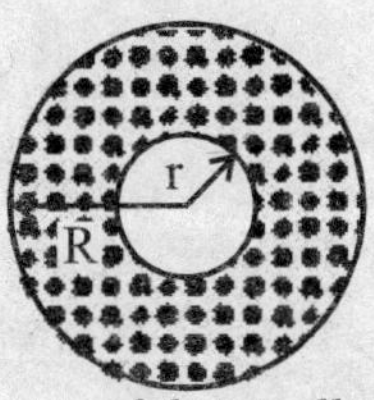

(ii) the perimeter of the smaller circle

$$= 2\pi R\left(\sqrt{\frac{1}{1+\frac{y}{x}}}\right),$$

(iii) the area of the smaller circle $= \pi R^2\left(\frac{1}{1+\frac{y}{x}}\right)$

and

(iv) the area of the shaded portion $= \pi R^2\left(\frac{1}{1+\frac{x}{y}}\right)$.

Where, R = radius of the larger circle.

Illustrative Example

Ex.: There are two concentric circles. Radius of the larger circle is 28 cm. Area of the smaller circle is $\frac{1}{3}$rd of the area between the two cirlces. Find the area and perimeter of the smaller circle.

Soln: Applying the above theorem, we have

Perimeter of the smaller circle = $2\pi \times 28 \times \sqrt{\frac{1}{1+3}}$

$$= 2\times\frac{22}{7}\times 28\times\frac{1}{2} = 88 \text{ cm}$$

Area of the smaller circle

$$= \frac{22}{7}\times 28\times 28\times\frac{1}{4} = 616 \text{ sq cm.}$$

Exercise

1. There are two concentric circles. Radius of the larger circle is 14 cm. Area of the smaller circle is $\frac{2}{5}$th of the area between the two cirlces. Find the area of the shaded portion.
   a) 440 sq cm b) 420 sq cm
   c) 220 sq cm d) 660 sq cm
2. There are two concentric circles. Radius of the larger circle is 7 cm. Area of the smaller circle is $\frac{1}{13}$th of the area between the two cirlces. Find the area of the smaller circle.
   a) 11 sq cm b) 22 sq cm c) 14 sq cm d) 28 sq cm
3. There are two concentric circles. Radius of the larger circle is 14 cm. Area of the smaller circle is $\frac{1}{3}$rd of the area between the two cirlces. Find the diameter of the smaller circle.
   a) 7 cm b) 14 cm c) 21 cm d) Data inadequate
4. There are two concentric circles. Radius of the larger circle is 35 cm. Area of the smaller circle is $\frac{4}{21}$ of the area between the two cirlces. Find the perimeter of the smaller circle.
   a) 44 cm b) 88 cm c) 66 cm d) Data inadequate

Answers

1. a 2. a
3. b; **Hint:** Radius of the smaller circle = 7 cm
   $\therefore$ diameter = 7 × 2 = 14 cm
4. b

Rule 53

Theorem: ***Length of a carpet 'd' m wide, required to cover the floor of a room which is x m long and y m broad, is given by*** $\left(\frac{xy}{d}\right)$ ***m. Or***

$$\textit{Length required} = \frac{\textit{Length of room} \times \textit{Breadth of room}}{\textit{Width of carpet}}$$

Illustrative Example

Ex.: How many metres of a carpet 75 cm wide will be required to cover the floor of a room which is 20 metres long and 12 metres broad?

Soln: Applying the above theorem, we have the

required length = $\frac{20\times 12}{0.75} = 320$ m.

Exercise

1. When 111 metres of carpet will cover a floor 18.5 metres by 7.5 metres, what is the width of the carpet?
   a) 1.25 metres b) 2 metres
   c) 2.75 metres d) 2.25 metres
2. How many metres of a carpet 60 cm wide will be required to cover the floor of a room which is 18 metres long and 15 metres broad?
   a) 320 m b) 360 m c) 420 m d) 450 m
3. How many metres of a carpet 40 cm wide will be required to cover the floor of a room which is 16 metres long and 10 metres broad?
   a) 400 m b) 380 m c) 350 m d) 325 m
4. How many metres of a carpet 1 m 4 cm wide will be required to cover the floor of a room which is 26 metres long and 20 metres broad?
   a) 525 m b) 450 m c) 500 m d) Data inadequate

Answers

1. a; **Hint:** $111 = \frac{18.5 \times 7.5}{x}$ $\therefore x = \frac{18.5 \times 7.5}{111} = 1.25$ metres

2. d 3. a 4. c

Rule 54

Theorem: ***A 'd' m wide carpet is used to cover the floor of a room which is x m long and y m broad. If the carpet is available at Rs A per metre, then the total amount required to cover the floor of the room is given by Rs*** $\left(A \times \frac{xy}{d}\right)$. ***Or***

Amount required = Rs

$$\left(Rate\ per\ metre \times \frac{length\ of\ room \times breadth\ of\ room}{width\ of\ carpet}\right)$$

Note: Length of the carpet

$$= \frac{xy}{d} \text{ or, } \frac{length\ of\ room \times breadth\ of\ room}{width\ of\ carpet}$$

Illustrative Example

Ex.: A 75 cm wide carpet is used to cover the floor of a room which is 20 metres long and 12 metres broad. What amount needs to be spent in carpeting the floor if the carpet is available at Rs 20 per metre?

Soln: Applying the above theorem, we have

the required answer $= 20 \times \frac{20 \times 12}{0.75} =$ Rs 6400.

Exercise

1. How many metres of carpet 75 cm wide will be required to cover a floor 27 m by 16 m, and what will be the cost at Rs 30 a metre.
   a) 576 m, Rs 17280 b) 767 m, Rs 20280
   c) 567 m, Rs 17010 d) Data inadequate
2. The cost of carpeting a room, 3 metres 2 dm broad with carpet at Rs 45 per sq m is Rs 1080. Find the length of the room.
   a) 7 m b) 7 m 5 dm c) 6 m 2 dm d) 7 m 2 dm
3. It costs Rs 15600 to carpet a room 7.5 metres wide with carpet 9 dm wide at Rs 234 per metre. What is the length of the room?
   a) 8 m b) 6 m c) 10 m d) 12 m
4. If it costs Rs 33750 to carpet a hall 25 metres by 15 metres with carpet at Rs 225 per metre, find the width of the carpet.
   a) 2.5 m b) 2 m c) 3 m d) 3.5 m

Answers

1. a

2. b; **Hint:** Area of the room $= \frac{1080}{45} = 24$ sq m

$\therefore$ length of the room $= \frac{24}{3.2} = 7.5$ m $= 7$ m 5 dm.

3. a; **Hint:** $\frac{7.5 \times x}{0.9} \times 234 = 15600$ $\therefore x = \frac{15600 \times 0.9}{7.5 \times 234} = 8$ m.

4. a; **Hint:** $33750 = \frac{25 \times 15}{x} \times 225$ $\therefore x = \frac{25 \times 15 \times 225}{33750} = 2.5$ m.

Rule 55

Theorem: ***Number of tiles, each measuring*** d_1 ***m ×*** d_2 ***m, required to pave a rectangular courtyard x m long and y m wide are given by*** $\left(\frac{x \times y}{d_1 \times d_2}\right)$.

Or

$$\textit{Number of tiles required} = \frac{length \times breadth\ of\ courtyard}{length \times breadth\ of\ each\ tile}$$

Illustrative Example

Ex.: How many paving stones each measuring 2.5 m × 2 m are required to pave a rectangular courtyard 30 m long and 16.5 m wide?

Soln: Applying the above theorem, we have the required answer $= \frac{30 \times 16.5}{2.5 \times 2} = 99$

Exercise

1. The dimensions of the floor of a rectangular hall are 4 m × 3 m. The floor of the hall is to be tiled fully with 8 cm × 6 cm rectangular tiles without breaking tiles to smaller sizes. The number of tiles required is:
   a) 4800 b) 2600 c) 2500 d) 2400
   [CDS Exam 1991]

2. How many paving stones, each measuring 10 dm by 9 dm are required to pave a verandah 60 m long and 6 m broad?
a) 600 b) 800 c) 400 d) 450

3. How many planks $10\frac{1}{2}$ m long and $3\frac{2}{3}$ dm broad will be required for the ground whose length is 42 m and breadth 12 m?
a) 160 b) 240 c) 60 d) 120

4. How many postage stamps 2 cm long and $1\frac{1}{2}$ cm wide will be required to cover a board of paper 3 dm long and 2 dm wide?
a) 300 b) 150 c) 250 d) 200

Answers

1. c; **Hint:** Area of the floor = $(400 \times 300)\ cm^2$

Area of one tiles = $(8 \times 6)\ cm^2$

Number of tiles = $\frac{400 \times 300}{8 \times 6} = 2500$

2. c 3. d 4. d

Rule 56

Theorem: ***Certain number of tiles, each measuring d_1 m × d_2 m, are required to pave a rectangular courtyard x m long and y m wide. If the tiles are available at Rs A per piece, then the amount needs to be spent is given by Rs*** $\left(A \times \frac{x \times y}{d_1 \times d_2}\right)$.

Or

Amount required

$$= \textit{price per tile} \times \frac{\textit{length} \times \textit{breadth of courtyard}}{\textit{length} \times \textit{breadth of each tile}}$$

Illustrative Example

Ex.: Certain number of paving stones each measuring 2.5 m × 2 m are required to pave a rectangular courtyard 30 m long and 16.5 m wide. What amount needs to be spent if the tiles of the aforesaid dimension are available at Re 1 per piece?

Soln: Applying the above theorem, we have

the required answer = $1 \times \frac{30 \times 16.5}{2.5 \times 2}$ = Rs 99

Exercise

1. Certain number of paving stones each measuring 5 m × 4 m are required to pave a rectangular courtyard 60 m long and 33 m wide. What amount needs to be spent if the tiles of the aforesaid dimension are available at Re 2 per piece?
a) Rs 99 b) Rs 198 c) Rs 96 c) Rs 192

2. Certain number of paving stones each measuring 4 m × 2 m are required to pave a rectangular courtyard 26 m long and 12 m wide. What amount needs to be spent if the tiles of the aforesaid dimension are available at Rs 5 per piece?
a) Rs 195 b) Rs 390 c) Rs 159 d) Rs 295

3. Certain number of paving stones each measuring 6 m × 5 m are required to pave a rectangular courtyard 20 m long and 15 m wide. What amount needs to be spent if the tiles of the aforesaid dimension are available at Rs 25 per piece?
a) Rs 250 b) Rs 150 c) Rs 125 d) Rs 160

Answers

1. b 2. a 3. a

Rule 57

Theorem: ***A room x m long and y m broad is to be paved with square tiles of equal sizes. The largest possible tile so that the tiles exactly fit is given by* "HCF of length and breadth of the room" *and the no. of tiles required are*** $\left[\frac{x \times y}{(HCF\ of\ x\ and\ y)^2}\right]$.

Illustrative Example

Ex.: A hall-room 39 m 10 cm long and 35 m 70 cm broad is to be paved with equal square tiles. Find the largest tile so that the tiles exactly fit and also find the number of tiles required.

Soln: Quicker Method:

Side of largest possible square tile = HCF of length and breadth of the room

= HCF of 39.10 and 35.70 m = 1.70 m

Also, number of tiles required

$$= \frac{\text{Length} \times \text{breadth of room}}{(\text{HCF of length and breadth of the room})^2}$$

$$= \frac{39.10 \times 35.70}{1.70 \times 1.70} = 483$$

Exercise

1. A rectangular courtyard 3.78 metres long and 5.25 metres broad is to be paved exactly with square tiles, all of the same size. What is the largest size of such a tile? Also find the number of tiles.
a) 21 cm, 450 b) 20 cm, 450
c) 25 cm, 500 d) Can't be determined

2. The length and breadth of a room are 10 m 75 cm and 8 m

25 cm respectively. The floor is to be paved with square tiles of the largest possible size. The size of the tiles is:

a) 25 cm × 25 cm b) 50 cm × 50 cm
c) 20 cm × 20 cm d) 30 cm × 30 cm

[Hotel Management 1991]

3. A room 7.3 metres by 7.1 metres is to be paved with equal square tiles. Find the least number of whole tiles which will exactly cover the floor.
a) 5183 b) 52
c) 5813 d) Can't be determined

4. A room 9.49 metres long and 9.23 metres broad is to be paved with equal square tiles. Find the biggest tile which will exactly fit and the number required.
a) 0.13 × 0.13 sq m, 5183
b) 13 × 13 sq m, 5183
c) 0.23 × 0.23 sq m, 5813
d) 0.13 × 0.13 sq m, 5813

Answers

1. a; **Hint:** Largest size of tile
= HCF of 378 cm and 525 cm = 21 cm

Number of tiles = $\left(\frac{378 \times 525}{21 \times 21}\right) = 450$

2. a; **Hint:** Largest possible size of the tile
= HCF of 1075 nd 825 = 25
∴ required answer = 25 cm × 25 cm

3. a; **Hint:** Required answer

$= \frac{7.3 \times 7.1}{(0.1)^2} = 5183$ [Since HCF of 7.3 and 7.1 is 0.1.]

4. a

Rule 58

Theorem: ***If a square hall x metres long is surrounded by a verandah (on the outside of the square hall) d metres wide, then the area of the verandah is given by 4d(x + d) sq metres.***

Illustrative Example

Ex.: A square field, 10 metres long, is surrounded by a path 2 metres wide. Find the area of the path.

Soln: Applying the above theorem,

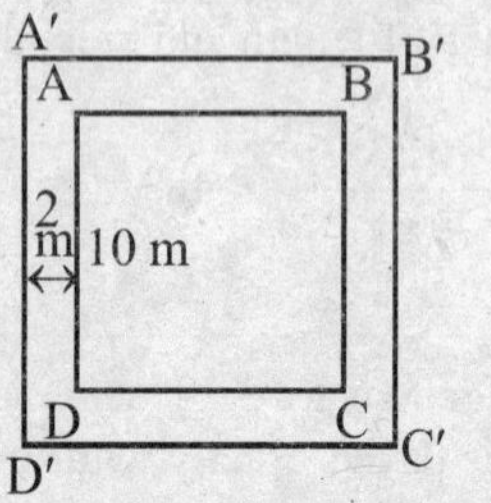

Area of the path = 4 × 2 (10 + 2) = 96 sq metres.

Exercise

1. A square field, 5 metres long, is surrounded by a path 1 metre wide. Find the area of the path.
a) 48 sq m b) 24 sq m
c) 16 sq m d) Data inadequate

2. A square field, 13 metres long, is surrounded by a path 3 metres wide. Find the area of the path.
a) 192 sq m b) 196 sq m c) 182 sq m d) None of these

3. A square field, 16 metres long, is surrounded by a path 4 metres wide. Find the area of the path.
a) 320 sq m b) 340 sq m c) 220 sq m d) None of these

4. A square field, 18 metres long, is surrounded by a path 2 metres wide. Find the area of the path.
a) 80 sq m b) 160 sq m c) 180 sq m d) 100 sq m

Answers

1. b 2. a 3. a 4. a

Rule 59

Theorem: ***If a square plot is x m long. It has a gravel path d m wide all round it on the inside, then the area of the path is given by 4d(x – d) sq m.***

Illustrative Example

Ex.: A square grass plot is 100 m long. It has a gravel path 2.5 metres wide all round it on the inside. Find the area of the path.

Soln: Applying the above theorem, we have the required answer = 4 × 2.5 × (100 – 2.5) = 975 sq metres.

Exercise

1. A path 2 metres wide, running all round a square garden has an rea of 7680 sq metres. Find the area of the part of the garden enclosed by the path.
a) 1.5 sq km b) 1.9 sq km
c) 2 sq km d) Can't be determined

2. A square grass plot is 22 m long. It has a gravel path 2 metres wide all round it on the inside. Find the area of the path.
a) 80 sq m b) 82 sq m
c) 96 sq m d) Data inadequate

3. A square grass plot is 24 m long. It has a gravel path 4 metres wide all round it on the inside. Find the area of the path.
a) 112 sq m b) 80 sq m c) 320 sq m d) None of these

4. A square grass plot is 31.5 m long. It has a gravel path 1.5 metres wide all round it on the inside. Find the area of the path.
a) 180 sq m b) 198 sq m c) 280 sq m d) 298 sq m

Answers

1. a; **Hint:** 4 × 2 × (x - 2) = 9680 or, $x - 2 = \frac{9680}{8} = 1210$
or, x = 1210 + 2 = 1212

∴ Area of the garden = $(1212)^2 = 1468944$ sq m

∴ required area = 1468944 - 9680 = 1459264 sq m
= 1.459264 sq km ≈ 1.5 sq km

2. a 3. c 4. a

Rule 60

Theorem: ***If a rectangular hall x m long and y m broad, is surrounded by a verandah (on the outside fo the rectangular hall) d m wide, then the area of the verandah is given by $2d[(x+y)+2d]\ m^2$.***

Or

Area of verandah = 2(width of verandah) × [length + breadth of room + 2 (width of verandah)]

Illustrative Example

Ex.: A rectangular hall 12 m long and 10 m broad, is surrounded by a verandah 2 metres wide. Find the area of the verandah.

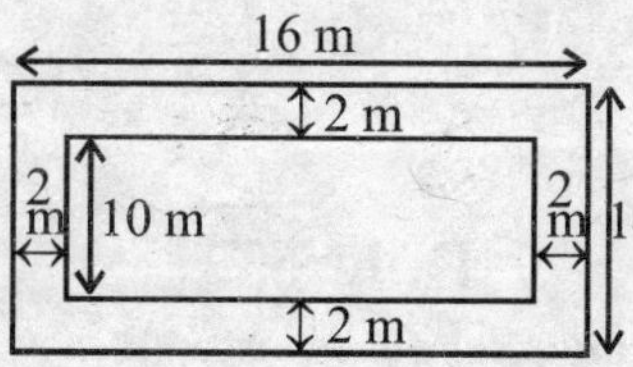

Soln: Since the verandah is outside the room, the above theorem will be applied.

Area of verandah $= 2\times2(10+12+2\times2) = 4\times26 = 104$ m^2

Exercise

1. A rectangular garden has 5 metres wide road outside around all the four sides. The area of the road is 600 square metres. What is the ratio between the length and the breadth of that plot?
   a) 3 : 2 b) 4 : 3 c) 5 : 4 d) Data inadequate
   (BSRB Calcutta PO - 1999)
2. The length and breadth of a rectangular field are in the ratio 7 : 4. A path 4 metres wide running all round outside it has an area of 416 square metres. Find the length and breadth of the field.
   a) 7 m, 4 m b) 28 m, 16 m
   c) 21 m, 12 m d) Data inadequate
3. A 5 m wide lawn is cultivated all along the outside of a rectangular plot measuring 90 m × 40 m. The total area of the lawn is:
   a) 1200 m^2 b) 1300 m^2 c) 1350 m^2 d) 1400 m^2
   (CDS 1991)
4. A rectangular room 10 m long and 8 m broad is surrounded by a verandah 2 metres wide. Find the area of the verandah.
   a) 89 sq m b) 88 sq m c) 86 sq m d) 98 sq m
5. A rectangular plot of grass 25 m 5 dm by 24 m 4 dm has a gravel walk 1.5 metres wide all-round it on the outside. Find the area of the walk in sq m.
   a) 158.7 sq m b) 160 sq m c) 168.7 sq m d) 157.8 sq m

Answers

1. d; **Hint:** $2\times5[(x+y)+(2\times5)] = 600$

$\therefore\ x+y = \dfrac{600-100}{10} = 50$

From this lone equation we cannot get the required ratio

2. b; **Hint:** Let the length and breadth be 7x and 4x

$2\times4[(7x+4x)+2\times4] = 416$ or, $88x = 416-64 = 352$

$\therefore\ x = 4$

$\therefore$ length $= 7\times4 = 28$ m and bradth $4\times4 = 16$ m.

3. d 4. b 5. a

Rule 61

Theorem: ***If a rectangular plot is 'x' m by 'y' m. It has a gravel path 'd' m wide all round it on the inside, then the area of the path is given by 2d(x + y – 2d) sq m.***

Illustrative Example

Ex.: A rectangular grassy plot is 112 m by 78 m. It has a gravel path 2.5 m wide all round it on the inside. Find the area of the path and the cost of constructing it at Rs 2 per square metre?

Soln: Since the path is inside the plot, the above theorem will be applied,

Area of the path $= 2\times2.5\times(112+78-2\times2.5)$
$= 5\times185 = 925$ sq m.

$\therefore$ cost of construction = rate × area
$= 2\times925 =$ Rs 1850

Exercise

1. A rectangular field is 125 m long and 68 m broad. A path of uniform width of 3 metres runs round the field inside it. Find the area of the path.
   a) 1122 sq m b) 1212 sq m
   c) 2211 sq m d) None of these
2. A footpath of uniform width runs round the inside of a rectangular field 38 metres long and 32 metres wide. If the area of the path be 600 sq metres, find its width.
   a) 6 metres b) 5 metres c) 4 metres d) 8 metres
3. A room 5 m × 4 m is to be carpeted leaving a margin of 25 cm from each wall. If the cost of the carpet is Rs 80 per sq m, the cost of carpeting the room will be:
   a) Rs 1440 b) Rs 1260 c) Rs 1228 d) Rs 1192
4. A rectangular court is 120 m long and 90 m broad and inside it a path of uniform width of 10 m runs round it. Find the cost of covering the path with flagstones at Rs 25 per sq m.
   a) Rs 95000 b) Rs 59000 c) Rs 9500 d) Rs 5900
5. Find the cost at Rs 20 per metre of carpeting a floor 47 m by 38 m with carpet 75 cm wide, so as to leave a margin of 1 m uncovered all round.
   a) Rs 45200 b) Rs 33200 c) Rs 43200 d) Rs 53200

Answers

1. a

2. b; **Hint:** $2d(38+32-2d) = 600$

or, $2d^2 - 70d + 300 = 0$

or, $d^2 - 35d + 150 = 0$

or, $d^2 - 30d - 5d + 150 = 0$

or, $d(d - 30) - 5(d - 30) = 0 \quad \therefore d = 30, 5$

$\therefore$ required answer = 5 metres

Note: This can also be solved by **Rule - 62.**

3. b; **Hint:** Area of the margin = $\left(5+4-2\times\frac{1}{4}\right)2\times\frac{1}{4} = 4.25$ sq m

Area of the carpet = $5 \times 4 - 4.25 = 15.75$ sq m

$\therefore$ Cost of carpeting = 80×15.75 = Rs 1260

4. a; **Hint:** Area of the path = $2 \times 10 \times [120 + 90 - 20]$

$= 20 \times 190 = 3800$ sq m

$\therefore$ required cost = 3800×25 = rs 95000

5. c; **Hint:** Area of the margin = $2 \times 1(47 + 38 - 2 \times 1)$

$= 166$ sq m

Area of the floor = $47 \times 38 = 1786$ sq m

$\therefore$ area of the carpet = $1786 - 166 = 1620$ sq m

$\therefore$ required cost = $20\times\frac{1620}{0.75}$ = Rs 43200 [See **Rule - 54**]

Rule 62

Theorem: ***If a path all around the inside of a rectangular park 'x' m by 'y' m occupies 'A' sq m, then the width of the path is given by*** $\left(\frac{(x+y)-\sqrt{(x+y)^2-4A}}{4}\right)$ ***m.***

Illustrative Example

Ex.: A path all around the inside of a rectangular park 37 m by 30 m occupies 570 sq m. Find the width of the path.

Soln: **Quicker Method - I:**

Area of path = 2 × width of path × [length + breadth of park – 2 × (width of path)]

$\Rightarrow 570 = 2 \times x \times [37 + 30 - 2x]$ (x is the width of path)

$\Rightarrow 570 = 134x - 4x^2 \Rightarrow 4x^2 - 134x + 570 = 0$

On solving this equation we get, $x = 5$ m.

Quicker Method - II:

Applying the above theorem, we have width of the path

$$= \frac{(37+30)-\sqrt{(37+30)^2-4\times570}}{4}$$

$$= \frac{67-\sqrt{4489-2280}}{4} = \frac{67-47}{4} = 5 \text{ m.}$$

Exercise

1. A carpet is laid on a rectangular floor of a drawing room in a house, measuring 8 metres by 5 metres. There is a boarder of constant width around the carpet and the area of the boarder is 12 square metres. What is the width of the boarder.
   a) 50 cm b) 1 m c) 120 cm d) 90 cm
2. A path all around the inside of a rectangular park 25 m by 20 m occupies 164 sq m. Find the width of the path.
   a) 1m b) 2 m c) 1.5 m d) 4 m
3. A path all around the inside of a rectangular park 30 m by 25 m occupies 204 sq m. Find the width of the path.
   a) 3m b) 2m c) 2.5 m d) 1 m

Answers

1. a; **Hint:** Required answer = $\frac{(8+5)-\sqrt{(8+5)^2-4\times12}}{4}$

$= \frac{13-\sqrt{169-48}}{4} = \frac{13-11}{4} = \frac{1}{2}$ m = 50 cm

2. b 3. b

Rule 63

Theorem: ***A rectangular garden is 'x' metres long and 'y' metres broad. It is to be provided with pavements 'd' metres wide all round it both on its outside as well as inside. Then the total area of the pavement is given by 4d(x + y) sq m.***

Illustrative Example

Ex.: A rectangular garden is 15 metres long and 10 metres broad. It has 3 metres wide pavements all around it both on its inside and outside. Find the total area of the pavements.

Soln: Applying the above theorem, we have the total area of the pavements = $4 \times 3(15 + 10) = 300$ sq m.

Exercise

1. A garden is 30 m long and 20 m broad. It has 1.5 m wide pavements all around it both on its inside and outside. Fnd the total area of the pavement.
   a) 300 sq m b) 150 sq m c) 600 sq m d) None of these
2. A garden is 18 m long and 12 m broad. It has 2 m wide pavements all around it both on its inside and outside. Find the total area of the pavement and the cost of construction of the pavement at the rate of Rs 3 per square metre.
   a) Rs 1440 b) Rs 960 c) Rs 480 d) Data inadequate
3. A garden is 35 m long and 25 m broad. It has 5 m wide pavements all around it both on its inside and outside. Find the total area of the pavement.
   a) 1500 sq m b) 800 sq m
   c) 1200 sq m d) Data inadequate
4. A garden is 20 m long and 15 m broad. It has 1.5 m wide pavements all around it both on its inside and outside. Find the total area of the pavement.
   a) 210 sq m b) 135 sq m c) 175 sq m d) 220 sq m

Answers

1. a 2. a 3. c 4. a

Rule 64

Theorem: ***A square garden is 'x' metres long. It is to be provided with pavements 'd' metres wide all round it both on its outside as well as inside. Then the total area of the pavement is given by (8dx) sq metres.***

Illustrative Example

Ex.: A square garden is 10 metres long. It has 3 metres wide pavements all round it both on its inside and outside. Find the total area of the pavements.

Soln: Applying the above theorem, we have
the total area of the pavements $= 8 \times 3 \times 10 = 240$ sq m.

Exercie

1. A square garden is 20 metres long. It has 2 metres wide pavements all round it both on its inside and outside. Find the total area of the pavements.
a) 320 sq m b) 325 sq m c) 240 sq m d) None of these
2. A square garden is 40 metres long. It has 1 metre wide pavements all round it both on its inside and outside. Find the total area of the pavements.
a) 320 sq m b) 160 sq m c) 420 sq m d) 230 sq m
3. A square garden is 15 metres long. It has 1.5 metres wide pavements all round it both on its inside and outside. Find the total area of the pavements.
a) 180 sq m b) 160 sq m
c) 240 sq m d) Data inadequate

Answers

1. a 2. a 3. a

Rule 65

Theorem: ***An oblong piece of ground measures x m by y m. From the centre of each side a path 'd' m wide goes across to the centre of the opposite side.***

I. Area of the path = d(x + y – d)
= (width of path) (length + breadth of park – width of path)

II. Area of the park minus the path = (x – d) (y – d)
= (length of park – width of path) × (breadth of park – width of the path)

Illustrative Example

Ex.: An oblong piece of ground measures 19 m 2.5 dm by 12 metres 5 dm. From the centre of each side a path 2 m wide goes across to the centre of the opposite side. What is the area of the path? Find the cost of paving these paths at the rate of Rs 1.32 per sq metre.

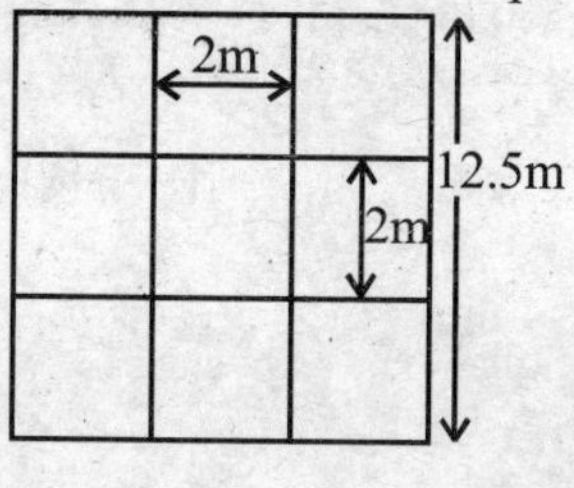

Soln: Quicker Method:
Now, for the given question,
Area of path $= 2 \times (19.25 + 12.5 - 2)$
$= 2 \times 29.75 = 59.5$ sqm.
$\therefore$ cost = rate × area = Rs (59.5 × 1.32) = Rs 78.54

Exercise

1. A rectangular lawn 60 metres by 40 metres has two roads each 5 metres wide, running in the middle of it, one parallel to length and the other parallel to the breadth. Find the cost of gravelling them at 60 paise per square metre.
a) Rs 285 b) Rs 385 c) Rs 275 d) Rs 475
2. A rectangular lawn 80 metres by 60 metres has two roads each 10 metres wide running in the middle of it, one parallel to the length and the other parallel to the breadth. Find the cost of gravelling them at Rs 30 per square metre.
a) Rs 3900 b) Rs 39000 c) Rs 3600 d) Rs 36000
3. The length and breadth of a rectangular field are 500 m and 400 m respectively. If two roads 10 m wide each are perpendicular to each other inside the field, what is the total area of the roads.
a) 8900 sq m b) 9800 sq m c) 9900 sq m d) 8000 sq m
4. A field is 19.25 m long and 12.5 m broad has two roads each 2 m wide running in the middle of it one parallel to length and the other parallel to breadth. Find the cost of gravelling them at Rs 1.32 per sq metre.
a) Rs 78 b) Rs 88.45 c) Rs 78.54 d) Rs 87.45
5. A field is 100 m long and 70 m wide. It has two roads each of same breadth. One road is parallel to the length and other parallel to breadth. If the cost of gravelling them at Rs 2 per sq metre is Rs 3200, find the breadth of the roads.
a) 5 m b) 10 m c) 12 m d) Data inadequate

Answers

1. a; **Hint:** Area of the path = 5 (60 + 40 - 5) = 475 sq m

$\therefore$ the cost of gravelling the path $= \dfrac{475 \times 60}{100} =$ Rs 285

2. b 3. a

4. c; **Hint:** Area of the path = 59.5 sq m
$\therefore$ the cost of gravellign the path = 59.5 × 1.32 = Rs 78.54

5. b; **Hint:** Area of the roads $= \dfrac{3200}{2} = 1600$ sq m

$d(100 + 70 - d) = 1600$ or, $d^2 - 17d + 1600 = 0$
By solving equation, we have, d = 160, 10
$\therefore$ required answer = 10 m.

Rule 66

Theorem: ***There is a square garden of side 'x' metres. From the centre of each side a path 'd' metres wide goes across to the centre of opposite side.***

I. Area of the path $= d(2x-d)$ ***sq metres.***

II. Area of the garden – minus the path $= (x-d)^2$ ***sq metres.***

Illustrative Example

Ex.: There is a square field of side 5 metres. A path 1 metre wide runs through the centre of the field, one each across its opposite sides. What is the total area of the path and the area of the remaining portion of the field?

Soln: Applying the above theorem, we have the total area of the path $1(2\times5-1)=9$ sq metres and the total area of the remaining portion of the field = $(5-1)^2 = 16$ sq metres.

Exercise

1. There is a square field of side 10 metres. A path 2 metres wide runs through the centre of the field, one each across its opposite sides. What is the total area of the path and the area of the remaining portion of the field?
   a) 36 sq m, 64 sq m b) 18 sq m, 32 sq m
   c) 32 sq m, 64 sq m c) 18 sq m, 64 sq m
2. There is a square field of side 18 metres. A path 3 metres wide runs through the centre of the field, one each across its opposite sides. What is the total area of the path and the area of the remaining portion of the field?
   a) 99 sq m, 225 sq m b) 98 sq m, 224 sq m
   c) 99 sq m, 441 sq m d) Data inadequte
3. There is a square field of side 21 metres. A path 1 metre wide runs through the centre of the field, one each across its opposite sides. What is the total area of the path and the area of the remaining portion of the field?
   a) 41 sq m, 400 sq m
   b) 82 sq m, 200 sq m
   c) 41 sq m, 200 sq m
   d) 82 sq m, 400 sq m

Answers

1. a 2. a 3. a

Rule 67

Theorem: ***If the sides of a rectangular field of 'A' sq m area are in the ratio a : b, then the sides are given by*** $\sqrt{A\times\frac{a}{b}}$ ***or*** $\sqrt{Area\times Ratio}$ ***and*** $\sqrt{A\times\frac{b}{a}}$ ***or*** $\sqrt{Area\times Inverse\ Ratio}$

Illustrative Example

Ex.: The sides of a rectangular field of 726 sq m are in the ratio of 3 : 2. Find the sides.

Soln: **Quicker Method:** Side = $\sqrt{\text{Area}\times\text{Ratio}}$

2nd side = $\sqrt{\text{Area}\times\text{Inverse Ratio}}$

$\therefore$ In the given question, first side = $\sqrt{726\times\frac{3}{2}}$

$= \sqrt{1089} = 33$ m

and the second side = $\sqrt{726\times\frac{2}{3}} = \sqrt{484} = 22$ m.

Exercise

1. The sides of a rectangular field of 128 sq m are in the ratio of 1 : 2. Find the sides.
   a) 16 m, 8 m b) 12 m, 6 m
   c) 14 m, 7 m d) Data inadequte
2. The sides of a rectangular field of 270 sq m are in the ratio of 6 : 5. Find the sides.
   a) 18 m, 15 m b) 24 m, 20 m
   c) 18 m, 12 m d) None of these
3. The sides of a rectangular field of 1125 sq m are in the ratio of 9 : 5. Find the sides.
   a) 45 m, 25 m b) 36 m, 25 m
   c) 54 m, 25 m d) None of these

Answers

1. a 2. a 3. a

Rule 68

Theorem: ***If the base and the height of a triangle are in the ratio x : y and the area of the triangle is A sq m, then the base is given by*** $\left(\sqrt{2\times A\times\frac{x}{y}}\right)$ ***m or*** $\sqrt{2\times Area\times Ratio}$ ***and the height is given by*** $\left(\sqrt{2\times A\times\frac{y}{x}}\right)$ ***m or*** $\sqrt{2\times Area\times Inverse\ Ratio}$.

Illustrative Example

Ex.: The base of a triangular field is three times its height. If the cost of cultivating the field at Rs 36.72 per hectare is Rs 495.72, find its base and height.

Soln: **Detail Method:** Area of the field

$$= \frac{\text{Rs } 495.72}{\text{Rs } 36.72} = \frac{27}{2} \text{ hectares.}$$

Also, area of the field $= \frac{1}{2}\times3\times\text{Height}\times\text{Height}$

$$= \frac{3}{2}(\text{Height})^2$$

$$\therefore \frac{3}{2}(\text{Height})^2 = \frac{27}{2} \text{ hectares.}$$

$\therefore$ (Height)$^2 = \frac{27}{2} \times \frac{2}{3}$ hectares

$= 9$ hectares $= 90000$ sq metres.

$\therefore$ Height $= \sqrt{90000}$ m $= 300$ m

Also, Base $= 3 \times$ Height $= 900$ m

Quicker Method: The ratio between base and height in above example is 3 : 1.

In such questions use the rule:

Base $= \sqrt{2 \times \text{Area} \times \text{Ratio}}$

Height $= \sqrt{2 \times \text{Area} \times \text{Inverse Ratio}}$

Now, ratio of base and height is 3 : 1. Hence, the ratio attached with base is 3, the ratio attached with height is 1.

$\therefore$ Base $= \sqrt{2 \times \frac{27}{2} \times \frac{3}{1}} = 900$ m.

Height $= \sqrt{2 \times \frac{27}{2} \times \frac{1}{3}} = 300$ m.

Exercise

1. The ratio of base and height of a triangular field is 3 : 2 and the area of the field is 108 sq m. Find its base and height.
   a) 18 m, 12 m b) 12 m, 8 m
   c) 21 m, 14 m d) Data inadequate
2. The ratio of base and height of a triangular field is 5 : 4 and the area of the field is 90 sq m. Find its base and height.
   a) 15 m, 12 m b) 20 m, 16 m
   c) 25 m, 20 m d) None of these
3. The ratio of base and height of a triangular field is 7 : 5 and the area of the field is 437.5 sq m. Find its base and height.
   a) 35 m, 25 m b) 21 m, 25 m
   c) 49 m, 35 m d) None of these

Answers

1\. a 2. a 3. a

Rule 69

Theorem: ***To find the area of a rhombus if one side and one diagonal are given.***

Area of a rhombus = diagonal $\times \sqrt{(side)^2 - \left(\frac{diagonal}{2}\right)^2}$

Illustrative Example

Ex.: Find the area of a rhombus one side of which measures 20 cm and one diagonal 24 cm.

Soln: Applying the above rule, we have the required area

$= 24 \times \sqrt{(20)^2 - \left(\frac{24}{2}\right)^2} = 24 \times \sqrt{400 - 144}$

$= 24 \times 16 = 384$ cm^2.

Exercise

1. The perimeter of a rhombus is 100 cm. If one of the diagonals measures 14 cm, what is the area of the rhombus?
   a) 336 sq cm b) 168 sq cm c) 504 sq cm d) 252 sq cm
2. Find the area of a rhombus one side of which measures 26 cm and one diagonal 20 cm.
   a) 480 sq cm b) 520 sq cm
   c) 840 sq cm d) Data inadequate
3. Find the area of a rhombus one side of which measures 40 cm and one diagonal 48 cm.
   a) 1536 sq cm b) 1436 sq cm
   c) 1636 sq cm d) Data inadequate

Answers

1\. a; **Hint:** Side $= \frac{100}{4} = 25$ cm.

Applying the given rule, we get the required answer $= 336$ sq cm.

2\. a 3. a

Rule 70

Theorem: ***To find the other diagonal of a rhombus, if perimeter of rhombus and one of its diagonals are given.***

Other diagonal = 2 $\times \sqrt{(side)^2 - \left(\frac{diagonal}{2}\right)^2}$ ***; where side*** $= \frac{Perimeter}{4}$

Illustrative Example

Ex.: The perimeter of a rhombus is 146 cm and one of its diagonal is 55 cm. Find the other diagonal and the area of the rhombus.

Soln: Applying the above rule,

one side of a rhombus $= \frac{146}{4} = 36.5$ cm

$\therefore$ other diagonal $= 2 \times \sqrt{(36.5)^2 - \left(\frac{55}{2}\right)^2} = 48$ cm

Now, area $= \frac{1}{2}$ (product of diagonals)

$= \frac{1}{2} \times 48 \times 55 = 1320$ sq cm.

Exercise

1. The perimeter of a rhombus is 144 cm and one of its diagonals is 18 cm, find the area of the rhombus and length of its other diagonal.
a) 627.4 sq cm (approx), 69.7 cm (approx)
b) 672.4 sq cm (approx), 67.9 cm (approx)
c) 527.4 sq cm (approx), 69.7 cm (approx)
d) None of these
2. The perimeter of a rhombus is 80 cm and one of its diagonal is 24 cm. Find the length of the other diagonal.
a) 32 cm b) 30 cm c) 64 cm d) Data inadquate
3. The perimeter of a rhombus is 26 cm and one of its diagonal is 20 cm. Find the length of the other diagonal.
a) 48 cm b) 38 cm c) 36 cm d) 64 cm

Answers

1. a 2. a 3. a

Problems on a Regular Polygon

Rule 71

To find the area of a regular polygon if length of its each side is given. Area of a regular polygon = $\left[\frac{n}{4}\cot\left(\frac{\pi}{n}\right)\right]\times(Side)^2$***. When 'n' = No. of sides. Now consider the following regular polygons.***

Hexagon: A hexagon has 6 sides.

(i) Area of a regular Hexagon $=\frac{6}{4}\cot\left(\frac{180}{6}\right)\times(Side)^2$

$$=\frac{3}{2}\cot 30^\circ\times(Side)^2=\frac{3\sqrt{3}}{2}\times(Side)^2$$

[Since Cot 30° = $\sqrt{3}$]

Illustrative Example

Ex.: Find the area of a regular hexagon whose side measures 9 cm.

Soln: Applying the above formula, we have

$$\text{area of a regular hexagon}=\frac{3\sqrt{3}a^2}{2}$$

Here $a=9$ cm.

$$\therefore \text{ area}=\frac{3\sqrt{3}\times 9^2}{2}\text{ sq cm}=210.4\text{ sq cm approx.}$$

(ii) If area of a regular hexagon is given, then

$$side=\sqrt{\left(\frac{2\times Area\ of\ a\ regular\ hexagon}{3\sqrt{3}}\right)}$$

Illustrative Example

Ex.: Find the side of a regular hexagon whose area is $24\sqrt{3}$ sq m.

Soln: Applying the above formula, we have

$$\text{Side}=\sqrt{\frac{2\times 24\sqrt{3}}{3\sqrt{3}}}=4\text{ m}$$

Octagon: ***An octagon has 8 sides.***

(i) Area of a regular Octagon

$$=\frac{8}{4}\cot\left(\frac{180}{8}\right)\times(Side)^2=2\left(\sqrt{2}+1\right)(Side)^2$$

Illustrative Example

Ex.: Find the area of an octagon whose side measures 6 m.

Soln: Applying the above formula, we have the area of an octagon

$$=2\left(\sqrt{2}+1\right)6^2=72\times 2.414=173.82\text{ sq m.}$$

(ii) If area of a regular octagon, is given, then

$$side\ of\ the\ regular\ octagon=\sqrt{\frac{Area}{2\left(\sqrt{2}+1\right)}}\ m$$

Illustrative Example

Ex.: Find to the nearest metre the side of a regular octagonal enclosure whose area is 1 hectare.

Soln: Area of a regular octagon = $2\left(1+\sqrt{2}\right)a^2$

Now, $2\left(1+\sqrt{2}\right)a^2=1$ hectare.

$$a^2=\frac{10000}{2\left(1+\sqrt{2}\right)}\text{ sq m.}\quad\text{or, } a^2=2071\text{ sq m approx.}$$

$\therefore$ a = 46 metres approx.

Exercise

1. Find the area of a regular hexagon whose side measures 8 cm.
a) 166.27 sq cm b) 156.27 sq cm
c) 166.72 sq cm d) 156.72 sq cm
2. Find the area of a regular octagon whose side measures 5 cm.
a) $\left(\sqrt{2}+1\right)10$ sq cm b) $50\left(\sqrt{2}+1\right)$ sq cm
c) $25\left(\sqrt{2}+1\right)$ sq cm d) None of these
3. The area of a regular octagon is 51 sq cm, find its side.
a) 3.25 cm b) 5.25 cm c) 4.25 cm d) 6.25 cm
4. Find to the nearest metre the side of a field which is in the form of a regular hexagon and measures 1 hectare in area.
a) 65 m b) 64 m c) 52 m d) 62 m

Answers

1. a 2. b 3. a
4. d; **Hint:** 1 hectare = 10000 sq m

Rule 72

Theorem: ***There is a regular polygon of 'n' sides. If length of each side is 'a' m, then the sum of the interior angles is given by $(n-2)\pi$; where, $n \geq 3$ and the value of each interior angle is $\left(\frac{n-2}{n}\right)\pi$.***

Illustrative Example

Ex.: Find the value of the sum of interior angles of a regular hexagon. Also find the value of each interior angle.

Soln: Applying the above theorem, we have

sum of the interior angles = $(6-2)\pi = 4\pi$ and

the value of each interior angle = $\left(\frac{6-2}{6}\right)\pi = \frac{2}{3}\pi$

Exercise

1. There is a regular polygon of 12 sides. Find the sum of interior angles and the value of each interior angle.

   a) $10\pi, \frac{5}{6}\pi$ b) $12\pi, \frac{8}{9}\pi$

   c) $6\pi, \frac{5}{6}\pi$ d) Data inadequate

2. There is a regular polygon of 8 sides. Find the sum of interior angles and the value of each interior angle.

   a) $6\pi, \frac{3}{4}\pi$ b) $8\pi, \frac{3}{4}\pi$

   c) $6\pi, \frac{5}{6}\pi$ d) Data inadequate

3. Find the value of the sum of interior angles of a regular pentagon. Also find the value of each interior angle.

   a) $3\pi, \frac{3}{5}\pi$ b) $2\pi, \frac{2}{5}\pi$

   c) $4\pi, \frac{4}{5}\pi$ d) Data inadequate

Answers

1. a 2. a 3. a

Rule 73

Theorem: ***There is a regular polygon of 'n' sides. If length of each side is 'a' metres, then the sum of the exterior angles is 2π and the value of each exterior angle is $\frac{2\pi}{n}$.***

Illustrative Example

Ex.: There is a regular polygon of 8 sides. Find the sum of the exterior angles and the value of each exterior angle.

Soln: Applying the above theorem, we have

sum of the exterior angles = 2π and the

value of each exterior angle = $\frac{2}{8}\pi = \frac{\pi}{4}$

Note: ***There is a regular polygon of 'n' sides and the length of each side is 'a' metres. Then the sum of each exterior and interior angle is given by π and the perimeter of the regular polygon is given by 'na' metres.***

Illustrative Example

Ex.: Find the perimeter of a regular pentagon whose each side measures 6 metres. Also, find the sum of each interior and exterior angles of the regular pentagon.

Soln: Applying the above theorem, we have the perimeter of a pentagon = $5 \times 6 = 30$ metres and the sum of each interior and exterior angle = π

Exercise

1. There is a regular hexagon. Find the value of each exterior angle.

   a) $\frac{\pi}{6}$ b) $\frac{\pi}{3}$ c) 2π d) None of these

2. There is a regular pentagon. Find the value of each exterior angle.

   a) $\frac{2\pi}{5}$ b) $\frac{\pi}{5}$ c) $\frac{\pi}{3}$ d) 2π

3. There is a regular polygon of 12 sides. Find the sum of the exterior angles and the value of each exterior angle.

   a) $2\pi, \frac{\pi}{6}$ b) $2\pi, \frac{\pi}{3}$ c) $2\pi, \frac{\pi}{2}$ d) Data inadequate

Answers

1. b 2. a 3. a

Rule 74

Theorem: ***If a room l metres long, b metres broad and c metres high has***

N windows $(a_1m \times b_1m, a_2m \times b_2, \ldots, a_nm \times b_nm)$ and

M doors $(x_1m \times y_1m, x_2m \times y_2m, \ldots, x_mm \times y_mm)$, then the cost of papering the walls with paper 'd' m wide at Rs X per metre is given by

$$Rs\ \frac{X}{d}[2(l+b)h - N(a_1b_1 + a_2b_2 + \ldots + a_nb_n) - M(x_1y_1 + x_2y_2 + \ldots + x_my_m)].$$

or, $\frac{\text{Cost of paper per metre}}{\text{Width of the paper}} \times$ Net area of the four walls

Illustrative Example

Ex.: A room 8 metres long, 6 metres broad and 3 metres

high has two windows $1\frac{1}{2}$ m × 1 m and a door 2 m × $1\frac{1}{2}$ m. Find the cost of papering the walls with paper 50 cm wide at 25 P per metre.

Soln: **Detail Method:**

Area of walls = 2 (8 + 6)3 = 84 sq m.

Area of two windows and door

$$= 2 \times 1\frac{1}{2} \times 1 + 2 \times 1\frac{1}{2} = 6 \text{ sq m.}$$

Area to be covered = 84 – 6 = 78 sq m.

$$\therefore \text{ length of paper} = \frac{78 \times 100}{50} \text{ m} = 156 \text{ m}$$

$$\therefore \text{ cost} = \text{Rs } \frac{156 \times 25}{100} = \text{Rs } 39$$

Quicker Method: Applying the above theorem, we have,

$X = 25 \text{ P} = \text{Rs } \frac{1}{4}$; $d = 50 \text{ cm} = \frac{1}{2}$ m;

n = 2 and m = 1

Now, required answer

$$= \frac{\frac{1}{4}}{\frac{1}{2}}\left[2(8+6)3 - 2\left(\frac{3}{2}\times 1\right) - 1\left(2\times\frac{3}{2}\right)\right]$$

$$= \frac{1}{2}[84-6] = \frac{1}{2}\times 78 = \text{Rs } 39$$

Exercise

1. The dimensions of a room are 12.5 metres by 9 metres by 7 metres. There are 2 doors and 4 windows in the room; each door measures 2.5 metres by 1.2 metres and each window 1.5 metres by 1 metre. Find the cost of painting the walls at Rs 3.50 per square metre.
   a) Rs 1011.50 b) Rs 1050.50
   c) Rs 1101.50 d) Can't be determined
2. Find the expense of papering a room whose length is 7.2 metres, breadth 6.3 metres and height 4 metres with paper half a metre wide at Rs 18 per metre, allowing for a door 1.8 m by 1.5 m and 4 windows each 1 m by 0.7 m.
   a) Rs 3690 b) Rs 3890 c) Rs 6390 d) Rs 6590
3. A room is 7.5 m long, 5.5 m broad and 5 m high. It has one door 1.6 m broad and 2.5 m high and two windows each 80 cm broad and 1.25 m high. What will be the expenditure in covering the walls by paper 40 cm broad at the rate of 75 paise per metre?
   a) Rs 232.50 b) Rs 230 c) Rs 233 d) Rs 233.50
4. A room is 7.5 m long, 4.5 m broad and 4 m high. It has two doors each 2.5 m long and 1.5 m broad and one window 1.50 m high and 0.8 m broad. What will be the expenditure in colouring the walls at Rs 1.50 per sq m.
   a) Rs 131 b) Rs 131.95
   c) Rs 130.95 d) Data inadequate

Answers

1. a; **Hint:** Net area = [2 (12.5 + 9) 7 - 2 (2.5 × 1.2) - 4 (1.5 × 1)] = 301 - 6 - 6 = 289
   ∴ cost of painting the walls = 289 × 3.50 = Rs 1011.50.
2. a; **Hint:** $\frac{18}{0.5}$ [2 × 4 (7.2 + 6.3) - 1 (1.8 × 1.5) - 4 (1 × 0.7)] = Rs 3690
3. a
4. c

Rule 75

Theorem: ***The radius of a circular wheel is r m. The no. of revolutions it will make in travelling 'd' km is given by*** $\left(\frac{d}{2\pi r}\right)$.

Or

No. of revolutions = $\frac{\text{Distance}}{2\pi r}$

Illustrative Example

Ex.: The radius of a circular wheel is $1\frac{3}{4}$ m. How many revolutions will it make in travelling 11 km?

Soln: **Detail Method:** Distance to be travelled = 11 km = 11000 m

Radius of the wheel = $1\frac{3}{4}$ m

$\therefore$ circumference of the wheel = $2 \times \frac{22}{7} \times 1\frac{3}{4}$ m = 11 m

∴ in travelling 11 m the wheel makes 1 revolution.

∴ in travelling 11000 m the wheel makes $\frac{1}{11} \times 11000$ revolutions, i.e. 1000 revolutions.

Quicker Method: Applying the above theorem,

$$\text{No. of revolutions} = \frac{\text{Distance}}{2\pi r} = \frac{11000}{2\times\frac{22}{7}\times\frac{7}{4}} = 1000.$$

Exercise

1. The diameter of the driving wheel of a bus is 140 cm. How many revolutions per minute must the wheel make in order to keep a speed of 66 km per hour?
   a) 250 b) 300 c) 200 d) 350
2. The diameter of a wheel is 2 cm. It rolls forward covering 10 revolutions. The distance travelled by it is:

a) 3.14 cm b) 62.8 cm c) 31.4 cm d) 125.6 cm
[Railway Recruitment Exam, 1990]

3. If the wheel of the engine of a train $4\frac{2}{7}$ metres in circumference makes seven revolutions in 4 seconds, the speed of the train in km/hr is:
a) 35 b) 32 c) 27 d) 20
(Clerk's Grade Exam 1991)

4. How far has a bicycle travelled when its driving wheel 35 cm in diameter, has made 6300 revolutions?
a) 6930 m b) 6390 m c) 6920 m d) 6830 m

5. The radius of a wheel is 42 cm. How many revolutions will it make in going 26.4 km?
a) 1000 revolutions b) 10000 revolutions
c) 5000 revolutions d) None of these

6. The driving wheel of a locomotive engine 2.1 m in radius makes 75 revolutions in one minute. Find the speed of the train in km/hr.
a) 59.4 km/hr b) 60 km/hr
c) 61.5 km/hr d) None of these

7. A carriage wheel makes 1000 revolutions in going over a distance of 3 km. Find its diameter.
a) $\frac{3}{2\pi}$ m b) $\frac{3}{\pi}$ m c) 3π m d) Data inadequate

Answers

1. a; **Hint:** Distance covered by wheel in 1 minute

$$=\frac{66\times1000\times100}{60}=110000\text{ cm}$$

$$\text{Circumference of wheel}=\left(2\times\frac{22}{7}\times70\right)=440\text{ cm}$$

$$\text{Number of revolutions in 1 minute}=\left(\frac{110000}{440}\right)=250.$$

2. b; **Hint:** Required distance $=\frac{22}{7}\times2\times10=62.8$ cm.

3. c; **Hint:** Distance covered in 4 seconds $=\frac{30}{7}\times7=30$ m

$\therefore$ speed of the engine per second

$$=\frac{30}{4}=\frac{30}{4}\times\frac{18}{5}=27\text{ km/hr}$$

4. a; **Hint:** Distance $=35\times\frac{22}{7}\times6300$ cm

$=693000$ cm $=6930$ m

5. b; **Hint:** Required revolutions $=\frac{2640000}{2\times\frac{22}{7}\times42}=10000.$

6. a; **Hint:** Distance travelled in one minute

$$=75\times2\times\frac{22}{7}\times2.1=990\text{ m}$$

$$\therefore\ \text{speed in m/sec}=\frac{990}{60}=\frac{990}{60}\times\frac{18}{5}=59.4\text{ km/hr.}$$

7. b; **Hint:** $1000=\frac{3\times1000}{\pi\times d}$ $\quad\therefore\ d=\frac{3}{\pi}$ m.

Rule 76

Theorem: ***The circumference of a circular garden is 'c' metres. Inside the garden a road of 'd' metres width runs round it. The area of the ring-shaped road is given by*** $d(c-\pi d)$ ***or*** $\pi d(2r-d)$ $[\because c=2\pi r]$, ***where r = radius of the circle.***

Area of ring-shaped road = width of ring (circumference of the circle − π × width of ring). Or

π × width of ring (2 × radius of the circle − width of ring)

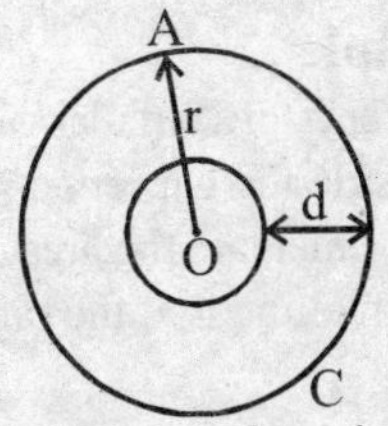

OAC is a circle of radius = r, there is pathway, inside the circle of width = d.

Illustrative Example

Ex.: The circumference of a circular garden is 1012 metres. Inside the garden, a road of 3.5 m width runs round it. Calculate the area of this road.

Soln: Applying the above theorem, we have

$$\text{area of the ring-shaped road}=3.5\left(1012-\frac{22}{7}\times3.5\right)$$

$$=3.5(1012-11)=3.5\times1001=3503.5\text{ sq m.}$$

Exercise

1. The circumference of a circular garden is 512 metres. Inside the garden, a road of 7 m width runs round it. Calculate the area of this road.
a) 4340 sq m b) 3430 sq m c) 3450 sq m d) 3550 sq m

2. The circumference of a circular garden is 644 metres. Inside the garden, a road of 14 m width runs round it. Calculate the area of this road.
a) 8400 sq m b) 4800 sq m c) 6400 sq m d) 7400 sq m

3. The circumference of a circular garden is 1215.4 metres. Inside the garden, a road of 4.9 m width runs round it. Calculate the area of this road.
a) 5800 sq m b) 8500 sq m c) 4800 sq m d) 6800 sq m

Answers

1. b 2. a 3. a

Rule 77

Theorem: ***The circumference of a circular garden is 'c' metres. Outside the garden, a road of 'd' m width runs round it. The area of the ring-shaped road is given by*** $d(c+\pi d)$ ***sq metres. Or*** $\pi d(2r+d)$ $[\because c = 2\pi r]$ ***where r = radius of the circle***

Area of ring-shaped road = width of ring (circumference + π × ***width of ring)***

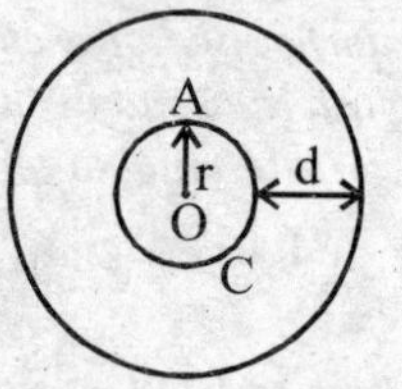

OAC is a circle of radius = r, there is pathway, outside the circle of width d.

Illustrative Example

Ex.: The circumference of a circular garden is 1012 m. Find the area. Outside the garden, a road of 3.5 m width runs round it. Calculate the area of this road and find the cost of gravelling it at the rate of 32 paise per sq m.

Soln: Quicker Method - I:

$$\text{Area} = \frac{(\text{circumference})^2}{4\pi} = \frac{(1012)^2}{4\times\frac{22}{7}} = 81466 \text{ sq cm}$$

Area of ring

$= \pi[(\text{width of ring})(2\times\text{inner radius} + \text{width of ring})]$

$$\text{Now, inner radius} = \sqrt{\frac{\text{Area}}{\pi}} = \sqrt{\frac{81466\times7}{22}} = 161\text{m}$$

$\therefore$ area of ring-shaped road

$$= \frac{22}{7}\times3.5\times(3.5+2\times161)$$

$$= \frac{22}{7}\times3.5\times(3.5+322) = 3580.5 \text{ sq m.}$$

$\therefore$ cost of gravelling $= 3580.5\times0.32 = 1145.76$ rupees.

Quicker Method - II: Applying the above theorem, we have

$$\text{the area of ring-shaped road} = 3.5\left(1012+\frac{22}{7}\times3.5\right)$$

$= 3.5(1012+11) = 3.5\times1023 = 3580.5$ sq m.

$\therefore$ cost of gravelling $= 3580.5\times0.32 = 1145.76$ rupees.

Note: If in the question, in place of circumference, radius is given, **Quicker Method I** will be applied.

Exercise

1. A circular grassy plot of land 70 metres in diameter has a path 7 metres wide running round it on the outside. How many stones 25 cm by 11 cm are needed to pave the path?
   a) 66100 b) 61600 c) 71600 d) 61700
2. A path of 4 metres width runs round a circular grassy plot whose circumference is $75\frac{3}{7}$ metres. Find
   (i) area of the fath
   a) 352 sq m b) 362 sq m c) 532 sq m d) 325 sq m
   (ii) the cost of gravelling the path at Rs 35 per sq metre
   a) Rs 12320 b) Rs 13220 c) Rs 12310 d) Rs 11320
   (iii) the cost of turfing the plot at Rs 21 per sq metre
   (a) Rs 7392 b) Rs 9504 c) Rs 9604 d) Rs 9732
3. The circumference of a circular garden is 165.6 m. Outside the garden, a road of 1.4 m width runs round it. Calculate the area of the road.
   a) 238 sq m b) 228 sq m
   c) 328 sq m d) None of these

Answers

1. b; **Hint:** Area of the path

$= (\pi\times70+\pi\times7)7 = 77\times22 = 1694$ sq m

$= (1694\times100\times100)$ sq cm

Area of the one stone $= (25\times11)$ sq cm

$$\text{Number of stones} = \frac{1694\times100\times100}{25\times11} = 61600$$

2. (i) a (ii) a

(iii) b; **Hint:** $2\pi r = \frac{528}{7}$ or, $r = \frac{528\times7}{7\times2\times22} = 12$ m

Or, area of the plot $= \pi r^2 = \frac{22}{7}\times12\times12$

$\therefore$ required cost $= \frac{22}{7}\times12\times12\times21 = \text{Rs } 9504.$

3. a

Rule 78

Theorem: ***A circular garden has ring-shaped road around it both on its inside and outside, each of width 'd' units. If 'r' is the radius of the garden, then the total area of the path is*** $(4\pi dr)$ ***sq units or 2Cd [where C = perimeter =*** $2\pi r$ ***]***

Illustrative Example

Ex.: A circular park of radius 25 metres has a path of width 3.5 metres all round it. Find the area of the path, if the garden has path both on its outside as well as inside.

Soln: Applying the above theorem, we have the

$$\text{required area} = 4\times\frac{22}{7}\times3.5\times25 = 1100 \text{ sq metres.}$$

Exercise

1. A circular park of radius 24 metres has a path of width 7 metres all round it. Find the area of the path, if the garden has path both on its outside as well as inside.
   a) 2110 sq m b) 2112 sq m
   c) 1221 sq m d) Data inadequate
2. A circular park of radius 15 metres has a path of width 1.4 metres all round it. Find the area of the path, if the garden has path both on its outside as well as inside.
   a) 264 sq m b) 254 sq m
   c) 284 sq m d) Data inadequate
3. A circular park of radius 18 metres has a path of width 4.9 metres all round it. Find the area of the path, if the garden has path both on its outside as well as inside.
   a) 1108.8 sq m b) 1106.8 sq m
   c) 1105.6 sq m d) 1104.8 sq m

Answers

1. b 2. a 3. a

Rule 79

Theorem: ***To find the area of the shaded portion of the following figure.***

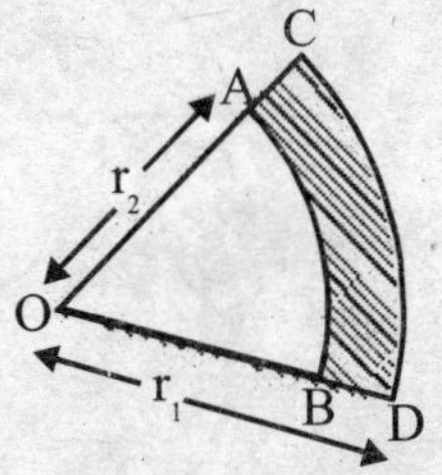

Area of the shaded portion ABCD $= \frac{\theta}{360}\pi\left(r_1^2 - r_2^2\right)$

Illustrative Example

Ex.: Find the area of the shaded portion

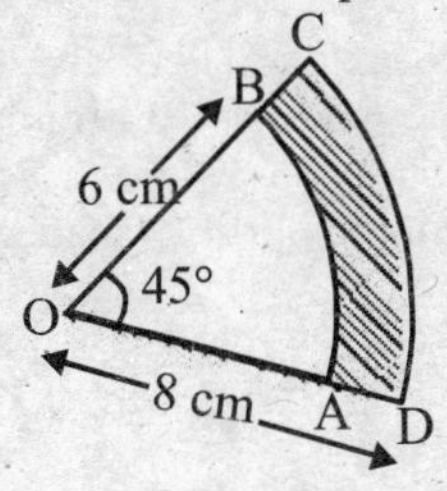

Soln: Detail Method: Area of sector $= \frac{\theta}{360} \times \pi r^2$

Area of sector AOB $= \frac{45}{360} \times \pi \times (6)^2 = \frac{9\pi}{2}$ sq m.

Area of sector DOC $= \frac{45}{360} \times \pi \times (8)^2 = 8\pi$ sq m.

Area of the shaded portion

$= 8\pi - \frac{9\pi}{2} = \pi\left(\frac{7}{2}\right) = \frac{22}{7} \times \frac{7}{2} = 11$ sq metres.

Quicker Method: Applying the above theorem, we have

the area of the shaded portion $= \frac{45}{360} \times \frac{22}{7}\left(8^2 - 6^2\right)$

$= \frac{1}{8} \times \frac{22}{7} \times 28 = 11$ sq m.

Exercise

1. Find the area of the shaded portion.

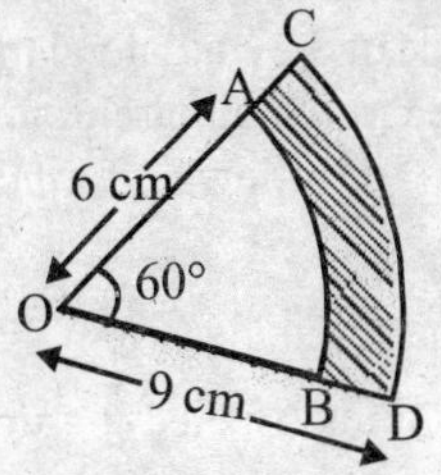

a) $23\frac{4}{7}$ sq m b) $27\frac{4}{7}$ sq m c) $24\frac{5}{7}$ sq m d) $23\frac{5}{7}$ sq m

2. Find the area of the shaded portion.

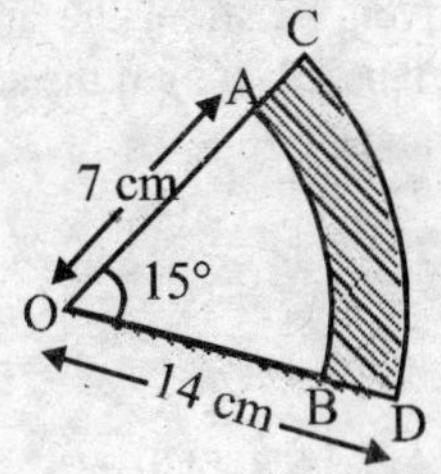

a) 19.25 sq m b) 18.75 sq m
c) 19.75 sq m d) Can't be determined

Answers

1. a 2. a

Rule 80

Theorem: ***There is an equilateral triangle of which each side is x m. With all the three corners as centres, circles are described each of radius $\frac{x}{2}$ m. The area common to all the circles and the triangle is $\frac{1}{8}\pi x^2$ or $\frac{1}{2}\pi$ (radius)^2 and the area of the remaining portion (shaded portion) of the triangle is $\left(\sqrt{3} - \frac{\pi}{2}\right)(\text{radius})^2$ or (0.162) (radius)^2 or, (0.0405)x^2.***

Illustrative Example

Ex.: There is an equilateral triangle of which each side is 2 m. With all the three corners as centres, circles are described each of radius 1 m. (i) Calculate the area common to all the circles and the triangle. (ii) Find the area of the remaining portion of the triangle.

(Take $\pi = 3.1416$)

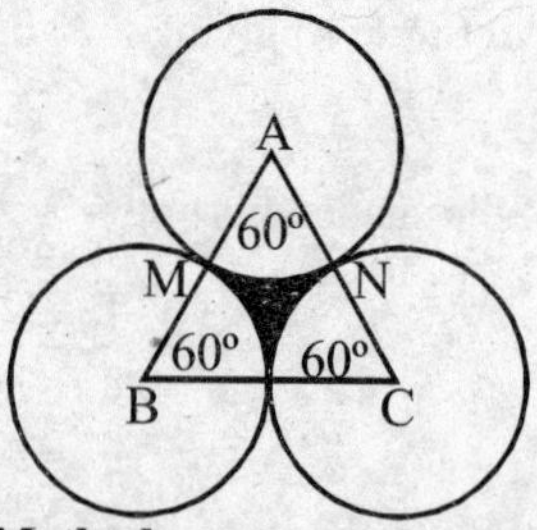

Soln: Quicker Method:
When the side of the equilateral triangle is double the radius of the circles, all circles touch each other and in such cases the following formula may be used:

Area of each sector $= \frac{1}{6} \times \pi \times (\text{radius})^2$

Area of remaining (shaded) portion

$= \left(\sqrt{3} - \frac{\pi}{2}\right)(\text{radius})^2 = (0.162)\,(\text{radius})^2$

(i) In this given question, the area common to all circles and triangle = sum of the area of three sectors AMN, BML and CLN

$= \frac{1}{6}\pi r^2 + \frac{1}{6}\pi r^2 + \frac{1}{6}\pi r^2 = \frac{1}{2}\pi r^2$

$= \frac{1}{2} \times \frac{22}{7} \times (1)^2 = 1.57$ sq m.

(ii) The area of the remaining portion of the triangle = The area of the shaded portion = $0.162 \times (1)^2 = 0.162$ sqm.

Exercise

1. An equilateral triangle has side 4 m. Three circles are drawn from the three vertices of the triangle, each of diameter equal to the side of the triangle. Find the area of the space inside the triangle which is not covered by the circles.
   a) 0.648 sq m b) 0.548 sq m
   c) 6.48 sq m d) Data inadequate
2. There is an equilateral triangle of which each side is 6 m. With all the three corners as centres, circles are described each of radius 3 m. Calculate the area common to all the circles and the triangle.
   (Take $\pi = 3.1416$)
   a) $\frac{9}{2}\pi$ sq m b) 3π sq m
   c) 4π sq m d) Data inadequate
3. There is an equilateral triangle of which each side is 4 m. With all the three corners as centres, circles are described each of radius 2 m. Calculate the area common to all the circles and the triangle.
   (Take $\pi = 3.1416$)
   a) 2π sq m b) $\frac{3}{2}\pi$ sq m c) $\frac{7}{2}\pi$ sq m d) 4π sq m

Answers

1. a 2. a 3. a

Rule 81

Theorem: ***The diameter of a coin is x cm. If four of these coins be placed on a table so that the rim of each touches that of the other two, then the area of the unoccupied space between them is*** $\left(\frac{4-\pi}{4}\right)x^2$ ***or*** $\frac{3}{14}x^2$ ***or*** $(0.215)x^2$ **sq cm**

and area of each sector is given by $\left(\frac{1}{16} \times \pi \times x^2\right)$ **sq cm.**

Illustrative Example

Ex.: The diameter of a coin is 1 cm. If four of these coins be placed on a table so that the rim of each touches that of the other two, find the area of the unoccupied space between them.

(Take $\pi = 3.1416$)

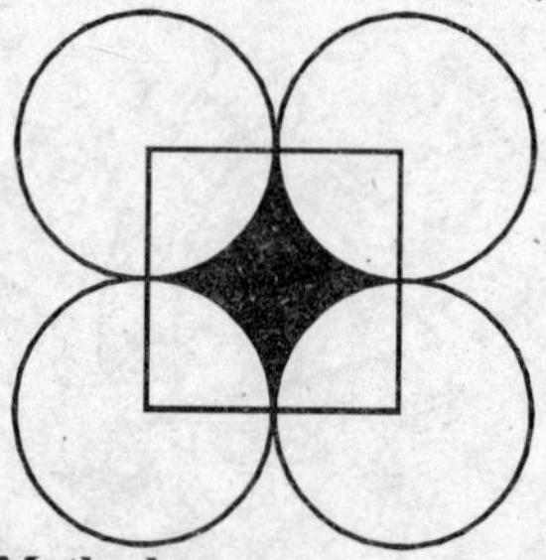

Soln: Quicker Method:
If the circles be placed in such a way that they touch each other, then the square's side is double the radius. In such cases the following formulae may be used:

Area of each sector $= \frac{1}{4} \times \pi \times (\text{radius})^2 = \frac{1}{16} \times \pi \times x^2$

Area of remaining portion (shaded part)

$= (4 - \pi)\,(\text{radius})^2 = \left(\frac{4-\pi}{4}\right)x^2 = (0.86)\,(\text{radius})^2$

Now, in the given question,
area of the unoccupied space

$= (0.86)\,(\text{radius})^2 = (0.215)x^2$

$= (0.86)\left(\frac{1}{2}\right)^2 = 0.215$ sq cm.

Exercise

1. Four circles are drawn from the four corners of a square.

The diameter of each circle is equal to the side of the square and hence the adjacent circles touch each other at the mid-point of the side of the square. If the side of the square is 7 cm, find the area of unoccupied space enclosed between the circles.

a) 38.5 sq cm b) 77 sq cm
c) 39.5 sq cm d) None of these

2. In the figure given below ABCD is a square and the circle are all congruent, each having its radius equal to 7 cm. Find the area of the shaded region in sq cm.

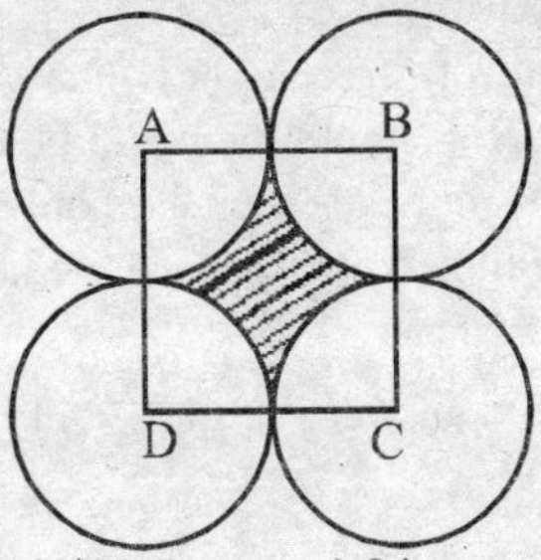

a) 42 sq cm b) 38.5 sq cm c) 84 sq cm d) 24 sq cm

3. The given figure represents a square of side 4 cm. At its 4 corners, circles of equal radii are drawn. What is the area of the shaded portion?

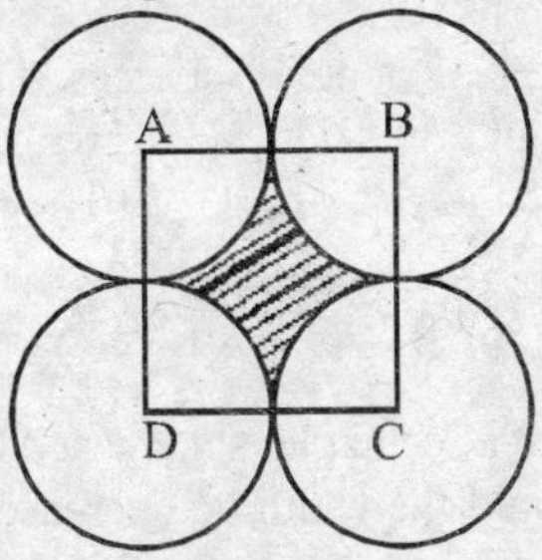

a) $3\frac{3}{7}$ sq cm b) $4\frac{4}{7}$ sq cm c) $3\frac{4}{7}$ sq cm d) $4\frac{3}{7}$ sq cm

Answers

1. a 2. a 3. a

Rule 82

Theorem: ***If the length of a rectangle is increased by x%, then the percentage decrease in width, to maintain the same area, is given by*** $\left(\frac{x}{100+x}\right)100$.

Illustrative Example

Ex.: The length of a rectangle is increased by 60%. By what per cent should the width be decreased to maintain the same area?

Soln: **Detail Method:** Let the length and breadth of the rectangle be *x* and *y*.

Then, its area = *xy*

New length = $x\left(\frac{160}{100}\right)=\frac{8x}{5}$

As the area remains the same, the new breadth of the rectangle $=\frac{xy}{\frac{8x}{5}}=\frac{5y}{8}$

$\therefore$ decrease in breadth $= y-\frac{5y}{8}=\frac{3y}{8}$

$\therefore$ % decrease in breadth $=\frac{3y\times100}{8\times y}=\frac{75}{2}=37\frac{1}{2}\%$

Quicker Method: You must have gone through similar examples in the chapter 'Percentage'. Applying the above theorem,

Required percentage decrease in breadth

$=60\left(\frac{100}{100+60}\right)=\frac{75}{2}=37\frac{1}{2}\%$.

Exercise

1. The length of a rectangle is increased by 20%. By what per cent should the width be decreased to maintain the same area?

a) $16\frac{2}{3}\%$ b) $26\frac{2}{3}\%$

c) $33\frac{1}{3}\%$ d) Data inadequate

2. The length of a rectangle is increased by 25%. By what per cent should the width be decreased to maintain the same area?

a) 20% b) 25% c) 16% d) 24%

3. The length of a rectangle is increased by 30%. By what per cent should the width be decreased to maintain the same area?

a) $\frac{100}{13}\%$ b) $\frac{200}{13}\%$ c) $\frac{300}{13}\%$ d) None of these

Answers

1. a 2. a 3. c

Rule 83

Theorem: ***If the length of a rectangle is increased by x%, then the percentage decrease in width, to reduce the area by y%, is given by*** $\left(\frac{x+y}{100+x}\right)\times100$.

Illustrative Example

Ex.: The length of a rectangle is increased by 20%. By what per cent should the width be decreased so that area of the rectangle decreases by 20%?

Soln: **Detail Method:** Let the length and width of the rectangle be *x* m and y m respectively.

Initial area $= x \times y = xy$ sq m.

Now, length is increased by 20%, therefore new length

$= x \times \frac{120}{100} = \frac{6x}{5}$ m.

Again, we suppose that the percentage decrease in width is d.

$\therefore$ new width $= y\left(\frac{100-d}{100}\right)$

New area $= xy\left[\frac{6}{5}\left(\frac{100-d}{100}\right)\right]$

Decrease in area $=$ Initial area $-$ New area

$= xy - xy\left[\frac{6}{5}\left(\frac{100-d}{100}\right)\right]$ or, $xy\left[1-\frac{600-6d}{500}\right]$

Percentage decrease in area

$= \frac{xy\left[1-\frac{600-6d}{500}\right]}{xy} \times 100$

As per the question, $\frac{xy\left[1-\frac{600-6d}{500}\right]}{xy} \times 100 = 20$

or, $\frac{500-600+6d}{5} = 20$ or, $6d = 200$

$\therefore d = \frac{200}{6} = 33\frac{1}{3}\%$

$\therefore$ required answer $= 33\frac{1}{3}\%$.

Quicker Method: Applying the above theorem,

Required answer

$= \frac{20+20}{100+20} \times 100 = \frac{400}{12} = \frac{100}{3} = 33\frac{1}{3}\%$.

Exercise

1. The length of a rectangle is increased by 25%. By what per cent should the width be decreased so that area of the rectangle decreases by 20%?
   a) 36% b) 30% c) 35% d) None of these
2. The length of a rectangle is increased by 50%. By what per cent should the width be decreased so that area of the rectangle decreases by 10%?
   a) 100% b) 20% c) 120% d) 125%
3. The length of a rectangle is increased by 25%. By what per cent should the width be decreased so that area of the rectangle decreases by 25%?
   a) 40% b) 20% c) 35% d) 36%

Answers

1. a 2. c 3. a

Rule 84

Theorem: ***If the length of a rectangle is increased by x%, then the percentage decrease in width, to increase the area by y%, is given by*** $\left(\frac{\text{Diff. in x and y}}{100+x}\right) \times 100$.

Illustrative Example

Ex.: The length of a rectangle is increased by 20%. By what per cent should the width be decreased so that area of the rectangle increases only by 4%.

Soln: Applying the above theorem, we have

the required answer $= \left(\frac{20-4}{100+20}\right) \times 100 = 13\frac{1}{3}\%$.

Exercise

1. The length of a rectangle is increased by 25%. By what per cent should the width be decreased so that area of the rectangle increases only by 5%.
   a) 12% b) 16% c) 18% d) 24%
2. The length of a rectangle is increased by 20%. By what per cent should the width be decreased so that area of the rectangle increases only by 10%.
   a) $8\frac{1}{3}\%$ b) 9% c) $8\frac{2}{3}\%$ d) $9\frac{2}{3}\%$
3. The length of a rectangle is increased by 50%. By what per cent should the width be decreased so that area of the rectangle increases only by 25%.
   a) $16\frac{2}{3}\%$ b) $18\frac{3}{4}\%$ c) $26\frac{2}{3}\%$ d) Data inadequate

Answers

1. b 2. a 3. a

Rule 85

Theorem: ***If the length of a rectangle is decreased by x%, then the percentage increase in width, to increase the area by y%, is given by*** $\left(\frac{x+y}{100-x}\right) \times 100$.

Illustrative Example

Ex.: The length of a rectangle is decreased by 20%. By what per cent should the width be increased, so that area of the rectangle increases by 20%?

Soln: Applying the above theorem, we have

the required answer

$$= \left(\frac{20+20}{100-20}\right)\times 100 = \frac{40}{80}\times 100 = 50\%$$

Exercise

1. The length of a rectangle is decreased by 25%. By what per cent should the width be increased, so that area of the rectangle increases by 25%?

   a) $66\frac{2}{3}\%$ b) $78\frac{2}{3}\%$ c) $26\frac{2}{3}\%$ d) $16\frac{2}{3}\%$

2. The length of a rectangle is decreased by 20%. By what per cent should the width be increased, so that area of the rectangle increases by 25%?

   a) $56\frac{1}{4}\%$ b) $58\frac{2}{3}\%$ c) $58\frac{1}{4}\%$ d) 58%

3. The length of a rectangle is decreased by 25%. By what per cent should the width be increased, so that area of the rectangle increases by 50%?

   a) 100% b) 175% c) 125% d) 80%

Answers

1. a 2. a 3. a

Rule 86

Theorem: ***If the length of a rectangle is decreased by x%, then the percentage increase in width, to maintain the same area, is given by*** $\left(\frac{x}{100-x}\right)100$.

Illustrative Example

Ex.: If the length of a rectangle is decreased by 20%, by what per cent should the width be increased to maintain the same area?

Soln: Apply the above rule, we have the required percentage increase in breadth

$$= 20\left(\frac{100}{100-20}\right) = 25\%$$

Note: To find the above formula, we have used the rule of fraction.

Exercise

1. If the length of a rectangle is decreased by 25%, by what per cent should the width be increased to maintain the same area?

   a) $33\frac{1}{3}\%$ b) $23\frac{1}{3}\%$ c) $16\frac{2}{3}\%$ d) None of these

2. If the length of a rectangle is decreased by 40%, by what per cent should the width be increased to maintain the same area?

   a) $66\frac{2}{3}\%$ b) $76\frac{1}{3}\%$ c) $56\frac{2}{3}\%$ d) 67%

3. If the length of a rectangle is decreased by 30%, by what per cent should the width be increased to maintain the same area?

   a) 40% b) $43\frac{6}{7}\%$

   c) $46\frac{2}{7}\%$ d) $42\frac{6}{7}\%$

Answers

1. a 2. a 3. d

Rule 87

Theorem: ***If length and breadth of a rectangle is increased x and y per cent respectively, then area is increased by*** $\left(x+y+\frac{xy}{100}\right)\%$.

Note: If any of the two measuring sides of rectangle is decreased then put negative value for that in the given formula.

Illustrative Examples

Ex. 1: If the length and the breadth of a rectangle is increased by 5% and 4% respectively, then by what per cent does the area of that rectangle increase?

Soln: **By Direct Formula:**

$$\% \text{ increase in area} = 5+4+\frac{5\times 4}{100} = 9+0.2 = 9.2\%$$

Ex. 2: If the length of a rectangle increases by 10% and the breadth of that rectangle decreases by 12%, then find the % change in area.

Soln: Since breadth decreases by y = –12, then % change in area

$$= 10+12+\frac{10\times(-12)}{100} = -2-1.2 = -3.2\%$$

Since there is –ve sign, the area decreases by 3.2%.

Exercise

1. If the height of a triangle is decreased by 40% and its base is increased by 40%. What will be the effect on its area?

   a) No change b) 16% increase
   c) 8% decrease d) 16% decrease

   [SBI PO Exam 1999]

2. If the length of a rectangle is increased by 20% and the breadth reduced by 20%, what will be the effect on its area?

   a) 4% increase b) 6% increase
   c) 4% decrease d) No change

   [BSRB Guwahati PO Exam 1999]

3. If the length of a rectangle is increased by $12\frac{1}{2}$% and the width increased by $6\frac{1}{4}$%, the area of the rectangle will:

a) increase by 19.53% b) decrease by $6\frac{1}{4}$%

c) increase by $19\frac{3}{4}$% d) increase by 19.92%

4. The length and breadth of a square are increased by 40% and 30% respectively. The area of resulting rectangle exceeds the area of the square by:
a) 42% b) 62% c) 82% d) None of these
[I. Tax & Central Excise 1988]

5. The length of a square is increased by 40% while breadth is decreased by 40%. The ratio of area of the resulting rectangle so formed to that of the original square is:
a) 25 : 21 b) 21 : 25 c) 16 : 15 d) 15 : 16
[I. Tax & Central Excise 1989]

6. If each of the dimensions of a rectangle is increased by 100%, its area is increased by:
a) 100% b) 200% c) 300% d) 400%

Answers

1. d; **Hint:** The given rule is applicable for any two dimensional figure. Hence,

the required effect = $\left(+40-40-\frac{40\times40}{100}\right)\% = -16\%$

ie the area will decrease by 16%.

2. c; **Hint:** Required effect = $\left(20-20-\frac{20\times20}{100}\right)\% = -4\%$

= 4% decrease.

3. a

4. c; **Hint:** In this case also the given rule will be applied.

5. b; **Hint:** Change in area = $\left(+40-40-\frac{40\times40}{100}\right)\% = -16\%$

ie area of the resulting rectangle is decreased by 16%.

$\therefore$ required ratio = $\frac{100-16}{100} = \frac{84}{100} = 21:25.$

6. c

Rule 88

Theorem: ***If all the measuring sides of any two dimensional figure is changed by x%, then its area changes by*** $\left(2x+\frac{x^2}{100}\right)\%$.

Illustrative Examples

Ex. 1: If radius of a circle is increased by 5%, find the percentage increase in its area.

Soln: By the theorem:

% increase in its area = $2\times5+\frac{5^2}{100}$

$= 10 + 0.25 = 10.25\%$

Ex. 2: If all the sides of a hexagon (six-sided figure) is increased by 2%, find the % increase in its area.

Soln: Required % increase $= 2\times2+\frac{2^2}{100} = 4+0.04 = 4.04\%$

Note: **1.** Whenever there is decrease, use –ve value for x. Whenever you get the –ve value, don't hesitate to say that there is decrease in the area.

2. If there is decrease in the above cases, find the percentage decrease in area, then answer for above two examples is 9.75% and 3.96% respectively.

Ex. 1: % decrease in its area

$= 2\times(-5)+\frac{(-5)^2}{100} = -10+0.25 = -9.75\%$

–ve sign shows that there is a decrease.

Ex. 2: % decrease in its area

$= 2\times(-2)+\frac{(-2)^2}{100} = -4+0.46 = -3.96\%$

–ve sign shows that there is a decrease.

Exercise

1. If the side of a square be increased by 50%, the per cent increase in area is:
a) 50 b) 100 c) 125 d) 150
(NDA Exam 1987)

2. Of the two square fields, the area of one is 1 hectare, while the other one is broader by 1%. The difference in areas is:
a) 101 m^2 b) 201 m^2 c) 100 m^2 d) 200 m^2

3. If radius of a circle is increased by 20%, find the percentage increase in its area.
a) 40% b) 41% c) 44% d) 43%

4. If radius of a circle is increased by 25%, find the percentage increase in its area.
a) $56\frac{1}{2}$% b) $66\frac{1}{4}$% c) $56\frac{1}{4}$% d) $56\frac{3}{4}$%

5. If all the sides of a hexagon (six-sided figure) is increased by 1%, find the % increase in its area.
a) 2% b) 2.1% c) 2.01% d) 3%

6. If all the sides of a octagon (eight-sided figure) is increased by 3%, find the % increase in its area.
a) 6% b) 6.3% c) 6.03% d) 6.09%

Answers

1. c

2. b; **Hint:** % increase $= 2\times1+\frac{1^2}{100}=\frac{201}{100}\%$

1 hectare = 10000 sq m

$\therefore$ increse in area $=\frac{201\times10000}{100\times100}=201$ sq m = defference in areas.

3. c 4. c 5. c 6. d

Rule 89

Theorem: ***If all the measuring sides of any two-dimensional figure are changed (increased or decreased) by x% then its perimeter also changes by the same, ie, x%.***

Illustrative Example

Ex.: If diameter of a circle is increased by 12%, find the % increase in its circumference.

Soln: Although diameter is rarely used as the measuring side of a circle, the above theorem holds good for it. Thus, by the theorem, % increase in circumference = 12%.

Exercise

1. If the radius of a circle is increased by 5% find the % increase in its cirucmference.
a) 10% b) 5.01% c) 10.25% d) 5%
2. If the side of a square is decreased by 25.5%, find the % decrease in its circumference.
a) 50% b) 25.5% c) 26% d) 26.5%
3. If the sides of an equilateral triangle is increased by 2%, find the % increase in its circumference.
a) 2% b) 4% c) 6% d) 5%

Answers

1. d 2. b 3. a

Rule 90

Theorem: ***If all sides of a quadrilateral are increased by x% then its corresponding diagonals also increased by x%.***

Illustrative Examples

Ex. 1: If the sides of a rectangle are increased each by 10%, find the percentage increase in its diagonals.

Soln: Required % increase in diagonals = 10%.

Ex. 2: If the length and the two diagonals of a rectangle are each increased by 9%, then find the % increase in its breadth.

Soln: From the above theorem it can be concluded that its breadth also increases by the same value, i.e. 9%.

Exercise

1. If the sides of a rectangle are increased each by 5%, find the percentage increase in its diagonals.
a) 10% b) 15% c) 2.5% d) 5%
2. If the sides of a rectangle are increased each by 6%, find the percentage increase in its diagonals.
a) 6% b) 6.5% c) 10% d) None of these
3. If the length and the two diagonals of a rectangle are each increased by 8%, then find the % increase in its breadth.
a) 8% b) 4% c) 16% d) No change
4. If the length and the two diagonals of a rectangle are each increased by 19%, then find the % increase in its breadth.
a) 9.5% b) 28%
c) 19% d) Can't be determined

Answers

1. d 2. a 3. a 4. c

Rule 91

Theorem: ***If a parallelogram, the length of whose sides are x cm and y cm, has one diagonal z cm long, then the length of the other diagonal is*** $\left[\sqrt{2\left(x^2+y^2-\frac{z^2}{2}\right)}\right]$ ***m.***

Illustrative Example

Ex.: A parallelogram, the length of whose sides are 12 cm and 8 cm, has one diagonal 10 cm long. Find the length of the other diagonal.

Soln: **Detail Method:**

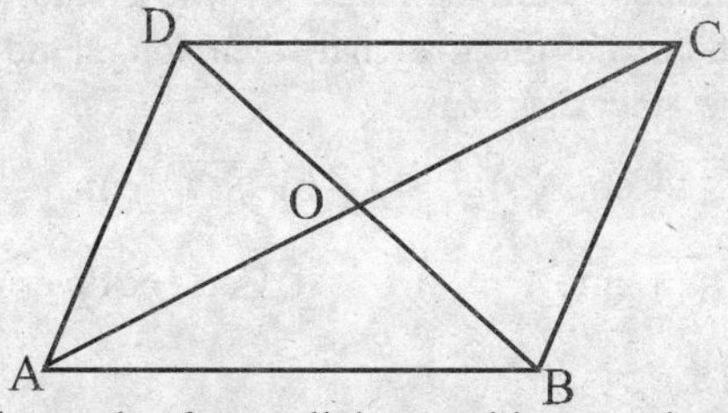

Diagonals of a parallelogram bisect each other.

Let BD = 10 cm. $\therefore$ OB = 5 cm

In triangle ABC, O is the mid-point of AC.

By a very important theorem in plane geometry, we have in triangle ABC

$$AB^2+BC^2=2\left(OB^2+AO^2\right)$$

$$\Rightarrow 12^2+8^2=2\left(5^2+AO^2\right)$$

$$\Rightarrow 144+64=50+2AO^2 \Rightarrow AO^2=79$$

$\therefore$ AO = 8.9 (approximately)

$\therefore$ the other diagonal = AC = 2AO = 2 × 8.9 = 17.8 cm.

Quicker Method:

By the above mentioned theorem, we have

$$AB^2+BC^2=2\left(OB^2+AO\right)^2$$

or, $2AO^2=AB^2+BC^2-2(OB)^2$

$\therefore AO = \sqrt{\frac{1}{2}\{AB^2 + BC^2 - 2(OB)^2\}}$

$\therefore$ Other diagonal

$= 2AO = 2\sqrt{\frac{1}{2}\{AB^2 + BC^2 - 2(OB)^2\}}$

or, Other diagonal $= \sqrt{2\{AB^2 + BC^2 - 2(OB)^2\}}$

$= \left(\sqrt{2\left(x^2 + y^2 - \frac{z^2}{2}\right)}\right)$ m

Thus, in this case, other diagonal

$= \sqrt{\{144 + 64 - 2 \times 25\}}$

$= \sqrt{2 \times 158} = \sqrt{316} = 17.8$ (approx.)

Exercise

1. A parallelogram, the length of whose sides are 15 cm and 10 cm, has one diagonal 12 cm long. Find the length of the other diagonal.
   a) $\sqrt{255}$ cm b) 17 cm c) $\sqrt{253}$ cm d) $\sqrt{506}$ cm
2. A parallelogram, the length of whose sides are 18 cm and 12 cm, has one diagonal 14 cm long. Find the length of the other diagonal.
   a) 27 cm b) 27.2 cm c) $\sqrt{640}$ cm d) 28.5 cm
3. A parallelogram, the length of whose sides are 16 cm and 8 cm, has one diagonal 10 cm long. Find the length of the other diagonal.
   a) $\sqrt{540}$ cm b) $\sqrt{270}$ cm
   c) $\sqrt{640}$ cm d) Can't be determined

Answers

1. d 2. b 3. a

Rule 92

Theorem: ***If a semi-circle is constructed on each side of a square of length x m, then the area of the whole figure is given by*** $\left[\frac{x^2}{2}(2+\pi)\right]$ ***sq m.***

Illustrative Example

Ex.: A semi-circle is constructed on each side of a square of length 2 m. Find the area of the whole figure.

Soln: **Detail Method:** Total area = Area of square + 4(Area of a semi-circle)

$= 2^2 + 4\left(\frac{1}{2}\pi r^2\right) = (4 + 2\pi)m^2$ (radius $= \frac{2}{2} = 1$)

Quicker Method: Applying the above theorem,

required area $= \frac{2^2}{2}(2+\pi) = (4 + 2\pi)$ sq m.

Exercise

1. A semi-circle is constructed on each side of a square of length 1 m. Find the area of the whole figure.
   a) $\left(1+\frac{\pi}{2}\right)$ m b) $\left(1-\frac{\pi}{2}\right)$ m
   c) $\frac{\pi}{2}$ m d) Can't be determined
2. A semi-circle is constructed on each side of a square of length 4 m. Find the area of the whole figure.
   a) $8(2+\pi)$ m b) $2(2+\pi)$ m
   c) 8π m d) $16(2+\pi)$ m
3. A semi-circle is constructed on each side of a square of length 6 m. Find the area of the whole figure.
   a) $18(2+\pi)$ m b) $16(2+\pi)$ m
   c) 16π m d) Can't be determined

Answers

1. a 2. a 3. a

Rule 93

Theorem: ***If the radius of a circle is decreased by 'x' metres, then the ratio of the area of the original circle to the reduced circle becomes a : b. The radius is given by*** $\left(\frac{x}{1-\sqrt{\frac{a}{b}}}\right)$ ***metres.***

Illustrative Example

Ex.: The area of a circle is halved when its radius is decreased by n. Find its radius.

Soln: By the question we have,

$\frac{\pi(r-n)^2}{\pi r^2} = \frac{1}{2}$ or, $r^2 = 2(r-n)^2$

or, $r^2 - \{\sqrt{2}(r-n)\}^2 = 0$

or, $\{r - \sqrt{2}(r-n)\}\{r + \sqrt{2}(r-n)\} = 0$

Since $r + \sqrt{2}(r-n) \neq 0$, we have

$r - \sqrt{2}(r-n) = 0$ or, $r(\sqrt{2}-1) = \sqrt{2}n$

$\therefore r = \frac{\sqrt{2}n}{\sqrt{2}-1}$

Quicker Method - I: We have, $\frac{\pi(r-n)^2}{\pi r^2}=\frac{1}{2}$

or, $\left\{\frac{\sqrt{2}(r-n)}{r}\right\}^2=1$ or, $\frac{\sqrt{2}(r-n)}{r}=1$

or, $r(\sqrt{2}-1)=\sqrt{2}n$ $\therefore r=\frac{\sqrt{2}n}{\sqrt{2}-1}$

Quicker Method - II: Applying the above theorem, we have

$$\text{radius}=\frac{n}{1-\frac{1}{\sqrt{2}}}=\frac{\sqrt{2}n}{\sqrt{2}-1}.$$

Exercise

1. The radius of a circle is decreased by 2 m, then the ratio of the area of the original circle to the reduced circle becomes 1 : 4. Find its radius.
   a) 4 m b) 6 m c) 3 m d) 2 m
2. The radius of a circle is decreased by 3 m, then the ratio of the area of the original circle to the reduced circle becomes 1 : 3. Find its radius.
   a) $\frac{\sqrt{3}}{\sqrt{3}-1}$ m b) $\frac{3\sqrt{3}}{1-\sqrt{3}}$ m
   c) $\frac{3\sqrt{3}}{\sqrt{3}-1}$ m d) Can't be determined
3. The radius of a circle is decreased by 4 m, then the ratio of the area of the original circle to the reduced circle becomes 4 : 9. Find its radius.
   a) 12 m b) 16 m c) 14 m d) 20 m

Answers

1. a 2. c 3. a

Rule 94

Theorem: ***If the area of a square is x sq units, then area of the circle formed by the same perimeter is given by*** $\frac{4x}{\pi}$ ***sq units. Or*** $\left(\frac{14}{11}x\right)$ ***sq units***

Illustrative Example

Ex.: A cord is in the form of a square enclosing an area of 22 cm². If the same cord is bent into a circle, then find the area of that circle.

Soln: **Detail Method:** Area of square = 22 cm²

$\therefore$ Perimeter of the square = $4\sqrt{22}$ cm

Now, this perimeter is the circumference of the circle.

$\therefore$ circumference of the circle

$$=2\pi r=4\sqrt{22} \quad \therefore r=\frac{2\sqrt{22}}{\pi}$$

$\therefore$ area of the circle

$$=\pi r^2=\pi\left(\frac{2\sqrt{22}}{\pi}\right)^2=\frac{\pi\times4\times22}{\pi^2}$$

$$=\frac{4\times22}{\pi}=\frac{4\times22\times7}{22}=28\text{ cm}^2$$

Quicker Method: Applying the above theorem, we have

$$\text{area of circle}=\frac{4\times22\times7}{22}=28\text{ cm}^2.$$

Exercise

1. A cord is in the form of a square enclosing an area of 2.2 sq m. If the same cord is bent into a circle, then find the area of that circle.
   a) 2.8 sq m b) 3.8 sq m c) 2.9 sq m d) 3 sq m
2. A cord is in the form of a square enclosing an area of 11 sq cm. If the same cord is bent into a circle, then find the area of that circle.
   a) 22 sq cm b) 1.4 sq cm c) 14 sq cm d) 28 sq cm
3. A cord is in the form of a square enclosing an area of 33 sq cm. If the same cord is bent into a circle, then find the area of that circle.
   a) 42 sq cm b) 48 sq cm c) 24 sq cm d) 32 sq cm
4. A cord is in the form of a square enclosing an area of 4.4 sq m. If the same cord is bent into a circle, then find the area of that circle.
   a) 5.6 sq m b) 56 sq m c) 0.56 sq m d) 6.5 sq m

Answers

1. a 2. c 3. a 4. a

Rule 95

Theorem: ***Two poles 'x' m and 'y' m high stand upright. If there feet be 'z' m apart, then the distance between their tops is*** $\left(\sqrt{z^2+(y-x)^2}\right)$ ***metres.***

Illustrative Example

Ex.: Two poles 15 m and 30 m high stand upright in a playground. If their feet be 36 m apart, find the distance between their tops.

Soln: **Detail Method:** From the figure it is required to find the length CD.

We have CA = LB = 15 m

$\therefore$ LD = BD – LB = 15 m

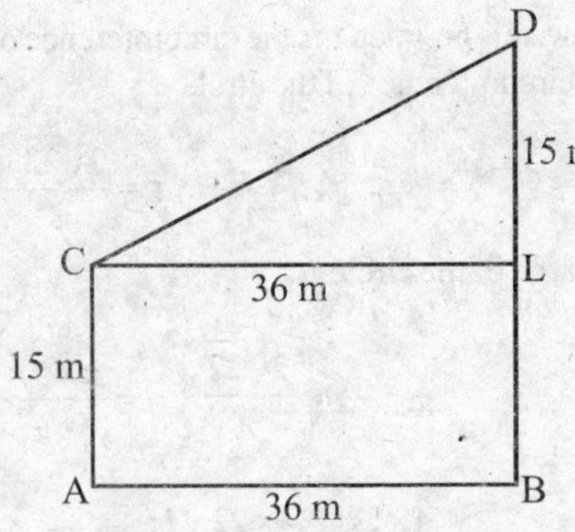

$\therefore$ CD $= \sqrt{CL^2 + DL^2} = \sqrt{36^2 + 15^2} = \sqrt{1521} = 39$ cm.

Quicker Method: Applying the above theorem, we have
the distance between their tops

$= \sqrt{36^2 + (30-15)^2} = \sqrt{1521} = 39$ cm

Exercise

1. Two poles 12 m and 18 m high stand upright in a play-ground. If their feet be 8 m apart, find the distance between their tops.
   a) 10 m b) 12 m
   c) 6 m d) Can't be determined
2. Two poles 7 m and 11 m high stand upright in a play-ground. If their feet be 3 m apart, find the distance between their tops.
   a) 8 m b) 6 m c) 5 m c) 9 m
3. Two poles 14 m and 32 m high stand upright in a play-ground. If their feet be 24 m apart, find the distance between their tops.
   a) 25 m b) 28 m c) 30 m d) 24 m

Answers

1. a 2. c 3. c

Rule 96

Theorem: ***Area of a square inscribed in a circle of radius r is $2r^2$ and side of a square inscribed in a circle of radius r is $\sqrt{2}r$.***

Note: Such a square is the largest quadrilateral inscribed in a circle.

Illustrative Example

Ex.: The circumference of a circle is 100 cm. Find the side of the square inscribed in the circle.

Soln: Circumference of the circle $= 2\pi r = 100$

$\therefore r = \frac{50}{\pi}$

$\therefore$ side of the inscribed square $= \sqrt{2}r = \sqrt{2} \times \frac{50}{\pi}$

Exercise

1. The circumference of a circle is 22 cm. Find the area of the square inscribed in the circle.
   a) 24 sq cm b) 24.2 sq cm
   c) 24.5 sq cm d) None of these
2. The circumference of a circle is 44 cm. Find the side of the square inscribed in the circle.
   a) $14\sqrt{2}$ cm b) $7\sqrt{2}$ cm
   c) 7 cm d) 14 cm
3. The circumference of a circle is 50 cm. Find the area of the square inscribed in the circle.
   a) $\frac{250\sqrt{2}}{\pi}$ sq cm b) $\frac{1250}{\pi^2}$ sq cm
   c) $\frac{50}{\pi}$ sq cm d) $\frac{2500}{\pi^2}$ sq cm

Answers

1. c 2. b 3. b

Rule 97

Theorem: ***The area of the largest triangle inscribed in a semi-circle of radius r is r^2.***

Illustrative Example

Ex.: The largest triangle is inscribed in a semi-circle of radius 14 cm. Find the area of the triangle.

Soln: Applying the above theorem, we have the required area $= (14)^2 = 196$ sq cm.

Exercise

1. The largest triangle is inscribed in a semi-circle of radius 4 cm. Find the area of the triangle.
   a) 16 sq cm b) 8 sq cm
   c) 12 sq cm d) Data inadequate
2. The largest triangle is inscribed in a semi-circle of radius 15 cm. Find the area of the triangle.
   a) 30 sq cm b) 225 sq cm
   c) 310 sq cm d) 350 sq cm
3. The largest triangle is inscribed in a semi-circle of radius 12 cm. Find the area of the triangle.
   a) 24 sq cm b) 144 sq cm
   c) 288 sq cm d) None of these

Answers

1. a 2. b 3. b

Rule 98

Theorem: ***If the largest triangle is inscribed in a semi-circle of radius r cm, then the area inside the semi-circle which is not occupied by the triangle is $\left(\frac{4}{7}r^2\right)$ sq cm.***

Illustrative Example

Ex.: The largest triangle is inscribed in a semi-circle of radius 14 cm. Find the area inside the semi-circle which is not occupied by the triangle.

Soln: **Detail Method:**

Such area = Area of semicircle – Area of such largest triangle

$$= \frac{\pi}{2}r^2 - r^2 = r^2\left(\frac{\pi}{2}-1\right) = 14^2 \times \frac{(22-14)}{14} = 112 \text{ cm}^2$$

Quicker Method: Applying the above theorem, we have

the required area $= \frac{4}{7}\times 14^2 = \frac{4}{7}\times 14\times 14 = 112$ sq cm.

Exercise

1. The largest triangle is inscribed in a semi-circle of radius 7 cm. Find the area inside the semi-circle which is not occupied by the triangle.
   a) 28 sq cm b) 35 sq cm
   c) 24 sq cm d) Data inadequate
2. The largest triangle is inscribed in a semi-circle of radius 21 cm. Find the area inside the semi-circle which is not occupied by the triangle.
   a) 352 sq cm b) 253 sq cm c) 252 sq cm d) 254 sq cm
3. The largest triangle is inscribed in a semi-circle of radius 28 cm. Find the area inside the semi-circle which is not occupied by the triangle.
   a) 448 sq cm b) 484 sq cm
   c) 844 sq cm d) None of these

Answers

1\. a 2. c 3. a

Rule 99

Theorem: ***The area of the largest circle that can be drawn in a square of side x is*** $\pi\left(\frac{x}{2}\right)^2$.

Illustrative Example

Ex.: Find the area of the largest circle that can be drawn in a square of side 14 cm.

Soln: By the formula, we have

the required area $= \pi\left(\frac{14}{2}\right)^2 = \frac{22}{7}\times 7^2 = 154$ cm².

Exercise

1. Find the area of the largest circle that can be drawn in a square of side 7 cm.
   a) 38.5 sq cm b) 42 sq cm c) 35 sq cm d) 42.5 sq cm
2. Find the area of the largest circle that can be drawn in a square of side 28 cm.
   a) 516 sq cm b) 616 sq cm
   c) 615 sq cm d) Data inadequate
3. Find the area of the largest circle that can be drawn in a square of side 2 m.
   a) π sq cm b) 2π sq cm
   c) $\frac{\pi}{2}$ sq cm d) Data inadequate
4. Find the area of the largest circle that can be drawn in a square of side 1 m.
   a) $\frac{\pi}{4}$ sq m b) π sq m c) 4π sq m d) $\frac{\pi}{2}$ sq m

Answers

1\. a 2. b 3. a 4. a

Rule 100

Theorem: ***To find the area of the quadrilateral when its any diagonal and the perpendiculars drawn on this diagonal from other two vertices are given.***

Area of the quadrilateral $= \frac{1}{2}\times$ ***any diagonal × (sum of perpendiculars drawn on diagonal from two vertices)***

Illustrative Example

Ex.: In a quadrilateral, the length of one of its diagonals is 23 cm and the perpendiculars drawn on this diagonal from other two vertices measure 17 cm and 7 cm respectively. Find the area of the quadrilateral.

Soln: In any quadrilateral,

Area of the quadrilateral $= \frac{1}{2}\times$ any diagonal × (sum of perpendi-culars drawn on diagonal from two vertices)

$$= \frac{1}{2}\times D\times(P_1+P_2)$$

$$= \frac{1}{2}\times 23\times(17+7) = 12\times 23 = 276 \text{ sq cm.}$$

Exercise

1. Find the area of a quadrilateral piece of ground one of whose diagonals is 50 metres long and the lengths of perpendiculars from the other two vertices are 29 and 21 metres respectively.
   a) 1250 sq m b) 1520 sq m
   c) 1230 sq m d) Can't be determined
2. In a quadrilateral, the length of one of its diagonals is 13 cm and the perpendiculars drawn on this diagonal from other two vertices measure 12 cm and 8 cm respectively. Find the area of the quadrilateral.
   a) 130 sq cm b) 135 sq cm c) 145 sq cm d) 144 sq cm
3. In a quadrilateral, the length of one of its diagonals is 16

cm and the perpendiculars drawn on this diagonal from other two vertices measure 15 cm and 7 cm respectively. Find the area of the quadrilateral.
a) 166 sq cm b) 146 sq cm
c) 176 sq cm d) 176 sq cm

4. In a quadrilateral, the length of one of its diagonals is 14 cm and the perpendiculars drawn on this diagonal from other two vertices measure 12 cm and 6 cm respectively. Find the area of the quadrilateral.
a) 126 sq cm b) 136 sq cm
c) 144 sq cm d) 124 sq cm

Answers

1. a 2. a 3. c 4. a

Rule 101

Theorem: ***The area of a circle circumscribing an equilateral triangle of side x is $\frac{\pi}{3}x^2$. (See figure)***

Illustrative Example

Ex.: The side of an equilateral triangle is 9 cm long. Find the area of the circle circumscribing the equilateral triangle.

Soln: Applying the above theorem, we have

area of the required circle $= \frac{\pi}{3} \times 9 \times 9 = 27\pi$ sq cm.

Exercise

1. The side of an equilateral triangle is 3 cm long. Find the area of the circle circumscribing the equilateral triangle.
a) 3π sq cm b) 4π sq cm
c) 6π sq cm d) 9π sq cm

2. The side of an equilateral triangle is 6 cm long. Find the area of the circle circumscribing the equilateral triangle.
a) 12π sq cm b) 16π sq cm
c) 24π sq cm d) Data inadequate

3. The side of an equilateral triangle is 12 cm long. Find the area of the circle circumscribing the equilateral triangle.
a) 36π sq cm b) 48π sq cm
c) 54π sq cm d) Data inadequate

Answers

1. a 2. a 3. b

Rule 102

Theorem: ***The area of a circle inscribed in an equilateral triangle of side x is $\frac{\pi}{12}x^2$. (See figure)***

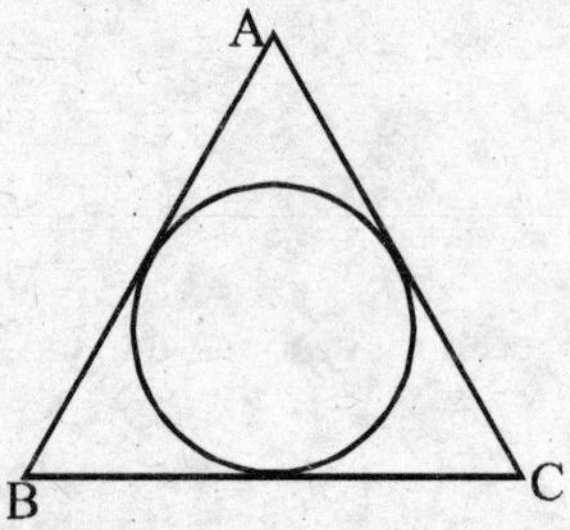

Illustrative Example

Ex.: The length of side of an equilateral triangle is 9 cm. Find the area of the circle inscribing the equilateral triangle.

Soln: Applying the above theorem, we have

the required area $= \frac{\pi}{12} \times 9 \times 9 = \frac{27\pi}{4}$ sq cm.

Exercise

1. The length of side of an equilateral triangle is 6 cm. Find the area of the circle inscribing the equilateral triangle.
a) 3π sq cm b) 4π sq cm
c) $\frac{3}{4}\pi$ sq cm d) Data inadequate

2. The length of side of an equilateral triangle is 4 cm. Find the area of the circle inscribing the equilateral triangle.
a) 4π sq cm b) $\frac{4}{9}\pi$ sq cm
c) $\frac{4}{3}\pi$ sq cm d) $\frac{3}{4}\pi$ sq cm

3. The length of side of an equilateral triangle is 12 cm. Find the area of the circle inscribing the equilateral triangle.
a) 12π sq cm b) 4π sq cm
c) 8π sq cm d) Data inadequate

4. The length of side of an equilateral triangle is 8 cm. Find the area of the circle inscribing the equilateral triangle.
a) $\frac{16}{3}\pi$ sq cm
b) 5π sq cm
c) 6π sq cm
d) Data inadequate

Answers

1. a 2. c 3. a 4. a

Rule 103

Theorem: ***Radius of a largest circle that can be drawn in an equilateral triangle of side x units is*** $\left(\frac{\sqrt{3}}{6}x\right)$ ***units.***

Illustrative Example

Ex.: There is an equilateral triangle of side 12 cm. Find the radius of the circle that can be drawn in the equilateral triangle.

Soln: Applying the above theorem, we have the

radius of the circle $= \frac{\sqrt{3}}{6} \times 12 = 2\sqrt{3}$ cm.

Exercise

1. There is an equilateral triangle of side 6 cm. Find the radius of the circle that can be drawn in the equilateral triangle.

   a) $4\sqrt{3}$ cm b) $\sqrt{3}$ cm

   c) $2\sqrt{3}$ cm d) Data inadequate

2. There is an equilateral triangle of side 4 cm. Find the radius of the circle that can be drawn in the equilateral triangle.

   a) $2\sqrt{3}$ cm b) $\sqrt{3}$ cm c) $\frac{2}{\sqrt{3}}$ cm d) $4\sqrt{3}$ cm

3. There is an equilateral triangle of side 18 cm. Find the radius of the circle that can be drawn in the equilateral triangle.

   a) $3\sqrt{3}$ cm b) $\frac{1}{\sqrt{3}}$ cm

   c) $\frac{2}{\sqrt{3}}$ cm d) Data inadequate

4. There is an equilateral triangle of side 24 cm. Find the radius of the circle that can be drawn in the equilateral triangle.

   a) $2\sqrt{3}$ cm b) $\frac{4}{\sqrt{3}}$ cm

   c) $4\sqrt{3}$ cm d) Data inadequate

Answers

1. b 2. c 3. a 4. c

Rule 104

Theorem: ***An equilateral triangle is circumscribed by a circle and another circle is inscribed in that triangle then the ratio of the areas of the two circles is 4 : 1. (See figure)***

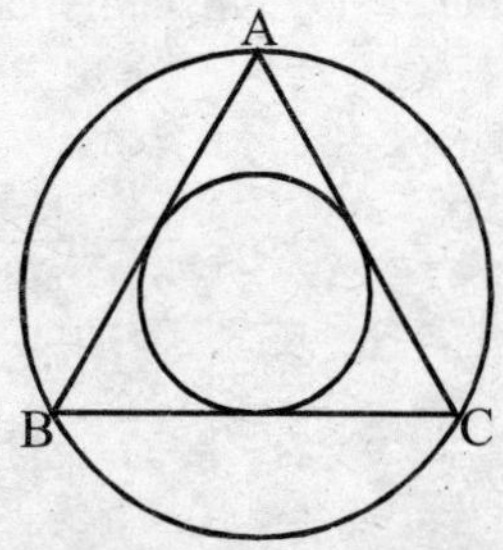

Illustrative Example

Ex.: The length of side of an equilateral triangle is 10 cm. Find the ratio of the areas of the circle circumscribing the triangle to the circle inscribing the triangle.

Soln: Applying the above theorem, we have the required ratio = 4 : 1.

Exercise

1. The length of side of an equilateral triangle is 12 cm. Find the ratio of the areas of the circle circumscribing the triangle to the circle inscribing the triangle.

   a) 4 : 1 b) 5 : 4 c) 2 : 1 d) 3 : 2

2. The length of side of an equilateral triangle is 15 cm. Find the ratio of the areas of the circle circumscribing the triangle to the circle inscribing the triangle.

   a) 7 : 4 b) 3 : 2 c) 4 : 1 d) Data inadequate

3. The length of side of an equilateral triangle is 25 cm. Find the ratio of the areas of the circle circumscribing the triangle to the circle inscribing the triangle.

   a) 4 : 1 b) 3 : 1 c) 5 : 2 d) 9 : 0

Answers

1. a 2. c 3. a

Rule 105

Theorem: ***If the area of a circle inscribed in an equilateral triangle is 'A' sq units, then the side of the equilateral triangle is*** $\sqrt{\frac{12A}{\pi}}$ ***units.***

Illustrative Example

Ex.: Area of a circle inscribed in an equilateral triangle is 616 sq cm. Find the side of the equilateral triangle.

Soln: **Detail Method:** According to the question,

$\pi r^2 = 616$, where r = radius of the circle

$\therefore r = \sqrt{\frac{616 \times 7}{22}} = 14$ cm

From geometry, we have,

AD = 3OD

ie Height of the equilateral triangle = 3x radius.

In right-angled-triangle ABD, $a^2 - \left(\frac{a}{2}\right)^2 = (42)^2$

where, a = side of the equilateral triangle

or, $\frac{4a^2 - a^2}{4} = (42)^2$

or, $3a^2 = (42)^2 \times 4$ or, $a = \sqrt{\frac{(42)^2 \times 4}{3}} = 28\sqrt{3}$ cm

Quicker Method: Applying the above theorem, we have

the required answer $= \sqrt{\frac{12 \times 616 \times 7}{22}} = \sqrt{12 \times 28 \times 7}$

$= \sqrt{28 \times 28 \times 3} = 28\sqrt{3}$ cm.

Exercise

1. Area of a circle inscribed in an equilateral triangle is 66 sq cm. Find the side of the equilateral triangle.
   a) $7\sqrt{6}$ cm b) $6\sqrt{7}$ cm c) 36 cm d) Data inadequate
2. Area of a circle inscribed in an equilateral triangle is 264 sq cm. Find the side of the equilateral triangle.
   a) $12\sqrt{7}$ cm b) 12 cm
   c) $14\sqrt{7}$ cm d) Data inadequate
3. Area of a circle inscribed in an equilateral triangle is 154 sq cm. Find the side of the equilateral triangle.
   a) $12\sqrt{3}$ cm b) $16\sqrt{3}$ cm c) $14\sqrt{7}$ cm d) $14\sqrt{3}$ cm

Answers

1. b 2. a 3. d

Rule 106

Theorem: ***There is a relation between the number of sides and the number of diagonals in a polygon. The relationship is given below. Number of diagonals*** $= \frac{n(n-3)}{2}$; ***where, n = no. of sides in the polygon.***

Illustrative Example

Ex.: Find the no. of diagonals of a hexagon.

Soln: Applying the above theorem,

for hexagon, there are $\frac{6(6-3)}{2} = 9$ diagonals.

Note: A 'hexagon' has six sides.

Exercise

1. Find the no. of diagonals of a pentagon.
   a) 5 b) 10 c) 15 d) 8
2. Find the no. of diagonals of a septagon.
   a) 14 b) 12 c) 16 d) 8
3. How many lines other than those shown in the figure are required to join each corner with another?

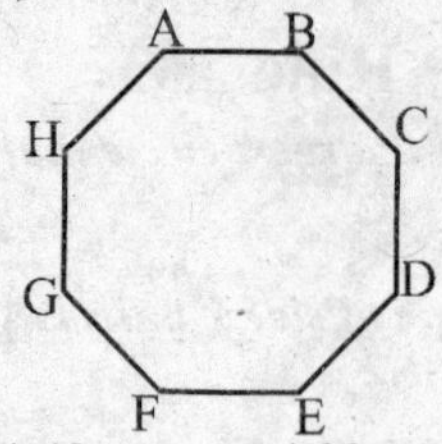

   a) 56 b) 48 c) 20 d) 32
   [Provident Fund Commissioner Exam, 2002]
4. Find the number of diagonals of a polygon having 12 sides.
   a) 66 b) 64 c) 54 d) 48

Answers

1. a 2. a 3. c; **Hint:** The given figure has 8 sides. 4. c

Rule 107

Theorem: ***The area of a rectangular plot is 'x' times its breadth. If the difference between the length and breadth is 'y' metres, then the breadth is given by (x - y) metres.***

Illustrative Example

Ex.: The area of a rectangular plot is 14 times its breadth. If the difference between the length and the breadth is 9 metres, what is its breadth?

Soln: **Detail Method:** Let l be the length and b be the breadth of the rectangular plot.

$l \times b = 14 \times b$ $\therefore l = 14$

Again, l - b = 9

or, 14 - b = 9 $\therefore b = 14 - 9 = 5$ metres.

Quicker Method: Applying the above rule, we have the required answer = 14 - 9 = 5 metres.

Note: If instead of difference, sum of the length and breadth is given, breadth is given by (y - x) m. Here x will be always less than y.

Exercise

1. Area of a rectangular plot is 15 times its breadth. If the difference between the length and the bredth is 10 metres, what is its breadth?
   a) 10 metres b) 5 metres
   c) 7.5 metres d) Data inadequate
   (BSRB Calcutta PO 1999)
2. Area of a rectangular plot is 12 times its breadth. If the difference between the length and the bredth is 8 metres, what is its breadth?
   a) 5 m b) 2 m c) 4 m d) 6 m
3. Area of a rectangular plot is 14 times its breadth. If the difference between the length and the bredth is 9 metres, what is its breadth?
   a) 5 m b) 4 m
   c) 6 m d) Can't be determined
4. Area of a rectangular plot is 18 times its breadth. If the difference between the length and the bredth is 6 metres, what is its breadth?

a) 12 m b) 14 m c) 10 m d) Data inadequate

Answers

1\. b 2. c 3. a 4. a

Miscellaneous

1\. The length and the breadth of the floor of a room is 20 ft and 10 ft respectively. Square tiles of 2 ft dimension having three different colours are placed on the floor. The first row of tiles on all sides is of black colour, out of the remaining one-third is of white colour and the remaining are of blue colour. How many blue-colour tiles are there?
a) 16 b) 32 c) 48 d) 24
[SBI PO Exam 2000]

2\. The perimeter of a rectangle is equal to the perimeter of a right-angled triangle of height 12 cm. If the base of the triangle is equal to the breath of the rectangle, what is the length of the rectangle?
a) 18 cm b) 24 cm c) 22 cm d) Data inadequate
[BSRB Mumbai PO 1998]

3\. The squared value of the diagonal of a rectangle is ($64 + B^2$) sq cm, where B is less than 8 cm. What is the breadth of that rectangle?
a) 6 cm b) 10 cm c) 8 cm d) Data inadequate
[BSRB Mumbai PO 1998]

4\. A rectangular plate is of 6 m breadth and 12 m length. Two apertures of 2 m diameter each and one aperture of 1 m diameter have been made with the help of a gas cutter. What is the area of the remaining portion of the plate?
a) 62.5 sq m b) 68.5 sq m
c) 64.5 sq m d) None of these
[Bank of Baroda PO 1999]

5\. The length and the breadth of a rectangle are in the ratio of 3 : 2 respectively. If the sides of the rectangle are extended on each side by 1 metre, the ratio of length to breadth becomes 10 : 7. Find the area of the original rectangle in square metres.
a) 256 b) 150 c) 280 d) None of these
[BSRB Guwahati PO 1999]

6\. The area of a right-angled triangle is two-third of the area of a rectangle. The base of the triangle is 80 per cent of the breadth of the rectangle. If the perimeter of the rectangle is 200 cm, what is the height of the triangle?
a) 20 cm b) 30 cm c) 15 cm d) Data inadequate
[BSRB Chennai PO 2000]

7\. Four sheets of 50 cm × 5 cm are to be arranged in such a manner that a square could be formed. What will be the area of inner part of the square so formed?
a) 2000 cm^2 b) 1600 cm^2
c) 1800 cm^2 d) None of these
[BSRB Bangalore PO 2000]

8\. The length of a rectangle is twice its breadth. If its length is decreased by 5 cm and breadth is incresed by 5 cm, the area of rectangle is increased by 75 cm^2. Therefore, the length of the rectangle is:
a) 20 cm b) 30 cm c) 40 cm d) 50 cm
[CDS Exam 1991]

9\. The area of a rectangle is thrice that of a square. Length of the rectangle is 40 cm and breadth of the rectangle is $\frac{3}{2}$ times that of the side of the square. The side of the square in cm is:
a) 60 b) 20 c) 30 d) 15

10\. The perimeters of both, a square and a rectangle are each equal to 48 m and the difference between their areas is 4 m^2. The breadth of the rectangle is:
a) 10 m b) 12 m c) 14 m d) None of these

11\. The expenses of carpeting a hall room were Rs 54000, but if the length had been 2 metres less than it was, the expenses would have been Rs 48000. What was the length?
a) 16 m b) 14 m c) 27 m d) 18 m

12\. A circular road runs round a circular garden. If the difference between the circumferences of the outer circle and the inner circle is 44 metres find the width of the road.
a) 7 m b) 14 m c) 8 m d) 16 m

Answers

1\. a; **Hint:**

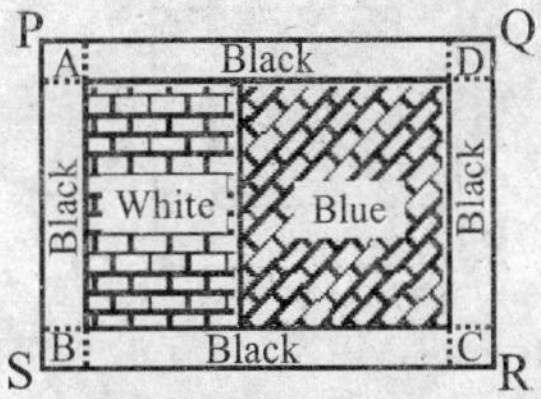

Area covered by black tiles
$= (20 + 20) \times 2 + (6 + 6) \times 2 = 80 + 24 = 104$ sq ft
Area of the floor PQRS $= 20 \times 10 = 200$ sq ft
$\therefore$ Remaining area $= 100 - 104 = 96$ sq ft

$\therefore$ Area covered by white tiles $= \frac{1}{3} \times 96 = 32$ sq ft

$\therefore$ Area covered by blue tiles $= 96 - 32 = 64$ sq ft

$\therefore$ No. of blue-colour tiles $= \frac{64}{2 \times 2} = 16$

2\. d

3\. a; **Hint:** $\text{Diagonal}^2 = 64 + B^2$ or, $10^2 = 64 + 6^2$

$\therefore B = 6$

4\. d; **Hint:** Reqd area $= 6 \times 12 - \left\{2 \times \pi\left(\frac{2}{2}\right)^2 + \pi\left(\frac{1}{2}\right)^2\right\}$

$$= 72 - \left(2\pi + \frac{\pi}{4}\right) = 72 - \frac{9\pi}{4} = 72 - \frac{9}{4} \times \frac{22}{7}$$

$$= 72 - \left(\frac{99}{14} = \right) 7.07 = 64.93 \text{ sq m}$$

5. d; **Hint:** Let the length and breadth be l and b respectively.

$\frac{l}{b} = \frac{3}{2}$ or, $l = \frac{3}{2}$ b .... (i)

$\frac{l+2}{b+2} = \frac{10}{7}$ or, $7l$ - 10b = 6 .... (ii)

From eq (i)

10.5b - 10b = 6 or, 0.5b = 6 or, b = 12 and l = 18

Area = $l \times b = 18 \times 12 = 216$ m^2

6. d; **Hint:** Let the base and height of triangle, and length and breadth of rectangle be L and h and L_1 and b_1 respectively. Then $\frac{1}{2} \times L \times h = \frac{2}{3} \times L_1 \times b_1$ .... (i)

$L = \frac{4}{5} b_1$ .... (ii) and $L_1 + b_1 = 100$ .... (iii)

In the above we have three equations and four unknowns. Hence the value of 'h' can't be determined.

7. d; **Hint:**

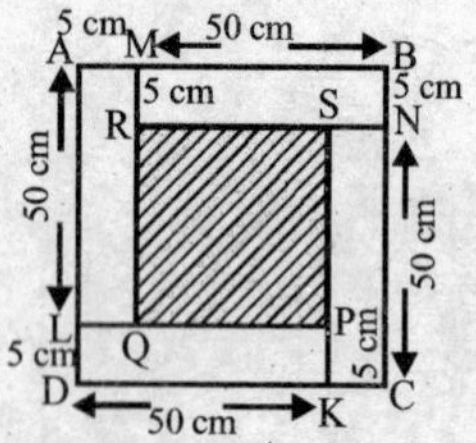

The four sheets are BMRN, AMQL, NSKC and DLPK

∴ Side of the new square sheet = 50 + 5 = 55 cm and side of the inner part of the square (55 - 10 =) 45 cm

Hence, area = $(45)^2$ = 2025 sq cm.

8. a; **Hint:** Let breadth = x cm and length = 2x cm.

Then, (2x - 5) (x + 5) - x × 2x = 75.

∴ $2x^2 + 5x - 25 - 2x^2 = 75$ or 5x = 50 or x = 10.

Hence, length = 20 cm.

9. b; **Hint:** Length of the rectangle = 40 cm.

Let the side of the square be x cm.

Then, breadth of the rectangle = $\frac{3}{2}$ x cm.

$40 \times \frac{3}{2} x = 3x^2 \Rightarrow x = 20$

∴ side of the square = 20 cm.

10. a; **Hint:** Let the length of rectangle = x metres & its breadth = y m.

Also, let the side of the square be z metres.

Then, 2 (x + y) = 4z = 48 ⇒ x + y = 24 and z = 12.

Also, z^2 - xy = 4 ⇒ xy = z^2 - 4 = 144 - 4 = 140.

So, $(x - y)^2 = (x + y)^2 - 4xy = 576 - 560 = 16$.

∴ x - y = 4 and x + y = 24. So, 2y = 20 or y = 10 m.

11. d; **Hint:** Let the length be xm and breadth be ym

Area = xy sq m

Cost of carpeting per sq m = Rs $\frac{54000}{xy}$ .... (i)

In the second case length is reduced by 2 m ie

Area = (x - 2) y sq m

Cost of carpeting per sq m = $\frac{48000}{(x-2)y}$ .... (ii)

Now, from the question, we have

$\frac{54000}{xy} = \frac{48000}{(x-2)y}$ ∴ x = 18 metres.

12. a; **Hint:** Let the radii be R_1 and R_2 of the outer and inner circles respectively.

Now, according to the question,

$2\pi R_1 - 2\pi R_2 = 44$

$R_1 - R_2 = \frac{44}{2 \times \pi} = 7$ m [$R_1 - R_2$ = width of the road.]

20 Elementary Mensuration - II

Rule 1

Cuboid

Theorem: ***To find volume of a cuboid if its length, breadth and height are given.***

Volume of a cuboid = length × breadth × height

Illustrative Example

Ex.: Find the volume of a cuboid 24 m long, 18 m broad and 16 m high.

Soln: Applying the above formula, we have volume of the cuboid = 24 × 18 × 16 = 6912 cubic metres.

Exercise

1. Find the volume of a cuboid 22 cm, by 12 cm, by 7.5 cm
   a) 1980 cu cm b) 1890 cu cm
   c) 1680 cu cm d) None of these
2. The area of a playground is 5600 sq. metres. Find the cost of covering it with gravel 1 cm deep, if the gravel costs Rs. 2.80 per cubic metre.
   a) Rs 166.70 b) Rs 146.80 c) Rs 186.50 d) Rs 156.80
3. Find the volume of a cuboid 90 metres by 50 metres by 75 cm.
   a) 3375 cu m b) 3735 cu m c) 3475 cu m d) 3875 cu m
4. A room 5 metres high is half as long again as it is broad and its volume is 480 cubic metres. Find the length and breadth of the room.
   a) 12 m, 8 m b) 9 m, 6 m
   c) 15 m, 10 m d) Data inadequate
5. A river 2 metres deep and 45 metres wide is flowing at the rate of 3 km per hour. Find how much water runs into the sea per minute.
   a) 5000 cu m b) 5400 cu m
   c) 4500 cu m d) None of these
6. How many lead shots each 0.3 cm in diameter can be made from a cuboid of dimension 9 cm by 11 cm by 12 cm?
   a) 84000 b) 86000 c) 85000 d) 48000
7. A wooden box of dimensions 8 m × 7 m × 6 m is to carry rectangular boxes of dimensions 8 cm × 7 cm × 6 cm. The maximum number of boxes that can be carried in the wooden box, is:
   a) 9800000 b) 7500000 c) 1000000 d) 1200000
   (CDS Exam, 1991)
8. A tank 3 m long, 2 m wide and 1.5 m deep is dug in a field 22 m long and 14 m wide. If the earth dug out is evenly spread out over the field, the level of the field will rise by nearly:
   a) 0.299 cm b) 0.29 mm c) 2.98 cm d) 4.15 cm
9. A rectangular tank measuring internally $37\frac{1}{3}$ metres in length, 12 metres in breadth and 8 metres in depth, is full of water. Find the weight of water in metric tons, given that one cubic metre of water weighs 1000 kilograms.
   a) 3584 metric tons b) 3684 metric tons
   c) 3485 metric tons d) 3884 metric tons
10. A room 3.3 metres high is half as long again as it is wide and its volume is $123\frac{3}{4}$ cub. metres. Find its length and breadth.
    a) 7.5 m, 6 m b) 8 m, 5 m c) 7.5 m, 5 m d) 8.5 m, 5 m
11. A rectangular block of stone measures 20 dm in length, 18 dm in breadth and 9 dm in height. What is its volume?
    a) 4230 cub dm b) 3240 cub dm
    c) 3328 cub dm d) None of these
12. A brick measures 22 cm by 10 cm by 7 cm. Find its volume.
    a) 1540 cub cm b) 1450 cub cm
    c) 1640 cm d) None of these
13. A piece of squared timber is 7 metres long and 0.1 metre both in width and thickness. What is its weight at the rate of 950 kg per cubic metres?
    a) 66 kg b) 67 kg c) 66.5 kg d) 68.5 kg
14. How many cubic metres of masonry are there in a wall 81 metres long, 4 metres high and 0.2 metre thick.
    a) 64.8 cub m b) 69 cub m c) 68 cub m d) 68.9 cub m
15. A tank contains 60000 cubic metres of water. If the length and breadth are 50 metres and 40 metres respectively,

find the depth.

a) 50 metres b) 25 metres c) 30 metres d) 20 metres

16. If 210 cubic metres of sand be thrown into a tank 12 metres long and 5 metres wide, find how much the water will rise?

a) 3.5 m b) 4 m c) 7 m d) Data inadequate

17. The area of playground is 4800 sq metres. Find the cost of covering it with gravel 1 cm deep, if the gravel costs Rs 480 per cubic metre.

a) Rs 23400 b) Rs 24300 c) Rs 23040 d) Rs 24030

18. A beam is 8 metres long, 0.5 m broad and 0.2 m thick. What is its cost at Rs 7000 per cubic metre?

a) Rs 5600 b) Rs 5800 c) Rs 6600 d) Rs 5400

19. What length must be cut off a straight plank 2.5 m broad and 0.025 m thick in order that it may contain 0.25 cubic metre?

a) 6 m b) 3 m c) 4 m d) 5 m

20. A rectangular tank is 50 metres long and 29 metres deep. If 1000 cubic metres of water be drawn off the tank, the level of the water in the tank goes down by 2 metres. How many cubic metres of water can the tank hold?

a) 14500 cub metres b) 12500 cub metres
c) 16500 cub metres d) 15500 cub metres

21. A river 10 metres deep and 200 metres wide is flowing at the rate of $4\frac{1}{2}$ km/hr. Find how many cubic m of water run into the sea per second.

a) 2500 cub metres b) 2000 cub metres
c) 2200 cub metres d) None of these

22. A swimming bath is 24 m long and 15 m broad. When a number of men dive into the bath, the height of the water rises by one cm. If the average amount of water displaced by one of the men be 0.1 cub m, how many men are there in the bath?

a) 32 b) 46 c) 42 d) 36

23. How many beams, each 4 metres long and measuring 20 cm by 12 cm at the end, can be cut from a piece of timber 12 metres long and 1 metre by 80 cm at its end?

a) 100 b) 150 c) 250 d) 125

24. A school room is to be built to accommodate 70 children, so as to allow 2.2 sq metres of floor and 11 cub metres of space for each child. If the room be 14 metres long, what must be its breadth and height?

a) 12 metres, 5.5 metres b) 11 metres, 5 metres
c) 13 metres, 6 metres d) 11 metres, 4 metres

25. A cistern is constructed to hold 200 litres, and the base of the cistern is a square metre. What is the depth of the cistern? A cubic metre is 1000 litres.

a) 50 cm b) 20 cm c) 25 cm d) 40 cm

26. Find the weight (to the nearest kilogram) of an iron rod of square section, 10 metres long and 2.3 cm broad. A cubic cm of iron weighs 7.207 grams.

a) 40 kg b) 39 kg c) 38 kg d) Data inadequate

27. A field is 500 metres long and 30 metres broad and a tank 50 metres long, 20 metres broad and 14 metres deep is dug in the field, and the earth taken out of it is spread evenly over the field. How much is the level of the field raised?

a) 0.5 m b) 1.5 m c) 1 m d) 2 m

Answers

1. a

2. d; **Hint:** Volume of gravel

$= \left(5600 \times \frac{1}{100}\right)$ cu metres. = 56 cu. m.

Cost of gravelling = Rs (56 × 2.80) = Rs 156.80.

3. a

4. a; **Hint:** Let breadth = x. Then, length = $\frac{3}{2}$ x metres.

or, $\frac{3}{2} x \times x \times 5 = 480$ or, x = 8

∴ breadth = 8 m and length = 12 m

5. c; **Hint:** Distance covered by water in 1 min

$= \left(\frac{3000}{60}\right) m = 50 m$

Water that runs in 1 min = (50 × 45 ×2) cu m = 4500 cu m

6. a; **Hint:** Number of lead shots = $\frac{\text{Volume of cuboid}}{\text{Volume of 1 lead shot}}$

$= \frac{9 \times 11 \times 12}{\frac{1}{6} \times \frac{22}{7} \times 0.3 \times 0.3 \times 0.3} = 84000$ [See **Rule-10**]

7. c; **Hint:** Number of boxes = $\frac{800 \times 700 \times 600}{8 \times 7 \times 6} = 1000000.$

8. c; **Hint:** Volume of earth dug out = $(3 \times 2 \times 1.5)\ m^3 = 9\ m^3$

Area over which earth is spread = $[(22 \times 14) - (3 \times 2)]\ m^2 = 302\ m^2$.

∴ Increase in level = $\frac{9}{302} m = \frac{9 \times 100}{302}$ cm = 2.98 cm.

9. a; **Hint:** volume of water = $37\frac{1}{3} \times 12 \times 8$ cub metres

Weight of water = $\frac{112}{3} \times 12 \times 8 \times 1000$ kg

= 3584000 kg = 3584 metric tons.

10. c; **Hint:** Length = $\frac{3}{2}$ × breadth; height = $3\frac{3}{10}$ metres

$\frac{3}{2} \times \text{breadth} \times \text{breadth} \times \frac{33}{10} = 123\frac{3}{4}$ cub metres

Or, $(\text{breadth})^2 = \frac{495}{4} \times \frac{2}{3} \times \frac{10}{33} = 25$ sq m

$\therefore$ breadth $= \sqrt{25} = 5$ m

$\therefore$ length $= \frac{3}{2} \times 5$ m $= 7.5$ m.

11. b 12. a

13. c; **Hint:** Volume of the timber $= 7 \times 0.1 \times 0.1 = 0.07$ cu m

$\therefore$ Weight of the timber $= 0.07 \times 950 = 66.5$ kg

14. a

15. c; **Hint:** Required depth $= \frac{60000}{50 \times 40} = 30$ metres

16. a; **Hint:** Let the initial height be h and the height after sand is thrown be H metres.
We have to find (H - h). According to the question, $12 \times 5 \times (H - h) = 210$

$\therefore H - h = \frac{210}{60} = \frac{7}{2} = 3.5$ metres

17. c; **Hint:** Cost of covering the playground $= 4800 \times 0.01 \times 480 =$ Rs 23040

18. a

19. c; **Hint:** Required length $= \frac{0.25}{2.5 \times 0.025} = 4$ metres.

20. a; **Hint:** $50 \times b \times 2 = 1000$ (See **Q. No. 16**)

$\therefore$ b = 10 m

$\therefore$ capacity of the tank $= 50 \times 10 \times 29 = 14500$ cub m
[Also see **Rule - 24**]

21. a; **Hint:** Speed of the river $= \frac{9}{2}$ km/hr $= \frac{9}{2} \times \frac{5}{18} = \frac{5}{4}$ m/sec

$\therefore$ required answer $= 10 \times 200 \times \frac{5}{4} = 2500$ cub m.

22. d; **Hint:** Let the no. of men be n. Now, from the question, we have
$24 \times 15 \times 0.01 = n \times 0.1$ [See **Q. No. - 16**]

$\therefore n = \frac{24 \times 15 \times 0.01}{0.1} = 36.$

23. a; **Hint:** Required no. of beams $= \frac{12 \times 1 \times 0.8}{4 \times 0.2 \times 0.12} = 100$

24. b; **Hint:** $14 \times b = 70 \times 2.2$

$\therefore b = \frac{70 \times 2.2}{14} = 11$ metres and $l \times b \times h = 70 \times 11$

$\therefore h = \frac{70 \times 11}{l \times b} = \frac{70 \times 11}{70 \times 2.2} = 5$ metres.

25. b; **Hint:** 1 sq m $\times$ depth $= \frac{200}{1000}$

$\therefore$ depth $= \frac{1}{5}$ m $= \frac{1}{5} \times 100 = 20$ cm.

26. c; **Hint:** Required answer

$= \frac{10 \times 100 \times 2.3 \times 2.3 \times 7 \times 7.207}{1000} = 38.125 \text{ kg} \approx 38$ kg.

27. c

Rule 2

Theorem: ***To find volume of a cuboid if its area of base or top, area of side face and area of other side face are given.***

Volume of the cuboid $= \sqrt{A_1 \times A_2 \times A_3}$

$= \sqrt{\text{area of base or top} \times \text{area of one face} \times \text{area of the other face}}$

Where, A_1 ***= area of base or top,***

A_2 ***= area of one side face and***

A_3 ***= area of other side face.***

Illustrative Examples

Ex. 1: Area of the base of a cuboid is 9 sq metres, area of side face and area of other side face are 16 sq metres and 25 sq metres respectively. Find the volume of the cuboid.

Soln: Applying the above formula, we have
the required answer $= \sqrt{9 \times 16 \times 25} = \sqrt{3600} = 60$ cu. metres.

Ex. 2: Find the volume of a cuboid whose area of base and two adjacent faces are 180 sq cm, 96 sq cm and 120 sq cm respectively.

Soln: We have, volume of a cuboid

$= \sqrt{\text{area of base} \times \text{area of one face} \times \text{area of the other face}}$

$= \sqrt{180 \times 96 \times 120} = 1440$ cu. cm.

Exercise

1. The area of a side of a box is 120 sq cm. The area of the other side of the box is 72 sq cm. If the area of the upper surface of the box is 60 sq cm then find the volume of the box.

a) 259200 cm^3 b) 86400 cm^3

c) 720 cm^3 d) Can't be determined

(BSRB Bangalore PO - 2000)

2. The area of a side of a box is 32 sq cm. The area of the other side of the box is 20 sq cm. If the area of the upper surface of the box is 10 sq cm then find the volume of the box.

a) 80 cm^3 b) 40 cm^3 c) 64 cm^3 d) 72 cm^3

3. The area of a side of a box is 30 sq cm. The area of the

other side of the box is 18 sq cm. If the area of the upper surface of the box is 15 sq cm then find the volume of the box.

a) 85 cm^3 b) 90 cm^3
c) 120 cm^3 d) None of these

Answers

1. c; **Hint:** Required answer

$= \sqrt{120 \times 72 \times 60} = \sqrt{72 \times 72 \times 100} = 720\ cm^3$

2. a 3. b

Rule 3

Theorem: ***To find the whole surface area of a cuboid if its length, breadth and height are given.***

Whole surface area of the cuboid = 2(lb + bh + lh)

Where, l = length, b = breadth and h = height of the cuboid.

Illustrative Example

Ex.: Find the surface area of a slab of stone measuring 4 metres in length, 2 metres in width and $\frac{1}{4}$ metre in thickness.

Soln: Applying the above formula, we have the

surface area = $2\left(4 \times 2 + 4 \times \frac{1}{4} + 2 \times \frac{1}{4}\right) = 19$ sq m

Exercise

1. Find the surface area of a cuboid 22 cm by 12 cm by 7.5 cm.
a) 1038 sq cm b) 1238 sq cm
c) 1138 sq cm d) 1308 sq cm
2. If the length, breadth and height of a cuboid are 2m, 2m and 1 m respectively, then its surfare area (in m^2) is:
a) 8 b) 12 c) 16 d) 24
[NDA Exam - 1990]
3. The area of the cardboard (in cm^2) needed to make a box of size 25 cm × 15 cm × 8 cm will be:
a) 390 b) 1000 c) 1390 d) 2780
4. If the length, breadth and height of a rectangular parallelopiped are in the ratio 6 : 5 : 4 and if total surface area is 33,300 m^2 then the length, breadth and height of a parallelopiped (in cm) respectivley are:
a) 90, 85, 60 b) 90, 75, 70 c) 85, 75, 60 d) 90, 75, 60
[NDA Exam - 1990]
5. A cuboid is 20 m × 10 m × 8 m. Find its length of diagonal, surface area and volume.
a) 23.75 m, 880 m^2, 1600 m^3
b) 27.35 m, 860 m^2, 1800 m^3
c) 23.75 m, 860 m^2, 1600 m^3
d) 23.75 m, 880 m^2, 1800 m^3
6. Find the volume and surface of a cuboid whose dimensions are 36 m, 12 m and 1 m.
a) 432 cu m, 960 sq m b) 532 cu m, 860 sq m
c) 532 cu m, 960 sq m d) None of these
7. A closed wooden box measures externally 45 cm long, 35 cm broad and 30 cm high, if the thickness of the wood is $2\frac{1}{2}$ cm, find the cost of painting the box inside at the rate of Rs 2 per square dm.
a) Rs 600 b) Rs 59 c) Rs 159 d) Rs 118

Answers

1. a 2. c
3. c; **Hint:** Area needed = 2 (lb + bh + lh)
$= 2[(25 \times 15) + (15 \times 8) + (25 \times 8)] = 1390\ cm^2$
4. d; **Hint:** Let the length, breadth and height be 6x, 5x and 4x metres respectivley.
Then, $2 \times [6x \times 5x + 5x \times 4x + 6x \times 4x] = 33300$
$\therefore\ 148x^2 = 33300$ or, $x^2 = 225$ or, $x = 15$
So, length = 90 m, breadth = 75 m and height = 60 m
5. a 6. a
7. d; **Hint:** Internal length = $45 - \frac{5}{2} \times 2 = 40$ cm

Internal breadth = $35 - \frac{5}{2} \times 2 = 30$ cm

Internal height = $30 - \frac{5}{2} \times 2 = 25$ cm

Internal surface area
$= 2[40 \times 30 + 40 \times 25 + 30 \times 25] = 5900$ sq cm

$\therefore$ required cost of painting = $\frac{5900 \times 2}{10 \times 10}$ = Rs 118.

Rule 4

Theorem: ***To find the diagonal of a cuboid if its length, breadth and height are given.***

Diagonal of cuboid = $\sqrt{l^2 + b^2 + h^2}$ ***; where l = length,***

b = breadth and

h = height of the cuboid.

Illustrative Example

Ex.: Find the length of diagonal of a cuboid 12 m long, 9 broad and 8 m high.

Soln: Applying the above formula, we have

diagonal = $\sqrt{12^2 + 9^2 + 8^2} = \sqrt{289} = 17$ m.

Exercise

1. Find the diagonal of a cuboid 22 cm, by 12 cm by 7.5 cm.

a) 684.25 cm b) 26.15 cm c) 25.14 cm d) 25.16 cm

2. Find the length of the longest pole that can be put in a room 10 metres by 8 metres by 5 metres.
a) 12 m b) 16 m c) 13 m d) 14 m
3. The diagonal of a cuboid 22 cm by 12 cm by 7.5 cm is:
a) 13.83 cm b) 6.04 cm c) 24.25 cm d) 26.15 cm
4. The length of the longest rod that can be placed in a room 30 m long, 24 m broad and 18 m high, is:
a) 30 m b) $15\sqrt{2}$ m c) 60 m d) $30\sqrt{2}$ m

[NDA Exam - 1991]

Answers

1. b
2. c; **Hint:** Length of longest pole
$= \text{diagonal} = \sqrt{10^2 + 8^2 + 5^2} = \sqrt{169} = 13$ m.
3. d
4. d; **Hint:** Length of the longest rod = length of diagonal
$= \sqrt{(l^2 + b^2 + h^2)} = \sqrt{(30)^2 + (24)^2 + (24)^2}$
$= \sqrt{1800} = 30\sqrt{2}$ m.

Rule 5

Theorem: ***To find total surface area of a cuboid if the sum of all three sides and diagonal are given. Total surface area = (Sum of all three sides)^2 – (Diagonal)^2***

Illustrative Example

Ex.: The sum of length, breadth and height of a cuboid is 25 cm and its diagonal is 15 cm long. Find the total surface area of the cuboid.

Soln: Applying the above theorem, we have the total surface area
$= (25)^2 - (15)^2 = 625 - 225 = 400$ sq cm.

Exercise

1. The sum of length, breadth and height of a cuboid is 5 cm and its diagonal is 4 cm long. Find the total surface area of the cuboid.
a) 9 sq cm b) 3 sq cm
c) 10 sq cm d) Data indequate
2. The sum of length, breadth and height of a cuboid is 26 cm and its diagonal is 14 cm long. Find the total surface area of the cuboid.
a) 840 sq cm b) 480 sq cm
c) 450 sq cm d) None of these
3. The sum of length, breadth and height of a cuboid is 12 cm and its diagonal is 8 cm long. Find the total surface area of the cuboid.
a) 60 sq cm b) 96 sq cm c) 90 sq cm d) 80 sq cm
4. The sum of length, breadth and height of a cuboid is 16 cm and its diagonal is 9 cm long. Find the total surface area of the cuboid.
a) 175 sq cm b) 170 sq cm c) 165 sq cm d) 185 sq cm

Answers

1. a 2. b 3. d 4. a

Rule 6

Theorem: ***If each edge (or side) of a cube is 'a' units then***
(i) volume of the cube = a^3 cubic units.
(ii) whole surface of the cube = $(6a^2)$ sq units.
(iii) diagonal of the cube = $(\sqrt{3}a)$ units.

Note: If diagonal of a cube is given, then the volume of the cube is given by $\left(\frac{\text{diagonal}}{\sqrt{3}}\right)^3$.

Illustrative Example

Ex.: Find the volume, surface area and the diagonal of a cube, each of whose sides measures 2 cm.

Soln: Applying the above theorem,

Volume $= a^3 = (2\times2\times2) = 8$ cm^3

Surface area $= 6a^2 = (6\times2\times2) = 24$ cm^2

Diagonal $= \sqrt{3}\times a = 2\sqrt{3}$ cm

Exercise

1. Find the volume of a cube each of whose sides measure 18 cm.
a) 5823 cu cm b) 8532 cu cm
c) 5832 cu cm d) None of these
2. Find the surface area and the diagonal of a cube, each of whose sides measure 18 cm.
a) 1944 sq cm, $18\sqrt{3}$ cm b) 1844 sq cm, $18\sqrt{3}$ cm
c) 1644 sq cm, $12\sqrt{3}$ cm d) None of these
3. A cube has a diagonal 17.32 cm long. Find the volume of the cube.
a) 1000 cu cm b) 1500 cu cm
c) 2000 eu cm d) 2500 cu cm
4. Find the volume and surface area of a cube, whose each edge measures 25 cm.
a) 15265 cu cm, 3750 sq cm b) 15625 cu cm, 3750 sq cm
c) 15625 cu cm, 3850 sq cm d) Data inadequate
5. Find the volume of a cube whose diagonal is $10\sqrt{3}$ metres.
a) 1000 cu m b) 1200 cu m c) 1500 cu m d) 1100 cu m
6. A certain cube of wood was bought for Rs 768. If the wood costs Rs 1500 per cubic metre, find the length of each edge of the cube.
a) 70 cm b) 80 cm
c) 90 cm d) Can't be determined

7. The length of diagonal of a cube is $(14\times\sqrt{3})$ cm. The surface area of the cube is:
a) 588 cm^2 b) 1176 cm^2 c) 33.908 cm^2 d) 294 cm^2
8. If each side of a cube is doubled, then its volume:
a) is doubled b) becomes 4 times
c) becomes 6 times d) becomes 8 times
9. The length of the longest rod that can fit in a cubical room of 4 m side, is:
a) 8.66 m b) 5.196 m c) 6.928 m d) 7.264 m
10. The surface area of a cube is 600 cm^2. The length of its diagonal is:
a) $10\sqrt{2}$ cm b) $10\sqrt{3}$ cm c) $\frac{10}{\sqrt{2}}$ cm d) $\frac{10}{\sqrt{3}}$ cm
11. The percentage increase in surface area of a cube when each side is doubled, is:
a) 25% b) 50% c) 150% d) 300%
12. A cubic metre of copper weighing 9000 kilograms is rolled into a square bar 9 metres long. An exact cube is cut off from the bar. How much does it weigh?
a) $333\frac{1}{3}$ kg b) $233\frac{1}{3}$ kg c) $334\frac{1}{3}$ kg d) $333\frac{1}{2}$ kg
13. Find the volume and surface of a cube whose edge is 15 metres.
a) 3375 cu m, 1350 sq m b) 3385 cu m, 1550 sq m
c) 3375 cu m, 1450 sq m d) 3835 cu m, 1350 sq m
14. The diagonal of a cube is $30\sqrt{3}$ metres. What is the solid content?
a) 27000 cu m b) 9000 cu m
c) 64000 cu m d) None of these
15. The three co-terminus edges of a rectangular solid are 36, 75 and 80 cm respectively. Find the edge of a cube which will be of the same capacity?
a) 70 cm b) 36 cm
c) 60 cm d) Data inadequate
16. A cube of metal each edge of which measures 5 cm, weighs 0.625 kg. What is the length of each edge of a cube of the same metal which weighs 40 kg?
a) 20 cm b) 25 cm c) 15 cm d) 30 cm

Answers

1. c 2. a

3. a; **Hint: See Note:** Required answer

$$=\left(\frac{17.32}{\sqrt{3}}\right)^3=\left(\frac{17.32}{1.732}\right)^3=(10)^3=1000\ cm^3$$

4. b 5. a

6. b; **Hint:** Volume of the cube

$$=\left(\frac{768}{1500}\times 100\times 100\right)=512000 \text{ cu cm}$$

$\therefore$ length of each edge

$=(512000)^{1/3}=(80\times 80\times 80)^{1/3}=80$ cm

7. b; **Hint:**

$\sqrt{a^2+a^2+a^2}=14\times\sqrt{3}\Rightarrow\sqrt{3}a=14\times\sqrt{3}\Rightarrow a=14$

$\therefore$ Surface area

$=6\times a^2=(6\times 14\times 14)\ cm^2=1176\ cm^2$.

8. d; **Hint:** Let original length of edge = a. Then, volume = a^3.

New edge = 2a; New volume = $(2a)^3=8a^3$.

$\therefore$ The volume becomes 8 times the original volume.

9. c; **Hint:** Diagonal = $4\sqrt{3}=6.928$ m.

10. b; **Hint:**

$6a^2=600\Rightarrow 3a^2=300\Rightarrow\sqrt{3a^2}=\sqrt{300}=10\sqrt{3}$

$\therefore$ Diagonal = $10\sqrt{3}$ cm.

11. d; **Hint:** Let the edge of the cube = a.

Then, surface area = $6a^2$.

New edge = 2a. So, new surface area = $6\times(2a)^2=24a^2$.

Increase = $\left(\frac{18a^2}{6a^2}\times 100\right)\% = 300\%$.

12. a; **Hint:** (Area of the square end) × 9 = vol = 1 cub metre

$\therefore$ side of the square end = $\sqrt{\frac{1}{9}}$ metres = $\frac{1}{3}$ metre

$\therefore$ Vol. of the cube = $\frac{1}{3}\times\frac{1}{3}\times\frac{1}{3}=\frac{1}{27}$ cub metre

$\therefore$ Weight of cube = $\frac{9000}{27}=333\frac{1}{3}$ kg.

[Also See **Rule - 25**]

13. a 14. a

15. c; **Hint:** Volume of the rectangular solid
= 36 × 75 × 80 = 216000 cu cm

$\therefore$ edge of the cube = $\sqrt[3]{216000}=60$ cm.

16. a; **Hint:** Volume of the cube = 5 × 5 × 5 = 125 cu cm
0.625 kg ≡ 125 cu cm

$\therefore$ 40 kg ≡ $\frac{125}{0.625}\times 40=8000$ cu cm.

$\therefore$ edge = $\sqrt[3]{8000}=20$ cm.

Cylinder

Rule 7

Theorem: ***If the radius of the base of a cylinder is 'r' units and its height (or length) is 'h' units, then volume of the cylinder is given by $\left(\pi r^2 h\right)$ cu. units. ie.***

Volume of the cylinder = Area of the base of cylinder × Height of the cylinder.

Illustrative Example

Ex.: Find the volume of a cylinder which has a height of 14 metres and a base of radius 3 metres.

Soln: Applying the above theorem,

$$\text{Volume} = \frac{22}{7} \times 3 \times 3 \times 14 = 396 \text{ cu. metres.}$$

Exercise

1. Find the volume of a cylinder of length 80 cm and the diameter of whose base is 7 cm.
   a) 8030 cu cm b) 3080 cu cm
   c) 3680 cu cm d) 3280 cu cm
2. Find the volume of an iron rod which is 7 cm long and whose diameter is 1 cm.
   a) 5.5 cu cm b) 6.5 cu cm c) 5 cu cm d) 11 cu cm
3. Water flows at 10 km per hour through a pipe with cross section a circle of radius 35 cm, into a cistern of dimensions 25 m by 12 m by 10 m. By how much will the water level rise in the cistern in 24 minutes?
   a) 15.13 m b) 5.13 m c) 4.13 m d) 6.13 m
4. A powder tin has a square base with side 8 cm and height 13 cm. Another is cylindrical with radius of its base 7 cm and height 15 cm. Find the difference in their capacities.
   a) 2130 cu cm b) 2310 cu cm
   c) 1478 cu cm d) 1468 cu cm
5. A metallic sphere of radius 21 cm is dropped into a cylindrical vessel, which is partially filled with water. The diameter of the vessel is 1.68 metres. If the sphere is completely submerged, find by how much the surface of water will rise.
   a) 1.75 cm b) 2 cm c) 2.25 cm d) 1.25 cm
6. A hollow garden roller 63 cm wide with a girth of 440 cm is made of iron 4 cm thick. The volume of iron is:
   a) 56372 cm^3 b) 58752 cm^3
   c) 54982 cm^3 d) 57636 cm^3
7. The radius of a wire is decreased to one third. If volume remains same, length will increase:
   a) 1 time b) 6 times c) 3 times d) 9 times
8. Find the volume of the cylinders in which
   (i) height = 6 cm, area of base = 5 sq m
   a) 30 cub cm b) 20 cub cm
   c) 15 cub cm d) 35 cub cm
   (ii) height = 7 metres, radius = 10 metres
   a) 2000 cub m b) 4400 cub m
   c) 2200 cub m d) 4000 cub m
9. The circumference of the base of a cylinder is 6 metres and its height is 44 metres. Find the volume.
   a) 126 cub m b) 128 cub m
   c) 136 cub m d) None of these
10. How many cubic metres of earth must be dug out to sink a well 35 metres deep and 4 metres in diameter?
    a) 220 cub m b) 660 cub m
    c) 440 cub m d) Can't be determined
11. The diameter of a cylindrical tank is 24.5 metres and depth 32 metres. How many metric tons of water will it hold? (One cubic metre of water weighs 1000 kg).
    a) 15062 metric tonnes b) 16092 metric tonnes
    c) 15092 metric tonnes d) 13062 metric tonnes
12. A cylindrical iron rod is 70 cm long, and the diameter of its end is 2 cm. What is its weight, reckoning a cubic cm of iron to weigh 10 grams?
    a) 4 kg b) 4.2 kg c) 2.2 kg d) Data inadequate
13. Find the height of the cylinder whose volume is 511 cub metres and the area of the base 36.5 sq metres.
    a) 14 m b) 22 m c) 21 m d) Data inadequate
14. A cylindrical vessel, whose base is 14 dm in diameter holds 2310 litres of water. Taking a litre of water to occupy 1000 cubic cm, what is the height of the vessel in dm?
    a) 150 dm b) 15 dm c) 1.5 dm d) Data inadequate
15. Find how many pieces of money $\frac{3}{4}$ cm in diameter and $\frac{1}{8}$ cm thick must be melted down to form a cube whose edge is 3 cm long?
    a) 820 (nearly) b) 480 (nearly)
    c) 489 (nearly) d) 889 (nearly)

Answers

1. b; **Hint:** Length = height of the cylinder = 80 cm and radius of base = 3.5 cm

$$\therefore \text{ volume} = \pi r^2 h = \left(\frac{22}{7} \times 3.5 \times 3.5 \times 80\right) = 3080 \text{ cu cm}$$

2. a; **Hint:** Required answer $= \frac{22}{7} \times 7 \times \frac{1}{2} \times \frac{1}{2} = \frac{11}{2} = 5.5$ cu cm

3. b; **Hint:** volume flown in 24 min

$$= \left(\frac{22}{7} \times \frac{35}{100} \times \frac{35}{100} \times \frac{10000}{60} \times 24\right) = 1540 \text{ cu m}$$

Initial volume of cistern = (25 × 10 × 10) = 3000 cu m

$$\text{New level} = \left(\frac{4540}{25 \times 12}\right) = 15.13 \text{ m}$$

$\therefore$ Rise in level = (15.13 - 10) = 5.13 m.

4. c; **Hint:** Required answer = $\frac{22}{7}\times7\times7\times15-8\times8\times13$

$= 2310 - 832 = 1478$ cu cm.

5. a; **Hint:** Volume of sphere

$=\left(\frac{4}{3}\times\frac{22}{7}\times21\times21\times21\right)=38808$ cu cm.

$\therefore \frac{22}{7}\times84\times84\times h=38808$

$\therefore$ h = 1.75 cm.

6. b; **Hint:** Circumference = 440 cm $\Rightarrow 2\pi r=440$

or, $r=\left(\frac{440}{2\times22}\times7\right)=70$ cm.

Inner radius = (70 - 4) cm = 66 cm
Volume of iron

$=\pi\left[(70)^2-(66)^2\right]\times63=\left(\frac{22}{7}\times136\times4\times63\right)=58752\ cm^3$

7. d; **Hint:** Let radius = R and length = h, volume = πR^2h

New radius = $\frac{1}{3}$R. Let new length = H

Volume = $\pi\times\left(\frac{1}{3}R\right)^2\times H=\frac{\pi R^2H}{9}$

$\therefore \pi R^2h=\frac{\pi R^2H}{9}$ or H = 9h.

8. (i) a (ii) c

9. a; **Hint:** $2\times\frac{22}{7}\times r=6 \quad \therefore r=\frac{6\times7}{22\times2}$

$\therefore$ volume = $\pi r^2h=\frac{22}{7}\times\frac{6\times7\times6\times7}{44\times44}\times44$

= 126 cub metres.

10. c; **Hint:** Required answer = $\frac{22}{7}\times2\times2\times35=440$ cub m.

11. c; **Hint:** Volume of the cylinder

$=\frac{22}{7}\times\frac{24.5\times24.5}{2\times2}\times32=15092$ cub metres

Since 1 cubic metre = 1000 kg
$\therefore$ 1 cubic metre = 1 metric ton
($\because$ 1000 kg = 1 metric ton)
$\therefore$ required answer = 15092 metric tonnes.

12. c; **Hint:** Volume of the iron rod

$=\frac{22}{7}\times1\times1\times70=220$ cub cm

$\therefore$ weight of the cylinder = $\frac{220\times10}{1000}=2.2$ kg

13. a; **Hint:** $\pi r^2h=511$ cub m and $\pi r^2=36.5$ sq m

$\therefore h=\frac{511}{36.5}=14$ m.

14. b; **Hint:** $\frac{22}{7}\times70\times70\times h=2310\times1000$ [$\because$ 1 dm = 10 cm]

$\therefore h=\frac{2310\times1000\times7}{22\times70\times70}=150$ cm = 15 dm.

15. c; **Hint:** Volume of one piece of money

$=\pi r^2h=\frac{22}{7}\times\frac{3}{8}\times\frac{3}{8}\times\frac{1}{8}$

$\therefore$ Required answer

$=\frac{3\times3\times3\times7\times8\times8\times8}{22\times3\times3\times1}=488.72\approx489$

Rule 8

Theorem: ***If the radius of the base of a cylinder is 'r' units and its height (or length) is 'h' units, then curved surface area of the cylinder is $(2\pi rh)$ sq units ie***
Surface area = circumference of the base × height

Illustrative Example

Ex.: Find the curved surface area of a cylinder which has a height of 14 metres and a base of radius 3 metres.

Soln: Applying the above theorem, we have the curved surface area = $2\times\frac{22}{7}\times3\times14=264$ sq metres.

Exercise

1. Find the curved surface area of a cylinder of length 80 cm and the diameter of whose base is 7 cm.
a) 1460 sq cm b) 1560 sq cm
c) 1760 sq cm d) 1960 sq cm

2. The area of the curved surface of a cylinder is 4400 cm^2 and the circumference of its base is 110 cm. Find the height and the volume of the cylinder.
a) 40 cm, 38500 cu cm b) 45 cm, 38500 cu cm
c) 40 cm, 38500 cu cm d) 45 cm, 38560 cu cm

3. How many cubic metres of the earth must be dug out to sink a well 21 metres deep and 6 metres in diameter? Find the cost of plastering the inner surface of the well at Rs 9.50 per sq metre.
a) 594 cu m, Rs 3762 b) 694 cu m, Rs 3672
c) 574 cu m, Rs 3752 d) None of these

4. A cylindrical tower is 5 m in diameter and 14 m high. The cost of white washing its curved surface at 50 paise per m^2 is:
a) Rs 90 b) Rs 97 c) Rs 100 d) Rs 110

5. The height of a cylinder is 14 cm and its curved surface is 264 cm^2. The radius of its base is:
a) 3 cm b) 4 cm
c) 2.4 cm d) 12.4 cm
6. Find the curved surface of the cylinders in which
(i) height = 10 m, circumference = 12 m
a) 120 sq m b) 140 sq m
c) 136 sq m d) 142 sq m
(ii) height = 14 m, radius = 10 m
a) 840 sq m b) 880 sq m
c) 780 sq m d) 980 sq m
7. The diameter of a cylindrical granite column is 1.5 metres and its height is 14 metres. Find the cost of polishing its curved surface at the rate of Rs 7.50 per sq metre.
a) Rs 495 b) Rs 595 c) Rs 695 d) Rs 395

Answers

1. c; **Hint:** Curved Surface = $2\pi rh = \left(2\times\frac{22}{7}\times 3.5\times 80\right)$
$= 1760$ sq cm.

2. a; **Hint:** height $= \frac{4400}{110} = 40$ cm

$2\pi r = 110$ cm or, $r = \frac{110\times 7}{44} = \frac{35}{2}$ cm

$\therefore$ Volume $= \pi r^2 h = \frac{22}{7}\times\frac{35}{2}\times\frac{35}{2}\times 40 = 38500$ cu cm.

3. a; **Hint:** Volume of the earth dug out = Volume of the well

$= \frac{22}{7}\times 3\times 3\times 21 = 594$ cu m

Curved surface area of the well

$= 2\times\frac{22}{7}\times 3\times 21 = 396$ sq m

$\therefore$ required cost = 396 × 9.50 = Rs 3762.

4. d; **Hint:** Curved surface

$= 2\pi rh = \left(2\times\frac{22}{7}\times\frac{5}{2}\times 14\right) = 220\ m^2$

Cost of white washing = Rs $\left(220\times\frac{1}{2}\right)$ = Rs 110

5. a; **Hint:** $2\times\frac{22}{7}\times r\times 14 = 264 \Rightarrow r = \frac{264}{88} = 3$ cm.

6. (i) a
(ii) b

7. a; **Hint:** Curved surface area $= \frac{2\times 22\times 1.5\times 14}{7\times 2} = 66$ sq m

$\therefore$ Cost of polishig = 66 × 7.50 = Rs 495.

Rule 9

Theorem: ***If the radius of the base of a cylinder is 'r' units and its height (or length) is 'h' units, then the total surface area of the cylinder is $\left(2\pi rh + 2\pi r^2\right)$ sq units.***

Or, Total surface area = $2\pi r(h+r)$ sq units = Circumference × (height + radius)

Illustrative Example

Ex.: Find the total surface area of a cylinder which has a height of 14 metres and a base of radius 3 metres.

Soln: Applying the above theorem, we have the

total surface area $= 2\times\frac{22}{7}\times 14\times 3 + \frac{22}{7}\times 2\times 3\times 3$

$= 264 + \frac{396}{7} = 320.57$ sq units.

Exercise

1. Find the total surface area of a cylinder of length 80 cm and the diameter of whose base is 7 cm.
a) 1837 sq cm b) 1687 sq cm
c) 1737 sq cm d) 1537 sq cm
2. Calculate the curved surface area, the total surface area and the volume of a cylinder with base radius 14 cm and height 60 cm.
a) 5280 sq cm, 6512 sq cm, 36960 cu cm
b) 2560 sq cm, 6152 sq cm, 36960 cu cm
c) 5280 sq cm, 6513 sq cm, 36960 cu cm
d) 5280 sq cm, 6152 sq cm, 39660 cu cm
3. A solid cylinder has a total surface area of 231 square cm. Its curved surface area is (2/3) of the total surface area. Find the volume of the cylinder.
a) 270 cu cm b) 269.5 cu cm
c) 256.5 cu cm d) 289.5 cu cm
4. The sum of the radius of the base and the height of a solid cylinder is 37 m. If the total surface area of the cylinder be 1628 sq m, find the volume.
a) 4620 cu m b) 4630 cu m c) 4520 cu m d) 4830 cu m
5. The ratio of total surface area to lateral surface area of a cylinder whose radius is 80 cm and height is 20 cm is:
a) 2 : 1 b) 3 : 1 c) 4 : 1 d) 5 : 1
6. If the diameter of the base of a closed right circular cylinder is equal to its height h, then its whole surface area is:
a) $2\pi h^2$ b) $\frac{4}{3}\pi h^2$ c) $\frac{3}{2}\pi h^2$ d) πh^2
[NDA Exam - 1990]

Answers

1. a; **Hint:** Total Surface = $\left(2\pi rh + 2\pi r^2\right)$

$= (1760) + \left(2\times\frac{22}{7}\times 3.5\times 3.5\right)$

$= (1760 + 77) = 1837$ sq cm

2. a

3. b; **Hint:** $2\pi rh + 2\pi r^2 = 231$ and $2\pi rh = \frac{2}{3}\times 231 = 154$

or, $2\pi r^2 = 77$ or, $\pi r^2 = \frac{77}{2}$ $\therefore r = \sqrt{\frac{77}{2}\times\frac{7}{22}} = \frac{7}{2}$ cm

or, $2\times\frac{22}{7}\times\frac{7}{2}\times h = 154$ $\therefore$ h = 7 cm

$\therefore$ Volume $= 2\pi r^2 h = \frac{22}{7}\times\frac{7}{2}\times\frac{7}{2}\times 7 = 269.5$ cu cm.

4. a; **Hint:** r + h = 37 and $2\pi r\ (r+h) = 1628$

or, $\pi r = \frac{1628}{74} = 22$

$\therefore$ r = 7 cm and h = 37 - 7 = 30 cm

$\therefore$ Volume $= \pi r^2 h = \frac{22}{7}\times 7\times 7\times 30 = 4620$ cu cm.

5. d; **Hint:** $\frac{\text{Total surface area}}{\text{Lateral surface area}} = \frac{2\pi rh + 2\pi r^2}{2\pi rh} = \frac{2\pi r(h+r)}{2\pi rh}$

$= \left(\frac{h+r}{h}\right) = \left(\frac{20+80}{20}\right) = \frac{5}{1} = 5:1$

Note: This result can be used as a general formula.

6. c; **Hint:** Radius $= \frac{1}{2}$ h and height = h

$\therefore$ Whole surface

$= 2\pi\times\left(\frac{1}{2}h\right)\times h + 2\pi\times\left(\frac{1}{2}h\right)^2 = \frac{3}{2}\pi h^2$.

Sphere

Rule 10

Theorem: ***If the radius of a sphere is r units, then volume of the sphere is*** $\left(\frac{4}{3}\pi r^3\right)$ ***cu units. If diameter is given, then volume of sphere becomes*** $\left(\frac{1}{6}\pi D^3\right)$ ***cubic units. [Where D = diameter]***

Illustrative Example

Ex.: Find the volume of a sphere of diameter 42 cm or radius 21 cm.

Soln: Applying the above theorem,

Case I: Volume of the sphere $= \frac{4}{3}\times\frac{22}{7}\times 21\times 21\times 21$

= 38808 cubic cm.

Case II: Volume of the sphere $= \frac{1}{6}\times\frac{22}{7}\times 42\times 42\times 42$

= 38808 cubic cm.

Exercise

1. Find the volume of a sphere whose radius is 2.1 metres.
   a) 38.808 cu m b) 388.08 cu m
   c) 68.808 cu m d) 38.88 cu m
2. A spherical shell of metal has an outer radius of 9 cm and inner radius of 8 cm. If the metal costs Rs 1.80 per cu cm, find the cost of shell.
   a) Rs 1630.80 b) Rs 1638.60 c) Rs 1636.80 d) Rs 1638.80
3. The largest sphere is carved out of a cube of side 7 cm. Find the volume of the sphere (Take $\pi = 3.14$).
   a) 179.6 cu cm b) 180.6 cu cm
   c) 176.9 cu cm d) 189.6 cu cm
4. A metal sphere of diameter 42 cm is dropped into a cylindrical vessel, which is partly filled with water. The diameter of the vessel is 1.68 metres. If the sphere is completely submerged, find by how much the surface of water will rise.
   a) 1.75 cm b) 2.75 cm c) 2 cm d) 1 cm
5. Find the weight of an iron shell, the external and internal diameters of which are 13 cm and 10 cm respectively, if 1 cu cm of iron weighs 8 gms.
   a) 6 kg b) 6.015 kg c) 5.016 kg d) 5.015 kg
6. A spherical ball of radius 3 cm is melted and recast into three spherical balls. The radii of two of these balls are 1.5 cm and 2 cm. Find the radius of the third ball.
   a) 5 cm b) 1.5 cm c) 3 cm d) 2.5 cm
7. Six spherical balls of radius r are melted and cast into a cylindrical rod of metal of same radius. The height of rod will be:
   a) 4 r b) 6 r c) 8 r d) 12 r
8. The amount of water contained in a sphere of radius 3 cm is poured into a cube of side 5 cm. The height upto which water will rise in the cube is:
   a) 2 cm b) $\frac{3\pi}{4}$ cm c) 0.5 π cm d) 1.44 π cm
9. The number of solid spheres of radius $\frac{1}{4}$ cm, which may be formed from a solid sphere of radius 4 cm is:
   a) 4096 b) 4964 c) 6904 d) 9640
10. How many bullets can be made out of a cube of lead whose edge measures 22 cm, each bullet being 2 cm in diameter?
    a) 5324 b) 2662 c) 1347 d) 2541
11. A cylindrical vessel 60 cm in diameter is partially filled with water. A sphere, 30 cm in diameter is gently dropped into the vessel. To what further height will water in the cylinder rise?

a) 15 cm b) 30 cm
c) 40 cm d) Can't be determined

12. Find the diameter of a sphere whose volume is $113\frac{1}{7}$ cub metres.
a) 6 m b) 8 m c) 3 m d) 4 m

13. How many bullets can be made out of a cube of lead whose edge measures 22 cm, each bullet being 2 cm in diameter?
a) 2341 b) 2641 c) 2541 d) 2451

Answers

1. a

2. c; **Hint:** Volume of metal = $\left\{\frac{4}{3}\pi(9)^3 - \frac{4}{3}\pi(8)^3\right\}$

$= \frac{4}{3}\pi(9-8)^3$ cu cm [Also see **Rule - 39]**

$= \left(\frac{4}{3}\times\frac{22}{7}\times 217\right) = \frac{2728}{3} =$ cu cm

Cost of metal = $\left(\frac{2728}{3}\times 1.80\right) =$ Rs 1636.80

3. a; **Hint:** Diameter of the sphere = 7 cm

$\therefore$ Volume of the sphere = $\frac{1}{6}\times\frac{22}{7}\times 7\times 7\times 7$

= 179.6 cu cm

4. a; **Hint:** Radius of the sphere = 21 cm

Volume of sphere = $\left(\frac{4}{3}\times\frac{22}{7}\times 21\times 21\times 21\right)$

= 38808 cu cm

Volume of water displaced by sphere = 38808 cu cm
Let the water rise by h cm

Then, $\frac{22}{7}\times 84\times 84\times h = 38808$ [See **Rule - 7]**

or, $h = \frac{38808\times 7}{22\times 84\times 84} = \frac{7}{4} = 1.75$ cm

5. c; **Hint:** Volume of iron

$= \frac{4}{3}\pi\left\{\left(\frac{13}{2}\right)^3 - (5)^3\right\} = \left(\frac{4}{3}\times\frac{22}{7}\times\frac{1197}{8}\right)$

= 627 cu cm [See **Rule - 39]**

$\therefore$ Weight of iron = $\left(\frac{627\times 8}{1000}\right) = 5.016$ kg

6. c; **Hint:** Volume of 3rd ball

$= \frac{4}{3}\pi(3)^3 - \left\{\frac{4}{3}\pi\left(\frac{3}{2}\right)^3 + \frac{4}{3}\pi(2)^3\right\} = \frac{125}{6}\pi$

$\therefore \frac{4}{3}\pi r^3 = \frac{125}{6}\pi \Rightarrow r = \frac{5}{2}$

= 2.5 cm

7. c; **Hint:** Let the height of the rod be h.

Then, $6\times\frac{4}{3}\pi r^3 = \pi r^2 h \Rightarrow h = 8r$

8. d; **Hint:** Let the required height be h cm

Then, $5\times 5\times h = \frac{4}{3}\pi\times(3)^3$

$\therefore h = \frac{36\pi}{25} = 1.44\ \pi$

9. a; **Hint:** Number of spheres

$$= \frac{\frac{4}{3}\times\pi\times 4\times 4\times 4}{\frac{4}{3}\times\pi\times\frac{1}{4}\times\frac{1}{4}\times\frac{1}{4}} = 4096$$

10. d; **Hint:** Volume of cube = $(22\times 22\times 22)\ cm^3$

[See **Rule - 6]**

Volume of 1 bullet = $\left(\frac{4}{3}\times\frac{22}{7}\times 1\times 1\times 1\right)cm^3$

$\therefore$ Number of bullets = $\left(\frac{22\times 22\times 22\times 3\times 7}{4\times 22}\right) = 2541$

11. c; **Hint:** Let H and h be the heights of water level before and after dropping the sphere into it.

Then, $\left[\pi\times(30)^2\times H\right] - \left[\pi\times(30)^2\times h\right] = \frac{4}{3}\pi\times(30)^3$

or $\pi\times 900\times(H-h) = \frac{4}{3}\pi\times 27000$

or, (H - h) = 40 cm.

12. a; **Hint:** $\frac{1}{6}\times\frac{22}{7}\times D^3 = \frac{792}{7}$

$\therefore D^3 = \frac{792}{7}\times\frac{7\times 6}{22} = 36\times 6 = 6\times 6\times 6$

$\therefore$ D = Diameter = $\sqrt{6\times 6\times 6} = 6$ m

13. c; **Hint:** Required answer = $\frac{22\times 22\times 22\times 3\times 7}{4\times 22\times 1\times 1\times 1} = 2541.$

Rule 11

Theorem: ***If the radius of a sphere is r units, then the surface area of the sphere is $(4\pi r^2)$ sq units. If in place of radius, diameter of the sphere is given, then the formula becomes as follows,***

Surface area of a sphere $= 4\pi r^2 = 4\pi\left(\frac{D}{2}\right)^2 = \pi D^2$ ***sq units.***

Illustrative Example

Ex.: Find the surface area of a sphere of diameter 42 cm.

Soln: Applying the above theorem, we have the surface area of a sphere

$= \frac{22}{7} \times 42 \times 42 = 5544$ sq units.

Exercise

1. Find the surface area of a sphere whose radius is 2.1 metres.
   a) 55.44 sq m b) 65.45 sq m
   c) 54.47 sq m d) None of these
2. Find the surface area of a sphere whose volume is 310464 cu cm.
   a) 22276 sq cm b) 22176 sq cm
   c) 12276 sq cm d) None of these
3. If a solid sphere of radius 10 cm is moulded into 8 spherical solid balls of equal radius, then surface area of each ball (in cm^2) is:
   a) 100 π b) 75 π c) 60 π d) 50 π
   [NDA Exam 1990]
4. If the volume of surface area of a sphere are numerically the same, then its radius is:
   a) 1 unit b) 2 units c) 3 untis d) 4 units
   (NDA Exam 1990)
5. Find the surfaces and volumes of the sphere having the following radii
   (i) 7 metres
   a) 616 sq m, $1437\frac{1}{3}$ cub m
   b) 661 sq m, $1347\frac{1}{3}$ cub m
   c) 616 sq m, $1447\frac{1}{3}$ cub m
   d) 616 sq m, $1377\frac{1}{3}$ cub m
   (ii) $3\frac{1}{2}$ mm
   a) 164 sq mm, $179\frac{1}{3}$ cub mm
   b) 154 sq mm, $179\frac{1}{3}$ cub mm
   c) 154 sq mm, $179\frac{2}{3}$ cub mm
   d) None of these
   (iii) 10.5 cm
   a) 1386 sq cm, 4851 cub cm
   b) 1256 sq cm, 4651 cub cm
   c) 1386 sq cm, 4651 cub cm
   d) 1386 sq cm, 4859 cub cm

Answers

1. a
2. b; **Hint:** $\frac{4}{3}\pi r^3 = 310464$

   or, $r^3 = \frac{310464 \times 3 \times 7}{4 \times 22} = 74088$ $\therefore$ r = 42 cm

   $\therefore$ surface area $= 4\pi r^2 = 4 \times \frac{22}{7} \times 42 \times 42 = 22176$ sq cm
3. a; **Hint:** Let r be the radius of each moulded sphere. Then,

   $8 \times \frac{4}{3}\pi r^3 = \frac{4}{3}\pi \times (10)^3 \Rightarrow (2r)^3 = (10)^3 \Rightarrow 2r = 10$

   $\therefore$ r = 5 cm.

   So, the surface area of each ball

   $= [4\pi \times (25)] = (100\pi)$ cm^2
4. c; **Hint:** Let r be the radius of the sphere.

   Then, $\frac{4}{3}\pi r^3 = 4\pi r^2 \Rightarrow r = 3$ units.
5. (i) a (ii) c (iii) a

Rule 12

Theorem: ***If the radius of a sphere is r units, then volume of a hemisphere*** $= \left(\frac{2}{3}\pi r^3\right)$ ***cu. units. If diameter is given, then volume of a hemisphere is given by*** $\left[\frac{\pi}{12}D^3\right]$ ***cu units.***

[Where, D = diameter of the sphere]

Illustrative Example

Ex.: Find the volume of a hemisphere of radius 21 cm.

Soln: Applying the above theorem, we have

Volume of hemisphere $= \left(\frac{2}{3} \times \frac{22}{7} \times 21 \times 21 \times 21\right)$

$= 19404$ cm^3

Exercise

1. A hemispherical bowl has inner diameter 42 cm. The quantity of liquid that the bowl can hold (in cm³) is:

a) $\frac{2}{3}\times\frac{22}{7}\times(21)^3$ b) $\frac{2}{3}\times\frac{7}{22}\times(21)^3$

c) $\frac{3}{2}\times\frac{22}{7}\times(21)^3$ d) $\frac{2}{3}\times\frac{7}{22}\times(42)^3$

[CDS Exam 1991]

2. How many litres of water will a hemispherical bowl contain whose radius is 2 metres? (1 litres = 1000 cub cm)

a) $1676\frac{19}{21}$ litres b) $1566\frac{19}{21}$ litres

c) $1686\frac{19}{21}$ litres d) None of these

3. Find the volume of a hemisphere of radius 7 cm.

a) $\frac{2056}{3}\text{cm}^3$ b) $\frac{2156}{7}\text{cm}^3$

c) $\frac{2156}{3}\text{cm}^3$ d) Data inadequate

4. Find the volume of a hemisphere of radius 14 cm.

a) $\frac{2}{3}\times\pi\times(7)^3\ \text{cm}^3$ b) $\frac{1}{12}\times\pi\times(28)^3\ \text{cm}^3$

c) $\frac{2}{3}\times\pi\times(6)^3\ \text{cm}^3$ d) $\frac{1}{12}\times\pi\times(12)^3\ \text{cm}^3$

Answers

1. c; **Hint:** (See the Illustrative Example) 2. a
3. c 4. b

Rule 13

Theorem: ***If the radius of a sphere is r units, then the curved surface area of the hemisphere is $(2\pi r^2)$ sq units.***
If in place of radius, diameter is given, then the curved surface area of the hemisphere becomes $\left(\frac{\pi}{2}D^2\right)$ sq units.

Illustrative Example

Ex.: Find the curved surface area of a hemisphere of radius 21 cm.

Soln: Applying the above theorem, we have the curved surface area = $2\pi r^2$

$$= \left(2\times\frac{22}{7}\times21\times21\right)\ \text{cm}^2 = 2772\ \text{cm}^2.$$

Exercise

1. Find the curved surface area of a hemisphere of radius 7 cm.

a) 308 sq cm b) 803 sq cm c) 306 sq cm d) 603 sq cm

2. Find the curved surface area of a hemisphere of radius 14 cm.

a) 1322 sq cm b) 1233 sq cm
c) 1232 sq cm d) 1122 sq cm

3. Find the curved surface area of a hemisphere of radius 28 cm.

a) 4928 sq cm b) 4298 sq cm
c) 4982 sq cm d) None of these

4. Find the curved surface area of a hemisphere of diameter 35 cm.

a) $\frac{\pi}{2}\times35\times35\ \text{cm}^2$ b) $2\times\pi\times35\times35\ \text{cm}^2$

c) $\frac{\pi}{2}\times(17.5)^2\ \text{cm}^2$ d) Data inadequate

Answers

1. a 2. c 3. a 4. a

Rule 14

Theorem: ***If the radius of a sphere is r units, then the whole surface area of the hemisphere is $(3\pi r^2)$ sq units. If in place of radius, diameter is given, then the whole surface area of the hemisphere is given by $\left(\frac{3}{4}\pi D^2\right)$ sq units.***

Illustrative Example

Ex.: Find the total surface area of a hemisphere of radius 21 cm.

Soln: Applying the above theorem, we have the total surface area

$$= 3\pi r^2 = \left(3\times\frac{22}{7}\times21\times21\right)\ \text{cm}^2 = 4158\ \text{cm}^2.$$

Exercise

1. Find the total surface area of a hemisphere of radius 7 cm.

a) 462 sq cm b) 642 sq cm
c) 468 sq cm d) Data inadequate

2. Find the total surface area of a hemisphere of radius 14 cm.

a) 1648 sq cm b) 1848 sq cm
c) 1448 sq cm d) Data inadequate

3. Find the total surface area of a hemisphere of diameter 16 cm

a) $\frac{2}{3}\times\pi\times(8)^2$ sq cm b) $\frac{3}{4}\times\pi\times(8)^2$ sq cm

c) $\frac{3}{4}\times\pi\times(16)^2$ sq cm d) None of these

Answers

1. a 2. b 3. c

Right Circular Cone

Rule 15

Theorem: ***To find the slant height of the right circular cone if radius of its base and height of the cone are given.***

Slant height (l) $= \left(\sqrt{h^2 + r^2}\right)$ ***units.***

Where h = height and r = radius of the base.

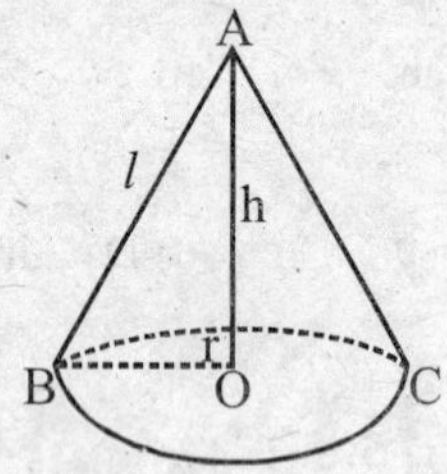

Illustrative Example

Ex.: Radius of the base of a right circular cone is 3 cm and height of the cone is 4 cm. Find the slant height of the cone.

Soln: Applying the above theorem, we have the slant height of the right circular cone

$= \sqrt{4^2 + 3^2} = 5$ cm.

Exercise

1. The height of a cone is 16 cm and the diameter of its base is 24 cm. Find its slant height.
   a) 10 cm b) 20 cm c) 15 cm d) 25 cm
2. The diameter of the base of a right circular cone is 4 cm and its perpendicular height is $2\sqrt{3}$ cm. The slant height of the cone is:
   a) 5 cm b) 4 cm c) $4\sqrt{3}$ cm d) 3 cm
3. The diameter of the base of a right circular cone is 6 cm and its perpendicular height is $3\sqrt{3}$ cm. Find the slant height of the cone.
   a) 6 cm b) $5\sqrt{3}$ cm c) $6\sqrt{3}$ cm d) $\sqrt{63}$ cm

Answers

1. b 2. b

3. a; **Hint:** $r = \frac{6}{2} = 3$ cm and $h = 3\sqrt{3}$ cm

$\therefore$ slant height (l)

$= \sqrt{h^2 + r^2} = \sqrt{\left(3\sqrt{3}\right)^2 + (3)^2} = \sqrt{27 + 9} = \sqrt{36} = 6$ cm.

Rule 16

Theorem: ***To find the volume of the right circular cone, if radius of the base and the height of the cone is given.***

Volume of the cone $= \left(\frac{1}{3}\pi r^2 h\right)$ ***cu. units.***

Illustrative Example

Ex.: Radius of the base of a right circular cone is 7 cm and the height of the cone is 3 cm. Find the volume of the cone.

Soln: Applying the above formula,

Volume of the cone $= \frac{1}{3} \times \frac{22}{7} \times 7 \times 7 \times 3 = 154$ cm^3

Exercise

1. The height of a cone is 16 cm and the diameter of its base is 24 cm. Find the volume of the cone.
   a) $\frac{16896}{7}$ cu cm b) $\frac{16876}{7}$ cu cm
   c) $\frac{19876}{7}$ cu cm d) None of these
2. Find the volume of a cone the diameter of whose base is 21 cm and the slant height is 37.5 cm.
   a) 4158 cu cm b) 4518 cu cm
   c) 4256 cu cm d) 4156 cu cm
3. A cone of height 7 cm and base radius 3 cm is carved from a rectangular block of wood 10 cm × 5 cm × 2 cm. Calculate the percentage of wood wasted.
   a) 66% b) 37% c) 67% d) 34%
4. From a solid right circular cylinder with height 12 cm and radius of the base 7 cm, a right circular cone of the same height and base is removed. Find the volume of the remaining solid.
   a) 1232 cu cm b) 1332 cu cm
   c) 1432 cu cm d) None of these
5. The radius and height of a right circular cone are in the ratio 5 : 12 and its volume is 2512 cu cm. Find the slant height, radius and curved surface area of the cone. (Take $\pi = 3.14$)
   a) 24 cm, 10 cm, 816.4 sq cm
   b) 22 cm, 11 cm, 816.4 sq cm
   c) 21 cm, 10 cm, 861.4 sq cm
   d) Can't be determined
6. If the height of a cone is increased by 100%, then its volume is increased by:
   a) 100% b) 200% c) 300% d) 400%
7. If a right circular cone of vertical height 24 cm has a volume of 1232 cm^3, then the area of its curved surface in cm^2 is:

a) 1254 b) 704 c) 550 d) 154

(NDA Exam 1990)

8. A right cylindrical vessel is full with water. How many right cones having same diameter and height as those of right cylinder will be needed to store that water?
a) 2 b) 3 c) 4 d) 5

[CDS Exam 1991]

9. A cylindrical piece of metal of radius 2 cm and height 6 cm is shaped into a cone of same radius. The height of cone is:
a) 18 cm b) 14 cm c) 12 cm d) 8 cm

(Railway Recruitment 1991)

10. Find the slant height of a cone whose volume is equal to 12936 cubic metres and the diameter of whose base is 42 metres.
a) 35 m b) 36 m c) 28 m d) None of these

11. The diameter of a cone is 21 cm. Its volume is 1848 cub cm. Find the perpendicular height of the cone.
a) 18 cm b) 14 cm c) 16 cm d) 20 cm

12. The volume of a cone is 616 cubic m. Its perpendicular height is 27 metres. find the radius of the base.
a) $4\frac{2}{3}$ m b) $4\frac{1}{3}$ m c) $4\frac{5}{6}$ m d) 5 m

13. From a solid right circular cylinder with height 10 cm and radius of the base 6 cm, a right circular cone of the same height and base is removed. Find the volume of the remaining solid.
a) $754\frac{5}{7}$ cub cm b) $754\frac{2}{7}$ cub cm
c) $756\frac{1}{7}$ cub cm d) Data inadequate

Answers

1. a

2. a; **Hint:** $h=\sqrt{(37.5)^2-(10.5)^2}=36$ cm.
[See **Rule - 15**]
Now, volume
$=\frac{1}{3}\pi r^2h=\frac{1}{3}\times\frac{22}{7}\times 10.5\times 10.5\times 36=4158$ cu cm

3. d; **Hint:** Volume of the rectangular block
$= 10\text{ cm}\times 5\text{ cm}\times 2\text{ cm}$ [See **Rule - 1**]
$= 100$ cu cm
Volume of the carved cone
$=\frac{1}{3}\times\frac{22}{7}\times 3\times 3\times 7=66$ cu cm
or, From 100 cu cm, volume carved = 66 cu cm
$\therefore$ required answer = 100 - 66 = 24%.

4. a; **Hint:** Remaining solid
$=\left(\frac{22}{7}\times 7\times 7\times 12-\frac{1}{3}\times\frac{22}{7}\times 7\times 7\times 12\right)=1232$ cu cm.

5. a; **Hint:** $\frac{1}{3}\times 3.14\times(25x^2)\times(12x)=2512\Rightarrow x=2$
$\therefore$ Radius = 10 cm & height = 24 cm
$\therefore$ $l=\sqrt{r^2+h^2}=\sqrt{(10)^2+(24)^2}=26$ cm
$\therefore$ Curved surface area
$=\pi rl=3.14\times 10\times 26=816.4$ sq cm

6. a; **Hint:** Let the height be h & radius be r.
New height = 2h
$\therefore$ Change in volume $=\left[\frac{\frac{1}{3}\pi r^2(2h)-\frac{1}{3}\pi r^2h}{\frac{1}{3}\pi r^2h}\times 100\right]$
= 100%

7. c; **Hint:**
$\frac{1}{3}\times\frac{22}{7}\times r^2\times 24=1232\Rightarrow r^2=\left(\frac{1232\times 3\times 7}{22\times 24}\right)\Rightarrow r=7$
Slant height $=\sqrt{(24)^2+(7)^2}=25$ cm
$\therefore$ Curved surface $=\left(\frac{22}{7}\times 7\times 25\right)=550\text{ cm}^2$

8. b; **Hint:** Volume of 1 cylinder = πr^2h
Volume of 1 cone $=\frac{1}{3}\pi r^2h$
Number of cones $=\frac{\pi r^2h}{\frac{1}{3}\pi r^2h}=3$

9. a; **Hint:** $\frac{1}{3}\pi\times(2)^2\times h=\pi\times(2)^2\times 6\Rightarrow h=18$ cm

10. a; **Hint:** $\frac{1}{3}\times\frac{22}{7}\times 21\times 21\times h=12936$
$h=\frac{12936\times 7\times 3}{21\times 21\times 21}=28$ m
$\therefore$ slant height (l) $=\sqrt{21^2+28^2}=\sqrt{1225}=35$ m.

11. c; **Hint:** $\frac{1}{3}\times\frac{22}{7}\times\frac{21}{2}\times\frac{21}{2}\times h=1848$
$\therefore$ $h=\frac{1848\times 3\times 7\times 2\times 2}{22\times 21\times 21}=16$ cm

12. a; **Hint:** $\frac{1}{3}\times\frac{22}{7}\times r^2\times 27 = 616$

or, $r^2 = \frac{616\times3\times7}{22\times27} = \frac{28\times3\times7}{27} = \frac{196}{9}$

$\therefore\ r = \sqrt{\frac{196}{9}} = \frac{14}{3} = 4\frac{2}{3}$ m

13. b; **Hint:** Required answer

$= \frac{22}{7}\times6\times6\times10 - \frac{1}{3}\times\frac{22}{7}\times6\times6\times10$

$= \frac{22}{7}\times6\times6\times10\times\frac{2}{3} = \frac{5280}{7} = 754\frac{2}{7}$ cub cm.

Rule 17

Theorem: ***To find the curved surface area of the cone,***

(i) if its slant height and radius of its base are given.

Curved surface area of the cone = πrl ***sq units.***

(ii) if its height and radius of its base are given,

Curved surface area of the cone = $\pi r\left(\sqrt{r^2+h^2}\right)$

sq units.

Illustrative Example

Ex.: Radius of the base of a right circular cone is 3 cm and the height of the cone is 4 cm. Find the curved surface area of the cone.

Soln: Applying the above formula,

Curved surface area of the cone $= \frac{22}{7}\times3\left(\sqrt{4^2+3^2}\right)$

$= \frac{22\times3\times5}{7} = 47\frac{1}{7}$ sq cm.

Exercise

1. The height of a cone is 16 cm and the diameter of its base is 24 cm. Find the area of curved surface of the cone.
   a) 754.28 sq cm b) 754.82 sq cm
   c) 774.28 sq cm d) None of these
2. It is required to make a hollow cone 24 cm high whose base radius is 7 cm. Find the area of the sheet metal required including the base.
   a) 467 sq cm b) 764 sq cm
   c) 704 sq cm d) None of these
3. The radius of base of a right circular cone is 6 cm and its slant height is 28 cm. The curved surface of the cone is:
   a) 268 cm^2 b) 658 cm^2 c) 462 cm^2 d) 528 cm^2
4. The area of the base of a right circular cone is 154 cm^2 and its height is 14 cm. The curved surface of cone is:
   a) $(154\times\sqrt{5})$ cm^2 b) 11 cm^2
   c) $(154\times\sqrt{7})$ cm^2 d) 5324 cm^2
5. Find the curved surface area of the cone when,
   (i) Slant height = 25 cm, diameter = 14 cm
   a) 450 sq cm b) 550 sq cm
   c) 660 sq cm d) 440 sq cm
   (ii) slant height = 17 dm, radius = 8 dm
   a) $427\frac{3}{7}$ sq dm b) $426\frac{3}{5}$ sq dm
   c) $427\frac{2}{7}$ sq dm d) Data inadequate
   (iii) perpendicular height = 36 dm, diameter = 30 dm
   a) $1838\frac{4}{7}$ sq dm b) $1828\frac{4}{7}$ sq dm
   c) $1818\frac{4}{7}$ sq dm d) None of these
6. The slant height of a conical tomb is $17\frac{1}{2}$ metres. If its diameter be 28 metres, find the cost of constructing it at Rs 135 per cubic metre and also find the cost of white-washing its slant surface at Rs 3.30 per square metre.
   a) Rs 261090, Rs 2541 b) Rs 291060, Rs 2741
   c) Rs 291060, Rs 2541 d) Rs None of these

Answers

1. a

2. c; **Hint:** Metal required = $(\pi rl + \pi r^2)$ sq cm

Where r = 7 cm and l = $\sqrt{(24)^2+(7)^2} = 25$ cm

$\therefore$ required answer = 550 + 154 = 704 sq cm.

3. d

4. a; **Hint:** $\pi r^2 = 154 \Rightarrow r^2 = \left(\frac{154}{22}\times7\right) \Rightarrow r = 7$

$\therefore$ slant height $= \sqrt{h^2+r^2} = \sqrt{(14)^2+(7)^2} = 7\sqrt{5}$ cm

Curved surface $= \left(\frac{22}{7}\times7\times7\sqrt{5}\right) = 154\sqrt{5}\ cm^2$

5. (i) b (ii) c (iii) a

6. c; **Hint:** Height of the cone $= \sqrt{\left(\frac{35}{2}\right)^2 - 14^2} = \frac{21}{2}$ m

Volume of the cone $= \frac{1}{3}\times\frac{22}{7}\times14\times14\times\frac{21}{2} = 2156$ cub m

$\therefore$ Cost of constructing the cone

$=2156\times 135=\text{Rs}\,291060$

Curved surface area of the cone

$=\pi rl=\frac{22}{7}\times 14\times\frac{35}{2}=770$ cub m

$\therefore$ Cost of white washing $=770\times 3.30=$ Rs 2541.

Rule 18

Theorem: ***To find the total surface area of the right circular cone.***

(i) if its slant height and radius of its base are given.

Total surface area of the cone

$=\left(\pi rl+\pi r^2\right)=\pi r(l+r)$ ***sq units.***

(ii) If its height and radius of its base are given.

Total surface area of the cone $=\pi r\left(\sqrt{h^2+r^2}+r\right)$

sq units.

Illustrative Example

Ex.: Radius of the base of a right circular cone is 3 cm and the height of the cone is 4 cm. Find the total surface area of the cone.

Soln: Applying the above formula,

$$\text{Total surface area}=\frac{22}{7}\times 3\left(\sqrt{4^2+3^2}+3\right)$$

$$=\frac{22\times 3\times 8}{7}=\frac{528}{7}=75\frac{3}{7}\text{ sq cm.}$$

Exercise

1. The height of a cone is 16 cm and the diameter of its base is 24 cm. Find the total surface area of the cone.
   a) 754.28 sq cm b) 452.57 sq cm
   c) 1206.85 sq cm d) None of these
2. Radius of the base of a right circular cone is 18 cm and the height of the cone is 24 cm. Find the total surface area of the cone.
   a) 864 π sq cm b) 468 π sq cm
   c) 854 π sq cm d) 485 π sq cm
3. Radius of the base of a right circular cone is 30 cm and the height of the cone is 40 cm. Find the total surface area of the cone.
   a) 240 π sq cm b) 2400 π sq cm
   c) 1600 π sq cm d) 1680 π sq cm

Answers

1. c 2. a 3. b

Frustum of a Right Circular Cone

Rule 19

Frustum: If a cone is cut by a plane parallel to the base so as to divide the cone into two parts as shown in the figure, lower part is called the frustum of the cone.

Frustum
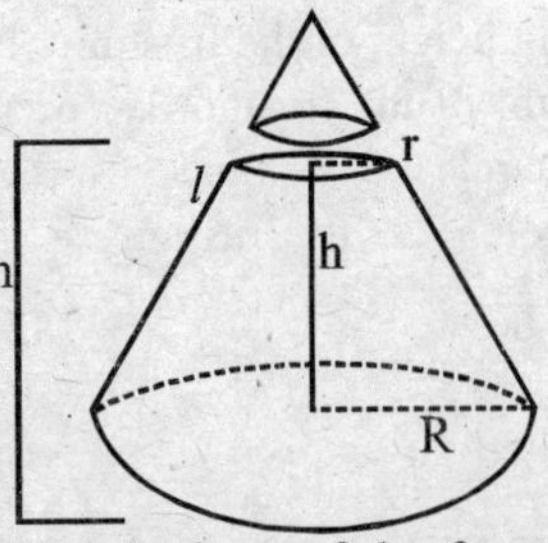

Let the radius of the base of the frustum = R, the radius of the top of the frustum = r, and slant height of the frustum = l units.

(i) Slant height $(l)=\sqrt{h^2+(R-r)^2}$ units.

(ii) Curved surface area $=\pi(R+r)l$ sq units.

(iii) Total surface area $=\pi\left[(R+r)l+r^2+R^2\right]$ sq units.

(iv) Volume of the frustum $=\frac{\pi h}{3}\left(r^2+R^2+rR\right)$ cu units.

Illustrative Example

Ex.: A frustum of a right circular cone has a diameter of base 10 cm of top 6 cm and a height of 5 cm. Find (i) slant height, (ii) curved surface area, (iii) total surface area and (iv) volume of the frustum.

Soln: Applying the above theorem,

Here, $r=\frac{6}{2}=3$ cm; $R=\frac{10}{2}=5$ cm; $h=5$ cm

(i) slant height $=\sqrt{h^2+(R-r)^2}$

$=\sqrt{5^2+(5-3)^2}=\sqrt{29}$ cm $=5.385$ cm

(ii) curved surface area

$=\pi(R+r)l=\frac{22}{7}\times 8\times 5.385=135.4$ sq cm.

(iii) Total surface area of the frustum

$=\pi\left[(R+r)l+r^2+R^2\right]$

$=\frac{22}{7}\left[8\times 5.385+(3)^2+5^2\right]$

$=\frac{22}{7}[43.08+9+25]=242.25$ sq cm.

(iv) Volume of the frustum $=\frac{\pi h}{3}\left(r^2+R^2+rR\right)$

$=\frac{22}{7}\times\frac{5}{3}\left[5^2+3^2+5\times 3\right]=256.67$ cu. cm.

Exercise

1. The slant height of the frustum of a cone is 20 cm and the height of the frustum is 16 cm. The radius of the smaller circle is 8 cm. Find
(i) the volume of the frustum

a) $\frac{72316}{7}$ cu cm b) $\frac{73216}{7}$ cu cm

c) $\frac{75216}{7}$ cu cm d) None of these

(ii) the total area of the surface of the frustum
a) 6424 sq cm b) 2464 sq cm
c) 4264 sq cm d) None of these

2. If the radii of the ends of a bucket 45 cm high are 28 cm and 7 cm, determine its capacity and the surface area.
a) 48510 cu cm, 5610 sq cm b) 48150 cu cm, 5610 sq cm
c) 48510 cu cm, 5510 sq cm d) 48510 cu cm, 6510 sq cm

3. A reservoir is in the shape of a frustum of a right circular cone. It is 8 m across at the top and 4 m across the bottom. It is 6 m deep. Its capacity is:
a) 176 m^3 b) 196 m^3 c) 200 m^3 d) 110 m^3
[CDS Exam 1999]

4. The circumference of one end of a frustum of a right circular cone is 48 cm and of the other end 34 cm, the height of the frustum is 10 cm, find its volume.
a) 1250 cub cm approx b) 1850 cub cm approx
c) 1350 cub cm approx d) 1360 cub cm approx

5. The radii of the ends of a frustum of a right circular cone are 33 cm and 27 cm, its slant height is 10 cm. Find its volume and whole surface.

a) 22704 cub cm, $7599\frac{3}{7}$ sq cm

b) 22407 cub cm, $7569\frac{3}{7}$ sq cm

c) 22704 cub cm, $7569\frac{3}{7}$ sq cm

d) Data inadequate

Answers

1. (i) b (ii) b

Hint: Let us denote the radius of smaller circle, radius of bigger circle, the height of frustum and the slant height by r, R, h and l respectively.
Then, r = 8 cm, h = 16 cm and l = 20 cm.

But $l = \sqrt{[h^2+(R-r)^2]} = \sqrt{[(16)^2+(R-8)^2]}$

or,

$20 = \sqrt{(R^2+320-16R)]}$ or, $R^2+320-16R = 400$

or, $R^2 - 16R - 80 = 0$ or $(R-20)(R+4) = 0$
$\therefore$ R = 20 or R = -4
But, R = -4 is not possible.
So, the radius of bigger circle is 20 cm.
Now,

(i) Volume of frustum $= \frac{\pi h}{3}(R^2+r^2+Rr)$

$= \left[\left(\frac{1}{3}\times\frac{22}{7}\times 16\right)(400+64+160)\right] = \frac{73216}{7}$ cu cm

(ii) Total area of surface of frustum $= \pi(R^2+r^2+Rl+rl)$

$= \frac{22}{7}(400+64+160+160) = 2464$ sq cm.

2. a; **Hint:** r = 7 cm, R = 28 cm and h = 45 cm

$l = \sqrt{h^2+(R-r)^2} = 49.6$ cm

Capacity = Volume of the frustum

$= \frac{1}{3}\pi h(R^2+r^2+Rr) = 48510$ cu cm

Surface area of the bucket

$= [\pi l(R+r)+\pi r^2] = 5610$ cu cm

3. a; **Hint:** Volume $= \frac{\pi h}{3}[R^2+r^2+Rr]$,

where R = 4 m, r = 2m, h = 6 m

$= \left\{\frac{22}{7}\times\frac{1}{3}\times 6\times(16+4+8)\right\} = 176\ m^2$

4. c; **Hint:** $2\pi R = 48$ $\therefore R = \frac{48\times 7}{2\times 22}$

$2\pi r = 34$ $\therefore r = \frac{34\times 7}{2\times 22}$

By applying the given rule find the volume.

5. a

Rule 20

Theorem: ***To find number of bricks when the dimensions of brick and wall are given.***

Required no. of bricks $= \frac{\textit{Volume of wall}}{\textit{Volume of one brick}}$

Illustrative Example

Ex.: A brick measures 20 cm by 10 cm by $7\frac{1}{2}$ cm. How many bricks will be required for a wall 25 m long, 2 m high and $\frac{3}{4}$ m thick?

Soln: Volume of wall = $25\times2\times\frac{3}{4}$ cu. m.

Volume of one brick

$=\frac{20}{100}\times\frac{10}{100}\times\frac{15}{200}=\frac{3}{2000}$ cu. m.

Reqd. number of bricks

$=\left(25\times2\times\frac{3}{4}\right)\div\frac{3}{2000}=25000$

Exercise

1. How many bricks will be required for a wall 8 metres long, 6 metres high and 22.5 cm thick, if each brick measures 25 cm by 11.25 cm by 6 cm?
   a) 6400 b) 3200 c) 64000 d) 86000
2. A wall of length 25 metres, width 60 cm and height 2 metres is to be constructed by using bricks, each of dimensions 20 cm by 12 cm by 8 cm. How many bricks will be needed?
   a) 16525 b) 15265 c) 15625 d) 15525
3. A brick measures 2.3 dm by 1.1 dm by 0.8 dm. How many bricks will be required for a wall 92 metres long, 4 metres high and 0.264 metre thick.
   a) 48000 b) 46000 c) 49000 d) 84000
4. The length of a room is 12 metres, width 8 metres, height 6 metres. How many boxes will it hold if each is allowed 1.5 cubic metres of space
   a) 356 b) 172 c) 256 d) 384
5. How many bricks 20 cm × 10 cm × 7.5 cm can be carried by a truck whose load is 5 metric tons? The bricks in question weigh 2500 kg per cubic metre.
   a) 1333 b) 1233 c) 1332 d) 1433

Answers

1. a; **Hint:** Volume of wall = (800 × 600 × 22.5) cu cm
   Volume of a brick = (25 × 11.25 × 6) cu cm
   ∴ Number of bricks
   $=\frac{\text{Volume of the wall}}{\text{Volume of a brick}}=\left(\frac{800\times600\times22.5}{25\times11.25\times6}\right)=6400$

2. c; **Hint:** Number of bricks $=\left(\frac{2500\times60\times200}{20\times12\times8}\right)=15625$

3. a; **Hint:** Required number $=\frac{92\times4\times0.264}{0.23\times0.11\times0.08}=48000$

4. d; **Hint:** Required answer $=\frac{12\times8\times6}{1.5}=384$

5. a; **Hint:** Required number of bricks
   $=\frac{5000}{2500}\times\frac{1}{0.2\times0.1\times0.075}=1333.3\approx1333.$
   [∵ 1 metric ton = 1000 kg]

Rule 21

Theorem: ***To find capacity, volume of material and weight of material of a closed box, when external dimensions (ie length, breadth and height) and thickness of material of which box is made, are given.***

(i) Capacity of box = (External length – 2 × thickness) × (External breadth – 2 × thickness) × (External height – 2 × thickness)

(ii) Volume of material = External Volume – Capacity

(iii) Weight of wood = Volume of wood × Density of wood.

Illustrative Example

Ex.: A closed wooden box measures externally 9 cm long, 7 cm broad, 6 cm high. If the thickness of the wood is half a cm, find (i) the capacity of the box and (ii) the weight supposing that one cubic cm. of wood weighs 0.9 gm.

Soln: **Quicker Method:**

In such cases,

Capacity = (external length – 2 × thickness) × (external breadth – 2 × thickness) × (external height – 2 × thickness)

Volume of material = External volume – Capacity

∴ in the given question,

Capacity = (9 – 2 × 0.5) (7 – 2 × 0.5) (6 – 2 × 0.5) = 8 × 6 × 5 = 240 cm^3.

∴ Volume of wood = external volume – capacity
= 9 × 7 × 6 – 240 = 138 cu.cm.

∴ Weight of wood = Volume of wood × density of wood = 138×0.9= 124.2 g.

Exercise

1. An open rectangular cistern when measured from out side is 1 m 35 cm long; 1 m 8 cm broad and 90 cm deep, and is made of iron 2.5 cm thick. Find (i) the capacity of the cistern, (ii) the volume of the iron used.
   a) 1171625 cu cm, 140575 cu cm
   b) 1711625 cu cm, 104575 cu cm
   c) 1171625 cu cm, 145075 cu cm
   d) None of these
2. Find the weight of a lead pipe 3.5 metres long, if the external diameter of the pipe is 2.4 cm and the thickness of the lead is 2 mm and 1 cc of lead weight 11.4 gm.
   a) 5.5 kg b) 5 kg c) 8 kg d) 10 kg
3. A closed rectangular box has inner dimensions 24 cm by 12 cm by 10 cm. Calculate its capacity and the area of tin foil needed to line its inner surface.
   a) 2680 cu cm, 1296 sq cm b) 2880 cu cm, 1396 sq cm
   c) 2880 cu cm, 1296 sq cm d) 2860 cu cm, 1296 sq cm
4. The dimensions of an open box are 52 cm, 40 cm and 29 cm. Its thickness is 2 cm. If 2 cm^3 of metal used in the box weight 0.5 gm, the weight of the box is:
   a) 8.56 kg b) 7.76 kg c) 7.756 kg d) 6.832 kg

5. A rectangular box whose external dimensions including the lid are 32, 27, 12 decimetres is made of wood 0.5 dm in thickness. What is the volume of wood in it?
   a) 1520 cu dm b) 1502 cu dm
   c) 1602 cu dm d) 1620 cu dm

Answers

1. a; **Hint:** (i) Capacity = (135 - 5) (108 - 5) (90 - 2.5) [Since cistern is open]
   = 130 × 103 × 87.5 = 1171625 cu cm
   (ii) Volume of iron = [(135 × 108 × 90) - (1171625) = 140575 cu cm
2. a; **Hint:** External radius of the pipe = 1.2 cm
   Internal radius of the pipe = (1.2 - 0.2) = 1 cm
   External volume
   $$=\left(\frac{22}{7}\times 1.2\times 1.2\times 3.5\times 100\right)=1584 \text{ cu cm}$$
   $$\text{Internal volume}=\left(\frac{22}{7}\times 1\times 1\times 3.5\times 100\right)=1100 \text{ cu cm}$$
   Volume of lead = (External Volume) - (Internal Volume)
   = (1584 - 1100) = 484 cu cm
   $$\text{Weight of the pipe}=\left(\frac{484\times 11.4}{1000}\right)=5.5176 \text{ kg.}$$
3. c; **Hint:** Capacity = Volume = (24 × 12 × 10) = 2880 cu cm
   Area of tin foil needed = Total surface area
   = [2 (24 × 12 + 12 × 10 + 24 × 10)] = 1296 sq cm.
4. d; **Hint:** Volume of metal
   $$=(52\times 40\times 29-48\times 36\times 27)=13664 \text{ cm}^3$$
   $$\therefore \text{ Weight of metal}=\left(13664\times 0.5\times \frac{1}{1000}\right)=6.832 \text{ kg.}$$
5. b; **Hint:** Required answer = 32 × 27 × 12 - 31 × 26 × 11
   = 10368 - 8866 = 1502 cu dm.

Rule 22

Theorem: ***To find the volume of a cube if the surface area of the cube is given.***

$$\textit{Volume of cube}=\left(\sqrt{\frac{\textit{Surface area}}{6}}\right)^3$$

Illustrative Example

Ex.: The surface of a cube is $30\frac{3}{8}$ sq metres. Find its volume.

Soln: Applying the above rule, we have

$$\text{volume}=\left(\sqrt{\frac{\frac{243}{8}}{6}}\right)^3=11\frac{25}{64} \text{ cu m.}$$

Exercise

1. The surface of a cube is 552.96 sq cm. Find the volume of the cube.
   a) 884.736 cu cm b) 848.736 cu cm
   c) 884.376 cu cm d) Data inadequate
2. Find the volume of a cube whose surface area is 150 square metres.
   a) 15625 cu cm b) 125 cu m
   c) 120 cu m d) 124 cu m
3. The surface of a cube is 1176 cm^2. The volume of this cube is
   a) 7056 cm^3 b) 4704 cm^3 c) 2744 cm^3 d) 3528 cm^3
4. The volume of a cube is 125 cm^3. The surface area of the cube is:
   a) 625 cm^2 b) 125 cm^2 c) 150 cm^2 d) 100 cm^2
5. The curved surface of a sphere is 1386 cm^2. Its volume is
   a) 2772 cm^3 b) 4158 cm^3 d) 4851 cm^3 d) 5544 cm^3
6. The volume of a sphere is $\frac{88}{21}(14)^3$ cm^3. The curved surface of this sphere is:
   a) 2424 cm^2 b) 2446 cm^2 c) 2464 cm^2 d) 2484 cm^2

Answers

1. a; **Hint:** $\text{Volume}=\left\{\sqrt{\left(\frac{552.96}{6}\right)}\right\}^3=\left[\sqrt{(92.16)}\right]^3=(9.6)^3$
   = 884.736 cu cm.
2. b 3. c
4. c; **Hint:** $125=\left(\sqrt{\frac{\text{Surface area}}{6}}\right)^3$
   or, $5=\sqrt{\frac{\text{Surface area}}{6}}$
   $\therefore$ Surface area = $5^2\times 6=150$ sq cm.
5. c
6. c; **Hint:** By direct formula, it is a time taking process. So solve this type of question by the method given below.
   $$\frac{4}{3}\pi r^3=\frac{88\times(14)^3}{21}\Rightarrow r^3=\left(\frac{88\times 14\times 14\times 14}{21}\times\frac{3}{4}\times\frac{7}{22}\right)$$
   $\therefore$ r = 14
   So, curved surface of the sphere
   $$=\left(4\times\frac{22}{7}\times 14\times 14\right)=2464\ cm^2.$$

Rule 23

Theorem: ***To find the volume of rain water at a place if the annual rainfall of that place is given.***
Volume of rain water = Height (or level) of water (ie Annual rainfall) × Base area (ie area of the place)

Illustrative Example

Ex.: The annual rainfall at a place is 43 cm. Find the weight in metric tonnes of the annual rainfall there on a hectare of land, taking the weight of water to be 1 metric tonne for 1 cubic metre.

Soln: **Quicker Method:**
Volume of water = height (level) of water × base area
In the given question, level of rainfall is 43 cm.

$\therefore$ volume of water $= \frac{43}{100}$ m × 10000 sq m
= 4300 cu m.

(As 1 hectare = 10,000 sq m).

$\therefore$ weight of water = 4300 × 1 = 4300 metric tonnes.

Exercise

1. In a shower 5 cm of rain falls. Find in cubic metres the volume of water that falls on 2 hectares of ground.
   a) 1000 cu cm b) 100 cu cm
   c) 1000 cu m d) None of these
2. The water in a rectangular reservoir having a base 80 metres by 60 metres is 6.5 metres deep. In what time can the water be emptied by a pipe of which the cross section is a square of side 20 cm, if the water runs through the pipe at the rate of 15 km per hour?
   a) 26 hrs b) 52 hrs c) 65 hrs d) 42 hrs
3. In a shower 2 cm of rain falls. Find in cubic metres the volume of water that falls on 6 hectares of ground.
   a) 12000 cub m b) 1200 cub m
   c) 1600 cub m d) Can't be determined

Answers

1. c; **Hint:** Depth of rain = 5 cm = $\frac{5}{100}$ ie $\frac{1}{20}$ metres.

Area of ground = 2 hectares = 20000 sq metres.

Volume of water = $20000 \times \frac{1}{20}$ = 1000 cu metres.

2. b; **Hint:** Volume of water = (80 × 60 × 6.5) = 31200 cu m

Area of cross section of the pipe

$= \left(\frac{20}{100} \times \frac{20}{100}\right) = \frac{1}{25}$ sq m

Volume of water emptied in 1 hour

$= \left(\frac{15 \times 1000 \times 1}{25}\right) = 600$ cu m

Time taken to empty the reservoir $= \frac{31200}{600} = 52$ hrs.

3. b; **Hint:** Required answer = 0.02 × 60000 = 1200 cub m

Rule 24

Theorem: ***A rectangular tank is 'l' metres long and 'h' metres deep. If 'x' cubic metres of water be drawn off the tank, the level of the water in the tank goes down by 'd' metres, then the amount of water (in cubic metres) the tank can hold is given by*** $\left(\frac{x \times h}{d}\right)$ ***cubic metres and the breadth of the tank is*** $\left(\frac{x}{ld}\right)$ ***metres.***

Illustrative Example

Ex.: A rectangular tank is 50 metres long and 29 metres deep. If 1000 cubic metres of water be drawn off the tank, the level of the water in the tank goes down by 2 metres. How many cubic metres of water can the tank hold? And also find the breadth of the tank.

Soln: **Detail Method:** Let the breadth of the tank be x metres.
Volume of the tank = 50 × 29 × x cubic metres.
From the question,
$50 \times 29 \times x = 1000 + 50 \times (29 - 2) \times x$
or, $x \times 50 \times 2 = 1000$ $\therefore x = 10$ metres.
$\therefore$ Breadth of the tank is 10 metres.
Volume of the tank = 50 × 29 × 10 = 14500 cubic metres.
Quicker Method: Applying the above theorem,

Volume of the tank $= \frac{1000 \times 29}{2} = 14500$ cub m.

Note: Looking at the above theorem, we may conclude that even if the length of the rectangular tank is not given. Volume of the tank can be calculated. To find breadth of the tank length is needed but not the height of the tank.

$\therefore$ Breadth of the tank $= \frac{1000}{50 \times 2} = 10$ metres.

Exercise

1. A rectangular tank is 15 metres long and 27 metres deep. If 450 cubic metres of water be drawn off the tank, the level of the water in the tank goes down by 3 metres. How many cubic metres of water can the tank hold? And also find the breadth of the tank.
   a) 4050 cub m, 9 m b) 4500 cub m, 10 m
   c) 4050 cub m, 10 m d) Data inadequate
2. A rectangular tank is 100 cm long and 58 cm deep. If 2000 cubic cm of water be drawn off the tank, the level of the water in the tank goes down by 4 cm. How many cubic metres of water can the tank hold? And also find the

breadth of the tank.

a) 48000 cub cm, 10 cm b) 58000 cub cm, 40 cm
c) 58000 cub cm, 5 cm d) Data inadequate

3. A rectangular tank is 20 metres long and 15 metres deep. If 500 cubic metres of water be drawn off the tank, the level of the water in the tank goes down by 2 metres. How many cubic metres of water can the tank hold? And also find the breadth of the tank.
a) 3750 cub m, 12.5 m b) 3570 cub m, 15.5 m
c) 3750 cub m, 12 m d) 3750 cub m, 15 m

4. A rectangular tank is 16 m deep. If 650 cubic metres of water be drawn off the tank, the level of the water in the tank goes down by 1 metre. How many cubic metres of water can the tank hold?
a) 10400 cub m b) 14000 cub m
c) 10500 cub m d) Data inadequate

Answers

1. c 2. c 3. a 4. a

Rule 25

Theorem: ***x cubic metres of copper weighing y kilograms is rolled into a square bar l metres long. An exact cube is cut off from the bar. Weight of the cube is given by*** $\left[\left(\sqrt{\frac{x}{l}}\right)^3 \times y\right]$ ***kg.***

Illustrative Example

Ex.: A cubic metre of copper weighing 9000 kilograms is rolled into a square bar 9 metres long. An exact cube is cut off from the bar. How much does it weigh?

Soln: Detail Method:

In this case a given volume of copper is rolled into a square bar (basically a cuboid with square base) of given length. Then an exact cube is cut off from this square bar. Obviously,. the exact cube should have the same dimensions as that of the square base of the square bar.

Now, given volume = 1 cu m.

= Area of square base × length

$\Rightarrow$ Area of square base × length = 1

$$\Rightarrow \text{Area of square base} = \frac{1}{\text{length}} = \frac{1}{9}$$

$$\therefore \text{side of square base} = \sqrt{\frac{1}{9}} = \frac{1}{3}$$

$\therefore$ Vol. of the cut off cube

$$= (\text{side of the square base})^3 = \left(\frac{1}{3}\right)^3 = \frac{1}{27}$$

$$\therefore \text{weight of cube} = \frac{1}{27} \times 9000 = 333.3 \text{ kg.}$$

Quicker Method:

In this type of question, use the formula:

Volume of cube cut off

$$= \left(\sqrt{\frac{\text{Volume of original solid}}{\text{length of the solid}}}\right)^3$$

$$\therefore \text{Volume} = \left(\sqrt{\frac{1}{9}}\right)^3 = \frac{1}{27}$$

$$\therefore \text{Weight} = \frac{9000}{27} = 333.3 \text{ kg.}$$

Note: We can also apply the above theorem directly.

$\therefore$ Weight of cube

$$= \left(\sqrt{\frac{1}{9}}\right)^3 \times 9000 = \frac{9000}{27} = 333.3 \text{ kg.}$$

Exercise

1. A cubic metre of copper weighing 8000 kilograms is rolled into a square bar 4 metres long. An exact cube is cut off from the bar. How much does it weigh?
a) 1000 kg b) 800 kg c) 500 kg d) 950 kg

2. A cubic metre of copper weighing 640 kilograms is rolled into a square bar 16 metres long. An exact cube is cut off from the bar. How much does it weigh?
a) 15 kg b) 32 kg c) 16 kg d) 10 kg

3. A cubic metre of copper weighing 2500 kilograms is rolled into a square bar 25 metres long. An exact cube is cut off from the bar. How much does it weigh?
a) 25 kg b) 20 kg c) 18 kg d) Data inadequate

4. A cubic metres of copper weighing 1600 kg is rolled into a square base 16 metres long. An exact cube is cut off from the bar. How much does it weigh?
a) 200 kg b) 600 kg c) 400 kg d) Data inadequate

Answers

1. a 2. d 3. b 4. a

Rule 26

Theorem: ***When many cubes integrate into one cube, the side of the new cube is given by side***

$= \sqrt[3]{\text{Sum of cubes of sides of all the cubes}}$

Illustrative Example

Ex.: Three cubes of metal whose edges are 3, 4 and 5 cm respectively are melted and formed into a single cube. If there be no loss of metal in the process find the side of the new cube.

Soln: Detail Method:

Volume of the first cube = $(3)^3 = 27$ cubic cm.

Volume of the second cube = $(4)^3 = 64$ cu. cm

Volume of third cube = $(5)^3 = 125$ cu. cm.

$\because$ Volume remains unchanged.

$\therefore$ Volume of the new cube = $27 + 64 + 125 = 216$ cu cm.

$\therefore$ Side of the new cube = $\sqrt[3]{216} = 6$ cm.

Quicker Method: Applying the above theorem, we have

side = $\sqrt[3]{3^3 + 4^3 + 5^3} = \sqrt[3]{27 + 64 + 125}$

$= \sqrt[3]{216} = 6$ cm.

Exercise

1. Three cubes of metal whose edges are 5, 6 and 7 cm respectively are melted and formed into a single cube. If there be no loss of metal in the process find the side of the new cube.
   a) $\sqrt{864}$ cm b) $\sqrt[3]{684}$ cm
   c) $\sqrt{684}$ cm d) None of these
2. Three cubes of metal whose edges are 30, 40 and 50 cm respectively are melted and formed into a single cube. If there be no loss of metal in the process find the side of the new cube.
   a) 60 cm b) 64 cm c) 90 cm d) 80 cm
3. Three cubes of metal whose edges are 2, 3 and 4 cm respectively are melted and formed into a single cube. If there be no loss of metal in the process find the side of the new cube.
   a) $\sqrt{99}$ cm b) $\sqrt[3]{99}$ cm
   c) $\sqrt{199}$ cm d) $\sqrt[3]{199}$ cm

Answers

1. b 2. a 3. b

Rule 27

Theorem: ***Total volume of a solid does not change even when its shape changes.***

$\therefore$ ***Old volume = New volume***

Illustrative Example

Ex.: A cubic metre of gold is extended by hammering so as to cover an area of 6 hectares. Find the thickness of the gold.

Soln: Applying the above theorem, we have

$\Rightarrow$ 1 cu m = 60000 × thickness

$\Rightarrow$ thickness = $\frac{1}{60000}$ m = 0.0017 cm.

Exercise

1. Half cubic metre of gold sheet is extended by hammering so as to cover an area of 1 hectare. Find the thickness of the gold.
   a) 0.05 cm b) 0.5 cm c) 0.005 cm d) 0.0005 cm
2. A metal sheet 27 cm long, 8 cm broad and 1 cm thick is melted into a cube. The difference between the surface area of two solids will be:
   a) 284 cm^2 b) 296 cm^2 c) 286 cm^2 d) 300 cm^2
3. A solid lead ball of 7 cm radius was melted and then drawn into a wire of 0.2 cm diameter. The length of wire will be:
   a) 458.43 m b) 457.33 m c) 468.26 m d) 437.29 m
4. The material of a cone is converted into the shape of a cylinder of equal radius. If the height of the cylinder is 5 cm, the height of cone is:
   a) 10 cm b) 15 cm c) 18 cm d) 20 cm
5. Two cubic metres of gold are extended by hammering so as to cover an area of twelve hectares. Find the thickness of gold.
   a) 0.017 cm b) 0.0017 cm c) 1.7 cm d) 0.17 cm
6. A cub cm of silver is drawn into a wire $\frac{1}{10}$ mm in diameter, find the length of the wire. ($\pi = 3.1416$)
   a) 128 metres b) 127.3 metres
   c) 129.3 metres d) 128.3 metres

Answers

1. c; **Hint:** Volume of Sheet

$= \frac{1}{2}$ cu m $= \left(\frac{1}{2} \times 100 \times 100 \times 100\right)$ cu cm

Area of Sheet = 1 hectare = 10000 sq metres

= (10000 × 100 × 100) sq cm.

Thickness

$= \frac{\text{Volume}}{\text{Area}} = \frac{1 \times 100 \times 100 \times 100}{2 \times 10000 \times 100 \times 100} = \frac{1}{200} = 0.005$ cm.

2. c; **Hint:** Volume of new cube formed

$= (27 \times 8 \times 1) = 216\ cm^3$

Edge of this cube = $(216)^{1/3} = (6 \times 6 \times 6)^{1/3} = 6$ cm

$\therefore$ Surface area of this cube = $6a^2 = 6 \times 6 \times 6 = 216\ cm^2$

Surface area of given cuboid = 2 (27 × 8 + 8 × 1 + 27 × 1)

$= 502\ cm^2$

$\therefore$ Difference between the surface areas = (502 - 216)

$= 286\ cm^2$.

3. b; **Hint:** Let the length of the wire be x cm.

Then, $\frac{4}{3} \times \pi \times 7 \times 7 \times 7 = \pi \times 0.1 \times 0.1 \times x$

$\therefore\ x = \frac{4}{3} \times \frac{7 \times 7 \times 7}{0.1 \times 0.1} = \frac{457333}{100} = 457.33$ m.

4. b; **Hint:** Let h be the height of the cone and r be the radius of the base of each of the cone and cylinder.

Then, $\pi r^2 \times 5 = \frac{1}{3}\pi r^2 h \Rightarrow h = 15$ cm.

5. b

6. b; **Hint:** $3.1416 \times \frac{0.01}{2} \times \frac{0.01}{2} \times h = 1$

$\therefore\ h = \frac{4}{3.1416 \times 0.01 \times 0.01} = 12732.365$ cm $= 127.3$ m.

Rule 28

Theorem: ***To find the number of possible cubes when disintegration of a cube into identical cubes.***

$$\textit{Number of cubes} = \left(\frac{\textit{Original length of side}}{\textit{New length of side}}\right)^3$$

Illustrative Example

Ex.: A cube of sides 3 cm is melted and smaller cubes of sides 1 cm each are formed. How many such cubes are possible?

Soln: Quicker Method:
In such questions use the above rule:

$$\text{Number possible} = \left(\frac{\text{original length of side}}{\text{new length of side}}\right)^3$$

$\therefore$ In this question, possible number of cubes

$$= \left(\frac{3}{1}\right)^3 = 27.$$

Exercise

1. The number of small cubes with edges of 10 cm that can be accommodated in a cubical box of 1 metre edge is:
   a) 100 b) 1000 c) 10 d) 10000
2. What number of 4 cm cubes can be cut from a 12 cm cube?
   a) 24 b) 25 c) 27 d) 64
3. A cube of sides 6 cm is melted and smaller cubes of sides 3 cm each are formed. How many such cubes are possible?
   a) 16 b) 8
   c) 27 d) Data inadequate
4. A cube of sides 15 cm is melted and smaller cubes of sides 5 cm each are formed. How many such cubes are possible?
   a) 27 b) 32 c) 29 d) 30

Answers

1. b; **Hint:** In the given question, it may be considered that the cubical box of 1 metre edge is disintegrated into numerous smaller cubes of 10 cm edge.
   $\therefore$ required number of cubes
   $$= \frac{100 \times 100 \times 100}{10 \times 10 \times 10} = 1000.$$

2. c 3. b 4. a

Rule 29

Theorem: ***A hollow cylindrical tube open at both ends is made of a thick metal. If the internal diameter or radius and length of the tube are given, then the volume of metal is given by*** **[π × height × (2 × Internal radius + thickness) × thickness] cu. units.**

Note: *In the given formula, we can write 2 × intenal radius = internal diameter.*

Illustrative Example

Ex.: A hollow cylindrical tube open at both ends is made of iron 2 cm thick. If the internal diameter be 50 cm and the length of the tube be 140 cm, find the volume of iron in it.

Soln: Applying the above theorem,
Here, internal diameter = 50 cm.

$\therefore$ internal radius $= \frac{50}{2} = 25$ cm

$$\text{Required volume} = \frac{22}{7} \times 140 \times (25 \times 2 + 2) \times 2$$

$$= \frac{22}{7} \times 140 \times 52 \times 2 = 45760 \text{ cu cm.}$$

Exercise

1. A hollow cylindrical tube open at both ends is made of iron 4 cm thick. If the internal diameter be 40 cm and the length of the tube be 144 cm, find the volume of iron in it.
   a) 25344 π b) 23544 π
   c) 26344 π d) None of these
2. A hollow cylindrical tube open at both ends is made of iron 1 cm thick. If the internal diameter be 20 cm and the length of the tube be 100 cm, find the volume of iron in it.
   a) 6000 cu cm
   b) 6600 cu cm
   c) 5600 cu cm
   d) None of these
3. A hollow cylindrical tube open at both ends is made of iron 2 cm thick. If the internal diameter be 33 cm and the length of the tube be 70 cm, find the volume of iron in it.
   a) 12400 cu cm b) 15400 cu cm
   c) 13800 cu cm d) 16400 cu cm

Answers

1. a 2. b 3. b

Rule 30

Theorem: ***A hollow cylindrical tube open at both ends is made of a thick metal. If the internal and external diameter or radius of the tube are given, then the volume of metal is given by π × height × $\left[(External\ radius)^2 - (Internal\ radius)^2\right]$ cu. cm.***

Illustrative Example

Ex.: A hollow cylindrical tube open at both ends is made of iron. If the external and internal radius of the tube are 25 cm and 23 cm respectively, find the volume of iron in it.

Soln: Applying the above theorem, we have volume of iron

$$= \frac{22}{7} \times 140 \times \left(25^2 - 23^2\right) = 42240 \text{ cu. cm.}$$

Exercise

1. The internal diameter of an iron pipe is 6 cm and the length is 2.8 metres. If the thickness of the metal be 5 mm and 1 cu cm of iron weighs 8 gm, find the weight of the pipe.
   a) 288.2 kg b) 22.88 kg
   c) 822.2 kg d) None of these
2. In a hollow cylinder made of iron, the volume of iron is 1430 cu cm. If the length of the cylinder be 35 cm and its external diameter be 14 cm, find the thickness of the cylinder.
   a) 1 cm b) 2 cm c) 1.5 cm d) 2.5 cm
3. If 1 cm^3 cast iron weighs 21 gm, then weight of a cast iron pipe of length 1 m with a bore of 3 cm and in which thickness of metal is 1 cm, is:
   a) 21 kg b) 24.2 kg c) 26.4 kg d) 18.6 kg
4. The radius of the inner surface of a leaden pipe is 1.5 dm, and the radius of the outer surface is 1.9 dm. If the pipe be melted and formed into a solid cylinder of the same length as before, find its radius.
   a) 1 dm b) 11.7 dm c) 1.17 dm d) 2.17 dm

Answers

1. b; **Hint:** Volume of the metal in the pipe

$$= \frac{22}{7} \times 280 \times \left[(3.5)^2 - (3)^2\right] = 2860 \text{ cu cm}$$

[External diameter = 3 + 0.5 = 3.5 cm]

$$\therefore \text{ Weight of the pipe} = \left(2860 \times \frac{8}{1000}\right) = 22.88 \text{ kg.}$$

2. a; **Hint:** $\frac{22}{7} \times 35 \times \left[7^2 - r^2\right] = 1430$

or, $5390 - 110r^2 = 1430$

or, $110r^2 = 5390 - 1430 = 3960$

or, $r^2 = \frac{3960}{110} = 36$ or, r = 6 cm.

$\therefore$ thickness of the cylinder = (7 - 6) = 1 cm.

3. c; **Hint:** External radius = 2.5 cm; Internal radius = 1.5 cm.

$\therefore$ Volume of metal

$$= \left[\pi \times (2.5)^2 \times 100 - \pi \times (1.5)^2 \times 100\right] cm^3$$

$$= \frac{22}{7} \times 100\left[(2.5)^2 - (1.5)^2\right] = \frac{8800}{7} cm^3$$

$$\text{Weight of the metal} = \left(\frac{8800}{7} \times 21 \times \frac{1}{1000}\right) = 26.4 \text{ kg.}$$

4. c; **Hint:** $\pi \times h \times (1.9^2 - 1.5^2) = \pi r^2 h$

or, $r^2 = 1.9^2 - 1.5^2 = 1.36$

$\therefore$ $r = \sqrt{1.36} = 1.661 \approx 1.17$ dm.

Rule 31

Theorem: ***A hollow cylindrical tube open at both ends is made of a thick metal. If the external diameter or radius and length of the tube are given, then the volume of metal is given by***

$[\pi \times height \times (2 \times outer\ radius - thickness) \times thickness]$ ***cu. units.***

Note: *In the given formula, we can write 2 × outer radius = outer diameter.*

Illustrative Example

Ex.: A hollow cylindrical tube open at both ends is made of iron 2 cm thick. If the external diameter be 50 cm and the length of the tube be 140 cm, find the volume of iron in it.

Soln: Applying the above theorem,
Here, external diameter = 50 cm

$$\therefore \text{ external radius} = \frac{50}{2} = 25 \text{ cm.}$$

$$\text{Required volume} = \frac{22}{7} \times 140(25 \times 2 - 2) \times 2$$

$$= \frac{22}{7} \times 140 \times 48 \times 2 = 42240 \text{ cu cm.}$$

Exercise

1. Find the volume of the material in a cylindrical tube in cubic dm, the radius of the outer surface being 10 dm the thickness 0.4 dm, and the height 9 dm.
   a) 22.76 cub m b) 221.67 cub m
   c) 221.76 cub m d) 220.67 cub m

2. Find to the nearest integer the inside volume of a hollow cylinder, (open at both ends), whose external diameter is 3.5 metres, thickness 0.05 metre and height 2.083 metres. (Take=3.1416)
a) 19 cub m b) 21 cub m c) 14 cub m d) 28 cub m
3. A hollow cylindrical tube open at both ends is made of iron 1 cm thick. If the external diameter be 22 cm and the length of the tube be 70 cm, find the volume of iron in it.
a) 4620 cu cm b) 4660 cu cm
c) 3620 cu cm d) 3820 cu cm

Answers

1. c 2. a 3. a

Rule 32

Theorem: ***If a rectangular sheet is rolled into a cylinder so that the one side becomes the height of the cylinder then the volume of the cylinder so formed is given by***

$$\frac{height \times (other\ side\ of\ the\ sheet)^2}{4\pi}.$$

Illustrative Example

Ex.: A rectangular sheet with dimension 22 m × 10 m is rolled into a cylinder so that the smaller side becomes the height of the cylinder. What is the volume of the cylinder so formed?

Soln: Applying the above theorem, we have

$$\text{Volume} = \frac{10 \times (22)^2}{4 \times \frac{22}{7}} = 385 \text{ cu m.}$$

Note: The height is 10 m since it is the smaller side. The other side is obviously 22 m.

Exercise

1. A rectangular sheet with dimension 11 m × 8 m is rolled into a cylinder so that the smaller side becomes the height of the cylinder. What is the volume of the cylinder so formed?
a) 77 cu m b) 87 cu m
c) 66 cu m d) Data inadequate
2. A rectangular sheet with dimension 33 cm × 24 cm is rolled into a cylinder so that the smaller side becomes the height of the cylinder. What is the volume of the cylinder so formed?
a) 2069 cu cm b) 2709 cu cm
c) 2079 cu cm d) 2089 cu cm
3. A rectangular sheet with dimension 44 m × 20 m is rolled into a cylinder so that the smaller side becomes the height of the cylinder. What is the volume of the cylinder so formed?
a) 3080 cu m b) 8030 cu m
c) 5060 cu m d) None of these

Answers

1. a 2. c 3. a

Cases of sphere changing shapes

Rule 33

Theorem: ***If a sphere of certain diameter or radius is drawn into a cylinder of certain diameter or radius, then the length or height of the cylinder is given by***

$$\frac{4 \times (radius\ of\ sphere)^3}{3 \times (radius\ of\ cylinder)^2}.$$

Illustrative Example

Ex.: A copper sphere of diameter 18 cm is drawn into a wire of diameter 4 mm. Find the length of the wire.

Soln: **Quicker Method:**
When a sphere is converted into a cylinder (**Note that wire is basically a cylinder**) the length of the wire is given by the rule:

$$\textbf{length of cylinder} = \frac{4 \times (\text{radius of sphere})^3}{3 \times (\text{radius of cylinder})^2}$$

$$\therefore \text{ In the given question, length} = \frac{4 \times (90)^3}{3 \times (2)^2}$$

$= 243000$ mm $= 24300$ cm.

Note: Sphere converted into a cylinder and cylinder converted into a sphere are one and the same thing.

Exercise

1. The diameter of a copper sphere is 6 cm. The sphere is melted and drawn into a long wire of uniform circular cross section. If the length of the wire is 36 cm, find its radius.
a) 1 cm b) 2 cm
c) $\sqrt{3}$ cm d) Can't be determined
2. A copper sphere of diameter 12 cm is drawn into a wire of diameter 2 cm. Find the length of the wire.
a) 288 cm b) 284 cm c) 286 cm d) None of these
3. A copper sphere of diameter 24 cm is drawn into a wire of diameter 4 cm. Find the length of the wire.
a) 576 cm b) 676 cm c) 756 cm d) 776 cm
4. A copper sphere of diameter 18 cm is drawn into a wire of diameter 6 cm. Find the length of the wire.
a) 108 cm b) 180 cm
c) 190 cm d) None of these

Answers

1. a; **Hint:** $36 = \frac{4}{3} \times \frac{3^2}{r^2}$ $\therefore$ r = 1 cm. 2. a
3. a 4. a

Rule 34

Theorem: ***A sphere is converted into a cylinder. If the length and the radius of the cylinder are given, then the radius of the sphere is given by***

$$\sqrt[3]{\frac{3}{4}(\textit{length of cylinder})(\textit{radius of cylinder})^2}$$

Illustrative Example

Ex.: A cylinder of radius 2 cm and height 15 cm is melted and the same mass is used to create a sphere. What will be the radius of the sphere?

Soln: Applying the above theorem, we have

the radius of sphere = $\sqrt[3]{\frac{3}{4}\times 15\times 2\times 2} = \sqrt[3]{45}$

Exercise

1. A cylinder of radius 9 cm and height 12 cm is melted and the same mass is used to create a sphere. What will be the radius of the sphere?
   a) 9 cm b) $\sqrt[3]{27}$ cm
   c) 8 cm d) 7 cm
2. A cylinder of radius 6 cm and height 8 cm is melted and the same mass is used to create a sphere. What will be the radius of the sphere?
   a) 7 cm b) 4 cm
   c) 6 cm d) $\sqrt[3]{36}$ cm
3. A cylinder of radius 12 cm and height 16 cm is melted and the same mass is used to create a sphere. What will be the radius of the sphere?
   a) 12 cm b) 10 cm c) 8 cm d) 6 cm
4. A cylinder of radius 15 cm and height 20 cm is melted and the same mass is used to create a sphere. What will be the radius of the sphere?
   a) 12 cm b) 18 cm
   c) 15 cm d) Data inadequate

Answers

1. a 2. c 3. a 4. c

Rule 35

Theorem: ***If a sphere of certain diameter or radius is drawn into a cylinder of certain height or length, then the radius of cylinder is given by*** $\sqrt{\frac{4\times(\textit{radius of sphere})^3}{3\times(\textit{length of cylinder})}}$.

Illustrative Example

Ex.: A copper sphere of 36 m diameter is drawn into a cylindrical wire of length 7.29 km. What is the radius of wire.

Soln: Applying the above theorem, we have

the radius of wire

$$= \sqrt{\frac{4\times\left(\frac{36}{2}\right)^3}{3\times 7.29\times 1000}} = \sqrt{\frac{4\times 18\times 18\times 18}{3\times 7290}} = 1.03 \text{ m.}$$

Exercise

1. A copper sphere of 12 m diameter is drawn into a cylindrical wire of length 1.8 km. What is the radius of wire.
   a) 4 m b) 40 cm c) 40 mm d) Data inadequate
2. A copper sphere of 15 m diameter is drawn into a cylindrical wire of length 4.5 km. What is the radius of wire.
   a) 1 m b) 2 m c) 0.1 m d) 0.1 cm
3. A copper sphere of 18 m diameter is drawn into a cylindrical wire of length 5.4 km. What is the radius of wire.
   a) 1 cm b) 1.2 cm c) 1.2 m d) 1 m

Answers

1. b 2. a 3. c

Rule 36

Theorem: ***If sphere is melted to form a cylinder whose height is n times its radius, then the ratio of radii of sphere to the cylinder is*** $\left(\frac{3}{4}\times n\right)^{1/3}$.

Illustrative Example

Ex.: A sphere is melted to form a cylinder whose height is $4\frac{1}{2}$ times its radius. What is the ratio of radii of sphere to the cylinder?

Soln: **Detail Method:** Let the radius of the sphere and cylinder be 'R' and 'r' respectively.

Volume of the cylinder

$$= \pi r^2 h = \pi r^2\left(\frac{9}{2}r\right)\ \left[\because h = \frac{9}{2}r\right] = \frac{9}{2}\pi r^3$$

Volume of the sphere = $\frac{4}{3}\pi R^3$

As per the question,

Volume of the sphere = Volume of the cylinder

or, $\frac{4}{3}\pi R^3 = \frac{9}{2}\pi r^3$ or, $\left(\frac{R}{r}\right)^3 = \frac{27}{8}$

$\therefore$ R : r = 3 : 2

Quicker Method: Applying the above theorem, we have

the required ratio = $\left(\frac{9}{2}\times\frac{3}{4}\right)^{1/3} = \left(\frac{27}{8}\right)^{1/3} = \frac{3}{2} = 3:2$

Exercise

1. A sphere is melted to form a cylinder whose height is 36 times its radius. What is the ratio of radii of sphere to the cylinder?
a) 3 : 2 b) 4 : 1 c) 9 : 1 d) 3 : 1
2. A sphere is melted to form a cylinder whose height is $15\frac{3}{16}$ times its radius. What is the ratio of radii of sphere to the cylinder?
a) 3 : 1 b) 7 : 4 c) 9 : 4 d) Data inadequate
4. A sphere is melted to form a cylinder whose height is $70\frac{5}{16}$ times its radius. What is the ratio of radii of sphere to the cylinder?
a) 5 : 4 b) 11 : 4 c) 15 : 4 d) 13 : 4

Answers

1. d 2. c 3. c

Rule 37

Theorem: ***If a cone, whose height is n times of its radius, is melted to form a sphere, assuming that there is no loss of material in this process, ratio of radius of the sphere to that of the cone is*** $\left(\frac{n}{4}\right)^{1/3}$.

Illustrative Example

Ex.: A cone, whose height is half of its radius, is melted to form a sphere. Find the ratio of radius of the sphere to that of the cone.

Soln: **Detail Method:** Let the raidus of sphere and cylinder be R and r respectively.

Volume of the cone $= \frac{1}{3}\pi r^2 h = \frac{1}{3}\pi r^2\left(\frac{r}{2}\right) = \frac{\pi}{3\times 2}r^2$

Volume of the sphere $= \frac{4}{3}\pi R^3$

As per the question,
Volume of the sphere = Volume of the cone

or, $\frac{4}{3}\pi R^3 = \frac{\pi}{3}\times\frac{r^3}{2}$

or, $\left(\frac{R}{r}\right)^3 = \frac{1}{8}$

$\therefore R : r = 1 : 2$

Quicker Method: Applying the above theorem,

Required ratio $= \left[\frac{1/2}{4}\right]^{1/3} = \left[\frac{1}{8}\right]^{1/3} = \frac{1}{2} = 1 : 2$

Exercise

1. A cone, whose height is $\frac{1}{16}$ of its radius, is melted to form a sphere. Find the ratio of radius of the sphere to that of the cone.
a) 1 : 3 b) 1 : 4 c) 2 : 3 d) 1 : 2
2. A cone, whose height is $13\frac{1}{2}$ times of its radius, is melted to form a sphere. Find the ratio of radius of the sphere to that of the cone.
a) 3 : 2 b) 2 : 3 c) 4 : 3 d) 3 : 1
3. A cone, whose height is 32 times of its radius, is melted to form a sphere. Find the ratio of radius of the sphere to that of the cone.
a) 1 : 2 b) 3 : 2 c) 2 : 1 d) None of these

Answers

1. b 2. a 3. c

Rule 38

Theorem: ***When one cylinder is converted into many small spheres, then the number of small spheres is given by the following formula,***

$$\textit{Number of small spheres} = \frac{\textit{Volume of cylinder}}{\textit{Volume of 1 sphere}}$$

Illustrative Example

Ex.: How many bullets can be made out of a lead cylinder 28 cm high and 6 cm radius, each bullet being 1.5 cm in diameter?

Soln: In this case, one cylinder is not converted into just one sphere but many spheres are being made. Here we will use the above formula:

$$\text{Number of bullets} = \frac{\text{volume of cylinder}}{\text{volume of 1 bullet}}$$

$$= \frac{\pi\times 6\times 6\times 28}{\frac{4}{3}\times\pi\times 0.75\times 0.75\times 0.75} = 1792.$$

Exercise

1. A cylinder of length 1 metre and diameter 15 cm is melted down and cast into spheres of diameter 5 cm. How many spheres can be made?
a) 270 b) 260 c) 290 d) 370
2. A cylinder of length 16 metres and diameter 4 cm is melted down and cast into spheres of diameter 1.6 cm. How many spheres can be made?
a) 475 b) 575 c) 375 d) 675
3. A cylinder of length 24 metres and diameter 5 cm is melted down and cast into spheres of diameter 2 cm. How many spheres can be made?
a) 450 b) 360 c) 480 d) 950

Answers

1. a 2. c 3. a

Rule 39

Theorem: ***If a sphere of radius R units has a spherical cavity of radius r units, then the volume of the spherical shell is given by*** $\left[\frac{4}{3}\pi\left(R^3 - r^3\right)\right]$ ***cubic units.***

Illustrative Example

Ex.: A sphere of radius 5 cm has a spherical cavity of radius 3 cm. Find the volume of the spherical shell.

Soln: Applying the above theorem,

Volume of the spherical shell $= \frac{4}{3} \times \frac{22}{7} \times \left[5^3 - 3^3\right]$

$= \frac{4}{3} \times \frac{22}{7} \times 98 = 410\frac{2}{3}$ cm³.

Exercise

1. A sphere of radius 5 cm has a spherical cavity of radius 4 cm. Find the volume of the spherical shell.
   a) 12 π b) $\frac{244}{3}\pi$ c) 144π d) Data inadequate
2. A sphere of radius 6 cm has a spherical cavity of radius 3 cm. Find the volume of the spherical shell.
   a) 253 π b) 255 π c) 252 π d) 152 π
3. A sphere of radius 7 cm has a spherical cavity of radius 5 cm. Find the volume of the spherical shell.
   a) $\frac{872}{3}\pi$ b) $\frac{874}{3}\pi$
   c) $\frac{782}{3}\pi$ d) Data inadequate

Answers

1. b 2. c 3. a

Rule 40

Theorem: ***The radius of a greatest sphere that can be carved out of a cone of radius r and height h is*** $\left(\frac{rh}{r+l}\right)$.

Where, l = slant height of the cone.

$= \sqrt{r^2 + h^2}$

Illustrative Example

Ex.: There is a cone of radius 3 metres and height 4 metres. Find the radius of the greatest sphere that can be carved out of that cone.

Soln: Applying the above theorem,

slant height $= \sqrt{3^2 + 4^2} = \sqrt{25} = 5$ and

radius of the greatest sphere

$= \frac{3 \times 4}{3+5} = \frac{12}{8} = 1\frac{1}{2} = 1.5$ metres.

Exercise

1. There is a cone of radius 18 cm and height 24 cm. Find the radius of the greatest sphere that can be carved out of that cone.
   a) 9 cm b) 12 cm
   c) 6 cm d) Can't be determined
2. There is a cone of radius 5 cm and height 12 cm. Find the radius of the greatest sphere that can be carved out of that cone.
   a) 3 cm b) $3\frac{1}{3}$ cm c) $4\frac{1}{3}$ cm d) $5\frac{1}{3}$ cm
3. There is a cone of radius 10 cm and height 24 cm. Find the radius of the greatest sphere that can be carved out of that cone.
   a) $6\frac{2}{3}$ cm b) 6 cm c) $6\frac{1}{3}$ d) Data inadequate
4. There is a cone of radius 9 cm and height 40 cm. Find the radius of the greatest sphere that can be carved out of that cone.
   a) $7\frac{1}{5}$ cm b) $8\frac{1}{5}$ cm
   c) $7\frac{2}{5}$ cm d) $8\frac{3}{5}$ cm
5. There is a cone of radius 7 cm and height 24 cm. Find the radius of the greatest sphere that can be carved out of that cone.
   a) $5\frac{1}{4}$ cm b) 6 cm
   c) $6\frac{2}{3}$ cm d) Data inadequate

Answers

1. a 2. b 3. a 4. a 5. a

Rule 41

Theorem: ***When a sphere disintegrates into many identical spheres, then the number of smaller identical spheres are given by*** $\left(\frac{\text{bigger radius}}{\text{smaller radius}}\right)^3$.

Illustrative Example

Ex.: Find the number of lead balls of diameter 1 cm each that can be made from a sphere of diameter 16 cm.

Soln: Detail Method:

$$\text{Number of balls} = \frac{\text{Volume of big sphere}}{\text{Volume of 1 small sphere}}$$

$$= \frac{\frac{4}{3}\pi \times 8 \times 8 \times 8}{\frac{4}{3}\pi \times 0.5 \times 0.5 \times 0.5} = 4096.$$

Quicker Method: Applying the above theorem, we have

$$\text{the required number} = \left(\frac{8}{0.5}\right)^3 = 16^3 = 4096$$

Exercise

1. Find the number of lead balls of diameter 2 cm each that can be made from a sphere of diameter 14 cm.
   a) 343 b) 243 c) 443 d) 433
2. Find the number of lead balls of diameter 1.5 cm each that can be made from a sphere of diameter 30 cm.
   a) 4000 b) 8000 c) 16000 d) Data inadequate
3. Find the number of lead balls of diameter 1 cm each that can be made from a sphere of diameter 18 cm.
   a) 5832 b) 8332 c) 8532 d) 5882

Answers

1. a 2. b 3. a

Rule 42

Theorem: ***If the ratio of surface areas of the two spheres are given, then the ratio of their volumes will be obtained from the following result.***
(Ratio of the surface areas)^3 = (Ratio of volumes)^2

Illustrative Example

Ex.: The curved surface areas of two spheres are in the ratio 1 : 4. Find the ratio of their volumes.

Soln: Applying the above theorem, we have
$(1:4)^3$ = (ratio of volumes)^2
or, $(1:64)$ = (ratio of volumes)^2
or, $\sqrt{1:64} = 1:8$ = ratio of volumes.

Exercise

1. The curved surface areas of two spheres are in the ratio 4 : 5. Find the ratio of their volumes.
   a) $8:5$ b) $8:5\sqrt{3}$ c) $8:5\sqrt{5}$ d) $5\sqrt{5}:8$
2. The curved surface areas of two spheres are in the ratio 1 : 3. Find the ratio of their volumes.
   a) $1:3\sqrt{2}$ b) $1:3\sqrt{3}$
   c) $1:3$ d) Can't be determined
3. The curved surface areas of two spheres are in the ratio 3 : 4. Find the ratio of their volumes.
   a) $3\sqrt{3}:4$ b) $3:4\sqrt{3}$ c) $3\sqrt{3}:8$ d) $8:3\sqrt{2}$
4. The curved surface areas of two spheres are in the ratio 1 : 2. Find the ratio of their volumes.
   a) $1:\sqrt{2}$ b) $2:1$ c) $2:2\sqrt{2}$ d) $1:2\sqrt{2}$

Answers

1. c 2. b 3. c 4) d

Rule 43

Theorem: ***If the ratio of the radii of two spheres are given, then the ratio of their surface areas will be obtained from the following result.***
(Ratio of radii)^2 = Ratio of surface areas.

Illustrative Example

Ex.: The radii of two spheres are in the ratio of 1 : 2. What is the ratio of their surface areas?

Soln: By the above theorem, we have,
(ratio of surface areas) = (ratio of radii)^2
$= (1:2)^2 = 1:4.$

Exercise

1. If the radii of two spheres are in ratio 1 : 4, then ratio of their surface areas will be
   a) 1 : 2 b) 1 : 4 c) 1 : 8 d) 1 : 16
   (NDA Exam 1990)
2. The radii of two spheres are in the ratio of 1 : 3. What is the ratio of their surface areas?
   a) 1 : 8 b) 1 : 9 c) 1 : 6 d) Data inadequate
3. The radii of two spheres are in the ratio of 2 : 5. What is the ratio of their surface areas?
   a) 4 : 25 b) 4 : 5 c) 6 : 25 d) 25 : 2
4. The radii of two spheres are in the ratio of 3 : 4. What is the ratio of their surface areas?
   a) 9 : 16 b) 16 : 25 c) 9 : 14 d) Data inadequate

Answers

1. d 2. b 3. a 4. a

Rule 44

Theorem: ***If the ratio of the radii of two spheres are given, then the ratio of their volumes will be obtained from the following result.***

$(Ratio\ of\ radii)^3$ = ***Ratio of volumes***

Illustrative Example

Ex.: The radii of two spheres are in the ratio of 1 : 2. What is the ratio of their volumes?

Soln: Applying the above theorem, we have
the ratio of volumes
$= (Ratio\ of\ radii)^3 = (1:2)^3 = 1:8$

Exercise

1. The radii of two spheres are in the ratio of 1 : 3. What is

the ratio of their volumes?
a) 1 : 9 b) 1 : 27 c) 1 : 18 d) Data inadequate

2. The radii of two spheres are in the ratio of 2 : 5. What is the ratio of their volumes?
a) 2 : 25 b) 4 : 25 c) 8 : 125 d) 8 : 25

3. The radii of two spheres are in the ratio of 4 : 9. What is the ratio of their volumes?
a) 16 : 81 b) 64 : 729
c) 81 : 64 d) Can't be determined

Answers

1. b 2. c 3. b

Two Cylinders

A. When volumes are equal

Rule 45

Theorem: ***If the ratio of the heights of two circular cylinders of equal volume are given, then the ratio of their radii is given by the following result.***

Ratio of radii = $\sqrt{\text{inverse ratio of heights}}$

Illustrative Example

Ex.: Two circular cylinders of equal volume have their heights in the ratio of 1 : 2. Ratio of their radii is?

Soln: Applying the above theorem, we have the ratio of radii

$= \sqrt{\text{inverse ratio of heights}} = \sqrt{2:1} = \sqrt{2}:1$.

Note: This formula is also applicable for two cones.

Exercise

1. Two circular cylinders of equal volume have their heights in the ratio of 1 : 4. Ratio of their radii is?
a) 2 : 1 b) 4 : 1 c) 16 : 1 d) Data inadequate

2. Two circular cylinders of equal volume have their heights in the ratio of 16 : 25. Ratio of their radii is?
a) 4 : 5 b) 5 : 4 c) 3 : 2 d) 2 : 3

3. Two circular cylinders of equal volume have their heights in the ratio of 4 : 9. Ratio of their radii is?
a) 2 : 3 b) 1 : 2 c) 2 : 1 d) 3 : 2

4. Two circular cylinders of equal volume have their heights in the ratio of 1 : 9. Ratio of their radii is?
a) 1 : 3 b) 3 : 1
c) 2 : 3 d) Data inadequate

Answers

1. a 2. b 3. d 4. b

Rule 46

Theorem: ***If the ratio of curved surface areas of two circular cylinders of equal volume are given, then the ratio of their heights is given by the following result.***

Ratio of curved surface areas = $\sqrt{\text{ratio of heights}}$.

Illustrative Example

Ex.: Two circular cylinders of equal volume have their heights in the ratio of 1 : 2. Find the ratio of their curved surface areas.

Soln: Applying the above theorem, we have the ratio of curved surface areas = $\sqrt{\text{Ratio of heights}}$.

$\therefore$ Ratio of curved surface areas = $\sqrt{1:2} = 1:\sqrt{2}$

Note: If in place of cylinders, cones are given, this formula is applicable.

Exercise

1. Two circular cylinders of equal volume have their heights in the ratio of 1 : 4. Find the ratio of their curved surface areas.
a) 1 : 2 b) 1 : 4 c) 2 : 1 d) 4 : 1

2. Two circular cylinders of equal volume have their heights in the ratio of 1 : 9. Find the ratio of their curved surface areas.
a) 1 : 3 b) 3 : 1 c) 2 : 1 d) 1 : 2

3. Two circular cylinders of equal volume have their heights in the ratio of 4 : 9. Find the ratio of their curved surface areas.
a) 9 : 4 b) 4 : 9 c) 2 : 3 d) 3 : 2

4. Two circular cylinders of equal volume have their heights in the ratio of 16 : 25. Find the ratio of their curved surface areas.
a) 4 : 5 b) 5 : 4 c) 16 : 5 d) 4 : 25

Answers

1. a 2. a 3. c 4. a

Rule 47

Theorem: ***If the ratio of radii of two circular cylinders of equal volume are given, then the ratio of their curved surface areas are given by the following result.***
Ratio of curved surface areas = inverse ratio of radii.

Illustrative Example

Ex.: Two circular cylinders of equal volume have their radii in the ratio of 1 : 2. Find the ratio of their curved surface areas.

Soln: Applying the above theorem, we have the ratio of curved surface areas
= inverse ratio of radii = 2 : 1

Note: If in place of cylinders, cones are given, this formula is applicable.

Exercise

1. Two circular cylinders of equal volume have their radii in the ratio of 1:3. Find the ratio of their curved surface areas.
a) 3 : 1 b) 1 : 3 c) 1 : 2 d) 9 : 1

2. Two circular cylinders of equal volume have their radii in the ratio of 2 : 5. Find the ratio of their curved surface areas.

a) 5 : 2 b) 25 : 4 d) 4 : 25 d) Data inadequate

3. Two circular cylinders of equal volume have their radii in the ratio of 4 : 9. Find the ratio of their curved surface areas.
   a) 9 : 4 b) 2 : 3 c) 3 : 2 d) 4 : 9

Answers

1. a 2. a 3. a

B. When radii are equal

Rule 48

Theorem: ***If the ratio of heights of two circular cylinders of equal radii are given then the ratio of their volumes are given by the following result.***
Ratio of volumes = Ratio of heights.

Illustrative Example

Ex.: Two circular cylinders of equal radii have their height in the ratio of 1 : 2. Find the ratio of their volumes.

Soln: Applying the above theorem, we have
the ratio of volumes = ratio of heights
$\therefore$ required answer = 1 : 2

Note: If in place of cylinders, cones are given, this formula is applicable.

Exercise

1. Two circular cylinders of equal radii have their height in the ratio of 1 : 3. Find the ratio of their volumes.
   a) 3 : 1 b) 1 : 9 c) 9 : 1 d) 1 : 3
2. Two circular cylinders of equal radii have their height in the ratio of 4 : 5. Find the ratio of their volumes.
   a) 4 : 5 b) 16 : 25 c) 25 : 16 d) 5 : 4
3. Two circular cylinders of equal radii have their height in the ratio of 3 : 7. Find the ratio of their volumes.
   a) 7 : 3 b) 3 : 7 c) 9 : 49 d) 49 : 9

Answers

1. d 2. a 3. b

Rule 49

Theorem: ***If the ratio of heights of two circular cylinders of equal radii are given then the ratio of their curved surface areas are given by the following result.***
Ratio of curved surface areas = Ratio of heights.

Illustrative Example

Ex.: Two circular cylinders of equal radii have their height in the ratio of 1 : 2. Find the ratio of their curved surface areas.

Soln: Applying the above theorem, we have the ratio of curved surface areas = 1 : 2.

Note: If in place of cylinders, cones are given, this formula is applicable.

Exercise

1. Two circular cylinders of equal radii have their height in the ratio of 1 : 9. Find the ratio of their curved surface areas.
   a) 1 : 3 b) 3 : 1 c) 1 : 9 d) Data inadequate
2. Two circular cylinders of equal radii have their height in the ratio of 1 : 4. Find the ratio of their curved surface areas.
   a) 4 : 1 b) 2 : 1 c) 1 : 2 d) 1 : 4
3. Two circular cylinders of equal radii have their height in the ratio of 4 : 9. Find the ratio of their curved surface areas.
   a) 2 : 3 b) 3 : 2 c) 9 : 4 d) 4 : 9

Answers

1. c 2. d 3. d

Rule 50

Theorem: ***If the ratio of volumes of two circular cylinders of equal radii are given then the ratio of their curved surface areas are given by the following result.***
Ratio of volumes = Ratio of curved surface areas.

Illustrative Example

Ex.: Two circular cylinders of equal radii have their volumes in the ratio of 3 : 1. Find the ratio of their curved surface areas.

Soln: Applying the above theorem, we have the ratio of curved surface areas = Ratio of volumes = 3 : 1.

Note: If in place of cylinders, cones are given, this formula is applicable.

Exercise

1. Two circular cylinders of equal radii have their volumes in the ratio of 2 : 1. Find the ratio of their curved surface areas.
   a) 1 : 2 b) 1 : 4 c) 4 : 1 d) 2 : 1
2. Two circular cylinders of equal radii have their volumes in the ratio of 9 : 4. Find the ratio of their curved surface areas.
   a) 3 : 2 b) 2 : 3 c) 36 : 81 d) 9 : 4
3. Two circular cylinders of equal radii have their volumes in the ratio of 4 : 7. Find the ratio of their curved surface areas.
   a) 16 : 49 b) 4 : 7 c) 2 : $\sqrt{7}$ d) Data inadequate

Answers

1. d 2. d 3. b

C. When heights are equal

Rule 51

Theorem: ***If the ratio of radii of two circular cylinders of equal heights are given, then the ratio of their volumes is given by the following result.***
Ratio of volumes = (Ratio of radii)^2

Illustrative Example

Ex.: Two circular cylinders of equal heights have their radii in the ratio of 2 : 3. Ratio of their volumes is ______.

Soln: Applying the above theorem, we have the ratio of volumes = (Ratio of radii)^2 = $(2:3)^2 = 4:9$

Note: If in place of cylinders, cones are given, this formula is applicable.

Exercise

1. Two circular cylinders of equal heights have their radii in the ratio of 1 : 3. Ratio of their volumes is ______.
   a) 3 : 1 b) 1 : $\sqrt{3}$ c) 9 : 1 d) 1 : 9
2. Two circular cylinders of equal heights have their radii in the ratio of 4 : 5. Ratio of their volumes is ______.
   a) 5 : 4 b) 2 : $\sqrt{5}$ c) 16 : 25 d) Data inadequate
3. Two circular cylinders of equal heights have their radii in the ratio of 2 : 1. Ratio of their volumes is ______.
   a) 4 : 1 b) 1 : 4 c) 1 : $\sqrt{2}$ d) $\sqrt{2}$: 1
4. Two circular cylinders of equal heights have their radii in the ratio of 3 : 4. Ratio of their volumes is ______.
   a) 9 : 16 b) 16 : 9 c) 3 : $\sqrt{2}$ d) $\sqrt{3}$: 2

Answers

1. d 2. c 3. a 4. a

Rule 52

Theorem: ***If the ratio of radii of two circular cylinders of equal heights are given, then the ratio of their curved surface areas is given by the following results.***
Ratio of curved surface areas = ratio of radii

Illustrative Example

Ex.: Two circular cylinders of equal heights have their radii in the ratio of 2 : 3. Find the ratio of their curved surface areas.

Soln: Applying the above theorem, we have the ratio of curved surface areas = ratio of radii = 2 : 3

Note: If in place of cylinders, cones are given, this formula is applicable.

Exercise

1. Two circular cylinders of equal heights have their radii in the ratio of 1 : 4. Find the ratio of their curved surface areas.
   a) 1 : 2 b) 2 : 1 c) 1 : 4 d) 4 : 1
2. Two circular cylinders of equal heights have their radii in the ratio of 2 : 5. Find the ratio of their curved surface areas.
   a) 5 : 2 b) 2 : 5 c) 3 : 2 d) 4 : 25
3. Two circular cylinders of equal heights have their radii in the ratio of 4 : 9. Find the ratio of their curved surface areas.
   a) 2 : 3 b) 3 : 2 c) 9 : 4 d) 4 : 9

Answers

1. c 2. b 3. d

Rule 53

Theorem: ***If the ratio of curved surface areas of two circular cylinders of equal heights are given, then the ratio of their volumes is given by the following results.***
Ratio of volumes = (Ratio of curved surface areas)^2

Illustrative Example

Ex.: Two circular cylinders of equal heights have their curved surface areas in the ratio of 2 : 3. Find the ratio of their volumes.

Soln: Applying the above theorem, we have,
Ratio of volumes = (Ratio of curved surface areas)^2
$= (2:3)^2 = 4:9$

Note: If in place of cylinders, cones are given, this formula is applicable.

Exercise

1. Two circular cylinders of equal heights have their curved surface areas in the ratio of 1 : 2. Find the ratio of their volumes.
   a) 1 : 4 b) 2 : 1 c) 4 : 1 d) 1 : $\sqrt{2}$
2. Two circular cylinders of equal heights have their curved surface areas in the ratio of 1 : 9. Find the ratio of their volumes.
   a) 1 : 81 b) 1 : 3 c) 3 : 1 d) Data inadequate
3. Two circular cylinders of equal heights have their curved surface areas in the ratio of 2 : 5. Find the ratio of their volumes.
   a) 2 : 5 b) 5 : 2 c) 4 : 25 d) 25 : 4

Answers

1. a 2. a 3. c

D. When curved surface areas are equal.

Rule 54

Theorem: ***If the ratio of radii of two circular cylinders of equal curved surface areas are given, then the ratio of volumes is calculated from the following result.***
Ratio of volumes = Ratio of radii

Illustrative Example

Ex.: Two circular cylinders of equal curved surface areas have their radii in the ratio of 3 : 4. Find the ratio of their volumes.

Soln: Applying the above theorem,
Ratio of volumes = Ratio of radii = 3 : 4

Note: If in place of cylinders, cones are given, this formula is applicable.

Exercise

1. Two circular cylinders of equal curved surface areas have their radii in the ratio of 3 : 5. Find the ratio of their

volumes.

a) 5 : 3 b) 9 : 25 c) 25 : 9 d) 3 : 5

2. Two circular cylinders of equal curved surface areas have their radii in the ratio of 1 : 4. Find the ratio of their volumes.

a) 1 : 4 b) 2 : 1 c) 4 : 1 d) 1 : 16

3. Two circular cylinders of equal curved surface areas have their radii in the ratio of 2 : 3. Find the ratio of their volumes.

a) 4 : 9 b) $\sqrt{2} : \sqrt{3}$ c) 2 : 3 d) Data inadequate

Answers

1. d 2. a 3. c

Rule 55

Theorem: ***If the ratio of heights of two circular cylinders of equal curved surface areas are given, then the ratio of their volumes is calculated from the following result.***
Ratio of volumes = inverse ratio of heights.

Illustrative Example

Ex.: Two circular cylinders of equal curved surface areas have their heights in the ratio of 3 : 4. Find the ratio of their volumes.

Soln: Applying the above theorem, we have,
Ratio of volumes

$= \text{Inverse ratio of heights} = \frac{1}{3} : \frac{1}{4} = 4 : 3$

Note: If in place of cylinders, cones are given, this formula is applicable.

Exercise

1. Two circular cylinders of equal curved surface areas have their heights in the ratio of 2 : 3. Find the ratio of their volumes.

a) 3 : 2 b) 4 : 9 c) 9 : 4 d) Data inadequate

2. Two circular cylinders of equal curved surface areas have their heights in the ratio of 1 : 2. Find the ratio of their volumes.

a) 2 : 1 b) 1 : 4 c) 4 : 1 d) 1 : $\sqrt{2}$

3. Two circular cylinders of equal curved surface areas have their heights in the ratio of 5 : 4. Find the ratio of their volumes.

a) $\sqrt{5}$: 2 b) 4 : 5 c) 2 : $\sqrt{5}$ d) Data inadequate

Answers

1. a 2. a 3. b

Rule 56

Theorem: ***If the ratio of heights of two circular cylinders of equal curved surface areas is given, then the ratio of their radii is given by the following result.***
Ratio of radii = Inverse ratio of heights.

Illustrative Example

Ex.: Two circular cylinders of equal curved surface areas have their heights in the ratio of 3 : 4. Find the ratio of their radii.

Soln: Applying the above theorem, we have

$\text{Ratio of radii} = \text{Inverse ratio of heights} = \frac{1}{3} : \frac{1}{4} = 4 : 3.$

Exercise

1. Two circular cylinders of equal curved surface areas have their heights in the ratio of 2 : 3. Find the ratio of their radii.

a) 3 : 2 b) 4 : 9 c) $\sqrt{3} : \sqrt{2}$ d) $\sqrt{2} : \sqrt{3}$

2. Two circular cylinders of equal curved surface areas have their heights in the ratio of 1 : 4. Find the ratio of their radii.

a) 4 : 1 b) 1 : 2 c) 2 : 1 d) Data inadequate

3. Two circular cylinders of equal curved surface areas have their heights in the ratio of 2 : 5. Find the ratio of their radii.

a) 5 : 2 b) 2 : 5 c) 4 : 25 d) $\sqrt{5} : \sqrt{2}$

Answers

1. a 2. a 3. a

Two Cones

Rule 57

Theorem: ***If the ratio of slant heights of two right circular cones of equal curves surface areas is given, then the ratio of their radii is given by the following result.***
Ratio of radii = inverse ratio of slant heights.

Illustrative Example

Ex.: Two right circular cones of equal curved surface areas have their heights in the ratio of 3 : 4. Find the ratio of their radii.

Soln: Applying the above theorem,
Ratio of radii

$= \text{Inverse ratio of slant heights} = \frac{1}{3} : \frac{1}{4} = 4 : 3$

Exercise

1. Two right circular cones of equal curved surface areas have their heights in the ratio of 1 : 2. Find the ratio of their radii.

a) 2 : 1 b) 4 : 1 c) 1 : $\sqrt{2}$ d) 1 : 2

2. Two right circular cones of equal curved surface areas have their heights in the ratio of 2 : 3. Find the ratio of their radii.

a) 3 : 2 b) 3 : 1

c) $\sqrt{3} : \sqrt{2}$ d) Data inadequate

3. Two right circular cones of equal curved surface areas

have their heights in the ratio of 4 : 9. Find the ratio of their radii.

a) 2 : 3 b) 9 : 4 c) 3 : 2 d) 4 : 9

Answers

1. a 2. a 3. b

Two Cubes

Rule 58

Theorem: ***If the ratio of sides of two cubes is given, then the ratio of their volumes is calculated from the following result.***

Ratio of volumes = (Ratio of sides)^3

Illustrative Example

Ex.: Sides of two cubes are in the ratio 2 : 3. Find the ratio of their volumes.

Soln: Applying the above theorem, we have

Ratio of volumes

$$= (Ratio\ of\ sides)^3 = (2:3)^3 = 8:27$$

Exercise

1. If the volumes of two cubes are in the ratio 8 : 1, the ratio of their edges is:
   a) 8 : 1 b) 4 : 1 c) 2 : 1 d) 3 : 1
2. If the volumes of two cubes are in the ratio 1 : 2, the ratio of their edges is:
   a) 1 : 4 b) 4 : 1 c) 8 : 1 d) 1 : 8
3. If the volumes of two cubes are in the ratio 3 : 4, the ratio of their edges is:
   a) 9 : 16 b) 27 : 64 c) $\sqrt{3}$: 2 d) Data inadequate
4. If the volumes of two cubes are in the ratio 1 : 3, the ratio of their edges is:
   a) 1 : 27 b) 27 : 1 c) 9 : 1 d) 1 : 9

Answers

1. c; **Hint:** Ratio of volumes = (Ratio of sides)^3

8 : 1 = (Ratio of sides)^3

$$\therefore \text{ Ratio of sides} = \left(\frac{8}{1}\right)^{1/3} = \frac{2}{1} = 2:1$$

2. d 3. b 4. a

Rule 59

Theorem: ***If the ratio of sides of two cubes is given, then the ratio of their surface areas is given by the following result.***

Ratio of surface areas $= (Ratio\ of\ sides)^2$

Illustrative Example

Ex.: Sides of two cubes are in the ratio of 2 : 3. Find the ratio of their surface areas.

Soln: Applying the above theorem, we have

Ratio of surface areas

$$= (Ratio\ of\ sides)^2 = (2:3)^2 = 4:9$$

Exercise

1. Sides of two cubes are in the ratio of 1 : 2. Find the ratio of their surface areas.
   a) 1 : $\sqrt{2}$ b) 1 : 4 c) 4 : 1 d) 2 : 1
2. Sides of two cubes are in the ratio of 3 : 5. Find the ratio of their surface areas.
   a) 9 : 25 b) 25 : 9 c) 5 : 3 d) Data inadequate
3. Sides of two cubes are in the ratio of 4 : 7. Find the ratio of their surface areas.
   a) 16 : 49 b) 2 : $\sqrt{7}$ c) $\sqrt{7}$: 2 d) 49 : 16

Answers

1. b 2. a 3. a

Rule 60

Theorem: ***If the ratio of volumes of two cubes is given, then the ratio of their surface areas is given from the following result.***

$$(ratio\ of\ surface\ areas)^3 = (ratio\ of\ volumes)^2$$

Illustrative Example

Ex.: Volumes of two cubes are in the ratio of 1 : 8. Find the ratio of their surface areas.

Soln: Applying the above theorem, we have

$$(Ratio\ of\ surface\ areas)^3 = (Ratio\ of\ volumes)^2$$

$$= (1:8)^2 = 1:64$$

$$\therefore \text{ Ratio of surface areas} = \sqrt[3]{1:64} = 1:4$$

Exercise

1. Surface areas of two cubes are in the ratio of 1 : 5. Find the ratio of their volumes.
   a) 1 : $5\sqrt{5}$ b) $5\sqrt{5}$: 1
   c) 5 : 1 d) 125 : 1
2. Volumes of two cubes are in the ratio of 1 : 9. Find the ratio of their surface areas.
   a) 1 : 3 b) 3 : 1
   c) 1 : $3(3)^{1/3}$ d) $3(3)^{1/3}$: 1
3. Volumes of two cubes are in the ratio of 1 : 4. Find the ratio of their surface areas.
   a) 1 : 2 b) 2 : 1
   c) 1 : $2(2)^{1/3}$ d) 1 : $2(2)^{1/2}$

Answers

1. a; **Hint:** $(\text{ratio of volumes})^2 = (1:5)^3 = 1:125$

$$\therefore \text{ ratio of volumes} = \sqrt{1:125} = 1:5\sqrt{5}$$

2. c 3. c

Rule 61

Theorem: ***If the ratio of heights (not slant height) and the ratio of diameters or radii of two right circular cones are given, then the ratio of their volumes can be calculated by the given formula.***

Ratio of volumes = $(Ratio\ of\ radii)^2 \times (ratio\ of\ heights)$

Illustrative Example

Ex.: If the heights of two cones are in the ratio 1 : 4 and their diameters in the ratio 4 : 5, what is the ratio of their volumes?

Soln: Applying the above theorem, we have

$$\text{Ratio of volumes} = (4:5)^2(1:4) = \frac{16}{25} \times \frac{1}{4} = 4:25$$

[$\because$ ratio of diameters = ratio of radii]

Note: This formula also holds good for cylinders.

Exercise

1. The radii of two cylinders are in the ratio of 2 : 3 and their heights are in the ratio 5 : 3. The ratio of their volumes is:
   a) 27 : 20 b) 20 : 27 c) 4 : 9 d) 9 : 4
   (Hotel Management 1991)
2. The radii of two cylinders are in ratio 2 : 3 and teir heights in ratio 5 : 3. Their volumes will be in ratio:
   a) 4 : 9 b) 27 : 20 c) 20 : 27 d) 9 : 4
   (Clerks' Grade Exam 1991)
3. If the heights of two cones are in the ratio 1 : 2 and their diameters in the ratio 2 : 3, what is the ratio of their volumes?
   a) 4 : 15 b) 2 : 9 c) 1 : 5 d) Data inadequate
4. If the heights of two cones are in the ratio 3 : 4 and their diameters in the ratio 8 : 3, what is the ratio of their volumes?
   a) 16 : 3 b) 3 : 16 c) 5 : 3 d) Data inadequate

Answers

1. b 2. c 3. b 4. a

Rule 62

Theorem: ***If the ratio of radii and the ratio of volumes of two right circular cones are given, then the ratio of their heights can be calculated by the following result.***
Ratio of heights = (inverse ratio of radii)² (ratio of volumes)

Illustrative Example

Ex.: If the radii of two cones are in the ratio 1 : 4 and their volumes in the ratio 4 : 5, what is the ratio of their heights?

Soln: Applying the above theorem, we have

$$\text{Ratio of heights} = \left(\frac{1}{1}:\frac{1}{4}\right)^2(4:5) = 16 \times \frac{4}{5} = 64:5$$

Note: This formula also holds for cylinders.

Exercise

1. If the volumes of two cones are in ratio 1 : 4 and their diameters are in ratio 4: 5, then the ratio of their heights is:
   a) 1 : 5 b) 5 : 4 c) 5 : 16 d) 25 : 64
   (NDA Exam 1990)
2. If the radii of two cones are in the ratio 1 : 2 and their volumes in the ratio 2 : 3, what is the ratio of their heights?
   a) 1 : 3 b) 4 : 3 c) 8 : 3 d) Data inadequate
3. If the radii of two cones are in the ratio 3 : 4 and their volumes in the ratio 9 : 14, what is the ratio of their heights?
   a) 8 : 5 b) 16 : 9 c) 8 : 7 d) 7 : 8

Answers

1. d; **Hint:** Ratio of diameters = Ratio of radii.

$$\therefore \text{ required answer} = \left(\frac{5}{4}\right)^2 \times \frac{1}{4} = 25:64.$$

2. c 3. c

Rule 63

Theorem: ***If the ratio of volumes and the ratio of heights of two right circular cones are given, then the ratio of their radii is given by the following result.***
ratio of radii

$= \sqrt{(ratio\ of\ volumes)(inverse\ ratio\ of\ heights)}$

Illustrative Example

Ex.: If the volumes of the two cones are in the ratio 4 : 1 and their heights in the ratio 4 : 9, what is the ratio of their radii?

Soln: Applying the above theorem,

$$\text{Ratio of radii} = \sqrt{(4:1)\left(\frac{1}{4}:\frac{1}{9}\right)} = \sqrt{(4:1)(9:4)} = 3:1$$

Note: This formula also holds good for cylinders.

Exercise

1. If the volumes of the two cones are in the ratio 9 : 1 and their heights in the ratio 9 : 16, what is the ratio of their radii?
   a) 5 : 1 b) 3 : 1 c) 4 : 1 d) 4 : 3
2. If the volumes of the two cones are in the ratio 16 : 9 and their heights in the ratio 4 : 9, what is the ratio of their radii?
   a) 2 : 1 b) 3 : 2 c) 4 : 3 d) Data inadequate
3. If the volumes of the two cones are in the ratio 25 : 16 and their heights in the ratio 4 : 9, what is the ratio of their radii?
   a) 15 : 8 b) 5 : 3
   c) 7 : 4 d) Can't be determined

Answers

1. c 2. a 3. a

Rule 64

Theorem: ***If the ratio of heights and the ratio of radii of two circular cylinders are given, then the ratio of their curved surface areas is given by (ratio of radii) (ratio of heights)***

Illustrative Example

Ex.: If the heights and the radii of two circular cylinders are in the ratio 2 : 3 and 1 : 2 respectively. Find the ratio of their curved surface areas.

Soln: Applying the above theorem, we have required ratio = (2 : 3) (1 : 2) = 1 : 3

Note: This formula also holds good for cones, just change to slant height instead of height.

Exercise

1. If the heights and the radii of two circular cylinders are in the ratio 3 : 4 and 4 : 5 respectively. Find the ratio of their curved surface areas.
   a) 4 : 5 b) 3 : 5 c) 3 : 4 d) Data inadequate
2. If the heights and the radii of two circular cylinders are in the ratio 2 : 3 and 4 : 5 respectively. Find the ratio of their curved surface areas.
   a) 8 : 15 b) 5 : 6 c) 5 : 8 d) 2 : 5
3. If the heights and the radii of two circular cylinders are in the ratio 1 : 3 and 1 : 5 respectively. Find the ratio of their curved surface areas.
   a) 1 : 15 b) 3 : 5 c) 1 : 3 d) 1 : 5

Answers

1. b 2. a 3. a

Rule 65

Theorem: ***If the ratio of radii and the ratio of curved surface areas of two circular cylinders are given then the ratio of their heights are given by (Ratio of curved surface areas) (Inverse ratio of radii).***

Illustrative Example

Ex.: If the radii and the curved surface areas of two circular cylinders are in the ratio 3 : 5 and 6 : 7 respectively. Find the ratio of their heights.

Soln: Applying the above theorem, we have the required ratio

$$=(6:7)\left(\frac{1}{3}:\frac{1}{5}\right)=(6:7)(5:3)=\frac{30}{21}=10:7$$

Note: This formula also holds good for right circular cones, just change to **slant heights** instead of heights.

Exercise

1. If the radii and the curved surface areas of two circular cylinders are in the ratio 2 : 3 and 4 : 5 respectively. Find the ratio of their heights.
   a) 6 : 5 b) 5 : 6 c) 3 : 5 d) 5 : 3
2. If the radii and the curved surface areas of two circular cylinders are in the ratio 3 : 4 and 5 : 6 respectively. Find the ratio of their heights.
   a) 10 : 9 b) 9 : 10 c) 5 : 8 d) 8 : 5
3. If the radii and the curved surface areas of two circular cylinders are in the ratio 4 : 5 and 6 : 7 respectively. Find the ratio of their heights.
   a) 14 : 15 b) 15 : 14 c) 35 : 24 d) 24 : 35

Answers

1. a 2. a 3. b

Rule 66

Theorem: ***If the ratio of heights and the ratio of curved surface areas of two circular cylinders are given, then the ratio of their radii is given by (Ratio of curved surface areas) (Inverse ratio of heights)***

Illustrative Example

Ex.: If the heights and the curved surface areas of two circular cylinders are in the ratio 1 : 3 and 4 : 5 respectively. Find the ratio of their radii.

Soln: Applying the above theorem, we have the

$$\text{required ratio} = (4:5)\left(1:\frac{1}{3}\right)=(4:5)(3:1)=12:5$$

Note: This formula also holds good for cones, just change to slant heights instead of heights.

Exercise

1. If the heights and the curved surface areas of two circular cylinders are in the ratio 1 : 2 and 4 : 7 respectively. Find the ratio of their radii.
   a) 7 : 8 b) 8 : 7 c) 2 : 7 d) 7 : 2
2. If the heights and the curved surface areas of two circular cylinders are in the ratio 2 : 3 and 8 : 9 respectively. Find the ratio of their radii.
   a) 3 : 4 b) 4 : 3 c) 5 : 6 d) 16 : 9
3. If the heights and the curved surface areas of two circular cylinders are in the ratio 3 : 4 and 5 : 8 respectively. Find the ratio of their radii.
   a) 5 : 6 b) 5 : 8 c) 6 : 5 d) 15 : 32

Answers

1. b 2. b 3. a

Rule 67

Theorem: ***x units of rain has fallen on a y square units of land. Assuming that R% of the raindrops could have been collected and contained in a pool having a*** x_1 ***units ×*** y_1

units base, the level, by which the water level in the pool would have increased, is given by $\frac{R}{100}\left(\frac{xy}{x_1y_1}\right)$ ***units.***

Illustrative Example

Ex.: Two cm of rain has fallen on a square km of land. Assuming that 50% of the raindrops could have been collected and contained in a pool having a 100 m × 10 m base, by what level would the water level in the pool have increased?

Soln: **Detail Method:** Volume of rain water = Area × height $= (1\text{ km})^2 \times 2\text{ cm} = (1000\text{ m})^2 \times 0.02\text{ m} = 20{,}000\text{ m}^3$

Quantity of collected water

$$= 50\%\text{ of }20{,}000\text{ m}^3 = \frac{1}{2}\times 20000 = 10000\text{ m}^3$$

∴ increased level in pool

$$= \frac{\text{Volume collected}}{\text{Base area of pool}} = \frac{10000}{10\times100} = 10\text{ m}$$

∴ the water level would be increased by 10 m.

Quicker Method: Applying the above theorem, we have the required answer

$$= \left[\frac{(1000)^2\times 0.02}{100\times10}\right]\frac{50}{100} = \frac{20000}{1000}\times\frac{1}{2} = 10\text{ m.}$$

Exercise

1. One cm of rain has fallen on 2 square km of land. Assuming that 25% of the raindrops could have been collected and contained in a pool having a 50 m × 5 m base, by what level would the water level in the pool have increased?
   a) 20 m b) 40 m c) 25 m d) Data inadequate
2. Two cm of rain has fallen on a square km of land. Assuming that 40% of the raindrops could have been collected and contained in a pool having a 200 m × 20 m base, by what level would the water level in the pool have increased?
   a) 2 m b) 1 m c) 4 m d) 1.5 m
3. 4 cm of rain has fallen on 2 square km of land. Assuming that 50% of the raindrops could have been collected and contained in a pool having a 100 m × 10 m base, by what level would the water level in the pool have increased?
   a) 40 m b) 60 m c) 80 m d) None of these

Answers

1. b 2. a 3. c

Rule 68

Theorem: ***If the radius of a cylinder becomes 'x' times and the height becomes 'y' times, then the ratio between the new volume and the previous volume is given by*** $\left(x^2y\right)$.

Illustrative Example

Ex.: If the radius of a cylinder is doubled and the height is halved, what is the ratio between the new volume and the previous volume?

Soln: **Detail Method:** Let the initial radius and height of the cylinder be r cm and h cm respectively.

Then $V_1 = \pi r^2 h$ and $V_2 = \pi(2r)^2\frac{h}{2} = 2\pi r^2 h$

$$\frac{\text{New volume}}{\text{Previous volume}} = \frac{2\pi r^2 h}{\pi r^2 h} = \frac{2}{1} = 2:1$$

Quicker Method: Applying the above theorem, we have

$x = 2$ and $y = \frac{1}{2}$

∴ required ratio $= (2)^2\times\frac{1}{2} = 2:1$

Note: This theorem also holds good for right circular cones.

Exercise

1. If the radius of a cylinder is trebled and the height is doubled, what is the ratio between the new volume and the previous volume?
   a) 8 : 1 b) 9 : 1 c) 9 : 2 d) 18 : 1
2. If the radius of a cylinder becomes 4 times and the height $\frac{3}{8}$ times, what is the ratio between the new volume and the previous volume?
   a) 6 : 1 b) 5 : 1 c) 4 : 1 d) 3 : 1
3. If the radius of a circular cone is halved and the height is doubled, what is the ratio between the new volume and the previous volume?
   a) 2 : 1 b) 1 : 2 c) 4 : 1 d) 1 : 4

Answers

1. d 2. a 3. b

Rule 69

Theorem: ***If the radius of a circular cylinder becomes x times and the height becomes y times, then the ratio between the new curved surface area and the previous curved surface area of the cylinder is given by (xy).***

Illustrative Example

Ex.: If the radius of a cylinder is doubled and the height is halved, what is the ratio between the new curved surface area and the previous curved surface area of the cylinder.

Soln: **Detail Method:** Let the initial radius and height of the cylinder be r cm and h cm respectively.

Then, curved surface area of the original cylinder

$= 2\pi rh$ and

curved surface area of the new cylinder

$$= 2\pi(2r)\times\frac{h}{2} = 2\pi rh$$

$\therefore$ required ratio

$$= \frac{\text{New curved surface area}}{\text{Previous curved surface area}} = \frac{2\pi rh}{2\pi rh} = 1:1$$

Quicker Method: Applying the above theorem,

Here, $x = 2$ and $y = \frac{1}{2}$

$\therefore$ Required ratio $= 2 \times \frac{1}{2} = 1:1$

Note: This formula also holds good for right circular cones. Just change to **slant height** instead of height.

Exercise

1. If the radius of a cylinder becomes 3 times and the height $\frac{1}{3}$ times,what is the ratio between the new curved surface area and the previous curved surface area of the cylinder?
   a) 1 : 1 b) 1 : 2
   c) 2 : 1 d) Can't be determined
2. If the radius of a cylinder is halved and the height is halved, what is the ratio between the new curved surface area and the previous curved surface area of the cylinder.
   a) 4 : 1 b) 1 : 4 c) 1 : 2 d) 2 : 1
3. If the radius of a right circular cone becomes 4 times and the slant height becomes $\frac{1}{3}$ times, what is the ratio between the new curved surface area and the previous curved surface area of the cone?
   a) 3 : 4 b) 2 : 3 c) 3 : 2 d) 4 : 3

Answers

1. a 2. b 3. d

Rule 70

Theorem: ***A well of 'D' m diameter or radius 'r' metre (here, $r = \frac{D}{2}$) is dug 'h' m deep. If the earth taken out has been spread all round it to a width of 'w' m to form a circular embankment, then the height of this embankment is given by $\left[\frac{r^2h}{w(w+D)}\right]$ metres.***

Illustrative Example

Ex.: A well of 11.2 m diameter is dug 8 m deep. The earth taken out has been spread all round it to a width of 7 m to form a circular embankment. Find the height of this embankment.

Soln: **Detail Method:** Volume of earth dug out

$$\pi r^2h = \frac{22}{7}\times\left(\frac{11.2}{2}\right)^2\times 8 = \frac{22}{7}\times 5.6\times 5.6\times 8 = 788.48 \text{ m}^3$$

Area of embankment $= \pi(5.6+7)^2 - \pi(5.6)^2$

$$= \pi\left[(5.6+7)^2 - (5.6)^2\right]$$

$$= \pi[(5.6+7-5.6)(5.6+5.6+7)]$$

$$= \frac{22}{7}\times 7\times 18.2 = 400.4 \text{ m}^2$$

$\therefore$ height of the embankment $= \frac{788.48}{400.4} = 1.97$ metres

Quicker Method: Applying the above theorem, we have
the height of embankment

$$= \frac{5.6\times 5.6\times 8}{7(7+11.2)} = \frac{6.4\times 5.6}{18.2} = 1.97 \text{ metres.}$$

Exercise

1. A well with 14 metres inside diameter is dug 8 metres deep. Earth taken out of it has been evenly spread all around it to a width of 21 metres to form an embankment. The height of embankment is:
   a) 43 cm b) 47.6 cm c) 53.3 cm d) 41 cm
2. A well with 10 metres inside diameter is dug 14 metres deep. Earth taken out of it has been spread all-round it to a width of 5 metres to form an embankment. Find the height of the embankment.
   a) $4\frac{2}{3}$ metres b) $4\frac{1}{3}$ metres
   c) 5 metres d) $5\frac{2}{3}$ metres
3. A well of 10 m diameter is dug 12 m deep. The earth taken out has been spread all round it to a width of 6 m to form a circular embankment. Find the height of this embankment.
   a) $3\frac{1}{8}$ m b) $3\frac{1}{4}$ m c) $3\frac{2}{3}$ m d) $3\frac{3}{8}$ m

Answers

1. c 2. a 3. a

Rule 71

Theorem: ***A right-angled triangle having base x metres and height equal to y metres is turned around the height, a right circular cone is formed. Then,***

(i) the volume of the cone = $\left[\frac{\pi}{3}x^2y\right]$ cubic metres and

(ii) the surface area of the cone

$$= \left[\pi x\left(\sqrt{x^2+y^2}\right)\right] \text{ sq metres.}$$

Illustrative Example

Ex.: A right-angled triangle having base 3 metres and height equal to 4 metres, is turned around the height. Find the volume of the cone thus formed. Also find the surface area.

Soln: Applying the above theorem, we have

Required volume

$$= \frac{\pi}{3}x^2y = \frac{\pi}{3}\times 3\times 3\times 4 = 12\pi \text{ cubic m.}$$

Required surface area $= \pi x\left(\sqrt{x^2+y^2}\right)$

$$= \pi\times 3\times\sqrt{3^2+4^2} = 15\pi \text{ sq metres.}$$

Exercise

1. A right-angled triangle having base 6 metres and height equal to 8 metres, is turned around the height. Find the volume of the cone thus formed. Also find the surface area.
   a) 96 π, 60 π b) 48 π, 60 π
   c) 96 π, 120 π d) 48 π, 30 π
2. A right-angled triangle having base 9 metres and height equal to 12 metres, is turned around the height. Find the volume of the cone thus formed. Also find the surface area.
   a) 324 π, 135 π b) 192 π, 120 π
   c) 342 π, 145 π d) None of these
3. A right-angled triangle having base 12 metres and height equal to 16 metres, is turned around the height. Find the volume of the cone thus formed. Also find the surface area.
   a) 768 π, 240 π b) 678 π, 240 π
   c) 668 π, 250 π d) None of these

Answers

1. a 2. a 3. a

Rule 72

Theorem: ***A right-angled triangle having base x metres and height equal to y metres is turned around the base, a right circular cone is formed. Then,***

(i) the volume of the cone $= \left[\frac{\pi}{3}xy^2\right]$ *cubic metres and*

(ii) the surface area of the cone $= \pi y\left(\sqrt{x^2+y^2}\right)$ *sq metres.*

Illustrative Example

Ex.: A right-angled triangle having base 3 metres and height equal to 4 metres, is turned around the base. Find the volume of the cone thus formed. Also find the surface area.

Soln: Applying the above theorem,

Volume of the cone $= \frac{\pi}{3}\times 3\times 4\times 4 = 16\pi$ cu. m.

Surface area of the cone $= \pi\times 4\times 5 = 20\pi$ sq m.

Note: The cone thus formed has height = the base of the triangle

Radius = Height of the triangle

Slant height = Hypotenuse of the triangle.

Exercise

1. A right-angled triangle having base 6 metres and height equal to 8 metres, is turned around the base. Find the volume of the cone thus formed. Also find the surface area.
   a) 128 π, 80 π b) 182 π, 90 π
   c) 128 π, 84 π d) 128 π, 48 π
2. A right-angled triangle having base 9 metres and height equal to 12 metres, is turned around the base. Find the volume of the cone thus formed. Also find the surface area.
   a) 432 π, 180 π b) 462 π, 190 π
   c) 432 π, 150 π d) None of these
3. A right-angled triangle having base 12 metres and height equal to 16 metres, is turned around the base. Find the volume of the cone thus formed. Also find the surface area.
   a) 1024 π, 320 π b) 1044 π, 320 π
   c) 1024 π, 420 π d) None of these

Answers

1. a 2. a 3. a

Rule 73

Theorem: ***If length, breadth and height of a cuboid is increased by x%, y% and z% respectively, then its volume is increased by*** $\left[x+y+z+\frac{xy+xz+yz}{100}+\frac{xyz}{(100)^2}\right]\%$.

Illustrative Example

Ex.: The length, breadth and height of a cuboid are increased by 5%, 10% and 15% respectively. Find the percentage increase in its volume.

Soln: Applying the above theorem,

Here, x = 5%, y = 10%, and z = 15%

Percentage increase in volume

$$= \left[5+10+15+\frac{(5\times 10)+(5\times 15)+(10\times 15)}{100}+\frac{5\times 10\times 15}{(100)^2}\right]$$

$$= 30+\frac{275}{100}+\frac{750}{(100)^2} = 30+2.75+0.075 = 32.825\%$$

Exercise

1. The length, breadth and height of a cuboid are increased by 10%, 15% and 20% respectively. Find the percentage increase in its volume.
   a) 45% b) 45.8% c) 51.5% d) 51.8%
2. The length, breadth and height of a cuboid are increased by 5%, 10% and 20% respectively. Find the percentage increase in its volume.
   a) 38.6% b) 36.8% c) 35% d) None of these
3. The length, breadth and height of a cuboid are increased by 10%, 10% and 20% respectively. Find the percentage increase in its volume.
   a) 40% b) 45% c) 45.2% d) 40.2%

Answers

1. d 2. a 3. c

Rule 74

Theorem: ***If length, breadth and height of a cuboid are made x, y, and z times respectively, then its volume is increased by*** $[(xyz-1)\times 100]$ ***per cent.***

Illustrative Example

Ex.: The length, breadth and height of a cuboid are made 3, 4 and 5 times respectively. Find the percentage increase in its volume.

Soln: Applying the above theorem,
Percentage increase in volume

$$= (3\times 4\times 5-1)100 = 5900\%$$

Note: 1. If any of the sides of a cuboid is made x times, then the percentage increase in its volume is $(x-1)\times 100$.

2. If any two of the sides of a cuboid are made x and y times, then the percentage increase in its volume is $[(xy-1)\times 100]$.

Exercise

1. The length, breadth and height of a cuboid are made 2, 3 and 4 times respectively. Find the percentage increase in its volume.
   a) 230% b) 23% c) 2300% d) Data inadequate
2. The length, breadth and height of a cuboid are made 2, 4 and 6 times respectively. Find the percentage increase in its volume.
   a) 4700% b) 47% c) 470% d) 4800%
3. If only length of a cuboid is made 4 times then, find the percentage increase in its volume.
   a) 6300% b) 640% c) 300% d) None of these
4. If the length and breadth of a cuboid are made 3 and 6 times respectively then, find the percentage increase in its volume.
   a) 1900% b) 1700% c) 170% d) 190%

Answers

1. c 2. a 3. c; **Hint:** See Note - 1 4. b

Rule 75

Theorem: ***If side of a cube is made x times, then***

(i) the percentage increase in its volume is $(x^3-1)\times 100$

and

(ii) the percentage increase in its total surface area is $(x^2-1)\times 100$

Illustrative Example

Ex.: Each edge of a cube is made 2 times. Find
(i) the percentage increase in its volume and
(ii) the percentage increase in its total surface area.

Soln: Applying the above theorem, we have
the percentage increase in its volume

$$= (2^3-1)100 = 700\% \text{ and}$$

the percentage increase in its total surface area

$$= (2^2-1)100 = 300\%.$$

Exercise

1. Each edge of a cube is made 4 times. Find
   (i) the percentage increase in its volume and
   (ii) the percentage increase in its total surface area.
   a) 6300%, 1700% b) 6400%, 1500%
   c) 6300%, 1500% d) None of these
2. Each edge of a cube is made 3 times. Find
   (i) the percentage increase in its volume and
   (ii) the percentage increase in its total surface area.
   a) 2600%, 800% b) 2700%, 900%
   c) 260%, 80% d) Data inadequate
3. Each edge of a cube is made $\frac{4}{3}$ times. Find
   (i) the percentage increase in its volume and
   (ii) the percentage increase in its total surface area.
   a) $\frac{3700}{27}\%, \frac{800}{9}\%$ b) $\frac{6400}{27}\%, 300\%$
   c) $\frac{3700}{27}\%, \frac{700}{9}\%$ d) $\frac{370}{27}\%, \frac{70}{9}\%$

Answers

1. c 2. a 3. c

Rule 76

Theorem: ***If side of a cube is increased by x%, then its volume increases by*** $\left[3x+\frac{3x^2}{100}+\frac{x^3}{100^2}\right]\%$ ***or***

$$\left[\left(1+\frac{x}{100}\right)^3-1\right]\times 100\%.$$

Illustrative Example

Ex.: Each edge of a cube is increased by 50%. What is the percentage increase in its volume?

Soln: From the theorem:

$$\% \text{ increase in volume} = 3 \times 50 + \frac{3(50)^2}{100} + \frac{(50)^3}{100^2}$$

$$= 150 + 75 + 12.5 = 237.5\%$$

Note: For decrease put the –ve value of x.

Exercise

1. Each edge of a cube is increased by 25%. What is the percentage increase in its volume?
   a) 95.3% b) 75% c) 90% d) 75.312%
2. Each edge of a cube is increased by 100%. What is the percentage increase in its volume?
   a) 300% b) 600% c) 650% d) 700%
3. Each edge of a cube is increased by 20%. What is the percentage increase in its volume?
   a) 60% b) 72.8% c) 60.8% d) 78.2%

Answers

1. a 2. d 3. b

Rule 77

Theorem: ***If side of a cube is increased by x%, then its surface area increases by*** $\left(2x + \frac{x^2}{100}\right)$ ***per cent.***

Illustrative Example

Ex.: Each edge of a cube is increased by 50%. What is the percentage increase in its volume.

Soln: Applying the above theorem,

$$\% \text{ increase in area} = 2 \times 50 + \frac{50 \times 50}{100} = 125\%$$

Note: **1.** For the area, we see that only two measuring sides are involved (as area has 2-dimensions). So, we use the above formula,

2. For decrease, put the –ve value of x.

Exercise

1. Each edge of a cube is increased by 10%. What is the percentage increase in its volume.
   a) 20% b) 21% c) 20.5% d) 22%
2. Each edge of a cube is increased by 20%. What is the percentage increase in its volume.
   a) 44% b) 44.5% c) 40% d) None of these
3. Each edge of a cube is increased by 25%. What is the percentage increase in its volume.
   a) 50% b) 31.25%
   c) 56.25% d) Data inadequate
4. Each edge of a cube is increased by 40%. What is the percentage increase in its volume.
   a) 80% b) 96% c) 88% d) 100.5%

Answers

1. b 2. a 3. c 4. b

Rule 78

Theorem: ***If the radius (or diameter) of a sphere or a hemisphere is changed by x% then its volume changes by*** $\left[3x + \frac{3x^2}{100} + \frac{x^3}{100^2}\right]$ ***% or*** $\left[\left(1 + \frac{x}{100}\right)^3 - 1\right] \times 100\%$.

Illustrative Example

Ex.: The diameter of a sphere is increased by 25 per cent. What is the per cent increase in its volume?

Soln: Applying the above theorem,

Percentage increase in volume

$$= \left[\left(1 + \frac{25}{100}\right)^3 - 1\right] \times 100$$

$$= \left[\frac{5 \times 5 \times 5 - 4 \times 4 \times 4}{4 \times 4 \times 4}\right] \times 100$$

$$= \frac{125 - 64}{64} \times 100 = 95.31\%$$

Note: We have used the word 'change' in place of increase or decrease in some cases. By this we conclude that if there is increase use the +ve value and if there is decrease then use the –ve value. If we get the answers +ve or –ve then there is respectively increase or decrease in the volume.

Exercise

1. The diameter of a sphere is increased by 25 per cent. What is the per cent increase in its volume?
   a) 95.3% b) 75% c) 90% d) 75.312%
2. The diameter of a sphere is increased by 20 per cent. What is the per cent increase in its volume?
   a) 60% b) 72.8% c) 60.8% d) 78.2%
3. The diameter of a sphere is increased by 40 per cent. What is the per cent increase in its volume?
   a) 160% b) 145.6% c) 174.4% d) None of these

Answers

1. a 2. b 3. c

Rule 79

Theorem: ***If the radius (or diameter) of a sphere or a hemisphere is changed by x% then its curved surface area changes by*** $\left[2x + \frac{x^2}{100}\right]$ ***per cent.***

Illustrative Example

Ex.: The diameter of a hemisphere is increased by 25%. What is the percentage increase in its curved surface area.

Soln: Applying the above theorem, we have
% increase in surface area

$$= 2 \times 25 + \frac{25^2}{100} = 50 + 6.25 = 56.25\%$$

Note: We have used the word 'change' in place of increase or decrease in some cases. By this we conclude that if there is increase use the +ve value and if there is decrease then use the –ve value. If we get the answers +ve or –ve then there is respectively increase or decrease in the volume.

Exercise

1. The diameter of a hemisphere is increased by 5%. What is the percentage increase in its curved surface area.
a) 10% b) 10.5% c) 10.75% d) 10.25%
2. The diameter of a hemisphere is increased by 3%. What is the percentage increase in its curved surface area.
a) 6% b) 6.9%
c) 6.09% d) Can't be determined
3. The diameter of a hemisphere is increased by 15%. What is the percentage increase in its curved surface area.
a) 30% b) 32.25% c) 34% d) 32.5%

Answers

1. d 2. c 3. b

Rule 80

Theorem: ***If height of a right circular cylinder is changed by x% and radius remains the same then its volume changes by x%.***

Illustrative Example

Ex.: The height of a cylinder is decreased by 25%. Keeping its radius unchanged. What is the % change in its volume?

Soln: Applying the above theorem, we have the percentage decrease in volume = 25%.

Note: **1.** This theorem also holds good for right-circular cones.

2. We have used the word 'change' in place of increase or decrease. By this we conclude that if there is increase use +ve value and if there is decrease then use the –ve value. If we get the answer +ve or –ve then there is respectively increase or decrease in the volume.

Exercise

1. The height of a cylinder is decreased by 4%. Keeping its radius unchanged. What is the % change in its volume?
a) 4% b) 2% c) 8% d) Data inadequate
2. The height of a cylinder is decreased by 8%. Keeping its radius unchanged. What is the % change in its volume?
a) 8% b) 4% c) 16% d) Data inadequate
3. The height of a cylinder is decreased by $4\frac{4}{5}$%. Keeping its radius unchanged. What is the % change in its volume?
a) 4.8% b) 4.6% c) 9.6% d) 9.2%

Answers

1. a 2. a 3. a

Rule 81

Theorem: ***If radius of a right circular cylinder is changed by x% and height remains the same the volume changes by***

$$\left[2x + \frac{x^2}{100}\right]\% \quad Or, \left[\left(1 + \frac{x}{100}\right)^2 - 1\right] \times 100\%.$$

Illustrative Example

Ex.: The radius of a cylinder is increased by 25%. Keeping its height unchanged. What is the percentage increase in its volume?

Soln: Applying the above theorem, we have

$$\text{required answer} = \left(2 \times 25 + \frac{25^2}{100}\right)\%$$

$$= 50 + \frac{625}{100} = 56.25\%$$

Note: **1.** This theorem also holds good for right-circular cones.

2. We have used the word 'change' in place of increase or decrease. By this we conclude that if there is increase use +ve value and if there is decrease then use the –ve value. If we get the answer +ve or –ve then there is respectively increase or decrease in the volume.

Exercise

1. The radius of a cylinder is increased by 30%. Keeping its height unchanged. What is the percentage increase in its volume?
a) 60% b) 69% c) 68.5% c) Data inadequate
2. The radius of a cylinder is increased by 35%. Keeping its height unchanged. What is the percentage increase in its volume?
a) 82.5% b) 82.25% c) 70% d) 80.5%
3. The radius of a cylinder is increased by 45%. Keeping its height unchanged. What is the percentage increase in its volume?
a) 110.25% b) 110.5% c) 90% d) 105.25%

Answers

1. b 2. b 3. a

Rule 82

Theorem: ***If radius of a right circular cylinder is changed by x% and height is changed by y% then volume changes***

$$by \left[2x + y + \frac{x^2 + 2xy}{100} + \frac{x^2 y}{100^2}\right]\%.$$

Illustrative Example

Ex.: The radius of a right circular cylinder is decreased by 10% but its height is increased by 15%. What is the percentage change in its volume?

Soln: Applying the above theorem,
Here, radius decreases by 10%.
$\therefore$ x = –10 [we put –ve sign because of the decrease in radius]
y = 15
Percentage change in volume

$$= \left[2\times(-10)+15+\frac{(-10)^2+2(-10)(15)}{100}+\frac{(-10)^2(15)}{(100)^2}\right]$$

$= -20+5-2+0.15 = -6.85\%$
–ve sign means decrease in volume
$\therefore$ Percentage decrease in volume = 6.85%

Note: **1.** This theorem also holds good for right-circular cones.
2. We have used the word 'change' in place of increase or decrease. By this we conclude that if there is increase use +ve value and if there is decrease then use the –ve value. If we get the answer +ve or –ve then there is respectively increase or decrease in the volume.

Exercise

1. The radius of a right circular cylinder is decreased by 5% but its height is increased by 10%. What is the percentage change in its volume?
a) 0.725% increase in volume
b) Remains unchanged
c) 0.725% decrease in volume
d) Data inadequate
2. The radius of a right circular cylinder is decreased by 10% but its height is increased by 20%. What is the percentage change in its volume?
a) 2.8% decrease in volume b) 3.2% increase in volume
c) Remains unchanged d) Data inadequate
3. The radius of a right-circular cone is increased by 15% but its height is decreased by 20%. What is the percentage change in its volume?
a) 5.8% increase in volume
b) 5.8% decrease in volume
c) Volume remains unchanged
d) None of these

Answers

1. c 2. a
3. a; **Hint:** Change in volume

$$= 2\times15+(-20)+\frac{(15)^2+2(15)(-20)}{100}+\frac{(15)^2(-20)}{(100)^2}$$

$= 30-20-3.75-0.75 = 5.8\%$
Since sign is +ve, hence there is a increase in its volume.

Rule 83

Theorem: ***If height and radius of a right circular cylinder both changes by x% then volume changes by***

$$\left[3x+\frac{3x^2}{100}+\frac{x^3}{100^2}\right]\%.$$

Illustrative Example

Ex.: Each of the radius and the height of a right circular cylinder is both increased by 10%. Find the % by which the volume increases.

Soln: Since all the three (two radius + one height) measuring sides increase by the same value, we use the formula

$$\%\text{ increase in volume} = 3\times10+\frac{3(10)^2}{100}+\frac{(10)^3}{100^2}$$

$= 30+3+0.1 = 33.1\%$

Note: **1.** This theorem also holds for right-circular cones.
2. We have used the word 'change' in place of increase or decrease. By this we conclude that if there is increase use +ve value and if there is decrease then use the –ve value. If we get the answer +ve or –ve then there is respectively increase or decrease in the volume.

Exercise

1. Each of the radius and the height of a right circular cylinder is both increased by 1%. Find the % by which the volume increases.
a) 3% b) 3.03% c) 3.3% d) Data inadequate
2. Each of the radius and the height of a right circular cylinder is both increased by 20%. Find the % by which the volume increases.
a) 62.8% b) 72.8% c) 72% d) 60%
3. Each of the radius and the height of a right circular cylinder is both increased by 12%. Find the % by which the volume increases.
a) 40.45% b) 40% c) 36.45% d) 45.45%

Answers

1. b 2. b 3. a

Rule 84

Theorem: ***If the radius of a right circular cylinder is changed by x% and height is changed by y% then curved surface area changes by*** $\left[x+y+\frac{xy}{100}\right]$ ***per cent.***

Illustrative Example

Ex.: The radius and height of a cylinder are increased by 10% and 20% respectively. Find the per cent increase in its curved surface area.

Soln: Applying the above theorem,
Per cent increase in surface area

$$= \left[10 + 20 + \frac{10 \times 20}{100}\right] = 32\%$$

Note: **1.** This theorem also holds good for right-circular cones. Just change to **slant height** instead of height.
2. We have used the word 'change' in place of increase or decrease. By this we conclude that if there is increase use +ve value and if there is decrease then use the –ve value. If we get the answer +ve or –ve then there is respectively increase or decrease in the volume.

Exercise

1. The radius and height of a cylinder are increased by 5% and 10% respectively. Find the per cent increase in its curved surface area.
   a) 15% b) 20.5% c) 15.5% d) None of these
2. The radius and height of a cylinder are increased by 15% and 20% respectively. Find the per cent increase in its curved surface area.
   a) 35% b) 38% c) 38.5% d) 35.8%
3. The radius and height of a cylinder are increased by 20% and 25% respectively. Find the per cent increase in its curved surface area.
   a) 45% b) 50% c) 56.5% d) Data inadequate
4. The radius of a right circular cone is increased by 25% and slant height is decreased by 30%. Find the change in curved surface area of the cone.
   a) 12.5% increase b) 12.5% decrease
   c) 62.5% increase d) Can't be determined

Answers

1. c 2. b 3. b
4. b; **Hint:** See **Note**
Required change in surface area

$$= 25 - 30 - \frac{25 \times 30}{100} = -12.5\% = \text{decrease in surface area.}$$

Rule 85

Theorem: ***If height and radius of a right-circular cone both change by x% then volume changes by*** $\left[3x + \frac{3x^2}{100} + \frac{x^3}{100^2}\right]\%$.

Illustrative Example

Ex.: Each of the radius and the height of a cone is increased by 20%. Then find the % increase in volume.

Soln: Applying the above theorem,

$$\% \text{ increase in volume} = 3 \times 20 + \frac{3(20)^2}{100} + \frac{(20)^3}{100^2}$$
$$= 60 + 12 + 0.8 = 72.8\%$$

Note: **1.** This theorem also holds good for right-circular cylinders.
2. We have used the word 'change' in place of increase or decrease. By this we conclude that if there is increase use +ve value and if there is decrease then use the –ve value. If we get the answer +ve or –ve then there is respectively increase or decrease in the volume.

Exercise

1. Each of the radius and the height of a cone is increased by 2%. Then find the % increase in volume.
   a) 6.1208% b) 6.1028%
   c) 6.0128% d) None of these
2. Each of the radius and the height of a cone is increased by 40%. Then find the % increase in volume.
   a) 168.4% b) 168% c) 174.4% d) Data inadequate
3. Each of the radius and the height of a cone is increased by 50%. Then find the % increase in volume.
   a) 237.5% b) 250% c) 225.25% d) 150%

Answers

1. a 2. c 3. a

Rule 86

Theorem: ***If two cubes each of edge a metres are joined to form a single cuboid, then the surface area of the new cuboid so formed is given by*** $(10a^2)$ ***sq metres.***

Illustrative Example

Ex.: Two cubes each of edge 10 cm are joined to form a single cuboid. What is the surface area of the new cuboid so formed?

Soln: **Detail Method:** Breadth and height of the new cuboid will remain as the edge of the cube but length of the cuboid will be doubled. Then for the cuboid;
length (l) = 2 × 10 = 20 cm.
breadth (b) = 10 cm
height (h) = 10 cm
∴ surface area of cuboid = $2[lb + bh + lh]$
$= 2[20 \times 10 + 10 \times 10 + 20 \times 10]$
$= 2[500] = 1000 \text{ cm}^2$

Quicker Method (Direct Formula):
Two cubes have 6 × 2 = 12 faces. When they are joined to form a cuboid, the two faces which are joined, vanish. And hence, we may say that the new cuboid has the same surface area as the total surface area of two cubes minus the two faces' area. That is, the formula comes as:

Surface area of new cuboid = $10a^2$
where a is the side of the cubes.
∴ in this case the total surface area of cuboid
$= 10 \times (10)^2 = 1000 \text{ cm}^2$

Exercise

1. Two cubes each of edge 2 cm are joined to form a single cuboid. What is the surface area of the new cuboid so formed?
   a) 40 sq cm b) 48 sq cm
   c) 36 sq cm d) Can't be determined
2. Two cubes each of edge 4 cm are joined to form a single cuboid. What is the surface area of the new cuboid so formed?
   a) 160 sq cm b) 192 sq cm c) 180 sq cm d) 152 sq cm
3. Two cubes each of edge 9 cm are joined to form a single cuboid. What is the surface area of the new cuboid so formed?
   a) 972 sq cm b) 810 sq cm c) 910 sq cm d) 820 sq cm

Answers

1. a 2. a 3. b

Rule 87

Theorem: ***If a circular wire of radius x units is cut and bent in the form of a rectangle whose sides are in the ratio of a : b, then the sides of the rectangle are given by*** $\left[\pi x\left(\frac{a}{a+b}\right)\right]$ ***units and*** $\left[\pi x\left(\frac{b}{a+b}\right)\right]$ ***units.***

Illustrative Example

Ex.: A circular wire of radius 42 cm is cut and bent in the form of a rectangle whose sides are in the ratio of 6 : 5. Find the smaller side of the rectangle.

Soln: **Detail Method:** Length of the wire = circumference of the circle

$$= 2\pi \times 42 = \frac{2\times 22\times 42}{7} = 264 \text{ cm}$$

Now, perimeter of the rectangle = 264 cm

Since, perimeter includes double the length and breadth, while finding the sides we divide by double the sum of ratio.

$$\text{Therefore, length} = \frac{264}{2(6+5)}\times 6 = 72 \text{ cm}$$

$$\text{and breadth} = \frac{264}{2(6+5)}\times 5 = 60 \text{ cm}$$

Quicker Method: Applying the above theorem,

$$\text{Sides} = \pi\times 42\times\left(\frac{6}{11}\right) = \frac{22}{7}\times 42\times\frac{6}{11} = 72 \text{ cm}$$

$$\text{and } \pi\times 42\times\frac{5}{11} = \frac{22}{7}\times 42\times\frac{5}{11} = 60 \text{ cm}$$

$\therefore$ Smaller side = 60 cm.

Exercise

1. A circular wire of radius 7 cm is cut and bent in the form of a rectangle whose sides are in the ratio of 4 : 7. Find the larger side of the rectangle.
   a) 14 cm b) 8 cm c) 12 cm d) 6 cm
2. A circular wire of radius 14 cm is cut and bent in the form of a rectangle whose sides are in the ratio of 9 : 2. Find the smaller side of the rectangle.
   a) 8 cm b) 36 cm c) 24 cm d) 18 cm
3. A circular wire of radius 21 cm is cut and bent in the form of a rectangle whose sides are in the ratio of 8 : 3. Find the sides of the rectangle.
   a) 48 cm, 18 cm b) 56 cm, 21 cm
   c) 40 cm, 15 cm d) Data inadequate

Answers

1. a 2. a 3. a

Rule 88

Theorem: ***A right circular cone is exactly fitted inside a cube in such a way that the edges of the base of the cone are touching the edges of one of the faces of the cube and the vertex is on the opposite face of the cube. If the volume of the cube is given, then the volume of the cone is given by*** $\left[\frac{\pi}{12}\times \text{volume of the cube}\right]$.

Illustrative Example

Ex.: A right circular cone is exactly fitted inside a cube in such a way that the edges of the base of the cone are touching the edges of one of the faces of the cube and the vertex is on the opposite face of the cube. If the volume of the cube is 343 cc, what **approximately** is the volume of the cone?

(Guwahati PO - 1999)

Soln: **Detail Method:** Edge of the cube = $\sqrt[3]{343} = 7$ cm

$\therefore$ radius of cone = 3.5 cm and height = 7 cm

$$\therefore \text{ volume of the cone} = \frac{1}{3}\pi r^2 h$$

$$= \frac{1}{3}\times\frac{22}{7}\times 3.5\times 3.5\times 7 = \frac{1}{3}\times 22\times 12.25 \approx 90 \text{ cc.}$$

Quicker Method: Applying the above theorem, we have the required answer

$$= \frac{\pi}{12}\times 343 = \frac{22\times 343}{7\times 12} = \frac{49\times 11}{6} \approx 90 \text{ cc.}$$

Exercise

1. A right circular cone is exactly fitted inside a cube in such a way that the edges of the base of the cone are

touching the edges of one of the faces of the cube and the vertex is on the opposite face of the cube. If the volume of the cube is 336 cc, what is the volume of the cone?

a) 88 cc b) 90 cc c) 66 cc d) 82 cc

2. A right circular cone is exactly fitted inside a cube in such a way that the edges of the base of the cone are touching the edges of one of the faces of the cube and the vertex is on the opposite face of the cube. If the volume of the cube is 126 cc, what is the volume of the cone?

a) 30 cc b) 27 cc
c) 33 cc d) 44 cc

3. A right circular cone is exactly fitted inside a cube in such a way that the edges of the base of the cone are touching the edges of one of the faces of the cube and the vertex is on the opposite face of the cube. If the volume of the cube is 470.4 cc, what is the volume of the cone?

a) 132.2 cc b) 123.2 cc
c) 125.3 cc d) None of these

4. A right circular cone is exactly fitted inside a cube in such a way that the edges of the base of the cone are touching the edges of one of the faces of the cube and the vertex is on the opposite face of the cube. If the volume of the cube is 1260 cc, what is the volume of the cone?

a) 330 cc b) 270 cc c) 440 cc d) 400 cc

5. A right circular cone is exactly fitted inside a cube in such a way that the edges of the base of the cone are touching the edges of one of the faces of the cube and the vertex is on the opposite face of the cube. If the volume of the cube is 1176 cc, what is the volume of the cone?

a) 308 cc b) 803 cc c) 380 cc d) 830 cc

Answers

1. a 2. c 3. b 4. a 5. a

Important Points to Remember

1. *If a cylinder, a hemisphere and a cone stand on the same base and have the same heights, then*
   (i) the ratio of their volumes = 3 : 2 : 1 and
   (ii) the ratio of their curve surface areas = $\sqrt{2}:\sqrt{2}:1$
2. *The ratio of the volumes of a cube to that of the sphere which will fit inside the cube is* $(6:\pi)$.
3. *If a cube of maximum volume (each corner touching the surface from inside is cut from a sphere, then the ratio of the volumes of the cube and the sphere is* $(2:\sqrt{3}\pi)$.
4. *The curved surface area of a sphere and that of a cylinder which circumscribes the sphere is the same.*

Miscellaneous

1. A hemisphere of lead of radius 7 cm is cast into a right circular cone of height 49 cm. Find the radius of the base.

a) 3.74 cm b) 3.47 cm c) 4.74 cm d) None of these

2. The radii of two cylinders are in the ratio 2 : 3 and their heights are in the ratio 5 : 3. Calculate the ratio of their volumes and the ratio of their curved surfaces.

a) 10 : 9, 6 : 5 b) 20 : 27, 10 : 9
c) 15 : 14, 5 : 3 d) None of these

3. A cone, a hemisphere and a cylinder stand on equal base and have the same height. Find the ratio of their volumes.

a) 2 : 3 : 4 b) 4 : 3 : 2
c) 1 : 2 : 3 d) Can't be determined

4. Sum of the length, width and depth of a cuboid is s and its diagonal is d. Its surface area is:

a) s^2 b) d^2 c) s^2-d^2 d) s^2+d^2

(CDS Exam 1989)

5. A cylinder and a cone have same height and same radius of the base. The ratio between the volumes of the cylinder and cone is:

a) 1 : 3 b) 3 : 1 c) 1 : 2 d) 2 : 1

6. A right cylinder and a right circular cone have same radius and same volume. The ratio of height of the cylinder of that of the cone is:

a) 3 : 5 b) 2 : 5 c) 3 : 1 d) 1 : 3

(CDS Exam 1991)

7. A right circular cone is cut off at the middle of its height and parallel to base. Call smaller cone thus formed A and remaining part B. Then:

a) Vol A < Vol B b) Vol A = Vol B
c) Vol A > Vol B d) Vol A = $\frac{\text{Vol B}}{2}$

(CDS Exam 1991)

8. A solid consists of a circular cylinder with exact fitting right circular cone placed on the top. The height of the cone is h. If total volume of the solid is 3 times and volume of cone, then the height of circular cylinder is:

a) 2h b) 4h c) $\frac{3h}{2}$ d) $\frac{2h}{3}$

(CDS Exam 1991)

9. The number of solid spheres, each of diameter 6 cm, that could be moulded to form a solid metal cylinder of height 45 cm and diameter 4 cm, is:

a) 3 b) 4 c) 5 d) 6

(NDA Exam 1990)

10. The areas of two spheres are in the ratio 1 : 4. The ratio of their volumes is:

a) 1 : 4 b) 1 : $2\sqrt{2}$ c) 1 : 8 d) 1 : 64

(CDS Exam 1991)

11. For a given volume, which of the following has minimum surface area?
a) cube b) cone c) sphere d) cylinder

Answers

1. a; **Hint:** Volume of hemisphere

$$=\left(\frac{1}{2}\times\frac{4}{3}\times\frac{22}{7}\times7\times7\times7\right)=\frac{2156}{3}\text{ cu cm}$$

Let the radius of the base of cone be r cm.

Then, $\frac{1}{3}\times\frac{22}{7}\times r^2\times49=\frac{2156}{3}$

or $r^2=\frac{2156\times7\times3}{3\times22\times49}=14$

$\therefore$ Radius $=\sqrt{(14)}=3.74$ cm

2. b; **Hint:** Let their radii be 2r, 3r and heights be 5h, 3h. Then,

$$\frac{V_1}{V_2}=\frac{\pi(2r)^2\times(5h)}{\pi(3r)^2\times(3h)}=\frac{20}{27}\ \&\ \frac{S_1}{S_2}=\frac{2\pi\times2r\times5h}{2\pi\times3r\times3h}=\frac{10}{9}.$$

3. c; **Hint:** Ratio of volumes $=\frac{1}{3}\pi r^3:\frac{2}{3}\pi r^3:\frac{3}{3}\pi r^3=1:2:3.$

4. c; **Hint:** $l+b+h=s$ and $\sqrt{l^2+b^2+h^2}=d.$

So, $\left(l^2+b^2+h^2\right)=d^2$

$\therefore\ (l+b+h)^2=s^2\Rightarrow\left(l^2+b^2+h^2\right)+2(lb+bh+lh)=s^2$

$\Rightarrow d^2+2(lb+bh+lh)=s^2$

$\Rightarrow 2(lb+bh+lh)=(s^2-d^2)$

$\therefore$ Surface area $=(s^2-d^2)$ [Also See **Rule - 5**]

5. b; **Hint:** $\frac{\text{Volume of the cylinder}}{\text{Volume of the cone}}=\frac{\pi r^2h}{\frac{1}{3}\pi r^2h}=\frac{3}{1}=3:1.$

6. d; **Hint:** Let r be radius of each. Let h and H be the heights of the cylinder and cone respectively. Then,

$$\pi r^2h=\frac{1}{3}\pi r^2H\Rightarrow\frac{h}{H}=\frac{1}{3}.$$

7. a; **Hint:** Let h be the height of the cone and r be the radius of its base.

Radius of base of A $=\frac{1}{2}r$, Height of A $=\frac{1}{2}h.$

Volume of A $=\frac{1}{3}\pi\times\left(\frac{1}{2}r\right)^2\times\left(\frac{1}{2}h\right)=\frac{1}{24}\pi r^2h$

Volume of B $=\left(\frac{1}{3}\pi r^2h-\frac{1}{24}\pi r^2h\right)=\frac{7}{24}\pi r^2h.$

$\therefore$ Vol A < Vol B

8. d; **Hint:** Let r be the radius of the base of each.

Then, volume of the cone $=\frac{1}{3}\pi r^2h$

Total volume $=\pi r^2h$

$\therefore$ Volume of cylinder $=\left(\pi r^2h-\frac{1}{3}\pi r^2h\right)=\frac{2}{3}\pi r^2h$

Let H be the height of the cylinder.

Then, $\pi r^2H=\frac{2}{3}\pi r^2h\Rightarrow H=\frac{2}{3}h.$

9. c; **Hint:** Volume of 1 sphere $=\frac{4}{3}\pi\times(3)^3=36\pi$

Volume of 1 cylinder $=\pi\times(2)^2\times45=180\pi$

$\therefore$ Number of spheres $=\frac{180\pi}{36\pi}=5.$

10. c; **Hint:** Let r and R be the radii of the spheres. Then,

$$\frac{4\pi r^2}{4\pi R^2}=\frac{1}{4}\Rightarrow\left(\frac{r}{R}\right)^2=\left(\frac{1}{2}\right)^2\Rightarrow\frac{r}{R}=\frac{1}{2}\Rightarrow\frac{r^3}{R^3}=\frac{1}{8}$$

$\therefore$ Ratio of volumes $=\frac{\frac{4}{3}\pi r^3}{\frac{4}{3}\pi R^3}=\frac{r^3}{R^3}=\frac{1}{8}$

11. c

21 Permutation & Combination

In this chapter, we include only rules (ie Quicker Methods) based on the questions asked in the various exams like, CAT, MAT, XLRI, FMS, Bank PO, AAO, Provident Fund, CET, UTI etc. For basics, please refer to '*Magical Book on Quicker Maths*'. Many aspirants find difficulty in understanding the basics of "Permutation and Combination". Therefore we advise you to go through all the rules discussed in the following pages and try to understand the 'detail method'. Still you are unable to understand, just mug the rules, apply to the appropriate questions and get the desired answers. Since weightage of this chapter is not much, only 1 or 2 questions are asked in the various competitive exams mentioned above, we again advise you to stick with these rules your purpose will be served.

Some Important Notations and Formulae

From the examination point of view the following few results are useful. Without going into details you should simply remember the results.

1. **Factorial Notations**

   The product of n consecutive positive integers beginning with 1 is denoted by n! or $\underline{|n}$ and read as factorial n.

   Thus according to the definition of $\underline{|n}$

   $\underline{|n} = 1 \times 2 \times 3 \times ... \times (n-1) \times n$

   $= n \times (n-1) \times (n-2) \times ... \times 3 \times 2 \times 1$

   For example,

   $\underline{|6} = 1 \times 2 \times 3 \times 4 \times 5 \times 6 = 6 \times 5 \times 4 \times 3 \times 2 \times 1 = 720$

2. **According to the definition of $\underline{|n}$**

   (a) $\underline{|n} = n \times (n-1) \times (n-2) \times ... \times 3 \times 2 \times 1$

   $= n\{(n-1) \times (n-2) \times ... \times 3 \times 2 \times 1\}$

   $= n(n-1)\{(n-2) \times ... \times 3 \times 2 \times 1\}$ and so on.

   $\therefore \underline{|n} = n\underline{|n-1} = n(n-1)\underline{|n-2}$

   $= n(n-1)(n-2)\underline{|n-3}$

   (b) If r and n are positive integers and r < n, then

   $$\frac{n!}{r!} = \frac{n\times(n-1)\times(n-2)\times...\times(r+1)\times r\times(r-1)\times...\times 3\times 2\times 1}{r\times(r-1)\times(r-2)\times...\times 3\times 2\times 1}$$

   $$= n(n-1)(n-2)...(r+1)$$

   $$\frac{n!}{(n-r)!} = n(n-1)(n-2)...(n-r+1)$$

   Caution: $\frac{n!}{r!} \neq \left(\frac{n}{r}\right)!$

   For example,

   $$\frac{8!}{4!} = 8\times 7\times 6\times 5 = 1680 \neq \left(\frac{8}{4}\right)! = 2! = 2$$

3. $${}^nP_r = \frac{n!}{(n-r)!}$$

   Where ${}^nP_r$ = number of permutations or arrangements of n different things taken r at a time.

4. $${}^nC_r = \frac{n!}{r!(n-r)!}$$

   Where ${}^nC_r$ = number of selections, or groups of n different things taken r at a time.

5. From (3) and (4), we have

   ${}^nP_r = r! \times {}^nC_r$

   Total number of arrangements = total no. of groups or selections × r!.

6. If ${}^nC_x = {}^nC_y$ then either x = y or x + y = n

7. Number of **permutations** of n things out of which P are alike and are of one type, q are alike and are of the other type, r are alike and are of another type and remaining [n – (p + q + r)] all are different $= \frac{n!}{p!\,q!\,r!}$.

8. Number of **selections** of r things (r ≤ n) out of n identical things is 1.

9. Total number of selections of zero or more things from n identical things = n + 1.

10. Total number of selections of zero or more things from

n different things

$$= {}^nC_0 + {}^nC_1 + {}^nC_2 + ... + {}^nC_n = 2^n.$$

11. Number of ways to distribute (or divide) n identical things among r persons where any persons may get any no. of things $= {}^{n+r-1}C_{r-1}$.

12. (a) ${}^nP_n = n!$

(b) $0! = 1$

(c) ${}^nC_r = {}^nC_{n-r} = \frac{n!}{r!(n-r)!}$

(d) ${}^nC_n = {}^nC_0 = 1$

13. Some Fundamental Principles of Counting

(a) Multiplication Rule

Suppose one starts his journey from place X and has to reach place Z via a different place Y.

For Y, there are three means of transport - bus, train and aeroplane - from X. From Y, the aeroplane service is not available for Z. Only either by a bus or by a train can one reach Z from Y. Also, there is no direct bus or train servie for Z from X. We want to know the maximum possible no. of ways by which one can reach Z from X.

For each means of transport from X to Y there are two means of transport for going from Y to Z. Thus, for going from X to Z via Y there will be 2 (firstly, by bus to Y and again by bus to Z; secondly, by bus to Y and thereafter by train to Z.)

+2 (firstly, by train to Y and thereafter by bus to Z; secondly, by train to Y and thereafter again by train to Z.)

+2 (firstly by aeroplane to Y and thereafter by bus to Z, secondly by aeroplane to Y and thereafter by train to Z.) = $3 \times 2 = 6$ possible ways.

We conclude:

If a work A can be done in *m* ways and another work B can be done in *n* ways and C is the final work which is done only when both A and B are done, then the no. of ways of doing the final work, C = m × n.

In the above example, suppose the work to reach Y from X = the work A $\rightarrow$ in m i.e. 3 ways. The work to reach Z from Y = the work B $\rightarrow$ in n i.e. 2 ways. Then the final work to reach Z from X = the final work C $\rightarrow$ in m × n, i.e. 3 × 2 = 6 ways.

(b) Addition Rule

Suppose there are 42 men and 16 women in a party. Each man shakes his hand only with all the men and each woman shakes her hand only with all the women. We have to find the maximum no. of handshakes that taken place at the party.

Case 1: Total no. of handshakes among the group of 42 men

$$= {}^{42}C_2 = \frac{42!}{2!(42-2)!} = \frac{42!}{2!\,40!} = \frac{42 \times 41 \times 40!}{2 \times 1 \times 40!}$$

$$= 21 \times 41 = 861$$

Case 2: Total no. of handshakes among the group of 16 women

$$= {}^{16}C_2 = \frac{16!}{2!(16-2)!} = \frac{16 \times 15 \times 14!}{2 \times 1 \times 14!} = 8 \times 15 = 120$$

$\therefore$ maximum no. of handshakes = 861 + 120 = 981.

I. Permutations

Rule 1

Problems based on direct application of the formula, ${}^nP_r = \frac{n!}{(n-r)!}$.

Working Rule

(i) ${}^nP_r = \frac{n!}{(n-r)!} = n(n-1)(n-2)...$ to r factors.

(ii) If LHS is the product of r consecutive integers, express RHS also as the product of r consecutive integers.

(iii) Factorise the RHS, find out the greatest factor and try with that factor. If greatest factor does not suit then try with greatest factor × least factor. Look at the example given below and try to understand the working rule.

Illustrative Example

Ex.: If ${}^nP_4 = 360$, find *n*.

Soln: Given ${}^nP_4 = 360$

$\therefore \frac{n!}{(n-4)!} = 360$

or, $n(n-1)(n-2)(n-3) = 360 = 6 \times 5 \times 4 \times 3$

$\therefore$ n = 6

[Here LHS is the product of 4 consecutive integers therefore, RHS ie 360 is to be expressed as the product of 4 consecutive integers.

$360 = 2 \times 2 \times 2 \times 3 \times 3 \times 5$, greatest of these factors is 5, therefore try with 5.

Integers just before and after 5 are 4 and 6. Both 4 and 6 are factors of 360. Thus we get four consecutive integers 6, 5, 4 and 3 whose product is 360.

If 5 does not suit, then try with 2 × 5 i.e. 10 etc.]

Exercise

1. If ${}^nP_3 = 9240$, find n.
2. If ${}^{10}P_r = 720$, find r.

Answers

1. **Hint:** Given, ${}^{n}P_3 = 9240 \quad \therefore \dfrac{n!}{(n-3)!} = 9240$

or, $n(n-1)(n-2) = 9240 = 22 \times 21 \times 20$

$\therefore$ n = 22

[Here $9240 = 2 \times 2 \times 2 \times 3 \times 5 \times 7 \times 11$, greatest of those factors is 11 but it does not serve our purpose, therefore try with 22]

2. **Hint:** Given, ${}^{10}P_r = 720 \quad \therefore \dfrac{10!}{(10-r)!} = 720$

$\therefore 10 \times 9 \times 8 \times \ldots$ to r factors $= 720 = 10 \times 9 \times 8$

$\therefore$ r = 3

Rule 2

Problems based on formation of numbers with digits when repetition of digits is not allowed.

Working Rule

(i) First of all decide of how many digits the required numbers will be.

(ii) Then fill up the places on which there are restrictions and then apply the formula ${}^{n}P_r$ for filling up the remaining places with remaining digits.

thousands place	hundreds place	tens place	units place

Illustrative Example

Ex.: How many numbers of four digits can be formed with the digits 1, 2, 3, 4, and 5? (if repetition of digits is not allowed).

Soln:

Here n = number of digits = 5

and r = number of places to be filled up = 4

$\therefore$ Required number $= {}^{5}P_4 = \dfrac{5!}{1!} = 5 \times 4 \times 3 \times 2 = 120$

Exercise

1. How many numbers between 400 and 1000 can be made with the digits 2, 3, 4, 5, 6 and 0?
a) 60 b) 70 c) 40 d) 120

2. Find the number of numbers between 300 and 3000 that can be formed with the digits 0, 1, 2, 3, 4 and 5, no digits being repeated in any number.
a) 90 b) 120 c) 160 d) 180

3. How many even numbers of four digits can be formed with the digits 0, 1, 2, 3, 4, 5 and 6; no digit being used more than once?
a) 300 b) 140 c) 120 d) 420

4. How many numbers of four digits greater than 2300 can be formed with the digits 0, 1, 2, 3, 4, 5 and 6; no digit being repeated in any number?
a) 480 b) 560 c) 660 d) 580

5. How many positive numbers can be formed by using any number of the digits 0, 1, 2, 3 and 4; no digit being repeated in any number?
a) 360 b) 260 c) 620 d) 280

Answers

1. a; **Hint:** [Here nothing has been given about repetition of digits, therefore, we will assume that repetition of digits is not allowed.]

Any number between 400 and 1000 must be of three digits only.

4 or 5 or 6		
↓ 3 ways	↓ ${}^{5}P_2$ ways	

Since the number should be greater than 400, therefore, hundreds place can be filled up by any one of the three digits 4, 5 and 6 in 3 ways.

Remaining two places can be filled up by remaining five digits in ${}^{5}P_2$ ways.

$\therefore$ Required number $= 3 \times {}^{5}P_2 = 3 \times \dfrac{5!}{3!} = 60$.

2. d; **Hint:** Any number between 300 and 3000 must be of 3 or 4 digits.

Case I: When number is of 3 digits.

3 or 4 or 5		
↓ 3 ways	↓ ${}^{5}P_2$ ways	

Hundreds place can be filled up by any one of the three digits 3, 4 and 5 in 3 ways.

Remaining two places can be filled up by remaining five digits in ${}^{5}P_2$ ways.

$\therefore$ Number of numbers formed in this case $= 3 \times {}^{5}P_2 = 3 \times \dfrac{5!}{3!} = 60$.

Case II: When number is of 4 digits

1 or 2			
↓ 2 ways	↓ ${}^{5}P_3$ ways		

Thousands place can be filled up by any one of the

two digits 1 and 2 in 2 ways and remaining three places can be filled up by remaining five digits in ${}^5P_3$ ways.

$\therefore$ Number of numbers formed in this case

$$= 2 \times {}^5P_3 = 2 \times \frac{5!}{2!} = 120$$

$\therefore$ Required number = 60 + 120 = 180

3. d; **Hint:** Each even number must have 0, 2, 4 or 6 in its units' place.

Here total number of digits = 7.

			0 or 2 or 4 or 6

[When 0 occurs at units place there is no restriction on other places and when 2 or 4 or 6 occurs at units place there is restriction on thousands place as 0 can not be put at thousands' place.]

Case I: When 0 occurs at units' place

			0

↓ ${}^6P_3$ ways ↓ 1 way

Units place can be filled up by 0 in 1 way and remaining three places can be filled up by remaining 6 digits in ${}^6P_3$ ways.

$\therefore$ Number of numbers formed in this case =

$1 \times {}^6P_3 = \frac{6!}{3!} = 120$

Case II: When 0 does not occur at units' place.

any one of remaining six digits except zero			2 or 4 or 6

↓ 5 ways ↓ ${}^5P_2$ ways ↓ 3 ways

Units' place can be filled up by any one of the three digits 2, 4 and 6 in 3 ways.

Thousands place can be filled up by any one of the remaining six digits except zero in 5 ways.

Remaining two places can be filled up by remaining five digits in ${}^5P_2$ ways.

$\therefore$ Number of numbers formed in this case

$$= 5 \times 3 \times {}^5P_2 = 15 \times \frac{5!}{3!} = 300$$

$\therefore$ Required number = 120 + 300 = 420

4. b; **Hint:** [Since number must be of four digits and greater than 2300, therefore any one of the five digits 2, 3, 4, 5 and 6 will occur at thousands' place. When any one of 3, 4, 5, 6 occurs at thousands' place the number will be definitely greater than 2300 but when 2 occurs at thousands' place there will be also restriction on hundreds' place to make the number greater than 2300.]

Case I: When 2 occurs at thousands' place:

2	3 or 4 or 5 or 6		

↓ 1 way ↓ 4 ways ↓ ${}^5P_2$ ways

Thousands' place can be filled up by 2 in 1 way and hundreds' place can be filled up by any one of the four digits 3, 4, 5 and 6 in 4 ways.

Remaining two places can be filled up by remaining five digits in ${}^5P_2$ ways.

$\therefore$ Number of numbers formed in this case

$$= 1 \times 4 \times {}^5P_2 = 4 \times \frac{5!}{3!} = 80$$

Case II: When anyone of 3, 4, 5 and 6 occurs at thousands' place:

3 or 4 or 5 or 6			

↓ 4 ways ↓ ${}^6P_3$ ways

Thousands' place can be filled up by any one of the four digits 3, 4, 5 and 6 in 4 ways and remaining three places can be filled up by remining six digits in ${}^6P_3$ ways.

$\therefore$ Number of numbers formed in this case

$$= 4 \times {}^6P_3 = 4 \times \frac{6!}{3!} = 480$$

$\therefore$ Required number = 80 + 480 = 560

5. b; **Hint: Case I:** When number is of five digits:

1 or 2 or 3 or 4				

↓ 4 ways ↓ ${}^4P_4$ ways

Ten thousands' place can be filled up by any one of the four digits 1, 2, 3 and 4 in 4 ways and the remaining four places can be filled up by the remaining four digits in ${}^4P_4$ ways.

$\therefore$ Number of numbers formed in this case = $4 \times {}^4P_4$

Case II: When number is of four digits.

1 or 2 or 3 or 4			

↓ 4 ways ↓ ${}^4P_3$ ways

$\therefore$ Number of numbers formed in this case $= 4 \times {}^4P_3$

Case III: When number is of three digits?

1 or 2 or 3 or 4		
↓ 4 ways	↓ ${}^4P_2$ ways	

$\therefore$ Number of numbers formed in this case $= 4\times {}^4P_2$

Case IV: When number is of two digits:

1 or 2 or 3 or 4	
↓ 4 ways	↓ ${}^4P_1$ ways

$\therefore$ Number of numbers formed in this case $= 4\times {}^4P_1$

Case V: When number is of one digit:

Number of positive numbers formed in this case = 4

$\therefore$ Required number

$$= 4\times {}^4P_4 + 4\times {}^4P_3 + 4\times {}^4P_2 + 4\times {}^4P_1 + 4$$

$$= 96 + 96 + 48 + 16 + 4 = 260$$

Rule 3

Number of permutations when repetition is allowed:

Working Rule:

Number of permutations of n different things taken r at a time when things can be repeated any number of times.

$= n \times n \times n$ .... r times $= n^r$.

Note: In such type of problems, you have to first determine as to which item can be repeated. And consider the value of repeated item as 'r' in the above formula.

Illustrative Example

Ex.: A gentleman has 6 friends to invite. In how many ways can he send invitation cards to them if he has three servants to carry the cards.

Soln:

[1st Servant, 2nd Servant] → same friend; [1st Friend, 2nd Friend] → same servant

Here we observe that invitation cards cannot be sent to the same friend by different servants but invitation cards may be sent to different friends by the same servant. Thus same servant may be repeated for different friends, therefore, we must start with friend. Invitation cards may be sent to each of the six friends by any one of the three servants in 3 ways.

$\therefore$ Required number $= 3 \times 3 \times 3 \times 3 \times 3 \times 3 = 3^6 = 729$.

Exercise

1. In how many ways 3 prizes can be given away to 7 boys when each boy is eligible for any of the prizes.

   a) 243 b) 343 c) 433 d) 2187

2. A telegraph has 5 arms and each arm is capable of 4 distinct positions, including the position of rest. What is the total number of signals that can be made?

   a) 1023 b) 1024 c) 3124 d) 3125

3. A letter lock consists of three rings each marked with 10 different letters. In how many ways it is possible to make an unsuccessful attempt to open the lock.

   a) 1000 b) 999 c) 1001 d) None of these

4. How many numbers greater than 1000 but not greater than 4000 can be formed with the digits 0, 1, 2, 3, 4 repetition of digits being allowed.

   a) 357 b) 375 c) 135 d) None of these

Answers

1. b; **Hint:**

[1st prize, 2nd prize] → same boy; [1st Boy, 2nd Boy] → same prize

Two prizes can be given to the same boy but two boys cannot get the same prize, therefore, we must start with prize.]

Each of the three prizes can be given away to any one of the 7 boys in 7 ways.

$\therefore$ Required number $7 \times 7 \times 7 = 343$.

2. a; **Hint:**

[1st arm, 2nd arm] → same position; [1st position, 2nd position] → same arm

Two arms may have same position but one arm cannot have two positions at a time, therefore, we must start with arm.]

Each of the 5 arms can have any one of the 4 positions in 4 ways. But all the 5 arms will be in rest position in $1 \times 1 \times 1 \times 1 \times 1 = 1$ way and in this case no signal will be made.

$\therefore$ Required number $= 4^5 - 1 = 1023$

3. b; **Hint:** Two rings may have same letter at a time but same ring cannot have two letters at a time, therefore, we must start with ring.

Each of the three rings can have any one of the 10 different letters in 10 ways.

$\therefore$ Total number of attempts $= 10 \times 10 \times 10 = 1000$

But out of these 1000 attempts only one attempt is successful.

$\therefore$ Required number of unsuccessful attempts

$$= 1000 - 1 = 999$$

4. b; **Hint:** We have to form a 4 digit number x such that $100 < x \le 4000$.

Clearly, number of such numbers = number of 4 digit numbers y such that $1000 \le y < 4000$. Any number greater than or equal to 1000 but less than 4000 must be of 4 digits and digits at its thousands place must be 1 or 2 or 3.

1 or 2 or 3	0 or 1 or 2 or 3 or 4		
↓	↓	↓	↓
3 ways	5 ways	5 ways	5 ways

Thousands' place can be filled up by any one of the three digits 1, 2 and 3 in 3 ways.

Hundreds', tens' and units' places can each be filled by any one of the five digits 0, 1, 2, 3 and 4 in 5 ways.

∴ Required number $= 3 \times 5 \times 5 \times 5 = 375$

Rule 4

Theorem: ***If there are two groups A and B consisting of 'm' and 'n' things respectively, then the number of ways in which no two of group B occur together are given by $\left({}^{m+1}P_n \times m!\right)$. Provided that $n \le m$.***

Illustrative Examples

Ex. 1: In how many ways can 7 I.A. and 5 I.Sc. students be seated in a row so that no two of the I. Sc. students may sit togther?

Soln: **Detail Method:** Here, there is no restriction on I.A. students, therefore, first we must fix the positions of 7 I.A. students.

× I.A. × I.A. × I.A. × I.A. × I.A. × I.A. × I.A. ×

Now 7 I.A. students can be seated in a row in 7! ways.

Now if I.Sc. students sit at the place (including the two ends) indicated by '×' then no two of the five I.Sc. students will come together.

Now there are 8 places for 5 I.Sc. students

∴ The five I.Sc. students can be seated in ${}^8P_5$ ways

∴ Required number of ways in which 7 I.A. students and 5 I.Sc. students can sit $= {}^8P_5 \times 7! = \frac{8!}{3!} \times 7!$

Quicker Method: Applying the above theorem, we have,

required answer $= {}^{7+1}P_5 \times 7! = \frac{8!}{3!} \times 7!$

Ex. 2: In a class of 10 students, there are 3 girls. In how many different ways can they be arranged in a row such that no two of the three girls are consecutive?

Soln: **Detail Method:** Number of girls = 3, number of boys = 7. Since there is no restriction on boys, therefore first of all fix the positions of the 7 boys.

Now 7 boys, can be arranged in a row in 7! ways.

× B × B × B × B × B × B × B ×

If the positions of girls are fixed at places (including the two ends) indicated by crosses, no two of three girls will be consecutive.

Now there are 8 places for 3 girls

∴ 3 girls can be arranged in ${}^8P_3$ ways

∴ Required number of ways $= {}^8P_3 \times 7! = \frac{8!}{5!} \times 7!$

Quicker Method: Applying the above theorem, we have

Required answer $= {}^{7+1}P_3 \times 7! = \frac{8!}{5!} \times 7!$

Exercise

1. In how many ways can 8 I.A. and 6 I.Sc. students be seated in a row so that no two of the I. Sc. students may sit togther?

   a) $\frac{9!8!}{3!}$ b) $\frac{8!7!}{2!}$ c) $\frac{9!8!}{4!}$ d) None of these

2. In how many ways can 6 I.A. and 4 I.Sc. students be seated in a row so that no two of the I. Sc. students may sit togther?

   a) $\frac{7!3!}{4!}$ b) $\frac{7!6!}{3!}$ c) $\frac{7!3!}{6!}$ d) Can't be determined

3. In a class of 12 students, there are 4 girls. In how many different ways can they be arranged in a row such that no two of the three girls are consecutive?

   a) $\frac{9!8!}{5!}$ b) $\frac{9!8!}{4!}$ c) $\frac{9!5!}{8!}$ d) $\frac{9!8!}{3!}$

4. In a class of 15 students, there are 5 girls. In how many different ways can they be arranged in a row such that no two of the three girls are consecutive?

   a) $\frac{11!10!}{5!}$ b) $\frac{11!10!}{6!}$ c) $\frac{11!10!}{8!}$ d) None of these

Answers

1. a 2. b 3. a 4. b

Rule 5

Theorem: ***The number of ways in which 'n' examination papers can be arranged so that the best and the worst papers never come together are given by $[(n-2) \times (n-1)!]$ ways.***

Illustrative Example

Ex.: In how many ways can 10 examination papers be arranged so that the best and the worst papers never come together.

Soln: **Detail Method:** The number of permutations of 10 papers when there is no restriction $= {}^{10}P_{10} = 10!$

When the best and the worst papers come together, regarding the two as one paper, we have only 9 papers.

These 9 papers can be arranged in ${}^9P_9 = 9!$ ways

But these two papers can be arranged among themselves in 2! ways.

$\therefore$ number of arrangements when the best and the worst papers do not come together

$= 10! - 9! \times 2! = 9!\,(10-2) = 8 \times 9!.$

Quicker Method: Applying the above theorem, we have,

the required number of ways $= (10-2) \times (10-1)!$

$= 8 \times 9! = 8 \times 9!$

Note: The number of ways in which 'n' books may be arranged on a shelf so that two particular books shall not be together is $[(n-2) \times (n-1)!]$

Exercise

1. In how many ways can 12 examination papers be arranged so that the best and the worst papers never come together.
   a) $10 \times 11!$ b) $12 \times 11!$ c) $10 \times 12!$ d) $10! \times 11!$
2. In how many ways can 15 examination papers be arranged so that the best and the worst papers never come together.
   a) $13! \times 14!$ b) $13 \times 10!$ c) $13 \times 14!$ d) None of these
3. Find the number of ways in which 21 books may be arranged on a shelf so that the oldest and the newest books never come together.
   a) $19! \times 20!$ b) $19 \times 21!$
   c) $19 \times 20!$ d) Can't be determined

Answers

1. a 2. c
3. c; **Hint:** See *Note.*

Rule 6

Theorem: ***There are 'm' boys and 'n' girls. The no. of ways in which they can be seated in a row so that all the girls do not sit together are given by*** $[(m+n)! - (m+1)! \times n!]$ ***ways.***

Note: This rule is different from the **Rule-I**. In Rule-I, "*no two occur together*" is given whereas in this rule "*all the girls do not sit together*" is given.

Illustrative Example

Ex.: There are 5 boys and 3 girls. In how many ways can they be seated in a row so that all the three girls do not sit together.

Soln: **Detail Method:** Total number of persons $= 5 + 3 = 8$

When there is no restriction they can be seated in a row in 8! ways.

But when all the three girls sit together, regarding the three girls as one persons, we have only $5 + 1 = 6$ persons.

These 6 persons can be arranged in a row in 6! ways.

But the three girls can be arranged among themselves in 3! ways

$\therefore$ number of ways when three girls are together

$= 6! \times 3!$

$\therefore$ Required number of ways in which all the three girls do not sit together $= 8! - 6! \times 3! = 6!\,(8 \times 7 - 6)$

$= 50 \times 6! = 36000.$

Quicker Method: Applying the above theorem, we have

the required no. of ways $= (5+3)! - (5+1)! \times 3!$

$= 8! - 6! \times 3! = 50 \times 6! = 36000.$

Note: There are 'm' boys and 'n' girls. The no. of ways in which they can be seated in a row so that all the boys do not sit together are given by $[(m+n)! - (n+1)! \times m!]$ ways.

Exercise

1. There are 3 boys and 2 girls. In how many ways can they be seated in a row so that all the three boys do not sit together.
   a) 72 b) 42 c) 172 d) 190
2. There are 8 boys and 4 girls. In how many ways can they be seated in a row so that all the girls do not sit together.
   a) $1320 \times 9!$ b) $1296 \times 9!$
   c) $1344 \times 9!$ d) $1296 \times 12!$
3. There are 9 boys and 5 girls. Find the no. of ways in which they can be seated in a row so that all the boys do not sit together.
   a) $240240 \times 9!$ b) $240240 \times 5!$
   c) $240234 \times 9!$ d) $240236 \times 9!$

Answers

1. a 2. b
3. c; **Hint:** See *Note.*

Rule 7

Theorem: ***The number of ways in which m boys and 'm' girls can be seated in a row so that boys and girls are alternate are given by 2(m! m!) ways.***

Illustrative Example

Ex.: In how many ways 4 boys and 4 girls can be seated in a row so that boys and girls are alternate?

Soln: **Detail Method:**

Case I: When a boy sits at the first place:

Possible arrangement will be of the from

B	G	B	G	B	G	B	G

Now there are four places namely 1st, 3rd, 5th and 7th for four boys, therefore, four boys can be seated in 4! ways. Again there are four places namely 2nd, 4th, 6th and 8th for four girls.

$\therefore$ four girls can be seated in 4! ways.

$\therefore$ Number of ways in this case $= 4!\,4!$

Case II: When a girl sits at the first place, possible arrangement will be of the form

G	B	G	B	G	B	G	B

∴ Number of arrangements in this case = 4! 4!

∴ Required number = 4! 4! + 4! 4! = 2(4! 4!) = 1152.

Quicker Method: Applying the above theorem, we have

the required answer = 2 (4! 4!) = 1152.

Exercise

1. In how many ways 3 boys and 3 girls can be seated in a row so that boys and girls are alternate?
   a) 9 b) 36 c) 72 d) Data inadequate
2. In how many ways 5 boys and 5 girls can be seated in a row so that boys and girls are alternate?
   a) 14400 b) 28800 c) 28000 d) None of these
3. In how many ways 2 boys and 2 girls can be seated in a row so that boys and girls are alternate?
   a) 4 b) 2 c) 8 d) 16

Answers

1. c 2. b 3. c

Rule 8

Theorem: ***The number of ways in which m boys and (m – 1) girls can be seated in a row so that they are alternate is given by [m! (m – 1)!] ways.***

Illustrative Example

Ex.: In how many ways 4 boys and 3 girls can be seated in a row so that they are alternate?

Soln: **Detail Method:** Possible arrangement will be of the form

B	G	B	G	B	G	B

There are four places namely 1st, 3rd, 5th and 7th for four boys.

∴ Four boys can be seated in 4! ways. Again there are three places namely 2nd, 4th and 6th for three girls.

∴ Three girls can be seated in 3! ways

∴ Requird number = 4! 3! = 144

Quicker Method: Applying the above theorem, we have

the required answer = 4! 3! = 144.

Exercise

1. In how many ways 10 boys and 9 girls can be seated in a row so that they are alternate?
   a) 10!9! b) 10!11! c) 9!11! d) Data inadequate
2. In how many ways 8 boys and 7 girls can be seated in a row so that they are alternate?
   a) 8! 7! b) 2(8! 7!) c) 8! 8! d) 8! 9!
3. In how many ways 5 boys and 4 girls can be seated in a row so that they are alternate?
   a) 2(5!4!) b) 4!4! c) 2(4!4!) d) 5!4!

Answers

1. a 2. a 3. d

Rule 9

Theorem: ***The number of ways in which 'm' persons of a particular group, caste, country etc. and 'm' persons of the other group, caste, community, country etc can be seated along a circle so that they are alternate, given by [m! (m – 1)!] ways.***

Illustrative Example

Ex.: In how many ways can 5 Indians and 5 Englishmen be seated along a circle so that they are alternate?

Soln: **Detail Method:** 5 Indians can be seated along a circle in 4! ways [See Note in **Rule - 10**].

If the Englishmen sit at the places indicated by cross '×' then Indians and Englishmen will be alternate.

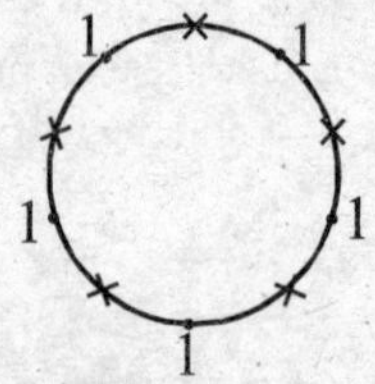

Now there are 5 places for 5 Englishmen.

∴ 5 Englishmen can be seated in 5! ways.

∴ Required number = 4! 5!.

Quicker Method: Applying the above theorem, we have,

the required number = 4! 5!.

Exercise

1. In how many ways can 4 Indians and 4 Englishmen be seated along a circle so that they are alternate?
   a) 2(4!3!) b) 4!4! c) 4!3! d) 4!5!
2. In how many ways can 6 Indians and 6 Englishmen be seated along a circle so that they are alternate?
   a) 6!5! b) 6!6! c) 2 × 6!5! d) None of these
3. In how many ways can 8 Indians and 8 Englishmen be seated along a circle so that they are alternate?
   a) 8!8! b) 8!9! c) 8!7! d) None of these

Answers

1. c 2. a 3. c

Rule 10

Theorem: ***A round table conference is to be held between 'n' delegates. The no. of ways in which they can be seated so that 'm' particular delegates always sit together are given by [(n – m)! × m!] ways.***

Illustrative Example

Ex.: A round table conference is to be held between 20

delegates of 20 countries. In how many ways can they be seated if two particular delegates are always to sit together?

Soln: **Detail Method:** Regarding two particular delegates who are to sit together as one person, we have only $18 + 1 = 19$ persons. These 19 persons can be seated at the round table in 18! ways. [See *Note*]

But two particular persons can be arranged among themselves in 2! ways.

$\therefore$ Required number $= 18! \times 2!$

Quicker Method: Applying the above theorem, we have

the required number $= (20 - 2)! \times 2! = 18! \times 2!$

Note: Always remember the following results based on the **Circular Permutations**.

1. Number of circular arrangements of 'n' different things = (n - 1)!.

2. When clockwise and anticlockwise arrangements are not different, number of circular arrangements of 'n' different things $= \frac{1}{2}(n-1)!$.

Exercise

1. A round table conference is to be held between 18 delegates of 3 countries. In how many ways can they be seated if two particular delegates are always to sit together?
   a) $15! \times 3!$ b) 18! 3!
   c) $15! \times 5!$ d) Data inadequate
2. A round table conference is to be held between 15 delegates of 2 countries. In how many ways can they be seated if two particular delegates are always to sit together?
   a) 13!×3! b) 15!2! c) 13!×2! d) 12!×2!
3. A round table conference is to be held between 16 delegates of 5 countries. In how many ways can they be seated if two particular delegates are always to sit together?
   a) $16! \times 5!$ b) $11! \times 5!$ c) $11! \times 4!$ d) None of these

Answers

1. a 2. c 3. b

Rule 11

Theorem: ***The number of ways in which 'n' different beads can be arranged to form a necklace are given by*** $\left[\frac{(n-1)!}{2}\right]$ ***ways.***

Illustrative Example

Ex.: Find the number of ways in which 6 different beads can be arranged to form a necklace.

Soln: **Detail Method:** Since in forming a necklace clockwise and anticlockwise arrangements are not different, therefore, 6 beads can be arranged to form a necklace in $\frac{(6-1)!}{2}$ ways $= \frac{5\times4\times3\times2\times1}{2} = 60$ ways

Quicker Method: Applying the above theorem, we have,

the required no. $= \frac{(6-1)!}{2} = \frac{5!}{2} = 60$ ways. [See *Note* of **Rule 10**].

Exercise

1. Find the number of ways in which 7 different beads can be arranged to form a necklace.
   a) $\frac{6!}{2}$ b) $\frac{5!}{2}$ c) $\frac{4!}{2}$ d) None of these
2. Find the number of ways in which 8 different beads can be arranged to form a necklace.
   a) $\frac{8!}{2}$ b) $\frac{7!}{2}$ c) $\frac{9!}{2}$ d) $\frac{6!}{2}$
3. Find the number of ways in which 12 different beads can be arranged to form a necklace.
   a) $\frac{11!}{2}$ b) $\frac{12!}{2}$ c) $\frac{10!}{2}$ d) None of these

Answers

1. a 2. b 3. a

Rule 12

(i) To find the number of permutations of n things taking all at a time when p of them are similar and of one type, q of them are similar and are of another type, r of them are similar and are of third type and the remaining $[n-(p+q+r)]$ are all different.

The required number of permutations $= \frac{n!}{p!\,q!\,r!}$

(ii) If a work A can be done in m ways and another work B can be done in n ways and C is the final work which is done only when both A and B are done, then the no. of ways of doing the final work C = m × n.

Illustrative Examples

Ex. 1: Find the number of permutations of the letters of the word "Pre-University".

Soln: There are 13 letters in the word Pre-University in which there are two e's two i's, two r's and 7 others are different letters.

$\therefore$ Required number of permutations $= \frac{13!}{2!\,2!\,2!}$

Ex. 2: How many different words can be formed with the letters of the word 'University'; so that all the vowels are together.

Soln: Total number of letters = 10; number of vowels = 4; i occurs twice.
Now, when 4 vowels are together, regarding the 4 vowels as one letter, we have only 6 + 1 = 7 letters.
Now these 7 letters can be arranged in 7! ways.
Since i occurs twice, therefore, four vowels can be arranged among themselves in $\frac{4!}{2!}$ ways.

$\therefore$ Required number $= 7! \times \frac{4!}{2!} = 60480$

Exercise

1. In how many ways can the letters of the word 'civilization' be rearranged?
   a) $\frac{12!}{4!}$ b) $\frac{12!}{4!}-1$ c) $\frac{13!}{5!}-1$ d) None of these
2. In how many ways can the letters of the word 'Director' be arranged so that the three vowels are never together?
   a) 1800 b) 18000 c) 16000 d) 1600
3. Find the number of rearrangements of the letters of the word 'Benevolent'. How many of them end in *l*?
   a) 302400, 30239 b) 302399, 30239
   c) 302399, 30240 d) None of these
4. How many words can be formed with the letters of the word 'Pataliputra' without changing the relative order of the vowels and consonants?
   a) 3600 b) 6300 c) 3900 d) 4600
5. How many different words can be formed with the letters of the word 'Pencil' when vowels occupy even places.
   a) 140 b) 147 c) 144 d) Can't be determined
6. How many different words can be formed with five given letters of which three are vowels and two are consonants, no two vowels being together in any word?
   a) 12 b) 16 c) 18 d) 10
7. Letters of the word DIRECTOR are arranged in such a way that all the vowels come together. Find out the total no. of ways for making such arrangement.
   a) 4320 b) 2720 c) 2160 d) 1120
   (SBI PO Exam 1999)
8. How many different letter arrangements can be made from the letters of the word RECOVER?
   a) 1210 b) 5040 c) 1260 d) 1200
   (SBI Associates PO 1999)

Answers

1. b; **Hint:** There are 12 letters in the word 'civilization' of which four are i's and others are different letters.

$\therefore$ Total number of permutations $= \frac{12!}{4!}$

But one word is civilization itself

$\therefore$ Required number of rearrangements $= \frac{12!}{4!} - 1$

2. b; **Hint:** Total number of letters = 8;
number of vowels = 3 r occurs twice.
Total number of arrangements when there is no restriction $= \frac{8!}{2!}$

When three vowels are together, regarding them as one letter, we have only 5 + 1 = 6 letters.

These six letters can be arranged in $\frac{6!}{2!}$ ways, since r occurs twice.
But the three vowels can be arranged among themselves in 3! ways.
Hence number of arrangements when the three vowels are together $= \frac{6!}{2!} \times 3!$

$\therefore$ Required number

$$= \frac{8!}{2!} - \frac{6!}{2!} \times 3! = \frac{6!}{2!}(8.7-6) = 18000$$

3. c; **Hint:** There are ten letters in the word benevolent of which three are e's and two are n's.

$\therefore$ Total number of arrangements $= \frac{10!}{3! \times 2!} = 302400$

But one word is benevolent itself

$\therefore$ Number of re-arrangements $= \frac{10!}{3!\,.\,2!} - 1 = 302399$

2nd part: When *l* is put in the end, number of remaining letters is 9 of which three are e's and two are n's

$\therefore$ number of words ending in $l = \frac{9!}{3!\,.\,2!} = 30240$

4. a; **Hint:** There are eleven letters in the word 'Pataliputra' and there are two p's, two t's, three a's and four other different letters.
Number of consonants = 6, number of vowels = 5
Since relative order of the vowels and consonants remains unchanged, therefore, vowels will occupy only vowel's place and consonants will occupy only consonant's place.
Now 6 consonants can be arranged among themselves in $\frac{6!}{2!\,2!}$ ways.

$[\because$ there are two p's and two t's$]$ and five vowels can be arranged among themselves in $\frac{5!}{3!}$ ways, since a occurs thrice

$\therefore$ Required number $= \frac{6!}{2!\,2!} \times \frac{5!}{3!} = 3600$

5. c; **Hint:** There are 6 letters in the word 'pencil' and no letter is repeated. There are two vowels e and i
Places are:

1st	2nd	3rd	4th	5th	6th

Even places are: 2nd, 4th and 6th
Now there are three even places for two vowels

$\therefore$ 3 vowels can be arranged in ${}^3P_3 = 3! = 6$ ways.

Four consonants can be arranged in remaining four places in $4! = 24$ ways

$\therefore$ Required number $= 6 \times 24 = 144$

6. a; **Hint:** Since there is no restriction on consonants, therefore, first of all we arrange the two consonants. Two consonants can be arranged in 2! ways.
Now if the vowels are put at the places (including the two ends) indicated by the '×' then no two vowels will come together

×	consonant	×	consonant	×

There are three places for three vowels and hence the three vowels can be arranged in these three places in

${}^3P_3 = 3!$ ways.

Hence number of words when no two vowels are together $= 2! \times 3! = 12$

7. c; **Hint:** Taking all vowels (IEO) as a single letter (since they come together) there are six letters with two 'R's.

Hence no. of arrangements $= \frac{6!}{2!} \times 3! = 2160.$

[3 vowels can be arranged in 3! ways among themselves, hence multiplied with 3!.]

8. c; **Hint:** Possible arrangements are $\frac{7!}{2!2!} = 1260$

[Division by 2 times 2! is because of the repetition of E and R]

Rule 13

To find the number of permutations, when certain things occur together, we do not have a general formula. But the following example will illustrate the concepts involved in this kind of questions.

Illustrative Example

Ex.: In how many ways can 8 Indians, 4 Americans and 4 Englishmen be seated in a row so that all persons of the same nationality sit together.

Soln: Regarding all persons of the same nationality as one person we have only three persons.
These three persons can be seated in a row in 3! ways. But 8 Indians can be arranged among themselves in 8! ways, 4 Americans can be arranged among themselves in 4! ways and 4 Englishmen can be arrranged among themselves in 4! ways.

$\therefore$ Required number $= 3!\,8!\,4!\,4!$

Exercise

1. There are 20 books of which 4 are single volume and the other are books of 8, 5 and 3 volumes respectively. In how many ways can all these books be arranged on a shelf so that volumes of the same book are not separated.
a) 7! 8! 5! 3! b) 7! 8! 4! 3!
c) 7! 6! 5! 3! d) None of these
2. A library has two books each having three copies and three other books each having two copies. In how many ways can all these books be arranged in a shelf so that copies of the same book are not separated.
a) 120 b) 180 c) 160 d) 140
3. 4 boys and 2 girls are to be seated in a row in such a way that two girls are always together. In how many different ways can they be seated?
a) 120 b) 720 c) 148 d) 240
(BSRB Guwahati PO 1999)
4. In how many different ways can the lettes of the word JUDGE be arranged so that the vowels always come together?
a) 48 b) 24 c) 120 d) 60
(SBI Bank PO 2001)

Answers

1. a; **Hint:** [Volumes of the same book are not to be separated ie all volumes of the same book are to be kept together.] Regarding all volumes of the same book as one book, we have only $4 + 1 + 1 + 1 = 7$ books.
These seven books can be arraned in 7! ways. Volumes of the book having 8 volumes can be arranged among themselves in 8! ways, volumes of the book having 5 volumes can be arranged among themselves in 5! ways.
And volumes of the book having 3 volumes can be arranged among themselves in 3! ways.
$\therefore$ Required number $= 7!\,8!\,5!\,3!$

2. a; **Hint:** Regarding all copies of the same book as one book, we have only 5 books.
These 5 books can be arranged in 5! ways. But all copies of the same book being identical can be arranged in only one way.
$\therefore$ Required number $= 5! \times 1 \times 1 \times 1 \times 1 \times 1 = 120$

3. d; **Hint:** Assume the 2 given students to be together ie (one). Now there are 5 students.
Possible ways of arranging them are $5! = 120$

Now they (two girls) can arrange themselves in 2! ways.
Hence, total ways = $120 \times 2 = 240$

4. a; **Hint:** Required number = 4!2! = 48.

II. Combinations

Rule 1

Problems based on direct application of the following formulae.

(i) ${}^nC_r = \frac{n!}{r!(n-r)!}$

(ii) ${}^nC_{r-1} + {}^nC_r = {}^{n+1}C_r$

(iii) If ${}^nC_x = {}^nC_y$ then either x = y or x + y = n

Illustrative Example

Ex.: If ${}^{15}C_{3r} = {}^{15}C_{r+3}$, find r.

Soln: We know that if ${}^nC_x = {}^nC_y$, then x = y

or, x + y = n

$\because {}^{15}C_{3r} = {}^{15}C_{r+3}$

$\therefore$ either 3r = r + 3, which gives r = $3/2$ which is not possible, since r is an integer.

or, 3r + r + 3 = 15, which gives r = 3. Hence r = 3.

Exercise

1. Find n, if ${}^nC_6 : {}^{n-3}C_3 = 33:4$

a) 11 b) 10 c) 13 d) 12

Answer

1. a; **Hint:** Given ${}^nC_6 : {}^{n-3}C_3 = 33:4$

$$\therefore \frac{n!}{6!(n-6)!} \times \frac{3!(n-3-3)!}{(n-3)!} = \frac{33}{4}$$

$$\text{or, } \frac{n!}{(n-3)!} \times \frac{3!}{6!} = \frac{33}{4}$$

$$\text{or, } \frac{n(n-1)(n-2)}{6.5.4} = \frac{33}{4}$$

or, $n(n-1)(n-2) = 6.5.33 = 11.3.3.2.5$

or, $n(n-1)(n-2) = 11.(3.3).(2.5) = 11.10.9$

$\therefore$ n = 11

Rule 2

Problems based on number of combinations.

(i) In simple cases

(ii) When certain things are included or excluded.

Illustrative Example

Ex.: Find the number of ways in which 5 identical balls can be distributed among 10 identical boxes, if not more than one ball can go into a box.

Soln: Number of boxes = 10 and number of balls = 5.
Now distributing 5 balls among 10 boxes, when not more than one ball can go into a box amounts to selecting boxes from among the 10 boxes. This can be done in ${}^{10}C_5$ ways.

$$\therefore \text{Required number of ways} = {}^{10}C_5 = \frac{10!}{5! \times 5!} = 252$$

Exercise

1. How many groups of 6 pesons can be formed from 8 men and 7 women?
a) 5000 b) 5005 c) 5050 d) None of these
2. There are 10 oranges in a basket. Find the no. of ways in which 3 oranges are chosen from the basket.
a) 125 b) 140 c) 110 d) 120
3. There are 25 students in a class. Find the number of ways in which a committee of 3 students is to be formed.
a) 2200 b) 2300 c) 2400 d) 3200

Answers

1. b; **Hint:** Total no. of persons = 8 + 7 = 15

$$\text{No. of groups} = {}^{15}C_6 = \frac{15!}{6!(15-6)!} = \frac{15!}{6!\,9!}$$

$$= \frac{15\times14\times13\times12\times11\times10}{6\times5\times4\times3\times2\times1} = 5005$$

2. d; **Hint:** Required number of ways = ${}^{10}C_3$

$$= \frac{10!}{3!\,(10-3)!} = \frac{10!}{3!\,7!} = \frac{10\times9\times8}{3\times2} = 120$$

3. b; **Hint:** Required no. of ways

$$= {}^{25}C_3 = \frac{25\times24\times23}{1\times2\times3} = 2300$$

Rule 3

Theorem: ***The number of triangles which can be formed by joining the angular points of a polygon of m sides as vertices are*** $\left[\frac{m(m-1)(m-2)}{6}\right]$.

Illustrative Example

Ex.: Find the no. of triangles formed by joining the vertices of a polygon of 12 sides.

Soln: **Detail Method:** A polygon of m sides will have m vertices. A triangle will be formed by joining any three vertices of the polygon.

$\therefore$ No. of triangles formed $= {}^mC_3 = \dfrac{m!}{3!\,(m-3)!}$

$$= \frac{m\times(m-1)\times(m-2)\times(m-3)!}{6\times(m-3)!} = \frac{m\times(m-1)\times(m-2)}{6}$$

Putting m = 12, we get

required no. of triangles $= \dfrac{12\times11\times10}{6} = 220$.

Quicker Method: Applying the above rule, we have the required no. of triangles $= \dfrac{12\times11\times10}{6} = 220$.

Exercise

1. Find the no. of triangles formed by joining the vertices of a hexagon.
   a) 15 b) 18 c) 20 d) 24
2. Find the no. of triangles formed by joining the vertices of a septagon.
   a) 42 b) 35 c) 32 d) 45
3. Find the no. of triangles formed by joining the vertices of a octagon.
   a) 56 b) 64 c) 42 d) 49

Answers

1. c 2. b 3. a

Rule 4

Theorem: ***The number of diagonals which can be formed by joining the vertices of a polygon of 'm' sides are*** $\left[\dfrac{m(m-3)}{2}\right]$.

Illustrative Example

Ex.: Find the no. of diagonals of a polygon of 12 sides.

Soln: **Detail Method:** A polygon of m sides will have m vertices. A diagonal or a side of the polygon will be formed by joining any two vertices of the polygon. No. of diagonals of the polygon + no. of sides of the polygon (= m) $= {}^mC_2$

$\therefore$ No. of diagonals of the polygon $= {}^mC_2 - m$

$$= \frac{m!}{2!\,(m-2)!} - m = \frac{m\times(m-1)}{2} - m$$

$$= \frac{m(m-1)-2m}{2} = \frac{m(m-3)}{2}$$

Putting m = 12, we get the reqd. no. of diagonals = $\dfrac{12\times9}{2} = 54$.

Quicker Method: Applying the above rule, we have the required no. of diagonals

$$= \frac{12\times(12-3)}{2} = \frac{12\times9}{2} = 54.$$

Exercise

1. Find the no. of diagonals of a hexagon.
   a) 9 b) 18 c) 12 d) 15
2. Find the no. of diagonals of a septagon.
   a) 16 b) 14 c) 12 d) 18
3. Find the no. of diagonals of a octagon.
   a) 20 b) 24 c) 16 d) 28
4. How many diagonals are there in a decagon?
   a) 35 b) 40 c) 49 d) 45

Answers

1. a 2. b 3. a

4. a; **Hint:** No. of sides of a decagon is 10.

$\therefore$ required no. of diagonals $= \dfrac{10\times(10-3)}{2} = 35$.

Rule 5

If there are 'm' horizontal and 'n' vertical lines, then the no. of different rectangles formed are given by $\left({}^mC_2 \times {}^nC_2\right)$.

Illustrative Examples

Ex. 1: In a chess board there are 9 vertical and 9 horizontal lines. Find the no. of rectangles formed in the chess board.

Soln: Applying the above rule, we have the total no. of rectangles $= {}^9C_2 \times {}^9C_2 = 36\times36 = 1296$.

Ex. 2:

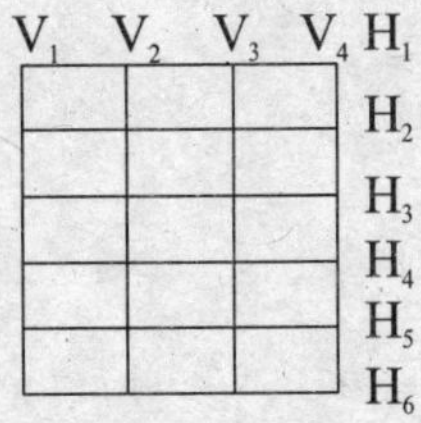

Count the total number of rectangles in the given figure.

Soln: Applying the above rule, we have the no. of required rectangles $= {}^6C_2 \times {}^4C_2 = 15\times6 = 90$.

Exercise

1. 10 parallel lines are intersected by 13 other parallel lines. Find the no. of parallelograms thus formed.
   a) 3150 b) 3510 c) 3610 d) 3501
2. ABCD is a rectangle. Count the no. of rectangles in the given figure.

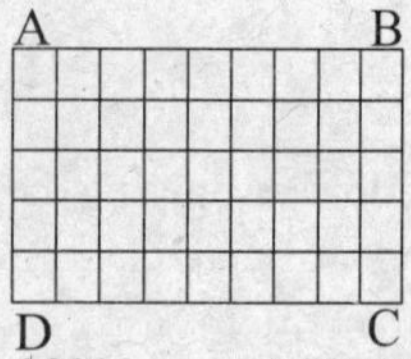

a) 37800 b) 38800 c) 38700 d) None of these

Answers

1. b; **Hint:** Required no. of parallelograms

$$= {}^{10}C_2 \times {}^{13}C_2$$

$$= \frac{10!}{2!(10-2)!} \times \frac{13}{2!(13-2)!}$$

$$= \frac{10 \times 9 \times 13 \times 12}{4} = 3510.$$

2. a

Rule 6

Theorem: ***In a party every person shakes hands with every other persons. If there was a total of H handshakes in the party, the no. of persons 'n' who were present in the party can be calculated from the equation given as*** $\frac{n(n-1)}{2} = H$.

Illustrative Example

Ex.: In a party every person shakes hands with every other persons. If there was a total of 210 handshakes in the party, find the no. of persons who were present in the party.

Soln: **Detail Method:** For each selection of two persons there will be one handshakes. So, no. of handshakes in the party = ${}^nC_2$, where n = no. of persons.

Now, ${}^nC_2 = 210$ (given)

or, $\frac{n \times (n-1)}{2} = 210$

or, $n \times (n-1) = 2 \times (2 \times 3 \times 5 \times 7) = 21 \times 20$

$\therefore n = 21$

Quicker Method: Applying the above theorem, we have,

the required no. of persons = $\frac{n(n-1)}{2} = 210$

$\therefore n = 21.$

Exercise

1. 8 men entered a lounge simultaneously. If each person shook hands with the other then find the total no. of handshakes.
   a) 16 b) 36 c) 56 d) 28
   (BSRB Bangalore PO 2000)
2. In a party every person shakes hands with every other persons. If there was a total of 45 handshakes in the party, find the no. of persons who were present in the party.
   a) 9 b) 10 c) 11 d) 12
3. In a party every person shakes hands with every other persons. If there was a total of 105 handshakes in the party, find the no. of persons who were present in the party.
   a) 15 b) 14 c) 16 d) 12
4. In a party every person shakes hands with every other persons. If there was a total of 120 handshakes in the party, find the no. of persons who were present in the party.
   a) 15 b) 18 c) 16 d) None of these
5. On the occasion of a certain meeting each member gave shakehand to the remaining members. If the total shakehands were 28, how many members were present for the meeting?
   a) 14 b) 7 c) 9 d) 8
   (NABARD 1999)

Answers

1. d; **Hint:** Applying the given rule, we have

required no. of handshakes = $\frac{8(8-1)}{2} = 28$.

2. b 3. a 4. c 5. d

Rule 7

Theorem: ***There are 'm' members in a delegation which is to be sent abroad. The total no. of members is 'n'. The no. of ways in which the selection can be made so that a particular 'r' members are always***

(i) ***included****, is given by* $\left({}^{n-r}C_{m-r}\right)$ *and*

(ii) ***excluded****, is given by* $\left({}^{n-r}C_m\right)$

Illustrative Example

Ex.: There are 5 members in a delegation which is to be sent abroad. The total no. of members is 10. In how many ways can the selection be made so that a particular member is always (i) included (ii) excluded?

Soln: **Detail Method:**

(i) Selection of one particular member can be done in $= {}^1C_1 = 1$ way. After the selection of the particular member, we are left with 9 members and for the delegation, we need 4 members more. So selection can be done in ${}^9C_4$ ways.

$\therefore$ required no. of ways of selection = ${}^1C_1 \times {}^9C_4$

$$= \frac{1 \times 9 \times 8 \times 7 \times 6}{24} = 126.$$

(ii) When one particular person has to be always excluded from the 5-member delegation, we are left with 10 – 1 = 9 persons. So selection can be done in $^9C_5$ ways.

$\therefore$ required no. of ways = $^9C_5 = 126$

Quicker Method:

(i) Applying the above theorem, we have

the required number = $^{10-1}C_{5-1} = ^9C_4 = 126$

(ii) Applying the above theorem, we have,

the required number = $^{10-1}C_5 = ^9C_5 = 126$

Exercise

1. There are 4 members in a delegation which is to be sent abroad. The total no. of members is 8. In how many ways can the selection be made so that a particular member is always (i) included (ii) excluded?
   a) 35, 35 b) 35, 40 c) 36, 32 d) None of these
2. There are 3 members in a delegation which is to be sent abroad. The total no. of members is 7. In how many ways can the selection be made so that a particular member is always (i) included (ii) excluded?
   a) 3, 20 b) 4, 21 c) 3, 18 d) 5, 20
3. There are 8 members in a delegation which is to be sent abroad. The total no. of members is 18. In how many ways can the selection be made so that 2 members are always (i) included (ii) excluded?
   a) 8800, 4920 b) 8008, 4290
   c) 8008, 4920 d) None of these

Answers

1. a 2. a 3. b

Rule 8

Theorem: ***There are 'n' points in a plane out of which 'm' points are collinear. The number of triangles formed by the points as vertices are given by*** $\left({}^nC_3 - {}^mC_3\right)$.

Illustrative Example

Ex.: There are 10 points in a plane out of which 4 are collinear. Find the number of triangles formed by the points as vertices.

Soln: Detail Method: For the time being let us suppose that the 10 points are such that no three of them are collinear. Now a triangle will be formed by any three of these ten points. Thus forming a triangle amounts to selecting any three of the 10 points.

Now 3 points can be selected out of 10 points in $^{10}C_3$ ways.

$\therefore$ number of triangles formed by 10 points when no 3 of them are collinear = $^{10}C_3$ ...... (i)

Similarly, the number of triangles formed by 4 points when no 3 of them are collinear = $^4C_3$ ...... (ii)

Now let the four points become collinear, then $^4C_3$ triangles formed by these 4 points vanish.

$\therefore$ Required number of triangles formed

$= {}^{10}C_3 - {}^4C_3 = 120 - 4 = 116$

Quicker Method: Applying the above theorem, we have,

the required no. of triangles = $^{10}C_3 - {}^4C_3 = 120 - 4 = 116$.

Exercise

1. There are 12 points in a plane out of which 5 are collinear. Find the number of triangles formed by the points as vertices.
   a) 200 b) 205 c) 210 d) 220
2. There are 18 points in a plane out of which 6 are collinear. Find the number of triangles formed by the points as vertices.
   a) 816 b) 796 c) 820 d) 790
3. There are 14 points in a plane out of which 4 are collinear. Find the number of triangles formed by the points as vertices.
   a) 360 b) 368 c) 364 d) None of these

Answers

1. c 2. b 3. a

Rule 9

Theorem: ***There are 'n' points in a plane out of which 'm' points are collinear. The number of straight lines formed by joining them are given by*** $\left({}^nC_2 - {}^mC_2 + 1\right)$.

Illustrative Example

Ex.: There are 10 points in a plane out of which 4 are collinear. Find the number of straight lines formed by joining them.

Soln: Detail Method: For the time being let us suppose that the 10 points are such that no three of the 'm' are collinear. Now a straight line will be formed by any two of these 10 points. Thus forming a straight line amounts to selecting two of the 10 points.

Now out of 10 points 2 can be selected in $^{10}C_2$ ways.

$\therefore$ number of straight lines formed by 10 points when no three of them are collinear = $^{10}C_2$. .... (i)

Similarly, the number of straight lines formed by 4 points when no 3 of them are collinear = $^4C_2$ ..... (ii)

Now let the four points become collinear, then ${}^4C_2$ straight lines formed by them will reduce to only one straight line.

$\therefore$ required number of lines formed

=

${}^{10}C_2 - {}^4C_2 + 1 = 45 - 6 + 1 = 40$

Quicker Method: Applying the above theorem, we have,

the required number of straight lines = ${}^{10}C_2 - {}^4C_2 + 1 = 45 - 6 + 1 = 40$.

Exercise

1. There are 12 points in a plane out of which 5 are collinear. Find the number of straight lines formed by joining them.
   a) 56 b) 57 c) 47 d) 46
2. There are 13 points in a plane out of which 4 are collinear. Find the number of straight lines formed by joining them.
   a) 73 b) 72 c) 70 d) 71
3. There are 8 points in a plane out of which 3 are collinear. Find the number of straight lines formed by joining them.
   a) 25 b) 26 c) 28 d) 29

Answers

1. b 2. a 3. b

Rule 10

Theorem: ***On a new year day every student of a class sends a card to every other student. If the postman delivers 'C' cards, then the number of students 'n' in the class can be calculated by the following equation. n(n – 1) = C.***

Illustrative Example

Ex.: On a new year day every student of a class sends a card to every other student. The postman delivers 600 cards. How many students are there in the class.

Soln: **Detail Method:** Let n be the number of students.

Now number of ways in which two students can be selected out of n students is ${}^nC_2$.

$\therefore$ number of pairs of students = ${}^nC_2$.

But for each pair of students, number of cards sent is 2 (since if there are two students A and B, A will send a card to B and B will send a card to A).

$\therefore$ for ${}^nC_2$ pairs, number of cards sent = $2 \times {}^nC_2$.

According to the question, $2 \times {}^nC_2 = 600$

or, $2 \times \frac{n(n-1)}{2!} = 600$

or, $n^2 - n - 600 = 0$

or, $(n-25)(n+24) = 0$

$\therefore$ n = 25, –24

But $n \neq -24$ $\therefore$ n = 25

Quicker Method: Applying the above theorem, we have,

the number of students in the class

$n(n-1) = 600$

or, $n^2 - n - 600 = 0$ $\therefore$ n = 25, –24

$\therefore$ required number of students = 25.

Exercise

1. On a new year day every student of a class sends a card to every other student. The postman delivers 552 cards. How many students are there in the class.
   a) 23 b) 24 c) 22 d) 33
2. On a new year day every student of a class sends a card to every other student. The postman delivers 1190 cards. How many students are there in the class.
   a) 35 b) 34 c) 33 d) 45
3. On a new year day every student of a class sends a card to every other student. The postman delivers 930 cards. How many students are there in the class.
   a) 30 b) 29 c) 41 d) 31
4. On a new year day every student of a class sends a card to other student. If the total no. of students are 51 then find how many cards did the post man deliver?
   a) 2550 b) 5250 c) 5220 d) 2530

Answers

1. b 2. a 3. d

4. a; **Hint:** Required answer = 51(51 – 1) = 51 × 50 = 2550 cards.

Rule 11

Theorem: ***If in an examination a minimum is to be secured in each of 'n' subjects for a pass, then the number of ways a student can fail is given by $(2^n - 1)$ ways.***

Illustrative Example

Ex.: In an examination a minimum is to be secured in each of 5 subjects for a pass. In how many ways can a student fail?

Soln: **Detail Method:** The student will fail if he fails in one or more subjects. Now, the students can fail in one or more subjects out of 5 subjects in

${}^5C_1 + {}^5C_2 + {}^5C_3 + {}^5C_4 + {}^5C_5$ ways

$= {}^5C_0 + {}^5C_1 + {}^5C_2 + {}^5C_3 + {}^5C_4 + {}^5C_5 - {}^5C_0$

$= 2^5 - 1 = 31$ ways

$\left[\because {}^nC_0 + {}^nC_1 + ... + {}^nC_n = (1+1)^n = 2^n\right]$

$\therefore$ Required number = 31

Quicker Method: Applying the above theorem, we have,

required number = $2^n - 1 = 2^5 - 1 = 31$ ways.

Exercise

1. In an examination a minimum is to be secured in each of 3 subjects for a pass. In how many ways can a student fail?
   a) 8 b) 9 c) 7 d) Data inadequate
2. In an examination a minimum is to be secured in each of 6 subjects for a pass. In how many ways can a student fail?
   a) 65 b) 63 c) 64 d) Can't be determined
3. In an examination a minimum is to be secured in each of 4 subjects for a pass. In how many ways can a student fail?
   a) 17 b) 26 c) 15 d) 31
4. In an examination a minimum is to be secured in each of 2 subjects for a pass. In how many ways can a student fail?
   a) 4 b) 2 c) 3 d) 5

Answers

1. c 2. b 3. c 4. c

Rule 12

Theorem: ***If there are 'n' questions in a question paper, then the no. of ways in which a student can solve one or more questions are given by $(2^n - 1)$ ways.***

Illustrative Example

Ex.: There are 6 questions in a question paper. In how many ways can a student solve one or more questions?

Soln: **Detail Method:** A student will solve one or more questions out of 6 questions in

$$^6C_1 + {}^6C_2 + {}^6C_3 + {}^6C_4 + {}^6C_5 + {}^6C_6 \text{ ways.}$$

$$= {}^6C_0 + {}^6C_1 + {}^6C_2 + {}^6C_3 + {}^6C_4 + {}^6C_5 + {}^6C_6 - {}^6C_0$$

$$= 2^6 - 1 = 64 - 1 = 63 \text{ ways.}$$

$\therefore$ required number = 63

Quicker Method: Applying the above theorem, we have

the required number = $2^6 - 1 = 64 - 1 = 63$.

Exercise

1. There are 7 questions in a question paper. In how many ways can a student solve one or more questions?
   a) 128 b) 63 c) 129 d) 127
2. There are 8 questions in a question paper. In how many ways can a student solve one or more questions?
   a) 256 b) 257 c) 127 d) 255
3. There are 5 questions in a question paper. In how many ways can a student solve one or more questions?
   a) 31 b) 32 c) 33 d) 30
4. There are 4 questions in a question paper. In how many ways can a student solve one or more questions?
   a) 16 b) 17 c) 31 d) 15

Answers

1. d 2. d 3. a 4. d

Rule 13

Theorem: ***From 'x' persons of a group A and 'y' persons from group B, the number of ways in which 'n' persons can be chosen to include exactly 'r' persons of group A and the rest of group B is given by $({}^xC_r \times {}^yC_{n-r})$ ways.***

Illustrative Example

Ex.: From 4 officers and 8 jawans in how many ways can 6 be chosen to include exactly one officer?

Soln: **Detail Method:**

No. of officers	No. of jawans	No. of ways
1	5	${}^4C_1 \times {}^8C_5$

$\therefore$ Required number = ${}^4C_1 \times {}^8C_5 = 4 \times 56 = 224$

Quicker Method: Applying the above theorem, we have,

the required answer = ${}^4C_1 \times {}^8C_{6-1}$

$= {}^4C_1 \times {}^8C_5 = 224$.

Exercise

1. From 5 officers and 7 jawans in how many ways can 4 be chosen to include exactly 2 officers?
   a) 210 b) 120 c) 200 d) 105
2. From 6 officers and 10 jawans in how many ways can 5 be chosen to include exactly 1 officer?
   a) 1290 b) 1160 c) 1260 d) None of these
3. From 8 officers and 12 jawans in how many ways can 7 be chosen to include exactly 3 officers?
   a) 27720 b) 27270 c) 26620 d) None of these

Answers

1. a 2. c 3. a

Rule 14

Theorem: ***In a basket there are certain number of fruits. Out of which, there are 'x' oranges, 'y' apples, 'z' mangoes and the remaining 'n' are of different kinds. Then the number of ways a person can make a selection of fruits from among the fruits in the basket are given by $[(x+1)(y+1)(z+1) \times 2^n - 1]$ ways.***

Note: Here we consider all fruits of the same type are identical.

Illustrative Example

Ex.: There are 4 oranges, 5 apples and 6 mangoes in a fruit basket. In how many ways can a person make a selection of fruits from among the fruits in the basket?

Soln: Detail Method:

Zero or more oranges can be selected out of 4 identical oranges in 4 + 1 = 5 ways.

Zero or more apples can be selected out of 5 identical apples in 5 + 1 = 6 ways.

Zero or more mangoes can be selected out of 6 identical mangoes in 6 + 1 = 7 ways.

∴ total number of selections when all the three types of fruits are selected (the number of any type of fruits may also be zero) $= 5 \times 6 \times 7 = 210$.

But in one of these selections number of each type of fruit is zero and hence this selection must be excluded.

∴ required number = 210 – 1 = 209.

Quicker Method: Applying the above theorem, we have,

the reqd number of ways $(4+1)(5+1)(6+1) \times 2^0 - 1$
$= 5 \times 6 \times 7 \times 1 - 1 = 210 - 1 = 209.$

Exercise

1. There are 5 oranges, 6 apples and 7 mangoes in a fruit basket. In how many ways can a person make a selection of fruits from among the fruits in the basket?
   a) 336 b) 337 c) 335 d) Can't be determined
2. There are 2 oranges, 3 apples and 4 mangoes in a fruit basket. In how many ways can a person make a selection of fruits from among the fruits in the basket?
   a) 61 b) 59 c) 60 d) 58
3. There are 5 oranges, 7 apples and 9 mangoes in a fruit basket. In how many ways can a person make a selection of fruits from among the fruits in the basket?
   a) 480 b) 481 c) 482 d) 479
4. In a basket, there are 4 oranges, 6 apples, 8 mangoes and the remaining 3 are of different kinds. In how many ways can a person make a selection of fruits from among the fruits in the basket?
   a) 2520 b) 2521 c) 2519 d) 2522

Answers

1. c 2. b 3. d 4. c

Rule 15

Theorem: ***The number of ways in which (n × m) different things can be divided equally among 'n' persons are given by*** $\frac{(nm)!}{(m!)^n}$.

Illustrative Example

Ex.: In how many ways 12 different things can be divided equally among 3 persons?

Soln: Detail Method:

Each person will get 4 things. Now first person can be given 4 things out of 12 different things in ${}^{12}C_4$ ways. Second person can be given 4 things out of remaining 8 things in ${}^{8}C_4$ ways. And third person can be given 4 things out of remaining 4 things in ${}^{4}C_4$ ways.

∴ Required number $= {}^{12}C_4 \times {}^{8}C_4 \times {}^{4}C_4$

$$\frac{12!}{4!\,8!} \times \frac{8!}{4!\,4!} \times 1 = \frac{12!}{(4!)^3}$$

Quicker Method: Applying the above theorem, we have,

the required number

$$= \frac{12!}{(4!)^3} \left[\begin{array}{l}\text{Since } n \times m = 12 \text{ and } n = 3 \\ \therefore m = \frac{12}{3} = 4\end{array}\right]$$

Note: If 'mn' different things are divided equally among 'n' groups, then the total no of different ways of distribution are given by $\frac{(nm)!}{(m!)^n \times (n!)}$

Exercise

1. In how many ways 12 different things can be divided in three sets each having 4 things.
   a) $\frac{12!}{(4!)^3 \times 3!}$ b) $\frac{12!}{(3!)^4 \times 4!}$ c) $\frac{12!}{(3!)^4 \times 3!}$ d) $\frac{12!}{(4!)^3 \times 4!}$
2. In how many ways 15 different things can be divided equally among 5 persons?
   a) $\frac{15!}{(3!)^5}$ b) $\frac{15!}{(5!)^3}$ c) $\frac{15!}{(5!)^2}$ d) $\frac{15!}{(3!)^3}$
3. In how many ways 18 different things can be divided equally among 6 persons?
   a) $\frac{18!}{(6!)^3}$ b) $\frac{18!}{(6!)^6}$ c) $\frac{18!}{(3!)^6}$ d) $\frac{18!}{(3!)^3}$
4. In how many ways 20 different things can be divided equally among 4 persons?
   a) $\frac{20!}{(4!)^5}$ b) $\frac{20!}{(5!)^4}$ c) $\frac{20!}{(5!)^5}$ d) None of these

Answers

1. a; **Hint:** See **Note**. Here, n × m = 12, n = 3 ∴ m = 4.
2. a 3. c 4. b

Rule 16

Theorem: ***The number of ways to distribute or divide 'n' identical things among 'r' persons when any person may get any number of things are given by*** $\left[{}^{n+r-1}C_{r-1}\right]$ ***ways.***

Illustrative Example

Ex.: In how many ways 20 apples can be divided among 5 boys.

Soln: Number of ways of dividing 'n' identical things among 'r' persons when any person may get any number of things = ${}^{n+r-1}C_{r-1}$ [From the above theorem]

Here, n = 20, and r = 5

∴ required number = ${}^{24}C_4 = \frac{24!}{4!\,20!} = 23 \times 22 \times 21 = 10626$.

Exercise

1. In how many ways 12 bananas can be divided among 4 girls.
   a) ${}^{12}C_4$ b) ${}^{13}C_3$ c) ${}^{15}C_3$ d) Data inadequate
2. In how many ways 16 oranges can be divided among 8 boys.
   a) ${}^{23}C_8$ b) ${}^{16}C_8$ c) ${}^{23}C_7$ d) None of these
3. In how many ways 13 apples can be divided among 5 students.
   a) ${}^{17}C_4$ b) ${}^{13}C_5$ c) ${}^{14}C_6$ d) ${}^{12}C_4$

Answers

1. c 2. c 3. a

Rule 17

Theorem: ***The number of quadrilaterals that can be formed by joining the vertices of a polygon of n sides are given by***

$\left[\frac{n(n-1)(n-2)(n-3)}{24}\right]$***; where n > 3.***

Illustrative Example

Ex.: Find the no. of quadrilaterals that can be formed by joining the vertices of an octagon.

Soln: Applying the above theorem,
an octagon has 8 sides, hence here, n = 8
∴ required number

$$= \frac{8(8-1)(8-2)(8-3)}{24} = \frac{8 \times 7 \times 6 \times 5}{24} = 70.$$

Exercise

1. Find the no. of quadrilaterals that can be formed by joining the vertices of an hexagon.
   a) 15 b) 20 c) 70 d) 16
2. Find the no. of quadrilaterals that can be formed by joining the vertices of an septagon.
   a) 45 b) 35 c) 42 d) 28
3. Find the no. of quadrilaterals that can be formed by joining the vertices of an decagon.
   a) 210 b) 200 c) 120 d) 160
4. Find the no. of quadrilaterals that can be formed by joining the vertices of a polygon of 12 sides.
   a) 495 b) 945 c) 545 d) 549

Answers

1. a 2. b
3. a; **Hint:** A decagon has 10 sides.
4. a

Rule 18

Theorem: ***If there are n points in a plane and no points are collinear, then the number of straight lines can be drawn using these 'n' points are given by*** $\left[\frac{n(n-1)}{2}\right]$.

Illustrative Example

Ex.: How many straight lines can be drawn with 16 points in a plane of which no points are collinear?

Soln: Applying the above theorem, we have the required number of straight lines

$$= \frac{16(16-1)}{2} = 15 \times 8 = 120.$$

Exercise

1. How many straight lines can be drawn with 15 points in a plane of which no points are collinear?
   a) 105 b) 120 c) 110 d) None of these
2. How many straight lines can be drawn with 18 points in a plane of which no points are collinear?
   a) 150 b) 153 c) 148 d) Can't be determined
3. How many straight lines can be drawn with 14 points in a plane of which no points are collinear?
   a) 90 b) 101 c) 91 d) 92

Answers

1. a 2. b 3. c

Rule 19

Theorem: ***If there are 'n' points in a plane and no three points are collinear, then the number of triangles formed with 'n' points are given by*** $\left[\frac{n(n-1)(n-2)}{6}\right]$.

Illustrative Example

Ex.: Find the no. of triangles that can be formed with 12 points in a plane of which no three points are collinear.

Soln: Applying the above theorem, we have the

required no. of triangles = $\frac{12 \times 11 \times 10}{6} = 220$.

Exercise

1. Find the no. of triangles that can be formed with 13 points in a plane of which no three points are collinear.

a) 276 b) 286 c) 296 d) Can't be determined

2. Find the no. of triangles that can be formed with 14 points in a plane of which no three points are collinear.
a) 346 b) 364 c) 384 d) 464
3. Find the no. of triangles that can be formed with 15 points in a plane of which no three points are collinear.
a) 454 b) 455 c) 544 d) 445

Answers

1. b 2. b 3. b

Rule 20

Theorem: ***n students appear in an examination. The number of ways the result of the examination can be announced are given by $(2)^n$.***

Illustrative Example

Ex.: 6 students appear in an examination. In how many ways can the result be announced?

Soln: **Detail Method:** Each student can either pass or fail in the examination. So, there exists 2 possibilities for each of the 6 students in the result.

$\therefore$ total number of ways for the result = $(2)^6 = 64$

Quicker Method: Applying the above theorem, we have

the required answer = $(2)^6 = 64$.

Exercise

1. 4 students appear in an examination. In how many ways can the result be announced?
a) 15 b) 16 c) 17 d) None of these
2. 5 students appear in an examination. In how many ways can the result be announced?
a) 32 b) 33 c) 31 d) Data inadequate
3. 7 students appear in an examination. In how many ways can the result be announced?
a) 126 b) 127 c) 129 d) 128

Answers

1. b 2. a 3. d

Rule 21

Theorem: ***'n' matches are to be played in a chess tournament. The number of ways in which their results can be decided are gien by $(3)^n$ ways.***

Illustrative Example

Ex.: 3 matches are to be played in a chess tournament. In how many ways can their results be decided?

Soln: The result of each of the 3 matches can be in three ways namely win, draw or loss.

$\therefore$ total no. of ways in which results of 3 matches can be decided = $(3)^3 = 27$.

Exercise

1. 2 matches are to be played in a chess tournament. In how many ways can their results be decided?
a) 27 b) 9 c) 8 d) None of these
2. 4 matches are to be played in a chess tournament. In how many ways can their results be decided?
a) 81 b) 16 c) 27 d) 64
3. 5 matches are to be played in a chess tournament. In how many ways can their results be decided?
a) 343 b) 243 c) 128 d) None of these

Answers

1. b 2. a 3. b

Rule 22

Theorem: ***A badminton tournament consists of 'n' matches.***

(i) The number of ways in which their results can be forecast are given by $(2)^n$ ways.

(ii) Total number of forecasts containing all correct results or all wrong results are given by 1.

Illustrative Example

Ex.: A badminton tournament consists of 3 matches.
(i) In how many ways can their results be forecast?
(ii) How many different forecasts can contain all wrong results.
(iii) How many different forecasts can contain all correct results?

Soln: Each badminton match can be decided in only 2 ways ie win or loss for a particular team.

$\therefore$ Total no. of ways the results of 3 matches can be forecast = $2^3 = 8$

Result of each match can be forecast wrong in only 1 way.

$\therefore$ Total no. of forecasts containing all wrong results

$= (1)^3 = 1$

Similarly, result of each match can be forecast correct in only 1 way.

$\therefore$ Total no. of forecasts containing all correct results

$= (1)^3 = 1$

Exercise

1. A badminton tournament consists of 4 matches. In how many ways can there results be forecast?
a) 16 b) 32 c) 15 d) 31
2. A badminton tournament consists of 5 matches. In how many ways can their results be forecast?
a) 32 b) 31 c) 33 d) 64
3. A badminton tournament consists of 7 matches. In how many ways can their results be forecast?
a) 64 b) 70 c) 128 d) 127

Answers

1. a 2. a 3. c

Rule 23

Theorem: ***The Indian hockey team is to play 'n' matches for the world cup.***

(i) The number of different forecasts that will contain all correct results is $(1)^n$ or 1.

(ii) The number of different forecasts that will contain all wrong results are $(2)^n$.

Illustrative Example

Ex.: The Indian hockey team is to play 4 matches for the world cup.
(i) How many different forecasts will contain all correct results?
(ii) How many different forecasts will contain all wrong results?

Soln: Result of each hockey match can be decided in 3 ways ie win, loss or draw.
(i) Only in 1 way we can correctly predict each match ie 'forecast' and the actual result are the same. Hence, total no. of ways to predict all 4 matches correctly = $(1)^4 = 1$.
(ii) For each match the prediction can go wrong in 2 ways. For example, prediction for any match is win but the actual result is either loss or draw.
$\therefore$ total no. of forecasts containing all wrong results = $(2)^4 = 16$.

Exercise

1. The Indian hockey team is to play 3 matches for the world cup. How many different forecasts will contain all correct results?
a) 8 b) 16 c) 1 d) 32
2. The Indian hockey team is to play 8 matches for the world cup.
(i) How many different forecasts will contain all correct results?
a) 8 b) 1 c) 256 d) None of these
(ii) How many different forecasts will contain all wrong results?
a) 128 b) 256 c) 512 d) None of these

Answers

1. c 2. (i) b; (ii) b

Miscellaneous

1. From 4 officers and 8 jawans in how many ways can be 6 chosen to include at least one officer.
a) 896 b) 986 c) 886 d) 996
2. From a group of 6 men and 4 women a committee of 4 persons is to be formed.
(SBI Bank PO Exam 2001)
(i) In how many different ways can it be done so that the committee has at least one woman?
a) 210 b) 225 c) 195 d) 185
(ii) In how many different ways can it be done so that the committee has at least 2 men?
a) 210 b) 225 c) 195 d) 185

Answers

1. a; **Hint: Detail Method:**

	No. of officers	**No. of jawans**	**No. of ways**
Case I	1	5	${}^4C_1 \times {}^8C_5 = 224$
Case II	2	4	${}^4C_2 \times {}^8C_4 = 420$
Case III	3	3	${}^4C_3 \times {}^8C_3 = 224$
Case IV	4	2	${}^4C_4 \times {}^8C_2 = 28$

$\therefore$ Required number = 224 + 420 + 224 + 28 = 896

Quicker Method: Applying the above theorem, we have,

$x = 4, y = 8$ and n = 6

The value of

$${}^xC_x \times {}^yC_{n-x} = {}^4C_4 \times {}^8C_{6-4} = {}^4C_4 \times {}^8C_2$$

Now, from the above theorem,

Required answer

$$= {}^4C_1 \times {}^8C_5 + {}^4C_2 \times {}^8C_4 + {}^4C_3 \times {}^8C_3 + {}^4C_4 \times {}^8C_2$$

$= 224 + 420 + 224 + 28 = 896$

2. (i) c; **Hint:** Required no. of ways

$$= {}^4C_1 \times {}^6C_3 + {}^4C_2 \times {}^6C_2 + {}^4C_3 \times {}^6C_1 + {}^4C_4$$

$$= 4 \times \frac{6\times5\times4}{1\times2\times3} + \frac{4\times3}{1\times2} \times \frac{6\times5}{1\times2} + \frac{4\times3\times2}{1\times2\times3} \times 6 + 1$$

$= 80 + 90 + 24 + 1 = 195$

(ii) d; **Hint:** Required no. of ways

$$= {}^6C_2 \times {}^4C_2 + {}^6C_3 \times {}^4C_1 + {}^6C_4$$

$$= \frac{6\times5}{1\times2} \times \frac{4\times3}{1\times2} = \frac{6\times5\times4}{1\times2\times3} \times 4 + \frac{6\times5\times4\times3}{1\times2\times3\times4}$$

$= 90 + 80 + 15 = 185$

22

Probability

In this chapter, only rules with illustrative examples have been given. For basics please consult "*Magical Book on Quicker Maths.*"

Rule 1

Theorem: ***Probability of an event (E) is denoted by P(E) and is defined as P(E)***

$$= \frac{n(E)}{n(S)} = \frac{no.of\ desired\ events}{total\ no.of\ events\ (ie\ no.of\ sample\ space)}$$

Illustrative Example

Ex.: A dice is thrown. What is the probability that the number shown on the dice is (i) an even no; (ii) an odd no.; (iii) a no. divisible by 2; (iv) a no. divisible by 3; (v) a no. less than 4; (vi) a no. less than or equal to 4; (vii) a no. greater than 6; (viii) a no. less than or equal to 6.

Soln: In all the above cases, S = {1, 2, 3, 4, 5, 6}, n(S) = 6.

(i) E (an even no.) = {2, 4, 6}, n(E) = 3

$$\therefore P(E) = \frac{n(E)}{n(S)} = \frac{3}{6} = \frac{1}{2}$$

(ii) E (an odd no.) = {1, 3, 5}, n(E) = 3

$$\therefore P(E) = \frac{n(E)}{n(S)} = \frac{3}{6} = \frac{1}{2}$$

(iii) E (a no. divisible by 2) = {2, 4, 6}, n(E) = 3

$$\therefore P(E) = \frac{3}{6} = \frac{1}{2}$$

(iv) E (a no. divisible by 3) = {3, 6}, n(E) = 2

$$\therefore P(E) = \frac{2}{6} = \frac{1}{3}$$

(v) E (a no. less than 4) = {1, 2, 3}, n(E) = 3

$$\therefore P(E) = \frac{3}{6} = \frac{1}{2}$$

(vi) E (a no. less than or equal to 4) = {1, 2, 3, 4} n(E) = 4

$$\therefore P(E) = \frac{4}{6} = \frac{2}{3}$$

(vii) E (a no. greater than 6) = {}, i.e. there is no number greater than 6 in the sample space

$$\therefore P(E) = \frac{0}{6} = 0$$

Probability of an impossible event = 0

(viii) E (a no. less than or equal to 6) = {1, 2, 3, 4, 5, 6}, n(E) = 6

$$\therefore P(E) = \frac{6}{6} = 1$$

Probability of a certain event = 1.

Note: $0 \le P(E) \le 1$

Exercise

1. In a simultaneous throw of two dice find the probability of getting a total of 8.

   a) $\frac{2}{9}$ b) $\frac{5}{36}$ c) $\frac{1}{6}$ d) Data inadequate

2. A coin is successively tossed two times. Find the probability of getting
   1) exatly one head 2) at least one head.

   a) $\frac{1}{2}, \frac{3}{4}$ b) $\frac{2}{3}, \frac{1}{4}$ c) $\frac{1}{4}, \frac{4}{5}$ d) $\frac{1}{2}, \frac{2}{3}$

3. In a box carrying one dozen of oranges, one third have become bad. If 3 oranges are taken out from the box at random, what is the probability that at least one orange out of the three oranges picked up is good?

   a) $\frac{1}{55}$ b) $\frac{54}{55}$ c) $\frac{45}{55}$ d) $\frac{3}{55}$

   (SBI Bank PO Exam 1999)

4. Out of 15 students studying in a class, 7 are from Maharashtra, 5 are from Karnataka and 3 are from Goa. Four students are to be selected at random. What are the chances that at least one is from Karnataka?

a) $\frac{12}{13}$ b) $\frac{11}{13}$ c) $\frac{10}{15}$ d) $\frac{1}{15}$

(BSRB Guwahati PO 1999)

5. The probability that a teacher will give one surprise test during any class meeting in a week is $\frac{1}{5}$. If a student is absent twice, what is the probability that he will miss at least one test.

a) $\frac{4}{15}$ b) $\frac{1}{15}$ c) $\frac{91}{25}$ d) $\frac{16}{125}$

(BSRB Mumbai PO Exam 1999)

6. In a throw of a coin, the probability of getting a head is _______.

a) $\frac{1}{2}$ b) $\frac{1}{4}$ c) 1 d) None of these

7. In a simultaneous throw of two coins, the probability of getting at least one head is _______.

a) $\frac{1}{2}$ b) $\frac{2}{3}$ c) $\frac{3}{4}$ d) $\frac{1}{3}$

8. Three unbiased coins are tossed, what is the probability of getting exactly two heads?

a) $\frac{1}{3}$ b) $\frac{3}{4}$ c) $\frac{2}{3}$ d) $\frac{3}{8}$

9. Three unbiased coins are tossed. What is the probability of getting at most 2 heads?

a) $\frac{1}{4}$ b) $\frac{3}{8}$ c) $\frac{7}{8}$ d) $\frac{1}{2}$

10. A fair coin is tossed 100 times. The probability of getting head an odd number of times is _______.

a) $\frac{1}{4}$ b) $\frac{2}{3}$ c) $\frac{1}{2}$ d) $\frac{3}{4}$

11. A bag contains 6 black balls and 8 white balls. One ball is drawn at random. What is the probability that the ball drawn is white?

a) $\frac{4}{7}$ b) $\frac{3}{4}$ c) $\frac{4}{5}$ d) $\frac{1}{8}$

12. A bag contains 8 red and 5 white balls. 2 balls are drawn at random. What is the probability that both are white?

a) $\frac{5}{16}$ b) $\frac{2}{13}$ c) $\frac{3}{26}$ d) $\frac{5}{39}$

13. A bag contains 5 blue and 4 black balls. Three balls are drawn at random. What is the probability that 2 are blue and 1 is black?

a) $\frac{1}{3}$ b) $\frac{2}{5}$ c) $\frac{1}{6}$ d) None of these

14. The odds in favour of an event are 3:5. The probability of occurrence of the event is _______.

a) $\frac{3}{5}$ b) $\frac{3}{8}$ c) $\frac{1}{3}$ d) $\frac{1}{5}$

15. The odds against the occurrence of an event are 5:4. The probability of its occurrence is _______.

a) $\frac{4}{5}$ b) $\frac{4}{9}$ c) $\frac{1}{5}$ d) $\frac{1}{4}$

16. In a lottery there are 20 prizes and 15 blanks. What is the probability of getting prize?

a) $\frac{1}{10}$ b) $\frac{2}{5}$ c) $\frac{4}{7}$ d) $\frac{2}{7}$

17. An urn contains 9 red, 7 white and 4 black balls. A ball is drawn at random. What is the probability that the ball drawn is not red?

a) $\frac{1}{11}$ b) $\frac{9}{20}$ c) $\frac{2}{11}$ d) $\frac{11}{20}$

18. What is the probability that a number selected from the numbers 1, 2, 3, 4, 5, ..., 16 is a prime number?

a) $\frac{1}{16}$ b) $\frac{5}{8}$ c) $\frac{3}{8}$ d) $\frac{7}{16}$

19. Ticket numbered 1 to 20 are mixed up and then a ticket is drawn at random. What is the probability that the ticket drawn bears a number which is a multiple of 3?

a) $\frac{3}{20}$ b) $\frac{3}{10}$ c) $\frac{2}{5}$ d) $\frac{1}{2}$

20. Ticket numbered 1 to 20 are mixed up and then a ticket is drawn at random. What is the probability that the ticket drawn bears a number which is a multiple of 3 or 7?

a) $\frac{1}{15}$ b) $\frac{1}{2}$ c) $\frac{2}{5}$ d) $\frac{7}{20}$

Answers

1. b; **Hint:** In a simultaneous throw of two dice
Sample space $= 6 \times 6 = 36$
Favourable cases are (2, 6) (3, 5) (4, 4) (5, 3) (6, 2)
So, the required probability $= \frac{5}{36}$

2. a; **Hint:** In tossing a coin 2 times the sample space is 4 ie (H, H), (H, T), (T, H), (T, T)
1) If A_1 denotes exactly one head
then, $A_1 = \{(H, T)(T, H)\}$ So, $P(A_1) = \frac{2}{4} = \frac{1}{2}$
2) If A denotes at least one head
then $A = \{(H, T), (T, H), (H, H)\}$ $\therefore P(A) = \frac{3}{4}$

3. c; **Hint:** $n(S) = {}^{12}C_3 = \frac{12\times11\times10}{3\times2} = 2\times11\times10 = 220$

No. of selection of 3 oranges out of the total 12 oranges $= {}^{12}C_3 = 2\times11\times10 = 220$

No. of selection of 3 bad oranges out of the total 4 bad oranges $= {}^4C_3 = 4$

$\therefore$ n(E) = no. of desired selection of oranges $= 220 - 4 = 216$

$\therefore\ P(E) = \frac{n(E)}{n(S)} = \frac{216}{220} = \frac{54}{55}$

4. b; **Hint:** Total possible ways of selecting 4 students out of 15 students $= {}^{15}C_4 = \frac{15\times14\times13\times12}{1\times2\times3\times4} = 1365$

The no. of ways of selecting 4 students in which no student belongs to Karnataka $= {}^{10}C_4$

$\therefore$ number of ways of selecting at least one student from Karnataka $= {}^{15}C_4 - {}^{10}C_4 = 1155$.

$\therefore$ Probability $= \frac{1155}{1365} = \frac{77}{91} = \frac{11}{13}$

5. b; **Hint:** The probability of absenting of the student in the class $= \frac{2}{6} = \frac{1}{3}$

$\therefore$ the probability of missing his test $= \frac{1}{5}\times\frac{1}{3} = \frac{1}{15}$.

6. a; **Hint:** Here S = {H, T} and E = {H}

$\therefore\ P(E) = \frac{n(E)}{n(S)} = \frac{1}{2}$

7. c; **Hint:** S = {HH, HT, TT, TH} and E = {HH, HT, TH}

$\therefore\ P(E) = \frac{n(E)}{n(S)} = \frac{3}{4}$

8. d; **Hint:** S = {HHH, HHT, HTH, THH, TTH, THT, HTT, TTT}and

E = Event of getting exactly two heads = {HHT, HTH, THH}

$\therefore\ P(E) = \frac{n(E)}{n(S)} = \frac{3}{8}$

9. c; **Hint:** n(S) = 8 [See hint of the Q. No. 8]

E = Event of getting 0, or 1 or 2 heads = {TTT, TTH, THT, HTT, HHT, HTH, THH}

or, n(E) = 7

$\therefore$ P(E) $= \frac{n(E)}{n(S)} = \frac{7}{8}$

10. c; **Hint:** $n(S) = 2^{100}$

n(E) = no. of favourable ways

$= {}^{100}C_1 + {}^{100}C_3 + ... + {}^{100}C_{99} = 2^{100-1} = 2^{99}$

$\left[\because {}^nC_1 + {}^nC_3 + {}^nC_5 + ... = 2^{n-1}\right]$

$\therefore\ P(E) = \frac{n(E)}{n(S)} = \frac{2^{99}}{2^{100}} = \frac{1}{2}$.

Note: The given case can be generalised as "*If a unbiased coin is tossed 'n' times, then the chance that the head will present itself an odd number of times is* $\frac{1}{2}$".

11. a; **Hint:** Total no. of balls = (6 + 8) = 14

No. of white balls = 8

$\therefore$ P(drawing a white ball) $= \frac{8}{14} = \frac{4}{7}$

12. d; **Hint:**

n(S) = Number of ways of drawing 2 balls out of 13

$= {}^{13}C_2 = \frac{13\times12}{2} = 78$

n(E) = No. of ways of drawing 2 balls out of 5

$= {}^5C_2 = \frac{5\times4}{2} = 10$

$\therefore$ P(E) $= \frac{n(E)}{n(S)} = \frac{10}{78} = \frac{5}{39}$

13. c; **Hint:** Let S be the sample space and E be the event of drawing 3 balls out of which 2 are blue and 1 is black. Then, n(S) = Number of ways of drawing 3 balls out of 9 $= {}^9C_3 = \frac{9\times8\times7}{3\times2\times1} = 84$

n(E) = Number of ways of drawing 2 balls out of 5 and 1 ball out of 4.

$= {}^5C_2 + {}^4C_1 = \left(\frac{5\times4}{2\times1} + 4\right) = 14$

$\therefore$ P(E) $= \frac{n(E)}{n(S)} = \frac{14}{84} = \frac{1}{6}$.

14. b; **Hint:** Number of cases favourable to E = 3

Total number of cases = (3 + 5) = 8

$\therefore$ P(E) $= \frac{3}{8}$.

15. b; **Hint:** Number of cases favourable to E = 4

Total number of cases = (5 + 4) = 9

$\therefore$ P(E) $= \frac{4}{9}$.

16. c; **Hint:** P(getting a prize) $= \frac{20}{20+15} = \frac{20}{35} = \frac{4}{7}$.

17. d; **Hint:** $P(\text{red}) = \frac{9}{9+7+4} = \frac{9}{20}$

$\therefore P(\text{not-red}) = \left(1 - \frac{9}{20}\right) = \frac{11}{20}$

18. c; **Hint:** S = {1, 2, 3, ..., 16} and E = {2, 3, 5, 7, 11, 13}

$\therefore P(E) = \frac{n(E)}{n(S)} = \frac{6}{16} = \frac{3}{8}$.

19. b; **Hint:** S = {1, 2, 3, ... 20} and E = {3, 6, 9, 12, 15, 18}

$\therefore P(E) = \frac{n(E)}{n(S)} = \frac{6}{20} = \frac{3}{10}$

20. c; **Hint:** Clearly, n(S) = 20 and E = {3, 6, 9, 12, 15, 18, 7, 14}

$\therefore P(E) = \frac{n(E)}{n(S)} = \frac{8}{20} = \frac{2}{5}$.

Rule 2

Problems based on leap year.

A leap year has 366 days, hence it has 52 complete weeks and 2 more days.

When the year is not a leap year, it has 52 complete weeks and 1 more day.

Illustrative Example

Ex.: (i) What is the chance that a leap year selected randomly will have 53 Sundays?

(ii) What is the chance, if the year selected is a not a leap year?

Soln: **(i)** A leap year has 366 days so it has 52 complete weeks and 2 more days. The two days can be {Sunday and Monday, Monday and Tuesday, Tuesday and Wednesday, Wednesday and Thursday, Thursday and Friday, Friday and Saturday, Saturday and Sunday}, i.e. n(S) = 7.

Out of these 7 cases, cases favourable for more Sundays are

{Sunday and Monday, Saturday and Sunday},

i.e. n(E) = 2

$\therefore P(E) = \frac{2}{7}$

(ii) When the year is not a leap year, it has 52 complete weeks and 1 more day that can be {Sunday, Monday, Tuesday, Wednesday, Thursday, Friday, Saturday}, n(S) = 7

Out of these 7 cases, cases favourable for one more Sunday is

{Sunday}, n(E) = 1

$\therefore P(E) = \frac{1}{7}$

Exercise

1. What is the probability that a leap year selected randomly will have 53 Mondays?

a) $\frac{2}{7}$ b) $\frac{5}{7}$ c) $\frac{1}{7}$ d) Data inadequate

2. What is the probability that an oridinary year has 53 Sundays?

a) $\frac{53}{365}$ b) $\frac{1}{7}$ c) $\frac{2}{7}$ d) $\frac{48}{53}$

Answers

1. a; **Hint:** A leap year has 366 days = 52 weeks + 2 days These 2 days can be (Sunday, Monday), (Monday, Tuesday), (Tuesday, Wednesday) ... or, (Saturday, Sunday). Out of these total 7 outcomes, there are 2 cases favourable to the desired event ie (Sunday, Monday) and (Monday, Tuesday)

$\therefore$ required probability $= \frac{2}{7}$.

2. b; **Hint:** An ordinary year has 365 days ie 52 weeks and 1 day. So the probability that this day is a Sunday is $\frac{1}{7}$.

Rule 3

Problems based on dice

Following chart will be helpful in solving the problems based on dice.

Chart: When two dice are thrown, we have,

S = {(1, 1), (1, 2), ..., (1, 6), (2, 1), (2, 2), ..., (2, 6), (3, 1), (3, 2), ..., (3, 6), (4, 1), ..., (4, 6), (5, 1), ..., (5, 6), (6, 1), ... (6, 6)}

n(S) = 6 × 6 = 36

Sum of the no. of the two dice		n(S)	Events	
(i)	(ii)		(i)	(ii)
2	12	1	{1, 1}	{6, 6}
3	11	2	{1, 2}, {2, 1}	{6, 5}, {5, 6}
4	10	3	{1, 3}, {3, 1}, {2, 2}	{6, 4}, {4, 6}, {5, 5}
5	9	4	{1, 4}, {4, 1}, {2, 3}, {3, 2}	{6, 3}, {3, 6}, {5, 4}, {4, 5}
6	8	5	{1, 5}, {5, 1}, {2, 4}, {4, 2}, {3, 3}	{6, 2}, {2, 6}, {5, 3}, {3, 5}, {4, 4}
7		6	{1, 6}, {6, 1}, {2, 5}, {5, 2}, {4, 3}, {3, 4}	

Illustrative Example

Ex.: When two dice are thrown, what is the probability that

(i) sum of numbers appeared is 6 and 7?

(ii) sum of numbers appeared ≤ 8?

(iii) sum of numbers is an odd no?

(iv) sum of numbers is a multiple of 3?

(v) numbers shown are equal?

(vi) the difference of the numbers is 2?

(vii) Sum of the numbers is at least 5.

Soln: (i) Use the above chart:

For 6, reqd probability $= \frac{n(E)}{n(S)} = \frac{5}{36}$

For 7, reqd probability $= \frac{6}{36} = \frac{1}{6}$

(ii) Desired sums of the numbers are 2, 3, 4, 5, 6, 7 and 8;

n(S) = 1 + 2 + 3 + 4 + 5 + 6 + 5 = 26

$\therefore$ reqd probability $= \frac{26}{36} = \frac{13}{18}$

(iii) Desired sums of the numbers are 3, 5, 7, 9 and 11;

n(S) = 2 + 4 + 6 + 4 + 2 = 18

$\therefore$ reqd probability $= \frac{18}{36} = \frac{1}{2}$

(iv) Desired sums of the numbers are 3, 6, 9 and 12;

n(S) = 2 + 5 + 4 + 1 = 12

$\therefore$ reqd probability $= \frac{12}{36} = \frac{1}{3}$

(v) Events = {1, 1}, {2, 2}, {3, 3}, {4, 4}, {5, 5}, {6, 6};

n(S) = 6

$\therefore P(E) = \frac{6}{36} = \frac{1}{6}$

(vi) Events = {3, 1}, {4, 2}, {5, 3}, {6, 4}, {4, 6}, {3, 5}, {2, 4}, {1, 3}

or n(S) = 8

$\therefore P(E) = \frac{8}{36} = \frac{2}{9}$

(vii) Events; either 2 or 3 or 4 or 5

n(E) = 1 + 2 + 3 + 4 = 10

n(S) = 36

$\therefore P(E) = \frac{n(E)}{n(S)} = \frac{10}{36} = \frac{5}{18}$.

Exercise

1. In a throw of a die, the probability of getting a prime number is

   a) $\frac{1}{3}$ b) $\frac{1}{2}$ c) $\frac{2}{3}$ d) $\frac{3}{4}$

2. In a simultaneous throw of two dice, what is the probability of getting a doublet?

   a) $\frac{1}{6}$ b) $\frac{1}{4}$ c) $\frac{3}{4}$ d) $\frac{2}{3}$

3. In a simultaneous throw of two dice, what is the probability of getting a total of 10 or 11?

   a) $\frac{7}{12}$ b) $\frac{5}{36}$ c) $\frac{1}{6}$ d) $\frac{1}{4}$

4. In a single throw of two dice what is the probability of not getting the same number on both the dice?

   a) $\frac{1}{6}$ b) $\frac{2}{3}$ c) $\frac{5}{6}$ d) $\frac{1}{3}$

Answers

1. b; **Hint:** Here S = {1, 2, 3, 4, 5, 6} and E = {2, 3, 5}

   $\therefore P(E) = \frac{3}{6} = \frac{1}{2}$.

2. a; **Hint:** In a simultaneous throw of two dice

   n(S) = 36

   Let E = {(1, 1), (2, 2), (3, 3), (4, 4), (5, 5), (6, 6)}

   $\therefore P(E) = \frac{6}{36} = \frac{1}{6}$.

3. b; **Hint:** Clearly, n(s) = 36

   Let (E) = {(4, 6), (5, 5), (6, 4), (5, 6), (6, 5)}

   $\therefore P(E) = \frac{5}{36}$

4. c; **Hint:** Clearly, n(S) = 36

   Let E = {(1, 1), (2, 2), (3, 3), (4, 4), (5, 5), (6, 6)}

   $\therefore P(E) = \frac{6}{36} = \frac{1}{6}$

   So, P(not - E) $= \left(1 - \frac{1}{6}\right) = \frac{5}{6}$

Rule 4

Problems based on cards

Following chart will be helpful to solve the problems based on cards.

Chart: A pack of cards has a total of 52 cards.

Red suit (26)		Black suit (26)	
Diamond (13)	Heart (13)	Spade (13)	Club (13)

The numbers in the brackets show the respective no. of cards in that category.

Each of Diamond, Heart, Spade and Club contains nine digit-cards 2, 3, 4, 5, 6, 7, 8, 9 and 10 (a total of 9 × 4 = 36 digit-cards) along with four Honour cards Ace, King, Queen and Jack (a total of 4 × 4 = 16 Honour cards).

Illustrative Examples

Ex. 1: A card is drawn from a pack of cards. What is the probability that it is

(i) a card of black suit?
(ii) a spade card?
(iii) an honours card of red suit?
(iv) an honours card of club?
(v) a card having the number less than 7?
(vi) a card having the number a multiple of 3?
(vii) a king or a queen?
(viii) a digit-card of heart?
(ix) a jack of black suit?

Soln:

For all the above cases $n(S) = {}^{52}C_1 = 52$

(i) $\frac{26}{52} = \frac{1}{2}\left[\text{or, } \frac{{}^{26}C_1}{{}^{52}C_1} = \frac{26}{52}\left(\because {}^nC_1 = n\right)\right]$

(ii) $\frac{13}{52} = \frac{1}{4}$ (iii) $\frac{4 \times 2}{52} = \frac{2}{13}$

(iv) $\frac{4}{52} = \frac{1}{13}$ (v) $\frac{5 \times 4}{52} = \frac{5}{13}$

(vi) $\frac{3 \times 4}{52} = \frac{3}{13}$

(vii) P(a king) = $\frac{4}{52} = \frac{1}{13}$; P(a queen) = $\frac{4}{52} = \frac{1}{13}$

$\therefore$ P(a king or a queen) = $\frac{1}{13} + \frac{1}{13} = \frac{2}{13}$

(viii) $\frac{9}{52}$ (ix) $\frac{2}{52} = \frac{1}{26}$

Ex. 2: From a pack of 52 cards, 2 cards are drawn at random. What is the probability that it has
(i) both the Aces? (ii) exactly one queen?
(iii) no honours card? (iv) no digit-card?
(v) One King and one Queen?

Soln: For all the above cases, $n(S) = {}^{52}C_2 = \frac{52 \times 51}{2} = 26 \times 51$

(i) Total no. of Aces = 4

$\therefore n(E) = {}^4C_2 = \frac{4 \times 3}{2} = 6$

$\therefore P(E) = \frac{6}{26 \times 51} = \frac{1}{221}$

(ii) Total no. of Queens = 4

Selection of 1 Queen card out of 4 can be done in ${}^4C_1 = 4$ ways. He can select the remaining 1 card from the remaining (52 – 4 =) 48 cards. Now, cards in ${}^{48}C_1 = 48$ ways.

$\therefore n(E) = 4 \times 48$ $\therefore P(E) = \frac{4 \times 48}{26 \times 51} = \frac{32}{221}$

(iii) Total no. of honours card = 16

To have no honours card, he has to select two cards out of the remaining 52 – 16 = 36 cards which he can do in ${}^{36}C_2 = \frac{36 \times 35}{2} = 18 \times 35$ ways

$\therefore P(E) = \frac{18 \times 35}{26 \times 51} = \frac{105}{221}$

(iv) $P(E) = \frac{{}^{16}C_2}{26 \times 51} = \frac{8 \times 15}{26 \times 51} = \frac{20}{221}$

(v) $n(E) = {}^4C_1 \times {}^4C_1 = 4 \times 4 = 16$

$\therefore P(E) = \frac{16}{26 \times 51} = \frac{8}{663}$

Ex. 3: From a pack of 52 cards, 3 cards are drawn. What is the probability that it has
(i) all three aces?
(ii) no queen?
(iii) one ace, one king and one queen?
(iv) one ace and two jacks?
(v) two digit-cards and one honours card of black suit?

Soln: For all the above cases, n(S)

$= {}^{52}C_3 = \frac{52 \times 51 \times 50}{3 \times 2} = 26 \times 17 \times 50$

(i) $n(E) = {}^4C_3 = 4$

$\therefore P(E) = \frac{4}{26 \times 17 \times 50} = \frac{1}{5525}$

(ii) $n(E) = {}^{48}C_3 = 8 \times 47 \times 46$

$\therefore P(E) = \frac{8 \times 47 \times 46}{26 \times 17 \times 50} = \frac{4324}{5525}$

(iii) $n(E) = {}^4C_1 \times {}^4C_1 \times {}^4C_1 = 4 \times 4 \times 4$

$\therefore P(E) = \frac{4 \times 4 \times 4}{26 \times 17 \times 50} = \frac{16}{5525}$

(iv) $n(E) = {}^4C_1 \times {}^4C_2 = 4 \times 6$

$\therefore P(E) = \frac{4 \times 6}{26 \times 17 \times 50} = \frac{6}{5525}$

(v) $n(E) = {}^{36}C_2 \times {}^8C_1 = 18 \times 35 \times 8$

$\therefore P(E) = \frac{18 \times 35 \times 8}{26 \times 17 \times 50} = \frac{252}{1105}$

Exercise

1. One card is drawn at random from a pack of 52 cards. What is the probability that the card drawn is a face card?

   a) $\frac{4}{13}$ b) $\frac{1}{4}$ c) $\frac{9}{52}$ d) $\frac{1}{13}$

2. One card is drawn at random from a pack of 52 cards. What is the probability that the card drawn is either a red card or a king?

   a) $\frac{6}{13}$ b) $\frac{1}{2}$ c) $\frac{7}{13}$ d) $\frac{27}{52}$

3. Two cards are drawn at random from a pack of 52 cards. What is the probability that the drawn cards are both aces?

   a) $\frac{1}{221}$ b) $\frac{2}{13}$ c) $\frac{3}{26}$ d) None of these

4. What is the probability of getting a king or a queen in a single drawn from a pack of 52 cards?

a) $\frac{1}{26}$ b) $\frac{1}{13}$ c) $\frac{2}{13}$ d) None of these

5. A card is drawn from a pack of 52 cards. A card is drawn at random. What is the probability that it is neither a heart nor a king?

a) $\frac{4}{13}$ b) $\frac{9}{13}$ c) $\frac{2}{13}$ d) $\frac{4}{13}$

6. A card is drawn at random from a pack of 52 cards. What is the probability that the card drawn is a spade or a king?

a) $\frac{4}{13}$ b) $\frac{3}{13}$ c) $\frac{2}{13}$ d) $\frac{1}{13}$

Answers

1. a; **Hint:** Clearly, n(S) = 52 and there are 16 face cards.

$\therefore$ P(E) $= \frac{16}{52} = \frac{4}{13}$

2. c; **Hint:** Clearly n(S) = 52. There are 26 red cards (including 2 kings) and there are 2 more kings.
Let (E) be the event of getting either a red card or a king.
Then, n(E) = 28

$\therefore$ P(E) $= \frac{n(E)}{n(S)} = \frac{28}{52} = \frac{7}{13}$.

3. a; **Hint:** n(S) = Number of ways of drawing 2 cards out of 52

$= {}^{52}C_2 = \frac{52 \times 51}{2 \times 1} = 1326$

n(E) = Number of ways of drawing 2 cards out of 4

$= {}^{4}C_2 = \frac{4 \times 3}{2} = 6$

$\therefore$ P(E) $= \frac{6}{1326} = \frac{1}{221}$.

4. c; **Hint:** Clearly, n(S)=52, there are 4 kings and 4 queens.

$\therefore$ P(E) $= \frac{n(E)}{n(S)} = \frac{8}{52} = \frac{2}{13}$.

5. b; **Hint:** There are 13 hearts and 3 more kings

$\therefore$ P(heart or a king) $= \frac{13+3}{52} = \frac{4}{13}$

$\therefore$ P(neither a heart nor a king) $= 1 - \frac{4}{13} = \frac{9}{13}$.

6. a; **Hint:** Let E and F be the event of getting a spade and that of getting a king respectively.
Then $E \cap F$ is the event of getting a king of spade
$\therefore$ n(E) = 13, n(F) = 4 and n(E $\cap$ F) = 1

So, P(E) $= \frac{13}{52} = \frac{1}{4}$, P(F) $= \frac{4}{52} = \frac{1}{13}$ and

P(E $\cap$ F) $= \frac{1}{52}$

$\therefore$ P(a spade or a king) = P(E $\cup$ F)
= P(E) + P(F) - P(E $\cap$ F)

$= \left(\frac{1}{4} + \frac{1}{13} - \frac{1}{52}\right) = \frac{4}{13}$.

Rule 5

Theorem: ***If a bag contains x red, y yellow and z green balls, 3 balls are drawn randomly, then the probability of the balls drawn contain balls of different colour is given by***

$$\left[\frac{6xyz}{(x+y+z)(x+y+z-1)(x+y+z-2)}\right]$$

or $\frac{{}^xC_1 \times {}^yC_1 \times {}^zC_1}{{}^{(x+y+z)}C_3}$

Illustrative Example

Ex: A bag contains 3 red, 5 yellow and 4 green balls. 3 balls are drawn randomly. What is the probability that the balls drawn contain balls of different colours?

Soln: **Detail Method:** Total no. of balls = 3 + 5 + 4 = 12

n(S) $= {}^{12}C_3 = \frac{12 \times 11 \times 10}{3 \times 2} = 220$

In order to have 3 different coloured balls, the selection of one ball of each colour is to be made.

n(E) $= {}^3C_1 \times {}^5C_1 \times {}^4C_1 = 3 \times 5 \times 4 = 60$

$\therefore$ P(E) $= \frac{60}{220} = \frac{3}{11}$

Quicker Method: Applying the above theorem, we have

the required answer $= \frac{6 \times 3 \times 5 \times 4}{12 \times 11 \times 10} = \frac{3}{11}$.

Exercise

1. A bag contains 4 red, 6 yellow and 5 green balls. 3 balls are drawn randomly. What is the probability that the balls drawn contain balls of different colours?

a) $\frac{24}{91}$ b) $\frac{14}{91}$ c) $\frac{13}{91}$ d) $\frac{21}{91}$

2. A bag contains 5 red, 7 yellow and 6 green balls. 3 balls are drawn randomly. What is the probability that the balls drawn contain balls of different colours?

a) $\frac{35}{138}$ b) $\frac{35}{136}$ c) $\frac{35}{134}$ d) $\frac{35}{163}$

3. A bag contains 6 red, 8 yellow and 4 green balls. 3 balls are drawn randomly. What is the probability that the balls drawn contain balls of different colours?

a) $\frac{4}{17}$ b) $\frac{3}{17}$ c) $\frac{6}{17}$ d) $\frac{8}{17}$

4. A bag contains 4 yellow, 5 red and 8 green balls. 3 balls are drawn randomly. What is the probability that the balls drawn contain balls of different colours?

a) $\frac{5}{17}$ b) $\frac{2}{17}$ c) $\frac{4}{17}$ d) $\frac{8}{17}$

Answers

1. a 2. b 3. a 4. c

Rule 6

Theorem: ***If a bag contains x red, y yellow and z green balls, 3 balls are drawn randomly, then the probability of the balls drawn contain***

(i) exactly 2 green balls is given by

$$\frac{3z(z-1)(x+y)}{(x+y+z)(x+y+z-1)(x+y+z-2)} \text{ or } \frac{{}^{(x+y)}C_1 \cdot {}^{z}C_2}{{}^{(x+y+z)}C_3}$$

(ii) exactly 2 yellow balls is given by

$$\frac{3y(y-1)(x+z)}{(x+y+z)(x+y+z-1)(x+y+z-2)} \text{ or } \frac{{}^{(x+z)}C_1 \cdot {}^{y}C_2}{{}^{(x+y+z)}C_3}$$

(iii) exactly 2 red balls is given by

$$\frac{3x(x-1)(y+z)}{(x+y+z)(x+y+z-1)(x+y+z-2)} \text{ or } \frac{{}^{(y+z)}C_1 \cdot {}^{x}C_2}{{}^{(x+y+z)}C_3}$$

Illustrative Example

Ex: A bag contains 3 red, 5 yellow and 4 green balls. 3 balls are drawn randomly. What is the probability that balls drawn contain exactly two green balls?

Sol: Detail Method:

Total no. of balls = 3 + 5 + 4 = 12

$n(S) = {}^{12}C_3 = \frac{12 \times 11 \times 10}{3 \times 2} = 220$

2 green balls can be selected from 4 green balls in ${}^{4}C_2$ ways and the rest one ball can be selected from the remaining (12 – 4) = 8 balls in ${}^{8}C_1$ ways.

$n(E) = {}^{4}C_2 \times {}^{8}C_1 = 6 \times 8 = 48$

$\therefore P(E) = \frac{48}{220} = \frac{12}{55}$

Quicker Method: Applying the above theorem, we have

the required answer

$$= \frac{3 \times 4 \times (4-1) \times 8}{12 \times 11 \times 10} = \frac{3 \times 4 \times 3 \times 8}{12 \times 11 \times 10} = \frac{12}{55}.$$

Exercise

1. A bag contains 4 red, 6 yellow and 5 green balls. 3 balls are drawn randomly.

(i) What is the probability that balls drawn contain exactly 2 green balls?

a) $\frac{20}{91}$ b) $\frac{15}{91}$ c) $\frac{10}{91}$ d) Data inadequate

(ii) What is the probability that balls drawn contain exactly 2 yellow balls?

a) $\frac{20}{91}$ b) $\frac{28}{91}$ c) $\frac{27}{91}$ d) $\frac{37}{91}$

(iii) What is the probability that balls drawn contain exactly 2 red balls?

a) $\frac{54}{455}$ b) $\frac{44}{455}$ c) $\frac{54}{91}$ d) None of these

2. A bag contains 5 red, 7 yellow and 6 green balls. 3 balls are drawn randomly.

(i) What is the probability that balls drawn contain exactly 2 green balls?

a) $\frac{14}{68}$ b) $\frac{13}{68}$ c) $\frac{15}{91}$ d) $\frac{15}{68}$

(ii) What is the probability that balls drawn contain exactly 2 yellow balls?

a) $\frac{77}{272}$ b) $\frac{76}{272}$ c) $\frac{77}{271}$ d) $\frac{76}{271}$

(iii) What is the probability that balls drawn contain exactly 2 red balls?

a) $\frac{55}{406}$ b) $\frac{54}{408}$ c) $\frac{55}{408}$ d) $\frac{55}{480}$

3. A bag contains 4 red, 5 yellow and 6 green balls. 3 balls are drawn randomly.

(i) What is the probability that balls drawn contain exactly 2 green balls?

a) $\frac{27}{91}$ b) $\frac{20}{91}$ c) $\frac{54}{91}$ d) None of these

(ii) What is the probability that balls drawn contain exactly 2 yellow balls?

a) $\frac{20}{91}$ b) $\frac{27}{91}$ c) $\frac{54}{455}$ d) Data inadequate

(iii) What is the probability that balls drawn contain exactly 2 red balls?

a) $\frac{20}{91}$ b) $\frac{19}{91}$ c) $\frac{12}{91}$ d) None of these

Answers

1. (i) a (ii) c (iii) a 2. (i) d (ii) a (iii) c
3. (i) a (ii) a (iii) c

Rule 7

Theorem: ***A bag contains x red, y yellow and z green balls. 3 balls are drawn randomly. The probability of the balls drawn contain***

*(i) no **yellow** ball is given by*

$$\left[\frac{(x+z)(x+z-1)(x+z-2)}{(x+y+z)(x+y+z-1)(x+y+z-2)}\right] \text{ or } \frac{^{(x+z)}C_3}{^{(x+y+z)}C_3}$$

*(ii) no **red** ball is given by*

$$\left[\frac{(y+z)(y+z-1)(y+z-2)}{(x+y+z)(x+y+z-1)(x+y+z-2)}\right] \text{ or } \frac{^{(y+z)}C_3}{^{(x+y+z)}C_3}$$

*(iii) no **green** ball is given by*

$$\left[\frac{(x+y)(x+y-1)(x+y-2)}{(x+y+z)(x+y+z-1)(x+y+z-2)}\right] \text{ or } \frac{^{(x+y)}C_3}{^{(x+y+z)}C_3}$$

Illustrative Example

Ex: A bag contains 3 red, 5 yellow and 4 green balls. 3 balls are drawn randomly. What is the probability that the balls drawn contain no yellow ball?

Soln: Detail Method: Total no. of balls = 3 + 5 + 4 = 12

$$n(S) = {}^{12}C_3 = \frac{12\times11\times10}{3\times2} = 220$$

3 balls can be selected from 3(red) + 4(green) = 7 balls in $^7C_3$ ways.

$$n(E) = {}^7C_3 = \frac{7\times6\times5}{3\times2} = 35$$

$$\therefore P(E) = \frac{35}{220} = \frac{7}{44}.$$

Quicker Method: Applying the above theorem, we have

$$\text{the required answer} = \frac{7\times6\times5}{12\times11\times10} = \frac{35}{220} = \frac{7}{44}.$$

Exercise

1. A bag contains 4 red, 5 yellow and 6 green balls. 3 balls are drawn randomly.
   (i) What is the probability that the balls drawn contain no yellow ball?
   a) $\frac{24}{91}$ b) $\frac{33}{91}$ c) $\frac{12}{65}$ d) Data inadequate
   (ii) What is the probability that the balls drawn contain no red ball?
   a) $\frac{12}{65}$ b) $\frac{33}{91}$ c) $\frac{24}{91}$ d) None of these
   (iii) What is the probability that the balls drawn contain no green balls?
   a) $\frac{12}{65}$ b) $\frac{33}{91}$ c) $\frac{24}{91}$ d) $\frac{1}{5}$
2. A bag contains 5 red, 6 yellow and 7 green balls. 3 balls are drawn randomly.
   (i) What is the probability that the balls drawn contain no yellow ball?
   a) $\frac{55}{204}$ b) $\frac{143}{408}$ c) $\frac{55}{272}$ d) $\frac{55}{208}$
   (ii) What is the probability that the balls drawn contain no red ball?
   a) $\frac{55}{282}$ b) $\frac{55}{272}$ c) $\frac{143}{408}$ d) $\frac{143}{406}$
   (iii) What is the probability that the balls drawn contain no green balls?
   a) $\frac{143}{408}$ b) $\frac{55}{272}$ c) $\frac{55}{204}$ d) $\frac{55}{208}$
3. A bag contains 3 red, 5 yellow and 7 green balls. 3 balls are drawn randomly.
   (i) What is the probability that the balls drawn contain no yellow ball?
   a) $\frac{24}{91}$ b) $\frac{25}{91}$ c) $\frac{44}{91}$ d) $\frac{8}{65}$
   (ii) What is the probability that the balls drawn contain no red ball?
   a) $\frac{8}{65}$ b) $\frac{44}{91}$ c) $\frac{24}{91}$ d) $\frac{45}{91}$
   (iii) What is the probability that the balls drawn contain no green balls?
   a) $\frac{44}{91}$ b) $\frac{24}{91}$ c) $\frac{18}{65}$ d) $\frac{8}{65}$.

Answers

1. (i) a (ii) b (iii) a 2. (i) a (ii) c (iii) b
3. (i) a (ii) b (iii) d

Rule 8

Theorem: ***There are 'x' boys and 'y' girls. If they sit in a row randomly, then the chance that***

(i) all the girls sit together is given by $\left[\frac{(x+1)!\,y!}{(x+y)!}\right]$.

(ii) all the boys sit together is given by $\left[\frac{(y+1)!\,x!}{(x+y)!}\right]$.

(iii) all the girls do not sit together is given by $\left[1-\frac{(x+1)!\,y!}{(x+y)!}\right]$.

(iv) all the boys do not sit together is given by $\left[1-\frac{(y+1)!\,x!}{(x+y)!}\right]$

(v) no two girls sit together (x > y) is given by $\frac{^{x+1}C_y\,x!\,y!}{(x+y)!}$.

Illustrative Example

Ex: There are 4 boys and 4 girls. They sit in a row randomly. What is the chance that all the girls do not sit together?

Soln: Detail Method:

Total no. of arrangements = n(S) = ${}^8P_8 = 8!$

Consider all the 4 girls as one, we have 4 boys + 1 girl = 5 persons. Which can be arranged in ${}^5P_5 = 5!$ ways.

But the 4 girls can also be arranged in ${}^4P_4 = 4!$ ways among themselves.

So, in 4! × 5! ways can the persons be arranged so that girls are together

$= \frac{4!5!}{8!}$ = P(girls are together) $= \frac{4\times3\times2}{8\times7\times6} = \frac{1}{14}$

∴ P(All girls are not together) = 1 - P

(All girls are together) $= 1-\frac{1}{14} = \frac{13}{14}$.

Quicker Method: Applying the above theorem, we have
the required answer

$= 1-\frac{5!4!}{8!} = 1-\frac{5!4!}{8!} = 1-\frac{1}{14} = \frac{13}{14}$.

Exercise

1. There are 5 boys and 4 girls. They sit in a row randomly.

(i) What is the chance that all the girls sit together?

a) $\frac{6!4!}{9!}$ b) $\frac{5!4!}{9!}$ c) $\frac{5!4!}{10!}$ d) None of these

(ii) What is the chance that all the boys sit together?

a) $\frac{6!4!}{9!}$ b) $\frac{5!5!}{9!}$ c) $\frac{5!5!}{10!}$ d) $\frac{6!4!}{10!}$

(iii) What is the chance that all the girls do not sit together?

a) $\frac{1}{21}$ b) $\frac{20}{21}$ c) $\frac{19}{21}$ d) $\frac{2}{21}$

(iv) What is the chance that all the boys do not sit together?

a) $\frac{5}{126}$ b) $\frac{6}{126}$ c) $\frac{121}{126}$ d) None of these

(v) What is the chance that the no two girls sit together?

a) $\frac{5}{42}$ b) $\frac{37}{42}$ c) $\frac{25}{42}$ d) Data inadequate

2. There are 6 boys and 5 girls. They sit in a row randomly.

(i) What is the chance that all the girls sit together?

a) $\frac{1}{66}$ b) $\frac{65}{66}$ c) $\frac{1}{77}$ d) $\frac{76}{77}$

(ii) What is the chance that all the boys sit together?

a) $\frac{1}{66}$ b) $\frac{1}{77}$ c) $\frac{76}{77}$ d) Data inadequate

(iii) What is the chance that all the girls do not sit together?

a) $\frac{65}{66}$ b) $\frac{1}{66}$ c) $\frac{7}{77}$ d) $\frac{76}{77}$

(iv) What is the chance that all the boys do not sit together?

a) $\frac{1}{77}$ b) $\frac{1}{66}$ c) $\frac{65}{66}$ d) $\frac{76}{77}$

(v) What is the chance that the no two girls sit together?

a) $\frac{21}{22}$ b) $\frac{3}{22}$ c) $\frac{19}{22}$ d) $\frac{1}{22}$

3. There are 5 boys and 5 girls. They sit in a row randomly.

(i) What is the chance that all the girls sit together?

a) $\frac{1}{42}$ b) $\frac{41}{42}$ c) $\frac{3}{35}$ d) None of these

(ii) What is the chance that all the boys sit together?

a) $\frac{3}{42}$ b) $\frac{41}{42}$ c) $\frac{3}{35}$ d) $\frac{1}{42}$

(iii) What is the chance that all the girls do not sit together?

a) $\frac{41}{42}$ b) $\frac{39}{42}$ c) $\frac{31}{42}$ d) None of these

(iv) What is the chance that all the boys do not sit together?

a) $\frac{3}{35}$ b) $\frac{41}{42}$ c) $\frac{1}{42}$ d) Data inadequate

Answers

1. (i) a (ii) b (iii) b (iv) c (v) a
2. (i) a (ii) b (iii) a (iv) d (v) d
3. (i) a (ii) d (iii) a (iv) b

Rule 9

Theorem: ***A box contains x black balls, y red balls and z green balls. 'r' balls are drawn from the box at random. The probability that all the balls are of the same colour is given by*** $\frac{{}^xC_r + {}^yC_r + {}^zC_r}{(x+y+z)_{C_r}}$ ***where r < x, y, z***

Illustrative Example

Case I: If r = 2; then the formula for required probability is given by $\left[\frac{x(x-1)+y(y-1)+z(z-1)}{(x+y+z)(x+y+z-1)}\right]$.

Ex: A box contains 4 black balls, 3 red balls and 5 green balls. 2 balls are drawn from the box at random. What is the probability that both the balls are of the same colour?

Soln: Detail Method:

Total no. of balls = 4 + 3 + 5 = 12

$$n(S) = {}^{12}C_2 = \frac{12 \times 11}{2} = 66$$

$$n(E) = {}^{4}C_2 + {}^{3}C_2 + {}^{5}C_2 = \frac{4 \times 3}{2} + \frac{3 \times 2}{2} + \frac{5 \times 4}{2}$$
$$= 6 + 3 + 10 = 19$$

$\therefore$ Required probability, $P(E) = \frac{n(E)}{n(S)} = \frac{19}{66}$

Quicker Method: Applying the above theorem,

$$\text{Required answer} = \frac{4(4-1)+3(3-1)+5(5-1)}{(4+3+5)(4+3+5-1)}$$

$$= \frac{12+6+20}{12 \times 11} = \frac{19}{66}$$

Note: The probability that both the balls are not of the same colour is given by 1 - P (Probability of the same colour)

Case II: If r = 3; then the formula for required probability is given by

$$\left[\frac{x(x-1)(x-2)+y(y-1)(y-2)+z(z-1)(z-2)}{(x+y+z)(x+y+z-1)(x+y+z-2)}\right].$$

Ex.: A box contains 5 green, 4 yellow and 3 white marbles. 3 marbles are drawn at random. What is the probability that they are not of the same colour?

(SBI Associates PO Exam 1999)

Soln: Detail Method:

Total no. of balls = 5 + 4 + 3 = 12

$$n(S) = {}^{12}C_3 = \frac{12 \times 11 \times 10}{1 \times 2 \times 3} = 220$$

i.e, 3 marbles out of 12 marbles can be drawn in 220 ways.

If all the three marbles are of the same colour, it can be done in

$${}^{5}C_3 + {}^{4}C_3 + {}^{3}C_3 = 10 + 4 + 1 = 15 \text{ ways}$$

Now, P(All the 3 marbles of the same colour) + P(all the 3 marbles are not of the same colour) = 1

$\therefore$ P(all the 3 marbles are not of the same colour)

$$= 1 - \frac{15}{220} = \frac{205}{220} = \frac{41}{44}.$$

Quicker Method: Applying the above theorem, we have,

the required answer

$$= 1 - \frac{5 \times 4 \times 3 + 4 \times 3 \times 2 + 3 \times 2 \times 1}{12 \times 11 \times 10}$$

$$= 1 - \frac{60+24+6}{12 \times 11 \times 10}$$

$$= 1 - \frac{90}{12 \times 11 \times 10} = 1 - \frac{3}{44} = \frac{41}{44}.$$

Note: The probability that all the balls are not of the same colour is given by 1 – P (Probability of the same colour).

Exercise

1. A box contains 5 black balls, 4 red balls and 6 green balls. 2 balls are drawn from the box at random. What is the probability that both the balls are of the same colour?

   a) $\frac{31}{105}$ b) $\frac{30}{150}$ c) $\frac{1}{105}$ d) $\frac{74}{105}$

2. A box contains 3 black balls, 5 red balls and 7 green balls. 2 balls are drawn from the box at random. What is the probability that both the balls are of the same colour?

   a) $\frac{74}{105}$ b) $\frac{34}{105}$ c) $\frac{71}{105}$ d) $\frac{31}{105}$

3. A box contains 4 black balls, 6 red balls and 8 green balls. 2 balls are drawn from the box at random. What is the probability that both the balls are not of the same colour?

   a) $\frac{49}{153}$ b) $\frac{50}{153}$ c) $\frac{103}{153}$ d) $\frac{104}{153}$

4. A box contains 4 green, 5 yellow and 6 white marbles. 3 marbles are drawn at random. What is the probability that all the three marbles are of the same colour?

   a) $\frac{102}{1365}$ b) $\frac{204}{1365}$ c) $\frac{102}{1635}$ d) None of these

5. A box contains 3 green, 5 yellow and 3 white marbles. 3 marbles are drawn at random. What is the probability that all the three marbles of the same colour?

   a) $\frac{51}{55}$ b) $\frac{4}{55}$ c) $\frac{3}{55}$ d) $\frac{52}{55}$

6. A box contains 4 green, 5 yellow and 4 white marbles. 3 marbles are drawn at random. What is the probability that all the three balls are not of the same colour?

   a) $\frac{9}{143}$ b) $\frac{134}{143}$ c) $\frac{8}{143}$ d) $\frac{135}{143}$

7. a box contains 4 black, 6 red and 8 green balls. 4 balls are drawn from the box at random. What is the probability that all the balls are of same colour?

   a) $\frac{47}{90}$ b) $\frac{43}{90}$ c) $\frac{49}{90}$ d) $\frac{41}{90}$

Answers

1\. a 2. b

3\. d; **Hint: See Note** Required probability $= 1 - \frac{49}{153} = \frac{104}{153}$.

4\. a 5. b 6. b

7\. b; **Hint:** Applying the given rule, we have the required probability

$$= \frac{{}^{4}C_4 + {}^{6}C_4 + {}^{8}C_4}{{}^{18}C_4} = \frac{1+15+70}{15 \times 4 \times 3} = \frac{43}{90}$$

Rule 10

Theorem: ***A bag contains 'x' red and 'y' black balls. If two draws of three balls each are made, the ball being replaced after the first draw, then the chance that the balls were red in the first draw and black in the second draw is given by***

$$\left[\frac{\{x(x-1)(x-2)\}\{y(y-1)(y-2)\}}{\{(x+y)(x+y-1)(x+y-2)\}^2}\right] \text{ or } \frac{{}^{x}C_3 \times {}^{y}C_3}{\left({}^{x+y}C_3\right)^2}.$$

Illustrative Example

Ex.: A bag a contains 5 red and 8 black balls. Two draws of three balls each are made, the ball being replaced after the first draw. What is the chance that the balls were red in the first draw and black in the second?

Soln: Detail Method:

Total no. of balls = 5 + 8 = 13

$$n(S) = {}^{13}C_3 = \frac{11\times12\times13}{1\times2\times3} = 286$$

Chance that the balls were red in first draw $= \frac{{}^{5}C_3}{{}^{13}C_3}$

and

Chance that the balls were black in the second draw

$= \frac{{}^{8}C_3}{{}^{13}C_3}$ [∵ balls are replaced after first draw]

Required probability $= \frac{{}^{5}C_3}{{}^{13}C_3} \times \frac{{}^{8}C_3}{{}^{13}C_3} = \frac{140}{20449}$

Note: In the above example, the two events are independent and can occur simultaneously. So, we used multiplication.

Quicker Method: Applying the above theorem, we have,

the required probability

$$= \frac{(5\times4\times3)\times(8\times7\times6)}{(13\times12\times11)^2} = \frac{20160}{2944656} = \frac{140}{20449}.$$

Exercise

1. A bag a contains 4 red and 7 black balls. Two draws of three balls each are made, the ball being replaced after the first draw. What is the chance that the balls were red in the first draw and black in the second?

   a) $\frac{28}{5445}$ b) $\frac{25}{5448}$ c) $\frac{28}{4554}$ d) $\frac{25}{4554}$

2. A bag a contains 5 red and 6 black balls. Two draws of three balls each are made, the ball being replaced after the first draw. What is the chance that the balls were red in the first draw and black in the second?

   a) $\frac{9}{1089}$ b) $\frac{8}{1089}$ c) $\frac{1081}{1089}$ d) $\frac{7}{1089}$

3. A bag a contains 7 red and 8 black balls. Two draws of three balls each are made, the ball being replaced after the first draw. What is the chance that the balls were red in the first draw and black in the second?

   a) $\frac{9}{845}$ b) $\frac{18}{845}$ c) $\frac{8}{845}$ d) None of these

Answers

1. a 2. b 3. c

Rule 11

Theorem: ***A bag contains x black and y white balls. A ball is drawn out of it and replaced in the bag. Then a ball is drawn again. The probability that***

(i) both the balls drawn were black is given by $\left(\frac{x}{x+y}\right)^2$.

(ii) both the balls drawn were white is given by $\left(\frac{y}{x+y}\right)^2$.

(iii) the first ball was white and the second black and vice versa is given by $\frac{xy}{(x+y)^2}$.

Illustrative Example

Ex: A bag contains 5 black and 7 white balls. A ball is drawn out of it and replaced in the bag. Then a ball is drawn again. What is the probability that (i) both the balls drawn were black; (ii) both were white; (iii) the first ball was white and the second black; (iv) the first ball was black and the second white?

Soln: The events are independent and capable of simultaneous occurrence. The rule of multiplication would be applied.

The probability that

(i) both the balls were black $= \frac{5}{12}\times\frac{5}{12} = \frac{25}{144}$

(ii) both the balls were white $= \frac{7}{12}\times\frac{7}{12} = \frac{49}{144}$

(iii) the first was white **and** the second black

$= \frac{7}{12}\times\frac{5}{12} = \frac{35}{144}$

(iv) the first was black **and** the second white

$= \frac{5}{12}\times\frac{7}{12} = \frac{35}{144}$

Note: From the above example we can see that how the quicker methods for such questions have been derived.

Exercise

1. A bag contains 4 black and 6 white balls. A ball is drawn out of it and replaced in the bag. Then a ball is drawn again. What is the probability that
(i) both the balls drawn were black;

a) $\frac{4}{25}$ b) $\frac{9}{25}$ c) $\frac{2}{5}$ d) $\frac{3}{5}$

(ii) both were white;

a) $\frac{3}{25}$ b) $\frac{3}{5}$ c) $\frac{2}{5}$ d) $\frac{9}{25}$

(iii) the first ball was white and the second black;

a) $\frac{6}{25}$ b) $\frac{21}{25}$ c) $\frac{19}{25}$ d) Data inadequate

2. A bag contains 6 black and 9 white balls. A ball is drawn out of it and replaced in the bag. Then a ball is drawn again. What is the probability that
(i) both the balls drawn were black;

a) $\frac{4}{25}$ b) $\frac{6}{25}$ c) $\frac{9}{25}$ d) None of these

(ii) both were white;

a) $\frac{9}{25}$ b) $\frac{16}{25}$ c) $\frac{21}{25}$ d) $\frac{4}{25}$

(iii) the first ball was white and the second black;

a) $\frac{6}{25}$ b) $\frac{19}{25}$ c) $\frac{4}{25}$ d) $\frac{9}{25}$

Answers

1. (i) a (ii) d (iii) a 2. (i) a (ii) a (iii) a

Rule 12

Theorem: A bag contains x red and y white balls. Four balls are drawn out one by one and not replaced. Then the probability that they are alternatively of different colours is given by $\left[\frac{2x(x-1)y(y-1)}{(x+y)(x+y-1)(x+y-2)(x+y-3)}\right]$.

Illustrative Example

Ex. 2: A bag contains 6 red and 3 white balls. Four balls are drawn out one by one and not replaced. What is the probability that they are alternatively of different colours?

Soln: **Detail Method:** Balls can be drawn alternately in the following order:
Red, White, Red, White OR White, Red, White, Red
If red ball is drawn first, the probability of drawing the balls alternatively

$$=\frac{6}{9}\times\frac{3}{8}\times\frac{5}{7}\times\frac{2}{6} \quad ..... (I)$$

If white ball is drawn first the probability of drawing the balls alternately

$$=\frac{6}{9}\times\frac{3}{8}\times\frac{5}{7}\times\frac{2}{6} \quad .... (II)$$

Required probability = (I) + (II) ..... (*)

$$=\frac{6}{9}\times\frac{3}{8}\times\frac{5}{7}\times\frac{2}{6}+\frac{3}{9}\times\frac{6}{8}\times\frac{2}{7}\times\frac{5}{6}=\frac{5}{84}+\frac{5}{84}=\frac{5}{42}.$$

Quicker Method: Applying the above theorem, we have the required probability

$$=\left[\frac{6\times3\times(6-1)(3-1)}{(6+3)(6+3-1)(6+3-2)(6+3-3)}\right]\times2$$

$$=\left[\frac{6\times3\times5\times2}{9\times8\times7\times6}\right]\times2=\frac{5}{42}.$$

Note: Wherever we find the word **AND** between two events, we use multiplication. Mark that both also means first **and** second. On the other hand, if the two events are joined with **OR**, we use addition as in the above example.

Exercise

1. A bag contains 6 red and 4 white balls. Four balls are drawn out one by one and not replaced. What is the probability that they are alternatively of different colours?

a) $\frac{2}{7}$ b) $\frac{4}{7}$ c) $\frac{1}{7}$ d) Data inadequate

2. A bag contains 8 red and 3 white balls. Four balls are drawn out one by one and not replaced. What is the probability that they are alternatively of different colours?

a) $\frac{7}{15}$ b) $\frac{7}{145}$ c) $\frac{14}{165}$ d) $\frac{7}{165}$

3. A bag contains 9 red and 7 white balls. Four balls are drawn out one by one and not replaced. What is the probability that they are alternatively of different colours?

a) $\frac{9}{65}$ b) $\frac{6}{65}$ c) $\frac{9}{130}$ d) $\frac{8}{130}$

4. A bag contains 5 red and 4 white balls. Four balls are drawn out one by one and not replaced. What is the probability that they are alternatively of different colours?

a) $\frac{5}{63}$ b) $\frac{10}{63}$ c) $\frac{15}{63}$ d) None of these

Answers

1. c 2. c 3. a 4. b

Rule 13

Theorem: ***A bag contains 'x' white and 'y' red balls. If two draws of one ball each are made without replacement, then the probability that one is red and the other white is given by*** $\left[\dfrac{2xy}{(x+y)(x+y-1)}\right]$.

Illustrative Example

Ex.: A bag contains 4 white and 6 red balls. Two draws of one ball each are made without replacement. What is the probability that one is red and other white?

Soln: **Detail Method:** Such problems can be very easily solved with the help of the rules of permutation and combination.

Two balls can be drawn out of 10 balls in $^{10}C_2$ or $\dfrac{10!}{2!8!}$ or $\dfrac{10\times 9}{2}$ or 45 ways.

One white ball can be drawn out of 4 white balls in $^4C_1$ or $\dfrac{4!}{1!3!}$ or 4 ways.

One red ball can be drawn out of 6 red balls in $^6C_1$ or 6 ways.

The total number of ways of drawing a white and a red ball are $^4C_1 \times {}^6C_1$ or $4 \times 6 = 24$.

The required probability would be

$$= \frac{\text{No. of cases favourable to the event}}{\text{Total no. of ways in which the event can happen}}$$

$$= \frac{24}{45} = \frac{8}{15}$$

Quicker Method: Applying the above theorem, we have

the required probability $= \dfrac{2\times 6\times 4}{10\times 9} = \dfrac{8}{15}$.

Note: The above theorem may be put as given below.

"*A bag contains 'x' white and 'y' red balls. If two balls are drawn in succession at random, then the probability that one of them is white and the other red, is given by* $\left[\dfrac{2xy}{(x+y)(x+y-1)}\right]$."

Exercise

1. A bag contains 8 white and 12 red balls. Two draws of one ball each are made without replacement. What is the probability that one is red and other white?

   a) $\frac{48}{95}$ b) $\frac{24}{95}$ c) $\frac{1}{19}$ d) None of these

2. A bag contains 5 white and 5 red balls. Two draws of one ball each are made without replacement. What is the probability that one is red and other white?

   a) $\frac{25}{81}$ b) $\frac{1}{3}$ c) $\frac{5}{9}$ d) $\frac{4}{9}$

3. A bag contains 4 white and 8 red balls. Two draws of one ball each are made without replacement. What is the probability that one is red and other white?

   a) $\frac{17}{33}$ b) $\frac{16}{33}$ c) $\frac{3}{11}$ d) Data inadequate

4. A bag contains 9 white and 3 red balls. Two balls are drawn in succession at random. What is the probability that one of them is white and the other red?

   a) $\frac{9}{22}$ b) $\frac{3}{22}$ c) $\frac{6}{22}$ d) $\frac{3}{11}$

Answers

1. a 2. c 3. b
4. a; **Hint:** See Note.

Rule 14

Theorem: ***A basket contains x_1 white and y_1 black balls. There is another basket which contains x_2 white and y_2 black balls if one ball is to be drawn from either of the two baskets, then the probability of drawing***

(i) a white ball is given by $\dfrac{1}{2}\left(\dfrac{x_1}{x_1+y_1}+\dfrac{x_2}{x_2+y_2}\right)$ ***and***

(ii) a black ball is given by $\dfrac{1}{2}\left(\dfrac{y_1}{x_1+y_1}+\dfrac{y_2}{x_2+y_2}\right)$.

Illustrative Example

Ex: A basket contains 3 white and 9 black balls. There is another basket which contains 6 white and 8 black balls. One ball is to be drawn from either of the two baskets. What is the probability of drawing a white ball?

Soln: **Detail Method:**

Since there are two baskets, each equally likely to be chosen, the probability of choosing either basket is $\frac{1}{2}$.

If the first basket is chosen, the probability of drawing a white ball $= \dfrac{1}{2}\times\dfrac{^3C_1}{^{12}C_1} = \dfrac{1}{2}\times\dfrac{3}{12} = \dfrac{1}{8}$

Similarly, if the second basket be chosen, the probability of drawing a white ball =

$\frac{1}{2}\times\frac{^6C_1}{^{14}C_1}=\frac{1}{2}\times\frac{6}{14}=\frac{3}{14}$

Since, the two events are mutually exclusive, we use addition, therefore, the probability of drawing a white ball from either basket is

$\frac{1}{8}+\frac{3}{14}=\frac{7+12}{56}=\frac{19}{56}$

Quicker Method: Applying the above theorem, we have,

the required probability $=\frac{1}{2}\left[\frac{3}{12}+\frac{6}{14}\right]=\frac{19}{56}$.

Exercise

1. A basket contains 4 white and 10 black balls. There is another basket which contains 5 white and 7 black balls. One ball is to be drawn from either of the two baskets.
(i) What is the probability of drawing a white ball?

a) $\frac{89}{168}$ b) $\frac{59}{84}$ c) $\frac{59}{168}$ d) $\frac{89}{84}$

(ii) What is the probability of drawing a black ball?

a) $\frac{119}{168}$ b) $\frac{89}{168}$ c) $\frac{59}{168}$ d) $\frac{109}{168}$

2. A basket contains 5 white and 9 black balls. There is another basket which contains 7 white and 7 black balls. One ball is to be drawn from either of the two baskets.
(i) What is the probability of drawing a white ball?

a) $\frac{4}{7}$ b) $\frac{6}{7}$ c) $\frac{3}{7}$ d) None of these

(ii) What is the probability of drawing a black ball?

a) $\frac{3}{7}$ b) $\frac{5}{7}$ c) $\frac{4}{7}$ d) $\frac{8}{15}$

3. A basket contains 6 white and 9 black balls. There is another basket which contains 8 white and 7 black balls. One ball is to be drawn from either of the two baskets.
(i) What is the probability of drawing a white ball?

a) $\frac{8}{15}$ b) $\frac{7}{15}$ c) $\frac{3}{5}$ d) $\frac{6}{15}$

(ii) What is the probability of drawing a black ball?

a) $\frac{3}{15}$ b) $\frac{8}{15}$ c) $\frac{7}{15}$ d) $\frac{9}{15}$

Answers

1. (i) c (ii) d 2. (i) c (ii) c
3. (i) b (ii) b

Rule 15

Theorem: ***A and B stand in a ring with 'x' other persons. If the arrangement of all the persons is at random, then the probability that there are exactly 'y' persons between A and B is given by*** $\left(\frac{2}{x+1}\right)$. ***Where*** $y<x$.

Illustrative Example

Ex: A and B stand in a ring with 10 other persons. If the arrangement of the 12 persons is at random, then the probability that there are exactly 3 persons between A and B is.

(Provident Fund Exam 2002)

Soln: Detail Method:

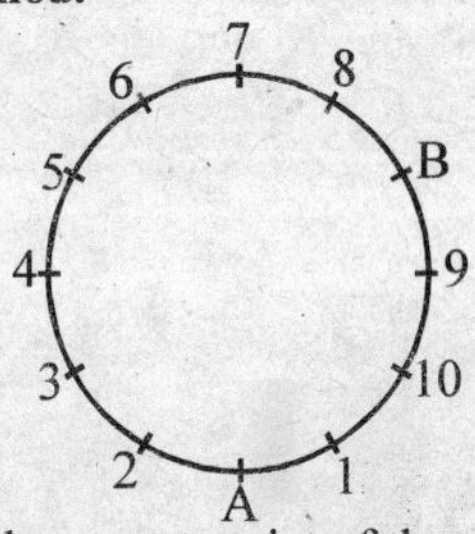

Let A stand on some point of the ring.
Then n(S) = the number of points on which B can stand = 11
If there be exactly 3 persons between A and B, then corresponding to any position occupied, B can take up only two position, the 4th place and the 8th place as counted from A.
Thus n(E) = 2

$\therefore$ P(E) $=\frac{n(E)}{n(S)}=\frac{2}{11}$

Quicker Method: Applying the above theorem, we have,

the required probability $=\frac{2}{10+1}=\frac{2}{11}$.

Exercise

1. A and B stand in a ring with 9 other persons. If the arrangement of the 11 persons is at random, then the probability that there are exactly 4 persons between A and B is.

a) $\frac{2}{11}$ b) $\frac{1}{5}$ c) $\frac{1}{10}$ d) $\frac{1}{11}$

2. A and B stand in a ring with 8 other persons. If the arrangement of the 10 persons is at random, then the probability that there are exactly 5 persons between A and B is.

a) $\frac{2}{7}$ b) $\frac{2}{9}$ c) $\frac{1}{9}$ d) $\frac{1}{7}$

3. A and B stand in a ring with 7 other persons. If the arrangement of the 9 persons is at random, then the probability that there are exactly 2 persons between A and B is

a) $\frac{1}{4}$ b) $\frac{3}{4}$ c) $\frac{2}{9}$ d) $\frac{3}{8}$

4. A and B stand in a ring with 11 other persons. If the arrangement of the 13 persons is at random, then the probability that there are exactly 3 persons between A and B is.

a) $\frac{1}{6}$ b) $\frac{2}{11}$ c) $\frac{3}{12}$ d) $\frac{4}{9}$

5. A and B stand in a ring with 14 other persons. If the arrangement of the 16 persons is at random, then the probability that there are exactly 6 persons between A and B is.

a) $\frac{2}{15}$ b) $\frac{1}{7}$ c) $\frac{3}{14}$ d) $\frac{3}{7}$

Answers

1. b 2. b 3. a 4. a 5. a

Rule 16

Theorem: ***If 'n' persons are seated at a around table then the probability that 'm' particular persons sit together is given by*** $\left[\frac{(n-m)!m!}{(n-1)!}\right]$.

Illustrative Example

Ex.: 10 persons are seated at a round table. What is the probability that two particular persons sit together?

Soln: **Detail Method:**

n(S) = no.of ways of sitting 10 persons at round table $=(10-1)!=9!$

Since 2 particular persons will be always together, then the no. of persons = 8 + 1 = 9

∴ 9 persons will be seated in (9 – 1)! = 8! ways at round table and 2 particular persons will be seated themselves in 2! ways.

∴ The number of ways in which two persons always sit together at round table = 8! × 2! = n(E)

$$\therefore P(E)=\frac{n(E)}{n(S)}=\frac{8!\times2!}{9!}=\frac{8!\times2}{9\times8!}=\frac{2}{9}$$

Quicker Method: Applying the above theorem, we have,

$$\text{the required probability}=\frac{(10-2)!2!}{(10-1)!}=\frac{8!2!}{9!}=\frac{2}{9}.$$

Exercise

1. 12 persons are seated at a round table. What is the probability that 4 particular persons sit together?

a) $\frac{4}{165}$ b) $\frac{8}{165}$ c) $\frac{9}{185}$ d) $\frac{4}{55}$

2. 8 persons are seated at a round table. What is the probability that 3 particular persons sit together?

a) $\frac{2}{7}$ b) $\frac{1}{7}$ c) $\frac{3}{14}$ d) None of these

3. 10 persons are seated at a round table. What is the probability that 3 particular persons sit together?

a) $\frac{1}{12}$ b) $\frac{2}{9}$ c) $\frac{1}{4}$ d) $\frac{1}{6}$

Answers

1. a 2. b 3. a

Rule 17

Theorem: If an event is repeated, under similar conditions, exactly 'n' times, then the probability that event happens exactly 'r' times is $\left[{}^nC_r\times p^r\times q^{n-r}\right]$, provided that

p = probability of happening and

q = probability of not happening ie p + q = 1.

Illustrative Example

Ex.: An unbiased coin is tossed 7 times, find the chance that exactly 5 times head will appear.

Soln: Here, n = 7, r = 5

p = probability of happening = $\frac{1}{2}$

q = probability of not-happening = $\frac{1}{2}$

∴ required probability

$$={}^7C_5\times\left(\frac{1}{2}\right)^5\times\left(\frac{1}{2}\right)^{7-5}=\frac{21}{128}.$$

Exercise

1. An unbiased coin is tossed 5 times, find the chance that exactly 3 times tail will appear.

a) $\frac{5}{16}$ b) $\frac{5}{32}$ c) $\frac{10}{64}$ d) None of these

2. An unbiased coin is tossed 9 times, find the chance that exactly 6 times head will appear.

a) $\frac{21}{128}$ b) $\frac{21}{256}$ c) $\frac{21}{64}$ d) Data inadequate

3. An unbiased coin is tossed 6 times, find the chance that exactly 4 times tail will appear.

a) $\frac{15}{32}$ b) $\frac{15}{128}$ c) $\frac{15}{64}$ d) $\frac{15}{256}$

Answers

1. a 2. a 3. c